CHAOS, CNN, MEMRISTORS AND BEYOND

A Festschrift for Leon Chua

CHAOS, CNN, MEMRISTORS AND BEYOND

A Festschrift for Leon Chua

Editors

Andrew Adamatzky

University of the West of England, UK

Guanrong Chen

City University of Hong Kong, P R China

World Scientific

NEW JERSEY · LONDON · SINGAPORE · BEIJING · SHANGHAI · HONG KONG · TAIPEI · CHENNAI

Published by

World Scientific Publishing Co. Pte. Ltd.
5 Toh Tuck Link, Singapore 596224
USA office: 27 Warren Street, Suite 401-402, Hackensack, NJ 07601
UK office: 57 Shelton Street, Covent Garden, London WC2H 9HE

British Library Cataloguing-in-Publication Data
A catalogue record for this book is available from the British Library.

CHAOS, CNN, MEMRISTORS AND BEYOND
A Festschrift for Leon Chua
With DVD-ROM, composed by Eleonora Bilotta

ISBN 978-981-4434-79-9

Typeset by Stallion Press
Email: enquiries@stallionpress.com

Printed by FuIsland Offset Printing (S) Pte Ltd Singapore

Dedicated to Leon O. Chua on the occasion of his 75th Birthday

PREFACE

This edited compendium is dedicated to Professor Leon O. Chua on the occasion of his 75th birthday this year.

Given the advanced evolution and extreme specialization in science and technology today, a scientist may take great pride in the body of his work if, during his career, he has had a significant impact upon just a single area of science or engineering. But the impact of Professor Chua's work is in not just one, but no less than three areas: nonlinear circuit theory [1], cellular neural networks [2] and solid-state memristors [3].

The significance and impact of Leon's scientific work are evidenced by the large number of awards and honors he has received over his career, the most important of which include the conferring of 13 *Doctor Honoris Causa* (Honorary Doctorate) degrees from the following prestigious institutions:

- École Polytechnique Fédérale de Lausanne, Switzerland (1983)
- University of Tokushima, Japan (1984)
- Technische Universität Dresden, Germany (1992)
- Technical University of Budapest, Hungary (1994)
- University of Santiago de Compostela, Spain (1995)
- University of Frankfurt, Germany (1996)
- Gheorghe Asachi Technical University of Iaşi, Romania (1997)
- University of Catania, Italy (2000)
- AGH University of Science and Technology, Cracow, Poland (2003)
- Dogus University, Istanbul Turkey (2005)
- Université du Sud, Toulon, France (2006)
- Université du Havre, France (2009)
- University of the West of England, UK (2011).

In addition, Leon has received numerous prestigious awards and prizes for his work, including

- The IEEE Browder J. Thompson Memorial Prize Award (1967),
- The IEEE Guillemin-Cauer Award (1972, 1985, 1989),
- The IEEE W. R. G. Baker Prize Paper Award (1973),
- The Frederick Emmons Award (1974),
- The Alexander von Humboldt Senior US Scientist Award (1982),
- The IEEE Centennial Medal (1985),
- The IEEE Neural Networks Pioneer Award (2000),
- The IEEE Third Millenium Medal (2000),
- The IEEE Circuits and Systems Society Golden Jubilee Medal (2000),
- The IEEE Gustav Robert Kirchhoff Award (2005),
- The Mac E. Van Valkenburg Award (1995 and 1998),
- The IEEE Circuits and Systems Society Vitold Belevitch Award (2007),

- The Guggenheim Fellowship (2010),
- Leverhulme Visiting Professor, UK (2010–2011).

Most recently, Leon was appointed Distinguished Affiliated Professor by the Technische Universitat Muenchen (TUM), Germany, which is the highest honor of its kind, along with a lifelong academic appointment based in the TUM Department of Electrical Engineering and Information Technology. This particular honor is awarded "only to those exceptional scientists and researchers of international prominence that have significantly shaped their own disciplines and also inspired other areas within the scientific community." Other honorees of this award include Nobel Laureates such as Professor Rudolph A. Marcus (Chemistry).

Regarding Leon's education and career background, he received his BSEE degree from the Mapúa Institute of Technology in the Philippines in 1959, the MSEE from the Massachusetts Institute of Technology in 1961, and the Ph.D. from the University of Illinois at Urbana-Champaign in 1964. He then started to work as Assistant Professor of Electrical Engineering at Purdue University and was promoted to Associate Professor there in 1967. Finally, he joined the University of California at Berkeley in 1970 and taught there until he retired recently.

Leon is an IEEE Life Fellow (1974) and a foreign member of the European Academy of Sciences (1997) as well as of the Hungarian Academy of Sciences (2006). He owns seven USA patents, and is the recipient of the top 15 most-cited authors in 2002 from all fields of engineering published during the ten-year period (1991–2001) according to the ISI Current Contents database. He is also the Founding Editor-in-Chief of the *International Journal of Bifurcation and Chaos*.

Regarding his personal background, Leon Ong Chua (Chinese name 蔡少棠) was born on 28 June 1936. He and his family moved to the United States in 1959. Leon and his wife Diana have four daughters — Amy, Michelle, Katrin and Cynthia — and 7 grandchildren.

To commemorate Leon's 75th birthday, this volume is a collection of essays and research articles celebrating Leon's great achievements in science and technology written by some of his closest collaborators, former students and long-term friends; these articles present scientific works in areas related to nonlinear circuits, cellular neural networks, and memristors.

We wish to express our sincerest appreciation to the staff of World Scientific Publishing Company for their assistance in producing this book. We especially thank World Scientific's Chairman and Editor-in-Chief, Dr. K. K. Phua for his strong and enthusiastic support of this work honoring the achievements of Professor Leon O. Chua.

Andrew Adamatzky (University of Western England, UK)

Guanrong Chen (City University of Hong Kong, Hong Kong)

Co-Editors

Winter, 2011

References

[1] Matsumoto, T. [1984] "A chaotic attractor from Chua's circuit," *IEEE Trans. Circuits Syst.* **CAS-31**(12) 1055–1058.

[2] Chua, L. O. [1988] "Cellular neural networks: Theory," *IEEE Trans. Circuits Syst.* **CAS-35**(10) 1257–1272.

[3] Chua, L. O. [1971] "Memristor — the missing circuit element," *IEEE Trans. Circuits Th.* **18**(5) 507–519.

CONTENTS

PART I

CELLULAR NONLINEAR NETWORKS, NONLINEAR CIRCUITS AND CELLULAR AUTOMATA

CHAPTER 1

GENEALOGY OF CHUA'S CIRCUIT

PETER KENNEDY

Department of Electrical and Electronic Engineering,
University College Cork, Ireland

Since its invention in 1983, Chua's circuit has become a reference circuit for studying bifurcations and chaos. This chapter plots the evolution of the circuit from the original simulations and experimental realization of a five-element topology with a three-segment nonlinear resistor to the latest three-element design comprising a capacitor, an inductor, and a memristor.

1. Introduction

In order to exhibit chaos, an autonomous circuit consisting of resistors, capacitors, and inductors must contain [15]:

- at least one nonlinear element,
- at least one locally active resistor, and
- at least three energy storage elements.

Chua's circuit, shown in Fig. 1, is one of the simplest electronic circuits that satisfies these criteria. In addition, this remarkable circuit is the *first* physical system for which the presence of chaos has been proven mathematically [3]. The circuit is readily constructed at low cost using standard electronic components [14] and exhibits a rich variety of bifurcations and chaos [7,8].

The circuit is described by three coupled ordinary differential equations:

$$\frac{dV_{C_1}}{dt} = \frac{G}{C_1}(V_{C_2} - V_{C_1}) - \frac{1}{C_1}f(V_{C_1}) \qquad (1)$$

$$\frac{dV_{C_2}}{dt} = \frac{1}{C_2}I_L - \frac{G}{C_2}(V_{C_2} - V_{C_1}) \qquad (2)$$

$$\frac{dI_L}{dt} = -\frac{1}{L}V_{C_2} \qquad (3)$$

The nonlinear resistor N_R has a symmetric piecewise-linear driving-point (DP) characteristic [6], as shown in Fig. 2. This characteristic is defined

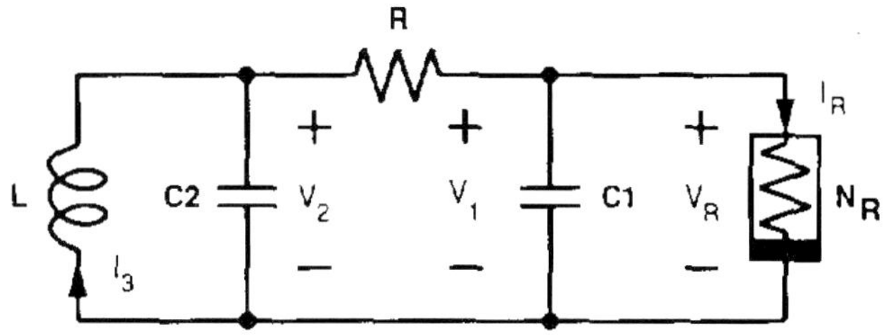

Fig. 1. Chua's five-element circuit consists of a linear inductor L, two linear capacitors (C_1, C_2), a linear resistor R, and a voltage-controlled nonlinear resistor N_R, called a "Chua diode" [14]

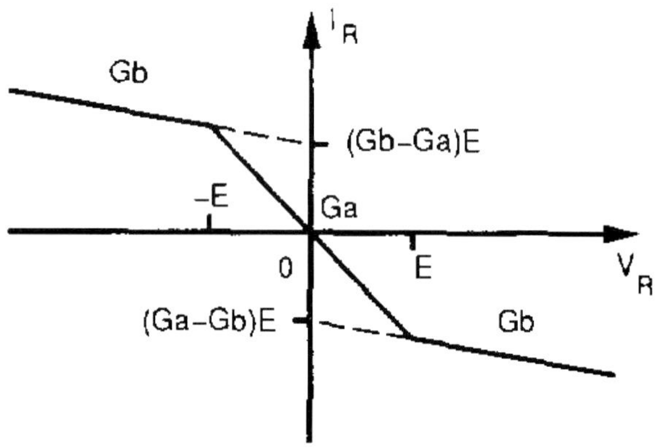

Fig. 2. The driving-point characteristic of the nonlinear resistor N_R in Chua's circuit has breakpoints at $\pm E$ and slopes G_a and G_b in the inner and outer regions, respectively.

analytically as follows:

$$f(V) = \begin{cases} G_bV + (G_b - G_a)E & \text{if } V < -E \\ G_aV & \text{if } -E \leq V \leq E \\ G_bV + (G_a - G_b)E & \text{if } V > E \end{cases} \tag{4}$$

where $E > 0$, $G_a < 0$, and $G_b < 0$.

2. History

While on a visit to Japan in October 1983, having witnessed a futile attempt at producing chaos in an electrical analog of Lorenz's equations, Prof. Leon Chua was prompted to develop a chaotic electronic circuit [4]. He realized that chaos could be produced in a piecewise-linear circuit if it possessed at least two unstable equilibrium points — one to provide stretching, and the other to fold trajectories. With this insight, he systematically identified those third-order piecewise-linear circuits containing a single voltage-controlled nonlinear resistor that could produce chaos. Specifying that the driving point characteristic of the voltage-controlled nonlinear resistor N_R should be chosen to yield at least two unstable equilibrium points, he invented the circuit shown in Fig. 1 [4].

Let the nonlinear resistor N_R in Chua's circuit have a symmetric piecewise-linear DP characteristic as shown in Fig. 2. Imagine that the values of the parameters are chosen such that the circuit possesses three equilibrium points (one at the origin where N_R has a locally negative slope or *conductance* G_a, and two in the outer regions where N_R has a locally negative conductance G_b) and that all three equilibrium points are *unstable*. There are three eigenvalues associated with each equilibrium point [15]. We assume that the equilibrium point at the origin has an *unstable* real eigenvalue and a *stable* pair of complex conjugate eigenvalues. In addition, suppose that the outer equilibrium point P_- has a *stable* real eigenvalue and an *unstable* complex conjugate pair. Two trajectories starting close together on the eigenplane in the outer region would be stretched apart as they spiralled away from the equilibrium point. If they could be *folded* back by the dynamics in the inner region and reinjected toward P_- along the stable real eigenvector, a *homoclinic* orbit would be produced, and chaos would result.

Figure 3 shows the structure of a spiral chaotic attractor in Chua's circuit [15]. The piecewise-linear state space is divided (by N_R) into an inner region (denoted D_0) and two outer regions (D_{-1} and D_1). There are equilibrium points in each of the three

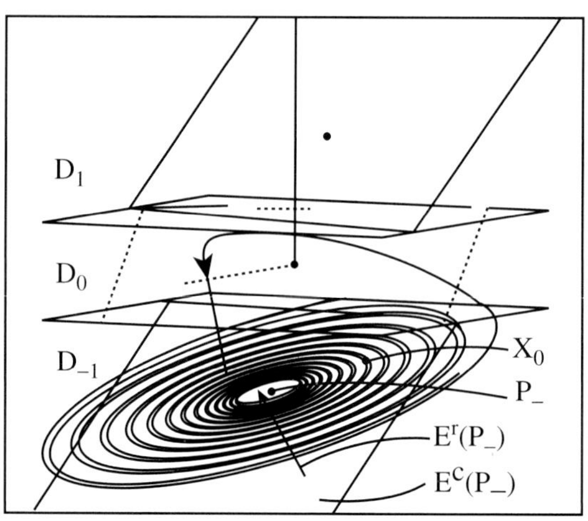

Fig. 3. Stretching and folding mechanism of chaos generation in Chua's circuit. A trajectory spirals away from the equilibrium point P_- along the eigenplane $E^c(P_-)$ until it enters the D_0 region, where it is folded back into D_{-1} and returns to the unstable eigenplane $E^c(P_-)$ close to P_- (reproduced from [15]).

regions. A homoclinic trajectory can be formed when the outer equilibrium points (P_- and P_+) have a stable real eigenvalue and an unstable complex conjugate pair of eigenvalues and the equilibrium point (0) at the origin has an unstable real eigenvalue and a stable complex conjugate pair. A trajectory starting from an initial state X_0 on the eigenplane $E^c(P_-)$ near P_- spirals away from the equilibrium point along the eigenplane $E^c(P_-)$ until it enters the D_0 region, where it is folded back into D_{-1} and returns to the unstable eigenplane $E^c(P_-)$ close to P_-. In this way, trajectories are repeatedly stretched apart (in D_{-1}) and folded (in D_0) to produce chaos [15].

Figure 4 shows the structure of a double scroll chaotic attractor in Chua's circuit. A trajectory starting from an initial state X_0 on the eigenplane $E^c(P_-)$ near P_- spirals away from the equilibrium point along the eigenplane $E^c(P_-)$ until it enters the D_0 region. If it enters the D_0 region below the eigenplane $E^c(0)$, it is reinjected into D_{-1}. However, if it enters the D_0 region above $E^c(0)$, it is injected into the D_1 region, where it spirals around P_+ with increasing amplitude until it returns to the D_0 region. In this way, a symmetric double scroll chaotic attractor is formed.

Soon after its conception, the rich dynamical behavior of Chua's circuit was confirmed by computer simulation [21]. Matsumoto used a fourth-order Runge-Kutta algorithm with a fixed step size to solve the system of equations, using the parameter values shown in Table 1. The simulations produced a complicated trajectory, as shown in Fig. 5.

Prof. Chua's doctoral student, Farhad Ayrom, had been working on synthesis techniques for piecewise-linear nonlinear resistors [5]. Leon phoned Farhad from Japan and asked him to synthesize a piecewise-linear resistor N_R with a prescribed driving-point characteristic. Using Ayrom's design, Zhong built the first physical realization of Chua's circuit the next day [1]. With the parameter values in Table 2, they recorded a double scroll attractor, as shown in Fig. 6 [27].

Since its invention in 1984, there has been an intensive effort to understand every aspect of the dynamics of Chua's circuit with a view to developing it as a paradigm for learning, understanding, and teaching about nonlinear dynamics and chaos [20]. That the circuit does in fact exhibit chaos in the sense of Shilnikov was proved by Chua *et al.* in 1986 [3]. By adding a linear resistor in series with

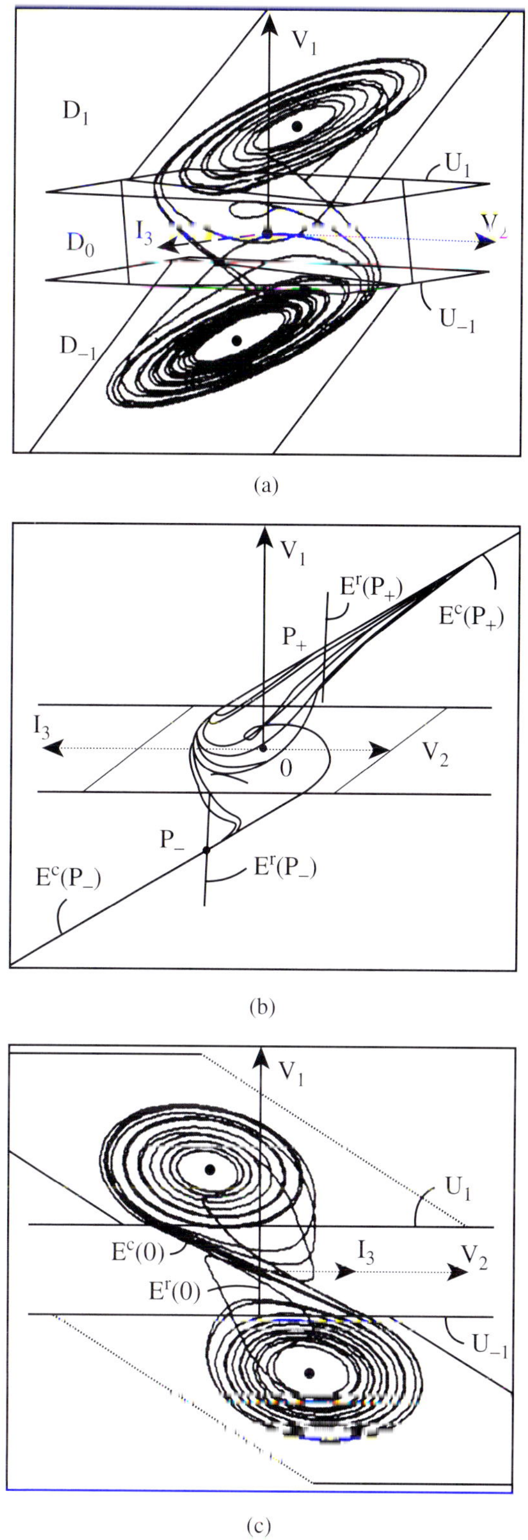

Fig. 4. Three views of a simulated double scroll attractor in Chua's circuit. (a) Reference view. (b) View along the edge of the outer complex eigenplanes $E^c(P_+)$ and $E^c(P_-)$; note how the trajectory in D_1 is flattened onto $E^c(P_+)$ and onto $E^c(P_-)$ in D_{-1}. (c) View along the edge of the complex eigenplane $E^c(0)$; a trajectory entering D_0 from D_1 above this plane returns to D_1 while one entering D_0 below $E^c(0)$ crosses to D_{-1} (reproduced from [15]).

Table 1. Parameter values used by Matsumoto in the first published simulation of Chua's circuit [21].

C_1 (F)	C_2 (F)	L (H)	R (Ω)	G_a (S)	G_b (S)	E (V)
1/10	2	1/7	1/0.7	−4	−0.1	1

the inductor, the circuit has been generalized to the canonical *Chua's oscillator* [7,8]. This augmented six-element circuit, shown in Fig. 7, is canonical in the sense that every continuous three-dimensional odd-symmetric piecewise-linear vector field may be mapped onto the circuit [7, 8]. Consequently, the circuit can exhibit *every dynamical behavior* that is known to be possible in a system described by a continuous odd-symmetric three-region piecewise-linear vector field. With the appropriate choice of parameters, the circuit can be made to follow the classic *period-doubling*, *intermittency*, and *torus breakdown* routes to chaos [7,8]. For a comprehensive bibliography of papers on Chua's circuit, we refer the reader to Chua's historical review paper [4].

The original five-element Chua's circuit quickly established itself as the first provably chaotic electronic circuit. As a result, it attracted huge interest in the electrical engineering community. With a solid underpinning theory and a robust practical implementation, Chua's circuit played an important role in the 1990s in the development of applications of chaos including chaos communications [10], cryptography, and chaos control [12]. When applications demanded high speed operation and integrated circuit implementation, appropriate variants were developed [9,25]. In 2008, a simplified four-element Chua was proposed [2]. Today, we have a three-element chaotic circuit which uses a memristor to realize simultaneously nonlinearity, local activity and memory [24].

In this chapter, we will discuss the six-, five-, four- and three-element chaotic circuits which have been inspired by Chua's original invention.

3. Five-element Chua's Circuit

As a circuit theorist, Leon Chua originally conceived of a five-element circuit that would exhibit chaos [4]. The circuit would require three dynamic elements (inductors and/or capacitors), a nonlinear resistor, and a linear resistor. The system described

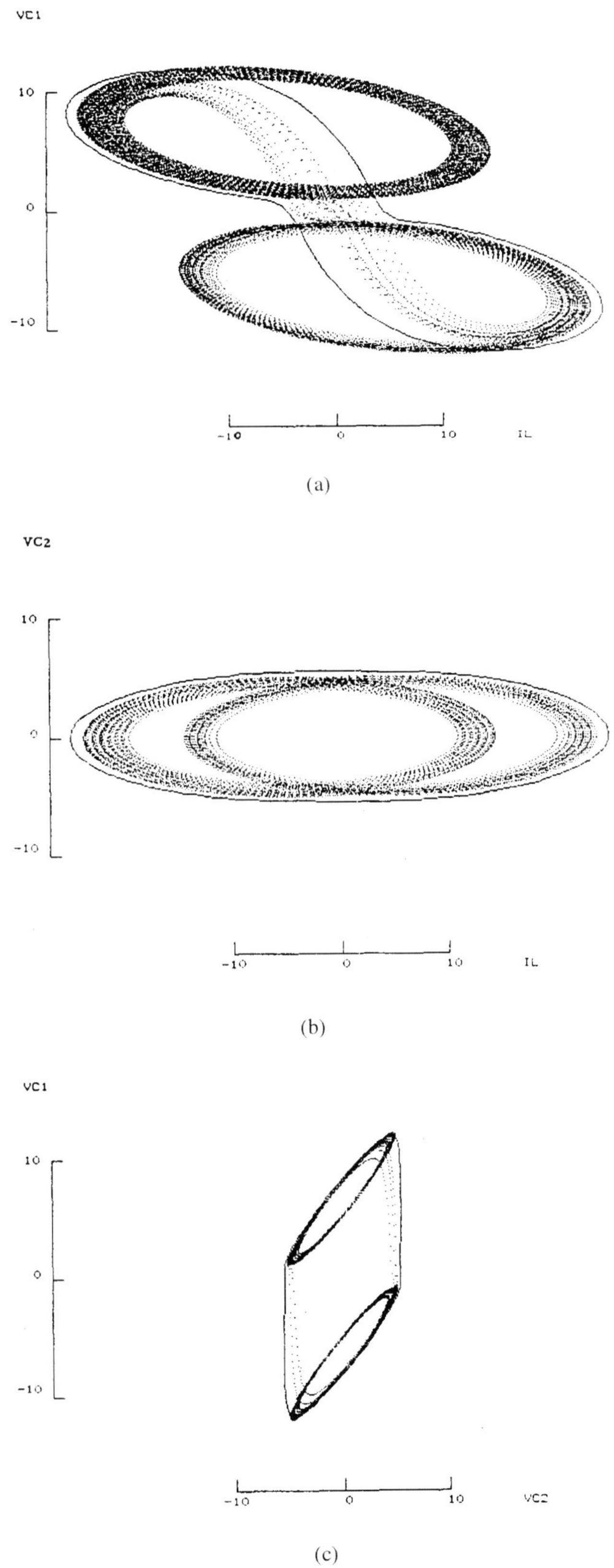

Fig. 5. Simulated trajectories in Chua's circuit with the parameters of Table 1. Projections on the (a) I_L–V_{C_1}, (b) I_L–V_{C_2}, and (c) V_{C_1}–V_{C_2} planes (reproduced from [21]).

Table 2. Parameter values used by Zhong and Ayrom in the first published experimental study of Chua's circuit [27].

C_1 (nF)	C_2 (nF)	L (mH)	R (kΩ)	G_a (mS)	G_b (mS)	E (V)
5.7	100	8	1.28	−0.77	−0.43	2.5

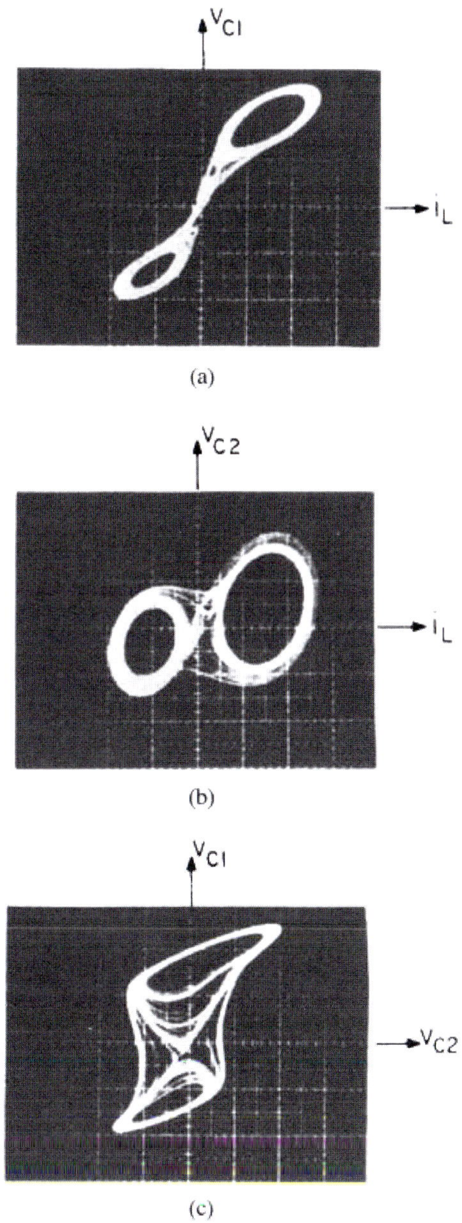

Fig. 6. Measured trajectory in Chua's circuit with the parameters of Table 2. Projections on the (a) I_L–V_{C_1}, (b) I_L–V_{C_2}, and (c) V_{C_2}–V_{C_1} planes (reproduced from [27]).

by the resulting circuit equations (1)–(4) can be implemented in many different ways, including analog and digital computers [19], as well as discrete and integrated [9] circuits.

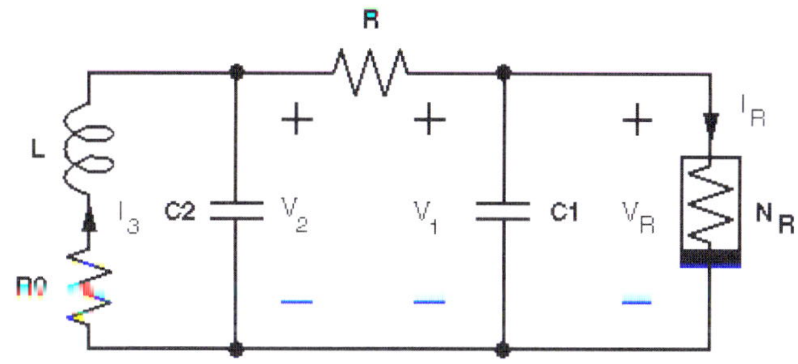

Fig. 7. In addition to the elements in Chua's original five-element circuit, the augmented six-element Chua's oscillator contains a linear resistor R_0 in series with the inductor L.

3.1. *Discrete circuit realization*

In the discrete circuit realization, there is a one-to-one mapping between the differential equations and circuit elements. Each state variable requires a dynamic element (an inductor or capacitor) and the algebraic relations between state variables are realized by resistive elements. Thus, the circuit has two capacitors (C_1 and C_2), an inductor (L), a linear resistor R and a nonlinear resistor N_R.

The linear capacitors and the resistor are implemented with physical capacitors and a resistor, respectively. The inductor can be implemented as a physical inductor or using a gyrator terminated in a capacitor; we will consider the inductor in detail in Sec. 3.5.

The nonlinearity in Chua's circuit is required to have at least two unstable equilibrium points [4]. Thus, the V–I characteristic must intersect the load line defined by the linear resistor R at least two points. The nonlinearity can have any number of different shapes. If one also requires symmetry then one can consider at least two classes of nonlinear function: piecewise-linear and cubic. Symmetry will yield *three* equilibrium points. Several subcircuits have been proposed to realize the required characteristic; these are discussed in Secs. 3.2 and 3.3.

3.2. *Piecewise-linear Chua diode*

The simplest symmetric piecewise-linear Chua diode is one which has a large negative slope G_a in its middle region and a smaller negative slope G_b in the outer regions, as shown in Fig. 2.

Practical resistors are *eventually passive*, meaning that they must lie in the first and third quadrants of the V–I plane for large values of current and voltage. In the case of a piecewise-linear

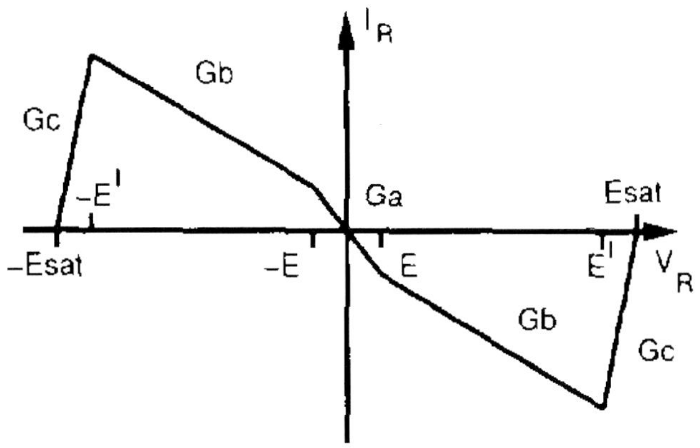

Fig. 8. The ideal three-segment driving point characteristic of the nonlinear resistor N_R in Chua's circuit has breakpoints at $\pm E$ and slopes G_a and G_b in the inner and outer regions, respectively. A practical implementation is eventually passive and therefore contains at least two additional segments that return the characteristic to the first and third quadrants for large values of voltage and current. We consider a five-segment case with breakpoints at $\pm E'$ and outermost slopes G_c.

resistor, this means that the characteristic must have two additional segments which return the characteristic to the first and third quadrants. Hence the characteristic has *five* segments and four breakpoints, two of which are required in order to produce the three equilibrium points, and two which satisfy the eventual passivity condition. Thus, a physical realization of the V–I characteristic shown in Fig. 2 has two additional breakpoints at $\pm E'$ and slopes G_c in the outermost regions, as shown in Fig. 8. We will call the outer breakpoints *parasitic* in the sense that they are not necessary for the operation of the circuit, but result from the implementation. The simplest eventually passive symmetric piecewise-linear nonlinearity has a slope G_a in its middle region, slope G_b in the outer regions, and slope G_c in the outermost regions, as shown in Fig. 8.

If the segments associated with eventual passivity are so far away from the inner breakpoints that the trajectory never reaches them, then the practical implementation is correctly modeled by the original Chua's circuit equations with a symmetric *three-segment* nonlinearity. If the attractor is too large, it may enter the outermost regions associated with eventual passivity; this distorts the attractor [28]. A further non-ideality due to implementation results from asymmetry in the breakpoints which causes the attractor to be asymmetric. This effect is noticeable in the early experimental realizations of Chua's circuit [27, 28] (see Fig. 6) but the problem was solved by Kennedy [14].

Negative resistance may be produced by using a Negative Impedance Converter (NIC) terminated in a resistor [5,13]. This turns a positive resistance into a negative resistance over the linear region of operation of the NIC. The NIC itself is eventually passive and therefore exhibits breakpoints associated with the extremities of the linear region of operation of the active element. The breakpoints in the Chua diode characteristic can be implemented in a number of ways including exploiting the eventual passivity of the NIC itself or by using resistor-diode networks.

3.2.1. *Op amp-based negative impedance converters*

In the first experimental implementation of Chua's circuit [27], the piecewise-linear nonlinear resistor was realized by connecting two NICs in parallel. One NIC provides the inner pair of breakpoints $\pm E$, while the other provides the outer pair $\pm E'$. The two-NIC implementation of Chua's circuit is shown in Fig. 9.

Op amp A_1 and resistors R_1–R_3 implement one negative resistor, op amp A_2 and resistor R_4–R_6 implement the other one. The slopes and breakpoints of the Chua diode are defined by:

$$G_a = -\frac{R_2}{R_1 R_3} - \frac{R_5}{R_4 R_6} \tag{5}$$

$$G_b = \frac{1}{R_3} - \frac{R_5}{R_4 R_6} \tag{6}$$

$$G_c = \frac{1}{R_3} + \frac{1}{R_6} \tag{7}$$

$$E = \frac{R_1}{R_1 + R_2} E_{sat} \tag{8}$$

$$E' = \frac{R_4}{R_4 + R_5} E_{sat} \tag{9}$$

Using the component values in Table 3, Zhong and Ayrom measured chaotic attractors. The corresponding values of the slopes and breakpoints of the Chua diode are summarized in Table 4.

Figure 6 shows experimentally-observed projections of a chaotic attractor. Note that this attractor should be symmetric but it is not; the bottom lobe of the double scroll attractor is smaller than the upper lobe. The reason for this is that the

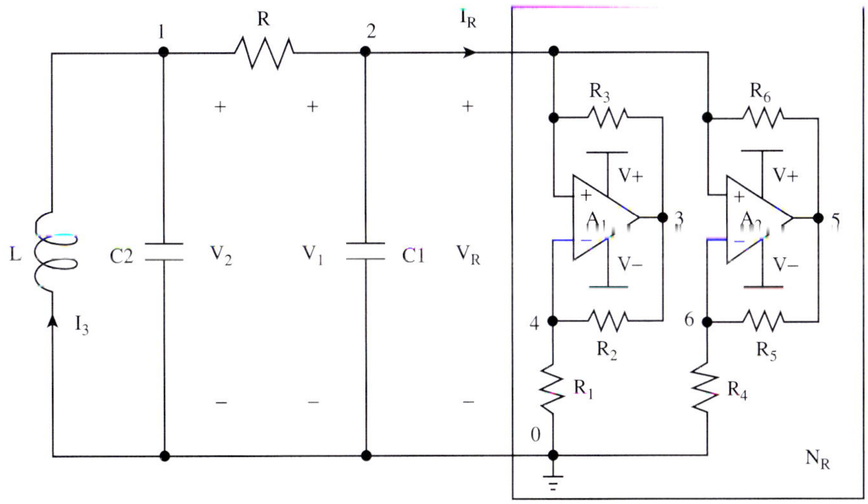

Fig. 9. Experimental realization of Chua's circuit using a two-NIC implementation of the Chua diode (reproduced from [15]).

Table 3. Component values in the first experimental implementation of Chua's circuit shown in Fig. 9 [27].

Element	Description	Value
C_1	Capacitor	5.7 nF
C_2	Capacitor	100 nF
L	Inductor	8 mH
R	Potentiometer	1.28 kΩ
A_1, A_2	Op amp (741)	
R_1	$\frac{1}{4}$W Resistor	128 Ω
R_2	$\frac{1}{4}$W Resistor	104 Ω
R_3	$\frac{1}{4}$W Resistor	5.36 kΩ
R_4	$\frac{1}{4}$W Resistor	5.43 kΩ
R_5	$\frac{1}{4}$W Resistor	3.67 kΩ
R_6	$\frac{1}{4}$W Resistor	1.09 kΩ

Table 4. Slopes and breakpoints of the Chua diode used by Zhong and Ayrom in the first experimental implementation of Chua's circuit [27].

G_a (mS)	G_b (mS)	E (V)	E' (V)	G_c (mS)
−0.77	−0.43	2.5	7.5	1.10

breakpoints are determined by the positive and negative saturation levels (E_{sat}^+ and E_{sat}^-) of the op amps. In a 741 op amp with power rails that are symmetrically placed about zero, the saturation levels are not symmetric. In particular, $E_{sat}^- < E_{sat}^+$; this makes the "bottom" lobe of the attractor smaller than the top.

The breakpoints in the op amp-based two-NIC Chua diode are determined by the saturation levels the op amps, which in turn depend on the internal structure of the op amp and the power supplies. If the saturation levels are asymmetric about zero, so are the breakpoints and the resulting attractors, as we have seen. Matsumoto *et al.* proposed an alternative implementation of the Chua diode [22] where an op amp-based NIC provides the underlying nonlinear resistance G_a and the outer breakpoints required by eventual passivity. Diodes and resistors are used to define the inner breakpoints and the slope G_b. Their circuit is shown in Fig. 10.

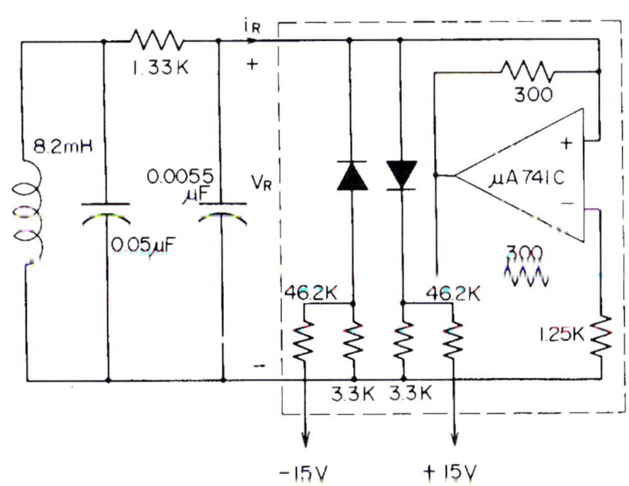

Fig. 10. Chua's circuit with a Chua diode comprising an op amp-based NIC to set the conductance G_a and a diode-resistor network to define the breakpoints (reproduced from [22].

Table 5. Slopes and breakpoints of the Chua diode in Fig. 10 used by Matsumoto *et al.* [22].

G_a (mS)	G_b (mS)	E (V)	E' (V)	G_c (mS)
−0.80	−0.48	1.0	10.9	3.66

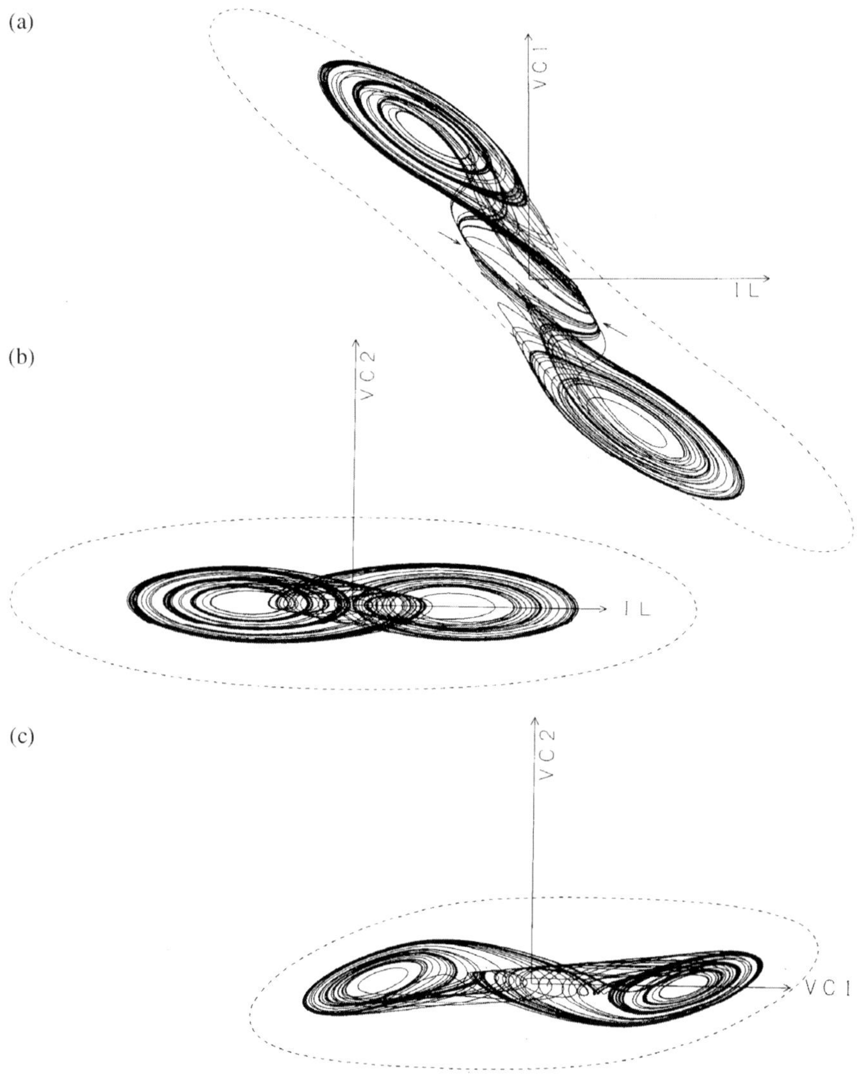

Fig. 11. Simulated trajectory in Matsumoto *et al.*'s realization on Chua's circuit with the parameters of Table 5 (reproduced from [22]). Projections on the (a) I_L–V_{C_1}, (b) I_L–V_{C_2}, and (c) V_{C_2}–V_{C_1} planes.

In this case, the inner breakpoints do not depend on the internal structure of the op amp.

With the parameter values shown in Fig. 10, the slopes and breakpoints of the Chua diode are as shown in Table 5. Figure 11 shows a simulated double scroll attractor in this circuit.

Kennedy [14] solved the asymmetric breakpoint problem with a two-NIC design which uses AD711 op amps instead of 741s. These provide symmetric saturation levels and thereby resolve the asymmetry problem. By setting $R_2 = R_3$ and $R_5 = R_6$, the design is further simplified. The corresponding design equations are:

$$G_a = -\frac{1}{R_1} - \frac{1}{R_4} \tag{10}$$

$$G_b = \frac{1}{R_3} - \frac{1}{R_4} \tag{11}$$

$$G_c = \frac{1}{R_3} + \frac{1}{R_6} \qquad (12)$$

$$E = \frac{R_1}{R_1 + R_2} E_{sat} \qquad (13)$$

$$E' = \frac{R_4}{R_4 + R_5} E_{sat} \qquad (14)$$

This implementation, shown in Fig. 9, has become the reference design for the Chua circuit with a piecewise-linear (PWL) Chua diode [14].

With the component values in Table 6, the corresponding values of the slopes and breakpoints in the Chua diode are summarized in Table 7.

Figure 12 shows an experimentally-observed projection of a double scroll chaotic attractor in Kennedy's robust implementation of Chua's circuit [15]. Note that the attractor does not enter the parasitic outer regions. Furthermore, due to the symmetric saturation levels of the AD711 op amp, the attractor is symmetric.

Figure 13 shows experimentally-observed projections of periodic and chaotic attractors in Kennedy's robust implementation of Chua's circuit [17].

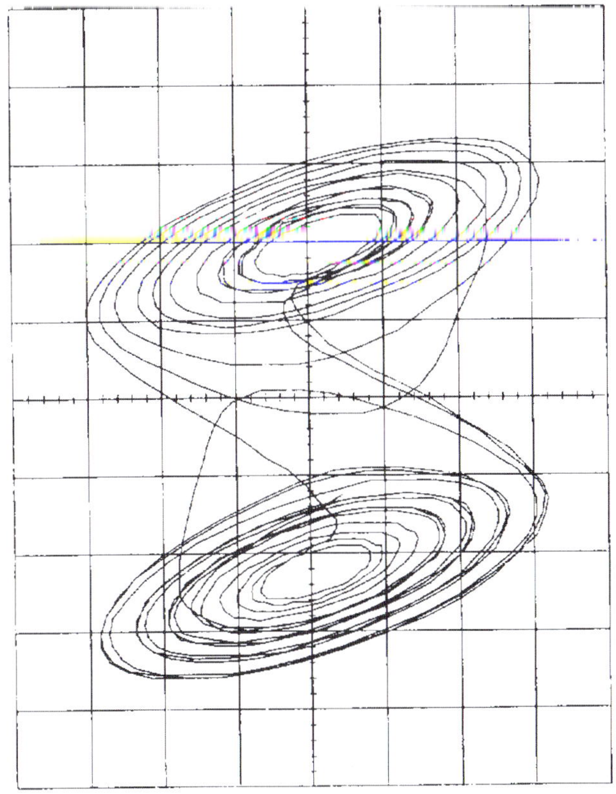

Fig. 12. Measured double scroll attractor in Kennedy's robust implementation of Chua's circuit with the parameters of Table 6 and $R = 1.52\,\mathrm{k\Omega}$ (reproduced from [15]). Horizontal axis: V_{C_2}; vertical axis: V_{C_1}.

Table 6. Component values for Kennedy's robust implementation of Chua's circuit [14].

Element	Description	Value
C_1	Capacitor	10 nF
C_2	Capacitor	100 nF
L	Inductor (TOKO type 10RB or equivalent)	18 mH
R	Potentiometer	2 kΩ
A_1, A_2	Op amp ($\frac{1}{2}$ AD712 or equivalent)	
R_1	$\frac{1}{4}$W Resistor	3.3 kΩ
R_2, R_3	$\frac{1}{4}$W Resistor	22 kΩ
R_4	$\frac{1}{4}$W Resistor	2.2 kΩ
R_5, R_6	$\frac{1}{4}$W Resistor	220 Ω

Table 7. Slopes and breakpoints of the Chua diode used by Kennedy in the robust implementation of Chua's circuit with $E_{sat} = 8.3$ V.

G_a (mS)	G_b (mS)	E (V)	E' (V)	G_c
−50/66	−9/22	1.08	7.61	101/22

3.2.2. *Current-mode negative impedance converters*

The performance of the op amp-based two-NIC implementation of the Chua diode is compromised by the relatively low bandwidth and slew rate limiting of the op amps. In order to improve the speed of the circuit, it is necessary to use components with higher bandwidth. One such circuit, shown in Fig. 14, uses current feedback op amps rather than traditional voltage op amps to realize the NICs [11]. The values of the dynamic elements are scaled by a factor of 10^{-2} compared to the values in Table 6, this makes the frequency of operation of this circuit 100 times higher.

In this case:

$$G_a = -\frac{1}{R_1} - \frac{1}{R_4} \qquad (15)$$

$$G_b = \frac{1}{R_3} - \frac{1}{R_4} \qquad (16)$$

$$G_c = \frac{1}{R_5} \qquad (17)$$

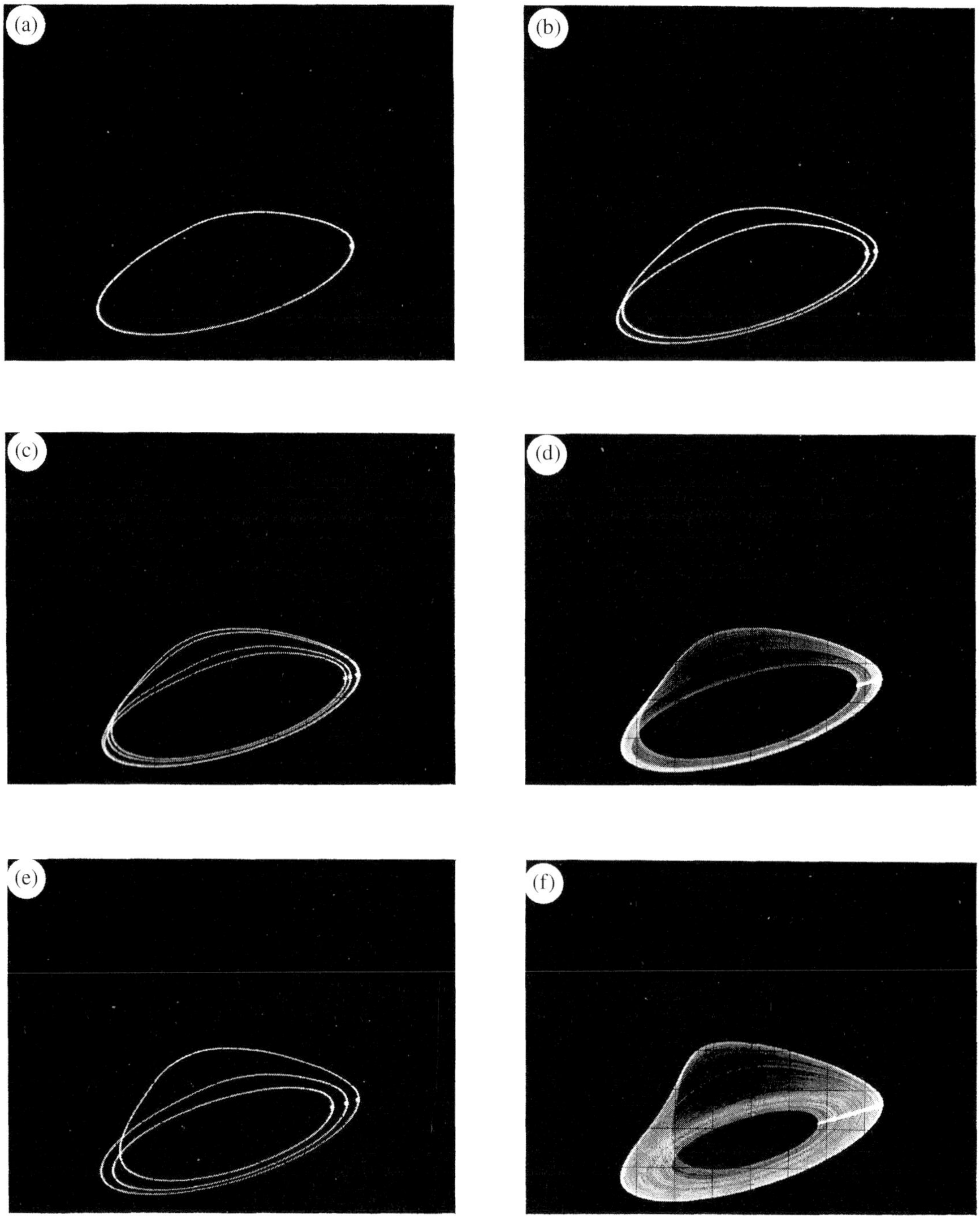

Fig. 13. Measured periodic and chaotic attractors in Kennedy's robust implementation of Chua's circuit with the parameters of Table 6 (reproduced from [17]). (a) Period-1 cycle; (b) period-2 cycle; (c) period-4 cycle; (d) spiral chaotic attractor; (e) period-3 cycle; (f) spiral chaotic attractor. Horizontal axis: V_{C_2}; vertical axis: V_{C_1}.

$$E = \frac{R_1}{R_1 + R_2} E_{sat} \tag{18}$$

$$E' = \frac{R_4}{R_4 + R_5} E_{sat} \tag{19}$$

With the component values shown in Table 8, the circuit produces the attractor shown in Fig. 15.

3.2.3. *Transistor-based negative impedance converter and diodes*

Another way to improve the bandwidth of the Chua diode is to replace the op amp-based NIC by a simpler transistor-based NIC. In 1986, Matsumoto *et al.* proposed an alternative implementation of the Chua diode, shown in Fig. 16, which uses

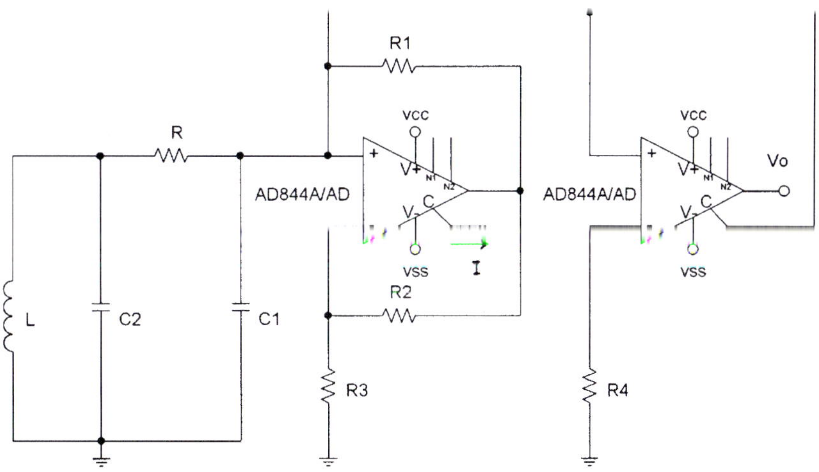

Fig. 14. High speed Chua's circuit using current feedback op amps in the two-NIC Chua diode (reproduced from [11]).

Table 8. Component values for Chua's circuit with current-mode two-NIC Chua diode.

Element	Description	Value
C_1	Capacitor	100 pF
C_2	Capacitor	1 nF
L	Inductor	180 μH
R	Potentiometer	1.6 kΩ
A_1, A_2	Op amp ($\frac{1}{2}$ AD844 or equivalent)	
R_1	$\frac{1}{4}$W Resistor	22 kΩ
R_2	$\frac{1}{4}$W Resistor	22 kΩ
R_3	$\frac{1}{4}$W Resistor	3.3 kΩ
R_4	$\frac{1}{4}$W Resistor	2.2 kΩ

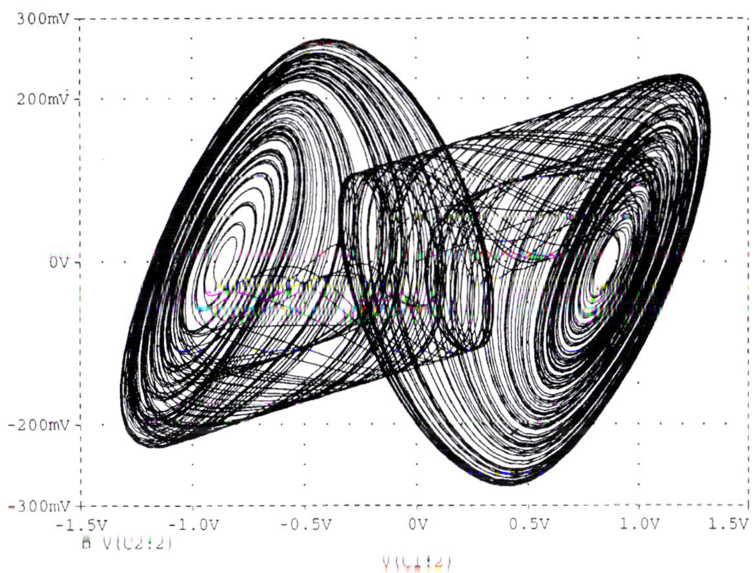

Fig. 15. PSpice simulation of V_{C_1}–V_{C_2} projection of chaotic attractor in the circuit of Fig. 14 with the parameter values in Table 8 (reproduced from [11]).

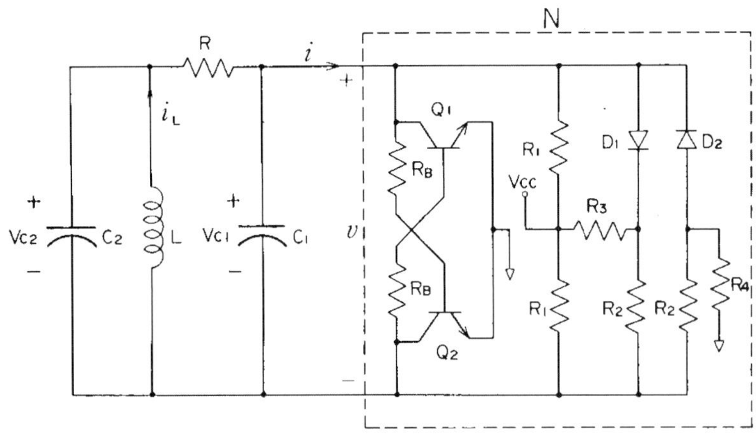

Fig. 16. The driving point characteristic of the nonlinear resistor N_R in Chua's circuit is realized using transistors, resistors, and diodes (reproduced from [23]).

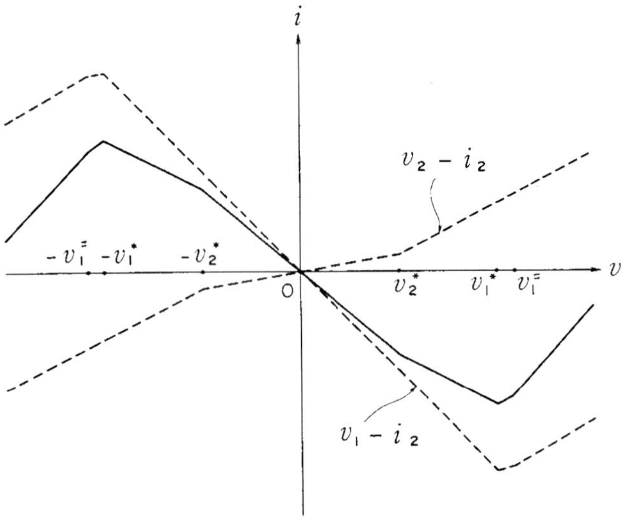

Fig. 17. The composite driving-point characteristic comprises two components due to the transistor-based NIC subcircuit (Q_1, Q_2, R_1 and R_B) and the diode and resistor subcircuit (D_1, D_2, R_2–R_4).

Table 9. Component value in the Chua's circuit of Fig. 16 with a transistor-based NIC.

Element	Description	Value
C_1	Capacitor	5.3 nF
C_2	Capacitor	47 nF
L	Inductor	6.8 mH
R	Potentiometer	1.21 kΩ
D_1, D_2	1S1588 Diode	
R_1	$\frac{1}{4}$W Resistor	1 kΩ
R_2	$\frac{1}{4}$W Resistor	3.3 kΩ
R_3	$\frac{1}{4}$W Resistor	88 kΩ
R_4	$\frac{1}{4}$W Resistor	39 kΩ
R_B	$\frac{1}{4}$W Resistor	56 kΩ
Q_1, Q_2	2SC1815 Transistor	

bipolar junction transistors to implement the NIC and diodes and resistors to set the breakpoints [23]. The transistor subcircuit consisting of Q_1, Q_2, R_1 and R_B implements a NIC. The positions of the two inner breakpoints are determined by the diodes D_1 and D_2 and the resistors R_2–R_4. The driving-point characteristic of the Chua diode is shown in Fig. 17.

With the component values in Table 9, the circuit produces the double scroll attractor shown in Fig. 18.

While this circuit works, it has a number of drawbacks. Firstly, the breakpoints depend on the thresholds of the diodes, which are not well-defined. Secondly, the diodes and transistors do not turn on abruptly, so the characteristic is not truly piecewise-linear.

Although Chua originally specified that the nonlinear resistor N_R should have a symmetric piecewise-linear driving-point characteristic, this is not strictly necessary [4]. All that is required is that the characteristic should produce at least two unstable equilibrium points with the appropriate stability properties when intersected by the load line of the linear resistor R. This can also be achieved by a cubic nonlinearity, as we shall discuss in Sec. 3.3.

3.3. *Cubic nonlinearity*

As mentioned in Sec. 3, the V–I characteristic of the nonlinear resistor N_R is required to intersect the

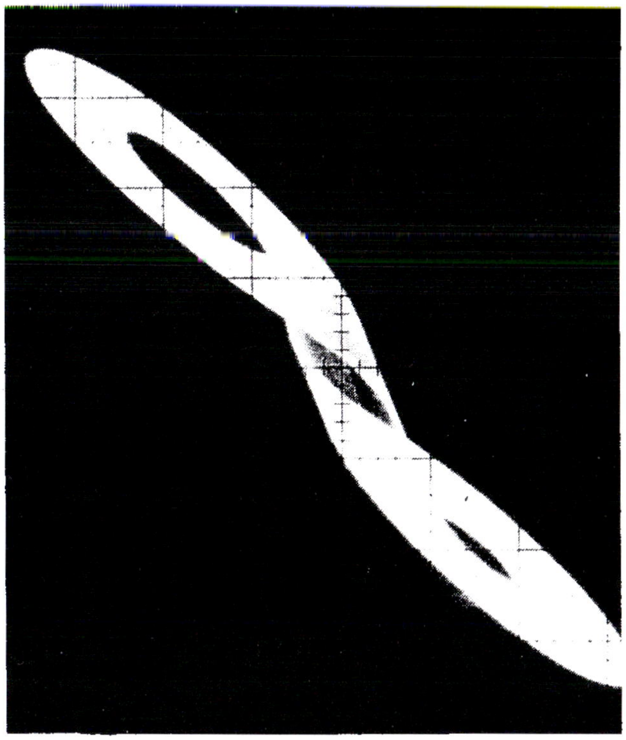

Fig. 18. I_L–V_{C_1} projection of chaotic attractor in the circuit of Fig. 16 with the parameter values in Table 9 (reproduced from [23]).

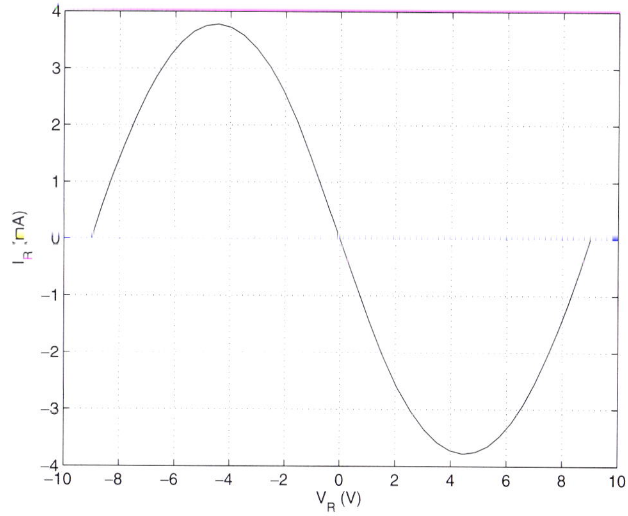

Fig. 19. Cubic driving-point characteristic with $V_{max} = 9$ V.

load line associated with the resistor R in at least two places. In addition, the resistor should be eventually passive, requiring that the characteristic lies in the first and third quadrants of the V–I plane. This can be achieved by means of a five-segment piecewise-linear characteristic as described thus far. It can also be realized with a cubic characteristic. In this case, the driving-point characteristic of N_R is defined by

$$f(V) = G_a V - \frac{G_a}{{V_{max}}^2} V^3 \tag{20}$$

where G_a is the slope of the characteristic at the origin and $\pm V_{max}$ are the outer zeroes, as shown in Fig. 19.

In 1985, Matsumoto *et al.* simulated a Chua's circuit with a cubic nonlinearity [22]. Their double scroll attractor is shown in Fig. 20. A decade later, Zhong presented an implementation of Chua's circuit using a cubic nonlinearity [26]. The nonlinear resistor, shown in Fig. 21, contains two analog multipliers, an op amp, and five resistors.

This implementation requires a significant amount of circuitry and suffers from limited speed of operation due to the compensation capacitor in the op amp. An approximately cubic nonlinearity can be realized more simply using four cross-coupled MOSFETs, as shown in Fig. 22 [25]. The measured V–I characteristic of this subcircuit is shown in Fig. 23.

Chua's circuit with a four-MOSFET approximately cubic nonlinear resistor is shown in Fig. 24. The resulting experimentally-measured double scroll attractor is shown in Fig. 25. This implementation of the nonlinear resistor has the advantage of very small parasitic dynamics, allowing it to behave resistively at relatively high frequencies. The circuit in Fig. 24 holds the record for the fastest reported Chua's circuit [25].

3.4. *Asymmetric nonlinearity*

Although most studies of Chua's circuit have considered the case when the nonlinear resistor has a symmetric driving-point characteristic, symmetry is not a necessary ingredient for complex behavior. In 1995, Kennedy studied the dynamics of Chua's circuit with a gross asymmetry and showed that the dynamics of the continuous two-region piecewise-linear Colpitts oscillator [16] may be mapped onto an equivalent asymmetric Chua's oscillator by choosing the breakpoints of an asymmetric Chua diode such that the resulting attractor is confined to two regions of the state space [18]. The dynamics of the circuits are equivalent in the sense that the vector field of the Colpitts oscillator is linearly conjugate to that of the asymmetric Chua's circuit.

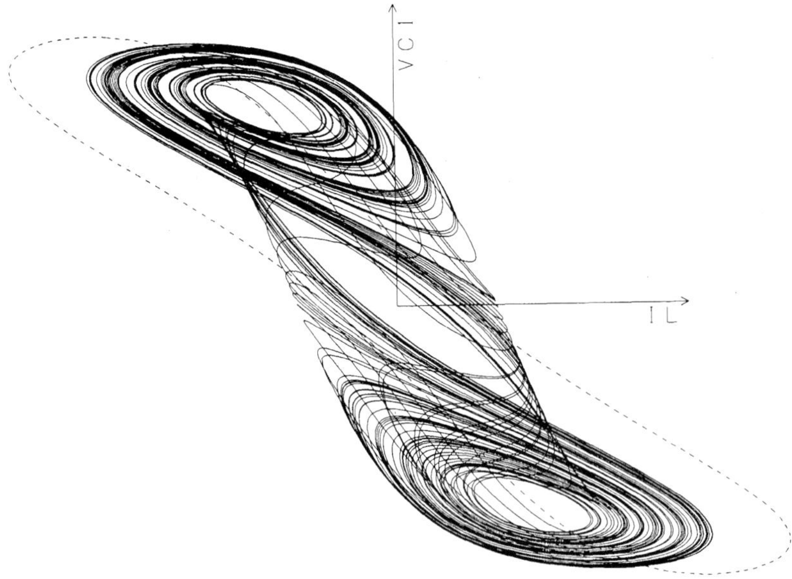

Fig. 20. I_L–V_{C_1} projection of chaotic attractor in Chua's circuit with a cubic nonlinearity (reproduced from [22]).

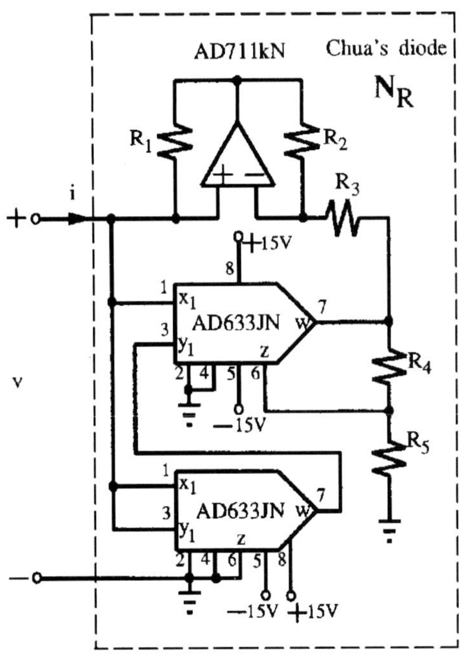

Fig. 21. Implementation of a nonlinear resistor with a cubic nonlinearity (reproduced from [26]).

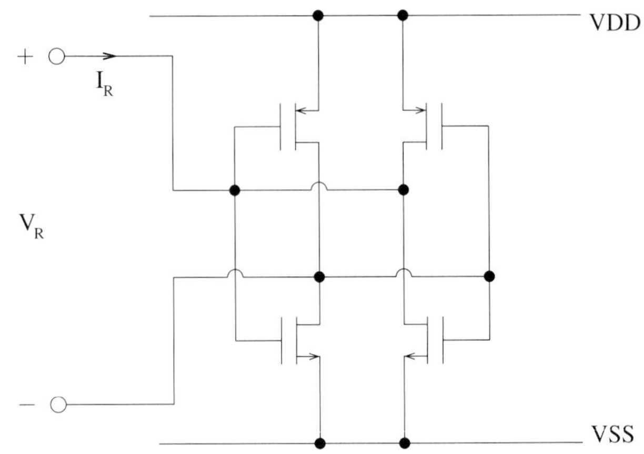

Fig. 22. Implementation of a nonlinear resistor with four cross-coupled MOSFETs (reproduced from [25]).

3.5. *Inductor*

The inductor in Chua's circuit can be realized by a physical inductor. Real inductors are wirewound components and have parasitic resistance associated with the windings of the coil. If this resistance is taken into account, then a physical realization of Chua's circuit using a real inductor has a resistor R_0 in series with the inductor. A problem with the early realizations of Chua's circuit was that the inductors had significant parasitic DC resistance which was not taken into account by the Chua's circuit equaitons. As a result, there was an unexplained difference between the simulated and experimentally measured behavior. The differences can be explained away by including the parasitic resistance in the simulations, namely by simulating Chua's oscillator rather than Chua's circuit equations. Chua's oscillator, shown in Fig. 7, is described by:

$$\frac{dV_{C_1}}{dt} = \frac{G}{C_1}(V_{C_2} - V_{C_1}) - \frac{1}{C_1}f(V_{C_1}) \tag{21}$$

$$\frac{dV_{C_2}}{dt} = \frac{1}{C_2}I_L - \frac{G}{C_2}(V_{C_2} - V_{C_1}) \tag{22}$$

$$\frac{dI_L}{dt} = -\frac{1}{L}V_{C_2} - \frac{R_0}{L}I_L \tag{23}$$

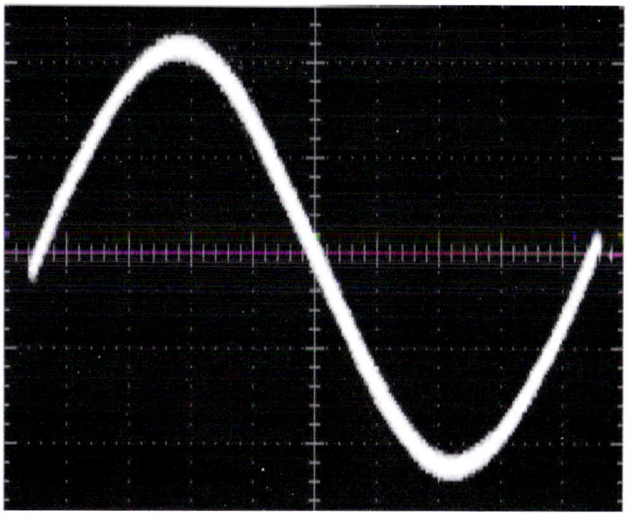

Fig. 23. Measured approximately cubic driving-point characteristic of the four-MOSFET Chua diode shown in Fig. 22 (reproduced from [25]).

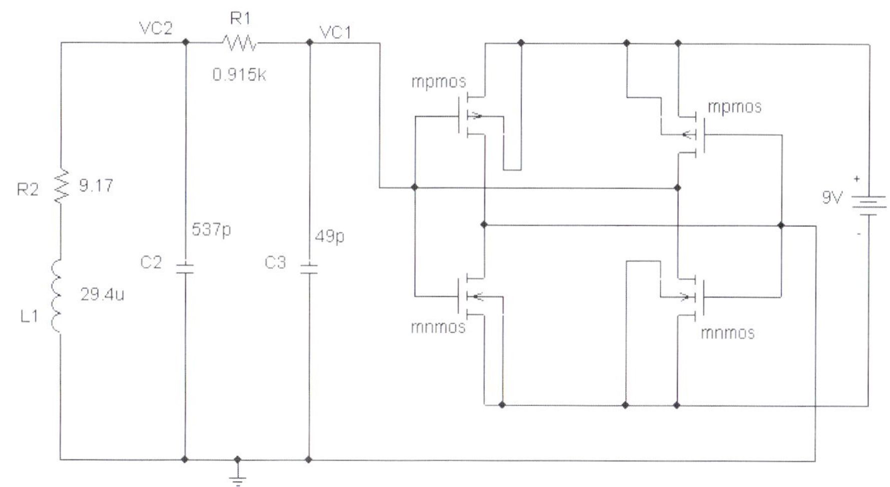

Fig. 24. Chua's circuit with a four-MOSFET approximately cubic nonlinear resistor (reproduced from [25]).

3.5.1. *Gyrator-based inductor*

In 1986, Prof. Leon Chua was preparing to give a keynote lecture entitled "The Sights and Sounds of Chaos" at a major international conference. He asked two of his doctoral students, Peter Kennedy and Ken Krieg, to build a portable system that would allow him to display trajectories from Chua's circuit on a large screen and simultaneously produce sound in the audio range. The solution was the Attractor Display System (ADS) — a system of galvanometers mounted at right angles that, when driven by X and Y signals from Chua's circuit, would deflect a laser beam, projecting a giant Lissajous figure on a distant screen. The problem was that the mirrors had to be operated open loop to obtain a sufficiently large image but their bandwidth was less than 500 Hz.

A Chua's circuit with a center frequency of several kHz could not be used. The challenge was to build a slow Chua's circuit. By scaling time, the sizes of the capacitances and inductances had to be increased. Finding a larger capacitance was relatively straightforward, but the required value of inductance would have necessitated a bulky inductor with a large parasitic resistance. The solution was to realize the inductance using a gyrator terminated with a capacitor. The resulting circuit is shown in Fig. 26. The diode clamp and the capacitor C_2 ensure that the inductor emulator operates correctly.

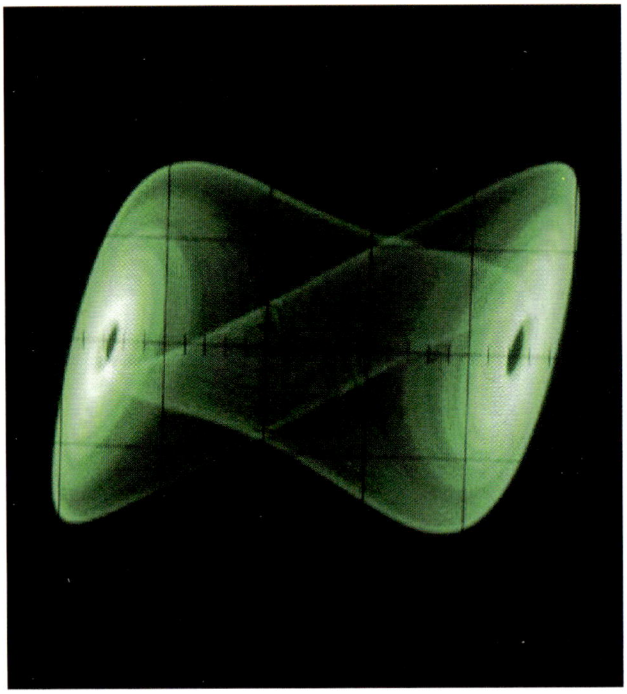

Fig. 25. Experimental double scroll attractor in Chua's circuit with an approximately cubic nonlinear resistor (reproduced from [25]).

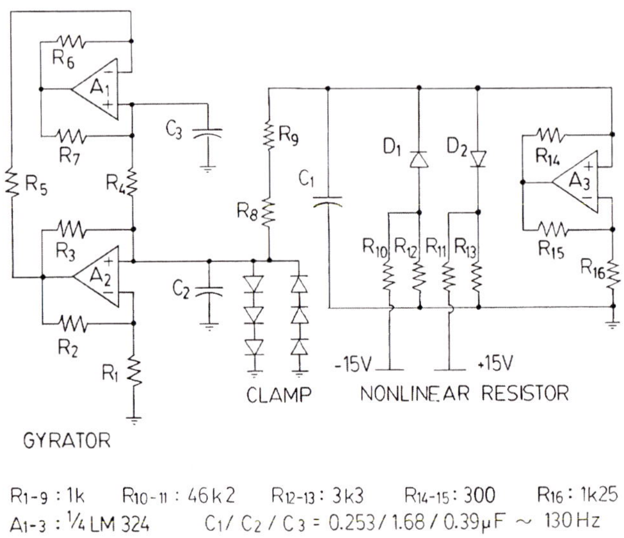

Fig. 26. Chua's circuit with a gyrator.

Two versions of the circuit with slightly different center frequencies were implemented in the ADS, with preset values of the resistors corresponding to period-1, period-2, spiral, double scroll and limit cycle attractors. Figure 27 shows the front and right side panels of the ADS. The front panel contains the controls for the two Chua's

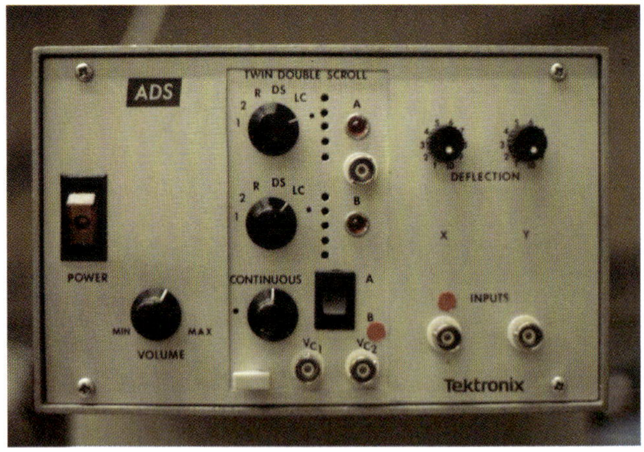

(a)

(b)

Fig. 27. (a) Front and (b) side panels of the Attractor Display System.

circuits. The left side panel contains a loudspeaker for the audio output; the right side panel has a perspex window through which the laser beam passes.

3.6. *Integrated circuit realization*

Cruz and Chua realized a Chua's circuit in integrated circuit form using a gyrator-based implementation of the inductor [9]. Figure 28 shows the circuit schematic. Operational Transconductance Ampliers OTA A and OTA B form the nonlinear resistor N_R. The gyrator comprising OTA C and OTA D terminated in capacitor C_3 emulates an inductance. Resistor R is connected externally between pins 1 and 2. The bias block between pins 4 and 5 provides the required bias voltage. Figure 29 shows a measured double scroll attractor in this circuit.

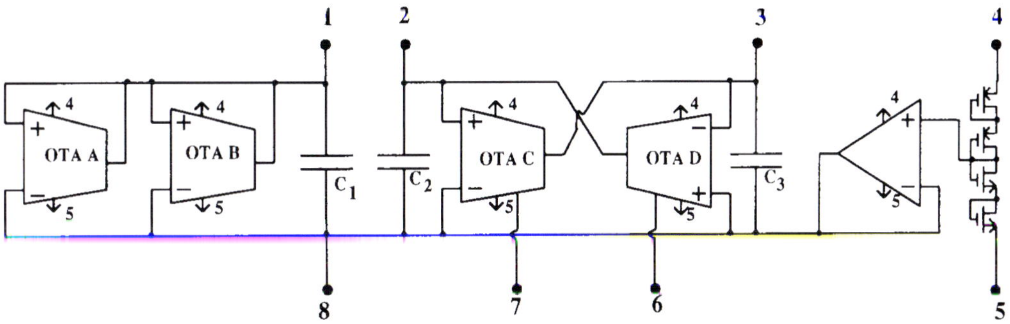

Fig. 28. Integrated circuit realization of Chua's circuit (reproduced from [9]).

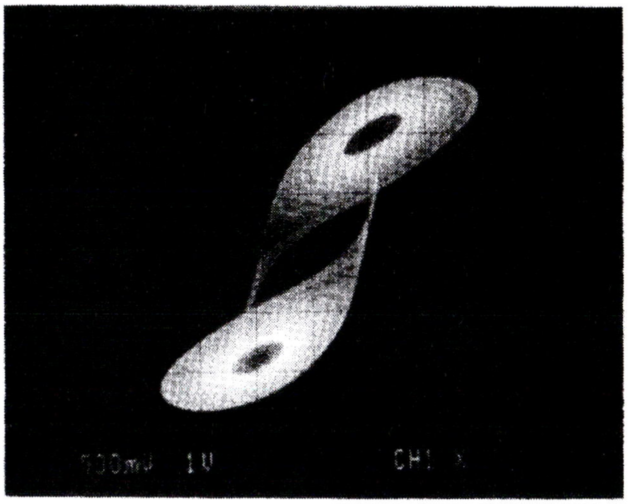

Fig. 29. Double scroll attractor in Cruz and Chua's integrated circuit realization of Chua's circuit (reproduced from [9]).

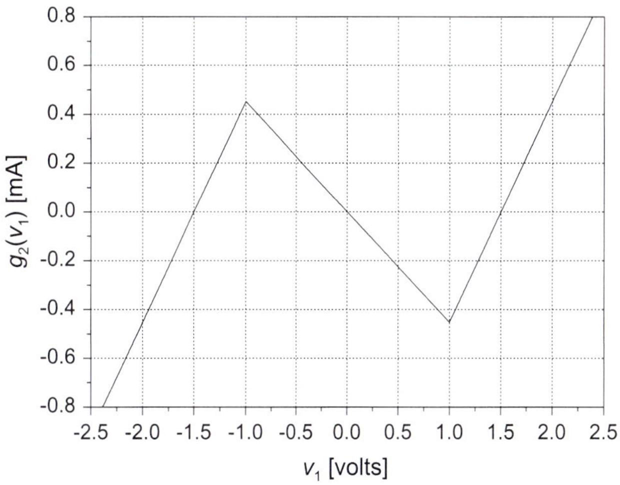

Fig. 31. DP characteristic in alternative Chua's circuit.

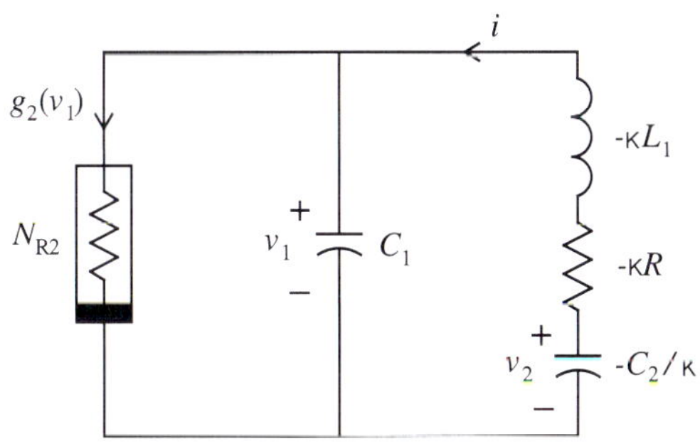

Fig. 30. Alternative configuration of the five-element Chua's circuit [2].

4. Four-element Chua's Circuit

The circuit shown in Fig. 30 contains the same circuit elements as the original five-element Chua's circuit, namely two capacitors, an inductor, a linear resistor and a piecewise-linear voltage-controlled nonlinear resistor. Three of the linear elements have negative values. Because of the difficulty in realizing negative circuit elements, this configuration, although proposed by Chua over two decades ago [4], was not implemented until 2007 [2].

The circuit in Fig. 30 is described by three ordinary differential equations:

$$\frac{dV_{C_1}}{dt} = \frac{1}{C_1} I_L - \frac{1}{C_1} f(V_{C_1}) \tag{24}$$

$$\frac{dV_{C_2}}{dt} = \frac{\kappa}{C_2} I_L \tag{25}$$

$$\frac{dI_L}{dt} = -\frac{R}{L} I_L - \frac{1}{\kappa L}(V_{C_2} - V_{C_1}) \tag{26}$$

where the voltage-controlled nonlinear resistor is the same Chua diode as in the original Chua's circuit, namely:

$$f(V) = \begin{cases} G_b V + (G_b - G_a)E & \text{if } V < -E \\ G_a V & \text{if } -E \le V \le E \\ G_b V + (G_a - G_b)E & \text{if } V > E \end{cases} \tag{27}$$

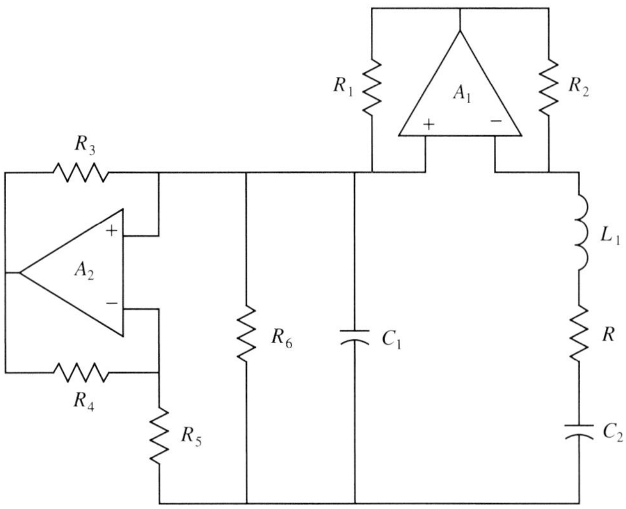

Fig. 32. Schematic realization of the physical implementation of the circuit shown in Fig. 30.

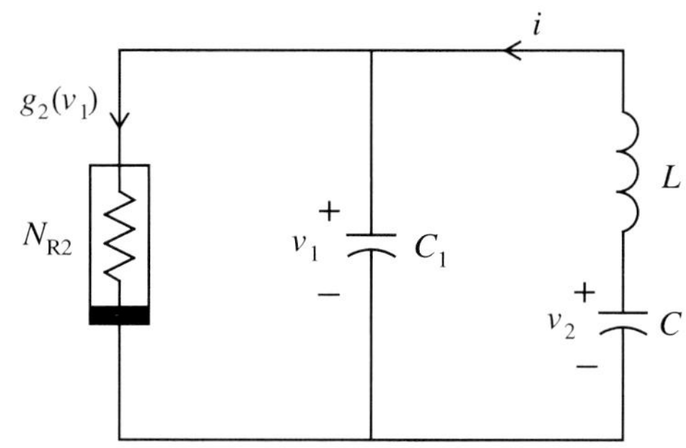

Fig. 33. A four-element equivalent Chua's circuit comprising a Chua diode, two capacitors and an inductor. Comparing to Fig. 30, $L = -\kappa L_1$ and $C = -C_2/\kappa$.

A practical realization of the circuit in Fig. 30 is shown in Fig. 32. Note that this implementation uses the same basic components as the original Chua's circuit shown in Fig. 9. Negative values of resistance, capacitance, and inductance are obtained by using a negative impedance converter consisting of the op amp A_1 and resistors R_1 and R_2. This subcircuit realizes:

$$\kappa = \frac{R_1}{R_2}. \tag{28}$$

The op amp A_2 and resistors R_3–R_6 implement the Chua diode, whose slopes and breakpoints are defined by:

$$G_a = -\frac{R_4}{R_3 R_5} + \frac{1}{R_6} \tag{29}$$

$$G_b = \frac{1}{R_3} + \frac{1}{R_6} \tag{30}$$

$$E = \frac{R_5}{R_4 + R_5} E_{sat}, \tag{31}$$

respectively. In the case of an LF351 op amp with $\pm$ 15 V power supplies, $E_{sat} \approx 13.5$ V.

A remarkable property of this circuit is that it can produce chaos even when $R = 0$. Thus, Chua's circuit is reduced to its simplest form, corresponding to a lossless LC network connected to a Chua diode, as shown in Fig. 33.

Table 10. Component values in the four-element Chua's circuit.

Element	Description	Value
C_1	Capacitor	33 nF
C_2	Capacitor	100 nF
L	Inductor	10 mH
R	$\frac{1}{4}$W Resistor	25 Ω, 0Ω
A_1, A_2	Op amp (LF351)	
R_1	$\frac{1}{4}$W Resistor	1 kΩ
R_2	$\frac{1}{4}$W Potentiometer	122 Ω
R_3	$\frac{1}{4}$W Resistor	10 kΩ
R_4	$\frac{1}{4}$W Resistor	9.8 kΩ
R_5	$\frac{1}{4}$W Resistor	780 Ω
R_6	$\frac{1}{4}$W Resistor	1.2 kΩ

Table 11. Parameter values used by Barboza and Chua in the four-element Chua's circuit [2].

G_a (mS)	G_b (mS)	E (V)
−0.42	0.93	1

Using the component values in Table 10, Barboza and Chua measured attractors in the alternative Chua's circuit configuration shown in Fig. 30 and the four-element circuit shown in Fig. 33 [2]. The corresponding values of the circuit parameters are summarized in Table 11; $R = 25\ \Omega$ in the first case and $R = 0\ \Omega$ in the second.

Figure 34 shows experimentally-observed chaotic attractors in the alternative Chua circuit and the four-element circuit, repectively [2].

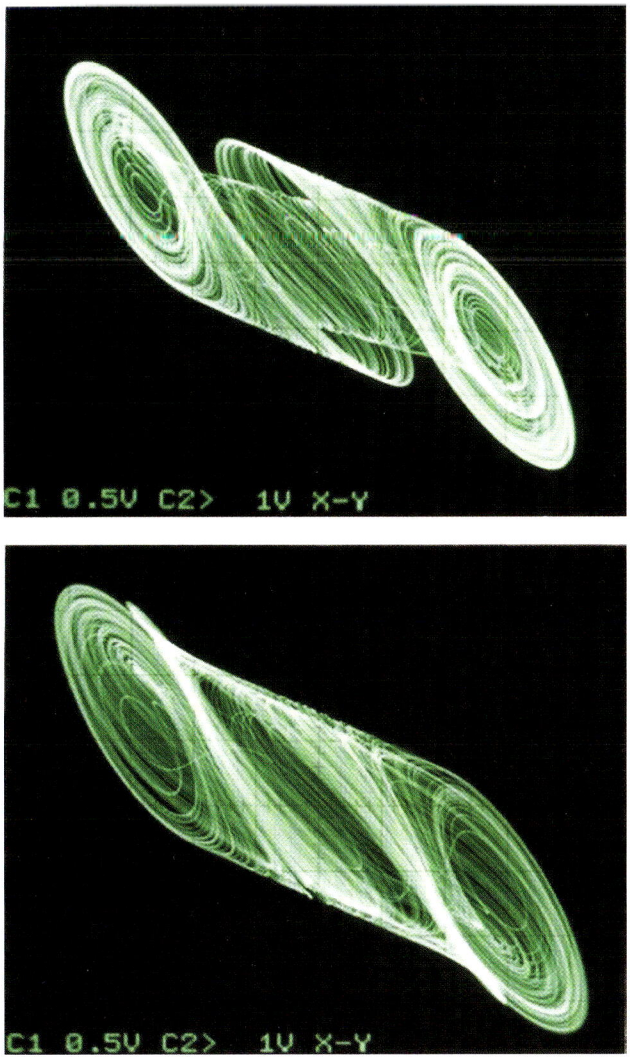

Fig. 34. Observed trajectories in the circuit of Fig. 32 with (top) $R = 25\ \Omega$ and (bottom) $R = 0\ \Omega$. These correspond to the alternative Chua's circuit (Fig. 30) and the four-element Chua's circuit (Fig. 33), respectively (reproduced from [2]).

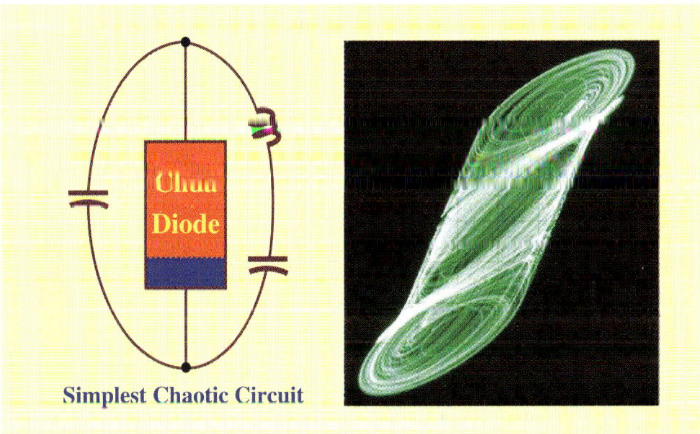

Fig. 35. The simplest Chua's circuit and its typical attractor.

5. Three-element Chaotic Circuit

We stated at the outset that, in order to exhibit chaos, an autonomous circuit consisting of resistors, capacitors, and inductors must contain

- at least one nonlinear element,
- at least one locally active resistor, and
- at least three energy storage elements [15].

The four-element Chua's circuit contains three energy storage elements and a locally active nonlinear resistor. A memristor can exhibit simultaneously the characteristics of a nonlinear element, a locally active resistor, and an energy storage element. Therefore, it should be possible to build a chaotic circuit using just *three* elements: a memristor and two energy storage elements. Muthuswamy and Chua reported this in 2010 [24]. Their "simplest chaotic circuit" is shown in Fig. 36.

This circuit is described by *two* ordinary differential equations:

$$\frac{dV_C}{dt} = \frac{1}{C} I_L \tag{32}$$

$$\frac{dI_L}{dt} = -\frac{1}{L}(V_C - V_M) \tag{33}$$

where V_M is the voltage across the memristor M, which is defined by:

$$V_M = \beta(X^2 - 1) I_M \tag{34}$$

$$\frac{dX}{dt} = I_M - \alpha X - I_M X. \tag{35}$$

The *memristance* $R(X)$ is given by:

$$R(X) = \beta(X^2 - 1). \tag{36}$$

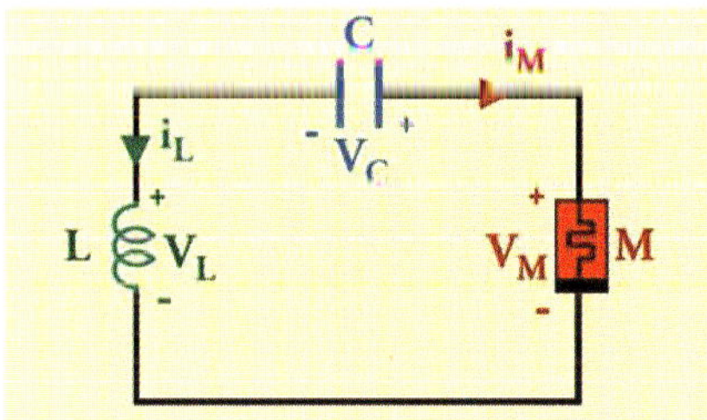

Fig. 36. Muthuswamy and Chua's "simplest chaotic circuit" contains two linear dynamic elements and a memristor (reproduced from [24]).

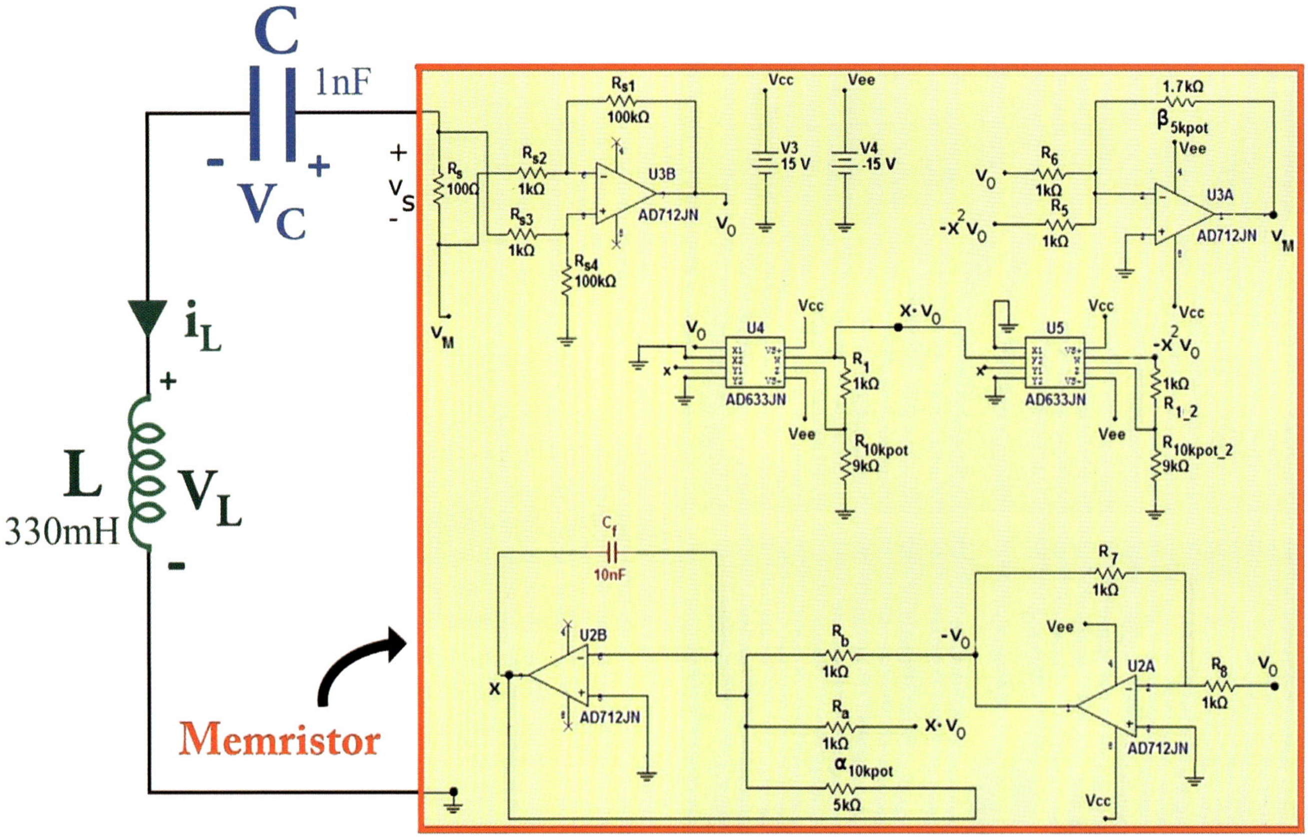

Fig. 37. Schematic of the three-element memristor-based chaotic circuit (reproduced from [24]).

Substituting (34) into (33) and recognizing that $I_M = -I_L$, one can show that the three-element chaotic circuit is described by *three* ordinary differential equations:

$$\frac{dV_C}{dt} = \frac{1}{C} I_L \tag{37}$$

$$\frac{dI_L}{dt} = -\frac{1}{L}(V_C + \beta(X^2 - 1)I_L) \tag{38}$$

$$\frac{dX}{dt} = -I_L - \alpha X + I_L X. \tag{39}$$

A practical realization of the circuit in Fig. 36 is shown in Fig. 37. Note that the memristor is emulated by an analog computer in this case. The op amp and analog multiplier subcircuits implement the memristance $R(X)$; the state variable X is stored on the capacitor C_f.

Using the component values indicated in Fig. 37, Muthuswamy and Chua measured a chaotic attractor in the circuit shown in Fig. 36 [24]. The corresponding values of the circuit parameters are summarized in Table 12. Figure 38 shows simulated and measured chaotic attractors in Muthuswamy and Chua's "simplest chaotic circuit."

Table 12. Parameter values used by Muthuswamy and Chua in the three-element chaotic circuit.

C (nF)	L (mH)	α	β
1	330	0.2	1.7

6. Summary

In this chapter, we have described six-, five-, four- and three-element chaotic circuits. Chua's original five-element design contained two linear capacitors, a linear inductor, a linear resistor, and a symmetric piecewise-linear voltage-controlled nonlinear resistor N_R. The circuit was first simulated by Matsumoto [21] and then realized experimentally by Zhong and Ayrom [27]. Chua *et al.* established the universality of the circuit by showing that the augmented six-element Chua's oscillator can reproduce all know dynamical behavior reported for third-order autonomous systems [7,8].

Elwakil and Kennedy increased the frequency of operation of Chua's circuit by a factor of 100 by replacing the voltage feedback op amps by current feedback op amps in the two-NIC realization

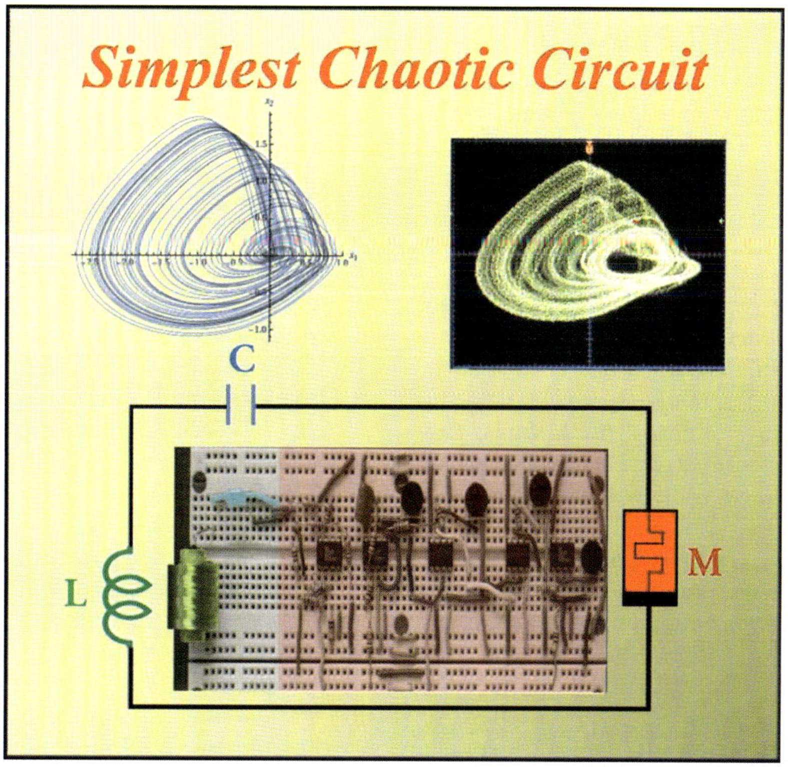

Fig. 38. Muthusawamy and Chua's simplest chaotic circuit and its typical attractor (reproduced from [24]).

of the Chua diode [11]. Kennedy also reduced the frequency of operation by replacing the inductor with a gyrator and capacitor; this is the slowest reported Chua's circuit. Cruz and Chua used the same gyrator-based structure in the first integrated circuit implementation of Chua's circuit [9].

In addition to the piecewise-linear Chua diode, Matsumoto *et al.* also simulated the circuit with a cubic nonlinearity [22]. A Chua's circuit with a cubic nonlinearity was first implemented by Zhong [26], and later simplified by O'Donoghue *et al.* to produce the fastest reported Chua's circuit [25]. Kennedy showed that the Colpitts oscillator with a piecewise-linear nonlinearity is equivalent to Chua's circuit with an asymmetric Chua diode [18].

Returning to a circuit topology first proposed by Chua, Barboza and Chua reported chaos in a simplified four-element version of Chua's circuit [2]; this contains two linear capacitors, a linear inductor, and a Chua diode. Muthuswamy and Chua showed that chaos could be produced in an autonomous circuit containing just three circuit elements: a linear capacitor, a linear inductor, and a memristor [24].

Key milestones in the development of Chua's circuit and its derivatives are summarized in Table 13.

Table 13. Milestones in the development of Chua's circuit.

1983	Chua invents five-element Chua's circuit [4]
1984	Matsumoto simulates Chua's circuit with PWL Chua diode [21]
1984	Zhong and Ayrom realize Chua's circuit in hardware [27]
1984	Matsumoto *et al.* simulate Chua's circuit with cubic nonlinearity [22]
1986	Kennedy implements "slow" Chua's circuit with gyrator-based inductor
1992	Kennedy builds robust Chua's circuit [14]
1993	Cruz and Chua report IC implementation of Chua's circuit [9]
1994	Zhong implements Chua's circuit with cubic nonlinearity [26]
1995	Kennedy demonstrates equivalence of Chua's circuit and Colpitts oscillator [18]
2004	O'Donoghue *et al.* build "fast" Chua's circuit with cubic nonlinearity [25]
2007	Barboza and Chua report four-element Chua's circuit [2]
2009	Muthuswamy and Chua build a three-element memristor-based chaotic circuit [24]

References

[1] F. Ayrom. Personal communication, October 2011.
[2] R. Barboza and L.O. Chua. The four-element Chua's circuit. *Int. J. Bif. Chaos*, 18(4):943–955, April 2008.

[3] L.O. Chua, M. Komuro, and T. Matsumoto. The double scroll family. *IEEE Trans. Circuits and Syst.*, CAS-33(11):1073–1118, November 1986.

[4] L.O. Chua. The genesis of Chua's circuit. *AEÜ*, 46(4):250–257, April 2008.

[5] L.O. Chua and F. Ayrom. Designing non-linear single op-amp circuits: A cookbook approach. *Int. J. Circuit Theory Appl.*, 13:235–268, July–September 1985.

[6] L.O. Chua, C.A. Desoer, and E.S. Kuh. *Linear and Nonlinear Circuits*. McGraw-Hill, 1987.

[7] L.O. Chua, C.W. Wu, A. Huang, and G.-Q. Zhong. A universal circuit for studying and generating chaos — part i: Routes to chaos. *IEEE Trans. Circuits and Systems — Part I: Fundamental Theory and Applications*, CAS-40(10):732–744, October 1993.

[8] L.O. Chua, C.W. Wu, A. Huang, and G.-Q. Zhong. A universal circuit for studying and generating chaos — part ii: Strange attractors. *IEEE Trans. Circuits and Systems — Part I: Fundamental Theory and Applications*, CAS-40(10):745–761, October 1993.

[9] J.M. Cruz and L.O. Chua. An IC chip of Chua's circuit. *IEEE Trans. Circuits and Systems — Part II: Analog and Digital Signal Processing*, 40(10):614–625, October 1993.

[10] H. Dedieu, M.P. Kennedy, and M. Hasler. Chaos shift keying: Modulation and demodulation of a chaotic carrier using self-synchronizing Chua's circuits. *IEEE Trans. Circuits and Systems — Part II: Analog and Digital Signal Processing, Special Issue on Chaos in Nonlinear Electronic Circuits — Part C: Applications*, 40(10):634–642, October 1993.

[11] A.S. Elwakil and M.P. Kennedy. Improved implementation of Chua's chaotic oscillator using current feedback op amp. *IEEE Trans. Circuits and Systems — Part I: Fundamental Theory and Applications*, CAS-47(1):76–79, January 2000.

[12] Z. Galias, C.A. Murphy, M.P. Kennedy, and M.J. Ogorzałek. Electronic chaos controller. *Chaos, Solitons & Fractals*, 8(9):1471–1484, 1997.

[13] M.P. Kennedy. Synthesis of continuous three-segment voltage-controlled piecewise-linear resistors for Chua's circuit family using operational amplifiers, diodes, and linear resistors. *Int. J. Circuit Theory Appl.*, 21(6):551–558, 1993.

[14] M.P. Kennedy. Robust op amp realization of Chua's circuit. *Frequenz*, 46(3–4):66–80, March–April 1992.

[15] M.P. Kennedy. Three steps to chaos part II: A Chua's circuit primer. *IEEE Trans. Circuits and Systems — Part I: Fundamental Theory and Applications, Special Issue on Chaos in Nonlinear Electronic Circuits — Part A: Tutorial and Reviews*, 40(10):657–674, October 1993.

[16] M.P. Kennedy. Chaos in the Colpitts oscillator. *IEEE Trans. Circuits and Systems — Part I: Fundamental Theory and Applications*, 41(11):771–774, November 1994.

[17] M.P. Kennedy. Experimental chaos from autonomous electronic circuits. *Phil. Trans. R. Soc. Lond. A*, 353(1701):13–32, 16 October 1995.

[18] M.P. Kennedy. On the relationship between the chaotic Colpitts oscillator and Chua's oscillator. *IEEE Trans. Circuits and Systems — Part I: Fundamental Theory and Applications*, CAS-42(6):376–379, June 1995.

[19] M.P. Kennedy, C.W. Wu, S. Pau, and J. Tow. Digital signal processor-based investigation of Chua's circuit family. *J. Circuits Syst. Computers*, 3(2):269–292, June 1993.

[20] R.N. Madan, editor. *Chua's Circuit: A Paradigm for Chaos*. World Scientific, Singapore, 1993.

[21] T. Matsumoto. A chaotic attractor from Chua's circuit. *IEEE Trans. Circuits and Syst.*, CAS-31(12):1055–1058, December 1984.

[22] T. Matsumoto, L.O. Chua, and M. Komuro. The double scroll. *IEEE Trans. Circuits and Syst.*, CAS-32(8):797–819, August 1985.

[23] T. Matsumoto, L.O. Chua, and K. Tokumasu. Double scroll via a two-transistor circuit. *IEEE Trans. Circuits and Syst.*, CAS-38(8):828–835, August 1986.

[24] B. Muthuswamy and L.O. Chua. Simplest chaotic circuit. *Int. J. Bif. Chaos*, 20(5):1567–1580, May 2010.

[25] K. O'Donoghue, M.P. Kennedy, P. Forbes, M. Qu, and S. Jones. A fast and simple implementation of Chua's oscillator with cubic-like nonlinearity. *Int. J. Bif. Chaos*, 15(9):2959–2971, September 2005.

[26] G.-Q. Zhong. Implementation of Chua's circuit with a cubic nonlinearity. *IEEE Trans. Circuits and Syst.*, CAS-41(12):934–941, December 1994.

[27] G.-Q. Zhong and F. Ayrom. Experimental confirmation of chaos from Chua's circuit. *Int. J. Circuit Theory Appl.*, 13:93–98, January 1985.

[28] G.-Q. Zhong and F. Ayrom. Periodicity and chaos in Chua's circuit. *IEEE Trans. Circuits and Syst.*, CAS-32(5):501–503, May 1985.

CHAPTER 2

IMPASSE POINTS, MUTATORS, AND OTHER CHUA CREATIONS

HYONGSUK KIM

Division of Electronics Engineering, Chonbuk National University, Korea

Most of the papers featured in this Festschrift are related to Professor Chua's more recent and active research areas, such as Cellular Neural/Nonlinear networks, Chua's circuit, and memristors. This chapter will attempt to identify some of his earlier works, ideas, and creations which are just as significant and in fact timeless.

1. Introduction

There is no formal "nonlinear circuit theory" until 1969 when Chua published what many researchers now regarded as a classic book entitled "Introduction to Nonlinear Network Theory" in 1969 [3].[a] This 949-page monumental book formally defines what is a nonlinear resistor, a nonlinear inductor, and a nonlinear capacitor, and develops the fundamental concepts, tools, and even symbols for developing a consistent, rigorous, yet elementary axiomatic theory of nonlinear circuits. For example, prior to Chua's book, standard textbooks on linear circuit analysis routinely treated the "controlled sources" just like independent voltage and current sources, thereby propagating numerous misconceptions. In [3], Chua showed that controlled sources are 2-ports that behave quite differently from independent sources, and should be distinguished by a symbol different from the traditional "small circles". The "diamond" symbol Chua proposed for controlled sources has since been adopted worldwide in elementary circuit theory textbooks.

Several fundamental concepts and tools first introduced and carefully illustrated in [3] have now been routinely applied. They include the concept of driving-point characteristic (DP plot), transfer characteristic (TC plot), and piecewise-linear analysis. The graphical method for deriving the DP and

[a] As a testimony of this statement, the reader will not find any used copy of Chua's book via Amazon.com, nor via any standard source. No one seems to be willing to let go his copy of [3]. This is as timeless a book as it can be!

TC plots of any 1-port made of an arbitrary series and parallel interconnection of 2-terminal nonlinear resistors have now been widely taught and exploited for its pedagogical and intuitive appeals. For example, the fundamental theorem in CNN which allows one to understand how a CNN template could give rise to a prescribed binary CNN cell is based entirely on the DP plot and piecewise-linear method developed in [3].

The above cited nonlinear circuit concepts and tools are all of original Chua vintage.

Another truly original and extremely powerful concept first developed in [3] is the concept of the "dynamic route", which has been exploited in numerous areas, including the CNN, and 1-dimensional cellular automata [16–20]. This powerful tool allows one to determine the dynamic behavior of any nonlinear circuit or system described by a first-order nonlinear differential equation. All nonlinear circuits containing one capacitor, or one inductor, and any interconnection of nonlinear resistors belong to this class. For example, both the CNN "cell" [15], and the basic "cell" for executing Wolfram's 1D cellular automata [16–20] belong to this class.

Another original and powerful contribution of the Chua's classic work [3] is the design and synthesis of nonlinear circuits with prescribed characteristics, using off-the-shelf circuit components. For example, it includes designing a 1-port with a prescribed DP plot, or a 2-port with a prescribed TC plot. Many practical large-signal electronic circuits can be understood, and designed via the principles developed in [3]. They include rectifiers, limiters, voltage compensating circuits, voltage regulators, flip flops, multivibrators, harmonic generators, frequency converters, wave-shaping circuits, etc. Indeed, virtually all practical nonlinear electronic circuits, circa 1969, were unified and explained via the concepts, techniques and tools developed in [3]. What is truly amazing about this seminal book on nonlinear circuit theory is that it has not aged, since the concepts treated are fundamentals, and fundamental concepts never die.

2. Impasse Points

Before Chua's 1964 Ph.D. thesis on nonlinear circuits, no one knew that, unlike "linear circuit theory" where it is always possible to write a system of n linear differential equations (called state

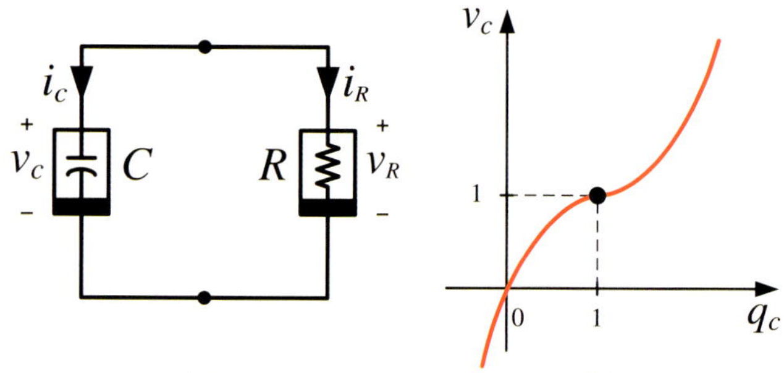

(a) (b)

Fig. 1. A 2-element RC network Л.

equations) for non-degenerate linear circuit, namely,

$$\frac{d\boldsymbol{x}}{dt} = A\boldsymbol{x} + B\boldsymbol{u}, \tag{1}$$

where $\boldsymbol{x}$ is an n-dimensional "state vector", $\boldsymbol{u}$ is an m dimensional "input vector", many nonlinear circuits do not have a system of continuously-differentiable state equations

$$\frac{d\boldsymbol{x}}{dt} = f(\boldsymbol{x}, \boldsymbol{u}) \tag{2}$$

where $f(\cdot)$ is a continuously-differentiable function of $\boldsymbol{x}$. This is because $f(\boldsymbol{x}, \boldsymbol{u})$ may be undefined (i.e., becomes infinite) at some isolated points $\boldsymbol{x}$, or $f(\boldsymbol{x}, \boldsymbol{u})$ may be as multiple values over some ranges of $\boldsymbol{x}$. To illustrate these subtle phenomena, we reproduce Examples 1, 2, 3, and 4 from pages 1060–1062 of [10] as follow:

Example 1. Consider the nonlinear RC network Л shown in Fig. 1(a). Let the element constitutive relations be described as follows:

$$\textit{Resistor } R: i_R = h(v_R), \quad v_R \in \mathrm{R} \tag{3}$$

$$\textit{Capacitor } C: v_R = g(q_C), \quad q_C \in \mathrm{R}. \tag{4}$$

If we choose the capacitor charge q_C as state variable, the state equation is given by

$$\dot{q}_C = -h(g(q_C)) \stackrel{\Delta}{=} f(q_C), \quad q_C \in \mathrm{R}. \tag{5}$$

Now suppose $g(\cdot)$ in (4) is *bijective* so that its inverse

$$q_C = \hat{q}_C(v_C), \quad v_C \in \mathrm{R} \tag{6}$$

exists. Then applying KCL and the chain rule, we obtain

$$C(v_C)\dot{v}_C = -h(v_C), \quad v_C \in \mathrm{R} \tag{7}$$

where $C(v_C) \stackrel{\Delta}{=} d\hat{q}_C(v_C)/dv_C$ is the incremental capacitance. Assuming that $C(v_C) \neq 0$ for all v_C, then (5) can be rewritten as

$$\dot{v}_C = -C^{-1}(v_C)h(v_C) \stackrel{\Delta}{=} f(v_C), \quad v_C \in \mathrm{R}. \tag{8}$$

Hence, (6) is also a well-defined state equation with the capacitor voltage v_C as the state variable, provided

$$C(v_C) \neq 0 \quad \textit{is defined for all} \quad v_C \in \mathrm{R}. \tag{9}$$

To illustrate the importance of (7), let R be a 1-Ω linear resistor and let C be described by [21]:

$$v_C = g(q_c) \stackrel{\Delta}{=} (q_c - 1)^3 + 1 \tag{10}$$

as shown in Fig. 1(b). Then (3) assumes the form

$$\dot{q}_C = -(q_C - 1)^3 - 1 \stackrel{\Delta}{=} f(q_C), \quad q_C \in \mathrm{R}. \tag{11}$$

Note that $g(\cdot)$ in (8) is bijective and has a continuous inverse function

$$q_C = \hat{q}_C(v_C) = 1 + (v_C - 1)^{1/3}. \tag{12}$$

Its incremental capacitance is given by

$$C(v_C) = \frac{1}{3}(v_C - 1)^{-2/3} \neq 0, \quad \textit{for all}, v_C \in \mathrm{R}. \tag{13}$$

Substituting (13) into (6) we obtain

$$\dot{v}_C = -3(v_C - 1)^{2/3} v_C \stackrel{\Delta}{=} f(v_C), \quad v_C \in \mathrm{R}. \tag{14}$$

Note that $f(v_C)$ is a continuously differentiable function which vanishes at $v_C = 1$. Hence $v_C = 1$ is a well-defined equilibrium point of (11) [3].

Observe, however, that $v_C = v_R = 1$ implies $i_C = -i_R = -1 \neq 0$. Hence we have arrived at a paradoxical situation where the capacitor current is not zero at equilibrium!

To resolve this paradox, note that $C(v_C)$ in (13) is undefined at $v_C = 1$; namely, $C(1) = \infty$. Consequently, (7) is violated. It follows, therefore, that the state equation which correctly describes Л is not (11) but rather (12):

$$\dot{v}_C = -3(v_C - 1)^{2/3} v_C \stackrel{\Delta}{=} f(v_C), \quad v_C \in \mathrm{R}., \quad v_C \neq 1. \tag{15}$$

The almost trivial distinction between (11) and (12) turns out to be crucially important in this example. Indeed, (11) can be interpreted as the state equation describing an infinite number of distinct networks (all sharing the same topology as Fig. 1) all of which differ from that defined by (8); namely; redefine R and C in Fig. 1(a) as follows:

$$\textit{Resistor}: i_R = 3C_O(v_R - 1)^{2/3} v_R \tag{16}$$

$$\textit{Capacitor}: q_C = C_O v_C \tag{17}$$

where C_O is any real number. Observe that at the equilibrium point $v_C = v_R = 1$, (16) implies $i_C = -i_R = 0$, as it should.

This example shows that while each network has a unique state equation (if it exists) with respect to a given state variable, each state equation represents infinitely many distinct networks.

The above example also demonstrates that even if a capacitor {resp., inductor} constitutive relation is *bijective*, it would still be preferable to choose the capacitor charge {resp., inductor flux} as a state variable. Such a choice not only obviates the need for checking additional conditions (such as (7)), but it offers certain additional numerical advantages [6: pp. 43–45, 431–432] over the capacitor voltage {resp:, inductor current}.

Example 2. Let R in Fig. 2(a) be a 1-Ω linear resistor and let C be a *charge-controlled capacitor*[b] described by

$$v_C = \hat{v}_C(q_C) \stackrel{\Delta}{=} \frac{1}{3}q_C^3 - 3q_C^2 + 8q_C, \quad q_C \in \mathrm{R} \tag{18}$$

as shown in Fig. 2(b). Then state equation (3) reduces to

$$\dot{q}_C = -\left[\frac{1}{3}q_C^3 - 3q_C^2 + 8q_C\right] \stackrel{\Delta}{=} f(q_C), \quad q_C \in \mathrm{R}. \tag{19}$$

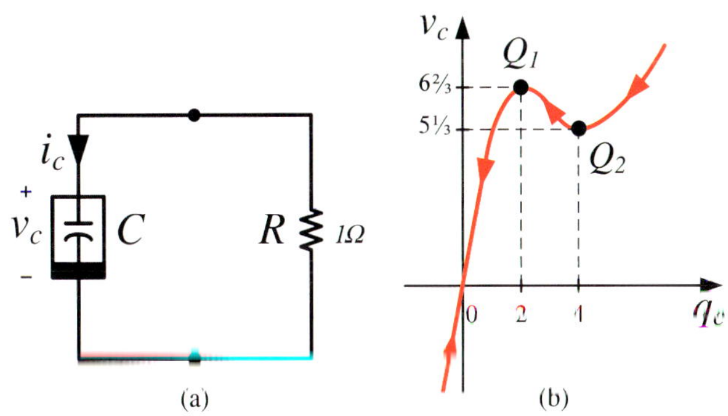

Fig. 2. The $q_C - v_C$ curve for charge-controlled capacitor C of Example 2. Here q_C is a global coordinate for this curve. The capacitor voltage v_C does not qualify as a global coordinate. Nor does it qualify as a "local" coordinate at Q_1 or Q_2.

[b] An $(n+1)$-terminal (or n-port) element described by $y = f(x)$, $x \in \mathrm{R}^n$ is said to be x-controlled if f is a continuous function for all $x \in \mathrm{R}^n$. In other words, "x-controlled" always means "x" is the independent variable of a function defined in the *entire* space R^n.

Note that C is not voltage-controlled. In particular, its incremental capacitance $C(v_C)$ is multivalued over the interval $[5\frac{1}{3}, 6\frac{2}{3}]$. If we choose v_C as state variable as in (6), then the state equation is defined only over 2 disjoint intervals $(-\infty, 5\frac{1}{3})$ and $(6\frac{2}{3}, \infty)$:

$$\dot{v}_C = -\frac{v_C}{C(v_C)},$$
$$-\infty < v_C < 5\frac{1}{3} \quad \text{and} \quad 6\frac{2}{3} < v_C < \infty. \tag{20}$$

Since $\dot{q}_C = -v_C$ implies $\dot{q}_C > 0$ whenever $v_C < 0$, and $\dot{q}_C < 0$ whenever $v_C > 0$, it follows that all solutions must tend toward the equilibrium point $q_C = 0$, as depicted by the "dynamic route" in Fig. 2(b) [3]. Hence, has a well-defined solution $\dot{q}_C(t)$ satisfying (14), in spite of the fact that the state equation (20) does not exist globally (i.e., in. R).

Since $v_C(t) = \hat{v}_C(q_C(t))$ is well defined, one might argue that (20) is also well defined so long as one picks the appropriate branch of the curve when numerically solving the state equation: the initial point specifies which branch applies at any time. This is in fact the way circuit simulation programs are designed, where the computation is carried out in a small neighborhood of an initial point [6]. Mathematically, v_C is called a local coordinate about the initial point $v_C(t_0)$. This argument is valid at all points in the curve in Fig. 2 except at Q_1 and Q_2. Note that given any neighborhood of either point, no matter how small, the $q_C - v_C$ curve remains a multivalued function of v_C. This difficulty is reflected in (20) in the value of $C(v_C)$ which tends to ∞ at precisely these 2 points. We conclude that v, does not qualify even as a local coordinate at Q_1 and Q_2.

Example 3. Let R in Fig. 3(a) be a 1-Ω linear resistor and let C be a *voltage-controlled capacitor* described by the "dual" of (13):

$$q_C = \hat{q}_C(v_C) \triangleq \frac{1}{3}v_C^3 - 3v_C^2 + v_C, \quad v_C \in \mathrm{R} \tag{21}$$

as shown in Fig. 3(b). Note that q_C does not qualify as a global coordinate for the $q_C - v_C$ curve, but

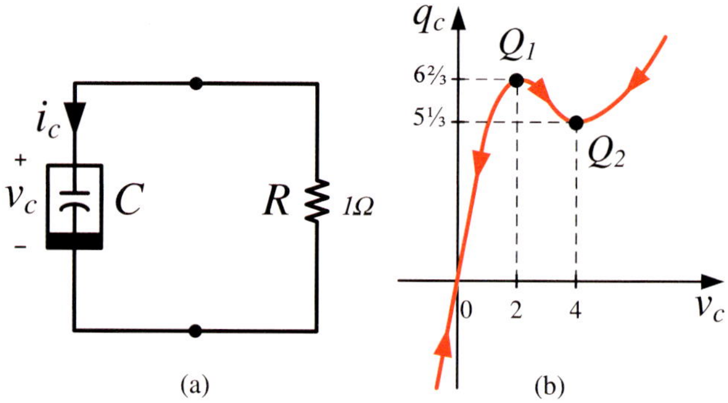

Fig. 3. The $q_C - v_C$ curve for voltage-controlled capacitor C of *Example 3*. Here v_C is a global coordinate for this curve. The capacitor charge q_C does not qualify as a global coordinate. Nor does it qualify as a local coordinate at Q_1 and Q_2.

v_C does. In fact, q_C does not even qualify as a local coordinate at Q_1 and Q_2 in Fig. 3(b). Hence, the state equation (3) does not exist globally.

Since v_C qualifies as a global coordinate for the $q_C - v_C$ curve, it is natural to write the state equation in the form of (6). Note however that since $C(v_C) = 0$ at Q_1 and Q_2, the state equation

$$\dot{v}_C = -\frac{v_C}{v_C^2 - 6v_C + 8} \triangleq f(v_C), \quad v_C \neq 2, 4 \tag{22}$$

is undefined at $v_C = 2$ and $v_C = 4$.

Unlike Fig. 2, however, the dynamic route (note that $\dot{q}_C < 0$ for all points with $v_C > 0$ is qualitatively quite different in Fig. 3. Here, the solution starting from a point on either side near Q_1 diverges from it {resp. Q_2 converges toward it}. One might dismiss this observation as no more than a simple example of an unstable equilibrium point of (15). Observe, however, that Q_1 and Q_2 are not equilibrium points since $\dot{v}_C$ does not tend to zero as $v_C \to 2$ or 4. In fact, $\dot{v}_C \to \infty$ as $v_C \to 4$ and $\dot{v}_C \to -\infty$ as $v_C \to 2$! This means that the solution starting at any point near Q_2 at $t = 0$ would arrive at Q_2 at some finite time $T < 0$. Likewise, the solution starting at any point near Q_1 at $t = 0$ would return to Q_1 at some *finite* time $T < 0$. Note that upon reaching either Q_2 or Q_1 in Fig. 3(b), the solution cannot be continued in forward or backward time, respectively. We call Q_1 and Q_2 impasse points[c] for this reason.

One could object to choosing either Q_1 or Q_2 as initial conditions on the ground that $f(v_C)$ in (16) is undefined at these points. However, the issue here

[c]We call a point x^* of an autonomous system $\dot{x} = f(x)$ an impasse point, if the solution ceases to exist after it reaches x^* at some "finite" time.

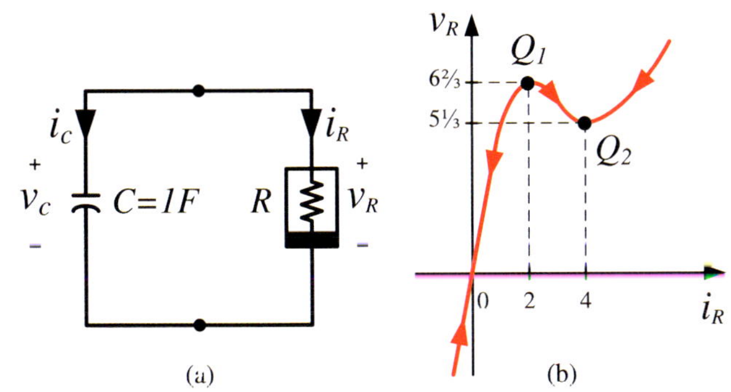

Fig. 4. The $i_R - v_R$ curve for current-controlled resistor R of Example 4. Here i_R is a global coordinate for this curve. The resistor voltage v_R does not qualify as a global coordinate. Nor does it qualify as a local coordinate Q_1 and Q_2.

is that starting in a small neighborhood of Q_2 any computer simulation would get us to Q_2 in *finite time* T, after which the computer will be unable to continue the integration routine since the solution ceases to exist.[d] Since all physical networks must have solution for all times $t \geq t_0$, any network which exhibits such impasse points is nonphysical and should be *remodeled*.

Example 4. Let C in Fig. 4(a) be a 1-F linear capacitor and let R be a *current-controlled resistor* described by

$$v_R = \hat{v}_R(i_R) \stackrel{\Delta}{=} \frac{1}{3}i_R^3 - 3i_R^2 + 8i_R, \quad i_R \in \mathrm{R} \quad (23)$$

as shown in Fig. 4(b). Since $\hat{v}(i_R)$ is not bijective, $h(v_R)$ in (3) is multivalued over the interval $5\frac{1}{3} < v_R < 6\frac{2}{3}$. Since both (3) and (6) are expressed in terms of $h(\cdot)$, *neither state equation exists globally* in this case. Note that this situation differs from those of Examples 2 and 3 where at least one state equation exists everywhere except for some isolated points.

To give a physical interpretation of this situation, observe that $\dot{v}_R = \dot{v}_C = i_C = -i_R$, implies $\dot{v}_R > 0$ whenever $i_R < 0$, and $\dot{v}_R < 0$ whenever $i_R > 0$. This gives rise to the dynamic route shown in Fig. 4(b). Again, we reach an impasse situation when the solution reaches Q_2 in some "finite" forward time $T < \infty$, or Q_1 in some "finite" backward time $T > -\infty$. (Note that Q_1 and Q_2 are *not* equilibrium points of either state equation (3) or (6), even if they are defined at Q_1 and Q_2.)

Hence, Q_1 and Q_2 are impasse points and we conclude that Fig. 4(a) represents a nonphysical circuit. To remodel this circuit, we need only insert a small series linear inductance representing the line inductance. It is easy to see that the resulting second-order circuit has a state equation defined globally in R^2 [3].

To uncover the source of the difficulty in Fig. 4(b), note that since $q_C = v_C = v_R$, neither q_C nor v_C qualifies as a global coordinate, or as a local coordinate at Q_1 and Q_2.

The above "impasse point" phenomenon is truly unexpected even by mathematicians, and has never been reported in the mathematical literature.

Yet, by invoking a "jump postulate", which Chua had justified by showing that the subsequent predictions agree with experimental observations [4], Chua had introduced a powerful and numerically much more efficient method for obtaining the correct result. Moreover, by applying his "dynamic route" technique, Chua was able to demonstrate, and explain the mechanism responsible for relaxation oscillations [11, 14]. This mechanism is illustrated in Example 1, page 1023 of [11], which we reproduce as Example 5 below:

Example 5. Consider the RL circuit shown in Fig.5(a), where the resistor is described by a non-monotonic voltage controlled $v_R - i_R$ curve in Fig. 6(b). Since $di_R(t)/dt = -v_R$, it follows that $di_R(t)/dt < 0$ whenever $v_R(t) > 0$, and $di_R(t)/dt > 0$ whenever $v_R(t) < 0$. The dynamic route [3] showing this behavior is indicated by arrowheads in Fig. 5(b). Note that for any initial current i_R other than at the origin, the solution reaches either Q_1 or Q_2 in finite time, beyond which the solution cannot be continued. Such points are called Impasse Points in [11]. The occurrence of impasse points in a network N always gives rise to a nonphysical situation which requires that N should be remodeled [11]. In this example, the simplest element (i.e., the lowest order element) that must be introduced to break the impasse is a linear capacitor, as shown in Fig. 5(c). The capacitance C is usually a small number (say a few *picofarads*) chosen to simulate the "stray capacitance" between the wires. The state equation for the remodeled circuit in Fig. 5(c) is well defined (provided $C \neq 0$)

[d]No existing circuit simulation program is capable of detecting this nonphysical situation. Clearly, whatever solution it generates after $t = T$ is plain garbage, which could be highly misleading to the unsuspecting user.

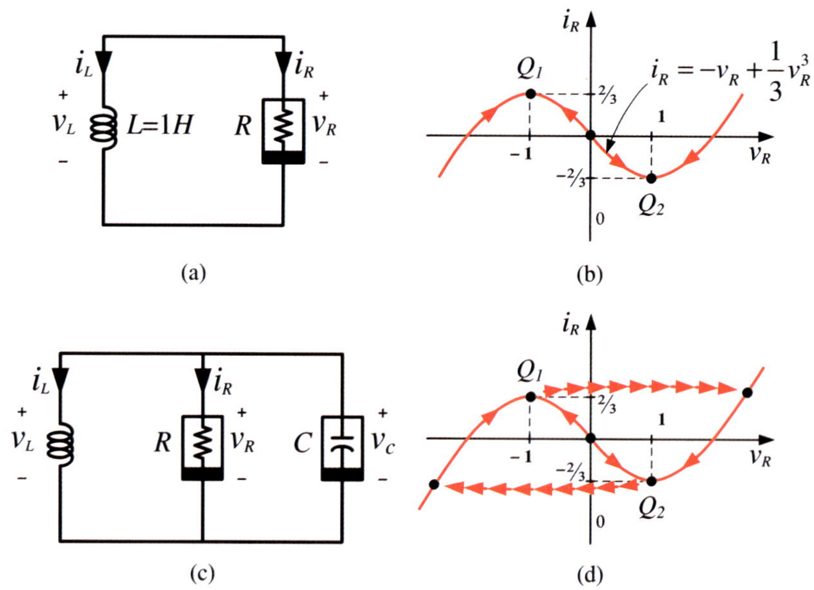

Fig. 5. Circuit for demonstrating the necessity of connecting a capacitor across a nonmonotonic voltage-controlled Resistor.

and has the dynamic route shown in Fig. 5(d). Note that the circuit functions as a *relaxation* oscillator [3].

Chua had conjectured that the action potential (spikes) emitted at the axon hillock of neurons operate by basically the same mechanism, namely, a stable equilibrium point situated near an unstable limit cycle, will execute a rapid transition to a much larger attractor, just like a relaxation oscillator, whenever a perturbation occurs outside of the small basin of attraction of the stable equilibrium point.

3. Mutators

The mutator is another highly original and fundamental creation of Chua [3] for synthesizing and building any 2-terminal nonlinear inductor and nonlinear capacitor from 2-terminal nonlinear resistors [3]. It has since been generalized to synthesize any memristor [4] as well as any $\alpha - \beta$ circuit elements [12]. The basic concept of the mutator is first presented in Section II of [2], which we reproduce below:

The mutator is an active, two-port linear network for transforming one type of nonlinear network element into another type. There are three classes of mutators-an *L-R mutator*, a *C-R mutator*, and an *L-C mutator*. An $L-R$ mutator is characterized by the property that if a nonlinear resistor with a $v-i$ curve Γ is connected across port two of this element, the resulting one-port network (seen across port one) becomes a nonlinear inductor in the sense that it can be characterized by a $\varphi - i$ curve, or by an $i-\varphi$ curve, which is identical to the original $v-i$ curve Γ. In the case where the $\varphi - i$ curve is identical to Γ, the element is called a *type 1 L-R mutator*. If the $i - \varphi$ curve is identical to Γ, the element is called a *type 2* $L-R$ *mutator*.[e] Conversely, if a nonlinear inductor with a $\varphi - i$ curve Γ (or $i - \varphi$ curve Γ), is connected across port one of a type 1 (or type 2) $L - R$ mutator, the resulting one-port network (seen across port two) becomes a nonlinear resistor whose $v - i$ curve is identical to Γ. In view of this two-way mutation property, the mutator is said to be a *reversible* two-port.

We will now express the above characterizations in the following precise mathematical terms:

Type I L-R Mutator: This two-port is defined by

$$\begin{aligned} v_1 &= \frac{dv_2}{dt} \\ i_1 &= -i_2 \end{aligned} \tag{24}$$

[e]In other words, the characteristic curve of the resulting inductor is simply obtained by relabeling the voltage and current variables of the $v - i$ curve Γ by the flux-linkage and current variables for a type 1 $L - R$ mutator, or by the current and flux-linkage variables for a type 2 $L - R$ mutator.

In the frequency domain, (24) is equivalent to the transmission ($ABCD$) matrix [25]

$$T_{LR_1}(p) = \begin{bmatrix} p & 0 \\ 0 & 1 \end{bmatrix} \tag{25}$$

where p is the complex frequency variable.

Type 2 L-R Mutator: This two-port is defined by

$$\begin{aligned} v_1 &= -\frac{di_2}{dt} \\ i_1 &= v_2 \end{aligned} \tag{26}$$

The above equations are easily obtained from the respective transformations and relationships $i = -i_2$, and $v = v_2$, where $\boldsymbol{i}$ and $\boldsymbol{v}$ are the current and voltage of the nonlinear resistor ***i-v*** curve. The symbols and relationships pertaining to an ***L-R*** mutator are tabulated in the upper portion of Table 1.

By an exactly analogous development, two types of ***C-R*** mutators for transforming a nonlinear resistor into a nonlinear capacitor, and vice versa,[f] can be defined. The symbols and relationships pertaining to a ***C-R*** mutator are tabulated in the middle portion of Table 1. Similarly, two types of ***L-C*** mutators can be defined for transforming a nonlinear capacitor into a nonlinear inductor, and vice versa. The symbols and relationships pertaining to an ***L-C*** mutator are tabulated in the lower section of Table 1.

3.1. *Realization of mutators*

Each of the mutators can be realized by either one or two controlled sources whose terminal variable is controlled in general by the time derivative or integral of the port variables. Several possible realizations are shown in Table 1 for each type of mutator. The controlled sources can be realized by an appropriate combination of active elements. Each of the basic realizations shown in Table 1 has been simulated by active networks with various degrees of success. The circuit shown in Fig. 6(a) represents a simple realization of a type 1 $L-R$ mutator corresponding to realization 4 in Table 1. A somewhat analogous realization for a type 1 $C-R$ mutator is shown in Fig. 6(b). Observe that both circuits require a voltage controlled current source. The six transistor circuit shown in Fig. 7 was designed specifically for this purpose.

3.2. *Experimental verification of mutators*

In order to investigate the performance of mutators, several circuits have been designed and verified experimentally.

Figure 8(a) shows the $v-i$ curve (right) of a typical nonlinear resistor (connected across port two of a type 1 $L-R$ mutator) and the $\varphi-i$ curve of the resulting inductor (left). In order to demonstrate that a mutator works in either direction, a nonlinear inductor with a $\varphi-i$ hysteresis curve as shown in Fig. 8(b) (left) is connected across port one of the same mutator and the resulting $v-i$ curve is shown on the right. Finally, Fig. 8(c) shows the mutation of a nonlinear resistor into a nonlinear capacitor having the same characteristic curve. Since a type 1 $L-C$ mutator is simply a gyrator, whose properties are now well known [25], no oscilloscope tracing is given for this type of mutator.

4. Other Chua Creations

While the above concepts of impasse points and mutators are concerned with specific applications, there is a lesser known aspect of Chua's research on nonlinear circuits which involves very general theories that formed the mathematical foundations nonlinear circuit theory. In particular, instead of inventing a new circuit with a specific purpose, Chua asked what circuit-theoretic and dynamical properties does certain broad classes of nonlinear circuit exhibit. These are "qualitative properties". For example, how does one characterize and generalize the circuit-theoretic properties of passivity and losslessness for nonlinear n-ports [1, 7, 9, 13].

Even more fundamental are a series of what are now regarded by mathematicians as of a foundational nature; namely, what class of dynamical nonlinear circuits can exhibit only a unique steady response for any input signal [8]? What class of autonomous dynamical nonlinear circuits have multiple equilibrium points [8]? This 1976 paper had provide an in-depth mathematical foundations of nonlinear circuits with multiple equilibrium points,

[f]Observe that a type 1 $L-C$ mutator is simply a gyrator [25], as might be expected. However, it is somewhat surprising to *find* another two port network element which also transforms a capacitor into an inductor, and vice versa.

Table 1. Characterization and realization of mutators.

	TYPE	SYMBOL AND CHARACTERIZATION	TRANSMISSION MATRIX $\begin{bmatrix} V_1 \\ I_1 \end{bmatrix} = \begin{bmatrix} T(p) \end{bmatrix} \begin{bmatrix} V_2 \\ -I_2 \end{bmatrix}$	BASIC REALIZATIONS USING CONTROLLED SOURCES
L-R MUTATOR	1	$(\varphi, i) \leftrightarrow (v, i)$ 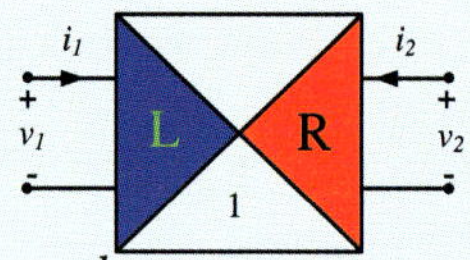 $v_1 = \frac{dv_2}{dt}$ $i_1 = -i_2$	$T_{LR1}(p) = \begin{bmatrix} p & 0 \\ 0 & 1 \end{bmatrix}$	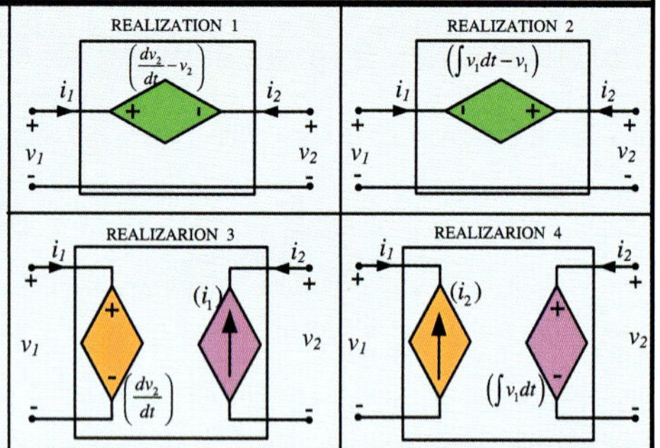
	2	$(\varphi, i) \leftrightarrow (i, v)$ 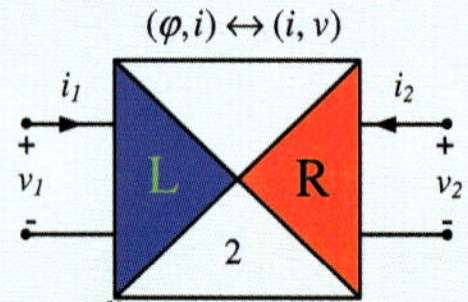 $v_1 = -\frac{di_2}{dt}$ $i_1 = v_2$	$T_{LR2}(p) = \begin{bmatrix} 0 & p \\ 1 & 0 \end{bmatrix}$	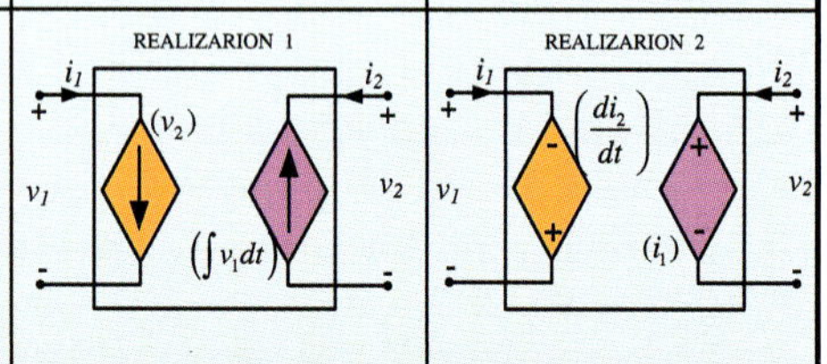
C-R MUTATOR	1	$(q, v) \leftrightarrow (i, v)$ 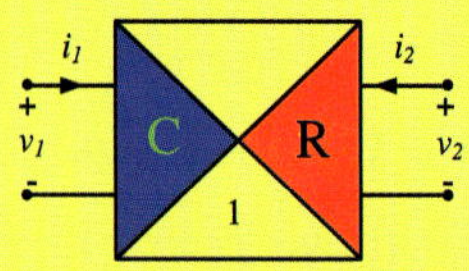 $v_1 = v_2$ $i_1 = -\frac{di_2}{dt}$	$T_{CR1}(p) = \begin{bmatrix} 1 & 0 \\ 0 & p \end{bmatrix}$	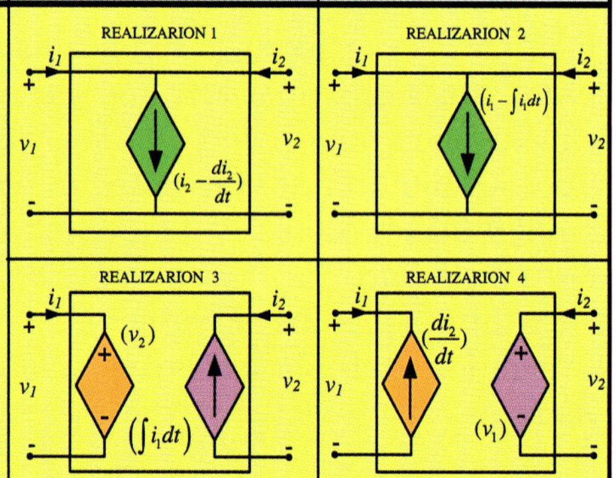
	2	$(q, v) \leftrightarrow (v, i)$ 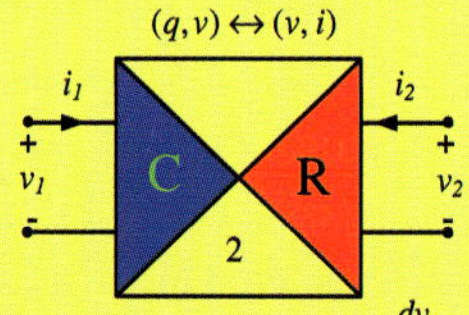 $v_1 = -i_2$ $i_1 = -\frac{dv_2}{dt}$	$T_{CR2}(p) = \begin{bmatrix} 0 & 1 \\ p & 0 \end{bmatrix}$	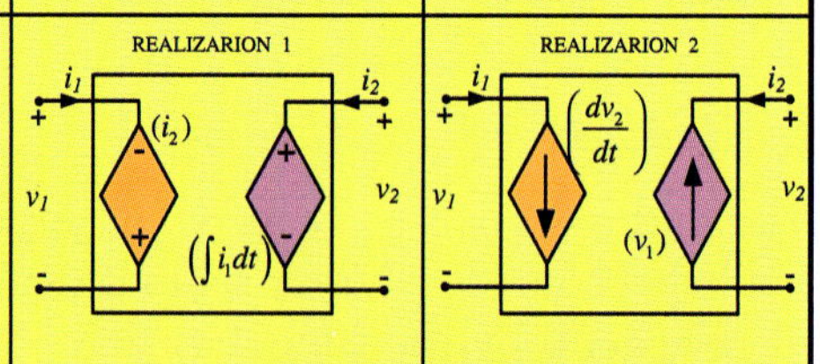
L-C MUTATOR	1	$(\varphi, i) \leftrightarrow (q, v)$ 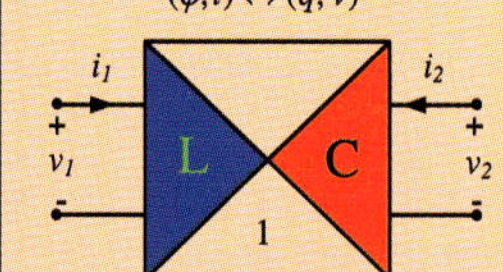 $v_1 = -i_2$ $i_1 = v_2$	$T_{LC1}(p) = \begin{bmatrix} 0 & 1 \\ 1 & 0 \end{bmatrix}$	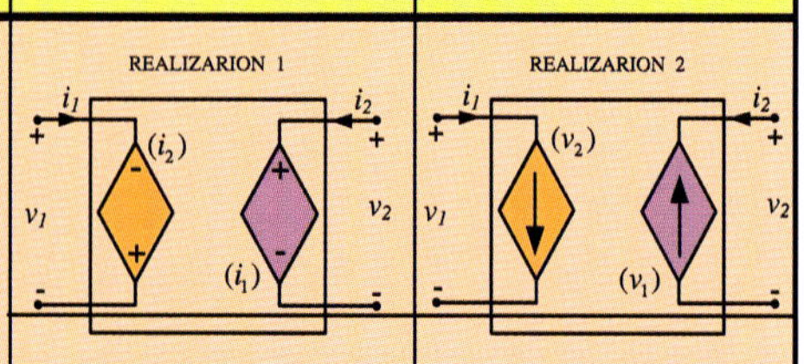
	2	$(\varphi, i) \leftrightarrow (v, q)$ 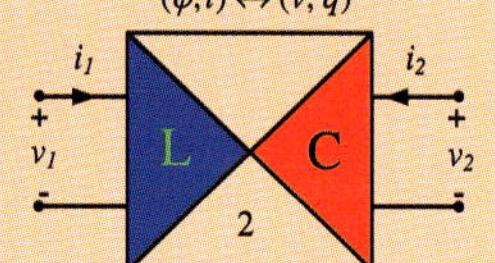 $v_1 = \frac{dv_2}{dt}$ $i_1 = -\int i_2 dt$	$T_{LC2}(p) = \begin{bmatrix} p & 0 \\ 0 & \frac{1}{p} \end{bmatrix}$	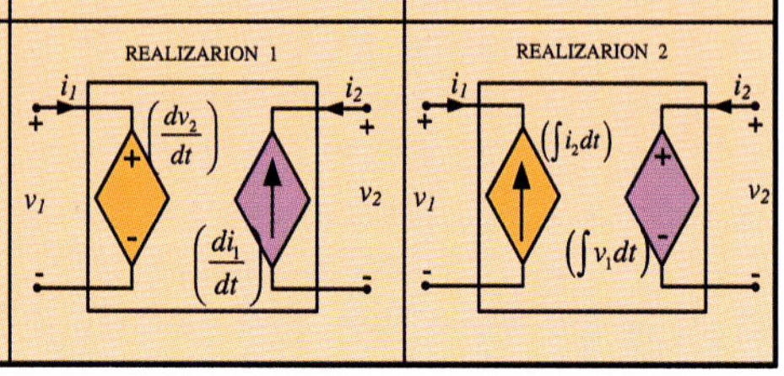

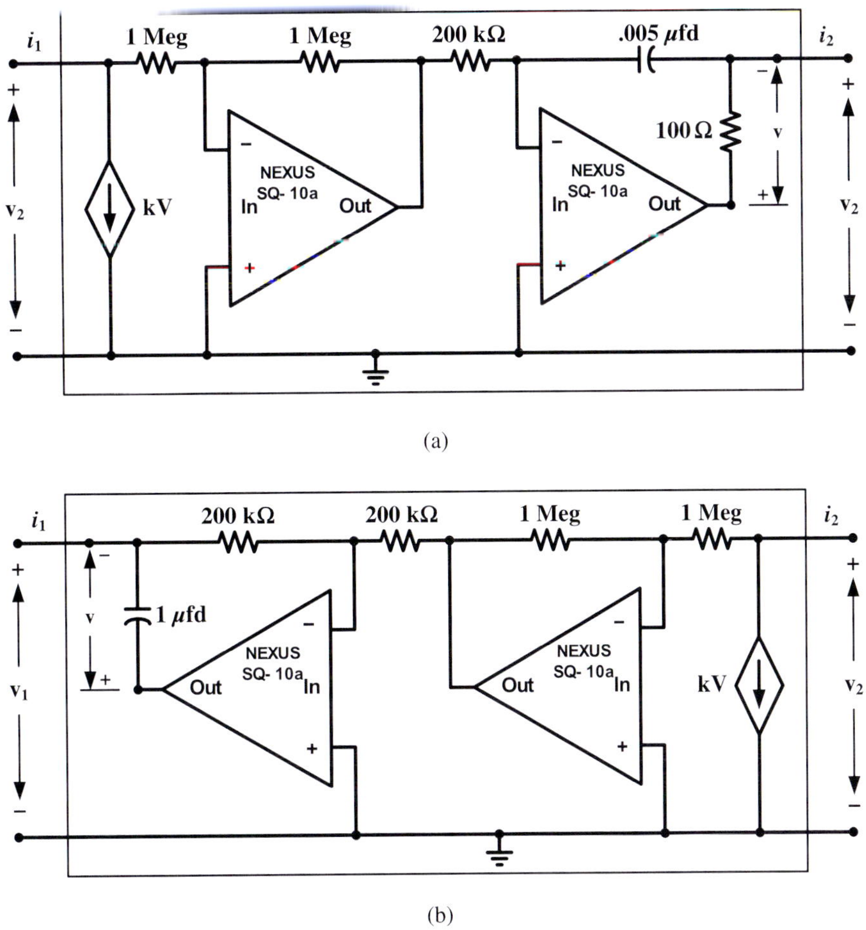

Fig. 6. Two inductorless circuits for realizing an $L-R$ mutator and a $C-R$ mutator.

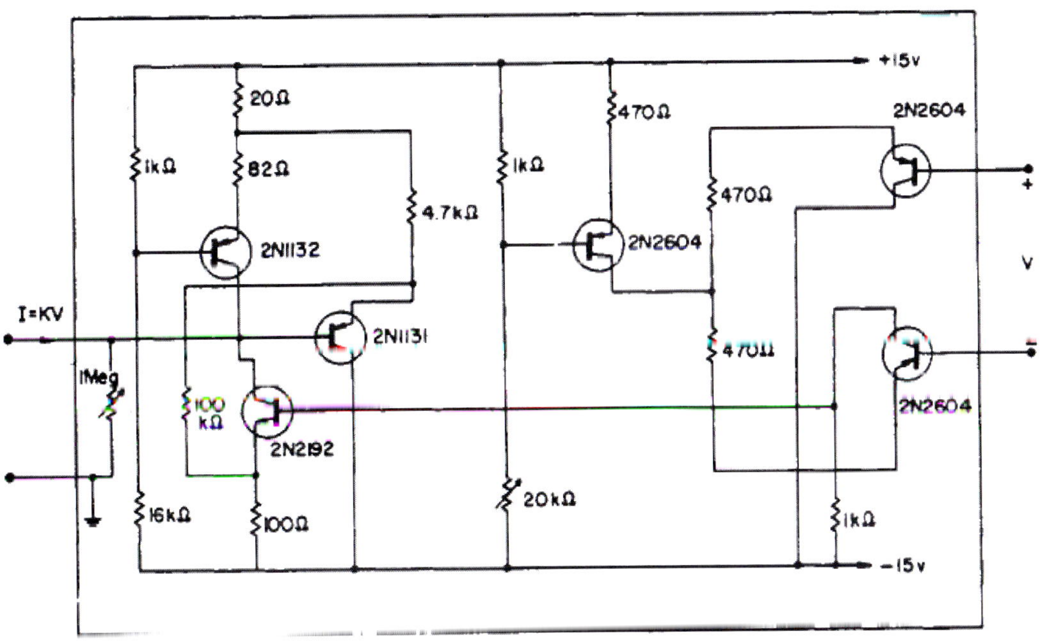

Fig. 7. An inductorless realization of a voltage controlled current source.

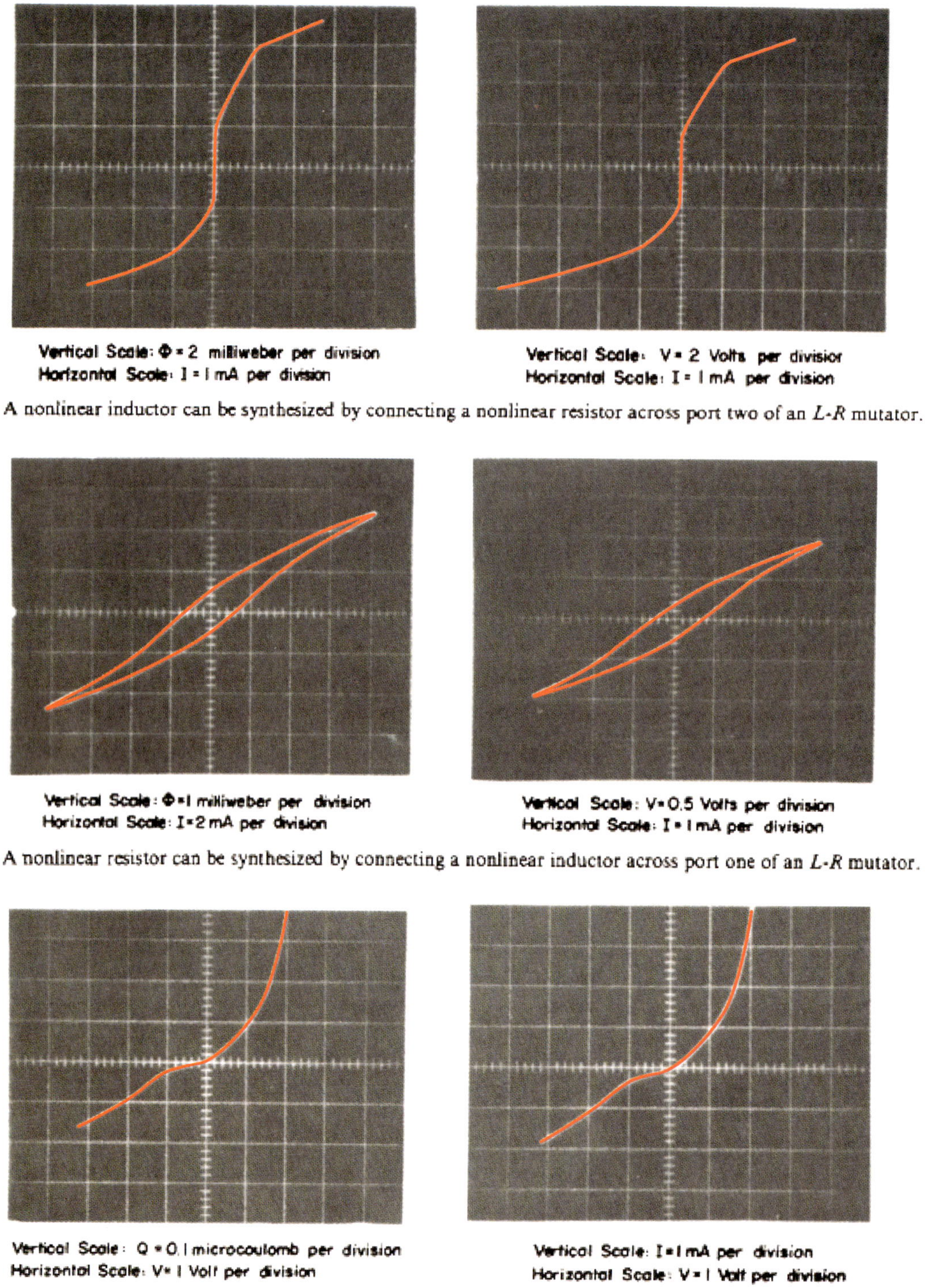

(a) A nonlinear inductor can be synthesized by connecting a nonlinear resistor across port two of an L-R mutator.

(b) A nonlinear resistor can be synthesized by connecting a nonlinear inductor across port one of an L-R mutator.

(c) A nonlinear capacitor can be synthesized by connecting a nonlinear resistor across port two of a C-R mutator.

Fig. 8. Oscilloscope tracings demonstrating the transformation of one type of nonlinear element into another type by a mutator. Retouched and colored from Fig. 3 from [Chua, 1968] for better readability.

and has in fact anticipated the era of neural networks launched by John Hopfield in 1982 [22, 23].

Yet another research subject that had predated and anticipated the recent active area of "synchronization of chaos" launched by Pecora and Carroll in [24] is a not-widely known yet fundamental and deep seminal paper entitled "Synchronization and Chaos" [26].

Acknowledgment

The author would like to thank Changju Yang for his assistance in the preparation of this manuscript.

References

[1] Brayton, R. K. Chua, L. O. Rhodes, J. D. & Spence, R. [1978], Modern network theory: an introduction, Georgi, Switzerland.

[2] Chua, L. O. [1968] "Synthesis of new nonlinear network elements," *Proceedings of the IEEE*, 56(8), 1325–1340.

[3] Chua, L. O., [1969] *Introduction to Nonlinear Network Theory*, McGraw-Hill, New York.

[4] Chua, L. O. & Alexander, G. R., [1971] "The effects of parasitic reactances on nonlinear networks," *IEEE Transactions on Circuit Theory*, 18(5), 520–532.

[5] Chua, L.O. [1971] "Memristor-the missing circuit element," *IEEE Trans. on Circuit Theory*, 18(5), 507–519.

[6] Chua, L. O. & Lin, P. M. [1975] *Computer-Aided Analysis of Electronic Circuits: Algorithms and Computational Techniques.*, Prentice Hall.

[7] Chua, L. O. & Green, D. N. [1976a] "Graph-theoretic properties of dynamic nonlinear networks," *IEEE Transactions on Circuits and Systems*, 23(5), 292–312.

[8] Chua, L. O. & Green, D. N. [1976b], "A qualitative analysis of the behavior of dynamic nonlinear networks: steady-state solutions of nonautonomous networks," *IEEE Trans. on Circuits and Systems*, 23(9), 530–535.

[9] Chua, L. O. [1977], "Qualitative properties of resistive networks containing multiterminal nonlinear elements: no gain properties," *IEEE Trans. on Circuits and Systems*, 24(3), 93–118.

[10] Chua, L.O., [1980a] "Dynamic nonlinear networks: state-of-the-art," *IEEE Trans. on Circuits and Systems*, 27(11), 1059–1087.

[11] Chua, L.O. [1980b] "Device modeling via basic nonlinear circuit elements," *IEEE Trans. on Circuits and Systems*, 27(11), 1014–1044.

[12] Chua, L. O. & Szeto, E.W. [1984] "Synthesis of higher order nonlinear circuit elements," *IEEE Transactions on Circuits and Systems*, 31(2), 231–235.

[13] Chua, L. O. [1984], "Nonlinear circuits," *IEEE Trans. on Circuits and Systems*, 31(1), 69–87.

[14] Chua, L. O. Desoer, C. A. & Kuh, E. S., [1987] *Linear and Nonlinear Circuits*, McGraw-Hill, New York

[15] Chua, L. O. & Yang L., [1988] "Cellular neural networks: applications," *IEEE Trans. on Circuits and Systems*, 35(10), 1257–1290.

[16] Chua, L. O., [2006] *A Nonlinear Dynamics Perspective of Wolfram's New Kind of Science* I, World Scientific Series on Nonlinear Science.

[17] Chua, L. O., [2007] *A Nonlinear Dynamics Perspective of Wolfram's New Kind of Science* II, World Scientific Series on Nonlinear Science.

[18] Chua, L. O., [2009] *A Nonlinear Dynamics Perspective of Wolfram's New Kind of Science* III, World Scientific Series on Nonlinear Science.

[19] Chua, L. O., [2011] *A Nonlinear Dynamics Perspective of Wolfram's New Kind of Science* IV, World Scientific Series on Nonlinear Science.

[20] Chua, L. O., [2012] *A Nonlinear Dynamics Perspective of Wolfram's New Kind of Science* V, World Scientific Series on Nonlinear Science.

[21] Göknar, I. C., [1980] "On the state description of nonlinear networks," *Proc. IEEE*, 68(6), 734–735.

[22] Hopfield, J. J. [1982] "Neural Networks and Physical Systems with Emergent Collective Computational Abilities," *Proc. of the National Academy of Sciences*, USA, 79, 2554–2558.

[23] Kennedy, M. P. & Chua, L. O. [1987] "Unifying the Tank and Hopfield linear programming circuit and the canonical nonlinear programming circuit of Chua and Lin," *IEEE Trans. on Circuits and Systems*, 34(2), 210–214.

[24] Pecora, L. M. & Carroll, T. [1990] "Synchronization in chaotic systems," *Physics Review Letters*, 64(8), 821–824.

[25] Su, K. L. [1965] *Active Network Synthesis*, McGraw-Hill, New York.

[26] Tang, Y. S. Mees, A. I. & Chua, L. O. [1983], "Synchronization and Chaos," *IEEE Trans. on Circuits and Systems*, 30(9), 620–626.

CHAPTER 3

CHUA'S LAGRANGIAN CIRCUIT ELEMENTS

ORLA FEELY

School of Electrical, Electronic and Communications Engineering
University College Dublin
Dublin 4, Ireland

The first part of Leon Chua's career, and the work that brought him to prominence within the international research community, saw him lay many of the rigorous foundations of nonlinear circuit theory. As part of these foundations, Chua included in his graduate courses in the University of California, Berkeley, a circuit theoretic consideration of the concept of the Lagrangian from classical mechanics. The aim of this review paper is to capture that presentation and include it as a further record of Chua's contributions to nonlinear circuits research and education.

1. Introduction

The first part of Leon Chua's career, and the work that brought him to prominence within the international research community, saw him lay many of the foundations of nonlinear circuit theory [2, 5]. This involved, among many other fundamental contributions, the definition of new circuit elements, most notably the memristor [3]. The explosion of interest in the memristor in recent years is a testament to the far-sighted and fundamental nature of Chua's work.

Another aspect of this work employed the concept of the Lagrangian from classical mechanics [1, 7] In Lagrangian mechanics, the Lagrangian is a function of a set of generalized coordinates η_i, velocities $\dot{\eta}_i$ (and possibly time) from which the dynamical equations of the system can be derived. The application of this concept in the context of circuit theory had been studied by a number of authors in the middle decades of the last century, and this work was significantly advanced by Chua in papers such as [4, 8]. Chua presented an introduction to the properties of circuit elements characterized by a Lagrangian in his graduate courses in the University of California, Berkeley, and in a number of keynote addresses, but did not include it in any of his textbooks. The aim of this review paper is to capture that presentation and include it as a further record of his contributions in nonlinear circuits research and education.

2. Chua's Presentation of Lagrangian Circuit Elements

A Lagrangian n-port is described by a C^1 Lagrangian function

$$L(q_1, q_2, \ldots q_m, \dot{q}_1, \dot{q}_2, \ldots \dot{q}_m, \phi_{m+1}, \ldots \phi_n, \dot{\phi}_{m+1}, \ldots \dot{\phi}_n) \quad (1)$$

for some partition $\{1, \ldots m\}$ and $\{m+1, \ldots n\}$ of the ports. q_j is the charge variable associated with port j, and its time derivative $\dot{q}_j$ is the port current. ϕ_j is the flux variable associated with port j, and its time derivative $\dot{\phi}_j$ is the port voltage. If $m = n$, the Lagrangian is a function of charge and current only, and if $m = 0$ it is a function of flux and voltage only. The constitutive relation of the n-port is derived from the Lagrangian by the equations

$$v_j = \frac{d}{dt}\left(\frac{\partial L}{\partial \dot{q}_j}\right) - \frac{\partial L}{\partial q_j} \quad j = 1, \ldots m$$

$$i_k = \frac{d}{dt}\left(\frac{\partial L}{\partial \dot{\phi}_k}\right) - \frac{\partial L}{\partial \phi_k} \quad k = m+1, \ldots n$$

Note that, although the theory is presented here in the context of n-ports, it can equally be applied to $(n+1)$-terminal elements.

Some examples of Lagrangian n-ports are as follows:

Example 1. A Lagrangian of the form $L(q_1)$ gives

$$v_1 = \frac{d}{dt}(0) - \frac{\partial L}{\partial q_1} = h(q_1).$$

This is the constitutive equation of a charge-controlled capacitor. In particular, if

$$L(q_1) = -\frac{q_1^2}{2C}$$

we have a linear capacitor defined by

$$v_1 = \frac{q_1}{C}.$$

Example 2. A Lagrangian of the form $L(\dot{q}_1) = L(i_1)$ gives

$$v_1 = \frac{d}{dt}\left(\frac{\partial L}{\partial i_1}\right) \quad \text{or} \quad \phi_1 = f(i_1).$$

This is the constitutive equation of a current-controlled inductor.

Building on these two basic examples we find examples such as the following:

Example 3. A Lagrangian of the form $L = f_1(q_1) + f_2(\dot{q}_1)$ gives

$$v_1 = \frac{d}{dt}\left(\frac{df_2}{d\dot{q}_1}\right) - \frac{\partial f_1}{\partial q_1},$$

the series combination of a current-controlled inductor and charge-controlled capacitor.

Example 4. A Lagrangian of the form $L(q_1, q_2, \ldots q_n)$ gives

$$v_1 = -\frac{\partial L}{\partial q_1} = f_1(q_1, q_2, \ldots q_n)$$
$$v_2 = -\frac{\partial L}{\partial q_2} = f_2(q_1, q_2, \ldots q_n)$$
$$\vdots$$
$$v_n = -\frac{\partial L}{\partial q_n} = f_n(q_1, q_2, \ldots q_n)$$

This is the constitutive equation of an n-port charge-controlled capacitor. Similarly, a Lagrangian $L(\phi_1, \phi_2, \ldots \phi_n)$ corresponds to an n-port flux-controlled inductor.

Further examples include the following:

Example 5. The Lagrangian $L = -nq_1\dot{\phi}_2$ represents an ideal transformer:

$$v_1 = n\dot{\phi}_2 = nv_2$$
$$i_2 = \frac{d}{dt}(-nq_1) = -ni_1$$

Example 6. The Lagrangian $L = G\phi_2\dot{\phi}_1$ represents an ideal gyrator:

$$i_1 = G\dot{\phi}_2 = Gv_2$$
$$i_2 = \frac{d}{dt}(-G\phi_1) = -Gv_1$$

Example 7. A Lagrangian of the form $L = Aq_1q_2\dot{q}_3$ gives

$$v_1 = -Aq_2i_3$$
$$v_2 = -Aq_1i_3$$
$$v_3 = \frac{d}{dt}(Aq_1q_2) \text{ i.e. } \phi_3 = Aq_1q_2$$

This is an example of the family of circuit elements known as *traditors* [6], defined more generally by a

Lagrangian $L = A\eta_1\eta_2\dots\eta_{n-1}\dot{\eta}_n$, where each $\eta_j = q_j$ or ϕ_j.

Of the three-port traditors, the Lagrangian $L = Aq_1q_2\dot{\phi}_3$ gives

$$\begin{aligned} v_1 &= -Aq_2v_3 \\ v_2 &= -Aq_1v_3 \\ q_3 &= Aq_1q_2 \end{aligned}$$

while the Lagrangian $L = A\phi_1\phi_2\dot{q}_3$ gives

$$\begin{aligned} i_1 &= -A\phi_2 i_3 \\ i_2 &= -A\phi_1 i_3 \\ \phi_3 &= A\phi_1\phi_2 \end{aligned}$$

The first of these is characterized by an algebraic relationship between the charge and voltage variables, and so represents a three-port capacitor, while the second is characterized by an algebraic relationship between flux and current, and so represents a three-port inductor. Most traditors, on the other hand, do not correspond to basic circuit elements.

The key principle here is the description of all of these elements by a scalar circuit-theoretic Lagrangian function. We see that this description can encompass a variety of basic circuit elements, along with others that do not arise as basic elements.

It may also be observed that all of the examples above are *lossless* circuit elements. An n-port representation N is lossless iff the total energy E entering N from $t = 0$ to any time τ depends only on $(q_1, q_2, \dots q_m, \dot{q}_1, \dot{q}_2, \dots \dot{q}_m, \phi_{m+1}, \dots \phi_n, \dot{\phi}_{m+1}, \dots \dot{\phi}_n)$ evaluated at time τ, i.e.

$$\int_0^{\tau} \left[\sum_{j=1}^{m} v_j(t)\frac{dq_j}{dt} + \sum_{k=m+1}^{n} i_k(t)\frac{d\phi_k}{dt} \right] dt = E(q_a(t), \phi_b(t), \dot{q}_a(t), \dot{\phi}_b(t))$$

By this definition, a network is lossless if the energy required to travel between two points of the state space is independent of the path taken. Once again, this concept is founded in classical mechanics.

There is a close relationship between the concept of the Lagrangian and that of energy transfer.

Lemma 1. *N is lossless iff there exists a C^1 function $E(q_a(t), \phi_b(t), \dot{q}_a(t), \dot{\phi}_b(t))$, the energy function, such that the following energy identity is satisfied:*

$$\begin{aligned} \sum_{j=1}^{m} v_j(t)\dot{q}_j + \sum_{k=m+1}^{n} i_k(t)\dot{\phi}_k &= \sum_{j=1}^{m} \left(\frac{\partial E}{\partial q_j}\dot{q}_j + \frac{\partial E}{\partial \dot{q}_j}\ddot{q}_j \right) \\ &+ \sum_{k=m+1}^{n} \left(\frac{\partial E}{\partial \phi_k}\dot{\phi}_k + \frac{\partial E}{\partial \dot{\phi}_k}\ddot{\phi}_k \right) \end{aligned}$$

Proof. Differentiate both sides of (3) with respect to time. □

Theorem 1. *Every Lagrangian n-port is lossless. In particular, its energy function is given by*

$$E(q_a(t), \phi_b(t), \dot{q}_a(t), \dot{\phi}_b(t)) = \sum_{j=1}^{m} \dot{q}_j(t)\frac{dL}{d\dot{q}_j} + \sum_{k=m+1}^{n} \dot{\phi}_k(t)\frac{dL}{d\dot{\phi}_k} - L \quad (2)$$

Proof. Substitute (2) into (4) and expand derivatives. □

Theorem 2. *If a Lagrangian n-port with energy function $E(q_a(t), \phi_b(t), \dot{q}_a(t), \dot{\phi}_b(t))$ given by (2) has ports 1, ... m terminated in short circuits and ports m+1, ... n terminated in open circuits, then $\frac{dE}{dt} = 0$ and $E(t) = E(0)\forall t \geq 0$.*

Proof. The terminating conditions imply that

$$\begin{aligned} \frac{d}{dt}\left(\frac{\partial L}{\partial \dot{q}_j}\right) &= \frac{\partial L}{\partial q_j} \quad j = 1, \dots m \\ \frac{d}{dt}\left(\frac{\partial L}{\partial \dot{\phi}_k}\right) &= \frac{\partial L}{\partial \phi_k} \quad k = m+1, \dots n \end{aligned}$$

It follows that

$$\begin{aligned} \frac{dL}{dt} &= \sum_{j=1}^{m} \left(\frac{\partial L}{\partial q_j}\dot{q}_j + \frac{\partial L}{\partial \dot{q}_j}\ddot{q}_j \right) \\ &+ \sum_{k=m+1}^{n} \left(\frac{\partial L}{\partial \phi_k}\dot{\phi}_k + \frac{\partial L}{\partial \dot{\phi}_k}\ddot{\phi}_k \right) \\ &= \sum_{j=1}^{m} \frac{d}{dt}\left(\dot{q}_j \frac{\partial L}{\partial \dot{q}_j} \right) + \sum_{k=m+1}^{n} \frac{d}{dt}\left(\dot{\phi}_k \frac{\partial L}{\partial \dot{\phi}_k} \right) \end{aligned}$$

and hence

$$\frac{d}{dt}\left(\sum_{j=1}^{m}\dot{q}_j(t)\frac{dL}{d\dot{q}_j}+\sum_{k=m+1}^{n}\dot{\phi}_k(t)\frac{dL}{d\dot{\phi}_k}-L\right)=0$$

□

It is possible to expand on the relationship between the Lagrangian and energy functions. If

$$E(q_a(t),\phi_b(t),\dot{q}_a(t),\dot{\phi}_b(t)) = E_0 + E_1 + E_2 + \cdots$$

and

$$L(q_a(t),\phi_b(t),\dot{q}_a(t),\dot{\phi}_b(t)) = L_0 + L_1 + L_2 + \cdots \tag{3}$$

where E_γ is the γth degree homogenous component of E expanded in ascending powers of $\dot{q}_j$ and $\dot{\phi}_k$, i.e. it is of the form

$$\sum_l a_l(q_a,\phi_b)\dot{q}_1^{\gamma_1}\ldots\dot{q}_m^{\gamma_m},\dot{\phi}_{m+1}^{\gamma_{m+1}}\ldots\dot{\phi}_n^{\gamma_n}$$

$$\gamma_1+\gamma_2+\ldots+\gamma_n=\gamma$$

with a corresponding definition for L_γ, we have the following theorem:

Theorem 3. *The γth degree homogeneous components of E and L are related by*

$$E_\gamma = (\gamma-1)L_\gamma, \quad \gamma = 1,\ldots n.$$

In particular, $E_1 = 0$.

Proof.

$$\begin{aligned}E_\gamma &= \sum_{j=1}^{m}\dot{q}_j\frac{dL_\gamma}{d\dot{q}_j}+\sum_{k=m+1}^{n}\dot{\phi}_k\frac{dL_\gamma}{d\dot{\phi}_k}-L_\gamma\\ &= (\gamma-1)L_\gamma\end{aligned}$$

□

An n-port is non-energic if the instantaneous power flow into it is always zero — it merely transfers power from one set of ports to another. The next theorem follows.

Theorem 4. *A Lagrangian n-port with L of the form (3) is non-energic iff $L = L_1$, i.e. each term in L can contain $\dot{q}_j$ and $\dot{\phi}_j$ to the first degree only.*

Proof. Follows from Theorem 4. □

Lagrangian 1-ports with $L = L_1$ are of the form $L(q_1,\dot{q}_1) = f(q_1)\dot{q}_1$ or $L(\phi_1,\dot{\phi}_1) = f(\phi_1)\dot{\phi}_1$, giving, respectively

$$v_1 = \frac{d}{dt}(f(q_1)) - f'(q_1)\dot{q}_1 = 0 \quad \text{or} \quad i_1 = 0,$$

the short circuit and the open circuit.

For $n = 2$ and $L = L_1$, the possibilities are:

$$L(q_1,q_2,\dot{q}_1,\dot{q}_2) = f_a(q_1,q_2)\dot{q}_1 + f_b(q_1,q_2)\dot{q}_2,$$

which gives

$$v_1 = \left(\frac{\partial f_a}{\partial q_2}-\frac{\partial f_b}{\partial q_1}\right)\dot{q}_2$$

$$v_2 = \left(\frac{\partial f_b}{\partial q_1}-\frac{\partial f_a}{\partial q_2}\right)\dot{q}_1$$

a non-linear gyrator;

$$L(\phi_1,\phi_2,\dot{\phi}_1,\dot{\phi}_2) = f_a(\phi_1,\phi_2)\dot{\phi}_1 + f_b(\phi_1,\phi_2)\dot{\phi}_2,$$

which again gives a nonlinear gyrator; and

$$L(q_1,\phi_2,\dot{q}_1,\dot{\phi}_2) = f_a(q_1,\phi_2)\dot{q}_1 + f_b(q_1,\phi_2)\dot{\phi}_2,$$

which gives

$$v_1 = \left(\frac{\partial f_a}{\partial \phi_2}-\frac{\partial f_b}{\partial q_1}\right)\dot{\phi}_2$$

$$i_2 = \left(\frac{\partial f_b}{\partial q_1}-\frac{\partial f_a}{\partial \phi_2}\right)\dot{q}_1$$

the equations of a nonlinear transformer.

3. Summary

This paper provides an introduction to Lagrangian circuit elements, following the presentation given by Leon Chua in graduate courses and invited lectures at various points over the course of his career. It highlights a number of aspects of his approach to nonlinear circuit theory: its roots in fundamental mathematical disciplines; the inclusion of elements beyond the basic set of circuit elements, based on the logical extension of existing classifications; and the pedagogical emphasis on relating new material and concepts to concepts with which the audience will already be familiar. The most spectacular example of the success of Chua's fundamental methods is the sudden flowering of interest in the memristor in recent years, following technological developments unimagined at the time of the device's inception. Lagrangian circuit elements, with their natural links to concepts of losslessness in circuits, may well experience a similar flowering of interest at some time in the future.

References

[1] Birkhoff, G. D. (1927) *Dynamical Systems* (American Mathematical Society).

[2] Chua, L. O. (1969). *Introduction to Nonlinear Network Theory* (McGraw-Hill).

[3] Chua, L. O. (1971). Memristor – the missing circuit element, *IEEE Transactions on Circuit Theory,* **18**, pp. 507–519.

[4] Chua, L. O. and McPherson, J. D. (1974). Explicit topological formulation of Lagrangian and Hamiltonian equations for nonlinear networks, *IEEE Transactions on Circuits and Systems*, **21**, pp. 277–286.

[5] Chua, L. O. (1984). Nonlinear circuits, *IEEE Transactions on Circuits and Systems* **31**, pp. 69–87.

[6] Duinker, S. (1959). Traditors — a new class of non-energic non-linear network elements, *Philips Research Reports*, **14**, pp. 29–51.

[7] Lanczos, C. (1949) *The Variational Principles of Mechanics*, (University of Toronto Press).

[8] Wyatt, J. L. and Chua, L. O. (1977). A theory of nonenergic n-ports, *International Journal of Circuit Theory and Applications*, **5**, pp. 181-208.

CHAPTER 4

FROM CNN DYNAMICS TO CELLULAR WAVE COMPUTERS

TAMAS ROSKA

Pazmany University and the Hungarian Academy of Sciences,
Prater 50/A, Budapest, 1083, Hungary
roska@ppke.hu

Embedded in a historical overview, the development of the Cellular Wave Computing paradigm is presented, starting from the standard CNN dynamics. The theoretical aspects, the physical implementation, the innovation process, as well as the biological relevance are discussed in details. Finally, the latest developments, the physical versus virtual cellular machines, as well as some open questions are presented.

1. Introduction

Until recently, Computers and Computing has been mainly considered as a discipline, technology, and industry where the processors with arithmetic and logic units and memories are *typically binary digital, and their spatial placement does not play a significant role* since the communication time between them in physical wires is negligible compared to the computation time of the processors. The wonderful and powerful body of principles and tools of computer science have been based on these assumptions, as well.

Today, we face three fundamental challenges:

- kilo- or mega- processor/core/thread chips are used where the geometric or *topographic address* of a digital computing cell (processor/core/thread) or storing cell (register/ caches, main memory, hard disk) plays a crucial role; moreover,

the *precedence of locality*, we call also a *cellular character*, is a must in designing algorithms and software;
- speed, power, area, bandwidth, and accuracy are traded of, the communication bandwidth between the processors and memories plays an important role in the computing power or computational complexity of a given architecture;
- at the interface between sensory arrays and processing arrays and in some cases the physical implementation of *arithmetic and logic array operations are made by mixed mode (analog and logic) circuits,* and not by pure digital; even, *in non-CMOS nanoscale* cell array architectures *the analog, including oscillatory, cells* as processors might play a crucial role.

In all these cases, spatial-temporal dynamics has a role not yet studied earlier (as a computational primitive, or as an activity pattern in a digital processor or memory array). Interestingly, the two pioneers, John von Neumann and Alan Turing proposed already the archetypes of these kinds of cellular architectures around 1952, proposing the cellular digital automata and the analog cellular dynamics, respectively [1,2]. These propositions, however, did not have a major effect in mainstream computer engineering, until recently.

During the Summer of 1988, I met again Professor Leon Chua in Budapest. We have told each other some interesting new findings. He mentioned the new Cellular Neural/nonlinear Network (CNN) principle, to be published soon [4]. Earlier, I was already aware of their new principle of optimization using nonlinear circuit array dynamics [3]. I was fascinated by the potential of this new idea (soon to become a paradigm) and told him about my new way of thinking about computing, combining analog spatial-temporal array dynamics and logic, inspired by the lateral asymmetry of the brain [5]. Soon, we started to combine these two principles to construct the CNN Universal Machine architecture [6], as a prototype for various Cellular Wave computers later materialized also in visual microprocessors [8], in other algorithmic frameworks, and in various products. This direction soon attracted some excellent scientists and many young talents. In 1990, we have organized the first IEEE International Workshop on Cellular Neural Networks and their Applications, CNNA 1990, Budapest (a biannual conference series, recently renamed as Cellular Nanoscale Networks and their Applications; Berkeley, 2010).

To form and use the CNN Universal Machine, as a universal machine on spatial-temporal flows, several fundamental theoretical results were needed. Later, the proof of concept in silicon became a major task. In addition to the two groups in Berkeley and Budapest, the groups of Professor Josef Nossek in Munich in the algorithmic and learning methods and Professor Ángel Rodriguez-Vázquez in Seville in making the breakthrough via mixed mode CMOS chip implementation (the first time to make analog spatial temporal actions as elementary instructions in an algorithmically programmed machine on a chip), and many other individuals contributed to the development of this new technology. Theoretical and practical tasks were solved, and a lot of new algorithmic principles and applications have been invented. Conference proceedings in various forms, special issues, journal papers, the first few monographs and textbooks marked the road of documenting this emerging paradigm.

Very soon, the biology relevance has been uncovered. The interaction pattern between the cells, called cloning template, and the notion of the receptive field in neurobiology became the two analogous organizational units to bridge the gap between neuroscience and computational primitives, leading to many solutions simply converting neurobiological anatomy into template architecture (the group of Professor József Hámori). The deeper insights in the retina studies by Professor Frank Werblin and his group generated a lot of new principles, including the “Bionic Eye”, moving towards mission critical applications as well. The motivation and advices of Drs. Larry Cooper and Joel Davis proved to be invaluable during the 90’s. The breakthrough in understanding the internal neurobiological architecture and signal processing operations of a mammalian retina changed the scenario in understanding the spatial-temporal processing in the retina (Nature, 2001, by Frank Werblin and Botond Roska [10]). After that, the direct biological relevance inspired many researchers to discover new neuromorphic or neuro-inspired CNN algorithms and applications. The modeling of some special functional mechanisms were developed as well, in particular the hyper acuity in space and time, elementary color processing and some illusions.

Fundamental insights have been developed later when the following questions arisen:

(1) PDEs can be modeled in discrete space (PDDE: partial differential difference equations) by single or multi-layer CNN architectures; is it possible, in the opposite way, to find a PDE for any template?
(2) What are the operators that can be implemented by both locally connected and fully connected networks?
(3) Is it possible to develop an analytic theory for binary Cellular Automata (the prophecy of John von Neumann)?
(4) Is it any relation between Gödel incompleteness and a binary input-output CNN?

After getting acquainted with the new type of spatial-temporal algorithms based on the CNN Universal Machine, several new type of actions have been implemented, as well as some new mathematical (morphology, wave metric, spatial-temporal chaos, etc.) and physical (true random pattern generation, different nonlinear PDE solutions, etc.) problems have been solved on this platform.

Meanwhile, after the turn of the century, it became clear that due to the power dissipation constraint one possibility is to increase the number of processors and memories on one single chip. Moreover, below about 60 nm, the wire delay becomes bigger than the gate delay; hence the precedence of locality is a must. Very soon, not only multi-core chips were made, however, hundreds and thousands of cores or processing cells have been placed on a chip, the so called many-core chips, and the cellular or multi-cellular character, resembling the basic CNN spatial-temporal dynamics or its local variants even in digital cores and memories became apparent. In addition, some digital CNN Universal Machine implementations have been successfully made on FPGAs or as ASIC chips directly attached to some sensor arrays, as well (e.g. the so called Xenon chip).

Meanwhile, textbooks had been published [7,9] and many special issues in major Journals dealt with the emerging Cellular Wave Computing paradigm [11].

A major class of open problems is now becoming obvious: what are the design principles and tools for such systems. Kilo-processor/core FPGAs and GPUs, as well as cellular supercomputers (e.g. CELL Multiprocessor) have been manufactured and the need for converting even non-topographic problems into topographic processing and memory cell arrays is a practical challenge. In this direction the computational framework of Physical versus the Virtual (multi) Cellular Machines has been introduced and the first results are already emerging.

In this paper in our review the focal points are: the new qualitative results, typical new actions, including bio-inspired ones, and various new type of physical implementations. Many other important results are missing in our review. The citation bibliography contains over 3000 papers and the author feels insecure whether some truly remarkable results have been omitted. Still, we do feel this exposition might help to get an overview about the emerging and fascinating area of cellular wave computing on topographic kilo- and mega-processor chips and systems.

2. Using Cellular Dynamics and Nonlinear Dynamical Circuits for Computation — A Prehistory

It is a remarkable historical fact that the two pioneers of digital computers, Alan Turing and John von Neumann, introduced the first two types of computer architectures that have the topographic, spatially distributed character, indeed a cellular (or "crystalline" geometric) array of elementary processing elements as the new way of computing. Interestingly, they had been made independently in about the same time.

Turing [1] used special second order continuous valued (analog) dynamic cells, inspired by chemical reactions, and applied a spatial interaction between the cells due to diffusion. The two state variables in a cell, called morphogen concentrations, were also influenced by the diffusion affect between the cells. The cells were placed topographically in a grid, a 1D grid or a 2D grid. Although, the equations were essentially linear, the practical nonlinearity inherent in any oscillator can be tacitly assumed. The reaction diffusion equations were actually partial differential difference equations (PDDE), their solutions were approximated by linear combinations of spatial-temporal waves. Turing discovered unusual phenomena in his simulations: spatial patterns depending on the local interactions between cells that are occurring in nature, however, impossible to define geometrically due to lack of any geometric regularity. Moreover, the emergence of asymmetry from symmetric structures was noticed due to instability and minor fluctuation at initial condition. In spite of the fact that Turing presented here a kind of a prototype case study only (later it turned out that his cells were locally active), the model showed a dramatically different kind of computation: using analog dynamics in spatially distributed processing cells with local interaction, and a spatial-temporal wave was the key computing mechanism.

John von Neumann, using a prototype by Stanislaw Ulam, designed a spatially distributed array of identical processing digital cells with local interactions to make self-reproducing automata [2]. Though all state variables and interactions were digital, the operational mechanism was spatial-temporal, a kind of waves. Again, the topographic, geometric address of a processing cell was important. Through a sequence of steps, due to the refractory period of his cells (reflecting the very simplified model of a neuron, understood at that time), the Cellular Automaton was able to reproduce a static pattern.

Interestingly, the need and practical possibility to use arrays of processors was far away in computer engineering for a while.

In 1973, an important principle, extending Maxwell's result for linear resistive networks to nonlinear resistive networks, the stationary principle [12], paved the way for a new way of thinking about computation as the optimum of a multivariable energy function. Namely, in any nonlinear resistive network, the stationary operating point is the minimum of a potential energy, the so called co-content function, as a sum of co-content functions of each resistor, actually being an integral of the two variables, the port current and voltage.

In the 1980's, several attempts had been made to use simple analog array processing cells (actually, dynamic circuits) for computing. The name "neural networks" was coined. A simple "neuron" model, as an amplifier with one time constant and output bipolar saturation nonlinearity, as well as a sum of weighted threshold function, has been considered. The typical network was a many input many output model, having all neuron outputs with weighted feedback connections to each other neuron input. The generic inputs were also connected with weights to all neuron inputs. Hence, having n neurons and n generic inputs, $2\,n^2$ weights defined the network. Placing inputs to the network, the settled outputs coded the input event. This, so called Hopfield network [13], had a learning rule to train the weights to perform some functionality, basically leading to some optimization of a cost function, typically a quadratic function of the state variables. Since this publication, 1984, many solutions have been proposed to solve various problems. In 1987, it was shown that, indeed, this optimization is nothing else than an application of the stationary principle for a nonlinear memoryless network. The name of the game was: solving a many-variable problem by a single circuit dynamics (with many analog processors/amplifiers).

The problem with these fully connected architectures was that practical implementations were feasible only in very limited size, and the full feedback led to instabilities (the stability criterion of symmetry cannot be ensured). The many learning algorithms though proved to be practical and useful, even if running on a single digital processor emulating the many processor/neuron network dynamics. Moreover, the neuromorphic character was mainly missing, and being one single network, its capability very limited. The term "neural" became an exaggeration in view of the great variety and complex behavior of real neurons.

3. The Standard CNN (Cellular Neural/Nonlinear Network) as the Practically Feasible Prototype Solution and Related Stability Issues

The seminal publication of the Cellular Neural/nonlinear Network (CNN) architecture in 1988 provided a different scenario [4]. The *topographic* array of simple cells ("neurons") with their important *geometric address*, as well as the precedence of locality (the cellular character) proved to be extremely important. In this architecture the robustness and stability is not restricted to the symmetry of interconnections [14].

This architecture provided for the physical, silicon and optical, implementation of kilocell/processor computing arrays. Note that due to the local feedback the indirect global interconnection is also achieved. That nature can be exploited in implementing not only local operators, however, global wave operators as well. In this way, complex PDEs, wave equations, even Navier Stokes equations for fluid dynamics can be represented by CNN operators in discretized space [15, 20, 21].

The cellular locally connected character is the basis for many neuromorphic models, literally converting real neural anatomy, their receptive fields, into cloning templates (local interaction patterns) [16,17]. This neuromorphic property can be used to generate CNN models for many sensory organs for vision, hearing, touching, etc. The key step is to convert the receptive field organization into a local interaction pattern defined by the cloning template of CNN, an obvious strength of this prototype architecture.

During the first few years of the 1990s, about a hundred various templates had been developed for different purposes, including visual, logic, PDE operators, as well as mathematical morphology, etc. [18].

Non-standard CNNs became important soon, by introducing nonlinear and delay templates, as well as space varying and adaptive templates [19].

Stability studies were important right from the beginning. Moreover, it turned out soon that analytic properties of CNN dynamics, depending on the templates, will be a vital condition to design stored programmable spatial-temporal computers. The issue is that under various initial and boundary conditions the feedback template defines the analytic properties, including stability and the maximal value of state variables.

The first challenge was to study various nonreciprocal CNN Templates, in particular when the feedback template (A template) is not centrally symmetric. The first result [14] showed that sign symmetry and positivity of the off diagonal elements in the template or the system matrix might lead to completely stable systems. This result was later generalized to many equivalent conditions [22]. Several other important results and analytic tools were published in [23–28].

A few non-stationary state patterns were soon discovered by some other template classes. A two cell chaotic behavior, and more general chaotic attractors were identified [29, 30]. An oscillating behavior was used to detect some patterns (e.g. edges) by the onset of oscillating cells [31], and an array of Chua's circuits were formed by a special template [32] and elementary spatial/temporal interaction, inspired by ant colonies was generated [33].

4. Inventing the Stored Programmable Spatial-temporal Computer: The CNN Universal Machine (CNN-UM) and the Cellular Wave Computer

The lesson learned from the practical limitations of any fixed program array processors with well defined cells and interconnections, as well as the need of the precedence of locality to make the architecture silicon-friendly, became soon clear. The role of stored programmability or algorithmic programmability in the digital computer was also clear from John von Neumann. To overcome the lack of stored programmability, another way of thinking was applied [5], combing analog/arithmetic array dynamics and logic. The new stored programmable spatial-temporal Machine, the CNN Universal Machine [6] became a fruitful paradigm. The basic idea is independent of technology, though silicon technology was considered as the main target platform.

The core of the cell array is a CNN architecture, each cell is extended by local analog/arithmetic and logic memories and a local communication and control unit connected to a Global Analog-and-logic Control Unit of the Global Analog-and-logic Program Unit. The latter contains also 3 registers for storing (and sending) the templates, the local logic operators, and the switch configurations defining and changing the local extended cell configuration (operating mode). Herewith, elementary instructions defined by spatial-temporal CNN dynamics can be used in a stored program algorithm. The optional local logic unit and local analog/arithmetic unit takes the values from the local memories, and places the results to them, respectively.

This architecture is a minimal architecture to make universal machine, universal in the logic universe, as a Turing Machine, and also as a nonlinear operator on topographic image flows. The algorithms running on this cellular wave computer can be defined formally as the so called alfa-recursive functions, as a counterpart of the mu-recursive functions for digital computers. The rigorous definition is as follows.

In the Cellular Wave Computer

- *Data are topographic flows* (*cell array* signals on a 1-, 2- or 3-dimensional grid, e.g. a visual, a tactile or an auditory flow, or the states of atoms in a molecular dynamics calculation).

Data type: *topographic flow*, Φ **(t)**, in R^n (1D), R^{nxm} (2D), or R^{nxmxp} (3D) as a function in time (continuous or, as a special case, discrete). In a 2D image flow

$$\Phi(t) : \varphi_{ij}(t)\}t \in [0, T]inR^1, \quad 1 \leq i \leq m1 \leq j \leq n$$

$\varphi_{ij}(t)$ are the *cell signals.*

For example, a 2D image flow could represent the input or the output image flow of a retina.

A nxm *Map* (e.g. an image or picture) P: at t = t∗, P = Φ (t∗)

if P is binary it is a *Mask* M

if

$$t = t0, t0 + \Delta t, t0 + 2\Delta t, \ldots\ldots t0 + k\Delta t$$

then we say that it is a *map sequence* (e.g. a video stream).

- *Instructions* are defined in space and time, typically as a spatial-temporal wave acting on the image flow data, the cell signals are *continuous* (real) valued (analog or digitally coded arithmetic) *and binary*; locally (cell by cell) stored in the cells.
- This local storage, providing for stored programmability in von Neumann sense, may be static or dynamic.
- The *protagonist elementary instruction*, Ψ (Φ), also called *wave instruction*, is defined as

$$\Phi \text{ output (t)} := \Psi(\Phi \text{ input(t)}, \text{ P}, \partial); \text{ t} \in [0, \text{T}]$$

where

Ψ: a function on image flows or image sequences
P: a map (image) defining initial state and or threshold (bias) map

∂: boundary conditions, ∂ (t) is a boundary input (might be connected to all cells in a boundary row or column)

T is the finite time interval

A scalar or logic valued functional γ on an image flow is used for branching instructions:

$$\text{q} := \gamma(\Phi \text{ input(t)}, \text{ P}, \partial);$$

For example, the so called global white functional on a binary mask M is logic 1, if the picture is full white, if at least one pixel is black it is logic 0. Another example, the maximum functional on a flow is defined by the highest scalar value at any cell signal value at any time. If the spatial-temporal instruction is of non-equilibrium type, this global output state can also be detected (Global Fluctuation, GF) or the different types of the non-equilibrium attractors can also be considered as global output parameters. Hence, GF could take values from a multivalued logic set.

We emphasize that the signal and instruction representation in this architecture may have various physical realizations, it might be analog, mixed mode, and digital using CMOS, or optical, etc.

- *As a simple special case* for the spatial-temporal dynamics, the simplest topographic wave instruction, is the standard Cellular Nonlinear Network (CNN) dynamics [4], defined for one layer of first order cells.

Having been defined the architecture and the formal description of the software (the alfa recursive function) a new world is opening for special architectural prototypes, as well as for a new kind of algorithmic thinking (like a spatial-temporal Rubik's cube).

5. Making the First Silicon Visual Microprocessors and its Computational Infrastructure — Other Physical Implementations

The next major challenge was to make the CNN Universal Machine architecture via CMOS Circuits. There was no precedence in this direction, never ever anybody designed a programmable spatial-temporal analog-and-logic dynamics algorithmically combined. In addition, optical on-chip sensors had been successfully integrated. After the first successful attempts, designing a 64 × 64 cell-processor chip (called ACE 4k), the full fledged 128x128 cell-processor was successfully made (ACE 16k) [34–37]. Next, a single chip, 2-layer CNN Universal Machine chip was made [38]. These chips paved the way, after making several applications on them, towards a commercially available visual microprocessor and vision system on a chip [40]. Several RISC architectures has also been developed, from a local shutter speed camera chip to a smart camera. The single chip EyeRIS system [40] represents the top achievement in this segment of the industry. Integration time with a speed of 20 ns on a smart camera chip means another speed record.

Other types of architectures have also been made using partially digital CMOS implementations [39, 41, 42]. The integration of the sensors onto the digital visual microprocessor chips was another challenge [43], recently a 3D integrated chip was designed, combining a sensory layer, a mixed/mode preprocessing layer, a special digital memory layer preparing the "foveal" processing in the fourth layer [44]. Optical implementation was also successful [45].

With the advent of the FPGA kilo-processor chips, a new physical implementation became possible. The first CNN Universal Machine on FPGA was, in a way a new kind of compiler, translating cellular spatial-temporal algorithms onto a FPGA [46]. Several other new directions will be studied later.

Using the analog/and/logic stored programmable chip, new vistas were found in experimenting on them. The test bed for studying various complex nonlinear phenomena led to the uncovering of spatial-temporal chaos, finding dynamic patterns of different cells, stationary, periodic and chaotic cells [47]. With another clever way of combining local noise with on chip algorithms, truly random patterns were generated [48].

6. Biological Relevance and Bio-inspiration

Following the gradual better and better understanding of the neuromorphic details of the retina, starting from the naïve "silicon retina" resistive grid model, a bio-inspired proposal was made [49], called the Bionic Eye. Though it was approximate in many aspects, the essence was practical. As it turned out by exploring the inner structure of the mammalian retina in 2001 [10], some fundamental modifications had to be made, especially reflecting the multi-channel output of a single retina. The next CNN model was more accurate [54], and many other refinements [55] and constructive combinations have been proposed.

In addition to vision, other topographic modalities were also studied. Moreover, the interesting phenomenon of hyper acuity was studied. The hyper acuity in space in electric fishes [50] and in time in barn owl [52] were successfully modeled via 1 D CNN models [51 and 53]. The hyper acuity in time model was used for implementing a high resolution time to digital conversion chip [56]. Diverse other areas, such as immune response related, color vision related, locomotion related, and tactile action related effects were also useful examples [57–60]. A receptive field ATLAS was compiled for diverse sensory phenomena in terms of CNN models [61].

7. Some Fundamental Theorems — More than PDE, Equivalence to Fully-connectedness, Analytic Theory of CA, Gödel Incompleteness

Right from the beginning, the types of universality of the CNN Universal Machine had been a main issue. The universality in the Turing sense, via the game of life, was proved [62], indeed using two layers with 2 templates. Later, other simpler constructions were also shown. The other type of universality is the functional richness of the cell by cell nonlinear operator, especially if delay templates are used as well.

For long, it was considered that the PDEs are the most complex operators in spatial-temporal description of dynamics systems. In 1987, Keener proved [71] that there exists a 1D special example for which for any finite spatial discretization there exist a range of parameters for which the qualitative behavior of the spatial-temporal response will be different compared to the spatially continuous case. This example triggered the fundamental question: is it true that for all Partial Differential Difference Equations (PDDE) defined by the CNN dynamics for each one of the cloning templates, there exists a PDE that is the limiting case in space, that is, the continuous space, version, with the same qualitative behavior? The detailed study led to the surprising conclusion: the answer is no [63]. Two types of equivalence were considered topological and numerical. It is quite interesting that Turing already studied the question of the relationship between PDE and its spatially discrete form [1], however, he did not noticed this possibility.

As mentioned before, due to the feedback template the whole network is indirectly interconnected via the direct local connections, still there were no firm results as to the equivalence of operators between the fully and locally connected dynamical systems. In [64] it was shown that for oscillator cells this equivalence is true, and in some cases a constructive proof is available that show that linear template is sufficient.

Cellular Automata were shown as special cases on CNN dynamics. In addition, several very interesting properties were discovered mostly by examples. John von Neumann's prophecy was still hanging in the air: there will be sometimes an analytic theory for binary automata [73]. That vision became true when Professor Leon Chua bridged the gap by applying a real valued representation of the binary state of a 1D Cellular Automaton with n cells [65]. The following sequence of seminal papers, e.g. [66–68] developed a brand new analytic theory of the binary universe as diverse Cellular Automata. Introducing a local memory in each cell (like in the CNN Universal Machine) showed additional properties [70].

What if we have not only an initial state, however, also an input to the Cellular Automaton, like in the CNN. This qeastion was studied and the first result was surprising [69]: much more complexity and even the capability of dynamic input detection.

Having been the deeper properties of Cellular Automata analyzed, special kind of attractors called Isles of Eden were found and their properties discovered. These Isles of Eden have been identified as special manifestations of Gödel's incompleteness theorem [72].

8. Prototype Spatial-temporal CNN Algorithms and Novel Applications

The invention of the CNN Universal Machine provided for the algorithmic combination of various spatial-temporal template dynamics, or put it in a different way, algorithms with spatial-temporal elementary instructions. Indeed, a new kind of algorithmic thinking has started forming a basis for different classes of alpha-recursive functions and related parts of a new kind of software library. Indeed, starting with the first software library, called Analogic CNN Program Library [18], a new edition has been made available almost every year, and a quite comprehensive "Software Library for Cellular Wave Computing Engines in an era of kilo-processor chips" was published and distributed with the IEEE CNNA-2010 conference Proceedings, also available via the internet [100]. Here, we will review three parts:

- Prototype functional CNN algorithms,
- Learning and adaptation methods, and
- Novel applications.

Prototype functional CNN algorithms

In some cases, a simpler version of CNN algorithms can be useful. Namely, in the so called discrete-time CNN (DTCNN) [74] the updating mechanism is performed by an iterative sequence of a given template with a memoryless (algebraic) CNN operator, actually calculating the spatial convolution and the bias term of the standard CNN equation [74]:

$$\begin{aligned} x_{ij}(n+1) &:= x_{ij}((n) + \Sigma Akl * ykl(n) \\ &\quad + \Sigma Bkl * ukl(n) \\ &\quad + z_{ij} y_{ij}(n+1) := f(x_{ij}(n=1)) \end{aligned}$$

where n is the time sequence, x and y are the state and output values of the cell and ij is the cell place and kl is a cell place in the neighborhood, f is a sigmoid function.

Many interesting applications have been performed by DT CNN that can be physically implemented in a simpler way.

As to the Boolean functions, using standard CNN dynamics, actually, practically all useful classes have been implemented [e.g. 75, 76, 77], including such functions as majority voting, parity check, and binary valued layout design algorithms.

Complex pattern calculation and skeleton generation even in a noisy environment are difficult tasks. Front propagation and active contour detection via nonlinear waves are key methods that were efficiently transformed into CNN wave dynamics operators [78].

Since the Turing patterns [1] it is a major challenge to form 2-dimensional patterns that are common in nature and are impossible to define in a geometric way. It turned out that a plethora of such patterns can be easily generated by autonomous CNN dynamics [79].

Some complex nonlinear analytic tools proved to be useful in analyzing CNN dynamics. For the continuous time case, the spectral approach was useful [80], and for the DT CNN the NLq theory was applied [81].

Exploring more complex spatial-temporal dynamics was possible via more complex cells, in particular using chaotic cells [82]. In addition to different wave dynamics, synchronization phenomena and related synchronization patterns had also been found [83].

A few typical more complex functions and special spatial filters had been invented. In particular, for binary and gray scale mathematical morphology [84], adaptive nonlinear spatial filters [85], separating signatures of various sound events [86], and computing optical flows [87].

A fundamental drift from classical problem setting emerged when the spatial/temporal wave nature of CNN dynamics was directly used in solving some complex tasks. In reticular, following the autowave metric principle of V.I. Krinsky et. al. [88], an autowave metric was developed by using a single CNN dynamic autowave for object classification [89,90,91].

Learning and adaptation methods

There have been several methods to "invent" or design or identify different templates to use standard or non-standard CNN dynamics as elementary instructions or combining them as subroutines for various tasks. A new direction was taken when learning methods had been used to identify and compute these templates [92, 93]. A new kind of adaptation was shown by introducing adaptive cells, instead of adapting the templates [94]. Introducing constraint diffusion led to a new kind of adaptation, as well [95].

Many other learning and adaptation techniques have been developed through the years with various combinations of template learning, cell adaptation, constraint diffusion techniques and their combination. A major step forward was represented by the systematic design procedure, the "art of CNN template design" [96].

Novel applications

The various kinds of applications started already in the early years [97], naturally, via simulations. Image compression applications [98], robust template design [101], detecting epilepsy seizures in brain electrical activity [105], multi-adaptive architecture for unmanned aerial vehicles (UAV) [102], implementing on/chip hyper acuity in time [106], and video authentication and copyright protection [99], are just a few of the many inventions. The latter problem partially influenced a new chip architecture (ACE 4k) to include a programmable boundary condition hosting the periodic boundary condition as well.

The software library [100] containing well over hundred templates, subroutines and applications influenced the invention of new algorithms. The emerging new capabilities of the cellular visual microprocessor chips provided for new applications, like ultra high frame rate applications going up to 50,000 frame per second [104].

The building of the stand alone ultra high speed cellular vision system [103] available in a lim ited number of pieces heralded the *beginning of a new phase in Cellular Wave Computing technology.* The spin-off companies, especially AnaFocus and Eutecus corporations, and hence the commercial availability of small stand alone visual systems with many-kilo-processor chips, both mixed mode (fully interfaced to outside digital systems) and cellular digital (like FPGAs), changed the technology scenario.

9. Physical and Virtual Cellular Machines with Kilo- and Mega-processor Chips and Related Topographic Algorithms

Since the beginning of our new century, silicon technology is shrinking towards the nanoscale domain. In addition to novel CMOS devices, the nanoscale CMOS technology is showing an astonishing march towards the Billion transistor era [107]. This means the mergence of kilo- and mega/processor chips. In this domain, in addition to the processor and memory array complexity, however, physical constraints force two major architectural and algorithmic constraints:

- *the precedence of locality*, that is the cellular character with a topographic architecture, where the *geometric address* of a processor or memory plays a crucial role due to the significant communication bandwidth/delay between the individual processors and memories, and
- the combined *optimization of an algorithm trading off 6 physical parameters*: speed, power, size/weight, energy dissipation, bandwidth/delay and accuracy.

As a consequence, the traditional computer science *notions and principles* as well as the traditional computer engineering *design methods and tools* are *no longer valid.*

This situation has already been identified in nanoscale device related research [108], and the need to go also beyond Boolen Logic was analyzed [109]. The fact that already kilo-processor FPGA chips pose new possibilities was shown in a complex application problem, computing compressible flow dynamics [110].

The need for cellular nanoscale wave computing and some novel principles and design methods has been recently published in an edited book [111]. The basic principles to bridge the gap between the physical and virtual machines, that are typically characterized by various multi-cellular architectures, has been formulated and two non-trivial application fields were shown [112]. The interfacing of the nanoscale electromagnetic field operated devices and the sub-micron scale CMOS chips is a technological challenge [113]. New design methods for computationally intensive 2D topographic problems had been presented in [114]. An efficient wave like solution of a wave problem was demonstrated in [115].

Implementing non-topographic problems on topographic architectures is another major algorithmic challenge. In [116] such a problem was solved for particle filters. A general problem is the geometric partition of program control, since there might not be possible for a global control. Such a problem was solved via local data driven methods [117].

Indeed, the new world of algorithm on topographic physical architectures is just opening for the computer science and computer engineering communities.

10. Conclusions and Major New Challenges

Motivated and sometimes required by the physical constraints and possibilities of semiconductor technology and new discoveries in neurobiology, respectively, the very notion of computers and computing has gradually changed during the last 20 years. Major new aspects are

- the need for the introduction and key role of a geometric or topographic address of a processor or a memory,
- the need for using continuous valued basic processing building blocks, and
- to use a new kind of algorithmic quality, the alpha recursive functions where elementary instructions are spatial-temporal dynamic processes

Nanoscale technologies, both CMOS and non-CMOS, both digital or analog-and-logic computing models need the spatial-temporal dynamic properties of algorithms. This fact leads to a new kind of computer engineering and computer science tha will help to design a realm of new algorithms for kilo- and mega- processor chip based computers. In addition, the non-CMOS nanoscale technologies require this kind of theory and design methods due to the use of new building blocks and interactions defined by physical processes.

Recent discoveries and related quantitative spatial-temporal nonlinear models in neurobiology also provide for new ideas, among others the role of synchronization in detecting events.

Acknowledgments

The ongoing support of the Hungarian Academy of Sciences, the Pazmany P. Catholic University, the Office of Naval Research, and a recent support from Intel Corporation are gratefully acknowledged. Comments from Dr. G. Pazienza and Professors P. Szolgay and Á. Zarándy are kindly acknowledged.

References

[1] A.M. Turing, The Chemical Basis of Morphogenesis, *Philosophical Transactions of the Royal Society of London. Series B, Biological Sciences*, Vol. 237, No. 641. pp. 37–72 Aug. 14, 1952

[2] J. von Neumann, The Theory of Automata: Construction, Reproduction, Homogeneity, Part II. in J.v. Neumann (edited and completed by A.W. Burks), *Theory of Self-Reproducing Automata*, University of Illinois Press, Urbana and London, 1966 (on cellular automata models: 1952–53)

[3] J. Kennedy and L. O. Chua, Circuit Theoretic Solutions for Neural Networks — an Old Approach to a New Problem, *Proceedings of the First International Conference on Neural Networks (ICNN)*, Vol. II, pp. 169–176, June 1987 (San Diego, CA).

[4] L. O. Chua and L. Yang, Cellular Neural Networks: Theory, *IEEE Trans. on Circuits and Systems, (CAS)*, Vol. 35. pp. 1257–1272, 1988

[5] T. Roska, Analog events and a dual computing structure using analog and digital circuits and operators, in *Discrete Event Systems: Models and Applications* (eds. P. Varaiya and A. B. Kurzhanski), New York, Springer, 1988

[6] T. Roska and L. O. Chua, The CNN Universal Machine — an Analogic Array Computer, *IEEE Trans. Circuits and Systems II*, Vol. 40, pp. 163–173, 1993

[7] L. O. Chua, CNN: A Paradigm for Complexity, World Scientific, Singapore, 1998

[8] T. Roska and Á. Rodriguez-Vázquez, Towards the visual microprocessor: VLSI Design and the use of Cellular Neural Network Universal Machines, John Wiley, 2000

[9] L. O. Chua and T. Roska, Cellular Neural Networks and Visual Computing, Cambridge University Press, Cambridge, UK, 2002

[10] B. Roska and F.S. Werblin, Vertical interactions across ten parallel, stacked representations in the-mammalian retina, *Nature*, Vol. 410, pp. 583–87, March 29, 2001

[11] T. Roska, Computational and Computer Complexity of Analogic Cellular Wave Computers, *Journal of Circuits, Systems and Computers*, Vol. 12, pp. 539–562, 2003

[12] L. O. Chua, "Stationary principles and potential functions for nonlinear networks, *J. Franklin Institute*, Vol. 296, pp. 91–114, Aug. 1973
[13] J.J. Hopfield, Neurons with graded response have collective computational properties like those of two-state neurons, *Proc. Natl. Acad. Sci. U.S.A.*, Vol. 81, pp. 3088–3092, 1984
[14] L.O. Chua and T. Roska, Stability of a Class of Nonreciprocal Cellular Neural Networks, *IEEE Trans. on Circuits and Systems, (CAS)*, Vol. 37, pp. 1520–1527, 1990
[15] P. Szolgay, G. Vörös and Gy. Erôoss, On the applications of the Cellular Neural Network (CNN) paradigm in mechanical vibrating systems, *IEEE Transactions on Circuits and Systems*, Vol. 40, pp. 222–227, 1993
[16] F.S. Werblin, Synaptic connections, receptive fields, and pattern of activity in the tiger salamander retina, *Investigative Ophthalmology and Visual Science, (Investigative Ophthalmology and Visual Science)*, Vol. 32, pp. 459–483, 1991
[17] T. Roska, K. Lotz, J. Hámori, E. Lábos and J. Takács, The CNN model in the visual system — Part 1: The CNN-retina and some direction- and length- selective mechanisms, *Research report of the Analogic (Dual) and Neural Computing Systems Laboratory, (DNS-8-1991)*, Budapest MTA SZTAKI, 1991
[18] Analogic CNN Program Library, Version 6.1, (eds. T. Roska and L. Kék), *Research report of the Analogic (Dual) and Neural Computing Systems Laboratory, (DNS-5-1994)*, Budapest MTA SZTAKI, 1994
[19] L.O. Chua and T. Roska, The CNN paradigm, *IEEE Trans. on Circuits and Systems I: Fundamental Theory and Applications, (CAS-I)*, Vol. 40, No. 3, pp. 147–156, 1993
[20] T. Roska, L.O. Chua, D.Wolf, T. Kozek, R. Tetzlaff, F. Puffer, Simulating nonlinear waves and partial differential equations via CNN — Part I. Basic Techniques, *IEEE Trans. on Circuits and Systems I: Fundamental Theory and Applications, (CAS-I)*, Vol. 42, No. 10, pp. 807–815, 1995
[21] T. Kozek, T. Roska, A Double Time-Scale CNN for Solving Two-dimensional Navier-Stokes Equations, *Int. J. Circuit Theory and Applications, (CTA)*, Vol. 24, pp. 49–56, 1996
[22] L. O. Chua and C. W. Wu, On the Universe of Stable Cellular Neural Networks, *Int. J. Circuit Theory and Applications, (CTA)*, Vol. 20. pp. 497–518, 1992
[23] T. Roska, C.W. Wu, M. Balsi, and L.O. Chua, Stability and dynamics of delay-type general and cellular neural networks, *IEEE Trans. on Circuits and Systems I: Fundamental Theory and Applications, (CAS-I)*, Vol. 39, No. 6, pp. 487–490, 1992
[24] P. Thiran, Influence of Boundary Conditions on the Behavior of Cellular Neural Networks, *IEEE Trans. on Circuits and Systems I: Fundamental Theory and Applications, (CAS-I)*, Vol. 40, No. 3, pp. 207–212, 1993
[25] P.P. Civalleri and M. Gilli, Some Dynamic Phenomena in Delayed Cellular Neural Networks, *Int. J. Circuit Theory and Applications, (CTA)*, Vol. 22. pp. 77–105, 1994
[26] M. Forti and A. Tesi, New Conditions for Global Stability of Neural Networks with Application to Linear and Quadratic Programming Problems, *IEEE Trans. on Circuits and Systems I: Fundamental Theory and Applications, (CAS-I)*, Vol. 42, No. 7, pp. 354–366, 1995
[27] S. Arik and V. Tavsanoglu, Equilibrium Analysis of Delayed CNN's, *IEEE Trans. on Circuits and Systems I: Fundamental Theory and Applications, (CAS-I)*, Vol. 45, No. 2, pp. 168–171, 1998
[28] M. Di Marco, A. Tesi, and M. Forti, Bifurcations and oscillatory behavior in a class of competitive Cellular Neural Networks, *Int. Journal of Bifurcation and Chaos, (Int. J. Bifurcation and Chaos)*, Vol. 10, No. 6, pp. 1267–1293, 2000
[29] F. Zou and J.A. Nossek, A chaotic attractor with cellular neural networks, *IEEE Trans. on Circuits and Systems, (CAS)*, Vol. 38. pp. 811–812, 1991
[30] F. Zou, J.A. Nossek, Bifurcation and Chaos in Cellular Neural Networks, *IEEE Trans. on Circuits and Systems I: Fundamental Theory and Applications, (CAS-I)*, Vol. 40, No. 3, pp. 166–173, 1993
[31] C-W. Wu, T. Roska, and L. O. Chua, Celullar Neural Networks Operating in Oscillatory Mode, *Memorandum UCB/ERL*, M94/5, Berkeley University of California at Berkeley, 1994
[32] P. Arena, S. Baglio, L. Fortuna and G. Manganaro, Chua's Circuit Can Be Generated by CNN Cells, *IEEE Trans. on Circuits and Systems I: Fundamental Theory and Applications, (CAS-I)*, Vol. 42, No. 2, pp. 123–125, 1995
[33] L. Nemes and T. Roska, A CNN Model of Oscillation and Chaos in Ant Colonies: A Case Study, *IEEE Trans. on Circuits and Systems I: Fundamental Theory and Applications, (CAS-I)*, Vol. 42, N. 10, pp. 741–745, 1995
[34] R. Domíniguez-Castro, S. Espejo, A. Rodriguez-Vázquez, R. Carmona, A CNN Universal Chip in CMOS Technology, *Proceedings of IEEE Int. Workshop on Cellular Neural Networks and Their Applications, (CNNA'94)*, pp. 91–96, Rome, 1994
[35] S. Espejo, R. Carmona, R. Domínguez-Castro and Á. Rodríguez-Vázquez, "A CNN Universal Chip in CMOS Technology", *International Journal*

of Circuit Theory and Applications, Vol. 24, pp. 93–109, 1996

[36] G. Linan, R. Domínguez-Castro, E. Roca, S. Espejo and Á. Rodríguez-Vázquez, A Mixed-signal 64x64 CNN Universal Chip, in [8], pp. 258–279, 2000

[37] T. Roska and Ángel Rodríguez-Vázquez, Toward Visual Microprocessors, Invited Paper, *Proceeding of the IEEE*, Special Issue on: Visual Perception: Technology & Tools, Vol. 90, pp. 1244–1257, 2002

[38] R. Carmona, F. Jiménez-Garrido, R. Domínguez-Castro, S. Espejo, T. Roska, Cs. Rekeczky, I. Petrás and Á. Rodríguez-Vázquez, A Bio-Inspired two-Layer Mixed-Signal Flexible Programmable Chip for Early Vision. IEEE Transactions on Neural Networks, Vol. 14, issue 5, pp. 1313–1336, ISSN 1045–9227, 2003

[39] P.J. Dudek, An Asynchronous Cellular Logic Network for Trigger-Wave Image Processing on Fine-Grain Massively Parallel Arrays, *IEEE Transactions on Circuits and Systems — II: Analog and digital signal processing*, Vol. 53, pp. 354–358, 2006

[40] www.anafocus.com

[41] P. Földesy, Á. Zarándy, Cs. Rekeczky, and T. Roska, Digital implementation of the cellular sensor-computers, *International Journal of Circuit Theory and Applications*, Vol. 34, No. 4, pp. 409–428, 2006

[42] M. Laiho, A.Paasio, J. Flak and K.A.I. Halonen, Template design for cellular nonlinear networks with 1-bit weights, *IEEE Transactions on Circuits and Systems I: Regular Papers*, Vol. 55, pp. 904–913, 2008

[43] Á. Zarándy, P. Földesy, R. Carmona, Cs. Rekeczky, J. A. Bean, and W. Porod, Cellular Multi-core Processor Carrier Chip for Nanoantenna Integration and Experiments, *in Cellular Nanoscale Sensory Wave Computing*, edited by C. Baatar, W. Porod and T. Roska, Springer, New York, ISBN: 978-1-4419-1010-3, 2010, pp. 147–162, 2010

[44] P. Földesy, R. Carmona-Galan, Á. Zarándy, Cs. Rekeczky, Á. http://www.scopus.com/authid/detail.url?authorId=35820042900 Rodríguez-Vzquez, T. http://www.scopus.com/authid/detail.url?authorId=35496244100Roska, Digital processor array implementation aspects of a 3D multi-layer vision architecture, *in Proceedings of the 12th International Workshop on Cellular Nanoscale Networks and their Applications, (IEEE CNNA2010)*, Berkeley (USA), pp. 329–332, 2010

[45] A. Ayoub, Sz. Tõkés, L. Orzó, Optical Cellular Wave Computer Implementation and Programming, *in Proceedings of the 10th IEEE International Workshop on Cellular Neural Networks and their Applications,* pp. 57–62, Istanbul, Turkey, 2006

[46] Z. Nagy, P. Szolgay, "Configurable Multi-layer CNN-UM Emulator on FPGA", *IEEE Transactions on Circuits and Systems I: Fundamental Theory and Applications*, Vol. 50, pp. 774–778, 2003

[47] I. Petrás, T. Roska, and L.O. Chua, New spatial-temporal patterns and the first programmable on-chip bifurcation test bed, *IEEE Transactions on circuits and Systems I*, Vol. 50, pp. 619–633, 2003

[48] M.Ercsey-Ravasz, T. Roska, Z. Néda, Stochastic simulation on the Cellular Wave Computers, *European Physics Journal*, Vol. 51. pp. 407–411, 2006

[49] F.S. Werblin, T. Roska, L.O. Chua, The analogic cellular neural network as a bionic eye, *Int. J. Circuit Theory and Applications, (CTA)*, Vol. 23, N. 6, pp. 541–569, 1995

[50] P. Baldi and W. Heiligenberg, How sensory maps could enhance resolution through ordered arragements of broadly tuned receivers, *Biol. Cybern.*, Vol. 59, pp. 313–318, 1988

[51] W. Heiligenberg and T. Roska, On biological sensory information processing principles relevant to cellular neural networks, in *Cellular Neural Networks*, (eds. T. Roska and J. Vandewalle), pp. 201–211, J.Wiley & Sons, Chichester, 1993

[52] M. Konishi, Neural mechanisms of auditory image formation, in: *The Cognitive Neurosciences*, Ed. M. S. Gazzaniga, MIT Press, pp. 269–277, 1995

[53] K. Lotz, L. Bölöni, T. Roska and J. Hámori, Hiperacuity in Time: a CNN model of a time-coding pathway of sound localization, *IEEE Trans. on Circuits and Systems I: Fundamental Theory and Applications, (CAS-I)*, Vol. 46, No. 8, pp. 994–1002, 1999

[54] D. Bálya, B. Roska, T. Roska, F.S. Werblin, A CNN Framework for Modeling Parallel Processing in a Mammalian Retina", *International Journal of Circuit Theory and Applications*, Vol. 30, pp. 363–393, 2002

[55] D. Bálya, Sudden Global Spatial-Temporal Change Detection and its Applications, *Journal of Circuits, Systems, and Computers (JCSC)*, Vol. 12, No. 6, pp. 845–856, 2003

[56] A Mozsáry, J-F Chung, T Roska, Function-in-layout: A demonstration with bio-inspired with bio-inspired hyperacuity chip, *International Journal of Circuit Theory and Applications*, Vol. 35, pp. 149–164, 2007

[57] Gy. Cserey, A. Falus, T. Roska, Immune Response Inspired Spatial-Temporal Target Detection Algorithms With CNN-UM, *International Journal of Circuit Theory and Applications,* Vol. 34, No. 1, pp. 21–47, 2006

[58] Á. Zarándy, L. Orzó, E. Grawes, and F. Werblin, CNN Based Models for Color Vision and Visual

Illusions, *IEEE Trans. on Circuits and Systems I*, Vol. 46, No. 2, pp. 229–238, 1999

[59] P. Arena, L. Fortuna and M. Branciforte, Reaction-Diffusion CNN Algorithms to Generate and Control Artificial Locomotion, *IEEE Trans. on Circuits and Systems I*, Vol. 46, No. 2, pp. 253–260, 1999

[60] A. Kis, F. Kovács, P. Szolgay, 3D Tactile sensor array processed by CNN-UM: a fast method for detecting and identifying slippage and twisting motion, *International Journal of Circuit Theory and Applications,* Vol. 34, No. 4, pp. 517–531, 2006

[61] V. Gál, J. Hámori, T. Roska, D. Bálya, Zs. Borostyánköi, M. Brendel, K. Lotz, L. Négyessy, L. Orzó, I. Petrás, Cs. Rekeczky, J. Takács, P. Venetiáner, Z. Vidnyánszky, Á. Zarándy, Receptive Field Atlas and Related CNN Models, *International Journal of Bifurcation and Chaos*, World Scientific Publishing Company, Vol. 14, No. 2, pp. 551–584, 2004

[62] L.O. Chua, T. Roska, and P.L. Venetianer, The CNN is Universal as the Turing Machine, *IEEE Trans. Circuits and Systems-I:Fundamental Theory and Applications,* Vol. 40, pp. 289–291, 1993

[63] M. Gilli, T. Roska, L. O. Chua and P. P. Civalleri, CNN Dynamics Represents a Broader Class than PDEs, *International Journal of Bifurcation and Chaos*, Vol. 12, pp. 2051–2068, 2002

[64] F. Corinto, M.Gilli, and T. Roska, On Full-Connectivity Properties of Locally Connected Oscillatory Networks, *IEEE Trans. on Circuits and Systems I*, Vol. 58, No. 5, pp. 1063–1075, 2011

[65] L.O. Chua, S. Yoon, R. Dogaru, A nonlinear dynamics perspective of Wolfram's new kind of science, Part I: Threshold of complexity, *Int. J. Bifurcation and Chaos* 12, 2655–2766, 2002

[66] L.O. Chua, K. Karacs, V.I. Sbitnev, J. Guan, and S. Jinwook, S. A nonlinear dynamics perspective of Wolfram's new kind of science. Part VIII: More isles of eden, *Int. J. Bifurcation and Chaos*, Vol. 17, 3741–3894, 2007

[67] L.O. Chua, G. Pazienza, L. Orzo, V. Sbitnev, and J. Shin, A nonlinear dynamics perspective of Wolfram's new kind of science. Part IX: Quasiergodicity, *Int. J. Bifurcation and Chaos*, Vol. 18, 2487–2642, 2008

[68] L. O. Chua and G. E. Pazienza, Nonlinear Dynamics Perspective of Wolfram's New Kind of Science, Part XIII: Bernoulli σ_τ-shift rules, *International Journal of Bifurcation and Chaos*, http://www.worldscinet.com/ijbc/mkt/archive.shtml?2010&20 Volume: 20, http://www.worldscinet.com/ijbc/20/2007/S0218127410207.html Issue: 7, pp. 1859–2003, 2010

[69] G. Pazienza and T. Roska, Dynamics of Cellular Automata with Input: Basis for a systematic study, *Int. J. Circuit Theory and Applications*, in press, 2011

[70] G.J. Martinez, A. Adamatzky, and R. Alonso-Sanz, How to make dull cellular automata complex by adding memory: Rule 126 case study, *Complexity*, Vol. 15, pp. 34–49, 2010

[71] J.P. Keener, J.P. Propagation and its failure in coupled systems of discrete excitable cells, *SIAM J. Appl. Math.* Vol. 47, pp. 556–572, 1987

[72] T. Roska, Isle of Eden in 1D Binary Cellular Automaton As a Manifestation of Gödel Incompleteness and a Proposal for a Bridge Between Analytical Results and Spatial-Temporal Logic Patterns, *International symposium on nonlinear theory and its applications. (NOLTA 2010)*, pp. 593–594, Krakow, 2010

[73] J. Von Neumann, Papers of John von Neumann on Computing and Computer Theory, ed. W. Aspray and A. Burks, (Section IV: Theory of Natural and Artificial Automata), The MIT Press and Tomash Publ., Los Angeles/San Francisco, 1987

[74] H. Harrer and J.A. Nossek, Discrete-time Cellular Neural Networks, *Int. J. Circuit Theory and Applications, (CTA)*, Vol. 20, pp. 453–467, 1992

[75] L. Nemes, L. O. Chua, T. Roska, Implementation of Arbitrary Boolean Functions on the CNN Universal Machine, *Int. J. Circuit Theory and Applications*, Vol. 26, pp. 593–610, 1998

[76] F Y. Chen, G. L. He and G. Chen, "ealization of Boolean functions via CNN: Mathematical theory, LSBF and template design, "IEEE Trans. on Circ. Syst.-I, Vol. 53, pp. 2203–2213, Oct. 2006

[77] P.L. Venetianer, P. Szolgay, K.R. Crounse, T. Roska, L.O. Chua, Analogue Combinatorics and Cellular Automata — Key Algorithms and Layout Design, *Int. J. Circuit Theory and Applications, (CTA)*, Vol. 24, pp. 145–164, 1996

[78] Cs. Rekeczky and L.O. Chua, "omputing with Front Propagation: Active Contour and Skeleton Models in Continuous-Time CNN" *Journal of VLSI Signal Processing*, Vol. 23, No.2/3, pp. 373–402, Kluwer, 1999

[79] P. Thiran, K.R. Crounse, L.O. Chua, and M. Hasler, Pattern Formation Properties of Autonomous Cellular Neural Networks, *IEEE Trans. on Circuits and Systems I: Fundamental Theory and Applications, (CAS-I)*, Vol. 42, No. 10, pp. 757–774, 1995

[80] P.P. Civalleri and M. Gilli, A Spectral Approach to the Study of Propagation Phenomena in CNNs, *Int. J. Circuit Theory and Applications, (CTA)*, Vol. 24, issue 1, pp. 37–48, 1996

[81] J.A. Suykens and J. Vandewalle, Discrete Time Interconnected Cellular Neural Networks within

NLq Theory, *Int. J. Circuit Theory and Applications, (CTA)*, Vol. 24, issue 1, pp. 25–36, 1996

[82] M.J. Ogorzalek, Z. Galias, W. Dabrowski and A. Dabrowski,Spatio-temporal Co-operative Phenomena in CNN Arrays Composed of Chaotic Circuits - Simulation Experiments, *Int. J. Circuit Theory and Applications, (CTA)*, Vol. 24, issue 3, pp. 261–268, 1996

[83] S. Jankowski, A. Londei, A. Lozowski and C. Masur, Synchronization and Control in a Cellular Neural Network of Chaotic Units by Local Pinnings, *Int. J. Circuit Theory and Applications, (CTA)*, Vol. 24. issue 3. pp. 275–282, 1996

[84] Á. Zarándy, A. Stoffels, T. Roska, and L. O. Chua, Implementation of Binary and Gray-Scale Mathematical Morphology on the CNN Universal Machine, *IEEE Trans. on Circuits and Systems I: Fundamental Theory and Applications, (CAS-I)*, Vol. 45, No. 2, pp. 163–168, 1998

[85] Cs. Rekeczky, T. Roska, and A. Ushida, CNN-based Difference-controlled Adaptive Nonlinear Image Filters, *Int. J. Circuit Theory and Applications, (CTA)*, Vol. 26, pp. 375–423, 1998

[86] J.A. Osuna and G.S. Moschytz, On the Separating Capability of Cellular Neural Networks, *Int. J. Circuit Theory and Applications, (CTA)*, Vol. 24, issue 3, pp. 253–260, 1996

[87] L. Török, Á. Zarándy, Analog VLSI, Array Processor Based, Bayesian, Multi-Scale Optical Flow, *International Journal of Circuit Theory and Applications*, Vol. 34, No. 1, pp. 49–75, 2006

[88] V.I. Krinsky, V.N. Biktașhev, and I.R. Efimov, Autowave principles for parallel image processing, *Physica D*, Vol. 49, pp. 247–253, 1991

[89] I. Szatmári, Cs. Rekeczky and T. Roska, A nonlinear wave metric and its CNN implementation for object classification, *Journal of VLSI Signal Processing* Vol. 23, No. 2/3, pp. 437–448, 1999

[90] I. Szatmári, Spatio-temporal Nonlinear Wave Metric for Binary and Gray-scale Object Comparison on Analogic Cellular Wave Computers, *Int. Journal of Functional Differential Equations*, Vol. 13, pp. 89–97, 2006

[91] I. Szatmári, Object comparison using PDE-based wave metric on cellular neural networks, *International Journal of Circuit Theory and Applications*, Vol. 34, No. 4, pp. 359–382, 2006

[92] F. Zou, S. Schwarz and J.A. Nossek, Cellular Neural Network Design Using a Learning Algorith, *Proceedings of IEEE Int. Workshop on Cellular Neural Networks and Their Applications, (CNNA'90)*, pp. 73–81, Budapest, 1990

[93] J. A. Nossek, Design and Learning with Cellular Neural Networks, *Int. J. Circuit Theory and Applications, (CTA)*, Vol. 24, issue 1, pp. 15–24, 1996

[94] R. Dogaru, L. O. Chua, and K. Crounse, "iramidal Cells: A novel class of adaptive coupling cells and their applications for cellular neural networks, *IEEE Trans. on Circuits and Systems I: Fundamental Theory and Applications, (CAS-I)*, Vol. 45, No. 10, pp. 1077–1090, 1998

[95] Cs. Rekeczky, CNN Architectures for Constrained Diffusion Based Locally Adaptive Image Processing, *International Journal of Circuit Theory and Applications*, Vol. 30, pp. 313–348, 2002

[96] Á. Zarándy, The Art of CNN Template Design, *Int. J. Circuit Theory and Applications*, Vol. 17, No. 1, pp. 5–24, guest ed: T. Roska, A. Rodriguez-Vazquez and J. Vandewalle, 1999

[97] T. Roska and J. Vandewalle (eds.), Cellular Neural Networks, J. Wiley, Chichester, New York, 1993

[98] M. Tanaka, Y. Tanji, M. Onishi, and M. Nakaguchi, Lossless image compression and reconstruction by cellular neural networks, *Proc. IEEE Cellular Neural Networks and Their Applications, (CNNA 2000)*, pp. 57–62, 2000

[99] M. Csapodi, J. Vandewalle and B. Preneel, CNN Algorithms for Video Authentication and Copyright Protection, *Journal of VLSI Signal Processing*, Vol. 23, No. 2/3, pp. 449–464, 1999

[100] Software Library for Cellular Wave Computing Engines in an era of kilo-processor chips, Version 3.1, edited by K. Karacs, Gy. Cserey, Á. Zarándy, P. Szolgay, Cs. Rekeczky, L. Kék, V. Szabó, G.Pazienza, and T. Roska, Report, Cellular Sensory and Wave Computing Laboratory of of the Computer and Automarion Research Institute of the Hungarian Academy of Sciences and the Jedlik Laboratories of the Pázmány P. Catholic University, Budapest, Hungary, 2010

[101] M. Haenggi and G. M. Moschytz, Analytic and VLSI Specific Design of Robust CNN Templates, *Journal of VLSI Signal Processing*, Vol. 23, No. 2/3, pp. 415–428, 1999

[102] Cs. Rekeczky, I. Szatmári, D. Bálya, G. Tímár, and Á. Zarándy, Cellular Multiadaptive Analogic Achitecture: A Computational Framework for UAV Applications, IEEE Transactions on Circuits and Systems — I.: Regular Papers, Vol. 51, No. 5, pp. 864–884, 2004

[103] Á. Zarándy, Cs. Rekeczky, Bi-i: A Standalone Ultra High Speed Cellular Vision System, *IEEE Circuits and Systems Magazine*, second quarter, pp. 36–45, 2005

[104] Á. Zarándy, R. Dominguez-Castro, S. Espejo, "ltra-high frame rate focal plane image sensor and processor" *IEEE Sensors Journal*, Vol. 2, pp. 559–565, 2002

[105] R. Kunz, R. Tetzlaff, and D. Wolf, "rain Electrical Activity in Epilepsy Characterization of the Spatio-temporal Dynamics with Cellular Neural

Networks Based on a Correlation Dimension Analysis" *IEEE International Symposium on Circuits and Systems* (ISCAS 00), Vol. pp. 2000

[106] A. Mozsáry, et. al. "unction-in-Layout: a Demonstration with Bio-Inspired Hyperacuity Chip" *International J. Circuit Theory and Applications*, Vol. 35, No. 2, pp. 149–164, 2007

[107] ITRS (International Technology Roadmap for Semiconductors) Roadmaps 2003, 2005, 2007, 2009

[108] W. Porod et al., "ioinspired Nano-Sensor Enhanced CNN Visual Computer,"n The Coevolution of Human Potential and Converging Technologies, Ed. M.C. Roco, C. Montemagno, *Annals of the NY Academy of Sciences*, Vol. 1013, pp. 92–109, 2004

[109] T. Roska, "ircuits, "omputers, and beyond Boolean Logic" *International Journal of Circuit Theory and Applications*, Vol. 35, pp. 485–496, 2007

[110] S. Kocsárdi, Z. Nagy, Á. Csík, P. Szolgay, Two-dimensional Compressible Flow Simulation on Emulated Digital CNN-UM, *International Journal of Circuit Theory and Applications*, Vol. 37, Issue 4, pp. 569–585, 2009

[111] C. Baatar, W. Porod, and T. Roska (Editors), Cellular Nanoscale Wave Computing, Springer, New York, ISBN: 978-1-4419-1010-3, 2010

[112] T. Roska, L.Belady, M.Ercsey-Ravasz, "ellular Wave Computing in Nanoscale via Million Processor Chips" in Cellular Nanoscale Sensory Wave Computing, edited by C. Baatar, W. Porod and T. Roska, Springer, New York, ISBN: 978-1-4419-1010-3, pp. 5–25, 2010

[113] Á. Zarándy, P. Földesy, R. Carmona, Cs. Rekeczky, J.A. Bean, and W. Porod, "Ellular Multi-core Processor Carrier Chip for Nanoantenna Integration and Experiments" Chapter 7 in Cellular Nanoscale Sensory Wave Computing, edited by C. Baatar, W. Porod and T. Roska, Springer, New York, ISBN: 978-1-4419-1010-3, pp. 147–162, 2010

[114] Á. Zarándy, Cs. Rekeczky, "ow-Power Processor Array Design Strategy for Solving Computationally Intensive 2D Topographic Problems" in Cellular Nanoscale Sensory Wave Computing, edited by C. Baatar, W. Porod and T. Roska, Springer, New York,ISBN: 978-1-4419-1010-3, pp. 215–245, 2010

[115] Z. Nagy, Zs. Vörösházi, P. Szolgay, "Emulated Digital CNN-UM Solution of Partial Differential Equations", *International Journal of Circuit Theory and Applications*, Vol. 34, No. 4, pp. 445–470, 2006

[116] A. Horváth and M. Rásonyi, "ast computation of particle filters on processor arrays" *Proceedings of the IEEE International Workshop on Cellular Nanoscale Networks and their Applications — Towards Megaprocessor Computing* (CNNA 2010, Berkeley), pp. 215–219, ISBN: 978-1-4244-6680-1, 2010

[117] Cs. Nemes, Z. Nagy, M. Ruszinkó, A. Kiss, P. Szolgay, "apping of High Performance Dataflow Graphs into Programmable Logic Devices" *Proceedings of the 2010 International Symposium on Nonlinear Theory and its Applications, NOLTA2010*, pp. 99–102, Krakow, Poland, September 5–8, 2010, ISBN: 978-4-88552-251-2

CHAPTER 5

CONTRIBUTIONS OF CNN TO BIO-ROBOTICS AND BRAIN SCIENCE

PAOLO ARENA* and LUCA PATANÉ

Dipartimento di Ingegneria Elettrica, Elettronica e Informatica,
Viale A. Doria 6 95125 Catania, Italy
**parena@dieei.unict.it*

The paradigm of Cellular Non-linear Networks is ubiquitously applied in different research fields. In this chapter the breakthrough in bio-robotics and brain science research is focused. In particular the implementation of CNN-based CPGs is discussed proposing different implementations on bio-inspired robots: hexapods, lamprey-like structures, crab-inspired platforms and others. Furthermore the locomotion control system has been extended following a bottom-up procedure to include higher cognitive capabilities. The CNN paradigm was applied to design and implement perceptual architecture inspired by the insect world. Reaction-Diffusion CNN systems have been used to model the arousal of behavioral solution optimized to the on-going environmental conditions. An approach based on Turing patterns is illustrated including experimental results on navigation control with a roving robot. Finally, interesting works based on the olfactory system of insects, modeled with the Winner-less Competition principle, are reported and discussed.

1. Introduction

Cellular Non-linear Networks are ubiquitous and their applications invest several different fields of research, from theoretical studies to hardware implementation, to industrial applications. The possibilities to realize CNNs in hardware both with discrete components and with VLSI devices is extremely important to realize control systems for robots. The extremely powerful sensory motor capabilities of living beings are largely due to the peculiarities of the neural architectures that acquire signals, process them, integrate with other sensory modalities and send the appropriate motor control commands to the muscle system. All these functions work concurrently and, although neuron main communication tools are spikes, their intrinsic dynamics is inherently analog, whereas the overall neural processing is peculiarly parallel and asynchronous. The somatotopic organization of sensory motor maps is an archetype of how living beings interact with outer world through proprioceptive and exteroceptive mechanisms. This Chapter will deal on how CNN largely contributed to a new generation of locomotion and perception systems embedded in Biologically inspired artificial machines.

CNNs can guarantee to a robotic system embedded solutions with parallel computation capabilities. These peculiar characteristics are extremely important in bio-robotics and in particular when multi-legged platforms are considered. The possibility to face with multiple degrees of freedoms in a robust and parallel way is a key point of CNNs and represents a significant contribution in this field. Locomotion in legged robots is by itself an hard task: under this perspective, analog CNN implementations of feedforward strategies based on Central Pattern Generators (CPGs) have been proposed and successfully applied to hexapod robots, crab-like structures and lamprey-like solutions. However CPGs can be enriched with sensory feedbacks coming from the environment to allow basic feedback control mechanisms like step-reflexes, steering control, searching reflexes, attitude control and others. A CNN-based CPG allows a distributed computation of the sensory information that can act at the level of single cells of the neural structure used as motor neurons for the robotic system. An implementation of these solutions was realized in a six-legged insect-like robot by using Reaction Diffusion (RD) Cellular Neural Networks [3]. Subsequently a new approach to the CPG implementation, based on the principle that the same group of neural cells can be reorganized by changing the synaptic connections, was proposed. Its corresponding CNN implementation is realized by using space variant templates and the mechanism was called Multi-template approach [4]. Additionally, such template values could be learned, in an unsupervised way, in order to provide a really flexible, environmentally mediated locomotion adaptation via a weighted combination of the fixed templates [7].

In the last years of research in using CNNs for Biorobotics, a question arose: is it possible to extend the CNNs paradigm to model higher level functionalities of the brain? Following this hypothesis a bottom-up approach was proposed to unravel the potentialities of CNNs in Brain Science. Particular attention was devoted to insects due to their simple brain that however allows the emergence of their complex adaptive behaviours.

Growing up from locomotion to cognition, a new general purpose perceptual control architecture, based on nonlinear neural lattices, was presented and applied to solve robot navigation tasks. Insects show the ability to react to certain stimuli with simple reflexes, using direct sensory-motor pathways, which can be considered as basic behaviors, inherited and pre-wired. Relevant brain centres, known as Mushroom Bodies (MB) and Central Complex (CX) were recently identified in insects: though their functional details are not yet fully understood, it is known that they provide secondary pathways allowing the emergence of cognitive behaviors [23, 104]. These are gained through the coordination of the basic abilities to satisfy the insect's needs. Taking inspiration from this idea, the proposed architecture modulates, through reinforcement learning mechanisms, a set of competitive and concurrent basic behaviors in order to complete the task assigned through a reward function. The core of the architecture is constituted by the so-called Representation layer, used to create a concise picture of the current environment situation, fusing together different stimuli for the emergence of perceptual states. These perceptual states are steady state solutions of lattices of Reaction-Diffusion Cellular Nonlinear Networks (RD-CNN), designed to show Turing patterns. The exploitation of the dynamics of the multiple equilibria of the network is emphasized through the adaptive shaping of the basins of attraction for each emerged pattern. Experimental campaigns on standard robotic platforms were performed to demonstrate the potentiality and the effectiveness of the approach. The robot, endowed with a Turing Pattern CNN based generator as perceptual controller was experimentally found able to cope with new situations, not encountered in the learning phase, which in any case had been performed in an unsupervised, reinforcement based way. The additional interesting aspect is that the same CNN cell, which is a standard two-layer Piecewise linear cell, can generate Turing

Patterns or spiking-like dynamics only changing its parameters. The overall architecture is therefore built around the same functional CNN cell, which, like the neuron in a brain structure, can give rise to completely different self organizing emergent dynamics if embedded in different neural lattices and if its parameters are slightly modulated. The main message to be drawn by CNN processing in Biorobotics is their analog, spatially distributed way of computation. In locomotion control, the whole set of actuators which concurrently move in the environment constitute the visible solution of the hidden spatial-temporal PDE-like processing performed by the CNN controller. Perception is also seen as the site where environment representation is incrementally formed, just like a morphogenetic pattern grows, under the control of endogenous and environmental stimuli. Moreover there is a topographically organized projection between motor neuron cell bodies and specific regions of the vertebrate spinal cord and higher brain centers, and this organization is maintained through a lot of living species. A wide variety of living beings reveal an extremely organised neural architectures CNN-based visual processors take inspiration from the retinotopic ganglion cells organization which is maintained within the appropriate area in the midbrain: this allowed to design and realize analog, digitally programmable VLSI chips able to implement extremely fast visual dynamics. Within the same perspective it was drawn by Neurobiologists.

A final, alternative way to face with the perception process using CNNs will be briefly addressed in the final part of the chapter. It will refer to the mathematical paradigm of the WinnerLess Competition (WLC), which was efficiently used to model olfactory processing in insects and mammalian. The sequence of saddle points crossed in the phase space of the multidimensional nonlinear system was used as a code for embedding the perceptual state of the robot. The large memory capacity shown by the WLC-CNN is able to contribute to the new perceptual framework for autonomous artificial agents, where the association between stimuli and sequences is learned through the experience.

The proposed examples lead to demonstrate that CNNs can be considered as an unifying framework for studying and implementing bio-inspired solution in robotics and for understanding basic brain functions.

2. CNN-based CPGs for Locomotion Control in Bio-robots

2.1. *Basis of locomotion*

It is often surprising how even simple beings possess a high level of adaptability while moving into various environments. Studies on how the control of locomotor activity is realized in a large category of animals reveal a hierarchical organization of the neural networks devoted to the locomotion control task. The key feature of these networks is that the pattern of neural activity is mapped into the pattern of locomotor activity. The generation of the rhythmic patterns of muscles activity takes place at the neural level called Central Pattern Generator (CPG). This structure has been revealed both in vertebrates (located in the spinal cord) and in invertebrates (located in relevant ganglia).

The CPG includes subnetworks of Command Neurons (CNs) and Local Pattern Generating Neurons (LPGNs). The former ones act as decisional centers: they receive stimuli from the sensorial organs or from higher level neurons. Therefore, at this level some important decisions such as escape reactions or the choice of the right locomotion pattern (i.e. swimming if the being is on the water or walking if it is on the terrain) take place. The LPGNs are responsible of coordination of the organs devoted to locomotion (cilia, legs). The neurons of this subsystem are called also motor-neurons, because they constitute the last neural level before movement actuators. In many biological cases this functional subdivision does not correspond to different neural networks: the same networks may act at different levels of the structure. It is also not so clearly established which are the mechanisms used to reorganize the LPGNs configuration in order to switch among locomotion patterns. Two mechanisms involved in this operation are focused by neurobiologists. In some cases it is necessary only to change the parameters of the synapses, while in other cases a strong reorganization of the network is required. In this latter case it is not rare that some synaptic connections are inhibited and other ones are excited towards neurons not yet involved in the former pattern generation.

The details of the mechanisms underlying rhythmic motions in animals have been a subject of research since last Century. Essentially two approaches were formalised: the first one considers the locomotion control as imposed from the high brain levels and in principle, does not need peripheral feedback. On the opposite side the other gives a mayor role to sensors, whose signals are strictly needed to initiate and modulate locomotion. In the case of some invertebrates, and in particular of the Stick insect, there is an evidence of no central coordination: legs are rather independent and only loosely connected to generate a coordinated motion: organisation among legs takes place at the level of the leg motor neurons, and no hierarchy is needed [44]. In the course of the last decades the former approach, referred to as the Central Pattern Generator paradigm, was the one that was found in the larger number of animal species taken into account. In fact, in many experiments on both invertebrates and vertebrates, basic rhythms are generated also in the case of fictive locomotion, i.e. in the absence of real movement of the locomotor organs, and so with no sensory feedback from these ones [79]. Of course, even if not needed, sensory feedback is involved in adapting the basic oscillations and their phases to environment conditions [90]. One classical example refers to the CPG induced initiation activity in Lampreys when they are disturbed mechanically moving their tail [101]. In general, there is evidence, from different points of view, that proprioceptive signals can entrain, modulate or even reset locomotion rhythmic activity [34, 81]. On the other hand, one of the main characteristics and advantages of the CPG approach, which makes it feasible for robotic implementation, is that only a reduced number of control signals are required from higher brain regions to initiate, modulate or even switch among complex patterns of locomotor activity [54,63]. According to this paradigm, (a few) descending pathways modulate or switch among different pre-programmed patterns of neural activity, which reside, for example, in the case of insects, within the thoracic ganglia. Scanning the literature, it is possible to find that mathematical models of CPG can be derived at different levels of abstraction, from the very low level of modelling the single neural compartments, to modelling networks of interconnected neurons (referred to as connectionist models) [61,62], to more abstract coupled oscillatory networks, and even to networks of intrinsically non-oscillating dynamical units, that, suitably entrained, give rise to oscillatory dynamics [37]. This last approach is called *Population dynamics or* neural field [92] approach, since the aim is to model the macroscopic, field generated dynamics of neural assemblies.

The CNN-based CPG approach lies rather on the boundary between the connectionist and the Neural Field approach, since, from the one hand, the network parameters are primarily taken into consideration and are indeed the main actors of the emergence of the overall spatial-temporal dynamics generated within the neural lattice, which, on the other hand, is analyzed and exploited taking into account the characteristics of the generated field.

The approach starts considering legged locomotion as a complex spatio-temporal non linear dynamical system and subsequently implementing the CPG mathematical model on large scale, digitally programmable analog processors, known as Cellular Neural Networks. One of the main advantages of such architectures is that they were conceived as being tightly related to a reference hardware relatively easy to be realised at the VLSI level. This gives the possibility to obtain real time locomotion controllers. On the other hand, most of the approaches to locomotion pattern generation and control stop at the software simulation level or at the high level hardware implementation, using finite state machines on digital micro controllers. Rarely attention is focussed on hardware architectures serving as paradigms for the topographic implementation of CPGs. Moreover, analog topographic approaches become appealing in view of controlling a large number of degrees of freedom, like in the biological case. Here the controller is topographically distributed along with the actuator and the sensor network, in order to reduce the delay among the stimulus and the consequent reaction and increase the efficiency when, for example in the case of escape reactions, speed is of primary importance.

In the proposed approach it is clear that rhythmic activity is the emerging solution of a spatially extended nonlinear dynamical system, solution which depends more on the connection parameters (the neuron synapses) than on the particular neuron model taken into account. This is in accordance with a number of relevant formal results (see for example [68]). According to the connectionist approach, the initial effort has been to design a proper cell dynamics, able to capture few essential features of the neural dynamics, but suitable to be implemented within the CNN paradigm. The following task was to develop suitable models for synaptic connections to lead to the emergence of proper travelling wave fronts in CNNs made of spiking neurons, useful for legged locomotion. Even if, in most cases, the CPG approach was considered, the CNN implementation of the locomotion controller still maintains interesting levels of adaptation, in terms of dynamic shape of the neural signals, phase and frequency control, that are indeed desirable properties on neural controllers [85, 86]. Such parameters could be modulated also from the sensory feedback, thus attaining interesting real time control properties.

2.2. *The CNN neuron model for CPG: a slow-fast controllable limit cycle*

Several biologically relevant neuron models were implemented under the CNN paradigm [9, 94]. The primary effort for the implementation of a biologically inspired neural network for locomotion control consisted in properly designing a cell model which should retain the main characteristics of a biological neuron and that could fit the main constraints of a CNN cell. The spiking dynamics should therefore be implemented in a PWL nonlinearity, like that one present in the standard CNN scheme, at the aim to exploit the large scale CNN hardware prototypes already available. We'll focus our attention to model neural dynamics directly related to a hardware large scale reference architecture. At this aim the following second order cell was designed:

$$\begin{aligned}\dot{x}_1 &= -x_1 + (1+\mu+\epsilon)\cdot y_1 - s_1\cdot y_2 + i_1;\\ \dot{x}_2 &= -x_2 + s_2\cdot y_1 + (1+\mu-\epsilon)\cdot y_2 + i_2;\\ y_i &= 0.5\cdot(\mid x_i+1\mid - \mid x_i-1\mid)\end{aligned}\qquad (1)$$

being $i = 1, 2$. It was demonstrated that such a cell, for a suitable selection of its parameters ($\mu = 0.7, s_1 = s_2 = 1, i_1 = -0.3$ and $i_2 = 0.3, \epsilon = 0$), is able to generate spiking dynamics, under the form of a stable slow-fast limit cycle [3], as it can be seen in Fig. 1.

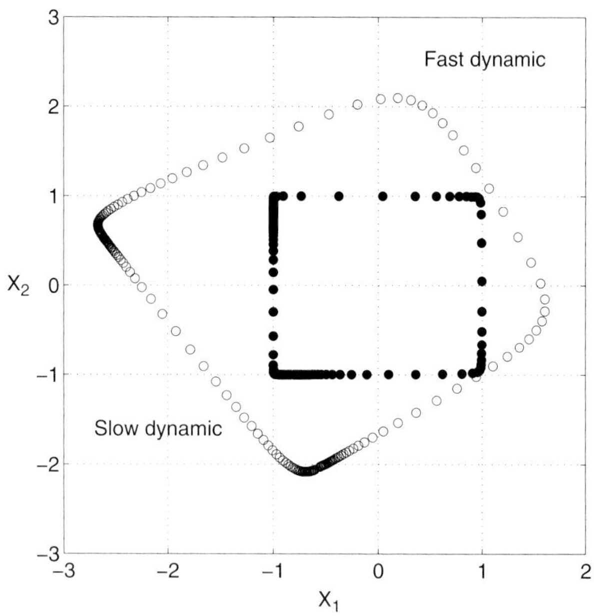

Fig. 1. The system behavior in the plane $x_1 - x_2$ (empty circles) and $y_1 - y_2$ (solid circles).

In view of the exploitation of this dynamics to produce the basic motion of a single leg in a bio-robot, it is clear that a suitable control of the dynamical characteristics of the cycle is reflected in the possibility to efficiently control the motion of each leg. Under this perspective, the stable limit cycle generated by this cell has a number of interesting properties, herewith summarised (for a detailed description see [47]):

- **Amplitude control**. The limit cycle arises in a PieceWise Linear dynamical system that, for the selected parameter set, possesses an unstable focus in the origin and a series of *virtual equilibrium points* [74]. These ones depend on the amplitude of the output saturation, giving the possibility to easily control the oscillation amplitude. For legged locomotion control, this aspect is relevant for example, to realise an efficient avoidance reflex. If only an *elevator reflex* is needed, a positive perturbation of the bias i_2 causes the basic cycle to be locally increased in amplitude. Also the *searching reflex* can be easily implemented. In this case we need to give rise to a disturbed limit cycle: the designed cell, with the addition of a sinusoidal term, can undergo chaotic trajectories, leading to the easy implementation of a reflex aimed at searching a contact point in the environment, after which to recover to the basic cycle [6].
- **Frequency control**. Different ways of frequency control can be implemented. In the case of slight changes, induced from the periphery, it was found that this non linear oscillator can be entrained by a weak periodic external signal: the oscillation frequency approaches that one of the forcing signal if the two frequencies are close enough, or otherwise synchronization still takes place, but the entrained oscillation approaches a fraction of the entrainment signal. A much larger frequency control can be obtained simply scaling the cell equations. When a hardware implementation is

considered, frequency modulation can be easily achieved through the switched capacitor technique [65]. This gives the possibility to easily control robots endowed with a number of different actuators, working at thoroughly different frequencies.

- **Oscillation period control**. In the case of this slow-fast limit cycle, a proper modulation of the bias term i_2, moves the nullclines within the phase portrait. There are critical values on i_2 such that the limit cycle collapses into a stable equilibrium point. Soon before the critical value, the slow part of the oscillator slows down. It means that the spiking characteristics of the signals are maintained, whereas the recovery time is slowed and the slow-fast ratio can be easily controlled. Being the slow part associated to the stance phase of the gait cycle, the period control is easily translated into a speed control in the locomotion pattern.

2.3. *CPG in a reaction-diffusion CNN structure*

The idea that inspired the RD-CNN approach for CPG, was to conceive locomotion as a complex spatial-temporal phenomenon, involving a large number of variables, whose dynamics is relevant in time and in space. Membrane-like dynamics are a classical example of this idea. Each node of the membrane cannot move alone: its dynamics depends heavily on that one imposed by its neighbors. This concept exactly matches with the electronic paradigm of the CNN, and is clearly related to the CPG, where each unit (leg or joint) motion is related to the motion of the neighboring units. Single activations (reaction), at the level of the single cell, imply energy dissipation (diffusion) which affects the neighbor dynamics, leading to the emergence of ordered spatial-temporal neural field dynamics over lattices of units, like the CNNs. The so-called *Reaction-Diffusion* paradigm was used in the last decades to model a huge quantity of natural phenomena in Chemistry, Morphogenesis, Neurobiology [77]. Typical emerging solutions of this complex dynamics are self-sustained oscillating phenomena, known as *autowaves* [67], as well as patterns arising in Morphogenesis, known as *Turing patterns* [97]. Both autowaves and Turing patterns were demonstrated to take place in CNN structures, called Reaction-Diffusion CNNs (RD-CNNs), and both phenomena were considered: the former to address locomotion, the latter to face with perception, as discussed in the next sections. From our perspective it has to be underlined that, from the one side, the neural firing possesses all the characteristics of an Autowave front [41], and on the other side, autowaves were demonstrated to take place in a RD-CNNs endowed with the cells (1) and with the following templates:

$$A_{11} = \begin{pmatrix} 0 & D_1 & 0 \\ D_1 & -4D_1 + \mu + \epsilon + 1 & D_1 \\ 0 & D_1 & 0 \end{pmatrix}; \tag{2}$$

$$A_{22} = \begin{pmatrix} 0 & D_2 & 0 \\ D_2 & -4D_2 + \mu - \epsilon + 1 & D_2 \\ 0 & D_2 & 0 \end{pmatrix}; \tag{3}$$

$$A_{12} = \begin{pmatrix} 0 & 0 & 0 \\ 0 & -s_1 & 0 \\ 0 & 0 & 0 \end{pmatrix}; A_{21} = \begin{pmatrix} 0 & 0 & 0 \\ 0 & s_2 & 0 \\ 0 & 0 & 0 \end{pmatrix}. \tag{4}$$

where $D_1 = 0.1$ and $D_2 = 0.1$ represent the synaptic diffusion coefficients and the other parameters are the same as specified in Sec. 2.2. These templates include both the dynamics of the single cell and the synaptic activity, modelled as a diffusion effect, coherently with the concept of electrical synapse. The so-called *Zero-Flux (Neumann)* boundary conditions were assumed. Moreover the robustness of this CNN model to parametric uncertainties and noise was assessed and a discrete component hardware implementation was also achieved [2].

Different RD-CNNs configurations were employed to work as CPGs for a number of Biorobots that were designed and built in our laboratory. It is to be emphasised that the focus is here more on the design and realization of the locomotion controller than on the robotic structure. On the other side, excellent work on the mechanical aspects of Biomimetic robots was already performed [70, 88]. Instead, in the present approach, the robot has to be viewed as the test bed in which the efficiency of the locomotion control can be appreciated, as the *visible solution* of the neural lattice where the nonlinear dynamics take place. The robot joints were, in most cases, implemented by using simple servomotors: so the trajectories shown by the cells are translated in sequences of position set points for the robot actuators. In Fig. 2 the circuit realization is depicted for a particular ring configuration of the

Fig. 2. A gallery of robots controlled by the RD-CNN CPG approach.

RD-CNN used as the CPG in a worm-like walking machine and in a hexapod robot [3], as well as in a lamprey-like robot [10]. These are also sketched in Fig. 2.

In the basic implementation referring to legged prototypes, each leg had two degrees of freedom, which span a 2D plane in the robot workspace in an elevator-protraction way, whereas they act as a flexor-extensor couple in each segment of the lamprey robot. So each joint in every structure was driven by one output variable of a CNN cell (see Eq. (1)). This hardware structure, joined to the simple implementation of local feedback strategies on each single cell, as outlined above, results in an efficient locomotion controller. In fact, besides the control strategies on each single cell outlined above, the overall dynamics in the neural lattice can be easily controlled in terms of speed of the travelling wave: this can be realised varying the number of cells in the ring, with the consequent variation of the slow part of the limit cycle for each cell [2]. Moreover, a transition among different gaits, typical of hexapods, can easily achieved by modulating the connections among the CPG and the legs, with a consequent variation of the phase among the moving legs, which varied from the slow (tetrapod) to the medium, to the fast (tripod) gait.

A typical characteristics of neural systems is *graceful degradation*: if some neurons die, the overall organization of the neural field dynamics is only slightly affected. This was experimentally observed in an hexapod robot: cutting a neuron out of the ring caused the partial loss of the phases among the oscillators, so locomotion control was affected, but nevertheless the structure was indeed able to move.

More sophisticated locomotion capabilities can be obtained by increasing the dexterity of the legs from two to tree joints. This was done in Rexabot III [5] and in the Gregor series of hexapod robots [20]. Here each leg was controlled by two identical cells of the type reported in Eq. (1), from which three controlling variables were extracted to drive the joint servomotors. In Rexabot III the problem of attitude control was also taken into consideration and solved with the design and implementation of another CNN structure, acting concurrently with the CNN-CPG, where each cell was a nonlinear PD controller, designed to maintain a set point in the roll-pitch plane. In this case the robot was controlled by the combination

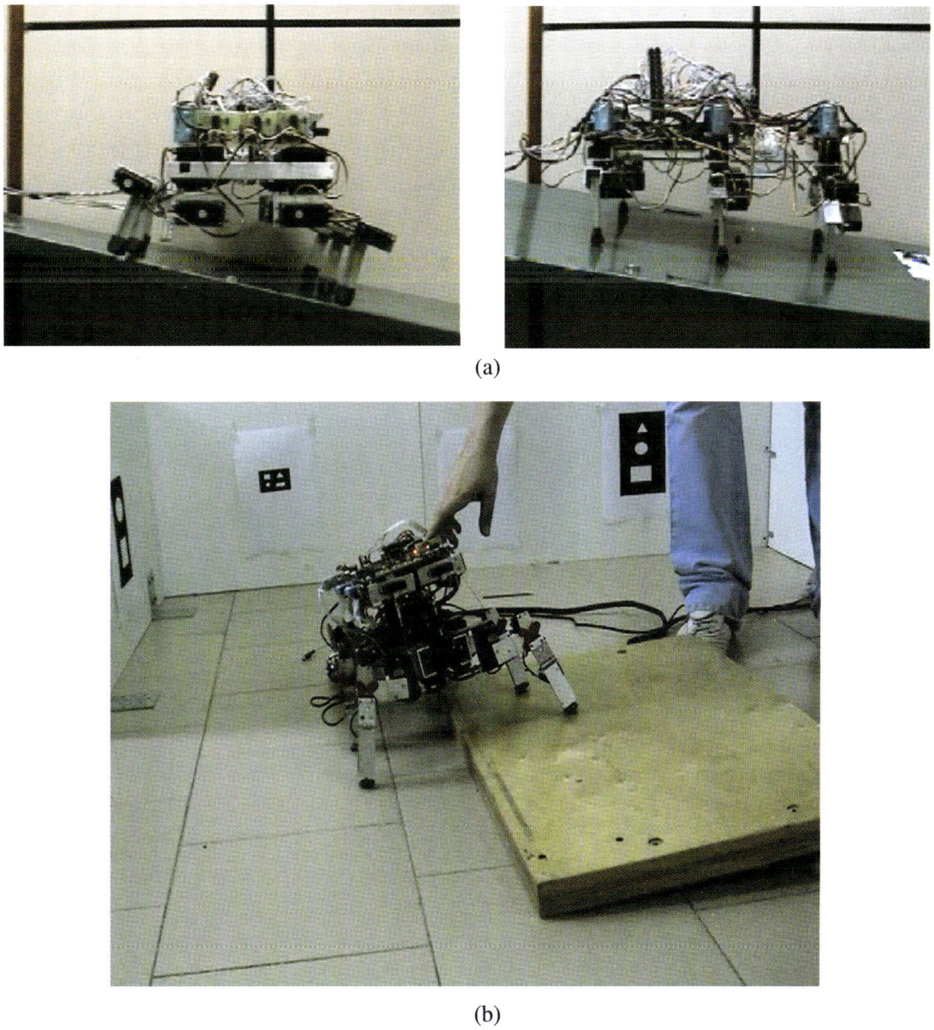

Fig. 3. Attitude control: (a) Rexabot III with the analogic attitude controller for roll and pitch; (b) Gregor III controlled through a reflex-based approach.

of the signals coming from the CPG and those ones coming from the attitude controller. A relevant characteristics of the approach is that the set points used for the attitude control come from only one double axis accelerometer, placed in the abdominal part of the robot. It derives that the feedback signal acts on the whole network controller. This causes each of the controllers to *cooperate* for compensating errors in the attitude that could not be caused by the corresponding leg. For example, if one of the legs is overloaded, or is broken, it was experimentally shown that the legs *self-organise* in such a way as to try to recover from overload or to attain the requested attitude even in presence of a broken leg. It is to be emphasised that this approach is analog, space distributed and characterised by a local connectivity, with the consequent characteristics of a real time efficiency and modularity that allows a direct extension to a larger number of actuators. In Fig. 3(a) Rexabot III is depicted in really sloped grounds, maintaining the horizontal attitude; moreover the Gregor robot controlled through a reflex-based strategy is also shown in Fig. 3(b) when facing with a uneven terrain.

Various locomotion patterns can be implemented using such a structure. The way to change the locomotion pattern is to use a different number of cells constituting the RD-CNN and to enroll or discard (by activating or inhibiting some synaptic inputs) some neurons in the structure to adapt the phase and the speed of the travelling wave front. Suitable commands (a few indeed) are therefore supposed to come down from the higher control regions to solve this task. The synchronization of the leg phases thus depends on the neural organization within the CPG. One advantage in terms of implementation is that the connection template values are constant and space invariant, and from the large scale VLSI implementation point of view, this architecture is feasible and chips implementing about one thousands of such cells, connected in a programmable RD-CNN structure have been made available recently [35].

2.4. *CPG in a multi-template-CNN*

In the approach discussed above, essentially the arousal of different patterns is a function of the number of neurons belonging to the CPG-CNN ring. Besides electrical synapses, modelled by means of diffusion connections, a simple model of chemical (excitatory or inhibitory) synapsis was introduced, to give the possibility to implement easily programmable CPG, without needing a variable number of neurons to generate different locomotion patterns. In fact in this case a minimal number of neurons (one for each leg) is required to generate a proper CPG: this was found at the expenses of the fact that now the templates are no longer space invariant, and that each locomotion pattern requires a proper template set. For this reason this approach was called *multi template* (MTA). A design method for a CPG following the MTA approach was given in [47], which allows to directly synthesize the connections (templates), given the leg number and the number of different, equally spaced phases, to be implemented into the structure. In Fig. 4 an hexapod robot controlled by the CPG-CNN implemented in discrete components is reported. In the Appendix all the details for the hardware implementation of the control system used for this robot are reported.

Fig. 4. The Hexapod robot controlled by the CPG-CNN implementing the multi template approach.

In addition to the possibility to design specific template set (specific programs) a strategy was introduced in [7] to simultaneously stimulate a given combination of these basic patterns (slow, medium and fast) to form a *continuous generalized* locomotion pattern. The degree of modulation of each basic pattern, subject to a suitable unsupervised and simple learning technique, was found useful to adapt the locomotion pattern to cover a large set of desired speed reference values.

The CNN-based CPG implemented with the Multi-template approach has been realized also in VLSI. The chip as presented in [8], was realized using a switched capacitor tecnique and has been used to control different multi-joint structures. The chip allows a frequency modulation of the output signals and has been tested on several robotic prototypes among which: a crab-like robot, a hexapod and a mini-hexapod structure, a sprawled hexapod named Gregor, a lamprey robot and a tiny legged structure actuated with piezo-electric materials (see Fig. 5). Taking for instance the crab-like structure, the neural structure and the corresponding stepping patterns for two different gaits are reported in Fig. 6.

3. A Brain for the Body: A CNN-based Spatio-temporal Approach

After finding in the CNN approach a suitable framework for the implementation of efficient real time locomotion controllers on biorobots, and also after experimentally verifying the efficiency in moving machines, the further problem that was faced consisted in giving to these structures the possibility to acquire some autonomous decision making capabilities. The challenging idea was to start to think at a robotic architecture that could be able in the near future, once assigned a proper mission, to find a way how to solve it, in spite of the disturbance coming from the environment: the robot should have the intrinsic basic capabilities (in terms of sensory-motor basic pathways) to face with a given problem, but has to properly use them. Under this perspective the attention now moves from learn locomotion to to cope with changing environment, but maintaining the same idea that the overall architecture should not discard the possibility to have a reference architecture for the hardware implementation.

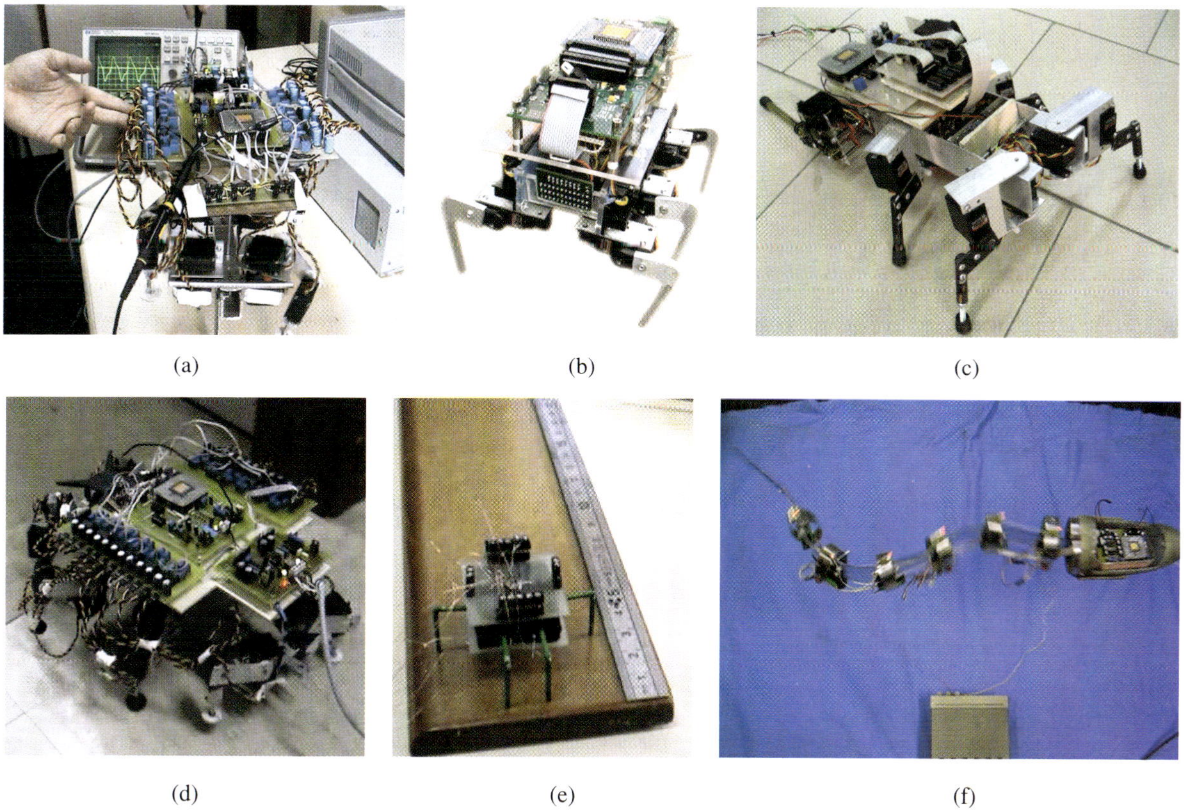

Fig. 5. Gallery of robots controlled through the VLSI CPG chip: (a) a 18 DoF hexapod robot, (b) a 12 DoF MiniHex, (c) a sprawled hexapod Gregor I, (d) a crab-like structure, (e) a piezoelectric actuated mini robot, (f) a lamprey-like robot.

An interesting idea came from the neurobiological studies performed on the behavioral control of the marine mollusc *Clione Limacina* [79]. Here it was found that suitable *command neurons*, placed in the cerebral ganglia, are able of impose or modulate the undergoing swimming pattern dynamics if particular conditions are met. In some cases, like in the case of Clione, it happens that some command neurons, for example those ones involved in the speed control, are non-spiking, i.e. they manifest long lasting plateau potential. From this biological evidence, a control structure was supposed, where a topological network of non-spiking neurons, constrained by signals coming from the sensors endowed within the robot, was able to generate combinations of plateau (low or high) signals. This, in a general 2D structure, can be considered as a bi-dimensional pattern, emerging from a lattice of neurons. The peculiarities of the emerged pattern represent the particular locomotion scheme to be executed at the level of the CPG. This concept reflects a common way on how animals interact with the environment: they, while moving in an environment and willing to proceed in a given direction, look around and decide to change behavior (for example steer), only if particular conditions are met, for example when they are too next to an obstacle. From a neural field point of view, it can be supposed that somewhere in the brain, one (or more than one) neural structure generates a neural field which is trapped within a given attractor, leading to the emergence of a particular stable pattern of activity. The respective basin of attraction is a function of the environment conditions. This is a common idea, deeply investigated by famous neuroscientists [48]. This would mean that physical navigation, for example, is reflected into a *mental* navigation within a landscape of different attractors, generated by a neural field. Behavioral variation is a consequence of an environmentally-modulated transition among the different basins of attraction. This approach was designed and experimentally tested in some robotic structures, where the perceptual core was a RD-CNN generating *Turing Patterns*. These ones are well known in developmental Biology after the work of Alan Turing [97]. They are solutions of Reaction Diffusion partial differential equations, and were demonstrated to take

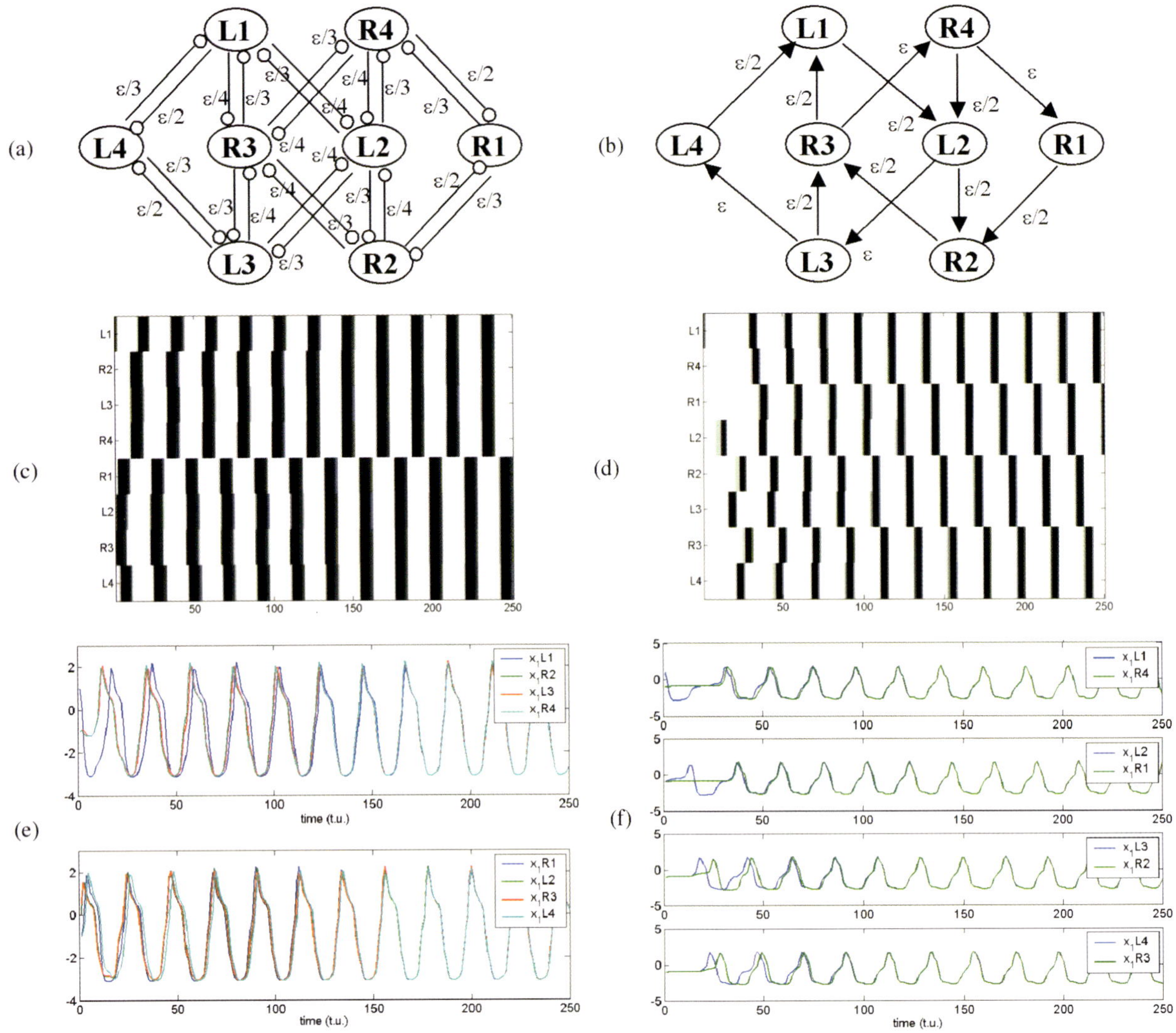

Fig. 6. MTA-CNN CPG for the crab-like robot: CPG structure, gait diagrams and waveforms for fast gait (a)(c)(e) and medium gait (b)(d)(f).

place in RD-CNNs [2]. More interestingly, it was shown that the same RD-CNN structure cell as in equation (1), with the same diffusive synapses for the CPG generation, but for another set of parameters ($\mu = -0.6, s_1 = 2, s_2 = 2.5, i_1 = 0$ and $i_2 = 0, \epsilon = 1.82, D_1 = 0.01, D_2 = 1$), was able to generate plateau potentials (i.e. a non spiking activity) leading to Turing pattern formation. As a consequence, the same basic architecture, joined to the programmability of the CNN structure, can be used to generate CPG locomotion activities as well as perceptual patterns. Following the paradigm known as Behavior-Based Robotics [25], in which the perceptual process is considered tightly interconnected with the agent behavioral needs, perception has been treated as an emerging complex phenomenon. Here a large amount of heterogeneous information is fused to create an abstract and concise internal representation of the surrounding environment, which at the same time takes into account the needs and the motivation of the agent [33], while the whole process is mediated through a behavior-dependent internal state [78].

Starting from these considerations and taking into account the latest results in Neurobiology [48] and the advancement in artificial cognitive system [99], we developed a general control architecture for implementing the sensing-perception-action cycle [73] to be potentially applied to different robotic platforms involved in several missions in cluttered environments. To this aim, we borrowed from the insect world neural structures responsible for both simple and complex behaviors.

3.1. *Control architecture*

As in insects, the proposed perceptual architecture is organized in various control levels consisting of functional blocks, acting either at the same level, as competitors, or at distinct hierarchical levels showing the capability to learn more complex, experience-based behaviors [104].

The control architecture is reported in Fig. 7. It consists of series of parallel sensory-motor pathways (i.e. basic behaviours) that are triggered and controlled by specific sensory events in a reflexive way, giving the knowledge baseline to the system which is thus able to react to low-level stimuli. Going up in the hierarchical scheme, two relevant centres of the insect brain are considered: the Mushroom Bodies (MB) and the Central Complex (CX). Both MB and CX are not yet well understood from a biological/neurogenetic point of view. However interesting studies [55, 59, 104] underlined how deeply these structures are involved in perceptual processes. In particular MBs are mainly devoted to the enhancement of causal relations arising among the basic behaviours, by exploiting the temporal correlation between sensory events; information storage and retrieval in the case of the olfaction sense; resolving contradictory cues through the visual sense by imposing continuation or adaptive termination of ongoing behaviour. CX is instead responsible of integration of visual information, storing and retrieving information on objects and their position in space, controlling the step length in order to approach or avoid such objects; motor control, landmark orientation and navigation, orientation storage and others. Some of these functionalities have been already developed creating a correlation-based anticipation layer.

These aspects were treated separately in previous papers by using causal Hebbian rule in an array

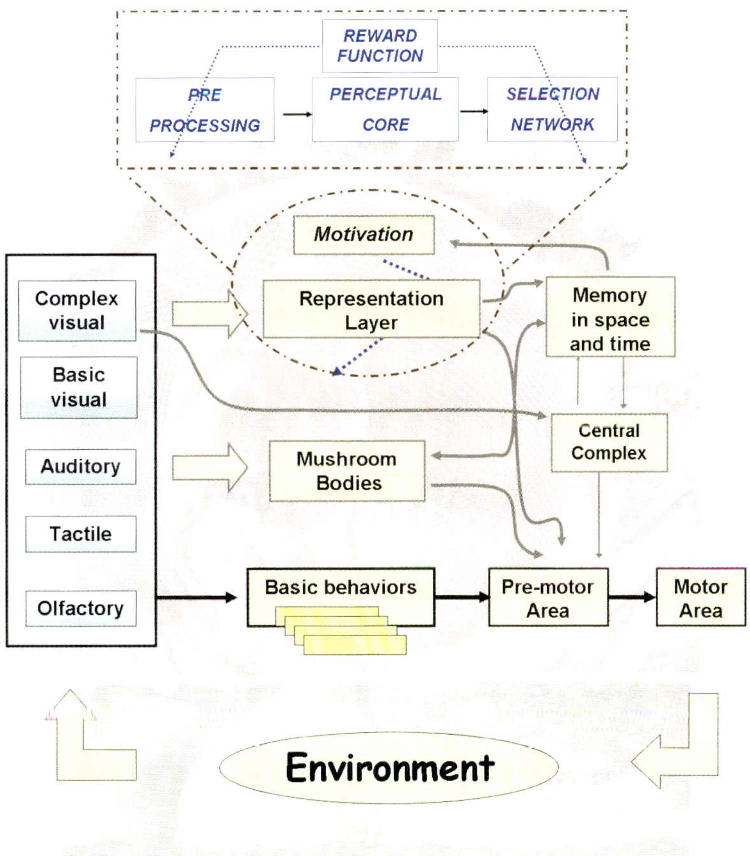

Fig. 7. Functional block diagram of the implemented control architecture. The interaction between the robot and the environment is realized by direct sensory-motor pathways, the *basic behaviors*, which are modulated by the representation layer. Mushroom Bodies (MBs) and Central Complex (CX) are relevant centers of the insect brain devoted to temporal correlation, information storage and retrieval, and other functionality summarized in a correlation layer. Finally the high level functions of the representation layer consists of a *preprocessing block*, a *perceptual core*, a *selection network*, while the *Reward function* drives the learning process.

of spiking neurons for anticipation [17], and in [23] where memory structures based on Recurrent Neural Networks were considered. In this chapter, for the sake of brevity, we briefly discuss only on our approach to modelling the high level the representation layer, where perception is formed, considering that the anticipation layer can be added within this architecture to further enhance the capabilities.

As depicted in Fig. 7 the control process can be divided into functional blocks: at the lowest level, we place the parallel pathways representing the basic behaviors, each one triggered by a specific sensor; at a higher level we introduce a representation layer that processes all the sensory information in order to define the final behavior. At the highest layer we introduce a lattice of non spiking neurons. This neural lattice shows distinct characteristics of complex dynamical systems. The emerging assemblies of neural states take on the meaning of percepts. These ones are then associated to suitable modulations of the basic behaviors. This modulation is performed through an unsupervised learning process which creates associations among sensory stimuli and patterns. In this way, at the end of the leaning stage, each pattern represents a particular behavior modulation, while its trained basin of attraction represents the set of all the environment conditions, as recorded through the sensors, leading to the emergence of that particular behavior modulation. The modulation parameters associated with each pattern are learned through a reinforcement learning: here the reinforcement signal is provided by a motivation layer implementing the degree of satisfaction of the robot. This depends on the local satisfaction of the single basic behaviors with the addition of other terms that reflect the robot mission. The presence of additional information into the motivation layer, not used by the basic behaviors can be exploited by the Representation layer in order to increase the robot performance.

Memory is of course distributed in the the whole architecture but a specific block has been also created (i.e. Memory in space and time in Fig. 7). This block develops a contextual layer, like in [100]. Here sequences of successful emerged patterns are memorized to be retrieved when needed. Details about this memory structure are reported in [12].

In such a way, as it happens in insects, the basic behaviors, which are often life-saving sensory-motor pathways, are progressively enriched with emergent capabilities which incrementally increase the animal skills. The main focus is therefore on the application of complex dynamics to learn a proper, complex, context-learned modulation of the basic skills. This process is the main characteristic of our approach which makes it different from the other control strategies, based on the subsumption architecture proposed by [32]. The latter in fact, uses a high level approach to design both the basic behaviors and the coordination block. In our strategy, complex dynamical systems are successfully used. Both architectures use a behavioral decomposition of the system to exploit parallel computation, although the Subsumption network makes a rigid hierarchy among the basic behaviors: the lower ones cannot influence the upper ones, while the latter can act on the former. In our scheme, taking inspiration from the insect brain organization, all the basic behaviors are sensory-motor pathways elicited by only one sensory modality and on the same hierarchical level: knowledge is incrementally built upon their modulation, giving importance to one or the other, depending on the context. Under this perspective the proposed architecture resembles the Motor Schemas, introduced by [25]. Turing Patterns in RD-CNN are hosted, in our architecture, within a layer here called *Representation Layer*. The Representation Layer in our architecture does not refer to a place where a predictive model of the body-environment interaction is learned. This area is rather a layer where the single sensory-motor modalities, constituted by the parallel sensory motor pathways, are modulated in a feedforward way, taking into account all the incoming sensory stimuli. This leads to the emergence of a contextually self organising activity, focused at modulating the basic behaviors.

All the sets of environmentally driven multisensory information, leading to one rewarding behavior modulation, are collected into a unique basin of attraction. It is represented by its steady state condition, depicted as a pattern. This pattern is a binary image, suitable for a very compact coding. It should be noted that the number of different patterns that are able to emerge from the neural RD lattice could be very high (on the order of some hundreds in a $4x4$ CNN). So the number of different behavior modulations could be as large as needed to cope with very complicated and cluttered environments. The result of the behavior modulation leads to a particular robot motion, at each time t. This is formalized with a final action $A_F(t)$ that consists of a variable turning movement (rotation) and a fixed-length forward movement. The main characteristics of the cognitive architecture are described in the following subsections.

3.1.1. *Sensory block*

To deal with the problem of autonomous navigation, the robot is provided with three distance sensors (covering the front, left and right hand side of the robot) for obstacle detection. Moreover, the robot receives information on the angle between the robot orientation and the direction of the robot-target and, in some simulations, also on the distance between the robot and the target. A graphic overview of the sensory apparatus is sketched in Fig. 8.

3.1.2. *Basic behaviors*

With *basic behaviors*, we refer to some "genetically" pre-wired reflexes, triggered by specific sensory events through direct sensory-motor pathways. Referring to crickets, these behaviours are: the capability showed by crickets to recover heading during walking, called *optomotor reflex* [30]; the female ability to follow the sound chirp emitted by a male, named *phonotaxis* [102]; and the ability to avoid obstacles, e.g. detected by the antennae.

At each time step t, the *optomotor reflex* tries to compensate for the previously executed rotation, as occurs in crickets that try to compensate leg asymmetry to maintain the heading during walk.

Even though a detailed neural network could be developed to carefully model the neural control system for such behavior (e.g. see [91]), in this work a very simple rule was adopted consisting of: $A_o(t) = -A_F(t-1)$, where $A_o(t)$ is the rotation triggered by the optomotor reflex at the time step t and $A_F(t-1)$ is the turn executed by the robot at the previous time step.

The *obstacle avoidance* behavior guides the robot in avoiding obstacles perceived using multiple distance sensors: $A_a(t) = f_a(d_F(t),d_L(t),d_R(t))$; here $A_a(t)$ is the rotation triggered by the obstacle avoidance, $f_a(\cdot)$ is a simplified version of the traditional potential field navigation algorithm [31] and $d_F(t)$, $d_L(t)$, $d_R(t)$ are the distances provided by the three distance sensors.

Finally, *phonotaxis* proposes a rotation $A_p(t)$, aiming to compensate for the phase between the robot heading and the robot-target direction: $A_p(t) = f_p(p(t))$, where $p(t)$ is the phase between the robot and the sound source. The function $f_p(\cdot)$, used in this application, is a simplified version of the model for phonotaxis behavior, reported in [60]. Table 1 summaries the model variables used to describe the system basic behaviours.

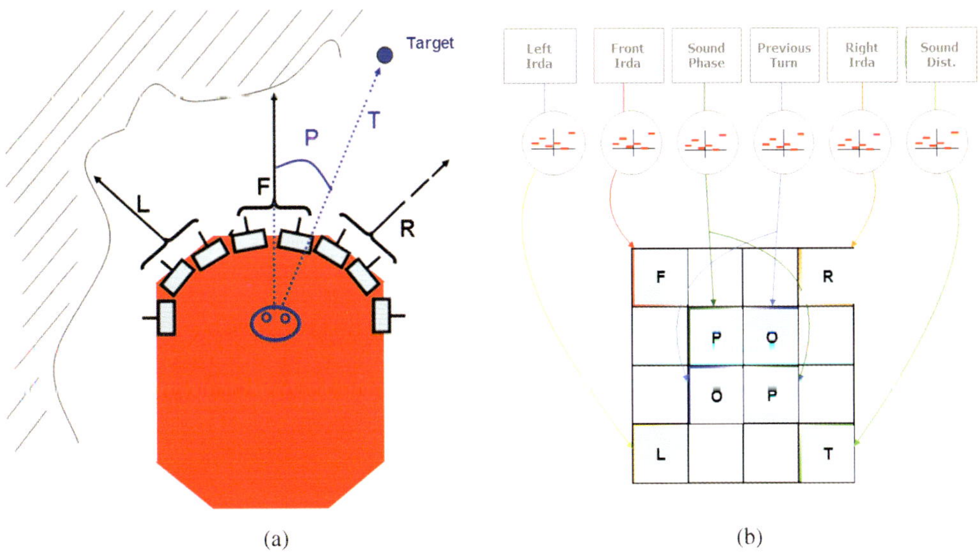

Fig. 8. (a) The robot acquire information from six sonar sensors grouped into three pairs (F:Front, L:Left, R:Right) and a target sensor providing the phase (P) and distance (T) between robot and the target. (b) Initialization for the first layer CNN cells in the representation layer. The corner cells are set by obstacle stimuli (Front, Left, Right obstacle distance sensors) and by the target distance sensor, if present. The central cells are set by the previous executed rotation (O) and by the angle between the robot heading and the direction robot-target.

Table 1. Summary of the model variables used to describe the basic behaviours and their meaning (see text for details).

$A_F(t)$	Final action
$A_o(t)$	Optomotor reflex
$A_p(t)$	Phonotaxis behaviour
$A_a(t)$	Obstacle avoidance behaviour
f_a	Potential field
f_p	Cricket inspired phonotaxis
$d_i(t)$	Distances robot-obstacle (i=F,R,L)
$p(t)$	Phase robot-target

3.1.3. *Representation layer*

The ability to interpret "situations" in terms of robot environment interaction (i.e. perception for action), is here considered as a *complex behavior*, growing up from the basic behaviors. The robot perceives using its sensory apparatus and processes at a cognitive level to optimize its behavior in relation to the mission assigned. The aim of the Representation layer, the highest control level within the whole cognitive process, is to achieve context dependent decisions. To this aim, all the available sensory modalities, each one separately being responsible of each single basic behavior, have to constitute the input to this layer. They are here incrementally transformed into environment *representations*, which lead to the modulation of the basic behaviors. These mechanisms are plastically modified by experience. In this work a CNN was designed to generate, on the basis of information coming from sensory events, Turing patterns as perceptual patterns. At the afferent (i.e. input) level, an unsupervised learning algorithm plastically shapes the basins of attraction of the Turing patterns in order to adjust the classification of the information with respect to the robot motivation.

The whole *representation layer* consists of a *preprocessing block*, a *perceptual core*, a *selection network* and a *motivation* layer, responsible for driving the learning process. The top side of Fig. 7 shows the main components of the representation layer.

3.1.4. *Preprocessing block*

The sensorial inputs, normalized in the range $[-1, 1]$, enter the preprocessing block: each stimulus is the input for a Sensing Neuron (SN) with piecewise linear activation function, made-up, in this case, of 10 amplitude-varying steps learned in an unsupervised way as briefly explained in Sec. 3.1.7. Finally, each output of the SNs sets the initial condition for a cell of the nonlinear dynamical system that realizes the perceptual core of the Representation layer.

3.1.5. *Perceptual core*

The creation of a concise representation of the environment is crucial for the cognitive process, since it is the result of the dynamic processing of the external stimuli.

To implement this feature, we use a nonlinear partial differential equation, discretised in space as a neural lattice made-up of second order CNN cells, connected by local diffusion. This constitutes a two-layers RD-CNN, able to generate Turing patterns [97]. The dimension of the network has been fixed to 4×4 on the basis of a previous work [12].

As shown in Fig. 8(b), the output of each SN sets the initial conditions for the state variable of two central cells or a corner cell, which have been proven to have higher control than the other cells [12]. The initial conditions for the state variables of the second layer are set around zero for all the cells.

To better understand how initial conditions influence the pattern emergence, we performed a simple experiment, where we set to zero the initial conditions for all the first layer cells except the top-right corner ($C(1;4)$) and bottom-left ($C(4;1)$) corner cells of a $4x4$ lattice, whose initial conditions have been varied in $[-1;1]$ range, mimicking two sensory inputs. The second layer cells are set to random values in the range $[0 : 005; 0 : 005]$. Fig. 9(a) shows the geometries of the basins of attraction for the 39 emerged patterns (represented by different colors), obtained by varying the initial conditions for the two cells above mentioned. At the end of the learning phase, executed with a simulation in which the two sensory inputs were associated to two distance-to-obstacle sensors, the number of basins of attraction and corresponding emerging patterns is tightly decreased to 15 (Fig. 9(b)). The effect of the unsupervised learning in the sensing block was to cluster the Turing patterns into a meaningful set of non redundant internal states, and to adapt the internal states (shape of basins of attraction) to the robot motivation.

To simplify the successive processing, we associate a simple integer code for each Turing pattern as already discussed in [12]. The code is obtained

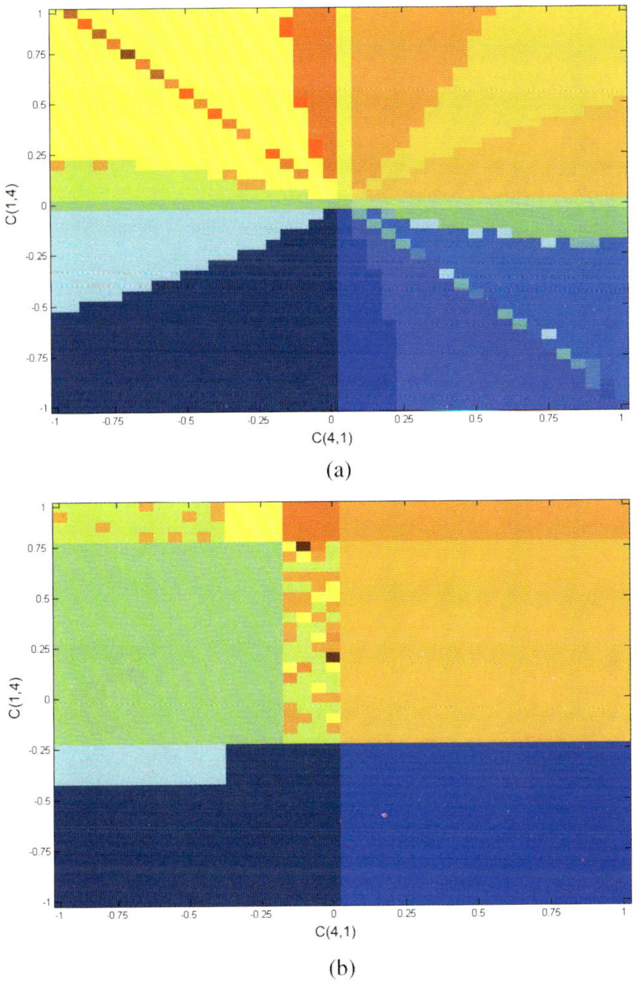

Fig. 9. (a) Initial basins of attraction for the 39 patterns and (b) final shape of the basins of attraction for the 15 patterns obtained with the introduction of the learned sensing neurons. The patterns emerged varying initial conditions for top-right corner cell (x-axis) and bottom-left corner cell (y-axis) of a 4×4 lattice in the range $[-1, 1]$.

converting the binary scheme of the pattern (i.e. sequence of black and white cells) into an integer value. This code is stored in a *Pattern Vector* at the first occurrence. Each element of the pattern vector contains the *Pattern Code* and the step of its last occurrence (*Occurrence Lag*). Once the external stimuli have been preprocessed, we reset the CNN, set again the initial conditions of the selected cells through the new outputs of the *SNs* (Fig. 8(b) and let the CNN re-evolve and generate a Turing pattern.

The use of Turing patterns as steady states of a dynamical system implies a form of sensor fusion, i.e. we synthesize heterogeneous sensory information into a single attractor. At each step, the information coming from sensors is fused to form a unique abstract and concise representation of the environment, as discussed in the Sec. 3.1.

3.1.6. *Selection network*

The *Selection Network* associates each element q of the pattern vector with a set of three parameters (k_o^q, k_a^q, k_p^q). At the first occurrence of the pattern q, they are randomly chosen in the range $[0, 1]$ with the constraint that: $k_o^q + k_a^q + k_p^q = 1$. Then, the parameters are modified under the effect of the learning process acting at the efferent (i.e. output) stage of the Representation layer as explained in the following. After completing the learning process and once the Turing pattern $q(t)$ has been generated at each time step t, the corresponding modulation parameters are selected and the behavior that emerges is the weighted sum of the actions suggested by the basic behaviors at that time: $A_F(t) = k_o^q \cdot A_o(t) + k_a^q \cdot A_a(t) + k_p^q \cdot A_p(t)$.

3.1.7. *Motivation layer and learning process*

The association between Turing patterns and modulation parameters is learned through a reward-based reinforcement learning implemented by a simplified Motor Map (MM) [12, 87], whereas the fitness of each action is evaluated by means of a Reward Function (RF), defined as follows:

$$RF(t) = \sum_i h_i \cdot RF_i(t) \tag{5}$$

where RF_i represents the degree of satisfaction related to the basic behavior i where $i = o, a, p$, indicating optomotor, avoidance and phonotaxis reflex, respectively:

$$\begin{aligned} RF_o(t) &= r_o(|A_F(t-1)|) \\ RF_a(t) &= \sum_i r_i(e^{d_i(t)}) \\ RF_p(t) &= r_p(|p(t)|) \end{aligned} \tag{6}$$

Here $A_F(t)$ is the action performed at time t, $d_i(t)$ is the distance between the robot and the obstacle detected by the sensor i ($i = Front(F), Right(R), Left(L)$) and $p(t)$ is the phase between the robot orientation and direction robot-target. The goodness of the behavior can be evaluated at each step via the function $DRF(t) = RF(t) - RF(t-1)$. A positive (negative) value for $DRF(t)$ indicates a successful (unsuccessful) behavior. Successful behaviors are followed by reinforcement,

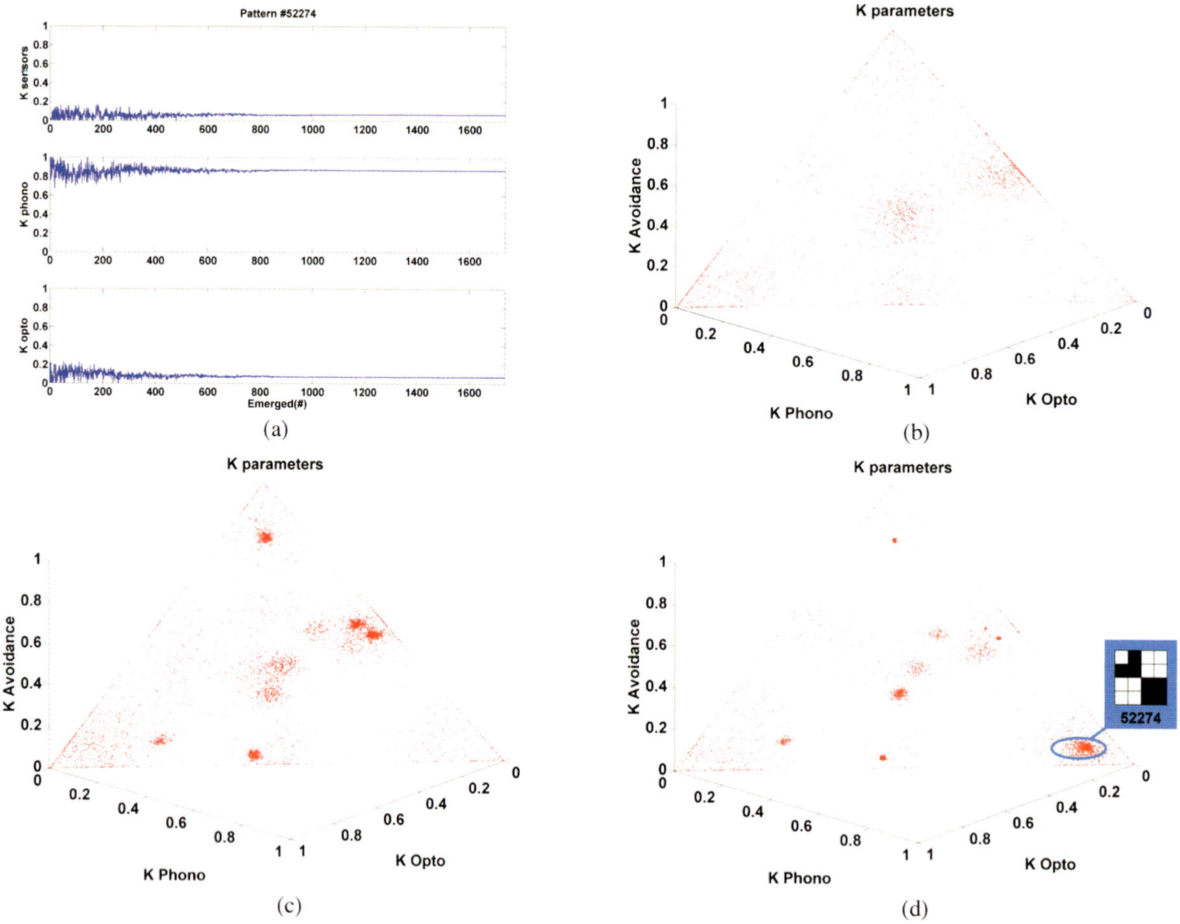

Fig. 10. (a) Evolution of the $k_i^f q_1$ for the emerged pattern ($q_1 = 52274$). The evolution of the modulation parameters is shown in (b–d) where the solutions adopted at the beginning of the learning stage (b), between 5000 and 10000 movements (c) and for the last 5000 movements are shown. In the last picture the region associated with the pattern 52274 is also indicated.

like in the Skinner's experiments [93] in order to maximize the RF. More details on the whole mathematical model are given in [12].

3.2. *Simulation results*

The software simulation environment, developed in $C++$, allows us to interface with the P3-AT robot and its simulation environment (MobileSim). The arena used consists of three rooms with six targets randomly placed.

3.2.1. *Learning phase*

As far as the simulated robot is concerned, the task assigned to the robot consists of aiming for a target and avoiding obstacles along the way. When the target is found it is switched off and another target appears in the arena. The learning phase lasts until one of the two following conditions occurs: either the a_q averaged on the last 1000 patterns drops below 10^{-4}, or after 5000 targets have been found. At the beginning of the learning phase, the robot randomly modulates the basic behaviors, due to the random initialization of the modulation parameters k_i^q ($i = a, o, p$), which determine the robot heading. During the learning process, the Motor Map-like algorithm corrects the parameters associated with each pattern. Fig. 10(a) shows the evolution of the $k_i^{q_1}$ for the pattern, i.e. $q_1 = 52274$. The modulation parameters associated to all the emerged patterns used in the first part, during and at the end of the learning phase (for a total number of 15000 actions divided in three blocks of 5000) is shown in Fig. 10(b)-(d).

This evolution is typical in several other simulations (for other results see [18]), taken into account and gives an idea of the clustering process that occurs during learning, where the basins of attraction of the Turing patterns that are associated to the modulation parameters, change in space and time leading to the emergence of specific rewarding behaviour combinations. For example it is interesting to notice that the pattern q_1 in this simulation is associated to a behaviour mainly guided by phonotaxis.

3.2.2. *Testing phase*

To evaluate the improvement of performance obtained during the learning process, we compared the result of the learned structure with other solutions: constant modulation parameters and randomly chosen modulation parameters. The constant behavior modulation parameters were chosen through a manual tuning aimed at optimizing the global performance of the robot. The parameters used in the following experiments are: $K_a = 0.35, K_p = 0.2, K_o = 0.05$. The randomly chosen modulation parameters gives an idea of the behaviour of the robot at the beginning of the learning phase when the behaviour modulation is initialized with random values.

The learning process leads to a significant reduction of actions needed to complete a sequence of targets searching, demonstrating the effectiveness of the control architecture and its capability to generalize the representations. Fig. 11 shows some examples of trajectories followed during the testing phase in case of fixed, random and learned modulation parameters.

To better understand the sequence of patterns (i.e. combination of behaviours) used during a target retrieving, an example is shown in Fig. 12. At the beginning of the trajectory the robot persists in the chosen behaviour combination until the environment conditions lead to a *jump* into the basin of attraction of another pattern, more adapt to represent the new situation. In this simulation about ten different behaviour modulations are used to fulfill the assigned goal.

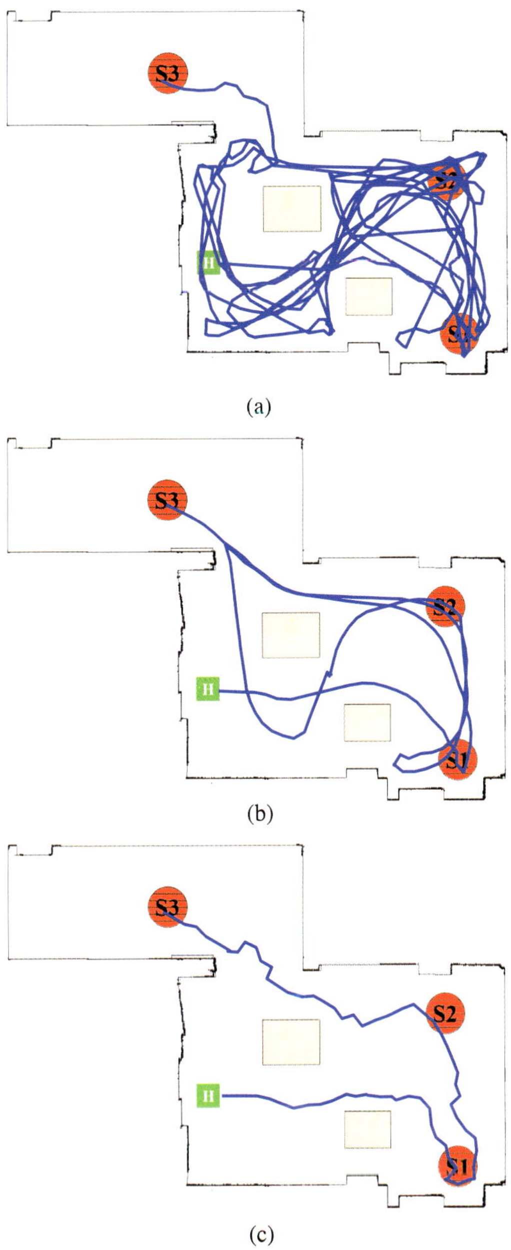

Fig. 11. Robot trajectories in the case of constant (a), randomly chosen (b) and learned (c) modulation parameters. The first two strategies (a–b) take a lot of time to reach the target and suffer from multiple collisions. From the learned modulation parameters, a very straightforward, although safe, behavior emerges.

3.2.3. *Experimental results*

To cross-check experimentally the promising results obtained in numerical simulations we use the roving robot P3-AT moving in a real environment. Fig. 13 shows the experimental set-up: the robot is placed in an environment with two targets. The trajectories followed by the robot are reported in Fig. 14. For the complete architecture the same behaviour modulation parameters obtained through the learning process in simulation were used for the tests with the real robot. The obtained performance are in line with the simulation results as reported in Table 2 for the path length and in Table 3 for the number of collisions.

It should be noted that when the robot is looking for the second target, after the first one is retrieved and switched off, the phono sensory system doesn't allow the detection of $S2$ because it is out of the sensor detection range (i.e. $[-90°, +90°]$). This is the reason for the long path followed to reach the second target also with the adaptive behaviour modulation.

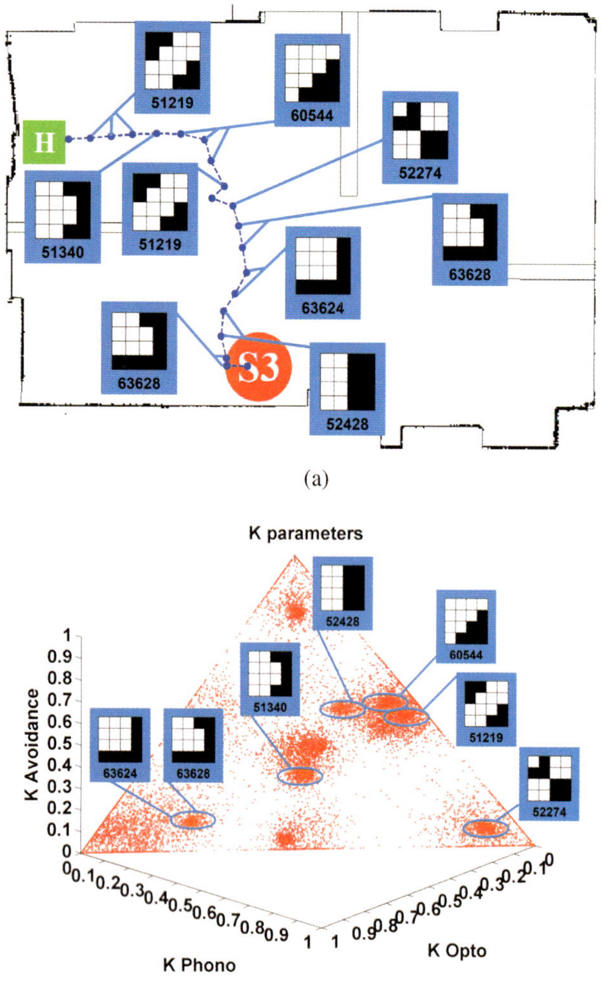

Fig. 12. (a) Sequence of patterns used during a target reaching task. Each pattern corresponds to a behaviour modulation (b).

Fig. 13. Experimental set-up. The rover P3-AT facing an environment with two targets.

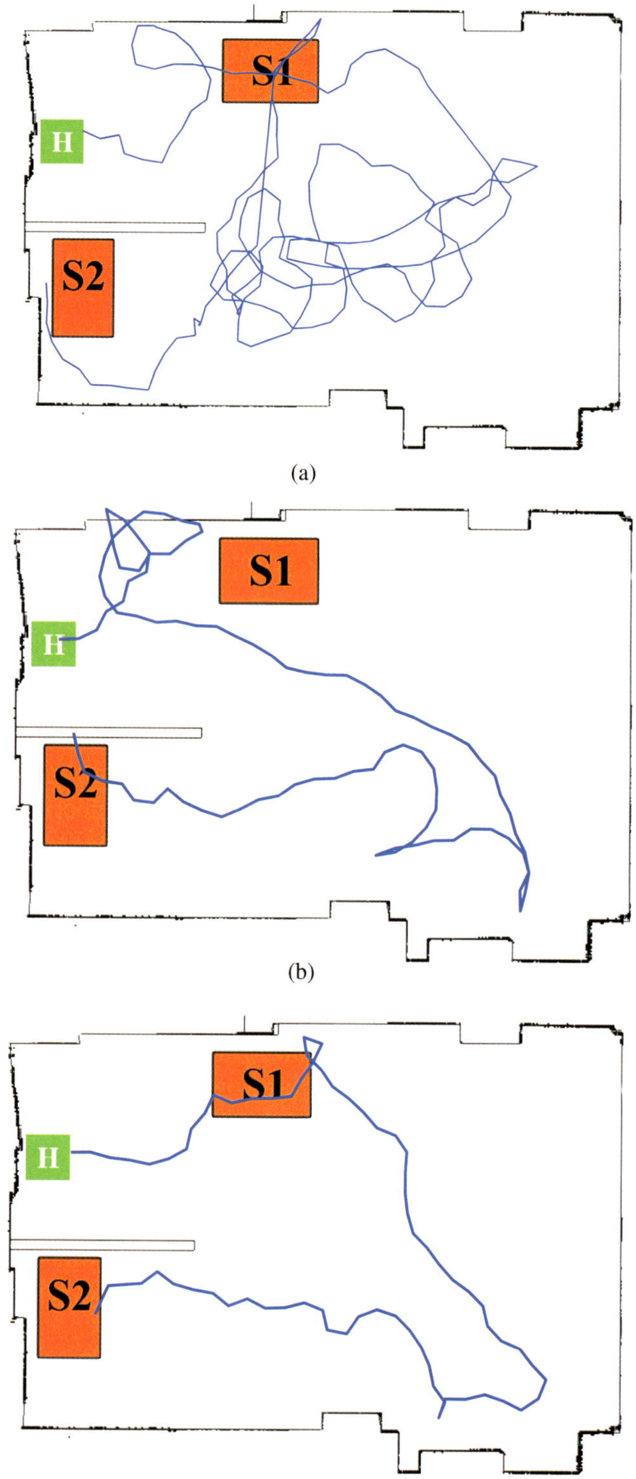

Fig. 14. Trajectories obtained with the real robot in the case of constant (a), randomly chosen (b) and learned (c) modulation parameters.

4. A Note on Winnerless Competition in CNNs

Going further in using the CNN paradigm to design and implement perceptual architecture, interesting works on the olfactory system of insects have been considered. Recent neurophysiological studies have demonstrated the ability of animals to discriminate between similar, but slightly different odors, which is crucial for the survival in a hostile environment. In particular experiments on the mitral cells of

Table 2. Experimental results, number of actions needed to retrieve three targets in a given order and improvement with respect to the worst case (i.e. Random modulation).

Sequence	Path length (m)		
	Random	Fixed	Learned
P	64	24	17.5
		Improvement (%)	
P	0	62.5	72.7

Table 3. Experimental results, number of collisions that occurs during the target retrieving process and improvement with respect to the worst case (i.e. Random modulation).

Sequence	Number of Collisions		
	Random	Fixed	Learned
P	11	6	2
		Improvement (%)	
P	0	45.4	81.8

the mammalian Olfactory Bulb [76] and on the projection neurons located within the insect Antennal Lobes [43] have shown that these neurons, involved in the second layer of sensation processing, are able to respond to different odors (or mixtures of odors) with repetitive and reproducible sequences of firing activity and quiescence [71]. Further, a deep analysis performed on the recorded activity of the 99 projection neurons of a locust antennal lobe [75] have shown how the transient dynamics of the neural activity is more effective in discriminate among different odors than the steady state condition which, nevertheless, is reached after a certain amount of time. These resulting spatial-temporal patterns of activity are then projected towards the higher level processing centres like the Mushroom Bodies and the Central Complex, in the insect brain [104].

To model the behavior shown in the olfactory system, a class of dynamical systems was proposed in [84]: the WinnerLess Competition (WLC) networks. These networks are able to transform a spatial distributed input into a spatial-temporal pattern realized by a quasi-periodic trajectory which pass near an heteroclinic contour connecting saddle fixed points or saddle limit cycles, which corresponds to the firing state of a neuron (or a subgroup of neurons). Such closed orbit represents a global attractor for the system [1] and, independently from the initial conditions, the trajectories tend asymptotically to it. It is to be emphasized that the results above mentioned, as also summarized in the following, were analytically formalized referring to three-species Lotka-Volterra systems (i.e. third order dynamics). The same results were generalized referring only to autonomous $n - species$ Lotka-Volterra systems, but, to the best of the author's knowledge, no analytical results were reported in literature for general $n - th$ order dynamics different from the Lotka-Volterra system.

Based on these considerations, given that we need such a non-autonomous high order dynamics to show the emergence of perceptual states, we referred in [16] to relevant neural models and for those one we'll build CNN networks, deriving relevant conditions on the network parameters in order to show the emergence of WLC relevant behavioral characteristics, useful to fulfill our classification task. More in details, referring to the classical WLC paradigm, the sequence of equilibria crossed by the trajectory depends on the stimulus and encodes the essential information relying on it. In this way, these main characteristics of WLC systems used for classification are:

- robustness to noise in the initial conditions;
- sensitivity to external stimuli;
- great number of possible classes.

Referring to the last feature, the advantage of this kind of approach with respect to coding with attractors in terms of capacity of the network is evident considering that in a multistable system, each class correspond to a basin of attraction of one of the fixed points. So, for a system of N neurons with N stable fixed points (or limit cycle), the maximum number of possible classes is N. In a WLC network, the use of a transient dynamics in place of the steady states equilibria greatly improves the number of possible classes: the number of sequences is potentially $e(N-1)!$ [84].

Starting from an analysis of the conditions for a Lotka-Volterra system to behave as a WLC network, given the lack of analytical WLC conditions, we investigate, in [16], the conditions for qualitatively reproducing the behavior of a WLC network by using locally connected cellular nonlinear networks (CNNs) in which the basic cells belong to three classes: the FitzHugh-Nagumo neuron [45], the Izhikevich neuron [64] and the single layer CNN standard cells [40]. The choice of these three models was made to exploit the advantage of the WLC for different computational perspectives: FitzHugh-Nagumo neurons can be used as a plausible model to reproduce many behaviors shown by the biological neurons, but introduce a high computational effort and can be used for computer simulation of a bio-inspired sensory system. Although showing many relevant neural characteristics, Izhikevich neurons are less biological plausible, but not so time consuming and can be implemented in dedicated digital hardware [69] with the advantage of behavior flexibility as a function of the parameters used. The single layer CNN standard cells have the advantage of a straightforward and already existing analog hardware implementation [35]: this guarantees high performance in terms of computation time. On the other hand, CNNs have widely been demonstrated to be a paradigm for the generation and also learning of complex spatial-temporal dynamics [106], some of them derived by biological signals [96], even in single layer cells [83]. Moreover, complex spatial-temporal dynamics were shown to take place also in digital hardware implementations [46]. For the proposed systems, we studied how the sequence of active states is sensitive to the incoming stimuli and independent from initial conditions, underlining the high capacity of the networks. We also present a compact code representing each different sequence in order to extract from the spatial-temporal pattern a compact information as a code for the class associated with the current stimulus.

Finally, in [16], we proposed an application demonstrating the possibility to use a WLC network in an artificial cognitive architecture in the challenging task of autonomous robot navigation where the ability to correctly interpret the incoming stimuli is crucial in presence of unpredictable and potentially hostile situations. In particular, the WLC network plays the role of perceptual core within an already developed control architecture discussed in the previous section, which models the sensing-perception-action loop, where, instead of using an RD-CNN, we use a WLC network, to generate the perceptual states, and consider, in place of the emerging Turing patterns, the spatiotemporal patterns coded by the sequence of activated neurons.

Also in this case the sequence constitutes a compact representation of the surrounding environment and is associated with an action by a selection network. To perform a given task, the robot is provided with no *a priori* knowledge about the action to be associated with each pattern and learns by trial and error. Learning is implemented by two mechanisms: an unsupervised learning acts at the sensing block allowing the system to modulate the "basins of attraction" of the sequences, while a simple reward-based reinforcement learning builds associations between spatiotemporal sequences and actions. All the details on the implementation and results are reported in [16].

5. Conclusions

CNNs are for sure a paradigm for complexity. Under this framework this chapter aims to survey a line of research focussed at using the CNN paradigm for the representation of spatial temporal neural dynamics in Biorobotics and Brain science. In particular the former aspects were dealt with referring to the capability of CNNs to act as central pattern generators for the locomotion control in legged and swimming robots. The design of the CNN was performed starting from the definition of the dynamic behavior of the single motor neuron (CNN cell) to the design of the proper templates to obtain different locomotion pattern. The suitability of the approach was tested in the most diverse situation in really working robot prototypes. The second part of the chapter focuses on how to add to these moving machines additional perceptual capabilities. Once again the capability of CNNs to represent and reproduce complex dynamics is exploited. Perception is here conceived as a complex phenomenon, in which environmental condition are considered and used to extract a compact and concise representation of the state of the surroundings, useful for taking on the action which increases a performance index, defined according to the robot mission. The architecture used as the site for the emergence of these perceptual states is once again a RD-CNN showing Turing Patterns. The hidden metaphor here is that, somewhere in the brain, the collection of all the information from the environment leads to the emergence of a perceptual state similar to the generation and evolution of a form in morphogenesis. More interestingly the same neuron used for the Central Pattern Generator implementation was used within the Turing pattern CNN generator, so as to indicate that very different dynamics can be obtained using the same basic dynamics (CNN neuron model) and the same paradigm (the RD-CNN). A last note was spent discussing the capability of CNNs to qualitatively reproduce the same dynamics as those ones show by the Winnerless Competition Paradigm, an archetype used to model neural information process in neural odor representation. The examples dealt with in this chapter are only a few cases out of all the potentialities offered by the CNN paradigm that will be more and more used in the future for the representation of neural activity, learning and memory for CNN based brain-like architectures to be embedded in autonomous robots. The inspiration drawn by studying CNNs and their relation to the biological world fostered a number of National and International research collaborations that made and are still contributing to make a breakthrough into the trans-disciplinary brain science and Biorobotics.

Acknowledgement

This work was supported by EU Project EMICAB, Grant N 270182.

Appendix: Hardware realization of a MTA CNN based locomotion controller

In this appendix a complete example of a walking hexapod is reported, employing the MTA approach. All the details of the hardware and mechanical implementation are reported to allow its reproduction. Some main guidelines are reported below, whereas the complete design can be downloaded from [24]. The authors acknowledge the contribution of Dr. M. Branciforte from ST Microelectronics for his contribution in the hardware design and realization of the robot. The whole MTA hexbot project was possible thanks to the Office of Naval research contribution.

MTA hexbot

The *MTA hexbot* is an autonomous six-legged walking robot. The structure of MTA hexbot consists of a body made of aluminum and six legs, each one with 2 degrees of freedom (DOF), actuated by twelve servomotors (Hitec high torque servos HS-945MG with metal gears). The robot is completely autonomous: all the electronics and power supply (three packages of five Ni-Mh batteries $2300mAh$) are on board. Provided this power supply the robot is able to carry a payload of 4kg and to show an autonomy of about $1h$ without payload. Some characteristics of the hexapod robot are given in Table 4.

Table 4. Characteristics of MTA hexbot.

dimensions	$31cm \times 25\,cm \times 20\,cm$
robot's weight	2250 g
servos' weight	750 g
weight of mechanical parts	600 g
batteries' weight	500 g
electronics' weight	400 g
payload	4 kg
autonomy	1 h

A CAD design of he mechanical structure of the robot is shown in Fig. 15.

A photo of the *MTA hexbot* is shown in Fig. 16. As regards the electronic part, it consists of 8 different circuits, a brief description of each board is given in the following (boards are labelled as in Fig. 16).

A — The CNN neurons. Each six second-order dynamical system is realized with OPAMPs. This circuit implements the equation (1) according to the design presented in [5, 74].

B — Output saturation. For each leg of the robot the two joints (α_{ij}-joint and β_{ij}-joint) are controlled by signals generated from the outputs of the motor-neuron ij associated to the leg. More precisely, the relationship between the motor-neuron outputs and the motor control signals can be described by the following equation:

$$\begin{cases} \alpha_{ij} = a_{2,ij} * y_{2,ij} + b_{2,ij} \\ \beta_{ij} = a_{1,ij} * f(y_{1,ij}) + b_{1,ij} \end{cases} \tag{7}$$

where α_{ij} and β_{ij} are used to control the motor position and $f(\cdot)$ is the following:

$$f(y) = \begin{cases} y & \text{if } y > k \\ k & \text{if } y \leq k \end{cases} \tag{8}$$

where k can be regulated by a trimmer from -1 to 0. This further saturation of the control signal prolongs the duration of the stance phase (corresponding to the low value of saturation of the control signal) and thus enhances the stability of the locomotion. The board B implements this nonlinearity $f(\cdot)$.

C — Motor settings. This circuit implements terms $a_{1,ij}$, $a_{2,ij}$, $b_{1,ij}$, $b_{2,ij}$ of equation (7) and thus provides the gains and offset used to modify the output variables for a proper control of the legs. The setup is made by using two groups of 12 trimmers (two for each

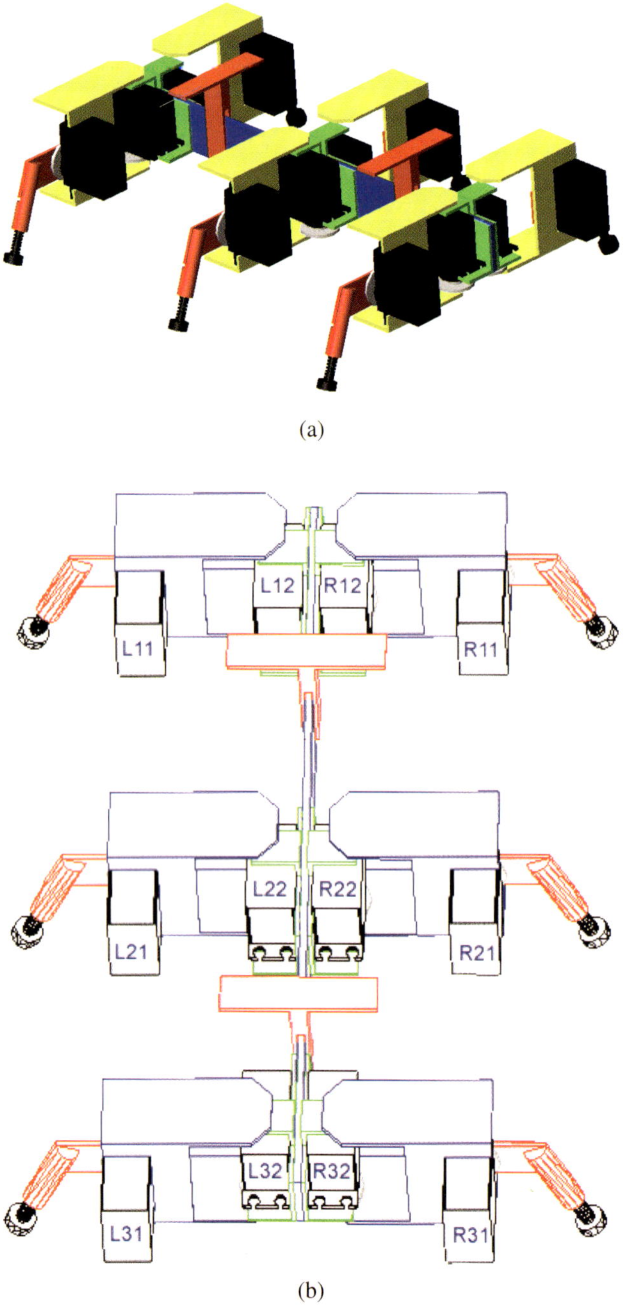

(a)

(b)

Fig. 15. CAD design of the mechanical structure.

Fig. 16. The autonomous six-legged walking robot driven by the CNN-based MTA control system.

DOF of each leg). These components set the parameters for a suitable control of the leg motion.

D — Connections between CNN neurons used to obtain fast, medium and slow gait. Analog switches digitally driven (MAX394) are used to chose the given set of templates.

E — Servomotor control. In order to generate the PWM signals needed to drive the servomotors, a Complex Programmable Logic Device (CPLD $XC9572$) is used. The CPLD with an external DAC ($DAC800$) generates a suitable digital ramp and provides the PWM signals by comparing this ramp with the analog signals α_{ij} and β_{ij} coming from the motor-neuron outputs. In this way PWM signals (i.e. signal with duty cycle proportional to the analog control signals) are generated to drive the robot servomotors. VHDL language is used to program the $XC9572$ that generates the pulses and controls the maximum range of duty cycle. It is worth to remark that the design of this board is strictly connected with the choice of the motors used to actuate the robot. The use of a microcontroller to generate PWM signals do not alter the generality of the analog approach. Moreover, a simple low-cost microcontroller can be used for this purpose since the effective computation takes place in the analog CNN.

F — The power supply. In order to obtain suitable dual power supply ($+5V$, $-5V$), two DC-DC converters ($ST755$, $ST777$) are used. The first one is an inverter used to generate the negative voltage ($-5V$), the second one is a step-up converter. The positive power supply is firstly decreased to $4.8V$ by two diodes, then is set to $+5V$ by the step-up converter. In this way if the battery voltage decreases the power supply remains stable.

G — An Infrared Remote Control (4-bit encoder $M145026$ on the remote command and a decoder $M145027$ on the robot) is used to pilot the robot. This allows to switch between the three different locomotion patterns and to change the motion direction turning left or right.

H — A PCB for connections between batteries, servomotors and power supply.

As it can be noticed the application of MTA-CNN to a hexapod robot offers the advantage of a simple circuitry and flexibility towards the implementation of new locomotion patterns.

A detailed description of the system including mechanical description of the robot and electronic parts design guidelines is reported in [24].

References

[1] Afraimovich, V. S., Rabinovich, M. I., Varona, P. (2004). Heteroclinic Contours in Neural Ensembles and the Winnerless Competition Principle, *Int. Jour. of Bif. and Chaos* 14(4), pp. 1195–1208.

[2] Arena, P., Branciforte, M., Fortuna, L. (1998). A CNN based experimental frame for patterns and autowaves, *Int. Jour. on Circuit Theory and Appls*, Vol. 26, pp. 635–650.

[3] Arena, P., Fortuna, L., Branciforte, M. (1999). Reaction-Diffusion CNN Algorithms to Generate and Control Artificial Locomotion, *IEEE Transaction on Circuits and Systems, I: Fundam. Theory and Applic.*, Vol. 46(2), pp. 259–266.

[4] Arena, P., Fortuna, L., Frasca, M. (2002a). Multi-Template Approach to realize Central Pattern Generators for Artificial Locomotion Control, *Int. J. Circ. Theor. Appl.* Vol. 30, pp. 441–458.

[5] Arena, P., Fortuna, L., Frasca, M. (2002b). Attitude control in walking hexapod robots: an analogic spatio-temporal approach, *Int. J. On Circuit Theory and Appls.*, Vol. 30, pp. 441–458.

[6] Arena, P., Fortuna, L., Frasca, M., Patané, L. (2003). Sensory feedback in CNN based Central Pattern Generators, *International Journal of Neural Systems*, Vol. 13(6), pp. 469–478.

[7] Arena, P., Fortuna, L., Frasca, M., Sicurella, G. (2004). An adaptive, self-organizing dynamical system for hierarchical control of bio-inspired locomotion, *IEEE Transactions on Systems, Man and Cybernetics, Part B*, Vol. 34(4), pp. 1823–1837.

[8] Arena, P., Fortuna, L., Frasca, M., Patané, L. (2005). A CNN-based chip for robot locomotion control. *IEEE Trans. On Circuits and Systems-I*, Vol. 52(9), pp. 1862–1871.

[9] Arena, P., Fortuna, L., Frasca, M., La Rosa, M. (2006a). Locally active Hindmarsh–Rose neurons, *Chaos, Solitons and Fractals*, Vol. 27(2), pp. 405–412.

[10] Arena, P., Fortuna, L., Frasca, M., Vagliasindi, G. (2006b). A wave-based CNN generator for the control and actuation of a lamprey-like robot, *Int. J. Bifurcation and Chaos*, Vol. 16(1), pp. 39–46.

[11] Arena, P., Bedia, M. G., Fortuna, L., Lombardo, D., Patanè, L., Velarde, M. G. (2006c). Spatio-temporal patterns in CNNs for classification: the winnerless competition principle, *Proceedings of CNNA 2006*.

[12] Arena P., Crucitti P., Fortuna L., Frasca M., Lombardo D., Patané L. (2007a). Turing patterns in RD-CNNs for the emergence of perceptual states in roving robots, *Int. J. of Bifurcation and Chaos*, 18(1), pp. 107–127.

[13] Arena, P., Fortuna, L., Lombardo, D., Patanè, L. and Velarde, M. G. (2007b). The WLC principle for action-oriented perception, *Proceedings of SPIE Europe 2007*.

[14] Arena, P., Fortuna, L., Lombardo, D., Patané, L. (2008a). Perception for action: Dynamic spatiotemporal patterns applied on a roving robot. *Adaptve Behavior*, Vol. 16(2/3), pp. 104–121.

[15] Arena, P., Fortuna, L., De Fiore, S., Patané, L. (2008b). Perception-action map learning in controlled multiscroll system applied to robot navigation. *Chaos*, Vol. 18, pp. 1–16.

[16] Arena, P., Fortuna, L., Lombardo, D., Patané, L., Velarde, M. (2008c). The WinnerLess Competition paradigm in Cellular Nonlinear Networks: Models and Applications. *Int. J. Circ. Theor. Appl.*, Vol. 37, pp. 505–528.

[17] Arena, P., Fortuna, L., Frasca, M., Patané, L. (2009a). Learning anticipation via spiking networks: application to navigation control. *IEEE Trans. on Neural Networks*, Vol. 20(2), pp. 202–216.

[18] Arena, P., De Fiore, S., Lombardo, D., Patané, L. (2009b). Emergence of perceptual states in nonlinear lattices: a new computational model for perception. In *Proc. of Int. Joint Conference on Neural Networks*.

[19] Arena P., De Fiore S., Patané L. (2009c). Web page available on-line: http://www.spark.dees.unict.it/

[20] Arena, P., Patané, L. (2011). Bio-inspired locomotion control of the hexapod robot Gregor III, In *Bio-inspired Computing and Networking*, Edited by Prof. Yang Xiao, and Prof. Fei Hu, Auerbach Publications, Taylor & Francis Group, CRC press, ISBN-10: 1420080326, ISBN-13: 978-1420080322.

[21] Arena, P., Patané, L. (2005). *Robot gallery*. Retrieved Dec 20, 2011, from EU Project SPARK I Web site: www.spark.diees.unict.it/Robot.html

[22] Arena, P., Patané, L. (2008). *Robot gallery*. Retrieved Dec 20, 2011, from EU Project SPARK II, Web site: www.spark2.diees.unict.it/Robot.html

[23] Arena, P., Patané, L.(eds) (2009). *Spatial Temporal Patterns for Action Oriented Perception in Roving Robots*. Cognitive Systems Monographs 1, Springer, ISBN: 978-3-540-88463-7.

[24] Arena, P., Patané, L. (2011). *CNN-based Hexapod robot*. Available on line at www.spark2.diees.unict.it/Download/Dissemination/MTAHexapod.pdf

[25] Arkin, R. C. (1991). *Behaviour Based Robotics*. MIT Press Cambridge.

[26] Arshavsky, Yu. I., Orlovsky, G. N., Panchin, Yu.V., & Pavlova, G.A. (1989). Control of locomotion in marine mollusc Clione limacina VII. Reexamination of type 12 interneurons. *Exp Brain Res.*, 78, 398–406.

[27] Balya, D., Roska, B., Roska, T., Werblin, F.S. (2002). A CNN framework for modeling parallel processing in a mammalian retina, *Int. J. Circuit Th. Appl.* Vol. 30, pp. 363–393.

[28] Balya, D. Petras, I., Rekeczky, C. (2004a). Pattern formation on the prototype complex-cell CNN-UM chip (CACE1k), *IEEE International Symposium on Circuits and Systems*, pp. 53–56

[29] Balya, D., Petras, I., Roska, T. (2004b). Implementing the Multilayer retinal model on the complex-cell cnn-um chip prototype, *International Journal of Bifurcation and Chaos*, Vol. 14(2), pp. 427–451.

[30] Böhm, H., Schildberger, K., & Huber, F. (1991). Visual and acoustic course control in the cricket Gryllus-bimaculatus. *Journal of Experimental Biology*, 159, 235–248.

[31] Borenstein, J., & Koren, Y. (1991). The vector field histogram fast obstacle avoidance for mobile robots. *IEEE Journal of Robotics and Automation*, 7, 278–288.

[32] Brooks, R. A. (1986). A robust layered control system for a mobile robot", *IEEE Jour. of Robotics and Automation*, 2(1), 14–23.

[33] Brooks, R. A. (1994). Intelligence without reason. In *Proc. of 12th Int. Conference on Artificial Intelligence*, eds. J. Mylopoulos and R. Reiter, San Mateo, CA.

[34] Büschges, A., Akay, T., Gabriel, J. P., Schmidt, J. (2008). Organizing network action for locomotion: Insights from studying insect walking, *Brain Research Reviews*, Vol. 57, pp. 162–171.

[35] Carmona, R., Jiménez-Garrido, F., Domínguez-Castro, R., Espejo, S., & Rodríguez-Vázquez, A. (2002). Bio-inspired analog VLSI design realizes rogrammable complex spatio-temporal dynamics on a single chip, *Proc. of Conf. Design Automation and Test in Europe*, pp. 362–366.

[36] Chauvet, E., Paullet, J. E., Previte, J. P., Walls, Z. (2002). A Lotka-Volterra Three-species Food Chain, *Mathematics Magazine* 75(4), pp. 243–255.

[37] Chen, Z., Iwasaki, T. (2008). Circulant Synthesis of Central Pattern Generators With Application to Control of Rectifier Systems, *IEEE Tran. on Automatic Control*, Vol. 53(1).

[38] Chiel, H. J., Beer, R. D., Gallagher, J. C. (1998). Evolution and analysis of model CPGs for walking I. Dynamical modules, *J. Computational Neuroscience*, Vol. 7(2) pp. 99-118.

[39] Chua, L.O., Desoer, C.A., Kuk, E.S. (1987). *Linear and nonlinear circuits*, McGraw-Hill, New York, NJ.

[40] Chua, L.O., Yang, L. (1988). Cellular neural networks: theory, *IEEE Trans. on Circuits and Systems*, Vol. 35, pp. 1257–1272.

[41] Chua, L.O., Hasler, M., Moschytz, G.S., Neirynck, J. (1995). Autonomous Cellular Neural Networks: A Unified Paradigm for Pattern Formation and Active Wave Propagation, *IEEE Trans. on Circuits and Systems — Part I*, Vol. 42(10), pp. 559–577.

[42] Cohen, A. H., Holmes, P. J., Rand, R. H. (1982). The nature of the coupling between segmental oscillators of the lamprey spinal generator for locomotion: a mathematical model, *J. Math. Biol.*, Vol. 3, pp. 345–369.

[43] Christensen, T., Hildebrand, J. (1987). Male-specific, sex pheromone-selective projection neurons in the antennal lobes of the moth *Manduca sexta*, *J. Comp. Physiol. A* 160, pp. 553–569.

[44] Dürr, V., Schmitz, J., Cruse, H. (2004). Behaviour-based modelling of hexapod locomotion: Linking biology and technical application, *Arthropod Struct Dev.*, Vol. 33(3), pp. 237–250.

[45] FitzHugh, R. (1961). Impulses and physiological states in theoretical models of nerve membrane, *Biophys.* 7(1).

[46] Földesy, P., Zarándy, Á., Rekeczky, C., Roska, T. Digital implementation of cellular sensor-computers, *International Journal of Circuit Theory and Applications* 34(4), pp. 409–428.

[47] Frasca, M., Arena, P., Fortuna L., (2004). Bio-Inspired emergent control of locomotion systems, *World Scientific*.

[48] Freeman, W. J. (2004). How and why brains create meaning from sensory information. *Int. Jour. of Bif. and Chaos*, 14, 515–530.

[49] Fukuoka, Y., Kimura, H., Cohen, A. H. (2003). Adaptive Dynamic Walking of a Quadruped Robot on Irregular Terrain based on Biological Concepts, *Int. Journal of Robotics Research*, Vol. 22(3–4), pp. 187–202.

[50] Freeman, W. J. (1987). Simulation of chaotic EEG patterns with a dynamic model of the olfactory system. *Biol. Cybern.*, 56, 139–150.

[51] Galan, R. C., Jimnez-Garrido, F., Domnguez-Castro, R., Espejo, S., Roska, T., Rekeczky, C.,

Petras, I., Rodrguez-Vzquez, A. (2003). A Bio-Inspired Two-Layer Mixed-Signal Flexible Programmable Chip for Early Vision, *IEEE Transaction on Neural Networks*, Vol. 14(5), pp. 1313–1336.

[52] Gnadt, W., & Grossberg, S. (2008). SOVEREIGN: An autonomous neural system for incrementally learning planned action sequences to navigate towards a rewarded goal. *Neural Networks* 21, 699–758.

[53] Goras, L., & Chua, L. (1995). Turing Patterns in CNNs — Part I, II. *IEEE Trans. Circuits and Systems — I*, 42, 602–626.

[54] Grillner, S., Georgopoulos, A. P., Jordan, L. M. (1997). Selection and initiation of motor behavior, In *P. S. G. Stein, S. Grillner, A. Selverston, D. G. Stuart (Eds.), Neurons, networks and motor behavior* MIT Press.

[55] Gronenberg, W.,& Lopez-Riquelme, G.O. (2004). Multisensory convergence in the mushroom bodies of ants and bees. *Acta. Biol. Hung.*, 55, 31–37.

[56] Harter, D. (2005) Evolving neurodynamics controllers for autonomous robots. *International Joint Conference on Neural Networks*, 137–142.

[57] Harter, D., & Kozma, R. (2005). Chaotic Neurodynamics for autonomous agents. *IEEE Trans. on Neural Networks*, 16(3), 565–579.

[58] Holmes, P., Full, R. J., Koditschek, D., Guckenheimer, J. (2006). The dynamics of legged locomotion: Models, analyses, and challenges, *SIAM Review*, Vol. 48(2), pp. 207–304.

[59] Homberg, U. (1987). Structure and functions of the central complex in insects. In *Arthropod brain, its evolution, development, structure and functions*, AP Gupta (Ed.), New York: Wiley, 347–367.

[60] Horchler, A., Reeve, R., Webb, B., & Quinn, R. (2004). Robot phonotaxis in the wild: a biologically inspired approach to outdoor sound localization. *Advanced Robotics*, 18(8), 801–816.

[61] Ijspeert, A. J., Hallam, J., Willshaw, D. (1999). Evolving swimming controllers for a simulated lamprey with inspiration from neurobiology, *Adaptive Behavior*, Vol. 7(2), pp. 151–172.

[62] Ijspeert, A. J. (2001). A connectionist central pattern generator for the aquatic and terrestrial gaits of a simulated salamander, *Biological Cybernetics*, Vol. 84(5), pp. 331–348.

[63] Ijspeert, J. (2008). Central pattern generators for locomotion control in animals and robots: A review, *Neural Networks*, Vol. 21, pp. 642–653.

[64] Izhikevich, E. M. (2003). Simple Model of Spiking Neurons, *IEEE Transactions on Neural Networks* 14, pp. 1569–1572.

[65] Johns, D. A., Martin, K. (1997). Analog Integrated Circuit Design, *John Wiley and Sons.*

[66] Kier, R. J., Ames, J. C., Beer, R. D., Harrison, R. R. (2006). Design and Implementation of Multipattern Generators in Analog VLSI, *IEEE Trans. on Neural Networks*, Vol. 17(4), pp. 1025–1038.

[67] Krinsky, V. I. (1984). Autowaves: Results, Problems, Outlooks, in *Self-Organization: Autowaves and Structures Far from Equilibrium*, Springer-Verlag, Berlin, pp. 9–18.

[68] Kuramoto, Y. (2003). Collective behavior of coupled phase oscillators, in *M. A. Arbib (Ed.), The handbook of brain theory and neural networks* MIT Press, pp. 223–226.

[69] La Rosa, M., Caruso, E., Fortuna, L., Frasca, M., Occhipinti, L., Rivolia, E. (2005). Neuronal dynamics on FPGA: Izhikevich's model, *Proceedings of SPIE 2005*, Sevilla, Spain.

[70] La Spina, G., Sfakiotakis, M., Tsakiris, D., Menciassi, A., Dario, P. (2007). Polychaetelike undulatory robotic locomotion in unstructured substrates, *IEEE Transactions on Robotics*, Vol. 23(6), pp. 1200–1212.

[71] Laurent, G., Wehr, M., Davidowitz, H. (1996). Temporal Representations of Odors in an Olfactory Network, *The Journal of Neuroscience* 16(12), pp. 3837–3847.

[72] Lee, Y. J., Lee, J., Kim, K.K., Kim, Y., Ayers, J. (2007). Low power CMOS electronic central pattern generator design for a biomimetic underwater robot, *Neurocomputing*, Vol. 71(1–3), pp. 284–296.

[73] Lynch, K. (1960). *The Image of the City.* MIT Press, Cambridge.

[74] Manganaro, G., Arena, P., Fortuna, L. (1999). Cellular Neural Networks: Chaos, Complexity, and VLSI Processing, Springer-Verlag, Berlin.

[75] Mazor, O., Laurent, G. (2005). Transient Dynamics versus Fixed Points in Odor Representations by Locust Antennal Lobe Projection Neurons, *Neuron*, 48, pp. 661–673, DOI: 10.1016/j.neuron.2005.09.032.

[76] Meredith, M. (1986). Patterned response to odor in mammalian olfactory bulb: the influence of intensity, *J. Neurophysiol* 56, pp. 572–597.

[77] Murray, J. D. (2002). *Mathematical Biology I: An Introduction (third edition).* Springer-Verlag, NJ.

[78] Nolfi, S. (2002). Power and Limits of Reactive Agents. *Neurocomputing*, 42(1), 119–145.

[79] Orlovsky, G. N., Deliagina, T. G., Grillner, S. (1999). Neuronal Control of Locomotion From Mollusc to Man, Oxford Univ. Press.

[80] Patané L. (2004). Bio-inspired Robots: from sensing toward perception, *PhD Thesis.*

[81] Pearson, K. G. (1995). Proprioceptive regulation of locomotion, *Current Opinion in Neurobiology*, Vol. 5, pp. 786–791.

[82] Pearson, G. K. (1993). Common principles of motor control in vertebrates and invertebrates, *Ann. Rev. Neurosci.* Vol. 16, pp. 265–97.

[83] Petràs, I. and Gilli, M. (2006). Complex dynamics in one-dimensional CNNs, *International Journal of Circuit Theory and Applications* 34(1), pp. 3–20.

[84] Rabinovich, M., Volkovskii, A., Lecanda, P., Huerta, R., Abarbanel, H. D. I., Laurent, G. (2001). Dynamical Encoding by Networks of Competing Neuron Groups: Winnerless Competition, *Phys.Rev.Lett.*.

[85] Righetti, L., Ijspeert, A. J. (2006). Programmable central pattern generators: An application to biped locomotion control, in *Proceedings of the 2006 IEEE international conference on robotics and automation.*

[86] Righetti, L., Ijspeert, A. (2008). Pattern generators with sensory feedback for the control of quadruped locomotion, in *Proceedings of the 2008 IEEE international conference on robotics and automation.*

[87] Ritter, H., Martinetz, T., Schulten, K. (1992). Neural Computation and Self-Organizing Maps — An Introduction, *Addison-Wesley, New York.*

[88] Ritzmann, R. E., Quinn, R. D., Fischer, M. S. (2004). Convergent evolution and locomotion through complex terrain by insects, vertebrates and robots, *Arthropod Structure and Development*, Vol. 33, pp. 361–379.

[89] Roska, B., Werblin, F. S. (2001). Vertical interactions across ten parallel stacked representations in mammalian retina, *Nature* 410, pp. 583–587.

[90] Rossignol, S., Dubuc, R., Gossard, J. P. (2006). Dynamic sensorimotor interactions in locomotion, *Physiological Reviews*, Vol. 86, pp. 89–154.

[91] Russo, P., Webb, B., Reeve, R., Arena, P., & Patané, L. (2005). A Cricket-Inspired Neural Network for FeedForward Compensation and Multisensory Integration. In *IEEE Conference on Decision and Control and European Control Conference 2005*, Sevilla, Spain.

[92] Schöner, G., Yang, W. Y., Kelso, J. A. S. (1990). A synergetic theory of quadrupedal gaits and gait transitions, *Journal of Theoretical Biology*, Vol. 142, pp. 359–361.

[93] Skinner, B. F. (1974). *About behaviorism.* Alfred Knopf, New York.

[94] Slavova, A., Zecca, P. (2007). Complex behavior of polynomial FitzHugh-Nagumo cellular neural network model, *Nonlinear Analysis: Real World Applications*, Vol. 8(4), pp. 1331–1340.

[95] Tenore, F., Etienne-Cummings, R., Lewis, M.A. (2004). A programmable array of silicon neurons for the control of legged locomotion, *Proc. 2004 International Symposium on Circuits and Systems, ISCAS 04*, pp. V-349–V-352.

[96] Tetzlaff, R., Niederhöfer, C., Fischer, P. (2006). Automated detection of a preseizure state: nonlinear EEG analysis in epilepsy by Cellular Nonlinear Networks and Volterra systems, *International Journal of Circuit Theory and Applications* 34(1), pp. 89–108.

[97] Turing, A. M. (1952). The chemical basis of morphogenesis, *Phil. Trans. R. Soc. London*, Vol. B327, pp. 37–72.

[98] Van Den Driessche, P., Zeeman, M. (1998). Three-dimensional competitive Lotka-Volterra systems with no periodic orbits, *Siam J. Appl.Math.* 58, pp. 227—234.

[99] Vernon, D., Metta, G., & Sandini, G. (2007). A survey of artificial cognitive system: Implications for the autonomous development of mental capabilities in computational agents. *IEEE Trans. Evolutionary Computation*, 11(2), 151—180.

[100] Verschure, P. F. M. J., Voegtlin, T., & Douglas, R. J. (2003). Environmentally mediated synergy between perception and behaviour in mobile robots. *Nature*, 425, 620–624.

[101] Viana Di Prisco, G., Wallén, P., Grillner, S. (1990). Synaptic effects of intraspinal stretch receptor neurons mediating movement-related feedback during locomotion, *Brain Research*, Vol. 530, pp. 161–166.

[102] Webb, B., & Scutt, T. (2000). A simple latency dependent spiking neuron model of cricket phonotaxis. *Biological Cybernetics*, 82(3), 247–269.

[103] Werblin, F. S. and Roska, B. M. (2004). Multichannel retinal processing: A tutorial of retinal function from photoreceptors to parallel feature abstractions mediated by the analog array processing capabilities of the retina, *Int. J. Bifurcation and Chaos* Vol. 14, pp. 843–852.

[104] Wessnitzer, J., & Webb, B. (2006). Multimodal sensory integration in insects — towards insect brain control architectures. *Bioinspiration and Biomimetics*, 1, 63–75.

[105] Williamson, M. M. (1998). Neural control of rhythmic arm movements, *Neural Networks*, Vol. 11, pp. 1379–1394.

[106] Xavier-de-Souza, S., Suykens, J. A. K., Vandewalle, J. (2006). Learning of spatiotemporal behaviour in cellular neural networks, *International Journal of Circuit Theory and Applications* 34(1), pp. 127–140.

CHAPTER 6

FROM RADIO-AMATEURS' ELECTRONICS TO TOROIDAL CHAOS

OTTO E. RÖSSLER

Division of Theoretical Chemistry, University of Tübingen, D-72076 Tübingen, Germany

CHRISTOPHE LETELLIER

CORIA UMR 6614 — Rouen University, F-76801 Saint-Etienne du Rouvray cedex France

On the occasion of Leon Chuas 75th birthday, we offer some new simulations and musings on toroidal chaos.

1. Introduction

Chaos is addictive. Its essence was recently rediscovered by the famous artist Olafur Eliasson who suspended a running desk-top fan from a long rope. The public was fascinated by the absolute unpredictability of the gadgets elegant motions.

One of the major — and perhaps most natural — roads into understanding chaos, as locally everywhere parallel hairlines, is the design of electronic systems while registering their activity on an oscilloscope. The electronic analog computer is itself an example. When you simulate a fast 2-variable triangle-shaped limit cycle on such a machine, then near the single sharp corner, the light point will tend to follow a highly irregular course that is different (shorter or longer) at every turn. We saw this in Tubingen jointly with Heinz Karfunkel in 1971 without realizing what it means (that there is a hidden third variable involved and that the "noise" we saw is not at all exogenous).

Chemical and biochemical (biological) systems do themselves consist of coupled smoothly varying variables. Life itself is an example. To model it electronically is in principle impossible because in nature, the state variables (concentrations) are at the same time the hardware. There is no big computer behind on which these variables are "simulated."

Radio amateurs have a much easier life. They can trust their hardware and bring it to producing beautiful sounds and music, and this at the speed of light across continents. Later, even pictures followed. Electrical engineering is the real home of chaos theory. The latter was invented by Poincaré in the context of celestial mechanics, but he already used an electronic system in his head, one feels.

The tool kit of astronomical interactions is somewhat less accessible mentally. The optimistic hope that eventually, very few controllable elements — capacitors, resistors, amplifiers — will suffice to get a beautiful nontrivial behavior out of their interaction was slow in coming. The fact that chemical soups and brains follow the same laws becomes fully comprehensible only to an electronics engineer.

Leon Chua is the person at the center. He did not "come from astronomy" like Poincaré, or from brain theory or "chemical soups" but plunged into the central heaven of electronic building elements directly. This made him stronger than other nonlinear scientists who thought of these elements only as analog tools to better understand their own objects of investigation. That these elements constitute heaven itself — the heaven of dynamics — no one saw more clearly than him.

His uniquely beautiful, large-format scientific journal,[a] with its primacy of color and simple clear drawings, reflects the powerful esthetic root of his chaos kingdom. No one else encouraged so many creative pictorial minds into taking themselves and their work seriously. No one forced more beautiful "double scrolls" into trajectories laid out in a clarity that removed any fear of the mind that this complexity could never be grasped in full. Hun Tun means "chaos" in Chinese mythology, but also the name of the person who cared most about all, without even having to open his eyes and ears and nose and mouth, because everything was already absolutely clear in his — the first emperor's — mind.

It is a privilege that we are allowed to offer some quasiperiodic — or no longer quasiperiodic but chaotic — motions on the occasion of Leon's birthday in this book that is an expression of his mind. The fact that there is a connection to human physiology — and breathing — in some of the simulations makes the connection to Hun Tun even more palpable.

2. The van der Pol Equation

The oldest example of complicated behavior found in an unconstrained (non-conservative) dynamical system is that found in the periodically forced van der Pol oscillator [1]. Cartwright & Littlewood's results were based on painstaking analytical arguments [2]. Is it, at the same time, also possible to understand their observations in terms of a so-called reinjection principle as we will detail below? The answer is yes. One can even say that Cartwright and Littlewood were the first to make use of the reinjection principle in order to find nontrivial behavior in a simple dynamical system.

In 1938, the Department of Scientific and Industrial Research issued a memorandum to the London Mathematical Society appealing for the assistance of pure mathematicians in solving the type of equations occurring in radio work, laying emphasis on the need to know how the frequencies of the periodic solutions varied with the parameters of the equation (see Cartwright [3] for a brief history of the earliest mathematical studies of this equation). At about the same time the van der Pol equation [4]

$$\ddot{v} - \epsilon(1 - v^2)\dot{v} + v = 0. \tag{1}$$

arose as a model of the oscillation of the potential in the cavity magnetron discovered at the end of 1939, cf. [5]. Such a device was used to produce the high-frequency oscillations required to develop a powerful radar. By then, the 1934 paper by van der Pol [6] already had a strong impact on radio-engineers [3, 7, 8]. Electronic experiments showed that the driven van der Pol equation has a periodic solution with a period which can be many times a high multiple of the period of the forcing term. Moreover, for some values of the amplitude of the forcing term, the driven van der Pol equation possesses two periodic solutions of different periods. Cartwright and Littlewood showed that the existence of two periodic solutions implies the existence of "bad" solutions which was "very difficult to rule out" [1]. These non-periodic orbits described as a "dicontinuously recurrent" motion would be called "chaotic" today. Later Norman Levinson (1912–1995) — who was Leon Chua's professor at MIT — used a piecewise linear form of the driven van der Pol equation to show the aperiodic character of some of its solutions [9]. It appears that Levinson's analysis led to Steven Smale's introduction of the Horseshoe map [10], a very important concept in chaos theory as pointed out in the first paper on chaos by the first author [11]. Following Levinson's approach,

[a] *International Journal of Bifurcation & Chaos.*

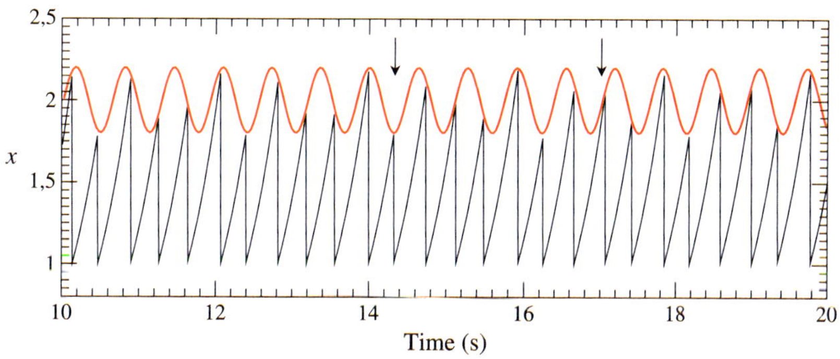

Fig. 1. Quasi-periodic regime in Eq. (6) obtained by numerical simulation. Parameters: $a = 0.2$, $\omega = \pi$ and $\mu = 1.8$. Time step: $\delta t = 0.001$ s.

Mark Levi investigated rotation numbers of periodic solutions to the driven van der Pol equation [12], thus enabling the topological analysis which was later applied to chaotic attractors.

Using Liénard's [13] familiar nonlinear transformation (cf. [14]), the periodically forced van der Pol oscillator

$$\ddot{x} - \mu\dot{x}(1 - x^2) + x = a\cos bt,$$

can be written in the form

$$\begin{cases} \epsilon\dot{x} = x - \dfrac{1}{3}x^3 - y \\ \dot{y} = x + a\cos bt, \end{cases} \tag{2}$$

where $\epsilon = \frac{1}{\mu}$. Cartwright and Littlewood considered Eq. (2) under the condition of very small values of ϵ. Later it was shown that such an equation can lead to chaotic behaviors [15].

3. A Related Prototype

The matter discussed in this section was mostly extracted from a textbook the first author wrote in the early 1980's but which was never published. With the help of Reimara Rössler, we found again the first draft of this textbook and we are preparing a publication of it, more than thirty years after its writing. As we will show in this chapter, Otto's views remain original even when they are considered after a so long time (says Christophe). In what follows, we preserved the content but not the writing.

The purpose was to provide a dynamical system producing a toroidal chaos governed by a mechanism simpler than the process underlying the van der Pol equation. Let us start from the two-dimensional system

$$\begin{cases} \dot{x} = \dfrac{x}{r}f(x) + \omega y \\ \dot{y} = -\omega x + \dfrac{y}{r}f(x) \end{cases} \tag{3}$$

in Cartesian coordinates which may be rewritten as the following two-dimensional system

$$\begin{cases} \dot{r} = \dfrac{1}{2}\dfrac{2x\,\mathrm{d}x + 2y\,\mathrm{d}y}{\sqrt{x^2+y^2}} = \left(\dfrac{x^2}{r} + \dfrac{y^2}{r}\right)\cdot f(r) \\ \qquad + \underbrace{\dfrac{\omega xy - \omega xy}{r}}_{=0} = r\cdot f(r) \\ \dot{\theta} = 1. \end{cases} \tag{4}$$

in polar coordinates. This system defines the evolution of the radius which is associated with a constant angular velocity. Using $f(r) = 1$, system (3) can be rewritten as the two-variable autonomous system

$$\begin{cases} \dot{x} = \dfrac{x}{r} + \omega y \\ \dot{y} = -\omega x + \dfrac{y}{r}, \end{cases} \tag{5}$$

which is an expansive system with a single fixed point which is an unstable focus. To prevent this system from diverging to infinity, a second condition is added to bound the radius. One of the simplest ways is to use, for example,

$$(r - 1) \bmod (1 + a\cos\pi bt).$$

System (4) thus becomes

$$\begin{cases} \dot{s} = s, \quad s \bmod (1 + a\cos\pi\omega t) \\ \dot{\theta} = 1. \end{cases} \tag{6}$$

For the simulation shown in Fig. 1, this meant that whenever $r = \sqrt{x^2+y^2}$ surpasses $2 + a\cos\pi\omega t$

from below, r is replaced with $r - (1 + a\cos\pi\omega t)$. The principle of modulation of the maximum value of the radius reached at each oscillation is shown in Fig. 1. These oscillations correspond to relaxation oscillations as introduced by van der Pol and Liénard.

System (6) corresponds to a particular linear combination of two periodic behaviors and, consequently, only quasi-periodic regimes are obtained. An example of them is shown in an $x - y$ plane projection of the phase space. The toroidal structure appears squeezed in this plane. A first-return map, built as a first-return to the maxima of the radius, clearly shows the quasi-periodic nature of the regime, whose strong signature is an annular map (Fig. 2b).

Systems (6) and (5) cannot produce more complex behaviors due their linear character. Chaotic behaviors can only be expected when a nonlinearly growing radius is involved. We thus introduced nonlinear terms in Eq. (5) to obtain the following equation

$$\begin{cases} \dot{x} = \mu \dfrac{x(1-y)}{r} + \omega y \\[2ex] \dot{y} = -\omega x + \mu \dfrac{y(1-x)}{r} \end{cases} \tag{7}$$

where $(r - 1)$ is $\mathrm{mod}(1 + a\cos\pi\omega t)$ as in the previous example. The behavior is now chaotic as evidenced by the first-return map built from the maxima of the radius (Fig. 3(b)). The map looks like a layered annular structure, thus confirming that the behavior is toroidal chaos. Better characterizing such toroidal chaos remains a great challenge. This apparently does not result from the emergence of a folding-in in a quasi-periodic regime. The stiff dynamics induced by the "mod ()" function squeezes the attractor onto itself, inducing some domains of the phase space (at the upper left corner of the return map, Fig. 3(b)) where the different trajectories cannot be distinguished, rendering the dynamics non-observable. The simplicity of the mechanism used to produce this behavior has an advantage: it is easy to understand; but the resulting dynamics is quite hard to fully characterize.

An "avatar" of Eq. (7) can be written in a slightly different way. The principle underlying the construction of this new model is the same as in the two previous examples but the equations are simpler

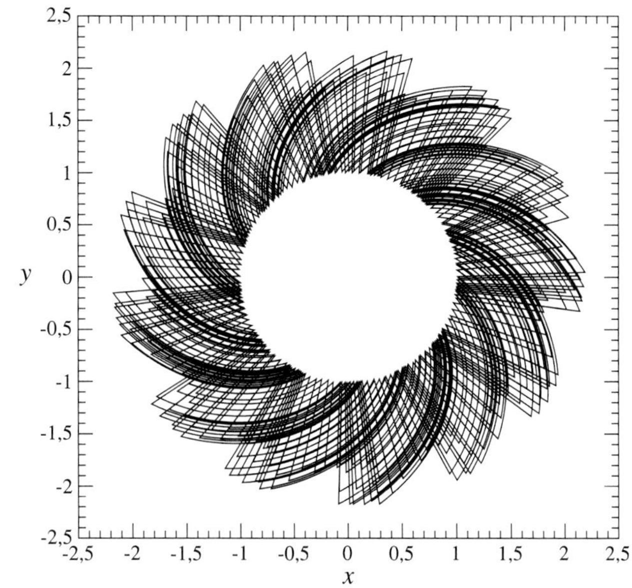

(a) Numerical simulation of Eq. (5)

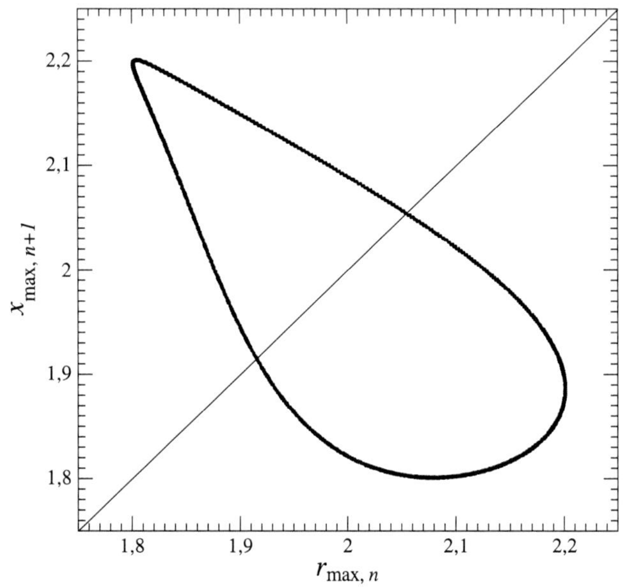

(b) The corresponding first-return map

Fig. 2. Quasi-periodic regime solution to Eq. (6), or equivalently to Eq. (5), obtained by numerical simulations. The first-return map is built as a return map to the maxima of radius $r_{\max}$. Parameters: $a = 0.2$ and $\omega = \pi/4$. Initial conditions: $x_0 = 1$ and $y_0 = 0$.

since there are no longer any fractional terms

$$\begin{cases} \dot{x} = \mu x(1-y) - \omega y \\ \dot{y} = \omega x + \mu y(1-x). \end{cases} \tag{8}$$

These equations with y as before produce the chaotic regimes shown in Fig. 4. It is far more difficult to show that this chaotic behavior has a toroidal character. In fact it has, because the first-return map is basically a closed loop with

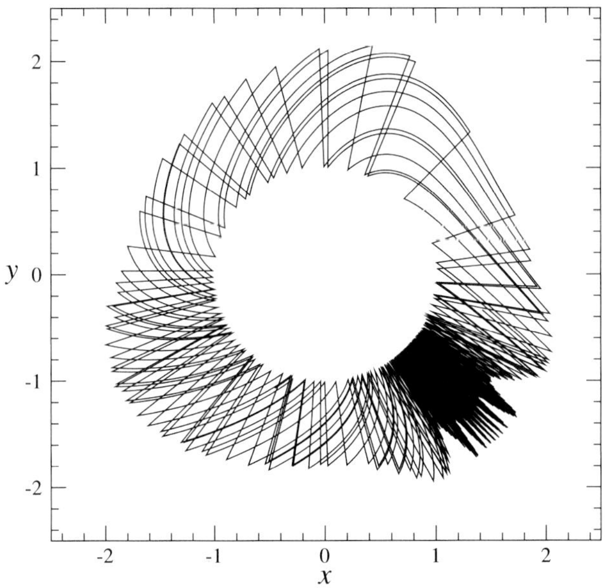

(a) Numerical simulation of Eq. (8)

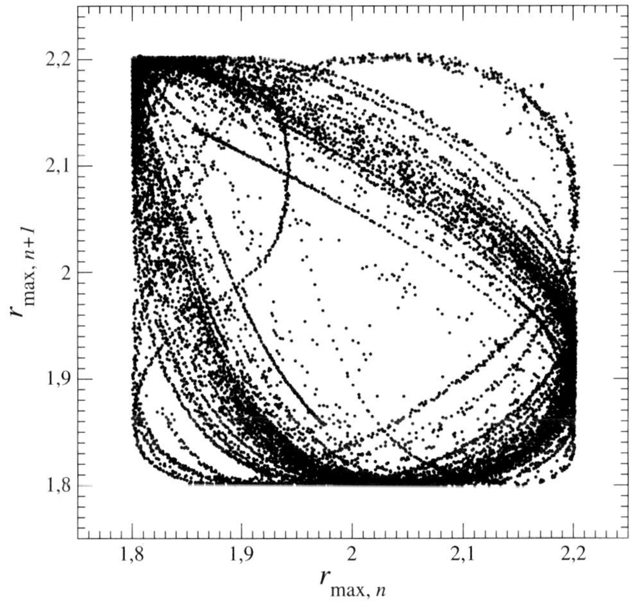

(b) The corresponding first-return map

Fig. 3. Chaos in Eqs. (6) and (8) obtained by numerical simulation. The first-return map is built on the maximum of radius $r_{\max}$. Parameters: $a = 0.2$, $\omega = \pi/4$ and $\mu = 1.1$. Initial conditions: $y_0 = 1$ and $z_0 = 0$.

some foldings (there are at least two foldings as exhibited by the dashed ellipse in Fig. 4b). Moreover, the first-return map should be viewed as "multiple looped"; this would mean that the torus is not trivial (is not a genus-one torus, but rather a knotted torus). The lack of a non-ambiguous phase space associated with this model — how is it possible to define a phase space of a system with a "modulo" function? — forbids any possibility to investigate in a deeper way the topology of this dynamics.

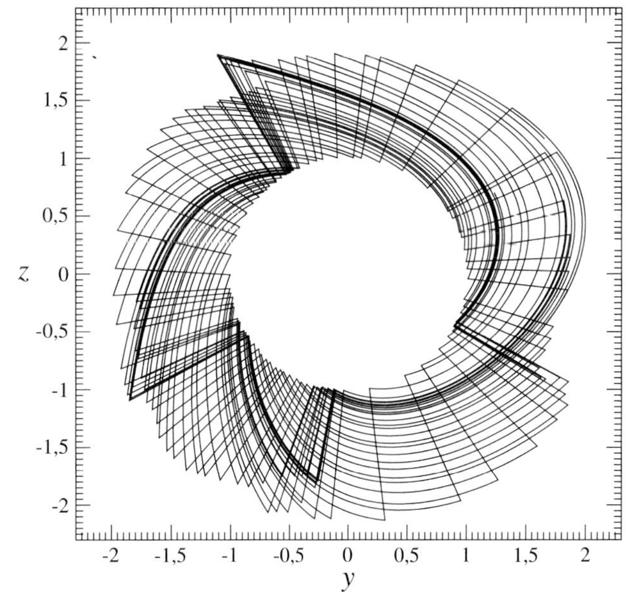

(a) Numerical simulation of Eq. (8)

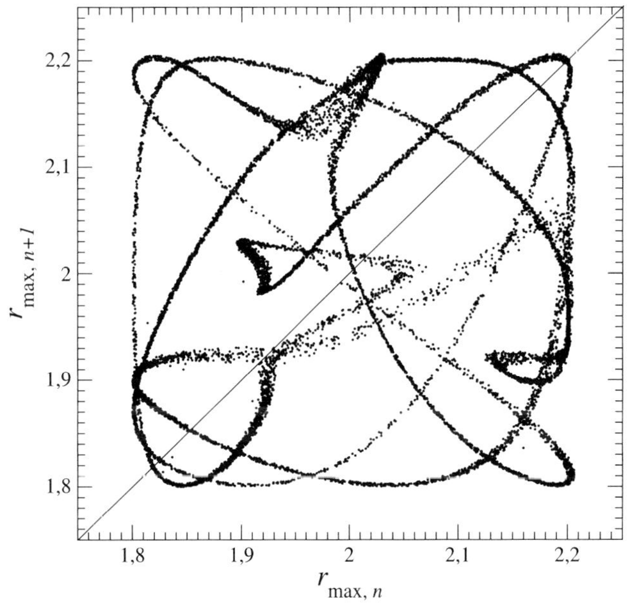

(b) The corresponding first-return map

Fig. 4. Chaos in Eqs. (6) and (8) obtained by numerical simulation. The first-return map is built on the maxima of the radius $r_{\max}$, Parameters: $a = 0.2$, $\omega = \pi$ and $\mu = 1.2$. Initial conditions: $y_0 = 1$ and $z_0 = 0$.

Some clues as to what should be attempted were proposed in an investigation of a spatio-temporal laser dynamics [19].

4. Conclusions

In this brief chapter, we proposed very simple models producing toroidal chaos. These models are illustrative in the sense that they help to figure out what general conditions are required for producing toroidal regimes. To produce a quasi-periodic regime, a linear radius growth with a periodic resetting to zero — whose period is incommensurable with the period of the radius growth — is enough. In order to produce toroidal chaos, a nonlinear function of the radius is required. Various types of toroidal chaos can be obtained according to the various parameter values chosen. These models are thus complex enough to produce many different sorts of toroidal chaos. A weakness of these models is that there is no natural phase space associated with them, a lack which forbids any topological analysis of these behaviors up until now.

References

[1] M. L. Cartwright & J. E. Littlewood, On nonlinear differential equations of the second order: I. The equation $\ddot{y} - k(1 - y^2)\dot{y} + y = b\lambda k \cos(\lambda t + \alpha)$, k large, *Journal of the London Mathematical Society*, **20**, 160–189 (1945).

[2] M. L. Cartwright, On non-linear differential equations of the second order, *Mathematical Proceedings of the Cambridge Philosophical Society*, **45**, 495–501 (1949).

[3] M. L. Cartwright, Nonlinear vibrations: a chapter in mathematical history, *The Mathematical Gazette*, **36** (316), 81–88 (1952).

[4] B. Van der Pol, Over "Relaxatie-trillingen", *Tijdschrift van het Nederlandsch Radiogenootschap*, **3**, 25–40 (1926).

[5] H. A. Boot & J. T. Randall, Historical notes on the cavity magnetron, *IEEE Transactions on electron devices*, **23** (7), 724–729 (1976).

[6] B. van der Pol, The nonlinear theory of electric oscillations, *Proceedings of the Institute of Radio Engineers*, **22** (9), 1051–1086 (1934).

[7] D. G. Tucker Forced oscillations in oscillator circuits and the synchronization of oscillators, *Journal of the Institute of Electrical Engineers*, **93** (1), 57–58 (1945).

[8] R. Adler, A study of locking phenomena in oscillators, *Proceedings of the Institute of Radio Engineers*, **34** (6), 351–357 (1946).

[9] N. Levinson, A second order differential equation with singular solutions, *The Annals of Mathematics*, **50** (1), 127–153 (1949).

[10] S. Smale, Differentiable dynamical systems, *Bulletin of the American Mathematical Society*, **73** (6), 747–817 (1967).

[11] O. E. Rössler, Chaotic behavior in simple reaction systems, *Zeitschrift für Naturfoschung A*, **31**, 259–264 (1976).

[12] M. Levi, Qualitative analysis of the periodically forced relaxation oscillations, *Memoirs of the American Mathematical Society*, **244**, 1–147 (1981).

[13] A. Liénard, Etudes des oscillations entretenues, *Revue Générale de l'Electricité*, **23**, 901–912, 946–954 (1928).

[14] N. Minorski, *Nonlinear Oscillations*, R. E. Krieger, Huntington, New York (1974).

[15] Y. Ueda & N. Akamatsu, Chaotically transitional phenomena in the forced negative-resistance oscillator, *IEEE Transactions on Circuit and Systems*, **28** (3), 217– (1981).

[16] O. E. Rössler, Chaotic oscillations: an example of hyperchaos. *Lectures in Applied Mathematics*, **17**, 141–156 (1979).

[17] O. E. Rössler, A synthetic approach to exotic kinetics, with examples, *Lecture Notes in Biomathematics*, **4**, 546–582 (1974).

[18] O. E. Rössler, Quasi-periodic oscillation in an abstract reaction system (abstract), *Biophysical Journal A*, **17**, 281 (1977).

[19] D. Amroun-Aliane, L. Pastur & C. Letellier Defects in spatio-temporal diagrams and their relations to phase coherence and lacks of observability, *Physical Review E*, **83**, 056212, 2011.

CHAPTER 7

ANALYZING THE DYNAMICS OF EXCITATORY NEURAL NETWORKS BY SYNAPTIC CELLULAR AUTOMATA

V. NEKORKIN*,†, A. DMITRICHEV*,† and D. KASATKIN*

**Institute of Applied Physics, Russian Academy of Sciences, 46 Ulyanov Str., Nizhny Novgorod, Russian Federation, 603950*

†Laboratory of Algorithms and Technologies for Network Analysis, The National Research University Higher School of Economics, 136, Rodionova Str., Nizhny Novgorod, Russian Federation, 603093

V. AFRAIMOVICH

Instituto de Investigacion en Comunicacion Optica, Universidad Autonoma de San Luis Potosi, Karakorum 1470, Lomas 4a 78220, San Luis Potosi, S.L.P., Mexico

An approach is presented for analyzing the dynamics of excitatory neural networks. It is based on the reduction of dynamics of networks to the dynamics of cellular automata (CA) on the graph of synaptic connections. We illustrate our approach on an example of network of Morris-Lecar neurons coupled by chemical synapses with short-term plasticity.

1. Introduction

Real neural networks are extremely complex systems [1]. They have a very complicated architecture of couplings, and individual neurons demonstrate nontrivial dynamics. Thus the study of spatio-temporal dynamics of such networks is often not a simple problem.

Recently in [2] an approach for the study of neural networks has been proposed. It consists in reducing a given network to a discrete model. In this model [3] each neuron is represented by a finite number of states and there are rules which determine how a neuron goes from one state to another. The approach is based essentially on a specific character of the dynamics of neurons. Indeed, the neurons are assumed to be relaxational. In this case each neuron can be in one of three consecutive phases: active phase, when a neuron can excite other neurons, refractory phase, when a neuron is non-responsive to the action of other neurons, and rest phase, when a neuron can be excited (activated) by other neurons. The duration of the refractory phase is assumed to be a multiple of the duration of the active phase. Thus the duration of the active phase is considered as a discretization time step. The interaction between neurons is taken into account in the following way. If a neuron receives an excitatory input, the excitation occurs instantaneously. On the other hand, if a neuron receives an inhibitory input, then it becomes excited due to the post-inhibitory rebound only when the inhibition ends. Moreover, the approach can be rigorously applied only to excitatory-inhibitory networks. In [4] the examples of application of this approach to specific biophysical models are presented.

Unlike [2, 4], we present here an approach for analyzing the dynamics of excitatory neural networks. It is based on the reduction of continuous dynamics of excitatory networks to a discrete one, in the form of a cellular automaton (CA) on the graph of connections. In the approach the main role is played by the dynamics of synapses but not by the specific features of neurons. In fact, the CA represents a network of synapses with a finite number of states which alternate each other according to some fixed rules. To determine the rules one needs to study only the responses of an individual synapse onto actions of neighboring (in graph of connections) synapses through corresponding neurons. As a result the numerical integration of the whole system of ordinary differential equations (ODEs) is not needed. Moreover, since the form of the neuron responses is not important, the approach is applicable to a broad set of networks including those consisting of neurons, which possess the neural excitability property (neurons of the class 2 excitability in the terminology of Izhikevich [5]).

The following is the scheme of an application of the approach for a given excitatory network consisting of N neurons:

(1) Single out the elementary clusters of the neural network. The cluster C_i $(i = 1, \dots, N)$ is defined as part of the network including the i-th synapse and j-th synapses coupled with the i−th one through j-th neurons.
(2) Based on the structure of each elementary cluster construct the graph of connections.
(3) Determine the responses of each cluster. In other words for each $i = 1, \dots, N$, find the responses of the i-th synapse caused by the different actions of the j-th synapse through the j-th neuron, $j \in K_i$. It is implied that a collective action of groups of several j-th synapses also has to be studied (if K_i consists of more than one element).
(4) Using the responses define the rules of the CA.
(5) Study dynamics of CA, find its attractors and spatio-temporal diagrams related to them.
(6) Describe regimes of spatio-temporal dynamics of the original neural network.

2. The Model

We demonstrate an example of an application of the approach for excitatory neural network with architecture shown on Fig. 1. To describe the dynamics of a neuron we use Morris-Lecar model [6]. Also we assume that the excitation among neurons is transferred by chemical synapses with short-term plasticity property [7]. The spatio-temporal dynamics of such network is given by following system of ODEs:

$$C\frac{dv_i}{dt} = -g_L(v_i - v_L) - g_{Ca}M_\infty(v_i)(v_i - v_{Ca}) - g_K n_i(v_i - v_K) + I_i^{ext} - g_{syn}s_i(v_i - v_{rev}), \quad (1)$$

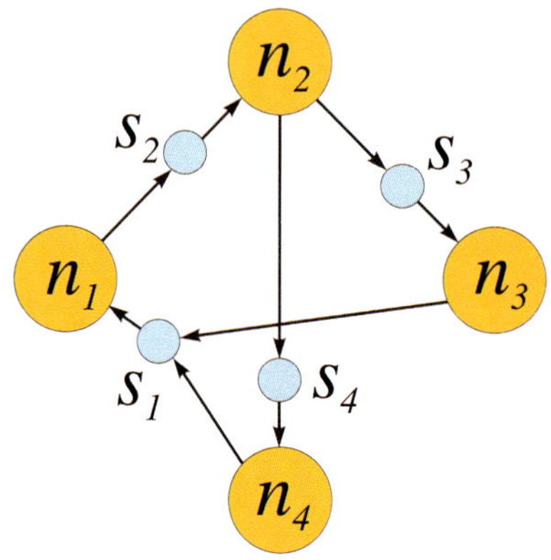

Fig. 1. The architecture of excitatory neural network. Here n_i and s_i signify the neurons and synapses, respectively.

$$\frac{dn_i}{dt} = \frac{n_\infty(v_i) - n_i}{\tau_n(v_i)}, \quad (2)$$

$$\frac{dr_i}{dt} = f_1(r_i) - s_i - k_1, \quad (3)$$

$$\frac{ds_i}{dt} = \varepsilon(f_{2i} - s_i - k_2). \quad (4)$$

The variables v_i, n_i characterize the membrane potential and the activation of the potassium ion channels of the i-th neuron ($i = 1, \ldots, 4$), C is the membrane capacitance. The terms at the RHS of Eq. (1) describe the currents flowing through the cell membrane of a neuron. The first three terms determine the leakage, the calcium and the potassium ionic currents, respectively, g_L, g_{Ca}, g_K are the maximal conductances, v_L, v_{Ca}, v_K are the equilibrium (reverse) potentials for the corresponding ion channels, and $M_\infty(v)$, $n_\infty(v)$, $\tau_n(v)$ are the stationary states of the activation levels and the characteristic relaxation time, respectively, depending on the membrane potential by the sigmoid law:

$$M_\infty(v) = 0.5\left[1 + \tanh\left(\frac{v - v^{(1)}}{v^{(2)}}\right)\right],$$

$$n_\infty(v) = 0.5\left[1 + \tanh\left(\frac{v - v^{(3)}}{v^{(4)}}\right)\right],$$

$$\tau_n(v) = \left[\phi\cosh\left(\frac{v - v^{(3)}}{2v^{(4)}}\right)\right]^{-1},$$

where $v^{(1)} = \quad 0.01$, $v^{(2)} - 0.15$, $v^{(3)} = 0$, $v^{(4)} = 0.3$, $\phi = 1$. The term I_i^{ext} is an external current. The last term at the RHS of Eq. (1) determines the synaptic current emerging as a result of action of other neurons on the i-th neuron by chemical synapse, where g_{syn} is the maximum synaptic conductance, v_{rev} is synaptic reversal potential and variable s_i corresponds to the fraction of open synaptic channels. To describe a short-term plasticity property, we introduced a phenomenological model [8] of synapse in the form of a system of Eqs. (3) and (4).

In Eq. (3), $f_1(r)$ is a piecewise linear (PWL) continuous function which resembles the shape of a fifth-degree polynomial:

$$f_1(r) = \begin{cases} -\gamma_1 r - A(1+\sigma_1), & r \leqslant -x_0 \\ \gamma_2 r, & -x_0 < r \leqslant x_0 \\ -\gamma_3 r + A(1+\sigma_3), & x_0 < r \leqslant x_1 \\ \gamma_4 r - A(1+\sigma_4), & x_1 < r \leqslant x_2 \\ -\gamma_5 r + A(1+\sigma_5), & r > x_2 \end{cases} \quad (5)$$

where

$$\sigma_1 = \frac{\gamma_1}{\gamma_2}, \quad \sigma_3 = \frac{\gamma_3}{\gamma_2}, \quad \sigma_4 = \frac{\gamma_4}{\gamma_2} + 2\frac{\gamma_4}{\gamma_3},$$

$$\sigma_5 = \frac{\gamma_5}{\gamma_2} + 2\gamma_5\left(\frac{1}{\gamma_4} + \frac{1}{\gamma_3}\right),$$

$$x_0 = \frac{A}{\gamma_2}, \quad x_1 = A\left(\frac{1}{\gamma_2} + \frac{2}{\gamma_3}\right),$$

$$x_2 = A\left(\frac{1}{\gamma_2} + \frac{2}{\gamma_3} + \frac{2}{\gamma_4}\right),$$

with $A = 2/3$, $\gamma_1 = 0.1$, $\gamma_2 = 1.136$, $\gamma_3 = 5$, $\gamma_4 = 41.113$, $\gamma_5 = 20$.

In Eq. (4), f_{2i} is a monotonically increasing continuous PWL function:

$$f_{2i} = \begin{cases} [\alpha + \mu\zeta_i\chi_i]r_i, & r_i < 0 \\ \beta r_i, & r_i \geqslant 0 \end{cases} \quad (6)$$

where $\chi_i = \sum_{j=1,\, j\neq i}^{N} H(v_j - \theta_{ji})$, $\zeta_i = H(\theta_{ref} - s_i)$ and $H(x) = 1/(1 + \exp(-x/1000))$ is a smooth approximation of the Heaviside step function. The term χ_i describes an action of presynaptic neurons on the synapse of ith neuron. Parameters θ_{ji} characterize the initial instants of activation of synaptic processes caused by input signals, i.e. they determine the synaptic thresholds. The term ζ_i accounts for the refractory property of a synapse. It is equal to 1 only while the synapse is not activated ($s_i < \theta_{ref}$).

It is assumed that due to short-term plasticity there are two types of synaptic responses (waveforms of synaptic conductance $s(t)$, see Fig. 2) — slow and fast ones which are distinguished by the characteristic rise times (T_2 and T_1) of the synaptic conductance. The characteristic times T_1 and T_2 are assumed to be in the following "resonance" relation $T_2 \simeq 2T_1$. Such behavior occurs because of the competition of presynaptic facilitation and postsynaptic desensitization processes. We reflect those features using the relaxational behavior of Eqs. (3)–(4) and a special design of nonlinear functions (5) and (6).

In all numerical simulations we use the following values of parameters and threshold constants: $\alpha = 0.2$, $\beta = 2$, $k_1 = -0.616$, $k_2 = -0.216$, $\varepsilon = 0.005$, $\mu = 2.3$, $\theta_{ref} = 0.01$, $C = 1.4$, $g_L = 0.1$, $g_{Ca} = 1.1$, $g_K = 2$, $v_L = -0.5$, $v_{Ca} = 1$, $v_K = -0.7$, $I_i^{ext} = 0.13$, $g_{syn} = 0.048$, $v_{rev} = 0.5$, $\theta_{12} = \theta_{23} = -0.009$, $\theta_{24} = -0.015$, $\theta_{31} = 0.01$, $\theta_{41} = 0.015$. To conform to the architecture, all other thresholds are set to 2.

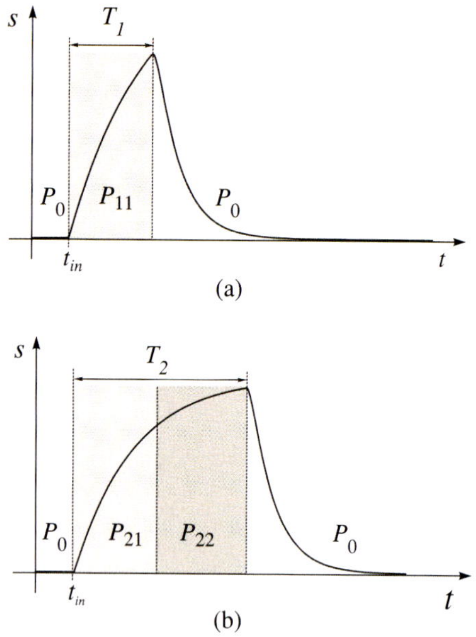

Fig. 2. Qualitative waveforms of synaptic conductance $s(t)$ corresponding to fast (a) and slow (b) synaptic responses ($T_2 \approx 2T_1$). Symbols P_0, P_{11}, P_{21}, P_{22} signify the discrete states of a synapse. See below, for details.

2.1. *The mechanism of formation of synaptic responses*

Let us briefly discuss the process of formation of synaptic responses. Consider the ith synaptic coupling starting from the case when $v_j < \theta_{ji}$ holds for all presynaptic neurons. In this case $\chi_i = 0$ and the system (3)–(6) has the phase portrait shown on Fig. 3(a). Point O_0 is globally stable. Note that s-coordinate of O_0 equals 0. Hence $\zeta_i = 1$ and the synapse can be activated. If at some instant of time the potentials of one or several neurons v_j exceed the threshold values θ_{ji}, then χ_i takes quickly some positive integer value. Numerical simulation of (3)–(6) shows that for $1 \leqslant \chi_i < \chi^*$ the system has a phase portrait shown on the Fig. 3(b) while for $\chi_i \geqslant \chi^*$ it has phase portrait shown on Fig. 3(c). In both cases the only attractor of the system is the limit cycle — either C_1 or C_2, which is established independently of initial conditions. Thus the dynamics of the synapse is governed by the alternating action of the trajectories from two phase planes.

If the exceeding of potentials v_j over values of θ_{ji} is multiple then a glued trajectory T_r (Fig. 4) appears on the (r_i, s_i)-plane. The dotted lines correspond to the pieces of trajectories of the Fig. 3(b) or Fig. 3(c) while continuous ones to trajectories of the Fig. 3(a). The process of formation of the glued trajectory T_r is ended when the inequalities $v_j(t) < \theta_{ji}$ hold for all j. The further dynamics is fully determined by the system (3)–(6) with the phase plane presented on Fig. 3(a). Two possibilities can occur. If the representing point of the trajectory lies on the right side of the separatrix W_1^s (Fig. 3(a)) or on the right side of the unstable manifold L^u (Fig. 3(a)),

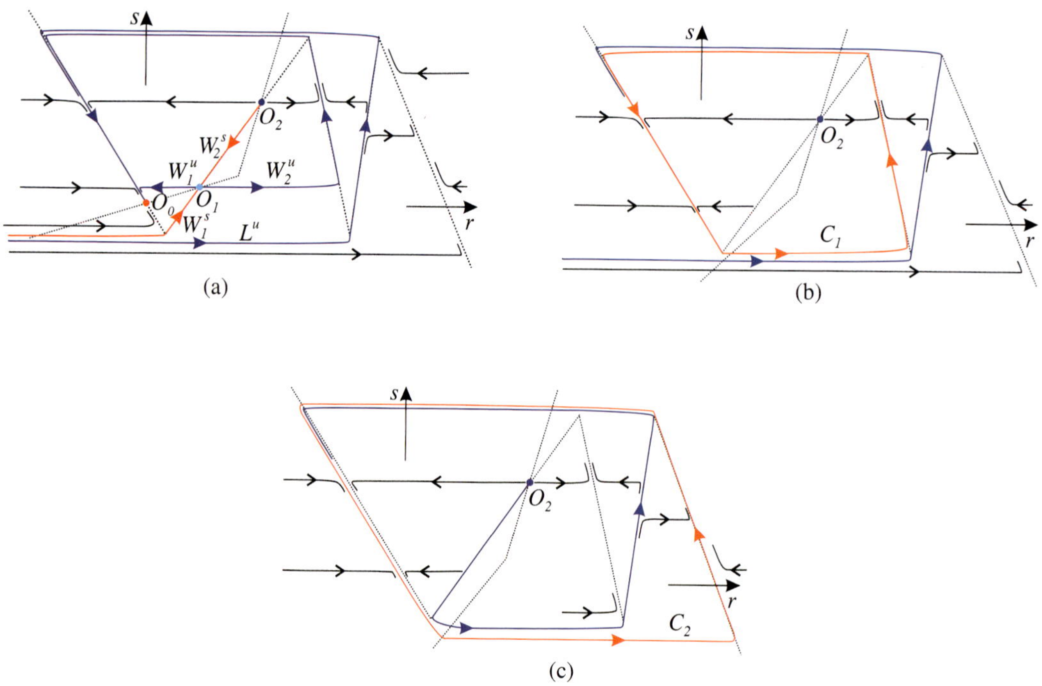

Fig. 3. The (r, s) phase planes of the system (3)–(6): $\chi_i = 0$ (a), $1 \leqslant \chi_i < \chi^*$ (b), $\chi_i \geqslant \chi^*$ (c).

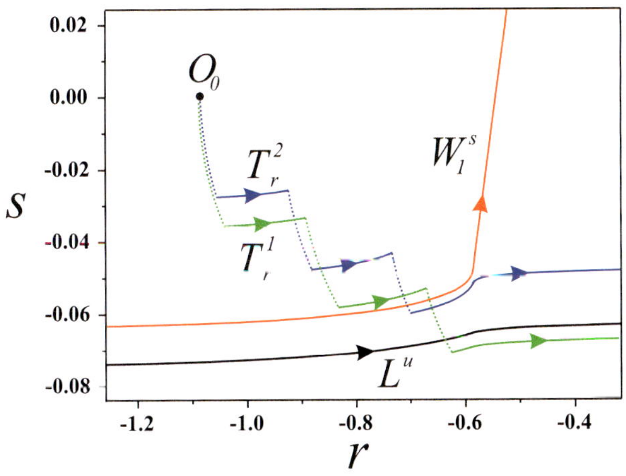

Fig. 4. The trajectory fragments formed in system (3)–(6) as a result of the processes of its activation ($v_j > \theta_{ji}$) and inactivation ($v_j < \theta_{ji}$) for weak (T_r^1) and strong (T_r^2) stimulations.

then the trajectory T_r comes to a vicinity of the equilibrium point O_0 according to the dynamics of the system (3)–(6). It is accompanied by a "powerful" impulsive change of the variable r, i.e. an activation of the synaptic interaction. Here the evolution of the variable s has a form typical for the synaptic current (Fig. 2). If the representing point on the (r_i, s_i)-plane lies on the left side of W_1^s or L^u the trajectory T_r again goes to O_0 but the relevant impulse response does not appear. In this case there is no activation of the synaptic coupling. The separatrix W_1^s and the unstable manifold L^u determine two thresholds of activation of the synaptic interaction. In order to go through the first threshold it is sufficient to have a comparatively weak pre-synaptic effect. On the other hand, for overcoming the second threshold the pre-synaptic action should be sufficiently strong. Moreover, overcoming of different thresholds leads to different efficiency of the activation of the synaptic interaction, i.e. characteristic times of the growth of the synaptic current. In the first case (weak pre-synaptic action) the growth of the synaptic current occurs slowly (slow synapses) (Fig. 2(b)). In the other case the strong presynaptic effect causes comparatively fast growth of the synaptic current (fast synapses) (Fig. 2(a)). Note that nonzero time which is necessary to get the representative point on the right side of W_1^s or L^u reflects the synaptic delay property.

Next, we will characterize fast and slow synapses by the length T of temporal interval that is equal to the time interval where $s(t)$ increases from 0 to its maximal value. We consider a "resonance" case when the value T_2 for the slow synapse and T_1 for the fast one satisfy $T_2 \simeq 2T_1$.

2.2. *Dynamics of a neuron under synaptic action*

Now we discuss basic properties of a neuron response to an excitatory synaptic action generated by the system of Eqs. (3)–(6). We consider the process of formation of the response on the example of Morris-Lecar neuron (the system of Eqs. (1)–(2)).

Under the action of the synaptic current $I_{syn}(t)$ the system (1)–(2) describing the i-th neuron may be treated as non-autonomous and, therefore, it has 3-dimensional generalized phase space. The system (3)–(6) has a small parameter ε and therefore, the evolution of the s_i-variable is slow with respect to the evolution of v_i and n_i. Hence, in the first approximation $s_i(t)$ may be considered as a parameter that is changing slowly. On the Fig. 5(a) a fragment of the bifurcation diagram of the system

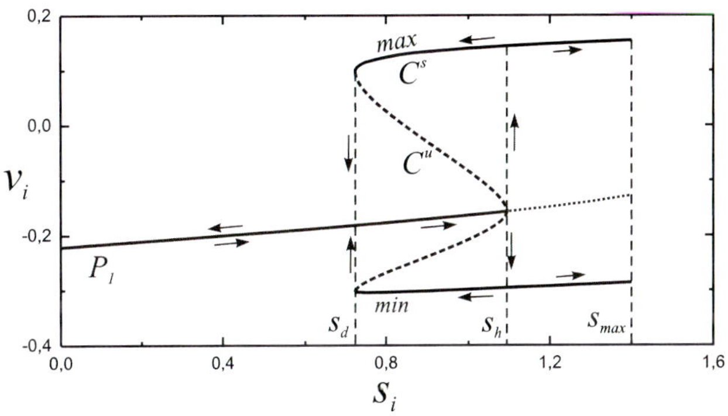

(a)

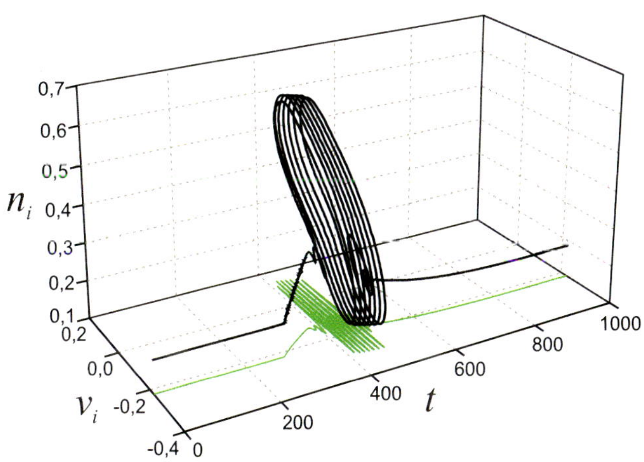

(b)

Fig. 5. (a) Bifurcation diagram. (b) A trajectory in the non-autonomous phase space.

(1)–(2) is presented. For $s_i < s_d \simeq 0.724$ the only equilibrium state P_1 exists. For $s_i = s_d$ the saddle-node bifurcation of a limit cycle holds, and two limit cycles — stable C^s and unstable C^u — appear on the phase plane. Further increasing of s_i leads to the shrinking of C^u to the equilibrium state P_1, and for $s_i = s_h \simeq 1.092$ the equilibrium state P_1 changes its stability due to the subcritical Andronov-Hopf bifurcation.

In the generalized phase space the stable equilibrium states $P_1(s_i)$ form a stable one dimensional integral manifold $M_{st}^s(0)$ and the stable limit cycles C^s form a stable two-dimensional integral manifold $M_c^s(0)$. According to the theorem of persistence of integral manifolds (see, e.g. [9]), there are integral manifolds $M_{st}^s(\varepsilon)$ and $M_c^s(\varepsilon)$ close to $M_{st}^s(0)$ and $M_c^s(0)$ correspondingly. Since initially the i-th neuron is at the phase of rest, the representative point in the generalized phase space is situated on the manifold $M_{st}^s(\varepsilon)$. Under a synaptic action the representative point starts moving along $M_{st}^s(\varepsilon)$ until it reaches the boundary of $M_{st}^s(\varepsilon)$. Due to a special choice of the parameter g_{syn} in the expression for the synaptic current, this piece of the itinerary of the representative point corresponds to the interval of the increasing of the variable $s_i(t)$ until its maximal value. After that the representative point passes rapidly into a vicinity of $M_c^s(\varepsilon)$ forming an oscillatory motion in the opposite direction (Fig. 5(a)). This piece of its itinerary corresponds to the decreasing part of the graph of $s_i(t)$ from the maximal value (Fig. 3). The oscillating process is over when the representative point reaches the boundary of the manifold $M_c^s(\varepsilon)$. As a result, a "transitive oscillating set" (see Fig. 5(b)) is forming in the generalized phase space. It corresponds to the generation of a collection of spikes in the i-th neuron. Because of our choice of parameters, the temporal intervals of the generation are relatively short with respect to the timescales of the synaptic response. Thus, in fact, the neuron can be in one of two states having different temporary scales: one of them corresponds to the slow monotonous change of the membrane potential v_i and another — to the fast generation of spikes (Fig. 5). In addition, these processes differ for the fast and slow synapses only by the duration of the slow phase of the change of v_i, and the spike oscillations in both cases are practically the same.

Though we considered here Morris-Lecar neuron similar behavior is typical for the other models of neurons.

3. Notes on Construction of Synaptic Cellular Automaton

Here we describe construction for an arbitrary neural network consisting of N neurons coupled by excitatory synapses. Let a synapse of the network has a number of inputs from presynaptic neurons and one output to postsynaptic one, i.e. the time course of synaptic conductance completely determines the dynamics of postsynaptic neuron. Thus to analyze the dynamics of whole neural network one can study only the dynamics of synapses layer. Next we define graph of connections, introduce discrete states of a synapse, and define a synaptic cellular automaton determined on the graph of connections.

3.1. *Definition of graph of connections*

Let us define an oriented graph $G =< Q_G, K_G >$. It has N vertices $Q_G = \{Q_1, Q_2, \ldots, Q_N\}$ which signify the synapses of the network. The set $K_G = \{K_1, K_2, \ldots, K_N\}$ reflects architecture of the connections in the network. Each K_i is a collection of indices such that $j \in K_i$ if and only if there is connection from j-th synapse of the network to i-th synapse through j-th neuron.

3.2. *Introducing discrete states of a synapse*

Let depending on the activity of presynaptic neurons and synaptic activation thresholds there are two types of synaptic responses (waveforms of synaptic conductance $s(t)$, see Fig. 2) — slow and fast ones which are distinguished by the characteristic rise times (T_2 and T_1) of the synaptic conductance. Then in the case of $T_2 = 2T_1$ the discrete states of a synapse can be introduced according to the behavior of the s-variable describing its dynamics. We say that the i-th synapse is in the state $\psi_i(s_i)$ that is equal to

- P_{11} if $s_i(t)$ undergoes the fast response and $s_i'(t) > 0$ (see Fig. 2(a));
- P_{21} if $s_i(t)$ undergoes the response of the slow type, $s_i'(t) > 0$ and $t_{in} \le t \le T_1 + t_{in}$, where $s_i(t_{in}) = 0$, $s_i'(t_{in}) > 0$, i.e. $s_i(t)$ starts growing at the instant of time t_{in} (see Fig. 2(b));
- P_{22} if $s_i(t)$ again undergoes the response of the slow type and $s_i'(t) > 0$, but $T_1 + t_{in} \le t \le 2T_1 + t_{in}$;
- P_0 if it is not in one of the previous states, i.e. either it is in the equilibrium state $s_i(t) = 0$ or $s_i'(t) < 0$ in both slow and fast cases.

Let us emphasize that exact resonance condition $T_2 = 2T_1$ is never realized precisely but numerical experiments show that even the approximate validity of this condition is sufficient to construct the cellular automaton which adequately describes the spatio-temporal order of activation of synapses and of firing of neurons of the network.

3.3. *Definition of CA rules*

Let us assume that neurons and synapses of the network possess the following properties:

- when a synapse is active it is non-sensitive to the presynaptic stimulation, i.e. it is in refractory state;
- a neuron is at rest in the absence of synaptic current (the conductance $s = 0$);
- a neuron generates spikes when the synaptic conductance overcomes some threshold value which is close to its maximum;
- the temporal intervals of generation of spikes are relatively short with respect to the characteristic timescales of evolution of the synaptic conductance;

These properties allow one to formulate the rules of evolution in time of ψ_i, i.e the rules of a CA:

(i) If i-th synapse at the instant of time t is in the state P_0 ($\psi_i(s_i(t)) = P_0$), and there is $j \in K_i$ such that $\psi_j(s_j(t)) = P_{11}$ or $\psi_j(s_j(t)) = P_{22}$ then depending on the values of synaptic activation thresholds (denote they by θ_{ji}) at the instant $t + T_1$ either it becomes excited (i.e. $\psi_i(s_i(t+T_1)) = P_{11}$ or $\psi_i(s_i(t+T_1)) = P_{21}$) or it stays in the rest (i.e. $\psi_i(s_i(t+T_1)) = P_0$).

(ii) If i-th synapse at the instant of time t is in the state P_{21}, i.e. $\psi_i(s_i(t)) = P_{21}$, then $\psi_i(s_i(t+T_1)) = P_{22}$ independently of other synapses. Moreover, P_{21}-synapse cannot make excited another synapse.

(iii) If $\psi_i(s_i(t)) = P_{11}$ or $\psi_i(s_i(t)) = P_{22}$, then $\psi_i(s_i(t+T_1)) = P_0$ independently of the states of other synapses.

(iv) Because of the existence of the refractory period, the synapses in the states P_{11}, P_{21}, P_{22} do not feel the action of other synapses.

These statements imply that for the fixed i the state $\psi_i(s_i(t+T_1))$ at the instant $t+T_1$ is uniquely determined by

- the state $\psi_i(s_i(t))$ at the previous instant of time t;
- the states of the j-th synapse, $j \in K_i$, at the instant of time t;
- the values of the threshold constants θ_{ji}, $j \in K_i$.

Table 1. Rules of the synaptic CA.

ψ_i	P_0	P_{11}	P_{21}	P_{22}
$\psi_j, j \in K_i$ $\forall j \in K_i$, $\psi_j = P_0$ or $\psi_j = P_{21}$	P_0	P_0	P_{22}	P_0
$\exists j \in K_i$, such that $\psi_j = P_{11}$ or $\psi_j = P_{22}$	P_0, P_{11} or P_{21} depending on the values of threshold constants θ_{ji}	P_0	P_{22}	P_0

Hence,

$$\psi_i(s_i(t+T_1)) = F_i(\psi_i(s_i(t)), \{\psi_j(s_j(t))\}, \{\theta_{ji}\}, j \in K_i),$$

where F_i is a function determined by the statements (i)–(iv). This relation defines a synaptic CA determined on the graph of connections G. In the Table 1 we summarize the rules of the CA. In what follows we identify $\psi_i(s_i(t))$ with the symbol ψ_i and $\psi_i(s_i(t+T_1))$ with the symbol $\overline{\psi_i}$.

Finally if we fix the values of thresholds θ_{ji} the rules of the CA takes the following form:

$$\overline{\psi_i} = F_i(\psi_i, \{\psi_j\}, j \in K_i),$$

where $\overline{\psi_i}, \psi_i \in \mathcal{P}$.

4. Analysis of Spatio-temporal Dynamics of the Network Based on CA Approach

Let us analyze the dynamics of neural network depicted on Fig. 1 using the CA approach according to the general scheme presented above (see Ref [8] as well).

At first we single out the elementary clusters of the network. It is easy to see from the architecture of the network that it contains only four elementary clusters: C_1-C_4, which are shown on Fig. 6(a). Based on the structure of each elementary cluster we construct corresponding graph of connections G. It is shown on Fig. 6(b).

Let us determine now the responses of each elementary cluster.

Cluster C_1 From the structure of this elementary cluster it follows that the state of the synapse s_1 depends on the states of the synapses s_3 and s_4 (i.e. $K_1 = \{3, 4\}$). Numerical experiments show that there are three different responses of the cluster. If

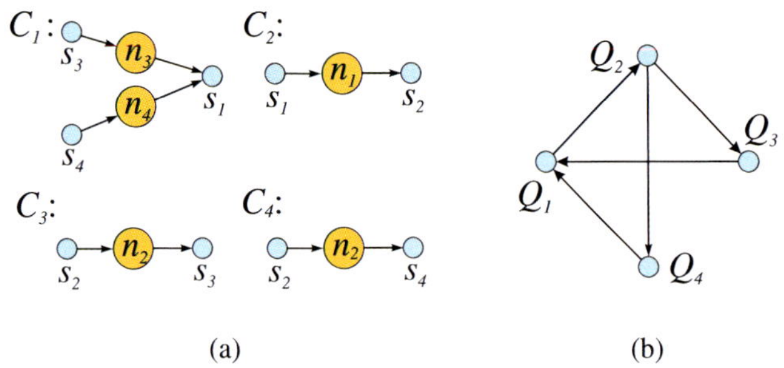

Fig. 6. (a) Elementary clusters of the network presented in Fig. 1 and (b) corresponding graph of connection G.

s_3 is the only active synapse in the cluster and it is either in the state P_{11} or in the state P_{22} then in next episode s_1 will always be activated with double time scale (go in the state P_{21}). Let now s_4 be the only active synapse in the cluster. Then in next episode s_1 will not be activated (stay in the state P_0) independently of the state of s_4. Finally, if s_3 and s_4 are the only active synapses in the cluster and they are in one of the states P_{11} or P_{22} then in next episode s_1 will always be activated with minimal time scale (go in the state P_{11}). From here we obtain the rules of the CA presented in Table 2.

Cluster C_2. From the structure of this elementary cluster it follows that the state of the synapse s_2 depends only on the state of the synapse s_1 (i.e. $K_2 = \{1\}$). There are two different responses of the cluster. If the synapse s_1 is in the state P_{11} (show the response with minimal time scale) then the synapse s_2 will be activated (if it is in the state P_0) with double time scale response.) On the other hand, if the synapse s_1 is in the state P_{22} (show the response with double time scale) then the synapse s_2 will be activated (if it is in the state P_0) with minimal time scale response. The corresponding rules are shown in Table 3.

Table 2. The rules $\overline{\psi_1} = F_1(\psi_1, \psi_3, \psi_4)$.

ψ_1	P_0	P_{11}	P_{21}	P_{22}
ψ_3/ψ_4				
P_0 /P_0	P_0	P_0	P_{22}	P_0
P_{11}/P_0	P_{21}	P_0	P_{22}	P_0
P_{21}/P_0	P_0	P_0	P_{22}	P_0
P_{22} /P_0	P_{21}	P_0	P_{22}	P_0
P_0 /P_{11}	P_0	P_0	P_{22}	P_0
P_{11}/P_{11}	P_{11}	P_0	P_{22}	P_0
P_{21}/P_{11}	P_0	P_0	P_{22}	P_0
P_{22}/P_{11}	P_{11}	P_0	P_{22}	P_0
P_0 /P_{21}	P_0	P_0	P_{22}	P_0
P_{11}/P_{21}	P_{21}	P_0	P_{22}	P_0
P_{21}/P_{21}	P_0	P_0	P_{22}	P_0
P_{22}/P_{21}	P_{21}	P_0	P_{22}	P_0
P_0 /P_{22}	P_0	P_0	P_{22}	P_0
P_{11}/P_{22}	P_{11}	P_0	P_{22}	P_0
P_{21}/P_{22}	P_0	P_0	P_{22}	P_0
P_{22}/P_{22}	P_{11}	P_0	P_{22}	P_0

Table 3. The rules $\overline{\psi_2} = F_2(\psi_2, \psi_1)$.

ψ_2	P_0	P_{11}	P_{21}	P_{22}
ψ_1				
P_0	P_0	P_0	P_{22}	P_0
P_{11}	P_{21}	P_0	P_{22}	P_0
P_{21}	P_0	P_0	P_{22}	P_0
P_{22}	P_{11}	P_0	P_{22}	P_0

Table 4. The rules $\overline{\psi_3} = F_3(\psi_3, \psi_2)$.

ψ_3	P_0	P_{11}	P_{21}	P_{22}
ψ_2				
P_0	P_0	P_0	P_{22}	P_0
P_{11}	P_{21}	P_0	P_{22}	P_0
P_{21}	P_0	P_0	P_{22}	P_0
P_{22}	P_{11}	P_0	P_{22}	P_0

Cluster C_3 The behavior of this cluster is similar to previous one. The state of the synapse s_3 depends on the state of the synapse s_2 (i.e. $K_3 = \{2\}$). If the synapse s_2 is in the state P_{11} (show the response with minimal time scale) then the synapse s_3 will be activated (if it is in the state P_0) with double time scale response. On the other hand, if the synapse s_2 is in the state P_{22} (show the response with double time scale) then the synapse s_3 will be activated (if it is in the state P_0) with minimal time scale response. The corresponding rules are shown in Table 4.

Cluster C_4 It is easy to see that the state of the synapse s_4 depends only on the state of the synapse s_2 (i.e. $K_4 = \{2\}$). It is established that the response of the synapse s_4 (while it is in the state P_0) to the action of the synapse s_2 is always of the minimal time scale T_1, independently of the state (P_{11} or P_{22}) of s_2. From here we obtain the rules of the CA presented in Table 5.

Tables 2–5 give us the full description of the rules $F_i, i = 1, \ldots, 4$, for the CA corresponding to the network. Using these rules one may

Table 5. The rules $\overline{\psi_4} = F_4(\psi_4, \psi_2)$.

ψ_4 / ψ_2	P_0	P_{11}	P_{21}	P_{22}
P_0	P_0	P_0	P_{22}	P_0
P_{11}	P_{11}	P_0	P_{22}	P_0
P_{21}	P_0	P_0	P_{22}	P_0
P_{22}	P_{11}	P_0	P_{22}	P_0

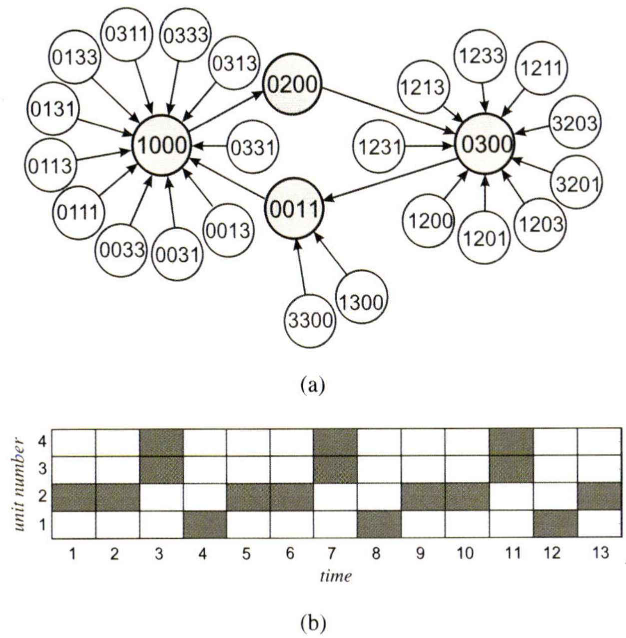

Fig. 7. A part of the basin of the periodic attractor A_1 (a); spatio-temporal order of synaptic activation corresponding to A_1 (b).

study the dynamics of the CA. The CA has three attractors: A_0, A_1 and A_2. They correspond to various scenarios of activation of synapses of the network and so define different firing patterns of neurons.

The attractor A_0 is the fixed point (0000). Here we denote the states of synapses by numbers: $P_0 \to 0$, $P_{11} \to 1$, $P_{21} \to 2$, $P_{22} \to 3$. The attractor A_0 is trivial — if all elements of the network is activated then all of them come to the rest state P_0 next episode.

The attractor A_1 is the periodic trajectory $(1000) \to (0200) \to (0300) \to (0011) \to (1000)$. A part of the basin of A_1 is shown on Fig. 7(a). Note that the periodic points in the attractor are marked by the grey color. Figure 7(b) depicts spatio-temporal order of synaptic activation determined by periodic attractor A_1. It is constructed as following. If i-th synapse in k-th episode is active, i.e. it is in one of the states P_{11}, P_{21} or P_{22}, corresponding rectangle is marked by the black color. On the other hand, if i-th synapse in k-th episode is silent, i.e. it is in the state P_0, corresponding rectangle is marked by the white color. From the symbolic sequence (and spatio-temporal diagram of synaptic activation as well) corresponding to A_1 it is easy to obtain the following order of firings of neurons: $1 \to 2 \to 3, 4$ and so on. Indeed, it is well known that a neuron fires the spikes when corresponding synaptic conductance overcomes some threshold value close to maximal conductance. In our case it corresponds to narrow time windows in the vicinity of transitions of synapses from the state P_{11} (or P_{22}) to P_0 (or similarly to the narrow regions in the vicinity of transitions from black to white rectangles on spatio-temporal diagram of synaptic activation).

The attractor A_2 is the periodic trajectory $(2000) \to (3000) \to (0100) \to (0021) \to (0030) \to (2000)$. A part of the basin of A_2 is shown on Fig. 8(a). Figure 8(b) depicts spatio-temporal order of synaptic activation determined by periodic attractor A_2. From the symbolic sequence corresponding to A_2 one can obtain the following order of firings of neurons: $1 \to 2 \to 4 \to 3$ and so on.

For the confirmation of the validity of these results we have carried out the numerical study of the whole system (1)–(6).

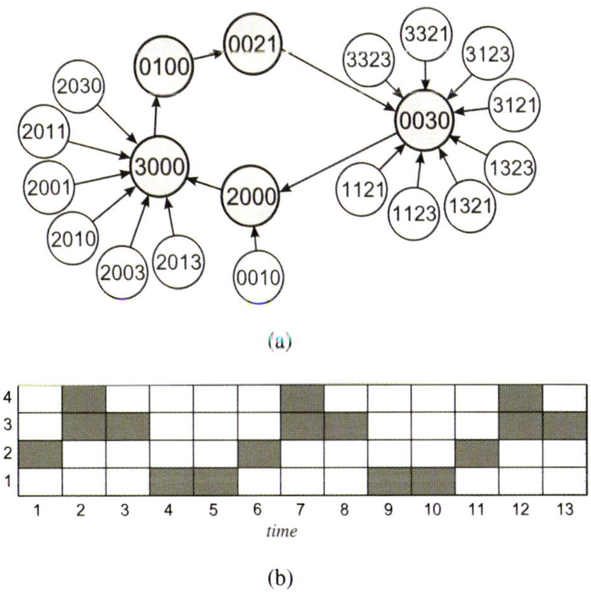

Fig. 8. A part of the basin of the periodic attractor A_2 (a); spatio-temporal order of synaptic activation corresponding to A_2 (b).

5. Numerical Analysis of Spatio-temporal Dynamics of the Network

Let us consider the network where the first neuron is subjected to time-varied external (applied) current in the form of a sequence of two square pulses ($I_1^{ext}(t) = 0.13 + I_{tsp}(t)$) with the same amplitude (A_{st}), the same duration ($T_{st} = 10$) and interpulse interval τ. Results of numerics show that depending on the characteristics of external stimulation there are two nontrivial spatio-temporal activity regimes of the network.

The first regime is shown on Fig. 9. Here Figure 9(a) depicts corresponding spatio-temporal diagram of synaptic activation and Fig. 9(b) — waveforms of synaptic conductances. The spatio-temporal diagram of synaptic activation is constructed as following. We considered the rectangles $D_i(a,b)$, $i = 1,2,3,4$ on the plane with the horizontal axis corresponding to the time and vertical axis corresponding to the synapse number. The height of $D_i(a,b)$ is 1, the width is $(b-a)$, the lower-left corner has coordinates (a,i), and it is black if the synapse i is active at the instant t ($s_i'(t) > 0$), $a < t < b$. It easy to see that the spatio-temporal order of synaptic activation obtained for the regime (see Fig. 9(b) as well) is in good correspondence with that for the attractor A_2 of the CA. Note, however, that it is not fully periodic. This is related to the fact that the resonance condition $T_2 = 2T_1$ is satisfied only with some (nonzero) accuracy. Nevertheless, first of all, the spacial order of activation of synapses is exactly the same as for the CA, and, secondly, temporal intervals of activation are approximately the same as for the CA up to the scaling factor T_1. On Fig. 9(c) waveforms of membrane potentials for different neurons are presented. We see that the temporal intervals of firing and the rest of neurons occur in the order prescribed by the order of synaptic activation. Note also that the number of spikes might be different for different neurons and different times of observation, since it depends on the state of full

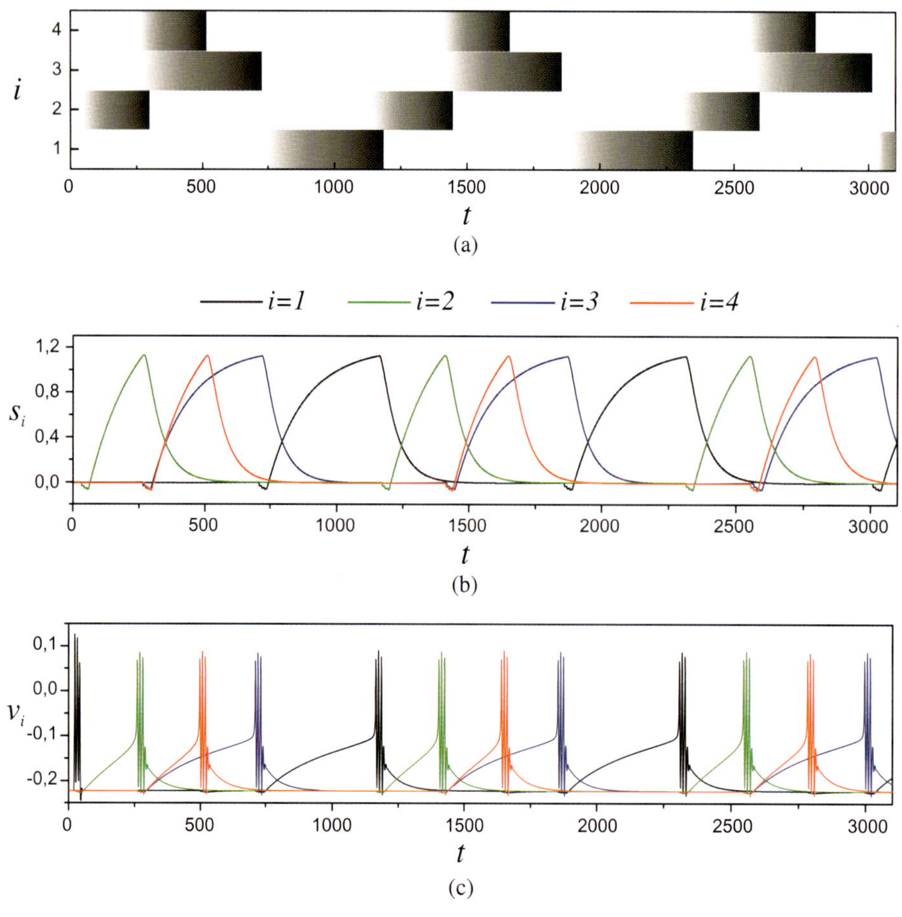

Fig. 9. Regime established in the network for the parameters $A_{st} = 0.05$, $\tau = 0.4$: (a) spatio-temporal diagram of synaptic activation; (b) realizations for synapses; (c) realizations for neurons.

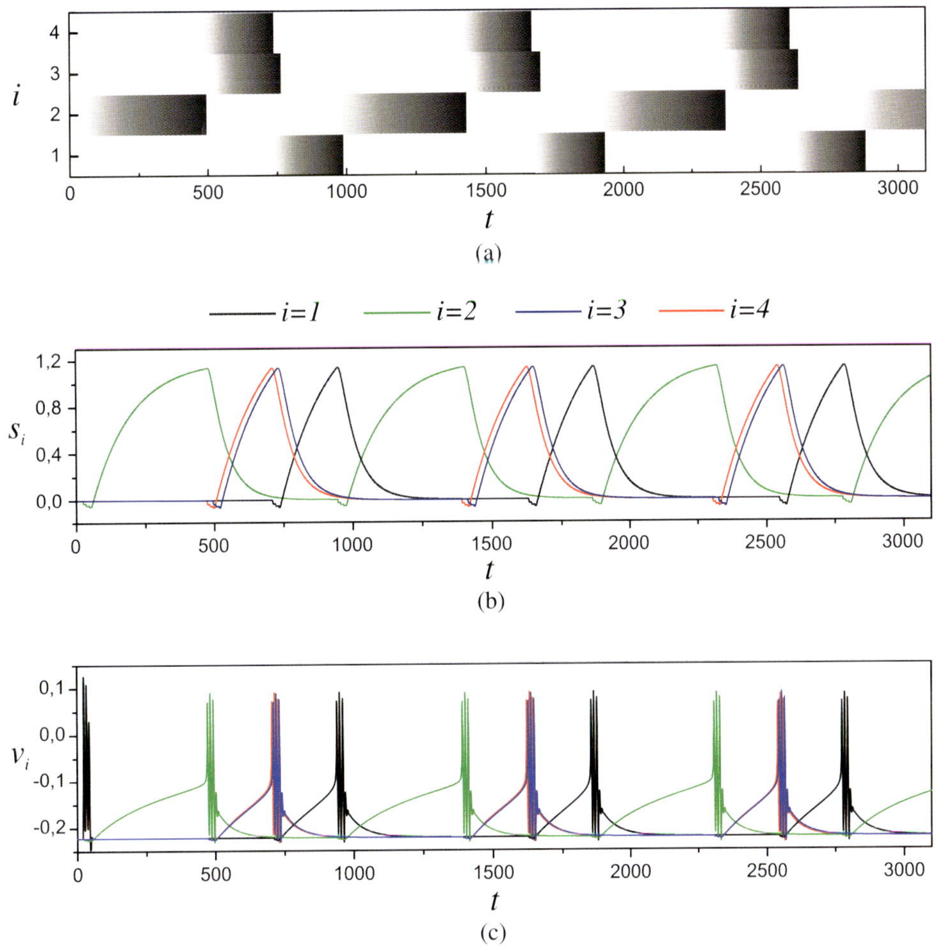

Fig. 10. Regime established in the network for the parameters $A_{st} = 0.05$, $\tau = 0.5$: (a) spatio-temporal diagram of synaptic activation; (b) realizations for synapses; (c) realizations for neurons.

network. Nevertheless, at an average the activation of the neuron network also occurs at the instants which are multiples of time scale T_1.

The second possible regime is shown on Fig. 10. We see that spatio-temporal order of synaptic activation (see 10(a),(b)) for the regime is in good correspondence with that for the attractor A_1 of the CA. From Fig. 10(c) one can also obtain that order of firing of neurons coincides fully with that obtained by means of the synaptic CA approach.

6. Conclusion

We have presented an example of the application of the synaptic CA approach to excitatory neural network consisting of four Morris-Lecar neurons coupled by chemical synapses with short-term plasticity. It has been shown that the corresponding CA has two nontrivial attractors — A_1 and A_2. They correspond to various scenarios of activation of synapses of the network and so define different firing patterns of neurons. The attractor A_1 defines the firing pattern $1 \to 2 \to 3, 4$ and so on, while the attractor A_2 — the firing pattern $1 \to 2 \to 4 \to 3$ and so on. Comparison of the results with direct numerical study of the original network has shown that the CA describes adequately the spatio-temporal dynamics (orders of synaptic activation and firing of neurons) of the network.

Acknowledgments

This work has been supported by Ministry of Education and Science of the Russian Federation under governmental contract No. 14.740.11.0348, the Russian Foundation for Basic Research under Grant Nos. 10-02-00643, 12-02-00526 and is partially supported by LATNA Laboratory, National Research University Higher School of Economics, Russian Federation Government Grant No. 11.G34.31.0057.

References

[1] E.R. Kandel, J.H. Schwartz, T.H. Jessell, *Principles of neuralscience.* (Prentice-Hall Int. Inc., London, 1991).

[2] D. Terman, S. Ahn, X. Wang, W. Just, Reducing neuronal networks to discrete dynamics, *Physica D* **237** (3), 324–338, (2008).

[3] W. Just, S. Ahn, D. Terman, Minimal attractors in digraph system models of neuronal networks, *Physica D* **237** (24), 3186–3196, (2008).

[4] S. Ahn, B.H. Smith, A. Borisyuk, D. Terman, Analyzing neuronal networks using discrete-time dynamics, *Physica D* **239** (9), 515–528, (2010).

[5] E. Izhikevich, *Dynamical systems in neuroscience.* (MIT Press, Cambridge, MA, 2007).

[6] C. Morris, H. Lecar, Voltage oscillations in the barnacle giant muscle fiber, *Biophys J.* **35** (1), 193–213, (1981).

[7] R.S. Zucker, W.G. Regehr, Short-term synaptic plasticity, *Annu. Rev. Physiol.* **64**, 355–405, (2002).

[8] V. I. Nekorkin, A. S. Dmitrichev, D. V. Kasatkin, V. S. Afraimovich, Relating the sequential dynamics of excitatory neural networks to synaptic cellular automata, *Chaos* **21** (4), 043124-1–13, (2011).

[9] M. V. Hirsh, C. C. Pugh, M. Shub, *Invariant manifolds. Lecture Notes in Mathematics.* (Springer-Verlag, 1977).

CHAPTER 8

DYNAMICAL SYSTEMS PERSPECTIVE OF WOLFRAM'S CELLULAR AUTOMATA

M. COURBAGE

Laboratoire Matière et Systèmes Complexes (MSC),
UMR 7057 CNRS and Université Paris 7 — Denis Diderot,
Case 7056, Bâtiment Condorcet, 10, rue Alice Domon et Léonie
Duquet 75205 Paris Cedex 13, France
maurice.courbage@univ-paris-diderot.fr

B. KAMIŃSKI

Faculty of Mathematics and Computer Science,
Nicolaus Copernicus University,
Chopina 12/18, 87-100 Torun, Poland
bkam@mat.uni.torun.pl

Leon Chua, following Wolfram, devoted a big effort to understand deeply the wealth of complexity of the rules of all elementary one-dimensional cellular automata from the point of view of the nonlinear dynamicist. Here we complete this point of view by a dynamical system perspective, extending them to the limit of infinite number of sites.

1. Chua Odyssey Through Cellular Automata

Many papers were written in the eighties on the general theory of one-dimensional cellular automata. But, the great number of the local rules and the many types of long-term behaviours required a special effort of classification in order to see more clearly in this jungle. Wolfram [21], has attempted such a dynamical classification and has established a first systematic study of all 256 rules for the range $r = 1$ and two-states alphabet, also called Elementary Cellular Automata (ECA). As in statistical mechanics of spin systems, the number of sites should tend to infinity.

Soon later, many papers from the community of topological and ergodic dynamical systems started to appear focusing on general and abstract properties, illustrated by case studies.

Sticking on a concrete point of view, Leon Chua [2] resumed a study, from the point of view of nonlinear dynamics, of each of the 256 rules, where the automata act on a finite ring. Using symmetries and introducing new criteria, he sifted each rule in order to understand globally this family.

Chua studied 1-D Wolfram's cellular automata on finite words (called "bit-strings" in the Chua language) on a circle $\mathbb{C}_l$ of length l, that is, the ring of integers modulo l, $(\mathbb{Z}/l\mathbb{Z})$. They are finite dynamical systems $\mathcal{X} = (X, f)$ such that the configuration space $X = S^{\mathbb{C}_l}$, $S = \{0, 1\}$ and $f : X \to X$ is a transformation defined by

$$(fx)_n = F(x_{n-r}, \ldots, x_{n+r}),$$
$$x = (x_i) \in X, \ n \in \mathbb{C}_l.$$

F is called the local rule and r is called the range of the cellular automaton map f. In the case $r = 1$, there are 256 possible rules numbered using the well known binary Wolfram numbering. Apart from f, consider the left shift transformation $\sigma : X \to X$ defined as follows.

$$(\sigma x)_n = x_{n+1}, \quad x \in X, \ n \in \mathbb{C}_l.$$

Studying the attractors of each rule, Chua observed that, in some large class of rules, the dynamical map f^n acts ultimately as a shift σ, or more generally as a k-step shift σ^k. He called such rules Bernoulli shifts. This investigation meets another investigation on the existence of traveling waves in cellular automata on $\mathbb{Z}$ in [5, 8, 9], where a method allowing the identification of those initial configurations $x \in S^{\mathbb{Z}}$ such that $f(x) = \sigma^k(x)$ showed that those configurations are spatially periodic for all k greater than r. When $k \leq r$ there are many non periodic traveling waves [20]. Using the isomorphism between $\{0, 1\}^{\mathbb{Z}}$ and the interval $[0, 1]$, Chua et al. associated to each cellular automaton a map of this interval into itself. Their explicit study of the basins of the Wolfram CA-maps is the most complete updated investigation providing an exhaustive classification including invariance and fractal properties, abundance of Isles of Eden in most rules, scale-free properties and some new empirical quasi-ergodicity.

2. Cellular Automata as Dynamical Systems

From now on, we extend the above definition substituting $\mathbb{C}_l$ by $\mathbb{Z}$. The CA-map is an example of a toy spatially extended system, since physical extended dynamical systems are given by PDE's in large space. However some notions here used may be generalized to Lattice Dynamical Systems.

The complex dynamics of spatial patterns of the extended dynamical system is very rich compared with the dynamics of systems with few degrees of freedom. Concepts of Lyapunov exponents and topological and Kolmogorov-Sinai entropy are reformulated in order to take into account the space-time interplay. One of these new concepts is the spatial density of entropy (cf. [3]) which is associated to the two parameters group of space translation and temporal evolution. In general, when this density of entropy is non zero, the entropy of the temporal evolution group is infinite.

The cellular automata dynamical systems are the simplest spatially extended toy systems on an infinite lattice with finite entropy. In order to distinguish the trivial CAs like shifts from those more complex CAs, Milnor ([15, 16]) introduced a concept of directional entropy, showing a class of CAs that unlike shifts, have positive entropy in many directions. Recently, a concept of directional Lyapunov exponents has also been introduced ([7]). Let us start with Lyapunov exponents. M. A. Shereshevsky has introduced a rigorous definition of the so-called Lyapunov exponents ([17]) after Wolfram ([21]). According to Wolfram these exponents aim to describe a velocity of propagation to the right or to the left of the "local perturbations" of initial patterns in one-dimensional CAs. In fact, modifying locally a spatial configuration, may lead either to an infinite expansion of the perturbation or to its containment. In the first case, to the instability of the initial configuration corresponds a positive Lyapunov exponent. Viewing a Lyapunov exponent in a space-time diagram is the best way to observe the intrinsic instability of such systems.

Let us first introduce some notations following Shereshevsky. For a given configuration $x = (x_i)$ denote by $W_s^+(x)$ the set of all configurations y obtained from x in modifying it only at the left of the site $s \in \mathbb{Z}$. In other words $y_i = x_i$ for all $i \geq s$, and arbitrary otherwise. $W_s^-(x)$ is similarly the set of perturbed configurations at the right of the site $(-s)$. Let us compute the evolution of a perturbed configuration $y \in W_0^+(x)$ at the left of the site 0 that will start to differ from x under the action of the CA-map, $f(y), f^2(y)$, etc. By using the formula

$$f(y)_0 = F(y_{-1}, y_0, y_1) \tag{1}$$

and the fact: $y_0 = x_0$, $y_1 = x_1$, we see that

$$f(y)_0 = F(y_{-1}, x_0, x_1) \tag{2}$$

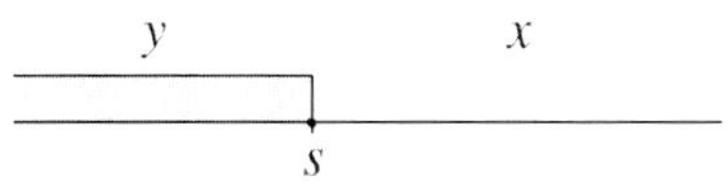

$y \in W_s^+(x)$

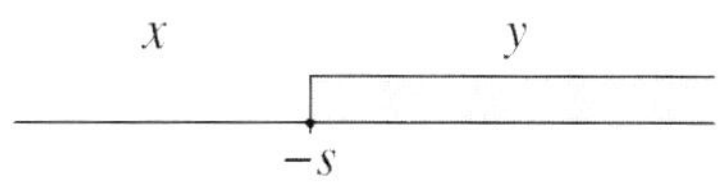

$y \in W_s^-(x)$

But as

$$f(x)_0 = F(x_{-1}, x_0, x_1) \tag{3}$$

and comparing (2) and (3) we see that $f(x_0)$ may be distinct from $f(y)_0$ if $y_{-1} \neq x_{-1}$. This however will depend on the rule F and the values of x_{-1}. If, for example, F does not depend explicitely of x_{-1}, comparing (2) and (3) shows that $f(x)_0 = f(y)_0$. However, F may depend on x without implying a distinction between $f(x)_0$ and $f(y)_0$ for some particular values of x. In the following example: $F(x_{-1}, x_0, x_1) = x_{-1}x_0x_1$ and for $x_0 = x_1 = 0$, we see that $f(y)_0 = f(x)_0$ whatever the value of y_{-1}. But if $x_0 = x_1 = 1$ then $f(y)_0 \neq f(x)_0$ as soon as $y_{-1} = x_{-1}$. In this case we say that for this rule and this x configuration, the y-perturbation of x at site -1 is propagated to the right. But how far can it spread? If we look at $f(y)_1$ and we use the same formulas (1)–(3) to compute $f(y)_1$, we see that $f(y)_1 = f(x)_1$ since $y_0 = x_0, y_1 = x_1, y_2 = x_2$, and consequently we have also $f(y)_i = f(x)_i$ for all $i \geq 1$. That is to say, that y perturbation propagates, in one iteration, to at most one site to the right.

Now take more and more iterations of y. For that, we use the following expression of $f^2(y)_i$:

$$\begin{aligned} f^2(y)_i &= F(f(y)_{i-1}, f(y)_i, f(y)_{i+1}) \\ &= F(F(y_{i-2}, y_{i-1}, y_i), F(y_{i-1}, y_i, y_{i+1}), \\ &\qquad F(y_i, y_{i+1}, y_{i+2})), \end{aligned} \tag{4}$$

and we apply it for $i = 0, 1, 2, \ldots$. From the definition of y, we see that

$$\begin{aligned} f^2(y)_0 &= F(F(y_{-2}, y_{-1}, y_0), F(y_{-1}, y_0, y_1), \\ &\qquad F(y_0, y_1, y_2)) \\ &= F(F(y_{-2}, y_{-1}, x_0), F(y_{-1}, x_0, x_1), \\ &\qquad F(x_0, x_1, x_2)). \end{aligned}$$

Comparing $f^2(y)_0$ with $f^2(x)_0$ given by

$$\begin{aligned} f^2(y)_0 = F(F(x_{-2}, x_{-1}, x_0), F(x_{-1}, x_0, x_1), \\ F(x_0, x_1, x_2)) \end{aligned} \tag{5}$$

we see here, and we saw in the case of $f(y)_0$, that the y perturbation at the sire -2 or -1, may propagate to the site 0 too. Now, with this second iteration, $f^2(y)_1$ may be now distinct from $f^2(x)_1$, i.e. from (3) we obtain:

$$\begin{aligned} f^2(y)_1 = F(F(y_{-1}, x_0, x_1), F(x_0, x_1, x_2), \\ F(x_1, x_2, x_3)) \end{aligned} \tag{6}$$

which can be distinct from $f^2(x)_1$ when $y_{-1} \neq x_{-1}$. Moreover, as above, looking at $f^2(y)_2$, $f^2(y)_3$, etc. the y perturbation is propagated, in two iterations, at the far right at most to the second site. It is possible that this perturbation propagates also to the left of 0, but we focus only on the propagation to the right of the (left) perturbations of x.

Therefore, doing more iterations, the perturbation propagates to the right step by step. The closest site to 0 never reached by any left perturbation y of x in n iterations is denoted by $\widetilde{\Lambda}_n^+(x)$. $\widetilde{\Lambda}_n^-(x)$ is defined in the same way for right perturbations of x that propagate to the left of the site 0.

It so happens that in this definition the site 0 plays a priviliged role while the rule is translation invariant regardless of the site position. In order to make the definition shift invariant a new exponent is defined as

$$\Lambda_n^{\pm}(x) = \sup_{j \in \mathbb{Z}} \widetilde{\Lambda}_n^{\pm}(\sigma^j x) \tag{7}$$

It is then defined the time average speed of propagation (by unit time) as:

$$\lambda^{\pm}(x) = \lim_{n \to \infty} \frac{1}{n} \Lambda_n^{\pm}(x) \tag{8}$$

This average speed was termed by Wolfram and Shereshevsky the left (in the case "−") and right (in the case "+") Lyapunov exponents. It can be proved using the Kingman subadditive ergodic theorem that the above limit indeed exists in the sense of the mathematical measure theory (i.e. almost everywhere). Later on, a slightly different concept of an average Lyapunov exponent was given in [19].

It is interesting to see the above definition in one example of CA defined the rule:

$$F(x_{-1}, x_0, x_1) = x_{-1} + x_0 \pmod 2.$$

The following scheme represents a space-time lattice, when the entering arrows to a site i at a time t correspond to the above rule.

It is easy to see that a perturbation $y_{-1} \neq x_{-1}$ will propagate after n iterations to site $n - 1$,

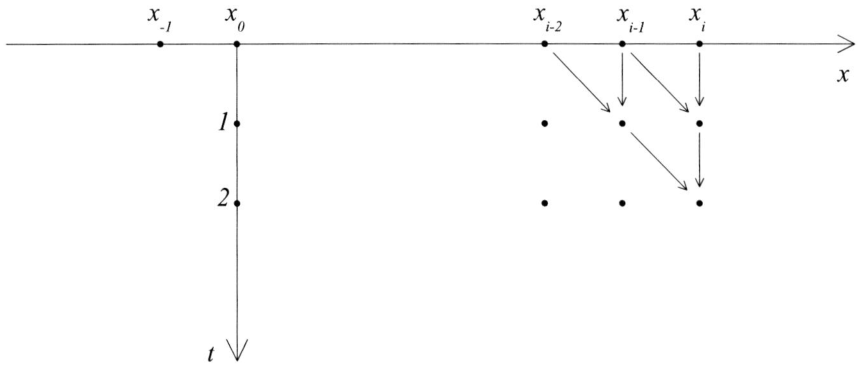

Scheme of the action of the automation as site i along time.

that is $f^n(y)_{(n-1)} \neq f^n(x)_{(n-1)}$. This implies that $\widetilde{\Lambda}_n^+(x) = n$. It is not the same for the left Lyapunov exponent which is obviously here vanishing. It follows then that $\Lambda_n^+(x) = n$ and so $\lambda^+(x) = 1$. In contrast, $\lambda^-(x) = 0$.

In this example, it is the one-to-one property of the map $x_{-1} \longrightarrow F(x_{-1}, x_0)$ which generates the maximal propagation to the right. In general a rule F such that $x_{-1} \longrightarrow F(x_{-1}, \overline{x_0}, \overline{x_1})$ is one-to-one for any value of $\overline{x_0}, \overline{x_1}$ is called by Hedlund [12] a left permutative rule. Right permutative rules are defined symetrically. L. Chua et al. [2] showed that there are 28 rules that are left or right permutative among them only 4 are left and right permutative. Hedlund [12] showed that any permutative rule defines a surjective CA. L. Chua and al. [2] introduced a class of rules called strongly quasi-ergodic by using the isomorphism between $\{0,1\}^{\mathbb{Z}}$ and the interval $[0,1]$, they associated to each cellular automaton a map of this interval into itself. Many properties of this map and its relation to permutativity and attractors of the CA-map have been given. Those properties have not yet been studied rigorously in the frame work of an infinite lattice $\mathbb{Z}$. A slightly different concept of Lyapunov exponent has been given by Tisseur in [19].

3. Instability, Invariant Measures and Entropy

We saw in the previous section how the Lyapunov exponents measure the instability of the initial conditions. A topological classification of the attractors of CA has been investigated by Kurka and others [1] on the infinite lattice $\mathbb{Z}$. In the finite case, the topology of an attractor may be complex as shown by Chua and al. [1], especially as the length of the lattice $L \longrightarrow \infty$. In this case, on account of instability of the initial conditions and complexity of attractors, only statistical knowledge may be reasonable for an understanding of the system.

Two statistical approaches to CA are possible. The first one corresponds to the description of long time behaviour of the CA-dynamics, when we start from a set of random initial conditions. Such a statistical description is reminiscent to the equilibrium micro-canonical ensemble in statistical mechanics. The second one corresponds to convergence to equilibrium under the CA-map of non-invariant measures. Concerning this problem, the only results to our knowledge in this respect are given by [4, 14] for some special CA-rule.

Concerning the first problem, there may be different stationary probability measures, invariant under the CA-map f and the shift σ, corresponding to disjoint attractors eventually. As in statistical mechanics, one considers a class of observables, we may conveniently identify them with the continuous functions, Φ, on X. Given an initial configuration x, a probability measure on the set X of all configurations is given by the equality of time averages to ensemble averages

$$\forall_\Phi \lim_{n\to\infty} \frac{1}{n} \sum_{k=0}^{n-1} \Phi(f^n x) = \int_X \Phi(x) d\mu(x).$$

It it exists, this measure μ depends on the initial condition x. We do not go here into a full discussion of the above ergodic property of the CA which may be very complicated, and not yet totally understood. It is therefore more practical, in studying the statistical property of a given CA-map f, to consider a family of shift invariant measures and to try to extract from them a subfamily of f-invariant measures.

A large class of shift invariant measures is known. One of the most remarkable is the uniform Bernoulli measure where the probability of any sequence

$$x_{i_1} = j_1, x_{i_2} = j_2, \ldots, x_{i_n} = j_n$$

is equal to $(\frac{1}{2})^n$. Let us call it ν. Nevertheless, this measure is f-invariant only for surjective CA-maps [12]. In the non-surjective case, the invariant measures are related to the geometric structure of the attractors that may be less simple than the Bernoulli measures, as computed using time averages in [11].

An important statistical quantity that characterizes the complexity of all trajectories of a dynamical system is the entropy.

Let us first start with the concept of topological entropy, so called because it depends on the concept of neighbourhood and topology of the configurations space. For Wolfram cellular automata a formal distance between two configurations $x = (x_i)$, $y = (y_i)$, $i \in \mathbb{Z}$, is given by

$$d(x, y) = \sum_{i=-\infty}^{+\infty} \frac{|x_i - y_i|}{2^{|i|}}.$$

That is, if m_0 is the radius of the window around zero where $x_i = y_i$, $|i| \leq m_0$, then all those y correspond to $\frac{1}{2^{m_0}}$ — neighbourhood of x.

For a dynamical system on a smooth surface, entropy characterizes how nearby trajectories may become distant and distinguishable. It is even more simple to illustrate this concept in cellular automata dynamical systems.

In the space-time diagram of a CA, let $R(m, t)$ be the rectangle

$$\{(i, \tau) \in \mathbb{Z} \times \mathbb{N}, -m \leq i \leq m, 0 \leq \tau \leq t\}.$$

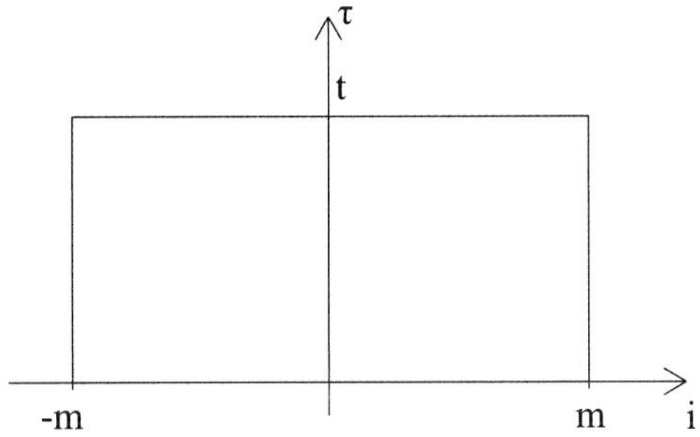

Two configurations x, y are said to be distinguished in $R(m, t)$, if for some $(i, \tau) \in R(m, t)$

$$(x(\tau))_i \neq (y(\tau))_i$$

where $x(\tau) = f^\tau(x)$.

Let $N(m, t)$ be the number of all such patterns generated by the CA-rule. As most of the CAs are not invertible, in general, starting from any initial conditions, the CA-patterns will end on some attractors. The scaling law of $N(m, t)$ as $t \longrightarrow \infty$ and $m \longrightarrow \infty$ defines the topological entropy by:

$$h(f) = \lim_{m \to \infty} \lim_{t \to \infty} \frac{\log N(m, t)}{t}.$$

That is

$$N(m, t) \sim K \cdot e^{h(f) \cdot t},$$

$m \longrightarrow \infty$, $t \longrightarrow \infty$, where K may depend on m. For example, for the shift CA-rule

$$N(m, t) = 2^{2m+t+2}$$

Note that, in this example, the limit $m \longrightarrow \infty$ is not important since the number of possible states by site is equal to 2, so that $h(f) = \log 2$. This is the maximal possible value of $h(f)$ among all rules. The scaling law of the increase of the number of patterns generated by the CA is related to the propagation of perturbations. In fact, two configurations that are initially identical on $\{-m, \ldots, m\}$ sites, will become distinguished in the space-time rectangle $R(m, t)$ under the action of the CA-map which propagate the perturbations from the outside of this interval. This enhances the number $N(m, t)$. The rigorous computation of entropy for CA is still unsolved, although there are examples of rules where this entropy is "undecidable" [13]. Nevertheless several computations were obtained in case studies especially for bipermutative CAs [6, 18].

Another point of view to the entropy is offered by the Shannon and Kolmogorov-Sinai (K-S) entropy defined in terms of f-invariant measures μ, $h_\mu(f)$. For a given CA-map as there may be several invariant measures, there are several (K-S)-entropies corresponding to a multitude of attractor. The topological entropy is an upper bound of those (K-S)-entropies. The K-S entropies corresponding to various ergodic measures provide a complete information about the CA-map, that can be completed by a multifractal analysis.

For any σ-ergodic or f-ergodic measure μ, Shereshevsky [17] showed the following inequality

$$h_\mu(f) \leq h_\mu(\sigma) \cdot (\lambda^+ + \lambda^-)$$

where $h_\mu(\sigma)$ is the K-S entropy of σ. This inequality become an equality for bipermutative CA-rules.

4. Cellular Automata as Space-time Dynamical Systems

Here we wish to understand the system under the joint action of the time dynamics (i.e. the map f) and the shift σ. The joint action of the shift and the CA-map, $\sigma^m f^n$, $m \in \mathbb{Z}, n \in \mathbb{N}$, is much more complex than the simple nonlinear dynamics of the CA-map. What are the consequences of the invariance of CA-dynamics under the spatial translation: are the chaotic properties invariant under translation? Is there a family of CA-maps for which the Lyapunov exponents and the entropy are also invariant under translation? Considering the CA-map action in a moving reference frame, is there an influence of the velocity of the frame on the growth of local perturbations ? These are some reasons to compute those quantities under the maps $\sigma^m f^n$. It is quite possible to define the Lyapunov exponents and the entropy of $\sigma^m f^n$, $\lambda^{\pm}(\sigma^m f^n)$, $h(\sigma^m f^n)$, as above for any CA. They represent the entropy for an observer moving on the axis Ox with a velocity $v = \frac{m}{n}$. Lyapunov exponents for a moving observer have been considered independently in the theory of coupled maps on lattice dynamical systems [10] in a different setting, in relation to open-flow fluids.

It is also possible to extend the above definitions from $(m,n) \in \mathbb{Z} \times \mathbb{N}$ to continuous parameters $(x,t) \in \mathbb{Z} \times \mathbb{R}^+$ allowing to consider the entropy in the continuous space-time, for an arbitrary velocity $v = \frac{x}{t}$ [6, 7]. We shall not go into these details. For the Wolfram elementary CA the computation of this entropy denoted $h(x,t)$ (or $h_\mu(x,t)$ in the metric case) is still an open question. Only in the permutative case we have partial answers in some cones of the space-time. The entropy is fully known in the bipermutative case. For the uniform Bernoulli measure we have the following results:

$$h_\mu(x,t) = \begin{cases} (x+t)\log k & t \le x, x > 0 \quad (I), \\ 2t \log k & -t \le x \le t \quad (II), \\ (t+|x|)\log k & t \le -x, x < 0 \ (III). \end{cases}$$

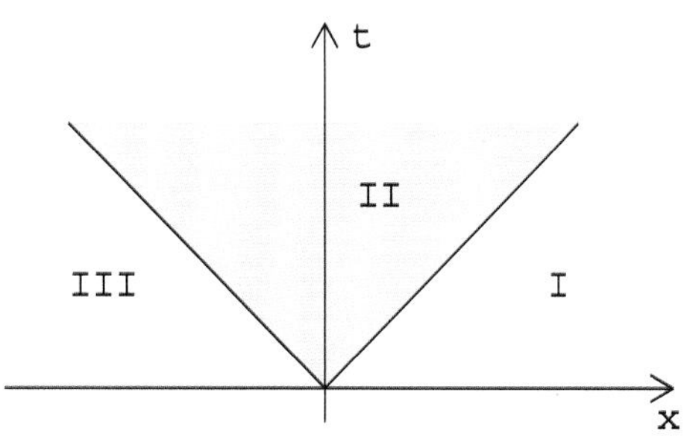

It is interesting to note that this entropy is a continuous function of (x,t), symmetric under the reflexion $x \longrightarrow -x$, independent on the from of the bipermutative map and also invariant under the spatial translation as soon as $v = \frac{|x|}{t} \le 1$. Only when $v > 1$, the spatial entropy start to change linearly in terms of x. It is possible to express the directional entropy in terms of the velocity v by using the homogeneity property of the entropy: $h_\mu(cx, ct) = ch_\mu(x,t)$, for $c > 0$. Then, $h_\mu(x,t) = th_\mu(v,1)$. The directional entropy by time unit as a function of the reference frame velocity v, $h_\mu(v,1)$, is then constant as the velocity has not reached the treshold of the interaction range, it increases linearly in terms of v as this treshold is exceeded. It would be interesting to compute the curves of $h_\mu(v,1)$, in all Wolfram CA to see exhaustively the influence of the dynamics on the directional entropy behaviour.

The directional Lyapunov exponents in the bipermutative maps have been also computed as follows:

$$\lambda^+_{(x,t)} = \begin{cases} 0 & \text{in } (I), \\ t-x & \text{in } (II), \\ t-x & \text{in } (III), \end{cases}$$

$$\lambda^-_{(x,t)} = \begin{cases} x+t & \text{in } (I), \\ x+t & \text{in } (II), \\ 0 & \text{in } (III). \end{cases}$$

Here, we can also make the same observations about the dependence of the directional Lyapunov exponents on the velocity since they have the same homogeneity property. A computation of the directional entropy in other example is given in [18]. We have the following equality for the bipermutative rules:

$$h_\mu(x,t) = h_\mu(\sigma)(\lambda^+_{(x,t)} + \lambda^-_{(x,t)}).$$

In general, for any invariant measure μ that is ergodic with respect to σ we have the Shereshevsky inequality [7]

$$h_\mu(x,t) \le h_\mu(\sigma)(\lambda^+_{(x,t)} + \lambda^-_{(x,t)}).$$

Several upper bounds have been obtained for the CA, they show a privileged role of the permutative rules. Among the 256 Wolfram rules, the maximal entropy and the maximal Lyapunov exponents are reached by the bipermutative ones. This means that these cellular automata maps are the most chaotic. However they are simple since they are surjective.

A detailed study of the space-time chaotic quantities may bring new surprises and especially concerning the existence of CA-maps for which they are translation invariant.

References

[1] F. Blanchard, P. Kurka, A. Maass, *Topological and measure-theoretic properties of one-dimensional cellular automata*, Phys. D 103 (1997), 86–99.

[2] L. Chua, *A Nonlinear Dynamics Perspective of Wolfram's New Kind of Science*, Volumes I-V, World Scientific 2006–2012.

[3] P. Collet, J.-P. Eckmann, *The definition and measurement of the topological entropy per unit volume in parabolic PDEs*. Nonlinearity 12 (1999), 451–473.

[4] M. Courbage, *On the approach to statistical equilibrium in an infinite-particle lattice dynamic model*. J. Math. Phys. 30 (1989), 1840–1850.

[5] M. Courbage, *On the Abundance of Traveling Waves in Infinite 1-D Cellular Automata*, Physica D 103 (1997), 133–144.

[6] M. Courbage, B. Kamiński, *On the directional entropy of $\mathbb{Z}^2$-actions generated by cellular automata*, Studia Math 153 (2002), 285–295.

[7] M. Courbage, B. Kamiński, *Space-time directional Lyapunov exponents for cellular automata*, J. Stat. Phys., 124 (2006), 1499–1509.

[8] M. Courbage, D. Mercier and S. Yasmineh, *Travelling waves and chaotic properties in cellular automata*, Chaos 9 (1999), 893–901.

[9] M. Courbage, D. Mercier and S. Yasmineh, *Wavelengths distributions of chaotic travelling waves in some cellular automata*, Physica D. 150 (2001), 63–83.

[10] R. J. Deissler, K. Kaneko, *Velocity-dependent Lyapunov exponents as a measure of chaos for open-flow systems*, Phys. Lett. A 119 (1987), No. 8, 397–402.

[11] P. A. Ferrari, A. Maass, S. Martnez, P. Ney, *Cesro mean distribution of group automata starting from measures with summable decay*, Ergodic Theory Dynam. Systems 20 (2000), 1657–1670.

[12] G. A. Hedlund, *Endomorphisms and automorphisms of the shift dynamical system*, Math. Systems Theory 3 (1969), 320–375.

[13] L. P. Hurd, J. Kari, K. Culik, *The topological entropy of cellular automata is uncomputable*, Ergodic Theory Dynam. Systems 12 (1992), 255–265.

[14] D. Lind. *Applications of ergodic theory and sofic systems to cellular automata*, Physica D 10 (1984), 36–44.

[15] J. Milnor, *Directional entropies of cellular automaton-maps*, in: Disordered Systems and Biological Organization, NATO Adv. Sci. Inst. Ser. F20 Springer, 1986, 113–115.

[16] J. Milnor, *On the entropy geometry of cellular automata*, Complex Systems 2 (1988), 357–386.

[17] M. A. Shereshevsky, *Lyapunov exponents for one-dimensional automata*, J. Nonlinear Sci., 2 (1992), 1–8.

[18] J. Smilie, *Properties of the directional entropy function for cellular automata*, Lectures Notes in Math. 1342 (1988), 689–705.

[19] P. Tisseur, *Cellular automata and Lyapunov exponents*, Nonlinearity, 13 (2000), 1547–1560.

[20] J. Urias, A. G. Salazar, E. Ugalde, A. Enciso, *Traveling patterns in cellular automata*, Chaos 6 (1996), 493–503.

[21] S. Wolfram, *Cellular Automata and Complexity*, Addison-Wesley Publishing Company, 1974.

CHAPTER 9

THE GENESIS OF CHUA'S CIRCUIT: CONNECTING SCIENCE, ART AND CREATIVITY

FRANCESCA BERTACCHINI*, ELEONORA BILOTTA†, GIUSEPPE LARIA‡ and PIETRO PANTANO§

Dipartimento di Linguistica, Universitá della Calabria,
Cubo17B, Ponte Bucci, Arcavacata di Rende (CS), Italy
**fbertacchini@unical.it*

The Chua's circuit has provided science with a real scientific and technological advancement in the field of dynamical systems. But the origin of the Chua's circuit, which has resulted in major areas of study within Chaos Theory, with thousands of printed scientific articles, is not widely known. In this chapter, we trace the fundamental stages of the Genesis of the Chua's circuit. To make the story more appealing for the new generations, a 3D movie was made (issued on DVD, with this volume). Besides presenting the main **characters** from the history of such system, it also introduces Lorenz, Poincare, Julia and Cantor as the precursors of the studies on chaos, which discuss the need for a physical system that demonstrates chaos, a phenomenon not yet understood, in a numerical, physical and experimental way. The story, enriched by the conversations of the above mentioned characters, was staged exactly as it occurred, derived from Professor Chua's famous article of 1983, *The Genesis of the Chua's circuit.* And the story continues with the authors contribution in expanding Chaos Theory, with the discovery of almost one thousand attractors, which have provided new important elements for the advancement of science in the field of nonlinear chaotic systems.

†Dipartimento di Fisica, Universitá della Calabria, Cubo31C, Ponte Bucci, Arcavacata di Rende (CS), Italy. Email: bilotta@unical.it.
‡Dipartimento di Linguistica, Universitá della Calabria. Email: giuseppe.laria@unical.it
§Dipartimento di Fisica, Universitá della Calabria. Email: piepa@unical.it.

1. Introduction

The focus of this chapter is the genesis of Chua's circuit, dedicated to Professor Leon Chua for his outstanding invention, and recounted by reconstructed animated characters in a 3D movie. This movie has been produced to make the topics of chaos and complexity accessible to the younger generation. To this end, advanced educational contexts of Edutainment (education plus entertainment), which involve the physical construction of Chua's circuit, its simulation using computer software and the creation of sounds, music and artworks using attractors have been developed.

In the 'sixties, E. Lorenz discovered chaos: its typical expression, represented by a strange attractor, was known as Lorenz's butterfly. However it was not possible to reproduce chaotic behavior in a laboratory; chaos was only a simulation obtainable by computer. Engineers and technicians did not consider chaos as important for their studies, even if the phenomenon attracted the attention of mathematicians and physicists. Moreover, the lack of tools to demonstrate the phenomenon experimentally, made chaos of scarce practical interest. As a result while chaos was a great scientific discovery it could not have a real impact. Lorenz's chaos was an abstraction of something that was not possible to observe in nature.

In early 'eighties, a group of Japanese scientists, led by professor Matsumoto, built a circuit based on Lorenz's equations and then carried out the first experiment to demonstrate chaos. Leon Chua was one of these scientists. While the experiment failed, on the same night a fundamental discovery, destined to change the history of contemporary science, was made. In fact, in his room, Professor Chua, after much thought about the causes of the failed experiment, realised how to reproduce chaos and invented his famous circuit.

On the basis of the reconstruction of the genesis of the circuit, as told by Professor Chua himself [1], and in the literature [2 8], we have produced a 3D movie, which faithfully reproduces this story, using virtual characters and environments. In this work, after a short introduction to the important scenes in the movie (in a DVD attached to the book), we describe the Gallery of Chua's attractors [9]. These have been developed starting from about 40 known attractor shapes and have been used creatively to make sounds, music, 3D images, jewels, and paintings. Moreover, these artistic tools make chaotic phenomena easy to understand, even for high school students, and so contribute to spreading scientific knowledge on chaos.

2. The Genesis of Chua's Circuit, the 3D Movie

Prologue.

Some scientists, such as Macbeth's witches who are able to foresee the future, or the *deus ex machina* of Greek drama, meet in an imaginary environment and talk about chaos. They are: Cantor, in a cloud of dust, brushing his clothes; Lorenz, introduced with a resentful tone and a flashy umbrella to protect himself from an unexpected hurricane, caused somewhere by the flapping wings of his butterflies; Poincaré, playing with a strange electronic game, which allows him to see the motion of three bodies, which are attracted to one another, according to Newton's law of gravity; Julia, with small fractal shapes, known as Douady rabbits, running around him, discoverer of the Julia set, which was later rediscovered by Mandelbrot in his fractal geometry. All the characters are waiting for the integration of chaos in a physical system, except Lorenz, who affirms that he has already discovered chaos. They wait for the demonstration that chaos is a real phenomenon and not just a purely mathematical abstraction, simulated by a computer (Fig. 1).

1983, Tokyo.

Professor Chua comes into a wonderful garden at the University of Waseda, Tokyo. Red maple leaves fill the environment. When he approaches, a branch forces him to lower his head. He contemplates the order and equilibrium of the universe and the strength and the majesty of nature. Then, smiling gently, he thinks that he is going to take part in an experiment on the physical demonstration of chaos, and that he is a scholar of chaos. After crossing a little bridge linking the garden to the institute, he looks at a stone: "I wonder if that is an immortal stone!", he thinks. In the play, this symbolic stone is associated with the Chua circuit, as an object that will continue to influence science progress. Professor Chua only has to do what needs to be done: to join the spirit of the stone. He enters the institute (Fig. 2).

An excited assistant, wearing a white gown, approaches Professor Chua and they go together to the experiment room. They walk along a hallway toward Professor Matsumoto's laboratory. On the left, large windows show the little lake, and

Fig. 1. Cantor, Julia, Poincaré and Lorenz (real and virtual) discuss and foresee the invention of a physical circuit for the mathematical, physical and numerical demonstration of chaos. There is an attractor and Lorenz affirms this to be the inventor of chaos. Each character has been represented with some characteristic element. In particular, Lorenz has an umbrella because he is always waiting for a possible hurricane, caused somewhere by the flapping of his butterflies. Images are arranged clockwise, while the central picture is one of the most salient points of the tale.

the ancient maple covered in leaves; many leaves have landed on the water, forming small elegant whirlpools. Many voices buzz in the laboratory, since many other university professors have been invited to witness the historic event. Professor Chua and Matsumoto's assistant come in. There, in an area of the laboratory, full of electronic instruments, the professors are eager to witness a live demonstration of, presumably, the first successful realization of the Lorenz Equations electronic circuit.

The circuit is perfectly built, with at a least a dozen components, and has many potentiometers and special capacitors (for fine tuning), to exploit the extreme sensitivity of the structure on which the circuit is based. The connections among the circuit elements have been made very precisely. Some assistants of Professor Matsumoto begin to work on the wonderful equipment. However, the circuit does not work and Matsumoto admits defeat. Chaos is still a mathematical abstraction, which can only be realised by numerical simulation (Fig. 3).

Professor Chua is also disappointed. He walks back to the garden toward the main road. Despondency overwhelms him and the beauty of the landscape does not make him feel better. The professor thinks that the failure of the experiment is not

Fig. 2. Waseda University, Professor Chua enters the garden and looks at the immortal stone. He knows that a great destiny awaits him. From a university corridor he observes a falling leaf that floats toward its destiny. In the image, Professor Chua's expression represents the prediction of the future.

due to Matsumoto's experimental skills. No, there must be something else. Maybe the failure is due to the lack of a critical nonlinear IC component, with a near-ideal characteristic and a sufficiently large dynamic range; namely, the analog multiplier. Yes, an accurate analogue multiplier is necessary. It is the key to building an autonomous chaotic circuit. Now in 1983 he sits on the same stone near the bridge, maybe the immortal stone, and picks up a stick to write in the sand. Professor Chua, acting as if reasoning it out himself says: "Let's recapitulate. We have only two autonomous systems, of ordinary differential equations which are generally accepted as being chaotic, the Lorenz equations (writing the formulas of the Lorenz equations in the sand), and the Roessler equations (writing these in the sand). And nonlinearity, in both systems, requires two variables, that is the product function. Only a demonstration that the circuit of Matsumoto-Lorenz works, could give Chaos its rightful place. How can we attract engineers and the non-linear circuit theorists to build circuits that exhibit chaos? How can we show its potential? How can we prove that some biological phenomena are chaotic?" (Fig. 4).

Outside the University, Chua takes a taxi and goes to his dormitory. After a light dinner, while he is going to bed, the feeling of failure still haunts him. Chua feels that the group is near success. He cannot resign himself to failure. Lorenz's butterfly

Fig. 3. At the centre of the scene there is the circuit, accurately built with many components, designed by Matsumoto's team. The experiment goes ahead but it fails. Everyone is disappointed.

is also there, even if it is only a numerical simulation. And there is also the Roessler attractor. Chua thinks," What is the essence of chaos? Maybe it is in the dynamic of these two instable focuses, passing through the three points of Lorenz? Or two in Roessler? Are at least two unstable equilibrium points present? And are these focuses fundamental? Can a simple nonlinear circuit be designed with these characteristics? Yes, maybe it is possible. Why haven't I thought about this before? At this point it is only a simple exercise in non-linear circuit theory. Yes. It is only necessary to examine all the possible circuits, and among them only eight are suitable. Let's reject all these non-chaotic circuits so there are only two possible chaotic circuits." In this way, in October 1983, in the dormitory of the University of Waseda, on napkins and the blank edges of used envelopes, in less than an hour, Chua draws the nearly illegible circuit diagrams and the piecewise-linear curve for the nonlinear resistor of Chua's circuit (Fig. 5).

The following day Chua rushes into Matsumoto's laboratory with the solution to the circuit problem. After making the suggested changes, Matsumoto obtains results. Professor Matsumoto goes on to simulate the circuit using a wide range of parameters, but his group cannot physically make it because Matsumoto has no experience in the synthesis of a device with non-linear characteristics (now called the Chua diode). However, the first step in the physical demonstration of chaos is achieved (Fig. 6).

Fig. 4. Chua comes out of the laboratory, walks and sits on the immortal stone, thinking about the failed experiment. He writes the equations of Lorenz and Roessler in the sand. Then he stands up and goes away.

Professor Chua is thinking in his office at the University of Waseda. Some months after Matsumoto's simulations of the Chua circuit, Chua is dissatisfied. In spite of his numerous suggestions, Matsumoto has not managed to build Chua's circuit, since his research group has never synthesized a prescribed non-monotonic $v - i$ characteristic.

Chua leaves for Berkeley and finds out that his circuit works. Ayrom and Zhong, Chua's collaborators at Berkeley University, have built the simplest electronic circuit exhibiting chaos, or rather, the phenomenon of bifurcation chaos. It will become a system of great importance, a paradigm in the study of complex phenomena in nonlinear circuits. Furthermore, if compared to Lorenz's attractor, Chua's circuit shows a much greater variety of complex phenomena. In this way, Zhong and Ayrom become the first researchers to experimentally document and observe the chaotic phenomena created by Chua's circuit (Fig. 7).

Garden of Waseda University, Tokyo. The story ends where it started, with the same characters commenting on Chua's discovery. Cantor, seated on the immortal stone, affirms that, thanks to Chua's circuit, the experimental demonstration of chaos dynamics has succeeded. This invention provides an important key in the interpretation of nature. Lorenz is disconsolate, thinking that his butterfly is not the only chaotic attractor. Chua's circuit demonstrates that chaos has many shapes. Poincaré suggests that maybe someone will find a physical demonstration of Lorenz's equations. Julia foretells

Fig. 5. Chua comes out of the University, takes a taxi and goes to the University dormitory. He sits on the bed and thinks about the possible characteristics of a circuit; then he suddenly stands up, takes a napkin from his pocket, writes the formula and draws a diagram of the chaotic circuit. At the centre of the image, there is the formula and diagram of Chua's circuit.

that Chua's circuit will allow the exploration of complex chaos dynamics and the following of infinite routes to chaos. And Julia, like an able magician, takes a topper and pulls out thousands of attractors, all produced from Chua's circuit; they are like doves flying in the sky and form a multidimensional universe of enchanting, unimaginable and surreally beautiful shapes and colourful patterns (Fig. 8).

3. The Gallery of Chua's Attractors

After the finding of about forty strange attractors deriving from Chua's circuit [2,3] and its generalizations, about a thousand of attractors have been find by [10–15]. Unlike Lorenz's attractor, which generates just a few kinds of dynamic patterns, Chua's circuit and systems derived from the circuit, can produce a large number of chaotic structures. In recent years, galleries of Chua attractors have been made, with 3D attractor images of extraordinary beauty, never previously seen in chaos literature. Basing our work on the existing literature, we have identified a set of basic attractors which we have used as the starting points for our investigation of the parameter space and dynamics of the physical Chua Oscillator. We have visualized these systems as points in a multidimensional space. This has allowed us to identify a series of model patterns that change only gradually when control parameters

Fig. 6. Chua rushes into Matsumoto's laboratory, waving the paper with the circuit formula. Chua and Matsumoto hug. At the centre of the image, there is the formula and diagram of Chua's circuit. Chua is at his desk when Matsumoto leaps for happiness: he has succeeded in the simulation of Chua's circuit and the realization of the first form of chaos. Chua is overjoyed with the results.

change. In fact, when these parameters are modified beyond a certain range, the models change, together with the dynamic phenomena they represent. The gallery contains about a thousand strange attractors, a small sample of the infinite range of chaotic phenomena the circuit is capable of producing (Fig. 9).

Moreover, a computational system able to produce shapes in the phase space [16], through discrete steps, can be associated to Chua's circuit. The shapes emerging from this system, depend on a series of parameters (parameter values distinguish particular machines that we have defined as Chua machines) leading to structures of extraordinary beauty and complexity, as well as highlighting bifurcation maps in multidimensional spaces individuating extremely suggestive patterns. In Fig. 9, some shapes which have emerged during the phases of Chua's circuit, and from its generalizations, are present. Figure 10 shows a collection of shapes emerging in Chua's machines.

4. Art from Chua's Circuit

The shapes emerging in Chua's circuit are beautiful and can inspire artistic creation. In Fig. 11 some

Fig. 7. After a period of time, Chua is thoughtful. Matsumoto cannot build the circuit. Chua writes a letter to his collaborators at Berkeley University, Ayrom and Zhong. Image of Berkeley. Ayrom and Zhong read the letter. Chua looks at the representation of chaos on the oscilloscope with Zhong and Ayrom. Chua's circuit, at the centre of the image, is built.

of these shapes have inspired the creation of flowers, realized by modifying the original shapes using photo retouching techniques.

Figure 12 shows a series of paintings presented in two exhibitions. The environment in which the paintings were exhibited can be virtually explored at the web address http://museomg.unical.it/scienar/. The second exhibition was in the context of different European conferences devoted to the connection between science and art.

In Fig. 13 we can observe a series of jewels created starting from the shapes emerging in Chua's circuit and appropriately adapted. Two of the pieces (the bracelet and the cornucopia) have been made by an artisan. The bracelet was made in gold and the cornucopia in silver.

The great creative possibilities linked to the circuit have persuaded us to start an international experimental program together with the international association of art, science and technology Leonardo (http://www.leonardo.info/). At the beginning, the project is only addressed at associated members but later it will be extended to all interested people. Its aim is the divulgation of Chua's attractors as tools for artistic, musical, and 3D creation.

Fig. 8. Cantor, Julia, Poincaré and Lorenz are again in the scene at Waseda University Scenography. Cantor sits on the immortal stone. Lorenz is nostalgic and opens his umbrella. Poincaré beckons to Lorenz while Julia, like an able magician, takes a topper and pulls out thousands of attractors. At the centre there is a collection of Chua attractors.

We can summarize the ESG's Science & Art route as follows:

- a series of theoretical research projects has led some researchers in the group to identify a variety of strange attractors, derived from Chua's circuit;
- these attractors have been represented as images, sounds and music by computational algorithms, embodied into software applications;
- from these applications, art forms such as 2D and 3D images, sound and musical compositions have been produced.

The group has also created a series of research products such as design elements, logos and other visual communication tools, new musical instruments, software environments for sound and music production and prototypes of video games. Another representation of the attractors, in interactive contexts, has allowed the construction of virtual imaginary environments and multimedia theatre performances [17–21].

It is important to note that the productions of Chua's circuit are not only artistically interesting from the visual point of view because the circuit is an interesting tool able to generate sounds and

Fig. 9. The Gallery of Chua's attractors.

Fig. 10. Shapes emerging in Chua's machines.

Fig. 11. Chua flowers, realized using photo retouching techniques.

Fig. 12. Real paintings inspired by Chua's attractors.

music [22]. Many images, shapes, movies, sounds, music and artistic creations are present at the following web address: http://chua.unical.it/.

5. Chaos is Simple and Everybody can Understand It

Chua's circuit is important not only in the field of science and for the artistic products it inspires, but also because it is an astonishing tool able to bring the topics of chaos and complexity (and science in general) nearer to the younger generation [23]. In [21], experiments with secondary school students are presented to show this. In ad-hoc laboratories, students have physically built the circuit, simulated it by computers, realized 3D attractors and produced sounds and music. In this way they have approached scientific topics outside the regular school curriculum and usually studied only by chaos specialists. In this way, they have understood that science is not difficult or boring and can be understood through entertainment and studied in a very interesting way.

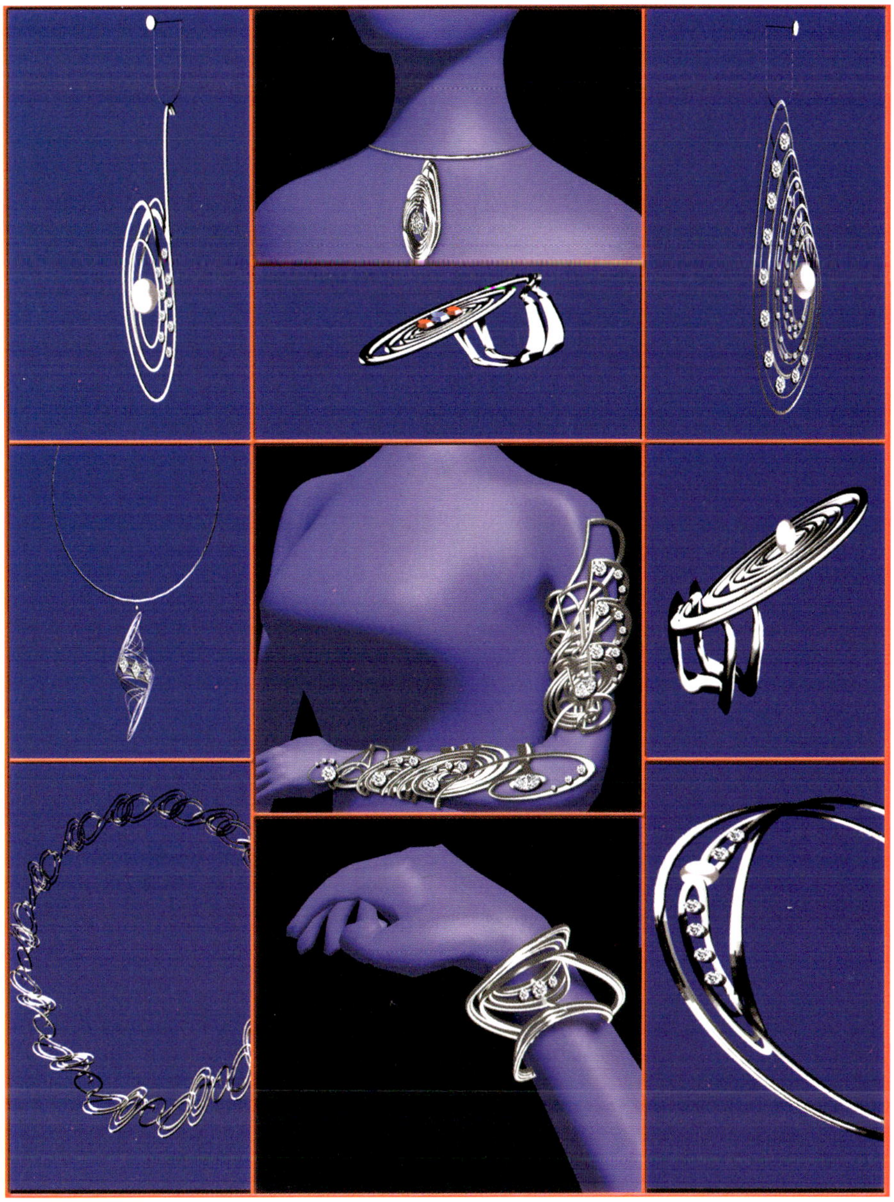

Fig. 13. Collection of jewels realized starting from the shape of Chua's attractors.

Experiments has been organized in theoretical lessons and laboratories. Students approached the chaotic phenomenon by creating 3D attractors, manipulating the control parameters of the simulated circuit and realizing images, sounds, and music (Fig. 14). Students have learned about many aspects of the contents of the experiment: the building and manipulation of circuit components, computer simulation, and the production of 3D attractors and music. All the students completed a learning questionnaire, providing positive results to the first empirical experiment. The materials used in this experimentation are available at the above mentioned web site.

6. Conclusions

The invention of Chua's circuit, and all Chua's work involves important aspects of contemporary science. In this work, we want to make a small contribution to the divulgation of the story of the genesis of Chua's circuit which sends a powerful message to the younger generation: science is beautiful because

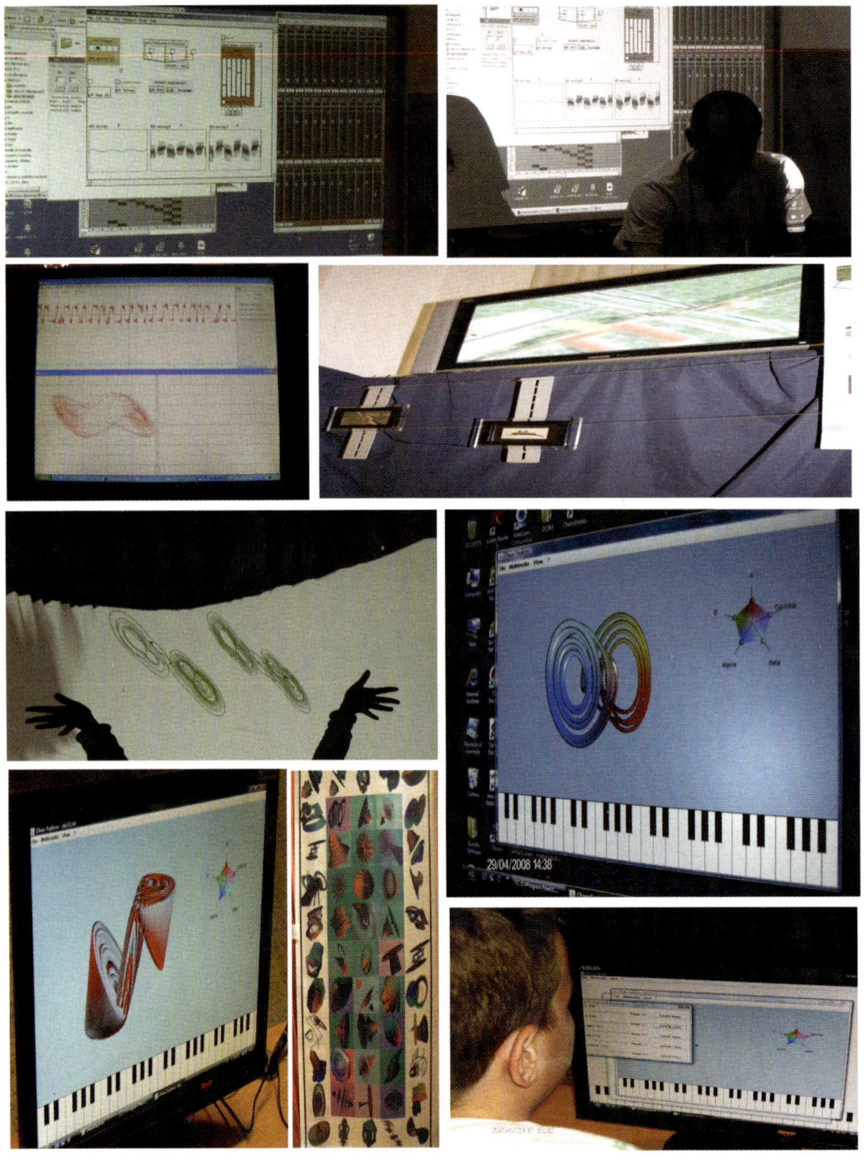

Fig. 14. The experiment.

it allows the understanding of nature. If science is also enjoyable and engaging, connecting us with our innate artistic side, we can appreciate the beauty of nature. Through non-conventional ways of study and research, the younger generation can better understand science. Our scientific path tries to connect two worlds, above all for promoting this message to future generations.

References

[1] L. O. Chua, The genesis of Chua's circuit, *Archiv fur Elektronik und Uebertragungstechnik*, **46**, 250–257, (1992).

[2] L. O. Chua, C. W. Wu, A. Huang & G. Q. Zhong, A universal circuit for studying and generating chaos-Part I: Routes to chaos, *IEEE Trans. Circuits Syst. I: Fund. Th. Appl.*, **40**, 732–744, (1993).

[3] L. O. Chua, C. W. Wu, A. Huang, & G. Q. Zhong, A universal circuit for studying and generating chaos. II. Strange attractors, *IEEE Trans. Circuits Syst. I: Fund. Th. Appl.*, **40**, 745–761, (1993).

[4] R. N. Madan, *Chua Circuit: A Paradigm for Chaos* (World Scientific, Singapore, 1993).

[5] L. O. Chua, Chua's Circuit 10 Years Later, *Int. J. Circuit Theory and Applications* **22**, 279–305, (1994).

[6] R. Barboza, & L. O. Chua, The Four-Element Chua's Circuit, *Int. J. Bifurcation and Chaos* **18**, 943–95, (2008).

[7] L. Fortuna, M. Frasca, & M. G. Xibilia, *Chua's Circuit Implementations: Yesterday, Today and Tomorrow* (World Scientific, Singapore 2008).

[8] R. Kilic, *A Pratical Guide for Studying Chua's Circuits* (World Scientific, Singapore, 2010).

[9] E. Bilotta, & P. Pantano, *A Gallery of Chua Attractors* (with Cd-Rom) (World Scientific, Singapore, 2008).

[10] E. Bilotta, P. Pantano & F. Stranges, A gallery of Chua attractors. Part I, *Int. J. Bifurcation and Chaos* **17**, 1–60, (2007).

[11] E. Bilotta, P. Pantano & F. Stranges, A gallery of Chua attractors. Part II, *Int. J. Bifurcation and Chaos* **17**, 293–380, (2007).

[12] E. Bilotta, P. Pantano & F. Stranges, A gallery of Chua attractors. Part III, *Int. J. Bifurcation and Chaos* **17**, 657–734, (2007).

[13] E. Bilotta, G. Di Blasi, S. Stranges & P. Pantano, A gallery of Chua attractors. Part IV, *Int. J. Bifurcation and Chaos* **17**, 1017–1078, (2007).

[14] E. Bilotta, G. Di Blasi, S. Stranges & P. Pantano, A gallery of Chua attractors. Part V, *Int. J. Bifurcation and Chaos* **17**, 1383–1511, (2007).

[15] E. Bilotta, G. Di Blasi, S. Stranges & P. Pantano, A gallery of Chua attractors. Part VI, *Int. J. Bifurcation and Chaos* **17**, 1801–1910, (2007).

[16] E. Bilotta & P. Pantano, Discrete Chaotic Dynamics from Chua's Oscillator: Chua Machines, *Int. J. Bifurcation and Chaos* **19**, 1–115, (2009).

[17] F. Bertacchini, E. Bilotta & P. Pantano, Educational Virtual Scenario for Learning Chaos and Complex Theories, *The International Journal of Virtual Reality*, **7**(2), 19–26, (2008).

[18] F. Bertacchini, E. L. Bossio, E. Bilotta, P. Pantano & Vena, S. Learning Chaos in an Interactive Virtual Museum, *VIWO 2009 Workshop*, 1–6, (2009).

[19] F. Bertacchini, E. L. Bossio, E. Bilotta & P. Pantano, Making Scientific Topics Simpler: A Web Site for Learning Chaos, *Proceedings of INTED 2010 Conference*, 5845–5853, (2010).

[20] F. Bertacchini, L. Gabriele & A. Tavernise, Educational Technologies. In *Bridging Educational Technologies and School Environment: implementations and findings from research studies.*, (Nova Publisher, 2011).

[21] E. Bilotta, E. L. Bossio & P. Pantano, Chaos at school: Chua's Circuit for Students in Junior and Senior High School, *Int. J. Bifurcation and Chaos*, **20**(1), 1–28, (2010).

[22] E. Bilotta, S. Gervasi & P. Pantano, Reading complexity in Chua's oscillator through music. Part I: A new way of understanding chaos, *Int. J. Bifurcation and Chaos* **15**, 253–382, (2005).

[23] G. Gandhi, G. Cserey, J. Zbrozek & T. Roska, Anyone can build Chua's circuit: Hands-onexperience with chaos theory for high school students, *Int. J. Bifurcation and Chaos*, **19**, (2009).

CHAPTER 10

NONLINEAR ELECTRONICS LABORATORY (NOEL): A REMINISCENCE

CHAI WAH WU

IBM Thomas J. Watson Research Center,
P.O. Box 218, Yorktown Heights, NY 10598, USA

The Nonlinear Electronics Laboratory (NOEL) at the University of California at Berkeley has been at the forefront of many breakthroughs in circuits and systems research over the years. I was fortunate enough to be part of this laboratory during the period 1990–1996. This account is a personal reminiscence of my time there and my thoughts on the impact of working at NOEL on my development as a research scientist.

1. Prelude

It would be appropriate to describe how I first get to be part of NOEL. When I received my Bachelor's degrees in Cognitive Science and Computer Engineering, I felt that I was not knowledgeable enough to make contributions to science and engineering.[a] I knew that I needed to attend graduate school to further my intellectual development. I applied to several schools and was accepted by the University of California at Berkeley (UCB) campus with a one-year fellowship[b] to pay for tuition and board thus allowing me to freely choose the research group and direction. During the spring of 1990 I was still contemplating what type of research I wanted be

[a]Although I was quite proud when as an undergraduate I found a way to reduce the number of gates in the Baugh-Wooley binary multiplier. But then I found out that this has already been done in Ref. [1] some 15 years ago.

[b]Thanks to the generosity of George C. Pardee and Helen N. Pardee, the 21st Governor of California and his wife. Speaking of California Governors, I had a chance to see the future 38th Governor of California, albeit from a distance (Sect. 5).

involved in when much to my surprise I received a thick envelope from Prof. Leon Chua. He had included several papers from his research group including the seminal papers on Cellular Neural Networks (CNN)[2, 3]. Also included in the envelope was a personal letter inviting me to work with him at NOEL on CNN. I was excited and humbled by the personal attention he showed to a newly minted university graduate. This is the rule rather than the exception, as over the years Leon welcomes both junior and senior researchers from all over the world and from a wide variety of disciplines (more on that later) to collaborate and work at NOEL. Leon was the only professor who had contacted me and I promptly accepted his offer. So I arrived in September 1990 at UC Berkeley without knowing much about how to do scientific research.

2. NOEL

During my time there, NOEL was located on the mezzanine of the second floor in Cory Hall, the Electrical Engineering and Computer Sciences building of the University of California at Berkeley. It is a happy coincidence that NOEL is LEON spelled backwards[c] since Leon Chua's leadership and vision define NOEL and in every part of the lab I see his involvement.

NOEL consisted of two sections, essentially one room plus half of another room connected via a doorway. The "half" room was separated from the other half (which are offices for graduate students used by other professors) by a wall of bookcases. Several of the bookcases were filled with reprints and research reports documenting the work that was done in the lab since the 1970's. It was quite interesting to see the evolution of research in nonlinear circuit theory, chaotic dynamics and Cellular Neural Networks by perusing these reports. There were several desks in the lab assigned to students and visitors. When I arrived at NOEL there was no desk available for me, and I was given a desk in the next office (i.e. the other "half" of the split room). Since all that separated my desk from NOEL were bookcases, I could talk to my colleagues in the lab simply by shouting. After a year or two I had my own desk in the lab proper, and I promptly covered it with books and papers (Fig. 2). The main area of the lab had an eclectic assortment of equipments. There were oscilloscopes, logic analyzers, power supplies and standard circuit building tools such as soldering irons, breadboards, wire wrap tools, and discrete components. But the lab also housed an anechoic chamber,[d]

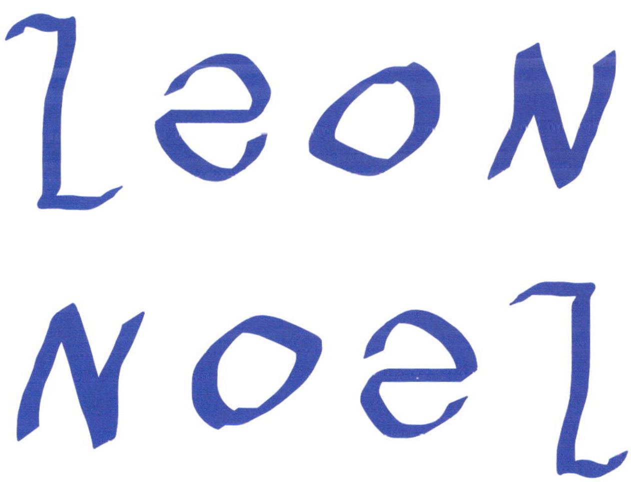

Fig. 1. An ambigram. The image appears the same when viewed upside down.

Fig. 2. My desk at NOEL. I am especially fond of the clock (the red device in the middle of the picture) that runs on juice, beer, fruits or vegetables. I found that yams work well. They don't spoil as easily and I can run the clock for weeks on a pair of tubers. The stickers on the bookshelf are from oranges and apples. Everytime I ate a fruit, I peel the sticker off the fruit and put in somewhere on the bookshelf. By the time I left NOEL, it was full of stickers.

[c]Fig. 1 is my feeble attempt at creating an ambigram to illustrate this fact. I have been fascinated with ambigrams ever since I read Douglas Hofstadter' book "Gödel, Escher and Bach".

[d]Large enough for a person to climb in and use as a sensory deprivation booth in a pinch when the stress of graduate school or research is too much to bear.

a high-voltage generator,[e] a canister of plant spores,[f] a collapsible full-frame camera and of course several computers and graphics workstations. With all the electronic equipments running, the lab could be quite hot sometimes. One year we decided to get a portable air conditioning unit and drilled a hole in the door to vent the hot air out.

3. Hours of Operation

We all had RFID card key access to Cory Hall and the lab door had a combination lock on it, and thus all of us could enter the lab at any time of the day. This provided us with a lot of flexibility in our schedule. For some reason, the senior members of the lab were more often there during the day and the junior members (i.e. students) tended to be at the lab during the night. My day usually started in late morning and continued late into the night. Luckily the hot dog stand across the street was open late,[g] so late-night snack attacks could be easily satisfied. Most of the time when I went home late at night there were still many people in the lab. However the streets were deserted and sometimes quite foggy and I tried to bike home as fast as I can. One time, I went a little too fast and got a ticket for running a stop light with my bicycle. Another time, while going downhill on Hearst Avenue the gravity assistance made my bicycle go too fast and when I hit a pothole (Fig. 3) I went flying off my bicycle, necessitating a trip to the emergency room.

Fig. 3. The pothole that turned my bicycle into a catapult.

During the day many of the visitors and students would work there and all the computers were usually occupied. During the night more computers were available and the lab was also quieter and that was when I did most of my work. Interestingly, many of my interactions with Leon occurred during the night (except when I was his Teaching Assistant for his circuits class) as he would come into the lab late at night or I would receive faxes and phone calls from him during the night hours. Apparently, both he and I get our best thinking done in the evening.

4. Research Directions

During my stay at NOEL, the lab's activities focused on two main areas of research: Cellular Neural Networks (CNN)[4] and chaotic dynamics. CNN's have emerged as a universal platform for massively parallel computation using an array of analog and/or digital processors that are connected locally in a regular fashion. The CNN research spans various areas: theory, algorithm design, hardware implementations and applications. This is a vibrant research area and an entire stack of technology needed to deploy CNN's in applications were developed, including theoretical analysis on performance limits, SDKs for writing algorithms, software emulators for rapid prototyping, VLSI implementation of various models of the CNN universal machine and complete systems for performing various high speed vision-related tasks. In addition to hosting many visitors around the world, there were also close collaboration between NOEL and various laboratories around the world. For instance, the Instituto de Microelectrónica de Sevilla (IMSE-CNM) at the University of Sevilla built several CNN processors in VLSI technology, and these were subsequently run on the hardware interface and development platform from the Hungarian Academy of Sciences, implementing many of the CNN algorithms developed at NOEL.

The research on chaotic dynamics focused on the theoretical analysis of chaotic circuits and systems and their possible applications in information technology. In particular, we study properties of chaotic systems such as Chua's circuit [5, 6] and

[e] This is a fun piece of equipment. In moments of boredom, we would use it to create a spark gap and punch tiny holes in paper.

[f] You may ask what the purpose of this was in the lab. The spores were used to show Chladni patterns by placing them on a vibrating plate. At specific frequencies and amplitudes we noticed that the spores looked like a boiling liquid.

[g] Last time I checked it was still in business.

chaotic digital filters[7, 8] and how to use them in various applications.

At any given time, there were researchers in NOEL proving theorems, writing code, running software simulations, designing integrated circuits, building electronic systems, running physical experiments and writing papers. The lab was always full of people and was a fertile ground for sprouting technical discussions and research projects among physicists, mathematicians, circuit designers, postdocs, students and professors.

I started my Ph.D. research working on CNN, particular on the stability of various types of CNNs [9–11]. I also created a maze-solving CNN that uses the dynamics of the network to find a solution to a maze. When Bert Shi was the Teaching Assistant (TA) for Leon's circuit theory class, he used this problem as an exercise problem for the class. As I will describe later on, this has profound consequences (for me). Sometime in the middle of my stay at NOEL, I changed my research focus to concentrate more on chaotic dynamics[12] and synchronization of chaos[13]. Even though this appeared to be a drastic change in research focus, the mathematical tools used (linear algebra, dynamical systems) are similar and the underlying theme of my research both in CNN and in chaotic dynamics was complex yet orderly behavior in a collection of interconnected systems. The mathematics department at UC Berkeley, located in Evans Hall not far from Cory Hall, is well known for its research faculty and their contributions to nonlinear and chaotic dynamics, with world renowned mathematicians such as Stephen Smale, Morris Hirsch and Charles Pugh on its staff. Being able to talk to them face to face was a boon to our research. In particular, Moe Hirsch's work on cooperative systems[14–16] shows that state trajectories of cooperative systems are eventually monotone, i.e. they preserve the natural partial ordering of the states over time and we used these results extensively in our work on CNN stability. I also had many interesting discussions with Charles Pugh who was my advisor for my Master's thesis in mathematics. The following coincidence is worth mentioning. On my first day of work at IBM I went to the security office to get my badge. I was quite surprised to see Charles Pugh standing in line at the same office, also waiting for his badge. It turned out he was visiting Mike Shub, who was my manager at IBM at the time, and Charles and I both arrived at IBM on the same day and went to the security office at around the same time!

Leon also had some of the progenitors of the burgeoning area of chaos research from the former Soviet Union visit the lab, including Leonid Shilnikov[h] and Alexander Sharkovsky (Fig. 4). Shilnikov[17] showed that under certain conditions

Fig. 4. From left to right: Vadim Anishchenko, Leonid Shilnikov and Alexander Sharkovsky.

[h]While writing this chapter, I learned with much sadness that Leonid Shilnikov has passed away in December 2011.

there exists countably infinite many periodic orbits near a homoclinic orbit of a 3-D dynamical system leading to chaos. Such Shilnikov bifurcations were found in Chua's circuit[18]. Sharkovsky[19][i] showed that the least periods of periodic orbits on a continuous map on an interval follow a specific ordering of the integers, now known as the Sharkovsky ordering. A consequence of Sharkovsky's theorem is that any continuous map on the unit interval with a period 3 orbit contains orbits of any period. While he was at NOEL he studied chaos in a Chua's circuit with a transmission line that is described by a functional differential equation[20]. I do not speak Russian and Alexander did not speak much English, so my conversation with Alexander sometimes resembled the conversation he had with Jim Yorke as described in Ref. [21]. Before he left, he gave me some nice picturesque postcards from the Ukraine (Fig. 5). Incidentally, my Master's thesis in mathematics focuses on extending another theorem by Sharkovsky[j] to star-shaped trees[22].

We developed the first chaotic communication system that is inherently error-free[23] and built an implementation using Chua's circuit as the chaos generator. In this setup, the chaotic signal is modulated by the information signal by embedding the information source into the dynamics of the chaotic circuit. The modulated chaotic signal is then sent to the receiver. At the receiver, an identical chaotic circuit demodulates the signal to recover the information signal. At the 1993 NOLTA conference in Hawaii, I was supposed to demo this at the end of Leon's talk. We built two identical Chua's circuits, one embedded in a receiver circuit and one embedded in a transmitter circuit. A low-amplitude music signal comprising of a selection from Mozart's Eine Kleine Nachtmusik was fed into the transmitter and is completely masked by the chaotic signal. This combined signal is then sent via a coax cable of several feet to the receiver circuit to be decoded. I have checked all the setup at NOEL and also right before the talk. Of course Murphy's Law kicked in and when I turned on the system in an auditorium full of people, the sound that came out of the receiver's speakers is . . . absolute silence. Since Chua's circuit contains an active nonlinear resistor,[k] it consumes power and this was delivered via a battery. It took several minutes of profuse sweating to realize that somehow I must have unplugged the battery to conserve power and neglected to plug it back in for the demo. Luckily the crowd was patient enough and eventually I was able to demo the system successfully. Ever since then, whenever something goes wrong, I would always check the power switch or power supply, even when I am writing software code. A video demoing the implementations we made around that time is available at `http://youtu.be/1o2z5KVuB7s`.

Fig. 5. Postcards from Kiev, Ukraine.

[i] `http://en.wikipedia.org/wiki/Sharkovskii's_theorem`.

[j] Which states that a point on the interval is a limit point of a continuous map if and only if every open interval containing that point contains at least 3 points of some trajectory of the map.

[k] A circuit with only passive elements simply cannot be chaotic.

5. Fun and Games

It was not always just work at NOEL. As children of the video game generation, many of the students unwound by playing computer games. The World Wide Web was just in its infancy, and somehow I became the person responsible for downloading games demos from an FTP server onto our computers for our enjoyment. Maciej Ogorzalek, who is a frequent visitor, would copy these games demos from the PC for his young sons. Since Maciej spent some time "testing" these games out in the lab, we were always wondering whether he was downloading the games for himself to play.

This was also long before Massive Multiplayer Online Games (MMOG) such as World of Warcraft existed. One popular game we played was Xtrek, a (primitive by today's standard) MMOG that ran on Unix workstations. On one occasion we sneaked into a conference room to play Xtrek by projecting it onto the big screen, making us feel like Captain Kirk on the bridge of the starship Enterprise. One time we were playing Xtrek late at night on the computers in NOEL when Leon walked in to tell us about some ideas he had. Luckily he was so excited to tell us what he was thinking about that he did not notice all the moving spaceships on our computer screens. How I wish I had a boss key[l] at the time although a spreadsheet would not have made much sense either for the research that we were doing.

We did many other activities together, including Hong Kong movie nights[m] and The Simpsons marathon. Bert Shi and I were both part of the UC Berkeley Wushu team. I participated in the annual UC Berkeley Wushu competition one year and placed third in the beginner category. I should mention that there were a total of 3 entrants in that category. One year we had a lab picnic at the Berkeley marina and Philippe Deregel volunteered to prepare crêpes for all of us. Since Philippe is French, he pronounced crêpe like a Frenchman would, which sounded to my unsophisticated ears very much like a word in English that is much less palatable.

When Arnold Schwarzenegger came to the UC Berkeley campus to film "Junior", Ann Chen and I went to see the movie being made and found out just how tedious movie making is. We sat there for several hours only to see a few takes being shot over and over again. Reading countless papers and writing pages of code don't seem so boring after that.

One highlight over the years was being invited to Leon's house for fine cheese, dinner and drinks.[n] In addition to walls of bookcases filled with books on a diverse set of topics, Leon's house also contained many interesting artifacts from around the world that he has collected over the years. Fig. 6 shows me attempting to blow a horn (albeit unsuccessfully). Leon is a fine wine connoisseur[o] and his wine cellar is always a delight to visit. He would show us his extensive French wine collection including bottles of wine of size Nebuchadnezzar (i.e. 15 liters!) and his homemade earthquake-resistant wine holders.

Fig. 6. Futile attempt to blow an alpine horn at Leon Chua's house.

[l]http://en.wikipedia.org/wiki/Boss_key.

[m]One time we were watching the Jackie Chan film "Wheels on Meals" and my Spanish roommate was amazed that Lola Forner, a former Miss Spain, can speak fluent Cantonese in that film.

[n]One Spring I mistakenly told Bert Shi the wrong date for changing the clock ahead one hour to accommodate for Daylight Saving Time and he arrived to dinner late.

[o]He is a Chevalier (Knight) of La Confrérie des Chevaliers du Tastevin.

Fig. 7. NOEL members on our way to the Grand Canyon from Las Vegas.

Speaking of libation, a few of us had a memorable night drinking Russian vodka with Andrei Shilnikov and Dmitry Turaev. Apparently they believed that the brain needs to be rebooted once and a while to properly do research and he invited us for a joint reboot session.

While we went on conferences, we also took the opportunity to do some sightseeing. I had many fond memories of our trips to Las Vegas, Seattle, Hawaii and Sevilla with various members of NOEL (Fig. 7). I still remember traveling around the south of Spain with Ken Crounse and got a parking ticket, but the policeman was kind enough to dismiss it because we were tourists. The streets were quite narrow and we knocked off the side mirror covers of our car on one of these narrow streets. On our trip to Hawaii to attend NOLTA, one of the planned attendees could not go and thus there was another airline ticket for another member of the lab to attend. Leon suggested that Bert Shi should go, but half a second later realized that Bert was driving us all to the airport, so Ken Crounse would go instead. Bert tells me how he drove the entire lab to the airport, waved goodbye to everyone and returned by himself to an empty lab.

6. Lasting Friendships

One of the benefits of working at NOEL that I have always appreciated is the large number of people from all over the world who would visit the lab. Fig. 8 shows a snapshot of some lab members at some point in time in the early 1990's. Leon has an uncanny ability to bring together a great group of researchers and working at NOEL with Leon is a conduit through which many of the lab members form collaborative relationships and many of us continue to collaborate with Leon to investigate Chua's circuits, chaotic dynamics, piecewise-linear vector fields, nonlinear circuit theory and CNN[p] in universities and research institutions around the world. Árpád and Ildikó Csurgay calls this extended network centered at NOEL Leon Chua's *Global Invisible College*.[q]

There was a great diversity in the experiences of the members of the lab. There were students in various stages of their academic programs,

[p] You can find a CNN world map circa 1998 in Ref. [24].

[q] Ref. [24], `http://www.eecs.berkeley.edu/~chua/gustav_award.pdf`.

Fig. 8. Some of the members of NOEL during the early 1990's.

Fig. 9. Commencement 1992 at UC Berkeley. Degrees of various kinds were handed out. The 4 students pictured received 1 Ph.D. degree, 1 M.S. degree and 2 B.S. degrees in total. From left to right: Stanley Pau, me, Michael Peter Kennedy, Leon Chua and Kevin Halle.

postdoctoral researchers, junior faculty members, and world-renowned researchers. There were also people who had done other things before turning their attention to science and engineering. For instance, Ray Brown had a successful career as a professional photographer before deciding to earn a doctorate in mathematics studying chaotic dynamics. Many of the friends I have made during this time I would see again in various conferences and workshops and many of us are still in contact today and reminisce about the time we spent in NOEL whenever we meet.

I have had the pleasure of meeting the following (surely incomplete) list of individuals during my

6-year stay at NOEL, a list that reads like a who's who in science and engineering:

- Vadim Anishchenko
- Marco Balsi
- Vladimir Belykh
- Ray Brown
- Ann Chen
- Ken Crounse
- Jose Cruz-Moreno
- Philippe Deregel
- Alexander Dmitriev
- Yuri Andreyev
- Tetsuro Endo
- Orla Feely
- Marco Gilli
- Kevin Halle
- Anshan Huang
- Makoto Itoh
- Claus Kahlert
- Michael Peter Kennedy
- Ljupco Kocarev
- Géza Kolumbán
- Tibor Kozek
- Sergei Kuznetsov
- Domine Leenaerts
- René Lozi
- Yuri Maistrenko
- Vladimir Maistrenko
- Gottfried Mayer-Kress
- George Moschytz
- Vicente Muñuzuri
- Vladimir Nekorkin
- Maciej Ogorzalek
- Ulli Parlitz
- Stanley Pau
- Ladislav Pivka
- Andras Radvanyi
- Csaba Rekeczky
- Tamas Roska
- Nikolai Rulkov
- M. Safanova
- Alain Shang
- Alexander Sharkovsky
- Bert Shi
- Andrey Shilnikov
- Leonid Shilnikov
- Krzysztof Slot
- Patrick Thiran
- Dmitry Turaev
- Akio Ushida
- Angel Rodriguez-Vazquez
- Joos Vandewalle
- Tao Yang
- Akos Zarandy
- Alexander Zheleznyak
- Guo-Qun Zhong

Even after I left NOEL, I would meet others who have worked at NOEL before or after me and invariably we have something in common to talk about.

6.1. *On a personal note*

A most happy outcome during this time was that I had the chance to meet Ann Chen, a student in Leon's circuit theory class. She saw the genuine joy and excitement Leon displayed while delivering his lectures and was also impressed by the intricate set of transparencies he used. By overlaying various transparencies in different orders, Leon was able to present his material dynamically, rivaling the animations we see today in lectures delivered via Powerpoint. She received an A+ in that class and was fascinated by CNN that Leon used in his class lectures. She was especially intrigued by the maze-solving CNN that she had to do as an exercise problem and wanted to work in NOEL on a senior project about CNNs. She would come to the lab quite often and a budding romance blossomed and she is now my best friend and wonderful wife. So in addition to being my mentor, friend and colleague, Leon was also a matchmaker!

7. Impact of NOEL on My Development as a Scientist

Another thing I especially appreciated during my time at NOEL is how Leon was involved in my growth intellectually as a researcher. When I wrote my first paper[9] with him shortly after arriving at Berkeley, he would have many comments and suggestions on every page of the manuscript. With time, he would let me be more independent in my research allowing me to pursue my own research paths that sometimes led to dead ends. In retrospect this pedagogical approach is very effective. When I first needed help he would provide more guidance but later on as I matured as a researcher he would let me have more independence and find my own ways and make my own mistakes. As a mentor, Leon always treated me like a friend. He is down to earth with an avuncular personality. In addition, he is a great motivator and has a superb work ethic that inspires others to strive for the same level of excellence. I try to use the same philosophy in mentoring others today.

Interdisciplinary research is essential today where many problems simply cannot be solved with expertise in only one technical area. The opportunities to interact with many people of different disciplines, experience and culture at NOEL trained me well for that. Today, I work with teams consisting of mathematicians, physicists, electrical engineers, computer scientists and software programmers on a daily basis.

Finally, doing research is akin to playing a chess game. There are many possible paths, but only a few (or only one) give you the result that you are looking for, and many paths lead you to wrong results. Since we have a limited amount of time, we can only pursue a subset of paths. When we first started in our careers we did not know which were the best paths to take and thus many of our trials led to dead ends. But as we mature intellectually and are more experienced, we start to know which paths to take and which ones to avoid. But sometimes we are overconfident in our skills and avoid paths that we feel are too trivial, are too operose[r] or are pointing to the wrong direction. But these paths could lead to new insights and perhaps the solution we are looking for. One lesson that Leon taught me is that sometimes we need to take an approach that initially looks unrewarding or has been tried before. We might be surprised where it could lead.

7.1. *Learning additional skills*

Since Leon was invited to give keynote lectures all over the world, he would ask us to help him create colorful computer graphics demonstrations. This provides me with the opportunities to learn additional skills. The lab had analog oscilloscopes and we need to take an image of the chaotic attractor generated from an electronic circuit shown on the oscilloscope screen as a phase portrait. Since the attractor is not periodic, the traces it made are not repeated and it took us some trial and error to find out the right brightness setting on the oscilloscope and the right exposure time on the film to obtain a good image of the attractor that is not overexposed or underexposed and reveal the salient features of the complex patterns in the attractor. In addition, we learned how to create 3D models and animation of strange attractors on an SGI workstation, shoot high resolution images on a full-frame camera (we still use film in those days) and produce video of experimental and simulation results. In NOEL we had a small set of professional-grade audio/video equipments to capture/edit video, insert/synchronize audio/soundtracks/narration, and record onto video tape. Since Leon's lectures are given in many different countries, we had the capabilities to record video in NTSC, PAL and SECAM formats. Alas, with the proliferation of digital video and desktop video editing software, such video production skills are becoming archaic and obsolete.

After Michael Peter Kennedy left NOEL, I became in charge of IT at NOEL, ordering computer equipment and ensuring that the computers were running smoothly for everyone. For serious problems, we would still rely on the IT staff at Cory Hall, but I learned to fix simple problems myself by poring over the various manuals in the Lab.[s]

As we are investigating complex phenomena that occurs in various areas of science and technology, we also learned how to mix various chemicals together to reproduce the Belousov-Zhabotinsky reaction[t] and generate circular wave patterns on a Petri dish.

8. Finally it is Time to Go

After I graduated in 1995, I remained as a postdoc at NOEL for the first part of 1996. When I received several job offers including an offer to join the Mathematical Sciences department in IBM Research as a Postdoctoral Fellow,[u] I asked Leon for his advice. Recall that only a few years earlier, IBM had major losses and had been written off for dead. Leon showed me an article in Business Week describing how Lou Gerstner is turning the company around[v] and he told me that it is a good company to work for. I followed his advice and I believe that was the right decision to make.

[r] Rather than admitting laziness, usually we appeal to Occam's Razor or the KISS (Keep it Simple, Stupid) principle.

[s] The Web and social media were not that well developed at the time, so finding answers to simple IT problems was more involved.

[t] http://en.wikipedia.org/wiki/Belousov-Zhabotinsky_reaction.

[u] http://www.research.ibm.com/goldstine/.

[v] http://www.businessweek.com/1995/44/b34481.htm.

I have had many wonderful memories at NOEL and I am grateful to have had the opportunity to be part of a wonderful team of people sharing the same appetence and passion to work with Leon at NOEL.

Acknowledgments

This anamnesis is constructed from my imperfect unaided memory along with contributions from friends and colleagues. Any errors are solely my responsibility.

References

[1] P. E. Blankenship, Comments on "a two's complement parallel array multiplication algorithm", *IEEE Transactions on Computers.* **23** (12), 1327, (1974).

[2] L. O. Chua and L. Yang, Cellular neural networks: Applications, *IEEE Transactions on Circuits and Systems.* **35**, 1273–1290, (1988).

[3] L. O. Chua and L. Yang, Cellular neural networks: Theory, *IEEE Transactions on Circuits and Systems.* **35**, 1257–1272, (1988).

[4] L. O. Chua and T. Roska, *Cellular Neural Networks and Visual Computing: Foundations and Applications.* (Cambridge University Press, 2005).

[5] T. Matsumoto, A chaotic attractor from Chua's circuit, *IEEE Trans. Circuits Syst.* **CAS-31**(12), 1055–1058, (1984).

[6] L. O. Chua, The genesis of Chua's circuit, *Archiv für Elektronik und Übertragungstechnik.* **46**(4), 250–257, (1992).

[7] L. O. Chua and T. Lin, Chaos in digital filters, *IEEE Transactions on circuits and systems.* **35**(6), 648–658 (June 1988).

[8] C. W. Wu and L. O. Chua, Properties of admissible symbolic sequences in a second order digital filter with overflow nonlinearity, *International Journal of Circuit Theory and Applications.* **21**(3), 299–307, (1993).

[9] L. O. Chua and C. W. Wu, On the universe of stable cellular neural networks, *International Journal of Circuit Theory and Applications.* **20**(5), 497–517, (1992).

[10] T. Roska, C. W. Wu, and L. O. Chua, Stability of cellular neural networks with dominant nonlinear and delay-type templates, *IEEE Transactions on Circuits and Systems-I: Fundamental Theory and Applications.* **40**(4), 270–272 (Apr., 1993).

[11] T. Roska, C. W. Wu, M. Balsi, and L. O. Chua, Stability and dynamics of delay-type general and cellular neural networks, *IEEE transactions on circuits and systems-I: Fundamental theory and applications.* **39**(6), 487–490 (June, 1992).

[12] R. N. Madan and C. W. Wu. Introduction to experimental chaos using Chua's circuit. In ed. R. N. Madan, *in Chua's circuit: A paradigm for chaos*, pp. 59–89. World Scientific, Singapore, (1993).

[13] C. W. Wu and L. O. Chua, Synchronization in an array of linearly coupled dynamical systems, *IEEE Transactions on Circuits and Systems–I: Fundamental Theory and Applications.* **42**(8), 430–447, (1995).

[14] M. W. Hirsch, Systems of differential equations that are competitive or cooperative II: Convergence almost everywhere, *SIAM J. Math. Anal.* **16**, 423–439, (1985).

[15] M. W. Hirsch. Convergence in neural nets. In eds. M. Caudill and C. Butler, *IEEE First International Conference on Neural Networks*, vol. 2, pp. 115–125, (1987).

[16] M. W. Hirsch, Convergent activation dynamics in continuous time networks, *Neural Networks.* **2**, 331–349, (1989).

[17] L. P. Shil'nikov, A case of the existence of a denumerable set of periodic motions, *Sov. Math. Dokl.* **6**, 163–166, (1965).

[18] L. P. Shilnikov, Chua's circuit: Rigorous results and future problems, *IEEE Transactions on Circuits and Systems-I.* **40**(10), 784–786, (1993).

[19] A. N. Sharkovskii, Coexistence of cycles of a continuous map of the line into itself, *Ukrainian Math. Z.* **16**, 61–71, (1964).

[20] A. N. Sharkovsky, Chaos from a time-delayed chua's circuit, *IEEE Transactions on Circuits and Systems-I.* **40**(10), 781–783, (1993).

[21] J. Gleick, *Chaos: Making a New Science.* (Penguin, 1988).

[22] C. W. Wu. Topological dynamics of continuous maps on the n-od. Master's thesis, University of California at Berkeley, (1994).

[23] C. W. Wu and L. O. Chua, A simple way to synchronize chaotic systems with applications to secure communication systems, *International Journal of Bifurcation and Chaos.* **3**(6), 1619–1627, (1993).

[24] K. K. Phua, Ed., *Ad Sexaginta Annos Prospere Celebrandos: Celebrating Six Decades of Achievement Dedicated to Leon O. Chua On The Occasion Of His Sixtieth Birthday.* (World Scientific, 1998).

CHAPTER 11

BURSTING IN CELLULAR AUTOMATA AND CARDIAC ARRHYTHMIAS

GIL BUB

Department of Physiology Anatomy and Genetics,
BHF Centre of Research Excellence,
University of Oxford, Oxford, UK

ALVIN SHRIER and LEON GLASS

Department of Physiology,
McGill University, Montreal, Canada

The mechanisms underlying the initiation and continuation of abnormal cardiac arrhythmias are incompletely understood. In this chapter, we summarize work that shows how simple cellular automata models of excitable media can display a range of interesting dynamical behavior including spontaneous bursts of reentrant spiral activity. Since the model incorporates basic physiological properties of excitability, heterogeneity, localized pacemakers, and fatigue in a schematic way, the model captures generic physiological dynamics that should be broadly observed in experimental and clinical settings as well as in more realistic mathematical models.

1. Introduction

Reentrant arrhythmias — where an excitatory wave travels in a circuit, repeatedly reactivating its cardiac substrate — are a leading cause of death in the developed world. The transition from normal sinus rhythm, where periodic wavefronts spread from a central pacemaking site to trigger cardiac contraction, to reentry has been the subject of numerous studies over the last century. Reentrant tachycardias are often self terminating, spontaneously converting back to a healthy sinus rhythm. These nonsustained ventricular tachycardias (NSVT) are common in heart disease patients (e.g. 33–80% of heart failure patients) [11, 39], but their incidence in the general population is unknown as the arrhythmic episode may be too brief to be noticed. However, the onsets of NSVTs are unpredictable, and termination of the arrhythmia is not guaranteed — sustained tachycardias can quickly evolve to a highly disorganized rhythm called fibrillation, which if in the ventricle is fatal unless treated within minutes. Predicting the appearance and duration of these ventricular tachycardias remains a challenging problem.

Repetitive episodes of NSVT can be broadly described as generating a bursting rhythm. Bursting, which can be defined as repeated periods of rapid activity interdispersed by periods of quiescence, is displayed by many different excitable cell systems. In general, there are two ways that bursts can be generated in groups of cells: one or more cells in a population have the property of endogenous bursting, or else the bursting arises from interactions between cells. Endogenous bursting cells have intrinsic conductances that produce slow potential waves, which activate voltage-sensitive conductances resulting in a spike train of activity. Bursting activity can also be generated inl networks of cells without endogenous bursters. Most network bursters depend on the activity of inhibitory connections on spontaneously firing cells: low frequency patterns are generated by alternate phases of rapid spontaneous firing and suppression of activity by activated neighbours [8, 27, 35]. Other variants use networks to average ionic currents or rely on a fixed delay excitation mechanism to generate bursting [28]. Bursting in cardiac disease states can also be thought of as either originating in from an endogenous bursting focus, or caused by emergent behaviour from large networks of coupled cells. Both focal and reentrant mechanisms have been reported during nonsustained atrial tachycardias, whereas reentrant mechanisms dominate in ventricular tachycardias and fibrillation. Further, diseases such as fibrosis and ischemia that promote reentrant arrhythmias introduce scars and regional changes in conductive properties of the muscle. The study of repetitive NSVT must therefore include an understanding of wave propagation through cardiac tissue with heterogeneities.

The propagation of electrical activity in cardiac muscle involves the interaction of different ion species across a variety of ion channels and diffusion of charge through a heterogeneous substrate with dynamically changing conductances. Despite complexity inherent in conduction at microscopic scales, the heart has been approximated as a continuous excitable media in many simulation studies. Excitable media can be broadly defined as spatially distributed systems that display undamped propagating waves which annihilate only when they reach a boundary or collide with other waves. Excitable media are most naturally represented as partial differential equations (PDEs). The number of differential equations required to represent a cell in the system may be large (see [14] for a recent review). For example, the Iyer-Mazhari-Winslow model has 67 variables and hundreds of parameters. The rational for developing such complicated models is the assumption that inclusion of all known components is needed to achieve an accurate simulation of the behaviour. Although this modelling approach is justified in many situations, the high dimensionality of such models hinders intuitive understanding of the underlying dynamics. Simplified low dimensional PDE models that encompasses many characteristics of excitable tissue including threshold behaviour, refractoriness, and dispersion effects have been developed; the two variable FitzHugh Nagumo model first formulated in 1961 continues to be used to simulate cardiac dynamics, and more recently, the Fenton-Karma three variable model [15] and its four variable counterparts [7] which also preserves cardiac tissue properties including action potential upstroke velocity and shape, different time scales for depolarization and repolarization, and action potential duration and conduction velocity restitution curves.

Excitable media can also be simulated using discrete time models with simple update rules. These models consists of a grid of points where each point takes on any of a small set of discrete states, and there is set of rules that determine future states of points in the grid based on the present state of the grid. Discrete models, also called cellular automatons (CAs), [32] have been applied to a variety of physical processes. The CAs states have values that reflect the functional characteristics of each grid point as opposed to the value of particular variables. Space is discretized by mapping the continuous effects of diffusion to simple rules based on neighbourhood interaction. Cellular automata models offer a level of abstraction over differential equation based models which allows them to represent many different natural systems. Because of their simplicity cellular automata models have been widely used. Moe, Rheinboldt and Abildskov [31] showed reentry in a simple cellular automata model at an early stage, and Greenberg and Hastings [21] systematized cellular automata models and demonstrated conditions leading to reentrant oscillations — see the original article for extensive references to the early literature.

Simple cellular automata models of excitable media have two advantages over differential equation models; they are more intuitively transparent, and computer simulations run far faster. Their simplicity come at a price, however. Radius of curvature and dispersion effects, which are common to all excitable media in nature, are absent in many simple CA models. Second generation CAs were formulated to account for these deficiencies. Models with thresholds greater than 1 (more than one neighbour must be active to induce a resting cell to the excited state) generate curvature effects if neighbourhood size or the number of excited states is greater than one [17]. Fast and Efimov [12] include dispersion effects by adding a weighing function (in which a cell's relative contribution to excitability decreases as a function of distance from the wavefront) and another state called the relative refractory period (where cells have a higher threshold than resting cells but can still be excited.) Gerhardt and colleagues [17] add dispersion by making threshold a decreasing function of time. Several methods have been proposed to remove anisotropic effects. Marcus and Hess [30] randomise the grid by leaving many (random) grid points empty in circular neighbourhoods. Kurrer and Shulten [29] select a random subset of a square neighbourhood as the active grid. Several second generation models define diffusion coefficients [18, 45] and use cellular automata to simulate continuous differential equation based excitable media models.

Several novel approaches which combine continuous and discontinuous dynamics have been developedcellular automata!hybrid approaches. Coupled map models, which are discrete in space but continuous in time, have been used to analytically derive spiral wave stability [26]. Hybrid models, which include both discrete switching between a cell's states (refractory, excitable, active) and continuous variables that can faithfully reproduce AP shape within each state, have been recently applied to the study of excitable media [47]. Hybrid automata preserve many mesoscopic properties of cardiac tissue, while being almost an order of magnitude more computationally efficient than their differential equation counterparts. Another computationally efficient approach are the use of scalar coupled map lattice models: here a single delay equation accurately reproduces action potential shape, and the models can be directly fit to real data of excitable waves, independent of the underlying cellular dynamics [22].

Reentrant waves of excitation in spatially extended systems such as the heart ventricle either travel around an unexcitable obstacle such as a scar or, in homogeneous substrates, form a characteristic spatiotemporal pattern called a spiral wave. Spiral waves (also called vortices and rotors) form naturally when a wavefront travels around a pivot point repeatedly re-exciting the substrate. A spiral is created as points further from the pivot have to travel farther to make a circuit, and therefore lag behind the wave at the center. Spiral waves are a general property of excitable media, and have been observed in the BZ reaction [46] intact [10] and cultured cardiac tissue, retinal [20] and neural [25, 38] preparations, and aggregating slime mold (Gerishch, 1986). Initiating a spiral wave involves generation of a wavefront with a free end, which can curve inward to re-excite tissue forming a wave with a spiral geometry. Free ends (also called broken wavefronts) can be generated in both homogeneous and heterogeneous excitable media.

Spiral waves can be generated in initially homogeneous media in several ways. Two sequential waves travelling perpendicularly to each other can generate a free end when the second wave collides with the refractory wake of the first. The resulting broken wave forms a spiral as the refractory substrate becomes excitable. This mechanism has been used to generate spirals in excised cardiac tissue [10]. Another mechanism is thought to play a significant role in systems displaying spontaneous oscillations leading to frequent target excitation patterns. Here, a region in the wake of travelling wave becomes excited, generating a wave moving outward from the excitation point. The excitation is blocked in the direction of first wave, which results in a semicircular wavefront with two free ends which curve inward to generate a pair of mirror-image spiral waves. This mechanism has been observed in spontaneously oscillating BZ reactions and slime mold preparations [33].

Spiral wave formation in heterogeneous media depends on how waves interact with unexcitable obstacles. Both the excitability of the tissue and the shape and size of the obstacle play a role on whether the wave detaches to generate a free end. There are two heuristic rules for predicting the effect of a planar wave interacting with a single obstacle. First, the obstacle must be greater in size than the width (sometimes termed wavelength, the product of wave speed and action potential duration) of the wavefront to induce a wavebreak. Second, the wave can detach from the obstacle generating a free end only if the curvature of the obstacle at the detachment point is greater than the inward propagation angle of the propagating free end, which in turn depends largely on excitability of the substrate. Situations with multiple obstacles or randomly distributed heterogeneities are more complex [4]. After a spiral wave is initiated, it can either persist, breakup into multiple wavelets, or die out. Spiral wave breakup is a complex phenomenon that has been dealt with in detail by several researchers (see [13] for an excellent review), however it is unlikely to play a role in bursting rhythms such as paroxysmal NSVT as it doesn't result in cessation of activity. Spiral wave termination occurs in homogeneous and heterogeneous media when the wave tip travels (a process called meander in homogeneous media, and drift in heterogeneous media) leading to a collision with an unexcitable boundary [12,34].

2. Network Bursting in Cultured Cardiac Monolayers

Heart cell monolayers are thin layers of tissue grown in culture dishes from embryonic or neonatal cardiac cells. Cardiac cells from very young animals have the capacity to form gap junctional connections with neighbouring cells in culture. After a few days in culture, embryonic cardiac cells are capable of supporting propagating waves of excitation over long distances. Cardiac monolayers were popular thirty years ago as model systems of two-dimensional conduction. More recently, the availability of potential mapping techniques have renewed interest in cultured monolayers, as they allow controlled environments for studying conduction on microscopic [36] and macroscopic [43] scales.

We measured calcium transients from dispersed monolayer cultures of embryonic chick heart cells as a function of age in culture, plating density, and gap junction connectivity. We found that while periodic target patterns predominate for well connected cardiac substrates, bursting patterns were frequently observed in relatively poorly connected monolayers (Fig. 1). Bursting rhythms were associated, in most

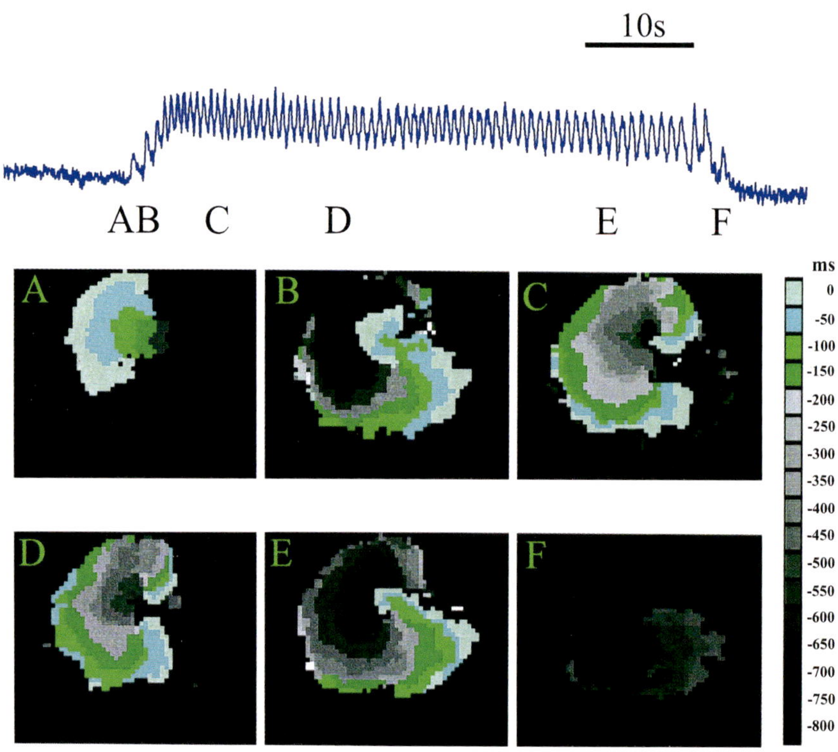

Fig. 1. An example of spiral wave driven bursting dynamics in the cultured cardiac monolayer. Top panel shows optically recorded calcium activity; Bottom panels show isochronal maps of initiation and termination (A-F) of a double armed spiral wave. Each panel is approximately 1 cm square. See for experimental details. (Reprinted from copyright PNAS).

cases, with the spontaneous onset and offset of spiral waves of activity. Spiral waves were initiated by a wave that propagates in a unidirectional fashion: waves blocked at a point near the site of initiation leading to unidirectional propagation and a wavefront free end. The free ends propagate back toward the initiation site, and, as more tissue is recruited into the wavefront, are able to pass through previously unexcitable regions leading to a spiral wave. The spiral wave's period increased over time (by up to 80%), and eventually (after on average 30 rotations) the spiral wave terminates. Wave termination appears to be due to a loss of excitability of the cardiac substrate, leading to the wave tip to collide with the border of the monolayer.

Although in most cases bursting was associated with clearly visible reentrant waves in our experimental system, in some cases bursting target waves were observed [6]. Since our imaging system has relatively low spatial resolution, we could not rule out that reentry occurred at smaller (cellular) scale. As cardiac cells tended in our hands to be either quiescent or beat periodically, we primarily focused our attention on the contribution of reentry to bursting rhythms. However, recent results from other groups strongly suggest that bursting is an intrinsic property of cardiac cells. Harada [23] showed, using an elegant experimental system which captured long term recordings of single cells and small cell clusters at high magnification, that cardiac cells display bursts of activity, and can even spontaneously transition between regular periodic activity bursting rhythms. Tong and Holden [42] showed that a cardiac cell model displays bursting dynamics for certain parameters. Finally, Himel [24] showed that pharmacological agents can induce bursting rhythms in the form early and delayed after depolarizations in isolated neonatal cardiac cells. These results suggest that communication between cardiac cells may not be relevant to the genesis of bursts in this system, and that spiral waves are the result of, but do not cause, bursting. However, it is difficult to reconcile this interpretation with the presence of cohesive waves in culture, the increasing period of rotation during bursts, or the prevalence of non bursting target waves in more well connected monolayer systems. Further studies, such as the use of higher resolution imaging equipment and simulations with populations of bursting cardiac cells, are needed to resolve this issue.

3. Simulating Network Bursting with CAs

Simulating the dynamics observed in the isotropic cell culture experimental system posed several challenges. First, the monolayer culture is heterogeneous at microscopic space scales, with cell-free patches and dense tissue areas. We assume that these heterogeneities play a critical role as the unidirectional propagation observed at initiation sites doesn't occur in homogeneous excitable systems. It's possible to use PDE and ODE models to simulate cardiac propagation at microscopic scales, however, it would be difficult to draw conclusions from a single simulation run since heterogeneities in our system are distributed throughout the tissue and do not have a single representative geometry. Furthermore, simulations at microscopic scales typically use many subcompartments per cell [40] and are therefore computationally expensive. Up-scaling cell-level simulations to macroscopic scales while running a sufficient number of geometries to gain confidence in the generalizability of the results is therefore impractical. Our group [3, 41] and others [1] have used a more simplified approach, consisting of a uniform reaction diffusion system with randomly distributed defects at simulation grid points. However, the physical interpretation of a single grid point defect in this case is difficult: in general, the underlying lattice spacing should be an order of magnitude smaller than the heterogeneities being simulated in order to avoid artefacts. Since we are primarily interested in how network activity generates bursting, we opted to use CAs as cellular activity is abstracted to a small number of essential states. The computational efficiency of CAs allow parameter sweeps and randomly generated geometries to be quickly simulated, and the simplicity of the formulation facilitates analysis of wave speed and propagation failure from first principles [3, 5].

A simple cellular automata model was developed which demonstrates that the addition of a limited number of rules can result in bursting behaviour in groups of spontaneously active, excitable, but nonendogenously bursting cells. Spontaneous activity is simulated by giving all grid points a probability of firing when excitable. The slowing and eventual annihilation of the spiral is modelled by adding a fatigue term. Fatigue refers to a phenomenon observed in cardiac tissue where periods of sustained rapid activity adversely affect conductive properties of the substrate. These effects are thought to be due to the gradual build-up of ion species such as sodium, potassium or calcium [2,9,28,44] which move across cell membranes during each action potential. Rapid activity overwhelms the cells ability to restore the ions concentration to steady state levels. The resulting imbalances are indirectly responsible for secondary effects such as hyperpolarisation or reduced cell coupling, which alter functional characteristics (such as period and conduction velocity) of the tissue. These long term effects are referred to as fatigue in conductive tissue [2] and overdrive suppression in pacemaking tissue [28]. Fatigue is simulated in the cellular automata models by decreasing a cell's ability to be stimulated by its neighbours as a function of its recent activity.

4. Bursting in a Greenberg-Hastings Model with Spontaneous Activity and Fatigue

We study a very simple cellular automaton modified from early proposals by Greenberg-Hastings [21] that we believe captures the effects of a heterogeneous cellular structure. The Greenberg-Hastings model is formulated as follows. Each site at time **t** is assigned a state, $u_j(t)$, where the subscript **j** refers to the cell number. The neighbourhood of a given site is determined using a square (Moore) metric of diameter **r**. The state is an integer: 0 is a rest state; states $1, 2, \ldots, E$ are excited states; states $E+1, E+2, \ldots, E+R$ are refractory states; state $(E+R+1)$ is identified with the rest state 0. The update rule for the state of a site is as follows: If $(1 <= u_i(t) <= E+R)$ then $u_i(t+1) = u_i(t) + 1$; and if $(u_i(t) = 0)$ then $u_i(t+1) = 0$ unless u_i is excited by its neighbours, or becomes spontaneously active with probability p_s.

A cell becomes excited if it is not refractory a sufficient number of excited cells is present in a defined neighbourhood around the cell. These models display bursting if each cell is given a probability of firing when $u_i(t) = 0$, and if excitability is modified by a fatigue term which increments for each cell every time it fires, and decays away exponentially at other times. The fatigue term in this formulation is added to the threshold for excitation.

The model shows initiation and annihilation of spirals similar to experimental observations. Impulses are initiated by the chance simultaneous firing of several sites in close proximity to each other, so that additional sites in the neighbourhood fire. Local heterogeneity prevents spread of excitation in all directions, resulting in unidirectional propagation. Two wave free ends double back to form doubled armed spiral waves. The build-up of fatigue eventually makes propagation impossible and the rotor is annihilated.

Spiral wave initiation in cellular automata depends on asymmetric initiation of a wave. Initiation of a wave occurs when a number of grid points greater than or equal to threshold K_{exci} are active within a neighbourhood. As the probability of initiation is low, we assume that at any initiation site only K_{exci} grid points are active, and that there are no grid points close to the initiation site that are refractory. If this is the case, unidirectional propagation can only occur if one or more of the initiating grid points become refractory before the target wave advances past the immediate neighbourhood of the initiation site. Consider a GH model with a Moore neighbourhood of size 1, $K_{exci} = 3, E = 5$, and $R = 6$. There are 56 $((8 \cdot 7 \cdot 6)/6!)$ different positions that three active grid points can have in a given 3×3 neighbourhood. Cells can have activation states ranging from 1 to 5 in each of these configurations. At least one grid point has a value of 1 at the moment of initiation. There are three possible locations for this grid point, giving $3 \times 5 \times 5 - 14 = 61$ different permutations (where 14 is the number of identical repeats) for the 56 spatial configurations at the initiation site. Of this $3416 (56 \times 61)$ different configurations, a certain portion give rise to unidirectional propagation and rotor waves, while the rest result in target patterns.

The probability that wave initiation results in a spiral can be determined by iterating the model for each of the 3416 possible configurations. As this is impractical, one spatial configuration (with 61 different configurations of activation state) is analysed in detail. The representative spatial configuration is chosen so that the results can be generalized to the remaining 55.

The two examples presented below demonstrate that different spatial configurations generate different patterns of activation even when the state of each grid point at wave initiation is similar. The numbers give the activation state (1 to 5) of each grid point. The symbol '+' denotes the extent of the neighbourhood at $t = 0$. Refractory grid points are represented as 0 and excitable grid points are left blank.

The first configuration gives rise to a target. The second results in unidirectional propagation that eventually evolves to a double armed rotor.

```
  t=0        t=5          t=10

                            1
                          12321
                         2345432
             121        135000531
  +++       13431       240000042
  141       25052      13500 00531
  +++       13431       240000042
             121        135000531
                         2345432
                          12321
                            1
```

Example 1.

```
  t=0        t=5          t=10

  ++4+       303         1350 0531
  1+1       15451        240000042
  +++        232         135000531
              1           2450542
                          1234321
                            121
```

Example 2.

Once initiated, the advancing wave is vulnerable to the effects of grid points becoming refractory only while the wave is close to (within one neighbourhood distance of) the initiating grid points. The configuration that results in the fastest propagating wave is more stable than others because the wave is in the neighbourhood of the initiating grid points for the least amount of time.

The configuration in which three grid points are activated in a row (as is shown in the first example) is more stable than other configurations as it gives rise to more rapidly propagating wave fronts than other configurations. Other configurations activate grid points in the center of the neighbourhood before the wave is able to advance outward. An example of this behaviour is seen by comparing the two activation patterns above. By t = 5, the propagating wave from the 'three in a row' activation pattern has advanced farther (in all directions) than the second disjoint configuration.

Analysis of the 'three in a row' configuration gives an upper bound on the number of targets for all possible spatial configurations as it is least susceptible to the effects of its initiating grid points becoming refractory. Of the 61 different state configurations, 47 gave rise to propagating waves, 24 of which were targets, and 23 of which gave rise to unidirectional propagation and rotors. Therefore, slightly less than half of all propagating waves seen in this system will give rise to rotors. The actual number is higher as other configurations generate proportionately more rotors than targets. For example, the configuration:

$$\begin{matrix} + & + & \mathbf{x} \\ + & + & + \\ \mathbf{x} & \mathbf{x} & + \end{matrix}$$

where **x** can be any activation state 1 through 5, generates only 35 propagating wavefronts (from a total of 61 different configurations) of which 31 give rise to rotors. Only one true target is generated (when all grid points fire at the same time), the remaining three start out as unidirectional propagating wavefronts, but collapse into targets.

5. Bursting in CAs with Heterogeneity in Cell Spacing

The bursting Greenberg-Hastings CA has several limitations when compared to real cardiac tissue. First, the CA displays artefacts associated with the underlying grid. These artefacts are apparent at neighbourhood sizes close to the base lattice spacing: simulations with small Moore neighbourhoods have planar wave fronts with sharp corners instead of smoothly curved waves. Although the use of larger neighbourhood sizes alleviate this problem, larger neighbourhoods average out asymmetries due to random activation, and therefore do not display unidirectional propagation and bursting. Second, the Greenberg-Hasting formalism doesn't simulate local changes in cell connectivity correctly. For example, unidirectional propagation caused by asymmetries in cell loading, which is observed in real cardiac tissue [36], cannot be replicated in a CA in which an excitable cell determines activity based only on the number of neighbouring active cells. Finally, experiments in which a control parameter is continuously varied are difficult to simulate: we observed subtle dynamic changes when continuously lowering connectivity of cardiac cell networks using pharmacological agents, but connectivity in the Greenberg-Hastings CA can only be changed by integer values. We incorporated several modifications to the original formulation that address these issues.

In the original Greenberg-Hastings model all cells were on a square lattice. We employ a method first used by Marcus and Hess [30] and introduce spatial heterogeneity in the cell density by perturbing the location of each cell. Given the original coordinates (x_0, y_0) with a base grid spacing of 1, the new coordinates are $(x_0 + e_x, y_0 + e_y)$, where e_x, e_y are uniformly distributed in the interval $(-0.5, 0.5)$. We also assign a random weight for a cell's influence on its neighbours. Each cell is assigned a random weight, S_i, uniformly distributed in the range $(0.5 < S_i < 1.5)$. The results are qualitatively the same for constant S_i. However, if each weight S_i is a whole number, the values of neighbourhood size where propagation fails cluster around the square root of whole numbers. Incorporating random values for S_i is a simple way to introduce a continuously varying number of neighbours with varying neighbourhood size. This approach allows analytic determination of wave speed as a continuous function of neighbourhood size, as well as an estimate of parameters that give rise to blocked wave propagation [4, 5]. The model generates smooth wavefronts for large neighbourhoods while displaying irregular fronts with local asymmetries for small neighbourhood sizes. Wave propagation in the revised model more closely mimic what is thought to occur in real tissue, where conduction is irregular at cellular space scales but is continuous at larger space scales [40].

The condition for a cell 'i' to become excited is as follows. If $u_i(t) = 0$, then $u_i(t+1) = 1$ if the sum of all excited cells divided by the sum of all nonexcited cells in a neighbourhood is over a threshold θ. The use of the ratio allows the model to simulate the effects of current sinks that are known to modulate propagation in cardiac systems [3]. The model can be used to simulate a classic cardiac experiment that demonstrates the interplay between local geometry and cell-cell connectivity. Figure 2 demonstrates the effect of a sudden expansion on propagation. Due to asymmetric cell loading at the expansion point, waves can propagate into the narrow channel from the expanse, but cannot propagate in the reverse direction. The wave can propagate in both directions if coupling is decreased by changing the neighbourhood size from 3 to 1.8. This behaviour cannot be simulated using a standard Greenberg-Hastings CA as the effects of asymmetric cell loading on an excitable cell are not accounted for.

In order to terminate bursting activity, we add a fatigue term similar to that used in the original formulation: the fatigue term of cell 'i' at time (t+1) is $\gamma_i(t + 1) = \gamma_i(t) + F_{add}$ if $u_i(t + 1) = 1$ and $\gamma_i(t + 1) = g\gamma_i(t) \times F_{sub}$ if $u_i(t + 1) \neq 1$. The fatigue term is added to the threshold θ. The model bursts as a function of network connectivity. The model displays target patterns for large neighbourhood sizes. Spirals do not form for these

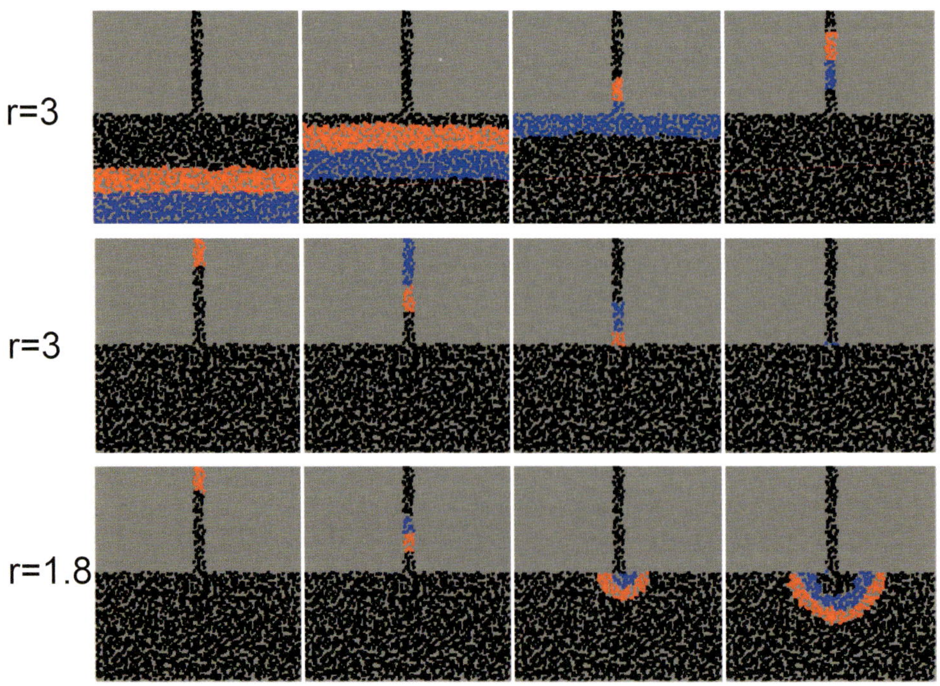

Fig. 2. Effect of geometry and connectivity on propagation. First row: a wave propagates from bottom of the grid to the top without blocking; Second row: a wave propagating from top of the grid to the bottom blocks at the expansion point; Third column: reducing r from 3 to 1.8 allows the wave to propagate through the expansion.

parameters because broken wave fronts seal at the initiation site and the wave does not develop wave breaks. The model displays reentrant dynamics for smaller neighbourhood sizes. Unidirectional block at the initiation site generates spiral waves. In contrast to the simple Greenberg-Hastings model used in earlier studies, reentrant activity is often initiated from a single active cell. Here, the mechanism for unidirectional propagation in the randomized CA is asymmetry in connectivity at the single cell level (see Fig. 2). The spiral tip tends to increase its meander as fatigue accumulates. The spiral tip eventually collides with the boundary of the simulation grid, terminating activity. This process repeats for some parameters, giving rise to a bursting rhythm. A parameter space map for the different dynamic regimes of this model is given in [5].

6. Network Bursting in CAs in the Absence of Reentry

The cellular automata model can also display triggered activity for large neighbourhood sizes. We explored the dynamics of the CA discussed in the last section, but here a small subset of cells beat with a fixed period randomly chosen within an interval instead of having all cells have a low probability of spontaneously firing [6]. Oscillators are phase reset when excited by neighbouring cells. Triggering in the model occurs when a wave acts to synchronise pacemakers, which increases the probability of subsequent synchronous activations. Pacemakers desynchronize over time because faster pacemakers fire before others and are subsequently refractory to phase resetting. We quantified the dynamics of the model by running the simulation for a large number of randomly assembled grids for different neighbourhood sizes. The potential for a grid to generate triggered bursting was measured by initiating a wave at one edge of the grid, and testing whether a triggered wave is generated after a short interval. The probability for generating spiral wave bursts is highest at $r = 1.65$ while the probability for triggered target wave generation is highest at $r = 2.1$ [6]. Triggered waves occur relatively infrequently: spontaneous reentry occurs in three times as many randomly generated simulation grids [6]. In contrast to spiral wave bursting, triggered wave bursting does not require addition of a fatigue term for spontaneous termination.

7. Conclusions

Initiation of bursting in CAs depends on the interplay between local differences in geometry and the connectivity of the network. Although cardiac tissue may be more accurately represented at the cell level with biophysically realistic continuous partial and ordinary differential equation models, the computational efficiency of CAs make them well suited for modelling the effects of randomly distributed heterogeneities, asymmetries in connectivity and variation in local spontaneous activity. The inherent stochasticity of paroxysmal cardiac arrhythmias such as nonsustained ventricular tachycardia may be a reflection of similar variability at microscopic space scales. If so, CAs can give valuable insights into arrhythmogenesis in diseased tissue.

Acknowledgments

Gil Bub acknowledges support from the UK Medical Research Council and the BHF Centre of Research Excellence, Oxford (RE/08/004).

Alvin Shrier and Leon Glass thank NSERC, MITACS (Mprime) and the Canadian Heart and Stroke Foundation for financial support.

References

[1] Alonso S, Barr M & Kapral R (2009). Effective medium approach for heterogeneous reaction-diffusion media. *J Chem Phys* **131**, 214102.

[2] Billette J, Metayer R & St-Vincent M (1988). Selective functional characteristics of rate-induced fatigue in rabbit atrioventriular node. *Circ Res* **68**, 790–799.

[3] Bub G, Shrier A & Glass L (2002). Spiral wave generation in heterogeneous excitable media. *Phys Rev Lett* **88(5)**, 058101 (1–4).

[4] Bub G & Shrier A (2002). Propagation through heterogeneous substrates in simple excitable media models. *Chaos* **12(3)**, 747–753.

[5] Bub G, Shrier A & Glass L (2005). Global organization of dynamics in oscillatory heterogeneous excitable media. *Phys Rev Lett* **94**, 028105 (1–5).

[6] Bub G, Tateno K, Shrier A & Glass L (2003). Spontaneous initiation and termination of complex rhythms in cardiac cell culture. *J Cardiovasc Electrophysiol* **14(10)**, s229–s236.

[7] Bueno-Orovio A, Cherry EM & Fenton FH (2008). Minimal model for human ventricular action potentials in tissue. *J Theor Biol* **253(3)**, 544–560.

[8] Chay TR (1990). Electrical bursting and intracellular Ca^{2+} oscillations in excitable cell models. *Biol Cybern* **63**, 15–23.

[9] Cohen J, Fozzard H & Shue SS (1982). Increase in intracellular sodium ion activity during stimulation in mammalian cardiac muscle. *Circ Res* **50**, 651–662.

[10] Davidenko JM, Kent PF, Chialvo DR, Michales DC & Jalife J (1990). Sustained vortex-like waves in normal isolated ventricular muscle. *Proc Natl Acad Sci USA* **87**, 8785–8789.

[11] Doval HC, Nul DR, Grancelli HO, Varini SD, Soifer S, Corrado G, Dubner S, Scapin O & Perrone S (1996). Nonsustained ventricular tachycardia in severe heart failure. *Circulation* **94**, 3198–3204.

[12] Fast VG & Etimov IR (1991). Stability of vortex rotation in an excitable cellular medium. *Physica D* **49**, 75–81.

[13] Fenton FH, Cherry EM, Hastings HM, Evans SJ (2002). Multiple mechanisms of spiral wave breakup in a model of cardiac electrical activity. *Chaos* **12(3)**, 852–889.

[14] Fenton F & Cherry EM (2008). Models of cardiac cell, *Scholarpedia* **3(8)**, 1868.

[15] Fenton F & Karma A (1998). Vortex dynamics in three-dimensional continuous myocardium with fiber rotation: Filament instability and fibrillation. *Chaos* **8**, 20–47.

[16] Fitz Hugh R (1961). Pulses and physiological states in theoretical models of nerve membrane. *Biophys J* **1**, 445.

[17] Gerhardt M, Schuster H & Tyson JJ (1990a). A cellular automaton model of excitable media including curvature and dispersion. *Science* **247**, 1563–1566.

[18] Gerhardt M, Schuster H & Tyson JJ (1990b). A cellular automaton modle of excitable media III. Fitting the Belousov-Zhabotinski reaction. *Physica D* **46(3)**, 416–426.

[19] Gerisch G. Chapter 6. in Current Topics in Developmental Biology, Vol. 3, eds. Moscona, A. Monroy, A. (Academic Press, New York), 1986.

[20] Gorelova NA & Bures J (1983). Spiral waves of spreading depression in the isolated chicken retina. *J Neurobiol* **14(5)**, 353–363.

[21] Greenberg JM & Hastings SP (1978). Spatial patterns for discrete models of diffusion in excitable media. *SIAM J Appl Math* **34**, 515–523.

[22] Guo Y, Zhao Y, Billings SA & Coca D (2010). Identification of excitable media using a scalar coupled map lattice model. *Int J Bifurc Chaos* **20(7)**, 2137–2150.

[23] Harada T, Yokogawa T, Miyaguchi T & Kori H (2009). Singular behavior of slow dynamics of single excititable cells. *Biophys J* **96**, 255–267.

[24] Himel HD, Garny A, Noble PJ, Wadgoangar R, Savarese J, Liu N, Bub G & El-Sherif N (2012). Electronic suppression of early afterdepolarizations in the neonatal rat ventricular myocyte monolayer. *J Physiol*, in preparation.

[25] Huang X, Xu W, Liang J, Takagaki K, Gao X & Wu J (2010). Spiral wave dynamics in neo-cortex. *Neuron* **68**, 978–990.

[26] Ito H & Glass L (1991). Spiral breakup in a new model of discrete excitable media. *Phys Rev Lett* **66**, 671–674.

[27] Izhikevich EM (2004). Which model to use for cortical spiking neurons? *IEEE T Neural Networ* **15(5)**, 1063–1070.

[28] Kunysz A & Glass L (1995). Overdrive suppression of spontaneously beating chick heart cell aggregates: Experiment and theory. *Am J Physiol* **269**, H1153–H1164.

[29] Kurrer C & Schulten K (1991). Propagation of chemical waves in discrete excitable media: Anisotropic and isotropic wave fronts. In Nonlinear Wave Processes in Excitable Media, ed. Holden A V, Markus M & Othmar HG, pp. 489–500, Plenum Press, New York.

[30] Marcus M & Hess B (1990). Isotropic cellular automaton for modelling excitable media. *Science* **347**, 56–58.

[31] Moe GK, Rheinboldt WC & Abildskov JA (1964). A computer model of atrial fibrillation. *Am Heart J* **67(2)**, 200–220.

[32] von Neuman J (1966). Theory of self-reproducing automata. University of Illinois Press, Urbana.

[33] Palsson E & Cox EC (1996). Origin and evolution of circular waves and spirals in Dictyostelium discoideum territories. *Proc Natl Acad Sci USA* **93**, 1151–1155.

[34] Pertsov AM & Ermakova EA (1988). Mechanism of drift of a spiral wave in an inhomogeneous medium. *Biofizika* **33**, 338–342.

[35] Rinzel J (1985). Bursting oscillations in an excitable membrane model. In Ordinary and Partial Differential Equations. Proceedings of the 8th Dundee Conference. Lecture Notes in Mathematics. ed. Sleeman BD & Jarvis RJ. 1151. Berlin: Springer, 304–316.

[36] Rohr S, Kucera JP, Fast VG & Kléber AG (1997). Paradoxical improvement of impulse conduction in cardiac tissue by partial cellular uncoupling. *Science* **275(5301)**, 841–844.

[37] Saxberg BEH & Cohen RJ (1991). Cellular automata models of cardiac conduction. In Theory of heart: Biomechanics, biophysics, and nonlinear dynamics of cardiac function. ed. L Glass L, Hunter P & McCulloch A. Springer-Verlag New York.

[38] Shibata J & Bures J (1974). Optimum topographical conditions for reverberating cortical spreading depression in rats. *J Neurobiol* **5**, 107–118.

[39] Singh SN, Fisher SG, Carson PE & Fletcher RD (1998). Prevalence and significance of nonsustained ventricular tachycardia in patients with premature ventricular contractions and heart failure treated

with vasodilator therapy. *J Am Coll Cardiol* **32**, 942–947.

[40] Spach MS & Heidlage JF (1995). The stochastic nature of cardiac propagation at a microscopic level: Electrical description of myocardial architecture and its application to conduction. *Circ Res* **76**, 366–380.

[41] Steinberg BE, Glass L, Shrier A & Bub G (2006). The role of heterogeneities and intercellular coupling in wave propagation in cardiac tissue. *Phil Trans R Soc A* **364(1842)**, 1299–1311.

[42] Tong WC & Holden AV (2005). Induced pacemaker activity in virtual mammalian ventricular cells. In FIMH2005, ed. Frangi AF, Radeva P, Santos A, Hernandez M. Lecture Notes in Computer Science, 3504 pp. 226–235, Springer.

[43] Tung L & Zhang Y (2006). Optical imaging of arrhythmias in tissue culture. *J Electrocardiol* **39**, s2–s6.

[44] Vasalle M (1977). The relationship among cardiac pacemakers: Overdrive suppression. *Circ Res* **41**, 269–277.

[45] Weimar J, Tyson JJ & Watson LT (1992). Diffusion and wave propagation in cellular automaton models of excitable media. *Physica D* **55**, 309–327.

[46] Winfree AT (1972). Spiral waves of activity. *Science* **175**, 634–636.

[47] Ye P, Entcheva E, Smolka SA & Grusu R (2008). Modelling excitable cells using cycle-linear hybrid automata. *IET Syst Biol* **2(1)**, 24–32.

CHAPTER 12

LOCAL ACTIVITY PRINCIPLE: THE CAUSE OF COMPLEXITY AND SYMMETRY BREAKING

KLAUS MAINZER

Technical University of München, Germany

The principle of local activity is precisely the missing concept to explain the emergence of complex patterns in a homogeneous medium. Leon O. Chua discovered and defined this principle in the theory of nonlinear electronic circuits in a mathematically rigorous way. The local principle can be generalized and proven at least for the class of nonlinear reaction-diffusion systems in physics, chemistry, biology and brain research. Recently, it was realized by memristors for nanoelectronic device applications in technical brains. In general, the emergence of complex patterns and structures is explained by symmetry breaking in homogeneous media. The principle of local activity is the cause of symmetry breaking in homogeneous media. We argue that the principle of local activity is really fundamental in science and can even be identified in quantum cosmology as symmetry breaking of local gauge symmetries generating the complexity of matter and forces in our universe. Finally, we consider applications in economic, financial, and social systems with the emergence of equilibrium states, symmetry breaking at critical points of phase transitions and risky acting at the edge of chaos. In any case, the driving causes of symmetry breaking and the emergence of complexity are locally active elements, cells, units, or agents.

1. The Local Activity Principle and the Emergence of Complexity

According to several prominent authors, including Stephen Hawking, a main part of 21st century science will be on complexity research. The intuitive idea is that global patterns and structures emerge from locally interacting elements like atoms in laser beams, molecules in chemical reactions, proteins in cells, cells in organs, neurons in brains, agents in markets etc. by self-organization [28, 30]. But what is the cause of self-organization? Complexity phenomena have been reported from many disciplines (e.g. biology, chemistry, ecology, physics, sociology, economy etc.) and analyzed from various perspectives such as Schrödinger's order from disorder [44], Prigogine's dissipative structure [42], Haken's synergetics [14], Langton's edge of chaos [21] etc. But concepts of complexity are often based on examples or metaphors only. We argue for a mathematically precise and rigorous definition of local activity as the cause of self-organizing complexity [10] which can be tested in an explicit and constructive manner.

Boltzmann's struggle in understanding the physical principles differenciating between living and non-living matter, Schrödinger's negative entropy in metabolisms, Prigogine's intuition of the instability of the inhomogeneous, and Haken's synergetics are in fact all direct manifestations of a fundamental principle of locality. It can be considered the complement of the second law of thermodynamics explaining the emergence of order from disorder instead of disorder from order, in a quantitative way, at least for reaction diffusion systems.

1.1. *The local activity principle in physical chemistry*

Self-organization is a well-known phenomenon in physical chemistry that can be observed in many particle systems of active media [40]. In mathematical terms, these systems are defined by reaction-diffusion equations for their individual components i with

$$\dot{x}_i = g_i(\{x_j\}) + D_i \nabla x_i$$

where $x_i(\overrightarrow{r}, t)$ are the local concentrations of i and $g_i(\{x_j\})$ are a set of nonlinear functions $g_i(x_i, \ldots, x_M)$ that describe the reaction rates for components i. The term D_i represents the diffusion constants of the different reacting components. ∇ is the second-order derivative operator of the spatial coordinate vector $\overrightarrow{r}$. Chemical reactions are assumed to take place in small volume elements of a homogeneous composition in which the concentrations change.

Active media are classified as bistable, excitable and oscillatory. In a bistable medium the equations $g_i(\{x_j\})$ have two states as the stationary solution. Large perturbations trigger transitions between these states which may result in trigger waves of, for example, flame propagation. They can also generate a large variety of patterns. A chemical example is the autocatalytic reaction scheme in which x is multiplied [49]:

$$A + 2\mathrm{x} \underset{k_1}{\overset{k_2}{\rightleftarrows}} 3\mathrm{x}$$

$$\mathrm{x} \underset{k_1}{\overset{k_2}{\rightleftarrows}} B$$

This simple scheme illustrates formation of spatio-temporal patterns by local finite-amplitude perturbations. It is a simplification of the autocatalytic model of the glycolysis cycle. A is a feed term and B an inert product. It can be shown that this system leads to the formation of local regions of concentration defined by sharp boundaries. These local regions provide a cell-like profile with multiplication and division behavior. It reminds us of self replication in protocellular systems. When the cell-like units exceed a particular dimension, the interior destabilizes because the necessary concentration gradient is not maintained. In this case, the cells divide. This kind of local activity by self replication was verified experimentally in a reaction system with ferrocyanide, sulfide and iodide that has a bistable stationary solution.

Different spatio-temporal patterns can emerge depending on the kinetics in this system. There are turbulent double rotating spirals, cellular, and target patterns. Reactions are represented by composition changes with a particular probability of grid positions, depending on the occupation of neighboring grid positions. In these cases, pattern formation is self-organized in the oscillatory medium of catalytic oscillators. Self-organizing patterns will only occur in specific reaction condition regimes.

1.2. *The local activity principle in discrete dynamics*

The assignment of reaction probabilities and different grid states as a function of the states of neighboring grid points can be considered a special case of cellular automata [32]. Cellular automata were originally designed by John von Neumann to model self-reproducing networks of cells in analogy to cellular reproduction through the replication of DNA. Cellular automata also model for an excitable medium how a two-dimensional automaton leads to target patterns or spiral waves. The simple mechanism that generates these target and spiral wave patterns illustrates that the key of self-organization is the local rules of interaction between neighboring sites.

The local activity principle of cellular automata leads us to Leon O. Chua who studied this principle in cellular automata (CA), cellular nonlinear networks (CNN), and electronic circuit theory [7]. These approaches allow a mathematical definition which can easily be translated into other non-electrical homogeneous media. The transistor is an example of a locally-active device, whereby a "small" (low-power) input signal can be converted into a "large" (high power) output signal at the expense of an energy supply (namely a battery). No radios, televisions, and computers can function without using locally-active devices such as transistors. Recently, the concept of locally-active memristors has been exploited for nanoelectronic device applications [5, 11], as well as for pinpointing the dynamical mechanisms for generating action potential in ion channels, such as the sodium and potassium channels in the celebrated Hodgkin-Huxley nerve membrane model.

For the formation of complex biological and chemical patterns, Schrödinger and Prigogine demanded nonlinear dynamics and an energy source as necessary conditions. But, for the exhibition of patterns in an electronic circuit (i.e. non-uniform voltage distributions), the claim for nonlinearity and energy source is too crude. In fact, no patterns can emerge from circuits with cells made of only batteries and nonlinear circuit elements which are not locally active.

In general, a spatially continuous or discrete medium made of identical cells interacting with all cells located within a neighborhood is said to manifest complexity iff the homogeneous medium can exhibit a non-homogeneous static or spatio-temporal pattern under homogeneous initial and boundary conditions.

The principle of local activity can be formulated mathematically in an axiomatic way without mentioning any circuit models. Moreover, any proposed unified theory on complexity should not be based on observations from a particular collection of examples and explained in terms that make sense only for a particular discipline, say chemistry. Rather it must be couched in discipline-free concepts, which means mathematics, being the only universal scientific language.

However, in order to keep physical intuition and motivation behind this concept, we start with a special class of Cellular Nonlinear Networks (CNN), namely the reaction-diffusion CNNs [7]. A CNN is any spatial arrangement of locally-coupled cells, where each cell is a dynamical system which has an input, an output, and a state evolving according to some prescribed dynamical laws, and whose dynamics are coupled only among the neighboring cells lying within some prescribed sphere of influence centered at the cell's location. In short: a CNN is defined by the state equations of isolated cells and the cell coupling laws (in addition to boundary and initial conditions).

In a reaction-diffusion CNN, each cell located at any 3-dimensional grid point with coordinates (α, β, γ) is defined by a state equation

$$\dot{x}_{\alpha\beta\gamma} = f(x_{\alpha\beta\gamma})$$

and a "diffusion" cell coupling law

$$i_{\alpha\beta\gamma} = D\nabla^2 x_{\alpha\beta\gamma}$$

where D is a constant diffusion matrix with diagonal elements (diffusion coefficients) associated with the state variables of $x_{\alpha\beta\gamma}$. The term $\nabla^2 x_{\alpha\beta\gamma}$ is the discrete Laplacian operator. If we add the cell coupling law to the cell state equation, we obtain the CNN reaction-diffusion equation

$$\dot{x}_{\alpha\beta\gamma} = f(x_{\alpha\beta\gamma}) + D\nabla^2 x_{\alpha\beta\gamma}.$$

1.3. *From discrete to continuous dynamics*

If the discrete Laplacian operator is replaced by its limiting continuum version, we get the standard reaction-diffusion partial differential equations (PDE). According to extensive numerical simulations, CNN reaction-diffusion equations and their associated PDE'S have similar qualitative behavior. For example, various spatio-temporal patterns from Prigogine's Brusselator and Oregonator PDE have been simulated by the corresponding reaction-diffusion CNNs.

A locally active CNN cell must behave exactly like a transistor: At a cell equilibrium point Q there exist a local (i.e. infinitesimal) input current signal such that by applying this small information-bearing signal about the DC operating point Q imposed by the battery, we can extract more local (infinitesimal) information-bearing energy $\delta\varepsilon(Q)$ at Q over some time interval, at the expense of the battery, which supplies the signal energy amplification. A CNN cell is said to be locally passive [8], if there is no cell equilibrium point where it is locally active. There are well-known sets of necessary and sufficient conditions for testing whether a cell is locally passive [10]. A violation of any one of these conditions would make a cell locally active. Thus, there are constructive procedures for testing local activity.

For reaction-diffusion CNN equations, the CNN parameters must be chosen so as to make its cells locally active, in order to make it possible for complex phenomena such as pattern formation to emerge. These parameters can be derived numerically from the set of local activity criteria. In a so-called linear stability analysis of, e.g. Prigogine's dissipative structures, at least one eigen value of the Jacobian matrix of $f(\cdot)$ associated with the kinetic laws of reaction-diffusion equations must have a positive real part for patterns to exist. But this is only one of four independent conditions each of which is sufficient for local activity. Another one can be satisfied with all eigenvalues having negative parts. This is the standard operating mode of most useful transistor amplifiers.

The key concept of this essay claims that a locally-active cell with only negative eigenvalues appears to be dead, when uncoupled because each is asymptotically stable, but together could generate patterns when they are coupled by a dissipative environment. This is counter-intuitive because dissipation should tend to damp out any non-uniformity! A mathematical theory which explains this perplexing phenomenon is what Turing [48] and Smale [46] had sought in vain. Prigogine was totally confused that he could only lament that a new term "instability of the homogeneous" need to be introduced. We claim the principle of local activity is precisely this missing concept and foundation. In particular it is the condition that the real part of the "complexity function", the cell impedance in circuit parlance, be negative at some frequency. It is this non-intuitive condition what we call "edge of chaos", which is a completely different concept from that advocated in the popular literature.

The fundamental new concept here is that if the cells are not locally active, no passive coupling circuits of any complexity can be found to create patterns in the resulting CNN. The universality of the concept of local activity is manifested clearly by its local activity criteria which are independent of the nature of the CNN cells. The only assumption we demand is that any system exhibiting complexity can be represented by a mathematical model which can be decomposed into elements and coupling laws.

2. The Local Activity Principle and Symmetry Breaking in the Universe

The emergence of complex patterns is explained by locally active units (elements, cells, agents etc.) in homogeneous media. In mathematical terms, a homogeneous state means symmetry. Thus, local activity leads to symmetry breaking of homogeneous and uniform states. The concept of locality, symmetry breaking and emergence of new structure is also fundamental in quantum cosmology. In order to understand cosmological applications of the local activity principle, a deeper analysis of quantum cosmology and quantum field theory is necessary.

In general, *principles of symmetry* play a central role in physics. The *invariance* and *covariance* properties of a system under specific symmetry transformations can either be related to the conservation laws of physics or be able to establish the structure of the fundamental physical interactions. This is the most essential aspect of symmetry as it concerns the basic principles of physics and the interactions themselves and not only the properties of geometric figures [25, 26].

2.1. *Mathematical definition of symmetry*

In *geometry*, figures or bodies are called symmetrical when they possess common measures or proportions. Thus the Platonic bodies can be rotated and turned at will without changing their regularity. Similarity transformations, for example, leave the geometric form of a figure unchanged, i.e. the proportional relationships of a circle, equilateral triangle, rectangle, etc. are retained, although the absolute dimensions of these figures can be enlarged or decreased. Therefore one can say that the form of a figure is determined by the similarity transformations that leave it unchanged (invariant).

In mathematics, a similarity transformation is an example of an automorphism.

In general an *automorphism* is the mapping of a set (e.g. points, numbers, functions) onto itself that leaves unchanged the structure of this set (e.g. proportional relations in Euclidean space). Automorphisms can also be characterized algebraically in this way:

(1) Identity I that maps every element of a set onto itself, is an automorphism.
(2) For every automorphism T an inverse automorphim T' can be given, with $T \cdot T' = T' \cdot T = I$.
(3) If S and T are automorphisms, then so is the successive application $S \cdot T$.

A set of elements with a composition that fulfils these three axioms is called a *group*. The symmetry of a mathematical structure is determined by the group of those automorphisms that let it unchanged (invariant).

Symmetry transformations can be classified in two classes: continuous and discrete transformations. Continuous transformations are in turn divided into *global* and *local* transformations. By definition, a symmetry transformation is said to be *continuous* if the set of parameters, which are necessary to describe the transformation, range over a continuous set of values. Examples of continuous transformations are the translation in space, the rotation around a given axis, and the translation in time. These symmetry transformations are global because once the transformation of a given point in space has been fixed, then the transformation at all other points in space is also fixed.

Basic principles of physics like linear momentum conservation, angular momentum conservation and energy result from the symmetry properties of the interactions under global space and time continuous transformations. According to Emmy Noether's theorem, a Lagrangian theory possesses N conserved quantities, if the theory (i.e. the Lagrangian function) is invariant under a N-parameter continuous transformation. Noether's theorem is not only a cornerstone of classical physics, but, by the correspondence principle, of quantum physics as well.

2.2. *Symmetries in the quantum world*

From a mathematical point of view, the state space of a quantum system is a Hilbert space of finite or countably-infinite dimension [1]. A *quantum state* is a one-dimensional subspace of the state space H. Any normalized vector in the one-dimensional subspace of a state can be used to represent this state, and is called a state vector. The original formulation of quantum mechanics assumed a one-to-one correspondence of one-dimensional subspaces of the state space with physical states, implying the unrestricted validity of the *superposition principle* for state vectors. This requirement is equivalent to the exclusion of *superselection rules*. A statement that selects some vectors, adding that they are physically unrealizable as state vectors is called a superselection rule. If there are superselection rules, then there exist subspaces of the state space that cannot be connected to each other by any observable. Not all self-adjoint operators on the state space are therefore *observables*.

Ignoring superselection rules, the states of a quantum system span a projective Hilbert space. Every vector ψ in the Hilbert space H determines a one-dimensional subspace, called the ray $\bar{\psi}$. The inner product of two rays $\bar{\psi}$ and $\bar{\phi}$ is defined by

$$\langle \bar{\psi} | \, \bar{\phi} \rangle = \frac{|\langle \psi | \;\, \phi \rangle|}{\|\psi\| \, \|\phi\|}.$$

The set of all rays in H is called the projective Hilbert space $\bar{H}$associated with the Hilbert space H. A symmetry transformation of quantum mechanics is an *automorphism* of the projective Hilbert space $\bar{H}$ associated with the state space H. Thus the symmetry of quantum mechanics is given by the *automorphism group* $Aut(\bar{H})$. A theorem of Eugene P. Wigner [54] asserts that the automorphism group $Aut(\bar{H})$ can be represented by the group of unitary operators acting on the state state space H. Let H_1 and H_2 be Hilbert spaces and F be a mapping from H_1 into H_2. Then F is called linear if

$$F(a\psi + b\psi) = a\; F\psi + b\; F\psi.$$

F is called antilinear iff

$$F(a\psi + b\psi) = a^*\; F\psi + b^*\; F\psi.$$

F is called isometric iff

$$\|F\psi\| = \|\psi\|$$

for all ψ and ϕ from H_1 and all complex numbers a and b. If the range of a linear isometric operator F: $H_1 \;\rightarrow\; H_2$ is the whole space H_2, then F is called *unitary*. An antiunitary operator F:

$H_1 \to H_2$ is an antilinear isometric operator having the range H_2. Wigner's theorem implies that two realizations of quantum mechanics whose state spaces are connected by a unitary or antiunitary transformation are from a logical point of view equivalent. Historically, the fact that symmetries in quantum mechanics are described by projective unitary representations has been known since Hermann Weyl's famous book on "Gruppentheorie und Quantenmechanik" [52]. Wigner published his theorem in his textbook (1931) without a full proof [54]. A complete proof was given by V. Bargmann [3].

2.3. *Global and local symmetries*

In quantum physics, all the properties of a system can be derived from the state or wave function associated with that system. The absolute phase of a wave function cannot be measured, and has no practical meaning, as it cancels out the calculations of the probability distribution. Only relative phases are measurable in an interference experiment. Therefore it is possible to change the *phase of a wave function* without leading to any observable effect. Formally a *phase transformation* of the wave function $\psi(x,t)$ can be written as

$$\psi(x,t) \to \psi'(x,t) = \mathrm{e}^{i\alpha}\psi(x,t)$$

with the parameter (phase) α of the transformation [27]. If α is *constant*, i.e. the same for all points in space-time, the equation expresses the fact that once a phase convention has been made at a given point in space-time, the same convention must be adopted at all other points. This is an example of a *global transformation* applied to the field $\psi\,(x,t)$. If $\alpha = \alpha(x,t)$is a *function of space and time*, then such a transformation will not leave invariant any equation of $\psi\,(x,t)$ with space or time derivatives. This is in particular true for the Schrödinger equation or any relativistic wave equation for a free particle.

In order to satisfy the *invariance under a local phase transformation* it is necessary to modify the equations in some way, which will no longer describe a free particle. Such modifications introduce additional terms, which describe the interaction of the particle with external fields and thereby generate the dynamics. That is the *gauge principle* or *principle of local symmetry* according to which the interactions are determined by invariance under local symmetry (phase or gauge) transformations. Historically, the principle of *gauge invariance* dates back to a (false) idea of Weyl who assumed a deeper dependence between the laws of matter and electromagnetism [51,53].

A *discrete symmetry transformation* is described by parameters ranging over a discrete set of values. Examples are symmetry operations that leave unaffected a crystal by reflections through planes, inversions with respect to a centre point and rotations around a given axis with angles $2\pi/n(n = 2,\ 3,\ 4$ or $6)$ corresponding to the periodicity of the crystal lattice. In elementary particle physics, there are three discrete transformations for interactions between leptons and quarks: the charge conjugation C, the parity transformation P, and the time reversal T. In a *charge conjugation operation*

$$C : q_\alpha \to -q_\alpha$$

All the particles of a system are replaced by their antiparticles and therefore all charges q_α change sign. The *parity transformation*

$$P : \mathbf{r} \to -\mathbf{r}$$

corresponds to a space inversion relative to a point. In a system of Cartesian coordinates, a point with coordinates (x, y, z) transforms into (-x, -y, -z) under the parity operation. The position vector $\mathbf{r}$ changes sign under a space inversion. The *time reversal operation*

$$T : t \to -t$$

corresponds to the inversion of the time variable t. The laws of physics are invariant with respect to T. Symmetry of time means that it is physically impos sible to distinguish between forward and backward moving in time.

Quantum theory of fields requires the invariance of the fields and interactions under the combined transformations of the three operations CPT. The CPT-theorem of quantum theory was proved by Wolfgang Pauli in 1957 [41]. If one of the three symmetries is violated, then, according to the *CPT-theorem*, one of the other two symmetries has also to be violated. For example, the violation of parity P requires that C or T be violated. If the invariance under the combination of two transformations holds, then the invariance under the third transformation must also hold. For example, the invariance under CP implies the invariance under T and vice-versa. The decay of Kaons is the only known example of time violation T which is enforced by a CP-violation. Further on, the CPT invariance implies that the masses and the lifetime of a particle is identical to those of antiparticles. CPT invariance has been empirically confirmed to very high precision.

2.4. *Local gauge symmetries and symmetry breaking*

Before 1956, it was assumed that *parity* was a fundamental symmetry of physical processes. In 1956, Tsung Dao Lee and Chin Ning Yang examined the question of whether processes driven by the weak interaction would distinguish left or right [22]. Their famous experiments performed in the beta decay of ^{60}Co, and in the weak decays of pions and muons,

$$\pi^+ \rightarrow \mu^+ + \nu_\mu \quad \text{and} \quad \mu^+ \rightarrow \mathrm{e}^+ + \nu_e + \bar{\nu}_\mu$$

not only provided the empirical support to the suggestions of Lee and Yang but also showed that parity violation was an universal property of the weak interaction.

The observation of *parity violation* was soon incorporated in the *theory of weak interaction* and is now a part of modern unified theory of electroweak interactions, the *Standard Model*. Actually, the fundamental physical forces of interaction can be characterized by *local gauge symmetries*. The unitary group $U(n)$ and the special unitary group $SU(n)$ refer to the unitary transformation of a n-dimensional complex coordinate space [25–27]. In the standard model, gravitation, electromagnetic, weak and strong interaction are represented by local Poincaré-, $U(1)$-, $SU(2)$-, and $SU(3)$-gauge groups. The research program of unified theories tries to unify the fundamental forces step by step in states of higher energy characterized by *unified local symmetries*.

In 1954, the *Yang-Mills theory* tried at first to unify proton and neutron by a *gauge theory of isospin-symmetry* [55]. But the Yang-Mills theory only forcasted massless gauge particles of interaction in contradiction to empirical observations. Later on, J. Goldstone [13] and P. Higgs [19] introduced the mechanism of *spontaneous symmetry breaking* (*Higgs' mechanism*) in order to give appropriate gauge particles the desired mass. The intuitive idea is that a symmetric theory can have asymmetric consequences. For example, the equations of a ball and the wheel of a roulette are symmetric with respect to the rotation axis, but the ball always keeps lying in an asymmetric position. In a first step, electromagnetic and weak forces could already be unified at very high energies in an accelerator ring. For energies of more than 100 Gigaelectron-Volts and distances less than 10^{-16} cm, there would be a perfect $U(1)$ x $SU(2)$ *symmetry*, in which the W^+ and Z^o field quanta would be exchanged as rapidly as the photon. Their transformations are described by the same symmetry group $U(1) \times SU(2)$. At a critical value of lower energy the symmetry spontaneously breaks apart into two partial symmetries $U(1)$ of electromagnetic force and $SU(2)$ of weak interaction. The gauge particles of weak interaction get their *mass* by the *Higgs' mechanism*, the photon of electromagnetic interaction remains massless.

After the successful unification of electromagnetic and weak interaction physicists try to realize the "*big*" unification of electromagnetic, weak and strong forces, and in a last step the "*superunification*" of all four forces. The symmetry of this last unification would be the "*holy grail*" of modern cosmology and physics. There are several research strategies of superunifications such as supergravity and superstring theories. Mathematically they are described by extensions of richer structures of local symmetries and their corresponding *gauge groups*. On the other hand the variety of elementary particles is realized by *spontaneous symmetry breaking*. The concept of local symmetry and symmetry breaking play an immense role in cosmology. During cosmic expansion and cooling temperature, the initial unified *supersymmetry* of all forces broke apart into the subsymmetries of physical interactions, and the corresponding elementary particles were generated in phase stages leading to more variety and complexity.

Spontaneous breaking of local symmetries (gauge groups) leads to the emergence of new matter and is obviously the most fundamental principle in nature [26]. Without the local activity of the Higgs' mechanism (or a corresponding local activity), there would be no matter and no existence. Even the beginning of cosmic expansion ('Big Bang') is assumed to be initiated by a kind of spontaneous symmetry breaking with respect to an equilibrium state of quantum vacuum [18]. In any case, our cosmological models assume a general *principle of local activity* which we have uncovered on several levels in nature.

The phases of cosmic expansion are determined by properties of *symmetry breaking*. For example, in the case of weak interaction, neutrinos occur only as a left-handed helix, but not as a right-handed one which means parity violation. This kind of *anti-symmetry* or *dissymmetry* seems also to be typical for *molecular structures of life* [25, 43]. Protein analysis shows that amino acids have an antisymmetrical carbon atom and occur only in the left-handed configuration. Weak interaction takes part in the chemical bonds. Thus, cosmic parity violation of weak interaction is assumed to cause the selection of chiral molecules. The reason is that the left-handed (L) and right-handed (D) examples of chiral molecules can be distinguished by a tiny parity violating energy difference ΔE_{pv}. The energetically stable examples (e.g., L-form of amino acids) are preserved. But, this assumption is only based on theoretical calculations (e.g., Hartree-Fock procedures in physical chemistry). We still lack exact measurements of experiments because of the tiny small parity violation energy difference ΔE_{pv} (e.g., $4 \cdot 10^{-14}$ (hc)cm^{-1} (H_2O_2), $1 \cdot 10^{-12}$ (hc)cm^{-1} (H_2S_2)), although there are proposed experiments with spectroscopic methods. In any case, symmetry breaking of weak interaction seems to be connected with the emergence of biomolecular structures of life.

In this article, we consider a hierarchy of pattern formation by a *local activity principle of symmetry breaking* from the emergence of elementary particles by the Higgs' mechanism and the emergence of biomolecules by parity violation up to the emergence of complex cellular structures by morphogenesis, the emergence of neural patterns in biological brains by ATP energetic processing, and finally the emergence of circuit patterns in electronic brains by transistors or memristors. The philosophical message of these principles of local activity and symmetry breaking is obvious: Without local activity leading to symmetry breaking in a homogeneous medium, no new patterns could emerge. In short:

Without local activity and symmetry breaking no existence! [7, 25, 26]

3. The Local Activity Principle and Symmetry Breaking in Economy and Society

3.1. *The local activity principle and social self-organization*

The self-organization of complex systems with locally active units can also be observed in social groups. An application of social dynamics is the emergence of behavioral patterns generated by car drivers as *locally active units*. In *automobile traffic systems*, a phase transition from non-jamming to jamming phases depends on the averaged car density as control parameter. The spontaneous emergence of chaotic patterns of traffic is a famous self-organizing effect of nonlinear interactions which can often not be reduced to single causes. At a critical value, fluctuations with fractal or self-similar features can be observed. The term *self-similarity* states that the time series of measured traffic flow looks the same on different time scales, at least from a qualitative point of view with small statistical deviations. This phenomenon is also called *fractality*. In the theory of complex systems, self-similarity is a (not sufficient) hint on chaotic dynamics. These signals can be used by guide systems of traffic [38].

In a political community, collective trends or majorities of opinions can be considered as *patterns* which are produced by mutual discussions and interaction of the locally active people in a more or less 'heated' situation. They can even be initiated by some few active people in a critical and unstable ('revolutionary') situation of the whole community. There may be a competition of opinions during heavy fluctuations. The essential point is that the winning opinion will dominate the collective (global) behavior of the people. In this case, stable modes are dominated ("enslaved") by some few unstable (locally active) modes. Therefore, Haken called this procedure the "slaving principle" [14, 15].

Thus, there is a kind of feedback: the collective order of a complex system is generated by the interactions of its locally active elements (*'self organization'*). After thermodynamic, genetic, and neural self-organization we also distinguish *social* and *economic self-organization*. On the one side, the behavior of the elements is dominated by the collective order. On the other side, people have their individual will to influence collective trends of society. Nevertheless, they are also driven by attractors of collective behavior.

3.2. *The local activity principle in innovation dynamics*

Sometimes, this approach is called "*econophysics*" (i.e. a combination of "economics" and "physics") [36, 39]. But, modeling self-organizing social systems is *no physicalism*, because their mathematical formalism of complex systems and nonlinear dynamics is independent of physical concepts and only considers economic and social data. Therefore, we prefer the term "*sociodynamics*". A forerunner of modern sociodynamics was the Austrian economist Joseph F. Schumpeter who analyzed the correlation of innovation dynamics and economic cycle theory. New ideas arise steadily. When enough ideas have accumulated, a cluster of innovations is introduced by entrepreneurship. They develop slowly at first, then accelerate as the methods are improved. A logistic curve characterizes the typical trajectory of an innovation. Some investment must precede the introduction of an innovation. Investment stimulates demand. Increased demand facilitates the spread of the innovation. Then, as all innovations are fully exploited, the process decelerates towards zero.

Schumpeter called this phenomenon the "swarming" of innovations. In his three-cycle model, the first short cycle relates to the stocks cycle and innovations play no role. The following longer cycle is related to innovations. Schumpeter recognized the significance of historical statistics and related the evidence of long waves to the fact that the most important innovations like steam, steel, railways, steamships, and electricity required 30 to 100 years to become completely integrated into the economy.

In general, Schumpeter described *economic evolution* as a technical progress in the form of "swarms" which were explained in a logistic framework. A technological swarm is assumed to shift the equilibrium to a new fixed point attractor in a cyclical way. The resulting new equilibrium is characterized by a higher real wage and higher consumption and output. Thus, Schumpeter's innovation dynamics can easily be interpreted in terms of sociodynamics with attractors. Innovation swarms at economical points of instability can be considered a global regime dominating long-term business cycles.

Historically, the great Depression of the 1930s inspired economic models of business cycles. However, from a mathematical point of view, the first models (for instance Hansen-Samuelson and Lundberg-Metzler) were linear and hence required exogenous shocks to explain their irregularity. The explanation by exogenous shocks has the great disadvantage that they are often arbitrary ad hoc hypotheses and hence can explain anything. The standard econometric methodology has argued in this tradition, although an intrinsic analysis of cycles has been possible since the mathematical discovery of strange attractors. The traditional linear models of the 1930s can easily be reformulated in the framework of nonlinear systems. From a methodological point of view, endogeneous nonlinear models with attractors seem to be more satisfactory. In this case, locally active agents generate global patterns of behaviour by nonlinear interactions. Nevertheless, linear models with exogenous shocks must be taken in earnest and tested in economics.

3.3. *The local activity principle in sociodynamics*

In contrast to physical, chemical, and biological systems, no equations of motion on the micro level are, in general, available for social systems. People are not atoms or molecules, but human beings with intentions, motivations, and emotions. In principle, their individual behavior and decision-making could be explained by their *brain dynamics*. Cognitive and emotional dynamics are determined by patterns of neural circuits characterizing individual thoughts, decisions, and motivations. But this is, until now, only a hypothesis, because the corresponding neural equations are not yet known. Furthermore, they would be too complex to solve them and predict the future behavior of people.

Therefore an alternative approach is suggested which gets along without microscopic equations, but nevertheless takes into account the local decisions and actions of individuals with probabilistic methods in order to derive the *macrodynamics of social systems* [50]. The modeling design consists of three steps: In the first step appropriate variables of social systems must be introduced to describe the states and attitudes of individuals. The second step defines the change of behavior by probabilistic phase transitions of individual states. The third step derives equations for the global dynamics of the system by stochastic methods.

In a society we can distinguish several sectors and sub-sectors that are denoted by variables. There are variables for material states, extensive and intensive personal states. The *sociconfiguration* of a social system is characterized by these material and personal macrovariables. They are measured by usual methods of public opinion research, sociology, or economics. Like in thermodynamics, there are intensive economic variables that are independent of the size of a system. Examples are prices, productivity, and the density of commodities. Extensive variables are proportional to the size of a system and concern, for example, the extent of production and investment, or the size and number of buildings.

Collective material variables are measurable. Their values are influenced by the individual activities of agents, which are often not directly measurable. The social and political atmosphere of a company is connected to socio-psychological processes, which are influenced by the attitudes, opinions, or actions of individuals and their subgroups. Thus, in order to introduce the socioconfiguration of collective personal variables, we must consider the states of individuals, expressed by their attitudes, opinions, or actions. Furthermore, there are subgroups with constant characteristics (e.g., sections or departments of a company or an institution), so that each individual is a member of one subgroup. The number of members of a certain state is a measurable macrovariable. The socioconfiguration of, e.g., a company is a set of macrovariables describing the distribution of attitudes, opinions, and actions among its subgroups at a particular time. The total macroconfiguration is given by the multiple of material configuration and socioconfiguration.

Their *probabilistic phase transitions* can be used for setting up the macroevolution equation of a social system. The probabilistic macrobehavior of a society is described by a probability distribution function over its possible socioconfigurations at a certain time. The distribution function $P(m, n; t)$ can be interpreted as the probability of finding a certain macroconfiguration of material configuration m and socioconfiguration n at time t. The evolution of the social system is the time-depending change of its probabilistic macrobehavior, i.e. the time derivative of the probability function $dP(m, n; t)/dt$. Thus, we get a stochastic nonlinear differential equation which is well-known in thermodynamics as *master equation*.

An example of social phase transitions and symmetry breaking is provided by world-wide *migration processes*. The behavior and the decisions of people to stay or to leave a region are illustrated by spatial distributions of populations and their change. The models may concern regional migration within a country, motivated by different economic and urban development, or even the dramatic worldwide migration between poor and rich countries in the age of globalization. The migration interaction of two human populations may cause several macro-phenomena, such as the emergence of a stable mixture, the emergence of two separated, but stable ghettos, or the emergence of a restless migration process. In numerical simulations and phase portraits of the migration dynamics, the macrophenomena can be identified with corresponding attractors.

3.4. *The local activity principle in financial dynamics*

Stability and welfare of our societies depend sensitively on the dynamics of international *financial markets*. We already mentioned that, in general, we do not know the microscopic motions of economic data and agents. Therefore, in 1900, the French mathematician L. Bachelier [2] considered the fluctuations of stock prices as statistical random walk (*Brownian motion*) before physicists like A. Einstein (1905) discovered it in the microscopic motion of small particles in fluids. Brownian motion does not only imply statistical stationarity of price increments and scaling of prices (i.e. invariance under displacement and change of scale), but also independence of price increments (knowing the past brings no knowledge about the future), continuity of price variation (a sample of Brownian motion is a continuous curve), rough evenness of price changes (normal Gaussian distribution or "white noise"), absence of clustering (no emergence of local patterns and structure) and absence of cyclic behavior. Further on, Gaussian distribution leads to the assumption of an *efficient market* and successful arbitraging with prices following a martingale: whether the past is known in full, in part, or not at all, price changes over all future time spans have zero as expectation.

Modern theory and practice of finance is still more or less based upon these fundamental beliefs. What makes an investment in the stock market risky is that there is a spread of possible outcomes. The usual measure of this spread is believed to be the standard deviation or variance in a bell-shaped (Gaussian) normal distribution. On this basis, H. Markowitz (1959) [37] suggested his well-known construction of *portfolios* in order to diversify away unique risk. A stock's contribution to the risk of a fully diversified portfolio depends on its sensitivity to market changes which is measured by a parameter 'beta'. Beta delivers the benchmarks for the expected risk premium which, since the mid-1960s, was calculated by the *capital asset pricing model* (CAPM). Bachelier did not only suggest the random walk of price changes, but also considered the effects of investing in options. A breakthrough in 1973 was the famous *Black-Scholes formula* calculating the present call option when there is a continuum of possible future stock prices on the basis of a normal (Gaussian) distribution [4]. Every day dealers on the options exchanges still use this formula to make their trades.

Brownian motion is mathematically more manageable than any alternative. But, unfortunately, it is an extremely poor approximation to financial reality. Since the end of the 1980s, we can observe financial crashes and turbulences deviating significantly from normal distribution. Investment portfolios collapsed and hedging with options à la Black-Scholes failed. From the viewpoint of dynamical systems, *patterns of time series* analysis illustrate the failures of traditional financial theory. While a record of Brownian motion changes looks like a kind of 'grass' with normal length, a record of actual price changes looks like an irregular alternation of quiet periods and bursts of volatility that stand out from the normal length of the grass. This feature demonstrates the apparent nonstationarity of the underlying rules. Further on, discontinuities appear as sharp peaks from the normally distributed Gaussian 'grass'. These peaks are not isolated but bunched together. Cyclic (but not periodic) behavior can be observed. Instability of the sample variance is expressed by a long-tailed distribution price changes. Last but not least, there is a long-term dependence of data [47].

Financial markets display some common properties with *fluid turbulence*. As for fluid turbulent fluctuations, financial fluctuations have intermittency at all scales. In fluid turbulence, a cascade of energy flux is known to occur from the large scale of injection to the small scales of dissipation. In the nonlinear and fractal approach of the financial system randomness can no longer be restricted to the 'normal' Gaussian distribution of price changes [34]. Non-Gaussian distributions with Levy- and Pareto-distributions are more appropriate to the wild turbulence of financial markets of today [12]. We must consider *degrees of randomness* [29, 31]. Gaussian distribution corresponds to a pattern of time series with a 'normal' length of 'grass' without extreme peaks. Therefore, it is called mild randomness which can be compared to a solid state of matter aggregation with low energy, stabile structure and defined volume. Wild randomness resembles the gas phase of matter with high energy, less structure and no defined volume. Slow randomness means the fluid state between the gas and solid state. From the viewpoint of time series, mild randomness corresponds to short- and long-run evenness. Slow randomness corresponds to short-run concentration and long-run evenness. Wild randomness corresponds to short- and long-run concentration [33, 35].

3.5. *The local activity principle and bounded rationality*

The rationality of human decision is bounded by the wild randomness of markets. Human cognitive capabilities are overwhelmed by the complexity of nonlinear systems they are forced to manage. Traditional mathematical decision theory assumed perfect rationality of economic agents (homo oeconomicus). Herbert Simon, Nobel Prize laureate of economics and one of the leading pioneers of systems science and artificial intelligence, introduced the principle of bounded rationality:

"The capacity of the human mind for formulating and solving complex problems is very small compared with the size of the problem whose solution is required for objectively rational behavior in the real world or even for a reasonable approximation to such objective rationality." [45]

Bounded rationality is not only given by the limitations of human knowledge, information, and time. It is not only the incompleteness of our knowledge and the simplification of our model. The constraints of short-term memory and of information storage in long-term memory are well-established. In stressful situations people are overwhelmed by a flood of information, which must be filtered under time pressure. People deviate from game-theoretically predicted equilibria. They act neither in the strict sense of the homo oeconomicus nor completely chaotically. They are *locally active agents* with *nonlinear effects of interactions* which cannot be predicted in the long run.

Therefore, we must refer to the real features of human information processing and decision making, which is characterized by emotional, subconscious, and kinds of affective and non rational factors. Even experts and managers often prefer to rely on rules of thumb and heuristics, which are based on intuitive feelings of former experience. Experience shows that human intuition does not only mean lack of information and the failure to make decisions. Our affective behavior and intuitive feeling are parts of our evolutionary heritage that enable us to make decisions when matters of survival are at stake. Therefore, we must know more about the factual microeconomic acting of people, their cognitive and emotional behavior, in order to understand macroeconomic trends and dynamics. This is the goal of experimental economics, observing, measuring, and analyzing the behavior of economic agents with methods of psychology, cognitive and social sciences at, e.g., stock markets or in situations of economic competition.

4. What can We Learn from the Local Activity Principle of Complex Systems?

What are the *human perspectives* of the *Local Activity Principle?* In the age of globalization, modern societies, economies, and information networks are highly dimensional systems with a complex nonlinear dynamics. From a methodological point of view, it is a challenge to improve and enlarge the instruments of modelling from low to high dimensional systems. Modern systems science offers an *interdisciplinary methodology* to understand typical features of self-organizing dynamics in nature and society [28]. The widespread presence of *power laws* has changed our evaluation of *extreme events* as rare and exceptional. They effect people and the environment in a sometimes dramatic way: societal disasters (pandemics), natural disasters (earth quakes, floods, cyclones), technical breakdown (power outages, chemical contaminations, nuclear disasters), or economic turbulence (collapse of banks, huge losses in the stock markets). Their high follow-up costs enforce us to take them serious and not as rare and negligible events.

In complex systems, people sometimes refuse responsibility because of impersonal "systemic effects" of their nonlinear dynamics. The *Local Activity Principle* demands to identify the locally active causes and responsible agents of these global effects. From a methodological point of view, we have significant hope that relations between power laws, causal networks, phase transitions, criticality and self-organization of complex systems can be exploited further.

As nonlinear models are applied in different fields of research, we gain general insights into the *predictable horizons* of oscillatory chemical reactions, fluctuations of species, populations, fluid turbulence, economic processes, and information dynamics. Obviously, nonlinear modelling with the Local Activity Principle explains the difficulties of the modern Pythias and Sibyls. The reason is that human societies are not complex systems of molecules or ants, but the result of highly *intentional acting beings* with a greater or lower degree of free will. A particular kind of *self-fulfilling prophecy* is the Oedipus effect in which people like the legendary Greek king try, in vain, to change their future as forecasted to them.

From a macroscopic point of view we may observe single individuals contributing with their *local activities* to the collective macrostate of society representing cultural, political, and economic order. Yet, macrostates of a society, of course, do not simply average over its parts. Its macrovariables strongly influence the individuals of the society by orientating their activities and by activating or deactivating their attitudes and capabilities. This kind of feedback is typical for complex dynamical systems. If the control parameters of the environmental conditions attain certain critical values due to internal or external interactions, the macrovariables may move into an unstable domain out of which highly divergent alternative paths are possible. Tiny unpredictable microfluctuations (e.g., local actions of very few influential people, scientific discoveries, new technologies) may decide which of the diverging paths in an unstable state of bifurcation society will follow.

Therefore, the paradigm of a centralized control must be given up by the insights in the self-organizing dynamics of highly dimensional systems. We act and decide under the conditions of *bounded rationality* and not with the Laplacian spirit of a totally informed homo oeconomicus. But, self-organization also leads to undesired effects. Cancer is a self-organizing process of growth. Turbulent financial markets are also out of control. Thus, we need a *balance* between *self-organization* and an appropriate *degree of control.* We need *global concepts of order* to realize *global governance.* Global crises like, e.g., the financial crisis of banks need *global response strategies* and international cooperation between nations.

Complexity management accepts the uncertainty that exists in the real world rather than ignoring it. The *Local Activity Principle* demands *local responsibility of acting people.* In short:

The Local Activity Principle leads to local responsibility in a global nonlinear world.

For example, managers and politicians must be obliged to personal accountability of risky investments, decisions, and actions. With respect to local responsibility, we can no longer accept that profits are individual, but wastages socialized. During a long evolution, cellular self-organization of organisms was embedded in a hierarchy of control processors, emerging in a learning process with changing challenges of environments. In engineering science, we should aim at self-organizing systems with *controlled emergence* of new appropriate features. By detecting global trends of complex dynamics, we have the chance of implementing favourite tendencies. By cooperation of locally active agents in complex systems we can make much more progress in choosing our next steps. *Cooperation in complex systems* supports deciding and acting for a sustainable future of a complex world.

References

[1] Audretsch, J., Mainzer, K. (eds.) (1996): *Wie viele Leben hat Schrödingers Katze? Zur Physik und Philosophie der Quantenmechanik.* Spektrum Akademischer Verlag: Heidelberg 2nd edition.

[2] Bachelier, L. (1900): Théorie de la spéculation. Dissertation. *Annales Scientifiques de l'École Normale Supérieure* 17, 21–86.

[3] Bargmann, V. (1954): On Unitary Ray Representation of Continuous Groups. *Annals of Mathematics* 59, 1–46.

[4] Black, F., Scholes, M. (1973): The pricing of options and corporate liabilities. In *Journal of Political Economy* 81, 637–654.

[5] Chua, L.O. (1971): Memristor – the missing circuit element. *IEEE Transactions on Circuit Theory* 18, 507–519.

[6] Chua, L.O., Yang, L. (1988): *IEEE Trans. Circuits Syst.* 35, 1257–1290.

[7] Chua, L.O. (1998): *CNN: A Paradigm for Complexity.* World Scientific: Singapore.

[8] Chua, L.O. (1999): Passivity and complexity. *IEEE Trans. Circuits Syst.*, vol. 46, No.1, 71–82.

[9] Chua, L.O., Roska, T. (2002): *Cellular Neural Networks and Visual Computing: Foundations and Applications.* Cambridge University Press: Cambridge.

[10] Chua, L.O. (2005): Local activity is the origin of complexity. *Intern. J. of Bifurcation and Chaos* 15, No. 11, 3435–3456.

[11] Chua, L.O. (2011): Resistance switching memories are memristors. *Applied Physics* A 102 4, 765–783.

[12] Embrechts, P., Klüppelberg, C., Mikosch, T (2003): *Modeling Extremal Events for Insurance and Finance.* Springer: Berlin.

[13] Goldstone, J. (1961): Field theories with 'Superconductor' solutions. *N. Cimento* 19, 154–164.

[14] Haken, H. (1983): *Synergetics. An Introduction*, Springer, New York 3rd edition.

[15] Haken, H., Mikhailov, A. (eds.) (1993): *Interdisciplinary Approaches to Nonlinear Complex Systems*, Springer: New York.

[16] Hayek, F.A. (1948). *Individualism and Economic Order.* The University of Chicago Press, Chicago.

[17] Hayek, F.A (1974): The pretence of knowledge, Nobel Lecture 1974. In *New Studies in Philosophy, Politics, Economics, and History of Ideas.* The University of Chicago Press: Chicago.

[18] Hawking, S.W., Hartle, J. B., Hertog, T. (2008): The No-Boundary Measure of the Universe. *Phys.Rev. Lett.* 100, 201301.

[19] Higgs, P.W. (1964): Broken Symmetries, Massless Particles, and Gauge Fields. *Phys. Letters* 12, 132–133.

[20] Imada, I. (2005): *Self-Organization and Society.* Springer: Tokyo.

[21] Langton, C.G. (1990), *Physica* 42, 12–37.

[22] Lee, T.D., Yang, C.N. (1956): Questions of Parity Conservation in Weak Interactions. *Phys. Rev.* 104, 254.

[23] Lorenz, H.W (1989): *Nonlinear Dynamical Economics and Chaotic Motion.* Springer, Berlin.

[24] Lux, T., Westerhoff, F (2009): Economics crisis. *Nature Physics* 5, 2–3.

[25] Mainzer, K. (1996): *Symmetries in Nature.* De Gruyter: Berlin (German: Symmetrien der Natur. De Gruyter: Berlin 1988).

[26] Mainzer, K. (2005a): *Symmetry and Complexity. The Spirit and Beauty of Nonlinear Science*, World Scientific: Singapore.

[27] Mainzer, K. (2005b): Symmetry and Complexity in Dynamical Systems. *European Review Academia Europaea* 13 suppl. 2, 29–48.

[28] Mainzer, K. (2007a): *Thinking in Complexity. The Computational Dynamics of Matter, Mind, and Mankind* Springer, Berlin 5th Edition.

[29] K. Mainzer, K. (2007b): *Der kreative Zufall. Wie das Neue in die Welt kommt* C.H. Beck Verlag, München.

[30] Mainzer, K. (ed.) (2009a): Complexity. *European Review* 17. Cambridge University Press: Cambridge.

[31] Mainzer, K. (2009b): Challenges of complexity in economics. *Evolutionary and Institutional Economics Review.* Japan Association for Evolutionary Economics 6 (1), 1–22.

[32] Mainzer, K., Chua, L.O. (2011): *The Universe as Automaton. From Simplicity and Symmetry to Complexity.* Springer: Berlin.

[33] Mandelbrot, B.B., Hudson, R.L. (2004): *The (mis)Behavior of Markets. A Fractal View of Risk, Ruin, and Reward* Basic Books, New York.

[34] Mandelbrot, B.B. (1997): *Fractals and Scaling in Finance. Discontinuity, Concentration, Risk*, Springer: New York.

[35] Mandelbrot, B.B. (1999): *Multifractals and 1/f Noise: Wild-Self-Affinity in Physics*. Springer: New York.

[36] Mantegna, R.N., H.E. Stanley, H.E. (2000): *An Introduction to Econophysics. Correlations and Complexity in Finance*. Cambridge University Press: Cambridge.

[37] Markowitz, H.H (1959): *Portfolio Selection: Efficient Diversification of Investments*. Yale University Press: New Haven.

[38] Mayinger, F. (ed.) (2001): *Mobility and Traffic in the 21st Century*. Springer: Berlin.

[39] McCauley, J.L (2004): *Dynamics of Markets. Econophysics and Finance*. Cambridge University Press: Cambridge.

[40] Mikhailov, A.S. (1994): *Foundations of Synergetics*. Springer: Berlin.

[41] Pauli, W. (1957): *Niels Bohr and the Development of Physics*. Pergamon Press: London.

[42] Prigogine, I. (1980): *From Being to Becoming*. Freeman: San Francisco.

[43] Quack, M. (1986): On the measurement of parity violating energy difference between enantiomers. *Chemical Physics Letters* 132 No. 2, 147153.

[44] Schrödinger, E. (1948): *What is Life?* Cambridge University Press: Cambridge.

[45] Simon, H. (1957): *Administrative Behavior: A Study of Decision-making Processes in Administrative Organizations*. Macmillian: New York, 198.

[46] Smale, S. (1974): *Lectures in Applied Mathematics*, Vol. 6 (American Mathematical Society) 15–26.

[47] Small, M. (2005): *Applied Nonlinear Time Series Analysis: Applications in Physics, Physiology and Finance*. World Scientific: Singapore.

[48] Turing, A.M. (1952): *Philos. Trans. Roy. Soc. London*, Series B 237, 37–72.

[49] Van Santen, R.A., Neurock, M. (Eds.) (2006): *Molecular Heterogeneous Catalysis. A Conceptual and Computational Approach*. Wiley-VCH: Weinheim.

[50] Weidlich, W (2002): *Sociodynamics. A Systematic Approach to Mathematical Modeling in the Social Sciences* Taylor and Francis: London.

[51] Weyl, H. (1918): Gravitation und Elektrizität. *Sitzungsberichte der Königlich Preußischen Akademie der Wissenschaften zu Berlin*, 465–480.

[52] Weyl, H. (1928): *Gruppentheorie und Quantenmechanik*. Hirzel: Leipzig (English trans. "The Theory of Groups and Quantum Mechanics", Dover, New York 1950).

[53] Weyl, H. (1929): Gravitation and the Electron. *Proc. Nat. Acad. Sci. (USA)* 15, 323–334.

[54] Wigner, E.P. (1931): *Gruppentheorie und ihre Anwendung auf die Quantenmechanik der Atomspektren*. Vieweg: Braunschweig (English trans. "Group Theory", Academic Press, New York 1959).

[55] Yang, C.N., Mills, R.L. (1954): Conservation of Isotopic Spin and Isotopic Gauge Invariance. *Phys. Rev.* 96, 191–195.

CHAPTER 13

EXPLORATIONS IN THE FOREST OF BIFURCATION TREES: ROUTE FROM CHUA'S CIRCUIT TO CHUA'S MEMRISTIVE OSCILLATOR

ŁUKASZ CZERWIŃSKI

Faculty of Electrical Engineering, Automatics,
Computer Science and Electronics,
AGH University of Science and Technology,
al.Mickiewicza 30, 30-059 Kraków, Poland

MACIEJ J. OGORZAŁEK

Department of Information Technologies,
Faculty of Physics, Astronomy and Applied Computer Science,
Jagiellonian University,
ul. Reymonta 4, Kraków, Poland

In this paper we study and compare bifurcation phenomena observed in the original Chua's circuit and Chua's oscillator with bifurcation trees constructed for three types of memristor oscillators derived from the original Chua's circuit by replacement of nonlinear elements by a memristive element. Bifurcation phenomena observed are very rich and include period doubling sequences, period adding sequences, coexisting attractors, attractor collisions, periodic windows.

1. Introduction

27 years have passed since the first observations of chaos in electronic circuits. This first chaotic attractor was discovered in Chua's circuit [19] — very simple circuit containing two linear capacitors, one linear inductor, one linear resistor and an active nonlinear (piecewise-linear) resistor called Chua's diode. Since then numerous theoretical and experimental studies were carried out by researchers in many laboratories around the world — chaos in electronic circuits became an independent area of research. Hundreds of publications have been

published exploring various aspects of nonlinear dynamics and chaos in Chua's circuit and its variants [5–11, 16–18]. A new impact on this area of research is linked with the real implementation of the memristor concept. Last three years have brought a wide-spread interest in applications of circuits containing memristors.

This is due to the fact that after a long period since this kind of elements have been proposed by L.O. Chua [12] there has been a breakthrough in technology permitting the fabrication of microelectronic memristive devices [21, 22]. In this paper we study and compare bifurcation phenomena observed in the original Chua's circuit, its variant called Chua's oscillator (with a series resistor added in series with the inductor) and three types of memristor oscillators derived from the original Chua's circuit by replacement of some elements by a suitable memristive element proposed by Itoh and Chua [15] . The circuit configurations considered are shown in Fig. 1.

2. Comparison of Circuit Equations

The dimension-less equations governing the prototype Chua's circuit read

$$\left.\begin{aligned} \frac{dx}{dt} &= \alpha(y - f(x)), \\ \frac{dy}{dt} &= x - y + z, \\ \frac{dz}{dt} &= -\beta y, \end{aligned}\right\} \tag{1}$$

while for the Chua's oscillator we add one term in the RHS of the first equation

$$\left.\begin{aligned} \frac{dx}{dt} &= \alpha(y - x - f(x)), \\ \frac{dy}{dt} &= x - y + z, \\ \frac{dz}{dt} &= -\beta y, \end{aligned}\right\} \tag{2}$$

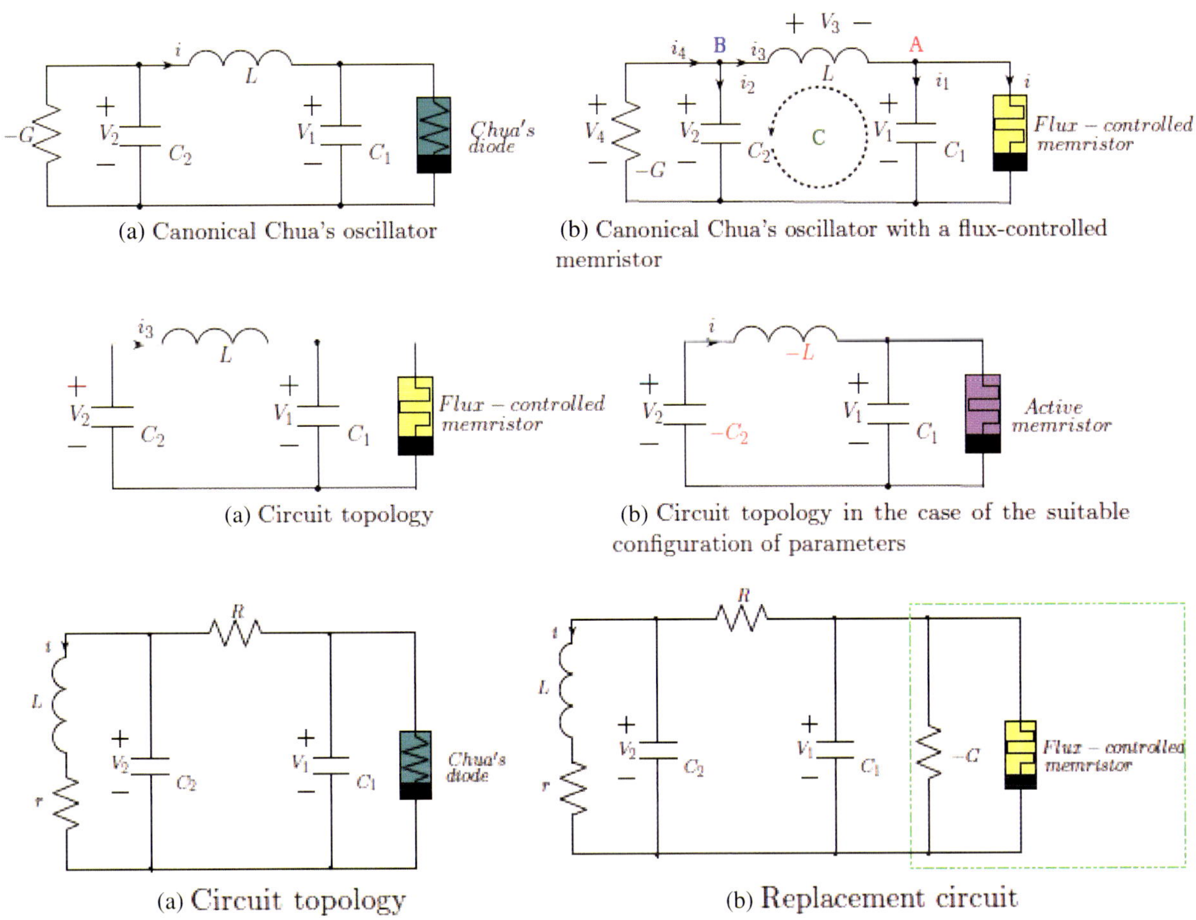

Fig. 1. Consider three circuit topologies as proposed by Itoh and Chua [15]. (Note: The circuit in the second row contains negative L and C elements thus is non-physical.

In both equations the nonlinear element is a three-segment piecewise linear function

$$i_R = f(v_R) = \begin{cases} m_1 v_R + m_1 - m_0 & \text{for } v_R \leq -1 \\ m_0 v_R & \text{for } -1 \leq v_R \leq 1 \\ m_1 v_R + m_0 - m_1 & \text{for } v_R \geq 1 \end{cases} \tag{3}$$

Adding to the canonical Chua's circuit a memristive element we introduce an additional variable in the state equations describing the dynamics of the circuit — thus we have a system of four nonlinear differential equations. It should be noted however that because of the fact that one of the state variables (voltage across the capacitor) is also the voltage across the memristor, one of the equations constitutes a constraining equation for system dynamics. A thorough discussion of the state equations of memristive oscillators can be found in [4]. The state equation for the first type of memristor oscillator circuit reads

$$\left.\begin{aligned} \frac{dv_1}{dt} &= \frac{1}{C_1}(i_3 - W(\varphi)v_1), \\ \frac{di_3}{dt} &= \frac{1}{L}(v_2 - v_1), \\ \frac{dv_2}{dt} &= \frac{1}{C_2}(-i_3 + Gv_2), \\ \frac{d\varphi}{dt} &= v_1. \end{aligned}\right\} \tag{4}$$

The parameters are chosen:

$L = 1$, $C_2 = 1$, $G = 0.65$, only C_1 can be varied in the interval $C_1 \in \langle 0, 0.22 \rangle$, and in normalized form

$$x = v_1, \quad y = i_3, \quad z = v_2, \quad w = \varphi,$$

$$\alpha = \frac{1}{C_1}, \quad \beta = \frac{1}{C_2}, \quad \gamma = \frac{G}{C_2}, \quad L = 1$$

$$\left.\begin{aligned} \frac{dx}{dt} &= \alpha(y - W(w)x), \\ \frac{dy}{dt} &= z - x, \\ \frac{dz}{dt} &= -\beta y + \gamma z, \\ \frac{dw}{dt} &= x. \end{aligned}\right\} \tag{5}$$

The second configuration in normal conditions is not able to generate chaotic solutions, but can behave as an oscillator when an active memristor is introduced. The circuit equations are given by

$$\left.\begin{aligned} \frac{dv_1}{dt} &= \frac{1}{C_1}(i_3 - W(\varphi)v_1), \\ \frac{di_3}{dt} &= \frac{1}{L}(v_2 - v_1), \\ \frac{dv_2}{dt} &= -\frac{1}{C_2}i_3, \\ \frac{d\varphi}{dt} &= v_1. \end{aligned}\right\} \tag{6}$$

The parameters are chosen:

$L = -1$ and $C_2 = -0.05$, only C_1 can be varied in the interval $C_1 \in \langle 0, 0.16 \rangle$. In this type of configuration, many nonlinear phenomena including different types of periodic orbits, period-adding bifurifications leading to chaos can be observed. Introducing new variables

$$x = v_1, \quad y = i_3, \quad z = v_2, \quad w = \varphi,$$

$$\alpha = \frac{1}{C_1}, \quad \beta = \frac{1}{C_2}, \quad \xi = \frac{1}{L},$$

the equation can be rewritten as follows:

$$\left.\begin{aligned} \frac{dx}{dt} &= \alpha(y - W(w)x), \\ \frac{dy}{dt} &= \xi(x + z), \\ \frac{dz}{dt} &= -\beta y, \\ \frac{dw}{dt} &= x. \end{aligned}\right\} \tag{7}$$

The dynamics of the third configuration are governed by the set of equations:

$$\left.\begin{aligned} \frac{dv_1}{dt} &= \frac{1}{C_1}\left(\frac{v_2 - v_1}{R} + Gv_1 = W(\varphi)v_1\right), \\ \frac{dv_2}{dt} &= \frac{1}{C_2}\left(\frac{v_1 - v_2}{R} - i\right), \\ \frac{di}{dt} &= \frac{1}{L}(v_2 - ri), \\ \frac{d\varphi}{dt} &= v_1. \end{aligned}\right\} \tag{8}$$

This circuit will be considered with the following values of parameters:

$L = 1$, $r = 0.35$, $G = 1.5$ and $C_1 = 0.1$, only C_1 can be varied in the interval $C_1 \in \langle 0.06, 0.08 \rangle$.

Substituting

$$x = v_1, \quad y = v_2, \quad z = -i, \quad w = \varphi, \quad \alpha - \frac{1}{C_1},$$
$$\beta = \frac{1}{L}, \quad \gamma = \frac{r}{L}, \quad \xi = G, \quad C_2 = 1, \quad R = 1$$

the above equation can be rewritten as follows:

$$\left.\begin{aligned} \frac{dx}{dt} &= \alpha(y - x + \xi x - W(w)x, \\ \frac{dy}{dt} &= x - y + z, \\ \frac{dz}{dt} &= -\beta y - \gamma z, \\ \frac{dw}{dt} &= x. \end{aligned}\right\} \quad (9)$$

In all considered types of circuits the nonlinear characteristic of the memristor was chosen as a three-segment piecewise linear one (Chua's diode type, comp. [15]). The slopes of the characteristic were chosen accordingly with the required property of local activity or passivity [15].

3. Bifurcation Analysis

Bifurcation analysis of the five circuits were carried out numerically. Matlab environment was used with the following integration routine parameters: simulation time = [0200], initial conditions ic = [0.010.010.01], numerical integration via the routine ode45 (with $InitialStep = 0.1$, $RelTol = 1e-9$, $AbsTol = 1e-9$, $MaxStep = 0.01$.

3.1. *Bifurcation trees for Chua's circuit and oscillator*

Chua's circuit and Chua's oscillator show very similar trajectory behavior (see Figs. 3 and 4) and what follows structure of bifurcation trees (compare Figs. 2–6). Bifurcations observed in Chua's circuit when parameter α was varied reveal two branches of the bifurcation corresponding to starting the experiments from two different (symmetric) initial conditions. Because of the perfect symmetry of the nonlinear characteristic we can observe coexisting symmetric trajectories in a wide range of system parameters — here we see periodic orbits of various length coming in pairs, each observed for an appropriate selection of initial conditions. Eventually the two attractors merge forming single chaotic attractor of the double scroll type. It can be noticed that the branches of the bifurcation tree (still corresponding to coexisting trajectories symmetric in state space) are not symmetric due to the selection of the variable selected in the display of the bifurcation tree. However the bifurcation points and the regions of existence of various types of attractor coincide. The structure of periodic orbits is extremely interesting and very rich — some very long trajectories can be seen in periodic windows e.g. period-12, period-23, period-32. The magnification of the parameter range in which period doubling sequence can be observed (Fig. 6) reveals the existence of a standard type period-3 window characteristic for many kinds of chaotic regimes observed in physical systems. Standard period doubling route to chaos is observed.

3.2. *Trajectories and bifurcations in Chua's memristive oscillators*

When testing the nonlinear behavior of the memristive oscillators first various types of trajectories have been observed together with their changes when varying the selected circuit parameter. In all simulation experiments we used the normalized circuit equations. In the circuit no.1 when changing the parameter β one could observe a period-adding sequence leading to chaos as shown in Fig. 7.

As in all the considered circuits the nonlinear function was selected as a piece-wise linear one with symmerty against the origin, in many cases one could observe coexisting periodic or chaotic attractors symmetric against the origin for the same parameter value. Such attractors when the bifurcation parameter is being varied often change in size, collide in the state space and form one chaotic attractor as shown in Fig. 8.

In the second type of oscillator we could observe an interesting type of heart-shaped periodic trajectories which undergo standard period doubling sequence of bifurcations (see Figs. 9(a)–9(d). The two symmetric chaotic attractors eventually collide and form one chaotic attractor as depicted in Figs. 9(e)–9(h).

3.3. *Bifurcation diagrams*

We constructed a number of bifurcation diagrams selecting various bifurcation parameters and also

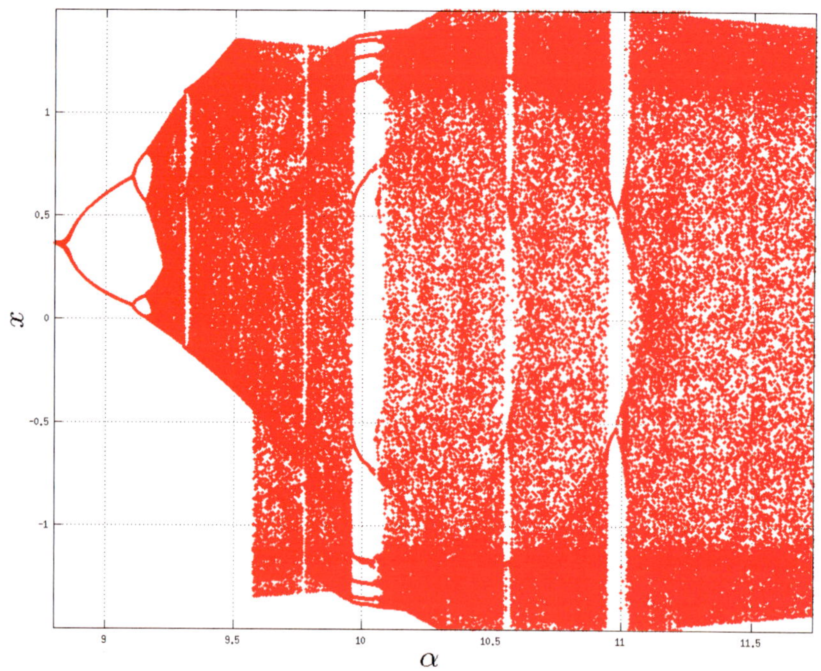

(a)

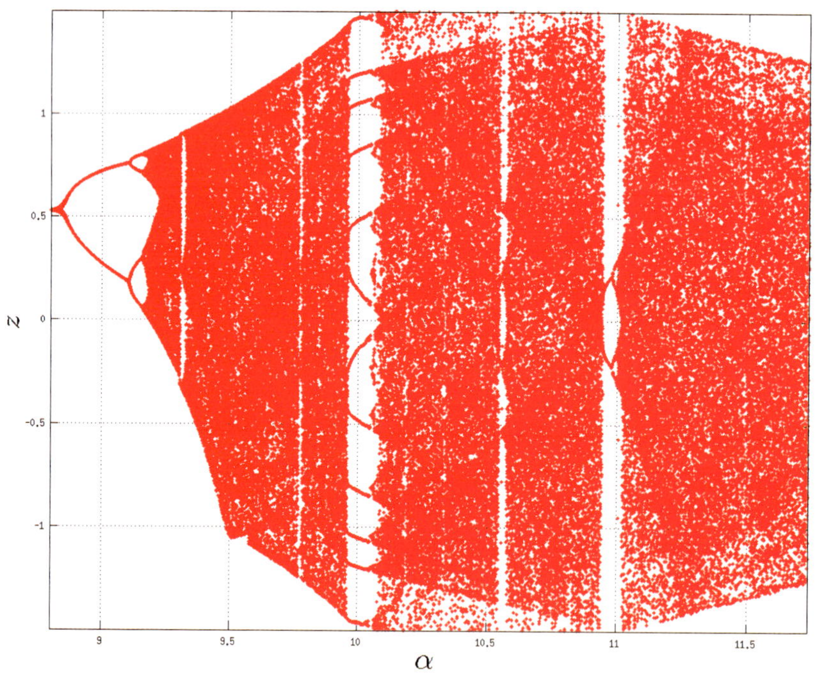

(b)

Fig. 2. Bifurcations observed in Chua's circuit when parameter α was varied. Two separate figures show two branches of the bifurcation tree followed when starting the experiments from two different (symmetric) initial conditions. It can be noticed that these branches are not symmetric due to the selection of the variable selected in the display of the bifurcation tree. However the bifurcation points and the regions of existence of various types of attractor coincide.

selecting different variables for the description of current stationary behavior. A few of the diagrams are shown below. For the construction we used the Matlab environment. All three types of memristor-based oscillators show a variety of bifurcation phenomena leading to chaotic behavior — here typical sequences are period addings and period doublings. A very rich variety of periodic

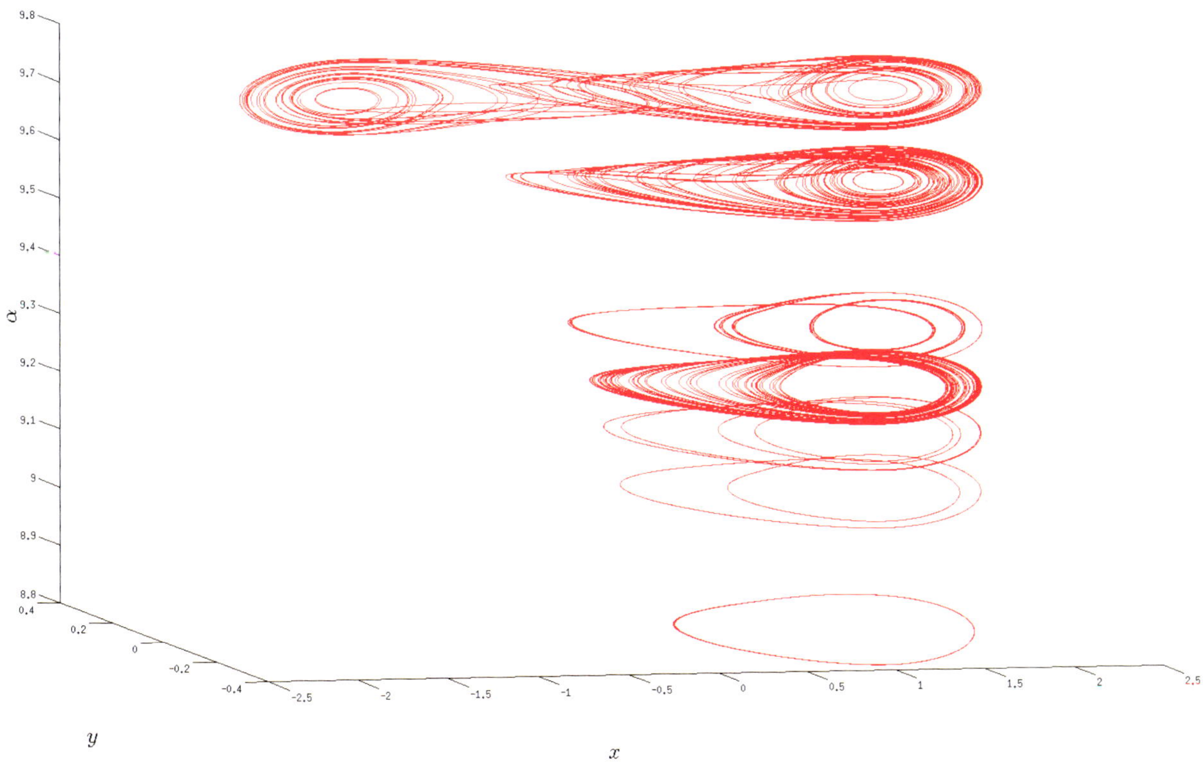

Fig. 3. Trajectories observed in successive period doubling bifurcations observed in Chua's circuit leading to fully developed double scroll attractor when parameter α was varied.

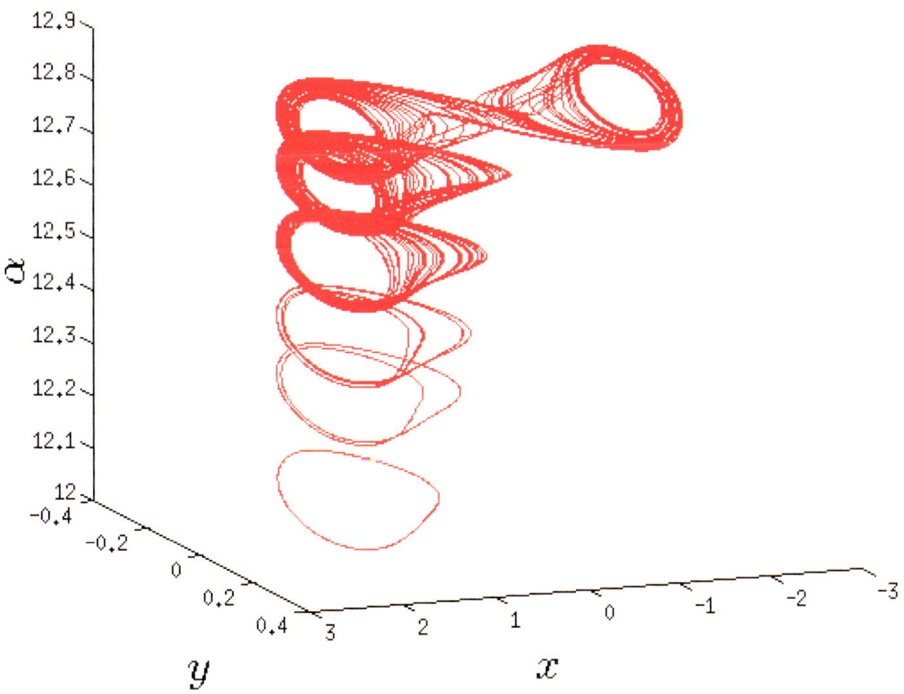

Fig. 4. Trajectories observed in successive period doubling bifurcations observed in Chua's oscillator leading to fully developed double scroll attractor when parameter α was varied — the trajectories observed in experiments show very close similarity to those observed in Chua's circuit depicted in Fig. 3.

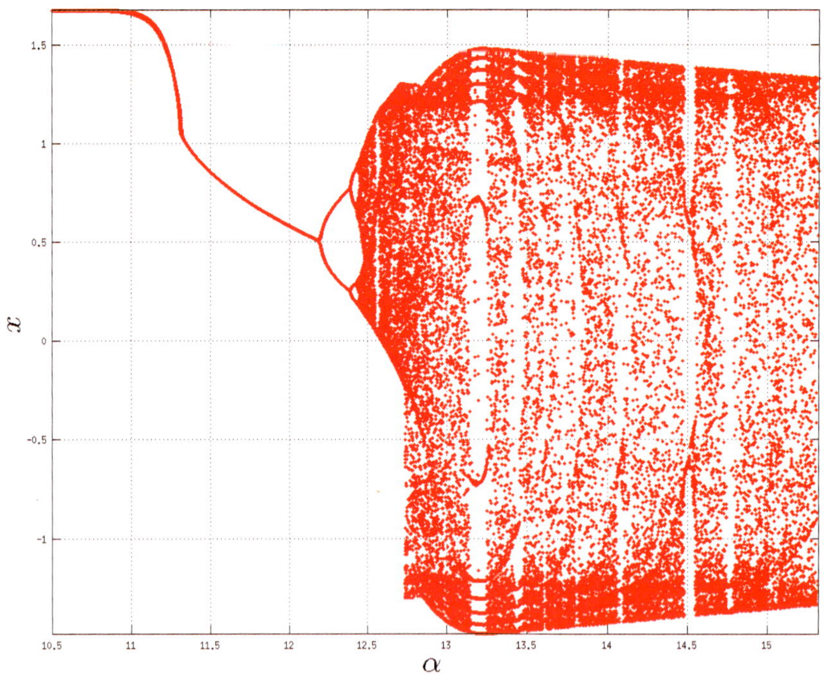

(a)

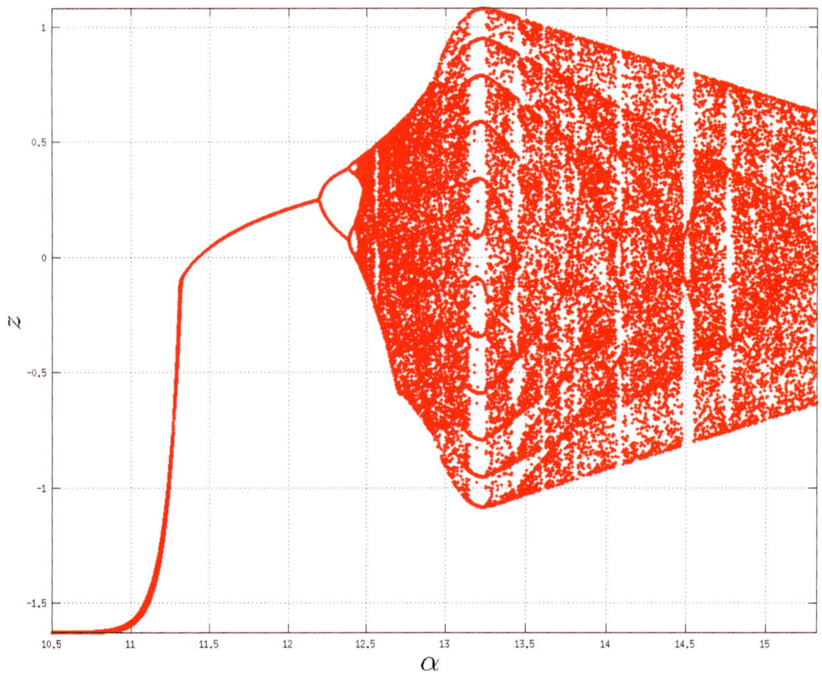

(b)

Fig. 5. Bifurcations observed in Chua's oscillator when parameter α was varied — two branches of the bifurcation tree can be distinguished — each observed when starting the experiment from different set of initial conditions. These branches are not symmetric due to the selection of the variable selected in the display of the bifurcation tree. However the bifurcation points and the regions of existence of various types of attractor coincide.

windows exist within the chaotic ranges in the parameter space. Also very high order periodic windows can be distinguished in the diagrams shown in Figs. 10–12.

In the bifurcation tree obtained when observing the variable w (controlling the memristive element) one can easily distinguish the region where two attractors coexist — there is a "missing"

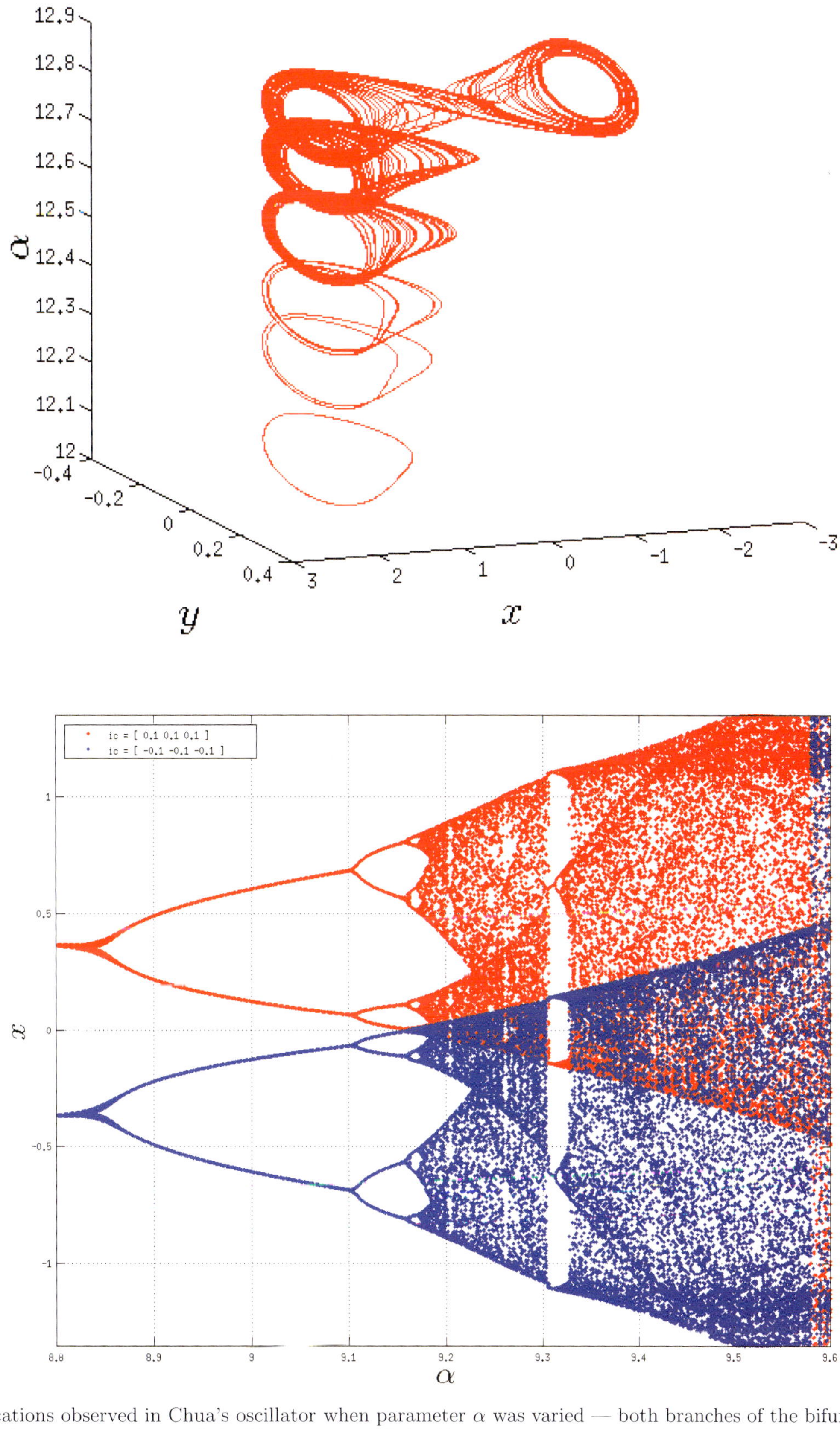

Fig. 6. Bifurcations observed in Chua's oscillator when parameter α was varied — both branches of the bifurcation tree and magnification of the region of period doublings — period doubling sequence is easily identified. Also the period-3 window within the chaos range is clearly visible.

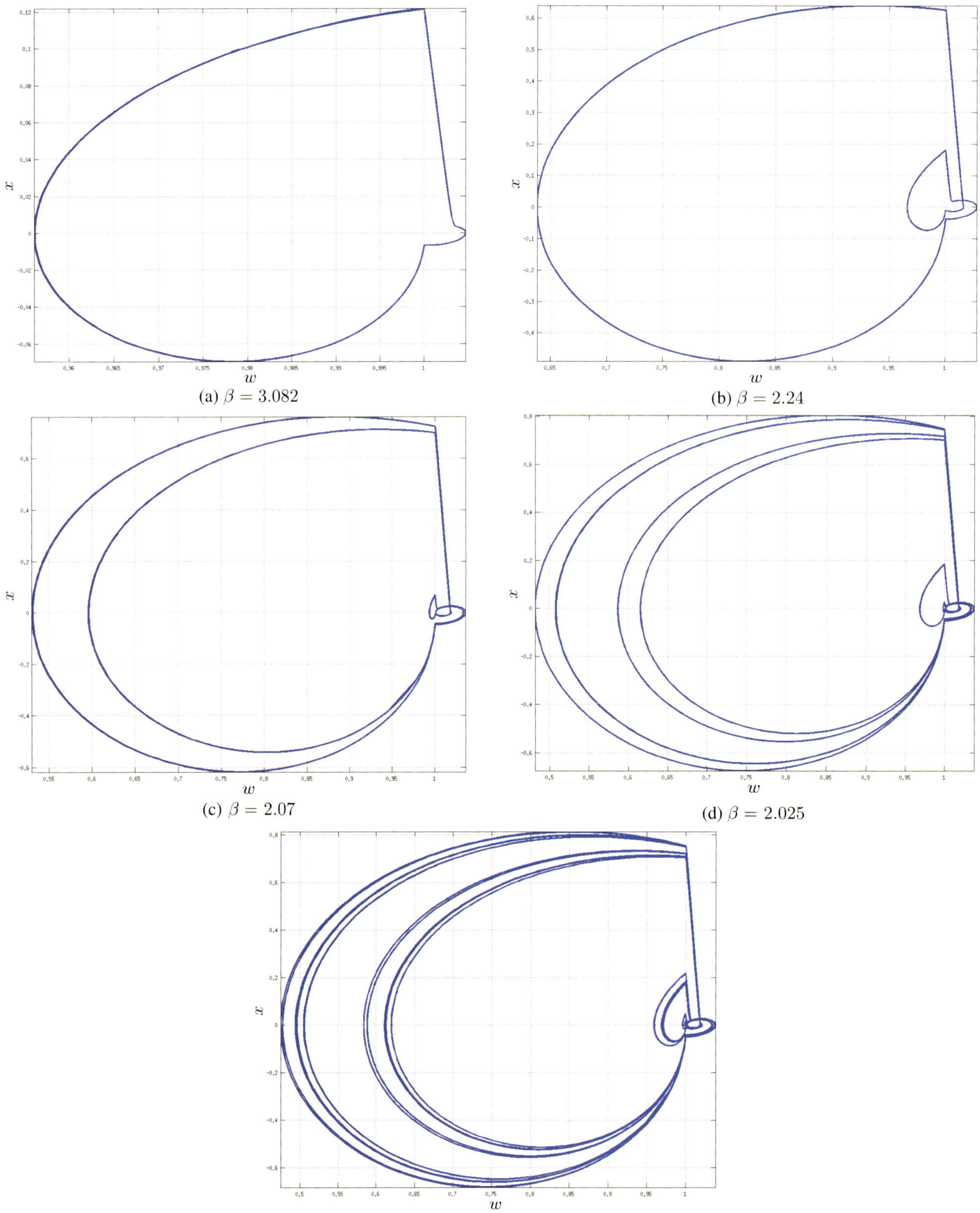

(a) $\beta = 3.082$

(b) $\beta = 2.24$

(c) $\beta = 2.07$

(d) $\beta = 2.025$

(e) $\beta = 2.017$

Fig. 7. Period adding bifurcations observed in circuit 1 when parameter β was varied. Two sizes of pieces of same orbit are visible — small-size rotation near the equilibrium point followed by large amplitude oscillation during one period of the trajectory. Different number of rotations of small and large amplitude in the successive orbits is also visible.

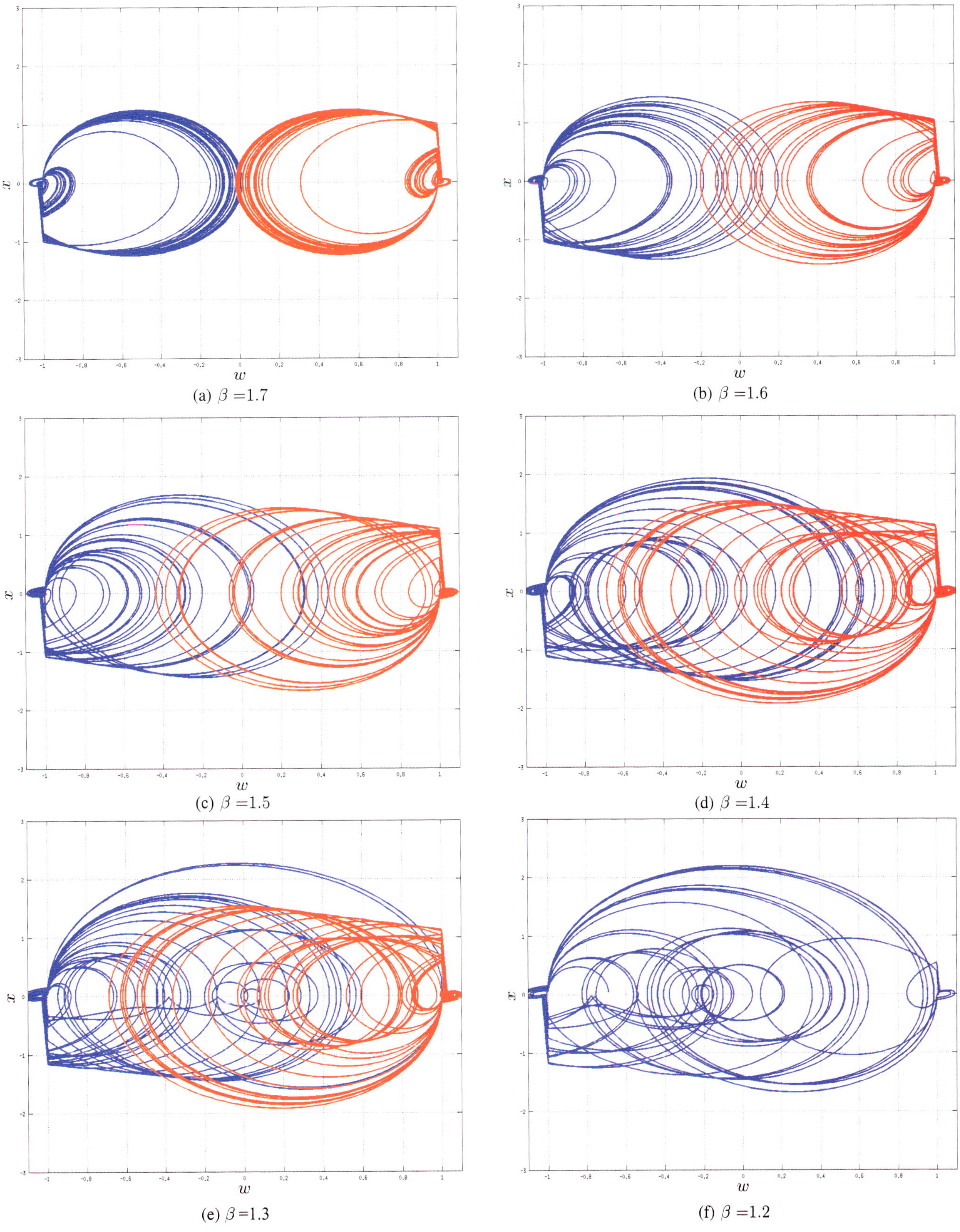

(a) β =1.7 (b) β =1.6 (c) β =1.5 (d) β =1.4 (e) β =1.3 (f) β =1.2

Fig. 8. Symmetry breaking — merging of attractors existing in the memristive circuit type 1.

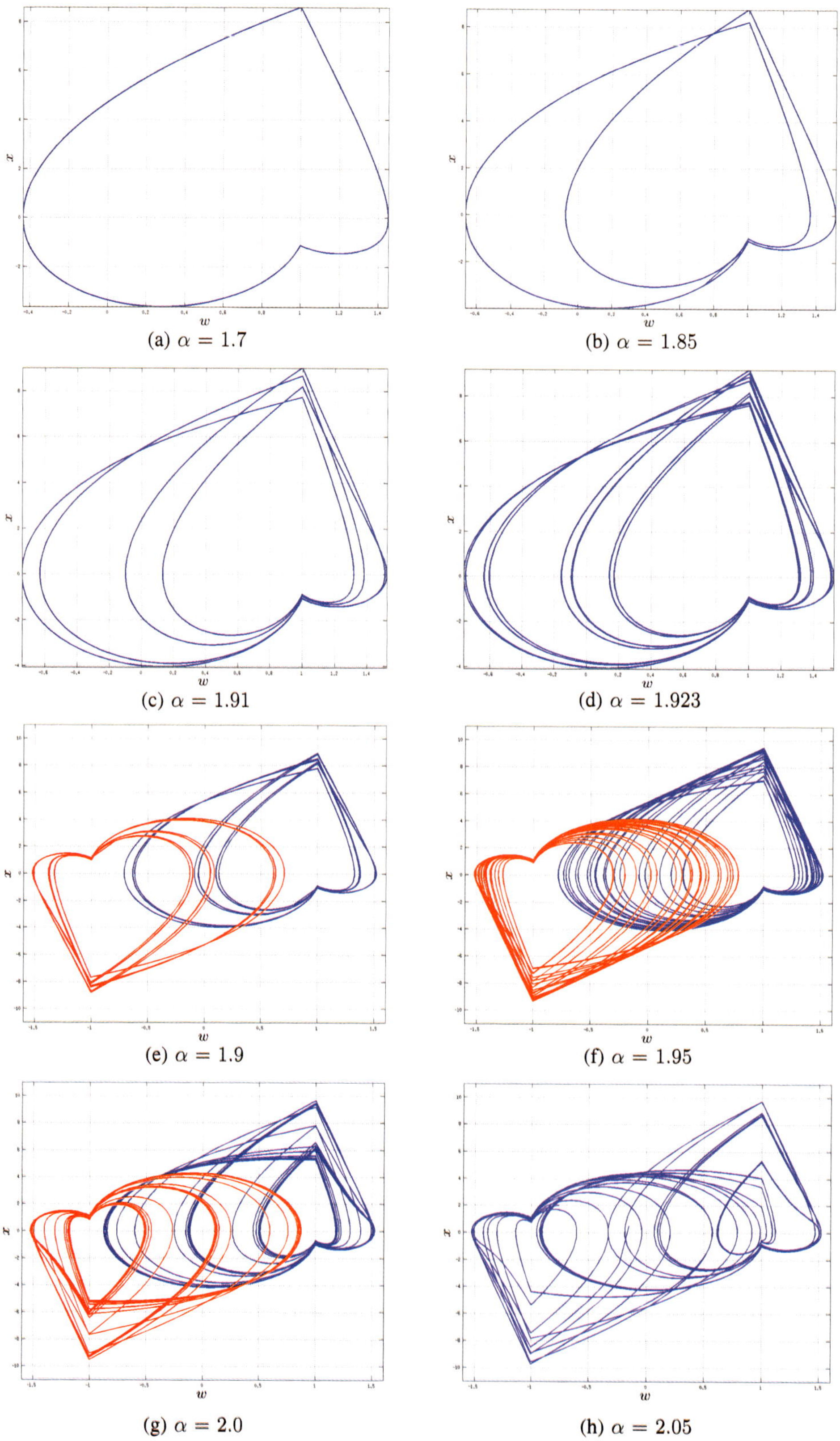

(a) $\alpha = 1.7$ (b) $\alpha = 1.85$

(c) $\alpha = 1.91$ (d) $\alpha = 1.923$

(e) $\alpha = 1.9$ (f) $\alpha = 1.95$

(g) $\alpha = 2.0$ (h) $\alpha = 2.05$

Fig. 9. Sequence of period doubling bifurcations of herat-shaped orbits observed in type 2 oscillator when parameter α is varied (a–d); Merging of the heart-shaped attractors when the bifurcation parameter α is being varied — two coexisting attractors (marked in blue and red) collide forming one attractor occupying the space of both predecessors (e–h).

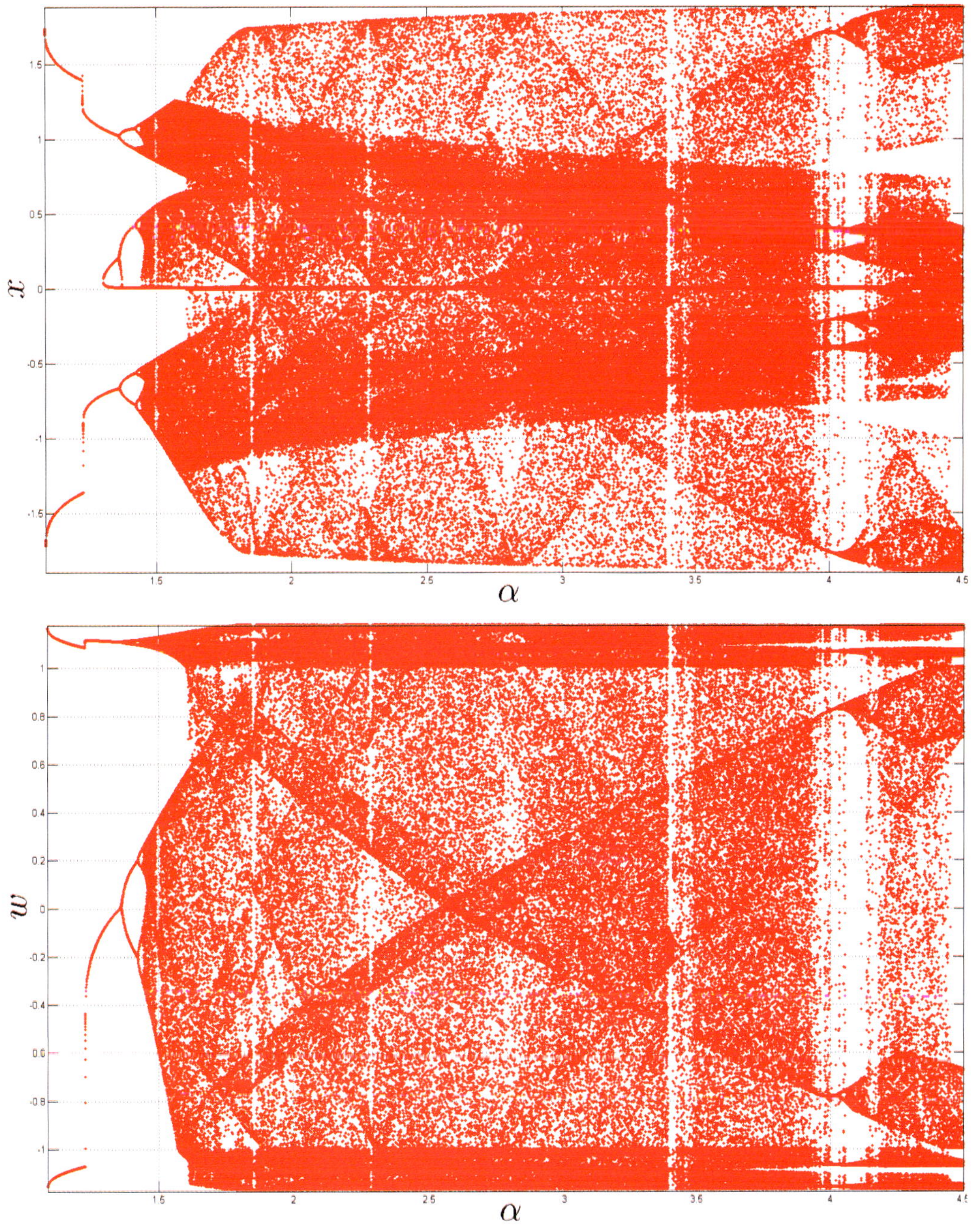

Fig. 10. Bifurcation trees for memristive oscillator type 1.

part in the diagram which could be only observed when starting the experiments from different initial conditions. Classic period doubling route to chaos is also clearly visible. Very interesting types of high-order periodic windows are also visible in both diagrams of Fig. 10. The bifurcation diagrams for this type of oscillator reveal also some so far not explained behaviors, namely first bifurcations observed for small values of parameter α suggest standard period doubling sequence but there are also some "bubbles" formed — possibly these are artefacts of the type of section/projection used and should require further investigation. Very interesting bifurcation structures can be seen in Fig. 11 constructed for oscillator type 2, namely secondary period doublings within the periodic windows are very clearly exemplified — period-8 window undergoing period doubling, period-15

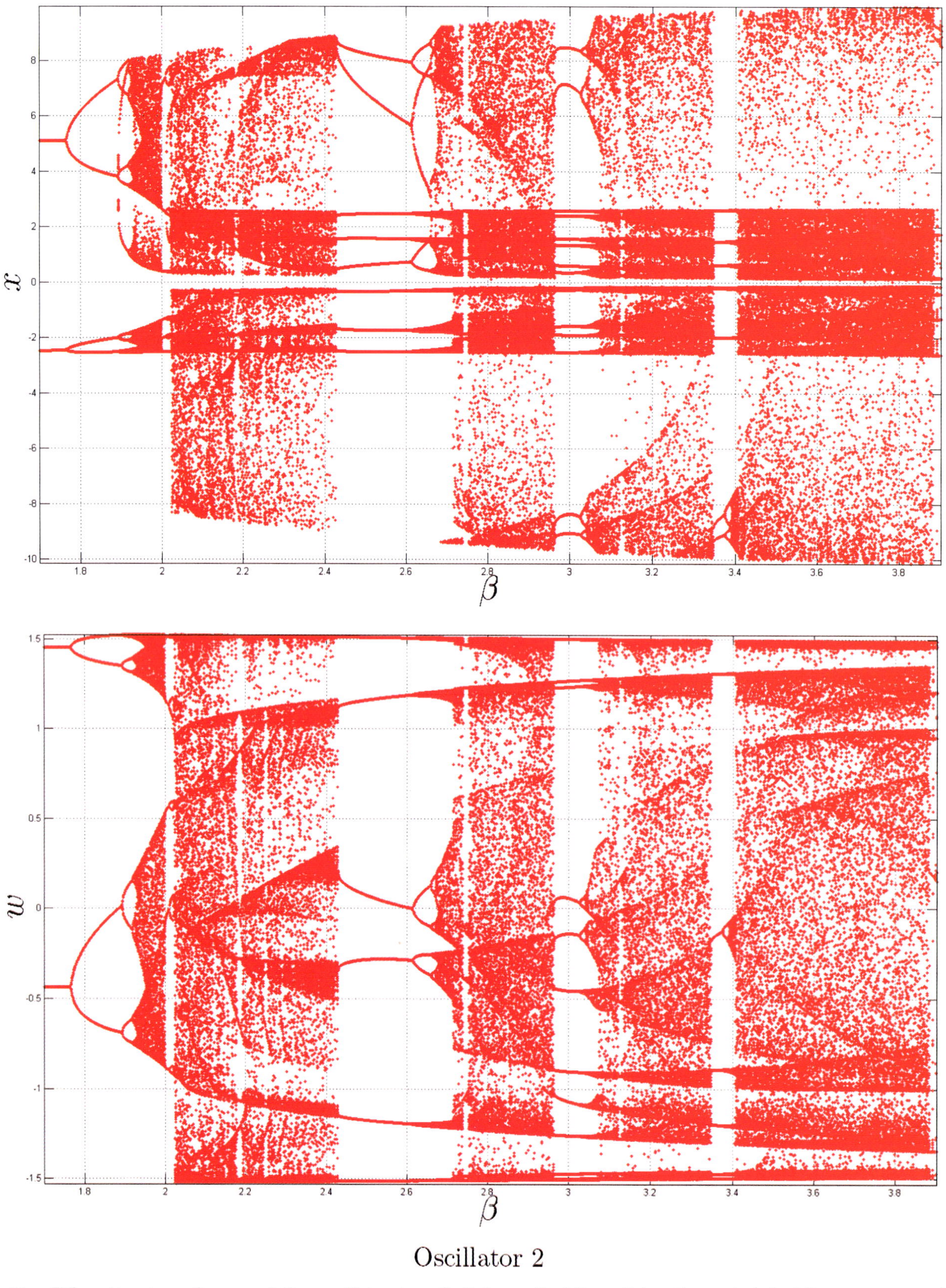

Fig. 11. Bifurcation trees for memristive oscillator type 2. It is worthwhile noticing the period doubling sequences of the periodic orbits in the periodic windows.

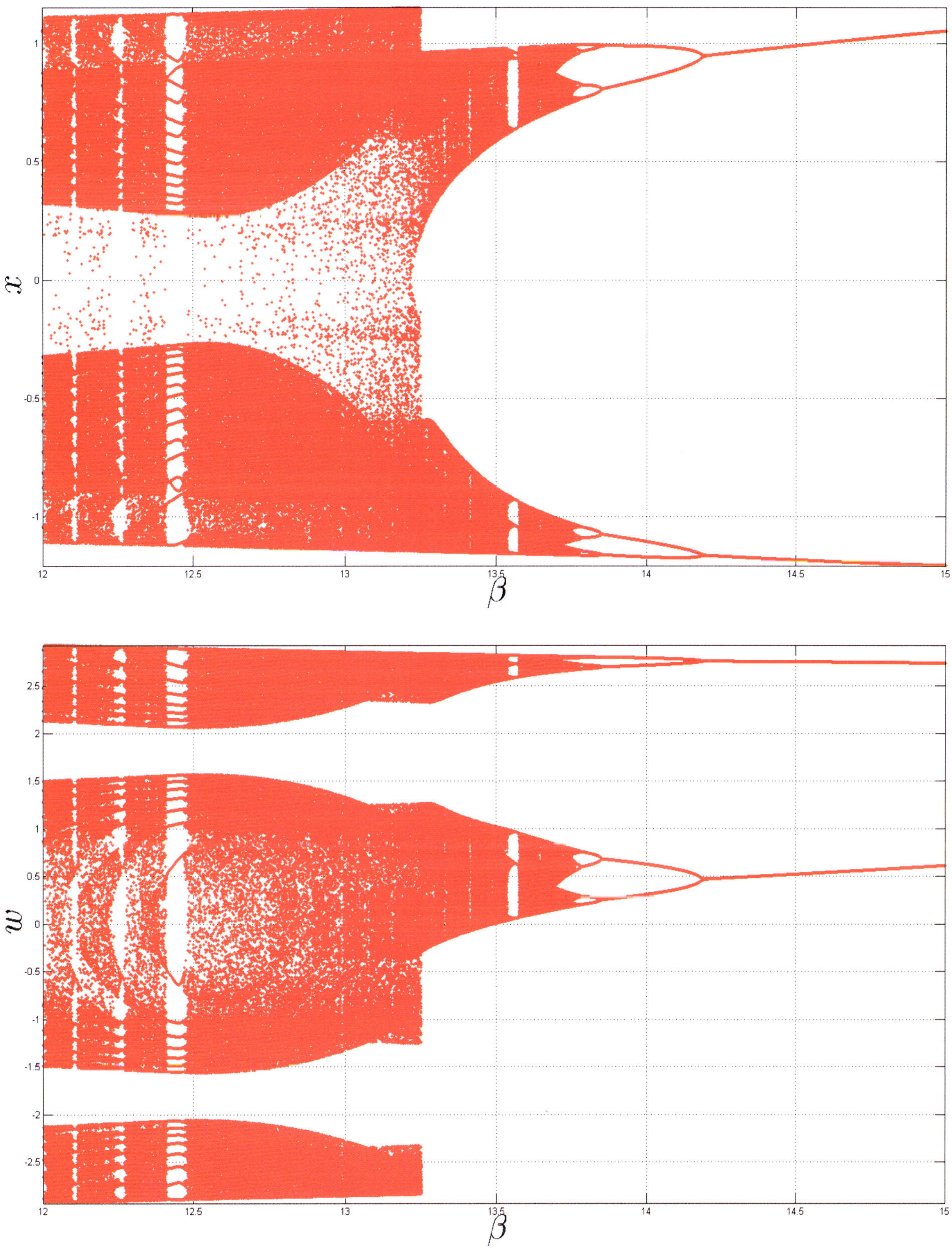

Oscillator 3

Fig. 12. Bifurcation trees for memristive oscillator type 3. Periodic windows of very high period are visible.

window undergoing period doubling sequence etc. In Fig. 12 we notice standard period-3 window and also a sequence of high-period windows starting with period-26 one.

One can notice also comparing the bifurcation trees obtained for the original Chua's circuit and Chua's oscillator with those for memristor-based oscillators that the extra degree of freedom introduced by the memristive element for the bifurcation trees become more developed and complicated. The period doubling sequences persist but within the chaotic range we have a very rich variety of periodic windows and changes of attractor size (jumps).

4. Conclusions

In this paper we reported a study of bifurcation phenomena observed in the family of circuits based on classic Chua's circuit using memristors. Finding the ranges of existence of various dynamic behaviors such as fixed points, periodic orbits, chaos ranges has not been reported before and is clearly important for understanding of circuit dynamics and finding possible applications. Memristor circuits have been already confirmed to have a variety of applications e.g. in memory circuits [3,14,20,23].

References

[1] Bao, B., Ma, Z., Xu, J., Liu, Z., Xu, Q., (2011) A Simple Memristor Chaotic Circuit with Complex Dynamics, Int. J. Bif. Chaos, Vol. 21, No. 8, pp. 2629–2645.

[2] Bilotta, E., Di Blasi, G., Stranges, F. and Pantano, P. (2007) A Gallery of Chua Attractors. Part VI. International Journal of Bifurcation and Chaos, 17: 1801–1910.

[3] Borghetti, J., Snider, G. S. , Kuekes, P.J., Yang, J.J., Stewart, D.R., Williams, R.S., (2010) Memristive switches enable 'stateful' logic operations via material implication, Nature 464 (7290): 873.

[4] Corinto, F., Ascoli, A., Gilli, M. (2011) Nonlinear dynamics of memristor oscillators, *IEEE Trans. CAS*, Part I, Vol. 58, No. 6, pp. 1323–1336.

[5] Chua, L. O. (1992) The Genesis of Chua's Circuit. Archiv fr Elektronik und Ubertragungstechnik, 46: 250–257.

[6] Chua, L. O. (1993) Global Unfolding of Chua's Circuit, IEICE Transactions on Fundamentals of Electronics, Communications, Computer Science, E76 A: 704 734.

[7] Chua, L. O. (1994) Chua's Circuit: An Overview Ten Years Later, Journal of Circuits, Systems and Computers, 4: 117–159.

[8] Chua, L. O. (2005) Local Activity is the Origin of Complexity, International Journal of Bifurcation and Chaos, 15: 3435–3456.

[9] Chua, L. O., Komuro, M., and Matsumoto, T. (1986) The double scroll family, IEEE Transactions on Circuits and Systems, 33: 1072–1118.

[10] Cruz, J. M. and Chua, L. O. (1993) An IC Chip of Chua's Circuit, IEEE Transactions on Circuits and Systems-II, 10: 596–613.

[11] Gandhi, G., Muthuswamy, B., and Roska, T. (2007) Chua's Circuit for High School Students, International Journal of Bifurcation and Chaos, 12.

[12] Chua, L. O. (1971) Memristor The Missing Circuit Element, *IEEE Transactions on Circuit Theory*, CT-18 (5): 507–519, 1971.

[13] Huang, A. S., Pivka, L., Wu, C. W, and Franz, M. (1996) Chua's Equation with Cubic Nonlinearity, International Journal of Bifurcation and Chaos, 12(A): 2175–2222.

[14] Eshraghian, K., Cho, K.R., Kavehei, O., Kang, S.K., Abbott, D., Kang, S.M., (2010) Memristor MOS Content Addressable Memory (MCAM): Hybrid Architecture for Future High Performance Search Engines", IEEE Transactions on Very Large Scale Integration (VLSI) Systems 1005: 3687.

[15] M. Itoh, M., Chua, L.O. (2008) Memristor Oscillators, *Int. J. Bif. Chaos*, Vol. 18, pp. 3183–3206.

[16] Kennedy, M. P. (1992) Robust Op Amp Realization of Chua's Circuit, Frequenz, 46: 66–80.

[17] Kennedy, M. P. (2005) Chua's Circuit, Encyclopedia of Nonlinear Science (Editor : A. Scott) 136–138.

[18] Madan, R. N. (1993) Chua's Circuit: A Paradigm for Chaos, World Scientific, Singapore.

[19] Matsumoto, T. (1984) A Chaotic attractor from Chua's Circuit, IEEE Transaction on Circuits and Systems, 31: 1055–1058.

[20] Snider G. *et al.*, (2011) From Synapses to Circuitry: Using Memristive Memory to Explore the Electronic Brain, IEEE Computer 44(2): 21–28, 2011.

[21] Strukov, D.B., Snider, G.S., Stewart, D.R., Williams, S.R. (2008) The missing memristor found, Nature 453(7191): 80–83, 2008.

[22] Tour, J.M., He, T. (2008), Electronics: The fourth element, Nature 453(7191): 42–43.

[23] Versace, M., Chandler, B., (2010) MoNETA: A Mind Made from Memristors, IEEE Spectrum 12: 30–37, 2010.

CHAPTER 14

CHUA'S NONLINEAR DYNAMICS PERSPECTIVE OF CELLULAR AUTOMATA

GIOVANNI E. PAZIENZA

Department of Mathematical Sciences and the Center for Large-Scale Integrated Optimization and Networks (CLION), The University of Memphis, TN, 38152 USA

Chua's 'Nonlinear Dynamics Perspective of Cellular Automata' represents a genuine breakthrough in this area and it has had a major impact on the recent scientific literature. His results have been accurately described in a series of fourteen papers appeared over the course of eight years but there is no compendious introduction to his work. Therefore, here for the first time, we present Chua's main ideas as well as a few unpublished results that have not been included in his previous papers. This overview illustrates the essence of Chua's work by using a clear terminology and a consistent notation, and it is aimed at those who want to approach this subject through a concise but thorough exposition.

1. Introduction

Decades ago John von Neumann and Stanislaw Ulam conjectured that simple dynamical systems interacting locally can generate complex dynamics [1]. That was the starting point of Cellular Automata (CA), a scientific success story that has been extensively studied ever since. Besides von Neumann and Ulam, the work by two other authors is being hailed as a milestone in this discipline: 'Conway's Game of Life' [2], which proved that even simple Cellular Automata can perform universal computation, and 'Wolfram's New Kind of Science' [3], which suggested that Cellular Automata can be used to define a novel approach to science. Because of his recent 'Nonlinear Dynamics Perspective of Cellular Automata', Chua can definitely be added to this prestigious list. His rigorous formalization of Cellular Automata in the framework of nonlinear dynamics has had a major impact in this area, proving a number of new results that are yet to be fully analyzed.

Chua presented his ideas in fourteen long papers [4–17] which are being edited also as a collection of volumes [18]. This monumental work may intimidate those who are new to this field and unfortunately there is no compendious study of Chua's approach to Cellular Automata. For this reason, we introduce the main concepts of the 'Nonlinear Dynamics Perspective of Cellular Automata' as well as propose a unified terminology and notation to amend the minor inconsistencies throughout Chua's papers, which are inevitable in such a voluminous study. Obviously, it is unrealistic to think of distilling into a book chapter all ideas contained in the original works and therefore we only present the main ideas behind this novel approach to Cellular Automata. Moreover, we reedited all Chua's tables and avoid redundant references throughout the text, also because often different aspects of the same topic were analyzed in different papers by Chua.

This chapter is structured as follows: in Sec. 2 we discuss the main features of elementary Cellular Automata, which are the CA model used by Chua; in Sec. 3, we study the dynamics of the single Cellular Automaton cell and prove that it can be used to define rigorously a complexity index for elementary CA rules; in Sec. 4, we focus on the global dynamics of the Cellular Automaton and present the original classification of elementary CA rules proposed by Chua; in Sec. 5, we formalize the link between Cellular Automata and Nonlinear Dynamics; finally, in Sec. 6 we give some final remarks on Chua's contribution and discuss how it can be fruitfully employed in the future.

2. The Tiny Universe of Globally-independent Elementary CA Rules

2.1. *Brief notes on elementary cellular automata*

In the most general definition, a Cellular Automaton is a lattice of coupled dynamical systems, called *cells*, whose states update synchronously according to a given rule $\mathcal{N}$. In general, each cell interacts only with a limited number of other cells included in its *radius of neighborhood.* If the state space of the Cellular Automaton is denoted by Σ, then any CA rule $\mathcal{N}$ induces a map:

$$\mathcal{M}_{\mathcal{N}} : \Sigma \to \Sigma \tag{1}$$

In general, the mapping of the state $\mathbf{x}^n$ onto the state $\mathbf{x}^{n+1}$, where the superscript refers to the iteration of the discrete-time system, performed by rule $\mathcal{N}$ can be summarized by the notation $\mathbf{x}^{n+1} = \mathcal{M}_{\mathcal{N}}(\mathbf{x}^n)$; similarly, $\mathbf{x}^{n+2} = \mathcal{M}^2_{\mathcal{N}}(\mathbf{x}^n) = \mathcal{M}_{\mathcal{N}}(\mathcal{M}_{\mathcal{N}}(\mathbf{x}^n))$ etc. Throughout this paper, italics and bold font will be used for the states of the cells and the Cellular Automaton, respectively.

There are countless models of Cellular Automata depending on the setting of its parameters (dimension of the lattice, number of states of the cells, radius of neighborhood etc.) Following the same approach as the majority of the authors, Chua studied the so-called *elementary* Cellular Automata that consist of a one-dimensional array of L binary cells, each interacting only with the two nearest neighbors. Therefore, the state of the cell in position i at iteration $n+1$ depends on the states of the cells in positions $i-1$, i, and $i+1$ at iteration n: $x_i^{n+1} = \mathcal{M}_{\mathcal{N}}(x_{i-1}^n, x_i^n, x_{i+1}^n)$. In this paper, we assume that the two cells in position 0 and $L-1$ are neighbors, thus $x_0^{n+1} = \mathcal{M}_{\mathcal{N}}(x_{L-1}^n, x_0^n, x_1^n)$ and $x_{L-1}^{n+1} = \mathcal{M}_{\mathcal{N}}(x_{L-2}^n, x_{L-1}^n, x_0^n)$. In other words, the state of an elementary Cellular Automaton is actually a ring because the first and the last cell of the one-dimensional array are 'glued together'. Such boundary conditions are called *periodic* and they are widely used in the scientific literature.

When we use the values ± 1 for the states of the binary cells of the Cellular Automaton,[a]

---

[a]This choice is arbitrary and numerous authors prefer to use 0 and 1 for the states of the Cellular Automaton cells. However, our choice is more convenient when the dynamics of the single cell is considered in the framework of nonlinear dynamics, as explained in Sec. 3.1.

Table 1. Truth tables of an elementary CA rule $\mathcal{N}$ and its globally-equivalent rules $\overline{\mathcal{N}}$, $\mathcal{N}^{\dagger}$, and $\mathcal{N}^{*}$. The output values β_k, $k \in [0,7]$, are binary, $\beta_k = \pm 1$.

x_{i-1}	x_i	x_{i+1}	$\mathcal{M}_{\mathcal{N}}$	$\mathcal{M}_{\overline{\mathcal{N}}}$	$\mathcal{M}_{\mathcal{N}^{\dagger}}$	$\mathcal{M}_{\mathcal{N}^{*}}$
−1	−1	−1	β_0	$\overline{\beta_7}$	β_0	$\overline{\beta_7}$
−1	−1	1	β_1	$\overline{\beta_6}$	β_4	$\overline{\beta_3}$
−1	1	−1	β_2	$\overline{\beta_5}$	β_2	$\overline{\beta_5}$
1	1	1	β_3	$\overline{\beta_4}$	β_6	$\overline{\beta_1}$
1	−1	−1	β_4	$\overline{\beta_3}$	β_1	$\overline{\beta_6}$
1	−1	1	β_5	$\overline{\beta_2}$	β_5	$\overline{\beta_2}$
1	1	−1	β_6	$\overline{\beta_1}$	β_3	$\overline{\beta_4}$
1	1	1	β_7	$\overline{\beta_0}$	β_7	$\overline{\beta_0}$

an elementary CA rule can be represented by means of a truth table, as shown in Table 1. There are $2^8 = 256$ possible combination of the eight values $\beta_k = \pm 1$, $k \in [0,7]$ in Table 1, each corresponding to one CA rule. The elementary CA rules can be numbered from 0 to 255 thanks to the following formula:

$$\mathcal{N} = (\beta_7\beta_6\beta_5\beta_4\beta_3\beta_2\beta_1\beta_0)_2 = \sum_{k=0}^{7}\left(2^k \cdot \frac{\beta_k + 1}{2}\right) \tag{2}$$

At first sight, elementary Cellular Automata may appear not complex enough to deserve to be studied; obviously, it is not the case. Actually, it is possible to prove that some elementary CA rules have the same computational complexity as the universal Turing machine [3]. The great advantage of studying such a simple model is that it is possible to formalize concepts that can be easily extended to more complex CA models. This is the reason why both Chua and Wolfram, alongside innumerable other authors, focused their attention mainly on this particular class.

2.2. *Global and quasi-globally equivalent rules*

It may happen that the dynamics of a rule $\mathcal{N}$ can be inferred from the one of a different rule $\mathcal{N}'$, and vice versa; in this case, the two rules are said to be *globally equivalent* according to the following definition:

Definition 14.1 [Global equivalence]. *Two CA rules $\mathcal{N}$ and $\mathcal{N}'$ are globally equivalent if there exists a transformation $T' : \Sigma \to \Sigma$ such that $\mathcal{M}^n_{\mathcal{N}}(\mathbf{x}^0) = T'\big(\mathcal{M}^n_{\mathcal{N}'}(T'(\mathbf{x}^0))\big)$, $\forall n \in \mathbb{N}$ and any initial state $\mathbf{x}^0$.*

By abuse of notation, we can write $\mathcal{N}' = T'(\mathcal{N})$. It is possible to prove that there are three such transformations. The first one is called *global complementation* $\overline{T}$ and it complements all cells of the state $\mathbf{x}$ of the Cellular Automaton; thus, $\overline{T}(\mathbf{x}) = \overline{\mathbf{x}}$, where $\overline{x_i} = -x_i$ $\forall i \in [0, L-1]$. Given a CA rule $\mathcal{N} = (\beta_7\beta_6\beta_5\beta_4\beta_3\beta_2\beta_1\beta_0)_2$, rule $\overline{\mathcal{N}} = \overline{T}(\mathcal{N})$ can be found directly from Def. 14.1 when $n = 1$. Therefore, $\mathcal{M}^1_{\mathcal{N}}(\mathbf{x}^0) = \overline{\mathcal{M}^1_{\overline{\mathcal{N}}}(\overline{\mathbf{x}^0})}$ which implies that $(x_{i-1}x_ix_{i+1}) \xrightarrow{\mathcal{N}} \beta_j \Rightarrow (\overline{x_{i-1}x_ix_{i+1}}) \xrightarrow{\overline{\mathcal{N}}} \overline{\beta_j}$, $\forall j \in [0,7]$. Applying this transformation to the eight rows of Table 1, we find that $\overline{\mathcal{N}} = \overline{(\beta_0\beta_1\beta_2\beta_3\beta_4\beta_5\beta_6\beta_7)}_2$.

The second transformation is called *left-right transformation* $T^{\dagger}$ and it mirrors the states of the Cellular Automaton; thus, $T^{\dagger}(\mathbf{x}) = \mathbf{x}^{\dagger}$, where $x_i^{\dagger} = x_{L-1-i}$ $\forall i \in [0, L-1]$. Given a CA rule $\mathcal{N} = (\beta_7\ \beta_6\beta_5\ \beta_4\beta_3\beta_2\beta_1\beta_0)_2$, rule $\mathcal{N}^{\dagger} = T^{\dagger}(\mathcal{N})$ can be found directly from Def. 14.1, when $n = 1$. Therefore, $\mathcal{M}^1_{\mathcal{N}}(\mathbf{x}^0) = (\mathcal{M}^1_{\mathcal{N}^{\dagger}}(\mathbf{x}^{0\dagger}))^{\dagger}$ which implies that $(x_{i-1}x_ix_{i+1}) \xrightarrow{\mathcal{N}} \beta_j \Rightarrow (x_{i+1}x_ix_{i-1}) \xrightarrow{\mathcal{N}^{\dagger}} \beta_j$, $\forall j \in [0,7]$. Applying this transformation to the eight rows of Table 1, we find that $\mathcal{N}^{\dagger} = (\beta_7\beta_3\beta_5\beta_1\beta_6\beta_2\beta_4\beta_0)_2$. The third transformation is called *left-right complementation* T^* and it is a combination of the previous two as it mirrors all states of the Cellular Automaton and complements the single cells; therefore, $T^*(\mathbf{x}) = \mathbf{x}^*$, where $x_i^* = \overline{x_{L-1-i}}$ $\forall i \in [0, L-1]$. Given a CA rule $\mathcal{N} = (\beta_7\beta_6\beta_5\beta_4\beta_3\beta_2\beta_1\beta_0)_2$, rule $\mathcal{N}^* = T^*(\mathcal{N})$ can be found directly from Def. 14.1, when $n = 1$. Therefore, $\mathcal{M}^1_{\mathcal{N}}(\mathbf{x}^0) = \big(\mathcal{M}^1_{\mathcal{N}^*}(\mathbf{x}^{0*})\big)^*$ which implies that $(x_{i-1}x_ix_{i+1}) \xrightarrow{\mathcal{N}} \beta_j \Rightarrow (\overline{x_{i+1}x_ix_{i-1}}) \xrightarrow{\mathcal{N}^*} \overline{\beta_j}$, $\forall j \in [0,7]$. Applying this transformation to the eight rows of Table 1, we find that $\mathcal{N}^* = (\overline{\beta_7\beta_3\beta_5\beta_1\beta_6\beta_2\beta_4\beta_0})_2$.

Some elementary CA rules are invariant with respect to some of these transformations. Specifically, 64 rules are fixed points of the left-right transformation $T^{\dagger}$, 16 rules are fixed points of the global complementation $\overline{T}$, and 16 rules are fixed points of the left-right complementation T^*. Remarkably, 8 rules are fixed points of all three transformations. To sum up, 176 rules are globally-equivalent to three other rules, 72 rules are globally-equivalent to one other rule, and eight rules are not globally-equivalent to any other rule. The list of all elementary CA rules and their globally-equivalent rules is displayed in Tables 2 and 3.

In conclusion, the 256 elementary CA rules can be partitioned into 88 equivalence classes, each made up of either one, two or four rules. For each

Table 2. Rules from 0 to 127 and their three globally-equivalent rules: rule $\mathcal{N}$ (first column), left-right transformation $\mathcal{N}^\dagger = T^\dagger(\mathcal{N})$ (second column), global complementation $\overline{\mathcal{N}} = \overline{T}(\mathcal{N})$ (third column), and left-right complementation $\mathcal{N}^* = T^*(\mathcal{N})$ (fourth column).

$\mathcal{N}$	$\mathcal{N}^\dagger$	$\overline{\mathcal{N}}$	$\mathcal{N}^*$	$\mathcal{N}$	$\mathcal{N}^\dagger$	$\overline{\mathcal{N}}$	$\mathcal{N}^*$	$\mathcal{N}$	$\mathcal{N}^\dagger$	$\overline{\mathcal{N}}$	$\mathcal{N}^*$	$\mathcal{N}$	$\mathcal{N}^\dagger$	$\overline{\mathcal{N}}$	$\mathcal{N}^*$
0	0	255	255	32	32	251	251	64	8	253	239	96	40	249	235
1	1	127	127	33	33	123	123	65	9	125	111	97	41	121	107
2	16	191	247	34	48	187	243	66	24	189	231	98	56	185	227
3	17	63	119	35	49	59	115	67	25	61	103	99	57	57	99
4	4	223	223	36	36	219	219	68	12	221	207	100	44	217	203
5	5	95	95	37	37	91	91	69	13	93	79	101	45	89	75
6	20	159	215	38	52	155	211	70	28	157	199	102	60	153	195
7	21	31	87	39	53	27	83	71	29	29	71	103	61	25	67
8	64	239	253	40	96	235	249	72	72	237	237	104	104	233	233
9	65	111	125	41	97	107	121	73	73	109	109	105	105	105	105
10	80	175	245	42	112	171	241	74	88	173	229	106	120	169	225
11	81	47	117	43	113	43	113	75	89	45	101	107	121	41	97
12	68	207	221	44	100	203	217	76	76	205	205	108	108	201	201
13	69	79	93	45	101	75	89	77	77	77	77	109	109	73	73
14	84	143	213	46	116	139	209	78	92	141	197	110	124	137	193
15	85	15	85	47	117	11	81	79	93	13	69	111	125	9	65
16	2	247	191	48	34	243	187	80	10	245	175	112	42	241	171
17	3	119	63	49	35	115	59	81	11	117	47	113	43	113	43
18	18	183	183	50	50	179	179	82	26	181	167	114	58	177	163
19	19	55	55	51	51	51	51	83	27	53	39	115	59	49	35
20	6	215	159	52	38	211	155	84	14	213	143	116	46	209	139
21	7	87	31	53	39	83	27	85	15	85	15	117	47	81	11
22	22	151	151	54	54	147	147	86	30	149	135	118	62	145	131
23	23	23	23	55	55	19	19	87	31	21	7	119	63	17	3
24	66	231	189	56	98	227	185	88	74	229	173	120	106	225	169
25	67	103	61	57	99	99	57	89	75	101	45	121	107	97	41
26	82	167	181	58	114	163	177	90	90	165	165	122	122	161	161
27	83	39	53	59	115	35	49	91	91	37	37	123	123	33	33
28	70	199	157	60	102	195	153	92	78	197	141	124	110	193	137
29	71	71	29	61	103	67	25	93	79	69	13	125	111	65	9
30	86	135	149	62	118	131	145	94	94	133	133	126	126	129	129
31	87	7	21	63	119	3	17	95	95	5	5	127	127	1	1

equivalence class, it is possible to select an arbitrary *representative specimen*; Chua's choice, which will be used throughout this paper, is reported in Table 4. Despite the limited number of globally-independent rules, elementary CA can display an extraordinary variety of behaviors: from the 'dull' rules 0 and 255 (these globally-equivalent rules map the whole state space Σ onto a single point) to the 'mythical' rule 110 which is capable of universal computation.

Globally-equivalent transformations are defined independently from the rule to which they are applied. Nevertheless, there is a weaker equivalence between rules, performed by a *quasi-global transformation* [19], that can be applied only to rules with a specific form. An example of quasi-global equivalent transformation is the *alternating-row complementation* $\widetilde{T}$ which complements the states of all cells only every other iteration, leaving them unchanged in the rest of the iterations. Therefore, rule $\widetilde{\mathcal{N}} = \widetilde{T}(\mathcal{N})$ can be defined as:

$$\mathcal{M}^n_{\widetilde{\mathcal{N}}}(\mathbf{x}) = \mathcal{M}^n_{\mathcal{N}}(\mathbf{x}) \Rightarrow \mathcal{M}^{n+1}_{\widetilde{\mathcal{N}}}(\mathbf{x}) = \overline{\mathcal{M}^{n+1}_{\mathcal{N}}(\mathbf{x})} \tag{3}$$

This formula implies that:

$$\mathcal{M}_{\widetilde{\mathcal{N}}}(\mathbf{x}) = \mathcal{M}_{\mathcal{N}}(\overline{\mathbf{x}}) \tag{4}$$

for a generic Cellular Automaton state $\mathbf{x}$. Nevertheless, only the elementary CA rules with the form $\mathcal{N} = (\overline{\beta_0}\ \overline{\beta_1}\ \overline{\beta_2}\ \overline{\beta_3}\beta_3\beta_2\beta_1\beta_0)_2$ have a quasi-global equivalent rule $\widetilde{\mathcal{N}}$, which has the form $\widetilde{\mathcal{N}} = (\beta_0\beta_1\beta_2\beta_3\overline{\beta_3}\ \overline{\beta_2}\ \overline{\beta_1}\ \overline{\beta_0})$; this implies that $\widetilde{\mathcal{N}} = 255 - \mathcal{N}$. There are only 16 quasi-globally equivalent rules — among which we find all eight rules with no globally-equivalent rule — and they can be grouped into eight pairs:

Table 3. Rules from 128 to 255 and their three globally-equivalent rules: rule $\mathcal{N}$ (first column), left-right transformation $\mathcal{N}^\dagger = T^\dagger(\mathcal{N})$ (second column), global complementation $\overline{\mathcal{N}} = \overline{T}(\mathcal{N})$ (third column), and left-right complementation $\mathcal{N}^* = T^*(\mathcal{N})$ (fourth column).

$\mathcal{N}$	$\mathcal{N}^\dagger$	$\overline{\mathcal{N}}$	$\mathcal{N}^*$	$\mathcal{N}$	$\mathcal{N}^\dagger$	$\overline{\mathcal{N}}$	$\mathcal{N}^*$	$\mathcal{N}$	$\mathcal{N}^\dagger$	$\overline{\mathcal{N}}$	$\mathcal{N}^*$	$\mathcal{N}$	$\mathcal{N}^\dagger$	$\overline{\mathcal{N}}$	$\mathcal{N}^*$
128	128	254	254	160	160	250	250	192	136	252	238	224	168	248	234
129	129	126	126	161	161	122	122	193	137	124	110	225	169	120	106
130	144	190	246	162	176	186	242	194	152	188	230	226	184	184	226
131	145	62	118	163	177	58	114	195	153	60	102	227	185	56	98
132	132	222	222	164	164	218	218	196	140	220	206	228	172	216	202
133	133	94	94	165	165	90	90	197	141	92	78	229	173	88	74
134	148	158	214	166	180	154	210	198	156	156	198	230	188	152	194
135	149	30	86	167	181	26	82	199	157	28	70	231	189	24	66
136	192	238	252	168	224	234	248	200	200	236	236	232	232	232	232
137	193	110	124	169	225	106	120	201	201	108	108	233	233	104	104
138	208	174	244	170	240	170	240	202	216	172	228	234	248	168	224
139	209	46	116	171	241	42	112	203	217	44	100	235	249	40	96
140	196	206	220	172	228	202	216	204	204	204	204	236	236	200	200
141	197	78	92	173	229	74	88	205	205	76	76	237	237	72	72
142	212	142	212	174	244	138	208	206	220	140	196	238	252	136	192
143	213	14	84	175	245	10	80	207	221	12	68	239	253	8	64
144	130	246	190	176	162	242	186	208	138	244	174	240	170	240	170
145	131	118	62	177	163	114	58	209	139	116	46	241	171	112	42
146	146	182	182	178	178	178	178	210	154	180	166	242	186	176	162
147	147	54	54	179	179	50	50	211	155	52	38	243	187	48	34
148	134	214	158	180	166	210	154	212	142	212	142	244	174	208	138
149	135	86	30	181	167	82	26	213	143	84	14	245	175	80	10
150	150	150	150	182	182	146	146	214	158	148	134	246	190	144	130
151	151	22	22	183	183	18	18	215	159	20	6	247	191	16	2
152	194	230	188	184	226	226	184	216	202	228	172	248	234	224	168
153	195	102	60	185	227	98	56	217	203	100	44	249	235	96	40
154	210	166	180	186	242	162	176	218	218	164	164	250	250	160	160
155	211	38	52	187	243	34	48	219	219	36	36	251	251	32	32
156	198	198	156	188	230	194	152	220	206	196	140	252	238	192	136
157	199	70	28	189	231	66	24	221	207	68	12	253	239	64	8
158	214	134	148	190	246	130	144	222	222	132	132	254	254	128	128
159	215	6	20	191	247	2	16	223	223	4	4	255	255	0	0

Table 4. Chua's choice for the representative specimen of the 88 equivalence classes in which the 256 elementary CA rules can be partitioned.

0	1	2	3	4	5	6	7
8	9	10	11	12	13	14	15
18	19	22	23	24	25	26	27
28	29	30	32	33	34	35	36
37	38	40	41	42	43	44	45
46	50	51	54	56	57	58	60
72	73	74	76	77	78	90	104
105	106	108	128	129	130	131	132
133	134	136	137	138	140	142	146
150	152	156	160	161	162	164	166
168	170	172	178	184	200	204	232

$\{15, 240\}$, $\{23, 232\}$, $\{43, 212\}$, $\{51, 204\}$, $\{77, 178\}$, $\{85, 170\}$, $\{105, 150\}$, $\{113, 142\}$. Not all these pairs are globally independent because $85 = T^\dagger(15)$, $240 = T^\dagger(170)$, $113 = T^\dagger(43)$, $212 = T^\dagger(142)$. Therefore, there are only six globally-independent pairs of quasi-globally equivalent rules; namely: $\{15, 240\}$, $\{23, 232\}$, $\{43, 212\}$, $\{51, 204\}$, $\{77, 178\}$, $\{105,150\}$.

3. Local Dynamics of Elementary CA Cells

3.1. *The discrete-time dynamics of single CA cell can be described by a continuous-time nonlinear dynamical system*

Cellular Automata are described as discrete-time dynamical systems; however, any computer used to simulate them is a continuous-time system with a very small, though non-zero, time scale. Therefore, it is meaningful to consider a single cell of the elementary Cellular Automaton as a continuous-time dynamical system with three binary ± 1 inputs (denoted by u_{i-1}, u_i, and u_{i+1}) which are the current state of the cell itself and its two nearest neighbors, and one binary ± 1 output (denoted by y_i) which is the next state of the cell. The inputs are not dependent on the continuous time t as we are analyzing what happens in the single iteration n. The internal state of the continuous-time system is denoted by $x_i(t)$ and it is a real-valued variable of time; even though the notation is very similar, it should not be confused with the discrete-time state of the cell x_i^n at the iteration n. The initial condition is set to be $x_i(0) = 0$. Our goal is proving that the discrete-time evolution from iteration n to iteration $n+1$ of the cell in position i, $x_i^n \xrightarrow{\mathcal{N}} x_i^{n+1}$, which is usually considered instantaneous, can be described by a dynamical system with the form

$$\frac{dx_i}{dt} = f(x_i(t), u_{i-1}, u_i, u_{i+1}) \tag{5a}$$

$$y_i(t) = g(x_i) \tag{5b}$$

where f and g are nonlinear functions, such that its attractors (ie, $y_i(+\infty)$) for all possible combinations of the input encode the truth table of any elementary CA rule. In other words, given a generic rule $\mathcal{N} = (\beta_7\beta_6\beta_5\beta_4\beta_3\beta_2\beta_1\beta_0)_2$, we desire that:

$$(u_{i-1}, u_i, u_{i+1}) = (000) \Rightarrow y_i = \beta_0 \tag{6a}$$

$$(u_{i-1}, u_i, u_{i+1}) = (001) \Rightarrow y_i = \beta_1 \tag{6b}$$

$$(u_{i-1}, u_i, u_{i+1}) = (010) \Rightarrow y_i = \beta_2 \tag{6c}$$

$$(u_{i-1}, u_i, u_{i+1}) = (011) \Rightarrow y_i = \beta_3 \tag{6d}$$

$$(u_{i-1}, u_i, u_{i+1}) = (100) \Rightarrow y_i = \beta_4 \tag{6e}$$

$$(u_{i-1}, u_i, u_{i+1}) = (101) \Rightarrow y_i = \beta_5 \tag{6f}$$

$$(u_{i-1}, u_i, u_{i+1}) = (110) \Rightarrow y_i = \beta_6 \tag{6g}$$

$$(u_{i-1}, u_i, u_{i+1}) = (111) \Rightarrow y_i = \beta_7 \tag{6h}$$

There are several choices for the functions f and g: Chua suggests the following scalar differential equation:

$$\frac{dx_i}{dt} = -x_i(t) + 2y_i(t) + (z_2 + c_1 \cdot |z_1 + c_0 \cdot |z_0 + b_0u_{i-1} + b_1u_i + b_2u_{i+1}||) \tag{7a}$$

$$y_i(t) = \frac{1}{2}(|x_i(t)+1| - |x_i(t)-1|) \tag{7b}$$

where the parameter vector $\mathcal{P} = \{b_0, b_1, b_2, c_0, c_1, z_0, z_1, z_2\} \in \mathbb{R}^8$; it has been proved that similar results can be achieved when f is a polynomial function [20]. Each realization of $\mathcal{P}$ defines one elementary CA rule; however, there are infinite realizations of $\mathcal{P}$ implementing a specific rule $\mathcal{N}$. It can be found experimentally that the choice of the values of the parameters is fairly robust for all 256 elementary CA rules. Even though we only discussed the case of elementary CA, in [4] it is proved that *any* rule of a binary Cellular Automaton, regardless of the spatial dimension and the neighborhood size, can be mapped into a nonlinear dynamical system whose attractors encode the associated truth table. The same paper conjectures that this result holds for Cellular Automata with an arbitrary number of states.

3.2. *A rigorous definition of the complexity index of elementary CA rules*

The attractors of the continuous-time dynamical system of Eqs. (7a) and (7b) can be written as:

$$y_i(+\infty) = \text{sgn } (z_2 + c_1 \cdot |(z_1 + c_0 \cdot |z_0 + b_0u_{i-1} + b_1u_i + b_2u_{i+1}|)|) \tag{8}$$

and, as already mentioned above, there exists a combination of the eight parameters to implement each of the 256 elementary CA rules. However, for some rules this equation can be simplified by removing one of the nested nonlinear functions, thus obtaining:

$$y_i(+\infty) = \text{sgn } (z_1 + c_0 \cdot |z_0 + b_0u_{i-1} + b_1u_i + b_2u_{i+1}|) \tag{9}$$

Remarkably, for some rules is is possible to simplify even more this equation by removing also the other nested nonlinear function, thus obtaining:

$$y_i(+\infty) = \text{sgn}(z_0 + b_0u_{i-1} + b_1u_i + b_2u_{i+1}) \tag{10}$$

The minimal number of nonlinear functions in the state equation necessary to implement a rule $\mathcal{N}$ can

Table 5. Parameters to obtain each of the 38 globally-independent rules with $\kappa = 1$ from Eq. (10).

Rule $\mathcal{N}$	z_0	b_0	b_1	b_2
0	-1	0	0	0
1	-2	-1	-1	-1
2	-2	-1	-1	1
3	-1	-1	-1	0
4	-2	-1	1	-1
5	-1	-1	0	-1
7	-1	-2	-1	-1
8	-2	-1	1	1
10	-1	-1	0	1
11	-1	-2	-1	1
12	-1	-1	1	0
13	-1	-2	1	-1
14	-1	-2	1	1

Rule $\mathcal{N}$	z_0	b_0	b_1	b_2
15	0	-1	0	0
19	-1	-1	-2	-1
23	0	-1	-1	-1
32	-2	1	-1	1
34	-1	0	-1	1
35	-1	-1	-2	1
42	-1	-1	-1	2
43	0	-1	-1	1
50	-1	1	-2	1
51	0	0	-1	0
76	-1	-1	2	-1
77	0	-1	1	-1
128	-2	1	1	1

Rule $\mathcal{N}$	z_0	b_0	b_1	b_2
136	-1	0	1	1
138	-1	-1	1	2
140	-1	-1	2	1
142	0	-1	1	1
160	-1	1	0	1
162	-1	1	-1	2
168	-1	1	1	2
170	0	0	0	1
178	0	1	-1	1
200	-1	1	2	1
204	0	0	1	0
232	0	1	1	1

Table 6. Parameters to obtain each of the 41 globally-independent rules with $\kappa = 2$ from Eq. (9).

Rule $\mathcal{N}$	z_1	c_0	z_0	b_0	b_1	b_2
6	1	-1	2	2	-1	-1
9	1	-1	2	2	1	-1
18	1	-1	2	1	2	1
22	1	-1	1	1	1	1
24	-2	1	0	1	-1	-1
25	3	-1	1	1	-2	4
26	3	-1	-1	2	-1	4
28	3	-1	-1	2	4	-1
30	2	-1	-1	-2	-1	-1
33	1	-1	1	-2	1	2
36	-2	1	0	1	-1	1
37	3	-1	-1	2	-1	-4
38	3	-1	-1	-1	2	4
40	1	-1	-1	-2	-2	1

Rule $\mathcal{N}$	z_1	c_0	z_0	b_0	b_1	b_2
41	1	-1	-1	-1	-1	1
44	3	-1	-1	-4	-2	1
45	2	-1	-1	-2	-1	1
54	2	-1	-1	-1	-2	-1
56	3	-1	1	2	4	-1
57	2	-1	-1	-1	-2	1
60	1	-1	0	1	1	0
72	1	-1	-1	-2	1	-2
73	1	-1	1	1	-1	1
74	3	-1	-1	2	1	4
90	1	-1	0	1	0	1
104	1	-1	1	-1	-1	-1
106	2	-1	-1	1	1	2
108	2	-1	1	-1	-2	-1

Rule $\mathcal{N}$	z_1	c_0	z_0	b_0	b_1	b_2
110	-2	1	-1	1	2	-3
122	3	-1	1	-2	-1	-2
126	2	-1	0	-1	-1	-1
130	1	-1	2	-1	1	-2
131	3	-1	1	4	-2	-1
132	1	-1	1	2	-1	-2
133	3	-1	1	-2	-1	4
134	1	-1	1	1	-1	-1
146	1	-1	-1	1	-1	1
152	3	-1	-1	1	4	-2
156	2	-1	-1	1	2	-1
164	-3	1	-1	-2	1	-2
166	2	-1	1	1	-1	-2

Table 7. Parameters to obtain each of the 41 globally-independent rules with $\kappa = 3$ from Eq. (8).

Rule $\mathcal{N}$	z_2	c_1	z_1	c_0	z_0	b_0	b_1	b_2
27	2	-1	-3	1	-3	-2	-4	1
29	2	-1	3	-1	-3	-2	1	-4
46	2	-1	3	-1	4	1	-2	-3
58	2	-1	3	-1	3	-1	2	-4
78	2	-1	-3	1	4	3	-1	2

Rule $\mathcal{N}$	z_2	c_1	z_1	c_0	z_0	b_0	b_1	b_2
105	1	-1	2	-1	1	-1	1	-1
150	3	-1	7	-1	-3	-4	-2	4
172	2	-1	3	-1	1	-3	-4	2
184	2	-1	-3	1	-4	1	-2	3

be used as a measure of its complexity. In particular, the rules that can be described by Eq. (10) have index of complexity $\kappa = 1$ (one nonlinear function); the rules that cannot be described by Eq. (10) but can be described by Eq. (9) have index of complexity $\kappa = 2$ (two nonlinear functions); finally, the rules that cannot be described by Eq. (9) but can be described by Eq. (8) have index of complexity $\kappa = 3$ (three nonlinear functions). Each rule has a unique index of complexity: among the 88 globally-independent elementary CA rules, 38 rules have $\kappa = 1$, 41 rules have $\kappa = 2$, and 9 rules have $\kappa = 3$. They are displayed in red, blue, and green, respectively, in Tables 2, 3, and 4. Observe that rule 110 has $\kappa = 2$: it is then natural to define such a value as the *threshold of complexity* for elementary CA rules. A similar notion was introduced by Wolfram in [3] but never rigorously formalized. For the reader's convenience, we summarized a realization of $\mathcal{P}$ for all rules with $\kappa = 1$ in Table 5, all rules with $\kappa = 2$ in Table 6, and all rules with $\kappa = 3$ in Table 7.

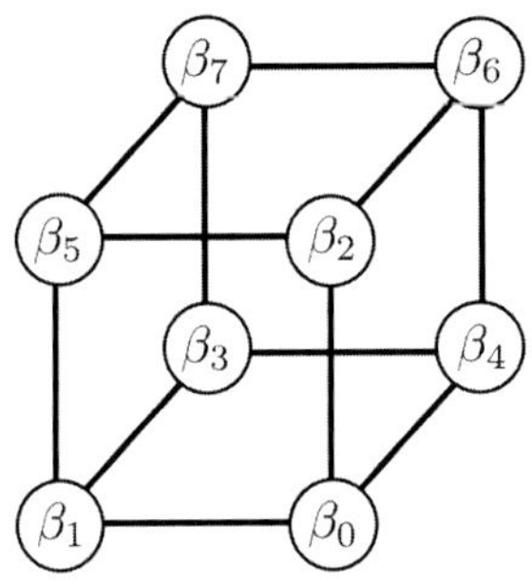

Fig. 1. Boolean cube for a generic elementary CA rule $\mathcal{N}$. Adjacent vertices map rows of the truth table with only one different input.

Furthermore, Chua proposed a more immediate way to find the index of complexity of an elementary CA rule $\mathcal{N}$ based on the so-called *Boolean cubes*. Given the truth table of $\mathcal{N}$ of Table 1, its Boolean cube can be constructed by mapping the next state for each of the eight inputs onto the vertices of a cube, where adjacent vertices rows of the truth table with only one different input (see Fig. 1). Remarkably, the index of complexity κ of an elementary CA rule corresponds to the minimal number of planes necessary to divide the vertices with different states (either $+1$ or -1) in its Boolean cube.

The 256 elementary CA rules can be partitioned according to the ratio of 1s and -1s in their correspondent truth table. For example, rule 1 and rule 254 have a single 1 and a single -1 in their truth table, respectively: then we say that both are 1/7 rules. Obviously, there are five such categories which we list in the following, including also into brackets the number of globally-independent rules belonging to each of them: 0/8 rules (1), 1/7 rules (6), 2/6 rules (18), 3/5 rules (34), and 4/4 rules (29). Within each category we can define a number of equivalence classes composed by those rules whose Boolean cubes (or those of any of the rules globally-equivalent to them) coincide when rotated. For instance, rule 1 and rule 2 are not globally equivalent, however their Boolean cubes coincide when rotated. By following this criterion, the 88 globally-independent elementary CA rules can be classified into 14 distinct equivalence classes [6] called *rotation groups* and denoted by the symbol $\mathcal{R}_j^i$, where the subscript j indicates the number of the category those rules belong (ie, 0 for 0/8 rules, 1 for 1/7 rules etc.) and the superscript i is a progressive number that distinguishes different equivalence classes having the same j. Observe that all and only the quasi-globally equivalent rules are contained in three rotation classes; namely, $\mathcal{R}_4^1$, $\mathcal{R}_4^2$, and $\mathcal{R}_4^7$.

4. Global Dynamics of Elementary Cellular Automata

4.1. *Periodic orbits: attractors and Isles of Eden*

Since the states of an elementary Cellular Automaton are binary arrays with length L, the total number of states is 2^L, when L is finite. Consequently, for any deterministic elementary CA rule, the temporal evolution starting from any initial state repeats itself after at most 2^L iterations (actually, this conservative estimate can be greatly reduced because different states may be equivalent under rotation). This observation allows us to define several fundamental concepts:

Definition 14.2. Given an initial state $\mathbf{x}^0$ and an elementary CA rule $\mathcal{N}$, t is the smallest non-negative integer for which there exists a positive integer l such that

$$\mathcal{M}_{\mathcal{N}}^l(\mathbf{x}^t) = \mathbf{x}^t \tag{11}$$

The parameter t is called transient length and the set of all states $\{\mathbf{x}^n, n \in [0, t-1]\}$ is called transient. The set of all states $\{\mathbf{x}^n, n \in [t, l+t-1]\}$ is called periodic orbit (or limit cycle) originating from the initial state $\mathbf{x}^0$ and l is called length of the orbit. The set of all transients of a given periodic orbit is called basin of attraction of the orbit.

The periodic orbits having a non-empty basin of attraction are called 'attractors'; otherwise, they are called 'Isles of Eden'. In general, an elementary CA rule has both attractors and Isles of Eden depending on the length of the Cellular Automaton and the initial state of the system. Nevertheless, some rules (called *conservative*) have only Isles of Eden, for any L and all initial states, and other rules (called *strictly-dissipative*) have no Isles of Eden at all. It is possible to prove that there are 6 conservative rules (4 of which are globally independent) and 28 strictly-dissipative rules (9 of which are globally independent). When L is finite, the basin of attraction of an attractor contains a finite number of states and some of which — called *Gardens of Eden* [21] — have no predecessors. Formally, a state $\mathbf{x}^n$ is a Garden of Eden if $\nexists\ \mathbf{x}^{n-1} \in \Sigma$ such that

Table 8. Rotation groups in which the 88 globally-independent elementary CA rules can be partitioned.

0/8 and 1/7 rules	
Boolean Cube	Rules
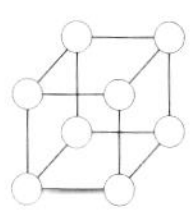	Group $\mathcal{R}_0^1$: 0
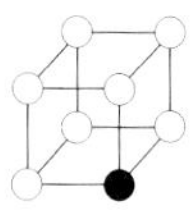	Group $\mathcal{R}_1^1$: 1, 2, 4, 8, 32, 128

2/6 rules	
Boolean Cube	Rules
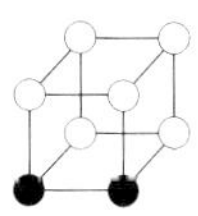	Group $\mathcal{R}_2^1$: 3, 5, 10 12, 34, 136 160
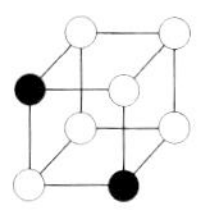	Group $\mathcal{R}_2^2$: 6, 9, 18 33, 40, 72 130, 132
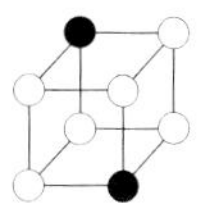	Group $\mathcal{R}_2^3$: 24, 36, 129

3/5 rules	
Boolean Cube	Rules
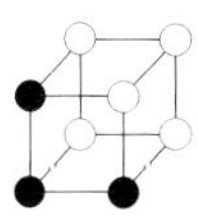	Group $\mathcal{R}_3^1$: 7, 11, 13, 14, 19, 35, 42, 50, 76, 138, 140, 162, 168, 200
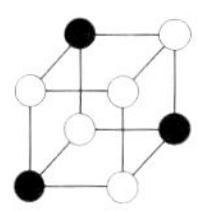	Group $\mathcal{R}_3^2$: 22, 41, 73 104, 134, 146
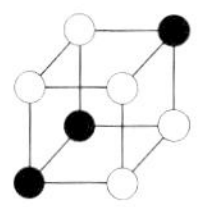	Group $\mathcal{R}_3^3$: 25, 26, 28 37, 38, 44, 56, 74, 131, 133, 137 152, 161, 164

4/4 rules	
Boolean Cube	Rules
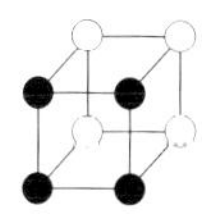	Group $\mathcal{R}_4^1$: 15, 51 170, 204
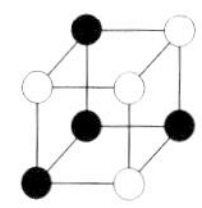	Group $\mathcal{R}_4^2$: 23, 43, 77 142, 178, 232
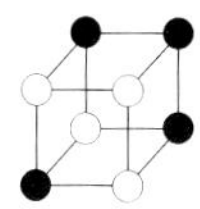	Group $\mathcal{RS}_4^3$: 30, 45, 54 57, 106, 108 156, 166
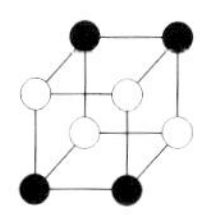	Group $\mathcal{R}_4^4$: 60, 90
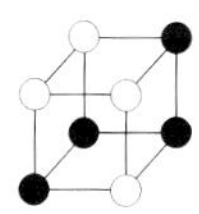	Group $\mathcal{R}_4^5$: 27, 29, 46 58, 78, 172 184
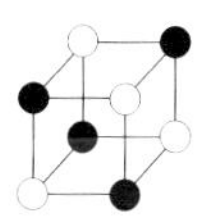	Group $\mathcal{R}_4^6$: 105, 150

$\mathbf{x}^n = \mathcal{M}_{\mathcal{N}}(\mathbf{x}^{n-1})$. An Isle of Eden have no Gardens of Eden, and its basin of attraction is empty, by definition.

Observe that a very similar approach, which was then used for an extensive study of Cellular Automata, was taken in [22].

A periodic orbit may contain two, or more, states that are equivalent under rotation. Therefore, it is necessary to introduce a measure of the number of states between a state of the periodic orbit and any rotated version of it.

Definition 14.3. The period τ of a periodic orbit is the smallest integer for which $\exists \sigma \in \mathbb{Z}$ such that $\mathbf{x}^\tau = \mathcal{S}^\sigma(\mathbf{x}^0)$, where $\mathcal{S}$ it the *right shift operator* and $\mathbf{x}^\tau = \mathcal{M}_{\mathcal{N}}^\tau(\mathbf{x}^0)$. This implies that $x_i^\tau = x_{i+\sigma}^0$ for any cell x_i^0 of $\mathbf{x}^0$, where the subscript refers to the position of the cell within the array.

The value of $|\sigma| \neq 0$ can be conveniently used to make a further distinction among periodic orbits: if $|\sigma| \neq 0$ then it is called Bernoulli-σ_τ periodic orbit; otherwise, it is called period-τ periodic orbit. For *period-τ* orbits $l = \tau$, ie, the length of the orbit is equal to its period; for Bernoulli-σ_τ orbits $l > \tau$, ie, the length of the orbit is greater than its period.

In the following, we will extensively make use of the concepts introduced above. Nevertheless, for consistency with Wolfram's works we will also visualize the evolution of elementary CA rules starting from a given initial state through the so-called

Table 9. Examples of space-time patterns of elementary CA rules and their classification according to Chua (C) and Wolfram (W): (a) Rule 0, C1 and W1; (b) Rule 51, C2 and W2; (c) Rule 62, C3 and W2; (d) Rule 170, C4 and W2; (e) Rule 90, C5 and W3; (f) Rule 110, C6 and W4.

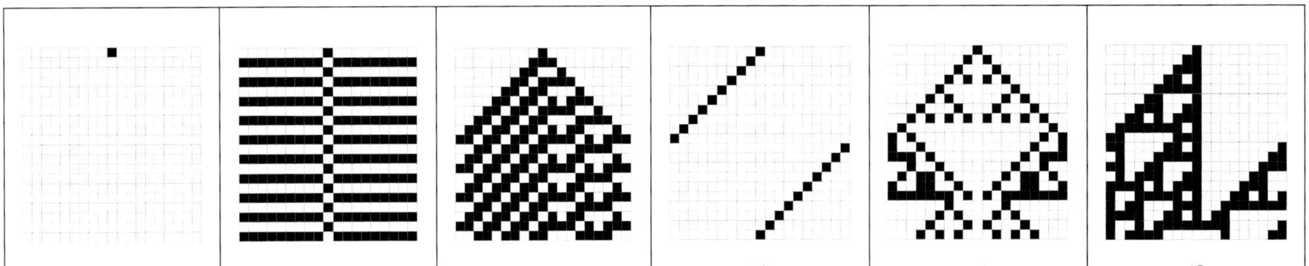

(a) (b) (c) (d) (e) (f)

Table 10. Distribution of the rules of the first four classes in Chua's classification according to the values of σ and τ of their robust periodic orbits. All these rules belong either to W1 or to W2 in Wolfram's classification.

	$\sigma = 0$	$\|\sigma\| = 1$	$\|\sigma\| = 2$
$\tau = 1$	C1	C4	
$\tau = 2$	C2	C4	C4
$\tau = 3$ and $\tau = 6$	C3		

space-time pattern. It is obtained by converting each state of the Cellular Automaton into a black and white string, where cells whose state is -1 (resp. 1) are in white (resp. black). All states are then stacked upon each other, with the initial state being on the first row. Examples of space-time patterns are in Tables 9 and 12.

4.2. *Classification of CA rules according to the robust periodic orbits*

In order to perform a complete analysis of the dynamics of a given elementary CA rule, it is necessary to study the evolution of all 2^L possible initial states,[b] but when L is very large such process is unfeasible. Therefore, numerous authors, including both Chua and Wolfram, limit their analysis to the so-called 'robust' periodic orbits, which are those obtained by using initial states chosen randomly in the state space Σ. Remarkably, the observation of the dynamics of a single robust periodic orbits is usually sufficient to make some general statements about an CA rule.

By performing a *qualitative* of their space-time patterns, Wolfram groups CA rules into four classes based on if the temporal evolution leads to:

- W1: a homogeneous state;
- W2: a set of separated simple stable or periodic structures;
- W3: a chaotic pattern;
- W4: complex localized structures, sometimes long-lived.

Conversely, by performing a *quantitative* of their periodic orbits, Chua groups CA rules into six classes depending on if the rule exhibits:

- C1: period-1 periodic orbits;
- C2: period-2 periodic orbits;
- C3: period-3 or period-6 periodic orbits;
- C4: Bernoulli σ_τ-shift periodic, where σ and τ do not depend on the initial bit string or the length L;
- C5: Bernoulli σ_τ-shift periodic orbits, where σ and τ depend on the initial bit string and/or the length L (bilateral rules);
- C6: Bernoulli σ_τ-shift periodic orbits, where σ and τ depend on the initial bit string and/or the length L (non-bilateral rules).

The difference between the rules of C5 and those of C6 is that the former are bilateral, which means that $\mathcal{N}^\dagger = \mathcal{N}$. The fact that a rule belongs to one of Chua's groups does not imply that all its periodic orbits have the same characteristics (ie, Bernoulli parameters σ and τ) as the robust ones. Actually, in the next section we will show that it is not the case. In Table 9 there is the space-time patterns of one rule for each of Chua's six groups. The rules of the first four groups in Chua's classification have a well-defined value of σ, and hence they can be

[b]This number can be actually be reduced approximately by a factor of L when strings obtained by rotating the initial state are not studied separately.

Table 11. Classification according to Chua of the 88 globally-independent elementary CA rules.

25 Period-1 rules					
0	4	8	12	13	32
28	40	44	72	76	77
78	104	128	132	136	140
160	164	168	172	200	204
232					

13 Period-2 rules			
1	5	19	23
28	29	33	37
50	51	108	156
178			

30 Bernoulli σ_τ-rules					
2	3	6	7	9	10
11	14	15	24	25	27
34	35	38	42	43	46
56	57	58	74	130	134
138	142	152	162	170	184

2 Period-3 rule			
131	133		

10 Complex Bernoulli-shift rules					
18	22	54	73	90	105
129	146	150	161		

8 Hyper Bernoulli-shift rules			
26	30	41	45
60	106	137	166

Table 12. Examples of non-robust periodic orbits with $\tau = 3$ for $L = 14$: (a) Rule 164, $\sigma - 1$; (b) Rule 37, $\sigma = 1$; (c) Rule 9, $\sigma = 2$; (d) Rule 25, $\sigma = 3$.

grouped according to their values of the Bernoulli parameters σ and τ too, as shown in Table 10. Observe that the maximum period for the robust periodic orbits of Bernoulli σ_τ rules is $\tau = 2$. An exhaustive analysis of all elementary CA rules of these four groups (performed in [16] and [17]) has led to the conclusion that each of them has *at most* two kinds of robust periodic orbits (i.e. different Bernoulli parameters σ and τ). In particular, rules 14, 43, 57, 142, 184 (all of them belonging to C4) have exactly two kinds of robust periodic orbits; all other rules have only one kind of robust periodic orbit. The complete classification according to Chua of the 88 globally-independent elementary CA rules is shown in Table 11.

Chua's and Wolfram's classifications can be easily related: in particular, rules of W1 are a proper subset of those of C1; rules of W2 can be in any of the groups from C1 to C4; rules of W3 and W4 can be either in C5 or in C6. At the same time, these two classifications are somehow complementary because Wolfram's classification is qualitative but it can be applied to any Cellular Automaton model regardless the number of states, spatial arrangement, neighborhood etc., whereas Chua's classification is quantitative but so far it has been applied only to elementary CA rules.

4.3. *Classification of CA rules according to the non-robust periodic orbits*

Generally, in addition to the robust periodic orbits, which are generated starting from a randomly chosen initial state, CA rules have also non-robust periodic orbits, which can be reached only from some specific initial states. When they exist, the density of non-robust periodic orbits tends to 0 as the size of the state space grows. We recall that the most complex rules — ie, those of groups C5 and C6 in Chua's classification — do not even have robust periodic orbits. The classification of elementary CA rules according to the properties of their robust periodic orbits may not reveal some important features [23, 24] and sometimes it can be even misleading [25]. For this reason, an accurate classification can be done only when non-robust periodic orbits are also taken into account.

How to define 'interesting' non-robust periodic orbits? We propose to look for orbits with period $\tau = 3$. The rational behind this choice is that the presence period-3 orbits in continuous real functions is an evidence of a very complex dynamical behavior, as proved in a classical work [26] of nonlinear dynamics. This result does not apply directly to elementary CA with a finite state space Σ (ie, in which the states are binary arrays with finite length L) since in this case Σ can be put into a one-to-one correspondence with rational numbers but not with real numbers, as explained in Sec. 5.1. Nevertheless, there are experimental evidences [27] that rules with periodic orbits with $\tau = 3$ exhibit considerably rich dynamics. Therefore, even though there is no rigorous result yet, it is meaningful looking for rules having non-robust periodic rules with $\tau = 3$. There are six such rules in the first four groups of Chua's classification; namely, rule 9, 25, 37, 74, 94, and 164. For instance, for $L = 14$ both rule 164 and rule 37 (belonging to C1 and C2, respectively) have a non-robust periodic orbit with $\tau = 3$ and $\sigma = 1$, as displayed in Table 12 (a) and (b), respectively. Similarly, also for $L = 14$ rule 9 and rule 25 (both belonging to C4) have non-robust periodic orbits with $\sigma = 2$ and $\sigma = 3$, respectively, and $\tau = 3$ (see Table 12 (c) and (d)).

These observations on the period of the non-robust periodic orbits suggest a new classification of the elementary CA rules of the first four Chua's groups. This classification contains only three classes: i) rules with robust periodic orbits only; ii) rules whose non-robust orbits have $\tau = 1$ and/or $\tau = 2$; iii) rules whose non-robust orbits have $\tau \geq 3$. Remarkably, this last class contains at least one rule from each of the first four Chua's groups.

5. Nonlinear Dynamics in Elementary Cellular Automata

5.1. *The states of elementary CA can be represented through rational numbers*

As already explained in Sec. 2.1, the states of an elementary Cellular Automaton are binary arrays with L cells; if L is finite, then the state space Σ is composed of 2^L states. In this case, to each state $\mathbf{x}^n = (x_0^n x_1^n \dots x_{L-1})$ of the Cellular Automaton we can associate a rational number q_n by using the function $\mathcal{Q} : \Sigma \to [0, 1)$ where:

$$q_n = \mathcal{Q}(\mathbf{x}^n) = \sum_{i=0}^{L-1} \left(\frac{x_i^n + 1}{2} \cdot 2^{-(i+1)} \right) \tag{12}$$

It is possible to prove [28] that if $L \to \infty$ then $\mathcal{Q}(\mathbf{x})$ tends to assume *all* rational values belonging to the interval $[0, 1)$ which are a dense subset of the real numbers. The mapping performed by an elementary CA rule $\mathcal{N}$, defined as $\mathcal{M}_{\mathcal{N}} : \Sigma \to \Sigma$, can be then associated to a function $f_{\mathcal{N}} : [0, 1) \to [0, 1)$, if $\forall \mathbf{x}^n \in \Sigma$, $\mathbf{x}^{n+1} = \mathcal{M}_{\mathcal{N}}(\mathbf{x}^n) \Leftrightarrow q_{n+1} = f_{\mathcal{N}}(q_n)$, where $q_{n+1} = \mathcal{Q}(\mathbf{x}^{n+1})$ and $q_n = \mathcal{Q}(\mathbf{x}^n)$. This formula establishes a one-to-one correspondence between elementary CA rules and functions defined of the set of rational numbers, and it is first fundamental step to link Nonlinear Dynamics and Cellular Automata.

5.2. *Time-τ characteristic function, time-τ return plot, and cobweb diagrams*

The time-τ characteristic function $\chi_{\mathcal{N}}^{\tau} : [0, 1) \to [0, 1)$ of rule $\mathcal{N}$ is a function defined as:

$$\mathcal{Q}\big(\mathcal{M}_{\mathcal{N}}^{\tau}(\mathbf{x})\big) = \chi_{\mathcal{N}}^{\tau}(\mathcal{Q}(\mathbf{x})), \quad \forall \text{ binary array } \mathbf{x} \text{ with length } L \tag{13}$$

The function $\chi_{\mathcal{N}}^{\tau}$ can be conveniently represented on a Cartesian coordinate system. For example, the time-1 characteristic function χ_{74}^{1} and time-2 characteristic function χ_{74}^{2} of rule 74 are shown in Fig. 2(a) and (b), respectively. Observe that if $\mathbf{x} = \mathcal{M}_{\mathcal{N}}^{\tau}(\mathbf{x})$, then $\mathcal{Q}(\mathbf{x})$ is a fixed point for $\chi_{\mathcal{N}}^{\tau}$. This implies that the graph of the time-τ characteristic function of a period-τ, ie, a rule in which the robust periodic orbits have period τ, will have most of the points lying on the diagonal line.

The special case in which $\tau = 1$ is particularly significant because it is an alternative way to define an elementary CA rule if the length L of the binary arrays is $L \geq 8$. In fact, when $L \geq 8$ at least one binary array, say $\mathbf{x}'$ among the 2^L possible contains all eight triplets of the truth table, from $(-1, -1, -1)$ to $(1, 1, 1)$, and hence $\chi_{\mathcal{N}}^{1}(\mathbf{x}')$ is a complete representation of the local rule $\mathcal{N}$ because it contains the information needed to derive the dynamic evolutions from any initial state.

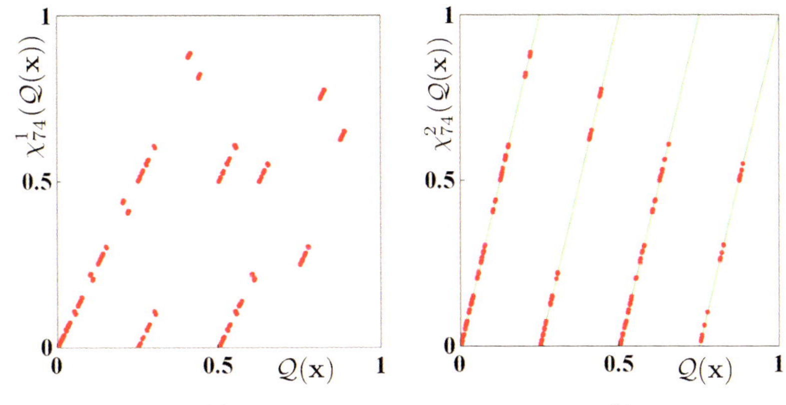

Fig. 2. (a) Time-1 χ^1_{74} and (b) time-2 characteristic function χ^2_{74} for rule 74 for $L = 8$. Observe that for this value of L all periodic orbits of rule 74 are Bernoulli-σ_τ with $\tau = 2$ and $\sigma = 2$; this is the reason why all points of its time-2 characteristic function lie on straight lines with slope 2.

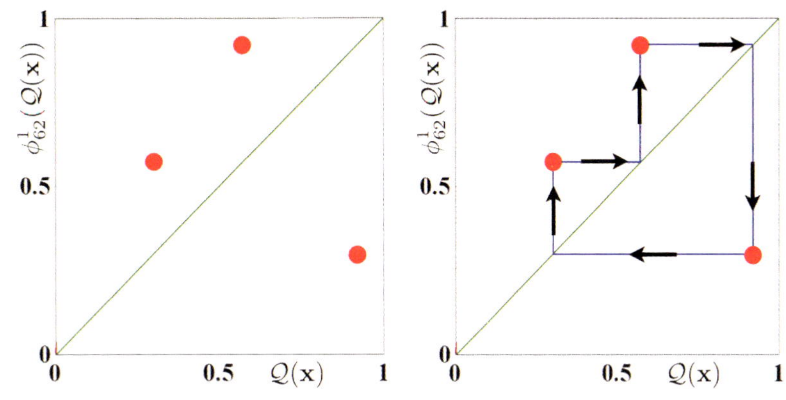

Fig. 3. (a) Time-1 return plot ϕ^1_{62} and cobweb plot (b) for rule 62 for $L = 15$ starting from the initial state $\mathbf{x}^0 = (1, -1, -1, -1, 1, 1, -1, 1, -1, 1, 1, -1, 1, -1, 1)$, where $\mathbf{x}^0 \xrightarrow{\mathcal{Q}} 0.552398681640625$.

Consequently, each rule has one, and only one, time-1 characteristic function but this is not true for higher values of τ: for example, the time-2 characteristic function for rule 105 is exactly the same as for rule 150 because they are quasi-globally equivalent under the alternating-row complementation $\widetilde{T}$ (see Sec. 2.2).

The time-τ characteristic function $\chi^\tau_{\mathcal{N}}$ represents the evolution at iteration τ of all possible states with length L. Its dual function is the time-τ return plot $\phi^\tau_{\mathcal{N}} : [0, 1) \to [0, 1)$ which represents the evolution of a specific state $\mathbf{x}^0$ at every τ iterations a CA rule $\mathcal{N}$. Hence, the time-τ return plot $\phi^\tau_{\mathcal{N}}$ is then defined as the collection of all states obtained by performing the $\mathcal{M}^\tau_{\mathcal{N}}$ mapping starting from an initial state $\mathbf{x}^0$ It can be represented on a Cartesian diagram system, such as done in Fig. 3(a) for $\mathcal{N} = 62$ and $\tau = 1$. This example contains only three points because we have chosen an initial state $\mathbf{x}^0$ belonging to a period-3 periodic orbit, ie, $\mathbf{x}^0 = \mathcal{M}^3_{62}(\mathbf{x}^0)$. In general, if $\mathbf{x}^0$ has period-τ, then the time-1 return plot contains τ points, and the time-τ return plot contains only one point. In this particular example we used $L = 15$ and $\mathbf{x}^0 = (1, -1, -1, -1, 1, 1, -1, 1, -1, 1, 1, -1, 1, -1, 1)$, $\mathbf{x}^1 = \mathcal{M}_{62}(\mathbf{x}^0) = (1, 1, 1, 1, -1, 1, 1, -1, -1, -1, 1, 1, -1, -1, -1)$ and $\mathbf{x}^2 = \mathcal{M}_{62}(\mathbf{x}^1) = (-1, 1, -1, 1, 1, -1, 1, 1, 1, 1, -1, 1, 1, 1, 1)$. Thanks to Eq. (12), we find that $\mathbf{x}^0 \xrightarrow{\mathcal{Q}} 0.552398681640625$, $\mathbf{x}^1 \xrightarrow{\mathcal{Q}} 0.961669921875$, and $\mathbf{x}^2 \xrightarrow{\mathcal{Q}} 0.358856201171875$. Obviously, the points of *any* time-τ return plot $\phi^\tau_{\mathcal{N}}$ of the local rule $\mathcal{N}$ are a subset of the time-τ

characteristic function $\chi_{\mathcal{N}}^{\tau}$ and the union of the time-τ return plots, for all possible $\mathbf{x}^0$, is equivalent to the time-1 characteristic function.

An alternative way to represent the dynamics of an elementary CA rule $\mathcal{N}$ is the Lameray diagram, also called cobweb plot [29]. It is similar to time-1 return plot, because it follows the evolution of a single binary array $\mathbf{x}^0$ throughout time, but with two main differences: first, we discard the transient and hence the circuit depicted will always be closed; second, we draw not only the points but also the lines connecting them (see Fig. 3(b)).

6. Final Comments on Chua's Original Contributions to the Study of Cellular Automata

The contribution given by Chua goes much beyond what has been presented in this paper, which has been rather focused on the fundamental notions of the 'Nonlinear Dynamics Perspective of Cellular Automata' in order to provide a synthetic, though comprehensive, overview of Chua's works. It is fair to say that some of the results included in Chua's papers had already been proved by other authors. Of course, Chua was aware of it, and this is why his papers have been been presented in the 'tutorial' section of the journals in which he published. Nevertheless, this circumstance is sometimes used to deny the originality of his approach. For this reason, we think that is useful to conclude this chapter by discussing what ideas among the many given by Chua represent a genuine breakthrough in this area.

The starting point of Chua's discourse is the formalization of the main characteristics of elementary CA rules, which we discussed in Sec. 2. This model of Cellular Automata is the most studied in the scientific literature, thus the majority of the concepts contained in that section, from the numbering of rules through Eq. (2) to the existence of globally-equivalent rules, are well-known. However, to the best of our knowledge, the definition of quasi-globally equivalent rules was introduced for the first time by Chua and we can expect that it will become part of the classical theory of Cellular Automata in the near future.

The study of the local dynamics of the single cell as a continuous-time nonlinear dynamical system, described in Sec. 3.1 is definitely among the most outstanding and original contributions by Chua. In fact, only by taking that approach it is possible to introduce rigorously the notion of index of complexity of an elementary CA rule and link it to the concept of Boolean cube. The discussion about the rotation classes of Boolean cubes, as displayed in Table 8, is also novel and not fully studied yet, for which we may expect that soon it will lead to new results.

Conversely, the global dynamics of Cellular Automata (see Sec. 4) has been the object of study of innumerable papers, and hence it is more difficult to realize what can be really considered original. We feel confident saying that even though other authors have analyzed the periodic orbits of elementary CA rules, probably nobody before Chua has proposed the rigorous classification in six classes shown in Table 11 as well as the analysis of non-robust periodic orbits in order to perform a more accurate study of the elementary CA rules.

Finally, most of the ideas of Sec. 5 are very similar to those of other authors, including Wolfram. Probably, the main merit by Chua has been the emphasis that has put on the importance of these tools to draw general conclusion about CA rules. For this purpose, he included a systematic study of the time-τ characteristic functions, the time-τ return plots, and the cobweb plots for *all* 256 elementary CA rules in his papers, which can be easily used as a reference even by unexperienced readers.

In conclusion, Chua's contribution to Cellular Automata has been fundamental and it definitely deserves the same credit as those by Ulam, von Neumann, Conway, and Wolfram. We can expect that an increasing number of researchers in the near future will be inspired by his works and possibly use them to provide further breakthroughs in this area.

References

[1] J. von Neumann, *The theory of self-reproducing automata.* Urbana, IL: University of Illinois Press, Urbana, 1966.

[2] E. Berlekamp, J. H. Conway, and R. K. Guy, *Winning ways for your mathematical plays.* New York: Academic Press, 1982.

[3] S. Wolfram, *A new kind of science* Wolfram Media, Inc, 2002.

[4] L. O. Chua, S. Yook, and R. Dogaru, "A nonlinear dynamics perspective of Wolfram's new kind of science. Part I: Threshold of complexity," *International Journal of Bifurcation and Chaos*, vol. 12, no. 12, pp. 2655–2766, 2002.

[5] L. O. Chua, V. I. Sbitnev, and S. Yook, "A nonlinear dynamics perspective of Wolfram's new kind of science. Part II: Universal neuron," *International*

Journal of Bifurcation and Chaos, vol. 13, no. 9, pp. 2377–2491, 2003.

[6] ——, "A nonlinear dynamics perspective of Wolfram's new kind of science. Part III: Predicting the unpredictable," *International Journal of Bifurcation and Chaos*, vol. 14, no. 11, pp. 3689–3820, 2004.

[7] ——, "A nonlinear dynamics perspective of Wolfram's new kind of science. Part IV: From Bernoulli shift to 1/f spectrum," *International Journal of Bifurcation and Chaos*, vol. 15, no. 4, pp. 1045–1183, 2005.

[8] L. O. Chua, V. I. Sbitnev, and S. Yoon, "A nonlinear dynamics perspective of Wolfram's new kind of science. part V: Fractals everywhere," *I. J. Bifurcation and Chaos*, vol. 15, no. 12, pp. 3701–3849, 2005.

[9] ——, "A nonlinear dynamics perspective of Wolfram's new kind of science. part VI: from time-reversible attractors to the arrow of time," *I. J. Bifurcation and Chaos*, vol. 16, no. 5, pp. 1097–1373, 2006.

[10] L. O. Chua, J. Guan, V. I. Sbitnev, and J. Shin, "A nonlinear dynamics perspective of Wolfram's new kind of science. Part VII: Isles of Eden," *International Journal of Bifurcation and Chaos*, vol. 17, no. 9, pp. 2839–3125, 2007.

[11] L. O. Chua, K. Karacs, V. I. Sbitnev, J. Guan, and J. Shin, "A nonlinear dynamics perspective of Wolfram's new kind of science. Part VIII: More isles of Eden," *International Journal of Bifurcation and Chaos*, vol. 17, no. 9, pp. 2839–3125, 2007.

[12] L. O. Chua, G. E. Pazienza, L. Orzo, V. I. Sbitnev, and J. Shin, "A nonlinear dynamics perspective of Wolfram's new kind of science. Part IX: Quasi-ergodicity," *International Journal of Bifurcation and Chaos*, 2008.

[13] L. O. Chua, G. E. Pazienza, and J. Shin, "A nonlinear dynamics perspective of Wolfram's new kind of science. Part X: Period-1 rules," *I. J. Bifurcation and Chaos*, vol. 19, no. 5, pp. 1425–1654, 2009.

[14] ——, "A nonlinear dynamics perspective of Wolfram's new kind of science. Part XI: Period-2 rules," *I. J. Bifurcation and Chaos*, vol. 19, no. 6, pp. 1751–1930, 2009.

[15] L. O. Chua and G. E. Pazienza, "A nonlinear dynamics perspective of Wolfram's new kind of science. Part XII: Period-3, period-6, and permutive rules," *I. J. Bifurcation and Chaos*, vol. 19, no. 12, pp. 3887–4038, 2009.

[16] ——, "A nonlinear dynamics perspective of Wolfram's new kind of science. Part XIV: More Bernoulli $\text{sigma}_{\text{tau}}$-shift rules," *I. J. Bifurcation and Chaos*, vol. 20, no. 8, pp. 2253–2425, 2010.

[17] ——, "A nonlinear dynamics perspective of Wolfram's new kind of science. Part XIII: Bernoulli $\text{sigma}_{\text{tau}}$-shift rules," *I. J. Bifurcation and Chaos*, vol. 20, no. 7, pp. 1859–2003, 2010.

[18] L. O. Chua, *A Nonlinear Dynamics Perspective of Wolframs New Kind of Science. Volumes I, II, III.* World Scientific, 2009.

[19] G. Pazienza and L. Chua, "Quasi global-equivalence in one-dimensional binary cellular automata," in *Circuit Theory and Design, 2009. ECCTD 2009. European Conference on*, aug. 2009, pp. 77–80.

[20] G. E. Pazienza and E. Gomez-Ramirez, "Complexity index of outer-totalistic binary cellular automata with arbitrary dimension and neighborhood," *International Journal of Bifurcation and Chaos*, 2012, to appear.

[21] E. F. Moore, "Machine models of self-reproduction," *Proc. Symp. Appl. Math.*, vol. 17, pp. 17–33, 1962.

[22] A. Wuensche and M. Lesser, *The global dynamics of cellular automata: an atlas of basin of attraction fields of one-dimensional cellular automata*, ser. Santa Fe Institute studies in the sciences of complexity: Reference volumes. Addison-Wesley, 1992.

[23] F. Ohi, "Chaotic properties of elementary CA rule 168," in *Automata 2008: EPSRC Workshop Cellular Automata Theory and Applications*, Bristol, UK, Jun. 12–14 2008.

[24] F. Chen, W. Jin, G. Chen, F. Chen, and L. Chen, "Chaos of elementary cellular automata rule 42 of Wolframs class II," *Chaos: An Interdisciplinary Journal of Nonlinear Science*, vol. 19, p. 013140, 2009.

[25] D. Eppstein. (2008) Gliders and Wolfram's classification. `http://www.ics.uci.edu/~eppstein/ca/wolfram.html`.

[26] T. Li and J. Yorke, "Period three implies chaos," *American mathematical monthly*, vol. 82, no. 10, pp. 985–992, 1975.

[27] G. E. Pazienza and M. Oswald, "Gardens of Eden: where Nonlinear Dynamics and Formal Languages meet," in *2010 International Symposium on Nonlinear Theory and its Applications (NOLTA 2010)*, 2010.

[28] I. Niven, *Irrational numbers.* The Mathematical association of America, 1967.

[29] K. T. Alligood, T. Sauer, and J. Yorke, *Chaos: An Introduction to Dynamical Systems.* Springer-Verlag NY, 1996.

CHAPTER 15

APPLICATION OF CNN TO BRAINLIKE COMPUTING

BERTRAM E. SHI

Department of Electronic and Computer Engineering
and
Division of Biomedical Engineering,
Hong Kong University of Science and Technology,
Kowloon, Hong Kong

We describe how a CNN architecture consisting of a network of four layer CNNs can be used to model the responses of visual cortical neurons. A key challenge is that visual cortical neurons are simultaneously selective along many stimulus dimensions, including retinal position, spatial frequency, orientation, temporal frequency, direction of motion, and binocular disparity. The ubiquity of intra-cortical feedback interconnections also suggests that that the neurons should operate in parallel and in continuous time. We discuss the modeling and implementation considerations that lead naturally to four layer CNN networks, and describe the realization of and experimental measurements from hardware implementations of this network architecture containing nearly 40,000 model neurons interacting via spikes in continuous time.

1. Introduction

Neurons in the visual pathway are organized into a large number of retinotopic maps, where each neuron responds only to a particular combination of image features at a certain retinal location. Moving from the retina to higher levels of visual processing, neurons become progressively more selective. Cells in the retina are selective along the stimulus dimensions of position, spatial frequency, temporal frequency. In the primary visual cortex (or V1), additional selectivity along the dimensions of orientation, direction of motion and binocular disparity emerges [12].

This chapter provides an overview of how the Cellular Neural (or Nonlinear) Network (CNN) can be used to construct neuromorphic models of neurons which are simultaneously selective along these six stimulus directions. Our approach uses different CNNs to compute the different retinotopic maps. A CNN consists of 2D arrays of locally interconnected processing elements, and thus is ideally suited to representing and computing retinotopic maps, since each processing element can be allocated to a different retinal position [8]. However, unlike the CNN Universal Machine [19], which computes a number of retinotopic maps by multiplexing the same CNN chip in time and storing the results in on-chip analog memories at each pixel, we propose an architecture consisting of a network of CNN chips that computes all retinotopic maps in parallel using separate chips to represent each map. Although this is more expensive in hardware, it has several advantages. First, this strategy enables feedback interactions between different retinotopic maps, which are an important aspect of the processing performed by the cortex. Second, this modular strategy enables us to expand both memory and processing power simultaneously, by adding more chips to the system, whereas the memory capability of a CNNUM chip is fixed at design time. Third, by allocating different chips for different functions, the design of each chip can be better optimized. The disadvantage of this application specific approach is increased design time.

Our motivation for this work arises from increasing evidence that by integrating complementary information about the environment, these maps account in part for the observed behavioral robustness of biological systems. For example, both motion parallax and binocular disparity can indicate depth discontinuities. A common engineering approach treats the two visual cues in isolation, solving separate structure from motion and structure from stereo problems, and later merging depth estimates in a subsequent step. However, biological systems appear to integrate information from different visual cues very early. For example, there are cells in V1 that are simultaneously selective to orientation, motion direction, speed and binocular disparity, which may be used to solve the binocular matching problem [10].

Indeed, the existence of cells simultaneously selective along multiple stimulus dimensions in the visual cortex is well established. Geisler and Albrecht [11] characterized large populations of cells in the primary visual cortex of both monkey and cat, and found that it was possible to characterize them accurately using a small set of descriptive functions with only a few parameters. This suggests that, although it is difficult to classify cortical cells into a small number of distinct classes, the cells may be conceptually organized according to a distribution of a relatively small number of cell properties.

The remainder of the chapter starts by describing energy models, which have been used to model the responses of neurons. This discussion serves to identify the required characteristics of the processing stages which we must take into account when building model neurons. Following that, we detail the architecture of our network of CNN chips that can be used to construct neurons tuned to respond to six different visual stimulus dimensions: retinal position, spatial frequency, temporal frequency orientation, direction of motion and binocular disparity. We describe the interactions between the CNN cells on each chip and present measured results from hardware realizations of this architecture. Due to its scope, the presentation is necessarily at a high level. References to more detailed descriptions are provided at appropriate points in the text.

2. Energy Models of Visual Cortical Neurons

The energy model has been used to model the selectivity and response of visual cortical neurons to orientation, disparity and motion. The basic components of the energy model is linear filtering by pairs of phase quadrature filters followed by squaring and summing of the two filter responses. The linear filter, whose kernel is often equated with the receptive field (RF) profile of the neuron, establishes the stimulus selectivity. Spatial filtering establishes orientation selectivity. Adding temporal filtering establishes selectivity to velocity and direction of motion. Linearly combining the spatio-temporal filter outputs from two eyes establishes selectivity to binocular disparity. Although inconsistencies between the energy model and biological data have been noted, this model is still a good first order approximation to the responses of visual cortical neurons, and thus is an appropriate starting point for constructing a neuromorphic model of them.

In the remainder of this section, we elaborate upon the energy model, and the different linear filtering stages which can be used to establish selectivity along the six dimensions of retinal position, spatial frequency, temporal frequency orientation, direction of motion and binocular disparity. We demonstrate that if we compute phase quadrature pairs of filter outputs simultaneously, then the required filtering steps are separable. In other words, the spatio-temporal binocular filters can be decomposed into the cascade of spatial, temporal and binocular filters. This decomposition is critical to the CNN implementation described in the next section.

2.1. *Retinal position, spatial frequency and orientation*

The spatial receptive field of V1 simple cells can be accurately modelled by a Gabor function [9, 13, 15], which accounts for neuronal selectivity along the dimensions of retinal position, spatial frequency and orientation. For vertical orientations, the Gabor function has the form

$$g_e(x,y) = \frac{1}{2\pi\sigma_x\sigma_y}\exp\left(-\frac{x^2}{2\sigma_x^2}-\frac{y^2}{2\sigma_y^2}\right) \times \cos(\Omega_x x+\phi) \quad (1)$$

The standard deviations of the Gaussian envelope, σ_x and σ_y, control the width of the receptive field and therefore the selectivity to spatial position. The parameter Ω_X determines the peak spatial frequency tuning, and ϕ is a phase shift parameter. By the uncertainty relationship, there is an inverse relationship between selectivity to spatial position and selectivity to spatial frequency. The subscript "e" indicates the even symmetry of the cosine function when $\phi = 0$. Other orientations can be obtained by rotating this function. Figure 1 shows Gabor functions with four different orientations.

The response of an orientation selective simple cell whose receptive field is centered at retinal position can be modelled by

$$r_s(x,y) = \left(\left|\iint g_e(x-x',y-y')I(x',y')dx'dy'\right|^+\right)^2 \quad (2)$$

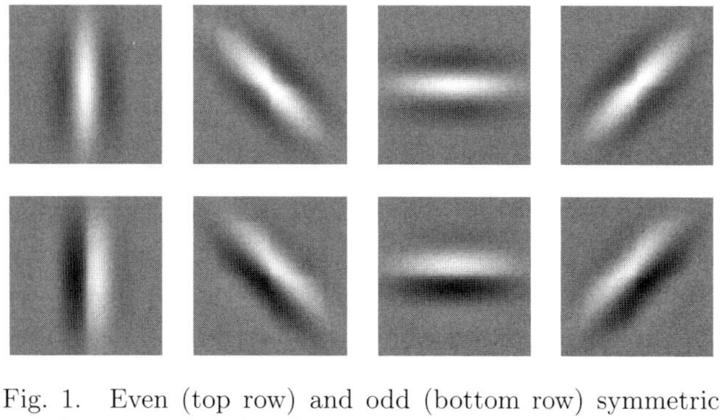

Fig. 1. Even (top row) and odd (bottom row) symmetric two-dimensional Gabor functions tuned to different orientations. White indicates positive values. Black indicates negative values.

where $I(x,y)$ represents the input image intensity and $|x|^+ = \max\{0,x\}$ is the positive half-wave rectifying nonlinearity. The combined operation of half-wave rectification and squaring is called *half-squaring*.

The response of a complex cell can be modelled as the sum of four simple cell responses whose phase parameters differ by $\pi/2$. Since filter kernels that differ in phase by π are equivalent, except for a change in sign, the response is equivalent to squaring and summing the non-rectified responses of two filters: one with kernel g_e and the other with kernel g_o, which is identical to g_e except that the cosine term is replaced by a sine term. The subscript "o" refers to the odd symmetry in x when $\phi = 0$. These two kernels are said to be in phase quadrature, since they differ in phase by 90 degrees. This sum of squared filter outputs is called the *orientation energy*, and is maximized for input stimuli whose orientation match the orientation of the filter kernels. Although the simple cells responses are phase sensitive, i.e. their response to a sine wave grating depends upon its phase, complex cell responses are largely phase invariant. Notationally, it is convenient to express this pair of filter kernels as a single complex valued filter kernel

$$g(x,y) = g_e(x,y) + jg_o(x,y) \quad (3)$$

$$= \frac{1}{2\pi\sigma_x\sigma_y}\exp\left(-\frac{x^2}{2\sigma_x^2}-\frac{y^2}{2\sigma_y^2}\right) \times \exp(j(\Omega_x x+\phi)) \quad (4)$$

The orientation energy is the squared modulus of the complex valued filter response.

2.2. *Temporal frequency and direction of motion*

By adding a temporal filtering stage to the Gabor spatial filters, we add selectivity for temporal frequency and direction of visual motion. These filters exploit the fact that rigid motion can be viewed as an orientation in the space-time domain [1, 24].

The temporal response of cortical cells can be modelled a function with the following form

$$h_e(t) = \frac{1}{\Gamma(\alpha)\tau^\alpha}\exp\left(-\frac{t}{\tau}\right)\cos(\Omega_t t+\phi_t)u(t) \quad (5)$$

where $u(t)$ is the unit step function which is 1 for and 0 otherwise, τ is the time constant of the envelope, α determines the degree of skewness, and Ω_t and ϕ_t determine the frequency and phase of the sinusoidal modulation [5]. For long time constants, the peak temporal tuning of this function is determined primarily by Ω_t.

The spatio-temporal receptive fields of cortical receptive fields can be modelled by combining the functions in (1) and (5) as follows:

$$f_e(x,y,t) = g_e(x,y)h_e(t) - g_o(x,y)h_o(t) \qquad (6)$$

where the function h_o is obtained from h_e by replacing the cosine by a sine. Motion selective simple cell responses can be modelled according to (2), where is g_e replaced by f_e, and the integral extends over time as well. Given the equations in (1) and (5), this simple cell would be tuned to horizontal motion with velocity $-\Omega_t/\Omega_x$. The sign of the velocity tuning (positive or negative) indicates the selectivity to direction of motion (right or left). Selectivity to motion in the opposite horizontal direction can be obtained by switching the sign of Ω_t or Ω_x. Selectivity non-horizontal directions of motion can be obtained by rotating the spatial receptive field profiles.

As in the case of orientation, motion tuned complex cell responses are obtained by summing four simple cell outputs. This sum is called the *motion energy*, and is equivalent to summing the squared (non-rectified) outputs of two phase quadrature linear filters, where the phase quadrature partner of (6) is

$$f_o(x,y,t) = g_e(x,y)h_o(t) + g_o(x,y)h_e(t) \qquad (7)$$

Notationally, it is convenient to express this pair of filters as a single complex valued filter kernel

$$f(x,y,t) = f_e(x,y,t) + jf_o(x,y,t) \qquad (8)$$

$$= g(x,y)h(t) \qquad (9)$$

where

$$h(t) = h_e(t) + jh_o(t) \qquad (10)$$

$$= \frac{1}{\Gamma(\alpha)\tau^\alpha}\exp\left(-\frac{t}{\tau}\right)$$

$$\times \exp(j(\Omega_t t + \phi_t))u(t) \qquad (11)$$

Notice that although the real valued kernels f_e and f_o are not separable in space and time, the complex kernel is.

2.3. *Binocular disparity*

For simplicity, we ignore the temporal filtering component of the responses of disparity tuned cortical neurons in the following discussion. However, this can be incorporated by adding a temporal filtering component to the monocular spatial filters discussed below, in a manner as discussed in the previous subsection [5].

Disparity selective simple cells in the visual cortex can be modelled by a linear binocular filter that sums the responses of two monocular filters modeled by Gabor functions, followed by a half squaring nonlinearity [2–4, 17]. The linear binocular filter output is

$$r_{lin} = \iint g_{eL}(x_L - x', y_L - y')I_L(x',y')dx'dy' + \iint g_{eR}(x_R - x', y_R - y')I_R(x',y')dx'dy' \qquad (12)$$

where $I_L(x,y)$ and $I_R(x,y)$ are the image intensities of the left and right eyes, and g_{eL} and g_{eR} are Gabor functions (1) whose parameters are identical except possibly for their phase values ϕ_L and ϕ_R. The monocular receptive field profiles of the left and right eyes are centered at locations (x_L, y_L) and (x_R, y_R). The binocular simple cell output is $r_s = (|r_{lin}|^+)^2$.

As in the case of orientation and motion tuning, complex cells are modelled by summing the outputs of four simple cell outputs. This sum is called the *binocular energy*, and can be obtained by summing the squared (non-rectified) outputs of two phase quadrature binocular filters, where the phase quadrature partner of (12) is obtained by each monocular Gabor function by its phase quadrature partner.

For monocular RFs tuned to vertical orientations, the resulting binocular filter will be selective for horizontal disparity. The disparity tuning of the cell is determined by the difference in the center locations $\Delta c = x_R - x_L$ and the difference in phase $\Delta\phi = \phi_R - \phi_L$. For a disparity of d, $I_R(x,y) \approx I_L(x-d,y)$. The binocular complex cell will respond maximally when the input has a disparity approximately

$$D_{pref} \approx \Delta c - \frac{\Delta\phi}{\Omega_x} \qquad (13)$$

3. CNN Architecture Modelling the Primary Visual Cortex

Figure 2 shows the architecture of a multi-chip CNN architecture that can implement neurons with the multidimensional selectivity observed in the primary visual cortex. The outputs of two silicon

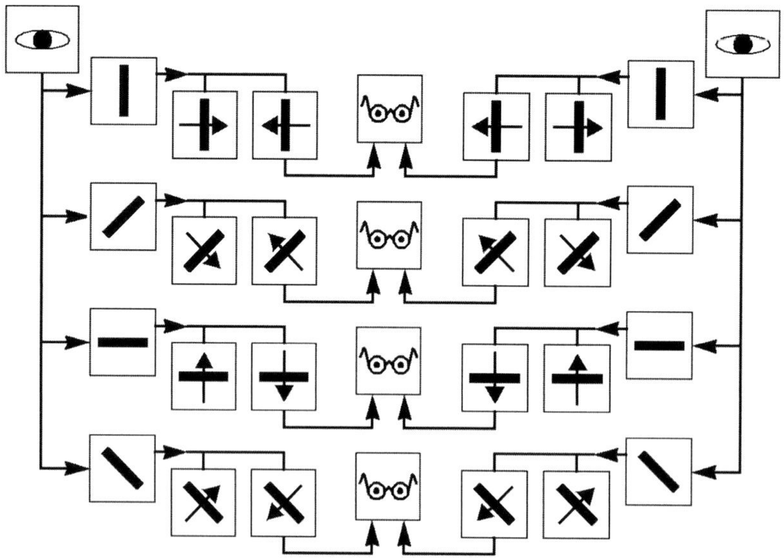

Fig. 2. Architecture of multi-chip network architecture implementing neurons selective along the six stimulus dimensions described in the text. Each block represents a separate chip containing a retinotopic array of neurons. The top blocks represent left and right eye silicon retina chips. Blocks with oriented bars are orientation selective spatial filtering chips. Blocks with oriented bars and arrows contain neurons selective for direction and speed of motion. These chips actually contain arrays of independent temporal filters, which when cascaded with the phase quadrature outputs of the orientation chip yield model neurons with the desired motion selectivity. Chips with eyeglasses combine filter outputs from the two eyes, resulting in binocular neurons selective for a particular combination of retinal position, spatial frequency, temporal frequency, orientation, direction of motion and binocular disparity.

retinae representing the left and right eyes are initially processed monocularly by a cascade of spatial and temporal filters before being combined binocularly. We can use CNNs to compute each of the required filter outputs. Thus, the architecture can be considered to be a network of CNNs. The outputs of the binocular stages are simultaneously selective along all six stimulus dimensions.

The first step of monocular processing is spatial frequency and orientation selective filtering. As discussed in the previous section, we must compute phase quadrature pairs of filter responses, which can be considered as the real and imaginary parts of a complex valued filter output. The spatial phase parameters is a free parameter, but it is only necessary to compute the pair of filter responses with $\phi = 0$, since quadrature pairs at other phase shifts can be computed by linear combinations of these two filter responses. In addition, different values of phase parameters only alter the phase properties of the simple cell responses. There is no difference between the responses of complex cells computed from the outputs of simple cells with different phase parameters, since the complex cell outputs are combined in such a way as to make their responses phase invariant. Since the responses of complex cells are much more robust to changes in the stimulus structure, it is likely that these complex cell outputs are what are used in further processing stages. Indeed, many biologically inspired algorithms for extracting disparity and motion information use complex cell outputs exclusively.

The requirement for phase quadrature responses motivated the development of two layer linear CNN architectures for Gabor-type filtering, where each layer represents one filter output at different spatial locations [20, 21]. A multi-layer CNN array for image filtering consists of a number of two-dimensional layers of discrete cells [8]. Each cell corresponds to one pixel in the image to be filtered, and has an associated state variable, input and output. The temporal evolution of the state at layer k is given by

$$\tau \dot{v}_k = \sum_l A_{kl} \star y_l + \sum_l B_{kl} \star u_l + \sum_l C_{kl} \star v_l \qquad (14)$$

where v_l, u_l and y_l are two dimensional arrays representing the state, input and output of layer l and τ is a positive time constant. The $\star$ indicates two dimensional correlation. The matrices A_{kl}, B_{kl} and C_{kl} are known as the output feedback, feedforward and state feedback cloning templates. The output is a memoryless nonlinear function of the state. If all of the A_{kl} are zero, we obtain a linear CNN.

Using only nearest neighbor connections, a two layer linear CNN can implement filters that are similar to Gabor filters with $\phi = 0$, except that the modulating function is not Gaussian, but decays like a Laplacian distribution. In one dimension, the filter kernel is given by

$$g(x) = \frac{\lambda}{2\alpha} \exp\left(-\frac{\lambda}{\alpha}|x|\right) \exp(j\Omega_x x) \tag{15}$$

While Gaussian functions are attractive theoretically, as they minimize the uncertainty relation, several studies have shown that the exact shape of the modulating function is not critical in real applications. It is possible to obtain closer approximations to Gaussian envelopes by using larger template radii.

A significant advantage of the two layer architecture is that it enables us to establish the spatially oscillatory RF profiles with a recurrent architecture with purely nearest neighbor connections between neurons. Similar RF profiles can be obtained with single layer architectures, but require second nearest neighbor connections, and would only represent one member of the pair [18].

Figure 3 shows a circuit implementing one pixel of the two layer linear CNN for Gabor-type filtering of one dimensional images. The voltages across the capacitors $v_e(n)$ and $v_o(n)$ represent the state variables of the two layers at pixel n. One way to

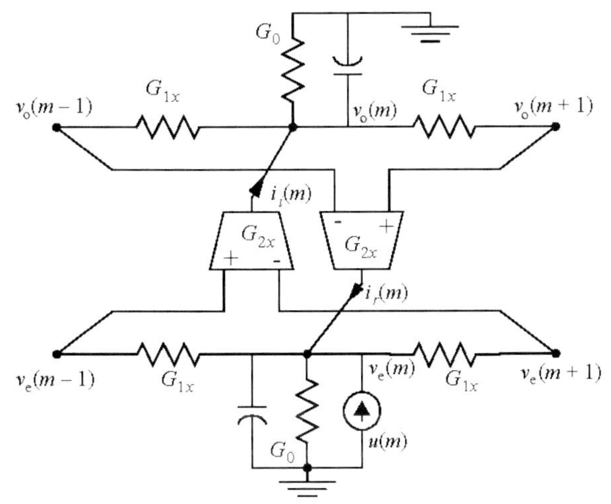

Fig. 3. The circuit architecture for one pixel in a one dimensional array. Resistors are labeled by their conductances. Trapezoidal blocks denote transconductance amplifiers labeled by their gains. Unit capacitance between each node and ground is assumed for modelling purposes.

define the CNN cloning template for this network is to define a complex valued state at each pixel by $v(n) = v_e(n) + jv_o(n)$. We obtain a single layer CNN with complex valued cloning templates. The nonzero cloning templates are given by

$$B = [0\ 1\ 0] \tag{16}$$

$$C = [(G_{1x} + jG_{2x})\ (-G_0 - 2G_{1x})\ (G_{1x} + jG_{2x})] \tag{17}$$

where G_0, G_{1x} and G_{2x} are real valued. Alternatively, we can define intra- and inter-layer templates between two real valued layers by

$$B_{ee} = B_{oo} = [0\ 1\ 0] \tag{18}$$

$$C_{ee} = C_{oo} = [G_{1x}\ (-G_0 - 2G_{1x})\ G_{1x}] \tag{19}$$

$$C_{eo} = -C_{oe} = [-G_{2x}\ 0\ G_{2x}] \tag{20}$$

For a real valued input image, we can set B_{oo} to zero. The filter parameters are related to the template coefficients by

$$\alpha^2 = \sqrt{G_{1x}^2 + G_{2x}^2} \tag{21}$$

$$\lambda^2 = G_0 + 2G_{1x} - 2\alpha^2 \tag{22}$$

$$\Omega_x = \arctan(G_{2x}/G_{1x}) \tag{23}$$

More recent network designs execute the same filtering operation, but contain four layers, where pairs of layers represent the positive and negative half wave rectified filter outputs, which we refer to as the ON and OFF components [6]. Although the four layer network requires twice as much hardware as the two layer network, this ON-OFF representation has the same advantages in our designs as it has in biology. In biology, the ON-OFF representation is energy efficient because quiescent signals correspond to small spike rates on both ON and OFF neurons. In our analog circuits, quiescent signals correspond to small currents in the ON and OFF circuits. Thus, this architecture automatically allocates power to active neurons, while neurons with little activity dissipate little power. Another advantage is that the ON-OFF signal representation includes the half-wave rectification operation that is part of the simple cell model.

Figure 4 shows the interaction between cells in the four layer network. We label the layers by $e+, e-, o+$ and $o-$, where the letter indicates the symmetry of the filter implemented by the layer (even or odd) and the $+$ or $-$ symbol indicates the polarity (ON or OFF). The non-zero cloning

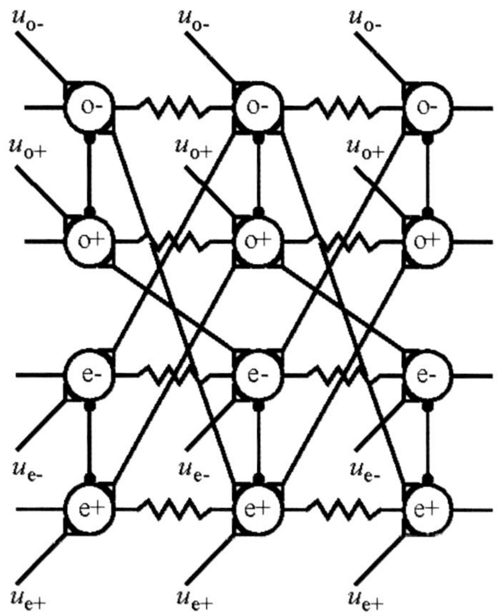

Fig. 4. Cell interconnections in a 1D version of the four layer network for Gabor-type filtering. The representation of the "e" and "o" layers of the two layer network is split across two layers representing the ON (+) and OFF (−) components. Interconnections terminating with filled circles are inhibitory. Interconnections terminating with open triangles are excitatory. Interconnections by resistors are implemented using the state feedback template. The mutually inhibitory interconnections between positive and negative layers of the same type are introduced by the output nonlinearity.

templates are given by

$$\begin{aligned} A_{e+o-} &= A_{e-o+} = A_{o+e+} \\ &= A_{o-e-} = [G_{2x}\ 0\ 0] \end{aligned} \tag{24}$$

$$\begin{aligned} A_{e+o+} &= A_{e-o-} = A_{o+e-} \\ &= A_{o-e+} = [0\ 0\ G_{2x}] \end{aligned} \tag{25}$$

$$\begin{aligned} B_{e+e+} &= B_{e-e-} = B_{o+o+} \\ &= B_{o-o-} = [0\ 1\ 0] \end{aligned} \tag{26}$$

$$\begin{aligned} C_{e+e+} &= C_{e-e-} = C_{o+o+} = C_{o-o-} \\ &= [G_{1x}\ (-G_0 - 2G_{1x})\ G_{1x}] \end{aligned} \tag{27}$$

Here we have assumed a complex valued input image where real and imaginary parts of the input are encoded using an ON-OFF representation, each associated with a corresponding CNN layer. Negative template coefficients in the two layer network are implemented using excitatory connections between ON and OFF layers. The outputs of the ON and OFF layers are equal to the difference between their states positive and negative half wave rectified

$$y_{e+} = |v_{e+} - v_{e-}|^{+} \tag{28}$$

$$y_{e-} = |v_{e+} - v_{e-}|^{-} \tag{29}$$

$$y_{o+} = |v_{o+} - v_{o-}|^{+} \tag{30}$$

$$y_{o-} = |v_{o+} - v_{o-}|^{-} \tag{31}$$

where $|x|^{+} = \max\{x, 0\}$ and $|x|^{-} = \min\{x, 0\}$. This output nonlinearity couples the ON and OFF layers by removing the common mode component in their states. This is important in ensuring the stability of the array, since all interlayer connections are excitatory.

Hardware implementations of retinotopic arrays of monocular neurons tuned to different orientations, as illustrated in the left and right columns of Fig. 2, have been reported [7]. Retinotopic arrays of neurons tuned to different orientations are obtained by fanning out the output of the silicon retina to different chips. Each chip is identical, containing a four layer network with arrays of 32×64 neurons in a 0.25 um CMOS process [6]. They are tuned to different orientations by different values of external bias voltages applied to each chip. The analog processing components of this network are implemented using transistors operating in weak inversion for maximum energy efficiency.

In this implementation, neuronal activity within the chips is spike-encoded using arrays of silicon integrate and fire neurons, and is communicated between chips using the Address Event Representation Protocol (AER), which is an asynchronous digital protocol that signals a spike within the neuron array on the chip by placing the address of the neuron that spiked on a digital bus [14]. The time the address appears on the bus directly encodes the spike time. Simultaneous spikes are handled by an arbiter that grants sequential access to the bus. Although this will introduce some jitter into the spike times, the magnitude of this jitter is on the order of tens of nanoseconds due to the high speed of modern electronic circuits, and is therefore negligible in comparison with the time constants of biological neurons. At the input side, an address appearing on the digital bus causes a spike (current pulse) to be delivered to input of the corresponding neuron in the array. The ON-OFF signal representation is advantageous when we use this spike-based communication scheme, since neurons will have low quiescent spike rates, which reduces the traffic on the digital bus. Fan-out of spikes from the silicon retina to the different chips is implemented using routing circuit contained on each Gabor-type filtering chip [16].

Figure 5 shows the multi-chip implementation of the CNN architecture for monocular orientation

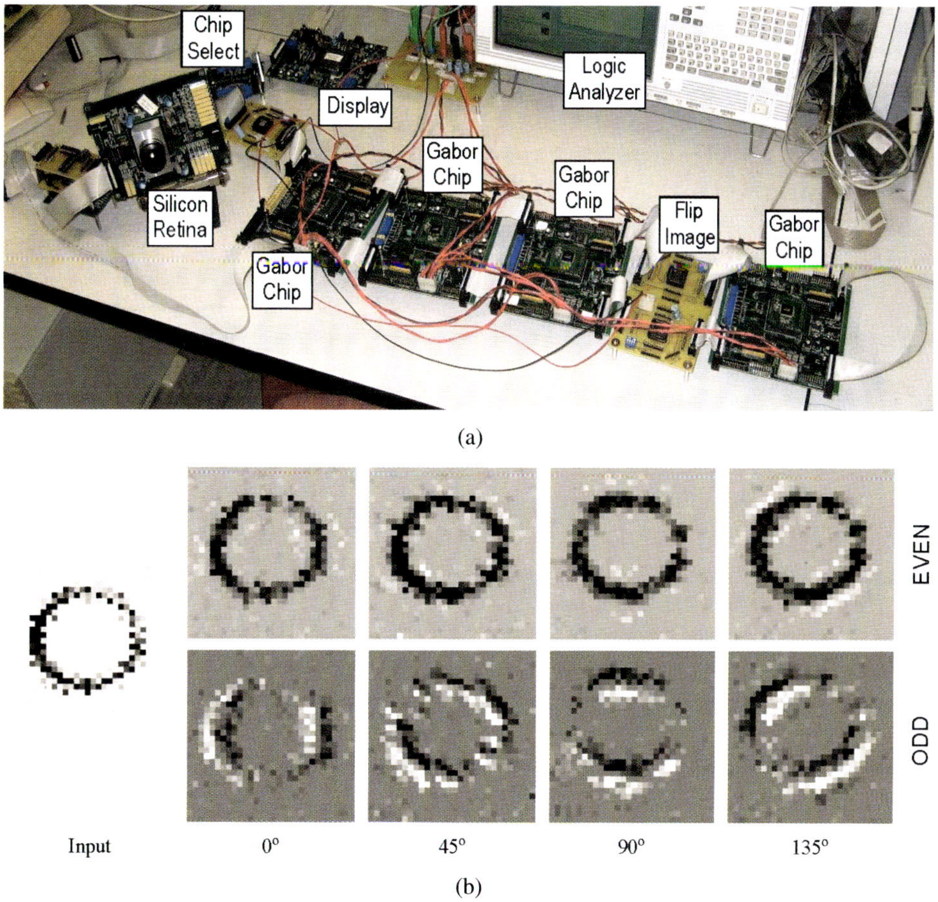

Fig. 5. (a) Photo of the multi-chip CNN architecture for monocular orientation selective filtering. Each "Gabor Chip" board is identical, but tuned to different orientations via different biasing voltages applied to the chip. Since implementation constraints restrict the range of tune to between 0° (vertical) and 90° (horizontal), we use a flip image board, which flips the output and input images of the fourth chip horizontally using address remapping. This enables us to implement neurons tuned to 135. The chip select board pass only spikes originating from a desired chip for display. The display board connects to a VGA monitor (not shown) for visualizing the chip outputs. The logic analyzer collects spikes for analysis. (b) Measured responses of the system to a black ring presented to the silicon retina. Image on the left shows the output of the silicon retina. The image intensity encodes the difference between the spike rates of the ON and OFF neurons at each position. The eight remaining images show the outputs of the neurons on the Gabor chips. The top row shows the difference between the responses of ON and OFF neurons with even symmetric receptive fields. The bottom row shows the difference between the responses of ON and OFF neurons with odd symmetric receptive fields. White corresponds to positive differential spike rates. Black corresponds to negative differential spike rates.

selective filtering, as well as measured data from the system. Input is provided by the silicon retina chip described in [25,26]. This implementation contains 35,648 model neurons spread across five chips. Neurons on different chips interact with each other in continuous time and communicate their activity via spikes transmitted via AER. Within each chip, neurons interact with each other in continuous time but via graded activity encoded as currents.

Because of the continuous time operation of the CNN arrays and the asynchronous communication protocol, we have been able to implement feedback interactions between neurons serving the same retinal location, but with different orientation tunings, even though these neurons reside on different chips [7]. These feedback interactions sharpen the orientation selectivity of the neurons.

The next stage in the monocular processing chain shown in Fig. 2 is temporal filtering to establish selectivity to temporal frequency and direction of motion. As described in the previous section, because we compute both phase quadrature outputs of the spatial filter simultaneously, the required spatio-temporal filter is separable in space and time. This implies that the required CNN array consists of decoupled cells, where each cell is constructed so that its temporal dynamics executes the required temporal filtering operation.

A four layer CNN cell can be used to implement a complex-valued temporal filter with kernel given by Eq. (11) where $\alpha = 1$ [22]. To facilitate integration into the multi-chip network, this cell circuit accepts and produces four spike encoded signals, which represent the ON and OFF components of the real and imaginary parts of the inputs and outputs. The four outputs of the spatial filtering chip at each pixel can be connected directly to the corresponding inputs of the array of CNN temporal filtering cells to implement the spatio-temporal filter with spatio-temporal receptive field given by Eq. (6).

Figure 4 shows the relationship between the inputs, states and outputs of the four layer temporal filtering cell. The relationship between inputs and outputs in this cell is identical to that used in the cells of the four layer orientation selective filtering array, as given by Eqs. (28)–(31). The non-zero cloning templates for the temporal filtering array are given by

$$A_{o+e+} = A_{e-o+} = A_{o-e-} = A_{e+o-} = [\beta] \tag{32}$$

$$B_{e+e+} = B_{e-e-} = B_{o+o+} = B_{o-o-} = [1] \tag{33}$$

$$\begin{aligned} C_{e+e+} &= C_{e-e-} = C_{o+o+} \\ &= C_{o-o-} = [-1] \end{aligned} \tag{34}$$

The value of β in the output feedback cloning template and the filter center frequency Ω_t in Eq. (5) are related by $\Omega_t = \beta/\tau$.

This cell circuit has been implemented in a 1.5μm CMOS process as described in [22]. The time constant τ and spatial frequency tuning Ω_t are adjustable via externally applied bias voltages. Internally, the spike trains are integrated to produce continuously varying currents, which are filtered using log-domain filter circuits constructed from CMOS transistors operating in weak inversion, where the low current levels enable us to obtain tunings and time constants on the order of tens of hertz, consistent with the temporal frequency tuning of real cortical cells.

The final step in the CNN architecture shown in Fig. 2 is to combine the monocular responses to create neurons selective to binocular disparity. According to (12), this can be done at each spatial location separately. Thus, the CNN array implementing disparity selective cells consists of a four layer array of cells which are decoupled both spatially and across layers. The CNN dynamics reduce to leaky integration at each layer and spatial position.

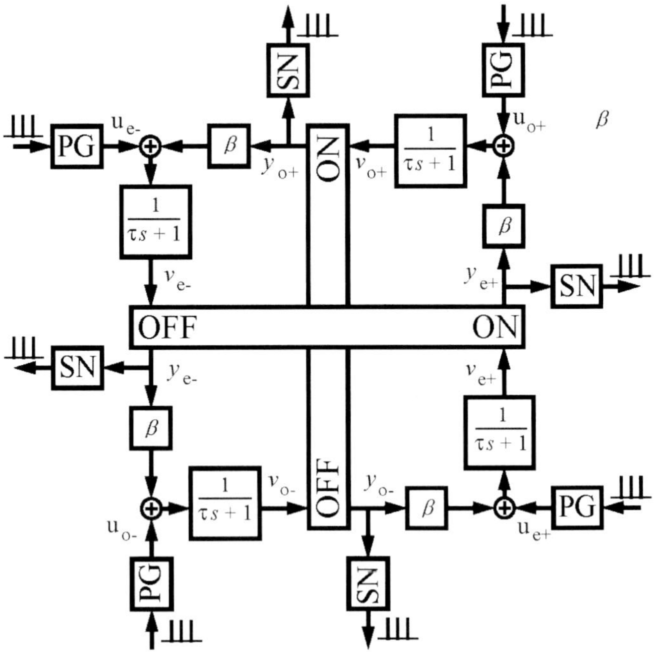

Fig. 6. The relationship between the inputs, states and outputs of the four layer CNN cell implementing the temporal filter used to image motion selectivity in cortical neurons. Internally, all signals are represented by currents. Externally, signals are encoded as voltage pulse trains (indicated by three spikes). The spiking neuron block (SN) converts currents to voltage pulses. The pulse generator block (PG) converts voltage pulses to current pulses. Blocks labelled $\frac{1}{\tau s+1}$ are leaky integrators.

The disparity selectivity of the cells is not adjusted by altering the dynamics of the CNN array, but by altering which outputs from the two eyes are combined. According to (13), the disparity tuning can be changed by introducing either a position shift or a phase shift between the two monocular receptive fields. Because neuron outputs from different positions and different phases are all encoded on a single AER bus, we can use address remapping circuits to select different combinations of monocular outputs, and thus alter the disparity tuning. A position shift can be implemented by adding a horizontal offset to the address of spikes coming from the left or right eye orientation selective filter chips. A phase shift can be implemented by changing the way in which the four filter outputs at each pixel from the left and right eye orientation selective chips are combined. To implement a zero phase shift, corresponding outputs from the left and right eye are combined at each layer, i.e. left eye e+ outputs are combined with right eye e+ outputs and so on. To implement a $-90°$ phase shift, left eye e+ (e$-$) outputs are combined with right eye o+ (o$-$) outputs

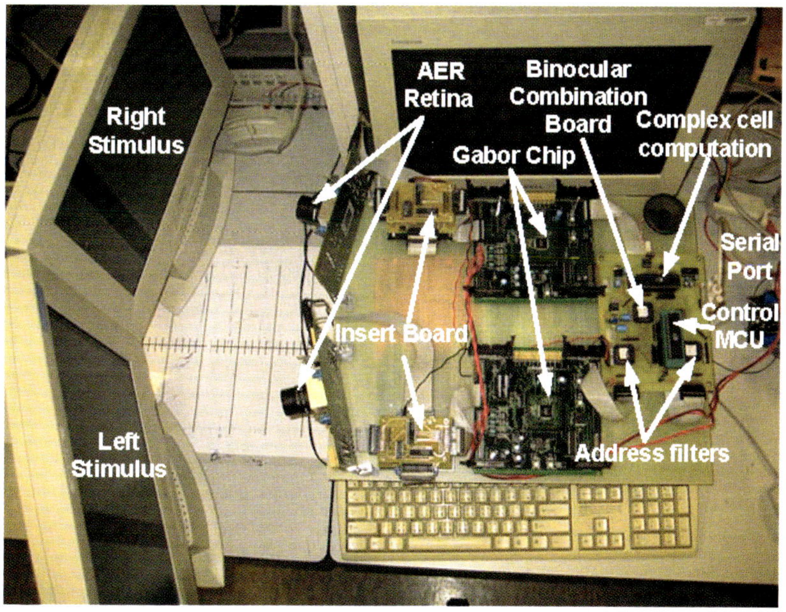

(a)

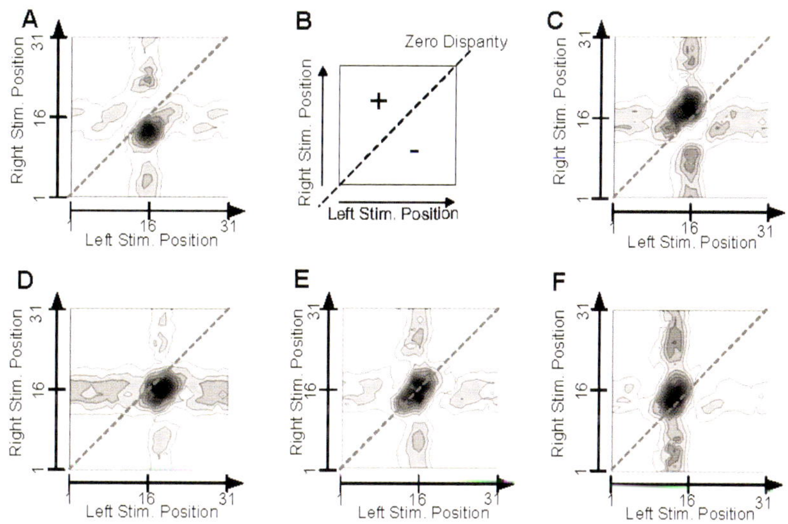

(b)

Fig. 7. (a) Photo of a multi-chip implementation of disparity tuned neurons. Two AER retinas provide analogs of left and right eye inputs to the system. The insert board converts the AER address format used by the silicon retina to that used by the Gabor-type chips. The left and right eye inputs are then processed by orientation selective neurons tuned to vertical orientations. The outputs of the orientation tuned neurons of the left and right eyes are then combined to yield neurons tuned to binocular disparities. (b) Measured binocular tuning curves shown as images where the gray level of pixel encodes the response of the complex cell to a pair of spatial bar stimuli focused onto the left and right retina chips. Each position indicates the firing rate for an input presented at a particular position in the left and right eyes. Darker gray levels indicate higher firing rates. (A, D) The tuning curves of neurons tuned to negative disparities by phase and position encoding. (B) The regions of the maps corresponding to different disparities. (C, F) The RF maps of neurons tuned to positive disparities by phase and position encoding. (E) The timomg curve of the neuron tuned to zero disparity, which is the same for both position and phase encoding.

and left eye o+ (o−) outputs are combined with right eye e− (o+) outputs.

In [23], we describe the construction of such spike remapping circuits and present measured results from neurons tuned to different disparities by both position and phase shifts. Figure 7 shows a photograph of the system, as well as measured disparity tuning curves in response to image presented to the system by focusing the outputs of two monitors onto silicon retinas representing the left and right eyes.

4. Summary

We have described a unified CNN architecture that can be used to build silicon neurons that mimic the multi-dimensional selectivity of visual cortical neurons along six stimulus dimensions. The feasibility of this CNN architecture has been validated through working hardware prototypes built using full custom mixed analog digital circuits. In order to support fully continuous time interactions between neurons and allow the possibility for real-time feedback, the architecture can be implemented by distributing computation across multiple chips, each containing a multi-layer CNN array and communicating via voltage pulses analogous to neural action potentials (spikes).

References

[1] Adelson and Bergen (1985). Spatiotemporal energy models for the perception of motion, *Journal Optical Society of America A*, Vol. 2, pp. 284–299.

[2] A. Anzai, I. Ohzawa, and R. D. Freeman (1999a). Neural mechanisms for processing binocular information I. Simple cells, *Journal of Neurophysiology*, Vol. 82, pp. 891–908.

[3] A. Anzai, I. Ohzawa, and R. D. Freeman (1999b). Neural mechanisms for processing binocular information II. Complex cells, *Journal of Neurophysiology*, Vol. 82, pp. 909–24.

[4] A. Anzai, I. Ohzawa, and R. D. Freeman (1999c). Neural mechanisms for encoding binocular disparity: receptive field position versus phase, *Journal of Neurophysiology*, Vol. 82, pp. 874–90.

[5] Y. Chen, Y. Wang, and N. Qian (2001). Modeling V1 disparity tuning to time-varying stimuli, *Journal of Neurophysiology*, Vol. 86, pp. 143–55.

[6] T. Y. W. Choi, B. E. Shi, and K. Boahen (2004). An ON-OFF Orientation Selective Address Event Representation Image Transceiver Chip, *IEEE Transactions on Circuits and Systems-I: Fundamental Theory and Applications*, Vol. 51, no. 2, Feb.

[7] T. Y. W. Choi, P. Merolla, J. Arthur, K. Boahen, and B. E. Shi (2005). Neuromorphic implementation of orientation hypercolumns. *IEEE Transactions on Circuits and Systems-I: Regular Papers*, Vol. 52, no. 6, pp. 1049–1060, Jun.

[8] L. O. Chua and L. Yang (1988). Cellular neural networks: Theory, *IEEE Transactions on Circuits and Systems*, Vol. 35, pp. 1257–1272, Oct.

[9] J. G. Daugman (1980), Two-dimensional spectral analysis of cortical receptive field profiles, *Vision Research*, Vol. 20, pp. 847–856.

[10] R. Van Ee and B. L. Anderson (2001). Motion direction, speed and orientation in binocular matching, *Nature*, Vol. 410, pp. 690–694.

[11] W. S. Geisler and D. G. Albrecht (1997). Visual cortex neurons in monkeys and cats: Detection, discrimination, and identification, *Visual Neuroscience*, Vol. 14, pp. 897–919.

[12] D. H. Hubel and T. N. Wiesel (2001). Receptive fields, binocular interaction and functional architecture in the cats visual cortex, *Journal of Physiology*, Vol. 160, pp. 106–154.

[13] J. P. Jones and L. A. Palmer (1987). An evaluation of the two-dimensional Gabor filter model of simple receptive fields in cat striate cortex, *Journal of Neuroscience*, Vol. 58, pp. 1233–1258.

[14] M. Mahowald (1994). *An Analog VLSI System for Stereoscopic Vision*. Kluwer Academic Publishers.

[15] S. Marceljia (1980). Mathematical description of the responses of simple cortical cells, *Journal Opt. Soc. Am.*, Vol. 70, pp. 1297–1300.

[16] P. Merolla, J. Arthur, B. Shi and K. Boahen (2007). Expandable networks for neuromorphic chips, *IEEE Transactions on Circuits and Systems–I: Regular Papers*, Vol. 54, no. 2, pp. 301–311, Feb.

[17] I. Ohzawa, G. C. DeAngelis, and R. D. Freeman (1990). Stereoscopic depth discrimination in the visual cortex: Neurons ideally suited as disparity detectors, *Science*, Vol. 249, pp. 1037–1041.

[18] L. Raffo (1995). Resistive network implementing maps of Gabor functions of any phase, *Electronics Letters*, Vol. 31, no. 22, pp. 1913–1914, Oct.

[19] T. Roska and L. O. Chua (1993). "The CNN Universal Machine: An analogic array computer," *IEEE Transactions on Circuits and Systems-II*, Vol. 40, no. 3, pp. 163–173.

[20] B. E. Shi (1998). Gabor-type filtering in space and time with cellular neural networks, *IEEE Trans. Circuits and Systems-I: Fundamental Theory and Applications*, Vol. 45, no. 2, pp. 121–132, Feb.

[21] B. E. Shi (2000). A low power orientation selective vision sensor, *IEEE Transactions on Circuits and Systems-II: Analog and Digital Signal Processing*, Vol. 47, pp. 435–440.

[22] B. E. Shi, E. K. C. Tsang and P. S. P. Au (2006). An ON-OFF temporal filter circuit for motion energy computations, *IEEE Transactions on Circuits and Systems-II: Express Briefs*, Vol. 53, no. 9, pp. 951–955, Sep.

[23] E. K. C. Tsang and B. E. Shi (2004). A preference for phase-based disparity in a neuromorphic implementation of the binocular energy model, *Neural Computation*, Vol. 16, no. 8, pp. 1579–1600, Aug.

[24] A. B. Watson and J. A. J. Ahumada (1985). Model of human visual-motion sensing, *Journal Optical Society of America A*, Vol. 2, pp. 322–342.

[25] K. A. Zaghloul and K. Boahen (2004a). Optic nerve signals in a neuromorphic chip I: Outer and inner retina models, *IEEE Transactions on Biomedical Engineering*, Vol. 51, no. 4, pp. 657–666, Apr.

[26] K. A. Zaghloul and K. Boahen (2004b). Optic nerve signals in a neuromorphic chip II: Testing and results, *IEEE Transactions on Biomedical Engineering*, Vol. 51, no. 4, pp. 667–675, Apr.

CHAPTER 16

IDEAL TURBULENCE PHENOMENON AND TRANSMISSION LINE WITH CHUA'S DIODE

E. Yu. ROMANENKO and A. N. SHARKOVSKY

Institute of Mathematics, National Academy of Sciences of Ukraine

Investigations of a lossless transmission line with Chua's diode allow us to understand the mathematical scenarios of how the transition from a simple behavior to an extremely complicated one may occur in parameter-distributed systems. We briefly present the concept of ideal turbulence — an original approach to modeling spatial-temporal turbulent processes, developed in our works on infinite-dimensional chaotic dynamics, and show that ideal turbulence is typical for idealized transmission lines of this kind. In particular, we specify parameters regions with the most evolved turbulence, when the statistical properties of vibrations are described by random processes.

1. Introduction

Mathematical research into electrical transmission lines has a reach history, originating in the mid-19th century from the work of J.C. Maxwell, W. Thomson (Lord Kelvin) and O. Heaviside, and continues to be relevant in our time. The number of modern-day works in this direction is very large (see, for example, [24, 25]). We and our colleagues have also been involved in this subject ("the most dense" in the 90s) — investigated complex vibrational regimes in a lossless line with Chua's diode (see [3,11,14,15,21,22]). Of importance was the fact that such idealized model admits a reduction to a one-dimensional map, which appear to be piecewise linear, the $v-i$ characteristic of Chua's diode being piecewise linear. This allowed for a fairly detailed analysis of the vibrations depending on the parameters and, in particular, the indication of parameter region in which the vibrations are extremely intricate — turbulent.

Over 20 years we have been actively engaged in the development of approaches to the study of spatal-temporal turbulent processes by the methods of dynamical systems theory, especially one-dimensional chaotic dynamics. These investigations have led to the concept of ideal turbulence — a certain mathematical idealization of real turbulence — and the fixation of the term '*ideal turbulence*' as a mathematical phenomenon (now this term is represented, in particular, in the "Encyclopedia of Nonlinear Science" [20]).

This paper is to look at our "old studies" on a transmission line with "the newly won positions" and describe in greater detail the vibrations in "the zone of turbulence". The paper consists of two parts. In the first we tried to clarify what is the ideal turbulence, in the form being laconic so to keep the mathematical rigor (so we had after all to adduce some cumbersome formulas). In the second part, we show that ideal turbulence is typical for the idealized transmission line with boundary nonlinearity and, in particular, highlight parameters regions which correspond to the most developed turbulence, when the statistical properties of the vibrations are described by random processes.

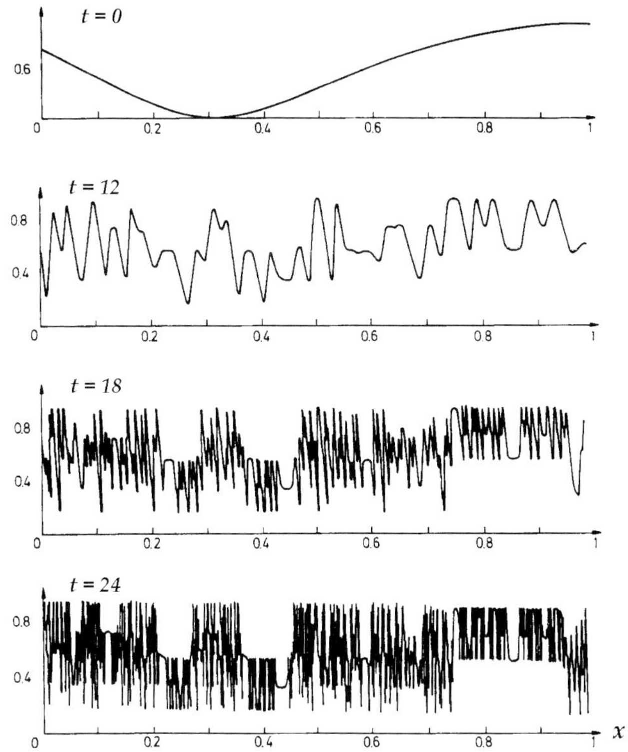

Fig. 1. Typical road to spatial-temporal chaos [3, 23].

2. Concept of Ideal Turbulence

The notion of *ideal turbulence* as spatial-temporal chaos in ideal media (without inherent resistance) was proposed one of the authors back in 1983; originally, it was named *dry turbulence* by analogy with von Neumann's dry water [12, 13]. Since then ideal turbulence is kept in the area of scientific interests of the authors. As a result a mathematically rigorous definition of ideal turbulence and approaches to its study have been elaborated (see our works given in the bibliography).

A meaningful mathematical definition of turbulence can be given for dynamical systems on spaces of smooth or piecewise smooth functions.

Let $\{\mathfrak{C}, S^t\}$ be an infinite-dimensional dynamical system, where $\mathfrak{C}$ is the space of continuous piecewise smooth functions $\phi : D \to E$, D and E are regions of Euclidean spaces, $t \in \mathbb{Z}^+$ or $t \in \mathbb{R}^+$. The trajectory $S^t[\phi]$ of $\phi \in \mathfrak{C}$ is understood as a sequence when $t \in \mathbb{Z}^+$ and as a directedness when $t \in \mathbb{R}^+$ (not to be confused with the orbit of ϕ).

The spatial-temporal chaotization in the dynamical system is perceived as an evolution of the functions $\phi \in \mathfrak{C}$ under S^t, which leads to that their behavior becomes more and more intricate and the limit states of the system are already impossible to describe with continuous functions (see Fig. 1). The attractor of the system, i.e. the set attracting all or almost all the trajectories, is not contained entirely in the phase space, and hence the system needs to be extended on a wider function space. The spaces of fractal and random functions are particularly appealing for use as such wider space.

For a space containing fractal functions we take the space of multivalent functions $\zeta : D \to E$, whose graphs are closed sets (such functions are called upper semicontinuous), and a metric is taken as follows

$$\rho^{\Delta}(\zeta_1, \zeta_2) = \sup_{\varepsilon>0} \min \left\{ \varepsilon, \sup_{x \in D} \Delta\left(V^{\varepsilon}_{\zeta_1}(x), V^{\varepsilon}_{\zeta_2}(x)\right) \right\},$$

where $\Delta(\cdot,\cdot)$ is for the Hausdorff distance between sets, $V^{\varepsilon}_{\zeta}(x) = \overline{\zeta(V_{\varepsilon}(x))}$, and $V_{\varepsilon}(x)$ is the ε-neighborhood of $x \subset D$.

The metric ρ^{Δ} compares the sets of values taken on by functions in the "optimal" neighborhoods of points from D and is equivalent to the metric given by the Hausdorff distance between the graphs of functions.

A function, random or deterministic, can be interpreted as a collection of its finite-dimensional distributions. For a deterministic function $\xi(x)$, all its distributions are defined uniquely by its one-dimensional distribution function $F_{\xi}(x, y) = \chi_{(-\infty, y]}(\xi(x))$ with $\chi_A(\cdot)$ being the indicator of the set A. For the space containing random and deterministic functions $\xi : D \to E$, a metric is conveniently chosen to compare the distributions of the functions in "optimal" neighborhoods of points from D. To obtain this metric, we have in the formula for ρ^{Δ} to replace the set $\zeta(V_{\varepsilon}(x))$ with the distributions of ξ in $V_\varepsilon(x)$. Thus we arrive at the metric

$$\rho^{\#}(\xi_1,\xi_2) = \sup_{\varepsilon>0} \min\left\{\varepsilon, \sum_{r=1}^{\infty} \frac{1}{2^r} \sup_{x\in D^r} \sup_{y\in E^r} \times \left| F^{r,\varepsilon}_{\xi_1}(x,y) - F^{r,\varepsilon}_{\xi_2}(x,y)\right|\right\},$$

where D^r and E^r are the direct products of r copies of D and E, and $F^{r,\varepsilon}_{\xi}(x,y)$ is the averaging of the r-dimensional distribution function $F^r_\xi(x,y)$ of ξ over ε-neighborhood of $x \in D^r$. The metric $\rho^{\#}$, that might appear to be too complicated, actually has a simple probability meaning and is closely linked to the notion of weak convergence.

Let $\mathfrak{C}^\Delta$ and $\mathfrak{C}^{\#}$ be the completions of the phase space $\mathfrak{C}$ respectively with upper semicontinuous functions via the metric ρ^Δ and with random and deterministic functions via the metric $\rho^{\#}$.

In dynamical systems theory, to describe the asymptotic behavior of trajectories, the notion of ω-limit set (i.e. the set of the limit points of a trajectory) is usually used. Reasoning from this, for an initial state $\phi \in \mathfrak{C}$, we will denote by $\omega_\Delta[\phi]$ and $\omega_{\#}[\phi]$ the sets of the limit points of the trajectory $S^t[\phi]$ respectively in the spaces $\mathfrak{C}^\Delta$ and $C^{\#}$. Finally, we will write $d(\cdot)$ for the Hausdorff dimension of the graph of a function.

Definition. The initial state $\phi \in \mathfrak{C}$ is said to generate *ideal turbulence* (IT) if the set $\omega_\Delta[\phi]$ contains at least one "point" which is a fractal function, i.e. if there exists $\zeta_* \in \omega_\Delta[\phi]$ such that $d(\zeta_*) > \dim D$. The initial state $\phi \in \mathfrak{C}$ is said to generate *stochastic turbulence* (SIT) if the set $\omega_{\#}[\phi]$ contains at least one "point" which is a random function (stochastic process). If there is a great many (in one sense or another, as the case requires) of initial states which generate IT or SIT, then IT or respectively SIT is said to occur in the dynamical system.

The majority of evolutionary boundary value problems (BVPs) for partial differential equations induce dynamical systems of shifts along solutions. It is reasonable to say that *there occurs IT or SIT in a BVP* if such a turbulence occurs in the dynamical system associated with the BVP.

The modern-day mathematical apparatus of dynamical systems theory (particularly, one-dimensional) allows one to investigate in detail the properties of ideal turbulence in the BVPs which are reducible to continuous argument difference equations or differential-difference equations. (see, for example, [16, 19, 23]). It has long been known that many BVPs for transmission lines fall into this class (see, for example, [1,4]), but an effective study of such BVPs by the "reduction method" became possible only in the last 20–30 years owing to the progress in continuous argument difference equations theory.

In this regard a transmission line with Chua's diode are among the representative examples.

3. Transmission Line with Chua's Diode

We consider the BVP for a lossless transmission line with Chua's diode

$$v_x = -Li_t, \quad i_x = -Cv_t, \quad 0 \le x \le l, \quad t \in \mathbb{R}^+, \tag{1}$$

$$v(0,t) = 0, \quad i(l,t) = G(v(l,t) - E - R\,i(l,t)). \tag{2}$$

Here $v(x,t)$ and $i(x,t)$ are the voltage and current along the line, L, C and R are the inductance, capacitance and resistance per unit length, E is the voltage across a voltage source (in series with the diode), and G is the piecewise linear function

$$G(u) = \begin{cases} m_0\,u, & |u| \le 1,\\ & m_0 < 0,\\ m_1\,u - (m_1 - m_0)\,\mathrm{sgn}\,u, & |u| \le 1,\\ & m_1 > 0, \end{cases} \tag{3}$$

that specifies the $v{-}i$ characteristic of the diode. We assume that the solutions of the BVP are continuous and piecewise smooth, which forces the initial conditions

$$v(x,0) = \varphi(x), \quad i(x,0) = \psi(x), \qquad \varphi,\psi : [0,l] \to \mathbb{R}, \tag{4}$$

to be likewise continuous and piecewise smooth, and be in agreement with the boundary conditions (2).

3.1. *Reducing to a difference equation*

At the condition $v(0,t) = 0$, the general solution of the BVP is of the form

$$v(x,t) = \alpha\left(t - \frac{x}{\nu}\right) - \alpha\left(t + \frac{x}{\nu}\right), \qquad i(x,t) = \frac{1}{Z}\left[\alpha\left(t - \frac{x}{\nu}\right) + \alpha\left(t + \frac{x}{\nu}\right)\right], \tag{5}$$

where $\nu = \sqrt{1/LC}$, $Z = \sqrt{L/C}$, and α is an arbitrary function. With the new variables

$$\beta(\tau) = \alpha(2l\tau/\nu), \quad \tau = (\nu t/l - 1)/2, \tag{6}$$

the combination of (5) and the second condition (2) yields the continuous argument difference equation

$$\beta(\tau + 1) = f(\beta(\tau)), \quad \tau \geq -1, \tag{7}$$

with the nonlinearity giving by the piecewise linear (single-valued or multivalent) map

$$f : z \mapsto A_s z - B_s \quad \text{for } z \in I_s, \ s = 0, \pm 1, \tag{8}$$

where

$$A_s = -1 + c_s, \quad B_s = \frac{c_s}{2}\left[E + s\left(1 - \frac{m_0}{m_s}\right)\right],$$

$$c_s = \frac{2Z}{\frac{1}{m_s} + R + Z}, \quad m_{-1} = m_1,$$

$$I_0 = \left\{z : \left|z - \frac{E}{2}\right| \leq |\delta|\right\},$$

$$I_{\pm 1} = \left\{z : \pm\left(z - \frac{E}{2}\right) > \delta\right\}, \quad \delta = \frac{m_0 Z}{c_0}.$$

The initial states $v_0(x)$ and $i_0(x)$ determine the initial condition for Eq. (7):

$$\beta(\tau)\Big|_{\tau \in [-1,0]} = \phi(\tau),$$

$$\phi(\tau) = \begin{cases} \dfrac{\varphi(-\theta) + Z\psi(-\theta)}{2}, \\ \qquad -1 \leq \tau < -\dfrac{1}{2}, \\ \dfrac{-\varphi(\theta) + Z\psi(\theta)}{2}, \\ \qquad -\dfrac{1}{2} \leq \tau < 0, \end{cases} \quad \theta = l(1 + 2\tau); \tag{9}$$

therewith $\phi(\tau)$ has the property: $\phi(0) = f(\phi(-1))$, which is a result of the consistency of the initial and boundary conditions for the BVP.

Equation (7) can be considered as the continuous family of the same maps (oscillators) $z_n \mapsto z_{n+1} = f(z_n)$, $z_0 = \phi(\tau)$, "placed" at all points of the interval $[-1, 0]$. Thus the dynamics of the lossless transmission line is reduced to the dynamics of a continuous family (continuous "lattice") of uncoupled oscillators.* Cumulative vibrations of these oscillators are described by the (infinite-dimensional) dynamical system

$$\mathbb{S} : \ \phi(t) \ \mapsto \ f(\phi(t)), \tag{10}$$

generated by Eq. (7) on the space of continuous piecewise smooth initial functions $\phi : [-1, 0] \to \mathbb{R}$ such that $\phi(0) = f(\phi(-1))$.

3.2. *Managing map*

It follows from the previous section that the behavior of solutions for the BVP (1), (2) is determined by the one-dimensional (1D) map f, which we will refer to as the *managing map*. As is shown in [22] the dynamics of f can prove to be intricate only in the following two cases:

$$(^*) \ \ -m_0 Z \leq E < 1 + m_0 R;$$

$$(^{**}) \ \ -m_0 R - \frac{m_0}{m_1} < E < -m_0 Z.$$

We restrict ourselves to the case (*). Then f is single-valued and has an invariant interval I that "takes up" all the points $z \in \mathbb{R}$: for any $z \in \mathbb{R}$, there exists $n > 0$ such that $f^n(z) \in I$. Moreover, f is equivalent on I to the two-segment map

$$g(z) = \begin{cases} p\,(z - a) + 1, & z \in [0,\, a], \\ q\,(1 - z), & z \in (a,\, 1], \end{cases} \quad \begin{array}{r} 0 < p < 1, \\ q > 1, \\ a = 1 - 1/q, \end{array} \tag{11}$$

with

$$p = \frac{-1 + (Z - R)m_1}{1 + (Z + R)m_1}, \quad q = \frac{1 - (Z - R)m_0}{1 + (Z + R)m_0}.$$

The two-parameter family of maps (11) has been studied closely enough, in particular, its main properties were given in the above-mentioned our works 90s. Here we discuss in more detail those needed for further.

Depending on the asymptotic dynamics of the trajectories of g the parameters region $P = \{(p, q) : 0 < p < 1, q < 1\}$ decomposes into the three

*Small perturbations, such as introducing losses into the transmission line, turns the uncoupled maps lattice into a coupled maps lattice.

subregions

$$P_0 = \{(p,q) : pq < 1\},$$
$$P_{IT} = \bigcup_{m>2} P_m,$$
$$P_m = \left\{(p,q) : \sum_{i=0}^{m-2} p^{-i} < q < p^{-m+1}\right\},$$
$$P_{ST} = P\backslash(\overline{P_0} \cup \overline{P_{IT}}),$$

each is open in P. The boundary between the subregions is advisable to represent in the form

$$G = \bigcup_{m\geq 2} G_m^- \cup \bigcup_{m\geq 3} G_m^+,$$
$$G_2^- = \{(p,q) : pq = 1\},$$
$$G_m^- = \{(p,q) : q = p^{-m+1}\} \cap \overline{P_m},$$
$$G_m^+ = \left\{(p,q) : \sum_{i=0}^{m-2} p^{-i} = q\right\} \cap \overline{P_m}, \quad m = 3, 4, \ldots$$

- Where $(p,q) \in P_0$, the map g has an attracting cycle of period 2, that attracts all the points from $[0,1]$ except the (repelling) fixed point $z = q/(1+q)$.
- Where $(p,q) \in P_m^+$, $m > 2$, the map g has an attracting cycle of period m; this cycle attracts all the points from $[0,1]$ except the set $\mathcal{D}(g)$ formed by the unstable points of g (and including the repelling cycles and their preimages) and is a cantor set of positive Hausdorff dimension ($0 < d_H\mathcal{D}(g) < 1$). The basin of the attracting cycle consists of a containable number of open intervals (which are the preimages of the m "main" intervals that contain this cycle).
- Where $(p,q) \in P_{ST}$, the map g has no attracting cycles but it has a cycle of intervals $J_1, \ldots, J_s$, $s = s(p,q)$, that "takes up" almost all the points from $[0,1]$. In this case, g is transitive on the set $J = \cup_{i=1}^s J_i$ (i.e. almost each trajectories is dense on J) and has a smooth invariant measure concentrated on J (in addition, g^s is probabilistically intermixing on J_k, $k = 1, \ldots, s$). When (p,q) is in the vicinity of G_m^-, then $s = 2m$. When (p,q) is in the vicinity of G_m^+, then $s = 1$ and, moreover, $J = [0,1]$; hence, the support of the smooth invariant measure consists of the unique interval $[0,1]$.

The boundary G is the site of bifurcations of two types.

- At the boundary G_m^-, $m \geq 2$, we observe a 'soft' doubling period bifurcation. The attracting cycle of period m, that exists when $(p,q) \in P_m$, transforms, when (p,q) gets on G_m^-, into a "taking up" cycle of intervals $J_1, \ldots, J_m$ of exclusive structure: $g^m(z)|_{J_i} = -z$, $i = 1, \ldots, m$; therewith the end-points of the intervals $J_1, \ldots, J_m$ are the preimages of the break-point $z = a$. Consequently, every interval J_i is the union of two adjacent intervals which transform one into another under g^m. When (p,q) moves away from G_m^- and passes in P_{ST}, the adjacent intervals "move apart" and produce a cycle of intervals with period $2m$. It is just the way in which the 'soft' doubling period bifurcations happen for piecewise linear maps.
- At the boundary G_m^+, $m \geq 3$, we deal with a 'hard' cycle birth bifurcation. Locally this cycle birth corresponds to a tangential bifurcation for smooth maps: the (repelling or semi-repelling) cycle of period m, produced during the transition of (p,q) from P_{ST} at G_m^+, splits into two cycles of period m, attracting and repelling, when (p,q) moves away from G_m^+ and passes in P_m.

However, the case $(p,q) \in G_m^+, m \geq 3$, is the first to give rise a cycle of period m, that has the so-called maximal type [2]: if $\gamma_1 < \gamma_2 < \ldots < \gamma_m$ are the points of the cycle, then $g(\gamma_i) = \gamma_{i+1}$, $i = 1, 2, \ldots, m-1$, and $g(\gamma_m) = \gamma_1$, which means

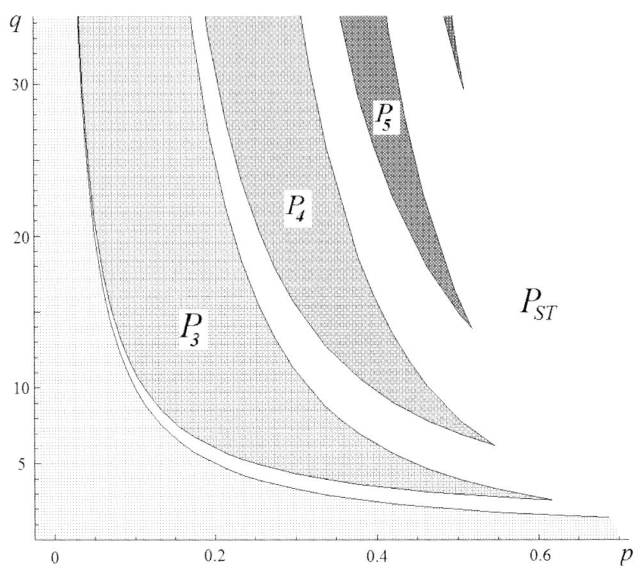

Fig. 2. Decomposition of the parameter region $P = \{(p,q) : 0 < p < 1, q > 1\}$.

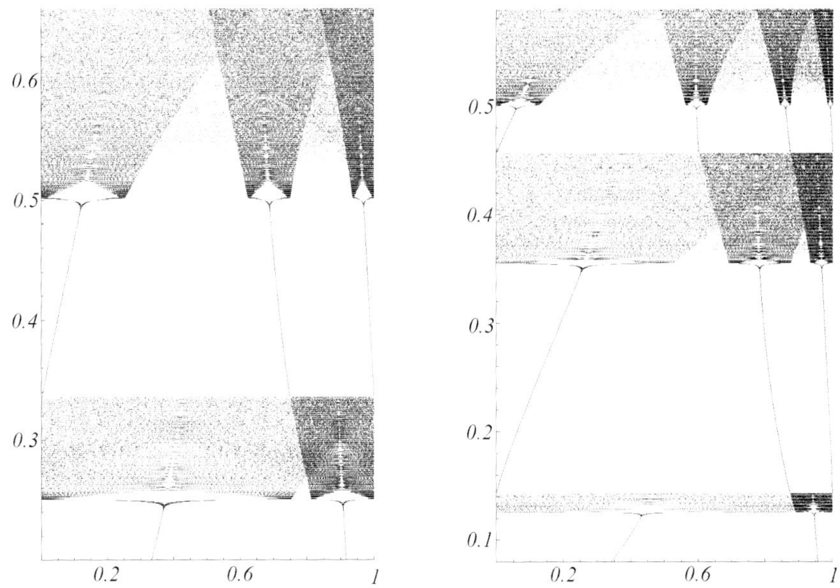

Fig. 3. Bifurcation diagram for the family (11) in the plane (z, p) with $q = 4$ (left) and $q = 8$ (right).

that all but one points of the cycle are placed on $[0, a]$ (the cycle includes, in particular, the points $z = 0, a, 1$.) Clearly the cycles of period m, that appear when (p, q) transfer into P_m, also are of maximal type. In families of unimodal maps, cycles of maximal type appear last among all the cycles of period m (when $(p, q) \in P_m$ and m is large, g has many cycles of period m; for instance, if m is a prime number, there are $(2^m - 2)/m$ such cycles, which is the largest number of cycles of period m, which is possible for the logistic map).

Just this feature of the cycles structure allows to observe in piecewise linear maps the period-adding phenomenon (or the period adding route to chaos), marked already in [5]. For the map (11) we see this phenomenon on the bifurcation diagrams in Fig. 2 and Fig. 3.

To make clear some details, it is useful to note certain differences in the bifurcations for the piecewise linear map (11) and for smooth maps by the example of the logistic map $x \mapsto \lambda x(1 - x)$, $3.83 < \lambda < 4$, when the latter already has a cycle with period 3.

With increasing the parameter λ in the logistic map, every cycle is born either as an attracting one (a soft period doubling bifurcation), or together with an attracting one (a hard tangential bifurcation). For $\lambda \approx 3.83$ a cycle of with period 3 appears and thereafter there arise successively cycles of maximal type with periods $4, 5, 6, \ldots$ The same sequence of cycles of maximal type realizes for (11), the only difference being in the following. In both the families, between the birth of a cycle of maximal type with period m and the birth of a cycle of maximal type with period $m+1$, a countable number of cycles with periods $> m + 1$ come into being. But for the logistic family, all these cycles appear, as noted, to be attracting or to be paired with a repelling one, and for the family (11), all the appearing cycles are, contrary, repelling (i.e., invisible). Thus, in the family (11) we observe the period-adding phenomenon — the successive appearance of attracting cycles, which is characterized by the increase of period of 1 only, and is alternated by "windows of chaos".

3.3. *Ideal turbulence from the lossless transmission line*

For any initial condition (4), there corresponds a unique solution

$$(v(x,t), i(x,t))_{\varphi,\psi} : [0, l] \times \mathbb{R}^+ \to \mathbb{R}^2$$

Hence, on the space of continuous, piecewise smooth initial vector-functions $(\varphi, \psi) : [0, l] \to \mathbb{R}^2$, the BVP induces the dynamical system

$$S^t : (\varphi(x), \psi(x)) \mapsto (v(x,t), i(x,t))_{\varphi,\psi}. \quad (12)$$

In order to understand which may be the asymptotic dynamics of the system (12), it is reasonable to consider in parallel with (12) the simpler discrete dynamical system (10), which is associated with (12) via the formulas (5)–(9) and includes the managing map f in an explicit form.

The properties of the model map g lead to the following theorem.

Theorem 1. *Let the conditions* (*) *be fulfilled. Then if* $(p,q) \notin P_0$ *every initial state* ϕ *other than a constant on any interval from* $[-1,0]$ *generates ideal turbulence in the system* (10); *therewith for each* $\zeta \in \omega_\Delta[\phi]$,

$$1 < d(\zeta) < 2 \quad \text{when } (p,q) \in P_{IT},$$
$$\text{and} \quad d(\zeta) = 2 \quad \text{when } (p,q) \in P_{ST}.$$

Moreover, if $(p,q) \in P_{ST}$, *every nonsingular initial state* ϕ *generates stochastic turbulence in the system* (10).

This theorem is a good example to understand why and how turbulence happens in the problems whose dynamics is specified by a 1D map.

- For ideal turbulence, the main factor is the complex topological structure of the set $\mathcal{D}(f)$ formed by the unstable points of f. Where $f: I \to I$ has a cycle of period other then the powers of 2, the Hausdorff dimension of $\mathcal{D}(f)$ is positive. The system (10) "inherits" this property. Schematically, this fact can be explained in the following way. Let $\mathcal{D}(f,\phi) = \phi^{-1}(\mathcal{D}(f))$. With the completion of the phase space via the metric ρ^Δ, the ω-limit set $\omega_\Delta[\phi]$ consists of functions $\zeta : [-1,0] \to 2^I$ such that $\zeta(\tau_*)$ is a singleton (i.e. a one-point set) when $\tau_* \notin \mathcal{D}(f,\phi)$, and a closed nongenerated interval when $\tau_* \in \mathcal{D}(f,\phi)$. In typical situations, $d(\zeta) = d_H\,\mathcal{D}(f,\phi) + 1$ and $d_H\,\mathcal{D}(f,\phi) > 0$, which implies $d(\zeta) > 1$.
- For stochastic turbulence, of fundamental importance is both the occurrence of a smooth invariant ergodic measure (*s.i.e.*-measure) for f and the fact that for interval maps, the availability of *s.i.e.*-measures is not extraordinary point (for ample classes of parameter-dependent maps, the parameter values at which the map possesses a *s.i.e.*-measure form a set of positive Lebesgue measure).

Where $f: I \to I$ possesses a *s.i.e.*-measure μ, almost every trajectory $f^n(z), n \in \mathbb{Z}^+$, restores μ: the residence time of the trajectory in a set $A \subset I$ is equal to $\mu(A)$. With the completion of the phase space via the metric $\rho^\#$, this 'temporal' stochasticity of f transforms into 'spatial-temporal' stochastization of the system (10). Most simply this fact is possible to explain when the support of μ consists of a single interval. In such a case, for all nonsingular[a] initial states $\phi(\tau)$, $\tau \in [-1,0]$, their ω-limit sets $\omega_\#[\phi]$ coincide with the same "point" — a stationary stochastic process $\xi^\#(\tau)$, $\tau \in [-1,0]$, with mutually independent values,[b] whose one-dimensional distribution function is of the form $F(y,\tau) = \mu((-\infty,y] \cap \operatorname{supp}\mu)$. With regard to Eq. (7), this means that, given $\varepsilon^* > 0$, for any solution $\beta(\tau)$ arising from a nonsingular initial function, and a however small $\sigma > 0$ there exists $\tau^* > 0$ with the property

$$\left|F_\beta^\varepsilon(y,\tau) - \mu((-\infty,y] \cap \operatorname{supp}\mu)\right| < \sigma$$
$$\text{for} \quad \tau > \tau^*, \ \ \varepsilon > \varepsilon^*, \ \ y \in \mathbb{R},$$

recall that $F_\beta^\varepsilon(y,\tau)$ is the ε-averaging of distribution function of the solution $\beta(\tau)$. It follows from the previous section that $\operatorname{supp}\mu = I$ in the vicinity of "subboundary" $\cup_{m\geq 3} G_m^+$.

We now return to the problem (1), (2) and its associated dynamical system (12). With Theorem 1 and the relationship between the systems (12) and (10), given by

$$S^t\,[(\varphi,\psi)]\,(x) = (\mathbb{S}^\kappa[\phi](\tau-\kappa) - \mathbb{S}^{\kappa+1}[\phi](\tau-\kappa),$$
$$\mathbb{S}^\kappa[\phi](\tau-\kappa) + \mathbb{S}^{\kappa+1}[\phi](\tau-\kappa)),$$
$$\tau = (\nu t/l - 1)/2, \phi(\tau) \text{ is as in (9)}, \kappa$$
$$\text{is the integer part of } \tau,$$

we have every reason to formulate the following theorem.

Theorem 2. *Let the conditions* (*) *be fulfilled and* $(p,q) \notin P_0$. *Then ideal turbulence occurs in the BVP* (1), (2). *Moreover, where* $(p,q) \in P_{ST}$, *there occurs stochastic turbulence in the BVP.*

The former assertion of Theorem 2 is virtually evident: for almost every initial state (φ,ψ) of the system (12), one can find $\zeta \in \omega_\Delta[(\varphi,\psi)]$ such that $1 < d(\zeta) \leq 3$.

---

[a]A function g is referred to as nonsingular, if mes $A = 0$ implies mes $g^{-1}(A) = 0$; here $g^{-1}(A)$ is the full preimage of the set A.

[b]For a process with independent values, its r-dimensional distribution function is the product of the r corresponding one-dimensional distribution functions.

But an accurate proof of the latter will require considerable efforts. Let we have the simplest situation: μ is a *s.i.e.*-measure μ of f and $\operatorname{supp}\mu$ consists of a single interval. Write $\lambda(w) = \mu((-\infty, w] \cap \operatorname{supp}\mu)$. The following statement is likely to be the case: given $\varepsilon^* > 0$, for any solution $(v(x,t), i(x,t))$ arising from a nonsingular initial state, and a however small $\sigma > 0$ there exists $t^* > 0$ with the property

$$\left|F_v^\varepsilon(y,(x,t)) - \int_{-\infty}^{+\infty}(1-\lambda(y-w))d\lambda(w)\right| < \sigma,$$
$$\left|F_i^\varepsilon(y,(x,t)) - \int_{-\infty}^{+\infty}\lambda(y-w)\,d\lambda(w)\right| < \sigma$$
$$\text{for } t > t^*,\ \ \varepsilon > \varepsilon^*,\ \ y \in \mathbb{R},$$

here $F_h^\varepsilon\big(y,(t,x)\big)$ is the ε-averaging of the distribution function of $h(x,t)$.

When the condition (**) hold, for the managing map f there also exists a model map, but it is three-segment. Analysis similar to that at the condition (*) leads to an analogous conclusion: there can happens ideal turbulence in the BVP. It is worth noting that at the condition (**), in the BVP we can observed a weak chaotization: the frequency of vibrations of the solutions increases with time but slowly, by a power law. We refer to such a phenomenon as *weak ideal turbulence* (WIT). Here, WIT resulted from the fact that the model map can possess, for example, cycles of period 1, 2 and 4 only.

References

[1] Brayton, R. K. (1967). Nonlinear oscillations in a distributed network, *Quart. Appl. Math.* **24**, pp. 289–301.

[2] Fedorenko V. V. and Sharkovsky A. N. (2010). On the coexistence of homoclinic and periodic trajectories, *Nonlinear Dynamics* (Russian Sci. J.) **6**, 1, pp. 207–217.

[3] Maistrenko Yu. L., Maistrenko V. L., Vikul S. I. and Chua L. O. (1995). Bifurcations of attracting cycles from time-delayed Chua's circuit, *Int. J. Bifurcation and Chaos* **5**, 3, pp. 653–671.

[4] Nagumo, J. and Shimura, M. (1961). Self-oscillation in a transmission line with a tunnel diode, *IEEE Int. J. Electron.* **49**, pp. 1281–1291.

[5] Pei L.-Q., Guo F., Wu S.-X. and Chua L. O. (1986). Experimental confirmation of the period adding route to chaos in a nonlinear circuit, *IEEE Trans. on Circuits and Systems* **34**, 4, pp. 438–441.

[6] Romanenko, E. Yu. (2010). Randomness in deterministic continuous time difference equations, *Int. J. Difference Equations and Appl.* **16**, 2–3, pp. 243–268.

[7] Romanenko, E. Yu. and Sharkovsky, A. N. (1996). From one-dimensional to infinite-dimensional dynamical systems: Ideal turbulence, *Ukrainian. Math. J.* **48**, 12, pp. 1604–1627.

[8] Romanenko, E. Yu. and Sharkovsky, A. N. (1999). From boundary value problems to difference equations: A method of investigation of chaotic vibrations, *Int. J. Bifurcation and Chaos* **9**, 7, pp. 1285–1306.

[9] Romanenko, E. Yu. and Sharkovsky, A. N. (2007). Dynamical systems and the modelling of turbulence, *Ukrainian Math. J.* **59**, 2, pp. 217–230.

[10] Romanenko, E. Yu. and Sharkovsky, A. N. (2010). Whether Chaos can be achieved in Piecewise Linear Boundary Value Problems, *J. Difference Equation and Dynamical Systems* **18**, 1–2, pp. 1–9.

[11] Severino, R., Sharkovsky, A., Sousa Ramos, J. and Vinagre, S. (2006). Topological invariants in a model of a time-delayed Chua's circuit, *Nonlinear Dynamics* **44**, pp. 81–90.

[12] Sharkovsky, A. N. (1983). "Dry" turbulence, in *Short Comm. Int. Congress Mathematicians* (1983) (Warszawa, Poland), **10**, 12, p. 4.

[13] Sharkovsky, A. N. (1984). "Dry" turbulence, in *Proc. Int. Workshop 'Nonlinear and Turbulent Processes in Physics'* (1984) (Gordon and Breach), **3**, pp. 1621–1626.

[14] Sharkovsky, A. N. (1993). Chaos from a time-delayed Chua's circuit, *IEEE Transactions on Circuits and Systems I* **40**, 10, pp. 781–783.

[15] Sharkovsky, A. N. (1994). Ideal turbulence in an idealized time-delayed Chua's circuit, *Int. J. Bifurcation and Chaos* **4**, 2, pp. 303–309.

[16] Sharkovsky, A. N. (2004). Difference equations and boundary value problems, in New Progress in Difference Equations *Proc. Int. Conf. 'Difference Equations and Appl.' (ICDEA-2001)* (Taylor and Francis, London), pp. 3–22.

[17] Sharkovsky, A. N. (2006). Ideal turbulence, *Nonlinear Dynamics* **44**, pp. 15–27.

[18] Sharkovsky, A. N. and Romanenko, E. Yu. (1992). Ideal turbulence: Attractors of deterministic systems may lie in the space of random fields, *Int. J. Bifurcation and Chaos* **2**, 1, pp. 31–36.

[19] Sharkovsky, A. N. and Romanenko, E. Yu. (2004). Difference equations and dynamical systems generated by some classes of boundary value problems, *Proc. Steklov Institute of Mathematics* (Moscow) **244**, pp. 264–279.

[20] Sharkovsky, A. N. and Romanenko, E. Yu. (2005). Turbulence: Ideal, in Alwyn Scott (ed.), *Encyclopedia of Nonlinear Science* (Routledge, New York-London), pp. 955–957.

[21] Sharkovsky, A. N., Deregel, Ph. and Chua, L. O. (1995). Dry turbulence and period-adding phenomena from a 1-D map, *Int. J. Bifurcation and Chaos* **5**, 5, pp. 1283–1302.

[22] Sharkovsky, A. N., Maistrenko, Yu. L., Deregel, Ph. and Chua, L. O. (1993). Dry turbulence from a time delayed Chua's circuit, *Systems and Computers* **3**, 2, pp. 645–668.

[23] Sharkovsky, A. N., Maistrenko, Yu. L. and Romanenko, E. Yu. (1986). *Difference Equations and their Applications* (Naukova dumka, Kiev) (in Russian). English transl.: (1993) Ser. "Mathematics and Its Applications" (Kluwer Academic Publishers, Dordrecht-Boston-London) **250**.

[24] Wang, L. and Ma, X. (2008). Analysis of spatiotemporal patterns in an infinite-dimensional electromagnetic system, *IEEE Transactions on Circuits and Systems I: Regular Papers* **55**, 4, pp. 1125–1132.

[25] Zou, J., Chen, F. and Ma, X. (2010). Analysis of complex spatiotemporal behaviors in a time-delayed Chua's circuit, *Int. J. Appl. Electromagnetics and Mechanics* **33**, 1–2, pp. 369–376.

CHAPTER 17

CHAOS IN ELECTRONIC CIRCUITS: CHUA'S CONTRIBUTION (1980–2000)

CHRISTOPHE LETELLIER

CORIA UMR 6614 — Rouen University
F-76801 Saint-Etienne du Rouvray cedex
France

This chapter is devoted to Leon Chua's contribution to the study of chaotic behaviors produced by electronic circuits. Obviously, we do not intend to be exhaustive. The chaotic attractors presented in this chapter were selected from a topological point of view among Chua and co-workers' papers. Mostly, the originality of the (apparent) structure of the chaotic attractors retained was the guideline. It is thus shown that Chua and co-workers were not only interested in three-dimensional chaos, among which is the well-known double-scroll attractor, but they also investigated higher-dimensional attractors. There are many examples of toroidal chaotic attractors which remain not so well studied from the topological point of view. The main difficulty with these attractors is that they are most often embedded within a space whose dimension is at least four. Chua and co-workers remarked that tori T^k with k greater than two require the "Poincaré section of a Poincaré section, a concept they named "second-order Poincaré section", a tool not too often studied today. This chapter thus reviews some chaotic attractors, some of them yet deserving further analysis.

1. Introduction

Electronic circuits can be considered as dynamical systems. There is a long tradition for this since major contributions in electronics using the emerging dynamical system theory can be dated back at least to the 1930s with the work by van der Pol [1] and Andronov and co-workers [2]; to only quote the most well known. Poincaré, who may be considered as one of the fathers of the dynamical system theory, investigated himself the equation governing a singing arc [3] as described by Ginoux & Letellier [4].

It is known that van der Pol and van der Mark remarked some "noise" reminiscent of a cornemuse [5] but no time series was then published. Before such a paper, William Duddell, who discovered the "musical arc" (or the singing arc as later named) published in 1901 [6] one time series which seems to be rather aperiodic [Fig. 1(a)]. In the same system, Blondel [7] observed aperiodic oscillations too [Fig. 1(b)]. Both of them were considering these oscillations as interesting but they did not have yet the theoretical tools required to investigate them carefully. Today we could investigate a little bit more deeply the nature of such oscillations, in particular to check whether they are chaotic or not.

Due to its simplicity, the van der Pol equation became widely studied. Using analytical approach, Cartwright and Littlewood [8], and Levinson [9] remarked that for certain parameter values, some aperiodic solutions can be found. But this is actually with the first computers that the chaotic solutions were actually identified. The emergence of chaos in electronic circuits during the late 70s and early 80s was reviewed by Letellier & Ginoux [10]. Unfortunately, in this review we forgot the important contribution by Igor Gumowski and Christian Mira (the "Toulouse Research Group") whose a short story was given in Ref. [11]. Stimulated by a paper by C. P. Pulkin [12] who showed that in one-dimensional noninvertible map infinitely many unstable cycles may lead to bounded complex iterated sequences, Gumowski and Mira studied the piecewise-linear map [13]

$$\begin{cases} x_{n+1} = (1-\lambda)x_n + y_n \\ y_{n+1} = y_n + f(x_n) \end{cases}$$

$$\text{where } f(x_n) = \begin{vmatrix} -2\lambda x_n - 0.9\lambda & x_n < -0.5 \\ -\dfrac{\lambda x_n}{5} & \text{if } |x_n| < 0.5 \\ -2\lambda x_n + 0.9\lambda & x_n > -0.5. \end{vmatrix} \tag{1}$$

For some λ-values, they obtained attractive limit set made of bounded cloud of points as shown in Fig. 2(a). This could have been the first now called "chaotic" solution to a piecewise-linear map reported with an explicit picture. By these times, Gumowski and Mira indicated these types of behaviors as "Pulkin phenomenon" [13]. This is only ten years later that first chaotic solutions were widely investigated in electronic circuits [10].

Our aim in this chapter is to provide a partial review of some Chua's contributions to chaos in electronic circuits. It should be clear that our choices were biased by our interest in the topological analysis of chaotic behaviors. We thus focused our attention on some different types of chaos Chua proposed with many co-workers.

2. Electronic Circuits as Dynamical Systems

The genesis of the simplest electronic circuit results from a deep know-how in nonlinear electronics.

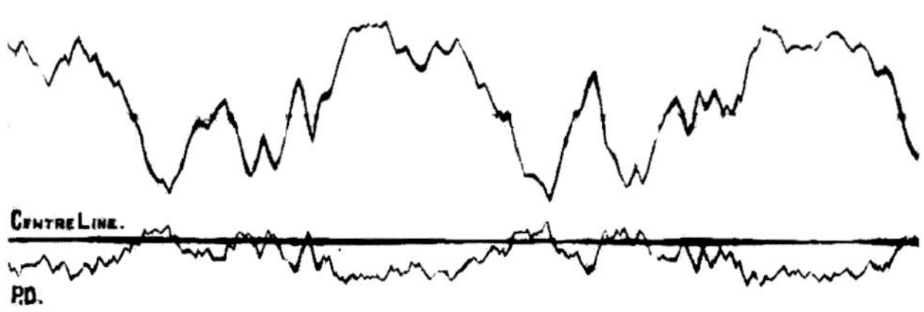

(a) Hissing regime observed by Duddell[6]

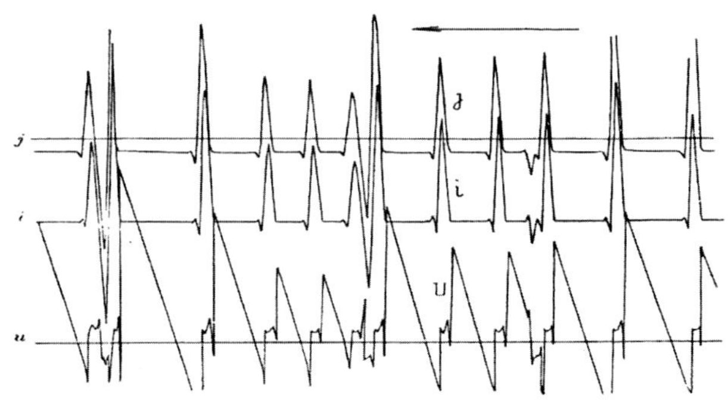

(b) Oscillations observed by Blondel[7]

Fig. 1. Aperiodic solution to a real implementation of a singing arc circuit.

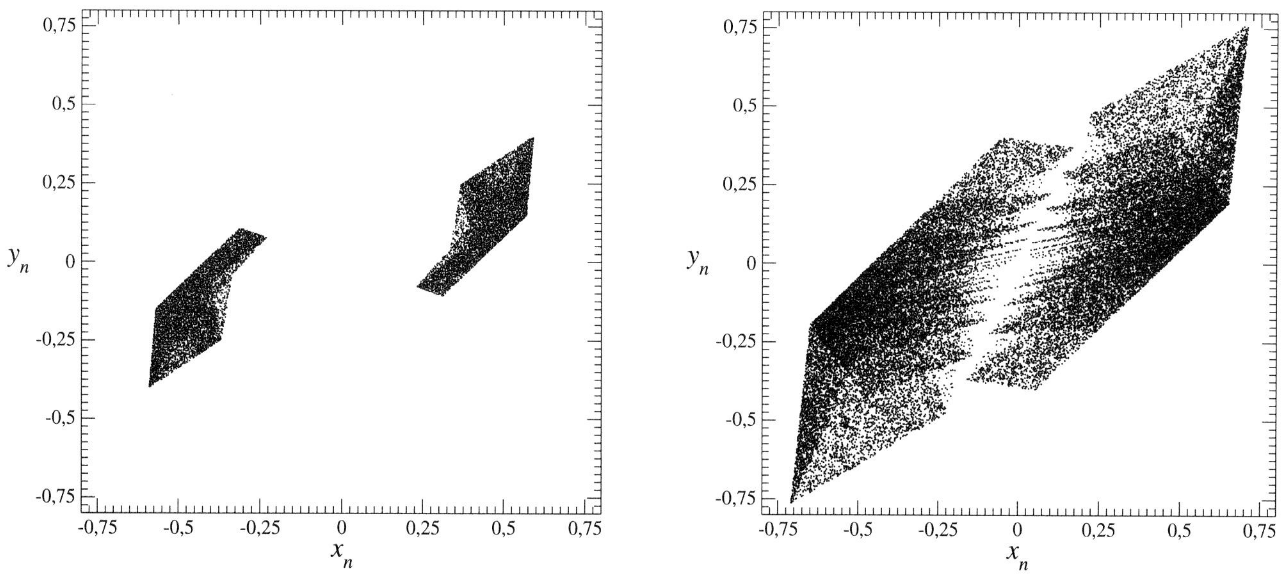

(a) $\lambda = 2.30$ (b) $\lambda = 2.39$

Fig. 2. Bounded-2 chaotic attractor solution to the piecewise-linear map (1) as (a) published in 1969 and (b) just before a boundary crisis.

Born June 28, 1936, Leon Chua was mainly interested in nonlinear electronic circuits and the dynamical system theory. His early papers discussed the synthesis of new elements for electronic circuits [14] and, in particular, in defining piecewise-linear or nonlinear resistors (see Figs. 6 and 8 in his 1968 paper [14]). Such aim leads Chua to discover the memristor in 1971 [15] which was recently designed by Hewlett Packard laboratories [16]. While he was a student at MIT (1962), Chua was studying the "qualitative theory" from books by Nemystskii & Stepanov [17] as well as by Lefschetz [18]. The three-volume Poincaré's book on celestial mechanics [19] was also studied. Thus Chua was developing a general theory for nonlinear electronic circuits with the desire to make a connection between the architecture of the circuit and its resulting dynamics. This is therefore a major problem, in a certain sense related to the still open question of the relationship between the algebraic structure of differential equations and the topology of the dynamics they produce.

2.1. *The inverse problem of the dynamical system theory*

Commonly, one investigates a given dynamical system by studying its fixed points, the nature of its asymptotic behaviors, etc. This is the *direct problem.* But sometimes, it is desirable to be able to write down some governing equations producing some prescribed features like the number and the nature of the fixed points, some given time series, etc. This is what is called the *inverse problem.* As far as we know, one of the first problem related to the nonlinear dynamical system theory attacked by Leon Chua was the inverse problem. In 1971, he published a paper where he proposed a way to write some governing equations which possess a finite number of prescribed singular points, as well as a prescribed set of eigenvalues associated with each singular points [20]. The background in nonlinear dynamical systems theory quoted in this 1971 paper is the book by Andronov, Vitt and Khaikin [2] and the book by Nemystskii and Stepanov [17].

Two years later, Chua provided a second contribution on this inverse problem with Douglas Green [21]. Their aim was then to find a class of differential equations having for solution a limit cycle with a prescribed frequency ω and a prescribed waveform. They treated the cases of autonomous and non-autonomous systems using Fourier decomposition and Chebichev polynomials associated with a least-square technique. One of their conclusion was that such a problem can be positively addressed but there is, in fact, an infinite number of different types of systems, all having the same limit cycles for solution.

This problem was addressed for chaotic solutions much later by Crutchfield and McNamara [22],

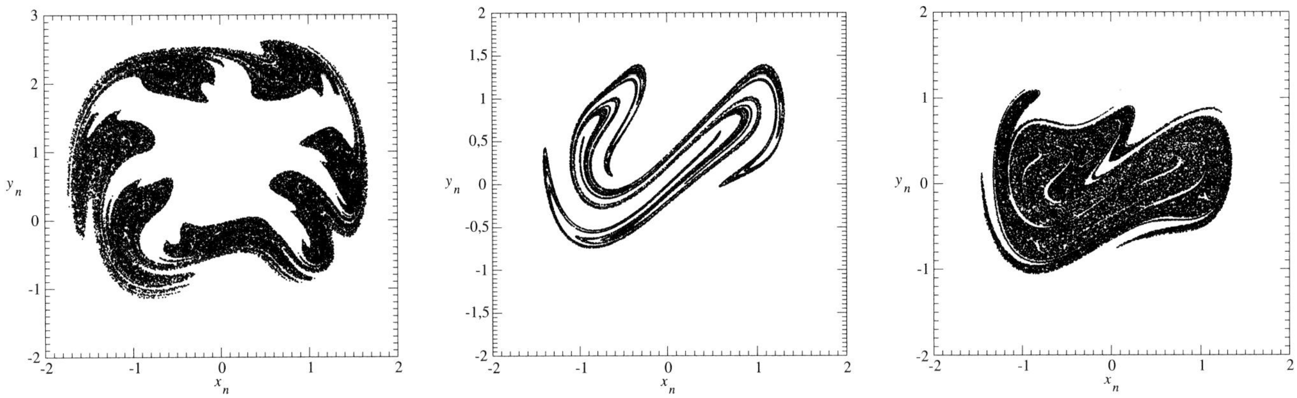

(a) $B = 1.0$, $\omega = 0.94$ (b) $B = 2.4$, $\omega = 0.70$ (c) $B = 2.0$, $\omega = 0.60$

Fig. 3. Poincaré sections of the chaotic attractors published by Ueda and Akamatsu in 1981. The parameter μ is set to 0.2.

once Takens had proposed his theorem for reconstructing the phase portrait from a single scalar time series [23]. This became what is called today *global modeling* whose a review was recently given by Aguirre and Letellier [24].

2.2. *Periodic and almost-periodic solutions*

Early analysis of solutions to nonlinear dynamical systems developed by Chua and co-workers were based on frequency domain analysis using, for instance, Volterra functional series [25]. Then the approach turned to use the Hopf theorem [26] for studying almost sinusoidal oscillations [27]. Electronic circuits were considered as dynamical systems and in 1990, Chua published a survey of the qualitative behavior of dynamic nonlinear (RLC) electronic circuits [28]. The motivation was to better understand the sometimes "bizarre" solutions observed for particular initial conditions. Among others, some solutions were investigated using "one-dimensional invariant sub-manifold", that is, using nullclines. Some relevant contributions to nonlinear dynamics with many applications in electronic circuits were quoted such as, for instance, Minorski's book [29] and Hayashi's book [30]. There is also one paper published by Gollub and co-workers about chaos in a system of coupled tunnel diode relaxation oscillators [31] where they remarked that "the nature of the coupling is much more important than the number of degrees of freedom in generating chaos". Two papers by Yoshisuke Ueda dealing with Chaos in the Duffing [32] and van der Pol [33] equations were also mentioned. These last two references are quite important in the sense that this is probably the first explicit reference to a chaotic attractor in an electronic circuit (Fig. 3) in Chua's published papers. In fact, in many books related to electronics like by Minorsky [29] and Hayashi [30], Duffing and van der Pol equations were treated as typical examples for producing nonlinear oscillations.

Ueda and Akamatsu used the van der Pol equation

$$\begin{cases} \dot{x} = y \\ \dot{y} = \mu(1 - x^2)y - x^3 + B\cos\omega t. \end{cases} \tag{2}$$

Note that in Ueda's papers, chaos is still presented as a "transitional phenomenon" by Ueda, apparently corresponding to the "bizarre" features mentioned by Chua. The qualification of "transitional" by Ueda may leave the impression that chaos was only an "artefact". Although that Ueda was not followed in his presentation of chaotic behaviors as "transitional phenomena", many others complained too that no proof for the chaotic nature of the solution to the Lorenz and the Rössler systems was provided [34–37]. Chua questioned himself too whether Lorenz's results [38] were actually asymptotic behaviors and not artefact. He was therefore pushed to develop some algorithms to be ensured that the numerical solutions he obtained were not artefact as we will see below.

Chua and Ushida therefore started by checking that the asymptotic behavior was actually reached when they were observing some quasi-periodic solutions [39]. The choice of quasi-periodic solutions was motivated by their relative simplicity and by the possibility to easily characterize them using a frequency domain analysis. To fulfill these criteria,

Chua and Ushida investigated Duffing's equation

$$\ddot{x} + 0.1\dot{x} + 2x + x^3 = 0.4\cos t + 0.4\cos 0.35t + 0.4\cos 0.155t. \quad (3)$$

This motivation was that, although describing a symmetrical mechanical pendulum, the Duffing system was often studied by Hayashi who used an electronic analog computer to make simulations [30]. It was also often quoted in books dealing with electronic circuits. Although deeply impressed by the Lorenz paper [38], Chua did not investigate the Lorenz system because it did not correspond to an electronic circuit by these times. In fact, the Lorenz system was only introduced in *IEEE* Journal in 1983 by Colin Sparrow [40]. The Duffing equation as investigated by Chua and Ushida can be rewritten as a forced second order differential equation

$$\begin{cases} \dot{x} = y \\ \dot{y} = -0.1y - 2x - x^3 \\ \qquad + A(\cos\omega_1 t + \cos\omega_2 t + \cos\omega_3 t) \end{cases} \quad (4)$$

where $A = 0.4$ and $\omega_1 = 1$, $\omega_2 = 0.35$ and $\omega_3 = 0.155$. The system is here forced with multiple input frequences [Fig. 4(a)]. Only an excerpt of a time series was proposed by Chua and his co-workers [39]. With all the conditions specified in the paper we obtained a period-211 limit cycle as exhibited by computing 50,000 points in the Poincaré section after the transient regime [Fig. 4(b)]. Thus, in spite of a phase portrait which could lead to conclude to a chaotic solution, the asymptotic behavior is periodic as stated in Chua and Ushida's paper.

2.2.1. *Early investigations of chaotic solutions*

The first occurrence of a chaotic attractor in Chua's works was the result of a collaboration with scientists involved in numerical studies, namely Jacques Neirynck (1931-) who spent few years in the Philips Laboratories and was in charge of Computer-Aided Design, and Philippe Verburgh (1957-) who was in the Circuit Theory Group of the Swiss Federeal Institute of Technology (Lausanne). One of their motivations was to demonstate that all properties for a chaotic motion was indeed observed, that is, sensitivity to initial conditions, hyperbolicity, existence of a strange attractor and a broadband frequency spectrum. This study was performed "through a long series of piecewise linear analysis using exact analytical formulas" [41].

The electronic circuit designed for this study was made of a single nonlinear element which is an inductor [Fig. 5(c)] and whose characteristic is the

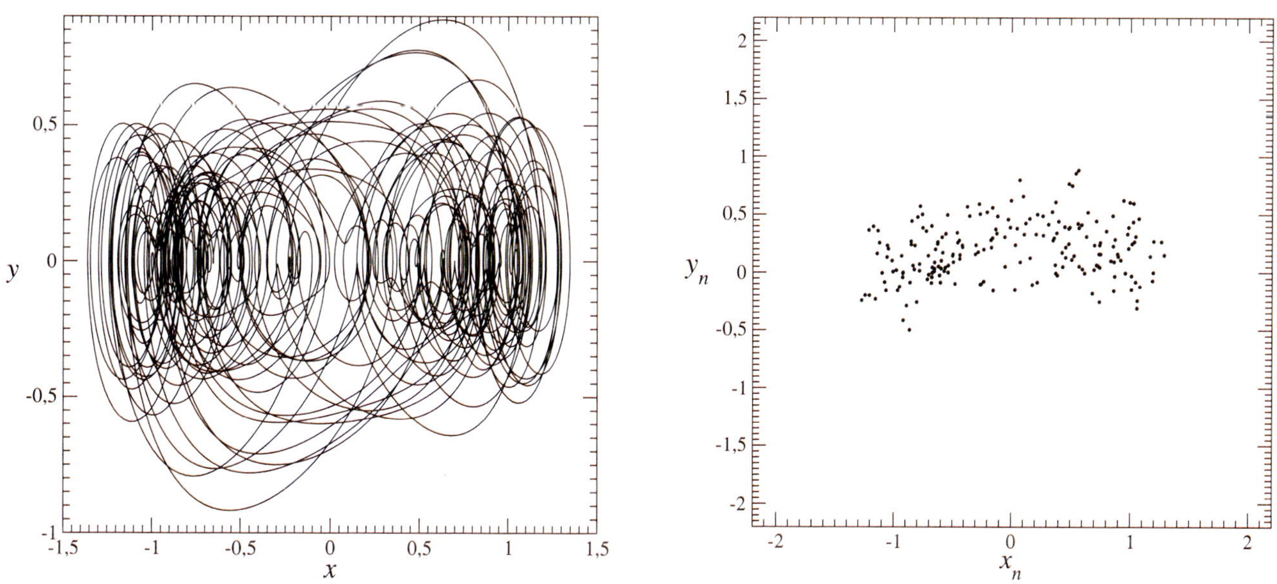

(a) Phase portrait

(b) Poincaré section

Fig. 4. Solution to the Duffing equation forced with multiple input frequencies as investigated by Chua and Ushida [39]. Initial conditions: $x_0 = 0.69667$ and $y_0 = -0.18304$. Only 211 different intersections with the Poincaré section based on frequency ω_1 were distinguished among the 50,000 intersections computed.

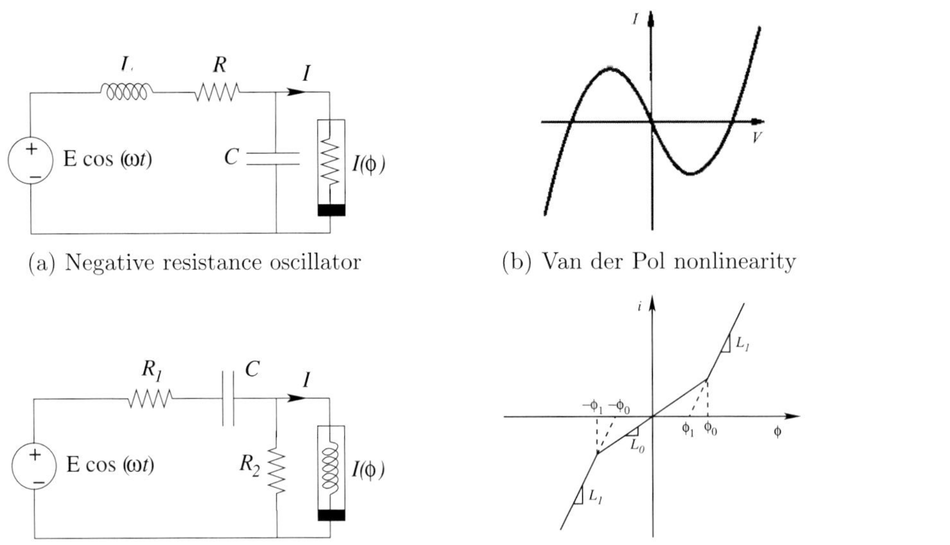

(a) Negative resistance oscillator

(b) Van der Pol nonlinearity

(c) Nonlinear resonant circuit

(d) Piecewise-linear function of the inductor

Fig. 5. (a) The negative resistance oscillator investigated by Ueda [33] and (b) the piecewise-linear resonant circuit investigated by Chua and co-workers in 1982 [41].

piecewise-linear function

$$I(\phi) = \begin{cases} \dfrac{(\phi - \phi_1)}{L_1} & \phi > \phi_0 \\ \dfrac{\phi}{L_0} & \text{for } |\phi| < \phi_0 \\ \dfrac{(\phi + \phi_1)}{L_1} & \phi < -\phi_0 \end{cases} \tag{5}$$

where $\phi_1 = \phi_0(1 - \frac{L_1}{L_0})$. The set of differential equations describing the circuit is

$$\begin{cases} \dot{\phi} = -\dfrac{R_1 R_2}{R_1 + R_2} I(\phi) - \dfrac{R_2}{R_1 + R_2}(V - E\cos\omega t) \\ \dot{V} = \dfrac{R_2}{C(R_1 + R_2)} I(\phi) - \dfrac{1}{C(R_1 + R_2)} \\ \quad \times (V - E\cos\omega t) \end{cases} \tag{6}$$

This is therefore a 2D driven dynamical system with a piecewise nonlinearity. This circuit differs from the circuit [Fig. 5(a)] used by Ueda [33] to investigate the van der Pol equation (2).

Chua and co-workers investigated the stability of periodic solutions and, finally, some chaotic behaviors as shown in Fig. 6. They noted the sensitivity to initial conditions and checked that it was a "*real phenomenon*" not caused by numerical instability. As they claimed, "these results represent the most reliable numerical confirmation to date of chaotic motion in a real physical circuit", that is, realistically modeled by ordinary differential equations [41].

To end their investigation, they computed a Poincaré section [Fig. 6(b)] and performed a spectral analysis. The signature of an attractive set was thus obtained according to the picture revealed by points in the Poincaré section: "These points seem to be all *attracted* toward certain well-defined curves in the state space." The Poincaré section shown in Fig. 6(b) is an example of the well-defined curve Chua and co-worker observed.

At this step, it is possible to draw the path followed by Chua and co-workers before investigating chaotic behaviors. (i) Chua introduced some nonlinear elements, (ii) he investigated the stability of periodic solutions, (iii) the evolution of asymptotic behaviors versus circuit parameters was investigated in terms of Hopf bifurcations, (iv) he developed algorithms to numerically integrate the equations (to be ensured that numerical instabilities were avoided) and (v) he investigated the dynamics of nonlinear circuits not satisfying any known stability theorem. This whole path was traced within the field of electronics without any major connection with the emerging "chaos world". Note that Chua should be regarded as a theoretician in electronics and was always focussed to develop a general theory for electronic circuits, dynamics included.

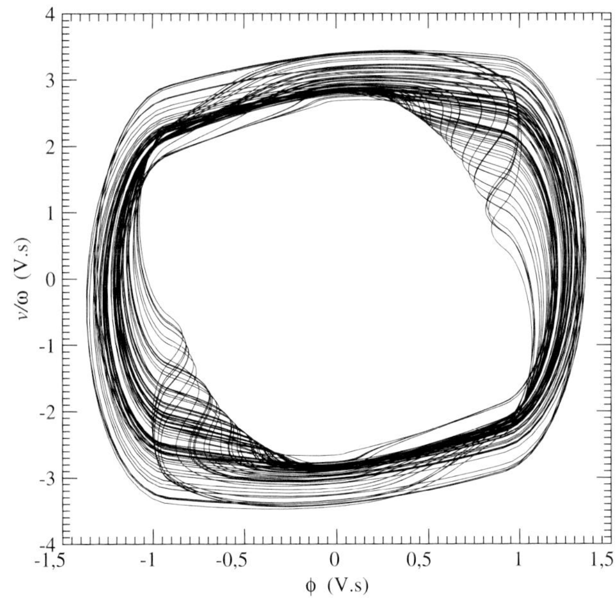

(a) Phase portrait

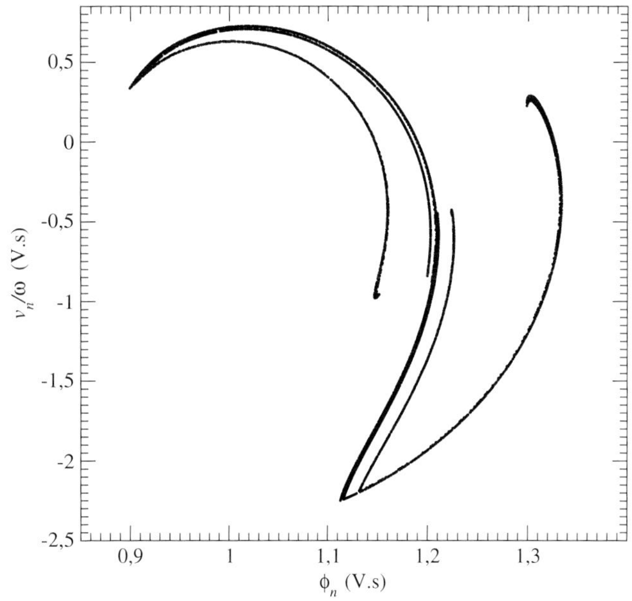

(b) Poincaré section

Fig. 6. Aperiodic solution to the driven piecewise-linear resonant circuit (6) for $E = 800\,\text{V}$, $R_1 = 50\ \Omega$, $R_2 = 50\,\text{k}\Omega$, $C = 250$ nF, $L_0 = 33.33$ H, $L_1 = 1.28$ H, $\phi_0 = 0.92$ V.s and $\frac{\omega}{2\pi} = 50\,\text{Hz}$.

After an investigation of periodic orbits (stability, symmetry) Chua and co-workers studied some chaotic behaviors. Although the sensitivity to initial conditions was often associated with a lack of reliability in numerical simulations, with their algorithm, Chua and co-workers were confident in their results [41]:

> "One may suspect this apparent instability to be caused by accumulating numerical errors at the boundaries rather than an intrinsic property of the true solution. To overcome such an objective we have checked the accuracy of the computed solutions [...]. It turns out that for this circuit (with $E = 800\,\text{V}$) the errors at boundaries do accumulate and thus *the solutions can be calculated with sufficient precision only for a certain length of time.* [...] Hence, the drifting apart of neighboring solutions [...] is a *real phenomenon not caused by numerical instability.*"

This conclusion was made possible because their

> "roundoff errors at different time steps within each *linear region* are independent of each other and do not get amplified over long time intervals. Hence, [their] piecewise-linear solutions are much more accurate and reliable compared to traditional numerical or analog (i.e. hardware) simulations."

The correctness of numerical simulations is one of the main advantages of piecewise-linear circuits because piecewise-linear model overcome this serious flaw that "long-term chaotic motions obtained by numerical integration diverge rapidly from the actual waveforms". In fact the shadowing lemma overcomes this lack of accuracy [42, 43].

2.3. *Relaxation oscillations*

In 1983 with Thomas S. Parker, Chua investigated relaxation oscillations using a piecewise-linear version of the van der Pol equation [44]. The electronic circuit (Fig. 7) was simplified compared to the one used by Ueda [Fig. 5(a)]. The governing equations

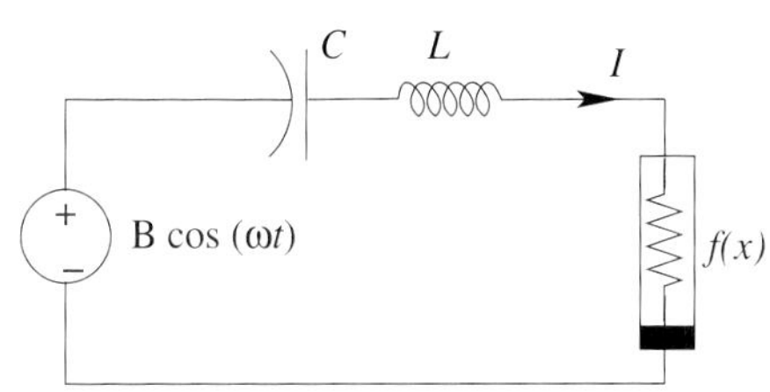

Fig. 7. The circuit producing relaxation oscillations investigated by Parker and Chua [44].

are

$$\begin{cases} \dot{x} = y \\ \dot{y} = \dfrac{1}{\epsilon}[-f(x)y - x + B\cos\omega t] \end{cases}$$

$$\text{where } f(x) = \begin{vmatrix} Rx + 2Rx_0 & & x < -x_0 \\ -Rx & \text{for} & |x| < x_0 \\ Rx - 2Rx_0 & & x > +x_0. \end{vmatrix} \tag{7}$$

The parameter values used for numerical simulations were $R = 1\ \Omega$, $x_0 = I_0 = 1\,\text{A}$ and $\omega = 1\,\text{rad.s}^{-1}$. Mostly, periodic solutions were investigated for providing "intuitive understanding" of relaxation oscillations in connection with the mathematical papers by Levinson [9] and Levi [45]. In particular, Levi "proved what types of behavior are possible and how this behavior is structured as B is varied." In order to do that, Levi used rotation numbers as introduced by Levinson. Using numerical simulations, Parker and Chua showed "how the different subharmonic solutions to the piecewise-linear relaxation oscillation equation are intimately related." It should be clear that B actually plays the role of an initial condition for system (7) when rewritten in the autonomous form

$$\begin{cases} \dot{x} = y \\ \dot{y} = \dfrac{1}{\epsilon}[-f(x)y - x + u] \\ \dot{u} = v \\ \dot{v} = -\omega^2 u. \end{cases} \tag{8}$$

Parameter B is thus "hidden" in the autonomous form as an initial condition $u_0 = B$. This is therefore the phase space which is investigated and not the parameter space [46]. Parker and Chua thus provided tables of rotation numbers in agreement with Levi's results.

In the other hand, Levinson found that " most of the solutions [...] are certainly not periodic. [...] Of the continuum of sequences there is a countable infinity of periodic sequences" [9]. This leads Parker and Chua to remark that for some ϵ and B values, Levinson showed that "given any (perhaps nonperiodic) sequence of component subharmonics, there exists a trajectory which follows that sequence." The existence of "essentially random behavior" was mentioned in the conclusion. When numerical simulations are performs, it can be observed that most of parameter values lead to limit cycle but we identified a quasi-periodic regime [Fig. 8(a)] and a metastable toroidal chaos [Fig. 8(b)] which ends with a period-2 limit cycle. Although chaotic attractors were not yet clearly investigated, plane projections of phase portraits were already widely used.

In another paper with Alistair Mees, Chua investigated chaos as a lack of synchronization [47] where chaos was defined by its properties: (i) sensitivity to initial conditions, (ii) non periodicity, (iii) broad band Fourier spectrum, and (iv) complicated recurrences. They investigated chaotic motion using a kind of symbolic dynamics validated by numerical simulations. Again, the analytical approach was preferred for providing a proof for chaotic motion. Three quotations

- T. Li & J. A. Yorke, A period-three implies chaos, *American Mathematics Monthly*, **82**, 985, 1975;
- R. May, Deterministic models with chaotic dynamics, *Nature*, **256**, 165–166, 1975.
- J. Guckenheimer, Symbolic dynamics and relaxation oscillations, *Physica D*, **1**, 227–235, 1980.

are remarkable in the sense that they are manifestations of the growing Chua's interest for "chaos theory".

With Tang and Mees, Chua then investigated an astable mutivibrator consisting of a capacitor connected across a resistive (memoryless) nonlinear circuit [48]. They investigated how the loss of synchronization implies chaos. In their circuit, they proved their model has the following features:

(1) There are no stable periodic solutions of any order.
(2) There is no almost periodic solution that is not periodic.
(3) There are unstable periodic solutions. If α is an integer, these exist for all possible periods.
(4) Any two trajectories either coincide exactly after a finite time, or are separated by order 1 infinitely often.

2.4. *The simplest nonautonomous circuit*

An extremely simple nonautonomous circuit consisting in a series connection of a linear resistor, a linear inductor, and a two-segment, piecewise-linear

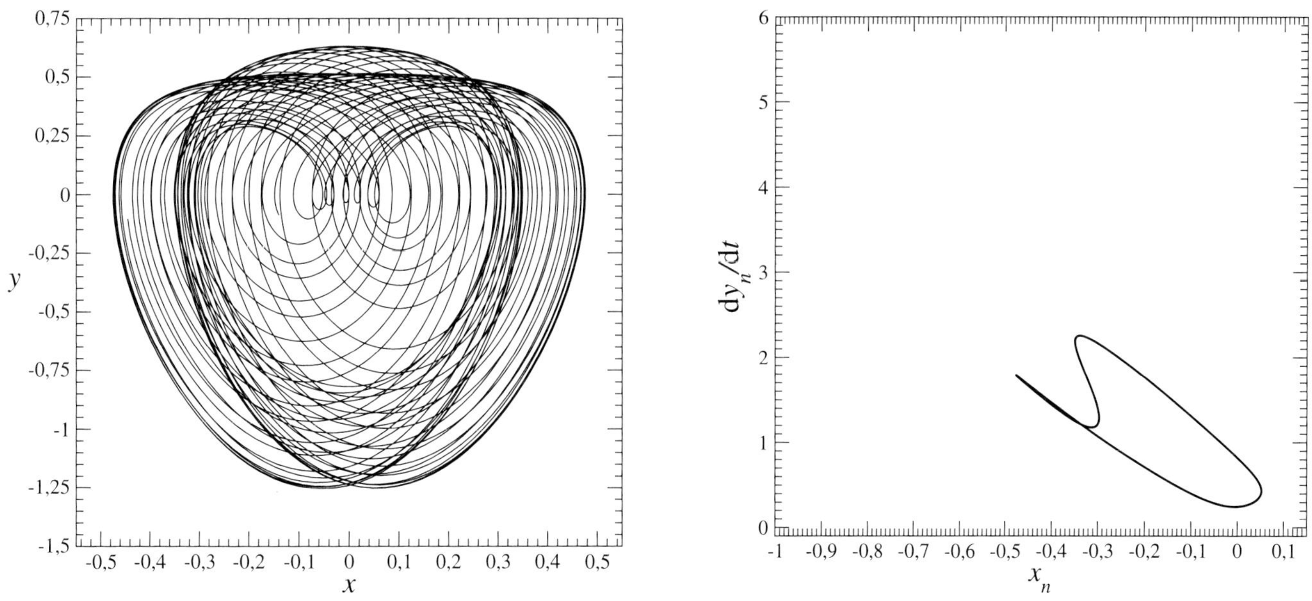

(a) $B = 0.18$: quasi-periodic regime

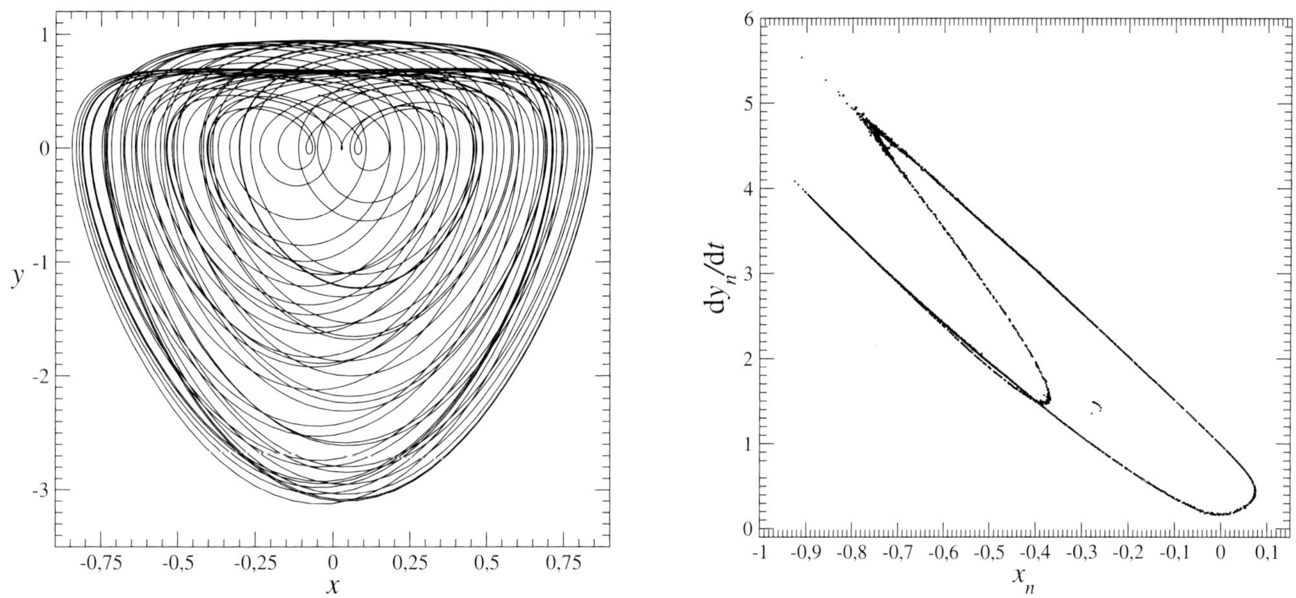

(b) $B = 0.15$: metastable toroidal chaos

Fig. 8. (a) Quasi-periodic regime and (b) metastable toroidal chaos solution to the piecewise-linear van der Pol equation (7). Parameter values: $\epsilon = 0.1825$, $\omega = 1$, $R = 1$, and $x_0 = 1$.

capacitor driven by a sinusoidal voltage source (Fig. 9) was investigated by Matsumoto, Chua and Tanaka [49]. A detailed study of the bifurcation diagram was then completed in a slightly different *R*-*L*-diode circuit [Fig. 9(b)] [50]. A two-dimensional map was proposed to reproduce "all essential features of the experimentally observed bifurcation phenomena."

In the *R*-*L* circuit as well in the two-dimensional map [50]

$$\begin{cases} x_{n+1} = y_n - 1 + \left| \begin{matrix} +\alpha_1 \, x_n & \text{if } x_n \geq 0 \\ -\alpha_2 \, x_n & \text{if } x_n < 0 \end{matrix} \right. \\ y_{n+1} = b \, x_n, \end{cases} \tag{9}$$

they observed that the ultimate scenario consists of three kinds of bifurcation: (i) saddle-node

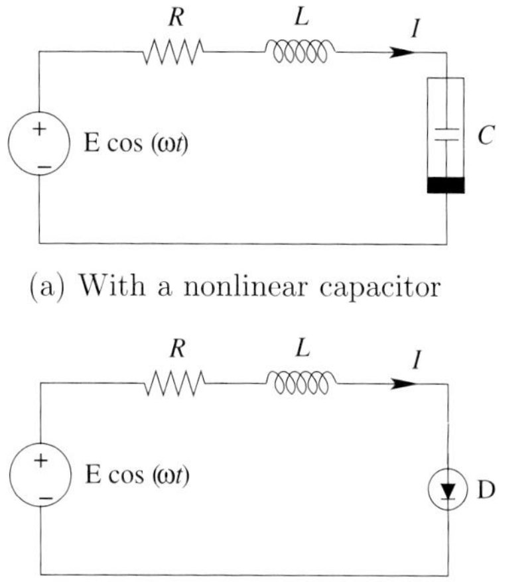

(a) With a nonlinear capacitor

(b) With a diode 3CC13

Fig. 9. Two simple nonautonomous circuits producing chaotic regimes, one with a nonlinear capacitor (a) as studied in Ref. [49], and one with a diode 3CC13 (b) as investigated in Ref. [50].

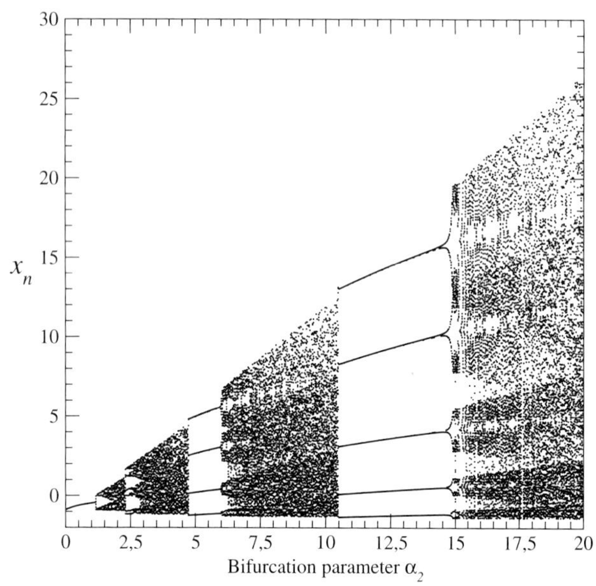

Fig. 10. Bifurcation diagram of the two-dimensional map (9) for reproducing the essential feature of the simplest nonautonomous chaotic electronic circuit. Parameter values: $b = -0.13$ and $\alpha_1 = 0.7$.

bifurcations, (ii) period-doubling cascades, and (iii) crisis. Using $b = -0.13$, and $\alpha_1 = 0.7$, a bifurcation diagram similar to the one observed in the experimental R-L-diode circuit was obtained (Fig. 10).

2.5. *Devil staircase in an experimental van der Pol-like circuit*

With Michael Kennedy, Chua examined a simple experimental circuit with a negative resistance device, as presented by van der Pol and van der Mark [51], which displays a transition to chaos through an alternating periodic-chaotic sequence, and described a nonlinear circuit model of the physical circuit which faithfully reproduces the observed period-adding phenomenon [52]. Van der Pol and van der Mark described an experiment in which by tuning the capacitor in a neon bulb R-C relaxation oscillator, driven by a sinusoidal voltage source, currents and voltages appear in the system which are whole submultiples of the driving frequency. To reproduce this, Kennedy and Chua considered a sinusoidally-driven neon bulb relaxation oscillator (Fig. 11). A high-voltage dc supply $E \approx 100$-V with large source resistance $R \approx 1$ MΩ was attached to a shunt connection of neon bulb and capacitor C, forming the basic relaxation oscillator. In series with the neon bulb was inserted a sinusoidal voltage source $E_0 \sin \omega t$ (a small current sense resistor R_s, was also inserted in series with the neon bulb to detect the current flowing in it).

The natural frequency of the undriven oscillator was set to 1 kHz by tuning capacitance C to C_0. Then a sinusoidal signal $E_0 \sin \omega t$ was applied as shown in Fig. 11. The resulting frequency of the voltage pulses measured across R, was recorded as the value of the bifurcation parameter C was brought gradually from C_0, to a much larger value. As the capacitance is increased, the system at first continues to oscillate at 1 kHz over a wide range of C, until the frequency suddenly drops to 1000/2 Hz, to maintain that value over a further range of capacitance. If C is increased still more, the frequency drops to 1000/3 Hz, then 1000/4 Hz, 1000/5 Hz,

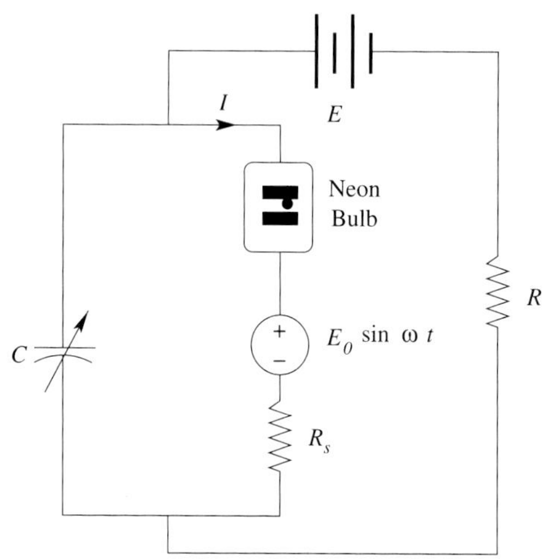

Fig. 11. Sinusoidally-driven neon bulb relaxation oscillator as investigated by Kennedy and Chua [52].

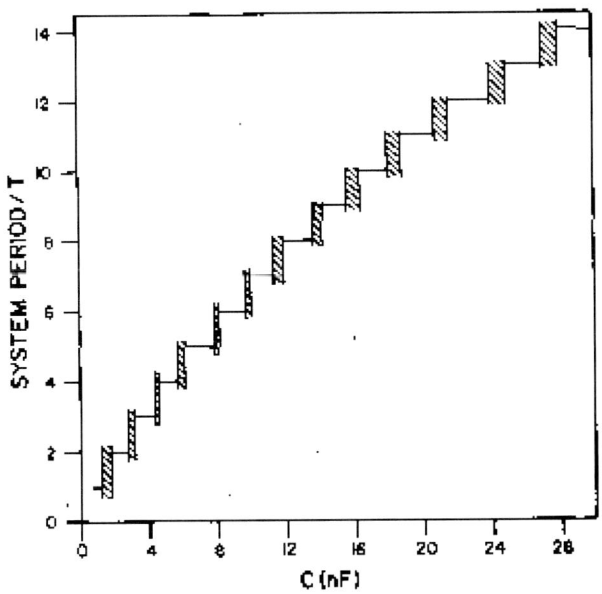

Fig. 12. Frequency of the oscillations produced by the sinusoidally-driven neon bulb oscillator investigated by Kennedy and Chua [52], and showing coarse staircase structure as observed by van der Pol and van der Mark [51].

and so on up to 1000/20 Hz (Fig. 12), exactly as reported by van der Pol and van der Mark in 1927 [51]. Such type of investigations was extended to other simple electronic circuits [53].

3. The Chua Circuit

3.1. *Discovery*

The first simple autonomous circuit with a chaotic attractor was proposed during a stay of Chua at Waseda University (Tokyo) from October 1983 to January 1984. Chua visited Takashi Matsumoto with whom he already published few papers [28, 54, 55]. As Chua wrote himself [56], Matsumoto was trying to perform an electronic circuit realization of the Lorenz equations. But, as Rössler failed in his trials to perform a chemical reaction with a Lorenz dynamics, Matsumoto failed in his attempt to design an electronic "Lorenz" circuit. In fact, a design for an electronic circuit which models the Lorenz equations was only proposed by F. N. H. Robinson in 1990 [57]. With the know-how Chua had in the use of piecewise-linear electronic elements, Chua quickly suggested a new circuit with a piecewise-linear resistor (Fig. 13). Chua was convinced that the use of a piecewise-linear function is convenient for realizing electronic circuits producing rich dynamics as well as facilitating analysis and programming. This is true for designing circuits, true for some type of analysis but has no deep consequence for computing. Very quickly, Matsumoto was able to produce a double-scroll attractor by using numerical integrations. Thus, the double-scroll attractor was observed in this electronic circuit on October 1983 [58], that is, at the beginning of Chua's stay at Waseda University.

But they apparently preferred to first complete the paper about the simplest non autonomous circuit [49] which was submitted on March 28, 1984. Then, Matsumoto sent a paper by fax to Chua who was back at the University of California at Berkeley. This paper was written by Matsumoto alone and was entitled "A chaotic attractor from Chua's circuit". This fax is dated on April 17, 1984 in Ref. [59] where it is quoted as a memorandum [60]. With Chua's agreement, Matsumoto submitted his paper on April 20, 1984 [61]. In this paper, Mastumoto presented himself the circuit consisting of "only one nonlinear element; a 3-segment R piecewise-linear

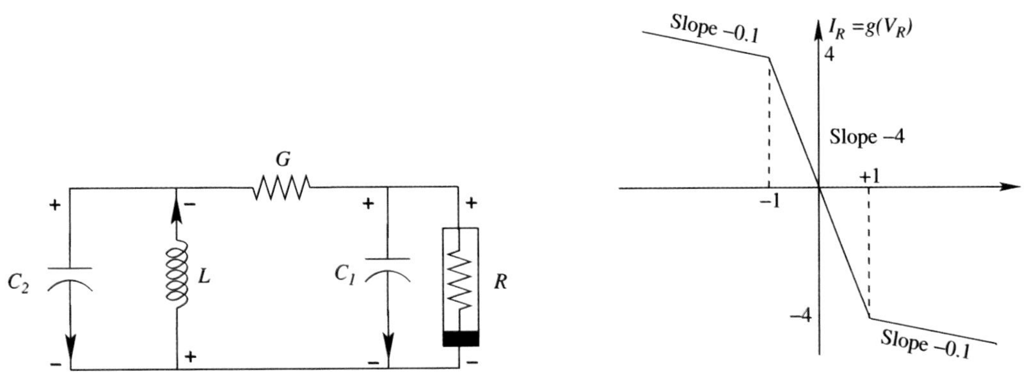

(a) Simplified Chua circuit (b) Characteristic of the nonlinear resistor

Fig. 13. Circuit with a nonlinear resistor proposed by Chua and simplified by Matsumoto.

resistor. It is a simplified version of a circuit suggested by Leon Chua." This was made without explicit reference to the original circuit. Matsumoto designated himself the circuit by Chua's name as seen from the title of the paper. The circuit investigated is described by the set of equations:

$$\begin{cases} C_1\dot{V}_{C_1} = G(V_{C_2} - V_{C_1}) - g(V_{C_1}) \\ C_2\dot{V}_{C_2} = G(V_{C_1} - V_{C_2}) + I_L \\ \dot{I}_L = -V_{C_2} \end{cases} \tag{10}$$

where V_{C_1}, V_{C_2} and I_L denote voltage across C_1, voltage across C_2, and current through L, respectively. Using notations adopted later [58], the piecewise-linear characteristics reads as:

$$g(V_{C_1}) = m_0 V_{C_1} + \frac{1}{2}(m_1 - m_0)|V_{C_1} + B_p| + \frac{1}{2}(m_0 - m_1)|V_{C_1} - B_p| \tag{11}$$

where $m_0 = -0.1$, $m_1 = -4$ and $B_p = 1$. From his numerical computations (Fig. 14), Matsumoto emphasized the role of phase portrait. He found the attractor structure different from the structure of the attractors solution to the Lorenz and Rössler equations, respectively. He wrote:

> "The origin of the Lorenz equations [...] is a saddle, i.e. all eigenvalues are real. Another difference is that (10) is symmetric with respect to the origin while the Lorenz equation is symmetric with respect to the z-axis. It is known that the Rössler attractor [62] does not contain any equilibrium. Hence, it is different from ours."

This claim was only supported by a fixed point analysis — and one wrong statement about the Rössler system — and some considerations about the symmetry inherent to the different systems.

With Chua's help (as it appeared in references quoted by Ref. [59]) and stimulated by Mastumoto's fax, F. Ayrom (University of California at Berkeley and, consequently, one of Chua's colleague) and one of his collaborator, Guo-Qun Zhong (Academia Sinica, Guangzhou, People's Republic of China), designed an electronic circuit with a 5-segment odd-symmetric V-I characteristic. It thus appears that the simplification mentioned by Matsumoto was to reduce the characteristic curve to three branches since the three central branches of the original characteristic can easily be approximated by a single branch as done by Matsumoto. The two-scroll chaotic attractor was thus observed by Zhong and Ayrom in a real electronic circuit. Their paper was submitted on July 20, 1984 and published in 1985, once Matsumoto's paper was published (December 1984). Note that the first chaotic

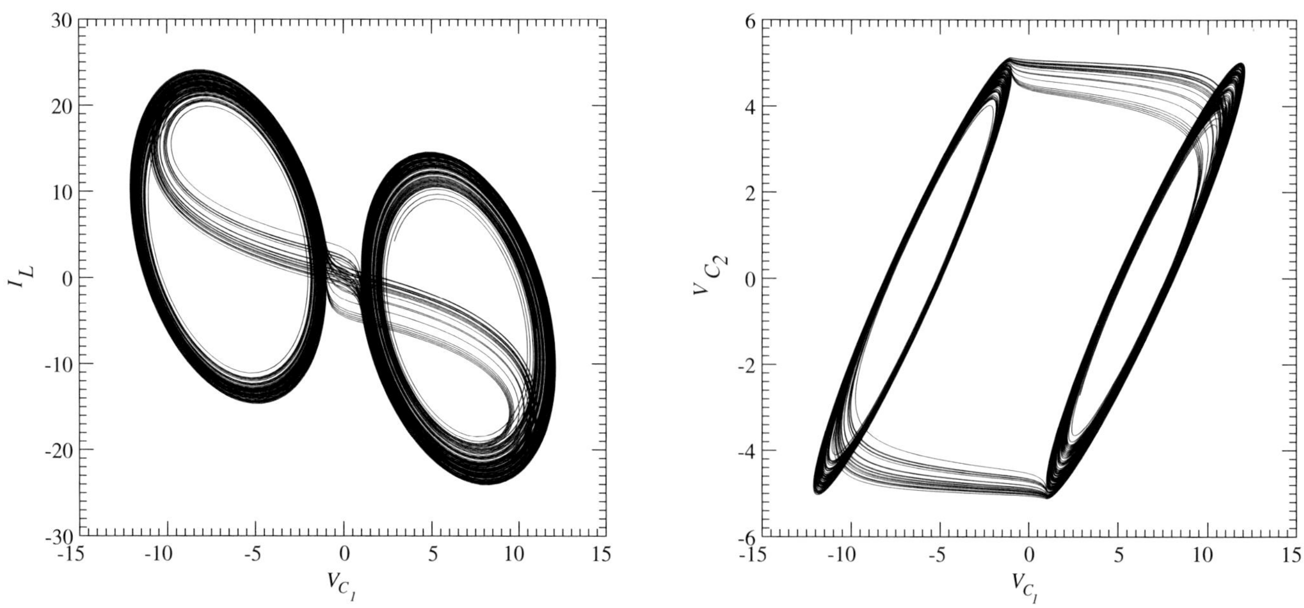

(a) V_{C_1}-I_L plane projection

(b) V_{C_1}-V_{C_2} plane projection

Fig. 14. Two-scroll chaotic attractor obtained by Matsumoto using a numerical integration. Parameter values: $G = 0.7$, $C_1 = 1/10$, $C_2 = 1/0.5$, $m_0 = -0.1$, $m_1 = -4$, $B_p = 1$ and $L = 1/7$.

attractor published and produced by an electronic circuit (with a phase portrait as a proof for the existence of a chaotic attractor) seems to be due to Pikovsky and Rabinovich [34]. It was followed by the report of a period-doubling cascade in a driven anharmonic oscillator by Paul S. Linsay [63].

Once Zhong and Ayrom's paper submitted, Matsumoto, Chua and Motomasa Komuro (Department of Mathematics, Tokyo Metropolitan University, Tokyo) submitted a paper dealing with the dynamics of the equations investigated by Matusmoto. Among the "friends" thanked in the first Matsumoto's paper [49], Chua and Komuro were the first two. The paper published in 1985 obviously appeared as an extended version of the 84 Matsumoto's paper. First, the parameter values were slightly changed to obtain a more esthetic attractor (Fig. 15). The breakthrough marked by the 1985 paper published by these three collaborators was clearly the deepness of the dynamical analysis, that is, Komuro's contribution. Komuro was highly involved in nonlinear dynamics [64, 65] and was aware about Williams' works on branched manifolds [66]. It was therefore natural for him to describe the structure of the attractor in the phase space with great details and in terms of a branched manifold (Fig. 16).

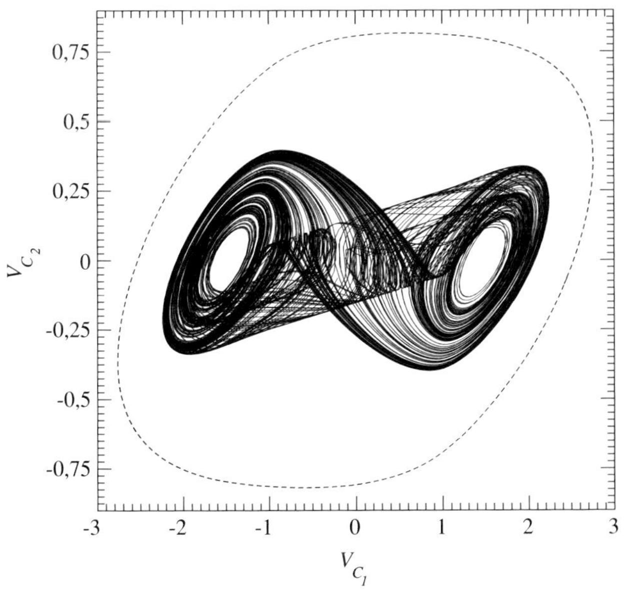

Fig. 15. Two-scroll chaotic attractor and its surrounding unstable limit cycle [58]. Parameter values: $G = 0.7, C_1 = 1/9, C_2 = 1, m_0 = -0.5, m_1 = -0.8, B_p = 1$ and $L = 1/7$. Initial conditions for the unstable limit cycle: $V_{C_1} = 2.532735, V_{C_2} = 1.285458 \cdot 10^{-3}$ and $I_L = -3.367482$.

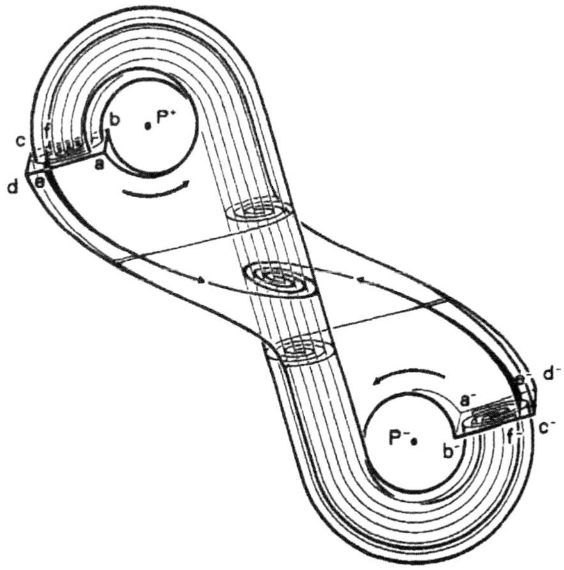

Fig. 16. Sketch of the geometrical structure of the chaotic attractor produced by the Chua circuit. From Ref. [58].

In this paper [58], the dimensionless form

$$\begin{cases} \dot{x} = \alpha[y - x - f(x)] \\ \dot{y} = x - y + z \\ \dot{z} = -\beta y \end{cases}$$

$$\text{where } f(x) = \begin{vmatrix} bx + a - b & & x \geq 1 \\ ax & \text{for} & |x| \leq 1 \\ bx - a + b & & x \leq -1 \end{vmatrix} \tag{12}$$

was introduced.

3.2. *The double scroll attractor*

Other topological analysis of the double scroll attractor were later performed in terms of template [67, 68]. In fact, Ghrist and Holmes showed that the Chua circuit (12) satisfies the theorem as follows [69, 70]:

Theorem: *Let $\dot{x} = f(x)$ for $x \in \mathbb{R}^3$ be a third order ODE which satisfies the following three conditions.*

(i) *There exists a fixed point p for the vector field, and the linearization Jf_p at p has eigenvalues $(-\lambda^s \pm i\omega, \lambda^u)$ with*

$$\lambda^u > \lambda^s > 0 \quad \omega \neq 0.$$

(ii) *f is equivariant under an order-2 rotation $\Psi \equiv \mathcal{R}_z(\pi)$ or an inversion symmetry $\Psi = \mathcal{P}$.*

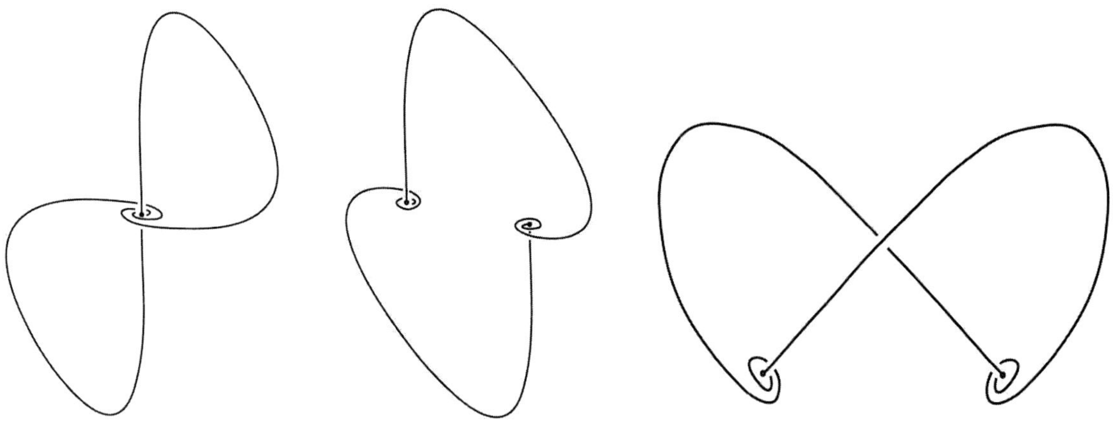

(a) Homoclinic (b) Heteroclinic (c) Heteroclinic

Fig. 17. Three symmetric homoclinic configurations as proposed by Ghrist and Holmes [69]. Chua and co-workers proved that the homoclinic connection (a) was realized in the Chua circuit (Fig. 9 in Ref. [71]).

(iii) *There exists a solution to the flow* ϕ_t *for which* $\lim_{t\to-\infty}\phi_t = p$ *and* $\lim_{t\to+\infty}\phi_t = \Psi(p)$.

Then, there exists a countable infinity of coupled suspended Smale Horseshoes in the flow in arbitrarily small tubular neighborhoods of the connections.
Adding the fourth requirement

(iv) *Either the homoclinic orbit* ζ, *and* $\Psi(\zeta)$ *form a pair of unknotted, unlinked loops joined at* p, *or the heteroclinic cycle formed by* ζ *and* $\Psi(\zeta)$ *is an unknotted loop as in Fig. 17.*

It was shown in Ref. [71] that the Chua circuit satisfies the four requirements which lead to the Corollary as follows [69]:

Corollary: *Given a third order system* $\dot{x} = f(x)$ *which has a pair of symmetric Shil'nikov connections (as per the four conditions (i) to (iv)) every knot and link type lives in an arbitrarily small tubular neighborhood of the Shil'nikov connections. This property persists under small* C^1 *perturbations.*

As a consequence, the Chua attractor is characterized by an universal template which contains all possible knots. This is not the case of the Lorenz system for which only degenerate heteroclinic orbits have been found by setting $b = 0$ [72, 73]. This is therefore another interest of the Chua circuit and its double-scroll attractor.

3.3. *A zoo of chaotic attractors*

Using the slighly different version of the Chua circuit

$$\begin{cases} \dot{x} = \kappa\alpha(y - x - \phi) \\ \dot{y} = \kappa(x - y + z) \\ \dot{z} = -\kappa(\beta y + \gamma z) \end{cases} \tag{13}$$

where the piecewise-linear function is

$$\phi = bx + \frac{(a-b)(|x+1| - |x-1|)}{2}, \tag{14}$$

Chua and co-workers obtained various chaotic attractors [74]. The most typical examples are shown in Fig. 18. The most suggestive plane projection was shown for each case according to the orientation of the attractor in the phase space. Let us start with two asymmetric funnel attractors [Fig. 18(a)] which can fuse into a single symmetric attractor through a merging attractor crisis when one of the parameters is varied. Such a symmetric attractor looks like the one shown in Fig. 18(b). This is not exactly a common double scroll in the sense that there are foldings in each scroll in the latter but not in the former. Depending on parameter values, the winding number in the funnel may vary but there is no crucial departure in the resulting topology (we precise that we focus here on the relevant main topological types of attractors). A very special attractor, here named the "cord" attractor in the view of the "cords" linking the two scrolls, is shown in Fig. 18(c). To our knowledge, this attractor was never reported elsewhere.

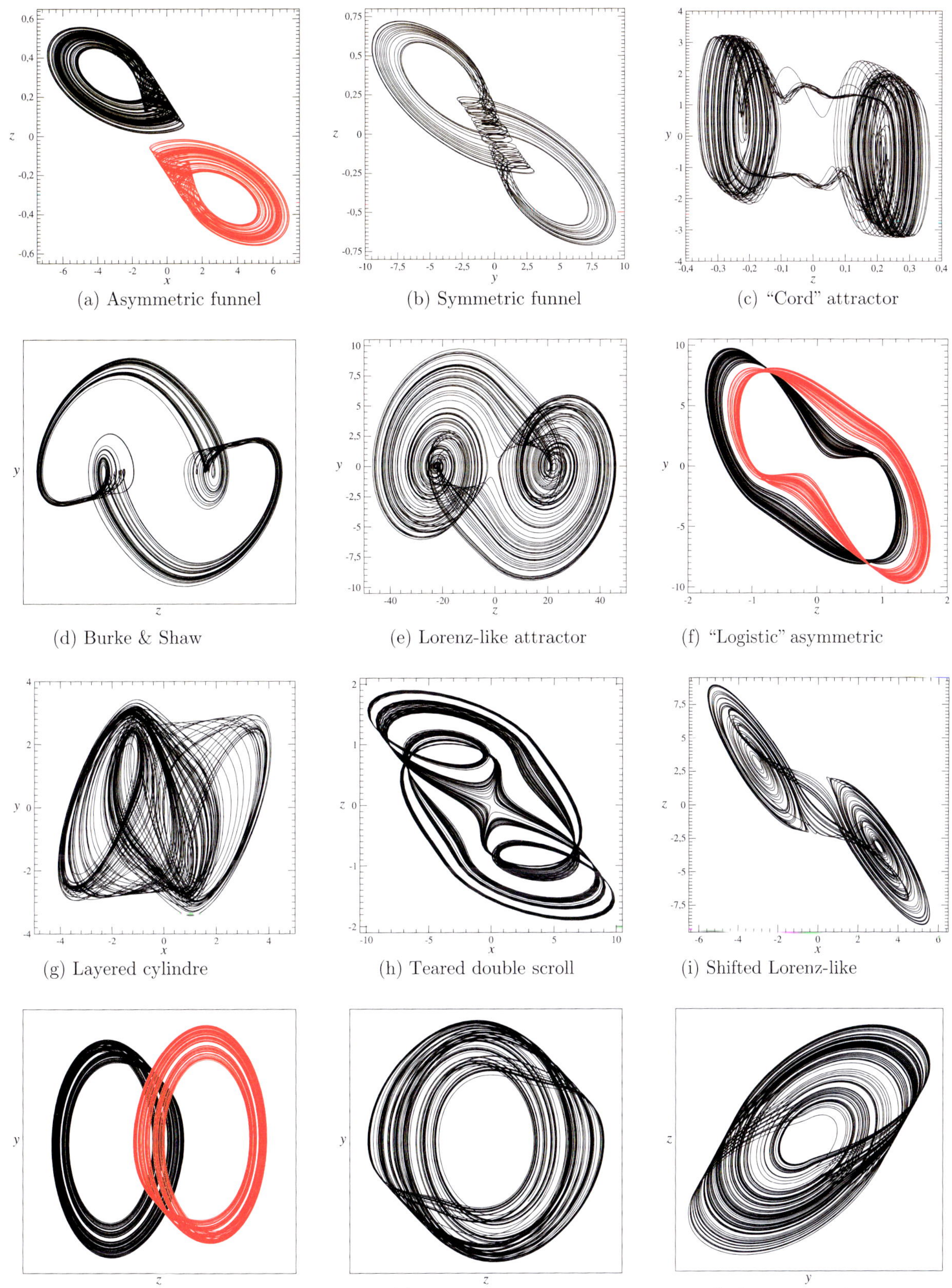

(a) Asymmetric funnel (b) Symmetric funnel (c) "Cord" attractor

(d) Burke & Shaw (e) Lorenz-like attractor (f) "Logistic" asymmetric

(g) Layered cylindre (h) Teared double scroll (i) Shifted Lorenz-like

(j) Spiral (k) Covered spiral (l) Covered spiral

Fig. 18. Zoology of chaotic attractors solution to the Chua circuit (13). Parameter values are reported in Table 1.

Table 1. Parameter values used to produced the attractors with system (13).

Fig.	α	β	γ	a	b	κ
18a	−1.5590535	0.0156453	0.1374556	−0.24385532	−0.0425189	−1
18b	−1.5590535	0.0156453	0.1574556	−0.2438532	−0.0425189	−1
18c	−1.3184010	0.0125741	0.1328593	−0.2241328	−0.0281101	−1
18d	−4.08685	−2.0	0	−1.142837	−0.7142858	+1
18e	−4.898979	−3.624135	−0.00118088	−2.501256	−0.9297201	+1
18f	−1.3635256	−0.0874054	−0.3114345	1.292150	−0.49717	−1
18g	3.7091002	24.0799705	−0.8592556	−2.7647222	0.1805569	+1
18h	−1.458906	−0.0930819	−0.3214346	1.218416	−0.5128436	−1
18i	143.1037	207.34198	−3.8767721	−0.855372	−1.09956	−1
18j	6.5792294	10.8976626	−0.0447440	−1.1819730	−0.6523354	+1
18k	8.4562218	12.0732335	0.0051631	−0.7056296	−1.1467573	−1
18l	−75.6	31.25	−3.125	−2.4	−0.98	−1
19	12.1414141	95.7977567	−0.8985829	−0.8557143	−1.1	−1
20	as specified	91.5210060	−0.7303376	−0.8557143	−1.1	−1
21	3.505	66.672752	−0.94779892	−0.855372	−1.09956	−1

The symmetric attractor shown in Fig. 18(d) looks like a Burke and Shaw attractor [75, 76]. Such an attractor is common in Lorenz-like systems [77,78]. In fact, there is also a Lorenz-like attractor [Fig. 18(e)] solution to the Chua circuit (13) but it has an inversion symmetry and not a rotation symmetry like the original Lorenz attractor. It thus corresponds to an embedding of the Lorenz attractor using variable x or y of the Lorenz system since, in that case, an inversion symmetry is recovered [79]. It is also possible to observe in the Chua circuit (13) two attractors obtained after a period-doubling cascade as observed in the Lorenz system for large R values [80]. These attractors have a Logistic-like first-return map to a Poincaré section. Nevertheless, these attractors are related by an inversion symmetry and not by a rotation symmetry as observed in the Lorenz system.

A "strange" attractor with a complex topology has also been observed [Fig. 18(g)]. Its complexity comes from the central zone, between the two scrolls. It will be shown below that such a structure is related to a toroidal structure. An attractor [Fig. 18(h)] similar to an attractor solution to a simple model for a pulsating star [81] was also found in the Chua circuit (13). This attractor has a very different structure from all the others already discussed. The topology of the similar attractor produced by the pulsating star model was investigated in Ref. [82]. The double scroll shown in Fig. 18(i) is interesting due to its central zone. The two scrolls are quite similar to the two scrolls of the common double scroll attractor but the way in which they are linked is rather unusual. It looks like a Lorenz attractor whose central part was mirrored. Note that such a special topology seems to be only observed in a system with an inversion symmetry.

The last three attractors [Figs. 18(j), 18(k), and 18(l)] are all related to the Rössler spiral attractor. The first ones [Fig. 18(j)] are two symmetry-related spiral attractors. They may become a single symmetric chaotic attractors through a merging attractor crisis. The case of the attractors shown in Fig. 18(k) looks like a two-fold cover of a spiral Rössler attractor [79]. It is thus a symmetric Rössler attractor with two foldings, one being the symmetric of the other under an inversion symmetry. A second example of such a two-fold cover was also obtained [Fig. 18(l)], but with a different orientation in the phase space compared to the attractor shown in Fig. 18(k). In spite of this, the last two attractors are topologically equivalent. The richness of the dynamics produced by the Chua circuit could result from the fact that it is characterized by an universal template as shown by Ghrist and Holmes [69].

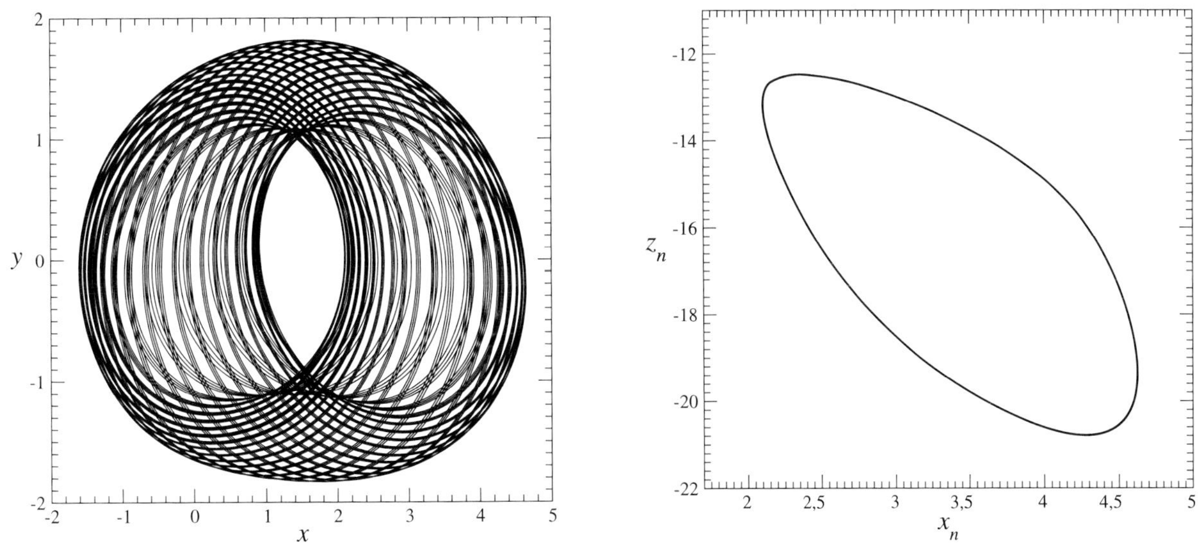

(a) Phase portrait (b) Poincaré section

Fig. 19. Quasi-periodic regime solution to Chua circuit (13). Parameter values: $C_1 = 0.0297\,\mu$F, $C_2 = 0.3606\,\mu$F, $G_a = 0.599$ mS, $G_b = 0.77$ mS, $R = -1/0.0007\,\Omega$, $R_0 = 13.4\,\Omega$, $L = 7.682$ mH.

4. Toroidal Chaos

Using the Chua circuit (13), Chua and co-workers [74] obtained a quasi-periodic regime [Fig. 19(a)] whose Poincaré section is a regular closed loop [Fig. 19(b)] as expected. More surprisingly, they observed a folded torus, that is, a toroidal chaotic attractor (Figs. 20). Toroidal chaos produced by three-dimensional autonomous system is quite rare, and only very few examples are known [83–85]. Varying the value of the capacitance C_1 allows to developed the toroidal chaotic regime [Figs. 20(b) and 20(c)]. Just before a boundary crisis leaving a single fixed point for attractor, the toroidal attractor [Fig. 20(d)] has a structure reminiscent of the double scroll attractor. When the capacitance C_1 is increased, another very strange toroidal structure is obtained [Fig. 20(a)] just before a second boundary crisis. These types of toroidal chaos are not yet fully understood and remain to be more carefully investigated. Let us note that it is possible to recover parameter values reported in Table 1 from the characteristics of the electronic components as follows: $\alpha = \frac{C_2}{C_1}$, $\beta = \frac{R^2 C_2}{L}$, $\gamma = \frac{R R_0 C_2}{L}$, $a = RG_a$, $b = RG_b$ and $\kappa = \pm 1$ if $RC_2 \gtrless 0$. The equivalent electronic circuit is explicited in [74].

Another toroidal chaotic attractor was also found by Chua and co-workers [74]. It is shown in Fig. 21(a). It is interesting in the sense that it has a single folding (bottom right of the Poincaré section as shown in Fig. 21(b)) as the attractor solution to the rather complicated model proposed by Bo Deng [83]. This attractor is therefore a good benchmark attractor to develop a topological analysis for three-dimensional toroidal chaotic attractors.

A "torus circuit" was investigated by Matsumoto and co-workers [86] using the governing equations

$$\begin{cases} \dot{x} = -\alpha f(y - x) \\ \dot{y} = -z - f(y - x) \\ \dot{z} = \beta z \end{cases} \tag{15}$$

where the piecewise-linear function reads as

$$f(u) = -au + \frac{a+b}{2}(|u+1| - |u-1|). \tag{16}$$

For some parameter values, a toroidal chaotic attractor was obtained (Fig. 22). It was shown that three mechanisms responsible for the breakdown of a torus T^2 are realized in the torus circuit. They are [87]

(i) a transition via a cascade of period-doubling bifurcation;
(ii) an abrupt transition to chaos in the homoclinic region via saddle-node bifurcation of a limit cycle;
(iii) a soft transition to chaos due to the loss of torus smoothness.

All these toroidal attractors are based on a torus T^2 structure. Higher-dimensional tori T^n $(n > 2)$

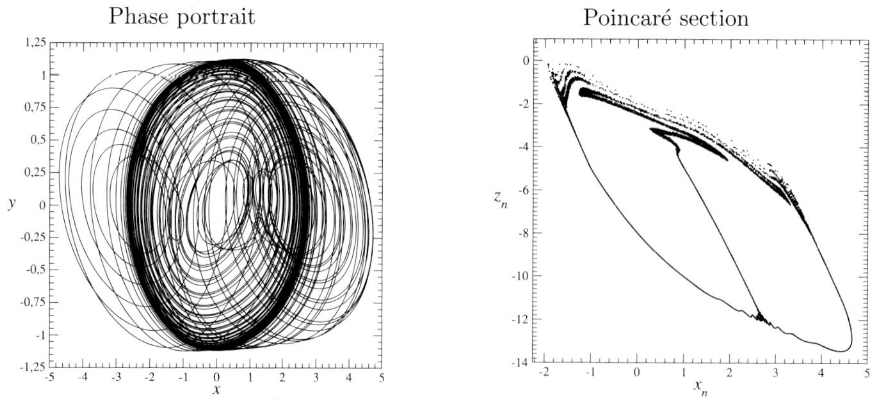

(a) $C_1 = 0.0164$ μF: Toroidal chaos

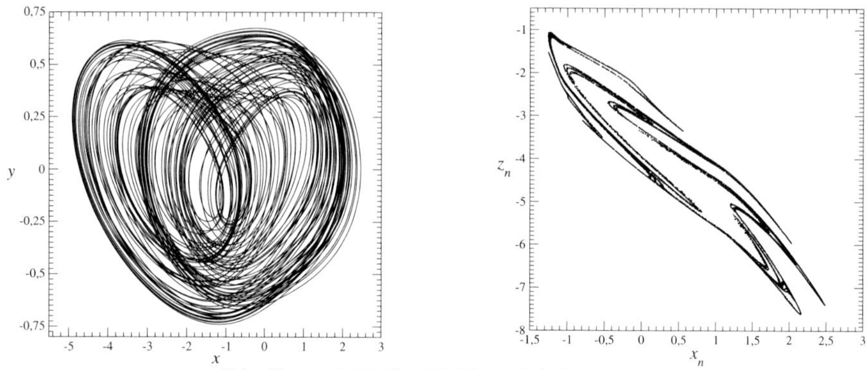

(b) $C_1 = 0.0105$ μF: Toroidal chaos

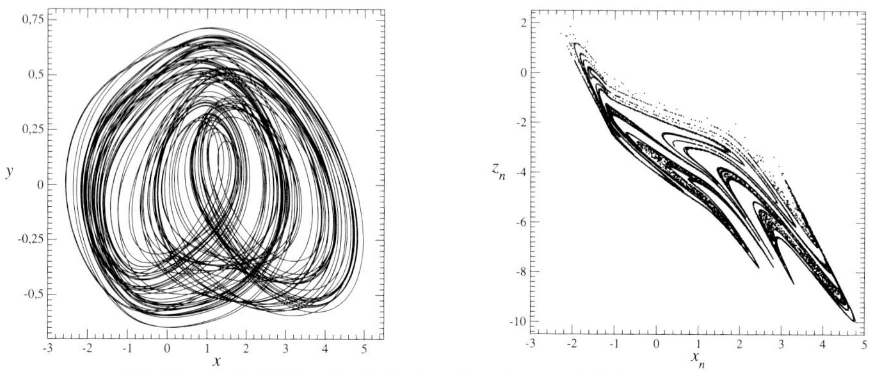

(c) $C_1 = 0.0100$ μF: Fully developed toroidal chaos

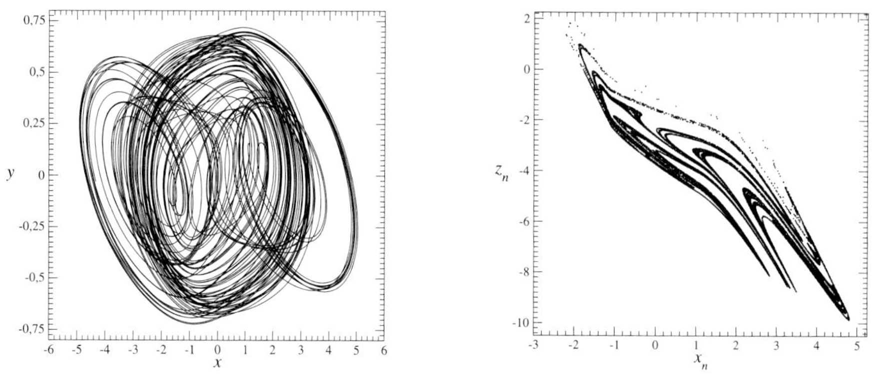

(d) $C_1 = 0.00946$ μF: Toroidal double scroll chaos

Fig. 20. Toroidal chaotic attractors solution to the Chua circuit (13). Parameter values: $C_2 = 0.3406$ μF, $G_a = 0.599$ mS, $G_b = 0.77$ mS, $R = -1/0.0007$ Ω, $R_0 = 11.4$ Ω, $L = 7.595$ mH and corresponding values of α, β, γ, a, b and κ are reported in Table 1.

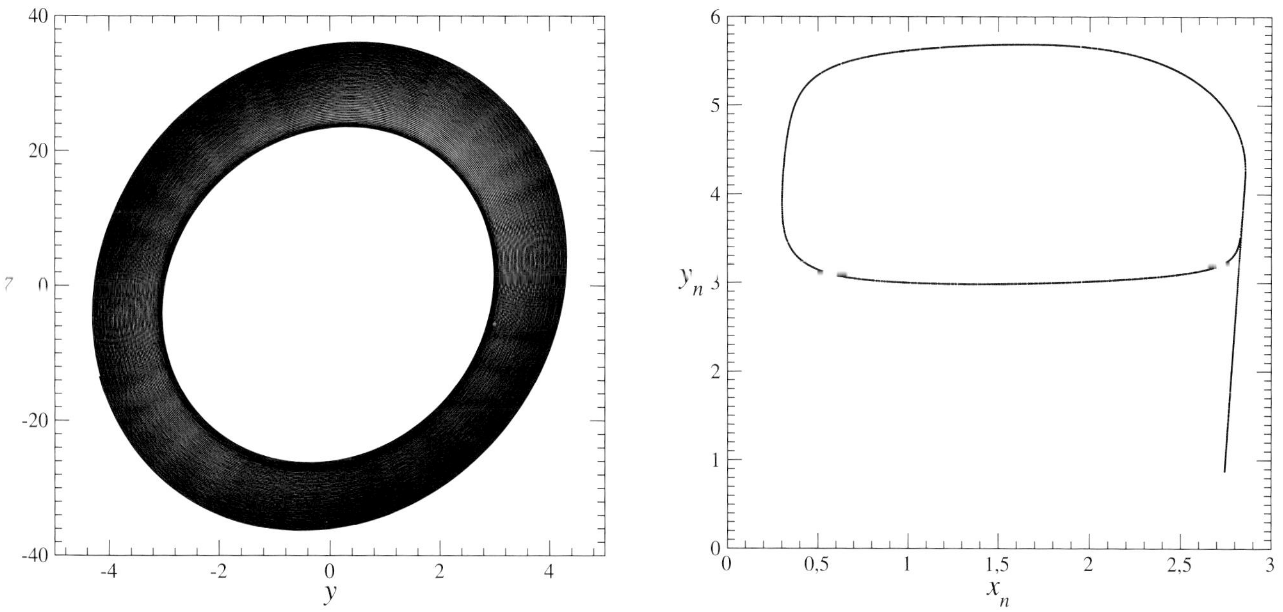

(a) Phase portrait

(b) Poincaré section

Fig. 21. Toroidal chaos solution to the Chua circuit (13). Parameter values are reported in Table 1.

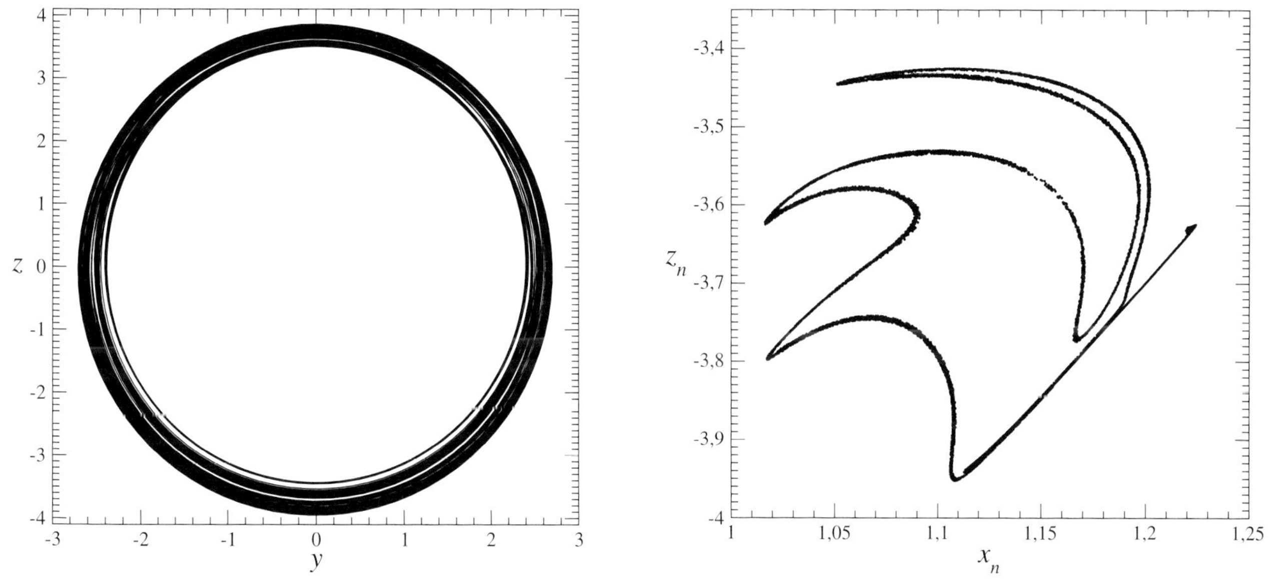

(a) Phase portrait

(b) Poincaré section

Fig. 22. Toroidal chaotic attractor solution to the "torus circuit" (15). Parameter values: $a = 0.07$, $b = 0.10$, $\alpha = 27.65$, and $\beta = 2.12$.

obviously require higher-dimensional systems to produce them. If there is no explicit nontrivial example in Chua's works of such torus, they were mentioned in the book written with Parker when they introduced "higher-order Poincaré maps" (Sec. 2.3 of that book [88]). The idea was to reduce the dimension of the limit set to investigate by sampling $(n-1)$ times the trajectory up to recover a two-dimensional object. For instance, a torus T^3 characterized by three independent frequences f_1, f_2, and f_3 would be investigated by using a second-order Poincaré map to perform a double sampling: the trajectory would be thus sampled at f_1 and f_2 simultaneously. While Parker and Chua provided a trivial example with a solution resulting from the sum of three sinusoids characterized by three incommensurable frequences,

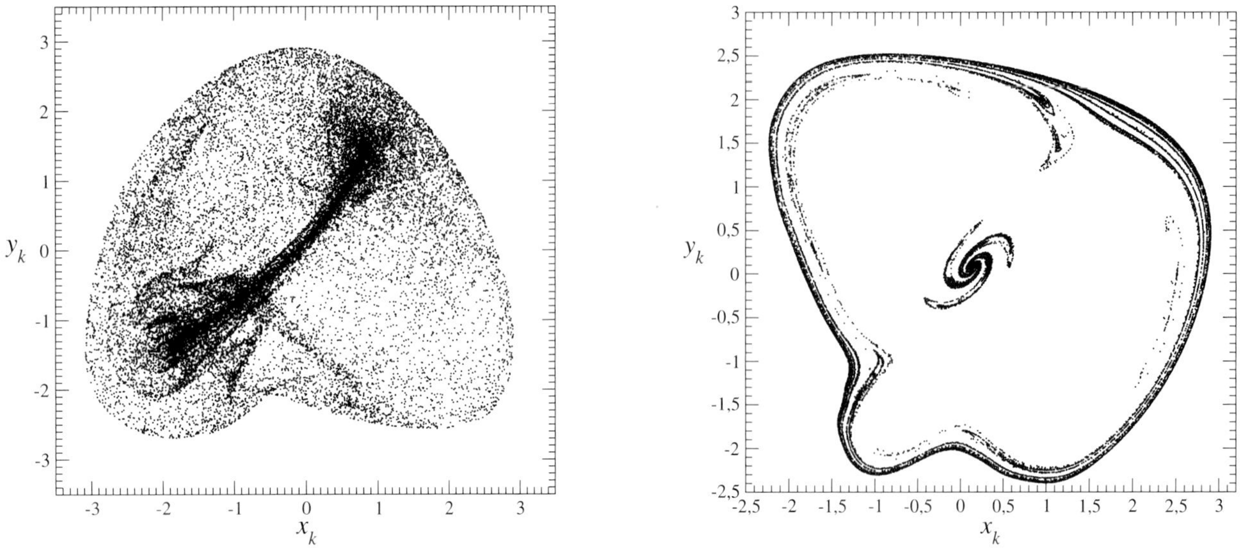

(a) First-order Poincaré map

(b) Second-order Poincaré map

Fig. 23. Toroidal chaos structured around a torus T^3 solution to map (17). Parameter values: $a = 0.28$, $b = c = 0.2$ and $d = 0.125$. The second-order Poincaré (b) map was obtained using $x_k \in [-\epsilon; +\epsilon]$ with $\epsilon = 0.005$.

let us exemplify this procedure with the three variables map

$$\begin{cases} x_{k+1} = -ay_k^2 + z_k \\ y_{k+1} = x_k + b + cx_k y_k + dz_k \\ z_{k+1} = y_k \end{cases} \tag{17}$$

which corresponds to the first-order Poincaré map of a torus T^3 [89]. With a single plane projection — or a first-order Poincaré map — (Fig. 23(a)), it is rather difficult to figure out that it corresponds to a toroidal chaos. But the toroidal structure characterized by an annular shape becomes obvious with a second-order Poincaré map computed by selecting points $x_k \in [-\epsilon; +\epsilon]$. The limit set is approached as $\epsilon \to 0$.

The advantage of a higher-order Poincaré map is that it allows easier identification of torus T^k. But, as mentioned by Parker and Chua, "the main drawback of the higher-order Poincaré map is that it requires an enormous amount of input data to produce a reasonable amount of output. [...] Another disadvantage is that, owing to the ϵ-neighborhood, the limit sets are fuzzy — a point becomes a segment and a circle becomes an annulus. This effect can be lessened by chosing a smaller value of ϵ, but this will increase the simulation time since more data points are required." Such a technique was recently used to investigate non trivial toroidal chaos in a laser system [90].

5. Lorenz-like Chaos in Electronic Circuits

As a solution to the Chua circuit (10), Philippe Bartissol and co-workers discovered another type of chaotic attractor (Fig. 24). They had to change drastically the parameter values which lead to a different circuit realization [91]. They named this attractor the "double Hook". In fact, this is nothing

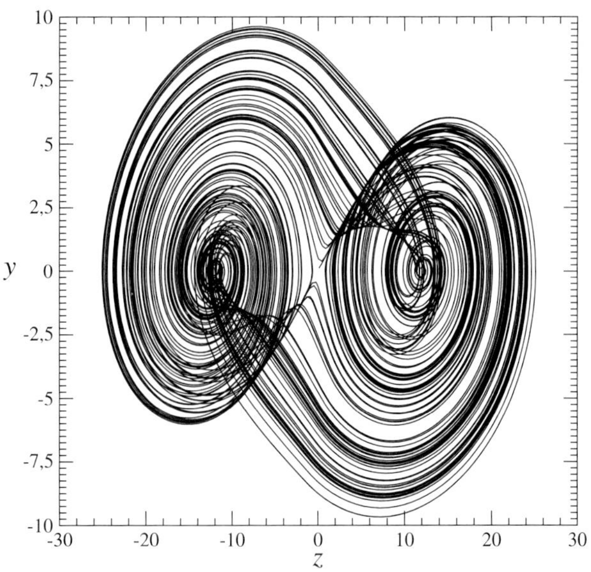

Fig. 24. The double hook attractor solution to the Chua circuit (10). Parameter values: $G = 0.539$, $L = -0.3005$, $B_p = 1$, $C_1 = -0.0647$, $C_2 = 0.3180$, $C_2 = 0.3170$, $m_0 = -0.5013$, and $m_1 = -1.3475$.

else than a Lorenz-like attractor (compare Fig. 24 to Fig. 18(e)). The dimensionless form (12) can be also used. In this case, parameter values for producing Fig. 24 are $\alpha = \frac{0.590}{0.120}$, $\beta = \frac{0.590}{0.162}$, $a = 1.5$, and $b = -0.07$.

One year later, Chua and co-workers designed a simple piecewise-linear circuit to reproduce Lorenz-like attractor [92]. Using $x = I_L$, $y = V_{C_1}$, and $z = V_{C_2}$, the governing equations arc [93]

$$\begin{cases} \dot{x} = \dfrac{y - Rx}{L} \\ \dot{y} = \dfrac{g_1(y-z) - g_2(y+z) - \gamma_2 y - x}{C_1} \\ \dot{z} = \dfrac{g_1(y-z) + g_2(y+z) + \gamma_3(z-E)}{C_2} \end{cases} \tag{18}$$

where g_i's are two piecewise-linear functions readings as

$$\begin{cases} g_1(u) = \dfrac{(m_0 - m_1)|u| + (m_0 + m_1)u}{2} \\ g_2(u) = \dfrac{(m_0 - m_1)|u| - (m_0 + m_1)u}{2}. \end{cases} \tag{19}$$

This system is equivariant under an order-2 rotation symmetry $\mathcal{R}_z(\pi)$. It has three fixed points, one designated by p^0 and located at $(x_0, y_0, z_0) = (0, 0, \frac{\gamma_3}{\gamma_3 - 2m_0})$, and two symmetry-related fixed points, $p^\pm$, around which the two scrolls take place as in the Lorenz system. A typical Lorenz-like attractor is shown in Fig. 25.

Chua and co-workers proved that

(i) there is a value of $\frac{1}{L} \in [0.0165; 0.048]$ for which there is a heteroclinic orbit (Fig. 26) connecting p^+ and p^-;
(ii) the Jacobian matrix at $p^\pm$ has eigenvalues $(-\lambda^s, \lambda^u \pm i\omega)$ where $\lambda^s > \lambda^u > 0$;
(iii) a Horseshoe is embedded in a neighborhood of the heteroclinic point.

As shown in Fig. 26, the heteroclinic orbit is unknotted as required by the Ghrist & Holmes theorem. Unfortunately, the eigenvalues at $p^\pm$ are not as required for applying this theorem and it cannot be proved that all possible knots are realized as periodic orbits solution to this Lorenz-like piecewise-linear circuit. Nevertheless, the chaotic nature of its attractor is proved in the sense of Shil'nikov theorem [93]. As shown by Ghrist [70], Lorenz-like systems cannot produce attractors containing every types of knots.

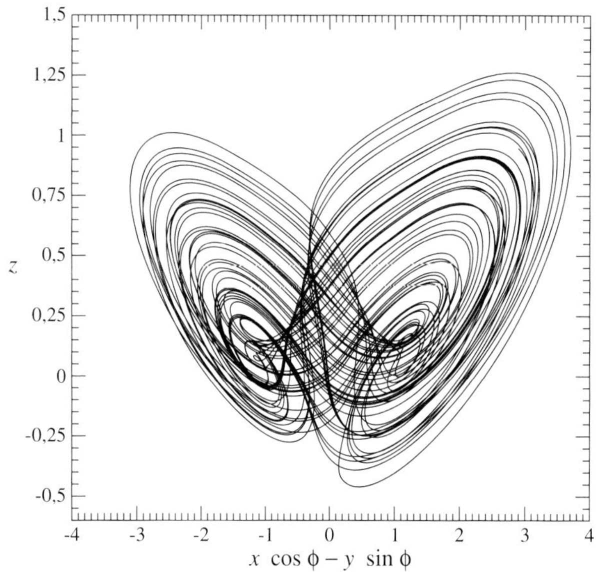

Fig. 25. A chaotic Lorenz-like attractor solution to the piecewise-linear circuit (18). Parameter values: $L = 1/0.047$ H, $C_1 = 1/0.145$ F, $C_2 = 1/0.0213$ F, $R = 1.8\ \Omega$, $\gamma_2 = 1.450\,\Omega^{-1}$, $\gamma_3 = 1.120\,\Omega^{-1}$, $m_0 = 1.48\,\Omega^{-1}$, $m_1 = 1.56\,\Omega^{-1}$, and $E = 1$ V.

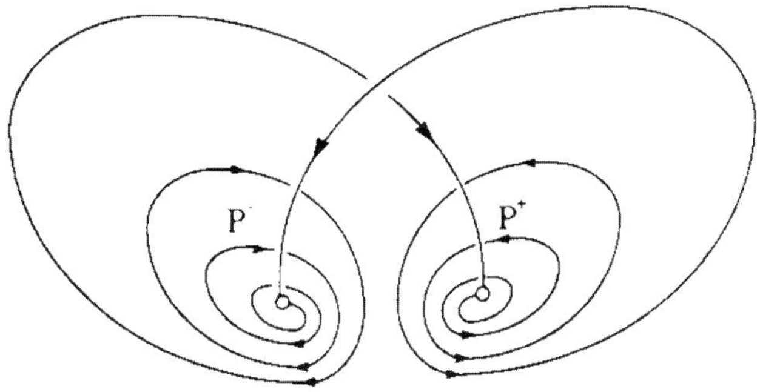

Fig. 26. Sketch of the heteroclinic orbit whose existence was proved by Tokunaga and co workers [93]. To compare with Fig. 17(c).

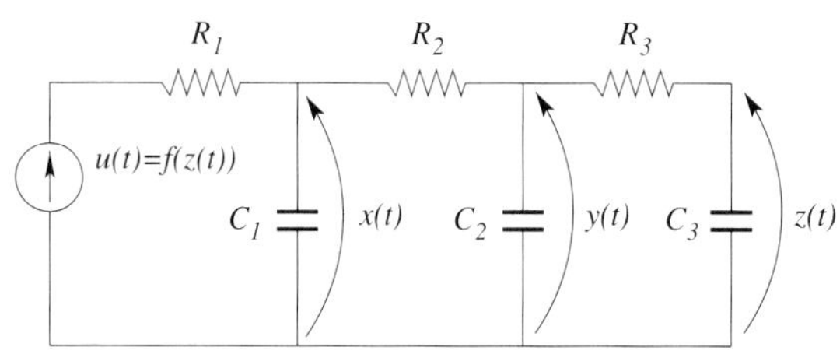

Fig. 27. Circuit diagram of the third-order R-C-ladder phase-shift oscillator.

6. A Symmetric Piecewise-linear Rössler Circuit

The R-C-oscillator circuit, consisting of a three-section R-C-ladder quadrupole and an inverting amplifier in the feedback path (Fig. 27), is one of

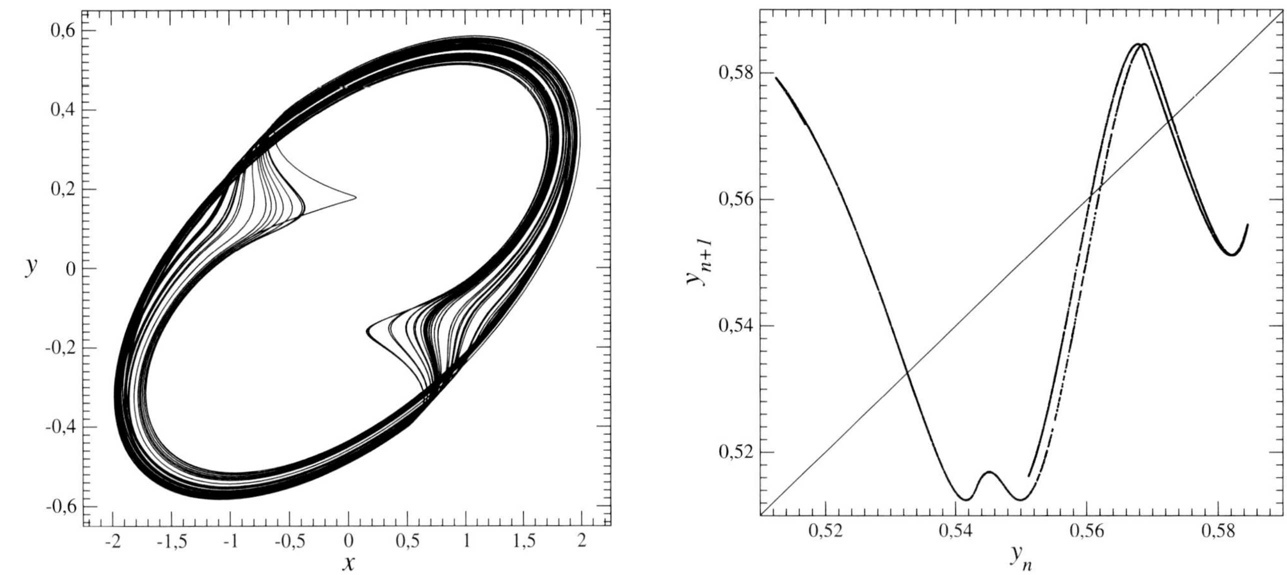

(a) Phase portrait (b) First-return map

Fig. 28. Symmetric Rossler-type chaotic attractor solution to the R-C oscillator consisting of a third-order R-C-ladder phase-shift circuit and a nonlinear amplifier in the feedback. Parameter values: $m_0 = 33.03$, $m_1 = 480$, and $\sigma = 0.2$.

the first examples of sine wave oscillators. Typical "engineer" procedure for the design of such oscillators has been based on the assumptions of linear operation of all elements used. Maciej Ogorzalek investigated how the nonlinearity introduced by active elements in the feedback path influences the dynamics of such a circuit. The governing equations of the circuit can be reduced to [94]

$$\begin{cases} \dot{x} = -2x + y + f(z) \\ \dot{y} = x - 2y + z \\ \dot{z} = y - z \end{cases}$$

$$\text{where } f(z) = \begin{vmatrix} +m_1 z + (m_0 + m_1)\sigma & z < -\sigma \\ -m_0 z & \text{if } |z| < \sigma \\ +m_1 z - (m_0 + m_1)\sigma & z > +\sigma. \end{vmatrix} \tag{20}$$

These system has an inversion symmetry, that is, it is left invariant under the coordinate transformation $(x, y, z) \mapsto (-x, -y, -z)$.

Varying slopes of the piecewise-linear characteristic, two simultaneaous period-doubling cascade are observed as route to chaos. They then lead to two co-existing asymmetric chaotic attractors which become a single symmetric Rössler-like attractor [Fig. 28(a)] through a merging attractor crisis. A first-return map to a Poincaré section [Fig. 28(b)] reveals the multimodal map with few differential extrema, characteristics of a symmetric Rössler-like attractor [79]. Two-fold cover of a spiral attractor is not unique as already shown in Figs. 18(k) and 18(l). The attractor shown in Fig. 28 is another example. The chaotic nature of the attractor was proved by applying the Shilnikov theorem [94].

7. Conclusion

We browsed many papers published by Chua and co-workers over two decades. We selected different types of attractors and presented them more or less in a chronological order. Chua is mostly known for the Chua circuit and the memristor but we showed that he contributed to chaos theory by providing many electronic circuits which can serve as specific experimental realizations producing the different types of chaotic attractors identified in the general framework of the nonlinear dynamical systems theory. Driven by his rigorous analytical approach, he can be considered as making the bridge between mathematical contributions by Levinson or Cartwright, for instance, and a general theory for electronic circuits.

References

[1] B. van der Pol, On "relaxation-oscillations," *The London, Edinburgh, and Dublin Philosophical Magazine and Journal of Science* **7**(2), 978–992 (1926).
[2] A. A. Andronov, A. A. Vitt & S. E. Khaikin, *Theory of Oscillators*, Addison-Wesley (1966).

[3] H. Poincaré, Sur la télégraphie sans fil, *Lumière électrique* **4**, 259–266, 291–297, 323–327, 355–359 & 387–393 (1908).

[4] J.-M. Ginoux & C. Letellier, "Van der Pol and the history of relaxation oscillations: toward the emergence of a concept," *Chaos*, submitted.

[5] B. Van der Pol & van der Mark, "The heartbeat considered as a relaxation oscillationsm and en electrical model of the heart," *The London, Edinburgh and Dublin Philosophical Magazine and Journal of Science* **7**(6), 763–775 (1928).

[6] W. Duddell, "On rapid variations in the current through the direct-current arc," *Journal of the Institution of Electrical Engineers* **30**(148), 232–267 (1900).

[7] A. Blondel, "Sur les phénomènes de l'arc chantant," *Éclairage Électrique* **44**(28), 41–58 and 81–104 (1905).

[8] M.-L. Cartwright & J. E. Littlewood, "On nonlinear differential equations of the second order: I. The equation $\ddot{y}+k(1-y^2)\dot{y}+y = b\lambda k\cos(\lambda t+a)$, k large," *Journal of the London Mathematical Society* **20**, 180–189 (1945).

[9] N. Levinson, "A second order differential equation with singular solutions," *Annals of Mathematics* **50**(1), 127–153 (1949).

[10] C. Letellier & J.-M. Ginoux, "Development of the nonlinear dynamical systems theory from radio engineering to electronics," *International Journal of Bifurcation & Chaos* **19**(7), 2131–2163 (2009).

[11] C. Mira, I. Gumowski and a Toulouse Research Group in the "prehistoric times of chaotic dynamics", *World Scientific Series in Nonlinear Science A* **39**, 95–198 (2000).

[12] C. P. Pulkin, "Oscillating iterated sequences (in Russian)," *Dokl. Akad. Nauk. SSSR* **76**(6), 1129–1132 (1950).

[13] I. Gumowski & C. Mira, "Sensitivity problems related to certain bifurcations in non-linear recurrence relations, *Automatica* **5**, 303–317 (1969).

[14] L. O. Chua, "Synthesis of new nonlinear network elements," *Proceedings of the IEEE* **56**(8), 1325–1243 (1968).

[15] L. O. Chua, "Memristor — the missing circuit element," *IEEE Transaction on Circuit Theory* **18**, 507–519 (1971).

[16] D. B. Strukov, G. S. Snider, D. R. Stewart & R. S. Williams, "The missing memristor found," *Nature* **453**, 80–83 (2008).

[17] V. V. Nemystskii & V. V. Stepanov, *Qualitative Theory of Differential Equations*, Princeton University Press (1960).

[18] S. Lefschetz, *Differential Equations: Geometric Theory* (1957), Reprinted by Dover (1977).

[19] H. Poincaré, *Les Méthodes Nouvelles de la Mécanique Céleste*, Gauthier-Vilars (1892–1899).

[20] L. O. Chua, "Synthesis of nonlinear systems with prescribed singularities," *IEEE Transactions on Circuit Theory* **18**(3), 375–382 (1971).

[21] L. O. Chua & D. Green, "Synthesis of nonlinear periodic systems," *IEEE Transactions on Circuit and Systems* **21**(2), 286–294 (1973).

[22] J. P. Crutchfield & B. S. McNamara, "Equations of motion from a data series," *Complex Systems* **1**, 417–452 (1987).

[23] F. Takens, "Detecting strange attractors in turbulence," *Lecture Notes in Mathematics* **898**, 366–381 (1981).

[24] L. A. Aguirre & C. Letellier, "Modeling nonlinear dynamics and chaos: A review," *Mathematical Problems in Engineering* **2009**, 238960 (2009).

[25] L. O. Chua & C.-Y. Ng, "Frequency domain analysis of nonlinear systems: general theory," *Electronic Circuits and Systems* **3**(4), 165–185 (1979).

[26] E. Hopf, "Bifurcation of a periodic solution from a stationary solution of a system of differential equations," *Berrichten der Mathematische Physikalische Klasses der Sachischen Acad. Wissenschoften* **94**, 3–22 (1942).

[27] A. I. Mees & L. O. Chua, "The Hopf bifurcation theorem and its applications to nonlinear oscillations in circuits and systems," *IEEE Transactions on Circuit and Systems* **26**(4), 235–254 (1979).

[28] L. O. Chua, "Dynamic nonlinear networks: State-of-the-art," *IEEE Transactions on Circuit and Systems* **27**(11), 1059–1087 (1980).

[29] N. Minorsky, *Nonlinear Oscillations*, Princeton (1962).

[30] C. Hayashi, *Nonlinear Oscillations in Physical Systems*, McGraw-Hill (1964).

[31] J. P. Gollub, T. O. Brunner & B. G. Danly, "Periodicity and chaos in coupled nonlinear oscillators," *Science* **200**, 48–50 (1978).

[32] Y. Ueda "Explosion of strange attractors exhibited by Duffing's equation," *Annals of the New York Academy of Sciences* **357**, 422–434 (1980).

[33] Y. Ueda & N. Akamatsu, "Chaotically transitional phenomena in the forced negative-resistance oscillator," *IEEE Transactions on Circuit and Systems* **28**(3), 217- (1981).

[34] A. S. Pikovskii & M. I. Rabinovich, "A simple autogenerator with stochastic behavior," *Sov. Phys. Dokl.* **23**(3), 183–185 (1978).

[35] C. T. Sparrow, "Chaos in a three-dimensional single loop feedback system with a piecewise linear feedback function," *Journal of Mathematical Analysis and Applications* **83**, 275–291 (1981).

[36] A. Arnéodo, P. Coullet & C. Tresser, "Possible new strange attractors with spiral structure," *Communications in Mathematical Physics* **79**, 573–579 (1981).

[37] R. F. Brockett, "On conditions leading to chaos in feedback systems," *Proceedings of the 21st IEEE Conference on Decision and Control* **2**, 932–936 (1982).

[38] E. N. Lorenz, "Deterministic nonperiodic flow," *Journal of the Atmospheric Sciences* **20**, 130–141 (1963).

[39] L. O. Chua & A. Ushida, "Algorithms for computing almost periodic steady-state response of nonlinear systems to multiple input frequencies," *IEEE Transactions on Circuit and Systems* **28**(10), 953–971 (1981).

[40] C. Sparrow, "An introduction to the Lorenz equations" *IEEE Transactions on Circuit and Systems* **30**(8), 533–542 (1981).

[41] L. O. Chua, M. Hasler, J. Neirynck & Ph. Verburgh, "Dynamics of a piecewise-linear resonant circuit," *IEEE Transactions on Circuit and Systems* **29**(8), 535–547 (1982).

[42] R. Bowen, "ω-limit sets for Axiom A diffeomorphisms," *Journal of Differential Equations* **18**, 333–339 (1975).

[43] S. Hammel, J. A. Yorke & C. Grebogi, "Do numerical orbits of chaotic dynamical processes represent true orbit?," *Journal of Complexity* **3**, 136–145 (1987).

[44] T. S. Parker & L. O. Chua, "A computer-assisted study of forced relaxation oscillations," *IEEE Transactions on Circuit and Systems* **30**(8), 518–533 (1983).

[45] M. Levi, "Qualitative analysis of the periodically forced relaxation oscillations," *Memoirs of the American Mathematical Society* **32**, 244 (1981).

[46] O. Ménard, C. Letellier, J. Maquet, L. Le Sceller & G. Gouesbet, "Analysis of a non synchronized sinusoidally driven dynamical system," *International Journal of Bifurcation & Chaos* **10**(7), 1759–1772 (2000).

[47] A. Mees & L. O. Chua, "Synchronization and chaos," *IEEE Transactions on Circuits & Systems* **30**(9), 620–626 (1983).

[48] Y. S. Tang, A. I. Mees & L. O. Chua, "Synchronization and chaos," *IEEE Transactions on Circuit and Systems* **30**(9), 620–626 (1983).

[49] T. Matsumoto, L. O. Chua & S. Tanaka, "Simplest chaotic non autonomous circuit," *Physical Review A* **30**(2), 1155–1157 (1984).

[50] S. Tanaka, T. Matsumoto & L. O. Chua, "Bifurcation scenario in a driven R-L-diode circuit," *Physica D*, **28**, 317–344 (1987).

[51] B. van der Pol & J. van der Mark, "Frequency demultiplication," *Nature* **120**(3019), 363–364 (1927).

[52] M. P. Kennedy & L. O. Chua, "Van der Pol and chaos," *IEEE Transactions on Circuit and Systems* **33**(10), 974–980 (1986).

[53] M. P. Kennedy & L. O. Chua, "The Devils staircase: The electrical engineers fractal," *IEEE Transactions on Circuit and Systems* **36**(8), 1133–1139 (1989).

[54] T. Matsumoto, L. O. Chua, H. Kawakami & S. Ichiraku, "Geometric-properties of dynamic nonlinear networks — Transversality, local-solvability and eventual passivity," *IEEE Transactions on Circuits & Systems* **28**(5), 406–428 (1981).

[55] T. Matsumoto, G. Ikegami & L. O. Chua, "Strong structural stability of resistive nonlinear n-ports," *IEEE Transactions on Circuits & Systems* **30**(4), 197–222 (1983).

[56] L. O. Chua, "A zoo of strange attractors from the canonical chuas circuits," *Proceedings of the 35th Midwest Symposium on Circuit and Systems* **2**, 916–926 (1992).

[57] F. N. H. Robinson, "Analogue electronic model of the Lorenz equations," *International Journal of Electronics* **68**(5), 803–819 (1990).

[58] T. Matsumoto, L. O. Chua & M. Komuro, "The double scroll," *IEEE Transactions on Circuits and Systems* **32**(8), 798–818 (1985).

[59] G.-Q. Zhong & F. Ayrom, "Experimental confirmation of chaos from Chua's circuit," *Circuit Theory and Applications* **13**(1), 93–98 (1985).

[60] T. Matsumoto, "A chaotic attractor from Chua's circuit," *Memorandum No. UCB/ERLM84/36*, University of California, Berkeley (April 1984).

[61] T. Matsumoto, "A chaotic attractor from Chua's circuit," *IEEE Transactions on Circuits and Systems* **31**(12), 1055–1058 (1984).

[62] O. E. Rössler, "An equation for continuous chaos," *Physics Letters A* **57**(5), 397–398 (1976).

[63] P. S. Linsay, "Period-doubling and chaotic behavior in a driven anharmonic oscillator," *Physical Review Letters* **47**(19), 1349–1352 (1981).

[64] M. Komuro, "On embeddings of subshifts of finite type," *Lecture Notes in Mathematics* **1021**, 299–301 (1983).

[65] M. Komuro, "Lorenz attractors do not have the pseudo-orbit tracing property," *Journal of the Mathematical Society of Japan* **37**(3), 489–514 (1985).

[66] R. F. Williams, "The structure of Lorenz attractors," *Publication de l'Institut des Hautes-Etudes Scientifiques* **50**, 73–99 (1979).

[67] L. Kocarev, Z. Tasev, D. Dimovski, "Topological description of a chaotic attractor with spiral structure," *Physics Letters A* **190**, 399–402 (1994).

[68] C. Letellier, G. Gouesbet & N. Rulkov, "Topological analysis of chaos in equivariant electronic circuits," *International Journal of Bifurcation & Chaos* **6**(12B), 2531–2555 (1996).

[69] R. W. Ghrist & P. J. Holmes, "An ODE whose solutions contain all knots and links," *International*

Journal of Bifurcation & Chaos **6**(5), 779–800 (1996).

[70] R. W. Ghrist, "Branched two-manifolds supporting all links," *Topology* **36**(2), 423–448 (1997).

[71] L. O. Chua, M. Komuro & T. Matsumoto, "The double scroll family," *IEEE Transactions on Circuit and Systems* **33**(11), 1073–1118 (1986).

[72] H. Kokubu & R. Roussarie, "Existence of a singularly degenerate heteroclinic cycle in the Lorenz system and its dynamical consequences: Part I," *Journal of Dynamics and Differential Equations* **16**(2), 513–557 (2004).

[73] Y. Liu & W. Pang, "Dynamics of the general Lorenz family," *Nonlinear Dynamics*, (Online First), DOI: 10.1007/s11071-011-0090-7 (2011).

[74] L. O. Chua, M. Komuro & T. Matsumoto, "A universal circuit for studying and generating chaos — Part II: Strange attractors," *IEEE Transactions on Circuit and Systems* **40**(10), 745–761 (1993).

[75] R. Shaw, "Strange attractor, chaotic behavior and information flow," *Zeitschrift fr Naturforschung A* **36**, 80–112 (1981).

[76] C. Letellier, P. Dutertre, J. Reizner & G. Gouesbet, "Evolution of multimodal map induced by an equivariant vector field," *Journal of Physics A* **29**, 5359–5373 (1996).

[77] C. Letellier, T. D. Tsankov G. Byrne & R. Gilmore, "Large-scale structural reorganization of strange attractors," *Physical Review E* **72**, 026212 (2005).

[78] C. Letellier, G. F. V. Amaral & L. A. Aguirre, "Insights into the algebraic structure of Lorenz-like systems using feedback circuit analysis and piecewise affine models," *Chaos* **17**, 023104 (2007).

[79] C. Letellier & R. Gilmore, "Covering dynamical systems: Two fold covers," *Physical Review E* **63**, 16206 (2001).

[80] O. E. Rössler,, "Horseshoe map in the Lorenz equation," *Physical Letters A* **60**(5), 392–394 (1977).

[81] M. Auvergne & A. Baglin, "A dynamical instability as a driving mechanism for stellar oscillations," *Astronomy & Astrophysics* **142**, 388–392 (1985).

[82] C. Letellier, *Caractérisation Topologique et Reconstruction des Attracteurs Etranges*, Ph'D Thesis, Rouen University, 1994.

[83] Bo Deng, "Constructing homoclinic orbits and chaotic attractors," *International Journal of Bifurcation & Chaos* **4**(4), 823–841 (1994).

[84] D. Li, "A three-scroll chaotic attractor," *Physics Letters A* **372**(4), 387–393 (2008).

[85] C. Letellier & R. Gilmore, "Poincaré sections for a new three-dimensional toroidal attractor," *Journal of Physics A* **42**, 015101 (2009).

[86] T. Matsumoto, L. O. Chua & R. Tokunaga, "Chaos via torus breakdown," *IEEE Transactions on Circuits and Systems* **34**(3), 240–253 (1987).

[87] V. S. Anishchenko, M. A. Safanova & L. O. Chua, "Confirmation of the Afraimovich-Shilnikov torus-breakdown theorem via a torus circuit," *IEEE Transactions on Circuit and Systems* **40**(10), 792–800 (1993).

[88] T. S. Parker & L. O. Chua, *Practical Numerical Algorithms for Chaotic Systems*, Springer-Verlag (1989).

[89] M. Klein, G. Baier & O. E. Rössler, "From n-tori to hyperchaos," *Chaos, Solitons & Fractals* **1**(2), 105–118 (1991).

[90] D. Amroun-Aliane, L. Pastur & C. Letellier, "Defects in spatiotemporal diagrams and their relations to phase coherence and lack of observability," *Physical Review E* **83**, 056212 (2011).

[91] Ph. Bartissol & L. O. Chua, "The double hook," *IEEE Transactions on Circuit and Systems* **35**(12), 1512–1522 (1988).

[92] R. Tokunaga, M. Komuro, T. Matsumoto & L. O. Chua, "Lorenz Attractor from an electrical circuit with uncoupled piecewise-linear continuous resistors," *International Journal of Control Theory and Applications* **17**(1), 71–85 (1989).

[93] R. Tokunaga, T. Matsumoto, L. O. Chua & S. Miyama, "The piecewise-linear Lorenz circuit is chaotic in the sense of Shilnikov," *IEEE Transactions on Circuits and Systems* **37**(6), 766–786 (1990).

[94] M. J. Ogorzalek, "Order and chaos in a third-order RC ladder network with nonlinear feedback," *IEEE Transactions on Circuit and Systems* **36**(9), 1221–547 (1989).

PART II

DYNAMICAL SYSTEMS AND CHAOS

CHAPTER 18

CONNECTIVITY OF JULIA SETS FOR SINGULARLY PERTURBED RATIONAL MAPS

ROBERT L. DEVANEY

Dept. of Mathematics, Boston University,
Boston, MA 02215, USA

ELIZABETH D. RUSSELL

Dept. of Mathematics, Western New England University,
Springfield, MA 01119, USA

In this paper we consider the family of rational maps of the form $F_\lambda(z) = z^n + \lambda/z^n$ where $n \geq 2$. It is known that there are two cases where the Julia sets of these maps are not connected. If the critical values of F_λ lie in the basin of ∞, then the Julia set is a Cantor set. And if the critical values lie in the preimage of the basin surrounding the pole at 0, then the Julia set is a Cantor set of concentric simple closed curves around the origin. We prove in this paper that, in all other cases, the Julia set of F_λ is a connected set.

1. Introduction

Our goal in this paper is to describe the connectivity properties of Julia sets of functions in the family of rational maps of the Riemann sphere given by

$$F_\lambda(z) = z^n + \frac{\lambda}{z^n}$$

where $n \geq 2$ and $\lambda \in \mathbb{C}$. These maps are called singular perturbations since, when $\lambda = 0$, the map is just $z \mapsto z^n$ and so the dynamics are completely understood. But when $\lambda \neq 0$, the degree of the map jumps from n to $2n$ and the dynamical behavior becomes much more complicated.

Despite the high degree of these maps, there are a number of similarities with complex polynomials of degree 2. For example, near ∞, the map is conjugate to $z \mapsto z^n$, so there is an immediate basin of attraction of ∞ which we denote by B_λ. Unlike polynomials, however, the full basin of ∞ is never a simply connected set. There is a pole at the origin and there are many parameters for which there is an open set surrounding 0 that is mapped n-to-one onto B_λ. If this set is disjoint from B_λ, we call this set the trap door and denote it by T_λ. If the orbit of any point eventually enters B_λ, it must do so by passing through the trap door.

In another analogy with quadratic polynomials, there is essentially only one free critical orbit for F_λ. One checks easily that there are $2n$ free critical points given by $c_\lambda = \lambda^{1/2n}$. (We call these points "free" critical points since ∞ is also a critical point, but it is fixed, and 0 is also a critical point, but 0 is mapped directly to ∞.) Despite the large number of critical points, there are only two critical values $v_\lambda = \pm 2\sqrt{\lambda}$. Exactly n of the critical points are mapped to one critical value and the other n are mapped to the other critical value. By symmetry, however, there is essentially only one free critical orbit, since, if n is odd, the orbits of $\pm v_\lambda$ behave symmetrically under $z \mapsto -z$, whereas if n is even, both $\pm v_\lambda$ are mapped to the same point by F_λ.

For quadratic polynomials, there is a well-known dichotomy: if the critical orbit escapes to ∞, the Julia set is a Cantor set, while if the critical orbit does not escape, the Julia set is a connected set. In analogy with this, there is a somewhat different situation for F_λ; in this case we have an "escape trichotomy." In [6], the following theorem was proved:

Theorem. *For the family* F_λ:

(1) *If one of the critical values lies in* B_λ, *then* $J(F_\lambda)$ *is a Cantor set and* $F_\lambda \,|\, J(F_\lambda)$ *is conjugate to the one-sided shift on* $2n$ *symbols.*
(2) *If one of the critical values lies in* T_λ *(and, by hypothesis,* T_λ *is disjoint from* B_λ*), then* $J(F_\lambda)$ *is a Cantor set of concentric simple closed curves (quasicircles).*
(3) *If one of the critical values lies in some preimage of* T_λ, *then* $J(F_\lambda)$ *is a Sierpiński curve.*

We remark that part two of this theorem was first proved by McMullen [7]. It turns out that this kind of Julia set does not occur when $n = 2$; indeed, the case $n = 2$ is very different from (and much more complicated than) the case $n > 2$. See [3].

A Sierpiński curve is any planar set that is homeomorphic to the well-known Sierpiński carpet fractal. The first such set to arise as a Julia set was found by Milnor and Tan Lei [9]. Since then, it has been shown that Sierpiński curves arise in many different ways as Julia sets [1, 2]. See Fig. 1 for examples of the three different cases arising in the family $z^4 + \lambda/z^4$.

In this paper, our goal is to strengthen the Escape Trichotomy Theorem by proving the following:

Theorem. *If the critical values of* F_λ *do not lie in either* B_λ *or* T_λ, *then the Julia set of* F_λ *is always a connected set.*

So, just as in the case of quadratic polynomials, for these rational maps, either the Julia set is connected or else it is disconnected and consists of uncountably many disjoint components. In this latter case, as mentioned above, either the Julia set is a Cantor set or else it is a Cantor set of simple closed curves. In Fig. 2, we display the parameter plane for the case $n = 4$. The external region is the Cantor set locus. The central disk is the region where the Julia set is a Cantor set of simple closed curves; this region is called the McMullen domain. The complement of the Cantor set locus and the McMullen domain is the connectedness locus since we shall show in this paper that all of these parameters are associated to maps whose Julia sets are connected sets. Those parameters drawn from the red disks in the connectedness locus are Sierpiński curve Julia sets alluded to in the escape trichotomy.

Dedication. This paper is dedicated to Leon Chua, who has made the International Journal of Bifurcation and Chaos into one of the most influential journals in the field of dynamical systems. The first author of this paper has enjoyed working with him on the editorial board of this journal for many years.

2. Proof of Connectivity

Our goal in this section is to prove that the Julia sets drawn from the connectedness locus are always connected sets. In order to prove this, we need to show that all of the Fatou components of F_λ are simply connected sets. There are thus only two ways that this could not happen; either there is a Herman ring or else there is an infinitely connected attracting or parabolic basin in the Fatou set. See [8]. If the Fatou set only has preimages of B_λ or simply connected attracting or parabolic basins or Siegel disks, then the Julia set is the complement of infinitely many open and disjoint disks and hence is a connected set.

We first show that there is no Herman ring in the Fatou set. As a remark, this was shown in [11], but the proof we give here is much simpler. This is the easiest case of the Theorem.

First suppose we have a periodic Herman ring. Recall that a Herman ring is an open annular region on which some iterate of F_λ is conjugate to an irrational rotation. At least one of the iterates of this ring must surround the origin, for if not, F_λ would map the interior component of each Herman ring to the corresponding interior component of its image. But then the family of iterates of F_λ would be normal on each such ring together with its bounded

$\lambda = 0.23$

$\lambda = 0.04$

$\lambda = -0.125i$

Fig. 1. Some Julia sets for $z^4 + \lambda/z^4$: if $\lambda = 0.04$, $J(F_\lambda)$ is a Cantor set; if $\lambda = 0.23$, $J(F_\lambda)$ is a Cantor set of circles; and if $\lambda = -0.125i$, $J(F_\lambda)$ is a Sierpiński curve.

interior component. Hence there would be no interior boundary of the Herman ring in the Julia set and the Fatou component would be the entire disk, not an annulus.

So let R be a Herman ring that surrounds the origin. We have that the Julia set of F_λ is symmetric under rotation by $2\pi/2n$, i.e., it has $2n$-fold symmetry. This follows from the fact that $F_\lambda(e^{\pi i/n}) = -F_\lambda(z)$ and, as we observed above, the orbits of $\pm z$ are either symmetric under $z \mapsto -z$ (when n is odd) or $F_\lambda(z) = F_\lambda(-z)$ when n is even.

So assume that we have a Herman ring R surrounding the origin. Then R must be symmetric under the rotation $z \mapsto e^{\pi i/n}z$. But then, by the above, the map cannot be one-to-one on R, so we have no such Herman ring. This eliminates the first possibility of a non-simply connected component in the Fatou set.

Now suppose that the map has a multiply connected attracting or parabolic basin. Then it is well known [8] that this domain must in fact be infinitely connected. As above, at least one iterate of this domain must surround the origin. To show that this cannot happen when λ is in the connectedness locus, we shall construct Cantor necklaces in the dynamical plane. The boundaries of these necklaces

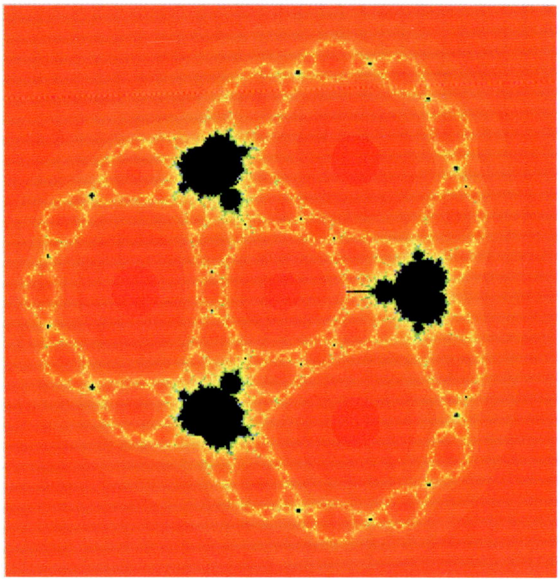

Fig. 2. The parameter plane when $n = 4$.

will be closed sets that connect the boundary of B_λ to the boundary of T_λ and lie in the Julia set. This will show that there is no such infinitely connected Fatou component surrounding the origin.

For simplicity, we shall concentrate on the case $n=2$; the cases where $n>2$ are handled in a similar manner; we will discuss the minor modifications for these cases at the end of this section. To define the Cantor necklace, we let Γ denote the Cantor middle-thirds set in the unit interval $[0,1]$. We regard this interval as a subset of the real axis in the plane. For each open interval of length $1/3^n$ removed from the unit interval in the construction of Γ, we replace this interval by an open disk of diameter $1/3^n$ centered at the midpoint of the removed interval. Thus the boundary of this open disk meets the Cantor set at the two endpoints of the removed interval. We call the resulting set the *Cantor middle-thirds necklace*. See Fig. 3. Any set homeomorphic to the Cantor middle-thirds necklace is called a *Cantor necklace*. We do not include the boundary of the open disks in the Cantor necklace for the following technical reason: it is sometimes difficult in practice to verify that these bounding curves are simple closed curves.

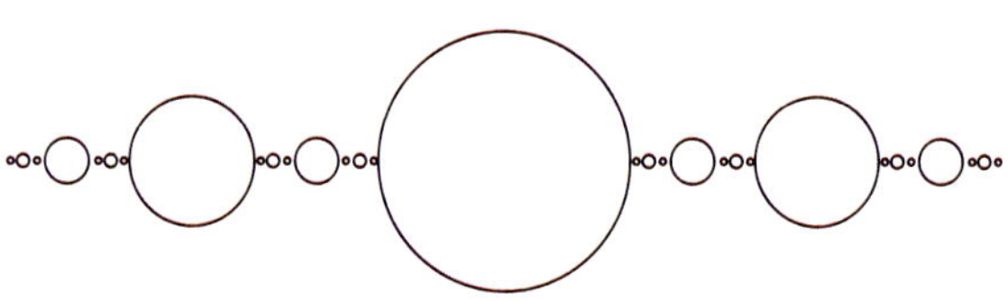

Fig. 3. The Cantor middle-thirds necklace.

Now let $\lambda = |\lambda|e^{i\eta}$. For the rest of this paper, we will consider only the case where $0 < \eta < 2\pi$. When λ is positive and real, it is known that F_λ is conjugate in the right half plane to a polynomial of the form z^2+c where $c \in \mathbb{R}^+$. See [4]. The boundary of this quadratic like Julia set then lies in the Julia set of F_λ and stretches from B_λ to T_λ and so no Fatou domain can surround the origin in this case. We remark that all of the results below concerning Cantor necklaces also hold when $\eta = 0$, but the arguments are slightly different.

When $|\lambda| < 1$, we have the following estimate of the size of B_λ.

Lemma. (The Escape Criterion). *Suppose* $|\lambda| < 1$ *and* $|z| \geq 2$. *Then* $z \in B_\lambda$, *so the Julia set is contained in the open disk* $|z| < 2$.

Proof. If $|z| \geq 2$, then we have

$$|F_\lambda(z)| \geq |z|^2 - \frac{|\lambda|}{|z|^2} \geq 2|z| - \frac{1}{4} > \frac{3}{2}|z|.$$

Inductively, we find

$$|F_\lambda^n(z)| \geq \left(\frac{3}{2}\right)^n |z|.$$

Therefore the orbit of any such z tends to ∞ so all points on or outside the circle of radius 2 lie in B_λ. □

As mentioned above, it is known that the Julia set is a Cantor set if $|\lambda| \geq 1$, so we also exclude this case for the remainder of this paper. Therefore we assume throughout that $\lambda = |\lambda|e^{i\eta}$ with $0 < \eta < 2\pi$ and $|\lambda| < 1$.

Recall that the critical points of F_λ are given by $\lambda^{1/4}$. Therefore one of the critical points of F_λ lies on the straight line through the origin given by $t\exp(i\eta/4)$ with $t > 0$. The image of this line lies along the straight line with argument $\theta = \eta/2$, and F_λ maps the line $t\exp(i\eta/4)$ with $t > 0$ in two-to-one fashion over the portion of this image straight line that extends from the critical value $2\sqrt{\lambda}$ whose argument is $\eta/2$ out to ∞. Note that the image of this line is disjoint from the line itself since we have assumed that $0 < \eta < 2\pi$. There is a second critical point of F_λ lying on the line with argument $\theta = \eta/4 - \pi/2$, and this line is mapped in two-to-one fashion to the opposite line $\theta = -\eta/2$ exactly as in the above case.

Let H_λ be the involution $\lambda^{1/2}/z$. Then we have $F_\lambda(H_\lambda(z)) = F_\lambda(z)$ so the Julia set of F_λ is

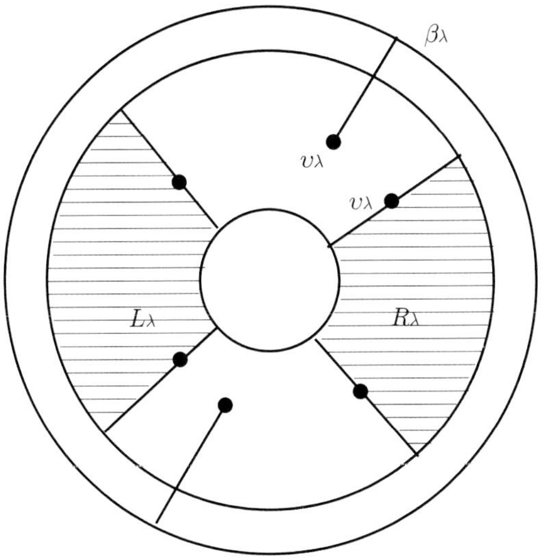

Fig. 4. R_λ and L_λ and their image under F_λ, which is the interior of β_λ minus the two segments connecting this circle to the critical values.

symmetric under H_λ. Note that H_λ interchanges B_λ and T_λ. By the escape criterion, we know that any point on or outside $r = 2$ is mapped closer to ∞. Let β_λ denote the image of this circle, so that $\beta_\lambda \subset B_\lambda$. Using the involution H_λ, there is a second circle, namely $r = |\lambda|^{1/2}/2$, that is also mapped two-to-one onto β_λ.

Consider the open region R_λ bounded by the rays $\theta = \eta/4$ and $\theta = \eta/4 - \pi/2$ and the two circular preimages of β_λ. The set R_λ is a quarter of an annulus. Let $L_\lambda = -R_\lambda$. We call R_λ (resp., L_λ) the right (resp., left) fundamental sector. These fundamental sectors are a pair of disjoint, open, simply connected regions in $\mathbb{C}$. Note that, for each λ, R_λ lies in the right half plane $\operatorname{Re} z > 0$, while L_λ lies in the left half plane. See Fig. 4.

Proposition. *F_λ maps each of the fundamental sectors in one-to-one fashion onto the open set $\mathcal{O}$ bounded by β_λ minus the portions of the two straight lines $\theta = \pm\eta/2$ extending from the critical values $\pm v_\lambda$ to β_λ. So the image of each of these fundamental sectors contains the closures of both R_λ and L_λ in its interior.*

Proof. The images of the straight rays bounding R_λ and L_λ are contained in the rays $\theta = \pm\eta/2$, both of which lie outside these sectors. The image of the outer circular boundary of each fundamental sector is exactly one-half of β_λ, while the inner boundary of each sector is mapped to the opposite half of β_λ. Hence each fundamental sector is mapped onto the open disk bounded by β_λ minus the two portions of the rays $\theta = \pm\eta/2$ lying beyond the critical values. This set is $\mathcal{O}$. By fourfold symmetry, this map must be one-to-one on each fundamental sector. □

Since F_λ maps the union of the fundamental sectors strictly over itself, most points in $R_\lambda \cup L_\lambda$ have orbits that leave this set at some iterate. Let Γ_λ be the set of points whose orbits remain for all iterations in $R_\lambda \cup L_\lambda$. Then we have:

Proposition. *The set Γ_λ is a Cantor set and $F_\lambda|\Gamma_\lambda$ is conjugate to the one-sided shift on two symbols.*

Proof. By the previous result, each of the fundamental sectors is mapped in one-to-one fashion onto the open region $\mathcal{O}$ that properly contains $R_\lambda \cup L_\lambda$ in $\mathbb{C}$. So we have a pair of well-defined inverses G_0 (resp., G_1) of F_λ that map $\mathcal{O}$ into R_λ (resp., L_λ). Standard arguments then show that these inverses are contractions in the Poincaré metric on $\mathcal{O}$. Then, for any one-sided sequence $(s_0 s_1 s_2 \ldots)$ of 0's and 1's, the set

$$\bigcap_{j=0}^{\infty} G_{s_0} \circ \ldots \circ G_{s_j}(\mathcal{O})$$

is a unique point and the map that takes the sequence $(s_0 s_1 s_2 \ldots)$ to this point defines a homeomorphism between the space of one-sided sequences of 0's and 1's endowed with the usual topology and Γ_λ. Hence Γ_λ is a Cantor set and we have that $F_\lambda \,|\, \Gamma_\lambda$ is conjugate to the one-sided shift on two symbols. □

Now suppose, in addition, that the critical values do not lie in B_λ. So B_λ is a simply connected open set. Our goal now is to construct a Cantor necklace in the dynamical plane. The Cantor set portion of the necklace will be the set Γ_λ constructed above, whereas the open disks will be certain of the preimages of the basin of ∞ that lie in R_λ and L_λ.

To construct the necklace, we have that F_λ is conjugate to z^2 in B_λ. Hence there are a pair of external rays in B_λ that correspond to the rays of angles 0 and π in the basin of ∞ for z^2. It is known [10] that the external ray of angle 0 limits on the fixed point p_λ that lies in $R_\lambda \cap \partial B_\lambda$ and the external ray of angle 1/2 limits on its preimage $-p_\lambda$ in L_λ. Let $\pm q_\lambda$ be the preimages of $-p_\lambda$ in L_λ and R_λ. Then there is a unique curve in T_λ passing through 0 and connecting q_λ to $-q_\lambda$ that is mapped onto the external ray connecting $-p_\lambda$ to ∞. Since T_λ is open and simply connected, we may define a homeomorphism that takes T_λ onto a disk centered at $1/2$ on the real line and having radius $1/6$. This is the central disk in the Cantor middle-thirds necklace. Moreover, we may arrange that this homeomorphism extends to the points $\pm q_\lambda$ in ∂T_λ so that the point $-q_\lambda$ is sent to $1/3$ and the point q_λ is sent to $2/3$.

Now consider the preimages of the trap door. Since the critical values do not lie in the trap door, there are four preimages of this set, but only two of them lie in the fundamental sectors by the fourfold symmetry. The other two lie in the complementary sectors. These give a pair of simply connected open sets that contain the preimages of $\pm q_\lambda$ in their boundaries; one of these sets lies in R_λ, the other lies in L_λ. These may be mapped homeomorphically to the open disks in the Cantor middle-thirds necklace whose diameter is $1/9$ with the property that an extension of this homeomorphism takes the preimages of $\pm q_\lambda$ to the corresponding endpoints of the Cantor middle thirds set. We then continue in this fashion by taking appropriate preimages of the trap door under compositions of G_0 and G_1 and defining a homeomorphism between these preimages and the appropriate disk in the Cantor middle-thirds necklace. Note that the symbolic dynamics described above dictates which preimages correspond to which disks. As above, we extend this homeomorphism to the two special boundary points that eventually map to p_λ. This defines the homeomorphism on a set whose image is then the union of the disks in the middle-thirds necklace together with the corresponding endpoints in their boundaries. By the symbolic dynamics, these preimages of p_λ are dense in the Cantor set Γ_λ, so we may extend this homeomorphism to all of the remaining points in the Cantor set. This then gives a homeomorphism between the points in Γ_λ together with the preimages of the trap door in $R_\lambda \cup L_\lambda$ and the Cantor middle thirds necklace. We have proved:

Proposition. *Suppose that the critical values of F_λ do not lie in B_λ or T_λ and that $0 < \operatorname{Arg}\lambda < 2\pi$. Then the set Γ_λ together with the preimages of B_λ under compositions of the maps G_0 and G_1 forms a Cantor necklace.*

Note that the boundary of the Cantor necklace consists of points that lie in either the Cantor set portion of the necklace or else on the boundaries of the preimages of the trap door. Thus the boundary of the necklace lies in the Julia set and therefore provides a closed set that connects the boundaries of T_λ and B_λ. This proves that there are no multiply connected Fatou components of F_λ and hence that the Julia set is connected when $\pm v_\lambda$ do not lie in B_λ or T_λ.

For the case $n > 2$, there is a similar escape criterion which is easy to compute and we also have the involution symmetry, now given by $H_\lambda(z) = \lambda^{1/n}/z$. The construction of the Cantor necklace is similar; the only difference is that we have $2n$ fundamental sectors in this case. Choosing the two that straddle the real axis then provides the region in which the necklace exists. We remark that, when $n > 2$ a much more complicated object known as a Cantor web may be constructed. See [5].

References

[1] Blanchard, P., Devaney, R. L., Look, D. M., Moreno Rocha, M., Seal, P., Siegmund, S., Uminsky, D. Sierpiński Carpets and Gaskets as Julia Sets of Rational Maps. In Dynamics on the Riemann Sphere. European Math Society (2006), 97–119.

[2] Devaney, R. L. A Myriad of Sierpiński Curve Julia Sets. In Difference Equations, Special Functions, and Orthogonal Polynomials, World Scientific (2007), 131–148.

[3] Devaney, R. L. Dynamics of $z^n + C/z^n$; Why the Case $n=2$ is Crazy. To appear in Conformal Dynamics and Hyperbolic Geometry, American Math Society, Contemporary Math.

[4] Devaney, R. L. Baby Mandelbrot Sets Adorned with Halos in Families of Rational Maps. In Complex Dynamics: Twenty-Five Years After the Appearance of the Mandelbrot Set. American Math Society, Contemporary Math, **396**, 2006, 51–62.

[5] Devaney, R. L. Cantor Webs in the Parameter and Dynamical Planes of Rational Maps. Fields Institute Communications **53** (2008), 105–123.

[6] Devaney, R. L., Look, D. M., and Uminsky, D. The Escape Trichotomy for Singularly Perturbed Rational Maps. Indiana University Mathematics Journal **54** (2005), 1621–1634.

[7] McMullen, C. The Classification of Conformal Dynamical Systems. Current Developments in Mathematics. International Press, Cambridge, MA, (1995) 323–360.

[8] Milnor, J. Dynamics in One Complex Variable. Third Edition. Annals of Mathematics Studies. Princeton University Press, (2006).

[9] Milnor, J. and Tan Lei. A "Sierpiński Carpet" as Julia Set. Appendix F in Geometry and Dynamics

of Quadratic Rational Maps. Experiment. Math. **2** (1993), 37–83.

[10] Petersen, C. and Ryd, G. Convergence of Rational Rays in Parameter Spaces, The Mandelbrot set: Theme and Variations, London Mathematical Society, Lecture Note Series 274, Cambridge University Press, 161–172, 2000.

[11] Xiao, Y. and Qiu, W. The Rational Maps $F_\lambda(z) = z^m + \lambda/z^d$ Have No Herman Rings. Proc. Indian Acad. Sci. **120** 2010, 403–407.

CHAPTER 19

STRUCTURAL TRANSFORMATIONS AND STABILITY OF DYNAMICAL NETWORKS

L. A. BUNIMOVICH and B. Z. WEBB

ABC Math Program and School of Mathematics,
Georgia Institute of Technology,
686 Cherry Street, Atlanta, GA 30332, USA

The existence of a steady state is one of the main properties of any real system. In the analysis of such systems most often the first step is to determine the system's fixed points and their stability. A recently developed approach in the study of dynamical networks allows for the simultaneous investigation of the individual network elements, their interactions, and the network's topology (structure). This general approach gives sufficient conditions under which a dynamical network has a globally attracting fixed point. It has been found that if one employs a new procedure of isospectral expansion to a dynamical network then these conditions can be weakened.

1. Introduction

The study of networks in nature, science, and technology is currently one of the most active areas of research. A good deal of this research however is focused on the structural or static features of these systems although most real networks are inherently dynamic [3, 7, 8, 10–12, 14]. That is, each network element has an associated state that changes with time.

To analyze this behavior we note that the dynamics of a network can be characterized by three key features; (i) the internal (local) dynamics of the network elements, (ii) the interactions between the network elements, and (iii) the structure of the network's graph of interactions. Features (i) and (ii) of a network are described by dynamical systems and can therefore be studied using methods from differential equations and dynamical systems theory. In fact, the internal dynamics and interactions of a network have been studied together for centuries. A popular example are reaction-diffusion equations where the reaction corresponds to the internal dynamics and the diffusion to the interactions of the system. However, what differentiates these systems from the present study of networks is their regular structure of interactions.

To describe the dynamics of a network with an irregular structure requires that we integrate this structure into our approach. However, the graph structure of a system is static and must therefore be studied using techniques outside standard dynamical systems theory.

Moreover, A major obstacle in understanding the dynamics of a high dimensional system is that the information needed to do so is spread throughout the various system components. Because of this it is tempting to find ways of simplifying the structure of a system while preserving some fundamental properties or features of the system's dynamics. One of the key dynamical quantities of a network is the spectrum associated with its graph structure.

To aid in understanding how a system's structure effects its dynamics the authors introduced the concept of an *isospectral graph transformation* in [5]. The main idea behind an *isospectral graph transformation* is that a graph can be modified in a number of ways that preserve its spectrum. As a special case, this includes reducing a graph to a smaller graph with less vertices. By so doing one is effectively concentrating the structural information contained in the graph.

By extending the notion of an isospectral graph transformation to a dynamical system it is also possible to modify the structure of almost any *discrete-time* dynamical system (F, X) with dynamics F and phase space X in a way that preserves the system's dynamics. As with isospectral graph reductions this is done by concentrating the information spread between the system components of (F, X). The resulting transformed system is denoted by $(\mathcal{X}_S F, X_S)$ and is called a *dynamical network expansion* of (F, X).

The importance of a *dynamical network expansion* is that it allows for better estimates (or at least no worse) of the untransformed system's global stability. A major result of [5], restated here, (see theorem 3) asserts that the global stability of a system (F, X) is better understood by studying one of its isospectral transforms $(\mathcal{X}_S F, X_S)$.

Importantly, the definition of a dynamical network is general enough that any dynamical system (F, X) where $F : X \to X$ is Lipschitz continuous and the dimension of X is greater than one can be considered to be a dynamical network. Therefore, the dynamics of a large number of dynamical systems can be analyzed in terms of their graph structure.

It is also worth mentioning that this idea of concentrating structural information by use of graph transformations has already lead to the introduction of new equivalence relations on graphs [5] and to improvements in the classical eigenvalues estimates of Geshgorin et al. [4].

This article is organized as follows. Section 2 introduces the mathematical framework for investigating the dynamics of a network. This section also provides sufficient conditions under which a dynamical network has a globally attracting fixed point (is strongly synchronizable). Section 3 then introduces the notion of an dynamical network expansion which is then used to gain improved estimates for whether a dynamical network has a global attractor. At each step in this paper examples are used to illustrate basic concepts and idea.

2. Networks as Dynamical Systems

As mentioned in the introduction, the dynamics of a network can be analyzed in terms of three key features; (i) the internal (local) dynamics of the network elements, (ii) the interactions between the network elements, and (iii) the topology or structure of the graph of interactions of the network.

To study the dynamics of networks our first task is to establish a mathematical framework for the investigation of networks as dynamical systems. This is done following the approach given in [1].

Let $i \in \mathcal{I} = \{1, \ldots, n\}$ and $T_i : X_i \to X_i$ be maps on the complete metric space (X_i, d) where

$$L_i = \sup_{x_i \neq y_i \in X_i} \frac{d(T_i(x_i), T_i(y_i))}{d(x_i, y_i)} < \infty. \tag{1}$$

Let (T, X) denote the direct product of the local systems (T_i, X_i) over $\mathcal{I}$ on the complete metric space (X, d_{max}) where for $\mathbf{x}, \mathbf{y} \in X$

$$d_{max}(\mathbf{x}, \mathbf{y}) = \max_{i \in \mathcal{I}} \{d(x_i, y_i)\}.$$

Definition 1. A map $F : X \to X$ is called an *interaction* if for every $j \in \mathcal{I}$ there exists a nonempty collection of indices $\mathcal{I}_j \subseteq \mathcal{I}$ and a continuous function

$$F_j : \bigoplus_{i \in \mathcal{I}_j} X_i \to X_j,$$

that satisfies the following Lipschitz condition for constants $\Lambda_{ij} \geq 0$:

$$d(F_j(\mathbf{x}|_{\mathcal{I}_j}), F_j(\mathbf{y}|_{\mathcal{I}_j})) \leq \sum_{i \in \mathcal{I}_j} \Lambda_{ij} d(x_i, y_i) \tag{2}$$

for all $\mathbf{x}, \mathbf{y} \in X$ where $\mathbf{x}|_{\mathcal{I}_j}$ is the restriction of $\mathbf{x} \in X$ to $\bigoplus_{i \in \mathcal{I}_j} X_i$. The (*interaction*) map F is defined as follows:

$$F(\mathbf{x})_j = F_j(\mathbf{x}|_{\mathcal{I}_j}), \quad j \in \mathcal{I}, \quad i \in \mathcal{I}_j.$$

The constants Λ_{ij} in definition 1 form the *Lipschitz matrix* $\Lambda \in \mathbb{R}^{n\times n}$ where the entry $\Lambda_{ij} = 0$ if $i \notin \mathcal{I}_j$.

Definition 2. The superposition $\mathcal{F} = F \circ T$ generates the dynamical system $(\mathcal{F}, X)$ which is a *dynamical network*.

To each dynamical network $(\mathcal{F}, X)$ there is an associated directed graph $\Gamma_{\mathcal{F}} = (V, E)$ called the network's *graph of interactions* with *vertex set* V and *edge set* E. For $V = \{v_1, \ldots, v_n\}$ we let e_{ij} denote the edge from vertex v_i to v_j.

Definition 3. Let $F : X \to X$ be an interaction. The graph $\Gamma_F = (V, E)$ with $V = \{v_1, \ldots, v_n\}$ and $E = \{e_{ij} : i \in \mathcal{I}_j,\ j \in \mathcal{I}\}$ is called the *graph of interactions* of F.

We note that each vertex $v_i \in V$ of the graph $\Gamma_F = (V, E, \omega)$ corresponds to the ith coordinate of the dynamical network $(\mathcal{F}, X)$. Moreover, there is an edge $e_{ij} \in E$ if and only if the jth coordinate of the interaction $F(\mathbf{x})$ depends on the ith coordinate of $\mathbf{x}$.

Remark 1. Suppose the interaction $F : X \to X$ is continuously differentiable and each $X_i \subseteq \mathbb{R}$. If DF is the matrix of first partial derivatives of F then the constants

$$\Lambda_{ij} = \max_{\mathbf{x}\in X} |(DF)_{ji}(\mathbf{x})|$$

satisfy condition (2) for the interaction F.

Example 1. Let $F : [0,1]^3 \to [0,1]^3$ be the interaction given by

$$F(\mathbf{x}) = \begin{bmatrix} 1 - x_1x_2/c \\ 1 - x_1x_3/c \\ 1 - x_2x_3/c \end{bmatrix} \quad \text{for } c \geq 1. \tag{3}$$

Using the Lipschitz constants $\Lambda_{ij} = \max_{\mathbf{x}\in X} |(DF)_{ji}(\mathbf{x})|$ the interaction F has the Lipschitz matrix

$$\Lambda = \begin{bmatrix} c^{-1} & c^{-1} & 0 \\ c^{-1} & 0 & c^{-1} \\ 0 & c^{-1} & c^{-1} \end{bmatrix}. \tag{4}$$

The graph of interactions Γ_F of F is the graph shown in Fig. 1.

2.1. *Stability of dynamical networks*

In this section we consider under what conditions a dynamical network can be shown to have a globally attracting fixed point.

Definition 4. The dynamical network $(\mathcal{F}, X)$ has a *globally attracting fixed point* $\tilde{\mathbf{x}} \in X$ if for any $\mathbf{x} \in X$,

$$\lim_{k\to\infty} d_{max}\big(\mathcal{F}^k(\mathbf{x}), \tilde{\mathbf{x}}\big) = 0.$$

If $(\mathcal{F}, X)$ has a globally attracting fixed point we say it is *globally stable*.

Recall that the constants L_i and Λ_{ij} come from (1) and (2) for the local systems (T, X) and interaction F respectively. For the dynamical network $(\mathcal{F}, X)$ we define the matrix $M_{\mathcal{F}} = \Lambda^T \cdot diag[L_1, \ldots, L_n]$. That is,

$$M_{\mathcal{F}} = \begin{pmatrix} \Lambda_{11}L_1 & \ldots & \Lambda_{n1}L_n \\ \vdots & \ddots & \vdots \\ \Lambda_{1n}L_1 & \cdots & \Lambda_{nn}L_n \end{pmatrix}.$$

We let $\rho(M_{\mathcal{F}})$ denote the *spectral radius* of the matrix Λ, i.e. if $\sigma(M_{\mathcal{F}})$ are the eigenvalues of Λ then

$$\rho(M_{\mathcal{F}}) = \max\{|\lambda| : \lambda \in \sigma(M_{\mathcal{F}})\}.$$

Theorem 1. *If $\rho(M_{\mathcal{F}}) < 1$ then the dynamical network $(\mathcal{F}, X)$ has a globally attracting fixed point.*

Example 2. Let $T_i : [0,1] \to [0,1]$ be the logistic map $T_i(x_i) = 4x_i(1 - x_i)$ for $1 \leq i \leq 3$. One can check that the constants $L_i = 4$ satisfies equation (1) for each local system T_i.

Consider the interaction $F : [0,1]^3 \to [0,1]^3$ given by (3) in example 1. Using the Lipschitz matrix Λ given by (4) the dynamical network $(\mathcal{F}, X)$ for $\mathcal{F} = F \circ T$ has the Lipshitz matrix

$$M_{\mathcal{F}} = \begin{bmatrix} 4/c & 4/c & 0 \\ 4/c & 0 & 4/c \\ 0 & 4/c & 4/c \end{bmatrix}.$$

As $\rho(M_{\mathcal{F}}) = 8/c$ then the dynamical network $(\mathcal{F}, X)$ has a globally attracting fixed point for any $c > 8$.

3. Dynamical Network Expansions

In this section we consider whether it is possible to transform a dynamical network while maintaining the dynamic properties of the network. The goal is to show (i) that such transformations exist and (ii) that these transforms allow for improved estimates on whether the original (untransformed) dynamical network has a unique global attractor.

As previously defined, a dynamical network $\mathcal{F} = F \circ T$ is the composition of the network's local

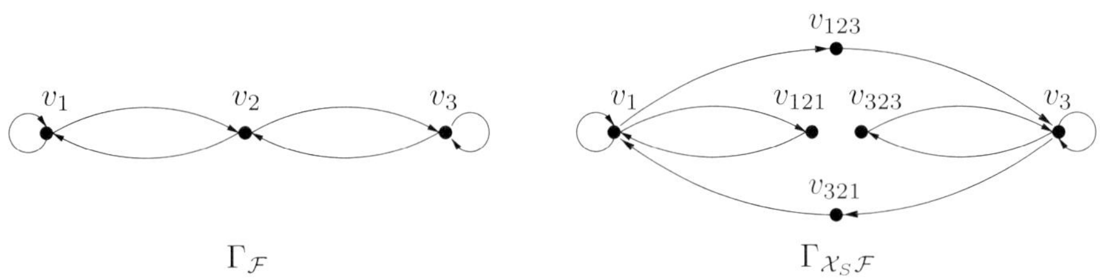

Fig. 1. The graph of interactions $\Gamma_{\mathcal{F}}$ and $\Gamma_{\mathcal{X}_S\mathcal{F}}$ of $(\mathcal{F}, X)$ and $(\mathcal{X}_S\mathcal{F}, X_S)$ in examples 1 and 3.

dynamics T and the interaction F. However, if the system has no local dynamics, i.e. $T = id$ is the identity map, then the dynamical network $\mathcal{F}$ is simply the interaction F.

Conversely, any composition $\mathcal{F} = F \circ T$ can be considered to be an interaction. Writing $\mathcal{F} = (F \circ T) \circ id$ the dynamical network $(\mathcal{F}, X)$ is simply the interaction $\mathcal{F} = F \circ T$ with no local dynamics.

Remark 2. If $(\mathcal{F}, X)$ is considered as a dynamical network with no local dynamics then $\mathcal{M}_{\mathcal{F}} = \Lambda$ for any constants Λ_{ij} satisfying (2) for $\mathcal{F}$. Hence, $(\mathcal{F}, X)$ has a globally attracting fixed point if $\rho(\Lambda) < 1$.

Suppose $(\mathcal{F}, X)$ is a dynamical network where $(\mathcal{F}, X)$ is considered as a dynamical network with no local dynamics. That is, the component

$$\mathcal{F}_j : \bigoplus_{i \in \mathcal{I}_j} X_i \to X \quad \text{for} \quad 1 \leq j \leq n.$$

Alternatively we write

$$\mathcal{F}_j(\mathbf{x}|_{\mathcal{I}_j}) = \mathcal{F}_j(x_{j_1}, \ldots, x_{j_m}), \quad \text{where}$$
$$\mathcal{I}_j = \{j_1, \ldots, j_m\}.$$

Note that if we *replace* the variable x_{j_i} of $\mathcal{F}_j$ by the function $f(y_1, \ldots, y_k)$ the result is the function

$$\mathcal{F}_j(x_{j_1}, \ldots, x_{j_{i-1}}, f(y_1, \ldots, y_k), x_{j_{i+1}}, \ldots, x_{j_m})$$

having variables $x_{j_1}, \ldots, x_{j_{\ell-1}}, y_1, \ldots, y_k, x_{j_{\ell+1}}, \ldots, x_{j_m}$. Additionally, if the sequence $\underline{\gamma} = \ell_1, \ldots, \ell_N$ let

$$\mathcal{F}_{j;\underline{\gamma}}(x_{j_1}, \ldots, x_{j_m}) = \mathcal{F}_j(x_{j_1,\underline{\gamma}}, \ldots, x_{j_m,\underline{\gamma}}).$$

That is, the variables of the function $\mathcal{F}_{j;\underline{\gamma}}$ are indexed by the sequences $j_i, \ell_1, \ldots, \ell_N$ for $1 \leq i \leq m$.

As $\mathcal{F}$ is an interaction the graph $\Gamma_{\mathcal{F}} = (V, E)$ where $e_{ij} \in E$ if x_i is a variable of $\mathcal{F}_j$. A *path* P in the graph $\Gamma_{\mathcal{F}}$ is an ordered sequence of distinct vertices $v_1, \ldots, v_m \in V$ such that $e_{i,i+1} \in E$ for $1 \leq i \leq m-1$. We call the vertices $v_2, \ldots, v_{m-1}$ of P the *interior* vertices of P. If the vertices v_1 and v_m are the same then P is a *cycle*.

Definition 5. For $\Gamma_{\mathcal{F}} = (V, E)$ let $S \subseteq V$. The set S is a *complete structural set* of $\Gamma_{\mathcal{F}}$ if

(i) each cycle of $\Gamma_{\mathcal{F}}$ contains a vertex in S; and
(ii) each vertex $v_k \in V$ belongs to a path $v_i, \ldots, v_j$ where $v_i, v_j \in S$.

Let $st_0(\Gamma_{\mathcal{F}})$ denote the set of all complete structural sets of the graph $\Gamma_{\mathcal{F}}$. Moreover, if the complete structural set $S = \{v_1, \ldots, v_m\}$ let $\mathcal{I}_S = \{1, \ldots, m\}$ denote the *index set* of S.

Let $(\mathcal{F}, X)$ be a dynamical network with graph of interactions $\Gamma_{\mathcal{F}} = (V, E)$ and suppose $S \in st_0(\Gamma_{\mathcal{F}})$. For $j \in \mathcal{I}_S$ let $\mathcal{F}_{\langle j,1\rangle}$ be the function

$$\mathcal{F}_j = \mathcal{F}_j(x_{j_1}, \ldots, x_{j_m})$$

in which each variable x_{j_ℓ} is replaced by $x_{j_\ell,j}$ if $j_\ell \notin \mathcal{I}_S$.

For $i > 1$ let $\mathcal{F}_{\langle j,i\rangle}$ be the function

$$\mathcal{F}_{\langle j,i-1\rangle} = \mathcal{F}_{\langle j,i-1\rangle}(x_{\underline{\gamma}_1}, \ldots, x_{\underline{\gamma}_t})$$

in which each $x_{\underline{\gamma}_\ell} = x_{\ell_1,\ldots,\ell_N}$ is replaced by the function $\mathcal{F}_{\ell_1;\underline{\gamma}_\ell}$ if $\ell_1 \notin \mathcal{I}_S$. If $\ell_1 \in \mathcal{I}_S$ for each $1 \leq \ell \leq t$ then define $\mathcal{X}_S\mathcal{F}_j = \mathcal{F}_{\langle j,i-1\rangle}$.

Definition 6. For $S \in st_0(\Gamma_{\mathcal{F}})$ let $\mathcal{B}_S(\Gamma_{\mathcal{F}})$ be the set of paths and cycles of $\Gamma_{\mathcal{F}}$ that begin and end with a vertex of S having no interior vertices in S. We say

$$\mathcal{A}_S(\mathcal{F}) = \{\ell_1, \ldots, \ell_N : v_{\ell_1}, \ldots, v_{\ell_N} \in \mathcal{B}_S(\Gamma_{\mathcal{F}}), \\ N > 2\} \tag{5}$$

is the set of *admissible sequences* of $\mathcal{F}$ with respect to S.

Let $\underline{\gamma} = \ell_1, \ldots, \ell_N \in \mathcal{A}_S(\mathcal{F})$. For $1 < i < |\underline{\gamma}| = N$ define the $N-2$ spaces

$$X_{i;\underline{\gamma}} = X_{\ell_1}.$$

Additionally, define the functions

$$\mathcal{X}_S\mathcal{F}_{i;\underline{\gamma}}(x_{i-1,\underline{\gamma}}) = x_{i-1,\underline{\gamma}}.$$

By way of notation we let

$$X_{N-1,\underline{\gamma}} = X_{\underline{\gamma}}, \; \mathcal{X}_S\mathcal{F}_{N-1,\underline{\gamma}} = \mathcal{X}_S\mathcal{F}_{\underline{\gamma}}, \text{ and } x_{1,\underline{\gamma}} = x_{\ell_1}.$$

Definition 7. Suppose $S \in st_0(\Gamma_{\mathcal{F}})$. Let

$$\mathcal{X}_S\mathcal{F} = \left(\bigoplus_{j\in\mathcal{I}_S} \mathcal{X}_S\mathcal{F}_j\right) \oplus \left(\bigoplus_{\substack{\underline{\gamma}\in\mathcal{A}_S(\mathcal{F})\\ 1<i<|\underline{\gamma}|}} \mathcal{X}_S\mathcal{F}_{i;\underline{\gamma}}\right).$$

and

$$X_S = \left(\bigoplus_{j\in\mathcal{I}_S} X_j\right) \oplus \left(\bigoplus_{\substack{\underline{\gamma}\in\mathcal{A}_S(\mathcal{F})\\ 1<i<|\underline{\gamma}|}} X_{i;\underline{\gamma}}\right).$$

For the function $\mathcal{X}_S\mathcal{F} : X_S \to X_S$ the dynamical network $(\mathcal{X}_S\mathcal{F}, X)$ is called the *dynamical network expansion* of $(\mathcal{F}, X)$ with respect to S.

Example 3. Consider the dynamical network $(\mathcal{F}, X)$ where

$$\mathcal{F}(\mathbf{x}) = \begin{bmatrix} \mathcal{F}_1(x_1, x_2) \\ \mathcal{F}_2(x_1, x_3) \\ \mathcal{F}_3(x_2, x_3) \end{bmatrix}.$$

The network has the graph of interactions $\Gamma_{\mathcal{F}} = (V, E)$ shown in Fig. 1 (left). Moreover, the set $S = \{v_1, v_3\}$ is a complete structural set of $\Gamma_{\mathcal{F}}$. To see this note that the cycles of $\Gamma_{\mathcal{F}}$ form the set

$$\{v_1, v_1; v_3, v_3; v_1, v_2, v_1; v_1, v_2, v_3; v_3, v_2, v_1; v_3, v_2, v_3\}$$

As (i) each cycle of $\Gamma_{\mathcal{F}}$ contains either v_1 or v_3 and (ii) each vertex in V belongs to the path v_1, v_2, v_3 then $S \in st_0(\Gamma_{\mathcal{F}})$. Hence, the expansion $(\mathcal{X}_S\mathcal{F}, X)$ is well defined.

To construct this expansion we first consider the components of $\mathcal{F}$ indexed by $\mathcal{I}_S = \{1, 3\}$. For $\mathcal{F}_1 = \mathcal{F}_1(x_1, x_2)$ note that the variable x_2 is indexed by an element not in $\mathcal{I}_S$. Hence, $\mathcal{F}_{\langle 1,1\rangle} = \mathcal{F}_1(x_1, x_{21})$. Replacing x_{21} by the function $\mathcal{F}_{2;21} = \mathcal{F}_2(x_{121}, x_{321})$ yields $\mathcal{F}_{\langle 1,2\rangle} = \mathcal{F}_1(x_1, \mathcal{F}_2(x_{121}, x_{321}))$. Since each variable of $\mathcal{F}_{\langle 1,2\rangle}$ is indexed by a sequence beginning with an element of $\mathcal{I}_S$ then

$$\mathcal{X}_S\mathcal{F}_1 = \mathcal{F}_1(x_1, \mathcal{F}_2(x_{121}, x_{321})).$$

Similarly, the function $\mathcal{X}_S\mathcal{F}_3$ can be shown to be

$$\mathcal{X}_S\mathcal{F}_3 = \mathcal{F}_3(\mathcal{F}_2(x_{123}, x_{323}), x_3).$$

Note that the set $\mathcal{B}_S(\Gamma_{\mathcal{F}}) = \{v_i, v_i; v_i, v_2, v_i; v_i, v_2, v_j : i, j \in \mathcal{I}_S\}$ implying

$$\mathcal{A}_S(\mathcal{F}) = \{121, 123, 321, 323\}.$$

Then for any $\underline{\gamma} \in \mathcal{A}_S(\mathcal{F})$ there is a single function $\mathcal{X}_S(\mathcal{F}_{2,\underline{\gamma}}$ corresponding to $\underline{\gamma}$ given by

$$\mathcal{X}_S\mathcal{F}_{\underline{\gamma}}(x_i) = x_i \quad \text{for} \quad \underline{\gamma} = i, 2, j.$$

By Definition 7 the expansion $\mathcal{X}_S\mathcal{F}$ is given by

$$\mathcal{X}_S\mathcal{F}(\mathbf{x}) = \begin{bmatrix} \mathcal{X}_S\mathcal{F}_1(x_1, x_{121}, x_{321}) \\ \mathcal{X}_S\mathcal{F}_3(x_{123}, x_{323}, x_3) \\ \mathcal{X}_S\mathcal{F}_{121}(x_1) \\ \mathcal{X}_S\mathcal{F}_{123}(x_1) \\ \mathcal{X}_S\mathcal{F}_{321}(x_3) \\ \mathcal{X}_S\mathcal{F}_{323}(x_3) \end{bmatrix}$$

$$= \begin{bmatrix} \mathcal{F}_1(x_1, \mathcal{F}_2(x_{121}, x_{321})) \\ \mathcal{F}_3(\mathcal{F}_2(x_{123}, x_{323}), x_3) \\ x_1 \\ x_1 \\ x_3 \\ x_3 \end{bmatrix} \tag{6}$$

where $X_S = (X_1 \oplus X_3) \oplus (X_{121} \oplus X_{321} \oplus X_{123} \oplus X_{323})$ The graph of interactions $\Gamma_{\mathcal{X}_S\mathcal{F}}$ is shown in Fig. 1 (right).

A natural question to ask is whether the expansion $(\mathcal{X}_S\mathcal{F}, X_S)$ and the initial (*real*) dynamical network $(\mathcal{F}, X)$ have similar dynamics.

Theorem 2. *Suppose $(\mathcal{X}_S\mathcal{F}, X_S)$ is a dynamical network expansion of $(\mathcal{F}, X)$. If $(\mathcal{X}_S\mathcal{F}, X_S)$ has a globally attracting fixed point then $(\mathcal{F}, X)$ also has a globally attracting fixed point.*

In particular, suppose $\mathbf{y}_S \in X_S$ is a globally attracting fixed point of $(\mathcal{X}_S\mathcal{F}, X_S)$. Then the globally attracting fixed point $\mathbf{y} \in X$ of $(\mathcal{F}, X)$ has the property

$$y_j = (\mathbf{y}_S)_j \quad \text{for} \quad j \in \mathcal{I}_S$$

from which the other coordinates of $\mathbf{y}$ can be easily deduced.

The following is an immediate corollary to Theorem 2 and Remark 36.1.

Corollary 1. *Let $(\mathcal{X}_S\mathcal{F}, X_S)$ be a dynamical network expansion of $(\mathcal{F}, X)$. Suppose the constants $\tilde{\Lambda}_{ij}$ satisfy (2) for $\mathcal{X}_S\mathcal{F}$. If $\rho(\tilde{\Lambda}) < 1$ then $(\mathcal{F}, X)$ has a globally attracting fixed point.*

The question then is whether there is any advantage in considering a dynamical network expansion over the original unexpanded network. As it turns out, dynamical network expansions always allow for better estimates (or at least no worse) of a dynamical network's global stability.

Theorem 3. (***Improved Stability Estimates for Dynamical Networks***) *Let $(\mathcal{X}_S\mathcal{F}, X_S)$ be an* *expansion of* $(\mathcal{F}, X)$. *Suppose there exist constants* Λ_{ij} *satisfying (2) for* $\mathcal{F}$. *Then there are constants* $\tilde{\Lambda}_{ij}$ *satisfying (2) for* $\mathcal{X}_S\mathcal{F}$ *such that* $\rho(\tilde{\Lambda}) \leq \rho(\Lambda)$.

Example 4. Let $(\mathcal{F}, X)$ be the dynamical network given in Example 2 once again considered as a network without local dynamics. That is,

$$\mathcal{F}(\mathbf{x}) = \begin{bmatrix} 1 - T_1(x_1)T_2(x_2)/\alpha \\ 1 - T_1(x_1)T_3(x_3)/\alpha \\ 1 - T_2(x_2)T_3(x_3)/\alpha \end{bmatrix} \quad \text{for } \alpha \geq 1$$

where $T_i(x_i) = 4x_i(1 - x_i)$ and $X = [0, 1]^3$.

Note that the system $(\mathcal{F}, X)$ has the same form as the dynamical network in Example 3. Hence, for $S = \{v_1, v_3\}$ the expansion $(\mathcal{X}_S\mathcal{F}, X_S)$ given by Eq. (6) is

$$\mathcal{X}_S\mathcal{F}(\mathbf{x}) = \begin{bmatrix} \mathcal{X}_S\mathcal{F}_1 \\ \mathcal{X}_S\mathcal{F}_3 \\ \mathcal{X}_S\mathcal{F}_{121} \\ \mathcal{X}_S\mathcal{F}_{123} \\ \mathcal{X}_S\mathcal{F}_{321} \\ \mathcal{X}_S\mathcal{F}_{323} \end{bmatrix} = \begin{bmatrix} 1 - T_1(x_1)T_2(1 - T_1(x_{121}) \\ \quad T_3(x_{321})/c)/c \\ 1 - T_2(1 - T_1(x_{123})T_3(x_{323})/c) \\ \quad T_3(x_3)/c \\ x_1 \\ x_1 \\ x_3 \\ x_3 \end{bmatrix} \tag{7}$$

Letting $\tilde{\Lambda}_{ij} = \max_{\mathbf{x} \in X_S} |(D\mathcal{F}_S)_{ji}(\mathbf{x})|$ one can compute that

$$\tilde{\Lambda} = \begin{bmatrix} \frac{16(c-1)}{c^3} & 0 & 1 & 1 & 0 & 0 \\ 0 & \frac{16(c-1)}{c^3} & 0 & 0 & 1 & 1 \\ \frac{16}{c^3} & 0 & 0 & 0 & 0 & 0 \\ 0 & \frac{16}{c^3} & 0 & 0 & 0 & 0 \\ \frac{16}{c^3} & 0 & 0 & 0 & 0 & 0 \\ 0 & \frac{16}{c^3} & 0 & 0 & 0 & 0 \end{bmatrix}$$

where the rows (and columns) of $\tilde{\Lambda}$ are numbered as in Eq. (7). The spectral radius of $\tilde{\Lambda}$ is then given by

$$\rho(\tilde{\Lambda}) = \frac{8c - 8 + 4\sqrt{4 + 2c(c^3 + 2c - 4)}}{c^3}$$

As $\rho(\tilde{\Lambda}) < 1$ for $c > 6.75$ then Theorem 3 implies $(\mathcal{F}, X)$ has a globally attracting fixed point for $c > 6.75$. Note that this is better than the estimate gained by direct (standard) analysis of $(\mathcal{F}, X)$ in Example 2.

Importantly, we note that the method of using network expansions generalizes the method used in [1, 2] for determining whether a network has a unique global attractor. As a final observation, in this section, we note that it is possible to sequentially expand a dynamical network and thereby sequentially improve ones estimate of whether the original untransformed network has a globally attracting fixed point.

4. Concluding Remarks

The large majority of real world networks are dynamic. Yet most studies consider either only the structure or the dynamics of these networks. As the spectrum is an important dynamical aspect of a network, isospectral graph transformations provide a way of studying the interplay between the topology of a network and its dynamics.

As a tool for investigating networks, isospectral transformations are quite flexible. Isospectral transformations can be used to reduce or expand the size of a network, considered as a graph, while either modifying or maintaining the edge weights of the network [5]. These *isospectral graph reductions* and *isospectral graph expansion* have the following useful properties and applications.

Isospectral graph reductions allow one the ability to uniquely reduce a network to any subsets of its vertex set while preserving the network spectrum. This in turn allows for the introduction of new equivalence relations on the space of all networks where two networks are equivalent if they can be reduced to the same network via some rule.

From the point of view of applications, such rules can be designed by experimentalist to compare different networks modulo some specific network structure. However, this requires experts in a particular field (biologist, physicist, etc.) to determine an appropriate set of network elements over which to reduce the network.

Isospectral graph expansions can be used to enlarge a network while preserving its sets of edge weights. As a general dynamical system has an associated graph structure (i.e. can be considered as a dynamical network) it is also possible to use an analogous procedure to expand the dynamical system while preserving its dynamics. This can be used to establish whether the system has a globally attracting fixed point.

Acknowledgments

The work of L. B. was partially supported by the NSF grant DMS-0900945.

References

[1] V. Afraimovich and L. Bunimovich. Dynamical networks: interplay of topology, interactions, and local dynamics, *Nonlinearity* **20** (2007) 1761–1771.

[2] Afraimovich V, Bunimovich L, and Moreno S 2010 Dynamical networks: Continuous time and general discrete time models *Regular and Chaotic Dynamics* **15** 129–147.

[3] Albert R and Barabási A-L 2002 Statistical mechanics of complex networks *Rev. Mod. Phys.* **74** 47–97.

[4] L. A. Bunimovich, B. Z. Webb, Isospectral Graph Transformations, Spectral Equivalence, and Global Stability of Dynamical Networks, *Nonlinearity* **25** (2012) 211–254.

[5] L. A. Bunimovich, B. Z. Webb, Isospectral Graph Reductions and Improved Estimates of Matrices' Spectra, *Linear Algebra and its Applications* **437**(7) (2012) 1429–1457.

[6] Chazottes J-R and Fernandes B (ed) 2005 Dynamics of Coupled Map Lattices and Related Spatially Extended Systems *Lect. Notes in Physics* Vol 671, (Berlin: Springer).

[7] Dorogovtsev S and Mendes J 2003 Evolution of Networks: From Biological Networks to the Internet and WWW Oxford: Oxford Univ. Press.

[8] Faloutsos M, Faloutsos P and Faloutsos C 1999 On power-law relationship of the internet topology *ACMSIGCOMM,'99, Comput. Commun. Rev.* **29** 251–263.

[9] Gershgorin S 1931 Über die Abgrenzung der Eigenwerte einer Matrix, *Izv. Akad. Nauk SSSR Ser. Mat.* **1** 749–754.

[10] Newman M, Barabási A-L and Watts D (ed) 2006 The Structure of Dynamic Networks (Princeton: Princeton Univ. Press).

[11] Porter M, Onnela J and Mucha P 2009 Communities in Networks *AMS Notices* **56** 1082–1097, 1164–1166.

[12] Strogatz S 2003 Sync: The Emerging Science of Spontaneous Order (New York: Hyperion).

[13] Varga R 2004 Gershgorin and His Circles (Germany: Springer-Verlag Berlin Heidelberg).

[14] Watts D 1999 Small Worlds: The Dynamics of Networks Between Order and Randomness (Princeton: Princeton Univ. Press).

CHAPTER 20

CHUA'S TIME

ARTURO BUSCARINO, LUIGI FORTUNA and MATTIA FRASCA

DIEEI, University of Catania, Italy

We discuss the relationship between the idea of time of the philosopher Bergson and the concept of time recurrence in chaotic systems. By taking into account the Chua's circuit, we saw that the "Chua's time", i.e., the average recurrence time of trajectories in the Chua's circuit, is able to qualitatively represent the features of the Bergon's time. Numerical and experimental results are presented.

1. Introduction

Henri Bergson (1859–1941) was one of the most influential philosophers of the beginning of the 20th century. A central theme of his thought is the problem of time [1, 2]. He refused the idea of an homogeneous and reversible, quantitative and measurable 'scientific' time, since in his conception what is measured is not the time interval itself, but only a portion of space. He stated that the time interval has not physical meaning, but it is the single consciousness which registers the real duration of time, which is qualitative and heterogeneous, not measurable and irreversible. Time according to Bergson is therefore a continuous flow of unique states of the consciousness. In this sense the concept of Bergson's time includes that of strange attractors before the discovery of these by scientists [3].

Bergson's idea of time is that the duration is not the sequence of instants, since this would imply that only the present could exist and the past could not perpetrate into the present, thus avoiding the possibility of evolution and of a real duration. Bergson retained that the duration is the continuous progress and growth of the past, which preserves itself entirely, so that the past is always present in the human consciousness. The man actions and thoughts are thus dictated by the whole history of the past, even if the man clearly and distinctly perceives only a part of the past. The consequence of this conception of the past is that the consciousness cannot pass two times through the same identical state. Even if the external circumstances are the same, the person which experiences them is not the same, since he/she is in continuous evolution thanks to the irreversibility of the duration and the continuous growth of the experience. The idea of Bergson about time is therefore that it is dynamic and not static. In this sense, the idea of Bergson is anticipated by the thought of St. Augustine, who conceived the time as a dynamic reality that is an extension of the soul itself.

Bergson's view on time can find analogies in the dynamical behavior occurring in chaotic systems. In the definition of strange attractors [4, 5], one of the key elements beyond the topologically transitivity and sensitive dependence on initial conditions, is the fact that points in the attractor are dense. This concept seems to find an analogy in the idea of Bergson that the consciousness cannot pass two times through the same identical state. Starting from these considerations and taking into account the importance of recurrences and related methods [6, 7] for investigating chaotic systems, we study the recurrence times of the Chua's circuit and introduce a characteristic index showing the matching between Bergson's conception and uniqueness of states in the chaotic Chua's circuit.

This Chapter is organized as follows. In Sec. 2 the definition of the characteristic index for recurrence is discussed. In Sec. 3 the experimental setup is briefly described. In Sec. 4 the results of the calculation of this index on the Chua's circuit are reported. In Sec. 5 conclusions are drawn.

2. The Average Recurrence Time

Let us consider a chaotic dynamical system defined by

$$\dot{\mathbf{x}} = f(\mathbf{x}) \tag{1}$$

with $\mathbf{x} \in \mathbb{R}^n$ and $f : \mathbb{R}^n \to \mathbb{R}^n$. Let us consider a time series $\mathbf{x}(t_i)$ $(i = 1, \ldots, N)$ extracted by the solution of system (1) with initial conditions $\mathbf{x}(t_0)$. Let us then take into account a general point $\bar{\mathbf{x}}_i = \mathbf{x}(t_i)$ of the trajectory. We start from this point and look when the trajectory comes sufficiently close to $\bar{\mathbf{x}}_i$. More rigorously, let us consider the following condition:

$$\|\mathbf{x}(t_j) - \bar{\mathbf{x}}_i\| < r \tag{2}$$

where $\|\cdot\|$ is the Euclidean norm and $r > 0$. This relationship will be in general satisfied for some $j = 1, \ldots, M$, from which recurrence times can be defined as:

$$\tau_{ij}(r) = \begin{cases} t_j - t_i & \text{if } j = 1 \\ t_j - t_{j-1} & \text{if } j = 2, \ldots, M \end{cases} \tag{3}$$

We then define the average recurrence time as a function of the radius r as follows:

$$\tau(r) = \frac{1}{N}\sum_{i=1}^{N} \frac{1}{M}\sum_{j=1}^{M} \tau_{ij}(r) \tag{4}$$

This parameter is characteristic for each given chaotic attractor. We calculated it, numerically and experimentally, for the Chua's circuit [8, 9] with respect to different values of the parameter α.

3. Experimental Setup

The Chua's circuit represents one of the first examples of an autonomous physical system exhibiting chaos [8]. The circuit was deliberately designed to exhibit aperiodic irregular waveforms. The circuit can be described by the following dimensionless equations:

$$\begin{cases} \dot{x} = \alpha[y - x - h(x)] \\ \dot{y} = x - y + z \\ \dot{z} = -\beta y - \gamma z \end{cases} \tag{5}$$

with $h(x) = m_1 x + 0.5(m_0 - m_1)(|x+1| - |x-1|)$. For $\alpha = 9, \beta = 14.286, \gamma = 0, m_0 = -1/7, m_1 = 2/7$ the so-called double scroll chaotic attractor is obtained. By varying the parameter α different behaviors can be obtained. In particular, in the following we have considered a single scroll attractor obtained for $\alpha = 8.6$, two different limit cycles obtained for $\alpha = 7.2$ and $\alpha = 8.1$ and another double scroll attractor obtained for $\alpha = 10$.

The analysis carried out in this Chapter is based on numerical integration of Eqs. (5), but also on real experiments conducted on a Chua's circuit. After his invention in 1983 [8], many different implementations of the Chua's circuit have been proposed [9]. In this Chapter, we used the implementation based on Cellular Neural/Nonlinear Networks (CNNs), which

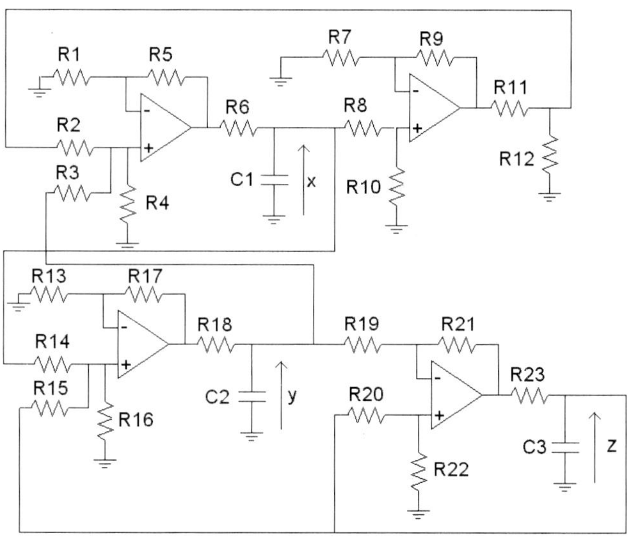

Fig. 1. Electrical scheme of Chua's circuit implementation based on Cellular Neural/Nonlinear Networks.

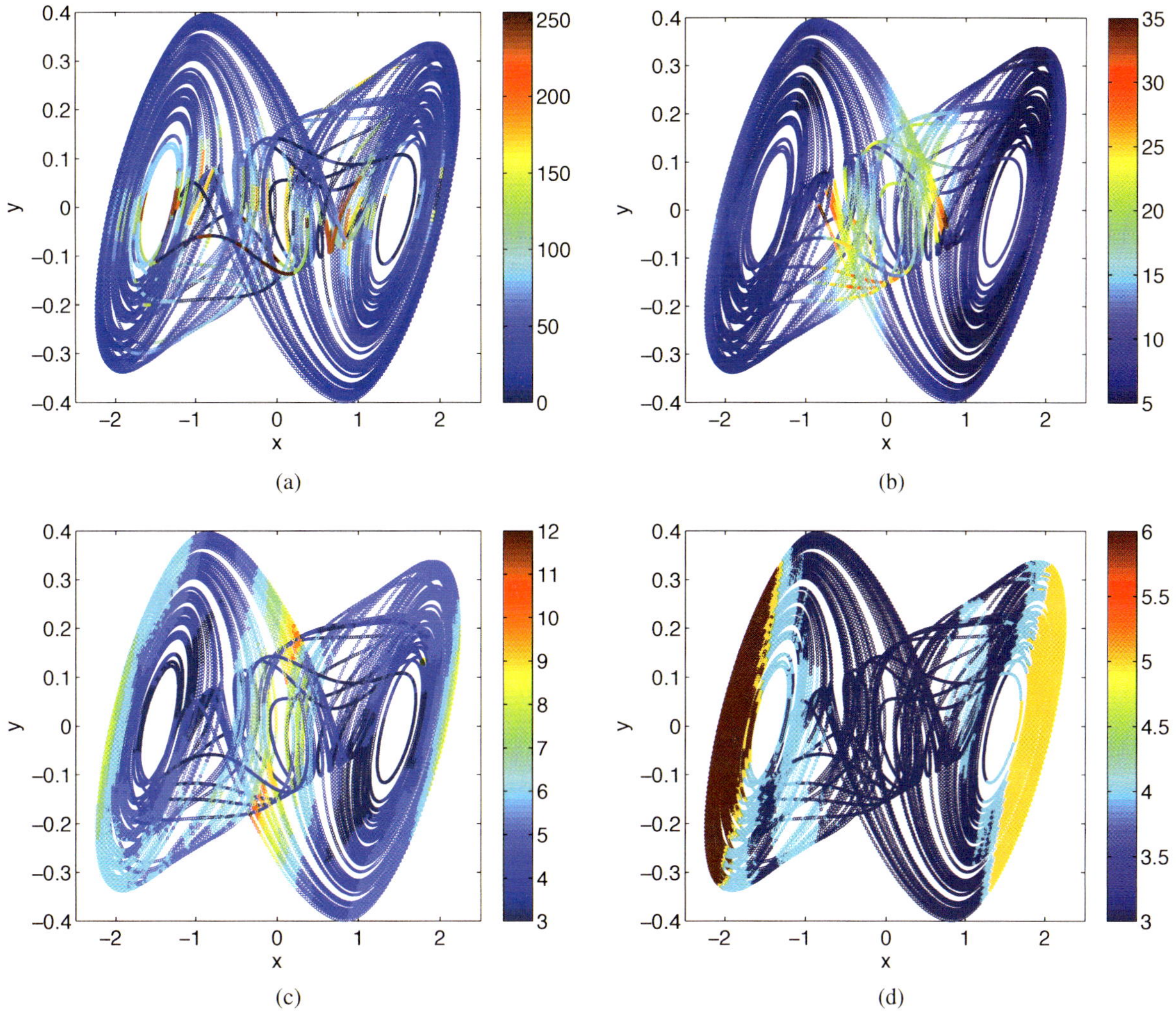

Fig. 2. Projection of the double scroll attractor ($\alpha = 9$) on the $x - y$ plane where color codes the first recurrence time $\tau_{i1}(r, x_i)$ for each calculated point $\bar{x}_i$ of the Chua's circuit attractor. Different values of r have been taken into account: (a) $r = 0.02$, (b) $r = 0.1$, (c) $r = 0.2$, (d) $r = 0.4$.

notably also represents a link between two of the inventions by Chua, the chaotic circuit and CNNs.

The circuit diagram is shown in Fig. 1. It is essentially an implementation based on operational amplifiers and easily to reproduce in an electronics lab. The circuit parameters are linked to those appearing in the mathematical model of Eqs. (5) by the following relationships:

$$\begin{aligned} \alpha &= \frac{R_5 R_{18}}{R_3 R_6} \\ \beta &= \frac{R_{21}}{R_{19}} \\ m_0 &= \frac{R_3}{R_5} - \frac{R_3 R_{21}}{R_2 R_{19}} \frac{R_{23}}{R_{22} + R_{23}} \\ m_1 &= \frac{R_3}{R_5} \end{aligned} \tag{6}$$

while the nonlinearity is given by:

$$\begin{aligned} h(x_1) = {} & \frac{R_3}{R_5} x_1 \\ & - \frac{R_{11}}{R_{11} + R_{12}} \frac{R_9}{R_8} \frac{R_3}{R_2} (|x_1 + 1| - |x_1 - 1|) \end{aligned} \tag{7}$$

Furthermore, $\frac{1}{C_1 R_{18}} = \frac{1}{C_2 R_{18}} = \frac{1}{C_3 R_{23}} = K$ represents a scaling factor of the circuit. The complete list of the circuit components for implementing several dynamical attractors is reported in [9].

4. Average Recurrence Time of the Chua's Circuit

The analysis based on the average recurrence time has been carried out both for the mathematical

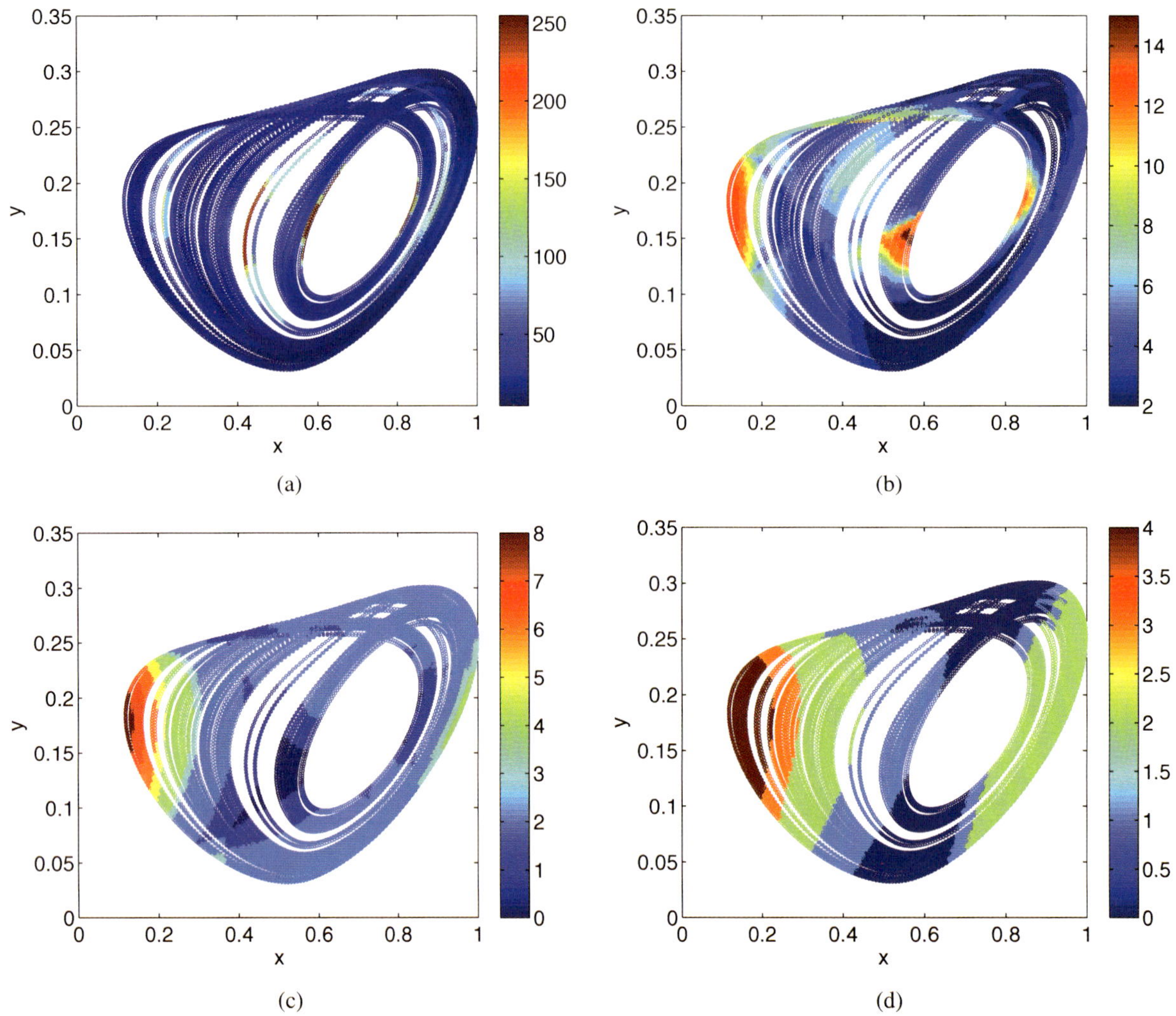

(a) (b) (c) (d)

Fig. 3. Projection of the single scroll attractor on the $x - y$ plane where color codes the first recurrence time $\tau_{i1}(r, \bar{x}_i)$ for each calculated point $\bar{x}_i$ of the Chua's circuit attractor. Different values of r have been taken into account: (a) $r = 0.02$, (b) $r = 0.1$, (c) $r = 0.2$, (d) $r = 0.4$.

model of the Chua's circuit as in Eqs. (5) and for the circuit described in Sec. 3.

The mathematical model of the Chua's circuit (5) has been numerically integrated for $T = 300\,s$ with a fixed integration step $\Delta t = 10\,ms$, thus obtaining $N = 30000$ points. For each point $\bar{x}_i$ in the trajectory the first recurrence time $\tau_{i1}(r)$ has been calculated according to Eq. (3). To explicit the dependence (index i) on the starting point $\bar{x}_i$ of the trajectory, let us now rewrite it as $\tau_{i1}(r) = \tau_{i1}(r, \bar{x}_i)$.

We first studied the first recurrence times for the Chua's double scroll attractor at $\alpha = 9$. The values of $\tau_{i1}(r, \bar{x}_i)$ have been then reported in a color-coded graph, where the position represents the initial starting point $\bar{x}_i$ and the color codes the value of $\tau_{i1}(r, \bar{x}_i)$. Different plots corresponding to different values of r are shown in Fig. 2.

When $r = 0.02$ (Fig. 2) the points in the periphery of the attractor recur after a long time while the inner points have lower recurrence times. Recurrence times are long and distributed along the attractor without particular correlation to their position in the phase plane. Increasing r, the values of $\tau_{i1}(r, \bar{x}_i)$ sensibly reduce. Increasing the value of the threshold r to $r = 0.1$ (Fig. 2) and $r = 0.2$ (Fig. 2), it is more evident the tendency of points in the outer region of the attractor to recur with longer times. Finally, for $r = 0.4$ (Fig. 2) the points far from the axis origin are those with higher recurrence times, even if these values are sensibly decreased with respect to smaller r. Fig. 3 shows another example of the analysis of the first recurrence times. In particular, it refers to the single scroll attractor for $\alpha = 8.6$. The analysis of the first recurrence

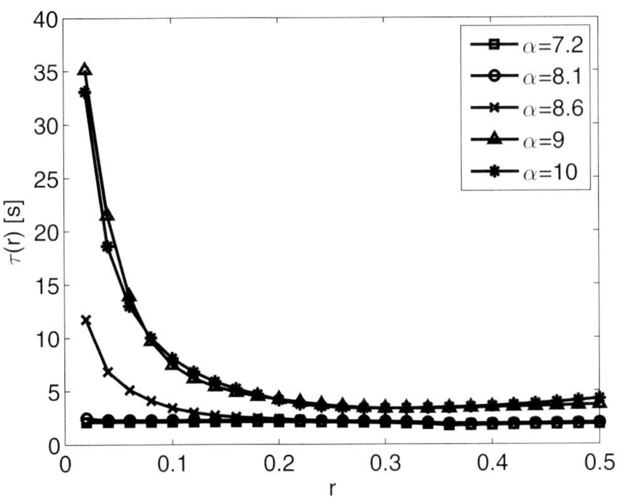

Fig. 4. Average recurrence time $\tau(r)$ with respect to the radius r for different values of α.

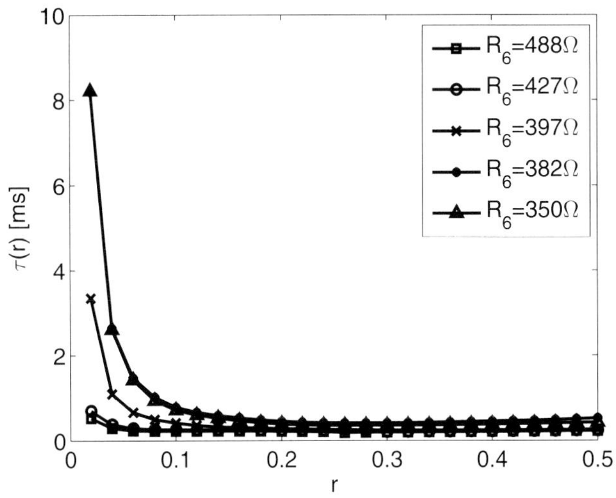

Fig. 5. Experimental results: average recurrence time $\tau(r)$ with respect to the radius r for different values of R_6.

times in the single scroll attractor confirms that, in general, decreasing the threshold r leads to longer times for the first recurrence, and also shows that the first recurrence times are not homogeneous for each point of the attractor.

The parameter $\tau(r)$, defined in Eq. (4) and representing the average recurrence time, is now analyzed. This is shown in Fig. 4 with respect to different values of r and for different values of α. It can be noticed that the trend of $\tau(r)$ is similar for the three chaotic attractors, and as $r \to 0$, the average recurrence time $\tau(r)$ of a chaotic state tends to infinity, while obviously this does not occur for limit cycle attractors. Therefore, when $r \to 0$, chaos allows to obtain a quantitative representation of the uniqueness of the states according to Bergson's view: the time to return back to the same state (or in other terms the duration of time) in fact becomes not measurable.

The same analysis was repeated for experimental data. Data were obtained by acquiring the three state variables of the Chua's circuit in Fig. 1 by using a data acquisition board (National Instruments AT-MIO 1620) with a sampling frequency of $f_s = 100kHz$. They have been then analyzed with the same routines used for the numerically integrated trajectories. Three qualitatively different dynamical behaviors have been obtained by varying the value of the variable resistor R_6. In particular, for $R_6 = 488\Omega$ and $R_6 = 427\Omega$ two limit cycles approximatively corresponding to those obtained for $\alpha = 7.2$ and $\alpha = 8.1$ are generated. For $R_6 = 397\Omega$ (approximatively corresponding to $\alpha = 8.6$) the single scroll attractor is obtained. For $R_6 = 382\Omega$ and $R_6 = 350\Omega$ (approximatively corresponding to $\alpha = 9$ and $\alpha = 10$) two double scroll attractors are obtained. The results are reported in Fig. 5 showing the trend of $\tau(r)$ vs. r. For limit cycle attractors, $\tau(r)$ is roughly constant in the limit of data accuracy. For chaotic attractors, $\tau(r)$ increases as r decreases, thus confirming the analogy between Bergson's time and what can be defined as the *Chua's time*, i.e., the recurrence time of the states of the Chua's circuit.

5. Conclusions

This Chapter starts from a series of considerations made by the philosopher Henri Bergson on the concept of time. Bergson's view on time is that the real duration of time, qualitative, heterogeneous, irreversible, is registered by the consciousness. The consciousness is in fact always influenced by the whole past, so that, even if the external circumstances may be the same, the person which experiences them is not the same and so are their thoughts and actions, since the evolution of his/her consciousness is continuous and irreversible. The consequence of this is that the consciousness cannot pass two times through the same identical state, and that each point of the continuous flow of the consciousness is unique.

Thinking at chaotic systems, many of these concepts expressed by Bergson can found an analogy. In fact, strange chaotic attractors are characterized by points which are dense in the phase plane. Thus, looking at a chaotic trajectory, one may have the impression that it returns back to a given point, but the point reached is not exactly the same. So, according to Bergson's thought each point is unique. This can be quantified by taking into account the average recurrence time of the trajectory to a given point. To this aim, we defined in this work the *Chua's time*, as the average recurrence time of a chaotic trajectory generated by the Chua's circuit in chaotic regime. Since the chaotic trajectory will never exactly return to the given point, the Chua's time depends on how large is the neighborhood of the point for which we consider that the trajectory approximately returned to it. This time tends to infinity as far as the neighborhood size is restricted. This represents a mathematical representation of the concept of uniqueness of the states by Bergson. Moreover, since chaotic systems are very sensitive to initial conditions, two trajectories starting from nearby points diverge after a characteristic time ruled by the maximum Lyapunov exponent. This means that the effect of the past on the consciousness according to Bergson is always present and may be very important, i.e., it may lead to totally different actions or decisions.

References

[1] Bergson, H., [1907], *L'Évolution créatrice*, Paris: Les Presses universitaires de France.

[2] Bergson, H., [1889], *Essai sur les donnes immdiates de la conscience*, Paris Alcan.

[3] Lorenz, E. N., [1963], "Deterministic Nonperiodic Flow", *Journal of the Atmospheric Sciences* **20**(2), pp. 130–141.

[4] Strogatz, S. H., [2000], *Nonlinear Dynamics and Chaos*, Cambridge, Massachussetts, Perseus Publishing.

[5] Sprott, J. C., [2003], *Chaos and Time-Series Analysis*, Oxford University Press.

[6] Eckmann, J. P., Kamphorst, S. O., Ruelle, D., [1987], "Recurrence Plots of Dynamical Systems", *Europhysics Letters* 4, pp. 973–977.

[7] Marwan, N., Romano, M. C., Thiel M., Kurths, J., [2007], "Recurrence Plots for the Analysis of Complex Systems", *Physics Reports* 438, pp. 237–329.

[8] Madan, R. N., *Chua's Circuit: A Paradigm for Chaos*, World Scientific Series on Nonlinear Sciences, Series B, Vol. 1, 1993.

[9] Fortuna, L., Frasca, M., Xibilia, M.G., [2009], *Chua's Circuit Implementations: Yesterday, Today, Tomorrow*, World Scientific.

[10] Rössler, O. E., [1976], "An equation for continuous chaos", *Physics Letters A*, 57(5), pp. 397–398.

CHAPTER 21

CHAOTIC NEURAL NETWORKS AND BEYOND

KAZUYUKI AIHARA[*,‡], TAIJI YAMADA[†] and MAKITO OKU[*]

[*]*Institute of Industrial Science, University of Tokyo, 4-6-1 Komaba, Meguro-ku, Tokyo 153-8505, Japan*

[†]*AIHARA Electrical Engineering Co., Ltd.*

A chaotic neuron model which is closely related to deterministic chaos observed experimentally with squid giant axons is explained, and used to construct a chaotic neural network model. Further, such a chaotic neural network is extended to different chaotic models such as a large-scale memory relation network, a locally connected network, a vector-valued network, and a quaternionic-valued neuron.

1. Introduction

Professor Leon Chua has been making great contributions to dynamical systems theory and its applications, including chaotic electrical circuits such as Chua's circuit (e.g., Chua *et al.* [1]), cellular neural networks [2], and memristors [3] that may realize the implementation of plastic circuits similar to biological neural networks. Thus, his achievements cover many disciplines such as nonlinear science, engineering, and neuroscience.

In this chapter, we present another approach that combines chaotic dynamics with neural networks, namely the modelling of chaotic neural networks [4, 5].

The brain is a large-scale nonlinear network composed of neurons and synapses. A seminal mathematical model of a neuron was proposed by McCulloch and Pitts in 1943 [6]. This model uses binary output values (0 and 1, that is, non-firing and firing, respectively), with discrete-time updating, which makes its operation similar to that of a logic element.

The actual behavior of biological neurons is far more complicated and dynamic than that of

[‡]aihara@sat.t.u-tokyo.ac.jp

the McCulloch–Pitts model. In particular, strange attractors with chaotic dynamics and their different bifurcations, or routes to chaos are observed in nerve membranes [5, 7–10] both experimentally (in squid giant axons) and numerically (in nerve equations like the Hodgkin-Huxley equations [11]). In this chapter, we introduce a neural network model composed of neurons with such chaotic neurodynamics [4, 5] and its extended models.

2. A Model of Chaotic Neural Networks

As a simple neuron model that can reproduce chaotic dynamics observed in squid giant axons, a chaotic neuron model has been proposed [4, 5, 12]. This model is based on Caianiello's neuronic equation [13] and the Nagumo–Sato model [14], and is represented as follows:

$$y(t+1) = f\left[A(t) - \alpha\sum_{d=0}^{t} k^d y(t-d) - \theta\right], \quad (1)$$

where $y(t+1)$ is the output of the neuron at $t+1$ and has a continuous value between 0 and 1; t is a discrete time step ($t = 0, 1, 2, \ldots$); f is the sigmoidal output function, e.g., $f\{x(t)\} = 1/\{1 + \exp(-x(t)/\epsilon)\}$ with the steepness parameter ϵ; $A(t)$ is the input strength at t; k is the decay factor; α is the refractory scaling parameter; and θ is the threshold value for firing. The term $\alpha\sum_{d=0}^{t} k^d y(t-d)$ in Eq. (1) represents the accumulation of refractory effects due to a past firing sequence, each effect of which decays exponentially with factor k and is linearly superimposed [14].

By defining the internal state $x(t)$ as $x(t) = A(t) - \alpha\sum_{d=0}^{t} k^d y(t-d) - \theta$, the dynamics of a single chaotic neuron receiving temporally constant stimulation $A(t) = A$ can be described by the following one-dimensional map [4]:

$$x(t+1) = F(x(t)) \equiv kx(t) - \alpha f\{x(t)\} + a, \quad (2)$$

where $a = (A - \theta)(1 - k)$.

Figure 1 shows the graph of the chaotic neuron map, which is bimodal and similar to return plots of the experimental data obtained from squid giant axons [15, 16]. Here, the bimodal map is a piecewise monotone continuous map with two turning points P_1 and P_2 [17]. The map of the chaotic neuron in Fig. 1 has $(+ - +)$ signs marking the slopes on the three monotonic intervals. As the decay factor k in Eq. (2) usually takes a value in

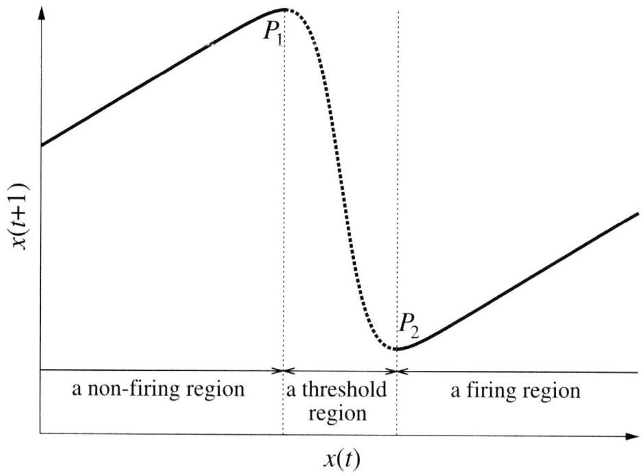

Fig. 1. The graph of the chaotic neuron map.

$0 < k < 1$ to characterize the monotonous decay, the derivatives in the two (+)-regions, which correspond to non-firing and firing, are also between 0 and 1. The derivatives in the middle (−)-region, on the other hand, can be less than −1 if the steepness parameter ϵ is sufficiently small [4]. This middle (−)-branch describes a continuous type of threshold separatrix peculiar to excitable dynamics in nerve membranes [18]; it generates sensitive dependence on initial conditions because of its steep slope of less than −1. Thus, the chaotic neuron model provides a new kind of bimodal map related to excitable nerve dynamics, one that is different from conventional bimodal maps such as cubic maps and circle maps.

Figure 2 shows an example of 1/3-entrained periodic response in the chaotic neuronal map and a squid giant axon, while Fig. 3 shows an example of chaotic response. Figure 4 shows (a) the bifurcation diagram of the internal state $x(t)$, (b) the average firing rate ρ, and (c) the Lyapunov exponent λ when the value of the bifurcation parameter a in Eq. (2) is changed from 0 to 1. Here, the average firing rate ρ and the Lyapunov exponent λ are defined as follows:

$$\rho = \lim_{n\to\infty} \frac{1}{n}\sum_{t=0}^{n-1} y(t), \quad (3)$$

$$\lambda = \lim_{n\to\infty} \frac{1}{n}\sum_{t=0}^{n-1} \ln\left|\frac{dF(x(t))}{dx(t)}\right|. \quad (4)$$

The dynamics of the ith chaotic neuron integrated in a chaotic neural network with M external neurons and N constituent neurons that

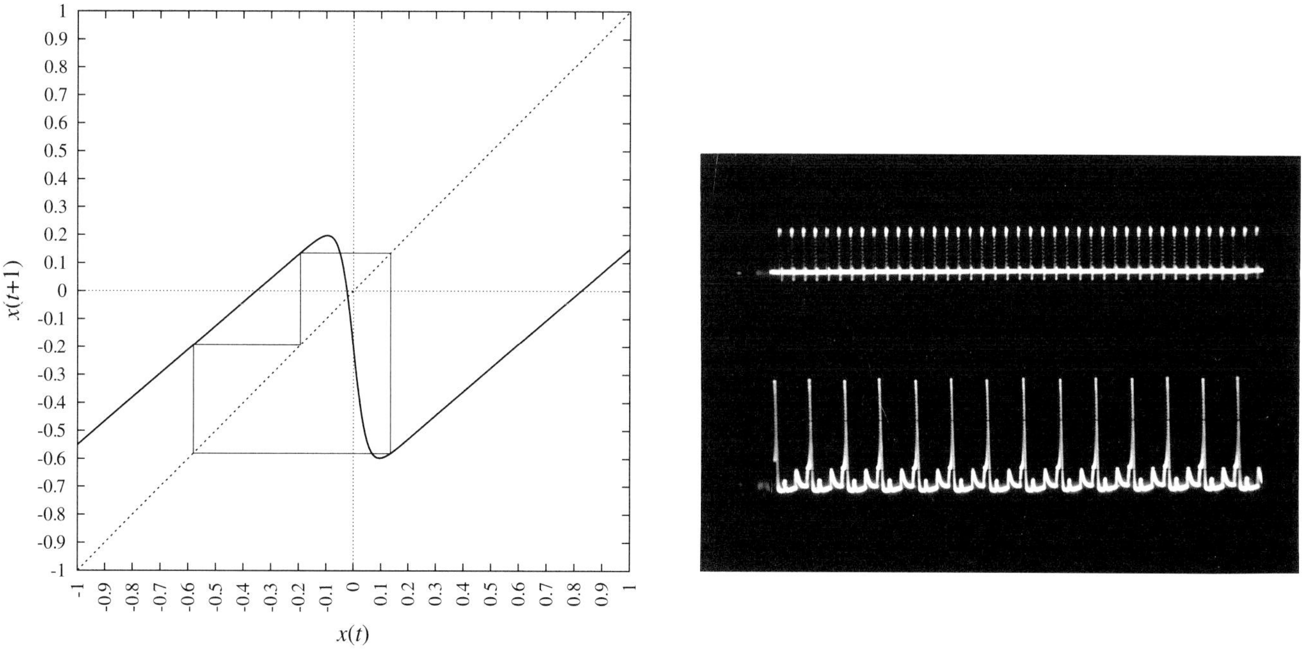

(a) (b)

Fig. 2. 1/3–entrained periodic response (a) in the chaotic neuronal map with $k = 0.85$, $\alpha = 1$, $a = 0.3$ and $\epsilon = 0.025$, and (b) in a squid giant axon initially in the resting state and stimulated by periodic pulses with pulse width 0.5 ms, pulse interval 5.0 ms, and amplitude $A = 7.6$ μA, or 1.297 times the threshold current.

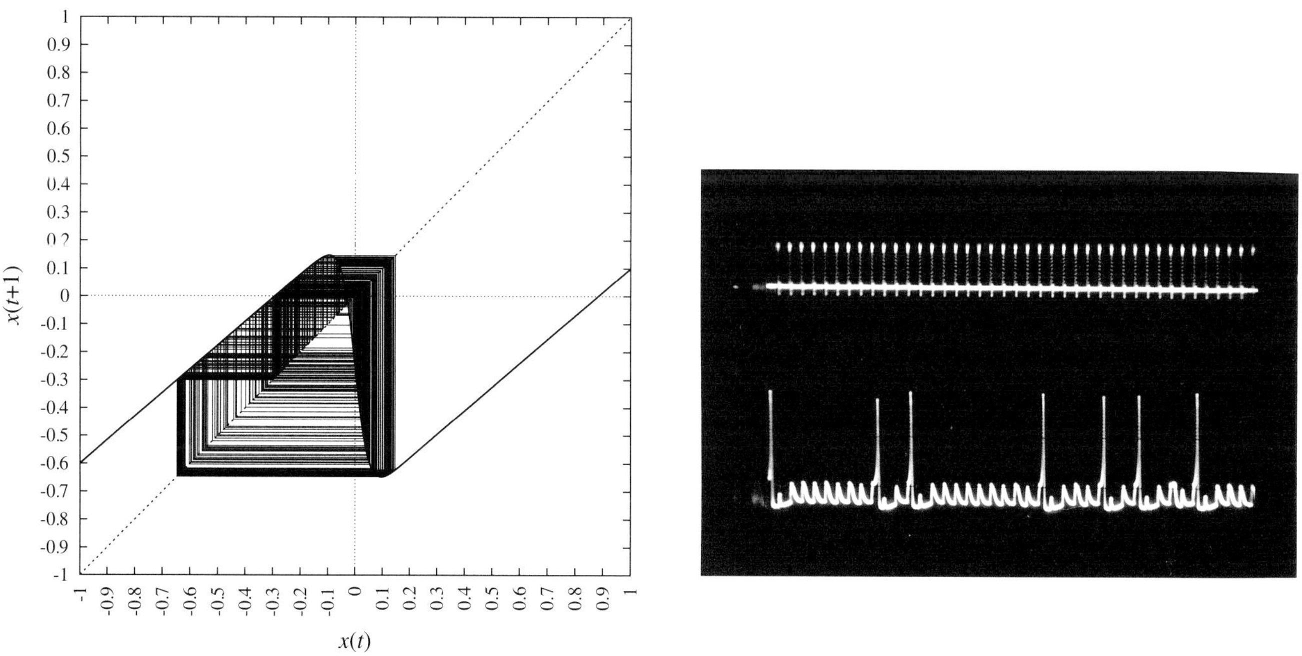

(a) (b)

Fig. 3. Chaotic response with a low firing rate (a) in the chaotic neuronal map with $k = 0.85$, $\alpha = 1$, $a = 0.25$ and $\epsilon = 0.025$, and (b) in a squid giant axon initially in the resting state and stimulated by periodic pulses with pulse width 0.5 ms, pulse interval 5.0 ms, and amplitude $A = 7.52$ μA, or 1.28 times the threshold current.

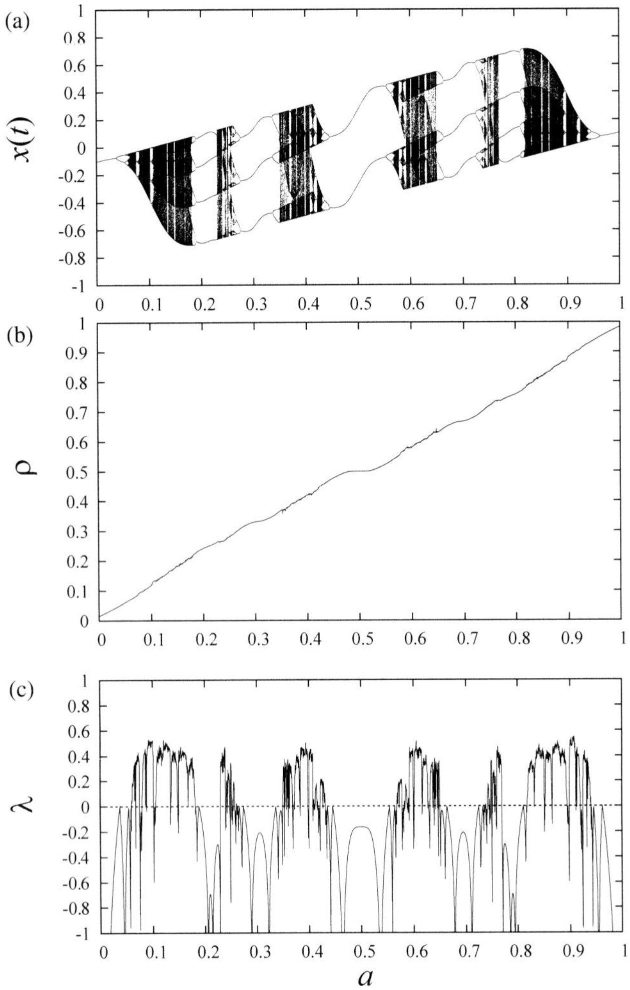

Fig. 4. Response characteristics including (a) the bifurcation diagram, (b) the average firing rate ρ, and (c) the Lyapunov exponent λ in the chaotic neuronal map where $k = 0.85$, $\alpha = 1$, and $\epsilon = 0.025$.

are mutually connected is generally described as follows [4, 19, 20]:

$$y_i(t+1) = f_i\left\{\sum_{j=1}^{M} v_{ij}\sum_{d=0}^{t} k_e^d A_j(t-d) + \sum_{j=1}^{N} w_{ij}\sum_{d=0}^{t} k_f^d h_j\{y_j(t-d)\} - \alpha\sum_{d=0}^{t} k_r^d g_i\{y_i(t-d)\} - \Theta_i\right\}, \quad (5)$$

where $y_i(t+1)$ is the output with a continuous value between 0 and 1 at discrete time $t+1$; f_i is the continuous output function; $A_j(t-d)$ is the strength of the input from the jth external neuron at discrete time $t-d$; v_{ij} and w_{ij} are synaptic weights from the jth external neuron and from the jth constituent neuron, respectively; k_e, k_f, and k_r are the decay factors for the external inputs, the feedback inputs, and the refractoriness, respectively ($0 \leq |k_e|, |k_f|, |k_r| \leq 1$); h_j and g_i are the transfer functions of the axon in the jth neuron for propagating action potentials and the refractory function that represents the strength of the refractory effects caused by the output y_i, respectively, where both h_j and g_i are usually assumed to be the identity functions for the sake of simplicity, i.e., $h_j(y_j) = y_j$ and $g_i(y_i) = y_i$; and α and Θ_i are the refractory scaling parameter and the resting threshold value, respectively.

By defining three internal state variables ξ_i, η_i, and ζ_i (which correspond to the externally applied spatio-temporal inputs from neurons outside the network, the spatio-temporal feedback inputs from the constituent neurons within the network, and the refractoriness, respectively), as $\xi_i(t+1) = \sum_{j=1}^{M} v_{ij}\sum_{d=0}^{t} k_e^d A_j(t-d)$, $\eta_i(t+1) = \sum_{j=1}^{N} w_{ij}\sum_{d=0}^{t} k_f^d y_j(t-d)$, and $\zeta_i(t+1) = -\alpha \sum_{d=0}^{t} k_r^d y_i(t-d) - \Theta_i$, we obtain the following difference equations for chaotic neural networks:

$$\xi_i(t+1) = k_e\xi_i(t) + \sum_{j=1}^{M} v_{ij}A_j(t), \quad (6)$$

$$\eta_i(t+1) = k_f\eta_i(t) + \sum_{j=1}^{N} w_{ij}y_j(t), \quad (7)$$

$$\zeta_i(t+1) = k_r\zeta_i(t) - \alpha y_i(t) - \theta_i, \quad (8)$$

where $\theta_i \equiv \Theta_i(1-k_r)$. The network dynamics is calculated by simultaneous Eqs. (6)–(8) for $i = 1, 2, \ldots, N$, whereas the output $y_i(t+1)$ of the ith neuron is obtained by transforming the internal states to the output value through the output function f_i as $y_i(t+1) = f_i\{\xi_i(t+1) + \eta_i(t+1) + \zeta_i(t+1)\}$.

3. Large-scale Chaotic Neural Networks with Memory Relation Network Structure

The chaotic neural network produces spatio-temporal chaotic dynamics of associative memory [19, 20]. Recently, we have developed simulation techniques and visualization methods for large-scale chaotic neural networks [21–23]. The computational cost associated with large-scale simulations can be reduced through the use of parallel computing and sparse connectivity; it then becomes possible to simulate networks composed of more than one million chaotic neurons, which become more than two millions-dimensional chaotic systems. Furthermore, we have also developed several methods for converting color images into binary codes so that the chaotic neural networks can memorize them. These methods can be used to visualize the network states by mapping them to their corresponding color images.

Such a large-scale chaotic neural network also has a large memory capacity. Thus, many patterns can be memorized by the network. The memory patterns we have analysed had associative relations among them, and formed complex network structure [24, 25]. Here, each memory pattern is considered a node in the network, and the associative relations among these patterns correspond to edges.

To introduce this kind of memory relation network (MRN) structure into the chaotic neural network model, we use hetero-associative connections [26–29]. The original auto-associative connections represent memory patterns themselves, whereas the hetero-associative connections specify the relations among these memory patterns. In addition, signal propagation through hetero-associative connections can involve some delay, as described in the Kleinfeld model [27].

The model equations are modified as follows [30]:

$$\eta_i(t+1) = k_f\,\eta_i(t) + \sum_{j=1}^{N} w_{ij} y_j(t) + \beta \sum_{j=1}^{N} u_{ij} y_j(t-\tau), \tag{9}$$

$$\zeta_i(t+1) = k_r\,\zeta_i(t) - \alpha y_i(t) + a_i, \tag{10}$$

$$y_i(t+1) = f\{\eta_i(t+1) + \zeta_i(t+1)\}, \tag{11}$$

where u_{ij} denotes the synaptic weight of the hetero-associative connection from the jth neuron to the ith one, $\beta \geq 0$ denotes the scaling parameter of these connections, $\tau \in \boldsymbol{N}$ denotes the synaptic delay, and a_i denotes the sum of the threshold and the temporally constant external inputs to the ith neuron.

In the present model, the memory patterns used are binary ones $\{s_1^k, \ldots, s_N^k\}$ ($k = 1, \ldots, K$, $s_i^k \in \{-1, 1\}$). For simplicity, let us assume that each memory pattern contains an equal number of 1's and −1's. Then, the synaptic weights w_{ij} and u_{ij} are determined as follows:

$$w_{ij} = \frac{1}{K} \sum_{k=1}^{K} s_i^k s_j^k, \tag{12}$$

$$u_{ij} = \frac{1}{|S|} \sum_{e_{kl} \in S} s_i^k s_j^l, \tag{13}$$

where S is the edge set of the MRN and e_{kl} is the edge from node l to node k. To reduce the computational cost, we remove some connections such that each neuron receives inputs from only L neurons at random.

To visualize the network state, we use the color images shown in Fig. 5(a) as the memory patterns. The MRN of these patterns includes "one-to-many associations" [31], i.e., a node in the MRN has multiple outgoing edges.

Figure 5(b) shows the actual transition frequencies among the memory patterns. The observed transitions are divided into two types based on whether or not the transition is consistent with the structure of the MRN. We can see that most of the observed transitions are consistent with the given MRN. Furthermore, most of the edges in the MRN are realized at least once; only a small number of edges are unrealized.

Figure 6 shows a typical time series for the decoded chaotic neural network output; many stored images appear successively. The mechanism of chaotic transitions can be explained as follows. Note that the strength of the hetero-associative connections β is small; therefore, the auto-associative connections dominate the dynamics, especially when a particular memory pattern is almost accurately retrieved. However, the network state becomes disordered again due to the accumulated refractoriness, where the influence of the auto-associative connections may be canceled out. At that point, the hetero-associative connections are effective because of the synaptic delay; therefore, they can lead the network dynamics to the next memory pattern. Thus, the hetero-associative connections induce nearly equal biases to all the candidate memory patterns for the next transition, and the chaotic dynamics chooses one memory pattern among them.

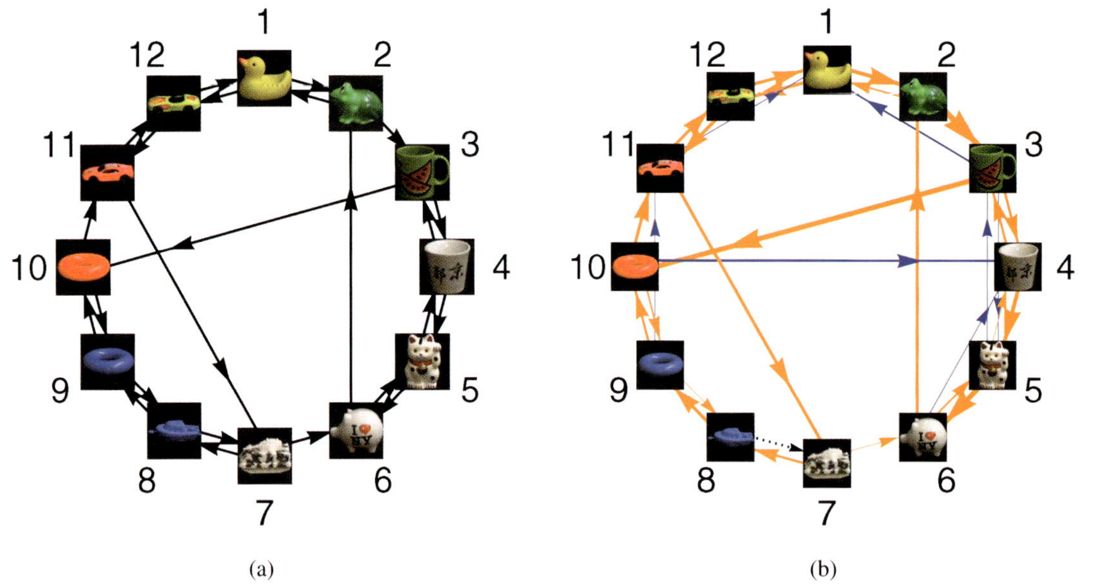

Fig. 5. (a) The 12 color images (128 × 128 pixels, 24-bit RGB mode) stored in the chaotic neural network, and the MRN of these memory patterns. All images were sampled from the Columbia Object Image Library (COIL-100) [32]. (b) Transition frequencies among the memory patterns observed over 2000 steps. The size of the arrows is proportional to the frequency. Orange, blue, and dotted arrows indicate consistent, inconsistent, and unrealized transitions, respectively. Self-loops are omitted.

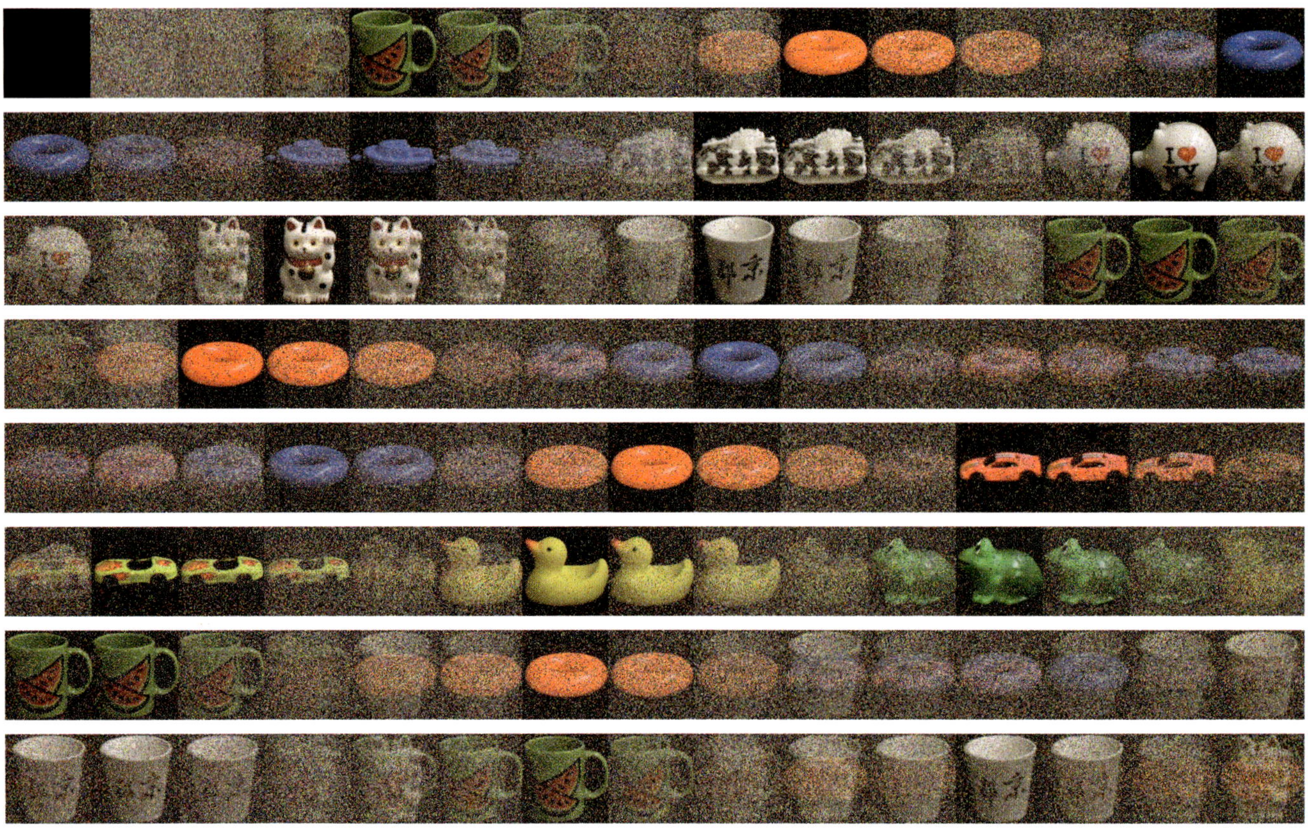

Fig. 6. An example of the time series of the decoded chaotic neural network output displayed at 4-step intervals, where $k_f = 0.8$, $k_r = 0.9$, $\alpha = 12$, $\epsilon = 0.015$, $\beta = 0.1$, $\tau = 10$, and $L = 480$. a_i follows uniform distribution in $[2, 4]$. Perturbation is also applied intermittently to facilitate chaotic transitions among the memory patterns. For more details, see [30].

4. Traveling Waves in Locally Connected Chaotic Neural Networks and Vector-valued Chaotic Neural Networks

The dynamics of a chaotic neural network heavily depends on the network connectivity. We have already considered the case of a sparse, globally connected network in Sec. 3. On the other hand, the cellular neural network [2] and other models [33,34] use a locally connected network architecture in which each neuron is connected only to its neighborhood.

A locally connected chaotic neural network can be constructed as follows [35]. Let us assume that the neurons are arranged in a two-dimensional lattice, and let (l_i, m_i) denote the position of neuron i with $l_i, m_i \in \boldsymbol{N}$. Its neighborhood N_i is defined as $N_i = \{(l, m) \mid \max(|l - l_i|, |m - m_i|) \leq d \text{ for } l, m \in \boldsymbol{N}\}$. Within the area N_i, we select L neurons at random and connect them to neuron i. All other connections are removed by setting the synaptic weights of these connections to 0.

The network shows rich spatio-temporal dynamics characterized by traveling waves. Starting from a random state, the network is initially divided into many small clusters of different memory patterns. These clusters merge to become bigger ones over time, and then the boundaries between the clusters start moving, as shown in Fig. 7. Various kinds of traveling waves such as plane waves, target waves, rotating waves, and spiral waves can be generated locally in space and temporally. The network state continues to change until it eventually converges to a globally coherent state such as a global plane wave with a constant period.

The waves discussed above correspond to moving boundaries of local cell assemblies. Here, cell assemblies refer to particular patterns of active and inactive neurons; these cell assemblies are distinguished by their different colors in Fig. 7. Movement of the boundaries between the assemblies results in the alternation of local activity patterns. It should be noted that the waves disappear if the neurons have no refractoriness. Therefore, the

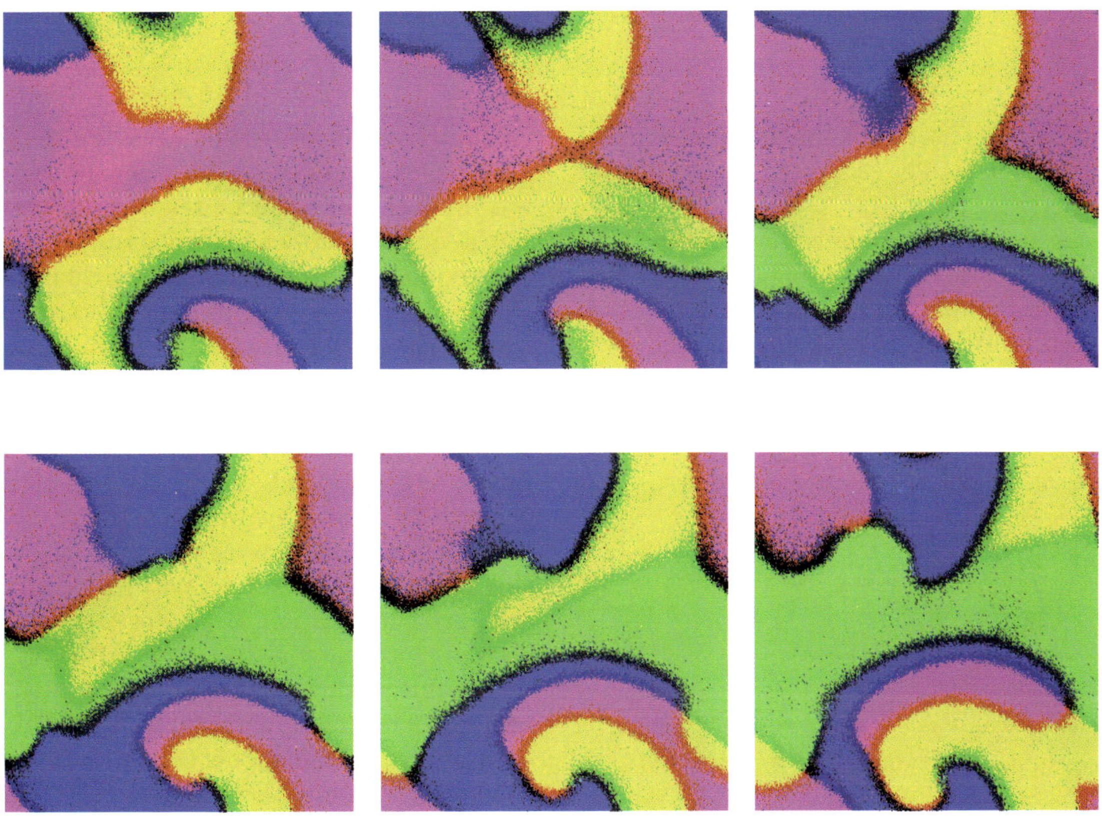

Fig. 7. Spatio-temporal patterns in the locally connected chaotic neural network at 10-step intervals. Four uniformly colored images are used as memory patterns. $k_f = 0.7$, $k_r = 0.9$, $\epsilon = 0.015$, $\alpha = 20$, $a = 6.4$, $d = 10$, and $L = 400$. For more details, see [35].

Fig. 8. Spatio-temporal patterns in the approximation model at 10-step intervals. $k_f = 0$, $k_r = 0.98$, $\epsilon = 1$, $\alpha = 0.2$, $a = 0$, $Q = 4$, and $w = 1$. The network size is 40×40.

refractoriness plays a key role in the emergence of traveling waves in locally connected chaotic neural networks.

We can also approximate the locally connected chaotic neural network using a vector-valued chaotic neural network. The vector-valued chaotic neural network model is a natural extension of the chaotic neural network model, in which neurons transmit vector-valued signals $\{y_i^1, \ldots, y_i^Q\}$ rather than scalar-valued outputs. The dynamics of the approximation model is described as follows:

$$\eta_i^p(t+1) = k_f \eta_i^p(t) + \sum_{j=1}^{N} \sum_{q=1}^{Q} w_{ij}^{pq} y_j^q(t), \tag{14}$$

$$\zeta_i^p(t+1) = k_r \zeta_i^p(t) - \alpha y_i^p(t) + a_i^p, \tag{15}$$

$$y_i^p(t+1) = \frac{\exp((\eta_i^p(t+1) + \zeta_i^p(t+1))/\epsilon)}{\sum_{q=1}^{Q} \exp((\eta_i^q(t+1) + \zeta_i^q(t+1))/\epsilon)}, \quad (i = 1, \ldots, N, \quad p = 1, \ldots, Q). \tag{16}$$

Each vector-valued chaotic neuron represents a small group of scalar-valued neurons in the locally connected chaotic neural network. Thus, the output $\{y_i^1, \ldots, y_i^Q\}$ indicates which memory pattern is retrieved within the local area. Here, Q corresponds to the number of memory patterns. $w_{ij}^{pq} = w > 0$, if vector-valued neurons i and j are neighbors and $p = q$. Otherwise, $w_{ij}^{pq} = 0$. In this study, we use the Moore neighborhood; that is, each vector-valued neuron has eight neighbors. We also set $a_i^p = a$, for simplicity.

Figure 8 shows a typical time series obtained from the approximation model. The model exhibits dynamics qualitatively similar to those of the locally connected chaotic neural network as shown in Fig. 7. For example, spiral waves can be observed in this time series, as well as patterns similar to target waves. In addition, these waves disappear if the vector-valued neurons lose their refractoriness.

In this section, we have demonstrated that the vector-valued chaotic neural network model can serve as an approximation model for the locally connected chaotic neural networks. In the next section, we discuss another extension of the chaotic neuron model in which a neuron transmits quaternionic-valued output rather than real-valued output.

5. A Quaternionic-valued Chaotic Neuron Model

Traditional neuron models using discrete and real-valued numbers have been extended to those that use complex-valued numbers [36–39]. Generally speaking, it is also possible to use quaternions and others.

A quaternionic-valued version of the chaotic neuron model in Eq. (2) is described as follows:

$$x(t+1) = Kx(t) - \alpha f\{x(t)\} + a, \tag{17}$$

where $x, K, a \in \boldsymbol{H}$ and $\alpha \in \boldsymbol{R}$. Referring to Georgiou and Koutsougeras [37], we define the output function $f(x)$ as follows:

$$f(x) = \frac{x}{\beta + |x|}, \tag{18}$$

where $x, \beta \in \boldsymbol{H}$.

Some examples of attractors in the quaternionic-valued chaotic neuron are shown in Fig. 9. It is our future problem to analyse the nonlinear

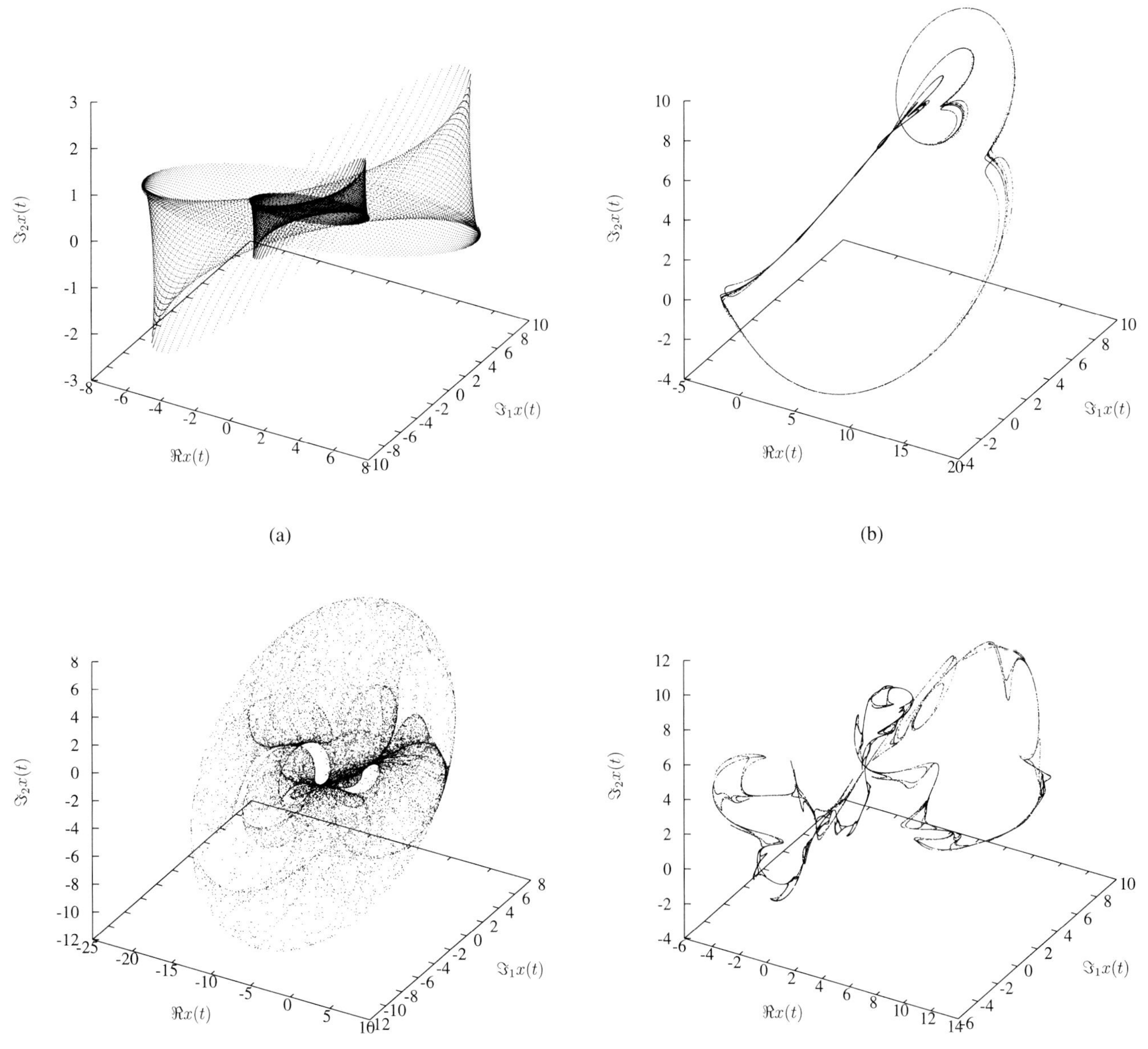

Fig. 9. Attractors of the quaternionic-valued chaotic neuron model with (a) $K = (0.9, 0, 0, 0)$, $\alpha = 10$, $\beta = (-1, -1, -1, -1)$, $a = (0, 0, 0.1, 0)$, $x(0) = (-1, 0, 0, 0)$, (b) $K = (0.9, 0, 0, 0)$, $\alpha = 10$, $\beta = (-1, 0.5, 0.5, 0.5)$, $a = (5, 5, 5, 5)$, $x(0) = (-1, 0, 0, 0)$, (c) $K = (0.9, 0, 0, 0)$, $\alpha = 10$, $\beta = (-1, 0.5, 0.5, 0.5)$, $a = (-6, -2, -2, -2)$, $x(0) = (-1, 0, 0, 0)$, and (d) $K = (0.82, 0.3, 0.15, 0)$, $\alpha = 10$, $\beta = (-1, 0.5, 0.5, 0.5)$, $a = (4, 0, 4, 4)$, $x(0) = (-1, 0, 0, 0)$.

dynamics of the quaternionic-valued chaotic neuron model and its networks in detail, and further extend the model to generalizations of quaternions, such as Clifford algebra.

6. Conclusions

We have presented a model of chaotic neural networks and its several extensions. Although we explained only memory related dynamics as an application, the models of chaotic neural networks have many other applications [40]. In particular, combinatorial optimization is one important application. Optimization dynamics of the chaotic neural networks is clarified both numerically [41, 42] and mathematically [43–45], and practical algorithms are also developed [46–49] (see Ikeguchi *et al.* [50] for a recent review). The formulation of the chaotic neural networks is a natural method to generate transitive dynamics among fixed point pseudo-attractors. Such transitive dynamics is realized by destabilization of fixed point attractors due to refractory effects, or equivalently self-recurrent inhibitory connections peculiar to chaotic neurons. It should be noted that transitive dynamics may be related to higher-brain functions in the prefrontal cortex [51].

Moreover, models of chaotic neural network have been implemented on electronic circuits [52–54]. Our future aim is to analyse spatio-temporal dynamics of extended chaotic neural networks in detail, consider their relations with the brain dynamics, and implement their electronic circuits for practical applications.

Acknowledgments

This research was partially supported by the Aihara Innovative Mathematical Modelling Project, the Japan Society for the Promotion of Science (JSPS) through the "Funding Program for World-Leading Innovative R&D on Science and Technology (FIRST Program)," initiated by the Council for Science and Technology Policy (CSTP). MO was supported by the Japan Society for the Promotion of Science, a Grant-in-Aid for JSPS Fellows (21·937).

References

[1] L. O. Chua, C. W. Wu, A. Huang, and G.-Q. Zhong, A universal circuit for studying and generating chaos. I. routes to chaos, *IEEE Transactions on Circuits and Systems I: Fundamental Theory and Applications.* **40**(10), 732–744 (oct, 1993). ISSN 1057-7122. doi: 10.1109/81.246149.

[2] L. O. Chua and L. Yang, Cellular neural networks: Theory, *IEEE Trans. Circuits Syst.* **35**(10), 1257–1272, (1988).

[3] L. Chua, Memristor-the missing circuit element, *IEEE Transactions on Circuit Theory.* **18**(5), 507–519 (sep, 1971). ISSN 0018-9324. doi: 10.1109/TCT.1971.1083337.

[4] K. Aihara, T. Takabe, and M. Toyoda, Chaotic neural networks, *Physics Letters A.* **144**, 333–340, (1990).

[5] K. Aihara, Chaos in neurons, *Scholarpedia.* **3** (5), 1786, (2008). URL `http://www.scholarpedia.org/article/Chaos_in_neurons`.

[6] W. S. McCulloch and W. H. Pitts, A logical calculus of the ideas immanent in neural nets, *Bulletin of Mathematical Biophysics.* **5**, 115–133, (1943).

[7] G. Matsumoto, K. Aihara, M. Ichikawa, and A. Tasaki, Periodic and nonperiodic responses of membrane potentials in squid giant axons during sinusoidal current stimulation, *Journal of Theoretical Neurobiology.* **3**, 1–14, (1984).

[8] K. Aihara and G. Matsumoto. Chaotic oscillations and bifurcations in squid giant axons. In ed. A. V. Holden, *Chaos*, pp. 257–269. Manchester and Princeton University Presses, Manchester and Princeton, (1986).

[9] K. Aihara, T. Numajiri, G. Matsumoto, and M. Kotani, Structures of attractors in periodically forced neural oscillators, *Physics Letters A.* **116**(7), 313–317, (1986). ISSN 0375-9601. doi: DOI: 10.1016/0375-9601(86)90578-5.

[10] G. Matsumoto, K. Aihara, Y. Hanyu, N. Takahashi, S. Yoshizawa, and J. Nagumo, Chaos and phase locking in normal squid axons, *Physics Letters A.* **123**(4), 162–166, (1987). ISSN 0375-9601. doi: DOI: 10.1016/0375-9601(87)90696-7.

[11] A. L. Hodgkin and A. F. Huxley, A quantitative description of membrane current and its application to conduction and excitation in nerve, *The Journal of Physiology.* **117**(4), 500–544, (1952).

[12] K. Aihara and G. Matsumoto. Forced oscillations and route to chaos in the Hodgkin-Huxley axons and squid giant axons. In eds. A. V. Holden and L. Olsen, *Chaos in Biological Systems*, pp. 121–131. Plenum Publishing, New York, (1987).

[13] E. R. Caianiello, Outline of a theory of thought-processes and thinking machines, *Journal of Theoretical Biology.* **2**, 204–235, (1961).

[14] J. Nagumo and S. Sato, On a response characteristics of a mathematical neuron model, *Kybernetik.* **10**, 155–164, (1972).

[15] A. Mees, K. Aihara, M. Adachi, K. Judd, T. Ikeguchi, and G. Matsumoto, Deterministic

prediction and chaos in squid axon response, *Physics Letters A*. **169**(1–2), 41–45, (1992). ISSN 0375-9601. doi: DOI: 10.1016/0375-9601(92) 90802-S.

[16] K. Aihara. Chaos in axons. In ed. M. Arbib, *The Handbook of Brain Theory and Neural Networks, 1st edition*, pp. 183–185. The MIT Press, Cambridge, Massachusetts, (1995).

[17] W. de Melo and S. van Strien, *One-Dimensional Dynamics*. (Springer-Verlag, 1993).

[18] R. FitzHugh. Mathematical models of excitation and propagation in nerve. In ed. H. Schwan, *Biological Engineering*, chapter 1, pp. 1–85. McGraw–Hill, NY, (1969).

[19] K. Aihara. Chaotic neural networks. In *Bifurcation Phenomena in Nonlinear Systems and Theory of Dynamical Systems*, vol. 710, *RIMS Kokyuroku*, pp. 145–163. Kyoto University, (1989).

[20] M. Adachi and K. Aihara, Associative dynamics in a chaotic neural network, *Neural Networks*. **10**(1), 83–98, (1997). ISSN 0893-6080. doi: DOI: 10.1016/S0893-6080(96)00061-5.

[21] M. Oku, K. Iwayama, K. Tokuda, H. Suzuki, and K. Aihara, Parallel simulation of a large-scale chaotic neural network, *IEICE Tech. Rep.* **109**(269), 55–59 (November, 2009). (in Japanese).

[22] M. Oku and K. Aihara. Associative dynamics of color images in a chaotic neural network. In *Proc. NOLTA'10*, pp. 465–468, (2010).

[23] M. Oku and K. Aihara, Associative dynamics of color images in a large-scale chaotic neural network, *NOLTA, IEICE*. **2**(4), 508–521 (October, 2011). doi: 10.1587/nolta.2.508.

[24] A. M. Collins and M. R. Quillian, Retrieval time from semantic memory, *J. Verb. Learn. Verb. Behav.* **8**(2), 240–247, (1969). ISSN 0022-5371.

[25] Y. Miyashita, Cognitive memory: cellular and network machineries and their top-down control., *Science*. **306**(5695), 435–440 (Oct, 2004). doi: 10.1126/science.1101864.

[26] S. Amari, Learning patterns and pattern sequences by self-organizing nets of threshold elements, *IEEE Transactions on Computers*. **c-21**(11), 1197–1206, (1972).

[27] D. Kleinfeld, Sequential state generation by model neural networks., *Proc. Natl. Acad. Sci. U.S.A.* **83**(24), 9469–9473 (Dec, 1986).

[28] H. Sompolinsky and I. Kanter, Temporal association in asymmetric neural networks, *Phys. Rev. Lett.* **57**(22), 2861–2864 (Dec, 1986).

[29] B. Kosko, Bidirectional associative memories, *IEEE Trans. Syst. Man Cybern.* **18**(1), 49–60, (1988).

[30] M. Oku and K. Aihara. Associative dynamics with structured relations among memories in a chaotic neural network. In *Proc. NOLTA'11*, pp. 366–369, (2011).

[31] Y. Osana, M. Hattori, and M. Hagiwara. Chaotic bidirectional associative memory. In *Proc. ICNN'96*, Vol. 2, pp. 816–821, (1996).

[32] S. A. Nene, S. K. Nayar, and H. Murase. Columbia Object Image Library (COIL-100). Technical Report CUCS-006-96, Department of Computer Science, Columbia University, (1996).

[33] Y. Shinohara, T. Kanamaru, H. Suzuki, T. Horita, and K. Aihara, Array-enhanced coherence resonance and forced dynamics in coupled FitzHugh-Nagumo neurons with noise, *Phys. Rev. E*. **65**(5 Pt 1), 051906 (May, 2002).

[34] N. F. Rulkov, I. Timofeev, and M. Bazhenov, Oscillations in large-scale cortical networks: map-based model, *J. Comput. Neurosci.* **17**(2), 203–223, (2004). doi: 10.1023/B:JCNS.0000037683.55688.7e.

[35] M. Oku and K. Aihara. Traveling waves in locally connected chaotic neural networks and their phenomenological modeling. In *Proc. ICCN'11*, (2011). (in press).

[36] A. Hirose, Ed., *Complex Valued Neural Networks — Theories and Applications*. (World Scientific, 2003).

[37] G. M. Georgiou and C. Koutsougeras, Complex domain backpropagation, *IEEE Transactions on Circuits and Systems II: Analog and Digital Signal Processing*. **39**(5), 330–334 (May, 1992). ISSN 1057-7130. doi: 10.1109/82.142037.

[38] I. Nemoto and K. Saito, A complex-valued version of Nagumo-Sato model of a single neuron and its behavior, *Neural Networks*. **15**(7), 833–853, (2002). ISSN 0893-6080. doi: DOI: 10.1016/S0893-6080(02)00066-7.

[39] G. Tanaka and K. Aihara, Complex-valued multistate associative memory with nonlinear multilevel functions for gray-level image reconstruction, *IEEE Transactions on Neural Networks*. **20**(9), 1463–1473 (Sept., 2009). ISSN 1045-9227. doi: 10.1109/TNN.2009.2025500.

[40] K. Aihara, Chaos engineering and its application to parallel distributed processing with chaotic neural networks, *Proceedings of the IEEE*. **90**, 919–930, (2002).

[41] T. Yamada and K. Aihara, Nonlinear neurodynamics and combinatorial optimization in chaotic neural networks, *Journal of Intelligent and Fuzzy Systems*. **5**, 53–68, (1997).

[42] I. Tokuda, T. Nagashima, and K. Aihara, Global bifurcation structure of chaotic neural networks and its application to traveling salesman problems, *Neural Networks*. **10**(9), 1673–1690, (1997). ISSN 0893-6080. doi: DOI: 10.1016/S0893-6080(97)00023-3.

[43] L. Chen and K. Aihara, Chaos and asymptotical stability in discrete-time neural networks, *Physica D: Nonlinear Phenomena*. **104**(3–4), 286–325, (1997). ISSN 0167-2789. doi: DOI: 10.1016/S0167-2789(96)00302-8.

[44] L. Chen and K. Aihara, Global searching ability of chaotic neural networks, *IEEE Transactions on Circuits and Systems I: Fundamental Theory and Applications.* **46**(8), 974–993, (1999).

[45] L. Chen and K. Aihara, Strange attractors in chaotic neural networks, *IEEE Transactions on Circuits and Systems I: Fundamental Theory and Applications.* **47**(10), 1455–1468, (2000).

[46] L. Chen and K. Aihara, Chaotic simulated annealing by a neural network model with transient chaos, *Neural Networks.* **8**(6), 915–930, (1995). ISSN 0893-6080. doi: DOI: 10.1016/0893-6080(95)00033-V.

[47] M. Hasegawa, T. Ikeguchi, and K. Aihara, Combination of chaotic neurodynamics with the 2-opt algorithm to solve traveling salesman problems, *Phys. Rev. Lett.* **79**(12), 2344–2347 (Sep, 1997). doi: 10.1103/PhysRevLett.79.2344.

[48] I. Tokuda, K. Aihara, and T. Nagashima, Adaptive annealing for chaotic optimization, *Phys. Rev. E.* **58**(4), 5157–5160 (Oct, 1998). doi:10.1103/PhysRev E.58.5157.

[49] M. Hasegawa, T. Ikeguchi, and K. Aihara, Exponential and chaotic neurodynamical tabu searches for quadratic assignment problems, *Control and Cybernetics.* **29**(3), 774–788, (2000).

[50] T. Ikeguchi, M. Hasegawa, T. Kimura, T. Matsuura, and K. Aihara. Theory and applications of chaotic optimization methods. In eds. N. Nedjah, L. Santos Coelho, V. Cocco Mariani, and L. d. Macedo Mourelle, *Innovative Computing Methods and their Applications to Engineering Problems*, Series: Studies in Computational Intelligence, chapter 8. Springer, (2011).

[51] K. Sakamoto, H. Mushiake, N. Saito, K. Aihara, M. Yano, and J. Tanji, Discharge synchrony during the transition of behavioral goal representations encoded by discharge rates of prefrontal neurons, *Cerebral Cortex.* **18**(9), 2036–2045, (2008). doi: 10.1093/cercor/bhm234.

[52] Y. Horio and K. Aihara, Neuron-synapse IC chip-set for large-scale chaotic neural networks, *IEEE Transactions on Neural Networks.* **14**(5), 1393–1404 (sept., 2003).

[53] Y. Horio, T. Ikeguchi, and K. Aihara, A mixed analog/digital chaotic neuro-computer system for quadratic assignment problems, *Neural Networks.* **18**(5–6), 505–513, (2005). ISSN 0893-6080. doi: DOI: 10.1016/j.neunet.2005.06.022. IJCNN 2005.

[54] Y. Horio and K. Aihara, Analog computation through high-dimensional physical chaotic neuro-dynamics, *Physica D: Nonlinear Phenomena.* **237**(9), 1215–1225, (2008). ISSN 0167-2789. doi: DOI: 10.1016/j.physd.2008.01.030. Novel Computing Paradigms: Quo Vadis?

CHAPTER 22

CHAOTIC NEOCORTICAL DYNAMICS

WALTER J. FREEMAN

Department of Molecular & Cell Biology MC 3206,
University of California at Berkeley, CA 94720, USA

The first step of the sensory systems is to construct the meaning of the information they receive from the senses. They do this by generating random noise and then filtering the noise with adaptive filters. We simulate the operation with the solutions of matrices of ordinary differential equations that predict subcritical Hopf bifurcations between point and limit cycle attractors. The second step is integration of the outputs from the several sensory systems into a multisensory percept, called a gestalt, which in the third step is consolidated and stored as knowledge. Simulation of the second step requires use of landscapes of non-convergent *chaotic* attractors. This is not deterministic chaos, which is much too brittle owing to the infinite sensitivity to initial conditions. It is a hybrid form we call *stochastic chaos*, which is stabilized by additive noise modeled on noise sources in the sensory systems. Thus bifurcation and chaos theory provides tools for succinct empirical models of cortical dynamics performing the most basic cognitive operations: generalization, abstraction, and categorization in constructing knowledge. The descriptions are in a form that is suitable for more advanced modeling using analog VLSI, neuropercolation from random graph theory, non-equilibrium dissipative thermodynamics, and macroscopic many-body physics. This review concludes with a summary of the applications of stochastic chaos in pattern classification and some prescriptions for neurobiologists on what to look for in large-scale anatomical formations.

1. Introduction

1.1. *Brain hierarchy of levels: Microscopic, mesoscopic, macroscopic*

This essay addresses two paradoxes stemming from the 1950s. On the one hand [76] in his preface to information theory famously wrote: "The fundamental problem of communication is that of reproducing at one point either exactly or approximately a message selected at another point. Frequently the messages have meaning; that is they refer to or are correlated according to some system with certain physical or conceptual entities. These semantic aspects of communication are irrelevant to the engineering problem (p. 384)". Yet the main role of the sensory cortices is to construct the meaning of sensory information, so that semantics is at the core of neurodynamics. On the other hand the ceaseless, aperiodic, unpredictable activity we observe in cortex invites modeling with chaotic dynamics in order to exploit the capacity of chaotic dynamics to create enormous quantities of new information [78]. Yet deterministic chaotic systems are stationary, autonomous, and noise-free, whereas brains are drenched in endogenous and environmental noise; they are continually evolving through successive bifurcations; and they are necessarily engaged with the environment in the constant extraction of energy for oxidative metabolism and the dissipation of heat [51]. What are the uses of chaos theory, and how is it that it has become so useful?

Within the decade von Neumann [82] wrote: "Whatever the language of the brain is, it cannot fail to differ considerably what we consciously and explicitly consider as mathematics .... Brains lack the arithmetic and logical depth that characterize our computations. ... We require exquisite numerical precision over many logical steps to achieve what brains accomplish in very few short steps" (pp. 80–81). Models of brain functions using the tools of computational neural networks and rule-driven symbol manipulation require many steps between input and output. Recent advances in nonlinear brain dynamics have emphasized the capabilities of brains for exceedingly rapid acquisition and filtering of relevant sensory information through all of the senses, followed by synthesis of the meanings of the information [9]. Cortices condense the meanings into various forms of knowledge, and express them in natural languages. All these operations transcend the capabilities of contemporary machine intelligence. The aim of this essay is to review progress in using chaos theory to answer the questions in qualitative terms, how brains construct knowledge and meaning from sensory information.

Any explanation that strives to meet von Neumann's challenge must rely on connections among very large numbers of neurons in many areas of the brain. Topographic mapping by parallel arrays of axons provides sensory input and motor output with the fine structure that enables brains to discriminate precisely small changes in input as fine as single sensory receptors and to focus control of motor output even to single motor units. In contrast, the steps between microscopic sensory inputs and microscopic motor outputs require wide divergence and convergence of axonal and dendritic connections to encompass very many neurons. Then the first of the "few short steps" in brain function requires transition from sparse high-intensity of *microscopic* sensory information carried by very few neurons to low-intensity of perceptual information carried by *mesoscopic* assemblies of neurons [10]. The requisite cognitive operations are generalization, abstraction and categorization.

This step taken by the sensory cortices is the construction of the meaning of the information. We postulate that the meaning of the information exists in the interrelations of the enormous number of neurons carrying each a fragment of information, which constitutes the knowledge brains create from information [11]. This knowledge must be stored and later mobilized in the serial processes for categorizing sensory information and then deciding on courses of action in response to the input. Mobilized knowledge has the form of bursts of oscillatory electroencephalographic (EEG) and electrocorticographic (ECoG) potentials carrying spatial patterns of amplitude modulation (AM) (Fig. 1(a)), which emerge from chaotic oscillations of background activity [45] through episodic bifurcations (phase transitions) by which the cortex jumps from chaotic disorder into phase-locked limit cycle activity [13] and back again.

The penultimate step of taking action requires the reverse operations of speciation and concretization of the low density broadly distributed firing [39, 40], which occur by focusing of neural activity into high-intensity but sparse firing of neurons in strategic mesoscopic assemblies. The assemblies activate microscopic networks of neurons that integrate tactical motor commands with proprioceptive

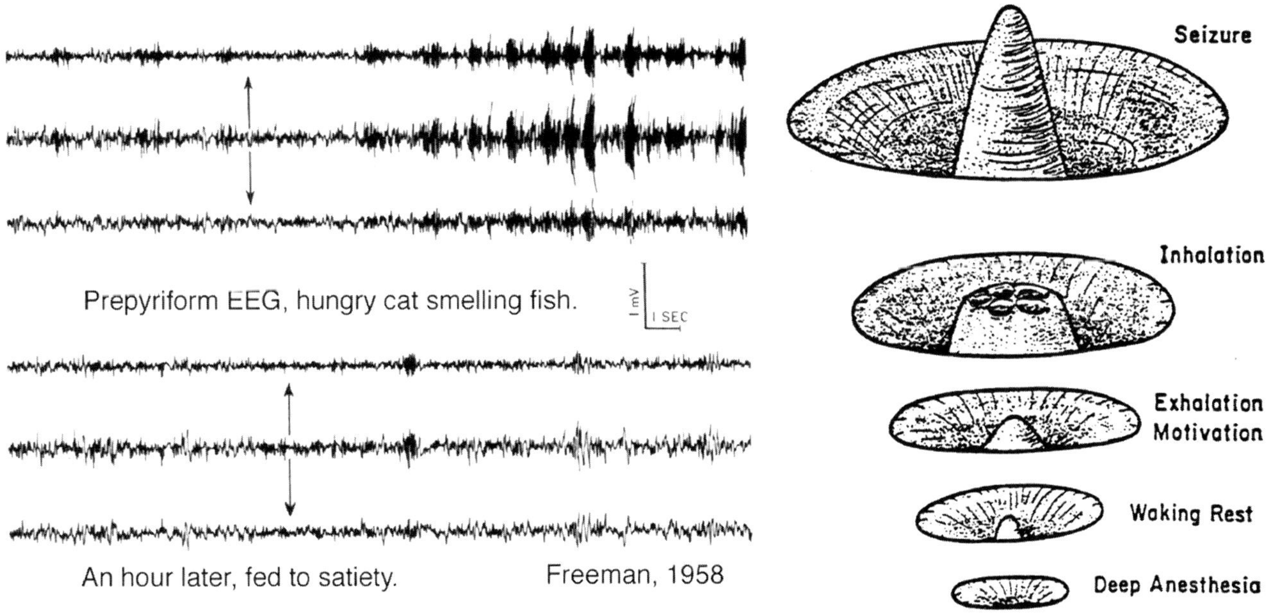

(a) (b)

Fig. 1. (a) The bipolar ECoG recording came from an electrode pair straddling the prepyriform cortical dipole field, first when the subject was habituated to a closed box. The upper signals were from the cat when given a whiff of fish when it was hungry, and the lower signals were when it had been fed to satiety. From Fig. 2, p. 115 in Freeman (1960). (b) A phase portrait of the olfactory system was constructed from recordings of the ECoG in the range of designated states. In the aroused state ('motivation') control of the ECoG repeatedly shifted between a non-zero point attractor during exhalation and a landscape of limit cycle attractors emergent during inhalation, each having its basin of attraction in the landscape. When the phase transition took place, the control of cortex was shifted from a point attractor to one of the limit cycle attractors representing one of the Bayesian priors that sensitized the bulb to a particular expected odorant stimulus. From Fig. 12, p. 145 in Freeman (1987).

information, which is the sensory feedback from muscle receptor neurons firing rapidly and precisely as tasks are performed. The output of the relatively few motor neurons is by pulse trains that convey information at high specificity [36, 37]. We conceive von Neumann's "few short steps" as state transitions between these three levels: microscopic — based in microelectrode recordings; mesoscopic — based in intracranial arrays to record the electrocorticographic (ECoG) recordings; and macroscopic — based in scalp arrays to capture the electroencephalographic (EEG) and functional magnetic resonance imaging (fMRI) recordings [42].

1.2. *Brain state space, state variables, state transitions*

For functional connectivity we further conceive that at each of the postulated "very few short steps", the operation must result in neural activity having a spatiotemporal pattern. The pattern must be observed in fluctuations of potentials — electric (EEG), magnetic (MEG), hemodynamic (fMRI, BOLD) — that are coherent over some contiguous collection, be they neurons, hypercolumns or modules. Each pattern must be stable long enough to be identified and measured, and it must recur sufficiently reliably in multiple trials of the action-perception cycle (Merleau-Ponty, 1942/1963, reduced by behaviorists to the stimulus-response reflex arc), in order to be collected in measurement space and categorized. At the microscopic level the patterns are observed in axon action potentials, that is, pulse trains of neurons in networks and measured as rates, intervals, and coincidences. At the mesoscopic level the patterns are observed in extracellular dendritic potentials of populations of neurons in areas of cortex and measured by digitizing the amplitudes. Depth electrodes serve to measure local field potentials (LFP); electrodes on cortical surfaces yield electrocorticograms (ECoG) (Fig. 1(a)). At the macroscopic level the space-time patterns are observed with EEG and MEG scalp potentials, and diverse patterns of cerebral blood flow revealed by functional magnetic resonance imaging (fMRI) and related techniques. For evolving cortical populations the critical measures

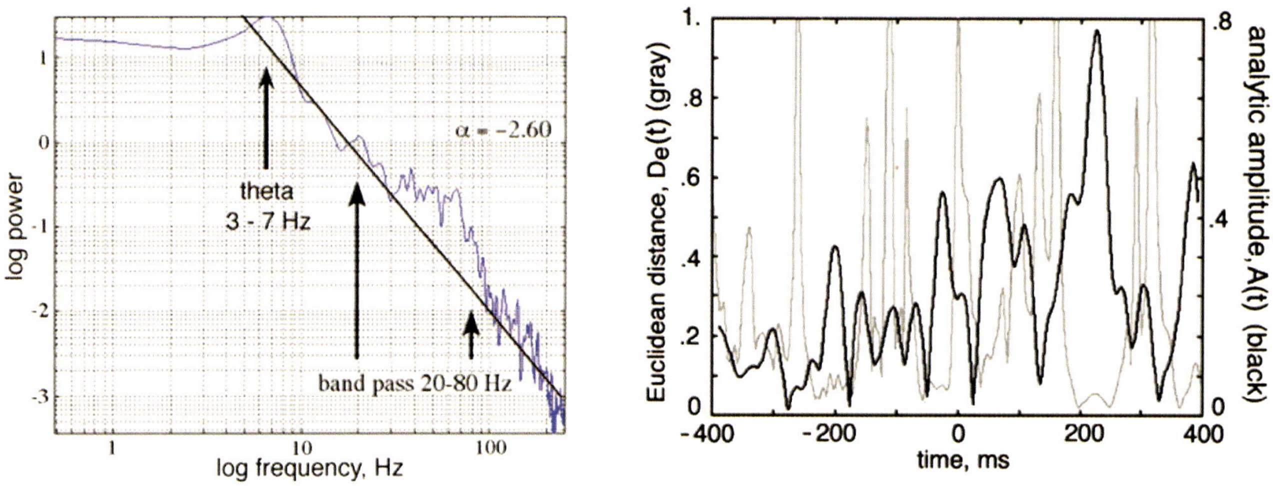

(a) (b)

Fig. 2. (a) The temporal power spectral density (PSD_T) had 1/f form in rest state but excess power above the trend line in theta and beta-gamma ranges during active states, implying increased order. (Fig. A1.02, A) (b) A(t) (black) and $1/\mathrm{D}_e(t)$ (gray) from the Hilbert transform on trials with CS (onset at 0 ms) showed intermittent frames with decreased rate of pattern change (gray curve), increased amplitude (black curve), and near-invariant instantaneous frequency (constant rate of increase in analytic phase), which imply increased order during frames between state transitions (Freeman, 2004b, Fig. 1.02). $\mathrm{D}_e(t)$ in (b) shows that ECoG power in each burst is concentrated in a narrow frequency band, so that frequency modulation across several bursts explains the broad peak in (a).

of functional connectivity are the magnitudes of correlation of activity between in well-defined frequency bands of the signals (Fig. 2) and the distances between cortical sites of recording, signifying the correlation lengths within spatially coherent patterns.

We make multichannel observations of spatial patterns simultaneously by using arrays of sensors and aggregate them in n-space, where n is the number of microscopic pulse trains, mesoscopic LFP/ECoG channels, macroscopic EEG/MEG channels, and fMRI voxels. Each train, channel and voxel contributes the coordinate axis of a dimension needed to create the n-space that defines an observable finite projection of the infinite brain state space. The range of observed values for all measures defines the magnitude of the n-space. Each measure that is simultaneously sampled at regular time intervals yields a state variable giving a point on its axis. The set of measurements at each time step determines an n-dimensional point in n-space. That point designates a spatial pattern to which all channels contribute, so the point and pattern represent the observed instantaneous state of the cortex.

A sequence of measurements gives a set of points forming a trajectory in state space that signifies a changing pattern. The Euclidean distance in n-space between successive samples in time gives the rate of change in the pattern. When the pattern is nearly fixed, the set of points forms a tight cluster. When the trajectory in n-space is projected into 2-space for graphic display, the trajectory appears as a line or curve with segments between and within clusters at the times of sampling. Each cluster represents the pattern of a transiently stable state. The trajectory between two clusters indicates a state transition. In dynamical terms each cluster manifests an attractor, and a set of clusters may represent a sequence of states that is variously described as an *attractor landscape* [78], a *metastable state* [29, 58, 59], a *chaotic itinerant trajectory* [81], the *connectome* forming the structural core of cerebral cortex [79], the *Bayesian brain* minimizing free energy [53], the *testable taxonomy* of unconscious, preconscious and conscious states [27], among others. The vector that defines each point is designated as an "order parameter" [54], because the spatial AM pattern it represents describes the specific order of the instantaneous cortical state [50].

The high degree of order is imposed by the strength of interaction (synaptic feedback gain) among all of the neurons comprising a population. The order parameter is a vector that is evaluated by the pattern. The rate of change in the order parameter is given by the Euclidean distance in n-space between successive points, $\mathrm{D}_e(t)$ (Fig. 2(b)). It is an empirical scalar index of disorder [37]; a low value signifies a cluster, and a high value signifies noise or a state transition through noise. The reciprocal, $1/D_e(t)$, serves as an index of persisting order. A cluster with small increments reveals high probability of convergence of the trajectory to an attractor. A large step may reveal a state transition by a jump across the separatrix between two basins of attraction. Haken [54] concluded that state transitions are essential to the mechanisms of information transfer between hierarchical levels. Each state transition by bifurcation provides one of the "very few short steps" by which brains mediate the action-perception cycle.

2. Relevant Properties of Connectivity in Evolving Scale-free Networks

2.1. *Types of probability distributions of connections*

The distributions of numbers of connection as a function of the distances of transmission can be estimated by measuring the lengths of axons. The predicted distributions are best compared in log-log coordinates. In the original formulation of random graphs by Erdös and Renyí [28] the probability of connections is uniform with distance (Fig. 3(a), lower line). In random cellular networks the connections are restricted to nearest or next-nearest neighbor nodes (e.g., [26]). In "small-world" graphs/networks a low percentage of the local connections is replaced with uniformly distributed long connections [84], which dramatically reduces the depth by bridging across many nodes. Small-world graphs/networks are related to cortical connectivity in the reported distributions of the lengths of axon projections (Fig. 3(b)), which display predominantly local connections and increasingly sparse connections with increasing distance from the cell bodies (Fig. 3(a), dashed curve). The power-law distribution gives a straight line, indicating that the patterns in EEG and ECoG activity are fractal and self-similar over wide dynamic ranges.

2.2. *Power-law connectivity distributions and scale-free networks*

Special significance is attached to the power-law distribution of connection distances, owing to the introduction of self-similarity at different scales of observation and measurement. The significance is found in the dynamical properties endowed by scale-free connectivity [20, 25]. The point raised here is that distributions of cortical connection lengths appearing to be exponential as commonly inferred (the convex dashed curve in Fig. 3(a)) may actually be power-law (the slanted line) because of three experimental limitations on determinations of distributions of axonal lengths. First, for short distances the observations using light microscopy

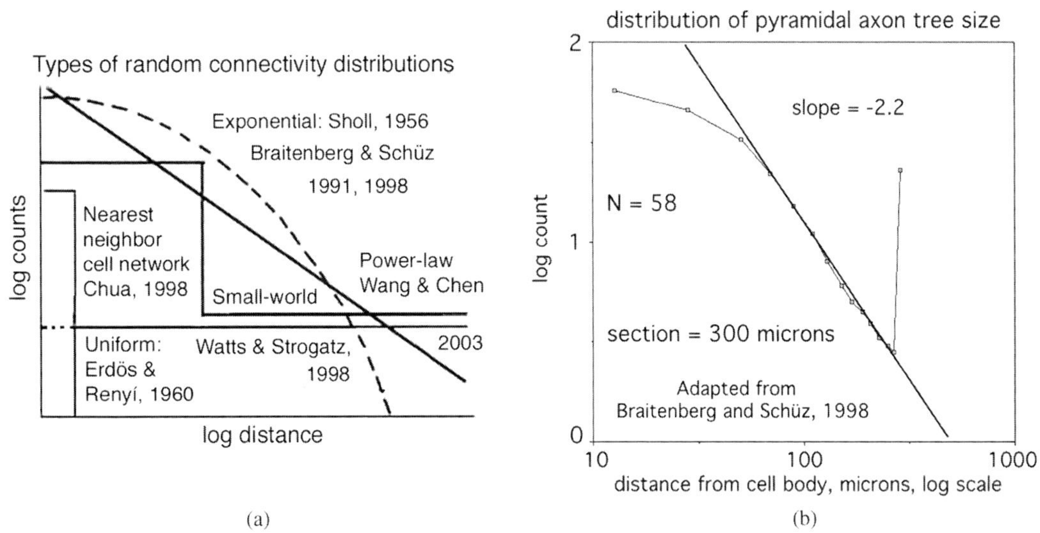

Fig. 3. (a) Distributions of the numbers of connections as a function of distances are schematized in log-log coordinates. The straight line indicates 1/f. (b) An example is shown of the distribution of measurements of lengths of axons that were made in histological sections of Golgi preparations of mouse cortex. The vertical line shows the count of axons that left the tissue section and couldn't be followed to the axon ends. The data were re-plotted in log-log coordinates in accord with their fractal properties. Adapted from [19].

omit most unmyelinated axons, which in electron micrographs substantially outnumber the myelinated axons. Second, the observations of axon lengths in Golgi preparations are made in tissue sections of limited thickness (e.g., 300 microns, Fig. 3(b)) in small mammals (here mouse). At the long end continuity is very difficult to follow in serial sections, leading to a deficit in the numbers of long connections. On the short end electron micrographs show that many unmyelinated terminal branches are smaller than the wavelengths used in optical microcopy, so they cannot be counted and can account for discrepancy in observed and power-law the distributions. Taking into account these experimental limitations, the data that have exponential distributions in semi-log coordinates may be consistent with power-law distributions in log-log coordinates, meaning that the distributions are fractal and support scale-free, self-similar cortical activity. The theory offers a prediction for anatomists that the power-law model should be rigorously tested. A method to test functional transmission distance is considered in the next section.

3. Relevant Properties of Activity in Evolving Cortex

3.1. *Origin, stabilization, and roles of background activity*

Cortical neurons continually fire in "spontaneous" background activity, which is ubiquitous in all parts of functioning brains. Pulse trains yield autocorrelation functions and interval histograms that conform to a Poisson process with a dead time [32, 36, 37]. The genesis of background activity has been modeled in detail for a population of interactive excitatory neurons in the olfactory bulb (a KIe set, [47]), in which the activity is sustained by the mutual synaptic excitation among excitatory neurons. The structural connectivity is sufficiently high that for each pulse emitted a neuron receives back on average more than one pulse, but owing to its refractory periods it might not respond. Hence the average magnitude of output is constrained for every neuron; it cannot remain silent, and it cannot fire above an average maximal rate. In physiological terms the background activity is everywhere locally stabilized homeostatically. The level of output is continuously adjusted by neuromodulation from brain stem histaminergic neurons that control the degree of arousal [38].

In terms of nonlinear dynamics a non-zero point attractor everywhere locally controls the mean firing level of every neuron, which guarantees life-long viability through exercise. Upon linearization at the operating point, the transfer function of the local population is shown to have a pole on the negative real axis close to the origin of the complex plane. The stability is demonstrated by electrical single shock stimulation of an afferent path. The impulse response is the probability of firing, which rises exponentially and decays at a rate proportional to response amplitude, which can be fitted with a sum of two exponential terms given by a 2nd order ordinary differential equation (ODE). The decay rate extrapolates to zero at zero input, proving that the population is governed by a point attractor that would give a pole at the origin in the absence of input. The increase in decay rate above zero at rest and in active behavioral states is due to the extrinsic background excitatory input on which the electrically excited test input is superimposed.

The conclusion is that the background activity is driven by mutual excitation (positive feedback among excitatory cells) that is self-stabilized by the refractory periods, so that the cortex is governed by a non-zero point attractor (Fig. 4). This attractor sustains a self-regulated state of criticality, in which the cortex is poised to jump from a receiving state to a transmitting state, but only when a conditioned stimulus is received (Fig. 1(a)). The power spectral density, PSD, in the receiving state is linear in coordinates of log power vs. log frequency, $1/\mathrm{f}^{\alpha}$, with a power-law exponent, $2 < \alpha < 4$. The form is determined by the background pulse input (white noise, $1/\mathrm{f}^0$) after it has been filtered by the dendritic integration approximated by a 2nd order ODE. The output would be brown noise ($1/\mathrm{f}^2$), but the refractory periods preventing firing remove power from the entire spectrum (Fig. 2(a)) in proportion to the square of frequency [52]. The result is "black noise" [75], which is characteristic of chaotic systems.

Populations of excitatory neurons are modeled by KIe sets, while populations of inhibitory neurons are modeled by KIi sets [2]. Both interactions are modeled by positive feedback loops, each subdivided into a receiving subset and a continually renewed transmitting subset and modeled with a 4th order ODE. The interaction of a KIe set with a KIi set gives negative feedback, which is modeled with an 8th order ODE (Chapter 5, [8, 32, 36, 37]). The role of inhibition is to generate oscillations in dendritic synaptic potentials and

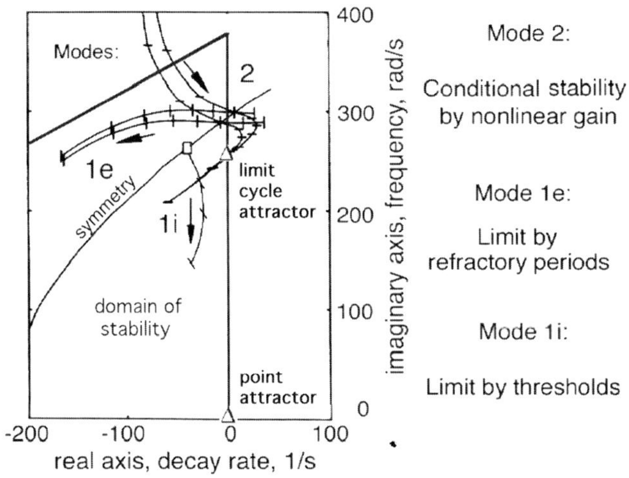

Fig. 4. The diagram displays root loci of the critical complex conjugate pole in the upper half of the complex plane closest to the imaginary axis and therefore dominating the impulse response. The root loci are plotted by repeated solutions of linearized ODEs fitted to impulse responses of the cortex. The arrows show the direction of displacement with increasing amplitude of activity. Mode 1e shows impulse responses with excitatory bias, and Mode 1i with inhibitory bias, both in explicit symmetry breaking. Mode 2 shows the tendency to spontaneous symmetry breaking on approach to the limit cycle attractor on the imaginary axis with increasing amplitude, which for the linear approximation represents a singularity. The state transition begins when the sign of the real part of the complex pole changes from −− to +. From Fig. 6.30 on p. 388, with equations for the root loci in Chapter 6, Freeman (1975/2004).

their accompanying axonal pulse probability waves. The oscillations cover a spectrum (Fig. 2(a)) from near zero to 3 Hz (delta), 3–7 Hz (theta), 7–12 Hz (alpha), beta (12–30 Hz), gamma (30–80 Hz), and epsilon (80–250 Hz). The latter is a new term to label the chatter of neural firing in the short loops of mutual excitation, by which positive feedback gives pulse intervals of 3–6 ms [32, 36, 37]. Cortical neurons require the excitatory bias to become re-excited on release from inhibition. In the rest state with minimal sensory input the EEG and ECOG oscillatory state variables have nearly Gaussian amplitude histograms, relatively flat autocorrelation functions, and broad spatial and temporal frequency distributions with power spectral densities (temporal, PSD_T, and spatial, PSD_X) conforming to power-law distributions in log-log coordinates (1/f).

The brain state at rest is symmetric, and the PSD reveals only the black noise. In the active state the symmetry is broken, bursts of oscillation emerge in the EEG and ECoG (Fig. 1(a)), and peaks appear above the 1/f trend line in the PSD (Fig. 2(a)). Linearization of the dynamics at the rest operating point yields transfer functions that again reveal the afore-mentioned pole at the origin of the complex plane. They also reveal an array of complex conjugate poles to the left of the imaginary axis and of complex zeroes to the right of the axis having values that are constrained by the value of the pole at the origin (Chapter 5 in [32, 36, 37]). Calculation of the root loci of the poles in the complex plane predicts the existence of a limit cycle attractor for the KII set, which in a piecewise linear approximation in the small-signal range is represented by a pair of complex conjugate poles on the imaginary axis (Chapter 6 in [32, 36, 37]). This discovery leads to the hypothesis that the sudden emergence of a burst of activity in the gamma range occurs when control of the cortex temporarily transfers from the non-zero point attractor to the limit cycle attractor (Fig. 4) in a transition that resembles a subcritical Hopf bifurcation.

3.2. *Cortical oscillations in response to input without and with phase transitions*

Input to cortex can change its activity in either of two ways. Perturbation by impulse input does not change the state of the cortex, because the cortex relaxes back to its prestimulus state while in doing so revealing its characteristic frequencies [2]. In this case the comparison of the perturbing input function with measures of the output functions define in terms of dynamics the operations by the cortex on the input to give the output. This approach is the basis for calculation of mutual information, cross-correlation, and transfer functions by piecewise linearization [32, 36, 37]. Particularly useful is the impulse input with electrical shocks, light flashes, auditory clicks, etc., giving "evoked potentials" that document the relaxation to the initial condition without change in state. Alternatively an input to cortex induces a phase transition revealed by an expanding oscillation [77]. The difference is crucial for understanding cortical dynamics. Both processes are initiated by input. Both can result in spatially coherent oscillation at a frequency in the beta-gamma range. Both oscillations are modulated spatially in both the amplitude and phase of the common carrier wave. In the relaxation process the amplitude and phase modulation (AM and PM) are imposed by the stimulus input in explicit symmetry breaking. In the phase transition the AM and PM patterns emerge from the cortical connectivity that has been shaped by prior learning in spontaneous symmetry breaking [50, 51] at extreme neural density resembling condensation of a gas to a liquid.

Demonstration of the relaxation process is by excitation of the olfactory bulb with single-shock afferent electrical stimulation of the olfactory nerve, which evokes a transient oscillation (an "evoked potential") having the form of a damped cosine at the characteristic frequency of the bulbar negative feedback loop (Fig. 4, Mode 1e). When recorded from an array of electrodes on the surface, the ECoG oscillation at the common frequency has a spatial phase gradient across the bulb with slope and direction that correspond to the slow conduction velocity of the parallel unmyelinated axons forming the afferent nerve. The location of peak amplitude is determined by the stimulus site; when the stimulating electrode is moved to new locations in the parallel sheet of axons, the location of peak AM likewise shifts.

In contrast, inhalation triggers a volley of pulses from olfactory receptors in the nose that cause a sustained excitation of bulbar neurons. The result is the emergence of burst of activity called a wave packet having a characteristic carrier frequency that is shared throughout the bulb. The spatial pattern of amplitude modulation (the AM pattern) does not conform to the specific triggering input but is instead contingent upon learning. The spatial pattern in the phase of the carrier wave (the PM pattern) is not a plane wave sweeping across the bulb like the phase gradient of the evoked potential. It is a radiating wave having a phase gradient with the shape of a cone. The direction and slope of the cone are determined not by the conduction velocities of the afferent axons but instead by the conduction velocities of intrabulbar axon collaterals [43]. These properties show that the wave packet in response to a conditioned stimulus is formed by a phase transition. The burst of oscillation is not like the ringing of a bell when struck; it is like the ignition of an explosion (Fig. 4, Mode 2). In a distributed medium as in formation of a snowflake or raindrop that phase transition does not take place everywhere simultaneously. It begins at a site of nucleation and spreads radially at an intrinsically determined velocity. Thus in the bulb the PM patterns are constrained by the conduction velocity of mitral collateral axons, and the reproducible AM pattern are determined by intrabulbar connectivity. A classifiable pattern emerges by spontaneous symmetry breaking only if the input contains a learned stimulus.

All of the primary sensory areas of neocortex share these properties. As in the bulb the AM patterned oscillations of neocortical wave packets in the beta-gamma range (20–80 Hz) result from a phase transition only if the subject has been conditioned to respond to the stimuli, including the control background air with which it is familiarized. The oscillation evolves in a relatively narrow range, giving a peak in the PSD_T from a carrier wave with an AM pattern that is classifiable with respect to CS, and a PM pattern that manifests the location of the site of nucleation and the phase velocity. In neocortex the phase velocities [44] conform to the conduction velocities of corticocortical axons [68, 80]. A sequence of amplitude patterns tends to repeat in the theta range (3–7 Hz) but with differing carrier frequencies in successive patterns. The near-invariance of frequency holds only within frames lasting $\sim$0.1 s but with modulation of the frequency from each pattern to the next, so the PSD_T computed over epochs $\geq$1 s has a relatively broad peak (Fig. 2(a)) spanning the high beta and low gamma ranges.

Most significantly, the AM pattern lacks invariance with respect to its fixed CS. When the CS is held constant but its meaning is changed, the AM pattern changes irreversibly [48]. The same result holds for all sensory areas of neocortex as well [17]. We conclude that the first of von Neumann's "very few short steps" in the sensory systems is the transduction of sensory energy in various forms into information-bearing action potentials, and that the second step is the phase transition on destabilization of the sensory cortices, which leads to the construction of an activity pattern that retrieves the knowledge about a CS that is stored in synaptic weights. The several sensory cortices send their AM patterns into the limbic system, where in von Neumann's third step they are integrated into a multisensory percept (gestalt). This review is restricted to his second step.

3.3. *Bifurcations, phase transitions and chaos*

The olfactory system has multiple components. The three largest parts suffice to model the system dynamics: the olfactory bulb (OB), prepyriform cortex (PC) and anterior nucleus (AON) [2]. When surgically isolated from each other and stimulated by impulse input, the OB, PC and AON reveal point attractors by giving damped cosines at step onset, each with its characteristic frequency. The three frequencies are incommensurate [61], so the parts cannot by synchronized at any one frequency. The parts are coupled by feedback pathways with delays, some of which are excitatory and tend continually to increase the system activity by positive feedback, while others are inhibitory and tend to sustain system oscillations by negative feedback. All interactions are limited by the thresholds and refractory periods of the neurons. A dynamical model of the system consists of three interconnected KII sets: KIIob, KIIpc, and KIIaon, with a bias-generating KIe set at the input, forming a KIII set. The solutions to the matrix of ODE suffice to simulate the background activity, the impulse responses of the component populations, the induced bursts of transient limit cycle oscillation (Fig. 1(a)) [3], and genesis of simple and complex partial epileptic seizure spikes (Fig. 1(b)) [33], from which we infer that the noise-free model has multiple Lyapunov exponents, most of which are negative or zero, and at least one of which is positive.

The most glaring deficiency is the failure to model the 1/f power-law form of the PSD. The reason is clear, that the continuum of the real 1/f form implies an infinite dimensional system, whereas the K-sets are rational approximations for implementation of low-dimensional systems on digital platforms. The parameters of the approximations can be tuned to approach the 1/f form [4], but in doing so the model becomes progressively more brittle, essentially by the addition of more basins of attraction. In effect the attractor landscape begins to resemble rabbit fur, owing to the proliferation of attractors, resulting in attractor crowding. The basins of attraction approach the size of the digitizing step, so that well within a fraction of a second the system placed in the basin of a chaotic attractor steps our into a neighboring basin of a point or limit cycle attractor. Empirically the remedy is provided by additive noise [4] in two forms. One form is low-level random numbers applied to each input channel in the KIe array uncorrelated between channels; the other is a single channel of low level random numbers given to the KIIaon. In effect the KIII landscape is smoothed from rabbit fur to alligator skin, enabling a manageable number of limit cycle attractors to coexist (Fig. 1(b)). The additive noise solves the problem of the infinite sensitivity of chaotic systems to changes in the initial conditions and ensures metastability in both brains and brain models [8]. In effect the ODE as implemented on digital platforms become stochastic difference equations. Multiplicative noise also plays a role in the homeostatic regulation of input gain to compensate for wide fluctuations in the intensity of sensory input. This janitorial operator is essential for signal normalization and range compression. It has been documented in studies that deserve more recognition than they have received [5, 31].

Clearly brains are not deterministic, and attempts to evaluate correlation dimensions and Lyapunov exponents using EEG and ECoG are certain to be confounded by noise [73]. The modulation of deterministic chaos by additive noise [46] creates a tool, stochastic chaos [7]. The formalism encourages the search for nonconvergent solutions for matrices of ODE to complement point and limit cycle attractors, taking advantage of the theories of metastability [58, 59], large-scale networks [22], and chaotic itinerancy [81]. The principal barrier to exploitation is the excessive run time required for numerical solutions of ODE, which might some day be circumvented by implementation of K-sets in analog VLSI [72] and in random graph theory using neuropercolation theory [42, 65].

4. Conclusions: Guidance by Chaos Theory of Further Biological Experimentation

4.1. *Applications of noise-modulated deterministic chaos*

Even though brain dynamics does not conform to low-dimensional chaos, the theory provides the guidance needed to construct and evaluate matrices of ODE with which to simulate brain functions in the cognitive realms of perception and choice. The non-convergent solutions of the ODE do suffice to simulate a wide range of spontaneous, evoked, and induced wave forms including the genesis of simple and complex partial epileptic seizures (Fig. 1(b)). More importantly when the KIII set is expanded into multiple 2-D layers of the several interactive KIe and KIi layers, the model performs pattern classification with exceptionally high levels of success, beginning with olfactory stimuli such as kinds of tea, [85]), and extending to quality control of machine screws in industry [86], Japanese vowel sounds [77], EEG patterns [61], face recognition [66], all 36 alphanumeric characters with an 8×8 array [66], and EEG patterns in seizure [74] and hypoxia [55], outperforming conventional neural networks in requiring small training sets (typically 5 to 10 examples), one-step convergence by phase transition rather than asymptotic approach by gradient descent, and memory capacity increasing geometrically with the number of nodes [8,57,65]. Further uses of multisensory pattern classification are achieved by conjoining three KIII sets to form a KIV set, which simulates the dynamics of goal-directed behavior of primitive vertebrates by constructing multisensory percepts (gestalts) for making decisions [63] and guiding navigation [62]. The full-scale implementation in a prototype of a Martian rover SSR2K [60] pioneered the field of intentional robotics.

4.2. *Neuroanatomy and neuroembryology*

In terms of hierarchical dynamics the allocortex is better described not as three-layered but as a single functional layer, and neocortex not as six-layered but as having two functional layers. The localized groups in Layers III to IV constitute modules for specialized information processing. The embedding neuropil in Layers I, II, V and VI (Fig. 6) forms a continuum over each cerebral hemisphere. Unlike allocortical neuropil, which has discontinuities in connection densities that segregate major areas from each other and from neocortex, the embedding neuropil of neocortex in each hemisphere forms a unified organ that enables the radial spread of activity by synaptic and diffusion-mediated communication. Evidence for this continuity is provided the spreading depression of Leão [23] from any site in the neocortex in each hemisphere but not to the other hemisphere or outside the neocortex. However, the mechanism of spreading depression is non-synaptic, nearest-neighbor. New data are needed to verify the hypothesis that connection distances have power-law distributions (Fig. 2), and to map the axonal distributions to provide the kernels needed to model the spatial integral transforms performed by divergent and convergent pathways. At present the best-documented example of a non-topographic, divergent-convergent projection is the lateral olfactory tract from the olfactory bulb to the prepyriform cortex [32,36,37].

Most analyses of anatomical data have emphasized point-to-point networks, as in the classic meta-study by Fellemin and Van Essen [29] of connectional architecture in visual cortex. These linkages are very important for the organization of the vectorial order parameters of mesoscopic and macroscopic AM patterns; however, they do not address the connectivity required for the long-range coherence and rapid phase transitions revealed by carrier waves. Braitenberg and Schüz [19] wrote: "A recent hypothesis by Miller [71], based on differences in both connectivity patterns and spontaneous activity between upper and lower layers, assigns to the upper layers the role of a neuronal 'library', storing most of the information encoded by assemblies ... while the lower layers are assumed to *catalyze* the process of assembly formation" (p. 150). Long cortico-cortical connections are not randomly distributed within the territory that they innervate. Instead they are distributed in patches with high local connection density, with intervening regions having few connections [69]. Kaas [57] wrote: "Generalizing from cats and monkeys it appears that the evolutionary advance in brain organization is marked by increases in the numbers of unimodal sensory fields, not by increases in multimodal association cortex as traditionally thought" (p. 147). Therefore Layers III-IV provide highly textured modules, while the embedding neuropil in Layers I, II, V and VI provides not only the connectivity needed for rapid integration of macroscopic modular activity in the encompassed clusters but also the uniformity and unity of neocortical anatomy that has often been noted by neuroanatomists. What is needed is the measurement and statistical analysis of the holistic organization of long connections in the manner pursued by Bok [18], and explication of the neuroembryological rules that govern formation of integrative systems, the developmental processes by which scale-free connectivity emerges in neocortex. Gulyas has demonstrated the ability to generate scale-free networks using local information in bottom-up construction. Close collaboration between network theorists and neuroembryologists will be needed to advance in this direction.

4.3. *Brain imaging combining structural and functional MRI, EEG and MEG*

The inference that foci seen with fMRI are signs of coordinated modules may best be tested by combining hemodynamic imaging with EEG and

MEG [22,67]. Preliminary data indicate that the necessary brain-wide macroscopic wave packets with beta carrier frequencies are found in human volunteers engaged in cognitive tasks [74]. Quantitative assays with a 64×1 linear array at 3 mm spacing on the scalp have shown (Freeman, Burke and Holmes, 2003) that re-synchronization of both beta and gamma activity recurs at rates in the theta and alpha ranges, commonly over correlation distances on the scalp of several cm and occasionally over the full length of the array, 19 cm. EEG data from intracranial arrays support the inference that cortical dynamics is scale-free, so that brains ranging 10^4 in mass from mice to whales have the capacity for widespread phase transitions that rapidly and repeatedly re-synchronize beta and gamma oscillations carrying AM patterns correlated with cognitive processes. Wang and Chen (2002) in studies of the temporal dynamics have shown that above a certain threshold of connection density, a scale-free network can undergo an abrupt phase transition and resynchronize globally no matter how large its diameter. Use of the Hilbert transform to calculate measures of beta-gamma analytic power is optimal, because the state variable thereby derived as an index for power is more closely related to the demands for metabolic energy than are the amplitudes of low frequencies (alpha, theta, delta) [67].

References

[1] Published in the International Journal of Bifurcation and Chaos.

[2] Freeman, WJ (1992) Tutorial in Neurobiology: From Single Neurons to Brain Chaos. Int J Bifurc Chaos 2: 451–482.

[3] Kay LM, Shimoide K, Freeman WJ (1995) Comparison of EEG time series from rat olfactory system with model composed of nonlinear coupled oscillators. Int J Bifurc Chaos 5: 849–858.

[4] Chang H-J, Freeman WJ (1998) Biologically modeled noise stabilizing neurodynamics for pattern recognition. Int J Bifurc Chaos 8(2), 321–345.

[5] Chang H-J, Freeman WJ (1999) Local homeostasis stabilizes a model of the olfactory system globally in respect to perturbations by input during pattern classification. Int J Bifurc Chaos 8: 2107–2123.

[6] Freeman WJ (1999) Noise-induced first-order phase transitions in chaotic brain activity. Int J Bifurc Chaos 9: 2215–2218.

[7] Freeman WJ (2000) Characteristics of the synchronization of brain activity imposed by finite conduction velocities of axons. Special Issue on Synchronization, Kurtz J (ed.) Int J Bifurc Chaos 10: 2307–2322.

[8] Kozma R, Freeman WJ (2001) Chaotic resonance: Methods and applications for robust classification of noisy and variable patterns. Int J Bifurc Chaos 10: 2307–2322. (Cover Figure).

[9] Freeman, W.J. (2003a) A neurobiological theory of meaning in perception. Part 1. Information and meaning in nonconvergent and nonlocal brain dynamics. Int J Bifurc Chaos 13: 2493–2511. http://repositories.cdlib.org/postprints/3348 http://repositories.cdlib.org/postprints/2355

[10] Freeman, W.J. (2003b) A neurobiological theory of meaning in perception. Part 2. Spatial patterns of phase in gamma EEG from primary sensory cortices reveal the properties of mesoscopic wave packets. Int J Bifurc Chaos 13: 2513–2535. http://repositories.cdlib.org/postprints/3343

[11] Freeman WJ, Gaál G, Jornten R (2003) A neurobiological theory of meaning in perception. Part 3. Multiple cortical areas synchronize without loss of local autonomy. Int J Bifurc Chaos 13: 2845–2856. http://repositories.cdlib.org/postprints/3344

[12] Freeman WJ, Burke BC (2003) A neurobiological theory of meaning in perception. Part 4. Multicortical patterns of amplitude modulation in gamma EEG. Int J Bifurc Chaos 13: 2857–2866. http://repositories.cdlib.org/postprints/3345

[13] Freeman WJ, Rogers LJ (2003) A neurobiological theory of meaning in perception. Part 5. Multicortical patterns of phase modulation in gamma EEG. Int J Bifurc Chaos 13: 2867–2887. http://repositories.cdlib.org/postprints/3346

[14] Freeman W.J. (2004) How and why brains create meaning from sensory information. Int J Bifurc Chaos 14: 513–530. http://repositories.cdlib.org/postprints/1689

[15] Li X, Li G, Wang L, Freeman WJ (2006) A study on a bionic pattern classifier based on olfactory neural system. Int J Bifurc Chaos 16(8): 2425–2434.

[16] Published in other Journals

[17] Barrie JM, Freeman WJ, Lenhart M (1996) Modulation by discriminative training of spatial patterns of gamma EEG amplitude and phase in neocortex of rabbits. J Neurophysiol 76: 520–539.

[18] Bok ST (1959) Histonomy of the Cerebral Cortex. Amsterdam: Elsevier.

[19] Braitenberg V, Schüz A (1998) Cortex: Statistics and Geometry of Neuronal Connectivity, 2nd ed. Berlin: Springer–Verlag.

[20] Breakspear M (2004) Dynamic connectivity in Neural systems: Theoretical and empirical considerations. Neuroinformatics 2(2): 205–225.

[21] Bressler SL, Kelso JAS (2001) Cortical coordination dynamics and cognition. Trends Cogn Sci 5: 2–36.

[22] Bressler SL, Menon V (2010) Large-scale networks in cognition: emerging methods and principles. Trends Cogn Sci 14: 277–290.

[23] Bures J, Buresová O, Kriváneк J (1974) The Mechanism and Applications of Leão's Spreading Depression of Electroencephalographic Activity. New York: Academic Press.

[24] Chang H-J, Freeman WJ, Burke BC (1998) Optimization of olfactory model in software to give 1/f power spectra reveals numerical instabilities in solutions governed by aperiodic (chaotic) attractors. Neural Networks (11: 449–466).

[25] Chen Q, Shi D (2004) The modeling of scale-free networks. Physica A 333: 240–248.

[26] Chua LO (1998) CNN. A Paradigm for Complexity. Singapore: World Scientific.

[27] Dehaene S, Changeux JP, Naccache L, Sackur J, Sergent C (2006) Conscious, preconscious, and subliminal processing: a testable taxonomy. Trends Cogn Sci 10: 204–211.

[28] Erdös P, Renyí A (1960) On the evolution of random graphs. Publ. Math. Inst. Hung. Acad. Sci. 5: 17–61.

[29] Fellemin DJ, Van Essen DC (1991) Distributed hierarchical processing in the primate cerebral cortex. Cerebral Cortex 1: 1–47.

[30] Freeman WJ (1960) Correlation of electrical activity of prepyriform cortex and behavior in cat. J Neurophysiol 23: 111–131.

[31] Freeman WJ (1974) Attenuation of transmission through glomeruli of olfactory bulb on paired shock stimulation. Brain Research 65: 77–90.

[32] Freeman WJ (1975) Mass Action in the Nervous System. New York: Academic Press. Electronic version 2004: http://sulcus.berkeley.edu/MANSWWW/MANSWWW.html

[33] Freeman WJ (1986) Petit mal seizure spikes in olfactory bulb and cortex caused by runaway inhibition after exhaustion of excitation. Brain Research Reviews 11:259–284.

[34] Freeman WJ (1987) Simulation of chaotic EEG patterns with a dynamic model *of* the olfactory system. Biol Cybern 56: 139–50. Fig 12.

[35] Freeman WJ (2000) A proposed name for aperiodic brain activity: Stochastic chaos. Neural Networks 13: 11–13.

[36] Freeman WJ (2004a) Origin, structure, and role of background EEG activity. Part 1. Analytic amplitude. Clin Neurophysiol 115: 2077–2088. http://repositories.cdlib.org/postprints/1006

[37] Freeman WJ (2004b) Origin, structure, and role of background EEG activity. Part 2. Analytic phase. Clin Neurophysiol 115: 2089–2107. http://repositories.cdlib.org/postprints/1486.

[38] Freeman WJ (2005) NDN, volume transmission, and self-organization in brain dynamics. J Integrative Neuroscience 4 (4): 407–421.

[39] Freeman WJ (2006) Definitions of state variables and state space for brain-computer interface. Part 1. Multiple hierarchical levels of brain function. Cogn Neurodyn 1(1): 3–14. http://dx.doi.org/10.1007/s11571-006-9001-x http://dx.doi.org/10.1007/s11571-006-9001-x http://repositories.cdlib.org/post prints/33712007

[40] Freeman WJ (2007) Definitions of state variables and state space for brain-computer interface. Part 2. Extraction and classification of feature vectors. Cogn Neurodyn 1(2): 85–96. http://www.springerlink.com/content/w7737h5tg6551888/ http://www.springerlink.com/content/w7737h5tg6551888/

[41] http://repositories.cdlib.org/postprints/3372

[42] Freeman WJ, Ahlfors SM, Menon V (2009) Combining EEG, MEG and fMRI signals to characterize mesoscopic patterns of brain activity related to cognition. Int J Psychophysiol 73(1): 43–52. http://repositories.cdlib.org/postprints/3386

[43] Freeman WJ, Baird B (1987) Relation of olfactory EEG to behavior: Spatial analysis: Behav Neurosci 101: 393–408.

[44] Freeman WJ, Barrie JM (2000) Analysis of spatial patterns of phase in neocortical gamma EEGs in rabbit. J Neurophysiol 84: 1266–1278.

[45] Freeman WJ, Burke BC, Holmes MD (2003) Aperiodic phase re-setting in scalp EEG of beta-gamma oscillations by state transitions at alpha-theta rates. Human Brain Mapping 19(4): 248–272. http://repositories.cdlib.org/postprints/3347

[46] Freeman WJ, Chang H-J, Burke BC, Rose PA, Badler J (1997) Taming chaos: Stabilization of aperiodic attractors by noise. IEEE Trans Circuits Systems 44:989–996.

[47] Freeman WJ, Erwin H (2008) Freeman K-set. Scholarpedia, 3(2): 3238 http://www.scholarpedia.org/article/Freeman_K-set

[48] Freeman WJ, Grajski, K.A. (1987) Relation of olfactory EEG to behavior: Factor analysis: Behav Neurosci 101: 766–777.

[49] Freeman WJ, Kozma R, Bollobás B, Riordan O (2009) Chapter 7. Scale-free cortical planar network, in: Handbook of Large-Scale Random Networks. Series: Bolyai Mathematical Studies, Bollobás B, Kozma R, Miklös D (Eds.), New York: Springer, Vol. 18, pp. 277–324. http://www.springer.com/math/numbers/book/978-3-540-69394-9

[50] Freeman WJ, Vitiello G (2006) Nonlinear brain dynamics as macroscopic manifestation of underlying many-body field dynamics. Physics of Life Reviews 3: 93–118. http://dx.doi.org/10.1016/j.plrev.2006.02.001 http://dx.doi.org/10.1016/

j.plrev.2006.02.001, http://repositories.cdlib.org/postprints/1515

[51] Freeman WJ, Vitiello G (2010) Vortices in brain waves. International Journal Modern Physics B, Vol: 24 (17): pp. 3269–3295. http://dx.doi.org/10.1142/S0217979210056025 http://dx.doi.org/10.1142/S0217979210056025

[52] Freeman WJ, Zhai J (2009) Simulated power spectral density (PSD) of background electrocorticogram (ECoG). Cognitive Neurodynamics 3(1): 97–103. http://dx.doi.org/10.1007/s11571-008-9064-y http://dx.doi.org/10.1007/s11571-008-9064-y

[53] Friston K (2009) The free-energy principle: a rough guide to the brain? Trends Cogn Sci 13(7): 293–301 doi:10.1016/j.tics.2009.04.005

[54] Haken H (1983) Synergetics: An Introduction. Berlin: Springer-Verlag.

[55] Hu M, Li J, Li G, Tang X, Freeman WJ (2006) Normal and hypoxia EEG recognition based on a chaotic olfactory model. LNCS Vol 3973: 554–559, doi: 10.1007/11760191

[56] Ilin R, Kozma R, Werbos PJ (2008) Beyond backpropagation and feedforward models: a practical training tool for a more efficient universal approximator. IEEE Trans Neur Netw 19(3), 929–937.

[57] Kaas JH (1987) The organization of neocortex in mammals: implications for theories of brain function. Ann Rev Psychol 38: 129–152.

[58] Kelso JAS (1995) Dynamic Patterns: The Self Organization of Brain and Behavior. Cambridge MA: MIT Press.

[59] Kelso JAS, Tognoli E (2006) Metastability in the brain. Neural Networks IJCNN'06: 363–368. doi: http://dx.doi.org/10.1109/IJCNN.2006.24670410.1109/IJCNN.2006.246704

[60] Kozma R (2003) On the constructive role of noise in stabilizing itinerant trajectories on chaotic dynamical systems. Chaos 11(3): 1078–1090.

[61] Kozma R, Freeman WJ (2002) Classification of EEG patterns using nonlinear dynamics and identifying chaotic phase transitions. Neurocomputing 44: 1107–1112.

[62] Kozma R, Freeman WJ (2003) Basic principles of the KIV model and its application to the navigation problem. J Integr Neurosci 2(1): 125–146.

[63] Kozma R, Freeman WJ, Erdí P (2003) The KIV model — nonlinear spatiotemporal dynamics of the primordial vertebrate forebrain. Neurocomputing 52–54: 819–826.

[64] Kozma R, Huntsberger T, Aghazarian H, Tunstel E, Ilin R, Freeman WJ (2008) Implementing intentional robotics principles using SSR2K platform. Advanced Robotics 22(12): 1309–1327.

[65] Kozma R, Puljic M, Bollobas, B, Balister, P, Freeman, WJ (2005) Phase transitions in the neuropercolation model of neural populations with mMixed local and non-local interactions. Biol. Cybernetics, 92(6), 367–379.

[66] Li G, Li L, Wang W, Freeman WJ (2004) Face recognition using a neural network simulating olfactory systems. LNCS Vol 3972: 93–97.

[67] Logothetis NK (2008) What we can do and what we cannot do with fMRI. *Nature* 453: 869–878. doi:10.1038/nature06976

[68] Lohmann H, Roerig B (1994) Long-range horizontal connections between supragranular pyramidal cells in the extrastriate visual cortex of the rat. J Comp Neurol 344: 543–558.

[69] Malach R (1994) Cortical columns as devices for maximizing neuronal diversity. TINS 17: 101–104.

[70] Merleau-Ponty M (1942/1963) The Structure of Behavior. Fischer AL (trans) Boston: Beacon.

[71] Miller R (1996) Neural assemblies and laminar interactions in the cerebral cortex. Biol Cybern 75: 253–261.

[72] Principe, J.C., Tavares, V.G., Harris, J.G. and Freeman, W.J. (2001) Design and implementation of a biologically realistic olfactory cortex in analog VLSI. Proceedings IEEE 89: 1030–1051.

[73] Rapp P (1993) Chaos in the neurosciences: Cautionary tales from the frontier. Biologist 40: 89–94.

[74] Ruiz Y, Li G, Freeman WF, Gonzalez E. (2009) Detecting stable phase structures on EEG signals to classify brain activity amplitude patterns. J Zhejiang Univ 10(10): 1483–1491.

[75] Schroeder M (1991) Fractals, chaos, power laws. Minutes from an infinite paradise. WH Freeman, San Francisco.

[76] Shannon CE (1948) A mathematical theory of communication. Bell System Tech. J. **27**: 379–623.

[77] Shimoide K, Freeman WJ (1995) Dynamic neural network derived from the olfactory system with examples of applications. IEICE Transaction Fundamentals E-78A: 869–884.

[78] Skarda, C.A., Freeman WJ (1987) How brains make chaos in order to make sense of the world. Behavioral and Brain Sciences 10: 161–195.

[79] Sporns O, Tononi G, Kötter R (2005) The human connectome: A structural description of the human brain. PLoS Comput Biol 1: 245–251. doi:10.1371/journal.pcbi.0010042

[80] Swadlow HA (1994) Efferent neurons and suspected interneurons in motor cortex of the awake rabbit: axonal properties, sensory receptive fields and subthreshold synaptic inputs. J Neurophysiol 71: 437–453.

[81] Tsuda I (2001) Towards an interpretation of dynamic neural activity in terms of chaotic dynamical systems. Behav Brain Sci 24:793–810.

[82] von Neumann J (1958) The Computer and the Brain. New Haven CT: Yale UP.

[83] Wang XF, Chen GR (2003) Complex networks: small-world, scale-free and beyond. IEEE Trans Circuits Syst, 31: 6–20.

[84] Watts DJ, Strogatz SH (1998) Collective dynamics of 'small-world' networks. Nature 393: 440–442.

[85] Yang X, Fu J, Lou Z, Wang L, Li G Freeman, WJ (2006) Tea classification based on artificial olfaction using bionic olfactory neural network. In: Wang X, Zhang L, Sun Y, Kim K-B, Lu W, Wu Y, Li B, Wang J, Wang W Wang, H (Eds.): Advances in Neural Networks — ISNN 2006. Lecture Notes in Computer Science. Heidelberg: Springer, pp. 343–348.

[86] Yao Y, Freeman WJ, Burke B, Yang Q (1991) Pattern recognition by a distributed neural network: An industrial application. Neural Networks 4: 103–121.

CHAPTER 23

NONLINEAR DYNAMICS OF A CLASS OF PIECEWISE LINEAR SYSTEMS

M. LAKSHMANAN

Centre for Nonlinear Dynamics, School of Physics,
Bharathidasan University, Tiruchirapalli 620024, India
lakshman@cnld.bdu.ac.in

K. MURALI

Department of Physics, Anna University, Chennai 600025, India
kmurali@annauniv.edu

Piecewise linear dynamical systems constitute an interesting class of nonlinear dynamical systems exhibiting many kinds of bifurcations and chaos, synchronizations and spatiotemporal patterns. They can be relatively easily designed through electronic circuits and the corresponding dynamical equations can be numerically analysed and theoretically investigated. The ubiquitous Chua's circuit, Murali-Lakshmanan-Chua (MLC) circuit, memristor based circuits, and negative conductance forced LCR circuits, as individual and coupled oscillators, are some of the well studied piecewise linear dynamical systems. More interestingly time-delay can be introduced in such circuits to study plethora of novel effects in complex systems, including networks. In this article, recent progress in understanding the nonlinear dynamics of such systems, especially of MLC circuits and its variants, will be briefly reviewed.

1. Introduction

Professor Leon Chua's multifaceted research interests and contributions have been a great source of inspiration for researchers in nonlinear dynamics and circuit theory around the world for the past five decades or so. Study of the nonlinear dynamics of piecewise-linear systems and circuits are essentially the outcome of the invention of Chua's circuit by Leon Chua in 1983 [4, 6, 21]. The nonlinear dynamics community has been indeed greatly influenced by Chua. Our group (specializing in theoretical physics) at Tiruchirapalli, a remote southern Indian town, was greatly motivated by a series of suggestions and encouragement by Leon Chua since early 1990 to supplement our theoretical analysis of nonlinear systems by taking up experimental works based on nonlinear electronic circuits. The present article essentially reviews our efforts to understand the behaviour of nonlinear dynamical systems through piecewise-linear systems, circuits, coupled systems and arrays, starting from MLC circuits, as a grateful appreciation to Professor Chua on his seventy fifth birthday.

In the field of nonlinear dynamics, the study of nonlinear electronic circuits is a convenient framework to undertake systematic exploration of fundamental mechanisms underlying the onset of bifurcations and chaos [12, 21, 24, 31]. Over the past three decades, piecewise linear circuits, either of autonomous or of nonautonomous nature, have become simple yet powerful models for studying various bifurcation structures and chaos phenomenon. The piecewise linear circuits usually consist of at least one circuit element whose transfer characteristics is piecewise linear in nature. These circuits are natural extensions of their linear counterparts, yet they exhibit a repertoire of nonlinear dynamical phenomena. From a practical point of view, these circuits are quite easy to construct, analyze, observe and even to prove the admittance of various bifurcations and chaos therein [6].

A variety of nonlinear electronic circuits consisting of piecewise linear circuit elements which exhibit chaos have been investigated in the recent literature [20]. Among them, the most prominent one is the autonomous, classic, Chua's circuit [6], which is represented by a system of three autonomous first order coupled ODEs. The system admits a variety of dynamical phenomena, including limit cycles, homoclinic and heteroclinic orbits, strange attractors, etc. Further, its higher dimensional version, namely the driven Chua's circuit [26], which is fourth order in nature, has been investigated as a black box exhibiting a rich variety of bifurcations and chaos phenomenon, ranging from period-doubling route to chaos to quasiperiodicity, intermittency, period adding sequences and so on.

Besides the above, even simpler chaotic circuits, namely the Murali-Lakshmanan-Chua (MLC) circuit and the variant MLC circuits, consisting of the piecewise linear resistor (Chua's diode) as the only nonlinear element, have been suggested to exhibit a wide variety of bifurcations and chaos phenomenon. In recent times several other piecewise linear circuits have also been suggested to study quasiperiodic orbits, strange nonchaotic attractors, hyperchaotic attractors and different kinds of synchronisations: These include the forced series LCR circuit with negative conductance [45], Inaba-Mori circuit [13, 14] and piecewise linear time delay systems [33]. Coupled arrays of such systems are also of considerable interest from the point of view of various types of synchronizations.

In this article, we will review briefly the bifurcations, chaos and synchronization phenomena exhibited by some of the above piecewise linear circuits using exact solutions, numerical analysis and experimental results. A preliminary version of these results has appeared earlier in Ref. [22]. Here we present a more upto date review of the progress. To start with, in Sec. 2, we will present the nonautonomous MLC circuit as a system of two coupled first order ordinary differential equations with a piecewise linearity representing the current-voltage characteristic of the nonlinear resistor, namely, the Chua's diode. This set of equations can also be rewritten as a single second order nonautonomous linear differential equation with discontinuous right hand side. Obtaining the exact solutions sector wise, and matching the solutions across the boundaries, we present the bifurcation and chaos scenario by varying the control parameters. The results are corroborated by direct numerical integration and experiments. In Sec. 3, the dynamics of a variant of the MLC circuit which belongs to a family of MLC circuit classification is discussed. Next, in Sec. 4, the spatio-temporal dynamics of coupled MLC circuits has been considered. Following this, in Sec. 5, nonlinear dynamics of certain simple electronic circuits consisting memristor, in particular memristor based MLC circuit, has been studied. Then, in Sec. 6, we point out how quasiperiodically forced piecewise linear systems are natural candidates to exhibit strange nonchaotic attractors of various types. Next, the bifurcation and chaos scenario in a scalar piecewise linear time delayed system is briefly described in Sec. 7. In Sec. 8, the mechanism of chaotic synchronization of coupled time-delay circuits and systems has been considered. Finally, in Sec. 9, we summarize our results.

2. The Basic MLC Circuit and its Chaotic Dynamics

The basic Murali-Lakshmanan-Chua (MLC) circuit [27], shown schematically in Fig. 1(a), consists of a capacitance (C), an inductance (L), a resistor (R), all of which are linear circuit elements, an external periodic forcing $f(t) = F \sin \omega t$ and only one nonlinear element, namely, the Chua's diode (N_R) which satisfies the piecewise linear current-voltage characteristic [Fig. 1(c)],

$$i_R = h(v_R), \tag{1}$$

where

$$h(v_R) = \begin{cases} G_b v_R + (G_b - G_a)B_p, & v_R \le -B_p \\ G_a v_R, & -B_p \le v_R \le B_p \\ G_b v_R + (G_a - G_b)B_p & v_R \ge B_p. \end{cases} \tag{2}$$

Essentially the circuit in Fig. 1(a) is a standard forced LCR circuit to which a nonlinear resistor, N_R, is connected in parallel. A current sensing resistor R_s is also included in the circuit for measurement purpose. Applying Kirchoff's laws for currents and voltages to the circuit in Fig. 1(a), one can write down the state equations as a system of two coupled first order piecewise linear differential equations,

$$C\frac{dv_R}{dt} = i_L - h(v_R), \tag{3a}$$

$$L\frac{di_L}{dt} = -Ri_L - R_s i_L - v_R + F \sin \omega t. \tag{3b}$$

Here i_L is the current along the inductor L and $v_R = v$ is the voltage across the capacitor C, while $h(v_R)$ is the piecewise linear function defined by equation (2).

Introducing the scaled variables

$$x = v/B_p, \quad y = i_L/GB_p, \quad t' = tG/C, \\ \omega' = \omega C/G, \quad G = 1/R, \tag{4}$$

Eq. (3) can be rewritten (after dropping the primes) as

$$\dot{x} = y - h(x), \quad (\cdot) = \frac{d}{dt} \tag{5a}$$

$$\dot{y} = -\beta(1+\nu)y - \beta x + f \sin(\omega t), \tag{5b}$$

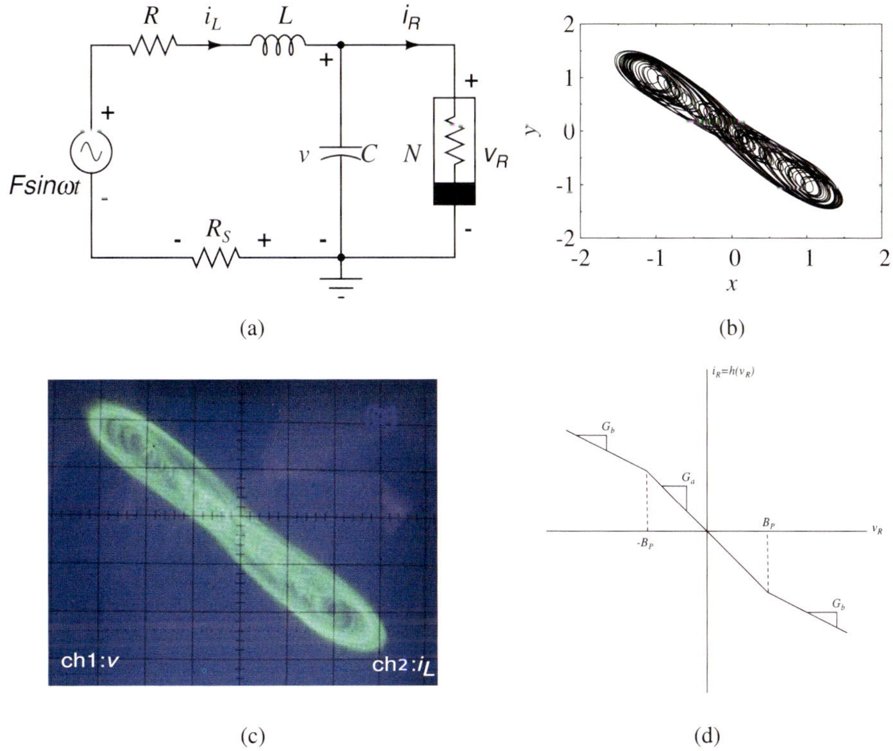

Fig. 1. (a) Experimental realization of the MLC circuit, (b) its typical double scroll chaotic attractor in the $(x-y)$ plane (numerical), (c) double scroll chaotic attractor in the $(v-i_L)$ plane (experimental) and (d) $(i_R - v_R)$ characteristic of Chua's diode.

where $\beta = C/LG^2$, $\nu = GR_s$ and $f = F\beta/B_p$. Here

$$h(x) = \begin{cases} bx + a - b, & x \geq 1 \\ ax, & |x| \leq 1 \\ bx - a + b & x \leq -1 \end{cases} \tag{6}$$

The parameters are $a = G_a/G$, $b = G_b/G$. Now the dynamics of (5) depends on the six parameters ν, β, a, b, ω and f.

2.1. *Explicit analytic solutions and period doubling route to chaos*

The dynamical equations (5) can be equivalently written as a second order linear differential equation with discontinuous right hand side:

$$\ddot{x} + \beta(1+\nu)\dot{x} + \beta x - f\sin\omega t = \Delta_i(t, x, \dot{x}), \quad i = 1, 2, 3, \tag{7}$$

where

$$\Delta_1 = -b\dot{x} - \beta b(1+\nu)x - \beta(1+\nu)(a-b), \quad \text{in } D_0\{x > +1\} \tag{8a}$$

$$\Delta_2 = -a\dot{x} - \beta a(1+\nu)x, \quad \text{in } D_{+1}\{-1 < x < +1\} \tag{8b}$$

$$\Delta_3 = -b\dot{x} - \beta b(1+\nu)x + \beta(1+\nu)(a-b), \quad \text{in } D_{-1}\{x < -1\}. \tag{8c}$$

A nearest known system is the bilinear oscillator defined by the equation (see for example, [1])

$$\ddot{u} + 2\zeta_i\dot{u} + k_i^2 u = \beta_i\cos\omega t + \Theta_i, \quad (\zeta_i, k_i, \beta_i, \Theta_i = \text{constants}), \tag{9}$$

where $i = 1$ if $u > 0$ and $i = 2$ if $u < 0$. Possible discontinuity induced bifurcations of equilibria and limit cycles of the above type of nonsliding systems have been reviewed recently [2].

The MLC circuit equations may be considered as an example of a nonsliding piecewise smooth system (PWS) with two discontinuities at $x = -1$ and $x = +1$. One way to study the dynamics of this system is to use the exact solutions of the underlying linear inhomogenous equations in each of the regions D_0, D_{+1} and D_{-1} and match them suitably across the boundaries at $x = \pm 1$ so as to obtain different kinds of dynamical behaviours as a function of one or more control parameters. In fact, the general solution in each of the three regions can be given as

$$x(t) = C^1_{0,\pm}\exp(\alpha_1 t) + C^2_{0,\pm}\exp(\alpha_2 t) + E_1 + E_2\sin\omega t + E_3\cos\omega t, \tag{10}$$

where $C_{0,\pm}$ are arbitrary constants in the corresponding regions $D_0, D_{\pm 1}$. The other parameters are given by

$$\alpha_1 = \frac{-A + \sqrt{A^2 - 4B}}{2}, \quad \alpha_2 = \frac{-A - \sqrt{A^2 - 4B}}{2}, \tag{11a}$$

where

$$A = \beta(1+\nu) + \mu, \; B = \beta(1 + \mu(1+\nu)), \quad \mu = a \text{ in } D_0 \quad \text{and} \quad \mu = b \text{ in } D_{\pm 1} \tag{11b}$$

and

$$E_1 = 0 \text{ in } D_0,$$
$$E_1 = \mp\beta(1+\nu)(a-b)/B \text{ in } D_{\pm 1},$$
$$E_2 = [f(B+\omega^2)/(\alpha_1^2+\omega^2)(\alpha_2^2+\omega^2)],$$
$$E_3 = [fA\omega/(\alpha_1^2+\omega^2)(\alpha_2^2+\omega^2)]. \tag{11c}$$

Knowing $x(t)$, $y(t)$ can be evaluated from Eq. (5) as

$$y(t) = \dot{x} + h(x). \tag{12}$$

Starting from the given initial conditions $x(0)$ and $y(0)$ in a given region, say D_0, the arbitrary constants C^1_0 and C^2_0 can be fixed. Then as the system evolves, the variable $x(t)$ may cross one of the boundaries, say $x = +1$ and enter into the region D_{+1} at a time $t = T_1$, provided $\dot{x}(T_1) > 0$. (If it crosses the boundary $x = -1$ into the region D_{-1}, then $\dot{x}(T_1) < 0$). Then the system will evolve in D_{+1} with the above solution with the arbitrary constants C^1_+ and C^2_+ fixed by the values of $x(T_1)$ and $y(T_1)$ from Eqs. (10) and (12). Then the trajectory can again come back to the region D_0, if $x(t)$ again crosses the boundary $x = +1$ at time $T = T_2 > T_1$, provided $\dot{x}(T_2) < 0$. Following the system again into D_0 with arbitrary constants C^1_0 and C^2_0 fixed by the values $x(T_2)$ and $y(T_2)$ through Eqs. (10) and (12), one can determine the time T_3 for the next crossing either into D_{-1} or D_{+1}. The procedure can be followed ad infinitum and the nature of the solution can be fixed. Following this procedure one can trace the entire trajectory starting from a given initial condition.

In Fig. 2, we present one parameter bifurcation diagrams which clearly show the presence of period-doubling bifurcations or inverse period doubling bifurcations followed by more complex bifurcations. A two parameter $(F - \omega)$ phase diagram

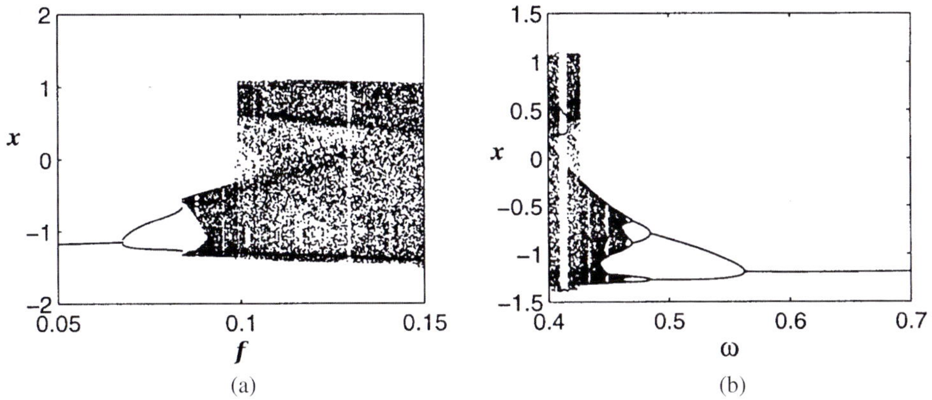

Fig. 2. One parameter bifurcation diagrams for (a) control parameter f (period doubling bifurcations) and (b) control parameter ω (inverse period doubling) for Eq. (5).

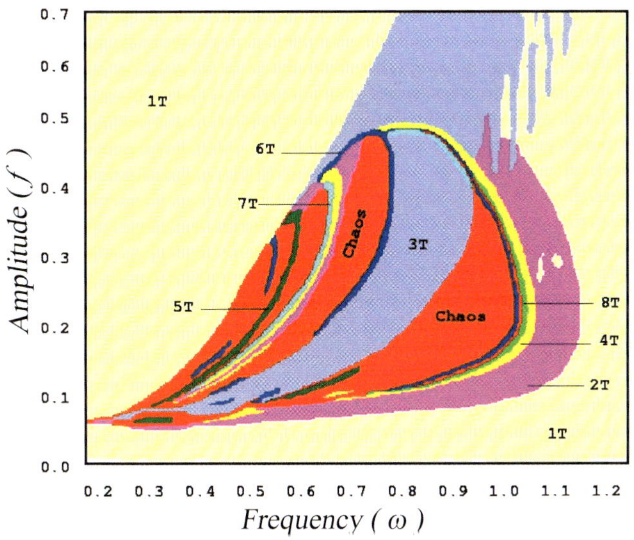

Fig. 3. Numerically obtained two parameter bifurcation diagram of the MLC circuit in the $(f - \omega)$ plane. Color: light yellow, period-1; magneta, period-2; yellow, period-4; light green, period-8; blue, period-3 window; green, period-5 window; violet, period-6 window and red, chaos.

is also presented in Fig. 3, where we identify a rich variety of bifurcations. These include

i) Period doubling bifurcations and chaos
ii) Period halving bifurcations
iii) Antimonotonicity
iv) Intermittency

and so on. For more details, see [44].

3. A Family of MLC Circuits

The family of MLC circuits, consisting of the basic MLC circuit and its different variants are shown in Fig. 4. These circuits exhibit very interesting dynamical phenomena including bifurcations and chaos. For example, we consider a variant of the MLC circuit [43, 44] discussed in Sec. 2. Instead of the series LCR circuit considered in Fig. 4(a), we now use parallel LCR circuit and include the nonlinear resistor, namely the Chua's diode (N_R) in parallel. The variant circuit is shown in Fig. 4(b).

The governing equations of motion of the variant of MLC circuit are

$$C\frac{dv_R}{dt} = \frac{1}{R}(F\sin(\Omega t) - v_R) - i_L - h(v_R), \tag{13a}$$

$$L\frac{di_L}{dt} = v_R, \tag{13b}$$

where $h(.)$ is the piecewise linear function given by Eq. (2).

As in the case of the MLC circuit discussed in Section 2, for the present circuit also the presence of a rich variety of bifurcations and different routes to chaos have been reported experimentally, confirmed numerically and proven mathematically [43, 44].

Rescaling the circuit equations (13) through the relations $x = \frac{v}{B_p}, y = \frac{i_L}{GB_p}$, $\omega = \frac{\Omega C}{G}, G = \frac{1}{R}$ and $t' = \frac{tG}{C}$ and then redefining t' as t, we have the dynamical equations

$$\begin{aligned} \dot{x} &= f\sin\omega t - x - y - h(x), \\ \dot{y} &= \beta x, \quad \cdot = \frac{d}{dt}, \end{aligned} \tag{14}$$

where $\beta = \frac{C}{LG^2}$, $f = \frac{F}{B_p}$.

The piecewise linear function $h(x)$ has again the piecewise behaviour as given by Eq. (6). Equations (14) can also be equivalently written as a single second order linear differential equation with

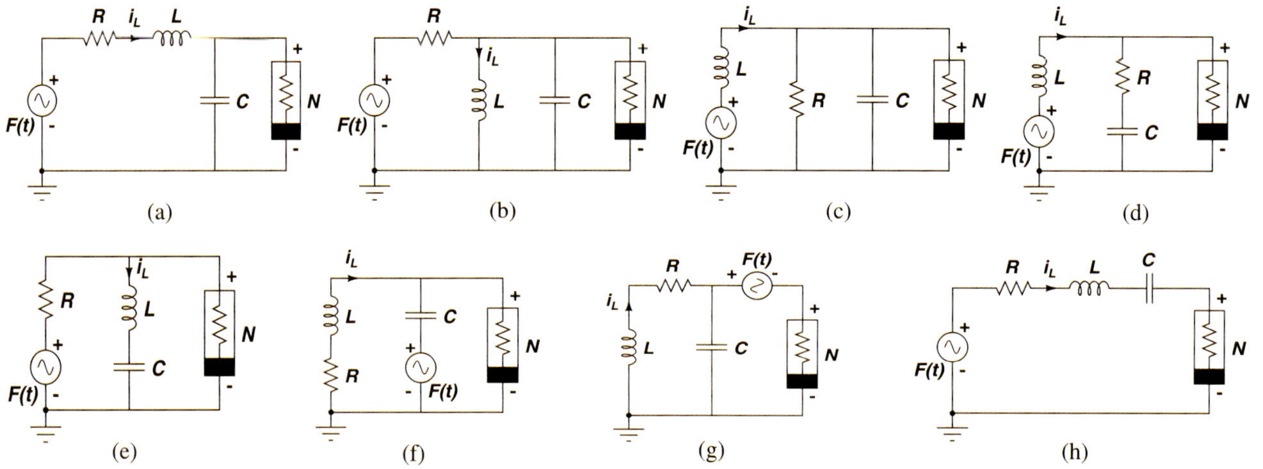

Fig. 4. (a–h) Family of MLC circuits.

discontinuities as in the case of the bilinear oscillator [1, 2]. In fact we can rewrite Eq. (14) as

$$\ddot{x} + \dot{x} + \beta x + \omega f \cos \omega t = h'(x)\dot{x}, \tag{15}$$

where

$$h'(x) = \begin{cases} b, & x > 1 \\ a, & -1 < x < 1 \\ b, & x < -1 \end{cases} \tag{16}$$

We can write down the explicit soloutions as in the previous section and match the solutions across the boundaries or carry out a numerical analysis of Eq. (14) or (15). Pspice simulations and experimental verifications have also been carried out in detail. A rich variety of bifurcations and chaos results in. These include the following:

i) Quasiperiodic and torus breakdown route to chaos
ii) Period-adding sequences and Farey sequences
iii) Period doubling and reverse period doubling bifurcations
iv) Remerging Feigenbaum trees, bubbling, band merging and antimonotonicity
v) Intermittency route to chaos
vi) Neimark bifurcation

To provide an idea of the varied kinds of dynamical phenomena admitted by this piecewise linear system (14), we present in Fig. 5 a two parameter bifurcation diagram in the amplitude-frequency $(f - \omega)$ space of the external force, while all the

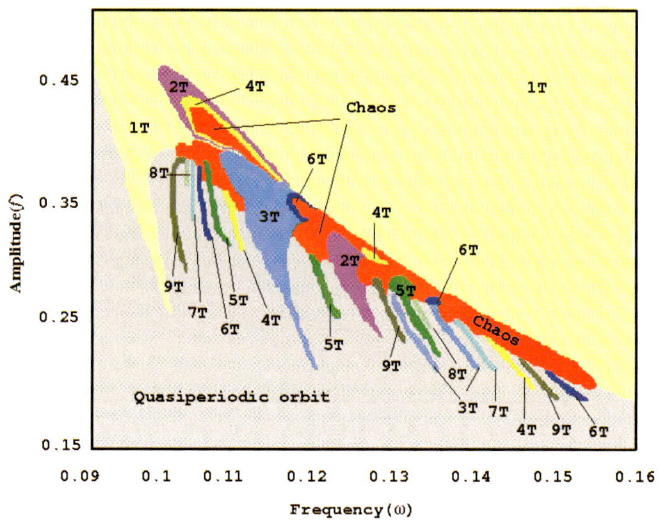

Fig. 5. Numerical two parameter phase diagram of variant MLC circuit (13b) in the $(f - \omega)$ plane. Colour: color code as given in Fig. 3.

other parameters are fixed. The diagram covers the transitions in the region of the external forcing frequency, $0.09 \le \omega \le 0.16$ and the forcing amplitude, $0.15 \le f \le 0.5$. Here also each coloured region denotes a particular type of steady state behaviour, identical to the phase diagram of the MLC circuit, see Fig. 3. Periodic windows are indicated by the corresponding numbers (such as 5T, 6T etc.). The numerical results indicated in Fig. 5 have also been confirmed with experimental results. More details may be obtained from references [43, 44].

In a similar way, the dynamics of the remaining circuits of the family given in Figs. 4 can be analysed. The details will be published elsewhere.

4. Spatiotemporal Dynamics of Coupled MLC Circuits

Coupled nonlinear oscillators in the form of arrays arise in many branches of science. These coupled arrays are often used to model very many spatially extended dynamical systems explaining reaction-diffusion process. One can visualize such an array as an assembly of a number of subsystems coupled to their neighbors, or as coupled nonlinear networks (CNNs) [5, 7]. For the past few years several investigations have been carried out to understand the spatiotemporal behaviours of coupled nonlinear oscillators and systems. The studies on these systems include the traveling wave phenomenon, Turing patterns, spatiotemporal chaos and synchronization [8, 18]. In this section we consider similar arrays of coupled nonlinear electronic circuits consisting of MLC circuit as a basic unit, which is essentially a two variable model.

4.1. *Arrays of MLC circuits*

We will now consider one and two-dimensional arrays of MLC circuits [28], where the intercell couplings are effected by linear resistors. Such resistive couplings are then equivalent to diffusive coupling.

4.1.1. *One-dimensional array*

Figure 6 shows a schematic representation of an one-dimensional chain of resistively coupled MLC circuits. Note that the individual cells correspond to isolated MLC circuits. Using the standard Kirchhoff's laws, the dynamics of the one-dimensional chain can be easily shown to be governed by the following system of equations, in terms of the rescaled variables,

$$\dot{x}_i = y_i - h(x_i) + D(x_{i+1} + x_{i-1} - 2x_i), \tag{17a}$$

$$\dot{y}_i = -\sigma y_i - \beta x_i + F\sin\omega t, \quad i = 1, 2, \ldots, N, \tag{17b}$$

where D is the diffusion coefficient, N is the chain length and $h(x)$ is the three segment piecewise linear function representing the current voltage characteristic of the Chua's diode,

$$h(x) = \begin{cases} \epsilon + m_2 x + (m_0 - m_1), & x \geq x_2 \\ \epsilon + m_0 x, & x_1 \leq x \leq x_2 \\ \epsilon + m_1 - (m_0 - m_1), & x \leq x_1 \end{cases} \tag{18}$$

where ϵ is the dc offset included for convenience. Naturally, depending on the choice of the slopes m_0, m_1 and m_2, one can fix the characteristic curve of the Chua's diode. For typical forms of $h(x)$, the coupled MLC circuit exhibits interesting dynamics such as wave propagation and its failure, Turing patterns, spatio-temporal chaos and so on. In the following discussion, we will fix N at $N = 100$.

4.1.2. *Two-dimensional array*

As in the case of the one-dimensional array introduced above, one can also consider a two-dimensional square array with each cell in the array being coupled to four of its nearest neighbours, for example, with linear resistors. The state equations can be now written in dimensionless form as

$$\begin{aligned} \dot{x}_{i,j} &= y_{i,j} - h(x_{i,j}) + D_1(x_{i+1,j} \\ &\quad + x_{i-1,j} + x_{i,j+1} + x_{i,j-1} - 4x_{i,j}) \\ &\equiv f(x_{i,j}, y_{i,j}), \\ \dot{y}_{i,j} &= -\sigma y_{i,j} - \beta x_{i,j} + D_2(y_{i+1,j} + y_{i-1,j} \\ &\quad + y_{i,j+1} + y_{i,j-1} - 4y_{i,j}) + F\sin\omega t \\ &\equiv g(x_{i,j}, y_{i,j}), \quad i, j = 1, 2, \ldots, N. \end{aligned} \tag{19}$$

This two-dimensional array has $N \times N$ cells arranged in a square lattice. In our numerical study we will again take $N = 100$.

In this analysis we fix the parameter at $\{\beta, \sigma, m_0, m_1, m_2, \epsilon, F\} = \{1.0, 1.0, -2.25, 1.5, .25, 0, 0\}$ so that the system admits bistability which is a necessary condition to observe a wave front. Carrying out

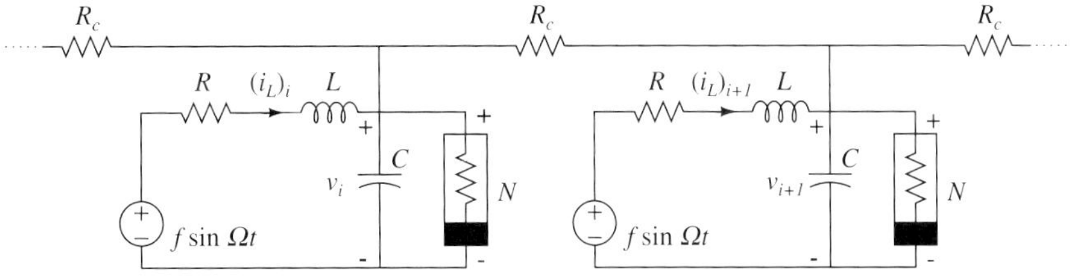

Fig. 6. Circuit realization of a one-dimensional array of coupled MLC circuits.

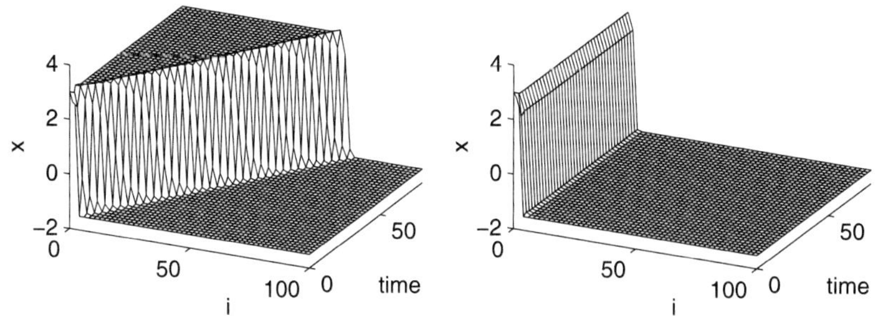

Fig. 7. Space-time plot (a) showing propagation of wave front in one-dimensional array (100 cells) of MLC circuits for $D = 2.0$ and (b) propagation failure for $D = 0.4$.

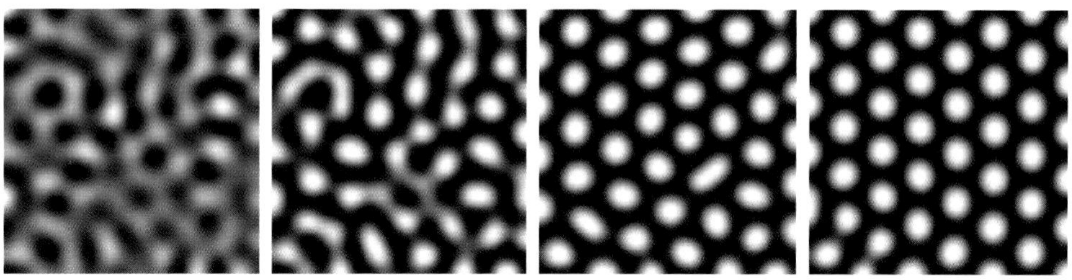

Fig. 8. The spontaneous formation of Turing patterns in an array of 100×100 oscillators for the parameters $\beta = 0.8$, $\sigma = 1$, $m_0 = -0.5$, $m_1 = 0.5$, $m_2 = 0.5$, $\epsilon = 0.1$, $F = 0.0$, $\omega = 0.75$, $D_1 = 1$ and $D_2 = 10$ in Eqs. (19) at various time units (a) $T = 50$, (b) $T = 100$, (c) $T = 500$ and (d) $T = 2000$.

a numerical integration of equations (17) with $N = 100$ and with the diffusion coefficients chosen at a higher value, $D = 2.0$, a motion of the wave front towards right (see Fig. 7) is observed, that is a travelling wave front is found. However, below a critical value of the diffusion coefficient ($D = D_c$) a failure in the propagation has been observed, which in the present case turns out to be $D = D_c = 0.4$. This means that the initiated wavefront is unable to move as time progresses and Fig. 7(b) shows the propagation failure for $D = 0.4$.

Another interesting dynamical phenomenon in the coupled arrays is the formation of Turing patterns. For this purpose, we have fixed the parameters for the two-dimensional model (19) as $\{\beta, \sigma, m_0, m_1, m_2, \ \epsilon, F\} = \{0.8, 0.92, 0.1, -0.5, 0.5, 0.5, \ 0\}$. The numerical simulations have been carried out using an array of size 100×100 and random initial conditions near the steady states have been chosen for the x and y variables. Figures 8(a)–(d) show how the diffusion driven instability leads to stable hexagonal pattern (Fig. 8(d)) after passing through intermediate stages (Figs. 8(a)–8(c)).

5. Dynamics of a Hyperchaotic Memristive MLC Circuit

In 1971 Leon Chua, while trying to place on a firm mathematical footing the interconnections between the three basic circuit elements known at that time, namely the resistor, capacitor and inductor, felt the need for a fourth basic circuit element, based on symmetry considerations, and predicted its existence [3]. He even postulated that this element should not only have a resistive property but should also be endowed with a memory nature. In order to justify its dual properties, he named this hypothetical element as memristor. Some three and a half decades later, Strutkov and his co-workers at Hewlett-Packard Labs reported the existence of this elusive element in nano-scale devices [41]. Ever since this discovery, much research has been going on to utilise the memristor in constructing nonlinear circuits [15, 16, 29, 48, 51, 52].

As memristance is a fundamental circuit quantity, Chua inferred that no combination of passive devices can reproduce the properties of a memristor. Nevertheless he theorised that, in principle, memristors with almost any $\phi - q$ characteristic

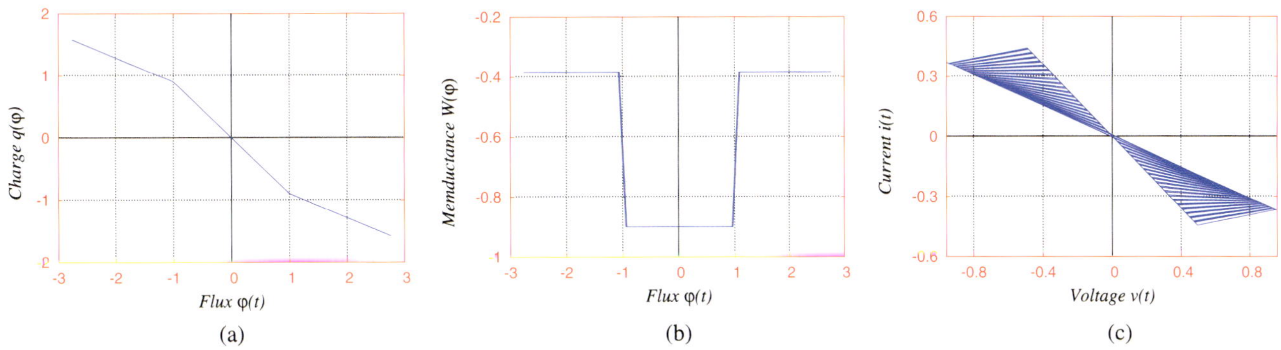

Fig. 9. (a) The characteristic curve of the memristor in the $(\phi - q)$ plane, (b) the variation of the memductance $W(\phi)$ as a function of the flux $\phi(t)$ across the memristor, (c) the $(v - i)$ characteristic of the memristor showing switching between the bi-stable memductance states in the normalized voltage versus current $(x - y)$ plane.

can be synthesised using analog circuits [3]. Itoh and Chua had proposed theoretically a three segment piecewise linear memristor with characteristic curves similar to those of Chua's diode [17], albeit with a difference. While the characteristic curve of the Chua's diode lies in the $(v - i)$ plane, that of the proposed memristor falls in the $(\phi - q)$ plane. The piecewise linear relationship for a Chua type flux controlled memristor is given as

$$q(\phi) = b\phi + 0.5(a - b)[(|\phi + 1|) - (|\phi - 1|)], \quad (20)$$

where a and b are the slopes in the inner and outer segments of the characteristic curve. This characteristic curve of the memristor in the $(\phi - q)$ plane is shown in Fig. 9(a). The current through the memristor is given by

$$i(t) = W(\phi)v(t), \quad (21)$$

where $W(\phi)$ is the memductance of the memristor. It is so called because it has the units of conductance (℧) and is given by

$$W(\phi) = \frac{dq(\phi)}{d\phi} = \begin{cases} a, & |\phi| \leq 1 \\ b, & |\phi| > 1 \end{cases} \quad (22)$$

where a, b are as defined earlier. From Eq. (22) we find that the memristor has a bistable nature (ie) it switches from a high conductance ON state to a low conductance OFF state as shown in Fig. 9(b). This bistable nature causes the current and voltage to vary with time as shown in Fig. 9(c).

A memristive MLC circuit can be constructed by replacing the Chua's diode of an ordinary MLC circuit with an active three segment piecewise linear flux controlled memristor described above, as its nonlinearity. This circuit is shown in Fig. 10.

Applying Kirchoff's laws, the circuit equations for the memristive MLC circuit given by Fig. 10 are written as

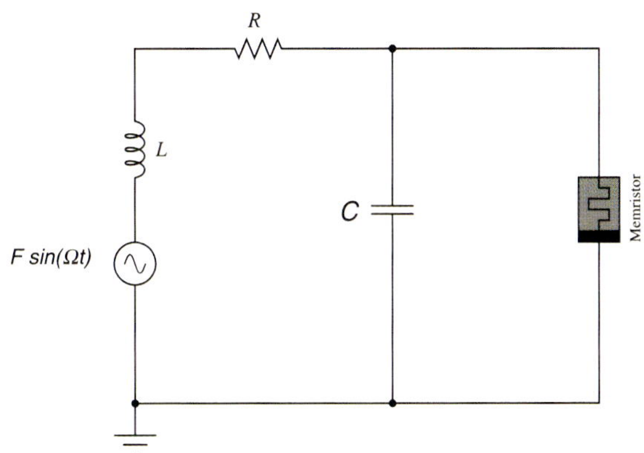

Fig. 10. Driven memristive MLC circuit.

$$C\frac{dv_1}{dt} = i - W(\phi)v,$$
$$L\frac{di}{dt} = -v - ri + F\sin(\Omega t),$$
$$\frac{d\phi}{dt} = v. \quad (23)$$

The normalized form of these circuit equations (23) are

$$\dot{x} = y - W(z)x,$$
$$\dot{y} = -\beta(x + y) + f\sin(\omega t),$$
$$\dot{z} = x. \quad (24)$$

Here $W(z)$ is the normalized value of the memductance of the memristor, given as

$$W(z) = \frac{dq(z)}{dz} = \begin{cases} a, & |z| \leq 1 \\ b, & |z| > 1 \end{cases} \quad (25)$$

where a, b are now the normalized values of the quantities mentioned earlier. The rescaling relationships used for normalization are

$$x = v,\ y = \frac{i}{G},\ z = \frac{G\phi}{C},\ G = \frac{1}{R},\ \beta = \frac{C}{LG^2},$$
$$\omega = \frac{\Omega C}{G} = \frac{2\pi\nu C}{G},\ \tau = \frac{Gt}{C},\ f = F\beta. \qquad (26)$$

The bistability nature of the memristor makes the memristive MLC circuit a first order piecewise-smooth continuous system of the Flippov type [1,2], due to the discontinuity is in $W(z)$ at $z \pm 1$. Referring to the memductance characteristic of the memristor in Fig. 9(b), we find that the memristor switches states at $z = +1$ and at $z = -1$ either from a more conductive ON state to a less conductive OFF state or vice versa depending on how we proceed across the discontinuity boundary. Consequently the phase space can be divided into three subspaces S_1, S_2 and S_3 due to the presence of two switching manifolds $\Sigma_{1,2}$ and $\Sigma_{2,3}$ defined by the zero set of the smooth functions $H_1(\mathbf{x}, \mu) = (1 - z), z = +1$ and $H_2(\mathbf{x}, \mu) = (1 + z), z = -1$. The memristive MLC circuit can be written as a set of smooth ODE's

$$\dot{\mathbf{x}}(t) = \left\{\begin{array}{l} F_1(\mathbf{x}, \mu), H_1(\mathbf{x}, \mu) \geq 0, \\ F_2(\mathbf{x}, \mu), H_1(\mathbf{x}, \mu) < 0, \\ F_2'(\mathbf{x}, \mu), H_2(\mathbf{x}, \mu) \geq 0, \\ F_3(\mathbf{x}, \mu), H_2(\mathbf{x}, \mu) < 0, \end{array}\right\}.$$

with the conditions that

$$\begin{aligned} F_1(\mathbf{x}, \mu) &= F_3(\mathbf{x}, \mu), \\ F_2(\mathbf{x}, \mu) &= F_2'(\mathbf{x}, \mu). \end{aligned} \qquad (27)$$

Here

$$F_1(\mathbf{x}, \mu) = F_3(\mathbf{x}, \mu) = \left\{\begin{array}{l} y - W_1(z)x \\ -\beta(x + y) - f sin(\omega t) \\ x \end{array}\right\}. \qquad (28)$$

and

$$F_2(\mathbf{x}, \mu) = F_2'(\mathbf{x}, \mu) = \left\{\begin{array}{l} y - W_2(z)x \\ -\beta(x + y) - f sin(\omega t) \\ x, \end{array}\right. \qquad (29)$$

where $W_1(z) = b$ and $W_2(z) = a$.

Grazing bifurcations are found to arise in the memrestive MLC Circuit at the switching manifolds [15, 16], leading to quasiperiodic, chaotic and hyperchaotic behaviours. The phase portrait, Poincaré map and the power spectrum for the hyperchaotic attractor are shown in Figs. 11(a), 11(b) and 11(c). The time plot for the $y(t)$ variable for an extended range is given in Fig. 11(d). The modulation envelopes isolated using Poincaré mapping of peak points are given in Fig. 11(e) and the power spectrum of the upper modulation envelope

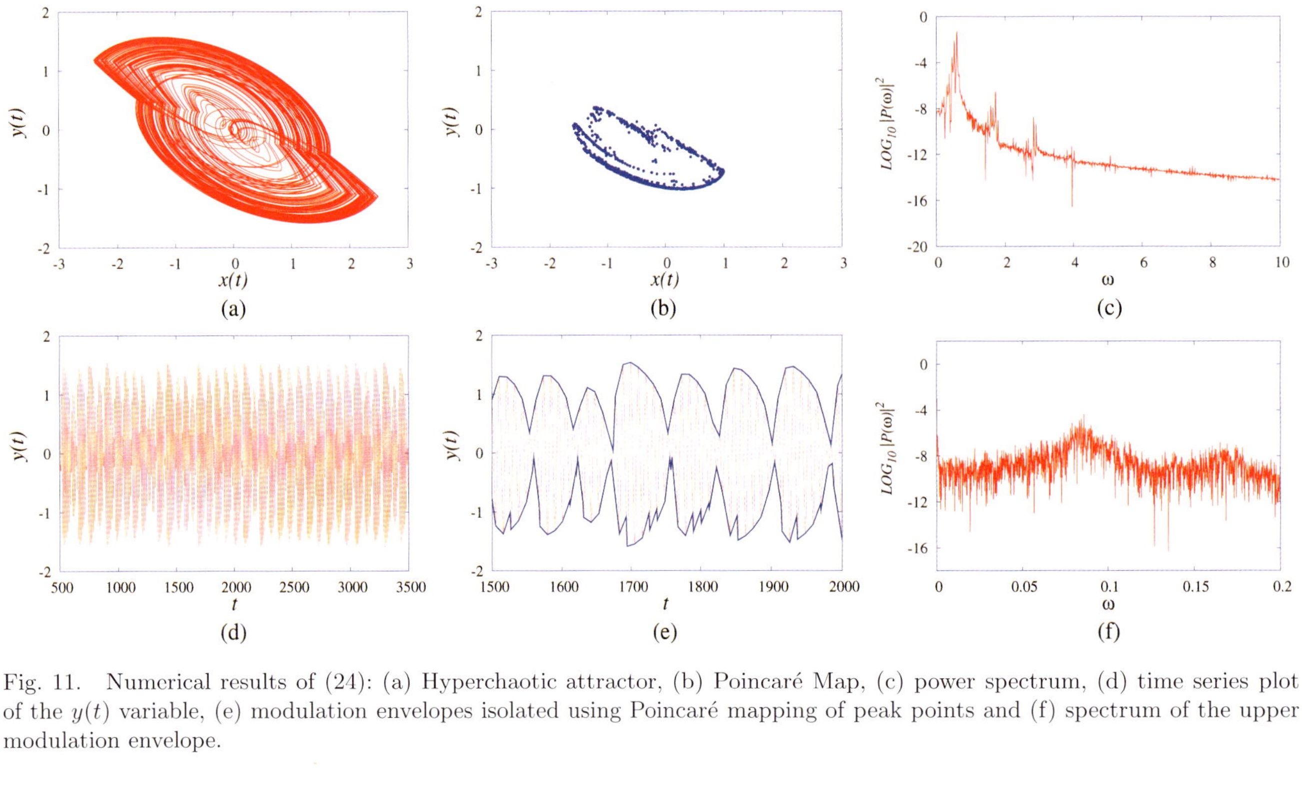

Fig. 11. Numerical results of (24): (a) Hyperchaotic attractor, (b) Poincaré Map, (c) power spectrum, (d) time series plot of the $y(t)$ variable, (e) modulation envelopes isolated using Poincaré mapping of peak points and (f) spectrum of the upper modulation envelope.

is given in Fig. 11(f). The Lyapunov exponents for the hyperchaotic attractor shown in Fig. 11(a) are $\lambda_1 = +0.043874, \lambda_2 = +0.014648, \lambda_3 = -0.108702$ and $\lambda_4 = 0.000000$. These diagrams as well as the Lyapunov exponents have been drawn/calculated after proper discontinuity mappings have been made at the switching manifolds.

6. Piecewise Linear Circuits and Strange Nonchaotic Attractors

In recent years, much attention has been given to a class of quasiperiodically forced dissipative dynamical systems which typically exhibits an interesting class of attractors which are not chaotic but yet display strange geometric properties. These are called strange nonchaotic attractors (SNAs), and were reported for the first time by Grebogi, Ott, Pelikan and Yorke [9, 11]. SNAs, although somewhat exotic, are not at all rare. They are regarded as structures in between regularity and chaos. They are geometrically strange as evidenced by their fractal nature, which is common to all chaotic systems. However, they are nonchaotic in a dynamical sense because they do not show sensitivity with respect to changes in initial conditions (as evidenced by negative or zero Lyapunov exponents) just like regular systems.

There are several mechanisms and routes by which SNAs can arise which have all been realized in a variety of dynamical systems [9, 32, 50]. These are included in Table. 1.

Table 1. Routes and mechanisms for the onset of various SNAs.

Type of route	Mechanism
Heagy-Hammel	Collision of period-doubled torus with its unstable parent
Gradual fractilization	Increased wrinkling of torus without any interaction with nearby periodic orbits
On-off intermittency	Loss of transverse stability of torus
Type-I intermittency	Due to saddle-node bifurcation, a torus is replaced by SNA
Type-III intermittency	Subharmonic instability
Homoclinic collision	Homoclinic collisions of the quasiperiodic orbits

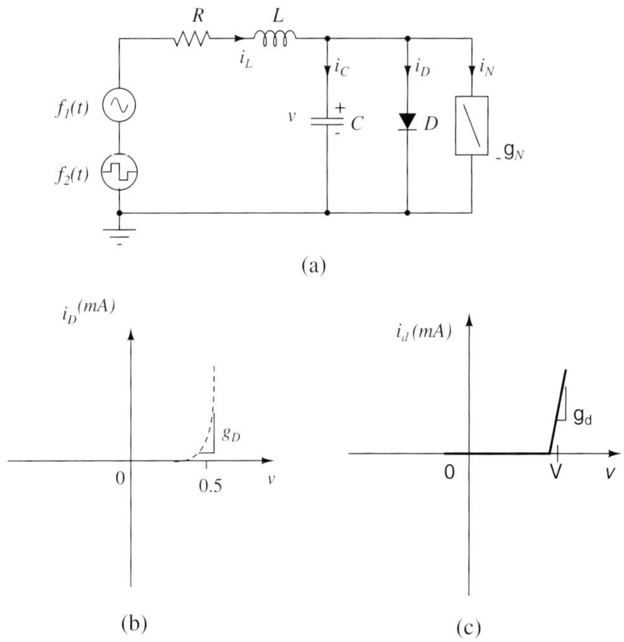

Fig. 12. (a) Circuit realization of a simple nonautonomous circuit. Here D is the *p-n junction* diode and g_N is negative conductance. The external emfs are $f_1(t) = E_{f1}\sin(\omega_{f1}t)$ and $f_2(t) = E_{f2}\text{sgn}[\sin(\omega_{f2}t)]$. The values of the circuit elements are fixed as $L = 50.3$ mH, $C = 10.35$ nF, $R = 1900$ ohms, $E_{f2} = 400$ mV and $\omega_{f2} = 17033$ Hz. The forcing amplitude E_{f1} and its frequency ω_{f1} are choosen as the control parameters, (b) $i - v$ characteristics of the *p-n junction* diode and (c) two segment piecewise-linear function.

It is very interesting to know that most of the above types of SNAs and even newer ones can be identified in a single dynamical system which is a piecewise linear system acted upon by quasiperiodic forcings. The circuit (Fig. 12) consists of a series LCR network, forced by a sinusoidal and a nonsinusoidal (square wave) voltage generator $f_1(t)$ and $f_2(t)$ (HP 33120A series), respectively. Two extra components, namely a *p-n junction* diode (D) and a linear negative conductor g_N, are connected in parallel to the forced series LCR circuit. The negative conductor used here is a standard op-amp based negative impedance converter (NIC). The diode operates as a nonlinear conductance, limiting the amplitude of the oscillator. In Fig. 12(a), v, i_L and i_D denote the voltage across the capacitor C, the current through the inductor L and the current through the diode D, respectively. The actual $v - i$ characteristic of the diode (Fig. 12(b)) is approximated by the usual two segment piecewise-linear function (Fig. 12(c)) which facilitates numerical analysis considerably. The state equations governing the circuit (Fig. 12) are a set of two first-order nonautonomous differential equations

$$C\frac{dv}{dt} = i_L - i_D + g_N v, \quad (30a)$$

$$L\frac{di_L}{dt} = -Ri_L - v + E_{f1}\sin(\omega_{f1}t) + E_{f2}sgn(\sin(\omega_{f2}t)), \quad (30b)$$

where

$$i_D(v) = \begin{cases} g_D(v - V), & v \geq V, \\ 0, & v < V. \end{cases} \quad (30c)$$

Here g_D is the slope of the characteristic curve of the diode, E_{f1} and E_{f2} are the amplitudes, ω_{f1} and ω_{f2} are the angular frequencies of the forcing functions $f_1(t) = E_{f1}\sin(\omega_{f1}t)$ and $f_2(t) = E_{f2}sgn(\sin(\omega_{f2}t))$, respectively. In the absence of $f_2(t)$, the circuit (Fig. 12(a)) has been shown to exhibit chaos and also strong chaos not only through the familiar period-doubling route but also via torus breakdown followed by period-doubling bifurcations [45].

We have fixed the actual experimental values of the inductance, L, capacitance, C, external frequency ω_{f2} and forcing strength E_{f2} of the square wave as $50.3mH$, $10.35nF$, $17033Hz$ and $400mV$, respectively, while we vary the amplitude E_{f1} and the frequency ω_{f1} of the sinusoidal force as control parameters. In order to study the dynamics of the circuit in detail, Eq. (30) can be converted into a convenient normalized form for numerical analysis by using the following rescaled variables and parameters $\tau = t/\sqrt{LC}, x = v/V$, $y = (i_L/V)(\sqrt{L/C}), E_1 = E_{f1}/V$, $E_2 = E_{f2}/V, \omega_1 = \omega_{f1}\sqrt{LC}, \omega_2 = \omega_{f2}\sqrt{LC}, a = R\sqrt{C/L}, b = g_N\sqrt{L/C}$, and $c = g_D\sqrt{L/C}$. The parameter space (E_1, ω_1) is scanned to distinguish the different dynamical behaviors exhibited by normalized form of Eqs. (30) in the range $E_1 \in (0.4, 1.1)$ and $\omega_1 \in (1.075, 1.275)$, which is shown in Fig. 13. It has also been confirmed that these dynamical behaviors are also exhibited by the circuit shown in Fig. 12 for the corresponding values of the parameters E_{f1} and ω_{f1}. Transitions from quasiperiodic attractor to SNA and subsequently to chaotic attractor can occur on increasing the value of the amplitude of the sinusoidal force E_1 for fixed value of its frequency ω_1. Strange nonchaotic attractors created through different mechanisms are found to occur for different values of the frequency ω_1 of the sinusoidal force. Now, we will outline the range of values of the frequency

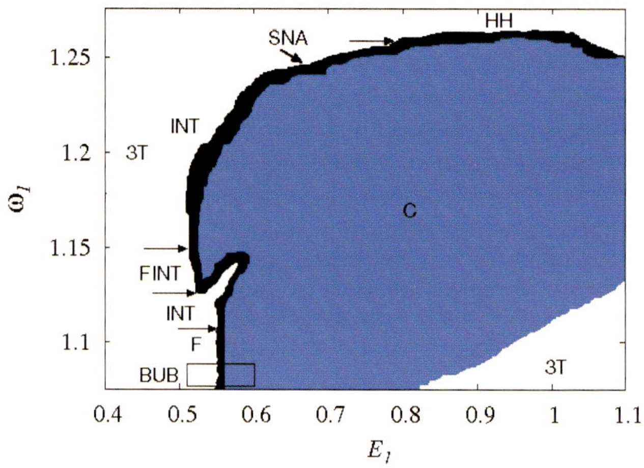

Fig. 13. Numerical phase diagram in the $(E_1 - \omega_1)$ plane for the circuit given in Fig. 12, represented by the normalized form of Eqs. (30). **3T** correspond to period-3 torus, **F**, **BUB**, **FINT**, **INT**, and **HH** denote the birth of SNAs through gradual fractalization, bubbling, fractalization followed by intermittency, intermittency and Heagy-Hammel routes, respectively. **C** corresponds to the chaotic attractor. Arrows indicate the transition regions between two different types of routes to SNA.

ω_1 for which SNAs arise from quasiperiodic attractors through different mechanisms on increasing the value of the amplitude E_1 of the sinusoidal force [38].

Strange nonchaotic attractor created through the new route, namely bubbling route, is identified in the range of frequency $\omega_1 \in (1.085, 1.086)$, where bubbles appear in the strands of period-3 torus and then the bubbles get increasingly wrinkled in the range of the amplitude $E_1 \in (0.54, 0.55)$ of the sinusoidal forcing resulting in SNA, the detail results of which can be found in [35]. This phenomenon is named as bubbling transition to SNA and it is denoted as (**BUB**) in Fig. 13. Further increase in the value of E_1 ends up with chaotic behavior indicated as **C** in Fig. 13 (see Fig. 14 also).

The routes and births of SNAs can be characterised and confirmed both experimentally and numerically by using a variety of qualitative measures. These include maximal Lyapunov exponents and their variance, Poincare maps, Fourier amplitude spectra and spectral distribution functions as well as the distribution of the finite-time Lypunov exponents [9,32,35,46,50]. Piecewise linear systems with quasiperiodic forcings turn out to be ideal candidates to realize the birth of SNAs.

In fact we have observed the occurence of strange nonchaotic attractors during the transition from two-frequency quasiperiodicity to a chaotic

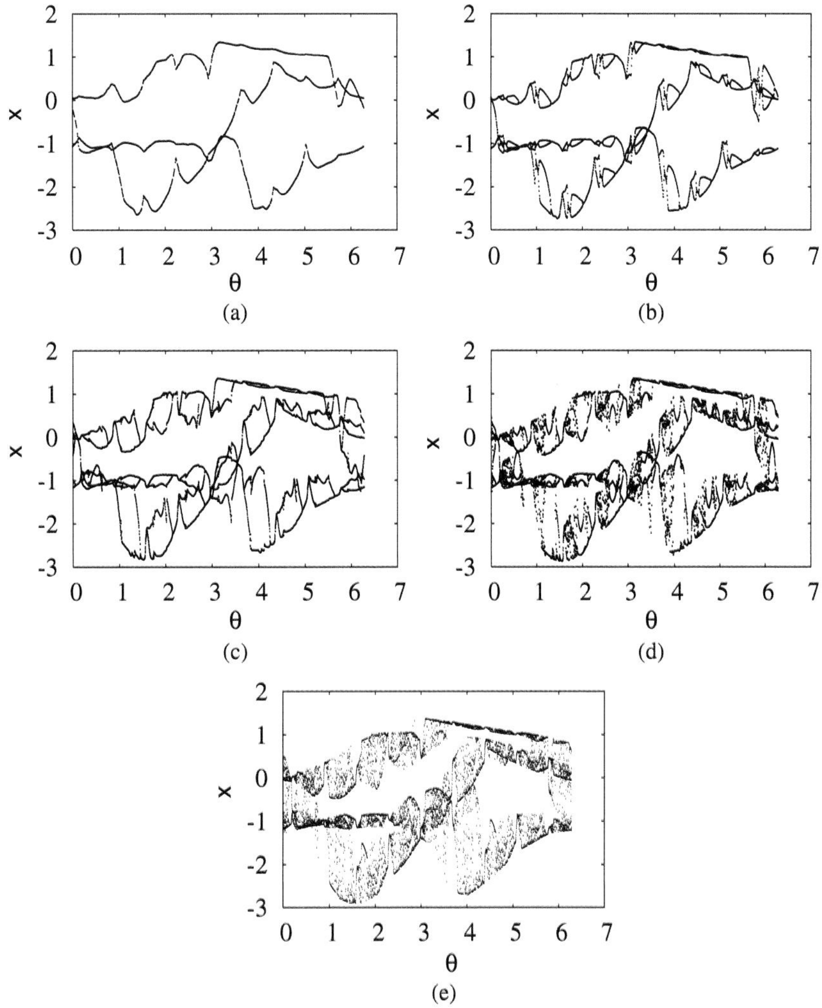

Fig. 14. Strands of the quasiperiodic attractor showing bubbling phenomenon in the $(x-\theta)$ plane as a function of the control parameter E_1. Here $\theta = \omega_2 t$. For details see [38].

attractor through a torus doubling bifurcation in a quasiperiodically forced MLC circuit also. The strange nonchaotic attractor arises when a a torus doubled attractor is interrupted by a subharmonic bifurcation, resulting in the inhibition of torus doubling sequence. This transition is shown to exhibit type III intermittent characteristic scaling behaviour (see [49]).

7. Bifurcation and Chaos in Scalar Piecewise Linear Time-delay Systems

Nonlinear time-delay systems are fascinating dynamical systems. Even scalar piecewise linear time-delay systems can exhibit very complex dynamical behaviour, essentially because they are effectively infinite dimensional in nature [23, 25, 30, 47]. For example, typically they can admit hyperchaotic attractors. In this section, we consider a simple delay differential equation (DDE) describing a time delayed feedback oscillator in the form

$$\frac{dx}{dt} = -ax(t) + bF[x(t-\tau)], \tag{31}$$

where a and b are positive scalar parameters, $x(t)$ is a dynamical variable, $F(x)$ is a nonlinear activation function and τ is the time delay. The activation function $F(x)$ in our study takes a symmetric piecewise linear form [39, 40],

$$F(x) = AF^* - Bx. \tag{32a}$$

Here

$$F^* = \begin{cases} -x^* & x < -x^*, \\ x & -x^* \le x \le x^*, \\ x^* & x > x^*, \end{cases} \tag{32b}$$

where A and B are positive parameters and x^* is the controllable threshold value. In our analysis, we

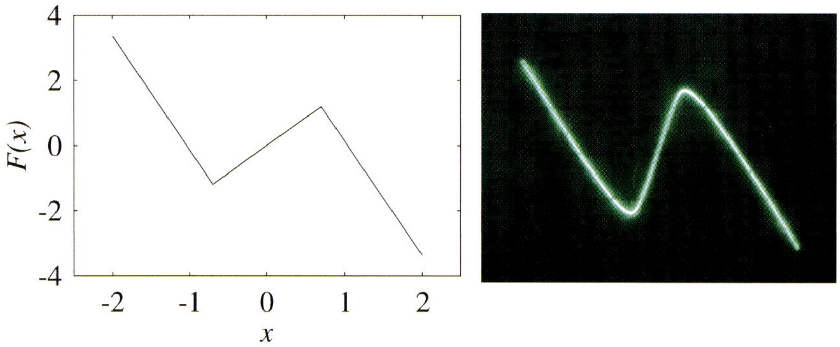

(a) (b)

Fig. 15. The nonlinear function $f(x)$. (a) Plot of the piecewise linear function $F(x)$ given by Eq. (32). (b) Measured characteristic curve of the nonlinear unit ND from Fig. 16, U_{in} vs U_{out}. Vertical scale $2V$/div., horizontal scale $1V$/div.

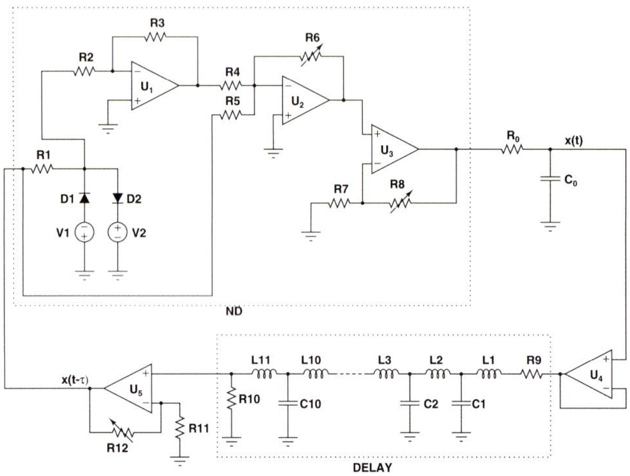

Fig. 16. Circuit diagram of the time delayed feedback oscillator with a nonlinear device unit (*ND*), a time delay unit (*DELAY*) and a lowpass first-order R_0C_0 filter.

chose $x^* = 0.7, A = 5.2, B = 3.5$. It may be noted that for $|x| > x^*$, the function $F(x)$ has a negative slope $-B$ and lies in all the four quadrants of the $(F - x)$ plane which is shown in Fig. 15(a). The figure reveals the piecewise linear nature of the function. Experimental implementation of the function $F(x)$ is shown in Fig. 15(b) in the form of V_{out} against V_{in} characteristic by the nonlinear device unit ND of Fig. 16.

This function $F(x)$ employs a threshold controller for flexibility. It efficiently implements a piecewise linear function. The control of this piecewise linear function facilitates controlling the shape of the attractors. Even for a small delay value this circuit system exhibits hyperchaos and can produce multi-scroll chaotic attractors by just introducing more number of threshold values, for example a square wave. In particular, this method is effective and simple to implement since we only need to monitor a single state variable and reset it if it exceeds the threshold and so has potential engineering applications for various chaos-based information systems.

7.1. *Experimental setup and results*

The system described by Eq. (31) with the nonlinear function $F(x)$ can be constructed using analog electronic devices. The circuit (Fig. 16) has a ring structure and comprises of a diode based nonlinear device unit with amplifying stages (U_2, U_3), a time delay unit with a buffer (U_4) and an amplifying stage (U_5).

To start with, Eq. (31) has been numerically integrated with the chosen nonlinear function $f(x)$ for the parameter values $a = 1.0, b = 1.2, \tau = 2.8, x^* = 0.7, A = 5.2$, and $B = 3.5$, with the initial condition $x = 0.9$ in the range $t \in (-\tau, 0)$. A one-band chaotic attractor is shown in Fig. 17a(i) for $\tau = 1.33$, while for $\tau = 2.8$ a double-band hyperchaotic attractor is obtained (Fig. 17a(ii)). The corresponding experimental results are shown in Figs. 17b(i) and 17b(ii) for the values of the parameter $R_0 = 5640\Omega$ (in this case $\tau = T_d/R_0C_0 \approx 1.331$) and $R_0 = 2680\Omega$ (now $\tau = T_d/R_0C_0 \approx 2.8022$), respectively. The experimental results are in good agreement with the numerical ones and also in their corresponding parameter values.

The scalar time delay equation admits a highly complex set of dynamical phenomena, including intricate bifurcations [39]. It also admits higher dimensional hyperchaotic attractors with several positive Lypunov exponents. In Fig. 18, we present the first seven maximal Lyapunov exponents for a range of time delays $\tau \in (2, 29)$, while the other parameters are fixed at $a = 1.0, b = 1.2$.

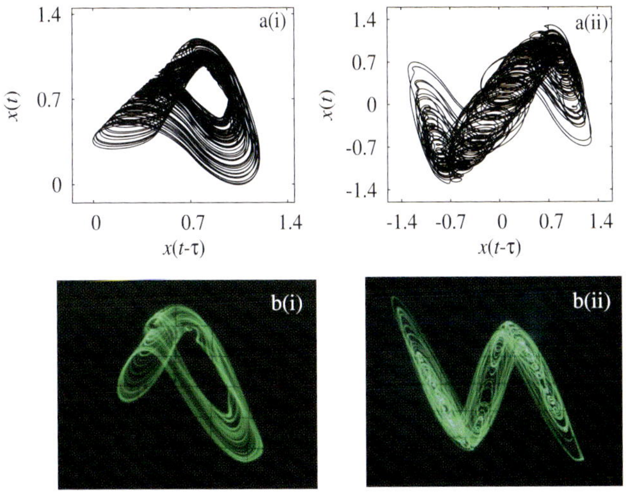

Fig. 17. (a) Phase portraits of chaotic attractors from Eqs. (31) and (32) for the parameters $a=1, b=1.2, A=5.2, B=3.5$ and $x^*=0.7$: (i) one-band chaos for $\tau=1.33$ and (ii) double-band chaos for $\tau=2.8$. (b) Phase portraits of chaotic attractors from the circuit (Fig. 16), $U(t-T_d)$ against $U(t)$, vertical scale $2V$/div., horizontal scale $0.2V$/div.: (i) one band chaos for $R_0 = 5640\,\Omega$ and (ii) double band chaos for $R_0 = 2680\ \Omega$.

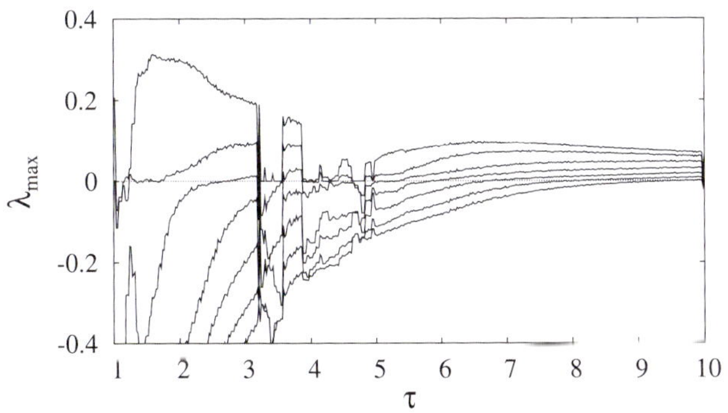

Fig. 18. The seven maximal Lyapunov exponents λ_{max} of the time-delay system (31) and (32) for the parameter values $a=1$, $b=1.2$, $x^*=0.7$, $A=5.2$, $B=3.5$ and $\tau \in (1,10)$.

Another piecewise system recently well studied by our group [33] is the five segment piecewise linear system, which is again of the form (1), where now

$$F(x(\tau)) = \begin{cases} 0, & x < -1.33, \\ -1.5x-2, & -1.33 < x < -0.8 \\ x, & -0.8 < x < 0.8, \\ 1.5x+2, & 0.8 < x < 1.33 \\ 0, & x > 1.33. \end{cases} \qquad (33)$$

The piecewise nature of the function $F(x)$ is shown in Fig. 19(a). The scalar time delay equation admits a highly complex set of dynamical phenomena, including intricate bifurcations. It also admits higher dimensional hyperchaotic attractors with several positive Lypunov exponents. In Fig. 19(c), we present the first ten maximal Lyapunov exponents for a range of time delays $\tau \in (2, 29)$, while the other parameters are fixed at $a = 1.0$, $b = 1.2$. A typical hyperchaotic attractor in the $x(t), x(t+\tau))$ pseudospace is shown in Fig. 19(b) for the value of $\tau = 15.0$. It is clear that piecewise linear time delay systems have very rich dynamical behaviour and can be taken as paradigmatic models to understand complex nonlinear systems.

8. Synchronization in Coupled Piecewise Linear Time-delay Circuits

Coupled chaotic systems can exhibit very many fascinating kinds of synchronizations, including complete, lag, anticipating, generalized and phase synchronizations [20,34,36,40]. Study of arrays and networks of such systems constitutes a challeng-

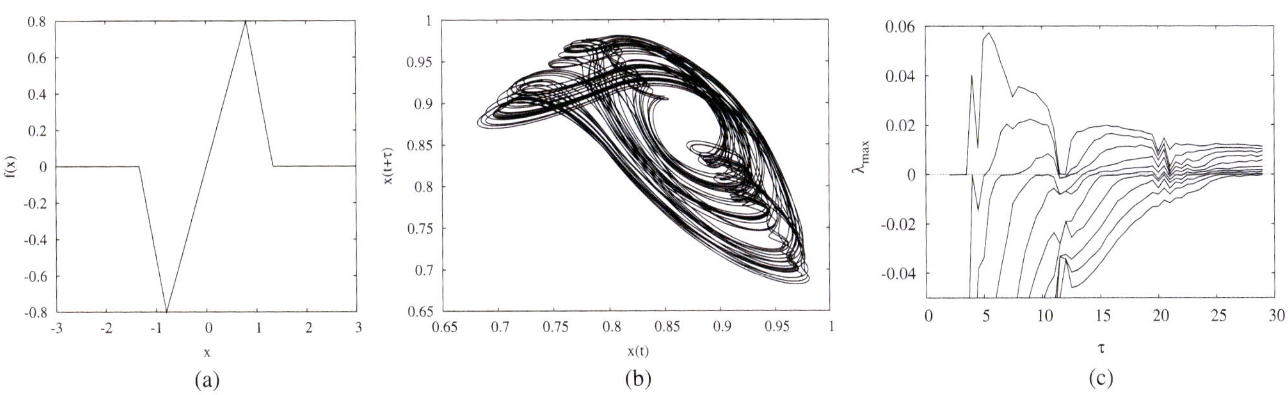

Fig. 19. (a) Piecewise nature of the function $F(x)$ given by Eq. (33). (b) A typical hyperchaotic attractor in the $x(t), x(t+\tau))$ pseudospace and (c) first ten maximal Lyapunov exponents.

ing area of study in nonlinear dynamics. Of particular interest in this connection is the effect of time delay in such systems. In this section, we will present some details on coupled piecewise linear time-delay systems and synchronizations exhibited by them.

8.1. *Synchronization transitions in coupled time-delay electronic circuits*

Now let us consider the following set of unidirectionally coupled first-order delay differential equations,

$$\frac{dx}{dt} = -a_1 x(t) + b_1 F[x(t-\tau_1)], \tag{34a}$$

$$\frac{dy}{dt} = -a_2 y(t) + b_2 F[y(t-\tau_1)] + b_3 F[x(t-\tau_2)], \tag{34b}$$

where $a_1 = a_2 > 0$ are positive constants, $b_1 \neq b_2$ contributes to the parameter mismatch resulting in coupled non-identical systems, b_3 is the coupling strength, τ_1 is the feedback delay and τ_2 is the coupling delay [40]. The nonlinear function $F(x)$ is of the same form as in Eq. (32).

Now to analog simulate the coupled time-delay systems (Eqs. (34)) and to demonstrate experimentally the existence of different types of synchronizations, a unidirectionally coupled time-delay electronic circuit is constructed as shown in the block diagram of Fig. 20. One of the electronic oscillator circuits is used as the drive system, while the other structurally identical circuit is used as the response system with some parameter mismatches. The drive voltage ($U_1(t)$) after the delay line in the drive system is fed back to the nonlinear part (ND_1) of the drive system and a fixed R_1C_1 filter with time delay to generate chaotic/hyperchaotic oscillations. Similarly, the response circuit with a nonlinear part (ND_2), a delay line and a fixed R_2C_2 filter is capable of generating chaotic/hyperchaotic oscillations. The signal after the nonlinear function of drive is used as the transmission signal, which is unidirectionally transmitted through the lowpass filter (R_3C_3), delay line and nonlinear part to the response circuit.

Experimental observations reveal transitions from anticipatory to lag synchronization via complete synchronization, when excitatory couplings are a function of the coupling delay (time τ_2) (see Fig. 21). Next, when we use an inhibitory coupling, $-b_3 f[x(t-\tau_2)]$, in Eq. (34b) instead of the excitatory coupling $+b_3 f[x(t-\tau_2)]$, we find the transition from inverse anticipatory to inverse lag synchronization via an inverse complete synchronization as a function of the coupling delay τ_2 (see Fig. 22).

8.2. *Experimental confirmation of chaotic phase synchronization*

We now consider the possibility of chaotic phase synchronization exhibited by coupled time delay systems. For this purpose, we consider the normalized evolution equation of the coupled time-delay electronic circuits with error coupling,

$$\begin{aligned} \dot{x} &= -ax(t) + b_1 F[x(t-\tau)], \\ \dot{y} &= -ay(t) + b_2 F[y(t-\tau)] + \varepsilon(x(t) - y(t)), \end{aligned} \tag{35}$$

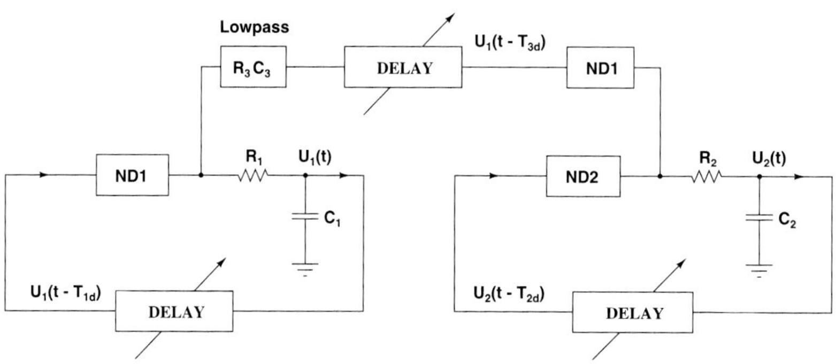

Fig. 20. Circuit block diagram of the coupled delayed feedback oscillator. Two delay oscillators are coupled through a nonlinear activation function (ND) but with a different time delay T_{3d} in the coupling with a low pass first-order R_3C_3 filter. $U_1(t)$ and $U_2(t)$ are the voltages across the capacitances C_1 and C_2, respectively. $U(t-T_{1d}) = U(t-T_{2d})$ are the voltages across the delay units of both the coupled oscillators.

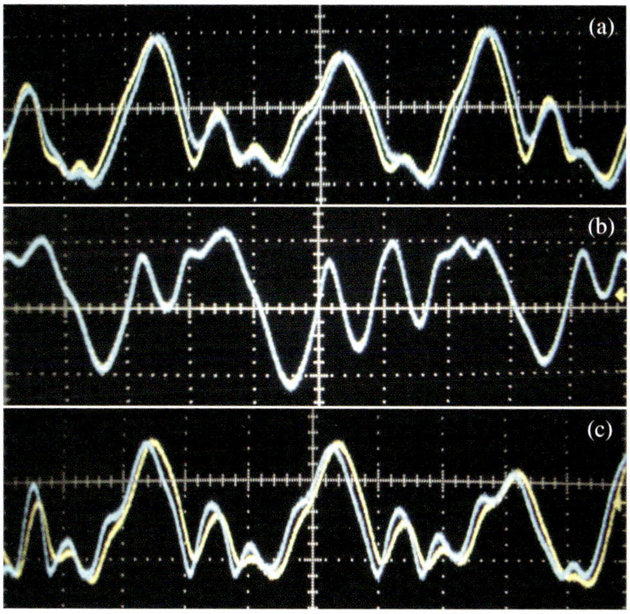

Fig. 21. Experimental time series plot (Eq. (34)) of the drive $x(t)$ (blue) and the response $y(t)$ (yellow), (a) anticipatory synchronization for $\tau_2 \approx 3.1007$, (b) complete synchronization for $\tau_2 \approx 2.8022$ and (c) lag synchronization for $\tau_2 = 2.5$; vertical scale $5V$/div., horizontal scale 1 ms.

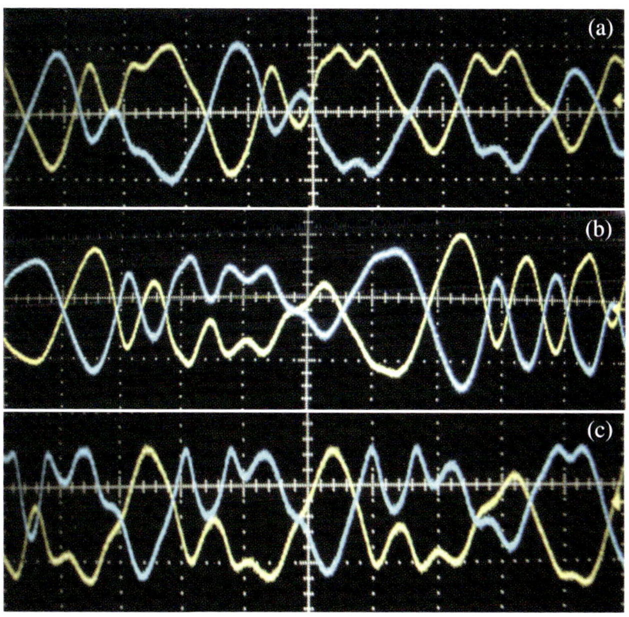

Fig. 22. Experimental time series plots of the drive $x(t)$ (blue) and the response $y(t)$ (yellow) (a) inverse anticipatory synchronization, (b) inverse complete synchronization and (c) inverse lag synchronization, see Eq. (34).

where a and b are positive parameters and τ is the time delay [36]. The function $F(x)$ is taken to be the symmetric piecewise linear function Eq. (32).

The coupled electronic circuit investigated here is shown in Fig. 23 as a block diagram. The individual time-delay units have a ring structure and comprise of a diode based nonlinear device unit (ND), a variable time-delay unit (DELAY) along with an integrator (R_0C_0) unit. The block diagram also contains a differential amplifier circuit (Λ) with a gain $\varepsilon = R_f/R$ used to find the difference between the two voltage signals $U1$ and $U2$. By changing the feedback resistance (R_f), the coupling parameter ε can be varied. We have fixed the same circuit parameter values of as in the above studies and demonstrate here the existence of chaotic phase synchronization (CPS) as a function of the coupling strength ε in both chaotic and hyperchaotic regimes for suitable values of the delay time τ.

Existence of CPS can now be demonstrated using the framework of localized sets in the chaotic regime. The sets obtained by sampling the time series of one of the systems whenever a maximum occurs in the other one are plotted along with the chaotic attractor of the same system for the delay time $\tau = 1.33$ experimentally in Fig. 24(i). The sets distributed over the entire attractor of both the drive [Figs. 24(a)i] and the response [Figs. 24(b)i] systems for the coupling strength $\varepsilon = 0.3$ indicate that the time-delay systems evolve independently. The sets that are localized on the chaotic attractor of both the drive [Figs. 24(c)i] and the response [Figs. 24(d)i] systems for the coupling strength $\varepsilon = 0.9$ correspond to a perfect locking of the phases of both circuits, indicating CPS.

8.3. *Zero-lag synchronization in coupled time-delay systems*

In this section we have considered the following three coupled piecewise linear time-delay systems with threshold nonlinear function,

$$
\begin{aligned}
\dot{x}_1(t) &= -\beta x_1(t) + \alpha F[x_1(t-\tau)] \\
&\quad + C[x_2(t-\tau) - 2x_1(t) + x_3(t-\tau)], \\
\dot{x}_2(t) &= -\beta x_2(t) + \alpha F[x_2(t-\tau)] \\
&\quad + C[x_1(t-\tau) - x_2(t)], \\
\dot{x}_3(t) &= -\beta x_3(t) + \alpha F[x_3(t-\tau)] \\
&\quad + C[x_1(t-\tau) - x_3(t)], \qquad (36)
\end{aligned}
$$

the parameter values are fixed as follows: $\beta = 1.0$, $\alpha = 1.2$, $\tau = 6.0$ and C is the coupling strength. The function $F(x)$ is taken to be a symmetric piecewise linear function Eq. (32). We note that in contrast

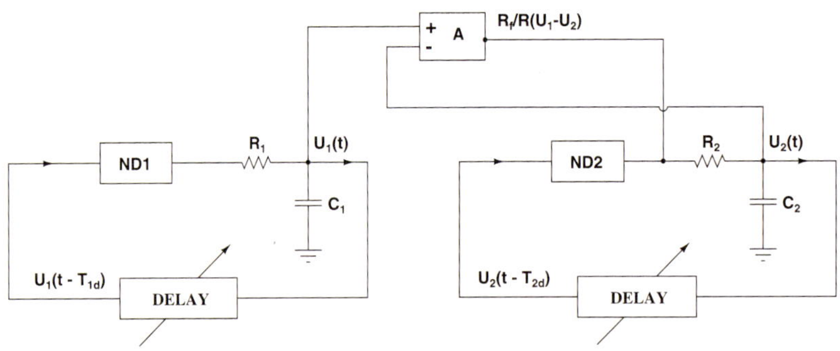

Fig. 23. Block diagram of the unidirectionally coupled time-delay electronic circuits with threshold nonlinearity.

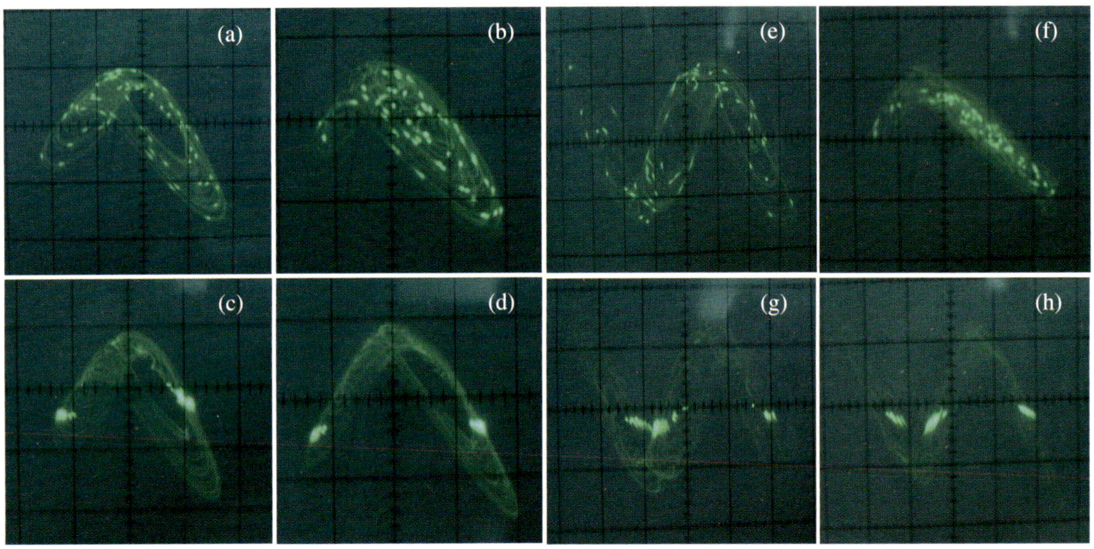

Fig. 24. Experimental characterization of CPS using the framework of localized sets in the chaotic regime (i) for $\tau = 1.33$ and in the hyperchaotic regime (ii) for $\tau = 6.0$. Sets in the drive and the response systems are distributed in (a) and (b) for $\varepsilon = 0.3$ indicating the asynchronous state and localized in (c) and (d) for $\varepsilon = 0.9$ indicating CPS, respectively.

to the previous example (34) and (35), the above three equations are delay coupled to one another. In particular, the first oscillator (drive system) is coupled to both the second and third oscillators, while the second and third oscillators are coupled to the first oscillator only. This system can be studied experimentally and it is found to exhibit phase and anti-phase synchronizations even with the delay coupling.

In the presence of weak coupling the oscillators 2 and 3 are in transition from nonsynchronized state to a phase synchronized (PS) state. These two oscillators are synchronized in an anti-phase state with the oscillator 1. The experimental snapshots of the time series of both the response systems (2 and 3) as seen from the oscilloscope are shown in Fig. 25(a) in the chaotic regime for the delay time $\tau = 1.33$ and the coupling strength $C = 1.12$ indicating the

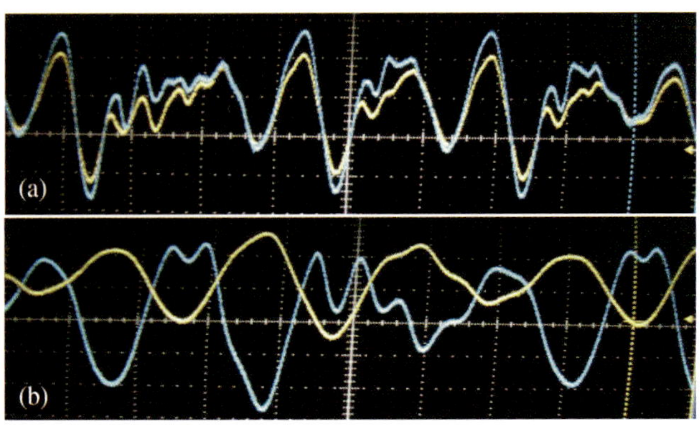

Fig. 25. Experimental results (a) Time series of the oscillators 2 ($x_2(t)$; yellow) and 3 ($x_3(t)$; blue) of Eq. (36) shows phase synchronization. (b) Transition to anti-phase synchronization in the oscillators 1 ($x_1(t)$; yellow) and 3 ($x_3(t)$; blue).

evolution of both systems in-phase with each other. The time series of drive and response (oscillators 1 and 2) are in anti-phase state with each other as shown in Fig. 25(b) for the same coupling.

8.4. Global phase synchronization in an array of piecewise linear time-delay systems

Even though the notion of CPS is well studied in low dimensional systems, there exist very little indepth studies in higher dimensional systems such as time-delay systems which are essentially infinite-dimensional in nature and often exhibit high-dimensional, highly non-phase-coherent hyperchaotic attractors with complex topological structure. Consequently, estimating phase explicitly to identify phase synchronization in such systems is quite difficult. Nevertheless, CPS has been recently demonstrated in two coupled piecewise linear and Mackey-Glass time-delay systems [34, 35, 42] by introducing a nonlinear transformation of the original dynamical variable to recast the original non-phase-coherent hyperchaotic attractors into smeared limit cycle-like attractors in order to facilitate the estimation of the phase variable using the available methods, and can be experimentally demonstrated as shown in the previous subsections. However, these investigations are carried out so far only in two coupled time-delay systems. In view of the widespread applications of CPS in ensembles of coupled oscillators, here we investigate the existence of global phase synchronization (GPS) in an array of unidirectionally coupled piecewise linear time-delay systems.

To verify GPS in a linear array we consider an array of unidirectionally coupled piecewise linear time-delay systems with free end boundary conditions represented by the following systems of coupled nonlinear first order delay differential equations,

$$\dot{x}_1 = -\beta x_1(t) + \alpha f(x_1(t-\tau)), \tag{37a}$$

$$\dot{x}_i = -\beta x_i(t) + \beta f(x_i(t-\tau)) + C[x_{i-1}(t) - x_i(t)], \tag{37b}$$

where $i = 2, 3, \ldots, N$. α, β are the system parameters, τ is the time-delay and C is the coupling strength. The nonlinear function $f(x)$ is chosen to be a piecewise linear function Eq. (33). We have fixed the parameter values at $\beta = 1.0$, $\alpha = 1.2$, $\tau = 15$, $p_1 = 0.8$ and $p_2 = 1.33$ and the nonlinear parameter α_i of the response systems in the array are randomly chosen in the range $\alpha_i \in (1.2, 1.4)$ so that all the systems are effectively nonidentical. For our simulation, we have fixed the total number of oscillators in the array at $N = 30$.

The uncoupled systems exhibit a highly non-phase coherent chaotic attractor for the chosen parameter values, see Fig. 19. Here we note that the phase calculated directly from the original non-phase coherent chaotic attractor cannot yield monotonically increasing behavior as it has several closed loops, which also contribute to the phase information, other than the main center of rotation of the major part of the trajectories. To overcome the problem a nonlinear transformation can be introduced so as to rescale the original non-phase coherent chaotic attractor into a smeared limit cycle-like attractor with a single center of rotation. The transformation is performed by introducing a new state variable [34, 35].

$$\Delta z(t+\tau) = x(t)x(t+\hat{\tau})/x(t+\tau). \tag{38}$$

where $\hat{\tau}$ is the optimal value of the delay to be chosen in order to avoid any additional center of rotation. We use the nonlinear transformation to recast the original non-phase coherent chaotic attractor of all $N = 30$ oscillators to smeared limit cycle-like attractors. We have fixed the value of the optimal time-delay $\hat{\tau}$ in Eq. (38) as $\hat{\tau} = 1.6$ for all the N oscillators. We have calculated the instantaneous phase of all the oscillators using the Poincaré section technique from their corresponding transformed attractors using the following equation

$$\phi_i(t) = 2\pi k + 2\pi \frac{t^i - t^i_k}{t^i_{k+1} - t^i_k}, \tag{39}$$

where t^i_k is the time of the kth crossing of the flow with the Poincaré section of the ith attractor.

To confirm the existence of GPS, we have calculated the average phase difference $\eta(t)$, defined as

$$\eta(t) = \frac{1}{N-1} \sum_{j=2}^{N} (\phi_1 - \phi_j). \tag{40}$$

The average phase difference ($\eta(t)$) for different values of C is shown in Fig. 26 as a function of time t. In the absence of the coupling ($C = 0.0$), the phases of all the oscillators evolve independently and hence their average phase difference increases linearly with time. Phase slips in the phase differences for small values of the coupling strength indicates that the oscillators are in the transition state to GPS. Further increase in the value of the coupling strength results in a strong boundedness of

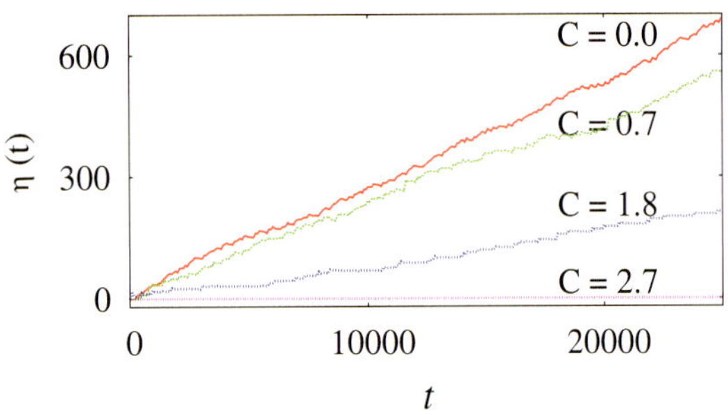

Fig. 26. The average phase difference ($\eta(t)$) for different values of the coupling C. For $C = 0.0$ the phase of all the systems are unbounded so the phase difference increases linearly with time but for $C = 2.7$ the phases of all the systems are bounded, showing a high-quality phase synchronization.

the phase of the oscillator. For sufficiently large coupling $C = 2.7$ the average phase difference of all the N oscillators becomes exactly zero showing a high quality GPS in the array.

9. Summary and Conclusions

In this article, we have given a brief overview of the nonlinear dynamics of a class of piecewise linear dynamical systems, modelling interesting nonlinear electronic circuits, describable by a system of two first order coupled nonlinear (piecewise linear) differential equations with external forcing as the constituent state equations. Equivalently, the dynamical equations can be rewritten as second order damped and forced linear differential equations with discontinuous right hand sides with one or two boundaries. These include the MLC circuit, variant MLC circuit, coupled MLC circuits, memristor based MLC circuit, forced negative conductance series LCR circuit with or without quasiperiodic forcing. We have also briefly considered the dynamics of certain piecewise linear time-delay systems and their coupled versions. All the above systems were shown to admit a very large variety of dynamical states, bifurcations and routes to chaos and synchronizations. The nonlinear dynamics underlying such piecewise linear systems continue to present challenging problems and fascinating results.

Acknowledgments

The authors are thankful to Mr. A. Ishaq Ahamed and Dr. K. Srinivasan for their help in preparing this article. The work of ML forms part of a Department of Science and Technology (DST), Government of India research project and a DST Ramanna Fellowship project. He is financially supported by a Department of Atomic Energy Raja Ramana Fellowship, which is gratefully acknowledged.

References

[1] Bernardo, M. di., Budd, C. J., Champneys, A. R., Kowalczyk, P., Nordmark, A. B., Tost, G. O. and Piirorinen, P. T. (2008). Bifurcations in Nonsmooth Dynamical Systems,*SIAM Rev.* **50**, pp. 629.

[2] Bernardo, M. di., Budd, C. J., Champneys, A. R. and Kowalczyk, P. (2008). *Piecewise Smooth Dynamical Systems: Theory and Applications*, (Springer-Verlag, London).

[3] Chua, L. O. (1971). Memristor – The missing circuit element, *IEEE Trans. Circuit Th. CT-V* **18**, pp. 507.

[4] Chua, L. O., Desoer, C. A. and Kuh, E. S. (1987). *Linear and Nonlinear Circuits*, (McGraw Hill, Singapore).

[5] Chua, L. O. and Yang, L. (1988). Cellular neural networks: Theory, *IEEE Trans. Circuits Syst.* **35**, pp. 1257.

[6] Chua, L. O. (1993). *IEICE Trans. Fund. E* **76**, pp. 704; Madan, R. N. Ed., (1996). *Chua's Circuit: A Paradigm for Chaos*, (World Scientific, Singapore).

[7] Chua, L. O. (1995). Special issue on Nonlinear waves, patterns and spatiotemporal chaos in dynamic arrays, *IEEE Trans. Circuits Syst.* **42**, pp. 557.

[8] Feingold, M., Gonzalez, D. L., Piro, O. and Viturro, H. (1988). Phase locking, period doubling and chaotic phenomena in externally driven excitable systems, *Phys. Rev. A* **37**, pp. 4060.

[9] Feudel, U., Kutznetzov, S. and Pikovsky, A. (2006). *Strange Nonchaotic Attractors: Dynamics between Order and Chaos in Quasiperiodically Forced Systems*, (World Scientific, Singapore).

[10] Filippov, A. F. (1988). *Differential Equations with Discontinuous Right Hand Sides*, (Kluwer Academic, Dordrecht).

[11] Grebogi, C., Ott, E., Pelikan, S. and Yorke, J. A. (1984). *Physica D*, **13**, pp. 961.

[12] Guckenheimer, J. and Holmes, P. (1983). *Nonlinear Oscillators, Dynamical Systems and Bifurcations of Vector Fields*, (Springer-Verlag, New York).

[13] Inaba, N. and Mori, S. (1991). Chaos via torus breakdown in a piecewise-linear forced vander Pol oscillator with a diode, *IEEE Trans. Circ. Syst-I* **38**, pp. 398.

[14] Inaba, N. and Mori, S. (1992). Folded torus in the forced Rayleigh oscillator with a diode pair, *IEEE Trans. Circ. Syst-I* **39**, pp. 402.

[15] Ishaq Ahamed, A., Srinivasan, K., Murali, K. and Lakshmanan, M. (2011). Observation of chaotic beats in a driven memristive Chua's circuits, *Int. J. Bifurcation and Chaos* **21**, pp. 737.

[16] Ishaq Ahamed, A., Srinivasan, K., Murali, K. and Lakshmanan, M. (2011). Chaotic and hyperchaotic beats in a memrestive MLC circuit (to be submitted to IJBC).

[17] Itoh, M. and Chua, L. O. (2008). Memristor oscillators, *Int. J. Bifurcation and Chaos* **18**, pp. 3183.

[18] Kocarev, L. and Parlitz, U. (1996). Synchronizing spatiotemporal chaos in coupled nonlinear oscillators, *Phys. Rev. Lett.* **77**, pp. 2206.

[19] Künze, M. (2000). *Non-Smooth Dynamical Systems, Lecture Notes in Math*, (Springer-Verlag, Berlin).

[20] Lakshmanan, M. and Murali, K. (1996). *Chaos in Nonlinear Oscillators: Controlling and Synchronisation*, (World Scientific, Singapore).

[21] Lakshmanan, M. and Rajasekar, S. (2003). *Nonlinear Dynamics: Integrability, Chaos and Patterns*, (Springer-Verlag, Berlin).

[22] Lakshmanan, M. (2009). Bifurcations and chaos in certain piecewise linear differential equations, *Dynamics of Continuous, Discrete and Impulsive Systems* **16**, pp. 891.

[23] Lakshmanan, M. and Senthilkumar, D. V. (2010). *Dynamics of Nonlinear Time-Delay Systems*, (Springer-Verlag, Berlin).

[24] Lichternberg, A. J. and Lieberman, M. A. (1983). *Regular and Stochastic Motion*, (Springer-Verlag, Berlin).

[25] Lu, H. and He, Z. (1996). Chaotic behaviour in first-order autonomous continuous-time systems with delay, *IEEE Trans. Circ. Syst-I.* **43**, pp. 700.

[26] Murali, K. and Lakshmanan, M. (1992). Effect of sinusoidal excitation on the Chua's circuit, *IEEE Trans. Circ. Syst-I.* **39**, pp. 264.

[27] Murali, K., Lakshmanan, M. and Chua, L. O. (1994a). The simplest dissipative nonautonomous chaotic circuit, *IEEE Trans. Circ. Syst-I.* **41**, pp. 462; (1994b). Bifurcation and chaos in the simplest dissipative non-autonomous circuit, *Int. J. Bifur. and Chaos* **4**, pp. 1511.

[28] Muruganandam, P., Murali, K. and Lakshmanan, M. (1999). Spatiotemporal dynamics of coupled array of Murali-Lakshmanan-Chua circuits, *Int. J. Bifurcation and Chaos* **9**, pp. 805.

[29] Muthuswamy, B. (2010). Implementing Memristor based chaotic circuits, *Int. J. Bifurcation and Chaos* **20**, pp. 1335.

[30] Namajūnas, A., Pyragas, K. and Tamaševičius, A. (1995). An electronic analog of the Mackey-Glass system, *Phys. Lett. A* **201**, pp. 42.

[31] Nayfeh, A. H. and Balachandran, B. (1995). *Applied Nonlinear Dynamics*, (John Wiley, New York).

[32] Prasad, A., Nandi, A. and Ramaswamy, R. (2007). Aperiodic nonchaotic attractors, strange and otherwise, *Int. J. Bifur. and Chaos* **17**, pp. 3397.

[33] Senthilkumar, D. V. and Lakshmanan, M. (2005a). Bifurcations and chaos in Time Delayed piecewise linear dynamical systems, *Int. J. Bifur. and Chaos* **15**, pp. 2895; (2005b). Transition from anticipatory to lag synchronisation via complete synchroniation in time-delay systems, *Phys. Rev. E* **71**, pp. 016211.

[34] Senthilkumar, D. V., Lakshmanan, M. and Kurths, J. (2008). Transition from phase to generalized synchronization in time-delay systems, *Chaos* **18**, pp. 023118; (2006). Phase synchronization in time-delay systems, *Phys. Rev. E* **74**, pp. 035205(R).

[35] Senthilkumar, D. V., Srinivasan, K., Thamilmaran, K. and Lakshmanan, M. (2008). Bubbling route to strange nonchaotic attractor in a nonlinear series LCR circuit with a nonsinusoidak force, *Phys. Rev. E* **78**, pp. 066211.

[36] Senthilkumar, D. V., Srinivasan, K., Murali, K., Lakshmanan, M. and Kurths, J. (2010). Experimental confirmation of chaotic phase synchronization in coupled time-delay electronic circuits, *Phys. Rev. E* **82**, pp. 065201(R).

[37] Sprott, J. C. (2010). *Elegant chaos: algebraically simple chaotic flows*, (World Scientific, Singapore).

[38] Srinivasan, K., Senthilkumar, D. V., Suresh, R., Thamilmaran, K. and Lakshmanan, M. (2009). Experimental realization of strange nonchaotic attractors in a nonlinear series LCR circuit with nonsinusoidal force, *Int. J. Bifurcation and Chaos* **19**, pp. 4131.

[39] Srinivasan, K., Raja Mohamed, I., Murali, K., Lakshmanan, M. and Sinha, S. (2011a). Design of time delayed chaotic circuit with threshold controller, *Int. J. Bifurcation and Chaos* **21**, pp. 725.

[40] Srinivasan, K., Senthilkumar, D. V., Murali, K., Lakshmanan, M. and Kurths, J. (2011b). Synchronization transitions in coupled time-delay electronic circuits with a threshold nonlinearity, *Chaos* **21**, pp. 023119.

[41] Strutkov, D. B., Snider, G. S., Stewart, D. R. and Williams, R. S. (2008). The missing memristor found, *Nature* **453**, pp. 80.

[42] Suresh, R., Senthilkumar, D. V., Lakshmanan, M. and Kurths, J. (2010). Global phase synchronization in an array of time-delay systems, *Phys. Rev. E* **82**, pp. 016215.

[43] Thamilmaran, K., Lakshmanan, M. and Murali, K. (2000). Rich variety of bifurcations and chaos in a variant of Murali-Lakshmanan-Chua circuit, *Int. J. Bifur. and Chaos* **10**, pp. 1781.

[44] Thamilmaran, K. and Lakshmanan, M. (2002). Classification of bifurcations and routes to chaos in a variant of Murali-Lakshmanan-Chua circuit, *Int. J. Bifur. and Chaos* **12**, pp. 783.

[45] Thamilmaran, K., Senthilkumar, D. V., Ishaq Ahamed, A. and Lakshmanan, M. (2005). Strong chaos in a forced negative conductance series LCR circuit, *Int. J. Bifur. and Chaos* **15**, pp. 637.

[46] Thamilmaran, K., Senthilkumar, D. V., Venkatesan, A. and Lakshmanan, M. (2006). Experimental realization of strange nonchaotic attractors in a quasiperiodically forced electronic circuit, *Phys. Rev. E* **74**, pp. 036205.

[47] Thangavel, P., Murali, K. and Lakshmanan, M. (1998). Bifurcation and controlling of chaotic delayed cellular neural networks, *Int. J. Bifur. and Chaos* **8**, pp. 2481.

[48] Tour, J. M. and He, Tao. (2008). The fourth element, *Nature* **453**, pp. 42.

[49] Venkatesan, A., Murali, K. and Lakshmanan, M. (1999). Birth of strange nonchaotic attractors through type III, *Phys. Lett. A* **259**, pp. 246.

[50] Venkatesan, A. and Lakshmanan, M. (2001). Interruption of torus doubling birfurcation and genesis of strange nonchaotic attractors in a quasiperiodically forced map: Mechanisms and their charaterizations, *Phys. Rev. E* **63**, pp. 026219.

[51] Wang, D., Zhao, H. and Yu, J. (2009). Chaos in memristor based Murali-Lakshmanan-Chua circuit, *Proc. IEEE*, **ICCCAS 2009**, pp. 958.

[52] Zhong, G. (1994). Implementation of Chua's Circuit with a Cubic Nonlinearity, *IEEE Trans. Circuits Syst.* **41**, pp. 934.

CHAPTER 24

CHAOTIC MATHEMATICAL CIRCUITRY

R. LOZI

Laboratoire J.A. Dieudonné — UMR CNRS 7351,
Université de Nice Sophia-Antipolis
Parc Valrose
06108 NICE Cedex 02
France
rlozi@unice.fr

Following the worldwide tradition of use of Chua's circuits for various purposes, we introduce the paradigm of chaotic mathematical circuitry which shows some similarity to the paradigm of electronic circuitry — the design of electronic circuit — especially in the frame of chaotic attractors. An electronic circuit is composed of individual electronic components, such as capacitors, diodes, inductors, resistors, transistors and connected by conductive wires. Recently, in 2009, three more components discovered by L. O. Chua have been added to the set of devices, namely: memristors, memcapacitors and meminductors. In the same way a mathematical circuit is composed of individual components we design: generators, couplers, samplers, mixers, reducers and cascaders, connected by streams of data. The combination of such mathematical components allows many new applications in chaotic cryptography, genetic algorithms in optimization or in control.

1. Introduction

Since the seminal work of E. N. Lorenz [16] in 1963, who discovered by accident the first chaotic strange attractor, chaotic dynamical systems have been fully studied. Fifty years after, only few chaotic attractors involving differential equations remain actively explored. Among them Chua's attractor is nowadays incredibly used, because both of its realizations: electronic circuit or system of differential equations can be combined for multiple purpose [8]. Following the first studies applying such combinations to crypted transmission, our aim is to build an analog of paradigm of electronic circuitry, which is the design of electronic circuit: the paradigm of chaotic mathematical circuitry, in order to improve the performance of well known chaotic attractors for application purpose (cryptography, generic algorithms in optimization, control, ...).

An electronic circuit is composed of individual electronic components, such as resistors, transistors, capacitors, inductors and diodes, connected by conductive wires through which electric current can flow. The combination of components and wires allows various simple and complex operations to be performed. In the same way a mathematical circuit is composed of individual components we introduce (generators, couplers, samplers, mixers, reducers and cascaders, ...) connected through streams of data. The combination of such mathematical components leads to several news applications such as improving the performance of well known chaotic attractors (Belykh [17] Lorenz, Rössler [28] ...) for application purpose (chaotic cryptography, genetic algorithms in optimization, control, ...).

In Sec. 2 we recall a first historic example of chaotic mathematical circuitry: the cascading of two identical receivers in Chua's circuit. In Sec. 3, we introduce the new paradigm of chaotic mathematical circuitry, some examples of applications of which are given in Sec. 4.

2. A First Historic Example of Chaotic Mathematical Circuitry: the Cascading of Two Identical Receivers in Chua's Circuit

2.1. *Chua's circuit*

In october 1983, visiting T. Matsumoto at Waseda University, L. O. Chua invented an electronic circuit (Figs. 1(a) and 1(b)) mimicking directly on an oscilloscope screen a chaotic signal (Fig. 1(c)).

Only two autonomous systems of ordinary differential equations were generally accepted then as being chaotic, the Lorenz equations [16] and the Rössler equations [28]. The nonlinearity in both

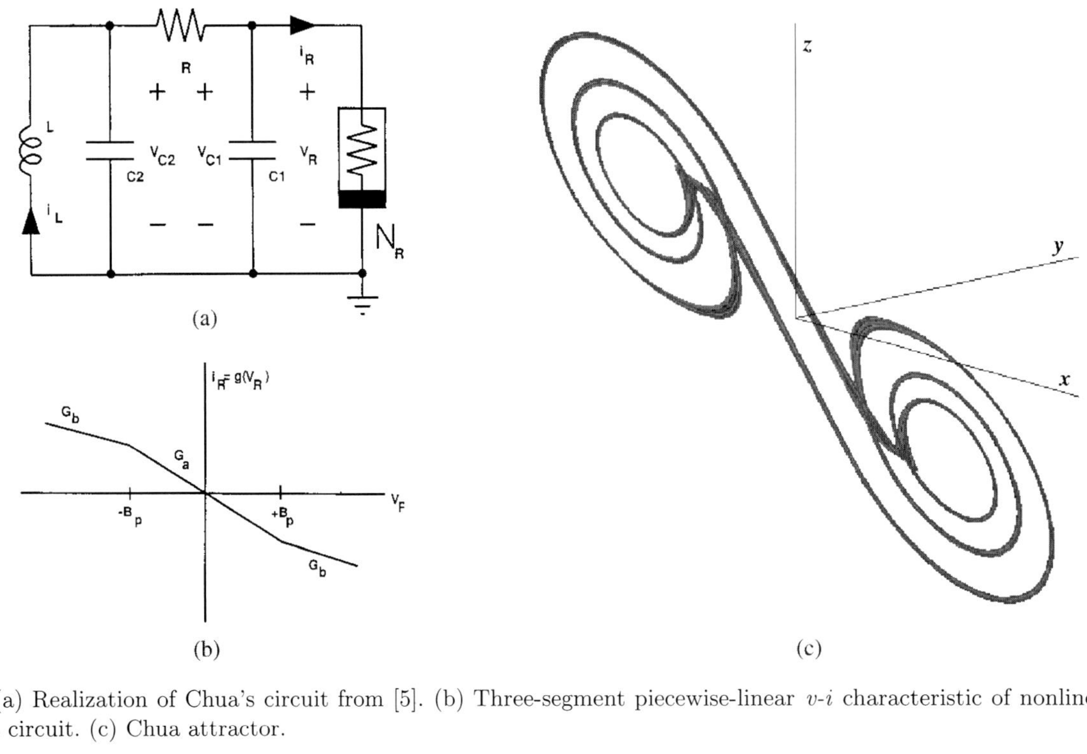

Fig. 1. (a) Realization of Chua's circuit from [5]. (b) Three-segment piecewise-linear v-i characteristic of nonlinear resistor in Chua's circuit. (c) Chua attractor.

systems is a the product function of two variables which is very difficult to build in electronic circuit. L. O. Chua [4] says, "The fault lies on the dearth of a critical nonlinear IC component with a near-ideal characteristic and a sufficiently large dynamic range; namely, the analog multiplier. Unfortunately, this component was the key to building an autonomous chaotic circuit in 1983." He adds, "Suddenly [in the evening of the precise day he attended an unsuccesful presentation of an electronic circuit realization of the Lorenz Equations] it dawned upon me that since the main mechanism which gives rise to chaos, in both the Lorenz and the Rössler Equations, is the presence of at least two unstable equilibrium points −3 for the Lorenz Equations and 2 for the Rössler Equations — it seems only prudent to design a simpler and more robust circuit [than that built by Matsumoto's team] having these attributes. Having identified this alternative approach and strategy, it becomes a simple exercise in elementary nonlinear circuit theory to enumerate systematically all such circuit candidates, of which there were only 8 of them, and then to systematically eliminate those that, for one reason or another, can not be chaotic."

The Chua's equations he laid down

$$\begin{cases} \dot{x} = \alpha(y - x - f(x)), \\ \dot{y} = x - y + z, \\ \dot{z} = -\beta y, \end{cases} \tag{1}$$

where

$$f(x) = bx + \frac{1}{2}(a - b)[|x + 1| - |x - 1|], \tag{2}$$

and

$$\alpha = 15.60, \quad \beta = 28.58, \quad a = -\frac{1}{7}, \quad b = \frac{2}{7}, \tag{3}$$

were soon numerically analyzed by T. Matsumoto [26]. The "nonlinear" characteristic which is in fact only piecewise linear allows some exact computations. Henceforth, L. O. Chua *et al.* [3] proved in the same breath that the mechanism of chaos exists in this attractor.

2.2. Secure communication via chaotic synchronization

2.2.1. *Synchronization*

The synchronization of two Chua's circuit was studied experimentaly height years later in 1992 [5], soon followed by its application to crypted transmission. The first laboratory demonstration of a secure communication system which uses a chaotic signal for masking purposes [27] and which exploits the chaotic synchronization techniques to recover the signal was reported in 1992 [15]. While the "transmitter" in this system is a direct implementation of the method proposed in Oppenheim *et al.* [27] the "receiver" differs from their computer simulation approach in that it actually contains two subsystems of the "chaotic" transmitter (Chua's circuit in that case).

2.2.2. *Single chaotic synchronization*

The mathematical translation of the dynamics of the circuit used by Kocarev *et al.* [15] for the experimental demonstration of secure communication is as follows: the basic building block is a Chua's circuit, the dynamics of which is given by the Chua's equations (1) and (2). Here $x(t)$ is used as noise-like "masking" signal. Let $s(t)$ be an information-bearing signal. The transmitted signal is $r(t) = x(t) + s(t)$, where the power level of $s(t)$ is assumed to be significantly lower than that of $x(t)$, in order to have the signal effectively hidden. The receiver consists of two subsytems.

The first one is driven by the transmitted signal $r(t)$:

$$\begin{cases} \dot{y_1} = r(t) - y_1 + z_1, \\ \dot{z_1} = -\beta y_1. \end{cases} \tag{4}$$

The second subsytem is driven by the signal $y_1(t)$:

$$\dot{x_2} = \alpha(y_1(t) - x_2 - f(x_2)). \tag{5}$$

Then $s(t)$ is recovered as

$$s_2(t) = r(t) - x_2(t) \approx s(t). \tag{6}$$

Actually the dynamics of the experimental set-up (see Fig. 2) is described by

$$\begin{cases} \dot{x_2} = \alpha(y_1(t) - x_2 - f(x_2)), \\ \dot{z_2} = -\beta y_1(t). \end{cases} \tag{7}$$

However, as long as we do not need $z_2(t)$ to recover $s_2(t)$, we continue to use Eq. (6) instead of Eq. (7) in the following improved system.

2.2.3. *Cascade chaotic synchronization*

Ten years after the invention of his ubiquitous real-world example of a chaotic system, leading some to declare it "a paradigm for chaos", and few months after the experimental demonstration of secure communication using the properties of this chaotic generator, Professor L. O. Chua was visiting us for one month, in May 1993, at the University of Nice.

In both implementations — electronic circuit realization (Fig. 3) or computer simulation (Fig. 2) — of the circuit used by Kocarev *et al.* [15] there is an inevitable error introduced by the signal $s(t)$. Then the claim was to enhance the performance of the chaotic masking technique by improving the convergence of the recovered signal $s_2(t)$ towards the information-bearing signal $s(t)$. Having in mind the knowledge of relaxation methods used in numerical analysis (in numerical mathematics, relaxation methods are iterative methods for solving systems of equations, including nonlinear systems), we proposed to iterate the process of recovering the signal, cascading a second identical receiver to the first one, i.e. introducing a second system of equations comparable to Eqs. (4)–(5) driven by $x_2(t)$ instead of $r(t)$, as displayed in Fig. 5. The second receiver also consists of two subsytems. The first one is driven by the signal $x_2(t)$ which is assumed to be more synchronized to $x(t)$ than the transmitted signal $r(t)$:

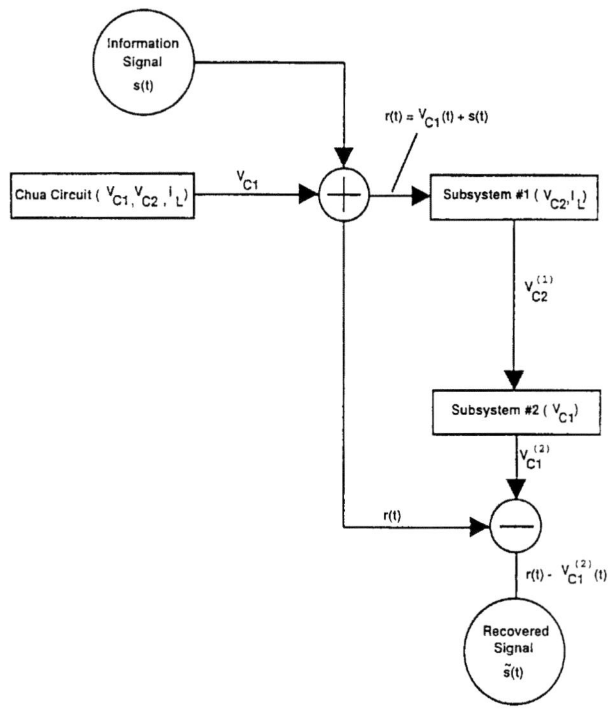

Fig. 2. Experimental set up. Block diagram of the system. It contains one Chua's circuit and two partial Chua's circuits, that is, subsystems #1 and #2 of Fig. 3, from [15].

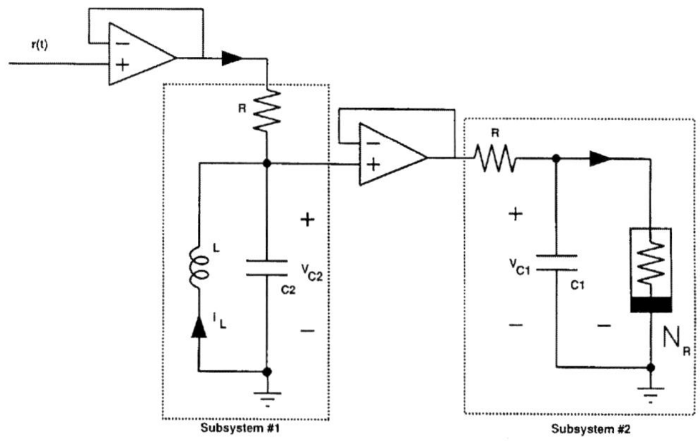

Fig. 3. Practical realization of the receiver. The first subsystem is a partial Chua's circuit consisting of the (v_{C_2}, i_L)-subsytem driven by the transmitted signal $r(t)$. The second subsystem is a partial Chua's circuit consisting of the (v_{C_1})-subsytem driven by the transmitted signal $v_{C_2}^{(1)}$. The triangular symbols are OpAmps which decoupled the systems, acting as the signal drive elements, from [15].

$$\begin{cases} \dot{y_3} = x_2(t) - y_3 + z_3 \\ \dot{z_3} = -\beta y_3 \end{cases} \tag{8}$$

The second subsystem of the second receiver is then driven by the signal $y_3(t)$ from Eq. (8):

$$\dot{x_4} = \alpha(y_3(t) - x_4 - f(x_4)) \tag{9}$$

Then $s(t)$ is recovered as

$$s_4(t) = r(t) - x_4(t) \approx s(t) \tag{10}$$

In practice we simply make two copies of the receiver as shown in Fig. 4. By identifying the symbols (V_{C_1}, V_{C_2}, I_l) in Chua's circuit (see Figs. 2 and 3) with (x, y, z), the electronic circuit implementation in Fig. 4 can be translated into the block diagram shown in Fig. 5. Although no two electronic circuits can be made perfectly identical in practice, this ideal situation can be approached with the help of the integrated circuit technology demonstrated in Delgado-Restituto & Rodriguez-Vasquez [6]. By fabricating several identical Chua's circuits on the *same* silicon chip, the resulting circuits are almost "clones" of each other. This technique has the additional security adavantage in that even if someone else has discovered the parameters (α, β) used in the system, integrating it into *another* silicon chip invariably introduces

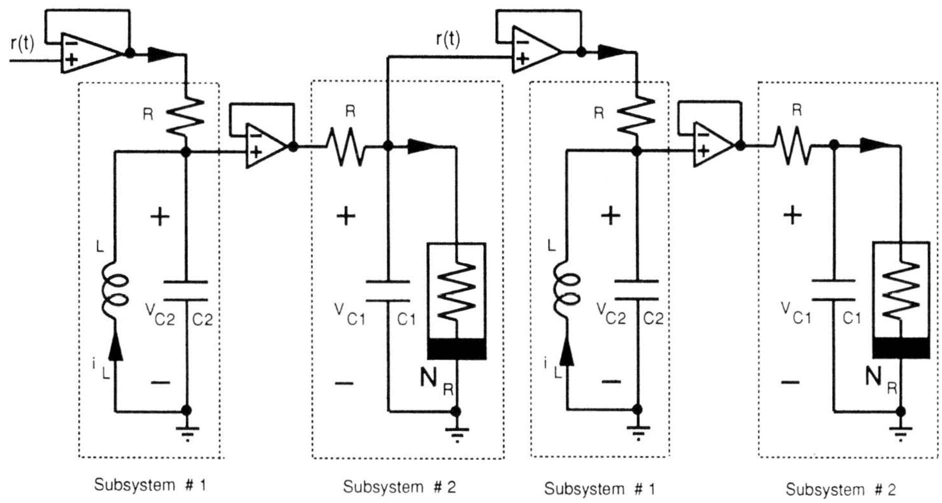

Fig. 4. Electronic circuit implementation of the two stage "receiver" consisting of two identical copies of the circuit given in Fig. 3.

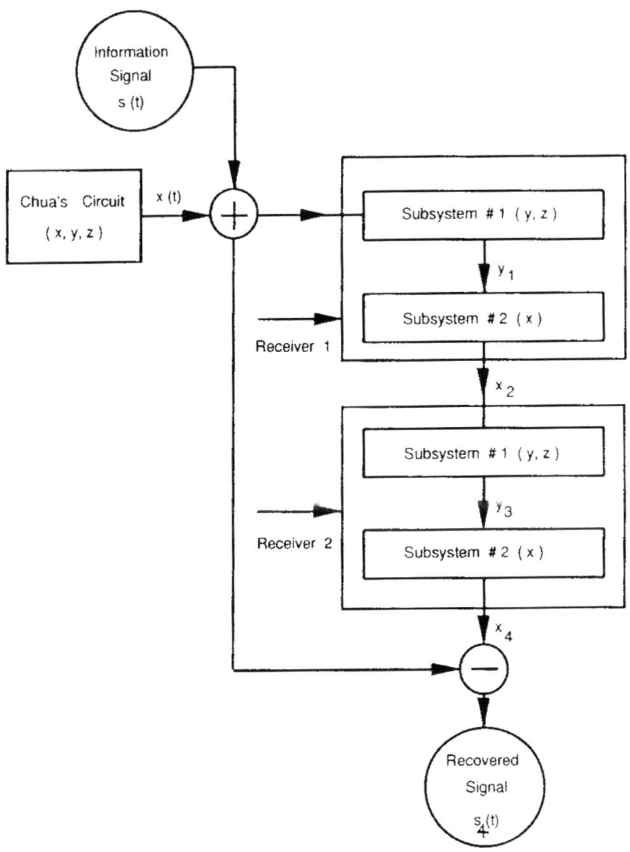

Fig. 5. Block diagram of the electronic circuit implemented in Fig. 4.

discrepancies due to the different processing parameters from different silicon "foundries". We have shown by computer experiments that by connecting two identical receivers, a significant amount of noise can be reduced, thereby allowing the recovery of a much higher quality signal [18].

2.2.4. *Numerical experiments*

Due to the limited extend of this chapter, we give only a sketch of results of numerical experiments. Assuming that the input (information-bearing) signal $s(t)$ is a single tone (sine wave) of amplitude $k \ll 1$:

$$s(t) = k \sin(\omega t) \quad \text{with } k > 0. \tag{11}$$

Therefore $s(t)$ has a power level significantly lower than that of $x(t)$.

Let us define the *errors*

$$\|e_s(k,\omega)\|_2 = \|s_2(t) - s(t)\|_2 \tag{12}$$

$$\|e_c(k,\omega)\|_2 = \|s_4(t) - s(t)\|_2 \tag{13}$$

where

$$\|f(t)\|_2 \triangleq \lim_{T\to\infty} \frac{1}{T}\left[\int_0^T f^2(t)dt\right]^{\frac{1}{2}} \tag{14}$$

denotes the quadratic norm of $f(t)$.

Extensive computer studies show that both $\|e_s(k,\omega)\|_2/\|s(t)\|_2$ and $\|e_c(k,\omega)\|_2/\|s(t)\|_2$ are independent of k, and that both norms $\|e_s(k,\omega)\|_2$ and $\|e_c(k,\omega)\|_2$ are decreasing in accordance to a power law with an exponent greater than 2 when ω is increasing. Moreover $\|e_c(k,\omega)\|_2$ is always less than $\|e_s(k,\omega)\|_2$. This shows that the cascade chaotic synchronization technique offers a good improvment over the single-stage chaotic synchronization results.

3. Mathematical Circuitry

Our aim is to build an analog of paradigm of electronic circuitry, which is the design of electronic circuit: the paradigm of mathematical circuitry and especially chaotic mathematical circuitry in order to improve the performance of well known chaotic attractors (Belykh [17] Lorenz [16] Rössler [28] ...) for application purpose (chaotic cryptography, genetic algorithms in optimization, control, emergence of ramdoness from chaos, ...).

An electronic circuit is composed of individual electronic components, such as capacitors, diodes, inductors, resistors, transistors and connected by conductive wires through which electric current can flow. Recently, in 2009, three more components discovered by L. O. Chua *et al.* [7] have been added: memristors, memcapacitors and meminductors (Fig. 6). The combination of components and wires allows various simple and complex operations to be performed: signals can be amplified, computations can be accomplished, and data can be moved from one place to another. In the same way a mathematical circuit is composed of individual components we introduce now: generators, couplers, samplers, mixers, reducers and cascaders, connected by streams of data. The combination of such mathematical components leads to several performing news applications, such as Chaotic Pseudo Random Number Generators (CPRNG), as we will show in the rest of this article.

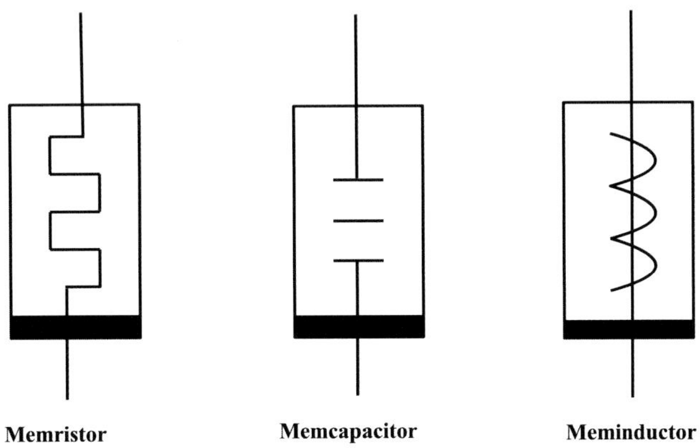

Fig. 6. Symbols of the three recently discovered circuit elements with memory.

3.1. *Generator*

Analog circuits are very commonly represented in schematic diagrams, in which wires are shown as lines, and each component has a unique symbol. We present in this section the symbols we design in order to draw mathematical schematic diagrams. The first symbols we describe, generator symbols, are, from a mathematical point of view, equivalent to a battery or a current generator in electronic circuit. However we consider that they generate a numerical signal (in one or more dimensions) rather than a voltage or an intensity variation (nonetheless, a voltage or intensity variation can be considered as a physical signal which can be discretized). This signal can be either continuous, as in Chua's circuit, or discrete as in Hénon map [13].

In the expanded symbol of continuous generator (see Fig. 7(a)), the solid line arrows coming out from the generator represent the three components of the signal $\underline{x}(t) = (x(t), y(t), z(t))$ (see Eq. (1)), the dashed line arrow points at λ which stands for the parameter value defined by Eq. (3), and the dot line arrow points at $\underline{x}_0 = \underline{x}(0)$, the given initial value of the signal. If there is no ambiguity on the nature of the generator used, the symbol can be simplified as in Fig. 7(b).

We need also to design chaotic circuitry for discrete signal. To this aim two generators can be considered: in dimension 2, the Hénon map

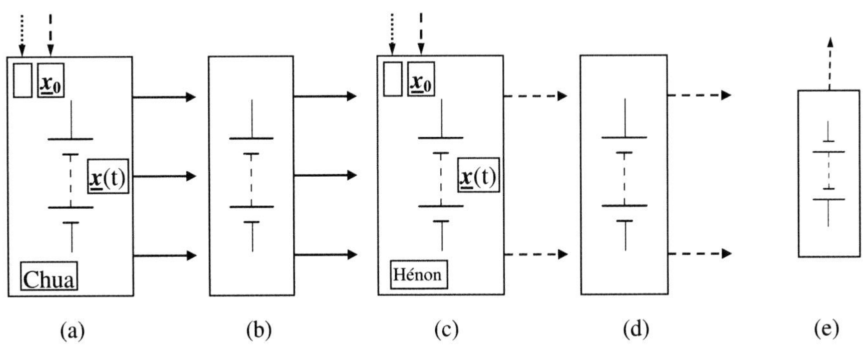

Fig. 7. Generators. (a) Continuous generator (Chua's circuit, expanded symbol). (b) Simplified continuous generator. (c) Discrete generator (Hénon map, expanded symbol). (d) Simplified discrete generator. (e) 1-dimensional logistic generator.

(see Figs. 7(c) and 7(d))

$$H_{a,b} : \begin{pmatrix} x \\ y \end{pmatrix} = \begin{pmatrix} y + 1 - ax^2 \\ bx \end{pmatrix} \tag{15}$$

with

$$a = 1.4, \quad b = 0.3, \tag{16}$$

which is associated to the dynamical system,

$$\begin{cases} x_{n+1} = y_n + 1 - ax_n^2 \\ y_{n+1} = bx_n \end{cases} \tag{17}$$

and the logistic map

$$f_r(x) = rx(1 - x) \tag{18}$$

linked to the one dimensional system which is chaotic when $r = 4$ (see Fig. 7(e)).

$$x_{n+1} = rx_n(1 - x_n) \tag{19}$$

Thereafter, another 1-dimensional chaotic generator, the symmetric tent map, will be, also, represented by the same symbol.

3.2. *Coupler*

Two years after the experimental study of the synchronization of two Chua's circuit, in 1994, experimental observation of hyperchaotic attractors in open and closed chain of Chua's circuit was reported [14]. The layout of the five identical coupled Chua's circuit forming a ring is displayed on Fig. 8, the state equations of this circuit are as follows

$$\begin{cases} C_1\dfrac{dv_{C_1}^{(1)}}{dt} = G\left(v_{C_2}^{(1)} - v_{C_1}^{(1)}\right) - f\left(v_{C_1}^{(1)}\right), \\ C_2\dfrac{dv_{C_2}^{(1)}}{dt} = G\left(v_{C_1}^{(1)} - v_{C_2}^{(1)}\right) + i_L^{(1)} + K_1\left(v_{C_2}^{(2)} - v_{C_2}^{(1)}\right), \\ L\dfrac{di_L^{(1)}}{dt} = -v_{C_2}^{(1)}, \\ C_1\dfrac{dv_{C_1}^{(2)}}{dt} = G\left(v_{C_2}^{(2)} - v_{C_1}^{(2)}\right) - f\left(v_{C_1}^{(2)}\right), \\ C_2\dfrac{dv_{C_2}^{(2)}}{dt} = G\left(v_{C_1}^{(2)} - v_{C_2}^{(2)}\right) + i_L^{(2)} + K_2\left(v_{C_2}^{(3)} - v_{C_2}^{(2)}\right), \\ L\dfrac{di_L^{(2)}}{dt} = -v_{C_2}^{(2)}, \\ \cdots \\ \cdots \\ \cdots \\ C_1\dfrac{dv_{C_1}^{(5)}}{dt} = G\left(v_{C_2}^{(5)} - v_{C_1}^{(5)}\right) - f\left(v_{C_1}^{(5)}\right), \\ C_2\dfrac{dv_{C_2}^{(5)}}{dt} = G\left(v_{C_1}^{(5)} - v_{C_2}^{(5)}\right) + i_L^{(5)} + K_5\left(v_{C_2}^{(1)} - v_{C_2}^{(5)}\right), \\ L\dfrac{di_L^{(2)}}{dt} = -v_{C_2}^{(5)}. \end{cases} \tag{20}$$

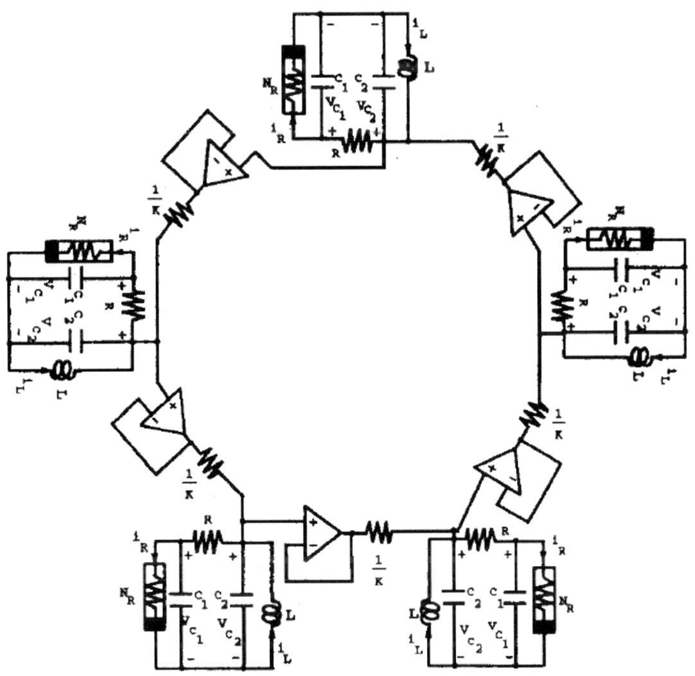

Fig. 8. Five identical coupled Chua's circuits forming a ring, from [14].

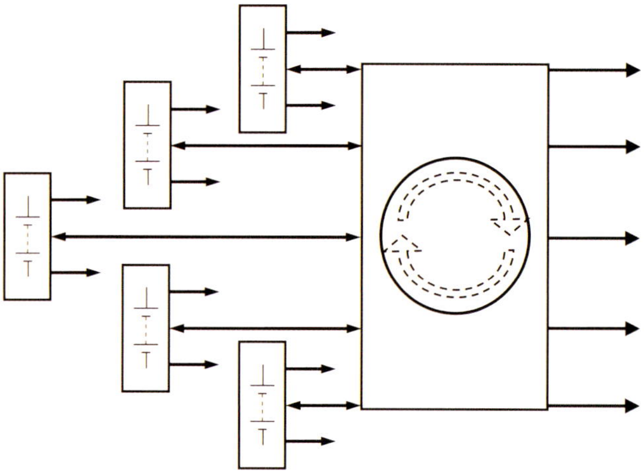

Fig. 9. Numerical circuit corresponding to electronic circuit of Fig. 8.

Each Chua's circuit (see Fig. 1(a)) contains three linear energy-storage elements (an inductor and two capacitors), a linear resitor, and a single nonlinear resistor, namely, Chua's diode with three segment linear characteristic defined by

$$f(v_R) = m_0 v_R + \frac{1}{2}(m_1 - m_0)[|v_R + B_p| - |v_R - B_p|] \tag{21}$$

where the slopes in the inner and the outer regions are m_0 and m_1, respectively, and $\pm B_p$ denotes the breakpoints (see Fig. 1(b)). Equation (21) is equivalent to Eq. (2). By identifying the symbols $(V_{C_1}^{(i)}, V_{C_2}^{(i)}, I_l^{(i)})$ in each Chua's circuit with (x^i, y^i, z^i), the state equations of the circuit can be translated into the differential equations (22) and the electronic circuit symbolized by the mathematical circuit of Fig. 9.

$$\begin{cases} \dot{x}^1 = \alpha(y^1 - x^1 - f(x^1)), \\ \dot{y}^1 = x^1 - y^1 + z^1 + k_1(y^2 - y^1), \\ \dot{z}^1 = -\beta y^1, \\ \dot{x}^2 = \alpha(y^2 - x^2 - f(x^2)), \\ \dot{y}^2 = x^2 - y^2 + z^2 + k_2(y^3 - y^2), \\ \dot{z}^2 = -\beta y^2, \\ \quad \cdots \\ \quad \cdots \\ \quad \cdots \\ \dot{x}^5 = \alpha(y^5 - x^5 - f(x^5)), \\ \dot{y}^5 = x^5 - y^5 + z^5 + k_5(y^1 - y^5), \\ \dot{z}^5 = -\beta y^5. \end{cases} \tag{22}$$

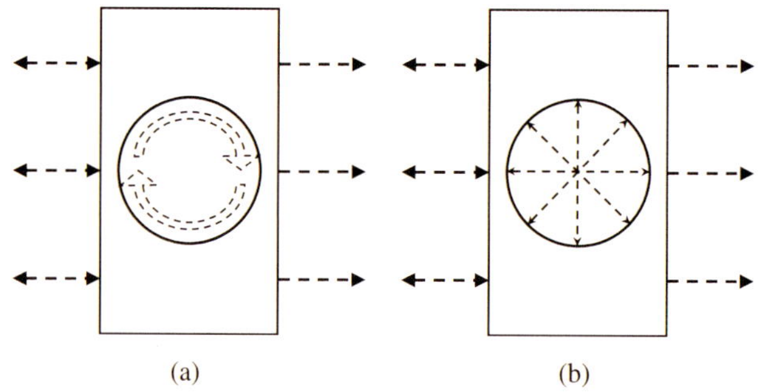

Fig. 10. (a) Ring coupler. (b) Full coupler.

In this figure the double arrows symbolize the coupling $k_i(y^{i+1} - y^i)$ of one Chua's circuit to the next one. In order to represent the coupling between mathematical equation, depending on the nature of the coupling, we can use both symbols: the ring coupler corresponding to the coupling of one generator to the next one (Fig. 10(a)), and the full coupler when the coupling involves more connections between the couplers (Fig. 10(b)).

Remark 1. In the rest of this article, we use solid line arrow for continuous signal $x(t)$, and dashed line arrow for discrete signal x_n.

It has been shown, a few years ago [20] that the ultra-weak coupling of several logistic maps (Eq. (18)) or symmetric tent maps

$$f_a(x) = 1 - a|x| \tag{23}$$

$$x_{n+1} = f_a(x_n) \tag{24}$$

when $a = 2$, allows the production of long series of chaotic numbers equally distributed over the interval $J = [-1, 1] \subset \mathbb{R}$.

When a dynamical system is realized on a computer using floating point numbers, the computation is of discretization, where finite arithmetic replaces continuum state space. For chaotic dynamical systems, the discretizations often have collapsing effects to a fixed point or to short cycles [25]. Instead, the ultra-weak coupling of logistic or symmetric tent map

$$\begin{cases} x_{n+1}^1 = (1 - 2\varepsilon_1) f\left(x_n^1\right) + \varepsilon_1 f\left(x_n^2\right) + \varepsilon_1 f\left(x_n^3\right) \\ x_{n+1}^2 = \varepsilon_2 f\left(x_n^1\right) + (1 - 2\varepsilon_2) f\left(x_n^2\right) + \varepsilon_2 f\left(x_n^3\right) \\ x_{n+1}^3 = \varepsilon_3 f\left(x_n^1\right) + \varepsilon_3 f\left(x_n^2\right) + (1 - 2\varepsilon_3) f\left(x_n^3\right) \end{cases} \tag{25}$$

symbolized by Fig. 11(a) gives rise to sterling model of generator of chaotic numbers with a uniform distribution of these numbers of the interval

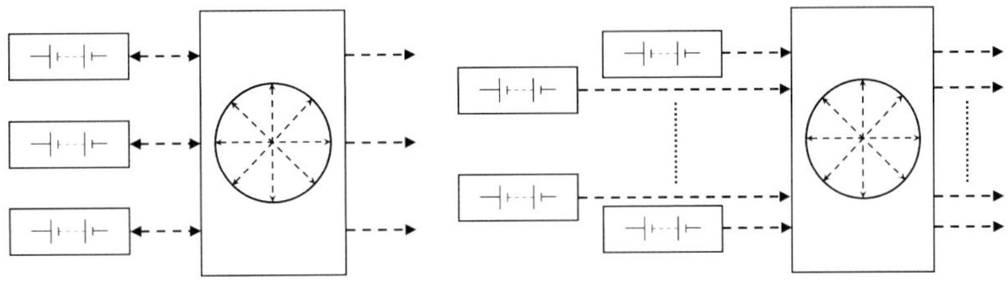

Fig. 11. (a) Circuit of ultraweak coupling of three 1-dimensional chaotic maps. (b) Circuit of ultraweak coupling of p 1-dimensional chaotic maps.

$[-1,1]$. Ultra-weak coupling means $\varepsilon_i \in [10^{-15}, 10^{-7}]$ for computations using double precision numbers [24]. The periodic solutions of such generator have period far greater than ten billions. Of course, more than three 1-dimensional generators can be coupled under the same process, enhancing the ergodic properties of the generator.

The considered system of the p-coupled dynamical systems is

$$X_{n+1} = F(X_n) = A.(\underline{f}(X_n)) \tag{26}$$

with

$$X_n = \begin{pmatrix} x_n^1 \\ \vdots \\ x_n^p \end{pmatrix} \quad \underline{f}(X_n) = \begin{pmatrix} f(x_n^1) \\ \vdots \\ f(x_n^p) \end{pmatrix} \tag{27}$$

and the coupling matrix A. F is a map of $J^p = [-1,1]^p \subset \mathbb{R}^p$ into itself. The mathematical circuit of this system is displayed in Fig. 11(b).

It is noteworthy that these families of very weakly coupled maps are more powerful than the usual formulas used to generate chaotic sequences

$$A = \begin{pmatrix} \varepsilon_{1,1} = 1 - \sum_{j=2}^{j=p} \varepsilon_{1,j} & \varepsilon_{1,2} & \cdots & \varepsilon_{1,p-1} & \varepsilon_{1,p} \\ \varepsilon_{2,1} & \varepsilon_{2,2} = 1 - \sum_{j=1, j\neq 2}^{j=p} \varepsilon_{2,j} & \cdots & \varepsilon_{2,p-1} & \varepsilon_{2,p} \\ \vdots & \ddots & & \vdots & \vdots \\ \vdots & & \ddots & \vdots & \vdots \\ \varepsilon_{p,1} & \cdots & \cdots & \varepsilon_{p,p-1} & \varepsilon_{p,p} = 1 - \sum_{j=1}^{j=p-1} \varepsilon_{p,j} \end{pmatrix} \tag{28}$$

mainly because only additions and multiplications are used in the computation process; no division being required. Moreover the computations are done using floating point or double precision numbers, allowing the use of the powerful Floating Point Unit (FPU) of the modern microprocessors. In addition, a large part of the computations can be parallelized taking advantage of the multicore microprocessors which are common in laptop computers, nowadays.

We display in Table 1 the discrepancies in quadratic norm (Eq. (14)) between the distribution of the iterated values x_n^1 and the Lebesgue measure vs. the number of iterates N_{iter}, for 2, 3 and 4-coupled symmetric tent maps. Computations are done using double precision numbers ($\sim$14–15 digits), $\varepsilon_{i,j} = i\varepsilon_1, j = 1, 4, \varepsilon_1 = 10^{-14}$, initial values: $x_0^1 = 0.330000013113$, $x_0^2 = 0.338756413113$, $x_0^3 = 0.331353442113$, $x_0^4 = 0.333213583113$.

Remark: the ring coupler of Fig. 10(a) corresponds to an hollow matrix, in which only one diagonal and one other coefficient are not empty. The full coupler of the same figure stands for a matrix more filled with non vanishing coefficients as in Eq. (28).

Table 1.

N_{iter}	2-coupled equation	3-coupled equation	4-coupled equation
10^5	0.100199	0.099820996	0.099610992
10^6	0.01006199	0.0098781898	0.01022057
10^7	0.0010442081	0.0010014581	0.0010055967
10^8	0.0001055816	9.8853067×10^{-5}	0.00010197872
10^9	1.567597×10^{-5}	1.0047459×10^{-5}	1.0326474×10^{-5}
10^{10}	7.3577797×10^{-6}	9.7251536×10^{-7}	9.9932242×10^{-7}
10^{11}	6.6338453×10^{-6}	1.0434293×10^{-7}	1.0070523×10^{-7}
10^{12}		1.116009×10^{-8}	9.6166733×10^{-9}
3×10^{12}		4.0443118×10^{-9}	3.2530773×10^{-9}

3.3. *Sampler*

However chaotic numbers are not pseudo-random numbers because the plot of the couples of any component (x_n^l, x_{n+1}^l) of iterated points (X_n, X_{n+1}) in the corresponding phase plane reveals the map f used as one-dimensional dynamical systems to generate them via Eq. (26). Nevertheless we have recently introduced a family of enhanced Chaotic Pseudo Random Number Generators (CPRNG) in order to compute very fast long series of pseudorandom numbers with desktop computer [21], [22], [23] and its properties have been analyzed [10], [11], [12]. This family is based on the previous ultra-weak coupling which is improved in order to conceal the chaotic genuine function.

In order to hide f of Eq. (26), in the phase space (x_n^l, x_{n+1}^l), two mechanisms are used. The pivotal idea of the first one mechanism is to sample chaotically the sequence $(x_0^l, x_1^l, x_2^l, \ldots, x_n^l, x_{n+1}^l, \ldots)$ generated by the l-th component x^l, selecting x_n^l every time the value x_n^m of the m-th component x^m, is strictly greater (or smaller) than a threshold $T \in J$, with $l \neq m$, for $1 \leq l, m \leq p$.

That is to say, to extract the subsequence $(x_{n_{(0)}}^l, x_{n_{(1)}}^l, x_{n_{(2)}}^l, \ldots, x_{n_{(q)}}^l, x_{n_{(q+1)}}^l, \ldots)$ denoted here $(\overline{x_0}, \overline{x_1}, \overline{x_2}, \ldots \overline{x_q}, \overline{x_{q+1}}, \ldots)$ from the original one, in the following way:

Given $1 \leq l, m \leq p$, $l \neq m$

$$\begin{cases} n_{(-1)} = -1 \\ \overline{x_q} = x_{n_{(q)}}^l, \\ \quad \text{with } n_{(q)} = \min_{r \in \mathbb{N}} \{r > n_{(q-1)} | x_r^m > T\} \end{cases} \tag{29}$$

The sequence $(\overline{x_0}, \overline{x_1}, \overline{x_2}, \ldots, \overline{x_q}, \overline{x_{q+1}}, \ldots)$ is then the sequence of chaotic pseudo-random numbers.

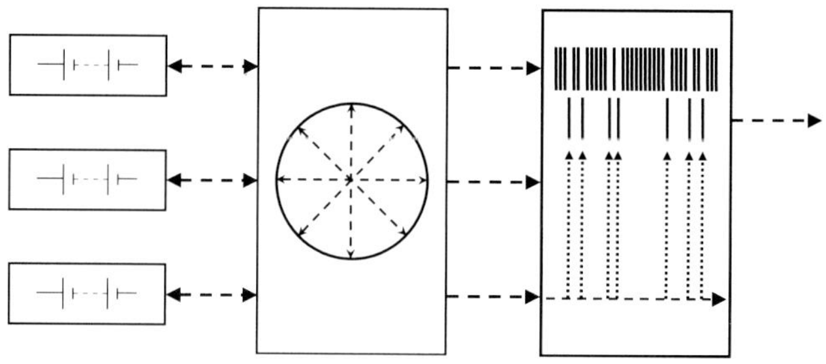

Fig. 12. Circuit of enhanced Chaotic Pseudo Random Number Generators (CPRNG) based on chaotic sampling.

The mathematical formula (29) can be best understood in algorithmic way. The pseudo-code, for computing iterates of (29) corresponding to N iterates of Eq. (26) is:

$X_0 = (x_0^1, x_0^2, \ldots, x_0^{p-1}, x_0^p) = seed$
$n = 0;\ q = 0;$
`do` { `while` $n < N$
  `do` { `while` $(x_n^m \leq T)$
    `compute`
      $(x_n^1, x_n^2, \ldots, x_n^{p-1}, x_n^p); n{+}{+}\}$
  `compute` $(x_n^1, x_n^2, \ldots, x_n^{p-1}, x_n^p)$;
  `then` $n(q) = n; \overline{x_q} = x_{n(q)}^1; n{+}{+}; q{+}{+}\}$

This chaotic sampling is possible due to the independence of each component of the iterated points X_n vs. the others [21]. We introduce the symbol on the right hand side of Fig. 12 in order to give a schematic representation of this sampling (also called subsampling) process.

3.4. *Mixer*

A second mechanism can improve the unpredictability of the pseudo-random sequence generated as above, using synergistically all the components of the vector X_n, instead of two. Given $p-1$ thresholds

$$T_1 < T_2 < \cdots < T_{p-1} \in J$$

and the corresponding partition of

$$J = \bigcup_{k=0}^{p-1} J_k$$

with $J_0 = [-1, T_1]$, $J_1 =]T_1, T_2[$, $J_k = [T_k, T_{k+1}[$ for $1 < k < p-1$ and $J_{p-1} = [T_{p-1}, 1[$, this simple mechanism is based on the chaotic mixing of the $p-1$ sequences $(x_0^1, x_1^1, x_2^1, \ldots, x_n^1, x_{n+1}^1, \ldots)$, $(x_0^2, x_1^2, x_2^2, \ldots, x_n^2, x_{n+1}^2, \ldots)$, $\ldots$, $(x_0^{p-1}, x_1^{p-1}, x_2^{p-1}, \ldots, x_n^{p-1}, x_{n+1}^{p-1}, \ldots)$, ...

using the last one $(x_0^p, x_1^p, x_2^p, \dots, x_n^p, x_{n+1}^p, \dots)$ in order to distribute the iterated points with respect to this given partition defining the subsequence $(\overline{x_0}, \overline{x_1}, \overline{x_2}, \dots, \overline{x_q}, \overline{x_{q+1}}, \dots)$ by

$$\begin{cases} n_{(-1)} = -1 \\ \overline{x_q} = x^k_{n_{(q)}}, \\ \quad \text{with } n_{(q)} = \min\limits_{1\le k\le p-1} \Big\{ s_k(q) \\ \quad = \min\limits_{r\in\mathbb{N}} \{ r_k > n_{(q-1)} \mid x^p_{r_k} \in J_k \} \Big\} \end{cases} \tag{30}$$

The pseudo-code, for computing the iterates of Eq. (30) corresponding to N iterates of Eq. (26) is

$X_0 = (x_0^1, x_0^2, \dots, x_0^{p-1}, x_0^p) = seed$
$n = 0;\ q = 0\ ;$
`do {` `while` $n < N$
  `do {while` $(x_n^p \in J_0)$ `compute`
    $(x_n^1, x_n^2, \dots, x_n^{p-1}, x_n^p); n{+}{+}\}$
  `compute` $(x_n^1, x_n^2, \dots, x_n^{p-1}, x_n^p)$
  `let` k `be such that` $x_n^p \in J_k$
  `then` $n(q) = n; \overline{x_q} = x^k_{n(q)}; n{+}{+}; q{+}{+}\}$

We introduce the symbol on the right hand side of Fig. 13 in order to give a schematic representation of the chaotic mixing process. For sake of simplicity we have only displayed a circuit with three 1-dimensional generators. However the mixing process runs better when more generators are coupled.

We display in Table 2 the discrepancies in quadratic norm (Eq. (14)) between the autocorrelation distribution of the iterated values and the Lebesgue measure vs. the number of subsampled or mixed iterates N_{iter}, for 4-coupled symmetric tent maps. Sampling: the first component x^1 is sampled by x^4 for the threshold value 0.998. Mixing: the three components x^1, x^2, x^3 are mixed and sampled by x^4 for the threshold values $T_1 = 0.998$, $T_2 = 0.9987$, $T_3 = 0.9994$. Computations are done using double precision numbers ($\sim 14 - 15$ digits), $\varepsilon_{i,j} = i\varepsilon_1$, $j = 1, 4$, $\varepsilon_1 = 10^{-14}$, initial values: $x_0^1 = 0.330$, $x_0^2 = 0.3387564$, $x_0^3 = 0.50492331$, $x_0^4 = 0.0$.

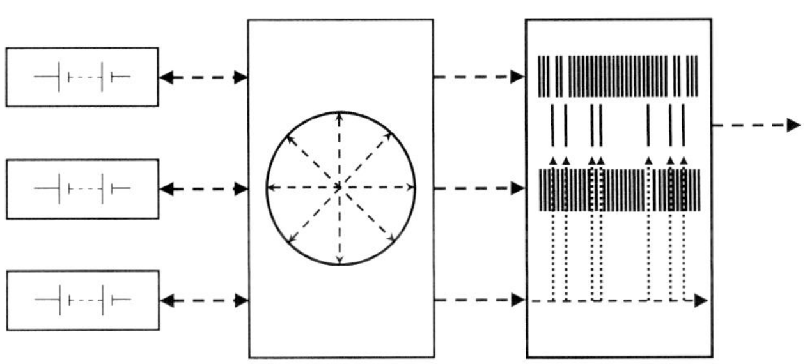

Fig. 13. Circuit of enhanced Chaotic Pseudo Random Number Generators (CPRNG) based on chaotic mixing.

Table 2.

N_{iter}	sampling of 4-coupled equation	mixing of 4-coupled equation
10^5	0.70947368	0.68924731
10^6	0.26570546	0.25881773
10^7	0.079871223	0.086706776
10^8	0.023190157	0.026815309
10^9	0.0071386288	0.0089111078
10^{10}	0.002493667	0.0027932033
10^{11}	0.00071561417	0.00085967214
10^{12}	0.00025442753	0.0002346851
10^{13}	0.000088445108	0.000073234736

We can say that the design of mathematical circuit including couplers, samplers or mixers allows the emergence of complexity in chaotic systems which leads to randomness [24].

3.5. *Reducer*

The accelerated development of modern data transaction applications such as telecommunications requires encoding techniques with higher standards of security.

Classically, these encoding sequences are obtained using Pseudo Random Number Generators (PRNG). Since the seminal work presented in Sec. 2, an efficient alternative, the chaotic-based generators (CPRNG) are used to achieve even higher demanding encryption standards. Indeed, the chaotic systems exhibit a plethora of properties which make them suitable to meet the above requirements.

The advantage to use chaotic systems lies in their extreme sensitivity to small parameter and initial conditions variations: in this way, as many different chaotic carriers as wanted can be generated. However, the appropriate selection of a chaotic map that satisfies cryptographic applications requirements is a huge problem. It has to be emphasized that all chaotic maps are not applicable, because the chaotic generator — which is deterministic — has to satisfy at the same time the criteria for closeness to random signals. Therefore many practical problems arise, from the choice of the chaotic generator and its parameters, to the chaotic properties verification after the quantisation. Ideally, for cryptographic applications and higher security, an everywhere dense chaotic attractor is required, so all chaotic signal samples will appear with the same probability. We have shown that highly efficient discrete-time chaotic generators can be obtained from coupling, sampling, and mixing quite simple models such as the logistic map or the symmetric tent map. We have built the corresponding mathematical circuit. Moreover there exist other combinations of such 1-dimensional chaotic attractor.

We introduce now, as an example, another combination which can directly provides random number without sampling or mixing, although it is possible to combine these processes with it.

The idea leading to this system is to confine on $J^p = [-1,1]^p \subset \mathbb{R}^p$, considered as a torus, a ring of p-coupled symmetric tent map (or logistic map) [9].

Consider the equation

$$\begin{cases} x_{n+1}^1 = 1 - 2\left|x_n^1\right| + k_1 x_n^2 \\ x_{n+1}^2 = 1 - 2\left|x_n^2\right| + k_2 x_n^3 \\ \vdots \\ x_{n+1}^p = 1 - 2\left|x_n^p\right| + k_p x_n^1 \end{cases} \tag{31}$$

where the parameters $k_i \in \{-1, 1\}$. In order to confine the variables x_{n+1}^i on J^p, we do, for every iteration the transform

$$if\ x_{n+1}^i < -1, \quad add\ 2 \tag{32}$$

$$if\ x_{n+1}^i > 1, \quad substract\ 2 \tag{33}$$

We desing a new symbol: the reducer (on the right hand side of Fig. 14) in order to give a schematic representation of the projection of the variable on the torus J^p. For sake of simplicity we have only displayed a circuit with three 1-dimensional generators. However this new pseudo-random number generator works better when more generators are coupled.

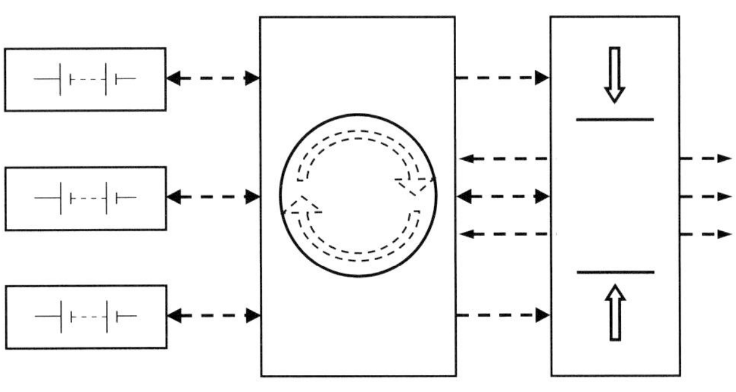

Fig. 14. Reducer for the circuit of Eq. (31), with $n = 3$.

To evaluate the random properties of these generators, a set of statistical based test known as NIST test developed by the National Institute of Standards and Technology have been used.

The random properties validation of a 4-dimensional system has been carried out. Additionally, the chaotic carrier output needs to be quantised and binarised (0 and 1) in order to be validated as being random using NIST tests. Therefore, different methods of binarisation (converting real signals to binary ones) have been implemented and compared.

A first 1−bit binarisation has been applied to the system (31) with $n = 4$, output: the results showed to be highly sensitive to the type of binarisation. Eventually, after testing several different methods, a 32−bit binarisation has been chosen as being the most suitable solution. Because the system is confined to the p−dimensional torus, 31 bits are assigned to represent the decimal part, and 1 bit to the sign. To illustrate the results, the NIST tests for the four dimensional system (31) with parameters $k_i = (-1)^{i+1}$ are shown in Fig. 15. The chosen conditions are:

Length of the original sequence: 10^8 bits, length of bit string: 10^6, quantity of bit strings: 100. The output of the system has been arbitrary chosen as being: $y = x_n^4$.

Furthermore, as the results show their independence from the initial conditions, every bit string in this test is the resulting sequence of a different randomly chosen initial condition.

The criterion for a successful test is that the p−value has to be superior to the significance level (0.01 for this case). For the present model, all tests were successful thus the sequence can be accepted as being random. As we said above, in order to improve even more the random properties of that random signal, two possible strategies are possible: increasing the system order or under-sampling the output signal, which is possible with the circuit of Fig. 14 in which a sampler is added on hand right side.

3.6. *Cascader*

Finally, turning back to the problem of chaotic masking, via the cascading of Chua's attractor,

RESULTS FOR THE UNIFORMITY OF P-VALUES AND THE PROPORTION OF PASSING SEQUENCES

generator is <data/lozi_4_bin32_alternes_CIrandom_y_x4_31250.txt>

C1	C2	C3	C4	C5	C6	C7	C8	C9	C10	P-VALUE	PROPORTION	STATISTICAL TEST
10	9	10	11	16	8	8	11	4	13	0.419021	100/100	Frequency
17	6	13	8	12	7	13	10	8	6	0.213309	100/100	BlockFrequency
8	9	13	8	9	11	11	9	12	10	0.978072	99/100	CumulativeSums
8	8	12	10	11	10	12	10	7	12	0.964295	99/100	CumulativeSums
6	11	15	13	7	12	5	7	7	17	0.075719	100/100	Runs
11	9	12	7	12	6	9	10	13	11	0.867692	99/100	LongestRun
7	10	16	12	9	8	11	10	12	5	0.494392	99/100	Rank
8	17	10	6	12	7	13	11	9	7	0.334538	99/100	FFT
12	7	8	8	6	9	9	9	18	14	0.213309	99/100	NonOverlappingTemplate
12	11	11	9	16	6	10	7	8	10	0.616305	99/100	OverlappingTemplate
11	9	8	6	10	14	9	13	8	12	0.779188	100/100	Universal
12	8	8	10	10	16	8	10	13	5	0.474986	99/100	ApproximateEntropy
9	7	9	5	8	4	5	10	8	4	0.452799	68/69	RandomExcursions
4	9	8	9	6	7	6	0	9	11	0.063482	69/69	RandomExcursionsVariant
15	6	12	11	12	8	13	5	9	9	0.437274	98/100	Serial
9	5	11	11	9	13	9	8	14	11	0.739918	99/100	LinearComplexity

Fig. 15. Successful result of the NIST tests.

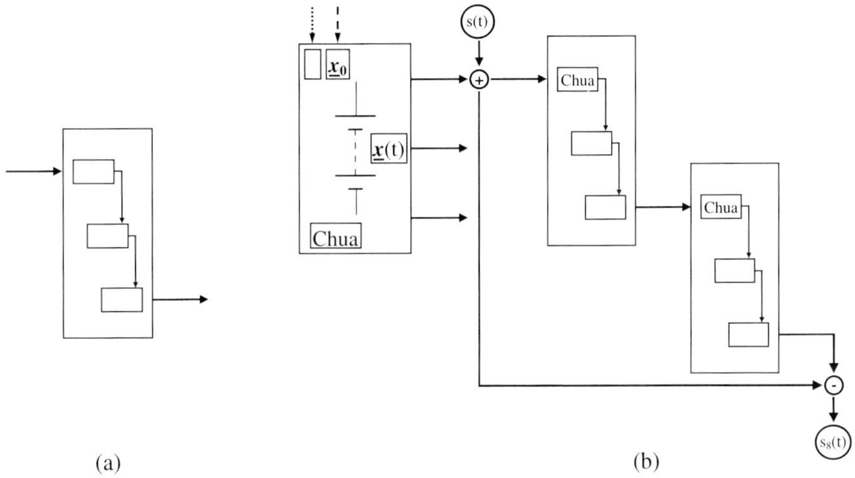

Fig. 16. (a) Cascader symbol. (b) Two cascading receivers combined.

the last symbol we design in order to schematise mathematically the cascading method is the cascader displayed on Fig. 16(a).

4. Mathematical Circuit Engineering

4.1. *Signal masking*

In the limited extend of this chapter, we give only two examples of the engineery of mathematical circuits we have introduced in the previous Section. The first one is an improvement of the cascading of two identical receivers of Chua's circuit. Albeit this improvement is rather from a numerical point of view than a practical one, it is given in order to illustrate more in deep the combination of circuit elements. It is possible to combine two cascading receiver as in Fig. 16(b).

The equations of this circuit are

$$\begin{cases} \dot{y_5} = x_4(t) - y_5 + z_5 \\ \dot{z_5} = -\beta y_5 \end{cases} \tag{34}$$

$$\dot{x_6} = \alpha(y_5(t) - x_6 - f(x_6)) \tag{35}$$

$$\begin{cases} \dot{y_7} = x_6(t) - y_7 + z_7 \\ \dot{z_7} = -\beta y_7 \end{cases} \tag{36}$$

$$\dot{x_8} = \alpha(y_7(t) - x_8 - f(x_8)) \tag{37}$$

Then $s(t)$ is recovered as

$$s_8(t) = r(t) - x_8(t) \approx s(t) \tag{38}$$

Numerical experiments show an improvement of the results [1], [19].

4.2. *Noise-resisting cryptography*

The second example which highlights the complex combination of circuit element belongs to the new

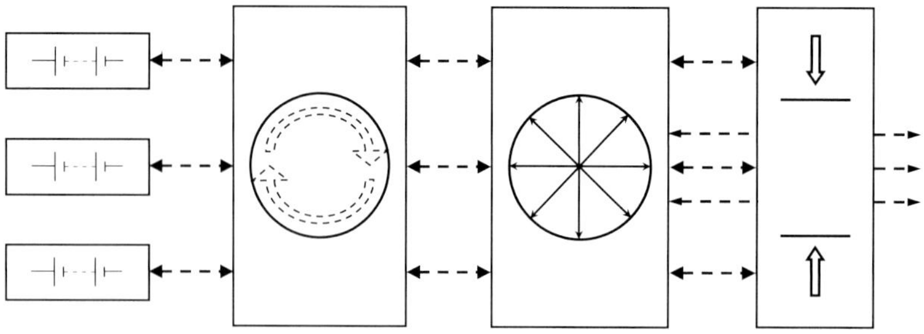

Fig. 17. Chaotic multistream PRNG (Cms-PNRG).

flourishing field of chaotic based cryptography. We proposed last year [2] a novel noise-resisting ciphering method resorting to a chaotic multi-stream pseudorandom number generator (denoted Cms-PRNG) described below. This Cms-PRNG co-generates an arbitrarily large number of uncorrelated chaotic sequences. These cogenerated sequences are actually used in several steps of the ciphering process. Noisy transmission conditions are considered, with realistic assumptions. The efficiency of the proposed method for ciphering and deciphering is illustrated through numerical simulations based on a Cms-PRNG involving ten coupled chaotic sequences.

The CPRNG decribed in Fig. 14, can be improved in order to generate uncorrelated sequences of pseudo-random numbers, possessing a large number of keys. This is simply obtained by adding a coupler as a keyer as in the circuit of Fig. 17 corresponding to Eq. (39) for 4-streams, or Eq. (40) for $p-$streams.

$$\begin{cases} x_{n+1}^1 = 1 - 2\left|x_n^1\right| + k_1(1 - \varepsilon_{1,3} - \varepsilon_{1,4})x_n^2 + \varepsilon_{1,3}x_n^3 + \varepsilon_{1,4}x_n^4 \\ x_{n+1}^2 = 1 - 2\left|x_n^2\right| + k_2(1 - \varepsilon_{2,4} - \varepsilon_{2,1})x_n^3 + \varepsilon_{2,4}x_n^4 + \varepsilon_{2,1}x_n^1 \\ x_{n+1}^3 = 1 - 2\left|x_n^3\right| + k_3(1 - \varepsilon_{3,1} - \varepsilon_{3,2})x_n^4 + \varepsilon_{3,1}x_n^1 + \varepsilon_{3,2}x_n^2 \\ x_{n+1}^4 = 1 - 2\left|x_n^4\right| + k_4(1 - \varepsilon_{4,2} - \varepsilon_{4,3})x_n^1 + \varepsilon_{4,2}x_n^2 + \varepsilon_{4,3}x_n^3 \end{cases} \tag{39}$$

$$\begin{cases} x_{n+1}^1 = 1 - 2\left|x_n^1\right| + k_1\left(\left(1 - \sum_{j=3}^{p}\varepsilon_{1,j}\right)x_n^2 + \sum_{j=3}^{p}\varepsilon_{1,j}x_n^j\right) \\ \vdots \\ x_{n+1}^m = 1 - 2\left|x_n^m\right| + k_m\left(\left(1 - \sum_{j=1, j\neq m, m+1}^{p}\varepsilon_{m,j}\right)x_n^{m+1} + \sum_{j=1, j\neq m, m+1}^{p}\varepsilon_{m,j}x_n^j\right) \\ \vdots \\ x_{n+1}^{p-1} = 1 - 2\left|x_n^{p-1}\right| + k_{p-1}\left(\left(1 - \sum_{j=1}^{p-2}\varepsilon_{p-1,j}\right)x_n^p + \sum_{j=1}^{p-2}\varepsilon_{p-1,j}x_n^j\right) \\ x_{n+1}^p = 1 - 2\left|x_n^p\right| + k_p\left(\left(1 - \sum_{j=1}^{p-2}\varepsilon_{p,j}\right)x_n^1 + \sum_{j=1}^{p-2}\varepsilon_{p,j}x_n^j\right) \end{cases} \tag{40}$$

The ring coupling which is expressed as a diagonal matrix in Eq. (31) is completed with numerous other non vanishing coefficients of the matrix used as secret keys in Eq. (40). Then the originality of the noise-resisting ciphering method introduced which uses this Cms-PRNG is twofold. First a novel ciphering method is proposed aimed at resisting to a noisy transmission channel. The main idea is to establish, between the transmitter and the receiver, a correspondence between the alphabet constituting the plain text and some intervals defining a partition of $[-1, 1]$. Some realistic assumption about the noise boundedness allows to restrict the bounds of the aforementioned intervals in order to precisely resist to the effects of the noise. An extra scrambling resorting to a co-generated chaotic sequence enhances the ciphering process. Then a new chaotic substitution method is developed: considering a chaotic carrier, belonging to the set of cogenerated and coupled pseudo-random chaotic sequences, the idea is to randomly/chaotically (in fact, this is determined by a second pseudo-random chaotic sequence) replace some elements of the carrier by a ciphered element (a letter here) of the message. At the receiver end, a copy of the Cms-PRNG, with the same parameters (hence we deal with a symmetrical ciphering method) allows to generate the necessary chaotic sequences and therefore to retrieve the initial message.

This process can be summarized in both circuits of Figs. 18, 19. Due again to the limited extend of this chapter, we cannot expand these figures in order to show the constituting symbols in each oval shaped region of the circuit. The originality of the method lies in the use of a chaotic pseudo-random number generator: several co-generated sequences can be used at different steps of the ciphering process, since they present the strong property of being uncorrelated. Each letter of the initial alphabet of the plain text is encoded as a subinterval of $[-1, 1]$. The bounds of each interval are defined in function of the known bound of the additive noise.

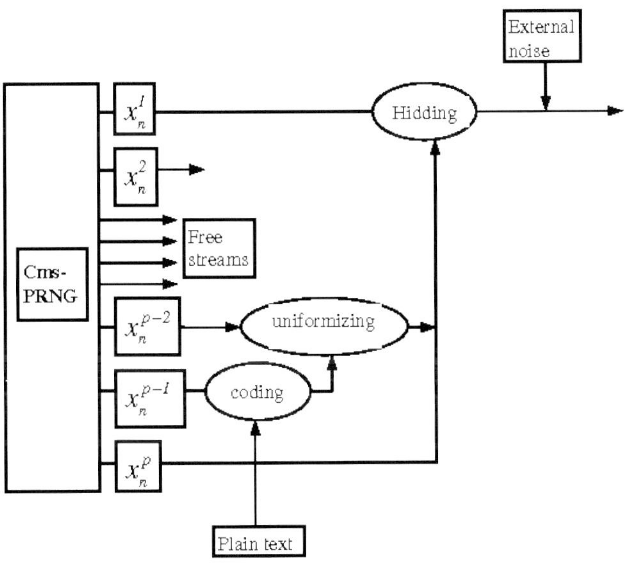

Fig. 18. Transmitter.

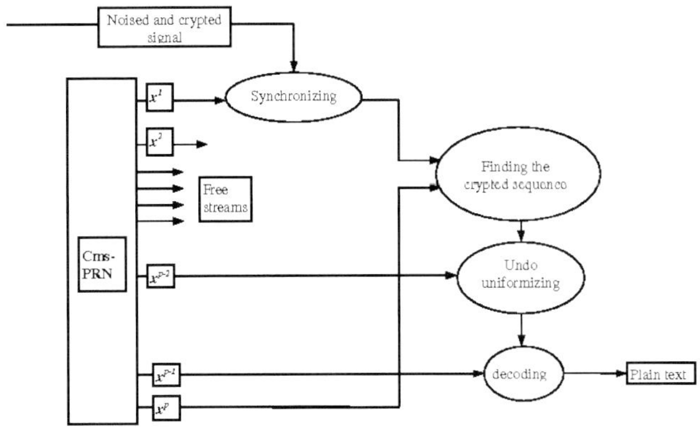

Fig. 19. Receiver.

A pseudo-random sequence is used to enhance the complexity of the ciphering. The transmission consists of a substitution technique inside a chaotic carrier, depending on another cogenerated sequence. The efficiency of the proposed scheme is illustrated on some numerical simulations [2]. As further work, some studies should be performed of several sets of unknown parameters, since with the considered Cms-PRNG with 10 states, the number of possible parameters amounts to 90 (the $\varepsilon_{i,j}$ and the k_i).

5. Conclusion

Following the worldwide tradition of use of Chua's circuits for various purposes, we have introduced the paradigm of chaotic mathematical circuitry which shows some similarity to the paradigm of electronic circuitry -the design of electronic circuits. This new paradigm allows, as an example, the building of new chaotic and random number generators. In this beginning of the third Millenium, the old tradition of design of electronic circuits is drastically revolutionized by the introduction by L. O. Chua of new components with memory, namely memristor, memcapacitor and meminductor. These devices are common at the nanoscale and their combination in circuits open up new functionnalities in electronics. Apart from the obvious use of these elements in nonvolatile memories, several applications can be already envisioned, especially in neuromorphic devices to stimulate learning, adaptative and spontaneous behavior [7].

Alongside to this electronic circuits revolution, the new theory of mathematical circuits allows many new applications in chaotic cryptography, genetic algorithms in optimization and in control,... Due to the versatility of the new components we introduce, the combined operation of these chaotic mathematical circuits is still largely unexplored. We hope our work will motivate experimental and theoretical investigations in this direction. For the 75th birthday of Professor L. O. Chua, one can highlight that his seminal work since more than forty years is fruitful exceeding his wide-ranging field of research. In this chapter (as suggested in the title of the anniversary book) we followed him from chaos to memristor and we went beyond.

References

[1] Aziz Alaoui, M. A. & Lozi, R., (1994), "Secure Communications via Chaotic Synchronization in Chua's Circuit: Numerical Analysis of The Errors of the Recovered Signal," in *Proc. Nonlinear Theory and its Application*, Kagoshima, Japan, 145–148.

[2] Cherrier, E. & Lozi, R., (2011), "Noise-resisting ciphering based on a chaotic multi-stream pseudo-random number generator," in *Proc. IEEE Conference Internet Technology and Secured Transactions (ICITST), 11–14 Dec. Abu Dhabi,* 91–96.

[3] Chua, L. O., Kumoro, M. & Matsumoto, T., (1984), "The Double Scroll Family," *IEEE Trans. Circuit and Systems*, **32** (11), 1055–1058.

[4] Chua, L. O., (1992), "The genesis of Chua's Circuit," *AEÜ*, **46** (4), 250–257.

[5] Chua, L. O., Kocarev, L., Eckert, K. & Itoh, M., (1992), "Experimental chaos synchronization in Chua's circuit," *Int. J. Bifurcation & Chaos*, **2** (3), 705–708.

[6] Delgado-Restituto, M. & Rodriguez-Vasquez, A. (1993), "A CMOS monolithic Chua's circuit," *J. Circuits, Systems and Computers*, **(3)** 2, 259–268.

[7] Di Ventra, M., Pershin, Y. V. & Chua, L. O., (2009), "Circuit elements with memory: memristors, memcapacitors, and meminductors," *Proceedings of IEEE*, **97** (10), 1717–1724.

[8] Duan, Z., Wang, J. and Huang, L. (2004), "Multi-input and multi-output nonlinear systems interconnected Chua's ciruits," *Int. J. Bifurcation & Chaos*, **14** (9), 3065–3081.

[9] Espinel, A., Taralova, I. & Lozi, R. (2011), "Dynamical and Statistical Analysis of a New Lozi Function for Random Numbers Generation," in *Proceeding or Physcon 2011, León, Spain, 5–8 September*, IPACS open Access Electronic Library.

[10] Hénaff, S., Taralova, I. & Lozi, R. (2009a), "Observers design for a new weakly coupled map function", in *Conference Proceedings of ICCSA 2009, 3rd International Conference on Complex Systems and Applications*, June 29–July 02, C. Bertelle, X. Liu, M.A. Aziz-Alaoui (Eds), 47–50.

[11] Hénaff, S., Taralova, I. & Lozi, R., (2009b), "Dynamical Analysis of a new statistically highly performant deterministic function for chaotic signals generation", *IPACS Open Access Electronic Library, Physics and Control 2009.*

[12] Hénaff, S., Taralova, I. & Lozi, R. (2010), "Exact and Asymptotic Synchronization of a new weakly coupled maps system", *Journal of Nonlinear Systems and Applications*, **1**, 3–4, 87–95.

[13] Hénon, M., (1976), "A Two-dimensional mapping with a strange attractor," *Commun. Math. Phys.*, **50**, 69–77.

[14] Kapitaniak, T., Chua, L. O. & Zhong, G.-H., (1994), "Experimental hyperchaos in coupled Chua's circuits," *IEEE Trans. Circuit and Systems*, **41** (7), 499–503.

[15] Kocarev, Lj., Hale, K.S., Chua, L.O. & Parlitz, U., (1992), "Experimental demonstration of secure communication via chaotic synchronization," *Int. J. Bifurcation & Chaos*, **2** (3), 709–713.

[16] Lorenz, E. N., (1963), "Deterministic nonperiodic flow," *J. Atmospheric Science*, **20**, 130–141.

[17] Loskutov, A., Yu. & Popkova, A., V., (2011), "Stabilization of chaotic oscillations in systems with a hyperbolic-type attractor," *JETP Letters*, **94** (1), 86–90.

[18] Lozi, R. & Chua, L.O., (1993), "Secure communications via chaotic synchronization II: noise reduction by cascading two identical receivers," *Int. J. Bifurcation & Chaos*, **3** (5), 1319–1325.

[19] Lozi, R., (1995), "Secure communications via chaotic synchronization in Chua's circuit and Bonhoeffer-Van der Pol equation: numerical analysis of the errors of the recovered signal," *ISCAS'95, IEEE International Symposium on Circuits and Systems, April 30 – May 3 1995,* Vol. 1, 684–687.

[20] Lozi, R., (2006),"Giga-periodic orbits for weakly coupled tent and logistic discretized maps," *Modern Mathematical Models, Methods and Algorithms for Real World Systems*, eds. A. H. Siddiqi, I. S. Duff & O. Christensen, (Anamaya Publishers, New Delhi, India), 80–124.

[21] Lozi, R., (2008a), "New Enhanced Chaotic Number Generators", *Indian Journal of Industrial and Applied Mathematics*, Vol. 1, no. 1, 1–23.

[22] Lozi, R., (2008b), "Chaotic Sampling, Very Weakly Coupling, and Chaotic Mixing: Three Simple Synergistic Mechanisms to Make New Families of Chaotic Pseudo Random Number Generators", in *6th EUROMECH Non Linear Dynamics Conference*, Saint-Petersburg, ENOC 2008, 30 June–4 July 2008, IPACS open Access Electronic Library, 1715–1724.

[23] Lozi, R. (2009), "Chaotic Pseudo Random Number Generators via Ultra Weak Coupling of Chaotic Maps and Double Threshold Sampling Sequences," in *Conference Proceedings of ICCSA 2009, 3rd*

International Conference on Complex Systems and Applications, June 29–July 02, C. Bertelle, X. Liu, M.A. Aziz-Alaoui, (Eds), 20–24.

[24] Lozi, R., (2011), "Complexity leads to randomness in chaotic systems" *Mathematics in Science and Technology: Mathematical Methods, Models and Algorithms in Science and Technology.* Proceedings of the Satellite Conference of ICM, 15–17 August 2010 Delhi, India (A.H. Siddiqi, R.C. Singh, P. Manchanda Eds.), World Scientific Publisher, Singapore, 93–125.

[25] Lozi, R., (2012), "Can we trust in numerical computation of chaotic solutions of Dynamical systems ?," in *Conference Proceedings of From Laser Dynamics to Topology of Chaos*, June 28–30, 2011, (Rouen), Ch. Letellier, (Eds), *In press.*

[26] Matsumoto, T., (1984), "A chaotic attractor from Chua's circuit," *IEEE Trans.*, **CAS-31** (12), 1055–1058.

[27] Oppenheim, A. L., Wornell, G. W., Isabelle, S. H. & Cuomo, K. M., (1992), "Signal processing in the context of chaotic signals," *Proc. 1992 IEEE ICASSP IV*, 117–120.

[28] Rössler, O. E., (1976), "Chaotic behavior in simple reaction system," *Zeitschrift für Naturforschung*, **A 31**, 259–264.

CHAPTER 25

CHUA'S EQUATION WAS PROVED TO BE CHAOTIC IN TWO YEARS, LORENZ EQUATION IN THIRTY SIX YEARS

BHARATHWAJ MUTHUSWAMY

Department of EECS, Milwaukee School of Engineering, USA

Although there are probably more publications on Chua's circuit than any other chaotic circuit, a tutorial with a historical emphasis is still lacking. Hence the goal of this chapter is to provide such a tutorial. This chapter will prove useful for a novice who is looking to understand the basics behind chaotic circuits without too much technical details. The chapter also includes a cookbook approach to a rigorous proof of chaos in piecewise-linear systems. The proof is a summary of the original piecewise-linear proof of chaos in Chua's circuit. The chapter concludes with a discussion of circuits derived from Chua's circuit.

1. Introduction

In this chapter, we study what is probably one of the most important contributions to chaos theory, Leon O. Chua's namesake Chua's circuit. This circuit is considered a paradigm because it is a fact that Chua's circuit is the first chaotic system in which chaos was systematically derived [5], physically confirmed [11] and rigorously analyzed [4]. The purpose of this chapter is to explore the systematic realization and rigorous analysis of chaos in Chua's circuit. Nevertheless, in order to understand why these facts are important, we must turn to a brief history of chaos [1].

Chaos was actually observed by Henri Poincaré in the late 1800s while working on the three body problem in astronomy. It was also observed by Van der Pol in the late 1920s in the very first electronic oscillator. However, in both these cases, chaotic behavior was not explicitly identified nor was it rigorously analyzed. For instance, Van der Pol dismissed chaos as "noise". Edward N. Lorenz published the first paper [10] that exclusively addressed the main features[a] of a chaotic system — sensitivity to initial conditions, aperiodicity and boundedness of trajectories. His 1963 paper was related to work he initially did in 1961 at the Travelers Insurance Company Weather Center in Hartford, Connecticut. Lorenz was shown a 7-equation model of fluid convective motion. Lorenz simplified this model into a 3-equation system of nonlinear Ordinary Differential Equations (ODEs), shown in Eqs. (1) through (3).

$$\dot{x} = -\sigma x + \sigma y \tag{1}$$

$$\dot{y} = -xz + \rho x - y \tag{2}$$

$$\dot{z} = xy - \beta z \tag{3}$$

The dot notation has been used to indicate a derivative with respect to time and will be employed throughout this chapter. For $\sigma = 10, \beta = \frac{8}{3}$, Lorenz observed via computer simulations that the system behaved "chaotically" whenever ρ exceeded the approximate critical value of 24.74. A phase plot $(x(t), y(t), z(t))$ obtained from numerical simulation of Eqs. (1) through (3) has been plotted in Fig. 1. The system equations have been numerically simulated and rendered using Mathematica 8 [20]. All calculations were also performed using Mathematica. The Mathematica source for this chapter can be found online at http://myweb.msoe.edu/muthuswamy/MuthuswamyWSPC2011ChuaBook-OnlineFiles/.

Although Lorenz's clever geometric arguments [10] implied that chaos in Eqs. (1) through (3) was not a mathematical anomaly, no one could reproduce chaos physically in the 1970s and early 1980s. Thus, in order to verify that steady-state chaotic behavior was not an artifact of computer simulations, there was a need for a physical chaotic system. One of the main reasons for the lack of a physical realization of early chaotic systems such as Lorenz and Rössler's equations [16] was the presence of multiplier terms in the nonlinearities. These terms were hard to realize physically, especially electronically since the analog multiplier did not become a reliable electronic component [5] till the late 1980s.

The lack of a physical realization would change in 1983, twenty years after Lorenz's paper, when

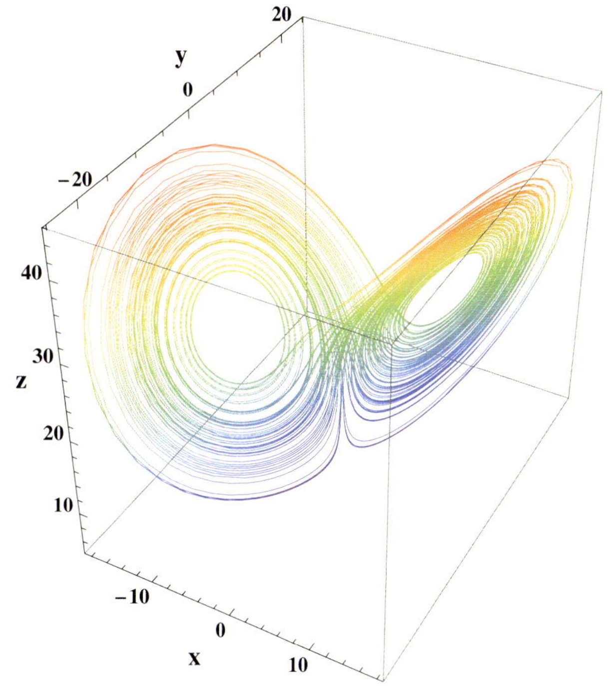

Fig. 1. The Lorenz butterfly. It is also called as a "chaotic attractor" because the underlying system exhibits chaotic behavior and nearby initial conditions are "attracted" into the butterfly. The attractor trajectory has been colored to highlight the unique shape in three dimensional space.

Leon O. Chua systematically derived Chua's circuit. Because of the systematic derivation, Chua's circuit utilized piecewise-linear nonlinearities as opposed to product terms. Hence physical realization of Chua's circuit and a rigorous proof of chaos would only take a total of two years [5]. In contrast the first electronic realization of Lorenz's system [18] was done, interestingly enough, using piecewise linear approximations to the product terms in 1989. Rigorous proof of chaos in the Lorenz system [19] would only be published in 1999, thirty six years after Lorenz's seminal work.

The rest of this chapter will study some of the most important properties of Chua's circuit. First we will systematically design and realize Chua's circuit [11]. Next we will emphasize why a rigorous proof of chaos is necessary and outline the approach for proving rigorous existence of chaos in Chua's circuit. This section will serve as a "cookbook" for the reader in proving rigorous existence of chaos in piecewise-linear systems. We emphasize piecewise-linear systems because Sec. 4 will provide

[a] We want to emphasize these features are not the definition of a chaotic system. To understand why no definition of chaos presently exists, please refer to [3].

a systematic tutorial on the Global Unfolding Theorem [6] by generalizing Chua's circuit to Chua's oscillator. This theorem will allow the reader to apply the rigorous proof of chaos from Sec. 3 to a much larger class of piecewise-linear systems. The chapter will conclude with suggestions for further exploring chaos.

2. The Design and Realization of Chua's Circuit

Based on the previous section, a brief problem statement [5] is given below.

> Design a physically realizable autonomous circuit exhibiting steady-state chaotic behavior.

We choose an autonomous chaotic circuit because this will eliminate the need for an external driving input such as a sinusoidal source. The requirement of an active circuit nonlinearity for chaos will be satisfied through operational amplifiers driven by DC sources (batteries).

Next we need to satisfy the Poincaré-Bendixson theorem, namely, we require at least three state variables for a continuous time autonomous dynamical system to be chaotic. In the case of electronic circuits, a natural choice of state variables are current and voltage. The typical linear passive inductor-capacitor or LC circuit in Fig. 2 is thus a good starting point because this circuit gives us two out of three state variables — the current through the inductor and the voltage across the capacitor.

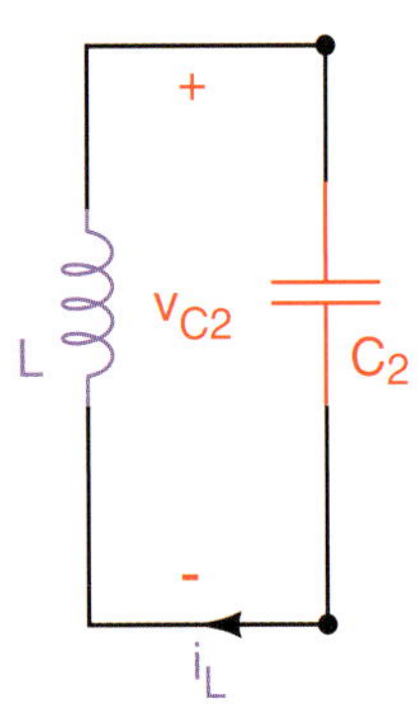

Fig. 2. The LC oscillator with the elements corresponding to different state variables highlighted in different color. We have two state variables, the current through inductor L given by $i_L(t)$ and the voltage across capacitor C_2 given by $v_{C2}(t)$. We will drop the explicit time reference since the independent variable in our differential equations is time.

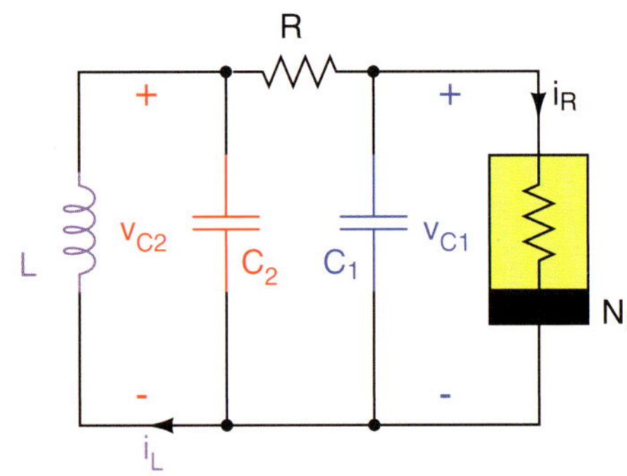

Fig. 3. Chua's circuit. The third state variable is the voltage v_{C1} across capacitor C_1. We have also added the linear passive resistor R and the nonlinear active resistor, N_R.

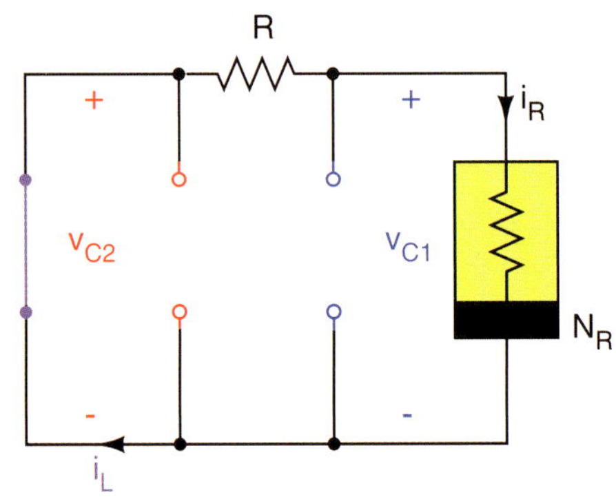

Fig. 4. Equilibrium or steady-state configuration of Chua's circuit. The dynamic elements have been replaced by their steady-state equivalents. The inductor is modeled by a short circuit and the capacitors are modeled by open-circuits.

In order to obtain the final state, we will choose a linear passive capacitor, since the selection of standard capacitor values significantly exceeds that of inductors. Since we have all three state variables, we will choose the nonlinear component as an active memoryless device. Hence we will need only one more component to complete the circuit because we need to separate the two voltage state variables across the capacitors. A linear passive resistor is chosen as the final component and the topology of Chua's circuit is shown in Fig. 3.

Note that the equilibrium configuration of circuit is very simple and is shown in Fig. 4. If we now choose the active nonlinearity as shown in Fig. 5, we get three equilibrium points and hence we can obtain chaos via Shilnikov's theorem [4].

Physical realization of the nonlinear resistor is usually done using operational amplifiers as shown in Fig. 6. Having designed the nonlinear function, Eqs. (4) through (6) are the dynamical equations

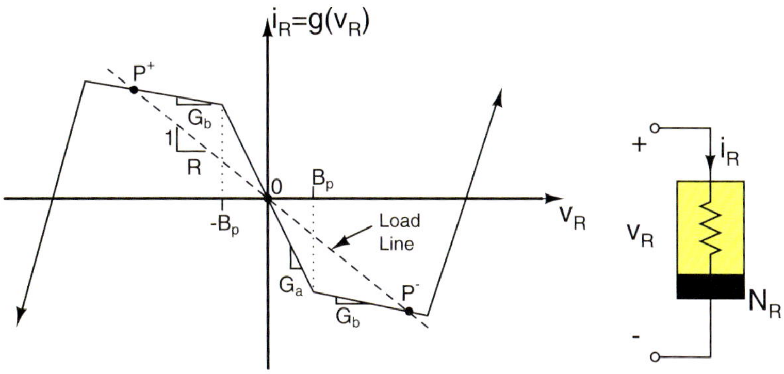

Fig. 5. The five segment piecewise-linear characteristic has three segments exclusively in the second and fourth quadrants. These correspond to the active region of the nonlinear resistor. In order to be physically realizable, notice that the nonlinear function eventually enters the first and third quadrants. Thus the nonlinear resistor is eventually passive. The linear resistor R has been plotted as a load and for particular values of R notice that we get three equilibrium points $P^+, 0, P^-$.

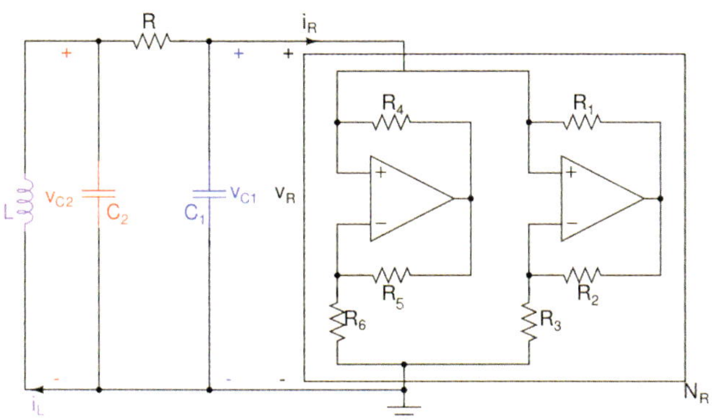

Fig. 6. Physical realization of Chua's circuit including the nonlinear resistor N_R. Parameter values [12] are $L = 7.14$ $mH, C_2 = 50$ $nF, C_1 = 5.56$ $nF, R = 1428$ Ω. Parameter values for the nonlinear resistor N_R [9] are $R_1 = R_2 = 220$ $\Omega, R_3 = 2.0$ $k\Omega, R_4 = R_5 = 22$ $k\Omega, R_6 = 3.3$ $k\Omega, B_p = 1$ V. These values for the N_R realization lead to $G_a = \frac{-1}{R_3} - \frac{1}{R_6}$ (since $R_1 = R_2$), $G_b = \frac{-1}{R_3} + \frac{1}{R_4}$ (since $R_4 = R_5$). AD712 or equivalent operational amplifiers can be used, power supply for the op-amps are $\pm 9\,V$. The op-amp power supplies have been omitted for clarity. There are a variety of other ways to realize N_R, such as cross-coupled [15] inverters. Also the inductor and capacitor values were chosen to match Matsumoto's original [12] paper. More realistic values [9] are $L = 18$ $mH, C_2 = 100$ $nF, C_1 = 10$ nF.

describing Chua's circuit in Fig. 6.

$$C_1 \frac{dv_{C1}}{dt} = \frac{v_{C2} - v_{C1}}{R} - g(v_{C1}) \tag{4}$$

$$C_2 \frac{dv_{C2}}{dt} = \frac{v_{C1} - v_{C2}}{R} + i_L \tag{5}$$

$$L \frac{di_L}{dt} = -v_{C2} \tag{6}$$

Eq. (7) defines a five-segment piecewise-linear function $g(\cdot)$ from Fig. 5, we emphasize that $G_a, G_b < 0$ in Eq.(7). Only the inner three segments in Fig. 5 are relevant for obtaining chaos in Chua's circuit. The outer two segments, or their nonlinear counterparts, will always be exhibited in any physical realization of the nonlinear resistor (also known in literature as the Chua diode). The outer two segments are a consequence of the law of conservation of energy. They play no role, however, in the state space region covered by Chua's attractor.

$$g(v_R) = G_b v_R + \frac{1}{2}(G_a - G_b)(|v_R + B_p| - |v_R - B_p|) \tag{7}$$

Usually Eqs. (4) through (7) are scaled to be dimensionless via the following substitutions.

$$x \triangleq v_{C1}/B_p \qquad y \triangleq v_{C2}/B_p \qquad z \triangleq i_L R/B_p, \tag{8}$$

$$\tau \triangleq t/(RC_2) \qquad a \triangleq G_a R \approx \frac{-8}{7} \qquad b \triangleq G_b R \approx \frac{-5}{7}, \tag{9}$$

$$\alpha \triangleq C_2/C_1 \approx 9 \quad \beta \triangleq C_2 R^2/L \approx 14\frac{2}{7} \tag{10}$$

Equations (11) through (14) are the dimensionless form of Chua's equation.

$$\dot{x} = \alpha(y - x - f(x)) \tag{11}$$

$$\dot{y} = x - y + z \tag{12}$$

$$\dot{z} = -\beta y \tag{13}$$

$$f(x) = bx + \frac{1}{2}(a - b)(|x + 1| - |x - 1|) \tag{14}$$

One further simplification for analysis is obtained by Eq. (15) from Eq. (14).

$$\begin{aligned} h(x) &\triangleq x + f(x) \\ &= m_1 x + \frac{1}{2}(m_0 - m_1)(|x+1| - |x-1|) \end{aligned} \quad (15)$$

Hence $m_0 = a+1, m_1 = b+1$. Thus the dimensionless form that will be used for a rigorous proof of chaos is given by Eqs. (16) through (19).

$$\dot{x} = \alpha(y - h(x)) \quad (16)$$

$$\dot{y} = x - y + z \quad (17)$$

$$\dot{z} = -\beta y \quad (18)$$

$$h(x) = m_1 x + \frac{1}{2}(m_0 - m_1)(|x+1| - |x-1|) \quad (19)$$

We will interchangeably use the words "Chua's circuit" and "Chua's system" to refer to Eqs. (16) through (19) from now on. From the parameter values in Fig. 6 we obtain $\{\alpha, \beta, m_0, m_1\} = \{9, 14\frac{2}{7}, -1/7, 2/7\}$ [4]. The classic Double-Scroll Chua's attractor obtained with these parameter values is shown in Fig. 7.

An useful quantity (or measure) for a chaotic attractor is the dimension. We are familiar with the concept of a dimension for everyday objects: a solid sphere is three dimensional, the surface of a sphere is two dimensional etc. In the case of chaotic attractors, many definitions of dimension exist, we will examine the Lyapunov dimension [12]. We can actually predict the range of the Lyapunov dimension without any computation. Figure 7 seems to imply that the chaotic attractor is three dimensional. However this attractor is not a solid object, it is more a "sheet-like" structure. So, is the dimension less than two? The answer is no because of the Poincaré-Bendixson theorem. Can the dimension be greater than three? The answer again is no because the system of equations for Chua's circuit is three dimensional.[b] Thus intuitively the Lyapunov dimension of the Double-Scroll must be between two and three. It turns out that the Lyapunov dimension of the Double-Scroll is approximately 2.13 [12].

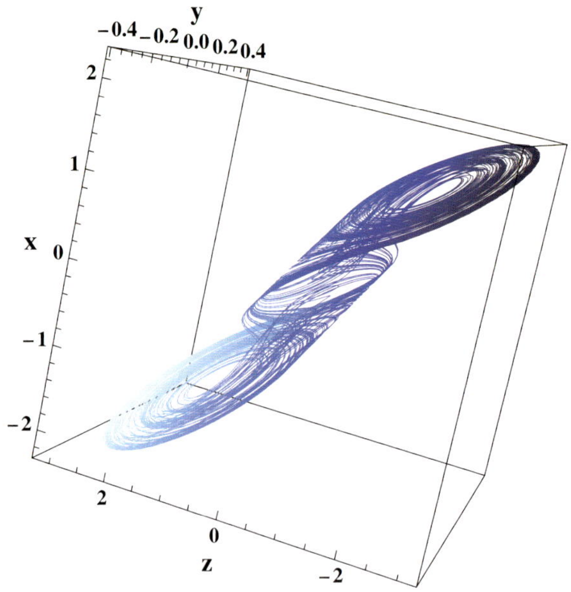

Fig. 7. The Double-Scroll Chua's attractor. Recall that Chua's circuit has three equilibrium points: $(P^+, 0, P^-)$. It turns out that the chaotic trajectory "wraps around" one of the outer equilibrium points (P^+, P^-) and then moves through the equilibrium point at the origin, to the other equilibrium point.

We conclude this section by discussing a subtle but very important concept that is often misunderstood: Chua's circuit is not an analog computer.

2.1. *Analog computer emulation of an ODE is not a physical circuit*

A common misconception is to equate an "analog computer" emulation of a system of ODEs to a physical circuit. The reasoning employed is that analog computers are made of physical circuit elements like op-amps, resistors, capacitors and inductors. However, this reasoning can be understood as erroneous by considering this fact: digital computers are also made of circuit elements. Yet no one calls a digital computer emulation of a system of ODEs as a physical circuit!

The importance of this concept can be further understood by considering the mass-damper-spring system in Fig. 8 and the two circuit realizations shown. One is the analog computer emulation and

[b]In the case of four (or more) autonomous nonlinear ordinary differential equations, we could have hyper (or higher-dimensional) chaotic attractors.

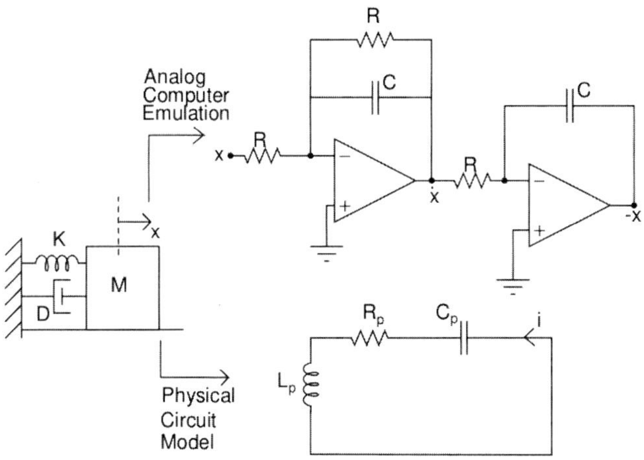

Fig. 8. The mass-damper-spring mechanical system and two electrical circuit versions. The voltages across the capacitor in the op-amp circuit correspond to the internal states of the mechanical system. However the analog computer version is an emulator whereas the LRC circuit is the actual physical electrical equivalent. Note that in deriving the analog computer, all mechanical parameters were assumed to have magnitude one. Moreover the analog circuit model actually requires one more inverting amplifier for producing x from $-x$.

the other is the LRC realization. It should be immediately apparent that the analog computer is simply a "signal flow" graph and cannot account for physical characteristics of the mass-damper-spring. For instance, the position and velocity of the mass is simply represented by the voltages across the two capacitors. Also, the transfer of energy between the mass and spring cannot be interpreted from the analog computer. This is because energy requires us to have the product of force waveform and the velocity waveform.[c] This is unavailable in the analog computer emulation[d].

In contrast, the LRC circuit is the true electrical analog of the mass-damper-spring. This is easily seen by simply comparing the differential equations. For the mass-damper-spring, we have Eq. (20).

$$M\ddot{x} + D\dot{x} + Kx = 0 \tag{20}$$

For the LRC circuit, we have Eq. (21)

$$L_p\ddot{q} + R_p\dot{q} + \frac{1}{C_p}q = 0 \tag{21}$$

Comparing Eqs. (20) and (21), we can see that the mass has a one-to-one correspondence to the inductor, the dampener corresponds to the resistor and the spring corresponds to $\frac{1}{C_p}$. Also note that the position is analogous to charge and velocity to current. Thus the LRC circuit is the electrical physical realization of the mass-damper-spring system.

By the same argument as above, Chua's circuit in Fig. 6 is not an analog computer model but a physical realization of Eqs. (16) through (18). The fact that Chua's circuit is a physical realization enables us to study interesting phenomenon such as local activity, that is essential [7] for chaos.

In concluding this section, note that the op-amp circuit is an "analog computer" simulation of the mechanical system, where the word "analog" is used, by tradition, to mean "continuous time" or "not digital". This should not be interpreted as an "analogy" of the mechanical mass-damper-spring system. Indeed, the word "analog computer" is just a non-technical jargon to mean any systematic collection of basic components, such as op amps, integrators, resistors, inductor, capacitors, etc., packed inside a compact cabinet, with externally accessible switches, and other extra features to allow one to quickly interconnect a continuous-time circuit such that its state equation is the desired one. One can also observe the time evolution of the state variable dynamics via an oscilloscope, in real time. In this sense, the op amp circuit in Fig. 8 for simulating the mass-damper-spring mechanical system is just a "poor man's analog computer", tailored to simulate the state equations of only one physical system.

3. The Need for a Rigorous Proof of Chaos

In the previous section, we examined the classic parameter values in Chua's circuit for obtaining Chua's attractor. If we can realize Chua's circuit and confirm chaos, why do we still need a rigorous proof? The answer is computer simulations have finite precision and experimental measurements have finite range for time or frequency. As a simple example, reconsider Chua's

[c]Scalar product of force and velocity gives instantaneous power. Time integral of instantaneous power gives work done or energy.

[d]Although the potential energy of the spring is analogous to the energy stored in the second capacitor and the kinetic energy of the mass is analogous to the energy stored in the first capacitor, this is not readily apparent.

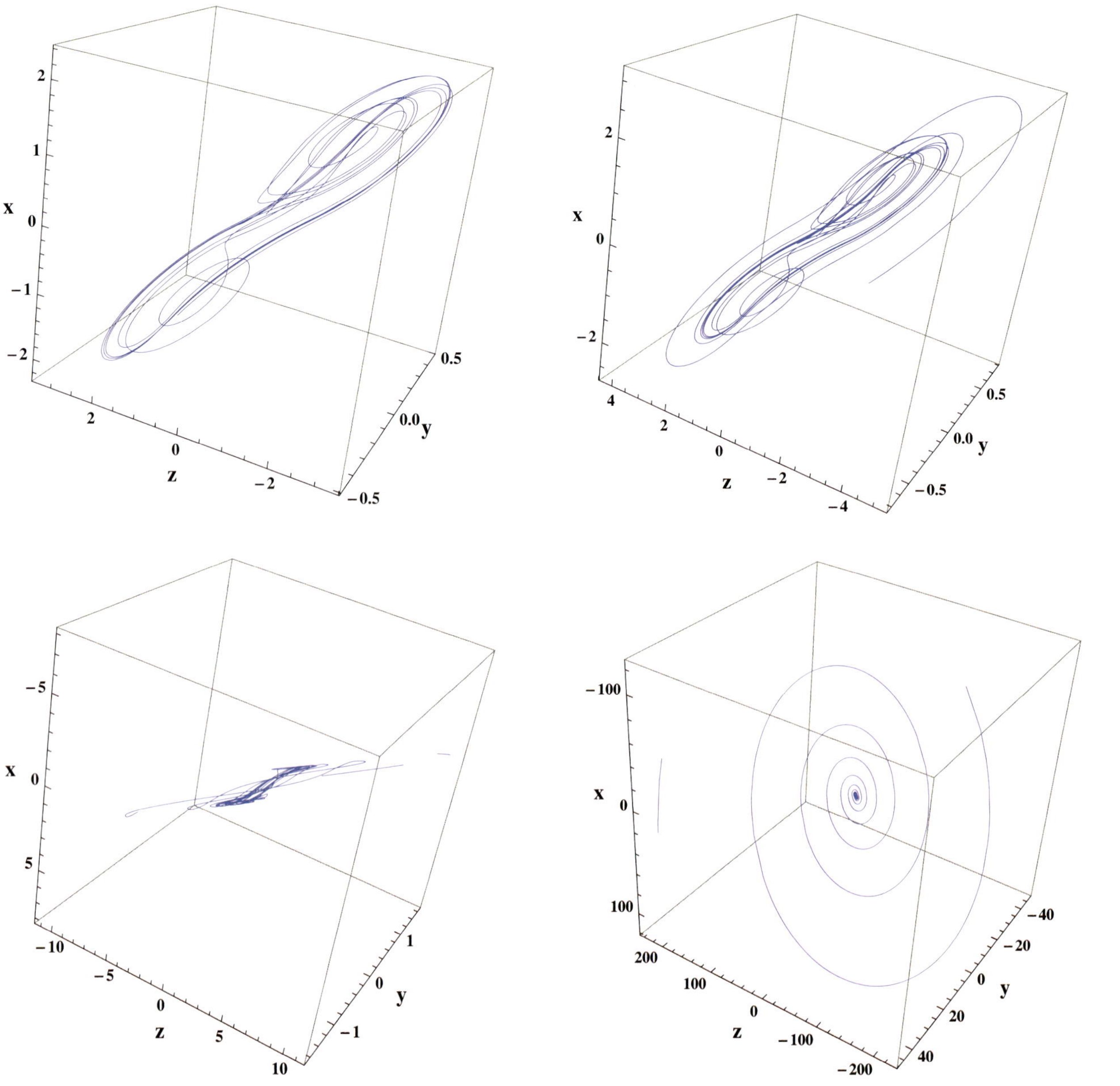

Fig. 9. Trajectory of Chua's circuit with $\alpha = 10.75$ for (a): t = 30 units, (b) t = 35 units, (c) t = 45 units and (d) t = 70 units. Notice how in each case the dimensions of our axes are increasing. The transient attractor is not even visible in (d). In other words Chua's circuit is unstable for this choice of parameter values. In the physical circuit we will quickly observe a large amplitude limit cycle due to amplifier saturation.

circuit with the parameter values $\{\alpha, \beta, m_0, m_1\} = \{10.75, 14\frac{2}{7}, -1/7, 2/7\}$. Note that we only changed the α value. A simulation of the system for four time intervals is shown in Fig. 9.

It is clear from Fig. 9 that we are observing chaos as a transient. If one were to consider only the first 30 time units, we would incorrectly conclude the system was exhibiting chaotic behavior. To address this issue a rigorous proof of chaos is necessary. We will now outline the approach for a rigorous proof of chaos in Chua's attractor.

3.1. *Piecewise-linear analysis of Chua's circuit*

The method employed in [4] was the very first rigorous proof of chaos in any physical, and not hypothetical, dynamical system. The essence of this approach is Shilnikov's theorem.

Theorem 1. *Let ξ be a continuous piecewise-linear vector field associated with a third-order autonomous system $\dot{\mathbf{x}} = \mathbf{f}(\mathbf{x}), \mathbf{x} \in \Re^3$. Assume the* *origin is an equilibrium point with a pair of complex conjugate eigenvalues* $\sigma \pm j\omega, (\sigma < 0, \omega \neq 0)$ *and a real eigenvalue* $\gamma > 0$ *satisfying* $|\sigma| < \gamma$. *If in addition,* ξ *has a homoclinic orbit through the origin, then* ξ *can be infinitesimally perturbed into a nearby vector field* ξ^* *with a countable set of horseshoes.*

The difficult part of Shilnikov's theorem is rigorously confirming the existence of a homoclinic orbit at the origin. Such an orbit is double-asymptotic at the origin, that is, $\mathbf{x}_h(t) \to \mathbf{0}$ as $t \to \pm\infty$. However, because of the piecewise-linear nature of the nonlinearity, we analyze the system in each linear region and then carefully "stitch" the trajectories at the boundaries. Since closed form solutions of trajectories and eigenspaces can be found for linear systems, we can plot the geometrical structure and typical trajectories of the piecewise-linear system. Fig. 10 shows a typical plot of eigenspaces for Chua's circuit.

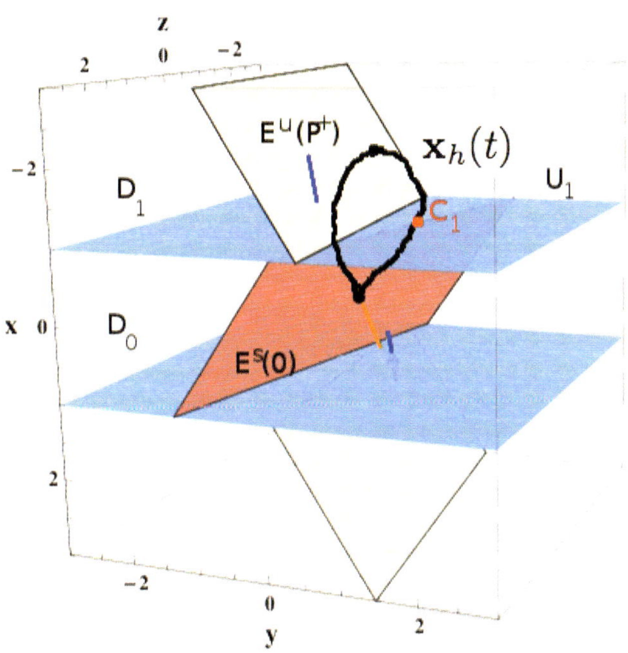

Fig. 10. Eigenspaces for the Chua system. The nonlinearity separates phase space into three distinct regions. $E^s(0)$ is the stable complex eigenspace associated with the origin. $E^u(P^+)$ is the unstable complex eigenspace associated with P^+. For clarity, only the complex eigenspaces are shown. A trajectory $\mathbf{x}_h(t)$ initially starts along the unstable real eigenspace associated with the origin. Once the trajectory passes into D_1, it is shown here to spiral tangentially to $E^u(P^+)$ before going through C_1. If there exists a C_1 such that it also lies on the line of intersection between $E^s(0)$ and U_1, we have found a homoclinic orbit $\mathbf{x}_h(t)$. We have exaggerated the homoclinic orbit by hand for clarity.

The detailed analysis is very involved and the reader is encouraged to go through the double scroll family paper [4] to understand the complete picture (such as rigorous analysis of bifurcation phenomenon). But there are three essential steps and we have included relevant equation, page numbers from the double scroll family paper [4] for correlating the "cookbook" approach in this chapter with Chua's original work.

(1) Determine the eigenvalues for each of the linear regions in Chua's system (Eqs. (16) through (19)). The transition matrix for region D_0 is

$$M_0 = \begin{pmatrix} -\alpha m_0 & \alpha & 0 \\ 1 & -1 & 1 \\ 0 & -\beta & 0 \end{pmatrix} \tag{22}$$

The transition matrix for region D_1 is

$$M_1 = \begin{pmatrix} -\alpha m_1 & \alpha & 0 \\ 1 & -1 & 1 \\ 0 & -\beta & 0 \end{pmatrix} \tag{23}$$

The eigenvalues for Eqs. (22) are defined as $\tilde{\gamma}_0, \tilde{\sigma}_0 \pm \tilde{\omega}_0$ and for (23) are defined as $\tilde{\gamma}_1, \tilde{\sigma}_1 \pm \tilde{\omega}_1$.

(2) Define necessary normalized eigenvalues (Eqs. (2.9) and (2.12), p. 1076 [4]) and strategic points (Eqs. (2.20) through (2.33), pp. 1077–1078 [4]).

$$\sigma_{0,1} = \frac{\tilde{\sigma}_{0,1}}{\tilde{\omega}_{0,1}} \qquad \gamma_{0,1} = \frac{\tilde{\gamma}_{0,1}}{\tilde{\omega}_{0,1}}$$
$$k_0 = \frac{-\tilde{\gamma}_0}{\tilde{\gamma}_1} \qquad k_1 = \frac{-\tilde{\gamma}_1}{\tilde{\gamma}_0} \tag{24}$$

$$p_0 = \sigma_0 + \frac{k_0}{\gamma_0}(\sigma_0^2 + 1) \quad p_1 = \sigma_1 + \frac{k_1}{\gamma_1}(\sigma_1^2 + 1) \tag{25}$$

$$Q_0 = (\sigma_0 - \gamma_0)^2 + 1 \qquad Q_1 = (\sigma_1 - \gamma_1)^2 + 1 \tag{26}$$

(3) Compute the following quantities and check if the given conditions are satisfied.

$$a_1 = \frac{-\sigma_1\gamma_1(p_1^2 + 1)}{Q_1} > 0 \tag{27}$$

$$\begin{aligned} a_2 = {} & Q_0^2 + \gamma_0^2(1 + p_0^2(2 + (\gamma_0 - \sigma_0)^2) \\ & + 2(\gamma_0 - \sigma_0)^2 + 2p_0(\sigma_0 - \gamma_0)) \\ & + Q_0\gamma_0(-3\gamma_0 + 3\sigma_0 \\ & + p_0(2 + p_0(\sigma_0 - \gamma_0)) > 0 \end{aligned} \tag{28}$$

$$a_3 = \sigma_1 \left(1 + \frac{2}{\sigma_1^2 - 1}\right) \geq \gamma_1 \quad (29)$$

$$a_4 = (1 + p_0^2) - (1 + \sigma_0^2) \times e^{2\sigma_0(\tan^{-1}(p_0) - \tan^{-1}(\sigma_0))} > 0 \quad (30)$$

$$0 < a_5 = (1 + p_0^2)e^{\sigma_0\left(\frac{\pi}{2} - \tan^{-1}(p_0)\right)} < 1 \quad (31)$$

$$a6 = \frac{\gamma_1(1 - \sigma_1(\sigma_1 - \gamma_1))}{Q_1} - \frac{(\sigma_1^2 + 1)\gamma_0 k_1}{(\sigma_0^2 + 1)\gamma_1 Q_1} \cdot (k_1\gamma_0(\sigma_1(\sigma_1 - \gamma_1) + 1) + 2\sigma_0\gamma_1(\sigma_1 - \gamma_1)) > 0 \quad (32)$$

$$0 < a7 = \frac{\sigma_1^2 + 1}{Q_1(\sigma_0^2 + 1)}\gamma_0 k_1(\gamma_0 k_1 - 2\sigma_0) < 1 \quad (33)$$

The conditions in Eqs. (27) through (30) check if the spiral trajectory in D_1 is sandwiched between two logarithmic spirals, for all time t. Thus one can ensure that the trajectory does not enter the stable eigenspace in D_1 (this would destroy the homoclinic orbit). These conditions also ensure that there is no "bouncing" behavior of the trajectory at the boundary U_1 in Fig. 10. Eq. (27) is Eq. (5.27) along with the condition in Eq. (5.24) on p. 1093 [4]. Eq. (28) can be obtained by using Eqs. (2.20) and (2.22) on p. 1077 [4] and then imposing the condition $\mathbf{OB_0} \cdot \mathbf{B_0A_0} > 0$ [4]. Eq. (29) is Eq. (5.36) on p. 1094 [4]. Eq. (30) is derived from Eqs. (5.37) and (5.39) on p. 1094 [4], along with the definition $\tan(\phi + \theta_0) = p_0$.

Equation (31) is to check that once the trajectory re-enters D_0, it does not intersect the plane $x = -1$. Eqs. (32) and (33) are to ensure that the point C_1 in Fig. 10 lies on the stable eigenspace $E^s(0)$ associated with the origin. These two conditions imply orbit $\mathbf{x}_h(t)$ in Fig. 10 does approach the origin as $t \to \infty$. Equation (31) is obtained from Eqs. (5.48) and (5.49) on p. 1094 [4]. Eq. (32) is obtained by imposing the condition in Eq. (5.54) on Eq. (5.57) (p. 1095 [4]). Eq. (33) is obtained by imposing the condition in Eq. (5.54) on Eq. (5.58) (p. 1095 [4]).

Note that all the expressions in Eqs. (27) through (33) are in closed form and hence we can calculate the result to the required precision. Also, the equations hold for corresponding trajectories from D_0 to D_{-1}.

We will now apply the technique to both prove and disprove chaos in Chua's circuit. In other words, one needs to be very careful in choosing the correct parameter values required for chaos. Not all parameter values for a dynamical system lead to chaotic behavior. Before we begin, note for both proofs below, the detailed calculations from Mathematica 8 (in the form of a pdf file) are available online at http://myweb.msoe.edu/muthuswamy/MuthuswamyWSPC2011ChuaBook-OnlineFiles/.

First we will prove that Chua's circuit is chaotic for parameter values $\{\alpha, \beta, m_0, m_1\} = \{9, 14\frac{2}{7}, -1/7, 2/7\}$.

Proof. Step one is to determine if the origin is an equilibrium point and is a hyperbolic saddle under Jacobi linearization. For our parameter values, the origin is an equilibrium point and the eigenvalues of linearized system at the origin are $\tilde{\gamma}_0 \approx 2.22, \tilde{\sigma}_0 \pm j\tilde{\omega}_0 \approx -0.97 \pm j2.71$. Thus Shilnikov's first condition is satisfied.

Now, we prove that there exists a homoclinic orbit based at the origin. One can compute and verify that all conditions in step (3) from the previous section are satisfied. Hence, there exists a homoclinic orbit at the origin. □

Now we will prove that Chua's circuit is not chaotic for parameter values $\{\alpha, \beta, m_0, m_1\} = \{10.75, 14\frac{2}{7}, -1/7, 2/7\}$. We only changed the α parameter.

Proof. Step one is to determine if the origin is an equilibrium point and is a hyperbolic saddle under Jacobi linearization. For our parameter values, the origin is an equilibrium point and the eigenvalues of linearized system at the origin are $\tilde{\gamma}_0 \approx 2.73, \tilde{\sigma}_0 \pm j\tilde{\omega}_0 \approx -1.10 \pm j2.61$. Thus Shilnikov's first condition is satisfied.

However we can determine that the last condition in step (3), Eq. (33), is not satisfied. In other words, there exists some $t > 0$ for which the point C_1 from Fig. 10 does not lie on the stable eigenspace associated with the origin. Note that we cannot determine t, we just know that a homoclinic orbit cannot exist for the given set of parameter values. □

Notice that our results for the value of $\alpha = 10.75$ agree with published [4] results on the "death of the Double-Scroll".

The piecewise-linear analysis technique in this section is especially useful since it can be applied to other chaotic systems via the Global Unfolding Theorem. A short tutorial on this important theorem is the topic of the next section.

4. The Global Unfolding Theorem

The piecewise-linear analysis technique from the previous section can be used to "globally unfold" Chua's circuit such that the resulting dynamical system is imbued with every possible qualitative dynamics of an extremely large family $\mathcal{C}$ of piecewise-linear dynamical systems [5]. This unfolding of Chua's circuit is obtained by simply adding a resistor R_0 in series with inductor L and is shown in Fig. 11. This new circuit is called Chua's oscillator.

In this section, we provide a short tutorial on the global unfolding theorem. First, the algorithm for global unfolding is shown below.

(1) Confirm that the system in question belongs to family $\mathcal{C}$ [5]. That is, the system defined by state equation

$$\dot{\mathbf{x}} = \mathbf{f}(\mathbf{x}), \mathbf{x} \in \Re^3 \tag{34}$$

has the following properties:

(1) $\mathbf{f}(\cdot)$ is continuous
(2) $\mathbf{f}(\cdot)$ is odd-symmetric (could also be with respect to a point other than the origin)
(3) $\Re^3$ is partitioned by 2 parallel boundary planes U_1 and U_{-1} into an inner region D_0 containing the origin and two outer regions D_1 and D_{-1}.

(2) Let (μ_1, μ_2, μ_3) and (ν_1, ν_2, ν_3) be the eigenvalues associated with the linear (D_0 region) and affine ($D_{1,-1}$ regions) respectively.

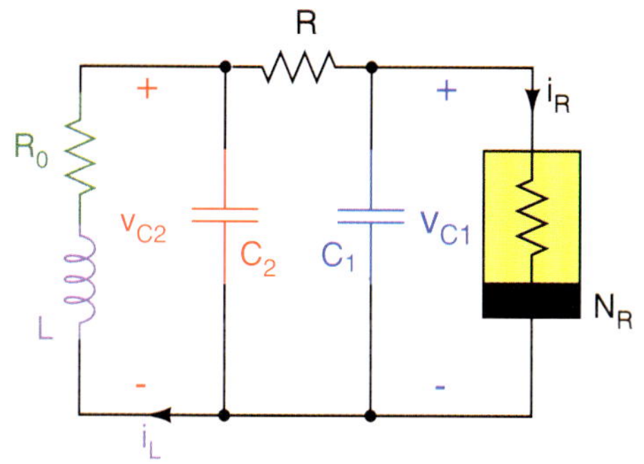

Fig. 11. Chua's oscillator obtained from Chua's circuit by adding linear passive resistor R_0. Note that R_0 is not the parasitic resistance associated with the inductor.

(3) Compute the equivalent eigenvalue parameters for Chua's oscillator from the following expressions. If $p_1 = q_1$, slightly perturb p_1 and q_1: $p_1 = p_1 + \delta, q_1 = q_1 - \delta$ [5].

$$p_1 = \mu_1 + \mu_2 + \mu_3 \qquad q_1 = \nu_1 + \nu_2 + \nu_3 \tag{35}$$

$$p_2 = \mu_1\mu_2 + \mu_2\mu_3 + \mu_3\mu_1 \quad q_2 = \nu_1\nu_2 + \nu_2\nu_3 + \nu_3\nu_1 \tag{36}$$

$$p_3 = \mu_1\mu_2\mu_3 \qquad q_3 = \nu_1\nu_2\nu_3 \tag{37}$$

(4) Compute the equivalent circuit parameters for Chua's oscillator from the following expressions [5].

$$k_1 \triangleq -p_3 + \frac{q_3 - p_3}{q_1 - p_1}\left(p_1 + \frac{p_2 - q_2}{q_1 - p_1}\right) \tag{38}$$

$$k_2 \triangleq p_2 - \left(\frac{q_3 - p_3}{q_1 - p_1}\right) + \left(\frac{p_2 - q_2}{q_1 - p_1}\right)\left(\frac{p_2 - q_2}{q_1 - p_1} + p_1\right) \tag{39}$$

$$k_3 \triangleq \left(\frac{p_2 - q_2}{q_1 - p_1}\right) - \frac{k_1}{k_2} \tag{40}$$

$$k_4 \triangleq -k_1k_3 + k_2\left(\frac{p_3 - q_3}{p_1 - q_1}\right) \tag{41}$$

$$C_1 = 1 \tag{42}$$

$$C_2 = -\frac{k_2}{k_3^2} \tag{43}$$

$$L = -\frac{k_3^2}{k_4} \tag{44}$$

$$R = -\frac{k_3}{k_2} \tag{45}$$

$$R_0 = -\frac{k_1k_3^2}{k_2k_4} \tag{46}$$

$$G_a = -p_1 - k_3 - \frac{k_1}{k_2} + \frac{k_2}{k_3} \tag{47}$$

$$G_b = -q_1 - k_3 - \frac{k_1}{k_2} + \frac{k_2}{k_3} \tag{48}$$

We will now apply the algorithm above to the system from [14] and thus obtain an attractor from Chua's circuit that is topologically conjugate to the attractor from [14].

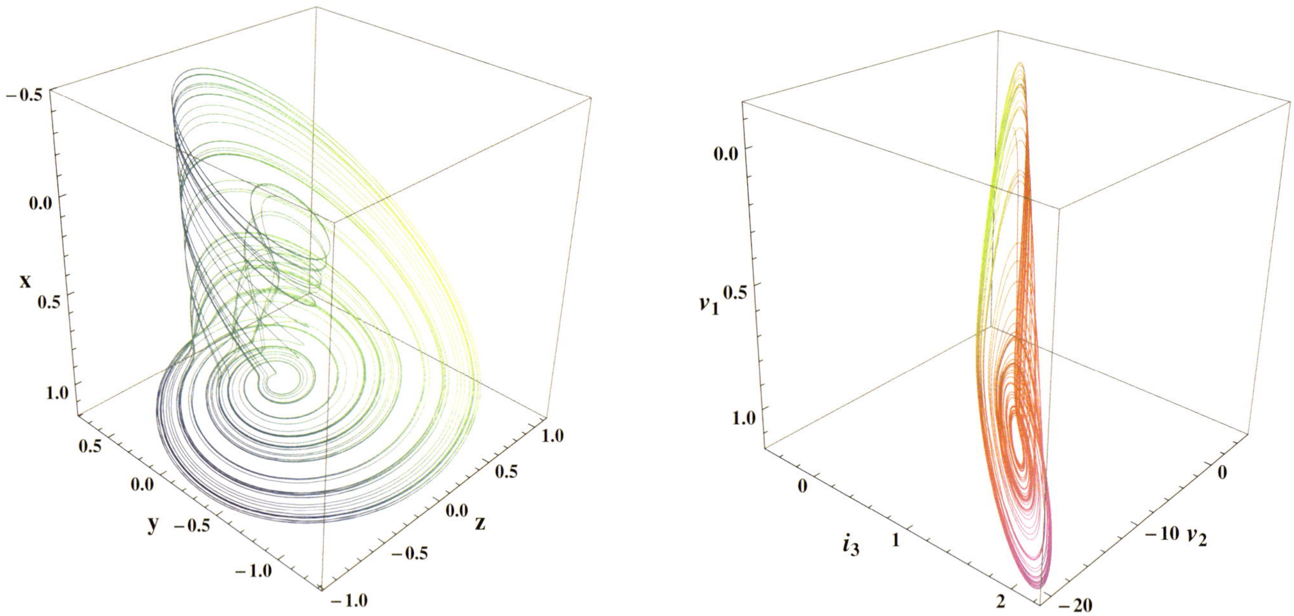

Fig. 12. (a) The original system from [14]. Parameter values for simulation are $a = 0.35, b = 1, \alpha = 0.5, \epsilon = 0.1$, initial conditions are $(0.1, 0.1, 0.1)$. (b) shows the linearly conjugate attractor for Chua's oscillator.

(1) The system of equations for the nonlinear inductor circuit [14] are given by Eqs. (49) through (52).

$$\dot{x} = -bz - bf(x) \tag{49}$$

$$\dot{y} = z \tag{50}$$

$$\dot{z} = (a-b)z - y - bf(x) \tag{51}$$

$$f(x) = \frac{x}{\epsilon} - \left(\alpha + \frac{1}{\epsilon}\right)\frac{|x+1| - |x+1|}{2} \tag{52}$$

Note that our system satisfies the conditions in step 1. Hence it can be mapped into Chua's oscillator.

(2) (μ_1, μ_2, μ_3) and (ν_1, ν_2, ν_3) are computed to be $(0.367, -0.283 + 1.131i, -0.283 - 1.31i)$ and $(-10.967, -0.133 + 0.946i, -0.133 - 0.946i)$.

(3) $(p_1, p_2, p_3, q_1, q_2, q_3)$ are computed to be $(-0.2, 1.15, 0.499, -10.7, -2, -10)$.

(4) Equivalent circuit parameters are thus $C_1 = 1$, $C_2 = -0.0326$, $L = -2.76$, $R = -10.116$, $R_0 = 9.2055$, $G_a = 0.5989$, $G_b = 11.099$.

Figure 12 shows the results. Notice how the attractor from Chua's oscillator is a linear transformation of the original attractor from [14]. Once we have globally unfolded a dynamical system into Chua's oscillator, we can apply the piecewise-linear analysis technique from the previous section to rigorously prove the existence of chaos.

5. Conclusions

From the date of invention and realization of Chua's circuit, it took two years to publish a rigorous proof of chaos. In this chapter, we covered all the main ideas from this time period. This chapter has thus provided the reader with all the tools necessary for further study of chaos in piecewise-linear dynamical systems. Ever since Chua's seminal work, a variety of other chaotic circuits have been discovered. Sprott's "jerky dynamics" based chaotic circuits [17] are especially elegant. Other techniques for rigorous analysis of chaos also exist, such as ones involving topological horseshoe theory [21].

For future work, a particularly interesting concept to explore is the minimal number of circuit elements required for chaos. Referring to Chua's circuit in Fig. 3, one can ask if we can eliminate the linear resistor R. This would however reduce the number of states from three to two since capacitors C_1 and C_2 are now in parallel. But in 2008, 25 years after the invention of Chua's circuit, Barboza and Chua reduced the number of elements by re-arranging Chua's circuit as shown in Fig. 13 [2]. Note that in reality, the nonlinear resistor in Fig. 13 has only one region with negative slope. The original Chua's circuit had three regions with negative slope. Nevertheless, Barboza and Chua demonstrated the circuit in Fig. 13 is linearly conjugate to Chua's circuit.

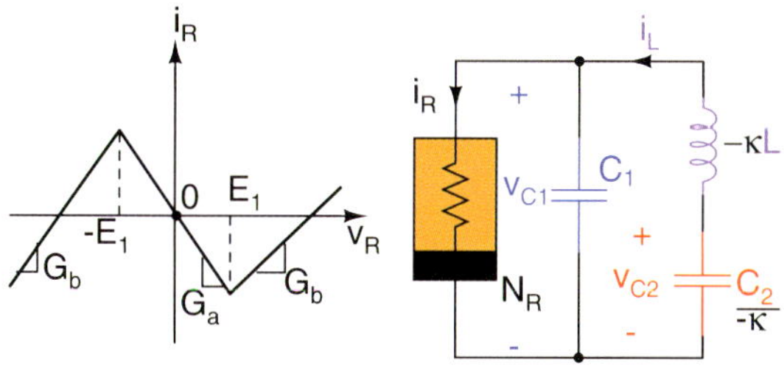

Fig. 13. The Four-Element Chua's circuit. The nonlinear resistor N_R has only one region of negative slope. Note that in the circuit $v_{C1} = v_R$. Capacitor C_2, C_1 and inductor L_1 are linear passive elements. C_2 and L_1 are made active ($-\kappa, \kappa > 0$) using an operational amplifier that acts as a current inverter.

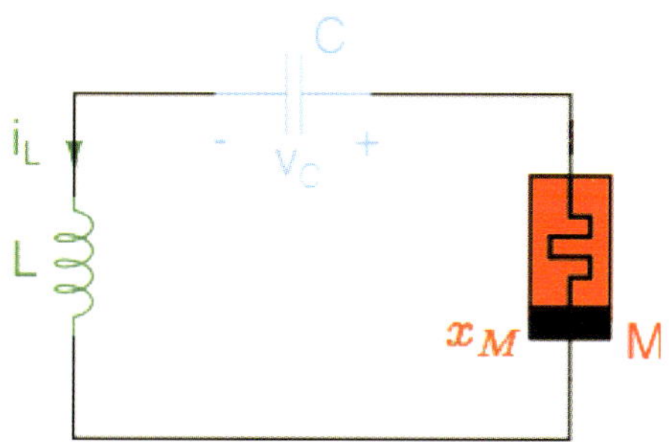

Fig. 14. The three element chaotic circuit. This circuit is simplest in the sense that we have only three elements and each element contributes to one of the three states required for chaos. Also the only nonlinear active element is the memristor, the capacitor and inductor are linear passive elements. Internal state of the memristor is denoted by x_M. The physical interpretation of this internal state depends on the memristor used.

Muthuswamy and Chua further eliminated the number of elements required for chaos by one when they systematically designed the circuit in Fig. 14 [13]. However the nonlinear characteristics of the memristor do not resemble the piecewise-linear nonlinearity in Chua's circuit. Although chaos has been rigorously proved [8] in the three element chaotic circuit, the nonlinear characteristics of the active memristor is not piecewise-linear. Thus a good candidate for future research would be to design a memristor with piecewise-linear characteristics to answer the question: does the three element circuit have a chaotic attractor that is topologically conjugate to the double scroll?

Acknowledgments

The author would like to thank Dr. Jean-Marc Ginoux, Dr. Jovan Jevtic and Dr. Gerald Thomas for insightful comments.

References

[1] Alligood, K. T., Sauer, T. D. and Yorke J. A. (1996). *Chaos: An Introduction to Dynamical Systems* (Springer).

[2] Barboza, R. and Chua, L. O. (2008). "The Four-Element Chua's Circuit", *International Journal of Bifurcation and Chaos*, Vol. 18(4). pp. 943–955.

[3] Brown, R. and Chua, L. O. (1996). "Clarifying Chaos: Examples and Counterexamples", *International Journal of Bifurcation and Chaos*, Vol. 6(2). pp. 219–249.

[4] Chua, L. O., Komuro, M. and Matsumoto, T. (1986). "The Double Scroll Family", *IEEE Transactions on Circuits and Systems*, Vol. 33(11). pp. 1072–1118.

[5] Chua, L. O. (1992). "The Genesis of Chua's Circuit", *Archiv fur Elektronik und Uebertragungstechnik*, Vol. 46(4). pp. 250–257.

[6] Chua, L. O. (1993). "Global Unfolding of Chua's Circuit", *IEICE Transactions on Fundamentals of Electronics, Communications and Computer Sciences*, Vol. E76-A(5). pp. 704–734.

[7] Chua, L. O. (2005). "Local Activity is the Origin of Complexity", *International Journal of Bifurcation and Chaos*, Vol. 15(11). pp. 3435–3456.

[8] Ginoux, J., Letellier, C. and Chua, L. O. (2010). "Topological Analysis of Chaotic Solution of a Three-Element Memristive Circuit", *International Journal of Bifurcation and Chaos*. Vol. 20(11). pp. 3819–3827.

[9] Kennedy, M. P. (1992). "Robust OP Amp Realization of Chua's Circuit", *Frequenz*, Vol. 46(3–4), pp. 66–80.

[10] Lorenz, E. N. (1963). "Deterministic Nonperiodic Flow", *Journal of Atmospheric Sciences*, Vol. 20, pp. 130–141.

[11] Matsumoto, T. (1984). "A Chaotic Attractor from Chua's Circuit", *IEEE Transactions on Circuits and Systems*, Vol. 31(12). pp. 1055–1058.

[12] Matsumoto, T., Chua, L. O. and Komuro, T. (1985). "The Double Scroll", *IEEE Transactions on Circuits and Systems*, Vol. 32(8). pp. 797–818.

[13] Muthuswamy, B. and Chua, L. O. (2010). "Simplest Chaotic Circuit", *International Journal of Bifurcation and Chaos*, Vol. 20(5). pp. 1567–1580.

[14] Nishio, Y., Mori, S. and Inaba, N. (1992). "Chaotic Phenomena In an Autonomous Circuit with a Nonlinear Negative Inductor", *Electronics and Communication in Japan*, Vol. 75(12). pp. 72–81.

[15] O'Donoghue, K., Kennedy, M. P., Forbes, P., Qu, M. and Jones, S. (2005). "A Fast and Simple Implementation of Chua's Oscillator with Cubic-Like Nonlinearity", *International Journal of Bifurcation and Chaos*, Vol. 15(9), pp. 2959–2971.

[16] Rössler, O. E. (1976). "An Equation for Continuous Chaos", *Physics Letters*, 57A(5). pp. 397–398.

[17] Sprott, J. C. (2000). "Simple chaotic systems and circuits", *American Journal of Physics*, 68(8), pp. 758–763.

[18] Tokunaga, R., Komuro, M., Matsumoto T. and Chua, L. O. (1989). "'Lorenz Attractor' from an Electrical Circuit with Uncoupled Continuous Piecewise-Linear Resistor", *International Journal of Circuit Theory and Applications*, Vol. 17(1). pp. 71–85.

[19] Tucker, W. (1999). *The Lorenz Attractor Exists*, Available from: <http://www2.math.uu.se/~warwick/main/thesis_2.1.html>. [13 November 2011].

[20] Wolfram Research. (2011). *Wolfram Research — Mathematica 8*. Available from: <http://www.wolfram.com>. [13 November 2011].

[21] Yang, X. S. (2009). "Topological Horseshoes and Computer Assisted Verification of Chaotic Dynamics", *International Journal of Bifurcation and Chaos*, Vol. 19(4). pp. 1127–1145.

CHAPTER 26

TOWARD A QUANTITATIVE FORMULATION OF EMERGENCE

G. NICOLIS

Interdisciplinary Center for Nonlinear Phenomena and Complex Systems, Université Libre de Bruxelles, Campus Plaine, CP 231 bd du Triomphe, 1050 Brussels, Belgium

A quantitative approach to emergence is developed, based on the property that in complex multivariate systems one can achieve under certain conditions a reduced closed-form description in terms of a limited number of new variables. The robustness of this description with respect to fluctuations is analyzed and the roles of the system size, the interaction range, the embedding dimensionality and the distance from criticalities are assessed.

1. Introduction

The work of Leon Chua opened new horizons in the fields of nonlinear dynamics and chaos, establishing connections between abstract dynamical systems and key real-world complex phenomena of concern, including the onset of spatial patterns, signal transmission, and information processing. As a tribute to his brilliant contributions I would like to convey in this short note some thoughts on the mathematical and physical foundations of emergence, one of the key concepts underlying the science of complex systems [1].

The essence of emergence is that complexity follows rules of its own transcending those governing the subunits constituting the system. On the other hand since the laws of nature are deterministic these rules are bound to follow in one way or the other from the interactions between these subunits, even though in most cases the precise relations linking them might be very difficult to determine and computationally very costly to implement. When dealing with emergence one is thus confronted at the outset with an apparent paradox, at the origin of its perception by part of the scientific community as an expression of ignorance or, at best, as a metaphor. A central task of complexity research is to show that emergence is in fact a well-defined, quantifiable observable property providing new ways to address and characterize systems composed of interacting subunits.

2. Quantifying Emergence

There are several instances, where switching from "elementary" to "higher" levels of description is accompanied by the appearance of novel, global scale features that one could be tempted to associate to emergence. These features are often brought out by averagings of some sort, in the spirit of statistical mechanics. A diffusion coefficient, a rate constant, or density and temperature for that matter, are novel concepts accompanying the passage from the microscopic to the macroscopic level of description. Likewise the concepts of entropy, dissipation, irreversibility, arise when the macroscopic level of description is further condensed into a thermodynamic description. In a similar vein, climate is a state of the atmosphere resulting from averaging over the short term variability of meteorological fields; the random-looking evolution of deterministic chaotic systems associated with the property of sensitivity to the initial conditions — a feature that stands at first sight in direct opposition to their defining character as deterministic systems — is best captured by switching from a pointwise description running on single trajectories to a coarse-grained description where one deals with ensembles of trajectories; and the very concept of attractor, central to the entire field of nonlinear dynamics, reflects the existence of regularities and structures not foreseen in the properties of the initial conditions and of the individual observables that become apparent once the evolution is embedded in phase space, an abstract concept that has little to do with our everyday perception of space-time events.

Can one then identify, in the light of the foregoing, emergence to the new collective features showing up in the course of switching to higher levels of description? The answer is, generally, in the negative. Indeed, if complexity theory is to be a predictive theory a much stronger property is needed, in which the new entities taking over at a given level determine the evolution of the system in a self-consistent manner, by talking to each other as it were, without the need to refer each time back and forth to what is going on at a lower level. We refer to this property as *closure*: given a set of n variables $\{x_i\}$ representing the state of the individual subunits of a complex multivariate system, there exist new variables $\{z_k\}$ whose number is much less than n, that obey to evolution laws whose structure rests solely on these variables. Alternatively, the system capitalizes on new structures arising through a process of self-organization and to which the aforementioned $\{z_k\}$ are associated and sets new rules supported entirely by these structures.

An example in which this mechanism is operating pertains to the normal form description of bifurcations. Let the time evolution of $\{x_i\}$ be given by the set of coupled equations

$$\frac{dx_i}{dt} = f_i(x_1, \ldots, x_n, \lambda) \quad i = 1, \ldots, n \tag{1}$$

where λ is a control parameter. Let λ_c be a critical value of λ beyond which new branches of solutions are bifurcating from a certain "reference" state. A major result of nonlinear dynamics [2, 3] is that for the description of the bifurcating branches as well as for transient behavior, a limited set of $\kappa \ll n$ quantities can be defined which obey a closed set of equations if the parameters lie close to their critical values λ_c. Therefore the original dynamics is effectively reduced into κ equations giving information on bifurcation, the remaining $n - \kappa$ equations being essentially irrelevant as far as bifurcation is concerned. We refer to the quantities that satisfy the bifurcation equations as *order parameters*.

A most important point is that the bifurcation equations have exactly the same form whatever the structure of the initial laws might be. In other words, the dynamics can be cast in a universal, *normal form* close to the bifurcation point.

Normal forms can be regarded as a new, "higher" level of description beyond that constituted by Eqs. (1). As an example, for a system operating close to a *pitchfork bifurcation*, the normal form describing the bifurcation of steady state solutions is

$$\frac{\partial z}{\partial t} = (\lambda - \lambda_c)z - uz^3 \tag{2}$$

where z is the unique order parameter and u is a combination of λ_c and the other control parameters present in the initial equations. Emergence corresponds here to the existence of non-trivial solutions for the order parameters z, inexistent prior to the transition across the critical point λ_c, which become subsequently the quantities that "perceive" the transition to complexity heralded by the occurrence of the bifurcation. The elementary interactions between the constituent subunits are still present and are even necessary. They are, however, by no means sufficient as they cannot capture properly some key elements coming into play. In short, we have a first quantitative measure of emergence within the framework of a deterministic description: the view of emergence as an expression of ignorance cannot be sustained.

3. Robustness Toward Fluctuations

Implicit in the concept of emergence as formulated above is the idea that the transitions leading to complex behaviors across bifurcation points are robust toward perturbations. Suppose that the variables $\{x_i\}$ in Eqs. (1) refer to a macroscopic-level description — the level at which complexity is perceived in the majority of situations of interest. A general class of the perturbations tending to modify the structure of the macroscopic evolution laws are then the intrinsic fluctuation, arising from the microscopic-level processes underlying the macroscopic behavior. To assess their effect one needs to resort to a probabilistic description, in which the principal quantity of interest is the probability density $P(x_1, \ldots, x_n, t)$ of the $\{x_i\}$'s. This quantity obeys to linear evolution equations such as the master equation or the Fokker-Planck equation [1, 4], in which the macroscopic-level complexity is reflected by the nonlinear dependence of certain coefficients on the $\{x_i\}$'s. Because of this, these equations generate an infinite hierarchy of evolution equations for the successive moments

$$\left\langle \prod_i x_i^{m_i} \right\rangle_t = \int dx_1 \cdots dx_n \; x_1^{m_1} \cdots x_n^{m_n} \; P(x_1, \ldots, x_n, t) \qquad (3)$$

in which the averages $\langle x_i \rangle$ are coupled to the second order variances $\langle \delta x_i \delta x_j \rangle = \langle x_i x_j \rangle - \langle x_i \rangle \langle x_j \rangle$ etc., the second order variances are coupled to the averages and to the third order variances, and so forth. This results in a set of equations of the form

$$\frac{d\langle x_i \rangle}{dt} = f_i(\langle x_1 \rangle, \ldots, \langle x_n \rangle, \lambda) + \frac{1}{2} \sum_{ij} \left(\frac{\partial^2 f_i}{\partial x_j \partial x_k} \right)_{\langle x_j \rangle \cdots} \langle \delta x_i \delta x_j \rangle + \cdots\cdots \qquad (4)$$

The question then arises, whether this infinite hierarchy can be truncated to a finite order and, if so, whether the resulting equations for the average values are identical to the macroscopic evolution laws, Eqs. (1). This property would in turn entail that the evolution of the macroscopic observables depends solely on the full set of these observables.

To secure the validity of the concept of emergence, it is thus necessary to determine the conditions under which the extra terms in Eq. (4) become negligible. One instance where these conditions are satisfied is in systems composed of a very large number n of subunits and functioning around a unique stable state (not necessarily time-independent) sufficiently far from a bifurcation point. The terms in question — which measure in fact, up to a proportionality factor, the strength of the stochastic forcing — are then of the order of $1/n$ and tend to zero as n gets very large, a limit referred to as the *thermodynamic limit*. One is then entitled to conclude that under these conditions a macroscopic-level (deterministic) description becomes an emergent property that springs into life starting from a full-scale probabilistic description.

Let now the system approach and cross a bifurcation point separating the unique "reference" stable state from a multiplicity of simultaneously stable states. To the observer, emergence of macroscopic-level complexity is tantamount to the existence of precisely such states. But to what extent can the above mechanism still operate in this range? On the one side, the mean value $\langle x \rangle$ in unlikely to remain representative since the probability density is expected to flatten and subsequently become multimodal. And on the other side, in systems of large spatial extent spatially inhomogeneous fluctuations may compromise the effective mixing implicit in the macroscopic "mean field" picture and lead eventually to segregation phenomena.

For any finite value of n a closed form global description limited to the macroscopic observables and valid for arbitrarily long times can no longer be guaranteed under the above conditions. An infinity of observables (the full set of moments) is necessary to describe adequately the system, and the states arising beyond bifurcation are by necessity metastable. In this sense emergence of complexity at the macroscopic level is to be understood in a weak sense, being limited both locally (in the vicinity of each of the probability peaks) and in time (the residence time around each peak prior to jumping to one of the other macroscopically stable states).

The situation becomes more clearcut in the limit of infinite size spatially extended systems of sufficiently high dimensionality, or in which the interactions between the subunits are long-ranged. As it turns out such systems are robust toward spatially inhomogeneous fluctuations in the sense that bifurcation persists and emergence of the different states beyond bifurcation is secured in a strong sense [5]. Technically, this shows up through the existence of more than one invariant probability densities satisfying appropriate regularity requirements, a property to be contrasted from the uniqueness of solutions of the (linear!) evolution equations of probability densities for systems involving a finite number of states [6].

4. Conclusions

We have seen how new levels of description following their own rules can be generated by the intrinsic dynamics underlying the evolution of a complex system. This allowed us to quantify the concept of emergence. The essential property sought here was *closure*, namely, the existence of an autonomous structurally stable set of laws governing the variables pertaining to the level of description considered.

It is important to realize that there are situations of considerable interest where the type of decoupling between levels of description envisioned here may be compromised. This happens, for instance, in nanoscale systems and in systems subjected to strong geometric or nonequilibrium constraints, like large biomolecules, catalytic surfaces, films, or membranes. Limitations in mobility tend then to compromise global-scale coherence and as a result bifurcations to self-organized states are suppressed [7]. Another important class pertains to phenomena associated with the occurrence of extreme values of the relevant variables such as earthquakes, floods, and financial crises. Fine scale properties, e.g. the tail of the probability distributions descriptive of the process begin here to matter. Still, universal laws such as fluctuation and large deviation type properties [8, 9] can be extracted pointing to an enlarged formulation of emergence, in which the deterministic and the probabilistic views become tightly intertwined.

References

[1] G. Nicolis and C. Nicolis, *Foundations of complex systems*, 2nd ed., World Scientific, Singapore (2012).
[2] J. Guckenheimer and Ph. Holmes, *Nonlinear oscillations, dynamical systems and bifurcations of vector fields*, Springer, Berlin (1983).
[3] G. Nicolis, *Introduction to nonlinear science*, Cambridge University Press, Cambridge (1995).
[4] C. Gardiner, *Handbook of stochastic methods*, Springer, Berlin (1983).
[5] D. Walgraef, G. Dewel and P. Borckmans, Nonequilibrium phase transitions and chemical instabilities, Adv. Chem. Phys. **49**, 311–355 (1982).
[6] M. Diakonova and R.S. MacKay, Mathematical examples of space-time phases, Bifurcation and Chaos **21**, 2297–2304 (2011).
[7] D. Ben-Avraham and S. Havlin, *Diffusion and reactions in fractals and disordered systems*, Cambridge University Press, Cambridge (2000).
[8] E. M. Sevick *et al.*, Fluctuation theorems, Ann. Rev. Phys. Chem. **59**, 603–633 (2008).
[9] H. Touchette, The large deviation approach to statistical mechanics, Phys. Rep. **478**, 1–69 (2009).

CHAPTER 27

CONTROLLED SYNCHRONIZATION OF CHAOTIC OSCILLATORS WITH HUYGENS' COUPLING*

J. PEÑA-RAMÍREZ, R. H. B. FEY AND H. NIJMEIJER

Department of Mechanical Engineering, Eindhoven University of Technology, P.O. Box 513, 5600 MB Eindhoven, The Netherlands

Oscillating systems are very common in, for example, biology, chemistry, physics, engineering, and many other fields. An intriguing phenomenon occurring in these systems is that two (or more) oscillating systems may synchronize, i.e. their rhythms are adjusted to a certain pattern. This work focuses on the synchronization problem of two chaotic oscillators with Huygens' coupling. A simplified version of Huygens' system of pendulum clocks is considered, where the clocks have been replaced by two chaotic oscillators. Two well known oscillators are used in the analysis, namely Duffing and van der Pol oscillators. The occurrence of synchronization in the oscillators is validated by means of experiments.

1. Introduction

Oscillating systems are very common in for example biology, chemistry, physics, engineering, and many other fields. Some examples of oscillatory systems are a flashing firefly, a pacemaker cell in the heart, and a mechanical pendulum. An intriguing phenomenon occurring in these systems is that two (or more) oscillating systems may synchronize, i.e. their rhythms are adjusted to a certain pattern.

The synchronization phenomenon was first observed by Christiaan Huygens around 1650 in a pair of pendulum clocks hanging from a wooden bar supported by two chairs. In a letter to his

*This paper is dedicated to Leon O. Chua on the occasion of his 75th birthday.

father, he provides a brilliant explanation about the "sympathy" of two pendulum clocks. He concluded that the synchronized motion of the pendula was due to the imperceptible motion of the coupling wooden bar, see e.g. [7], and [17]. Note that the phenomenon discovered by Huygens is particularly remarkable because the synchronization of the pendula is not induced by a control action. Rather it is a consequence of the interaction between the pendula through the coupling, i.e. the wooden bar.

Other systems, where synchronization occurs spontaneously, are biological systems, where it has been found that two or more cells, independently of their nature, or functioning, can show sympathetic behaviour, i.e. they synchronize by using signaling messengers (as for example light) as the coupling signal [9].

In other cases, the synchronization phenomenon is artificially induced by for instance a control law. This kind of synchronization is called, for obvious reasons, controlled synchronization [3]. As an example, the reader may consider the synchronization of wireless network terminals, which are of vital importance in communication systems [22].

Synchronized behavior can also be observed in chaotic systems. Due to their extreme sensitivity to initial conditions, these systems are not expected to synchronize in a natural way. However, it is possible to find suitable couplings/controllers such that two (or more) chaotic systems may synchronize, see e.g. [1,6,14].

Nowadays, there exist several couplings/controllers used to synchronize two chaotic systems, see e.g. [4]. As a matter of fact, these couplings can be used even in the case that the oscillators are not chaotic. A different situation occurs when the oscillators are linked through Huygens' coupling. In this case, the limit synchronizing behaviour of the system is not always the same. In fact, whether the oscillators are chaotic or not, has an influence in the steady state (synchronized) behaviour, as will be discussed in this manuscript.

This work focuses on the synchronization problem of two chaotic oscillators with Huygens' coupling. A simplified version of Huygens' system of pendulum clocks is considered, where the clocks have been replaced by two chaotic oscillators. The purpose is to study the existence of (in-phase and anti-phase) synchronized chaotic motion in the oscillators. Two well known oscillators are used in the analysis, namely Duffing and van der Pol oscillators. The occurrence of synchronization in the oscillators is validated by means of experiments.

The outline of this chapter is as follows. In Sec. 2, a simplified Huygens' model is presented and the experimental setup is briefly described. In Sec. 3, the synchronization of two Duffing oscillators with Huygens' coupling is analyzed. Then, in Sec. 4, it is shown that by driving the coupling bar with a periodic signal, it is possible to induce chaotic behaviour in the oscillators. A nonlinear controller is designed to achieve (in-phase and anti-phase) synchronization in the oscillators when they behave in a chaotic fashion. This is experimentally validated. Next, in Sec. 5, a similar analysis is conducted for a pair of van der Pol oscillators. Finally, a discussion and some conclusions are presented in Secs. 6 and 7, respectively.

2. A Generalized Huygens' Model

In the original setup of pendulum clocks used by Huygens in his discovery of synchronization, the coupling between the clocks was a wooden bar supported by two chairs, see Fig. 1(a). In its simplest

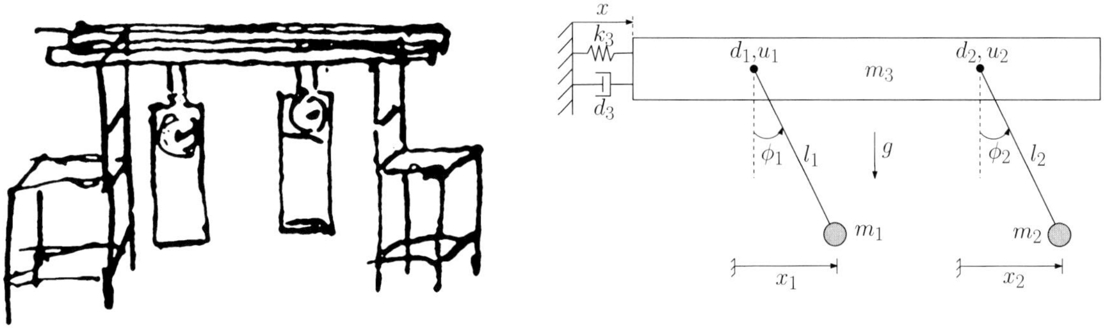

(a) (b)

Fig. 1. (a) Original Huygens drawing cf. [7,17]. (b) Simplified Huygens' model.

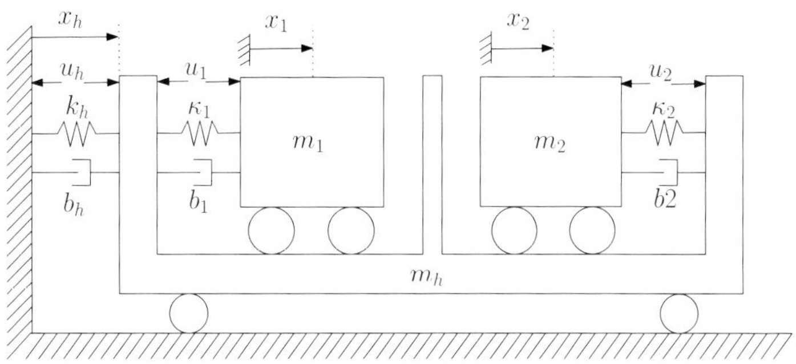

Fig. 2. Two mass-spring-damper oscillators with Huygens' coupling.

form, the wooden bar at the top of two chairs can be modeled by a rigid bar of one dof elastically attached to the fixed world. The pendulum clocks can be replaced by two damped/driven pendula as depicted in Fig. 1(b). This simplified model is able to mimic the in-phase and anti-phase synchronized motion of the pendula observed by Huygens 300 years ago, see e.g. [2, 12, 13, 19].

A further simplification of Huygens' system of Fig. 1 is the schematic model depicted in Fig. 2. In this model, the pendulum clocks are now replaced by two (actuated) mass-spring-damper oscillators and the wooden bar at the top of two chairs is again modeled by a one dof suspended rigid bar. Note that in this model rotational angles are replaced by translational displacements. The control inputs u_1, u_2 of the two oscillators can be used to guarantee steady-state oscillations and/or to modify the inherent dynamic properties of the system, like mass, stiffness, and damping, in a desired way.

As a matter of fact, the diagram of Fig. 2 schematically describes an experimental setup that has been constructed in order to study synchronization of coupled oscillators in general. By means of feedback control, it is possible to adjust the dynamical properties of the setup. Obviously, this opens the possibility of analyzing synchronous behaviour in a wide variety of dynamical systems. This experimental platform can be used to validate (robustness of) the theoretical findings. Moreover, the obtained experimental results may indicate new and interesting directions for research. For a detailed description of the setup, the reader is referred to [16].

The following questions are addressed in the remainder of this chapter: given the Huygens system (two pendulum clocks coupled through an elastic support), is it possible to replace the pendulum clocks by another type of second order nonlinear oscillator and still to observe the synchronized motion? In particular, what happens when the pendulum clocks are replaced by two chaotic oscillators? Since synchronization of the chaotic behaviour will not occur naturally, is it possible to induce the synchronized behaviour by means of a control action? The first question is addressed in the next section, whereas questions 2 and 3 are considered in Sec. 4 and Sec. 5.

3. Synchronization of Two Undriven/Undamped Duffing Oscillators with Huygens' Coupling

In this section, it is shown that two undriven and undamped nonlinear oscillators with Huygens' coupling can synchronize in a natural way. For the sake of illustration, a pair of Duffing oscillators is considered in the analysis. However, it should be stressed that at this point it is possible to consider any other pair of arbitrary second order nonlinear oscillators.

Consider the system depicted in Fig. 2 and assume that the undamped oscillators are identical, i.e. consider $m = m_1 = m_2$, $k_1 = k_2$, and $b_1 = b_2 = 0$. Furthermore, assume that the stiffness characteristic corresponding to both oscillators is given by:

$$k_i(\Delta x_i) = k\Delta x_i + k_3\Delta x_i^3, \quad \text{for} \quad i = 1, 2, \qquad (1)$$

where $\Delta x_i = x_i - x_h$, $k_3 \in \mathbb{R}^+$, and k can be either positive or negative. As stated before, the oscillators are unforced, so $u_1 = u_2 = 0$. Also assume that the coupling bar is free to oscillate as in Huygens' case, i.e. $u_h = 0$. Then, using Newton's second law, the (idealized) equations of motion of system of Fig. 2

become

$$\begin{aligned} m\ddot{x}_1 &= -k(x_1 - x_h) - k_3(x_1 - x_h)^3 \\ m\ddot{x}_2 &= -k(x_2 - x_h) - k_3(x_2 - x_h)^3 \\ m_h\ddot{x}_h &= -m_1\ddot{x}_1 - m_2\ddot{x}_2 - k_h x_h - b_h\dot{x}_h. \end{aligned} \tag{2}$$

System (2) describes the dynamics of two undamped and unforced Duffing oscillators coupled through an elastically supported and damped rigid bar. In order to prove the existence of a synchronized regime, the total energy of system (2) is computed

$$V_u = \frac{1}{2}m_h\dot{x}_h^2 + \frac{1}{2}k_h x_h^2 + \frac{m}{2}\sum_{i=1}^{2}\dot{x}_i^2 + \sum_{i=1}^{2}\int_0^{\Delta x_i} k_i(s)\mathrm{d}s. \tag{3}$$

Next, the time derivative of V is computed along the trajectories of (2). This yields

$$\dot{V}_u = -b_h\dot{x}_h^2 \leq 0. \tag{4}$$

The analysis continues by using LaSalle's invariance principle [8], from which it follows that all the system trajectories converge to a set where

$$x_1 = -x_2, \quad \dot{x}_1 = -\dot{x}_2, \quad \dot{x}_h = x_h = 0. \tag{5}$$

This shows that two undamped, undriven Duffing oscillators with Huygens' coupling will synchronize in anti-phase. Obviously, this result suggests that the coupling used by Huygens (a suspended bar) may be used to synchronize two arbitrary oscillators. In [21], an experimental analysis related to equation (2) has been presented. Other experimental results related to the phase synchronization of arbitrary oscillators with Huygens' coupling are provided in [15] and [20].

4. Synchronization of Two Chaotic Duffing Oscillators with Huygens' Coupling

The first objective of this section is to show that, by *driving* the coupling bar with an external periodic excitation and by *varying* its mass, it is possible to trigger the onset of chaos in two second order nonlinear oscillators coupled through a suspended rigid bar (i.e. with Huygens' coupling). When the oscillators are in a chaotic state, the synchronization phenomenon discussed in Sec. 3, will not occur spontaneously. Consequently, the second objective is to show that by using the well known

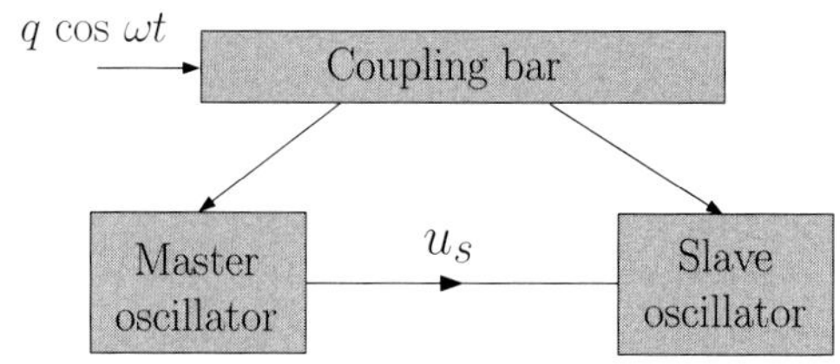

Fig. 3. Two oscillators with Huygens' coupling and master/slave configuration. The slave control input is indicated by u_s.

master/slave configuration cf. [10] it is possible to achieve (controlled) synchronization, i.e. the slave system will be forced to follow the chaotic dynamics of the master oscillator. Figure 3 visualizes this. However, a mutual synchronization scheme can be considered as an alternative [18].

Consider again the mechanical system of Fig. 2 and assume that (1) holds. Subindices $i = 1, 2$ are now replaced by the subindices $i = m, s$, which refer to the master oscillator and slave oscillator, respectively. Then, the equations of motion of the system are:

$$\ddot{x}_m = -\frac{b}{m}(\dot{x}_m - \dot{x}_h) - \frac{k}{m}(x_m - x_h) - \frac{k_3}{m}(x_m - x_h)^3 \tag{6}$$

$$\ddot{x}_s = -\frac{b}{m}(\dot{x}_s - \dot{x}_h) - \frac{k}{m}(x_s - x_h) - \frac{k_3}{m}(x_s - x_h)^3 + \frac{1}{m}u_s \tag{7}$$

$$\ddot{x}_h = -\frac{m}{m_h}(\ddot{x}_m + \ddot{x}_s) - \frac{k_h}{m_h}x_h - \frac{b_h}{m_h}\dot{x}_h + \frac{u_h}{m_h}, \tag{8}$$

where u_s is the slave control input to be designed. This set of equations again describes a pair of two identical Duffing oscillators with Huygens' coupling. Moreover, assume that the oscillators have negative nonlinear stiffness, i.e. $k < 0$. The rigid coupling bar is externally excited with the periodic force

$$u_h = q\cos\omega t, \tag{9}$$

where q is the amplitude of the force and $\omega = 2\pi f$ is the angular frequency of the external excitation.

The following parameter values are considered: $b = 0.0840$ $[\frac{\mathrm{Ns}}{\mathrm{m}}]$, $k = -0.2310$ $[\frac{\mathrm{N}}{\mathrm{m}}]$, $k_3 = 0.210$ $[\frac{\mathrm{N}}{\mathrm{m}^3}]$, $\omega = 1.2$ $[\frac{\mathrm{rad}}{\mathrm{s}}]$, $q = 89$ [N], $k_h = 388.71$ $[\frac{\mathrm{N}}{\mathrm{m}}]$, $b_h = 3.2656$ $[\frac{\mathrm{Ns}}{\mathrm{m}}]$, and $m = 0.210$ [kg]. The slave control

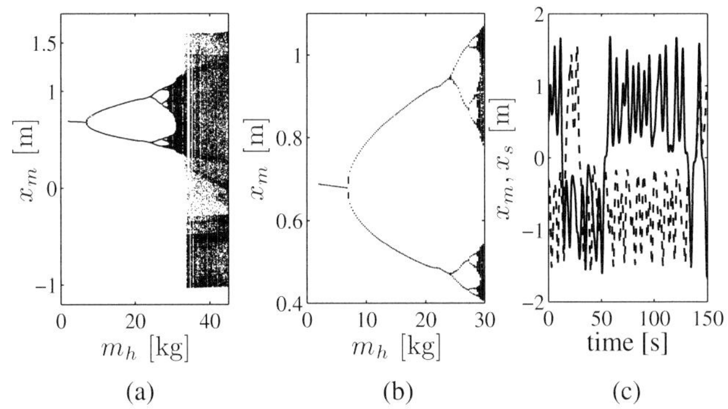

Fig. 4. (Left) Bifurcation diagram for x_m as a function of m_h. (Center) Enlargement of plot (a) in the interval $m_h = [4.1 - 30]$ [kg]. (Right) Chaotic trajectories of the oscillators.

input is not active yet, i.e. $u_s = 0$[N]. Note that for choices of k and k_3, oscillators (6)–(7) become driven Duffing oscillators with a double potential well. Different to other works, where chaotic behaviour of the Duffing equation is observed by varying the amplitude or frequency of the external periodic excitation, see e.g. [5], here, the mass m_h is considered as a bifurcation parameter taking values in the range $m_h \in [4.1, 45]$ [kg]. The bifurcation diagram is depicted in Figs. 4(a) and 4(b). A cascade of period-doubling bifurcations ultimately resulting in a chaotic state can be observed. Starting from $x_m(0) = 0.5$ and $x_s(0) = -0.3$, Fig. 4(c) shows chaotic trajectories of the oscillators for $m_h = 42$ [kg].

4.1. *Case A: in-phase synchronization*

When the synchronized motion of the oscillators is in-phase, the following equations should be satisfied:

$$e = x_s - x_m = 0 \quad \text{and} \quad \dot{e} = \dot{x}_s - \dot{x}_m = 0. \tag{10}$$

Writing down the error dynamics for system (6)–(8) yields:

$$\begin{aligned}\ddot{e} = \frac{1}{m}\big[&- b\dot{e} - ke - k_3e^3 - 3k_3\big(x_s x_m \\ &- x_h(x_s + x_m) + x_h^2\big)e + u_s\big].\end{aligned} \tag{11}$$

Similar to [11], the control input u_s is defined:

$$u_s = -k_p e - k_v \dot{e} + \eta, \tag{12}$$

where $k_p > 0$, $k_v > 0$, and η a is a compensation term to be designed. Substitution of this control in (11) yields the following closed-loop error dynamics

$$\begin{aligned}\ddot{e} = \frac{1}{m}[&-(b + k_v)\dot{e} - (k + k_p)e - k_3e^3 \\ &- f(x_s, x_m, x_h)e + \eta],\end{aligned} \tag{13}$$

where $f(x_s, x_m, x_h) = 3k_3(x_s x_m - x_h(x_s + x_m) + x_h^2)$.

Proposition 27.1. *If the gains k_p, k_v satisfy*

$$k_p > -k, \quad k_v > -b \tag{14}$$

and the compensation term η is designed such that

$$\eta = 3k_3(x_s x_m - x_h(x_s + x_m) + x_h^2)e, \tag{15}$$

then, system (13) is asymptotically stable, i.e.

$$\lim_{t \longrightarrow \infty} e = 0, \quad \lim_{t \longrightarrow \infty} \dot{e} = 0. \tag{16}$$

Proof. The closed-loop (11)–(12), with η as given in (15) is described by

$$\ddot{e} = \frac{1}{m}\big[- (b + k_v)\dot{e} - (k + k_p)e - k_3e^3\big]. \tag{17}$$

Now, consider the candidate Lyapunov function

$$\begin{aligned}V = \frac{1}{2}(\dot{e} + \alpha e)^2 + \frac{1}{2}\Bigg[&\left(\frac{k + k_p}{m}\right) \\ &+ \alpha\left(\frac{b + k_v}{m}\right) - \alpha^2\Bigg] e^2 + \frac{k_3}{4m}e^4\end{aligned} \tag{18}$$

where $\alpha \in \mathbb{R}^+$ is a constant. Function (18) is positive definite if α satisfies

$$\alpha < \frac{(b + k_v)}{m}. \tag{19}$$

The time derivative of (18) along the solutions of system (17) is given by

$$\dot{V} = -\left(\frac{b + k_v}{m} - \alpha\right)\dot{e}^2 - \frac{\alpha(k + k_p)}{m}e^2 - \frac{\alpha k_3}{m}e^4. \tag{20}$$

Next, assume that α satisfies (19). If the gains of the controller (12) are chosen according to (14) then the closed-loop system (17) is asymptotically stable and in consequence

$$\lim_{t \longrightarrow \infty} (x_s - x_m) = 0, \quad \text{and} \quad \lim_{t \longrightarrow \infty} (\dot{x}_s - \dot{x}_m) = 0. \tag{21}$$

□

4.2. *Case B: anti-phase synchronization*

It has been shown in Sec. 3 that when the coupling bar is free to oscillate, i.e. $u_h = 0$, the undamped oscillators will synchronize in anti-phase and as a consequence $x_h \longrightarrow 0$. However, in the present case the coupling bar is being excited with a periodic force and anti-phase synchronization in the same spirit of Huygens will not occur. Therefore, a control input is designed (in analogy with the in-phase synchronization case) to realize anti-phase synchronization of the slave oscillator with respect to the chaotic motion of the master oscillator.

The anti-phase synchronization errors are defined by

$$e_1 = x_s + x_m = 0 \quad \text{and} \quad e_2 = \dot{e}_1 = \dot{x}_s + \dot{x}_m = 0. \tag{22}$$

Consequently, the error dynamics of the coupled Duffing oscillators (6)–(7) are

$$\begin{aligned} \dot{e}_1 &= e_2 \\ \dot{e}_2 &= \frac{1}{m}[-be_2 - ke_1 - k_3 e_1^3 + \gamma(x_x, x_s, e_1) + u_s]. \end{aligned} \tag{23}$$

where

$$\begin{aligned} \gamma(\cdot) = -k_3\big[&3x_m x_s e_1 - 3x_h(x_m^2 + x_s^2) \\ &+ 3x_h^2 e_1 - 2x_h^3\big] + 2kx_h + 2b\dot{x}_h. \end{aligned} \tag{24}$$

As before, the control input u_s is defined as:

$$u_s = -\gamma(x_m, x_s, e_1) - k_p e_1 - k_v e_2. \tag{25}$$

Note that the closed-loop dynamics (23)–(25) coincides with the closed-loop dynamics (17) with $e = e_1 = (x_s + x_m)$ and $\dot{e} = e_2 = (\dot{x}_s + \dot{x}_m)$. Then, by following the same reasoning as used in Proposition 1, it is possible to show that system (23)–(25) with k_p and k_v satisfying (14) is asymptotically stable, i.e. the oscillators will asymptotically synchronize in anti-phase.

4.3. *Experimental validation*

The synchronized motion of the chaotic Duffing oscillators with Huygens' coupling is experimentally verified in this section. As a testbed, the electromechanical setup schematically depicted in Fig. 2 is used. For a detailed explanation about this setup, the reader is referred to [16]. By means of feedback control, the inherent mechanical properties are adjusted such that the dynamical behaviour of the setup is described by the set of equations (6)–(8). In all experiments, the following parameter values are considered: $b = 0.0840$ $[\frac{\text{Ns}}{\text{m}}]$, $k = -0.2310$ $[\frac{\text{N}}{\text{m}}]$, $k_3 = 1.6 \cdot 10^5$ $[\frac{\text{N}}{\text{m}^3}]$, $\omega = 1.2$ $[\frac{\text{rad}}{\text{s}}]$, $q = 0.225$ [N], $k_h = 388.71$ $[\frac{\text{N}}{\text{m}}]$, $b_h = 3.2656$ $[\frac{\text{Ns}}{\text{m}}]$, $m = 0.210$ [kg], and $m_h = 42$ [kg].

In the first experiment, the in-phase synchronized motion of oscillators (6)–(7) is investigated. The nonlinear feedback controller (12) is implemented for the slave oscillator (7). For this experiment, the oscillators are released from the initial conditions $x_m(0) = 5$ [mm], $x_s(0) = -3$ [mm], and $x_h(0) = \dot{x}_h(0) = \dot{x}_m(0) = \dot{x}_s(0) = 0$. The values of the gains in the controller are $k_p = 1.3$ $[\frac{\text{N}}{\text{m}}]$ and $k_v = 0.5$ $[\frac{\text{Ns}}{\text{m}}]$. With this choice, (14) is satisfied.

The corresponding experimental results are depicted in Fig. 5. The transient responses for the displacements, the synchronization error, and the control input are shown in Figs. 5(a), 5(c), and 5(e) respectively.

For $t \in [0, 20)$, the controller is not activated and the behaviour in the oscillators is uncorrelated as depicted in Fig. 5(a). At $t = 20$ [s], the controller is activated and as a consequence, the slave oscillator synchronizes in-phase with the chaotic trajectory of the master. This can be seen in Fig. 5(b), where the complete time series corresponding to the displacements of the oscillators is depicted and synchronized chaotic steady-state behaviour is achieved. The synchronization error $e = x_s - x_m$ obviously is large when the control action is not applied yet, but once the controller is activated, this error becomes practically zero as illustrated in Figs. 5(c) and 5(d) respectively. In theory, this error should become exactly zero. However, in practice the oscillators are not exactly equal and consequently in the experimental result there is a small error. Fig. 5(e) shows that the control effort stays very small once in-phase synchronization is realized. Fig. 5(f) presents a projection of the displacements corresponding to the oscillators onto the plane (x_m, x_s). The light grey part of the curve corresponds to the transient behaviour, whereas the dark grey part of the curve indicates the synchronized steady-state behaviour.

In a second experiment, anti-phase synchronization is investigated. In this case, the control input (25) is used with $k_p = 1.1$ $[\frac{\text{N}}{\text{m}}]$ and $k_v = 0.5$ $[\frac{\text{Ns}}{\text{m}}]$. In order to better illustrate the effect of the

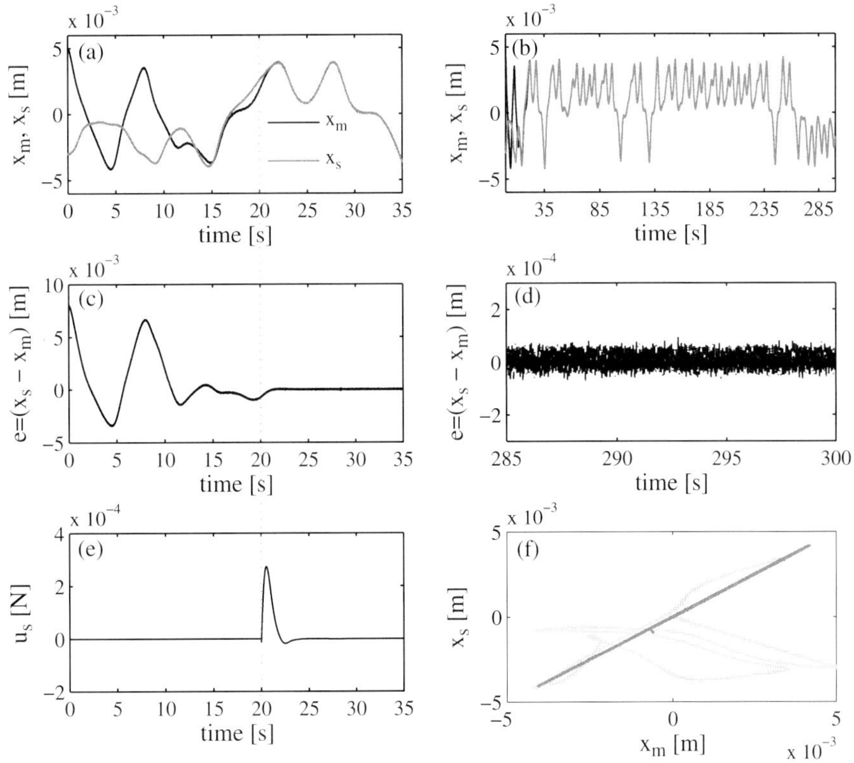

Fig. 5. Experimental result: oscillators synchronize in-phase.

control action, the controller is again activated at $t = 20$ [s]. After this time, the displacement and velocity of the slave oscillator synchronize in anti-phase with the trajectories of the master, so the control objective is achieved. Figure 6 summarizes the main results of this experiment. Plots 6(a) and 6(c) correspond to the initial transient responses, whereas plots 6(b), 6(d), and 6(e) show the steady-state behaviour for the displacements x_m and x_s of the oscillators, the synchronization error $e = x_m + x_s$, and the control input u_s, respectively. The anti-phase synchronization of the coupled Duffing oscillators becomes immediately clear by drawing the projection of the displacements corresponding to the oscillators onto the plane (x_m, x_s) as depicted in Fig. 6(f). As before, the light grey part of the curve corresponds to the transient behaviour, whereas the dark grey part of the curve indicates the synchronized steady-state behaviour.

5. Synchronization of Two Chaotic van der Pol Oscillators

In the previous section, a (slightly) modified version of Huygens' system was considered, where the pendulum clocks were replaced by two Duffing oscillators. This section is in the same spirit, in the sense that now the pendula are replaced by two van der Pol oscillators. In order to achieve (in-phase or anti-phase) synchronization, again a nonlinear control action is included to one of the oscillators.

Consider again the system depicted in Fig. 2 and assume identical oscillators, i.e. $m = m_1 = m_2$, $k = k_1 = k_2$, and $b = b_1 = b_2$. Furthermore, assume that the damping characteristic corresponding to both oscillators is given by

$$\beta_i(x_i, \dot{\Delta}_{xi}) = -b(1 - ax_i^2)\dot{\Delta}_{xi}, \quad i = m, s, \tag{26}$$

where $\dot{\Delta}_{xi} = (\dot{x}_i - \dot{x}_h)$. Note that subindices $i = 1, 2$ in Fig. 2 have been replaced by subindices m, s respectively.

For now, consider the case $u_m = u_h = 0$. Then, the equations of motion of the system are

$$\begin{aligned}\ddot{x}_m &= \frac{b}{m}(1 - ax_m^2)(\dot{x}_m - \dot{x}_h) - \frac{k}{m}(x_m - x_h)\\ \ddot{x}_s &= \frac{b}{m}(1 - ax_s^2)(\dot{x}_s - \dot{x}_h) - \frac{k}{m}(x_s - x_h) + \frac{1}{m}u_s\\ \ddot{x}_h &= -\frac{m}{m_h}(\ddot{x}_m + \ddot{x}_s) - \frac{k_h}{m_h}x_h - \frac{b_h}{m_h}\dot{x}_h.\end{aligned} \tag{27}$$

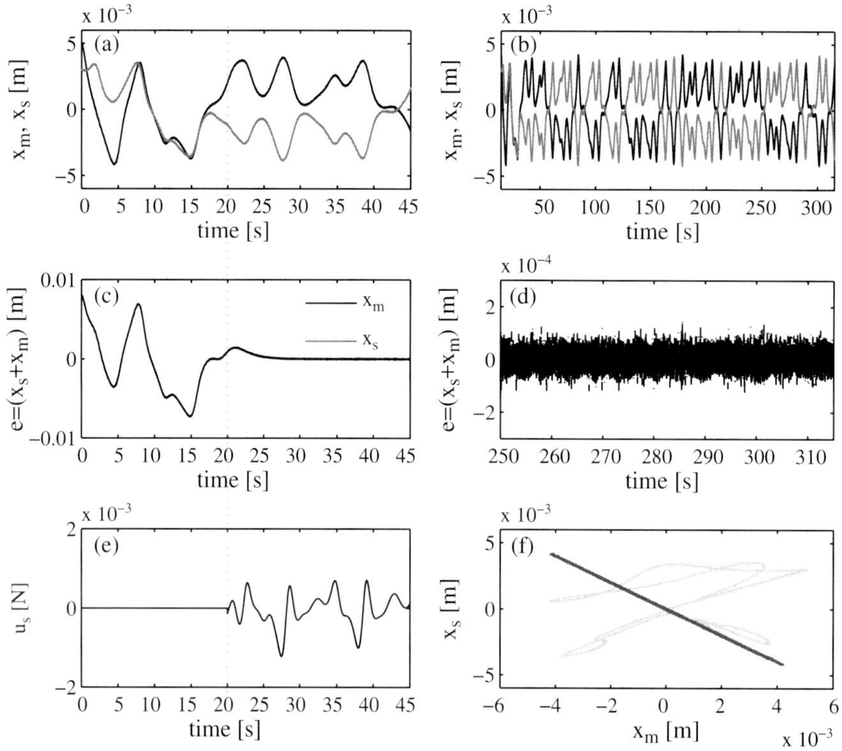

Fig. 6. Anti-phase synchronization is achieved in two chaotic Duffing oscillators with Huygens' coupling.

This set of equations describes a pair of van der Pol oscillators with Huygens' coupling. It has been found that this system shows chaotic behaviour for certain parameter values. Similar to the case of Duffing oscillators discussed before, the ratio $\frac{m}{m_h}$ is a key parameter in the onset of chaotic behaviour. Note, however, that now the coupling bar is not driven.

Also here, in the chaotic regime, the coupled van der Pol oscillators may show synchronized behaviour provided that a control action is applied to one of them. Next, the particular cases of in-phase and anti-phase synchronization are discussed.

5.1. *Case A: in-phase synchronization*

By defining in-phase synchronization errors like (10), the corresponding error dynamics for system (27) can be written as

$$\ddot{e} = \frac{1}{m}\big[b\dot{e} - ba\big(x_s^2\dot{x}_s - x_m^2\dot{x}_m - \dot{x}_h\big(x_s^2 - x_m^2\big)\big) - ke + u_s\big]. \tag{28}$$

In order to achieve in-phase synchronized motion in the chaotic oscillators, the following controller may be considered

$$u_s = -k_p e - k_v\dot{e} + \eta, \tag{29}$$

where $k_p > 0$ and $k_v > 0$ and η can be chosen as

$$\eta = ba\big(x_s^2\dot{x}_s - x_m^2\dot{x}_m - \dot{x}_h\big(x_s^2 - x_m^2\big)\big). \tag{30}$$

Following the approach used in Proposition 27.1, it can be proven that the controller (29)–(30) yields global asymptotic stability for system (28), i.e. the coupled oscillators asymptotically synchronize in-phase, provided that $k_v > b$ and $k_p > 0$.

5.2. *Case B: anti-phase synchronization*

If the two van der Pol oscillators with Huygens' coupling show in-phase synchronized chaotic behaviour, the coupling bar will vibrate. However, when anti-phase synchronization is induced in the oscillators, the oscillations in the coupling bar will die out, i.e. $x_h \to 0$ and the coupling bar comes to a standstill as in the original Huygens system, see also the third equation in (27). Therefore, if anti-phase is achieved, the originally chaotic motion

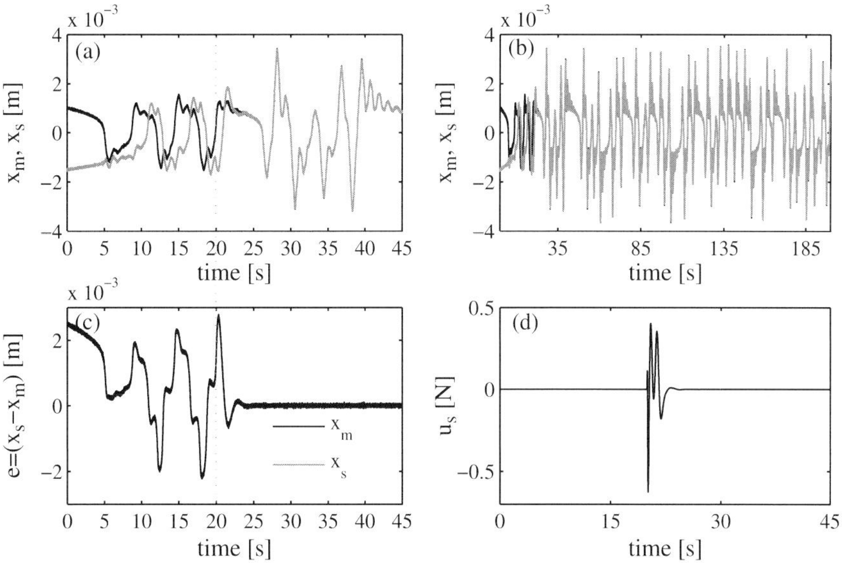

Fig. 7. In this experiment, two van der Pol oscillators with Huygens' coupling synchronize in-phase.

will change to periodic motion in both van der Pol oscillators, which are now uncoupled. To avoid this scenario and maintain chaotic motion for anti-phase synchronization, the external periodic excitation (9) is applied to the coupling bar.

The anti-phase synchronized motion in the chaotic oscillators is achieved by considering controller (29) with $e = (x_m + x_s)$, $\dot{e} = (\dot{x}_m + \dot{x}_s)$, and $\eta(\cdot)$ chosen as

$$\eta(\cdot) = ba\big(x_s^2\dot{x}_s + x_m^2\dot{x}_m - \dot{x}_h\big(x_s^2 + x_1^2\big)\big) - 2kx_h + 2b\dot{x}_h. \quad (31)$$

As before, the control gains in (25) are chosen according to $k_v > b$ and $k_p > 0$.

5.3. *Experimental results*

The synchronized chaotic motion of two van der Pol oscillators is validated by means of experiments. System (27) is implemented in the experimental setup with the following parameter values: $b = 18.3135$ $[\frac{\text{Ns}}{\text{m}}]$, $b_h = 3.2656$ $[\frac{\text{Ns}}{\text{m}}]$, $k = 6.315$ $[\frac{\text{N}}{\text{m}}]$, $k_h = 388.71$ $[\frac{\text{N}}{\text{m}}]$, $a = 55 \cdot 10^5$ $[\frac{1}{\text{m}^2}]$, $m = 4.210$ [kg], and $m_h = 5$ [kg].

In a first experiment, the oscillators are synchronized in-phase by applying to one of them the control input (29)–(30) with $k_p = 25$ $[\frac{\text{N}}{\text{m}}]$ and $k_v = (b + 10)$ $[\frac{\text{Ns}}{\text{m}}]$. The controller is activated at $t = 20$ $[s]$. The oscillators are released from the initial conditions $x_m(0) = 1$ [mm] and $x_s(0) = -1.5$ [mm]. The remaining initial conditions are set to zero. The obtained experimental results are presented in Fig. 7. When the controller is off, the chaotic trajectories of the oscillators are not synchronized and consequently the error $e = (x_s - x_m)$ is large, but once the controller is activated, this error becomes practically zero as depicted in Figs. 7(a) and 7(c). Although theoretically this error should asymptotically converge to zero, in practice the error signal keeps showing small vibrations caused by measurement noise and (slight) differences in the oscillators. The control effort is depicted in Fig. 7(d).

Next, the experiment is repeated, but now control law (29) with (31) is applied because it is desired to synchronize the oscillators in anti-phase. To better illustrate the influence of the control law, the controller is now switched on at $t = 115$ [s]. As can be seen from Fig. 8, when the control action is not active, the system behaves chaotically

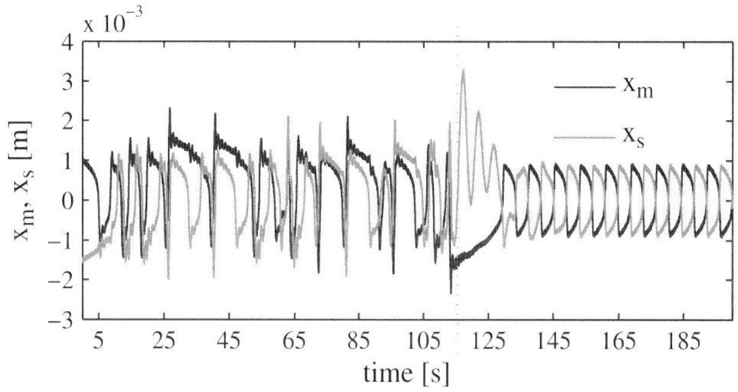

Fig. 8. Although the oscillators synchronize in anti-phase, the chaotic behaviour disappears.

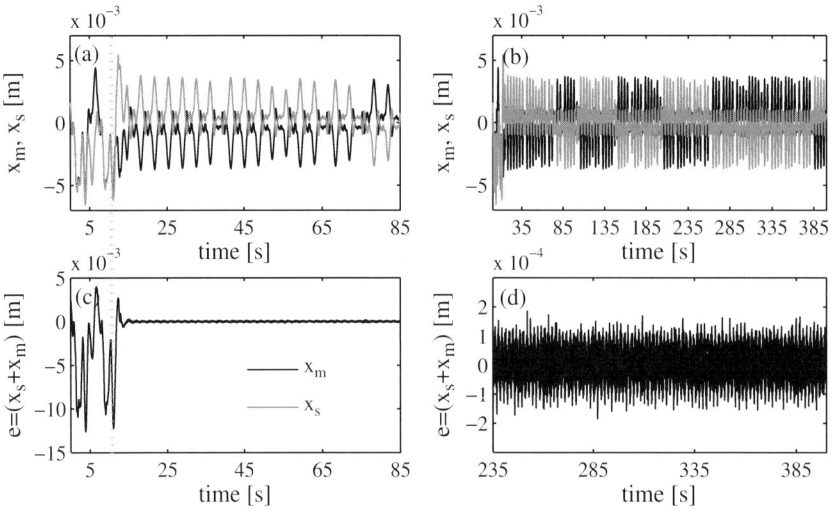

Fig. 9. In this experiment anti-phase synchronization is achieved.

and unsynchronized. When the controller is activated, the oscillators asymptotically synchronize in anti-phase but in a periodic regime. As discussed above, the chaotic behaviour disappears because the coupling bar comes to a standstill. The obtained periodic solution is identical to the solution of each individual oscillator uncoupled from the coupling bar.

To avoid the disappearance of the chaotic behaviour, an external periodic force is applied to the coupling bar. This force is given in (9) and is used in a third experiment, with $q = 0.5$ [N] and $\omega = 3.5$ $[\frac{\text{rad}}{\text{s}}]$. The remaining parameter values are identical to the first and second experiment. The non-zero initial conditions are $x_m(0) = 2.5$ [mm] and $x_s(0) = 2$ [mm]. The control action (29) with (31), with $k_p = 25$ $[\frac{\text{N}}{\text{m}}]$ and $k_v = 28$ $[\frac{\text{Ns}}{\text{m}}]$, is applied from $t = 10$ [s]. As observed in Fig. 9, the chaotic trajectories of the oscillators asymptotically synchronize in anti-phase.

6. Discussion

The contributions of the present work can be summarized as follows

- The mass of Huygens' coupling bar, which determines the coupling strength, has been considered as a bifurcation parameter leading to chaos. In the case of two Duffing oscillators, the onset of chaotic behaviour is induced by driving the coupling bar with a periodic excitation. In the case of two self-driven van der Pol oscillators, the chaotic behaviour appears 'spontaneously'.
- Huygens' coupling can be used to synchronize not only pendulum clocks as in the original experiment of Huygens, but also to synchronize other type of oscillators like Duffing oscillators as shown in Section 3. However, it is important to emphasize that if the oscillators behave chaotically, then Huygens' coupling will not lead to spontaneous synchronization and an additional control input should be added to one of the oscillators such that synchronized chaotic motion is induced in the oscillators.
- The experimental results presented here validate that it is possible to induce synchronized behaviour in a pair of chaotic oscillators with Huygens' coupling.

It should be noted that the nonlinear controllers used to synchronize the oscillators, see e.g. (12), require inclusion of a compensation term η in order to deal with the nonlinearities and/or to cancel the influence of the coupling bar, see Eq. (15). Sometimes, however, it is desired to have a controller that on the one hand is as simple as possible, but on the other hand should be robust enough to guarantee that the stability is not compromised. In other words, it should be assured that the error dynamics at least remains ultimately uniformly bounded. For these reasons, the following simple feedback controller may be used

$$u_s = -k_p e - k_v \dot{e}. \tag{32}$$

This control law has the advantage that it does not require knowledge of the parameters of the oscillator to be controlled. Computer simulations and

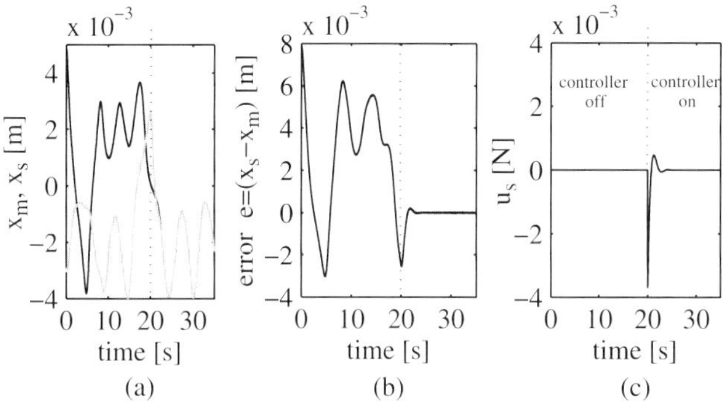

Fig. 10. In this experimental result, in-phase synchronization is induced in system (6)–(8) by using controller (32).

experimental results have revealed that in some cases, by using (32), the oscillators may (asymptotically) synchronize, whereas in other cases the so called practical synchronization will appear. The stability proof of this turns out to be involved but it will be worthwhile to address this problem also because experimental results and computers simulations have revealed that the control effort is small. This has several (beneficial) implications. First, the probability to have saturation effects in the actuators is considerably reduced. Secondly, small control signals result in low energy consumption. For the sake of illustration, an experimental result corresponding to system (6)–(8) with parameter values as given in Sec. 4.3 and with controller (32), with $k_p = 1.2$ and $k_k = 0.5$ is presented in Fig. 10.

7. Conclusions and Recommendations

In this work, it has been shown that two Duffing and two van der Pol oscillators with Huygens' coupling may experience chaotic behaviour. For the case of the Duffing oscillators, the coupling bar must be driven by a periodic force. The coupling strength — or rather the mass of the coupling bar — has been considered as the bifurcation parameter and for the case of the Duffing oscillators, a cascade of period-doubling bifurcations resulting ultimately in chaotic states has been found. In the chaotic regime, for the Duffing oscillators as well as the van der Pol oscillators, the oscillators do not synchronize spontaneously. This is not a surprise, because, although the oscillators are identical, there is extreme sensitivity to initial conditions. For this reason, a controller was added to one of the oscillators, i.e. the slave oscillator, in order to induce in-phase or anti-phase synchronization of chaotic motion. This principle has been experimentally validated. When Huygens' coupling is used to couple two arbitrary oscillators, the occurrence of spontaneous synchronization (in the same spirit of Huygens) depends on whether the oscillators have chaotic solutions or not.

For reasons mentioned in Sec. 6, it is interesting to perform a stability analysis of the (in-phase and/or anti-phase) synchronization error dynamics when the simple controller (32) is used.

Finally, it is worth saying that Huygens' system is an exciting and challenging nonlinear dynamical system, which complex dynamical behaviour is far from being completely understood. Further studies of this system will lead to unveiling more details about the complex yet intriguing synchronization phenomenon.

References

[1] Afraimovich, V. S., Verichev, N. N. and Rabinovich, M. I. (1986). Stochastic synchronization of oscillation in dissipative systems, *Radiophysics and Quantum Electronics* **29**, 9, pp. 795–803.

[2] Bennett, M., Schatz, M., Rockwood, H. and Wiesenfeld, K. (2002). Huygens' clocks, *Proceedings of the Royal Society of London A: Mathematical, Physical and Engineering Sciences* **458**, 2019, pp. 563–579.

[3] Blekhman, I. I., Fradkov, A. L., Nijmeijer, H. and Pogromsky, A. Yu. (1997). On self-synchronization and controlled synchronization, *Systems and Control Letters* **31**, pp. 299–305.

[4] Boccaletti, S., Kurths, J., Osipov, G., Valladares, D. L. and Zhou, C. S. (2002). The synchronization of chaotic systems, *Physics Reports* **366**, 1–2, pp. 1–101.

[5] Chen, G. and Dong, X. (1993). On feedback control of chaotic continuous-time systems, *IEEE Transactions on Circuits and Systems I: Fundamental Theory and Applications* **40**, 9, pp. 591–601.

[6] Fujisaka, H. and Yamada, T. (1983). Stability theory of synchronized motion in coupled-oscillator systems, *Progress of Theoretical Physics* **69**, 1, pp. 32–37.

[7] Huygens, C. (1660). in M. Nijhoff (ed.), *Oeuvres completes de Christiaan Huygens*, Vol. 17, pp. 156–189.

[8] Khalil, H. (2002). *Nonlinear Systems*, 3rd edn. (Prentice Hall, New Jersey).

[9] Kim, J. R., Shin, D., Jung, S. H., Heslop-Harrison, P. and Cho, K. H. (2010). A design principle underlying the synchronization of oscillations in cellular systems, *Journal of Cell Science* **123**, 4, pp. 537–543.

[10] Murali, K. and Lakshmanan, M. (1994). Drive-response scenario of chaos synchronization in identical nonlinear systems, *Physical Review E* **49**, 6, pp. 4882–4885.

[11] Nijmeijer, H. and Berghuis, H. (1995). On Lyapunov control of the Duffing equation, *IEEE Transactions on Circuits and Systems I: Fundamental Theory and Applications* **42**, 8, pp. 473–477.

[12] Oud, W. T., Nijmeijer, H. and Pogromsky, A. Yu (2006). A study of Huijgens' synchronization: experimental results, in K. Pettersen, J. Gravdahl and H. Nijmeijer (eds.), *Group Coordination and Cooperative Control*, *Lecture Notes in Control and Information Sciences*, Vol. 336 (Springer Berlin/Heidelberg), pp. 191–203.

[13] Pantaleone, J. (2002). Synchronization of metronomes, *American Journal of Physics* **70**, 10, pp. 992–1000.

[14] Pecora, L. M. and Carroll, T. L. (1990). Synchronization in chaotic systems, *Physical Review Letters* **64**, 8, pp. 821–824.

[15] Pena-Ramirez, J., Fey, R. H. B. and Nijmeijer, H. (2011). On phase synchronization of oscillators mounted on a movable support, in *Proceedings of the 7th EUROMECH Nonlinear Dynamics Conference (ENOC 2011)* (July 24–July 29, 2011, Rome, Italy).

[16] Pena-Ramirez, J., Fey, R. H. B. and Nijmeijer, H. (2012). An experimental study on synchronization of nonlinear oscillators with Huygens' coupling, *Nonlinear Theory and its Applications, IEICE* **3**, 2, pp. 128–142.

[17] Pikovsky, A., Rosenblum, M. and Kurths, J. (2001). *Synchronization. A universal concept in nonlinear sciences* (Cambridge University Press, Cambridge).

[18] Pogromsky, A. Y. (1998). Passivity based design of synchronizing systems, *International Journal of Bifurcation and Chaos* **8**, 2, pp. 295–319.

[19] Pogromsky, A. Y., Belykh, V. N. and Nijmeijer, H. (2003). Controlled synchronization of pendula, in *Proceedings 42nd IEEE International Conference on Decision and Control* (December 9–12, Maui, Hawaii, USA.), pp. 4381–4386.

[20] Pogromsky, A. Y., Rijlaarsdam, D. and Nijmeijer, H. (2010). Experimental Huygens synchronization of oscillators, in M. Thiel, J. Kurths, M. C. Romano, G. Karolyi and A. Moura (eds.), *Nonlinear Dynamics and Chaos: Advances and Perspectives*, *Understanding Complex Systems*, Vol. 33 (Springer Berlin/Heidelberg), pp. 195–209.

[21] Rijlaarsdam, D., Pogromsky, A. Yu and Nijmeijer, H. (2010). Synchronization between coupled oscillators: an experimental approach, in G. Leonov, H. Nijmeijer, A. Yu Pogromsky and A. Fradkov (eds.), *Dynamics and control of hybrid dynamical systems*, *B*, Vol. 14 (World Scientific), pp. 153–165.

[22] Wolf, B. J., Russell, H. B. and Wang, K. (2007). Synchronizing transmission schedules of partitioned ad hoc networks, in *Military Communications Conference. MILCOM 2007 IEEE* (October 29–October 31, Orlando, Fl, USA), pp. 1–7.

CHAPTER 28

USING TIME-DELAY FEEDBACK FOR CONTROL AND SYNCHRONIZATION OF DYNAMICAL SYSTEMS

KESTUTIS PYRAGAS,* VIKTORAS PYRAGAS and TATJANA PYRAGIENE

Semiconductor Physics Institute, Center for Physical Sciences and Technology, LT-01108 Vilnius, Lithuania

An application of time-delayed feedback signals in the field of chaos control and synchronization is discussed. In the first part of the chapter, we present an overview of delayed feedback control techniques for stabilization of unstable periodic orbits of chaotic systems. The second part is devoted to anticipating synchronization, where the time-delayed feedback in a slave system enables us to forecast the chaotic dynamics of the identical master system.

1. Introduction

Time-delay effects are very common in many natural and engineering systems. They may arise due to the finite propagation speed of information between the constituents of a complex system or due to processing and latency times. Unlike ordinary differential equations, delay systems are infinite-dimensional in nature and their analysis is much more complicated. The time-delay feedback is, in many cases, a source of instability. However, in some specially designed systems the time-delay feedback can manifest itself as a stabilizing factor.

In this chapter, we discuss two methods in the field of chaos control and synchronization where the time-delay feedback is used for the stabilization of some desired unstable states of chaotic systems. The first method is the delayed feedback control (DFC) algorithm in which the time-delay feedback is utilized for the stabilization of unstable periodic orbits (UPOs) of chaotic systems. The second method is

*Corresponding author.

the algorithm of anticipating synchronization. Here the time-delay feedback provides the stabilization of unstable anticipating manifold of the slave system driven by the identical master system. In both cases, we present the main ideas of the methods, their developments up to date and our recent achievements.

2. Delayed Feedback Control Algorithm

2.1. *Review of theoretical and experimental results*

The DFC algorithm has been invented 20 years ago [32] as a simple, robust, and efficient method to stabilize UPOs in chaotic systems. Nowadays, it becomes one of the most popular methods in the chaos control research [46]. The method allows a noninvasive stabilization of UPOs of dynamical systems in the sense that the control force vanishes when the target state is reached. The block diagram of the method is presented in Fig. 1. The DFC algorithm is reference-free and makes use of a control signal $K\Delta s(t)$ obtained from the difference $\Delta s(t) = s(t) - s(t-\tau)$ between the current state $s(t)$ of the system and the state of the system $s(t-\tau)$ delayed by one-period τ of the target orbit. The UPO may become stable under the appropriate choice of feedback strength K. Note that only the stability properties of the orbit are changed, while the orbit itself and its period remain unaltered. The controlled system can be treated as a black box, since the method does not require any exact knowledge of either the form of the periodic orbit or the system's equations. The method is particularly appealing for experimentalists, since one does not need to know anything about the target orbit beyond its period τ. The DFC algorithm is notably superior to other control methods in fast dynamical systems, since it does not require any real-time computer processing.

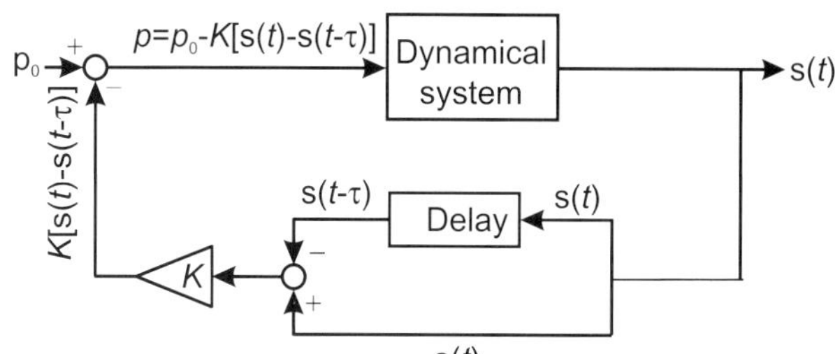

Fig. 1. Block diagram of the delayed feedback control method. $s(t)$ is an output variable, p is a control parameter, p_0 is its value at which the dynamical system possesses an unstable periodic orbit with a period τ, and K is the feedback gain.

The algorithm has been successfully applied to a number of real-world problems in diverse experimental systems, including electronic chaotic oscillators, mechanical pendulums, lasers, gas discharged systems, a current-driven ion acoustic instability, a chaotic Taylor-Couette flow, chemical systems, high-power ferromagnetic resonance, helicopter rotor blades, and cardiac systems (cf. [37] for a review). An important practical application of the DFC algorithm has been recently demonstrated in [58]. By successful implementation of the DFC method in an atomic force microscope the authors managed to stabilize cantilever oscillation and remove artifacts on a surface image. It is also possible to perform the DFC based bifurcation analysis for experiments [47, 48].

The DFC method has been verified for a large number of theoretical models from different fields. Examples include stabilization of high-speed semiconductor lasers, buck converters, chaotic systems with dry friction, excitable systems, car-following traffics, economic models, satellite attitude control, chaotic roll motion of a flooded ship in waves, ensembles of globally coupled oscillators, etc. (cf. [36]). More recently the DFC has been proposed to eliminate chaotic oscillations in robots with kinematical redundancy [22] and in the Internet [23].

A reach variety of modifications of the DFC has been suggested in order to improve its performance. Adaptive versions of the DFC with automatic adjustment of delay time [14, 20, 28] and control gain [4, 21] have been considered. For spatially extended systems, various modifications based on spatially filtered signals have been analyzed [2, 15]. The wave character of dynamics in some systems allows a simplification of the DFC algorithm by replacing the delay line with the spatially distributed detectors. In [24] such a simplification is reported for an ionization wave experiment. Due to dispersion relations, the delay in time is equivalent to the spatial displacement, and the control signal can be constructed without use of the delay line. Most important modification of the DFC has been proposed in [12]. The authors improved an original DFC scheme by using an information from many previous states of the system. This extended DFC (EDFC) scheme achieves stabilization of UPOs with a greater degree of instability [3, 33].

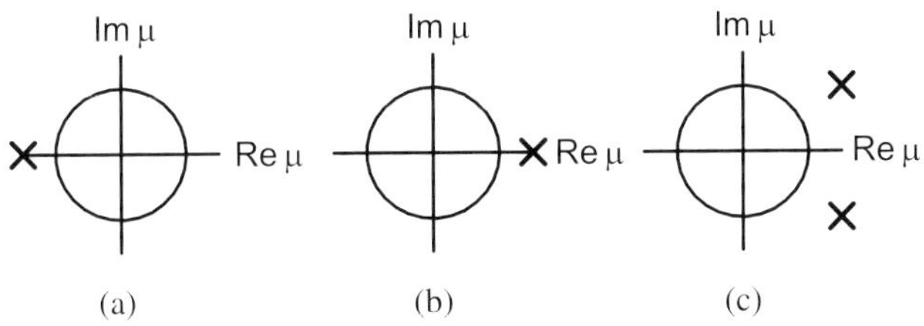

Fig. 2. Leading Floquet multipliers of unstable periodic orbits arising from different bifurcations: (a) period doubling, (b) tangent or subcritical Hopf, and (c) Nejmark-Sacker (discrete Hopf) bifurcations. The unit circle defines the region of stability.

The theory of DFC is difficult because the delayed feedback induces an infinite number of degrees of freedom. Even linear analysis of such systems is complicated due to the infinite number of Floquet exponents (FEs) characterizing the stability of controlled orbits. Nevertheless, some analytical approaches have been developed in vicinity to various bifurcations of periodic orbits, such as the period doubling bifurcation [18, 35], the subcritical Hopf bifurcation [34,37,38] and the Nejmark-Sacker (discrete Hopf) bifurcation [44]. The leading Floquet multipliers (FMs) of UPOs arising from these bifurcations are shown in Fig. 2. Note that all three types of orbits have different topological properties. The orbits with the negative real multiplier flip their neighborhood during one turn. The orbits with the complex conjugate pair of the multipliers have a finite torsion, and the orbits with the positive real multiplier have no torsion. Below we discuss separately the theoretical treatments of the DFC algorithm for these three different types of UPOs.

The orbits with the negative real multiplier [Fig. 2(a)] arise from a period-doubling bifurcation and are typical, for example, for the Rössler system. The mechanism of stabilization of such orbits by delayed feedback is well understood [18, 35]. In [35] there has been derived the relationship between the FEs of periodic orbits controlled by proportional and delayed feedback algorithms. Then it is enough to know the dependence of the Floquet exponents on the control gain for the proportional feedback control algorithm. Similar dependence for the EDFC is obtained from a transcendental equation and we do not need to solve the delay differential equations.

An analytical treatment of the delayed feedback controlled UPOs with the complex conjugate pair of leading FMs [Fig. 2(c)] has been developed in [44]. Such orbits arise from a Nejmark-Sacker (discrete Hopf) bifurcation and cannot appear in low-dimensional chaotic attractors. However, such orbits may appear in low-dimensional non-chaotic systems, e.g. in a weakly nonlinear van der Pol oscillator subjected to a periodic force and delayed feedback control. For this system, we have derived a simplified averaged equation and obtained analytical expressions defining the region of stabilization of the controlled system as well as an optimal value of the control gain.

The orbits with the positive real multiplier [Fig. 2(b)] come, for example, from a tangent or a subcritical Hopf bifurcation and are typical, for instance, for the Lorenz system. They satisfy the so-called odd number limitation, which states that any UPOs with an odd number of real Floquet multipliers greater than unity can never be stabilized by any DFC technique. During the last fifteen years, this limitation has been intensively discussed in the literature. The history of the limitation is dramatic. This statement was first proven by Ushio [52] for discrete time systems. [18] and [27] proved the same limitation for the continuous time DFC, and then this proof was extended for the EDFC [29]. It was commonly accepted that it is impossible to overcome this inherent limitation. However, [11] have shown by a simple example that this limitation does not hold in general for autonomous systems. Nevertheless, it remains valid for non-autonomous systems.

To overcome the odd number limitation a counterintuitive idea based on an unstable controller has been proposed [34, 37, 38]. By including an additional unstable mode into the control loop one artificially enlarges the set of real multipliers greater than unity to an even number. The performance of such an unstable controller has been numerically demonstrated for the Lorenz system [34] as a representative of systems with torsion-free UPOs. Later on, it has been realized that the unstable DFC allows an analytical treatment if a system is close to a subcritical Hopf bifurcation. First, such a treatment has been performed for a simple nonlinear second order electronic circuit modeling a subcritical Hopf bifurcation [38]. Then this approach has been extended for wider class of dynamical systems with an arbitrary large phase space dimension [37].

In the next two sections, we discuss in more details our recent achievements in the theory of the DFC. Section 2.2 is devoted to the modification of the DFC algorithm that improves it global properties. The ergodicity — an universal feature of chaotic systems is invoked in order to enlarge the basin of attraction of the stabilized orbit. In Sec. 2.3, we describe an adaptive modification of the DFC algorithm with the continuously varying time delay that provides an asymptotic convergence of the delay time to the period of the target UPO.

2.2. *Improving the global properties of the DFC*

The standard tool for discussing the control performance of DFC systems consists in linear stability analysis. However, even if such a local analysis predicts stable states, experimental success is not guaranteed because the control performance may strongly depend on initial conditions. The analysis of global properties of DFC systems, such as basins of attraction of stabilized orbits, is much more complicated problem. There exists virtually no systematic investigation of time-delayed feedback control beyond the linear regime. Numerical analysis of particular systems shows that the DFC algorithm can form incredibly complex basins of attraction [56,57].

To improve the global properties of the linear DFC algorithm several modifications have been proposed. A first heuristic idea has been suggested in the original paper [32]. It has been shown that limiting the size of the control force by a simple cutoff increases a basin of attraction of the stabilized orbit. This idea has proven itself in a number of chaotic systems, and now it is widely used in experimental implementations of DFC method. An alternative two-step DFC algorithm has been proposed in [50]. In [17], the authors have developed a theory which stated that depending on the type of transition at the control boundary there appear basins of attraction of different size; for a continuous transition the basin of attraction is large, while for a discontinuous transition it is small. Finally, a nonlinear DFC for systems close to a subcritical Hopf bifurcation has been proposed in [16]. It has been shown that the basin of attraction can be enlarged considerably by coupling control forces through the phase of the signal.

Unfortunately, the above-proposed nonlinear DFC schemes are not universal. In recent papers [41, 42], we have elaborated a DFC algorithm with the improved global properties by invoking an egodicity — the universal feature of chaotic systems. The ergodicity means the fact that a trajectory of any chaotic system visits a neighborhood of each periodic orbit with finite probability. The idea of using ergodicity in chaos control research was first formulated in the seminal paper by Ott, Grebogy, and Yorke (OGY) [30] and has been employed in their OGY control algorithm. However, a straightforward implementation of ergodicity in DFC schemes has not been considered. Our main motivation was to fill this gap and adapt the OGY ideas for DFC algorithm. The specific point of our problem is that the DFC force increases the phase space dimension of the closed-loop system. As well as in the OGY algorithm, we do not perturb the system until it comes in a small neighborhood of the desired orbit. Using a scalar observable, we have developed a technique which allows us to evaluate a moment when the state of the free time-delay system approaches the target orbit. At this moment, we activate the DFC force and stabilize the target. Note that the problem of evaluating the moment when the state of the free system approaches the target orbit has to be considered in an infinite-dimensional phase space, since the DFC force increases the phase space dimension.

To demonstrate the algorithm let us consider the Rössler equations [45] subjected to DFC force

$$\dot{x} = -y - z, \tag{1a}$$

$$\dot{y} = x + ay - K\Delta s(t), \tag{1b}$$

$$\dot{z} = b + z(x - c). \tag{1c}$$

Here x, y, z are the dynamic variables of the Rössler system. The parameters $a = 0.2$, $b = 0.2$ and $c = 5.7$ are chosen such that the system exhibits chaotic behavior. We suppose that $s(t) = y(t)$ is an observable and the DFC perturbation $K\Delta s(t)$ is applied only to the second equation of the Rössler system. Here K is the feedback gain and $\Delta s(t)$ denotes the difference of the observable between the current state, and the state delayed by the period τ of UPO:

$$\Delta s(t) = s(t) - s(t - \tau) = y(t) - y(t - \tau). \tag{2}$$

Numerical analysis shows that the basins of attraction of the period-1 and period-2 UPOs of the Rössler system are rather large. The straightforward application of the original DFC algorithm (without waiting until the system approaches the desired orbit) is successful for these orbits at any initial conditions placed on the strange attractor. However, this is not the case for the period-3 UPOs. The free ($K = 0$) Rössler system has two period-3 UPOs with slightly different periods $\tau = 17.51$ and $\tau = 17.60$. Here we concentrate on the stabilization of the orbit with the period $\tau = 17.51$. The original DFC algorithm is not able to stabilize this orbit for any initial conditions, and thus we need a modification.

The linear analysis of Eqs. (1) with the perturbation (2) shows that the target orbit is stable in the interval of control gain $0.05 < K < 0.34$. The optimal value of the feedback gain, which leads to the fastest convergence of nearby initial conditions towards the desired orbit is $K_{op} = 0.06$. Although the whole basin of attraction of the stabilized orbit may be very complex, it necessarily occupies some region close to the target orbit. This general feature of the basin of attraction results from linear stability of the orbit. Thus, the DFC algorithm should be successful if the initial conditions are in fair proximity to the target. On the other hand, due to the ergodicity, the free system ($K = 0$) should approach the target orbit as close as desired for any initial conditions placed on the strange attractor.

Our strategy is as follows. We do not perturb the system ($K = 0$) when its state is far away from the desired orbit and activate the control in the form of Eq. (2) with $K = K_{op}$ when the state approaches the target UPO. The main problem here is to define the moment when the state of the free system falls in a small neighborhood of the desired orbit. Having a scalar observable $s(t)$, the standard way to reconstruct the phase space of the system is to introduce delay coordinates. Using an $(M+1)$-dimensional delay-coordinate vector $\mathbf{r}(t) = [s(t), s(t-\delta t), \ldots, s(t-M\delta t)]$ with delay δt one can write inequalities (requirements)

$$|\mathbf{r}_j(t) - \mathbf{r}_j(t-\tau)| \equiv |\Delta s(t - j\delta t)| < \varepsilon, \quad j = 0, \ldots, M. \tag{3}$$

that guarantee that the state of the system at time t is close to the periodic orbit of the period τ in the reconstructed phase space. These inequalities mean that the absolute value of the difference $|\Delta s(t)|$ has to be small in $M + 1$ equally spaced points in the time interval $[t - \Delta t, t]$ with $\Delta t = M\delta t$. Alternatively, we may replace the above $M + 1$ inequalities (3) by one inequality for the averaged difference, $(1/M)\sum_{j=0}^{M} |\Delta s(t - j\delta t)| < \varepsilon$. For continuous-time systems with an infinite-dimensional phase space, we take the limit $\delta t \to 0$, $M \to \infty$. On the assumption that the product $M\delta t = \Delta t$ is finite, we obtain $(1/\Delta t)\int_{t-\Delta t}^{t} |\Delta s(t')|dt' < \varepsilon$. The latter condition defines the smallness of a moving average of the difference $|\Delta s(t)|$, where Δt is a window of moving average. The main advantage of such a condition is that the moving average can be simply estimated electronically. The simplest moving average filter can be designed as a first-order low-pass filter.

As a result, we come to the following modification of the DFC algorithm. To estimate the moving average of the difference $|\Delta s(t)|$ we introduce an auxiliary variable w that satisfies first order filter equation

$$\tau_w \dot{w} = |\Delta s(t)| - w. \tag{4}$$

An asymptotic solution of this equation is

$$w(t) = \frac{1}{\tau_w}\int_{-\infty}^{t} e^{(t'-t)/\tau_w} |\Delta s(t')|dt'. \tag{5}$$

Thus the variable w represents an exponentially weighted moving average of the difference $|\Delta s(t)|$. The characteristic window of the moving average is $\Delta t = \tau_w$. The smallness of this variable indicates the closeness of the system state to the target orbit in an infinite-dimensional phase space.

The control procedure is as follows. We start from initial conditions placed on the strange attractor of the free Rössler system and integrate Eqs. (1) and (4) for $K = 0$ as long as $w > \varepsilon$. As soon as the variable w becomes small, $w < \varepsilon$, we set $K = K_{op}$ and continue the integration of the system subjected to DFC. We assume that the control goal is achieved when w decreases to a given value $\varepsilon_F = 3 \times 10^{-2} \ll \varepsilon$. We repeat this procedure for 10^3 different initial conditions and plot in Fig. 3 histograms of time needed to achieve the control. For suitably chosen filter parameter $\tau_w = \tau/7$ and fairly small $\varepsilon < \varepsilon_c \approx 0.33$, the algorithm produces 100% success rate for any initial conditions. For $\varepsilon = \varepsilon_c$ the mean time of control amounts approximately 10 periods of UPO.

We emphasize that our modification is very simple. The closeness of the system state to the target orbit is estimated from the usual DFC signal (2) by the standard low-pass filter (4). Thus the algorithm allows a simple analogue implementation and preservers the main advantages of the original DFC technique. By further modification of this algorithm, one can select and stabilize a particular UPO among different UPOs of the same period [41, 42].

2.3. *Adaptive DFC with a state dependent time delay*

Experimental implementation of the DFC method requires a knowledge of delay time, which is equal to the period of actual UPO. However, for

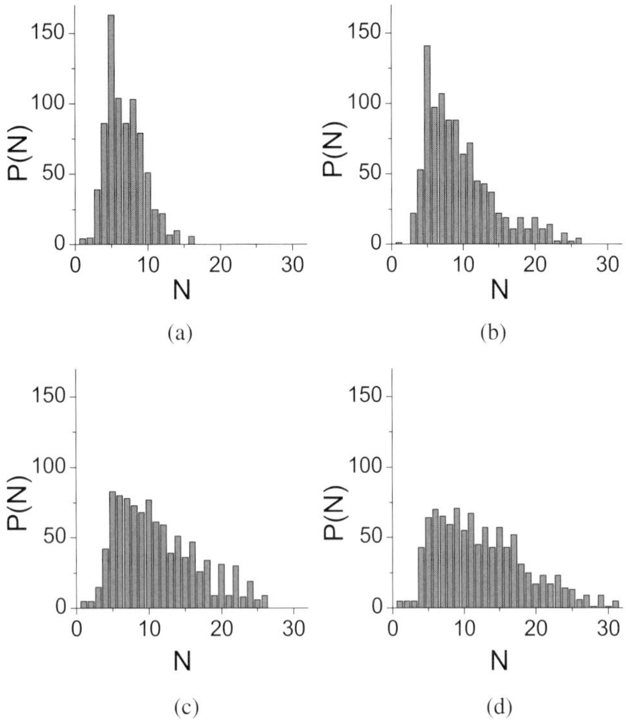

Fig. 3. Histograms of time needed to achieve control in the Rössler system for $K = K_{op} = 0.06$, $\tau = 17.51$, and $\tau_w = \tau/7$. N is the number of periods τ needed for stabilization of the target UPO and $P(N)$ is the number of successful stabilizations with the given time of control. The parameter ε and the mean number $\langle N \rangle$ of periods needed to achieve control are (a) $\varepsilon = 0.7$, $\langle N \rangle = 6.87$; (b) $\varepsilon = \varepsilon_c = 0.33$, $\langle N \rangle = 9.54$; (c) $\varepsilon = 0.25$, $\langle N \rangle = 11.25$; (d) $\varepsilon = 0.2$, $\langle N \rangle = 12.4$. In (a) only 79% of initial conditions are successful, while in (c)–(d) the 100% success rate is obtained.

autonomous systems, this period is not known a priori. Furthermore, in real experimental situation, the period of actual UPO may evolve due to the evolution of the system parameters under the effects of exogenous, unpredictable factors. In this context, adaptive control techniques with an automatic adjustment of the delay time are desired.

The problem of estimating the period of a target UPO from observed experiential data has been considered in several publications. In the original paper [32], it has been shown that the amplitude of DFC perturbation has a resonance-type dependence on the delay time, and the periods of UPOs can be extracted from the minima of this dependence. More sophisticated theoretical foundation for obtaining the UPO period from the observed control signal has been developed in [19]. The first adaptive technique employing online variation of the delay time has been proposed [20]. Here the delay time is adjusted in a discrete way according to the distance between successive maxima of the output signal. Another adaptive discrete-time technique based on the gradient-descent method has been proposed in [5].

Here we briefly outline our recently proposed [43] adaptive continuous-time DFC algorithm with a state-dependent delay, which is based on resonance dependence of the DFC perturbation on the delay time pointed out in [32]. We implement an adaptive strategy similar to that proposed in [49] to maintain a synchronous state in time-evolving complex networks. Note that in the last years the systems described by state dependent delay differential equations have become of particular interest. Recently, [13, 26] have demonstrated that the variation of the delay time can enhance the stability properties of the DFC.

Consider a general case of DFC controlled dynamical system:

$$\dot{\mathbf{X}}(t) = \mathbf{F}[\mathbf{X}(t), K\Delta s(t)]. \tag{6}$$

The first argument in the function $\mathbf{F}$ shows the dependence of the vector field on internal degrees of freedom, while the second argument denotes the dependence on control force $K\Delta s(t)$. Here K is the feedback gain, and the delayed difference is defined as $\Delta s(t) = s(t) - s(t-\tau) = g[\mathbf{X}(t)] - g[\mathbf{X}(t-\tau)]$, where $s(t) = g[\mathbf{X}(t)]$ is a measurable scalar signal that is a function of state vector variable $\mathbf{X}(t)$. Assume that the free system ($K = 0$) has an unstable periodic solution $\mathbf{X}(t) = \xi(t) = \xi(t-T)$ that we intend to stabilize by control perturbation $K\Delta s(t)$ with time-dependent delay time $\tau(t)$. Our aim is to construct an algorithm, which provides an asymptotic convergence of the delay $\tau(t)$ towards the period of desired UPO and stabilizes this UPO.

As a helpful tool for this purpose, we construct the following functional:

$$\bar{\Delta}(t) = \int_0^t \exp[-\nu(t-t')][s(t') - s(t' - \tau(t'))]^2 dt'. \tag{7}$$

This functional describes the running time-average of the quantity $[s(t) - s(t-\tau(t))]^2$. It estimates the averaged value of this quantity in the time interval $[t - 1/\nu, t]$. The parameter ν defines the window of averaging.

If we chose τ as a constant of time, we could scan the values of τ, and plot the dependence of the functional $\bar{\Delta}$ vs. τ. At the periods of UPOs it is exactly zero, i.e. we have resonances for $\tau = T_k$ (here T_k is the period of the k-th UPO).

We now determine the dynamics of the delay by the following equation:

$$\frac{d\tau}{dt} = -\beta \frac{d\Delta}{d\tau}. \tag{8}$$

Here Δ is the same as $\bar{\Delta}$ but after a substitution (approximation) $\tau(t') \to \tau(t)$. We choose the value of the parameter β in such a way as to satisfy $1/\beta \gg 1/\nu$, i.e. the delay time τ evolves much slower than the time of averaging. The form of Eq. (8) suggests that the functional Δ may be regarded as a potential in the space of delay τ. Since the dependence $\Delta = \Delta(\tau)$ has a form of parabolas in the neighborhoods of periods of UPOs, one may expect that the nearby solution of varied delay will converge towards the minima, $\tau \to T_k$, where T_k is the period of the desired UPO.

After denoting the gradient as $G(t) \equiv \frac{d\Delta}{d\tau}$, and differentiating it with respect to time, we can collect the above results into one system:

$$\dot{\mathbf{X}} = \mathbf{F}(\mathbf{X}, K[s(t) - s(t - \tau(t))]), \tag{9a}$$

$$\dot{\tau} = -\beta G, \tag{9b}$$

$$\dot{u} = \gamma[s(t - \tau(t)) - u], \tag{9c}$$

$$\dot{G} = -\nu G + 2\gamma[s(t) - s(t - \tau(t))][s(t - \tau(t)) - u]. \tag{9d}$$

Equation (9c) is introduced for an approximate estimation of the time-derivative, i.e. we set $\dot{s}(t - \tau(t)) \approx \gamma[s(t-\tau(t))-u]$ in Eq. (9d). Equations (9b)–9(d)) enable us to find the dynamics of the variable delay $\tau(t)$. They are valid if the parameters satisfy the following inequalities: $1/\gamma \ll 1/\nu \ll 1/\beta$.

In Fig. 4 we show the results of numerical simulations for the period-one UPO of the Rössler system, when the state variable $\mathbf{X} = (x, y, z)$ satisfies Eqs. 1. We see that the adaptive algorithm forces the variable delay to converge towards the period of UPO, although the initial delay $\tau_0 = 3$ differs considerably from the period $T \approx 5.88$ of the UPO.

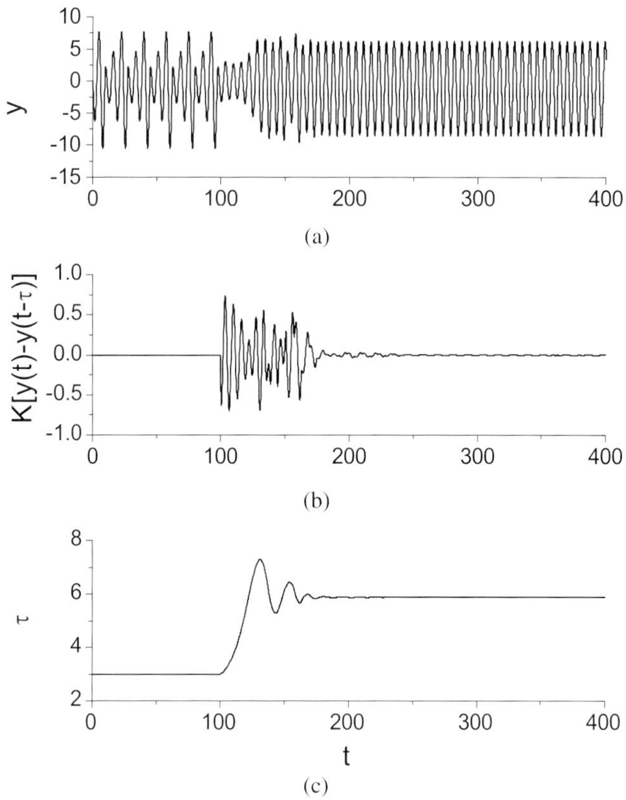

Fig. 4. Adaptive stabilization of the period-1 UPO of the Rössler system. (a) Dynamics of the measured signal $s(t) = y(t)$, (b) control force $K[y(t) - y(t - \tau)]$, and (c) state-dependent delay time $\tau(t)$. The controller parameters are: $\gamma = 5$, $\beta = 0.002$, and $\nu = 0.02$. The initial value of the delay time is $\tau_0 = 3$. For $t < t_c = 100.0$ the feedback force is switched off ($K = 0$), and for $t \geq t_c$ it is activated with an optimal value of the feedback gain $K = K_0 = 0.116$.

3. Anticipating Synchronization of Chaotic Systems

3.1. *Brief review*

Synchronization of oscillations is a phenomenon common to a large variety of nonlinear dynamical systems in physics, chemistry, and biology [31]. Whereas first investigation of the synchronization phenomenon goes back to the work by Huygens in 1665, the past decades have witnessed a considerable interest in the topic of synchronization of chaotic systems. Intuitively it would seem that chaos and synchronization are two mutually exclusive notions. However, it has been shown that synchronization can appear in chaotic systems in many different ways, including the identical, phase, lag, and anticipating synchronization. The latter type of synchronization introduced by Voss [53] is most counterintuitive.

In the case of anticipating synchronization one deals with two systems, a "master" and a "slave" which are coupled unidirectionally via a time-delay feedback in such a manner that the slave system predicts the behavior of the master system. The coupling scheme introduced by Voss is as follows:

$$\dot{\mathbf{r}}_1 = \mathbf{f}(\mathbf{r}_1) \tag{10a}$$

$$\dot{\mathbf{r}}_2 = \mathbf{f}(\mathbf{r}_2) + \mathbf{K}[\mathbf{r}_1 - \mathbf{r}_2(t - \tau)], \tag{10b}$$

where $\mathbf{r}_1(t)$ and $\mathbf{r}_2(t)$ are the dynamic vector variables of the master and slave system, respectively,

f is a nonlinear vector function, τ is a delay time, and **K** is a coupling matrix. It is easy to see that the anticipatory synchronization manifold $\mathbf{r}_2(t) = \mathbf{r}_1(t+\tau)$ is a solution of Eqs. (10). For appropriate choice of τ and **K**, this solution can be stable, i.e., the slave anticipates by an amount τ the output of the master. This phenomenon has been studied numerically for a variety of systems and justified experimentally in electronic circuits [55] and lasers [51].

Implementation of anticipating synchronization as a strategy for real-time forecasting of a given dynamics requires a design of coupling schemes with a possibly large anticipation time. The analysis performed in [53] shows that the scheme (10) with a diagonal matrix **K** is ineffective. Its maximum stably anticipation time is much shorter than the characteristic time scales of the system's dynamics. In order to enlarge the prediction time it was proposed to extend Eq. (10b) with a chain of N unidirectionally coupled slave systems [54]:

$$\dot{\mathbf{r}}_i = \mathbf{f}(\mathbf{r}_i) + \mathbf{K}[\mathbf{r}_{i-1} - \mathbf{r}_i(t-\tau)], \quad i = 2, \ldots, N+1. \tag{11}$$

Formally, the prediction time of this scheme is N times larger as compared to the scheme (10b). However, it was shown that the chain (11) is unstable to propagating perturbations, and this convective-like instability limits the number of slaves in the chain which can operate in a stable regime [25].

Recently, we addressed the question, whether it is possible to considerably prolong the prediction time via a suitable choice of the coupling matrix **K** and for typical low-dimensional chaotic systems, we gave a positive answer [39, 42]. In the next two sections, we briefly outline our ideas. We describe two algorithms of design **K** based on the scheme (10) with the single slave system. In Sec. 3.2, we show that using the idea of phase-lag compensating coupling (PLCC), the prediction time can be enlarged several times in comparison to the diagonal coupling. In the Sec. 3.3, we describe an extended PLCC (EPLCC) algorithm that allows us to prolong the anticipating time to the characteristic period of chaotic oscillations.

3.2. *Phase-lag compensating coupling*

First, we demonstrate a heuristic idea of the PLCC algorithm with the Rössler system [45], which is given by a three-dimensional (3D) vector variable $\mathbf{r} = [x, y, z]$ and vector field

$$\mathbf{f}(\mathbf{r}) = [-y - z, x + ay, b + z(x - c)]. \tag{12}$$

In the following, we set $a = 0.15$, $b = 0.2$, $c = 10$ and suppose that both **r** and **f** are the vectors columns. Although the Rössler system has two fixed points, the strange attractor originates from one of them, $\mathbf{r}_0 = [(c-s)/2, (s-c)/2a, (c-s)/2a]$, located close to the origin, where $s = (c^2 - 4ab)^{1/2}$. The fixed point is a saddle-focus with an unstable 2D manifold (an unstable spiral) almost coinciding with the (x, y) plane and a stable 1D manifold almost coinciding with the z axis. The phase point of the system spends most time in the (x, y) plane moving along the unstable spiral according to approximate equations $\dot{x} = -y$, $\dot{y} = x + ay$. Whenever x approaches a value $x \approx c$, the z variable comes into play. The phase point leaves for a short time the (x, y) plane and then returns to the origin via a stable z-axis manifold.

Taking into account such a topology of the strange attractor, we choose the coupling matrix as $\mathbf{K} = k\mathbf{Q}$, where k is a scalar parameter defining the coupling strength and

$$\mathbf{Q} = \begin{pmatrix} \cos\alpha & -\sin\alpha & 0 \\ \sin\alpha & \cos\alpha & 0 \\ 0 & 0 & 0 \end{pmatrix} \tag{13}$$

is a 3×3 matrix that projects the vector field onto the unstable (x, y) plane and rotates this projection by the angle $\alpha = \omega\tau$. Here ω is a frequency of the unstable spiral, which for the Rössler system is ≈ 1. The main advantage of such a choice consists in phase-lag compensation of the time-delay feedback term in Eq. (10b). When the system moves along the unstable spiral in the (x, y) plane, the vector $\mathbf{Qr}_2(t-\tau)$ is in-phase with the vector $\mathbf{r}_2(t)$ [cf. Fig. 5], and thus the term $\mathbf{Kr}_2(t-\tau)$ provides a correct negative feedback. We refer to this coupling law as a phase-lag compensating coupling (PLCC). In Fig. 6 we compare the effect of PLCC with the

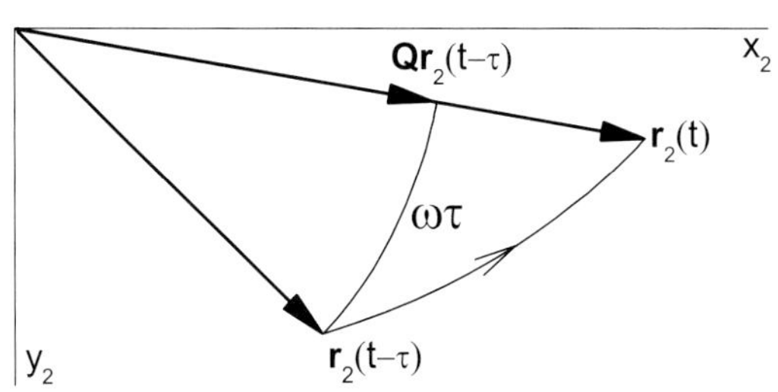

Fig. 5. Phase-lag compensation of delayed vector $\mathbf{r}_2(t-\tau)$.

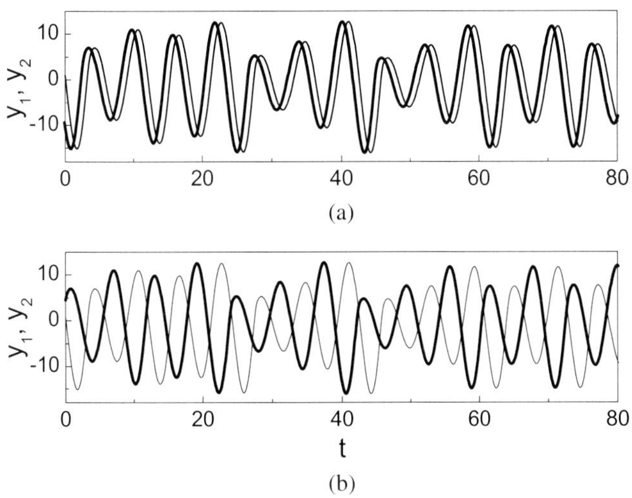

Fig. 6. Time series $y_1(t)$ of the master (thin line) and $y_2(t)$ of the slave (bold line) Rössler systems. (a) The diagonal coupling with $\mathbf{K} = 0.36\ \mathrm{diag}[1, 1, 1]$ and $\tau = 0.9$. (b) The PLCC with $\mathbf{K} = 0.18\ \mathbf{Q}$, $\omega = 1$, and $\tau = 3.6$. In both cases the coupling parameters are optimized in such a way as to attain the maximum stable anticipation time.

usual diagonal coupling, when $\mathbf{K} = k\,\mathrm{diag}[1, 1, 1]$. The time of reliable prediction for the PLCC is $\tau \approx 3.8$. It exceeds 4 times the maximum prediction time for the diagonal coupling. The characteristic period of chaotic oscillations in the Rössler system is ≈ 6. Thus, our algorithm allows us to make a prediction for more than a half of this period. We stress, that the PLCC enables forecasting of the *global* dynamics of the system, although the coupling matrix (13) takes into account only the *local* properties of the phase space. Indeed, the phase-lag compensation via rotation of the vector field is strongly valid only close to the fixed point.

We now consider a more precise and more general development of the above idea for any Rössler-type dynamical system. Suppose that a strange attractor of a 3D chaotic system $\dot{\mathbf{r}} = \mathbf{f}(\mathbf{r})$ originates from a saddle-focus fixed point $\mathbf{r}_0$ such that $\mathbf{f}(\mathbf{r}_0) = 0$. Generally, the unstable and stable manifolds of the fixed point may have arbitrary orientations in the phase space. To design the coupling for this general case we first shift coordinates to the fixed point and rewrite the governing equation in the form $\dot{\mathbf{R}} = \mathbf{J}\mathbf{R} + \mathbf{N}(\mathbf{R})$, where $\mathbf{J} = \partial\mathbf{f}/\partial\mathbf{r}|_{\mathbf{r}=\mathbf{r}_0}$ is the Jacobian matrix, $\mathbf{N}(\mathbf{R}) = \mathbf{f}(\mathbf{r}_0 + \mathbf{R}) - \mathbf{J}\mathbf{R}$ is a nonlinear function, and $\mathbf{R} = \mathbf{r} - \mathbf{r}_0$. Then Eq. (10b) for the slave system can be written as

$$\dot{\mathbf{R}}_2 = \mathbf{J}\mathbf{R}_2 + \mathbf{N}(\mathbf{R}_2) + \mathbf{K}[\mathbf{R}_1 - \mathbf{R}_2(t-\tau)]. \quad (14)$$

According to our assumptions, the Jacobian $\mathbf{J}$ has a pair of complex conjugate eigenvalues $\lambda_{1,2} = \gamma \pm i\omega$ with $\gamma > 0$, corresponding to 2D unstable spiral manifold, and a real negative eigenvalue λ_3, representing the stable 1D manifold. By a suitable change of variables, the Jacobian $\mathbf{J}$ can be transformed to Jordan normal form — that is,

$$\mathbf{E}^{-1}\mathbf{J}\mathbf{E} = \begin{pmatrix} \gamma & -\omega & 0 \\ \omega & \gamma & 0 \\ 0 & 0 & \lambda_3 \end{pmatrix}, \quad (15)$$

where $\mathbf{E}$ is the matrix of eigenvectors of $\mathbf{J}$. This transformation orients the unstable 2D manifold towards the (x, y) plane and the stable manifold towards the z axes. After such a transformation, we can apply the above theoretical arguments and use the coupling law $\mathbf{K} = k\mathbf{Q}$. This means that in the original (non-transformed) variables the coupling matrix has to be constructed as

$$\mathbf{K} = k\mathbf{E}\mathbf{Q}\mathbf{E}^{-1}. \quad (16)$$

Equation (16) gives a general algorithm of coupling design for typical chaotic systems. Application of the general coupling law (16) to the Rössler system does not advance significantly the forecasting algorithm in comparison to the above considered heuristic approach. This is because the Jacobian of the Rössler system is close to Jordan normal form and the matrix $\mathbf{E}\mathbf{Q}\mathbf{E}^{-1}$ does not differ significantly from the matrix $\mathbf{Q}$.

The general coupling law (16) allows us to extend the prediction horizon not only for the Rössler-type systems, but also for more complex double-scroll attractors. A typical example of such a system is the Chua circuit [6] defined by the flow

$$\mathbf{f}(\mathbf{r}) = [(g(y-x) - i_N)/C_1, (g(x-y) + z)/C_2, -y/L], \quad (17)$$

where x, y, and z represent, respectively, the tension across two capacitors and the current through the inductor, and $i_N = m_0 x + 0.5(m_1 - m_0)(|x + B_p| - |x - B_p|)$ is a piecewise-linear function. We choose the parameters $C_1 = 0.1$, $C_2 = 1$, $L = 1/5$, $m_0 = -0.5$, $m_1 = -0.8$, $B_p = 1$, and $g = 0.58$ such that the circuit operates in the double scroll regime. The double scroll attractor originates from two saddle-focus fixed points $\mathbf{r}_0^{\pm} = [\pm x_0, 0, \mp g x_0]$, where $x_0 = (m_0 - m_1)B_p/(g + m_0)$. These fixed points have identical eigenvalues and eigenvectors since their Jacobian matrixes coincide. Therefore, the same coupling matrix (16) is suited to the both fixed points, and we get correct phase-lag compensation when the trajectory twists along of one or

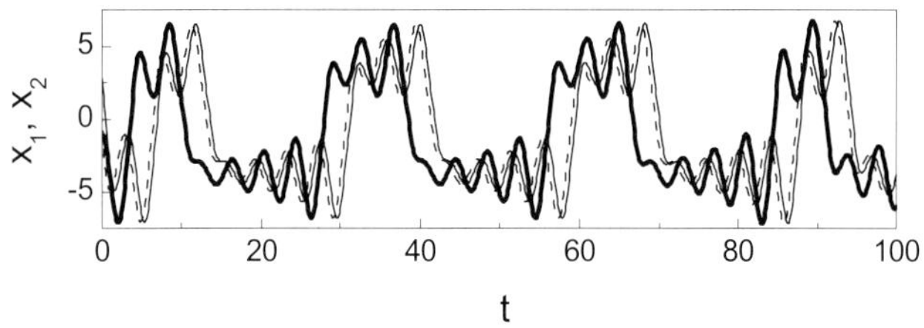

Fig. 7. Dynamics of the master (thin line) and the slave systems coupled by PLCC (bold line) and diagonal coupling (dashed line) for the Chua circuit. The parameters for PLCC are $k = 0.18$, $\omega = 1.5698$, $\tau = 3.3$ and for diagonal coupling $\mathbf{K} = 0.3$ diag[1,1,1] and $\tau = 0.5$. The parameters are chosen in such a way as to attain the maximum stable anticipation time.

other scroll of the strange attractor. In Fig. 7 we see that the maximum anticipation time obtained with the PLCC is 6.6 times larger as compared to the diagonal coupling.

The linear stability of the anticipating synchronization is determined by variational equation

$$\delta\dot{\mathbf{r}}_2 = \left.\frac{\partial \mathbf{f}}{\partial \mathbf{r}_2}\right|_{\mathbf{r}_2=\mathbf{r}_1(t+\tau)} \delta\mathbf{r}_2 - \mathbf{K}\delta\mathbf{r}_2(t-\tau), \quad (18)$$

where $\delta\mathbf{r}_2 = \mathbf{r}_2(t) - \mathbf{r}_1(t+\tau)$ is a transversal deviation from the synchronization manifold. The growth rates of this deviation define the transversal Lyapunov exponents (TLEs). A necessary condition for the synchronized regime to be stable is that the maximum TLE $\lambda_\perp$ is negative. Numerical analysis of the maximum TLE for the coupled Rössler systems shows that the interval of delay τ, where synchronization manifold is stable is considerably larger for the PLCC than that for the diagonal coupling [39]. These numerical results can be explained by a simple analytical model. If we suppose that the main contribution to the maximum TLE comes from the system's motion along the unstable spiral, then it is reasonable to consider anticipating synchronization for the mere spirals. Specifically, assume that dynamical system under consideration is described by two linear equations $\dot{x} = \gamma x - \omega y$, $\dot{y} = \omega x + \gamma y$, which define an unstable spiral with the positive increment γ and frequency ω. For the complex variable $Z = x + iy$, this system can be presented by a single equation $\dot{Z} = (\gamma + i\omega)Z$. Then equations for anticipating synchronization of two spirals take the form

$$\dot{Z}_1 = (\gamma + i\omega)Z_1 \quad (19a)$$

$$\dot{Z}_2 = (\gamma + i\omega)Z_2 + ke^{i\alpha}[Z_1 - Z_2(t-\tau)], \quad (19b)$$

where $ke^{i\alpha}$ is a complex coupling coefficient. By a suitable choice of the phase α we can model both the PLCC ($\alpha = \omega\tau$) and diagonal coupling ($\alpha = 0$). The solution of the master system is $Z_1 = Z_0 e^{(\gamma+i\omega)t}$, and the deviation $\Delta Z_2 = Z_2(t) - Z_0 e^{(\gamma+i\omega)(t+\tau)}$ from the anticipated state satisfies $\Delta\dot{Z}_2 = (\gamma + i\omega)\Delta Z_2 - ke^{i\alpha}\Delta Z_2(t-\tau)$. Substituting $\Delta Z_2 = Ce^{(\lambda+i\omega)t}$ we obtain the characteristic equation

$$\lambda = \gamma - ke^{i(\alpha-\omega\tau)}e^{-\lambda\tau}, \quad (20)$$

that defines the eigenvalues λ of the synchronized state in a rotating frame with the frequency ω. For PLCC, $\alpha = \omega\tau$ and solution of Eq. (20) is $\lambda = \gamma + W(-\tau k e^{-\gamma\tau})/\tau$, where W(z) is the Lambert W function, satisfying by definition $W(z)e^{W(z)} = z$ [10]. The necessary condition for stability of the synchronized state is: $\tau < 1/\gamma$. As expected, the limit for prediction horizon is determined by the inverse of the largest Lyapunov exponent, which for the spiral is $1/\gamma$. In Fig. 8, we compare stability regions in the (τ, k) plane for the PLCC ($\alpha = \omega\tau$) and diagonal coupling ($\alpha = 0$) for the values of the parameters $\omega = 0.997$ and $\gamma = 0.074$ the same as the eigenvalues of the fixed point of the Rössler system. We see that the stability region of the PLCC is considerably larger than that of the diagonal coupling. As shown in [39], these regions are in

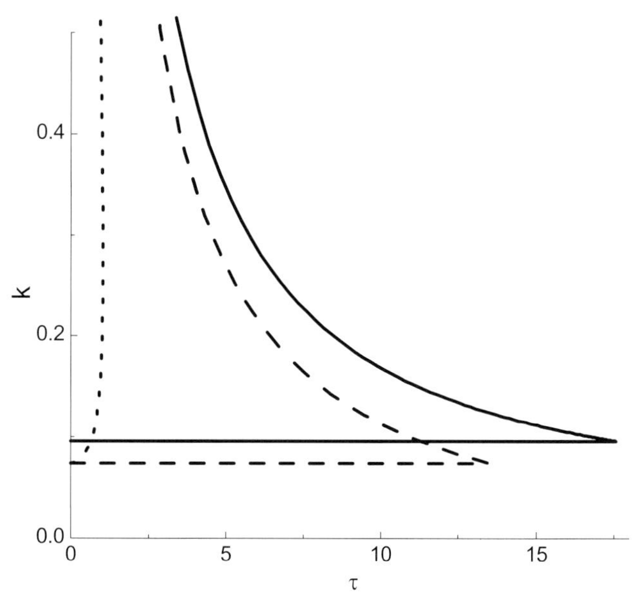

Fig. 8. $\tau - k$ stability diagrams for coupled spirals (19) and (21) with the parameters $\omega = 0.997$ and $\gamma = 0.074$ the same as the eigenvalues of the fixed point of the Rössler system. The dotted line bounds the region of stability for the diagonal coupling ($\alpha = 0$). The boundaries of stability for PLCC are depicted by dashed lines. The solid lines confine the region of stability of the synchronization manifold for the EPLCC at $R = -0.3$. The lower bound is determined by $k_1 = (1 - R)\gamma$. The upper bound is defined by Eq. (27).

approximate quantitative agreement with the corresponding regions of the Rössler system. Thus, the characteristic parameters of anticipating synchronization can be simply estimated from the linear Eqs. (19) that model the local dynamics of the chaotic system close to the fixed point.

3.3. *Extended phase-lag compensating coupling*

To further improve the PLCC algorithm, we employ an idea from the DFC theory. In the original DFC algorithm (cf. [32]), the coupling law is analogous to that of the anticipating synchronization. To improve the DFC, [12] proposed an extended version of the original DFC, which incorporates in the feedback loop an information from many previous states of the system. This modification, known as an extended DFC (EDFC), makes possible the stabilization of high-periodic orbits with large positive Lyapunov exponents (cf. [33]).

We use the idea of the EDFC to improve the performance of the PLCC algorithm [42]. We refer to this modification as an extended PLCC (EPLCC). The PLCC uses the single difference between the current state of the master system and the time-delayed state of the response system, whereas the EPLCC extends this perturbation to an infinite series of corresponding differences that are time-delayed by integer multiples of the delay τ. Again, we demonstrate the idea for two coupled unstable spirals

$$\dot{Z}_1 = (\gamma + \mathrm{i}\omega)Z_1, \tag{21a}$$

$$\dot{Z}_2 = (\gamma + \mathrm{i}\omega)Z_2 + kS(t). \tag{21b}$$

Now the EPLCC perturbation $S(t)$ is governed by

$$\begin{aligned} S(t) &= \mathrm{e}^{\mathrm{i}\alpha}[Z_1(t) - Z_2(t-\tau)] \\ &\quad + \sum_{n=1}^{\infty} R^n \mathrm{e}^{\mathrm{i}(n+1)\alpha}[Z_1(t-n\tau) \\ &\quad - Z_2(t-(n+1)\tau)], \end{aligned} \tag{22}$$

where R is a memory parameter that regulates the weight of the time-delayed information. We require $-1<R<1$ to guarantee the convergence of the infinite sum. The case of $R = 0$ corresponds to the PLCC considered in the previous section (cf. Eq. (19b)). Note that $S(t)$ vanishes for any value of R when the system moves along the synchronization manifold, since the equality $Z_1(t-n\tau) = Z_2(t-(n+1)\tau)$ holds on the manifold for all n. The phase lag of each term in the sum is compensated by the multiplier $\mathrm{e}^{\mathrm{i}(n+1)\alpha}$, and the amplitude is weighted by the multiplier R^n, such that the contribution of terms exponentially decreases with the increase of time delay. At the first look, Eq. (22) seems to be very complex. Fortunately, the infinite sum can be rewritten in an equivalent, recursive form

$$S(t) = \mathrm{e}^{\mathrm{i}\alpha}[Z_1(t) - Z_2(t-\tau) + RS(t-\tau)] \tag{23}$$

and implemented experimentally with the single delay line (cf. [12]). Now, the stability of the synchronized state is defined by variational equations

$$\dot{\Delta}(t) = (\gamma + \mathrm{i}\omega)\Delta(t) - kS(t-\tau), \tag{24a}$$

$$S(t) = \mathrm{e}^{\mathrm{i}\alpha}\left[\Delta(t) + RS(t-\tau)\right], \tag{24b}$$

where $\Delta(t) = Z_1(t) - Z_2(t-\tau)$ describes the deviation from the synchronization manifold. This leads to the characteristic equation

$$(\lambda - \gamma)(\mathrm{e}^{-\mathrm{i}(\alpha-\omega\tau)+\lambda\tau} - R) + k = 0. \tag{25}$$

For $R = 0$, it coincides with Eq. (20) analyzed in the previous section. Here, as well as in the previous section, we set $\alpha = \omega\tau$ and analyze the simplified version of Eq. (25)

$$(\lambda - \gamma)(\mathrm{e}^{\lambda\tau} - R) + k = 0. \tag{26}$$

Unfortunately, the solution of Eq. (26) for $R \neq 0$ cannot be expressed through the Lambert function. Although the analytical solution of this equation is not available, the boundaries of stability can be obtained analytically. The lower bound of stability $k_1 = \gamma(1-R)$ is deduced from Eq. (26) by substituting $\lambda = 0$. The upper bound of stability k_2 is defined by a Hopf bifurcation. Substituting $\lambda = \mathrm{i}\Omega$ in Eq. (26) and denoting $\Omega\tau = \beta$, we obtain the dependence k_2 on τ in a parametric form

$$\begin{aligned} \tau(\beta) &= \frac{\beta(\cos\beta - R)}{\gamma\sin\beta}, \\ k_2(\beta) &= \gamma(\cos\beta - R) + \frac{\beta\sin\beta}{\tau(\beta)}. \end{aligned} \tag{27}$$

These boundaries are depicted in Fig. 8 for $R = -0.3$. We see that the EPLCC extends the stability region of the synchronization manifold in the (τ, k)-plane. The maximum prediction time can be obtained from Eqs. (27) in the limit $\beta \to 0$. The necessary condition for stability of the synchronized state is

$$\tau < \tau_H = (1-R)/\gamma. \tag{28}$$

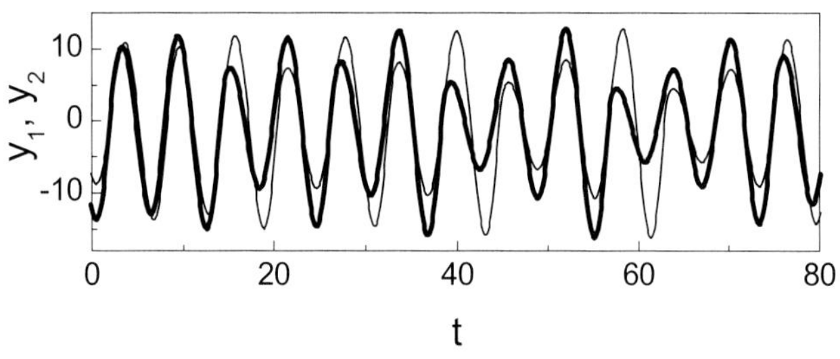

Fig. 9. Time series $y_1(t)$ of the master (thin line) and $y_2(t)$ of the slave (bold line) Rössler systems for the EPLCC. The parameters of the EPLCC are $\mathbf{K} = 0.18\ \mathbf{EQE^{-1}}$, $\omega = 0.997$, $R = -0.46$ and $\tau = 6.3$.

In comparing with the PLCC, we see that the EPLCC extends the prediction horizon by a factor $(1-R)$. This factor is greater that 1 only for negative values of the parameter R. When $R \to -1$, this factor reaches a maximum value equal to 2. Thus, for coupled spirals, the maximum prediction horizon of the EPLCC exceeds two times the maximum prediction horizon of the PLCC. Note that, in the case of negative values of R, the feedback law (22) describes a series with alternating signs. Only such a series advances the stability of anticipating synchronization.

Now we consider an application of the EPLCC algorithm for chaotic systems. We extend Eqs. (10) for the EPLCC as follows:

$$\dot{\mathbf{r}}_1 = \mathbf{f}(\mathbf{r}_1) \tag{29}$$

$$\dot{\mathbf{r}}_2 = \mathbf{f}(\mathbf{r}_2) + k\mathbf{S}(t), \tag{30}$$

where the perturbation $\mathbf{S}(t)$ is governed by the recursive equation

$$\mathbf{S}(t) = \mathbf{K}[\mathbf{r}_1(t) - \mathbf{r}_2(t-\tau) + R\mathbf{S}(t-\tau)] \tag{31}$$

and the coupling matrix $\mathbf{K}$ is defined by Eq. (16). In Fig. 9 we show the application of the EPLCC for the Rössler system. We see that the EPLCC allows us to prolong the prediction time to the value $\tau = 6.3$, which is equal to the characteristic period of chaotic oscillations of the system.

4. Conclusion

Summing up, we have presented a review of the delayed feedback control (DFC) methods, including experimental implementations, verifications for theoretical models and various modifications. We have also discussed the analytical theories developed for this method, which are connected with the bifurcations of target orbits. We stress that the research on the global properties of the DFC algorithm as well as adaptive modifications are still weakly developed and require further attention of investigators. In Secs. 2.2 and 2.3, we have presented our recent achievements in this field.

Another way of application of a time-delay feedback is the forecasting chaotic dynamics of a master system via anticipating synchronization with a slave system. In Sec. 3, we have presented two algorithms based on a phase-lag compensating coupling (PLCC) and extended PLCC that allowed us to achieve a long term anticipation with the prediction time equal approximately to the characteristic period of chaotic oscillations. We hope that our findings in this field will stimulate the search for appropriate coupling laws in other problems of anticipating synchronization, e.g., to enhance the predictability of chaotic systems with unknown dynamical models [8] or excitable systems driven by random forcing [7, 9]

Although the DFC and anticipating synchronization algorithms are similar in the sense that both involve a time-delay feedback, however, the DFC is much more effective. In its usual form, it provides a stabilization of UPOs whose periods are comparable with the characteristic period of chaotic oscillations. In contrast, the usual anticipating synchronization algorithm is able to stabilize the anticipatory manifold only for small time-delays. To prolong this time we have designed a special phase-lag compensating coupling. The difference in efficacy of these two algorithms is due to different role of the delayed feedback in these algorithms. The DFC law (2) involves both the proportional feedback term $y(t)$ and the delay feedback term $y(t-\tau)$. The stabilization of UPOs is mainly caused by the proportional feedback. The delay term here is necessary to set to zero the control force on the desired UPO. In contrast, the feedback law $\mathbf{K}[\mathbf{r}_1 - \mathbf{r}_2(t-\tau)]$ in the anticipating synchronization algorithm (10) does not contain the proportional feedback term. It involves only the delayed feedback term $\mathbf{r}_2(t-\tau)$ and only this term is responsible for the stabilization of the anticipatory manifold.

References

[1] Baba, N., Amann, A., Schöll, E. and Just, W. (2002). Giant improvement of time-delayed feedback control by spatio-temporal filtering, *Phys. Rev. Lett.* **89**, p. 074101.

[2] Bleich, M. E., Hochheiser, D., Moloney, J. V. and Socolar, J. E. (1997). Controlling extended systems

with spatially filtered, time-delayed feedback, *Phys. Rev. E* **55**, pp. 2119–2126.

[3] Bleich, M. E. and Socolar, J. E. S. (1996). Stability of periodic orbits controlled by time-delay feedback, *Phys. Lett. A* **210**, pp. 87–94.

[4] Boccaletti, S. and Arecchi, F. T. (1995). Adaptive control of chaos, *Europhys. Lett.* **31**, pp. 127–132.

[5] Chen, G. and Yu, X. (1999). On time-delayed feedback control of chaotic systems, *IEEE Trans. Circ. Syst. I* **46**, pp. 767–772.

[6] Chua, L. O., Komuro, M. and Matsumoto, T. (1986). The double scroll bifurcations, *IEEE Transactions on Circuits and Systems* **33**, pp. 1073–1118.

[7] Ciszak, M., Calvo, O., Masoller, C., Mirasso, C. R. and Toral, R. (2003). Anticipating the response of excitable systems driven by random forcing, *Phys. Rev. Lett.* **90**, p. 204102.

[8] Ciszak, M., Gutiérrez, J. M., Cofino, A. S., Mirasso, C., Toral, R. and Ortín, L. P. S. (2005). Approach to predictability via anticipated synchronization, *Phys. Rev. E* **72**, p. 046218.

[9] Ciszak, M., Marino, F., Toral, R. and Balle, S. (2004). Dynamical mechanism of anticipating synchronization in excitable systems, *Phys. Rev. Lett.* **93**, p. 114102.

[10] Corless, R. M., Gonnet, G. H., Hare, D. E. G., Jeffrey, D. J. and Knuth, D. E. (1996). On the lambert w function, *Advances in Computational Mathematics* **5**, pp. 329–359.

[11] Fiedler, B., Flunkert, V., Georgi, M., Hoevel, P. and Schöll, E. (2007). Refuting the odd-number limitation of time-delayed feedback control, *Phys. Rev. Lett.* **98**, p. 114101.

[12] Gauthier, D. J., Sukow, D. W., Concannon, H. M. and Socolar, J. E. S. (1994). Stabilzing unstable periodic orbits in a fast diode resonator using continuous time-delay autosynchronization, *Phys. Rev. E* **50**, pp. 2343–2346.

[13] Gjurchinovski, A. and Urumov, V. (2008). Stabilization of unstable steady states by variable-delay feedback control, *Europhys. Lett.* **84**, p. 40013.

[14] Herrmann, G. (2001). A robust delay adaptation scheme for Pyragas' chaos control method, *Phys. Lett. A* **287**, pp. 245–256.

[15] Hochheiser, D., Moloney, J. V. and Lega, J. (1997). Controlling optical turbulence, *Phys. Rev. A* **55**, pp. R4011–R4014.

[16] Höhne, K., Shirahama, H., Choe, C. U., Benner, H., Pyragas, K. and Just, W. (2007). Global properties in an experimental realization of time-delayed feedback control with an unstable control loop, *Phys. Rev. Lett.* **98**, p. 214102.

[17] Just, W., Benner, H. and von Loewenich, C. (2004). On global properties of time-delayed feedback control: weakly nonlinear analysis, *Physica D* **199**, pp. 33–44.

[18] Just, W., Bernard, T., Ostheimer, M., Reibold, E. and Benner, H. (1997). Mechanism of time-delayed feedback control, *Phys. Rev. Lett.* **78**, pp. 203–206.

[19] Just, W., Reckwerth, D., Mockel, J., Reibold, E. and Benner, H. (1998). Delayed feedback control of periodic orbits in autonomous systems, *Phys. Rev. Lett.* **81**, pp. 562–565.

[20] Kittel, A., Parisi, J. and Pyragas, K. (1995). Delayed feedback control of chaos by self-adapted delay time, *Phys. Lett. A* **198**, pp. 433–438.

[21] Lehnert, J., Hvel, P., Flunkert, V., Guzenko, P. Y., Fradkov, A. L. and Schöll, E. (2011). Adaptive tuning of feedback gain in time-delayed feedback control, *Chaos* **21**, p. 043111.

[22] Li, L., Liu, Z., Zhang, D. and Zhang, H. (2006). Controlling chaotic robots with kinematical redundancy, *Chaos* **16**, p. 013132.

[23] Liu, C. L. and Tian, Y. P. (2008). Eliminating oscillations in the internet by time-delayed feedback control, *Chaos, Solitons & Fractals* **35**, pp. 878–887.

[24] Mausbach, T., Klinger, T., Piel, A., Atipo, A., Pierre, T. and Bonhomme, G. (1997). Continuous control of ionization wave chaos by spatially derived feedback signals, *Phys. Lett. A* **228**, pp. 373–377.

[25] Mendoza, C., Boccaletti, S. and Politi, A. (2004). Convective instabilities of synchronization manifolds in spatially extended systems, *Phys. Rev. E* **69**, p. 047202.

[26] Michiels, W., Assche, V. V. and Niculescu, S. I. (2005). Stabilization of time-delay systems with a controlled time-varying delay and applications, *IEEE Trans. Autom. Control* **50**, pp. 493–504.

[27] Nakajima, H. (1997). On analytical properties of delayed feedback control of chaos, *Phys. Lett. A* **232**, pp. 207–210.

[28] Nakajima, H., Ito, H. and Ueda, Y. (1997). Automatic adjustment of delay time and feedback gain in delayed feedback control of chaos, *IEICE Transactions on Fundamentals of Electronics Communications and Computer Sciencies* **E80A**, pp. 1554–1559.

[29] Nakajima, H. and Ueda, Y. (1998). Limitation of generalized delayed feedback control, *Physica D* **111**, pp. 143–150.

[30] Ott, E., Grebogi, C. and Yorke, J. A. (1990). Controlling chaos, *Phys. Rev. Lett.* **64**, pp. 1196–1199.

[31] Pikovsky, A., Rosemblum, M. and Kurths, J. (2001). *Synchronization: A universal concept in nonlinear sciences* (Cambridge University Press, Cambridge, England).

[32] Pyragas, K. (1992). Continuous control of chaos by self-controlling feedback, *Phys. Lett. A* **170**, pp. 421–428.

[33] Pyragas, K. (1995). Control of chaos via extended delay feedback, *Phys. Lett. A* **206**, pp. 323–330.

[34] Pyragas, K. (2001). Control of chaos via an unstable delayed feedback controller, *Phys. Rev. Lett.* **86**, pp. 2265–2268.
[35] Pyragas, K. (2002). Analytical properties and optimization of time-delayed feedback control, *Phys. Rev. E* **66**, p. 026207.
[36] Pyragas, K. (2006). Delayed feedback control of chaos, *Philos. Trans. R. Soc. A* **364**, pp. 2309–2334.
[37] Pyragas, K. and Pyragas, V. (2006). Delayed feedback control of the Lorenz system: An analytical treatment at a subcritical hopf bifurcation, *Phys. Rev. E* **73**, p. 036215.
[38] Pyragas, K., Pyragas, V. and Benner, H. (2004). Delayed feedback control of dynamical systems at a subcritical hopf bifurcation, *Phys. Rev. E* **70**, p. 056222.
[39] Pyragas, K. and Pyragienė, T. (2008). Coupling design for a long-term anticipating synchronization of chaos, *Phys. Rev. E* **78**, p. 046217.
[40] Pyragas, K. and Pyragienė, T. (2010). Extending anticipation horizon of chaos synchronization schemes with time-delay coupling, *Phil. Trans. R. Soc. A* **368**, pp. 305–317.
[41] Pyragas, V. and Pyragas, K. (2009). Using ergodicity of chaotic systems for improving the global properties of the delayed feedback control method, *Phys. Rev. E* **80**, p. 067201.
[42] Pyragas, V. and Pyragas, K. (2010). Modification of delayed feedback control using ergodicity of chaotic systems, *Lith. J. Phys.* **50**, pp. 305–316.
[43] Pyragas, V. and Pyragas, K. (2011). Adaptive modification of the delayed feedback control algorithm with a continuously varying time delay, *Phys. Lett. A* **375**, pp. 3866–3871.
[44] Pyragienė, T. and Pyragas, K. (2005). Delayed feedback control of forced self-sustained oscillations, *Phys. Rev. E* **72**, p. 026203.
[45] Rössler, O. E. (1976). An equation for continuous chaos, *Phys. Lett. A* **57**, pp. 397–398.
[46] Schoell, E. and Schuster, H. G. (eds.) (2008). *Handbook of Chaos Control* (Wiley-VCH, Weinheim).
[47] Sieber, J., Gonzalez-Buelga, A., Neild, S. A., Wagg, D. J. and Krauskopf, B. (2008). Experimental continuation of periodic orbits through a fold, *Phys. Rev. Lett.* **100**, p. 244101.
[48] Sieber, J. and Krauskopf, B. (2008). Control based bifurcation analysis for experiments, *Nonlinear Dyn.* **51**, pp. 365–377.
[49] Sorrentino, F. and Ott, E. (2008). Adaptive synchronization of dynamics on evolving complex networks, *Phys. Rev. Lett.* **100**, p. 114101.
[50] Tamaševičius, A., Mikolaitis, G., Pyragas, V. and Pyragas, K. (2007). Delayed feedback control of periodic orbits without torsion in nonautonomous chaotic systems: Theory and experiment, *Phys. Rev. E* **76**, p. 026203.
[51] Tang, S. and Liu, J. M. (2003). Experimental verification of anticipated and retarded synchronization in chaotic semiconductor lasers, *Phys. Rev. Lett.* **90**, p. 194101.
[52] Ushio, T. (1996). Limitation of delayed feedback control in nonlinear discrete-time systems, *IEEE Trans. Circuits Syst. I* **43**, pp. 815–816.
[53] Voss, H. U. (2000). Anticipating chaotic synchronization, *Phys. Rev. E* **61**, pp. 5115–5119.
[54] Voss, H. U. (2001). Dynamic long-term anticipation of chaotic states, *Phys. Rev. Lett.* **87**, p. 014102.
[55] Voss, H. U. (2002). Real-time anticipation of chaotic states of an electronic circuit, *Int. J. Bifurc. Chaos Appl. Sci. Eng.* **12**, pp. 1619–1625.
[56] Yamasue, K. and Hikihara, T. (2004). Domain of attraction for stabilized orbits in time delayed feedback controlled duffing systems, *Phys. Rev. E* **69**, p. 056209.
[57] Yamasue, K. and Hikihara, T. (2006). Persistence of chaos in a time-delayed-feedback controlled duffing system, *Phys. Rev. E* **73**, p. 036209.
[58] Yamasue, K., Kobayashib, K., Yamada, H., Matsushige, K. and Hikihara, T. (2009). Controlling chaos in dynamic-mode atomic force microscope, *Phys. Lett. A* **373**, pp. 3140–3144.

CHAPTER 29

MODELS OF CATASTROPHIC EVENTS AND SUGGESTIONS TO FORETELL THEM

YVES POMEAU

Department of Mathematics, University of Arizona, Tucson, USA

MARTINE LE BERRE

Institut des Sciences Moléculaires d'Orsay ISMO-CNRS, Univ. Paris-Sud, Bat. 210, 91405 Orsay, France

Catastrophes of all kinds happen all the time in the real world. They can be roughly defined as short duration-large amplitude events, following and followed by long periods of "ripening" of the system under consideration. They are obviously related to a class of phenomena called sometime "relaxation" oscillations. This makes them *a priori* a good subject of theoretical investigation because of the widely separated time scales. This separation of scales requires the existence of a small dimensionless parameter in the formulation of the problem, giving hope of a "general" theory based upon this smallness. We propose first a generic saddle-node bifurcation model to describe fast transitions of finite amplitude resulting from the intrinsic dynamics of the system. In this model with an added external source of noise, the correlation time of the fluctuations increases before the transition and their spectrum drifts towards lower frequencies, following the classical pattern of slowing down at bifurcations points. Having in mind applications to earthquakes, we examine next, from the same point of view, a mathematical model of stick-slip sliding between two solids, as introduced by Dieterich and Ruina. This class of model has a dimensionless parameter which is very large for geophysical applications. An asymptotic analysis of the solutions shows a sharp difference between the two steps of the stick-slip dynamics: the stick (slow) part takes place along a slow manifold in phase space, but, in contrast with what happens at the slow crossing of a saddle-node bifurcation studied first, the transition between slow and fast dynamics occurs when the slow dynamics has a finite time singularity breaking the assumption of slowness of the adiabatic theory. With an external source of noise, the correlation time of the fluctuations *decreases* long before the fast transition, the opposite of what happens in the saddle-node model where this correlation time *increases*. In both cases, the fluctuations change long before the transition and this could be a way of detecting catastrophic events before they happen.

It is a great pleasure and honour for us to participate to this volume in honor of Professor Leon Chua, whose contributions to Nonlinear Science are so outstanding. We dedicate to him this piece on a domain of Nonlinear Science and bifurcation theory, namely the analysis of dynamical systems behaving with very different time scales. As we explain below, we believe it is one area where Nonlinear Science could help find new ways to improve our understanding of Nature and to forecast catastrophic events.

1. Introduction

Earthquakes, like volcanic eruptions as well as other physical phenomena and perhaps also some kinds of socio-economical "revolutions", show an abrupt transition from one state to another. We develop a first approach to this question by introducing an abstract model. In this model, one assumes that an earthquake results locally from the evolution of a time-dependent potential which distorts as time evolves. The catastrophic motion of geophysical plates is imaged as the rocking of a ball in a double-well potential when the two minima exchanges their relative stability properties. From the point of view of dynamical systems this model shows a "dynamical saddle-node" bifurcation. At the bifurcation a pair of fixed points, one locally stable, the other locally unstable, merge and vanish as a control parameter changes. This transition is intrinsic (not the result of an excitation from outside) and dynamical in the sense that, as the parameter changes slowly, the system quickly jumps within a time much shorter than the typical time of evolution of the external parameter. In earthquakes, this typical time of evolution is on geological scales of plate tectonics (typically hundred years for major earthquakes), although the duration of earthquakes is within the second to minute range. This saddle-node model is covered in Sec. 2 below.

Afterwards we examine another class of models, taken from the solid mechanics and geophysics literature which aimed at describing the relative slide of two plates subject to solid friction at their interface, one of them being pulled by an external spring driven at constant velocity.

In this second model, looked at in Sec. 3, and for realistic parameters values, there are relaxation oscillations of the stick-slip kind with widely different time scales, as in the case of the first model.

However the responses to an external noise source are completely different for the two models. The first one displays the classical slowing-down of the fluctuations near the transition, with a progressive increase of the correlation time starting noticeably before the transition itself. In the stick-slip model on the contrary, the response to the noise becomes more and more short-memory as the transition approaches. While they present different character, the response to noise changes long before the transition in both models and so could be used to detect catastrophes.

2. Dynamical Double-well Model: Saddle-node Bifurcation

In this section, we consider a saddle-node bifurcation in a "gradient flow", that is a damped dynamical system such that a coordinate $x(t)$ is solution of the equation of motion of the type

$$\frac{\mathrm{d}x}{\mathrm{d}t} = -\frac{\partial V}{\partial x}. \tag{1}$$

In this equation $V(x)$ is a potential, and the dynamics tends to lower the value of $V(x)$. In the geophysical context, the scalar variable x could be the displacement of one plate with respect to the one its slides over. In the following we ultimately consider the case of a time dependent double well potential $V(x,t)$, distorting with time, of the form

$$V_q(x,t) = -\frac{x^3}{3} - atx + \frac{x^4}{4}, \tag{2}$$

illustrated in Fig. 1(b), where the three curves display the evolution of the potential as time evolves (see caption). In Eq. (2) the constant parameter a gives the distortion rate, small value of a corresponding to a slow distortion of the potential. The solution of the dynamical model (1)–(2) illustrated

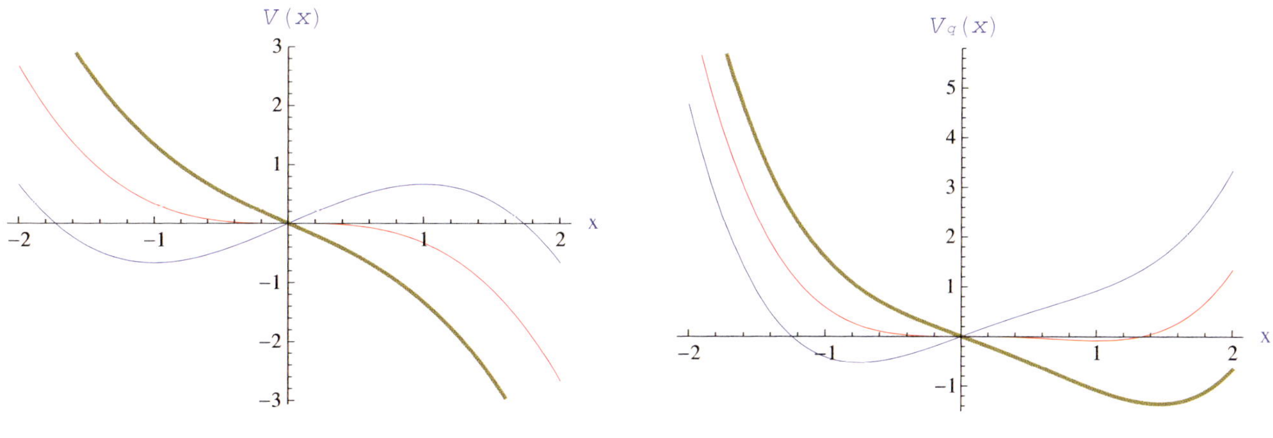

Fig. 1. (a) Cubic potential, (b) Quartic potential. The blue, red, and yellow curves correspond to the increasing time values $at = -1, 0, 1$ respectively.

in Fig. (3) displays a jump from negative to positive values occurring at a time value t_c named the transition time.

Note that the present approach differs from the catastrophe theory of Thom and Arnol'd [1] which studies how steady equilibria of equations like (1) change when some constant parameter of the potential $V(.)$ is modified. Below we consider a different kind of question, namely what happens to the solution of equation (1) when the potential becomes itself a slowly varying function of time, like the potential (2) with the time dependent parameter at. We show in particular that the pair of equilibrium points disappears by a saddle-node bifurcation at $t = 0$. While the question of rocking of the solution between two stationary steady state has been already studied [2] with various applications in mind, this question of the slow sweeping across bifurcations (by "sweeping" we mean crossing of a transition point with a time dependent parameter in the equation(s) of motion) has not been so much developed to the best of our knowledge. We emphasize the occurrence of an intermediate time scale in the case of a slow sweeping that has not been noticed before, although we believe it to be crucial for a strategy of foretelling catastrophes in the real world.

2.1. *Cubic potential just before the transition*

Before studying this quartic potential model, let us first investigate what happens in a close vicinity of the transition, for small values of x, where the potential reduces to a cubic expansion with respect to x. Assume first that $V(.)$ does not depend explicitely on time and takes the form

$$V(x) = -\left(\frac{1}{3}x^3 + bx\right), \tag{3}$$

with b real constant (for the moment).

For b negative $V(x)$ has two real extrema (*i.e.* the roots of $\frac{\partial V}{\partial x} = 0$), one $-\sqrt{-b}$ is a stable equilibrium, the other, $\sqrt{-b}$, is an unstable equilibrium. For $b = 0$ the two equilibria merge and disappear for b positive. This is the saddle-node bifurcation. The shape of $V(x)$ near $x = 0$ and for b small is universal: for a given smooth $V(x)$ showing this saddle-node bifurcation, one can always rescale time and external parameter to find the "local" problem in this form.

The extension to a time dependent control parameter b goes as follows. If b is a smooth function of time, one can assume that $b(t)$ crosses the critical value, *i.e.* zero in the present case, at time zero in such a way that $b(t) = at + \cdots$ with a non zero constant, and the dots being for higher terms in the Taylor expansion of $b(t)$. The time dependent cubic potential becomes

$$V_c(x, t) = -\left(\frac{1}{3}x^3 + atx\right), \tag{4}$$

evolving with time as illustrated in figure(1-a). With the rescaling $x \mapsto xa^{-1/3}$ and $t \mapsto ta^{1/3}$, the dynamical equation (1) corresponding to the cubic potential writes

$$\dot{x} = -(t + x^2), \tag{5}$$

which is the "universal" parameterless dynamical equation for a saddle-node bifurcation.

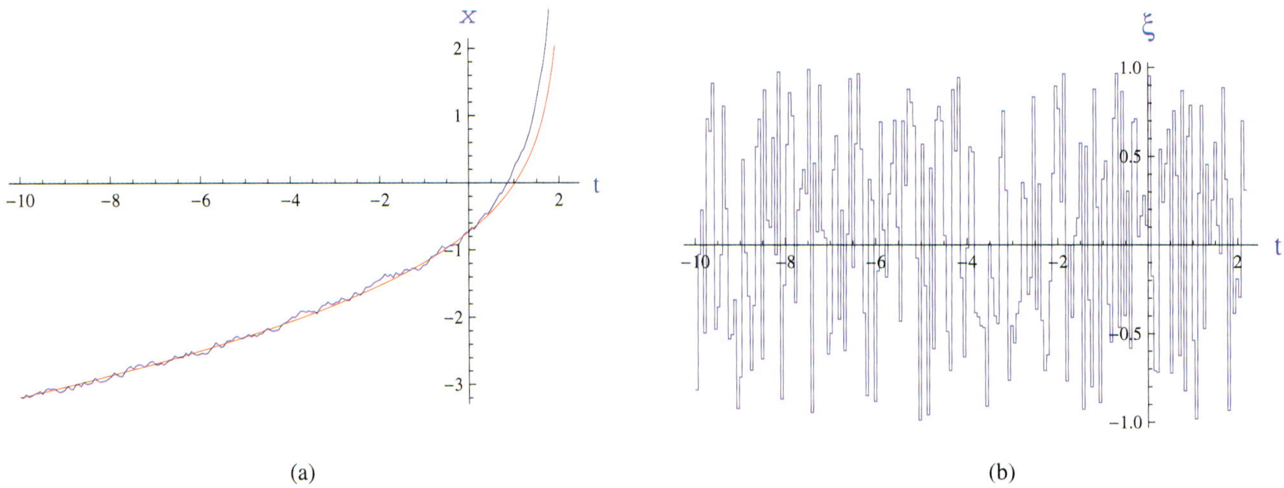

(a) (b)

Fig. 2. (a) Solutions of Eq. (12), with and without noise. $\epsilon_n = 0$ (smooth curve) and $\epsilon_n = 1$ (noisy curve). With small amplitude noise the two curves cannot be distinguished. (b) A single sample of short memory noise $f(t)$.

2.1.1. *Van der Pol oscillation relaxations*

We point out that this generic equation describes for example the transition from slow to fast motion in the van der Pol oscillator, as shown by Dorodnitsyn [3] and recalled here briefly. The van der Pol dynamical system

$$\ddot{x} + \eta(x^2 - 1)\dot{x} + x = 0, \tag{6}$$

is a prototype for relaxation oscillations (RO) with slow drift and fast jumps in saddle folding systems, in the limit of large non linearities (large positive η values). Using the Liénard transform $y = -\dot{x}/\eta - \int_0^x (x^2 - 1)$, then setting $t \mapsto t/\eta$, equation (6) writes

$$\begin{cases} \dot{y} = x \\ \dot{x} = -\eta^2 \left[y + \left(\frac{x^3}{3} - x \right) \right]. \end{cases} \tag{7}$$

which shows that the dominant component of the velocity field is $\dot{x}$ in the limit η large. As well known the trajectory in the plane (x, y) follows in the slow phases the cubic $y = x - x^3/3$, but becomes unstable at $x = \pm 1$, where it jumps along a line at y almost constant toward $x = \pm 2$ on the cubic. Close to the transition point ($x = -1$, $y = -2/3$), the equations of motion for $\delta x = x + 1$ small compared to one and $\delta y = y + 2/3$ write

$$\begin{cases} \delta\dot{y} = -1 \\ \delta\dot{x} = -\eta^2(\delta y - \delta x^2). \end{cases} \tag{8}$$

which have the obvious solution $\delta y = -t + t_0$, t_0 constant, taken as zero, and $\delta\dot{x} = \eta^2(t + \delta x^2)$. The constant η can be eliminated by taking as new time scale $t \mapsto \eta^{4/3} t$ and new variable $x \mapsto \delta x \eta^{2/3}$, which gives exactly the equation (5) introduced above to describe the dynamics near a saddle node bifurcation with a drifting parameter.

2.1.2. *Solution of the "universal" model*

We look for a solution of equation (5) transiting from the "stable" fixed point at "large" negative times to the rolling down one towards positive value of x at positive times. This solution behaves like $x(t) \approx -\sqrt{-t}$ at large negative times. The equation (5) is of the Riccati type and can be integrated by introducing the function $y(t)$ such that $x(t) = -\frac{y'}{y}$ where $y' = \frac{\mathrm{d}y}{\mathrm{d}t}$ and $y(t)$ is a solution of Airy's equation,

$$y'' + ty = 0. \tag{9}$$

The solution relevant with the given condition for $x(t)$ at $t \to -\infty$, is Airy's function $Ai(-t)$ which writes

$$Y(t) = Ai(-t) = \int_0^{+\infty} \cos\left(\frac{u^3}{3} - ut\right) \mathrm{d}u, \tag{10}$$

and leads to the blue curve $x(t)$ drawn on Fig. 2(a). The transition ends up when t becomes equal to the first zero of Airy's function $Ai(-t)$, corresponding to a divergence of the original function $x(t)$. Let t_c be this critical value of t, *i.e.* the smallest root of $Y(t) = 0$, a pure number, about $t_c \approx 2.338$, corresponding to a divergence of the solution $x(t)$ of the cubic potential model. Just before the transition, the behavior of $x(t)$ deduced from the Laurent expansion of $Y(t)$ close to t_c, writes

$$x(t) \approx \frac{1}{t_c - t} - \frac{t_c}{3}(t_c - t) \ldots, \tag{11}$$

in agreement with the numerical solution.

Actually we have to point out that the expression (11) is not of great interest since it looses its validity some time before the transition time t_c. Actually the "universal" dynamical equation (5) was proposed for x close to zero, that reduces the validity range of the generic model to time values $(t_c - t) \succeq 1$. Outside this range the cubic potential model must include higher order terms to avoid the divergence of the solution in a close vicinity of t_c, that is done in the next section dealing with the quartic potential.

In its domain of validity, $(t_c - t)$ larger than unity, we may explore the response of our system to a small external noise, and search whether the response to the noise changes qualitatively with time, and so could be a signal ahead of the transition.

Let us consider Eq. (5) with a small noise added, so that Eq. (5) is replaced by

$$\frac{\mathrm{d}x}{\mathrm{d}t} = x^2 + t + \epsilon_n f(t), \tag{12}$$

where $f(t)$ is a random function of time, and ϵ_n a small factor.

In the limit ϵ_n small, one can solve Eq. (12) by expansion in powers of ϵ_n,

$$x(t) = x_0(t) + \epsilon_n x_1(t) + \cdots \tag{13}$$

where

$$x_0(t) = -\frac{Y'(t)}{Y(t)}, \tag{14}$$

The linear response to the noise is

$$x_1(t) = \frac{1}{Y^2(t)} \int_{t_0}^{t} \mathrm{d}\tilde{t}\; f(\tilde{t})\; Y^2(\tilde{t}). \tag{15}$$

Because $Y^2(\tilde{t})$ tends rapidly to zero as $(\tilde{t})$ tends to minus infinity, one can take $t_0 = -\infty$ to get rid of the effect of the initial conditions. From expression (15) we may infer that the fluctuations increase as time approaches t_c because $Y(t_c) = 0$, but recall that our study here is limited to $t_c - t \succeq 1$. Therefore let us investigate the response to noise far away from t_c.

Let us take a delta-correlated (or white) noise, such that

$$\langle f(t_a) f(t_b) \rangle = \delta(t_a - t_b). \tag{16}$$

The correlation function of $x_1(t)$ is given by

$$\langle x_1(t) x_1(t') \rangle = \frac{1}{Y^2(t) Y^2(t')} \int_{-\infty}^{\inf(t,t')} \mathrm{d}\tilde{t} Y^4(\tilde{t}), \tag{17}$$

whose behavior for large negative values of both t and t', is derived from the asymptotic expression of Airy's function, $Ai(-t) \approx \frac{e^{-\frac{2}{3}(-t)^{3/2}}}{2\sqrt{\pi}(-t)^{1/4}}$. Setting $w = \frac{\tilde{t}}{t}$, and $F(w) = 1 - w^{3/2}$, the variance $\sigma^2 = \langle (x(t) - x_0(t))^2 \rangle$ of the fluctuations writes

$$\langle x_1(t)^2 \rangle \approx (-t) \int_1^{\infty} \frac{\mathrm{d}w}{w} e^{\frac{8}{3}(-t)^{3/2} F(w)}. \tag{18}$$

In the limit $(-t) \to \infty$ the integral is concentrated near $w = 1$ so that

$$\langle x_1(t)^2 \rangle \approx \frac{1}{4}(-t)^{-\frac{1}{2}}, \tag{19}$$

which shows that the variance of the fluctuations increases long before t_c, as time goes on.

2.2. *Dynamical saddle-node bifurcation toward a new steady state*

The unbounded growth of $x(t)$ in the "universal" model is a consequence of the *local* cubic form of $V(x)$ as expanded near $x = 0$, which is valid around $x = 0$ only, in obvious contradiction with the fact that $x(t)$ diverges. To suppress the divergence of $x(t)$ at the transition we add a stabilizing (positive) term to the potential $V(x)$ which becomes quartic,

$$V_q(x) = -\frac{x^3}{3} - bx + \frac{x^4}{4}, \tag{20}$$

as drawn in Fig. 1(b). Because of the growth of $V_q(x)$ at infinity, like x^4, the solution of the differential equation

$$\frac{\mathrm{d}x}{\mathrm{d}t} = -\frac{\partial V_q}{\partial x} = b + x^2 - x^3, \tag{21}$$

does not diverge at finite time. Equation (20) can be written in the given scaled form, provided the coefficient of x^4 is positive. For such a potential one parameter only remains. In Eq. (20) the coefficient is chosen as b, the one of the linear term. For $b = 0$ the dynamical system (21) is exactly at the saddle-node bifurcation, because at $b = x = 0$ the first and second derivative of $V_q(x)$ both vanish, but not the third one. Contrary to the case of the pure cubic potential, this system has always, that is for any value of b, a stable fixed point beyond the pair of fixed points collapsing at the saddle-node bifurcation. This makes it a fair candidate for describing the dynamical saddle-node bifurcation without blow-up.

As in the previous case, we shall look now at the case of a time dependent b, that will be taken as $b = at$ with a positive constant. Because of the rescaling of the cubic and quartic term, the parameter a cannot be eliminated (another possibility is to put a parameter in front of the cubic term). For the potential $V_q(x,t)$ of Eq. (2) we shall analyze the solution of the dynamical equation

$$\frac{\mathrm{d}x}{\mathrm{d}t} = at + x^2 - x^3, \tag{22}$$

tending at large times to the quasi-equilibrium point $x = (at)^{1/3}$, t being considered as a parameter, see Fig. 3. Moreover we consider the limit a small, which could describe a wide range of slip phenomena [4], as earthquakes where a is generally very small, of order 10^{-9} (see below).

2.2.1. *Three time ranges for small a*

In this limit we show first that there are three characteristic time intervals, depending how x is close to zero.

The long time scale is the average recurrence time of an earthquake at the same site along the fault. It is typically of order $t_b^{\text{phys}} \sim 200$ years. In our model it is the time needed for the potential $V_q(x,t)$ to change significantly, to move from a pair of fixed points to a saddle-node bifurcation. Because time enters in $V_q(x,t)$ through the combination (at), the adimensional time needed for a change of shape

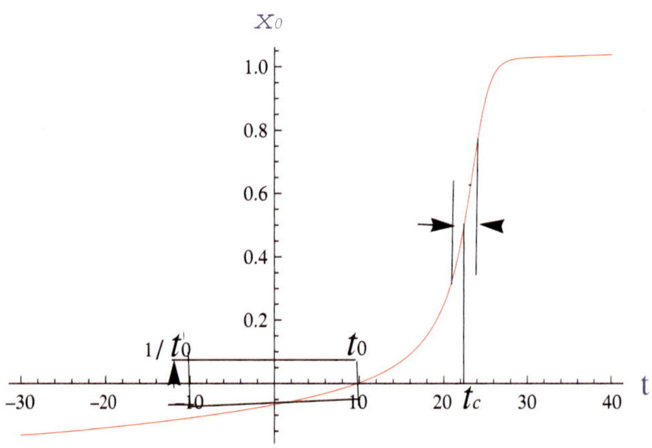

Fig. 3. Solution of Eq. (22) for $a = 10^{-3}$. The rectangle around the origin defines the region $-t_0 < t < t_0$ and $-1/t_0 < x < 1/t_0$, with $t_0 \sim a^{-1/3}$. The critical time is $t_c \sim 2.34\, t_0$. The two vertical lines inserted between the two arrows delimitate the large slope time duration, of order unity.

of V_q is of order

$$t_b \sim a^{-1}. \tag{23}$$

The short time t_{eqk} is of order unity in our model equation (22) as stated in the next paragraph. It is the duration of the abrupt change of the slope of the solution $x(t)$. This short time corresponds to the dynamic rupture duration, which is typically of order ten seconds for a magnitude 6 earthquake, $t_{\text{eqk}}^{\text{phys}} \sim 10s$ [5]. Therefore the ratio of these two time scales $t_{\text{eqk}}^{\text{phys}}/t_b^{\text{phys}} = t_{\text{eqk}}/t_b$, is as small as $a = 1.6\ 10^{-9}$ in the geophysical context.

There is another time scale, t_0, the time interval standing before the transition, and close to it, during which the potential is very flat, while the solution has not jumped. During this time, x and at are much smaller than unity, then the cubic term on the right-hand side of Eq. (22) is negligible. In this range one recovers the universal equation of the dynamical saddle-node bifurcation (5) by taking $X = xa^{-1/3}$ and $T = ta^{1/3}$, with the boundary condition $X(t) \approx -\sqrt{-T}$ at T tending to minus infinity. This property concerns the rectangular domain drawn in Fig. 2(c), where x is small, $x \sim a^{1/3}$, and t extends from $-a^{-1/3}$ to $t \sim a^{-1/3}$, located before the abrupt increase. Therefore the time extent of this domain introduces the intermediate time scale,

$$t_0 \sim a^{-1/3}, \tag{24}$$

long compared to unity (the adimensional short time scale t_{eqk} for the duration of a seismic rupture) and small compared to $t_b = a^{-1}$, the average recurrence time between earthquakes.

Let us prove that the short time is of order unity, by matching the solution $X(T)$ of the universal equation to the solution of Eq. (25) below, in the vicinity of the critical point $t_c(a) = a^{-1/3}$ t_c. Because $X(T)$ behaves like $(t_c - T)^{-1}$ before it diverges, it follows that the solution $x(t)$ behaves as $\approx (a^{-1/3}t_c - t)^{-1}$ for "large" values of $\delta t = t - a^{-1/3}t_c$ before the critical time. Using δt in this development as time variable, $x(\delta t)$ becomes of order one when δt becomes of order one too. When this happens, the term at in equation (22) is negligible, therefore the solution of this equation which can be matched with the solution near the bifurcation is the solution of the integrable equation

$$\frac{\mathrm{d}x(\delta t)}{\mathrm{d}(\delta t)} = x^2(\delta t) - x^3(\delta t), \tag{25}$$

with the asymptotic behavior for very large negative times $x(\delta t) \sim -1/\delta t$. This equation shows that the time scale for the earthquake rupture t_{eqk} is of order one, because it has no explicit dependence with respect to the small parameter a. This result is confirmed by the numerics: For a small we find that the rising time of $x(t)$ close to $t_c(a)$ (defined as the half-width of the slope $\frac{dx}{dt}$ solution of Eq. (22)) is $t_{\text{eqk}} \sim 2.5$, independent of a. From the observational point of view, the catastrophe takes place during this time t_{eqk} of order one, because the displacement is of order one then, compared to the displacement of order $a^{1/3}$ taking place during time $a^{-1/3}$ typical of the "universal" transition process. The two solutions match in the range $1 \ll (-\delta t) \ll a^{-1/3}$. Supposing that the physical fast time scale for earthquakes is $t_{\text{eqk}}^{\text{phys}} \sim 10s$, the intermediate time scale is $t_0^{\text{phys}} \sim a^{-1/3} t_{\text{eqk}}^{\text{phys}}$, which is a few hours for $a = 10^{-9}$.

2.2.2. *Precursor effects in the response to an external noise*

With a noise source added, the dynamical equation (1) becomes,

$$\frac{\mathrm{d}x}{\mathrm{d}t} = x^2 - x^3 + at + \epsilon_n f(t). \tag{26}$$

Actually the effective noise amplitude is not equal to ϵ_n close to the saddle-node, but depends on the value of the parameter a. Indeed for $|t| \leq t_0$, the cubic term in Eq. (27) is negligible, and the equation reduces to

$$\frac{\mathrm{d}x}{\mathrm{d}t} = x^2 + at + \epsilon_n f(t). \tag{27}$$

which may be written as $\mathrm{d}X/\mathrm{d}T = X^2 + T + \tilde{\epsilon}(a)f(t)$, by setting $X = xa^{-1/3}$, $T = ta^{1/3}$, and $\tilde{\epsilon}(a) = \epsilon_n a^{-2/3}$. Therefore the effective noise is larger than ϵ_n in the rectangular domain of Fig. 3.

Let us study the fluctuations of the solution $x(t)$ of Eq. (27). For a small noise input, the solution may be expanded in powers of ϵ_n as above. At first order it gives

$$\frac{\mathrm{d}x_1}{\mathrm{d}t} = \left[2x_0(t) - 3x_0^2(t)\right]x_1(t) + f(t), \tag{28}$$

whose solution is formally

$$x_1(t) = \int_{t_0}^{t} \mathrm{d}\tilde{t}\; f(\tilde{t}) \exp[g_0(t) - g_0(\tilde{t})], \tag{29}$$

where $g_0(t)$ is the time integral of the second derivative of the potential $-d^2V_q(x,t)/dx^2$, $g_0(t) = \int_{t_0}^{t} du[2x_0(u) - 3x_0^2(u)]$. The standard deviation $\sigma_{x_1}(t)$ has to be calculated numerically. We expect it to display the same behavior as for the cubic case in the whole domain where $|x(t)| \ll 1$, i.e. before the transition, and close to it, because the potential is cubic in this region. After the transition, we expect that the fluctuation decreases, because the solution without noise becomes quasi-steady. This is confirmed by the numerics: as for the cubic potential, the fluctuations strongly increase close to the critical time $t_c(a)$. More precisely the standard deviation σ_x of the fluctuations and the velocity $x_0'(t)$ evolve identically for small noise, getting a maximum when the catastrophe occurs, at $t_c(a)$. Therefore the variance of the signal fluctuations cannot be used as a precursor for predicting the transition. Such a correlation between the standard deviation of the fluctuations and the sudden change of the solution has been reported recently [6] where the GPS geodetic signal, which can be assimilated to our $x_0(t)$, is shown to be strongly correlated to the seismic signal (which we see as related to the fluctuations $x_1(t)$). Note that when the noise increases, the growth of the fluctuations occurs earlier and earlier, their maximum progressively shifting before t_c. This shift becomes visible only for an effective noise amplitude larger than unity, that is physically outside the range of noise values.

Consider the case of small effective noise, where the correlation function and the spectrum of the fluctuations $(x(t) - x_0(t))$ are well described by the correlation function and spectrum of $x_1(t)$, respectively. The calculation of these functions requires some care because the system is not in a statistically steady state. Therefore the spectral density of the fluctuations depends on time and the correlation function

$$\Gamma_{x_1}(t, \tau) = \langle x_1(t - \tau/2) x_1(t + \tau/2) \rangle, \tag{30}$$

depends both on t and on τ. A time dependent spectrum is formally defined by the (real) Wigner transform

$$\mathrm{S}_{x_1}(t, \nu) = \int_{-\infty}^{+\infty} \mathrm{d}\tau e^{-2i\pi\nu\tau} \times \langle x_1(t - \tau/2) x_1(t + \tau/2) \rangle, \tag{31}$$

that has to be modified for numerical applications, by introducing a slipping window. Choosing a Gaussian window function of width w, the numerical spectrum writes

$$\mathrm{S}_{x_1}(t, \nu) = \left\langle \left| \int_{t_a}^{t_b} \mathrm{d}\tau e^{-(\frac{t-\tau}{w})^2} x_1(\tau) e^{-2i\pi\nu\tau} \right|^2 \right\rangle, \tag{32}$$

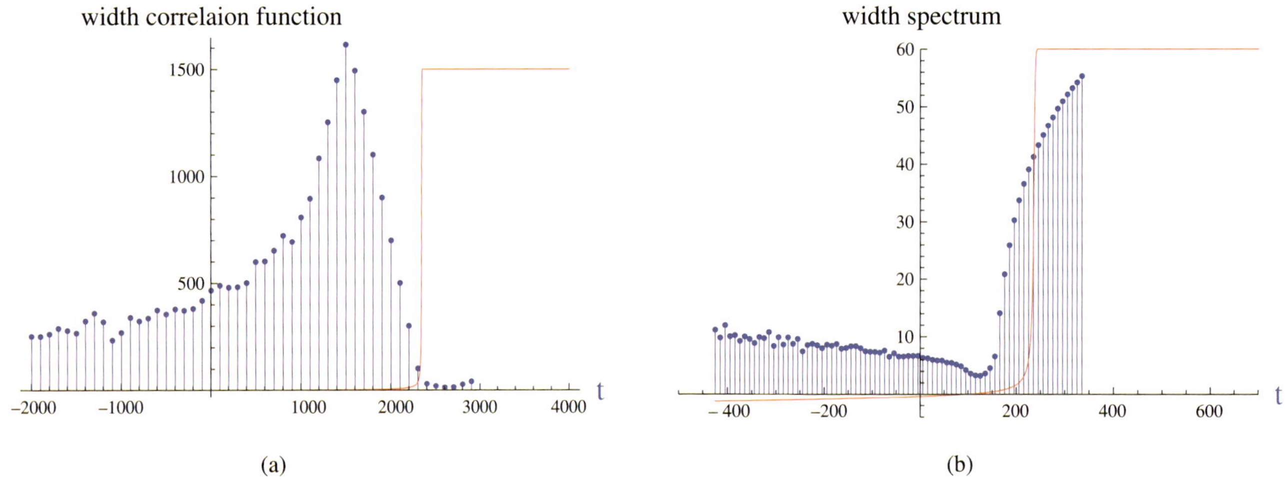

Fig. 4. (a) Width (in arbitrary units) of the correlation function of the fluctuation $x(t) - x_0(t)$ for $a = 10^{-9}$ ($t_0 = 10^3$, $t_c(a) \sim 2340$); (b) Spectral width (a. u.) defined as the peak width of the spectrum $S_{x_1}(t, \nu)$, Eq. (32).

where $t_{a,b}$ are the numerical integration time boundaries. The half-height half-width τ_x of the correlation function (and $\Delta\nu$ of the spectrum) are reported in Fig. 4. Both widths display the critical slowing down which is well-known for gradient flows near bifurcations. Consider first the left curve. The figure displays a strong increase of the correlation time τ_x of the fluctuation $(x(t) - x_0(t))$ in the intermediate time range, reaching its maximum value at time $t \sim 1.5\ t_0$. This sudden enhancement of the correlation time corresponds to the "slowing down" before the critical time, it is preceded by a slow growth of τ_x (not shown in the figure, increasing by a factor ten for $-100t_0 < t < 0$), and is followed by a rapid decrease before the critical time.

This rapid decrease is connected with the rapid growth of the variance close to t_c, and does not exist when the correlation function is defined as $\Gamma_{x_1}(t_1, \tau)/\sigma(t - \tau/2)\sigma(t + \tau/2)$ [1] where $\sigma(t)$ is the time dependent variance of the random function $x_1(t)$. In the case of earthquakes, this result leads to the following behavior: τ_x slowly grows during nearly a week, it strongly increases few hours before the catastrophe reaching its maximum about two hours before t_c, then it strongly decreases. Such a remarkable behavior should be used as precursor. Mathematically the increase of the correlation length *before* the catastrophe can be understood when looking at the formal expression (29). The second derivative of the potential vanishes at $t = t_0$, that leads to the flatness of $g_0(t)$ in the whole domain $0 < t < t_c(a)$, as shown in Fig. 5(a). Fig. 5(b) shows that the "precursor time" (see caption) is given by the relation

$$t_{prec} \sim 2t_0 \tag{33}$$

which corresponds to about 4 hours.

As for the spectral width, the result is just the opposite: it continuously decreases from large negative time, until the time $t \sim 1.5t_0$, where it suddenly grows. The slow decrease of $\Delta\nu$, corresponding to a slow shift of the spectrum towards low frequencies, is followed by a rapid spectral broadening at the end of the intermediate domain, before the transition time $t_c(a)$. As a increases, the shift towards low-frequencies remains visible with an appropriate time window, but the last stage (spectral broadening) becomes unnoticed. Note that the choice of the time extension w of the window function is important for such a study, here we have chosen $w \sim t_0$, because the maximum of the correlation time is of order t_0 before the catastrophe. The two stages of the spectral width change are both important, because they occur *before the transition.*

The growth of the fluctuations and their shift to lower frequencies can be understood as follows. As the transition approaches the potential $V(x,t)$ becomes flatter and flatter, making weaker and weaker the restoring force towards equilibrium. Therefore, at constant noise source, the amplitude of the fluctuations driven by this noise source will grow because the damping is less and less efficient. Moreover, the typical time scale for this damping will get larger and larger because of the decreasing stiffness of the potential, which will favor noise at lower and lower frequencies.

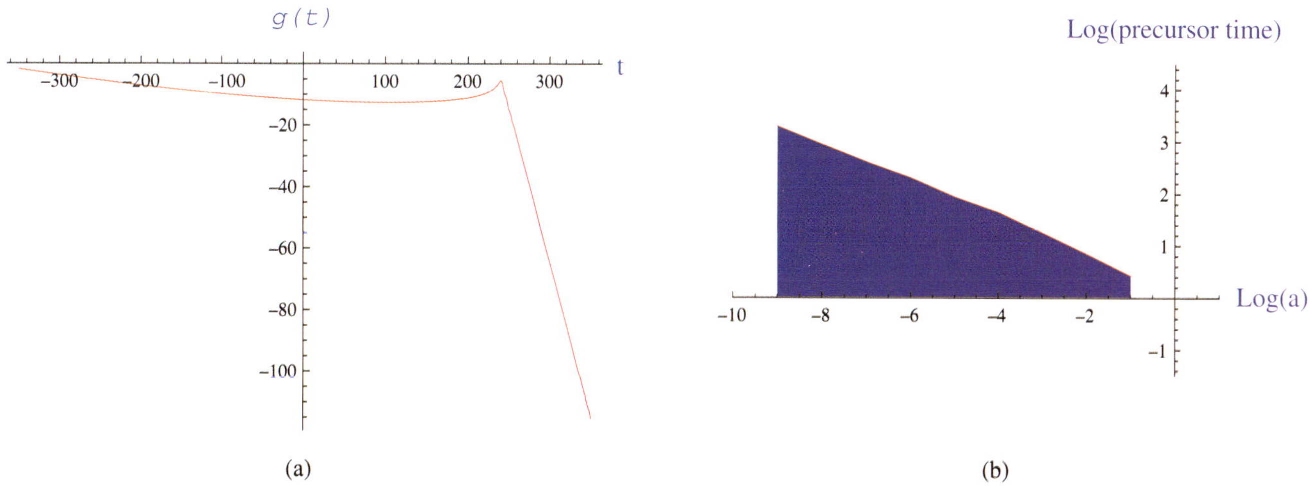

Fig. 5. (a) Flat shape of $g_0(t)$ before the critical time, for $a = 10^{-9}$, or $t_0 = 1000$; (b) "Precursor time" $t_{prec} = t_c(a) - t_{1/3}$ versus a, in Log scale ($t_{1/3}$ being the time where the width is $\frac{1}{3}$ of its maximum value, before the catastrophe time $t_c(a)$).

This result seems to agree with the observation made by R. Peters [7] who developed unconventional seismometers with output proportional to the ground acceleration. The author reports the observation of low-frequency components enhanced some hours before the Oakland (C.A.) earthquake on July 20th 2007. This result was illustrated by the author by plotting the cumulative spectral power versus the period (reciprocal frequency). In the Fig. 6(a), the cumulative spectra are derived from successive 12 hours time records. The blue curve, which corresponds to few hours before the critical time, displays low frequency build-up (in the range 10^{-2} to 10^{-1} Hz). For comparison we have drawn the cumulative spectra of the acceleration $\ddot{x}(t)$ in Fig. 6(b), before and at the critical time. Note that the low-frequency build up effect is less visible for the second order time derivative $d^2x(t)/dt^2$ than for $x(t)$, nevertheless the lower curves in figure (b) display clearly the enhancement of the energy in low frequency range, whereas the two upper curves just before and at the catastrophe display the vanishing of the low frequency energy, i.e. a shift of the spectrum towards high frequency range.

3. Model of Frictional Slipping

The model of fast transition separated by long intervals of quietness studied in the previous section could look a bit arbitrary because the saddle-node bifurcation is postulated, not derived from a physico-mathematical model describing, however approximately, a concrete problem. Below we try to apply related ideas to a model of frictional slipping, a priori pertinent for the earthquakes, assuming earthquakes are the slip episode of the stick-slip dynamics of two sliding solid plates.

Generally speaking, the stick-slip observed in the friction between solids makes it a priori a good example of relaxation dynamics. It is ubiquitous in real world phenomena like in geophysics and in engineering and the class of model proposed by Dieterich [8] and by Ruina [9] was designed to represent this physics. They use a "state" variable, denoted as θ later on, which depends on the physical state of the surfaces facing each other and describe memory effects of the response. This state variable has its own dynamics, in agreement with the observation of time dependent solid/solid friction.

Numerical solutions of this family of models (we shall refer to them as the DR model below) show well relaxation-like oscillations with two widely separated time scales. However we show that, in this family of models, the transition from slow to fast and conversely does not occur by an instability of the slow manifold (SM) similar to the one existing in the strongly nonlinear regime of the van der Pol equation. The stick-slip behavior in the DR model makes a new class of relaxation oscillations: here the slow-fast transition occurs because the slow dynamics becomes singular in a finite time, i.e. by a runaway of the slow dynamics, not by transverse instability of the SM. The DR model has been used in particular to analyze a simple physical

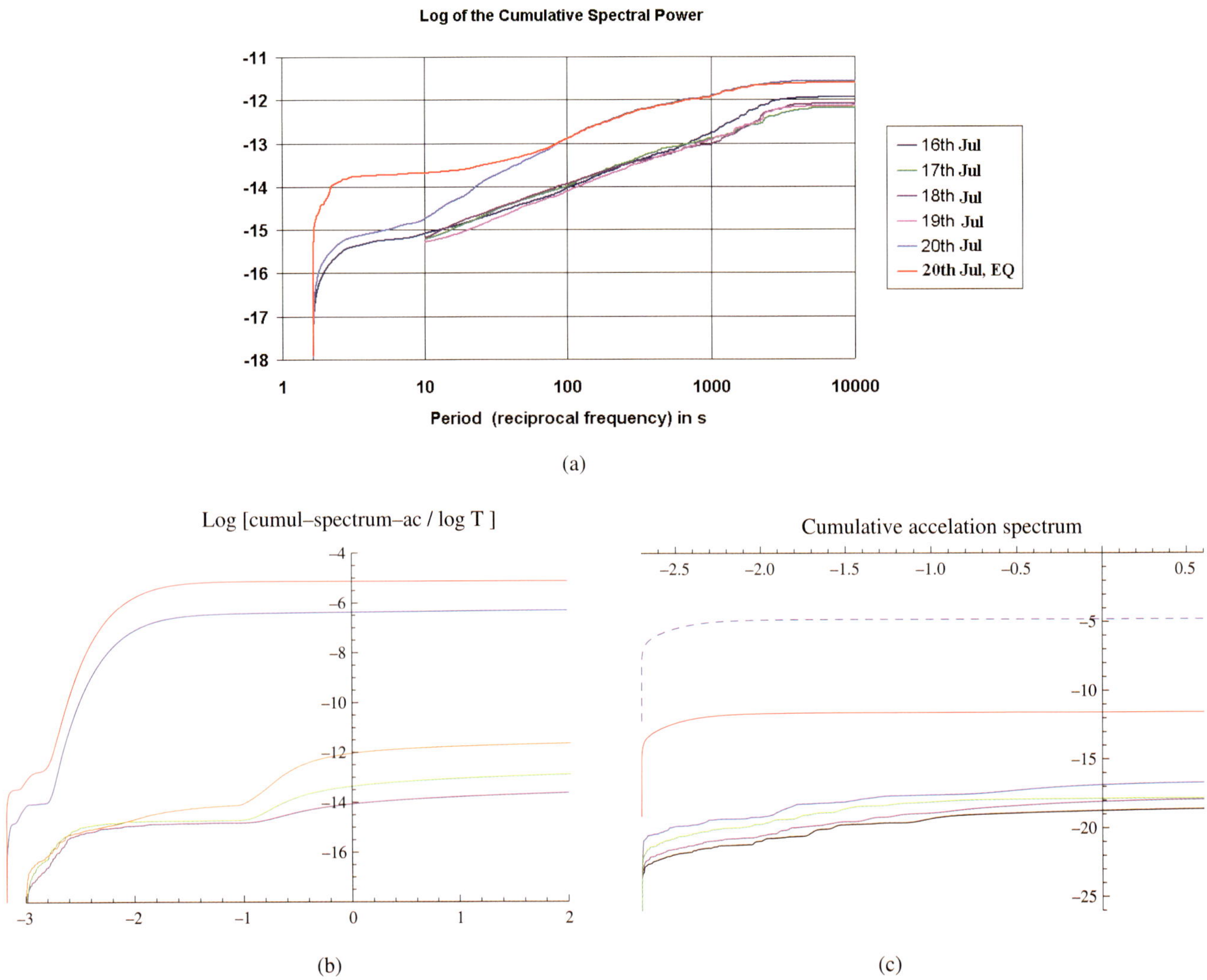

Fig. 6. Cumulative spectral power of the ground acceleration versus the inverse of the frequency, $T = 1/\nu$. (a) From R.D. Peters [7] data, during and before the Oakland earthquake. (b) Calculated with the saddle-node model with $a = 10^{-6}$,$w = t_0/2$, the three lower curves correspond to time sequences taken during the "critical slowing down" process, (around $t = -t_0$ purple, $t = 0$ green, $t = t_0$ orange), the two upper curves after the maximum of the slowing down process ($t = 1.5t_0$ blue, $t = 2t_0$ red). (c) Calculated with the DR model with $\gamma = 10^3$, $\epsilon = 0.72$, $\xi = 0.7$. From up to down, taking $t = 0$ as the critical time, curves are taken around the time values $t = 0$ (dashed), $t = -0.3t_0$ (red), $t = -0.5t_0$ (blue), $t = -t_0$ (green), $t = -2t_0$ (purple), and $-4.5t_0$ (black). Note that for strong earthquakes, the typical values of t_0 are ~ 2 hours for the saddle-node model, ~ 10 years for the DR model.

situation, a solid block sliding on a fixed surface, the block being driven by an external spring. This is believed to be a qualitative representation of the dynamics of two adjacent tectonic plates near a fault in geophysical context. While the evaluation of the different forces involved in the plate tectonic is still controversial, the irregularity of the seismic dynamics allows to expect the model also represent earthquakes when the sliding becomes much faster than the slow phase.

The DR equations of motion of a single sliding block coupled with rate-and-state dependent friction [9], [10], are given by

$$\dot{u} = v - v_0, \tag{34}$$

$$\dot{v} = -\frac{1}{M}(ku + \theta + A\ln(v/v_1)), \tag{35}$$

and

$$\dot{\theta} = -\frac{v}{D_c}(\theta + B\ln(v/v_1)), \tag{36}$$

The first equation relates u, the position of the block, to its speed v. The constant v_0 is the difference of speed between the point holding the spring and the surface the block is sliding on. The force due to the external spring is proportional to u, with a spring constant k, and M is the mass of the block. Moreover θ, a contribution to the friction force, depends on the state of the surfaces facing each other, a function of the history of sliding, as given by the solution of Eq. (36). The intermediate equation (35) is a way of writing the dynamics of the block under the effect of the pulling force proportional to u and the effect of an effective friction which includes the state dependent part proportional to θ. The quantity $(\theta + A\ln(v/v_1))$ on its right-hand side is the friction force plus a constant, absorbed into a constant displacement added to u. Even after addition of a constant friction (independent on v), the DR equations cannot be valid for all values of (u, θ, v). Because A is positive, the friction force is dominated at low speeds v by the logarithm, and so becomes very large positive, in the direction of v, a negative friction which is clearly nonphysical. Because the added constant is independent on the parameters used to write the equations above, one cannot tell, without knowing what is its value in a particular application, if the non physical regime of negative friction is reached or not for this case. We assume that, in the regime we consider, negative friction never happens. Note that in case of very low speed, Lapusta *et al.* [11] propose to change the above relation $v/v_1 = \exp(\tau_f)$ between the velocity v and the friction τ_f, by the relation $v/v_1 = 2\sinh(\tau_f)$, that changes $\ln(v)$ into $\ln(v + \sqrt{v^2+1})$.

The velocity v_1 and the coefficient A, B and D_c are phenomenological quantities derived in principle from experiments. One can absorb the ratio v_0/v_1 by adding another constant to u. Using the scaled quantities v/v_0, θ/A,u/D_c, v_0t/D_c, one transforms the above equations into the mathematically convenient form [12]

$$\dot{u} = v - 1, \tag{37}$$

$$\dot{v} = -\gamma^2\left(u + \frac{1}{\xi}(\theta + \ln(v))\right), \tag{38}$$

and

$$\dot{\theta} = -v(\theta + (1+\epsilon)\ln(v)). \tag{39}$$

This version of the DR equations keeps three dimensionless parameters, γ, ϵ and ξ, related to the quantities M, v_0 and to the other phenomenological parameters A, B and D_c derived in principle from experiments [9], [13]. The geophysical literature gives ϵ and ξ of order one, and γ large. To give an order of magnitude, γ is typically of order 10 for ice flows in Antarctica [14], where the sliding phase lasts about 20 minutes once a day, while for major earthquakes $\gamma \sim 10^8$ [10] with sliding phases of typically 5 seconds occurring once every 200 years.

When the parameter γ becomes very large and ϵ/ξ is above unity, the fixed point $(u = \theta = 0, v = 1)$ of equations (37)–(39) undergoes a Poincare–Andronov bifurcation, leading to a stable periodic solution, see Fig. 7(a). At threshold the period is

$$\tau_c = \frac{2\pi}{\sqrt{\xi}}. \tag{40}$$

The period τ increases noticeably with the control parameter, as illustrated by curves (b-c) in Fig. 7, it reaches a linear dependence with respect to ϵ for given γ, and increases nearly as $5\log(\gamma)$ for given ϵ.

In the case of earthquakes, the typical parameter values [10] $v_0 = 30$ mm/year, $D_c/v_0 = 2.7$ years, $\xi = 0.8, \epsilon = 1, \gamma = 10^8$ lead to a period equal to 75 years, or $\tau = 28$ in units of Eq. (37)–(39). More generally, considering earthquakes separated by 30 to 200 years, it gives the dimensionless period

$$\tau \sim 10 - 70, \tag{41}$$

which is the range of period values considered below in our numerical calculations.

The striking point is the stick-slip character of the solutions with almost motionless long intervals (stick step) interrupted by short bursts of rapid displacement (slip step). In this limit the stiffness of the DR equations makes them hard to solve numerically, so that most numerical simulations are done with γ of order unity, unrealistic for earthquake modeling. Therefore it is important to carry an asymptotic analysis of the solutions of this set in the large γ limit, what is done below.

Moreover we check our theoretical results by a systematic numerical investigation of the solutions for γ in the range 10-10^4, plus some calculations for larger values of γ (until 10^7). This was performed on Mathematica, by the "StiffnessSwitching" method using a pair of extrapolation methods as the default. The nonstiff solver uses the "ExplicitModifiedMidpoint" base method, the stiff solver uses the "LinearlyImplicitEuler" base method.

3.1. *Slow and fast regimes*

3.1.1. *Stick step*

Consider first the stick regime, where the acceleration of the motion is small,

$$\dot{v} \ll \gamma^2, \tag{42}$$

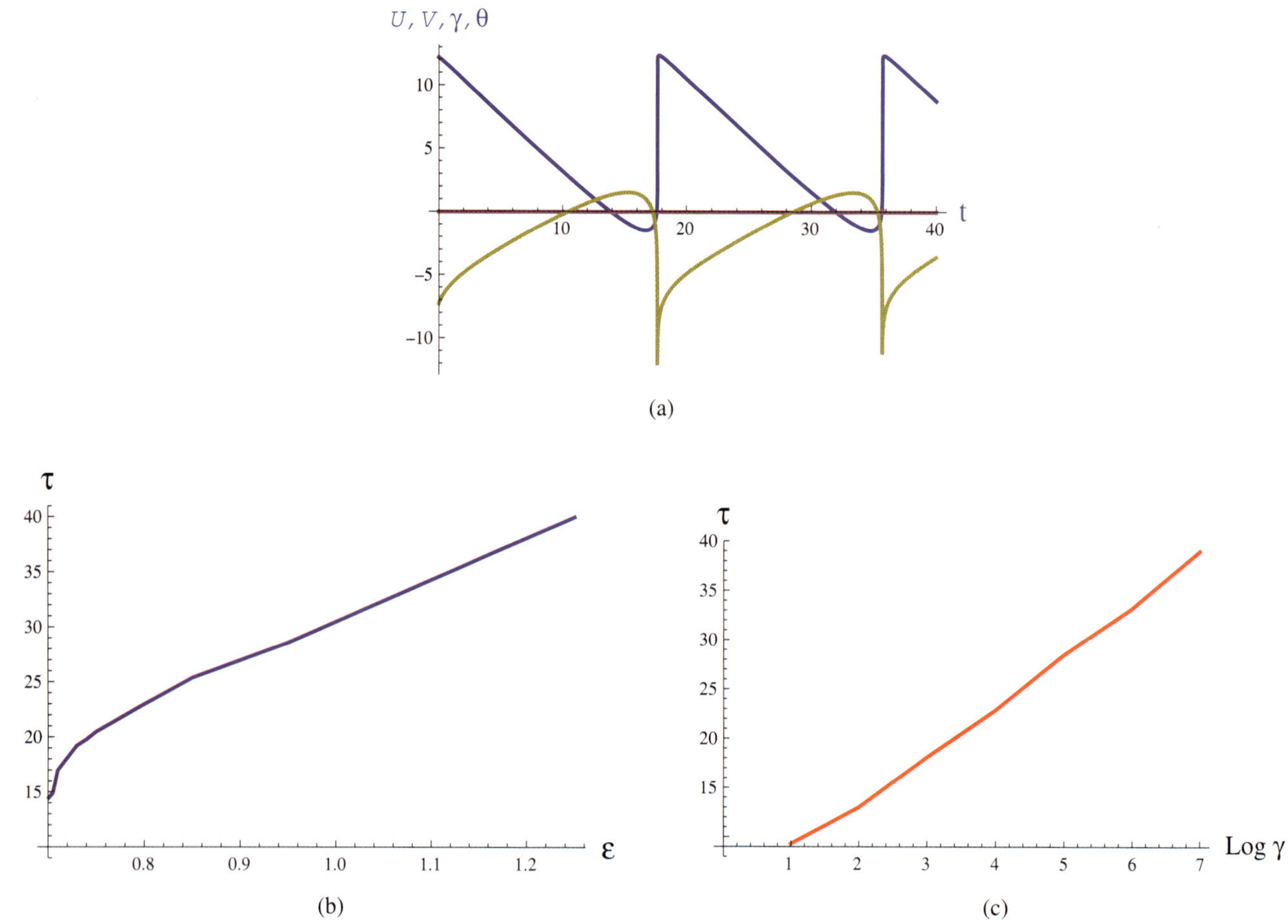

Fig. 7. (a) Periodic solution of Eqs. (37)–(39) for $\gamma = 10^3$, $\xi = 0.7$, $\epsilon = 0.72$, $u(t)$ (blue), $v(t)$ (purple) and $\theta(t)$ (yellow). (b–c) Period of the limit cycle for $\xi = 0.7$, in (b) as function of ϵ for $\gamma = 10^3$; in (c) as function of $\log_{10}(\gamma)$ for $\epsilon = 0.72$.

although there is always some sliding as illustrated in Fig. 7(a). The time duration of this step is very close to the period τ of the limit cycle.

The velocity v and the acceleration $\dot{v}$ are smaller than one in the main part of the sticking stage, they both becomes unity at time nearly equal to $t_c - 1$. This defines the intermediate time scale t_0, the time interval separating the instant where $v(t) = 1$ and the catastrophe time t_c where the velocity is maximum. For the whole range of parameters that we investigated numerically, we have found the relation

$$t_0 \cong 1, \tag{43}$$

within the units of Eqs. (37)–(39). After reaching the value $v = 1$ and $\dot{v} = 1$ the velocity and the acceleration increase. At the end of the stick step, in the range defined by $1 \ll v \ll \gamma$ and by Eq. (42), the solution is analytically tractable, see the Sec. 3.1.3. Note that in the case of earthquake, the intermediate time scale corresponds to few years (2.7 years in [10]), that makes the DR model completely different from the saddle-node model where the intermediate stage lasts a few hours only [15].

During the whole sticking episode defined by equation (42), the coefficient of γ^2 on the right-hand side of Eq. (38) can be set to zero, that gives the relation

$$\xi u + \theta + \ln(v) = 0, \tag{44}$$

defining a surface S of equation $v(u, \theta) = e^{-(\xi u + \theta)}$ in the phase space, (u, v, θ). Putting the relation (44) in Eqs. (37) and (39), one get a set of two first order ODE's for u and θ in which the large parameter γ has disappeared,

$$\begin{cases} \dot{u} = e^{-(\xi u + \theta)} - 1 \\ \dot{\theta} - e^{-(\xi u + \theta)}[\epsilon\theta + (1 + \epsilon)\xi u], \end{cases} \tag{45}$$

defining a *2D flow* (two dimensional).

From the point of view of the 3D dynamical system we started from, the 2D flow defined by equations (45) takes place on the surface (manifold) S, a reduction of the original 3D flow done by Gu *et al.* [16] in the same limit. The "stick" phase of the 3D dynamics fulfills Eqs. (42), (44) and (45).

Let us show that in the 3D flow, the SM is attracting. The surface S splits the 3D phase space (u, v, θ) into two subspaces, the one for which $(\xi u + \theta + \ln(v))$ is positive and the one for which it is negative. In the large γ limit, the dominant component of the time derivative of the 3D vector (u, v, θ) is $\dot{v}$. The negative sign on the r.h.s. of Eq. (38), ensures that the SM is attractive from both subspaces, because taking v very large positive one sees that $\dot{v}$, the dominant component of the "velocity", defined by the set (37), (38) and (39) is directed toward v negative, and conversely for $(\xi u + \theta + \ln(v))$ negative. This shows that the slow manifold is everywhere attracting in this limit γ large. Therefore the classical "scenario" for van der Pol-type relaxation oscillation with an ejection out of a repelling SM cannot hold for DR equations. We show later that in the DR case, the 3D flow escapes from S because the 2D flow trajectory goes to infinity in *finite time*, see Figs. 9 and 10.

3.1.2. *Slip step*

During the slip, both v and $\dot{v}$ are "large", of order γ and γ^2 respectively. Neglecting (-1) on the right-hand side of Eq. (37) allows to eliminate the large parameter γ from the equations by rescaling and addition. This is done by putting $v = \gamma V(T)$, $t = T/\gamma$, $u = U(T) + c_u \ln(\gamma)$ and $\theta = \Theta(T) + c_\theta \ln(\gamma)$. The quantities denoted as T, V, Θ and U are of order 1 with respect to γ as well as the constants c_θ, c_u. Setting to zero the part independent of T in the equations derived from (38) and (39) one finds two algebraic equations for c_θ and c_u with the solution

$$\begin{cases} c_\theta = -(1+\epsilon) \\ c_u = \frac{\epsilon}{\xi}, \end{cases} \tag{46}$$

The functions $U(T)$,$V(T)$ and $\Theta(T)$ are solutions of a set of three ODE's free of the large parameter γ

$$\begin{cases} U_{,T} = V \\ V_{,T} = -\left[U + \frac{1}{\xi}(\Theta + \ln(V))\right] \\ \Theta_{,T} = -V\left[\Theta + (1+\epsilon)\ln(V)\right], \end{cases} \tag{47}$$

where $X_{,T}$ is for $\frac{\mathrm{d}X}{\mathrm{d}T}$.

This predicts that the slip lasts a (short) time δt_{eqk} of order $1/\gamma$ although the sticking lasts a time independent of γ. This scaling law is in excellent agreement with the numerical study of the full DR equations for γ large. We observed that, as a function of t, $v(t)$ peaks with a time half-width

$$\delta t_{eqk} \sim 2.5/\gamma, \tag{48}$$

in a wide range of parameters, as illustrated in Fig. 8.

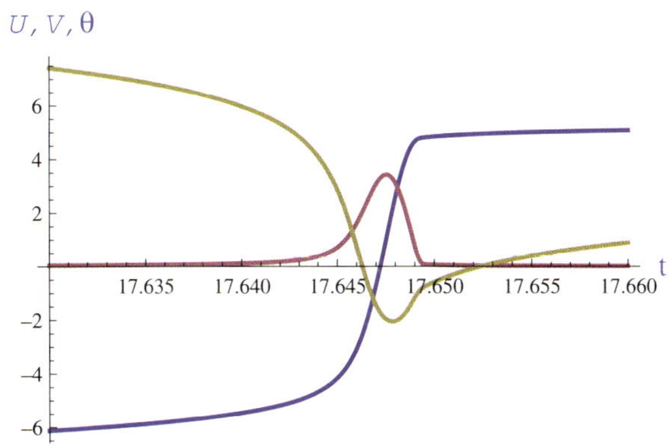

Fig. 8. Fast step. Part of the solution shown in Fig. (7), using the same color code close to the catastrophe, in units of Eqs. (47).

3.1.3. *Matching slow and fast solutions*

There remains to interpolate between the stick and slip parts of the dynamics. To the best of our knowledge, this matching has not been done before for DR-like equations, in relation with the idea of a finite time singularity of the slow dynamics. According to matching theory, the solution in the interpolation domains has to satisfy the set of equations common to the two phases, slow and fast,

$$\dot{u} = v, \tag{49}$$

together with Eqs. (39) and (44). The integral curves of (39) and (49) are solutions of the single linear ODE

$$\frac{\mathrm{d}\theta}{\mathrm{d}u} = \epsilon\theta + (1+\epsilon)\xi u, \tag{50}$$

that gives

$$\theta(u) = ce^{\epsilon u} - \frac{(1+\epsilon)\xi}{\epsilon^2}(1+\epsilon u), \tag{51}$$

with c arbitrary constant. In our problem this expression agrees well with the numerics, with c calculated from a particular value of (u, θ) belonging to the matching region, see the discussion below concerning the Figs. 9(a), 9(b) and (8). The general time dependent problem in the matching region may be written as $\dot{u} = \exp(\Phi(u))$ with

$$\Phi(u) = -ce^{\epsilon u} + \frac{\xi}{\epsilon}u + \frac{(1+\epsilon)\xi}{\epsilon^2}. \tag{52}$$

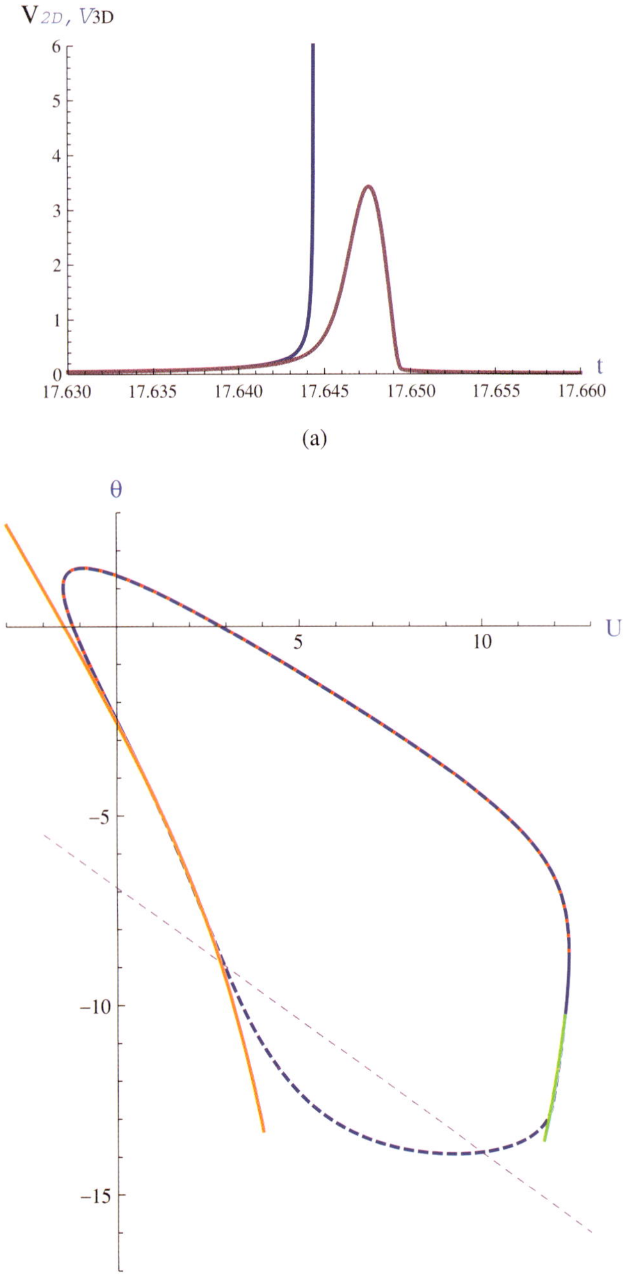

Fig. 9. (a) $v(t)/\gamma$ for the 2D and 3D flows (blue and purple curves respectively) (b) Phase space (u, θ). The 3D limit cycle (closed dashed blue curve) is superimposed with the 2D flow (solid portion) on the SM. The orange and green portions correspond to the two branches of matching solutions, Eq. (51).

The solution reads

$$t - t_0 = \int_{u(t_0)}^{u} du' \exp(-\Phi(u')), \qquad (53)$$

where $u(t_0) = u_0$. It yields a singularity at finite time in the forward time direction if $\int_{u_0}^{\infty} du' \exp(-\Phi(u'))$ is a converging integral, which requires c to be negative. This occurs at the end of the stick interval, as illustrated in Fig. 9(a), but their is no finite time singularity in the full DR system. There the divergence of the singularity is replaced by a large back and forth swing ending with a slowing down and a return to the SM. A physical consequence of this behavior is that, before the catastrophe u is negative and positive at the end of it, changing the strain from dilative to compressive. This does not seem to be what happens in real earthquakes, where the strain remains dilative after the quake and so can generate secondary quakes weeks after a major slide. This points toward a large effect of friction during the quake itself. In the DR model this means that ξ should be rather small to enhance friction with respect to inertia: a large ξ means that inertia is dominant, which means that an almost complete back and forth swing is needed to return to a small velocity at the end of the fast phase. By decreasing ξ one increases the effect of friction during the fast phase. This makes possible to end it with u still positive (as observed) whilst the friction term $\ln(v)/\xi$ is increased in Eq. (30). This puts physical constraints on the DR model to represent large earthquakes, particularly on the value of ξ (in real life it is constrained by the ratio of the strain before and after the quake, a measurable quantity in principle).

Note that the solution (52) was derived within the adiabatic approximation (42), which limits its validity to a domain we are going to precise. On the surface S, the time derivative of $v = \exp(-\xi u - \theta)$ writes $\dot{v} = -e^{-2(\xi u + \theta)}[\xi + \epsilon\theta + (1+\epsilon)\xi u]$ by using Eq. (49) and the second equation (45). Therefore in the space (u, θ) the straight line of Cartesian equation

$$\epsilon\theta(u) = -(1+\epsilon)\xi u - \xi. \qquad (54)$$

splits the SM into a part such that trajectories crossing this line tends to still lower values of $\dot{v}$, and so follow even more closely the SM, although when the crossing is in the other direction, $\dot{v}$ tends to increase, as well as v itself (always positive) and so ends up with the finite time divergence just described. A rough estimate of condition (42) limits the 2D flow to the domain located above the line

$$\theta(u) = -\xi u - \ln(\gamma). \qquad (55)$$

These results are illustrated in Fig. 9(b). In this figure the closed curve (blue-dashed) displays the limit cycle (3D flow) evolving counterclockwise in the phase space (u, θ). Superimposed onto this curve, we plot the stick episode (solid red portion), and the matching analytical solutions (51) (orange and green curves). The adiabatic approximation domain lies above the dashed straight line, Eq. (55). The two pieces of analytical solutions fit well the two intermediate regimes, they are calculated with two distinct values of c (derived from Eq. (51) and using values of (u, θ) before and after the catastrophe respectively). The orange curve which belongs to the SM, matches the stick regime and the fast one, then it diverges on the surface S, while the limit cycle escapes from S. This happens when the adiabatic approximation fails. The escape from the SM of the 3D flow in the fast regime is visible on Fig. 10.

The physics of the oscillation can be understood as follows. The acceleration of the block at the end of the stick phase is unstable because it lowers the friction and so feeds itself. The transition to the "sliding" fast regime occurs when the acceleration becomes big enough to make the inertia of the block relevant. In this fast regime the dynamical system gets one more dimension (in other terms the effect of inertia increases by one the number of coupled ODE's). If one neglects in this fast regime the friction, the dynamics is the one of a harmonic oscillator making a half swing while its velocity keeps the same sign. When this oscillator has slowed down enough to yield back its energy to the potential energy of the spring, its velocity returns to small values and friction becomes so large that the SM is reached again. This could explain that, in earthquakes, a finite fraction of the initial elastic energy is not dissipated in the event itself, because part of it remains as potential energy after the large scale pendulum-like motion. In this model, the spring is under dilative strain before the slip episode and under compressive strain at the end of it.

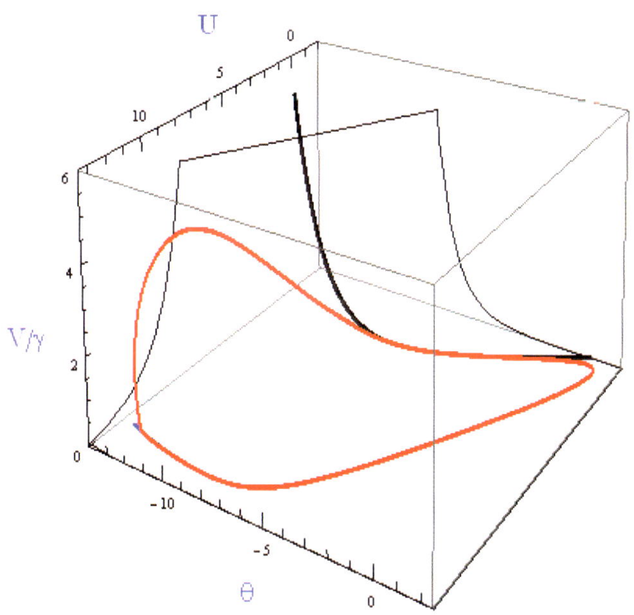

Fig. 10. 3D view of the trajectory (solid red curve) leaving and landing out and in the SM (light blue surface), the black part diverging on S displays the 2D flow in its "non slow" regime.

3.2. *Lyapunov analysis along the trajectory*

We have noted above that the SM seems to be everywhere attractive, taking argument of the negative sign on the right hand side of Eq. (38). This statement is invalid for the close vicinity of the SM, more precisely at distance of order $1/\gamma^2$ from the SM. In the very slow part of the trajectory, we may investigate how attractive is the SM by performing the linear stability analysis of the flow. Along a peculiar trajectory $(u_0(t), \ln(v_0(t)), \theta_0(t))$ the Jacobian matrix is

$$\begin{pmatrix} 0 & 1 & 0 \\ -\gamma^2 & -\dfrac{\gamma^2}{\xi v_0} & -\dfrac{\gamma^2}{\xi} \\ 0 & -A & -v_0 \end{pmatrix}, \tag{56}$$

where

$$A = \theta_0 + (1+\epsilon)(1+\ln(v_0)). \tag{57}$$

The eigenvalues of the matrix (56) are solutions of the equation

$$\lambda^3 + \left(\frac{\gamma^2}{\xi v_0} + v_0\right)\lambda^2 + \gamma^2\left(\frac{1-A}{\xi} + 1\right)\lambda + \gamma^2 v_0 = 0. \tag{58}$$

In the large γ limit, the two eigenvalues are

$$\lambda(t) = -B(t) \pm \sqrt{B(t)^2 - \xi v_0(t)^2}, \tag{59}$$

with

$$B(t) = \frac{1}{2}v_0(1 - A + \xi). \tag{60}$$

The three eigenvalues are drawn as functions of time in Fig. 11, for the flow corresponding to Fig. 7(a), or 9(b) or (10). One eigenvalue (curve a) is negative all along the trajectory. The two other eigenvalues are real in the first part of the trajectory, and become complex conjugate at time about $t_c - 4.6$, see Fig. 7(b). The real parts are the red and blue curves in (b), the imaginary parts are drawn in Fig. 7(c). The real part crosses zero at time $t \sim t_c - 2.3$, with an eigenfrequency ω of order unity.

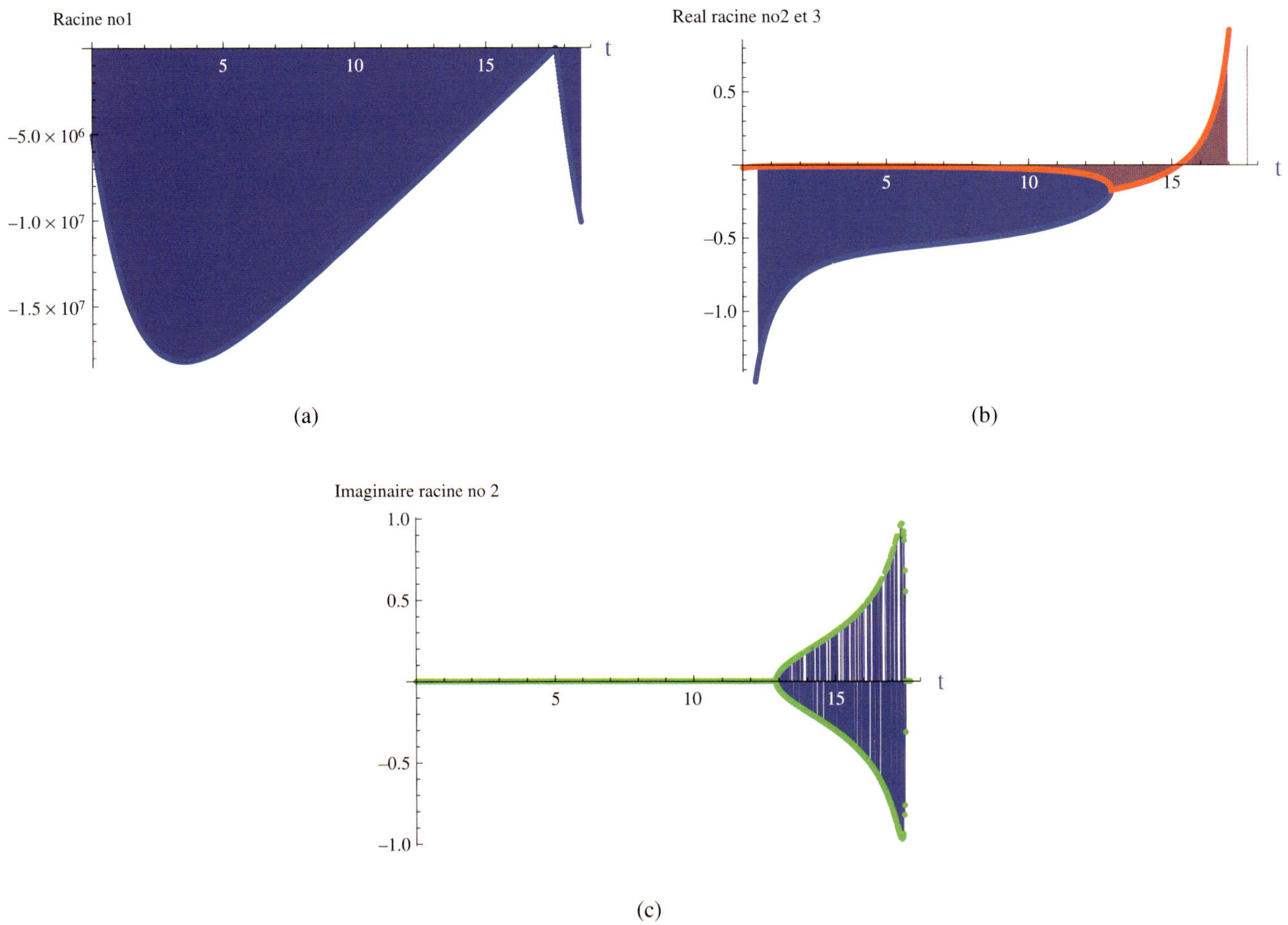

Fig. 11. Local stability along the trajectory, for the solution drawn in Fig. 7(a). (a) Negative real solution of Eq. (58), (b)–(c) two other solutions, real parts in (b) and imaginary parts in (c). The real part of the two complex conjugate exponents crosses the zero real value at time $t_c - 2.3t_0$, here $t_c = 17.6475$.

This occurs inside the "very slow" stick regime ending near $t \sim t_c - t_0$ where $v = 1$. At this time, the local stability analysis loses physical meaning, because the motion cannot be considered as steady during the time interval $2\pi/\omega$.

3.3. *Response to noise*

In view of forecasting catastrophes we have studied the response of the DR system to an external white noise, as we did for the saddle-node bifurcation in the potential model [15]. We solved the set of stochastic equations

$$\begin{cases} \dot{u} = v - 1 + \eta_u f_u(t) \\ \dot{v} = -\gamma^2 \left(u + \dfrac{1}{\xi}(\theta + \ln(v)) \right) + \eta_v f_v(t) \\ \dot{\theta} = -v(\theta + (1+\epsilon)\ln(v)) + \eta_\theta f_\theta(t), \end{cases} \tag{61}$$

where f_u, f_v, f_θ are three independent noise functions with short memory time (of order δt_{eqk}) and $\eta_u, \eta_v, \eta_\theta$ are amplitudes smaller than γ^2. We have calculated the correlation functions of $x(t)$, x standing for any function u, v, or θ,

$$\Gamma_x(t, t-t') = \langle x(t)x(t-t')\rangle - \langle x(t)\rangle\langle x(t-t')\rangle, \tag{62}$$

which are functions depending on the delay t' but also on the time t, and the standard deviations

$$\sigma_x(t) = \Gamma_x(t,t). \tag{63}$$

The results are very different from those of the potential model. First because here the standard deviation σ_v of the response dramatically increases with time before the event, much more than the velocity $v(t)$ itself, that could be seen as a precursor. This contrasts with the saddle-node case, where the standard deviation of the response grows nearly like $v(t)$ and therefore cannot be used as precursor. The two functions $\sigma_v(t)$ and $v(t)$, are drawn in Fig. 12(a) in Log scale, showing that the growth of σ_v is three

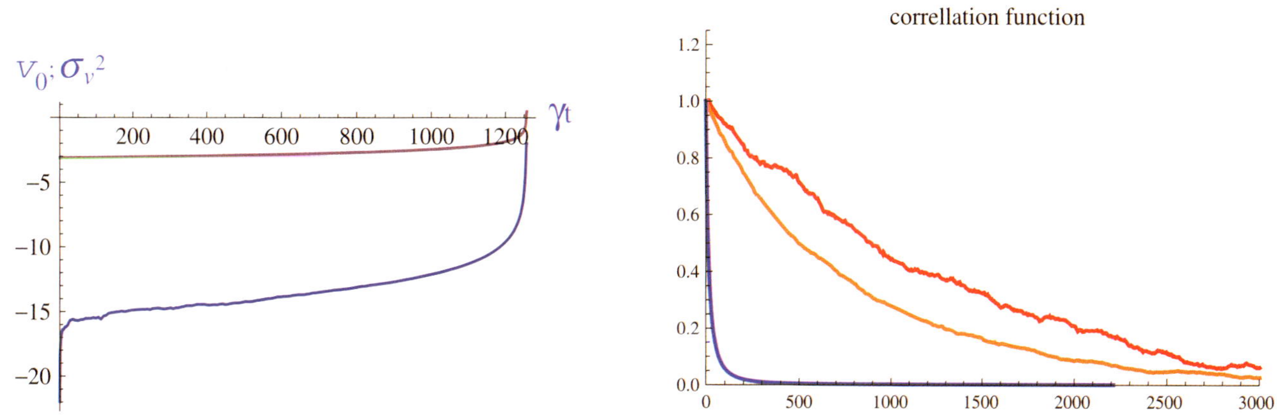

(a) (b)

Fig. 12. Response to noise. (a) Comparison of the increase of velocity and its standard deviation, in $\log_{10}$ scale in the last part of the stick regime, versus γt, until $tc - 0.1$, (b) correlation function $\Gamma_u(t, t - t')$ as function of t' for three time values $t = t_c - 2.3$ (red), $t = t_c - 1$ (orange), $t = tc - 30/\gamma$ (blue), for same parameters as in Fig. 7.

orders of magnitude larger than the growth of $v(t)$, on the time interval $(t_f - t_0, t_f)$, where the final time $t_f = t_c - 0.1$ corresponds to about two months before the event for typical earthquake case.

Secondly the correlation time τ_x (defined as the the width of correlation function) decreases very strongly during the large time interval $\sim 2t_0$ before the event, see Fig. 12(b). This could be understood as a sort of speed-up, contrary to what happens in the potential model where the correlation time becomes maximum shortly before t_c, in agreement with the well-known critical slowing-down phenomenon (or critical opalescence in spatial systems).

This striking result appears when we compare Figs. 13 and 4(b) which display the correlation time τ_x of the fluctuating part of the displacement for the DR and saddle-node model respectively. For the saddle-node model τ_x increases slowly before it get enhanced (critical slowing down) at time $\sim t_c - t_0$ (t_0 being the intermediate time scale in this system[15]). For the DR system we find that the correlation time of the response to noise decreases very slowly until the singularity (speed-up). The duration of the slow decrease of τ_x is of order few t_0, that is a noticeable part of the total period for the DR model. In summary, while the maximum of τ_x occurs before the catastrophe in both cases, the time interval is very large for the DR model, while is short for the saddle-node model, because we have

$$t_0/\delta t_{eqk} \sim \gamma, \tag{64}$$

and

$$t_0/\delta t_{eqk} \sim a^{-\frac{1}{3}}, \tag{65}$$

respectively. Therefore the effects (rapid increase of τ_x /slowing-down for the saddle-node model versus slow decrease of τ_x/ speed-up for the DR model) seem opposite while they both result from the fact that the memory time of the response shifted from t_c by a time interval of order t_0.

The two relations (64) and (65) explain why the response to noise is changed much earlier in the DR model than in the saddle-node model. Using parameters pertinent for earthquake, the precursor time t_0 is about few hours for the saddle-node model,

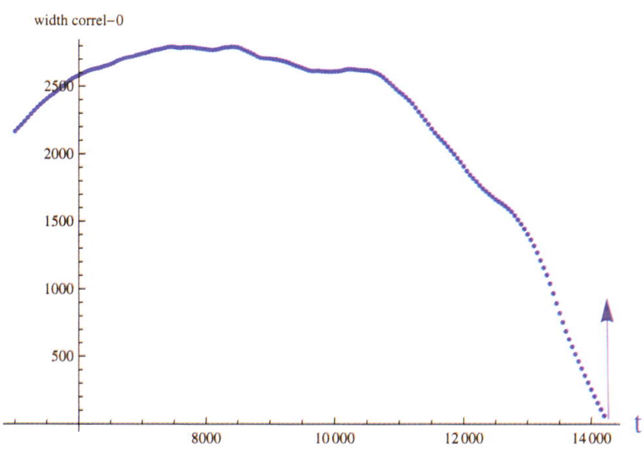

Fig. 13. Correlation time for the DR model with $\gamma = 10^3$, $\xi = 0.7$, $\epsilon = 0.72$ versus γt during half of the stick regime. The arrow indicates the catastrophe time.

while it is about few years for the DR one. Both results are related to the linear stability of the SM since the memory time of the response to external noise is maximum when the temporal integral of the largest stability exponent is close to zero, that gives a "precursor time" of order t_0 in both cases.

3.4. *Power law friction*

Finally it is worth questioning the relevance of this idea of transition for other mathematical models. In other terms how "generic" is this kind of transition? One can say first that it persists (as a way of going from slow to fast and conversely for the full dynamics) in the DR equations by changing the numerical values of the parameters. Another indication that it is a generic scenario is to check that it is still there in (slightly) modified equations.

We changed the logarithm $\ln(v)$ in the DR equations (38) and (39), by the function

$$g_\alpha(v) = \frac{v^\alpha - 1}{\alpha}, \tag{66}$$

with α positive.

Solving the modified system of equations

$$\begin{cases} \dot{u} = v - 1 \\ \dot{v} = -\gamma^2 \left(u + \dfrac{1}{\xi}(\theta + g_\alpha(v)) \right) \\ \dot{\theta} = -v(\theta + (1+\epsilon) g_\alpha(v)), \end{cases} \tag{67}$$

we found stable periodic solutions, with slow and fast steps. More precisely this appears above a certain threshold (close to the Poincaré–Andronov bifurcation the limit cycle has ordinary behavior). The curves (a) and (b) of Fig. 14 display slow-fast

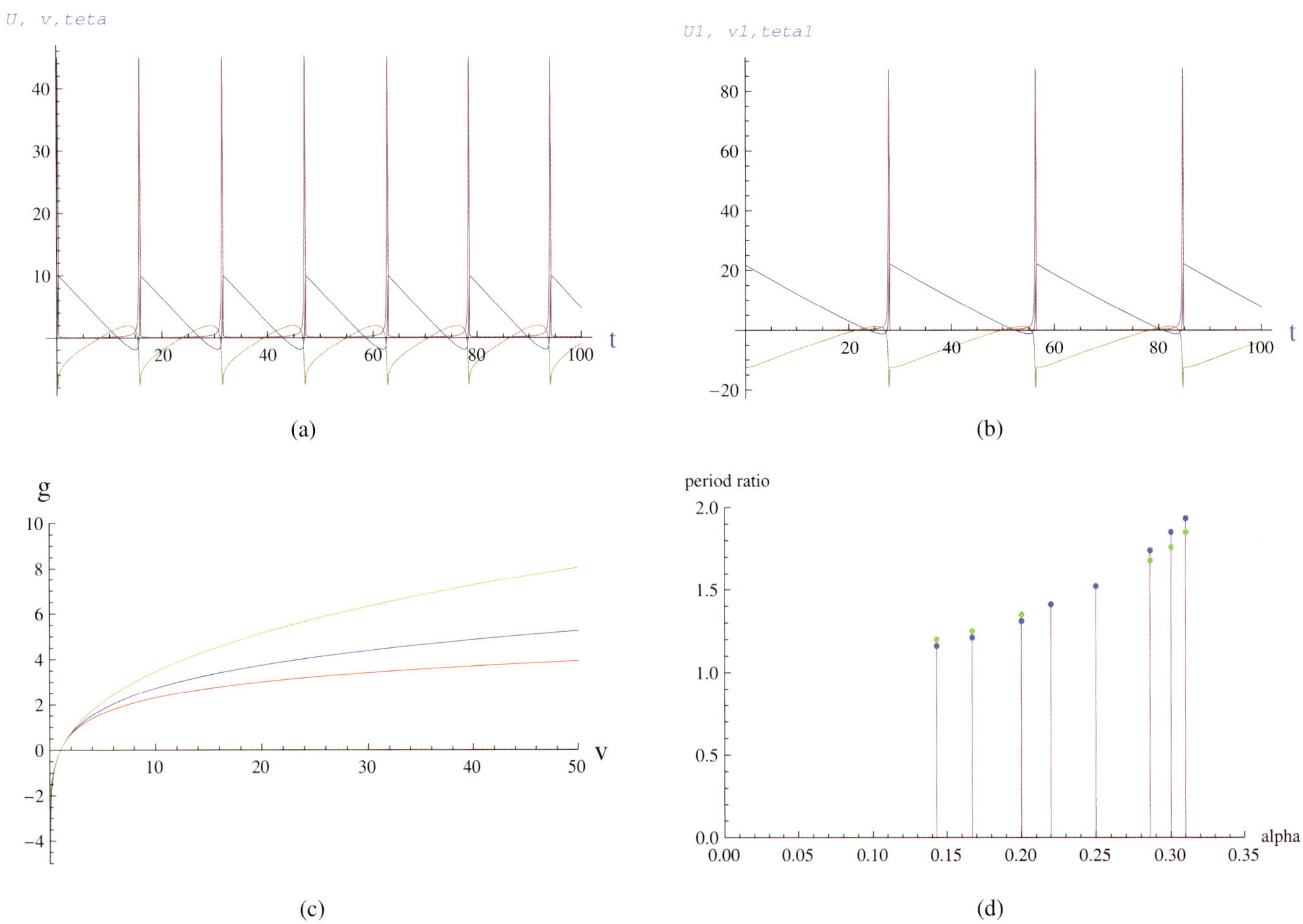

Fig. 14. Solution of the DR equations with $\ln(v)$ friction (a), and their modified form with a power law friction (b) (Eqs. (67) with $\alpha = 0.31$) for the set of parameters $\gamma = 10, \epsilon = 1, \xi = 0.7$. (c) $\ln(v)$ (red curve), and $g_\alpha(v)$ (blue and green curves for $\alpha = 0.143$ and 0.333 respectively. (d) The ratio of the limit cycle periods (blue points) and amplitudes (green points) for the power law and ln(v) cases, as function of α.

limit cycles obtained for the same parameter values, with the $\ln(v)$ friction and the power law friction, respectively (DR) equations and Eqs. (67) respectively). In both cases we observe that the period of the limit cycle increases with ϵ/ξ for a given value of γ. Moreover in the power law case we find that the period of the limit cycle increases with α, see Fig. 14(c). This occurs because the nonlinearity increases with the parameter α. The intermediate time (interval between the instant where $v(t) = 1$ and t_c) is found to be nearly equal to unity,

$$t_0 \simeq 1, \tag{68}$$

The Jacobian matrix is

$$\begin{pmatrix} 0 & 1 & 0 \\ -\gamma^2 & -\dfrac{\gamma^2}{\xi g_0'} & -\dfrac{\gamma^2}{\xi} \\ 0 & -A(\alpha) & -v_0 \end{pmatrix}, \tag{69}$$

where $f_0' = v_0^{\alpha-1}$, $A(\alpha) = \theta_0 + (1+\epsilon)(v_0^\alpha(\alpha+1) - 1/\alpha)$, and $\theta_0(t), v_0(t)$ are the dynamical variables along the trajectory, supposed to evolve very slowly with respect to the inverse of the imaginary part of any eigenvalue of the matrix (local stability). We found that the eigenvalues of matrix (69) and (56) display similar behavior with one negative real exponent, and two others alternately real and complex. For example for $\gamma = 10^3$, $\xi = 0.7$ and $\epsilon = 0.78$ we observe that the real part of the complex conjugate exponents cross zero at a time $t_c - 2.45$, which is very similar to the result obtained with the $\ln(v)$ friction term, cf Fig. 11.

We have also studied the response to noise in the case of a power-law friction (Eqs. (67) plus noise). The Fig. 15 displays a progressive speed-up, the correlation time decreasing during an interval of order t_0 before t_c, as in the case of the $\ln(v)$ friction law (compare with Fig. 13).

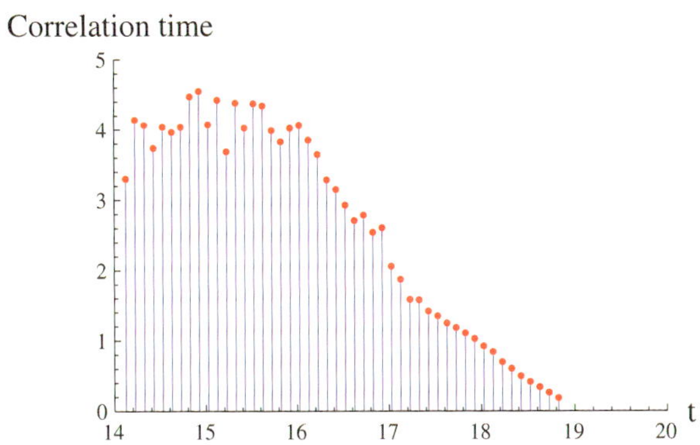

Fig. 15. Half-height half-width of the correlation function, as a function of time for $\alpha = 0.15, \gamma = 10^3, \epsilon = 0.78, \xi = 0.7$.

4. Discussion and Perspectives

We have presented two models of "relaxation dynamics" with similar looking behavior: long periods of "quiet" slow motion are interrupted by fast transitions, followed again by slow phases, etc. However the two models are deeply different, both because they show different mechanisms of slow to fast and reverse transitions, but also because they are objectively different in their response to an external source of noise. The saddle-node like model shows the standard "critical slowing down" with time correlations becoming more and more long ranged before the transition, something well-known in many areas. Here the correlation time reaches a maximum within a short time interval before the transition, of order t_0. On the contrary the DR-inspired model shows a very slow squeezing of the correlation time of the response as the transition approaches, the opposite of the dynamical saddle-node transition. In both cases the maximum of the correlation time occurs at time $t_c - t_0$, which is delayed from the catastrophe time t_c by the time scale t_0, but in the DR model the intermediate time t_0 is much longer than in the saddle-node model, relations (64)–(65). Therefore the response to noise of the two models can be summarized as follows: a slowing down which persist practically until t_c for the saddle-node model, and a very slow "speed-up" behavior until t_c for the DR-model. The sharp difference between the two models could perhaps be shown in real life conditions, and could be a way of proving (or disproving) the validity of some models of stick-slip behavior with laboratory experiments of solid friction.

We note that the geophysics literature describes many phenomena arising during the days or hours preceding strong earthquakes, that seems better in favor of the saddle-node model involving the short intermediate time. These phenomena have been attributed [17] to the emergence of electrical currents at the earth surface, arising from hard crystalline rocks near dip-slip faults, and caused by charge carriers, stress activated in the rocks deep below. Moreover the observation of clear build up of low-frequency components in the ground acceleration spectrum some hours before the Oakland earthquake [7] is in agreement with the prediction of the saddle-node model, whereas the DR model predicts enhancement of high frequency components long before the catastrophe, see Fig. 6.

We take this opportunity to present Professor Leon Chua our best wishes for many more years of fruitful research on our common topic of investigation, non linear phenomena.

Acknowledgments

Paul Clavin, Jean-Yves Le Mouël, Clément Narteau and Randall Peters are greatly acknowledged for interesting discussions, and Patrice Fromy for his technical help.

References

[1] Peters, R. D., Le Berre, M. and Pomeau, Y. (2012). Prediction of catastrophes: An experimental model, *Phys. Rev. E* **86**, 026207; and arXiv:1204.1551.

[2] Bouchaud, J-P. and Cont, R. (1998). A Langevin approach to stock market fluctuations and crashes, *Eur. Phys. J. B* **6**, pp. 543–550.

[3] Dorodnitsyn, A. A. (1947). Asymptotic solution of the van der Pol equation, *Priklad Matem. i Mekhan* **11**, pp. 313–328,(in Russian), English Translation: (1953) *Trans. Am. Math. Soc.* **88**, pp. 1–24. The authors thank V. Hakim for pointing out the relevance of this work.

[4] Peng, Z. and Gomberg, J. (2010). An integrated perspective of the continuum between earthquakes and slow-slip phenomena, *Nature Geoscience* **3**, pp. 599–607.

[5] Scholz, C. (1990). *The mechanism of earthquakes and faulting*, 2nd edn. 2001 (Cambridge University Press).

[6] Brudzinski, M. R. and Allen, R. M. (2007). Segmentation in episodic tremor and slip all along Cascadia, *Geology* **35**, pp. 907–910.

[7] Peters, R. D. (2007). Precursor warnings of structural catastrophe through observation of a seismic bounce, http://physics.mercer.edu/hpage/CSP/bounce.html.

[8] Dieterich, J. H. (1978). Time-dependent friction and the mechanics of stick-slip, *Pure appl. Geophys.* **116**, pp. 790–805; (1979) Modelling of rock friction 1. Experimental results and constitutive equations, 2. Simulations and presismic slip *J. Geophys. Res* **84** (1979) pp. 2161–2168 and pp. 2169–2175.

[9] Ruina, A. L. (1983). Slip instability and state variable friction laws, *J. of Geophysical Res.* **88**, pp. 10359–10370.

[10] Rice, J. R. and Tse, S. T. (1986). Dynamic motion of a single degree of freedom system following a rate and state dependent friction law, *J. of Geophys. Res.* **91**, pp. 521–530.

[11] Lapusta, N., Rice, J.R., Ben-Zion, Y. and Zheng, G. (2000). Elastodynamic analysis for slow tectonic loading with spontaneous rupture episodes on faults with rate- and state-dependent friction, *Journal of Geophysical Research* **105**, pp. 23765–23789.

[12] Erickson, B.A. (2010). *Complexity in the Nonlinear Dieterich-Ruina Friction Law*, PhD thesis, University of California, Santa Barbara (USA); Erickson, B. A., Birnir, B. and D. Lavallee, D. (2011) Periodicity, chaos and localization in a Burridge-Knopoff model of an earthquake with rate-and-state friction, *Geophysical Journal International*, **187**, pp. 178–198; Madariaga, R. (1998) Unpublished Notes.

[13] Rice, J. R. and Ruina, A. L. (1983). Stability of Steady Frictional Slipping, *J. of Applied Mechanics* **50**, pp. 343–349.

[14] Sergienko, O. V., Macayeal, D. R. and Bindschadler, R. A. (2009). Stick-slip behavior of ice streams: modeling investigations, *Annals of Glaciology* **50**, pp. 87–94.

[15] Pomeau, Y. and Le Berre, M. (2011) http://arxiv.org/abs/1102.5637. 1); Pomeau, Y., Le Berre, M., Le Mouël, J-L., Narteau, C. and Fromy, P. (2011) Foretelling catastrophes?, in Comptes-rendus de la 14e Rencontre du Non Linéaire, édités par C. Josserand, M. Lefranc et C. Letellier, pp. 135–144 , Non-Linéaire Publications, Saint-Etienne du Rouvray. ISBN:978-2-9538596-0-7.

[16] Gu, J. C., Rice, J. R., Ruina, A. M. and Tse, S. T. (1984). Slip motion and stability of a single degree of freedom elastic system with rate and state dependent friction, *J. Mech. Phys. Solids* **32**, pp. 167–196.

[17] Freund, F. T., Takeuchi, A. and Lau, B.W. (2006). Electric currents streaming out stressed igneous rocks- A step towards understanding pre-earthquake low-frequency E. M. emissions, *Physics and Chemistry of the Earth* **31**, pp. 389–396.

CHAPTER 30

SYNCHRONIZATION PROPENSITY IN NETWORKS OF DYNAMICAL SYSTEMS: A PURELY TOPOLOGICAL INDICATOR

STEFANO FASANI

DEI, Politecnico di Milano, Milano, Italy

SERGIO RINALDI

DEI, Politecnico di Milano, Milano, Italy

EEP, International Institute for Applied Systems Analysis, Laxenburg, Austria

Synchronization in networks of identical dynamical systems is enhanced by the number of manifolds in which synchrony of groups of systems is conserved or reinforced. Since the number of these invariant manifolds depends only on the coupling architecture of the network, it can be proposed as a purely topological indicator of synchronization propensity. The proposal is empirically validated through the detailed study of an ecological application.

1. Introduction

All properties of networks of N interacting dynamical systems depend, in general, upon network topology, coupling strength and local dynamics, and the most challenging problem is to identify the dependence upon topology [5,15]. This is what we do here by suggesting a purely topological indicator for estimating the propensity of the network to synchronize, namely to have all systems or groups of them behaving in unison at least intermittently if not permanently [12]. This indicator is particularly useful for detecting the impact of changes in the coupling architecture: for example, it could be used to establish if the randomization of a regular network (see Fig. 1) enhances or reduces the chances of synchronization, independently upon local dynamics and coupling strength. Our analysis differs from previous ones [3, 10, 17] which refer only to the special case in which all systems behave in unison (complete synchronization). The proposed topological indicator is based on the number S of partitions (here called synchronous) of the nodes of the network satisfying a special topological relationship.

The reason for suggesting this indicator is purely theoretical and based on the fact that all synchronous partitions are in one-to-one

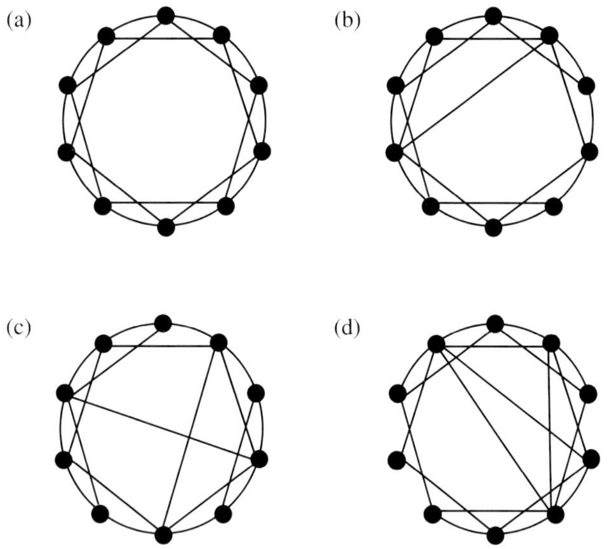

Fig. 1. Networks with 10 identical systems and different coupling architectures. Networks (b), (c), (d) are obtained from (a) by randomly rewiring $m = 1, 2, 3$ connections.

correspondence with invariant manifolds in state space in which groups of systems behave in unison (this is often called cluster or concurrent synchrony). Properties of these invariant manifolds have been discussed in [1, 4, 11] yet without stressing the relationship with network topology. If the state of the network is in one of these manifolds, synchronization can not be lost (and can actually be strengthened), while if the state of the network approaches the manifold and remains close to it for a while then synchronization shows up only temporarily. Thus, the greater is S, the higher are the chances to be in (or be attracted by) a permanent or intermittent synchronous regime.

This study suggests, in a sense, what might be intuitively expected, namely that regular networks (for which S turns out to be high) can more easily support synchrony than other networks. However, the problem is not as simple as one could imagine, because perceiving if S is high or low by a simple inspection of the coupling architecture is very difficult, if not impossible.

2. Synchronous Partitions

Consider a network of N identical dynamical systems described by

$$\dot{x}_h(t) = f(x_h(t)) + D \sum_{k\in\mathcal{N}} c_{hk}(x_k(t) - x_h(t)), \quad h \in \mathcal{N} \qquad (1)$$

where $\mathcal{N} = \{1, 2, \ldots, N\}$, is the set of systems, x_h is the n-dimensional state vector of system h, $f(\cdot)$ is the function describing the local dynamics $\dot{x} = f(x)$ of each system when isolated, $c_{hk} = c_{kh}$ is equal to 1 when systems h and k are connected and 0 otherwise, and D is a non-zero $n \times n$ matrix. When D is diagonal, model (1) allows one to deal with the majority of applications involving diffusion of energy, populations, and matters [5, 12, 15]. In such a case the diagonal elements of D are standard diffusion rates.

System (1) is usually associated with an undirected graph in which the set of nodes is $\mathcal{N}$ and the arcs are systems connections.

We now consider partitions $\Pi = \{\pi_1, \pi_2, \ldots, \pi_p\}$ of the set $\mathcal{N}$ where π_i is a subset of the nodes of the graph, i.e. a group of systems in the network. Among all possible partitions, S of them, here called *synchronous partitions*, satisfy the following topological condition:

$$n_{h_1\pi_j} = n_{h_2\pi_j} \quad \forall h_1, h_2 \in \pi_i \quad \forall (i,j) : i \neq j \qquad (2)$$

where $n_{h_l\pi_j}$ indicates the number of arcs connecting a node h_l of a set π_i with a disjoint set of nodes π_j. In words, a partition $\Pi = \{\pi_1, \pi_2, \ldots, \pi_p\}$ is synchronous if for all pairs of groups of nodes, all nodes of the first group are connected with the same number of arcs to the second group. For example, condition (2) is satisfied in the partition of Fig. 2(a) but not in that of Fig. 2(b).

We can now prove the following

Theorem 1. *If $\Pi = \{\pi_1, \pi_2, \ldots, \pi_p\}$ is a synchronous partition, the manifold*

$$x_{h_1} = x_{h_2} \quad \forall h_1, h_2 \in \pi_i \quad \forall i \in \{1, 2, \ldots, p\} \qquad (3)$$

is an invariant manifold of system (1).

Proof. Consider a generic point ($x_h = x^{(i)}$ $\forall h \in \pi_i$ $\forall i$) of manifold (3) as initial state at time t_0 of system (1). Thus, for $h \in \pi_i$ we have

$$\begin{aligned}\dot{x}_h(t_0) &= f(x^{(i)}) + D \sum_{k\in\mathcal{N}-\pi_i} c_{hk}(x_k(t_0) - x^{(i)}) \\ &= f(x^{(i)}) + \sum_{j\neq i} n_{h\pi_j} D(x^{(j)} - x^{(i)}).\end{aligned}$$

But condition (2) says that $n_{h\pi_j}$ for $h \in \pi_i$ does not depend on h, so that $\dot{x}_h(t_0)$ is the same for all $h \in \pi_i$. This means that the solution of (1) remains in manifold (3) which is therefore invariant. □

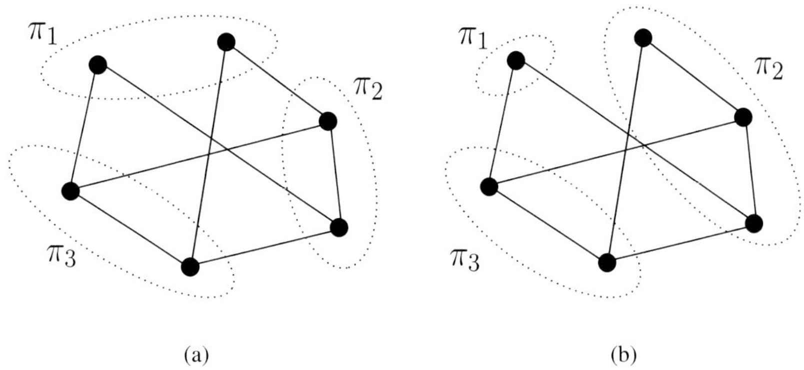

Fig. 2. The partition in (a) is synchronous while the partition in (b) is not synchronous.

In words, Theorem 1 says that if the state of the network belongs at a given time to manifold (3), characterized by the synchrony of groups of systems, then this synchrony is conserved forever and possibly enhanced because the states of two different groups of systems might asymptotically converge one to each other. This is why partitions satisfying condition (2) have been called synchronous.

In principle, there could exist manifolds preserving synchrony different from those pointed out by Theorem 1. Fortunately, this possibility is ruled out by the following.

Theorem 2. *If a manifold with groups of synchronous systems of the form* (3), *where* $\Pi = \{\pi_1, \pi_2, \ldots, \pi_p\}$ *is a partition of* $\mathcal{N}$, *is invariant, then* Π *is a synchronous partition.*

Proof. When the system starts at time t_0 in the invariant manifold (3), the state vectors of all the systems of the same group π_i are identical, so that two different systems h_1 and h_2 belonging to the same group π_i are described by

$$\dot{x}_{h_1}(t_0) = f(x^{(i)}) + D\sum_{j\neq i} n_{h_1\pi_j}(x^{(j)} - x^{(i)})$$

$$\dot{x}_{h_2}(t_0) = f(x^{(i)}) + D\sum_{j\neq i} n_{h_2\pi_j}(x^{(j)} - x^{(i)})$$

But $\dot{x}_{h_1}(t_0)$ must be equal to $\dot{x}_{h_2}(t_0)$ because manifold (3) is invariant, so that

$$\sum_{j\neq i}(n_{h_1\pi_j} - n_{h_2\pi_j})(Dx^{(j)} - Dx^{(i)}) = 0.$$

Since this condition must hold for any arbitrary vector $(Dx^{(j)} - Dx^{(i)})$ in the range of D, which is at least one-dimensional because $D \neq 0$, we must have $n_{h_1\pi_j} - n_{h_2\pi_j} = 0\ \forall j \neq i$ which is, indeed, condition (2). □

Thus, S represents the number of manifolds in which synchrony is preserved (if not enhanced). S can be computed by an algorithm that checks if the topological condition (2) is satisfied for each partition. A Matlab version of this algorithm that takes advantage of an incremental construction of the possible partitions using a tree structure and a prune strategy which prematurely discards families of partitions that early violate condition (2) is available on request.

3. Topological Indicators of Synchronization Propensity

All forms of synchrony of system 1 depend on network topology (matrix $[c_{hk}]$), local dynamics (function f) and dispersal (matrix D). Functions that depend only on topology and capture the probability that a particular form of synchrony is present, when local dynamics and dispersal belong to a suitably defined admissible class, are called *topological indicators of synchronization propensity*. In this section we first review the three known topological indicators of synchronization propensity and then present a new one based on the notion of synchronous partitions discussed in the previous section.

The first topological indicator denoted by (a) in the following, has been proposed by Wu and Chua (see [17]) and refers to complete synchrony, namely to the case where all systems in the network behave in unison and return to this peculiar collective behavior after any small perturbation. The topological indicator (a) is theoretically justified when the admissible local dynamics are characterized by Master Stability Functions [10], which are negative above a threshold value $\underline{\varepsilon}$. In fact, under these conditions, the completely synchronous regime is stable for small perturbations provided

$$\lambda_2 > \frac{\underline{\varepsilon}}{d} \tag{4}$$

where $d = \sqrt{d_1^2 + \ldots + d_n^2}$ and λ_2 is the minimum non-zero eigenvalue of the connectivity matrix [10] (We recall that the connectivity matrix is a zero row sum matrix in which g_{ij}, $i \neq j$, is equal to -1 if systems i and j are coupled and equal to 0 otherwise, whereas g_{ii} is the degree of system i, namely the number of systems directly coupled with i). Since λ_2 depends on topology, while $\underline{\varepsilon}$ and d depend on local dynamics and dispersal and are therefore randomly selected within their admissible sets, we can summarize condition (4) by saying that topologies with higher values of λ_2 have higher probabilities of giving rise to complete synchronization in (1).

The second topological indicator, denoted by (b), also refers to local stability of the completely synchronous state, but the set of admissible local dynamics is not constrained to the functions f with Master Stability Functions negative above a threshold $\underline{\varepsilon}$. Indeed, the property required for the Master Stability Functions is weaker, namely to be negative in an interval $(\underline{\varepsilon}, \overline{\varepsilon})$. Under these assumptions, it can be proved [10] that the stability condition of the completely synchronous state is

$$\frac{\lambda_2}{\lambda_N} > \frac{\underline{\varepsilon}}{\overline{\varepsilon}} \tag{5}$$

where λ_N is the maximum eigenvalue of the connectivity matrix. Thus, $\underline{\varepsilon}/\overline{\varepsilon}$ being a random variable selected within the admissible sets, we can conclude that topologies with higher values of λ_2/λ_N have higher chances to give rise to complete synchrony.

The third topological indicator of synchronization propensity, denoted by (c), still refers to the completely synchronous state, but pretends that the network returns to it after any perturbation and not only after small perturbations. This property (global stability) can be studied using suitable Liapunov functions, as done in [3], and the final result is that the completely synchronous state has higher chances to be globally stable if the inverse of the highest sum of the lengths of all shortest paths in the network passing through the same arc is higher.

The fourth topological indicator of synchronization propensity, denoted by (d) in the following, is the one we propose in this paper. It is radically different from the three other indicators (a), (b) and (c) because it does not refer only to complete synchrony but also to weak forms of synchrony, like partial (cluster) synchrony and intermittent synchrony. Since the probability of being in a state of weak synchrony certainly increases with the number of manifolds where partial synchrony is preserved (if not enhanced), on the basis of the two theorems reported in the previous section we can propose the number S of synchronous partitions as topological indicator of synchronization propensity.

In order to compare the four indicators we have just described, we can consider the regular network of Fig. 1(a) and the families of networks obtained from it by randomly rewiring $m = 1, 2, 3$ connections, with the aim of determining if this randomization increases or decreases the chances of synchronization. In the following section we will go into more details on this problem by making reference to a specific ecological application.

For the network in Fig. 1(a), corresponding to $m = 0$, the four indicators can be easily computed (the value of the indicator (d) computed with our algorithm is $S = 64$). But the four indicators can also be computed for the less regular networks shown in Fig. 1 (for example, our algorithm gives $S = 7, 5, 3$, respectively, for the networks with $m = 1, 2, 3$ in Fig. 1). Computing the four indicators for many randomly generated networks with the same m one obtains the four curves reported in Fig. 3. The curves (a), (b), (c) show that randomization has only a small impact on synchronization, while curve (d) predicts just the opposite, namely a dramatic drop of synchrony with randomization. However, these conflicting results can be easily justified if one recalls that the topological indicator (d) refers to all forms of synchrony while the indicators (a), (b), and (c) refer only to the strongest form of synchrony.

4. An Ecological Application

Spatially extended ecological systems are often composed of N very similar, if not identical, patches in each one of which n species dynamically interact. Some of the patches are connected by channels through which individuals of different species can migrate at various rates. Thus, Eq. (1) can be used to model such ecosystems.

If some patches are synchronous, even intermittently, then all populations of the same species of those patches reach their minimum density at the same time. This is a highly risky situation because if an environmental shock of large spatial scale occurs exactly at that time, there are high chances that all individuals of the species in those patches die. For this reason high probabilities of synchronization are usually perceived by ecologists as high probabilities of extinction. This justifies studies in which relationships are established between probabilities of synchronization and topological characteristics of the network.

We consider from now on a spatially extended ecosystem composed of n species living in 10 patches located on a circle. Migration of all species is possible from each patch to its four neighbors (two on the left and two on the right), as shown in Fig. 1(a). Networks of this kind mimic aquatic or terrestrial ecosystems distributed around islands or lakes. Moreover, we assume to be interested in knowing if randomization of the migration channels in the form shown in Fig. 1 is beneficial or not in terms of risk of extinction. The problem can be immediately solved by using the topological indicator (d) described in the previous section because in this application we are interested in all forms of synchrony. The answer is therefore given by the curve (d) of Fig. 3 which allows one to conclude that randomization reduces very effectively the risk of extinction.

In order to verify if this is indeed the case, we assume that there are only two species, namely a prey x_1 and a predator x_2, described by the most

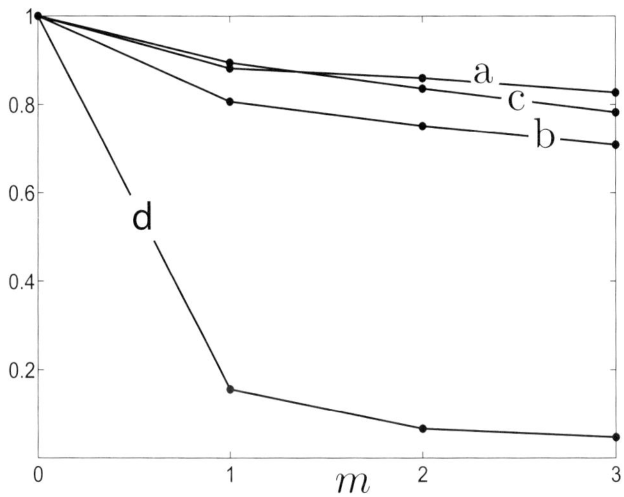

Fig. 3. Topological indicators of synchronization propensity for the networks of Fig. 1. Each curve (normalized to 1 for $m = 0$) represents the mean values of 100 randomly generated networks for every m.

standard model used in ecology [14]:

$$\begin{aligned} \dot{x}_1 &= f_1(x_1, x_2) = rx_1\left(1 - \frac{x_1}{K}\right) - \frac{ax_1}{b + x_1}x_2 \\ \dot{x}_2 &= f_2(x_1, x_2) = e\frac{ax_1}{b + x_1}x_2 - zx_2 \end{aligned} \tag{6}$$

where r and K are net growth rate and carrying capacity of the prey, $ax_1/(b + x_1)$ is the Holling type II functional response of the predator and e and z are predator efficiency and mortality. Thus, model (1) is completely specified because $f(\cdot)$ is given by (6), the matrix D is a 2×2 diagonal matrix with migration rates d_1 and d_2 on its diagonal and the elements c_{hk} are specified by the graphs of Fig. 1.

In order to estimate the synchronization propensity in the network we can use a pragmatic approach somehow similar to those used by field ecologists [8,9]. Here we closely follow [7] who have studied, through extensive simulations, the dependence of synchrony on various factors in the networks of Fig. 1 with model (6). More precisely, we integrate Eq. (1) for a given initial condition and determine, through correlation analysis, how many are the groups of patches (p) that can be roughly considered to be synchronous after the system has settled on one of its numerous attractors (limit cycles, tori and strange attractors). Then, we repeat this procedure for 100 randomly generated initial conditions, as well as for many different graphs with the same value of m, thus estimating the distribution of the number p of synchronous groups of patches for each m. Since $p = 1$ and $p = 10$ indicate completely synchronous and asynchronous regimes respectively, we extract from each distribution the probability P that $p \leq 5$, i.e. that the groups of synchronous patches are at most 5 and consider this as the probability of synchronization. Finally, in order to check the robustness of our results we have repeated all computations for more than 10 parameter settings of model (6) and for various values of the migration rates. All the computed probabilities decline with m, as shown in the four examples reported in Fig. 4, and therefore confirm the prediction of our topological indicator (see curve (d)) in Fig. 3). The curves in Fig. 4 compare favorably with the analysis performed by Holland and Hastings (see Figs. 11(a), 10(b), 3, 9(a) in [7]).

It is worth noticing, however, that in this specific application millions of long simulations of 20 ODE's are needed to obtain the same conclusions

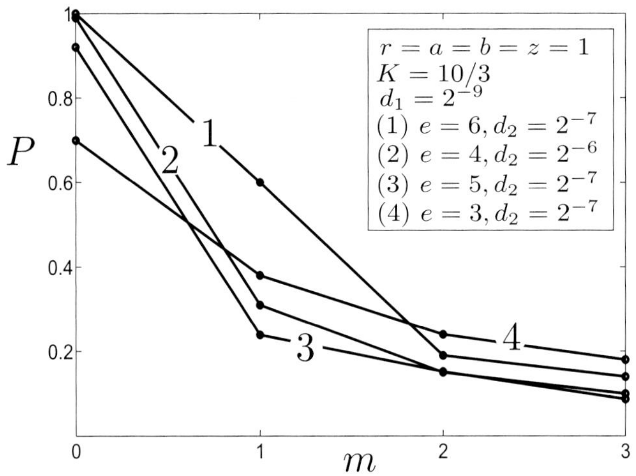

Fig. 4. Probability P that at most 5 groups of patches are synchronous in the networks of Fig. 1.

that can be obtained in a few minutes with our indicator. Moreover, thanks to our theorems we can state that the results are valid not only for other prey-predator communities but also for any other model.

5. Concluding Remarks

The results presented in this paper are certainly worth to be extended to more complex networks, in particular those described by directed graphs, and to different types of coupling rules like periodic or aperiodic couplings (e.g. [2,3,13]), which are important when migration or diffusion can occur only in specific seasons or during rare and random events. Moreover, since high values of our topological indicator are not necessarily correlated with high chances of complete synchronization, it would make sense to try to identify more complex indicators that might also include the propensity to complete synchronization. Finally, our indicator could be used to identify the coupling architectures that are most promising for controlling epidemics in social networks, since this problem has been shown to be critically influenced by synchrony [6].

References

[1] Belykh, I., Belykh, V. and Hasler, M. (2000). Hierarchy and stability of partially synchronous oscillations of diffusively coupled dynamical systems, *Physical Review E* **62**, 5, pp. 6332–6345.

[2] Belykh, I., Belykh, V. and Hasler, M. (2004). Blinking model and synchronization in small-world networks with a time-varying coupling, *Physica D* **195**, 1-2, pp. 188–206.

[3] Belykh, I., Hasler, M., Lauret, M. and Nijmeijer, H. (2005). Synchronization and graph topology. *I. J. Bifurcation and Chaos* **15**, 11, pp. 3423–3433.

[4] Belykh, V., Osipov, G., Petrov, V., Suykens, J. and Vandewalle, J. (2008). Cluster synchronization in oscillatory networks. *Chaos* **18**, 3, p. 037106.

[5] Boccaletti, S., Latora, V., Moreno, Y., Chavez, M. and Hwang, D. (2006). Complex networks: Structure and dynamics, *Physics Reports* **424**, 4–5, pp. 175–308.

[6] Earn, D. J., Levin, S. A. and Rohani, P. (2000). Coherence and conservation, *Science* **290**, pp. 1360–1364.

[7] Holland, M. D. and Hastings, A. (2008). Strong effect of dispersal network structure on ecological dynamics, *Nature* **456**, pp. 792–794.

[8] Koenig, W. D. (1999). Spatial autocorrelation of ecological phenomena, *Trends in Ecology and Evolution* **14**, pp. 22–26.

[9] Liebhold, A. M., Koenig, W. D. and Bjornstad, O. (2004). Spatial synchrony in population dynamics, *Annual Review of Ecology, Evolution and Systematics* **35**, pp. 467–490.

[10] Pecora, L. M. and Carroll, T. L. (1998). Master stability functions for synchronized coupled systems, *Physical Review Letters* **80**, 10, pp. 2109–2112.

[11] Pham, Q.-C. and Slotine, J.-J. (2007). Stable concurrent synchronization in dynamic system networks, *Neural Network* **20**, 1, pp. 62–77.

[12] Pikovsky, A., Rosenblum, M. and Kurths, J. (2003). *Synchronization : A Universal Concept in Nonlinear Sciences* (Cambridge University Press).

[13] Rinaldi, S. (2009). Synchrony in slow-fast metacommunities, *I. J. Bifurcation and Chaos* **19**, pp. 2447–2453.

[14] Rosenzweig, M. L. and MacArthur, R. H. (1963). Graphical representation and stability conditions of predator-prey interactions, *The American Naturalist* **97**, 895, pp. 209–223.

[15] Strogatz, S. H. (2001). Exploring complex networks, *Nature* **410**, 6825, pp. 268–276.

[16] Wu, C. W. (2007). *Synchronization in Complex Networks of Nonlinear Dynamical Systems* (World Scientific).

[17] Wu, C. W. and Chua, L. O. (1996). On a conjecture regarding the synchronization in an array of linearly coupled dynamical systems, *IEEE Transactions on Circuits and Systems I: Fundamental Theory and Applications* **43**, 2, pp. 161–165.

CHAPTER 31

FURTHER PROGRESS IN PARTIAL CONTROL OF CHAOTIC SYSTEMS

JUAN SABUCO and MIGUEL SANJUAN

Departmento de Fisica, Universidad Rey Juan Carlos, Tulipan, s/n 28933, Mostoles, Madrid, Spain

SAMUEL ZAMBRANO

In-Vivo Chromatin and Transcription Group, San Raffaele Scientific Institute, Milan, Italy

The Partial Control technique is a new paradigm in the control of chaotic systems in the presence of noise. This new approach of control allows one to keep trajectories of a dynamical system inside a region from which nearly all the trajectories diverge. Its main advantage is that this goal is achieved even if the corrections applied to the trajectories (control) are smaller than the action of environmental disturbances (noise) on the dynamics. This is a rather counterintuitive result, that is achieved thanks to what we call safe sets. Here, we study the use of the Partial Control technique in one of the most famous chaotic maps, the Hžnon map, and the deep relationship between the safe sets and the sets of points with different escape times, the escape time sets. We also show how it is possible to find certain extended safe sets that can be used instead of the safe sets in the Partial Control technique. We also discuss briefly the development of a new algorithm to find these new safe sets in any dynamical system with a chaotic behaviour.

1. Introduction

Some dynamical systems do not present a chaotic attractor but, instead, they present a nonattractive set in phase space where the dynamics is chaotic: a chaotic saddle. In fact, it is common to find that for a system with a chaotic attractor, by varying one of the system's parameters, a boundary crisis might occur, by which the chaotic set becomes nonattractive and nearly all trajectories escape from it. This kind of behavior is known as transient chaos [2, 8, 16].

In dynamical systems with transient chaos a complete analysis is possible for those showing a horseshoe map. It is the horseshoe which allows us to bound the location of the chaotic saddle in the phase space and thus the chaotic and escaping behavior.

In many real applications it might be desirable to control all the orbits starting close to the chaotic saddle while preserving the chaotic behavior. In mechanics, for example, preservation of transient chaos can prevent the appearance of undesired resonances. In lasers, it has been shown that maintaining transient chaos can help to avoid undesired intensity peaks. In engineering, it is known that the thermal pulse combustor operates chaotically, but when one tries to achieve high efficiency this can destroy the chaos and to cause the engine to flameout. In population dynamics the transition from transient chaos to periodicity is usually related with pathological situations (extinctions). These are examples that can be found in the techniques that have been proposed in recent years to achieve the goal of preserving transient chaos [3, 4, 6, 7, 14, 15, 18]. But two important issues must be addressed to solve this problem: The first one is the repulsive nature of the chaotic saddle, the second one is the environmental noise present in many physical situations, that typically makes the orbits escape even faster (although in some cases noise can slow down the escape process, see, [5]).

The partial control technique [20, 21] which is based in the application of small perturbations that will always keep the orbits inside the region Q, close to the chaotic saddle, was recently proposed to control transient chaos. The remarkable achievement of the partial control technique is that it allows us to control the system even when the amplitude of the corrections applied to the trajectories (the *control*) is smaller than the maximum deviation of the trajectories from their deterministic path due to the presence of environmental noise (the *noise*). The basic ingredient in order to obtain this somehow counterintuitive result is the use of certain sets referred to as *safe sets* [23] in Q. These sets have certain particular geometrical properties that are related to the typical stretching and folding action of the horseshoe-like mapping of Q. This advantageous control technique has been applied to well-known physical models [1, 17, 19, 20, 22, 23].

The main goal of this chapter is to develop a way of improving the partial control technique using what we call extended safe sets [11], instead of the safe sets. The extended safe sets are sets that we build making use of the escape times sets, and as we will show they also allow to keep trajectories bounded, with control smaller than noise offering some advantages over the safe sets. We will also explore in this chapter how does perform the partial control technique using safe sets, and extended safe sets and we will show which one is the best choice in different scenarios.

We also describe here an algorithm [12, 13] to find safe sets automatically: given the region in phase space from which trajectories escape and the value of the noise and control, safe sets are readily found. Basically we choose an initial phase space region and by iterations the algorithm eliminates those parts that do not hold with the condition required to be a safe set to finally obtain the desired safe set. We call it Iterative Sculpting Algorithm, as an analogy to removing material as in sculpting a statue.

This chapter is organized as follows: In Sec. 2 we describe the problem that we want to solve with our control strategy and we describe the system that we use in our explorations: the Hénon map. In Sec. 3 we review the main concepts of the safe sets. In Sec. 4 we define the escape time sets and explore their relation with the safe sets, and in Sec. 5 we show the conditions that extended safe sets need to fulfill. In Sec. 7 we review an algorithm that we have proposed recently to find safe sets in more general situations, i.e., situations where it is difficult to find horseshoe maps or when they are topologically very complex. Section 6 provides a numerical exploration of our control technique and a comparison between the results obtained with extended safe sets and safe sets. In Sec. 8 we draw the main conclusions of this work.

2. Problem Statement

We consider that the dynamics of the system considered is given by a map $\mathbf{p}_{n+1} = \mathbf{f}(\mathbf{p}_n)$, that can also be a Poincaré map, where $p_n \in \mathbb{R}^2$. We assume that the map f acts on a square Q like a horseshoe like-map, for details see [21]. This implies that nearly all the trajectories inside Q (except a zero measure set) escape from it after some iterations. On the other hand, the behavior inside the square Q is erratic due to the existence of this zero measure nonattractive set, the chaotic saddle.

As we said before, we consider systems with this kind of escaping dynamics $p_{n+1} = f(p_n)$ and also affected by noise. This is modeled here by adding at each iteration a random perturbation $\xi_n \in \mathbb{R}^2$ that we refer to as the *noise*, that is bounded by the constant ξ_0, $\|\xi_n\| \leq \xi_0$. Thus, the system to be controlled is $p_{n+1} = f(p_n) + \xi_n$. The effect of noise is that *all* the trajectories inside the square Q will now escape from it under iterations. In order to test our results we use here a noise with a uniform probability distribution but the control technique has to work for any other kind of distributions.

The ubiquity of this type of dynamical systems suggests that there are situations in which it might be desirable to control the system in order to keep the trajectories in the region where the horseshoe is defined (and thus far from an undesired attractor), although we might not need to determine exactly where the trajectory will go in Q. We call this type of control *partial control* of the system, and with this purpose, we can design a control strategy based on applying an accurately chosen *control* $\mathbf{u}_n$ each iteration, that we assume also bounded by a positive constant $||\mathbf{u}_n|| \leq u_0$, in such a way that the global dynamics is given by

$$\begin{cases} \mathbf{q}_{n+1} = \mathbf{f}(\mathbf{p}_n) + \boldsymbol{\xi}_n \\ \mathbf{p}_{n+1} = \mathbf{q}_{n+1} + \mathbf{u}_n, \end{cases} \tag{1}$$

so the control $\mathbf{u}_n$ depends on $\mathbf{p}_n$ and $\boldsymbol{\xi}_n$, as in other paradigmatic control methods [10].

A question that arises naturally here is, which is the value of u_0 needed to control the system, given the value of ξ_0. If the control is very strong, such that $u_0 > \xi_0$, it will not be difficult to find a strategy allowing to avoid escapes from the region Q. If $u_0 = \xi_0$, it might be possible. Remarkably, the partial control technique that we describe below allow us to achieve this goal even if the control is smaller than the noise, that is, if $u_0 < \xi_0$. As it will become clear later, this is due to the existence of certain sets inside the square Q: the *safe sets*.

An example of this type of dynamical system is the Hénon map *with an adequate choice of parameters*. The Hénon map defined as

$$\begin{cases} x_{n+1} = a - by_n - x_n^2 \\ y_{n+1} = x_n \end{cases} \tag{2}$$

is a paradigmatic system in nonlinear dynamics and for that reason we have chosen it, from now on, to show how the partial technique works.

We are interested here in the situation where $a = 6$ and $b = 0.4$. This is due to the fact that the Hénon map f acts like a horseshoe map on the square $Q \equiv [-4, 4] \times [-4, 4]$, as shown in Fig. 1.

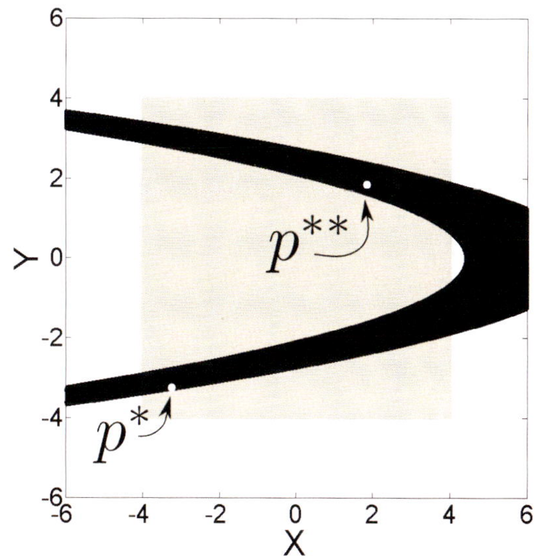

Fig. 1. The square $Q = [-4, 4] \times [-4, 4]$ (grey) and its image under the Hénon map $x_{n+1} = 6 - 0.4y_n - x_n^2$, $y_n = x_n$ (black). These are the parameters used in the chapter, for which the map acts like a horseshoe in Q. The two saddle fixed points p^* and p^{**} are shown.

The first implementation of the partial control technique required that the considered map acts on a square like a horseshoe map, i.e., that it satisfies the Conley-Moser conditions [21]. For these values of the parameters, the Hénon map satisfies these conditions, so we use it here both to illustrate and to numerically test our results.

In Fig. 1 we also show the two fixed points of this horseshoe-like map, p^* and p^{**}, the former will play later an important role (remember that every horseshoe map has associated two fixed points). The chaotic saddle responsible for transient chaos is shown in Fig. 2 and has been computed using the DYNAMICS software [9]. Due to the horseshoe mapping, this set is topologically equivalent to an intersection of two Cantor sets of vertical and horizontal lines, as expected. Thus, *nearly all* the points inside the square (except a zero measure set, the chaotic saddle and its stable manifold) escape from it under iterations. The dynamics inside the set is chaotic, but we have transient chaos due to its nonattracting nature for a typical trajectory starting inside Q.

3. Safe Sets

The partial control technique was originally defined, using safe sets, as the target of the control perturbations. We offer here a small review of their

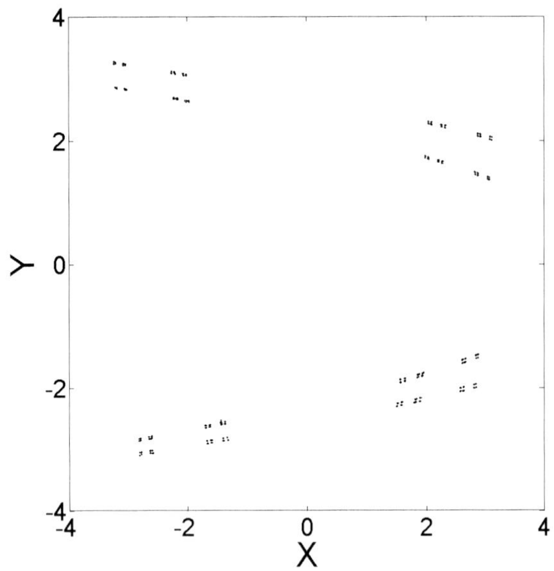

Fig. 2. Chaotic saddle for the Hénon map $x_{n+1} = 6 - 0.4y_n - x_n^2$, $y_n = x_n$.

main properties that would allow us to introduce the extended safe sets in a natural way.

Basically the safe sets are a set of curves that are placed in the square where is located the horseshoe map. We can see the form of different S^k safe sets in the Hénon map with escaping dynamics in Fig. 3. As we can see in that figure it is possible to group these curves using a specific order that we denote as k.

Bearing this graphical idea of safe sets in mind we can already define the main properties for the curves that are part of S^k (the safe set of order k.):

- S^k consist of 2^k vertical curves.
- Any vertical curve of S^k has two *adjacent* vertical curves of S^{k+1} closer to it than any other curve of S^k.
- The maximum distance between any of the 2^k curves of S^k and its two adjacent curves of S^{k+1}, denoted as δ_k, goes to zero as $k \to \infty$.
- All the curves of a S^k set can be grouped in pairs. Each pair would be composed of the two closest curves.

The algorithm to find the adequate safe set in a system with escapes make use of the inverse of the horseshoe map. The procedure consists of making backward iterations (applying the inverse map) of a straight line centered in the square as we can see in Fig. 4. Making these iterations and computing their intersections with the square, we easily obtain all the curves that are part of a particular S^k, where the order of the safe set corresponds to the number of iterations. Thus, it is possible to obtain the adequate safe set S^k that makes possible to avoid escapes even with a control smaller than the noise.

To properly quantify the properties of the partial control technique we need to define three parameters to establish the condition of a control smaller than the noise for a particular S^k:

- The first is the curve, ζ_i, that represents the middle distance in a particular pair of curves for a given S^k set.
- The second is the distance, $\delta_{max(i)}$, from the two points of that pair whose distance to ζ_i is maximum.
- The third is the distance, $\delta_{min(i)}$, from the two points of that pair whose distance to ζ_i is minimum.

We can visualize these parameters in Fig. 5 where we have plotted all of them for the two pairs of curves present in the S^2 set.

If we label as δ^*_{max} as the largest $\delta_{max(i)}$ among all the pairs of curves, then it is easy to understand that the condition to have a control smaller than the noise is that

$$\xi_0 > \delta^*_{max}. \tag{3}$$

When this condition is satisfied, we can also compute the maximum control present in the system if we consider δ^*_{min} as the smallest of all the δ_{min} among all the pairs of curves. Then we have that

$$u_0 = max\{\delta^*_{max}, \xi_0 - \delta^*_{min}\}. \tag{4}$$

4. Escape Time Sets

In order to define what is a extended safe set, it is necessary to review the idea of escape time of a trajectory starting in a region Q where exists a escaping dynamics. A point has a *escape time* n if the number of iterations that are needed to leave Q is n.

Using the idea of escape time it is possible to group all the points in the square Q in what we call T^n sets. We denote as T^n sets the sets of points in a square with a horseshoe dynamics, that stay inside the square under n iterations or more, that is:

$$T^n = \{p \in Q / f^n(p) \in Q\}. \tag{5}$$

This means that all the points with escape time n, $n+1$, $n+2\dots$ etc lie inside T^n. For example, T^3 would be composed of all the points with escape time 3, 4, 5 and so forth into infinity.

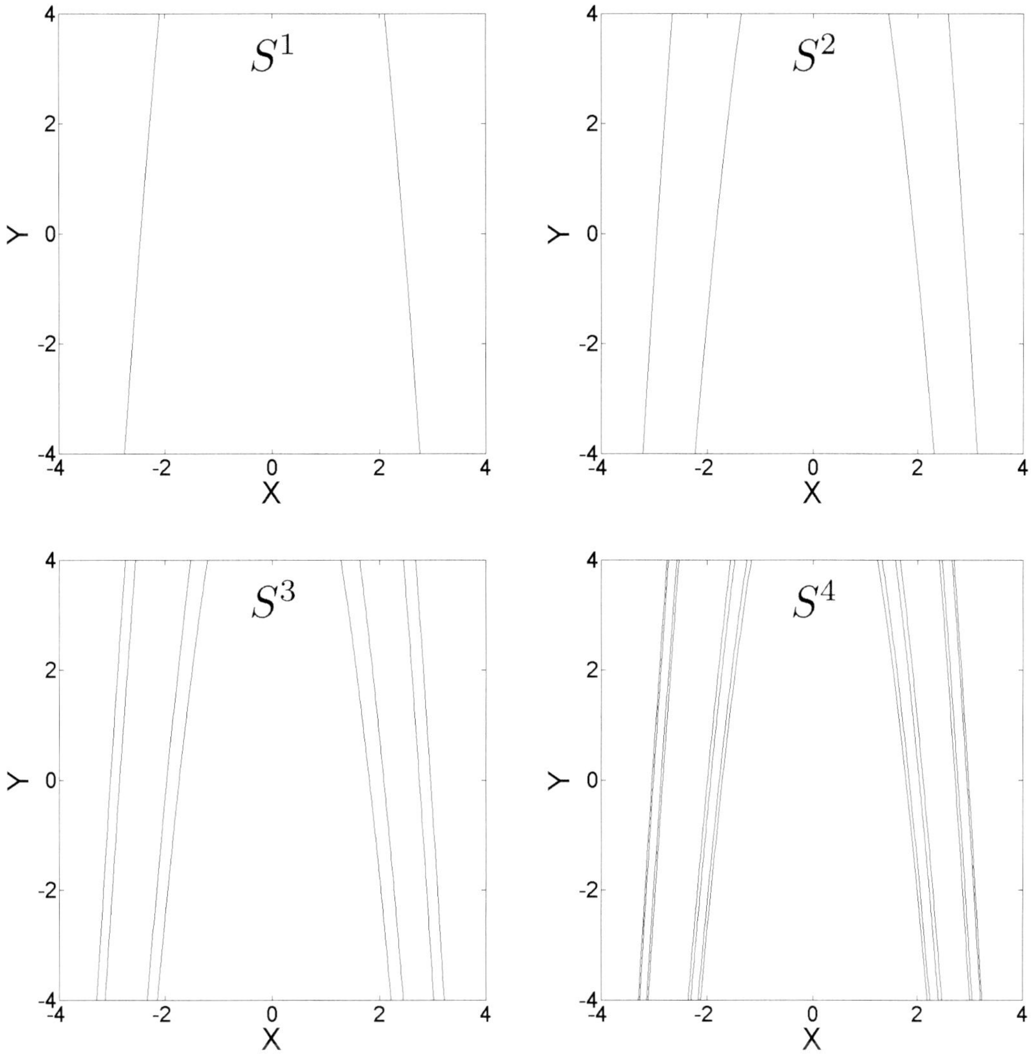

Fig. 3. The sets S^1, S^2, S^3 and S^4 that are the result of computing the corresponding preimage of S^0, that would be the vertical segment splitting the square into two equal rectangles. The set S^1 consists of a pairs of curves, the set S^2 has two pairs of curves, the set S^3 consists of four pairs of curves, and so on. Their geometrical properties are key for application of the partial control technique.

The appearance of the T^n sets, for a horseshoe-like map, are quite similar to curved strips as we can see in Fig. 6 for the T^3 set. With this graphical idea of the T^n sets it is possible to establish the following properties for the T^n sets:

- The T^n set consist of 2^n vertical strips.
- Inside each T^n we have all the sets of higher order, i.e $T^{n'>n} \subset T^n$.
- As the order n increases the width of the strips of the size of T^n sets decreases.
- All the strips of a T^n set can be grouped in pairs. Each pair would be composed of the two closest strips.

We also know that the safe sets of order n (S^n) are by definition inside the corresponding T^n set. This is due to the fact that on the one hand S^0 has escape time $n = 0$ and that on the other hand any S^n set is computed using n preimages of S^0. So it is quite obvious that, for example, S^1 has escape time $n = 1$ because in one iteration all their points are mapped to S^0 that still stays in Q. For S^2 we have that in one iteration we go to S^1 and in two iterations to S^0 so it has escape time $n = 2$. Thus it is clear that

$$S^n \subset T^n. \tag{6}$$

Considering this, it might be possible that some of the points inside T^n (not only those of S^n) could be used as "extended safe sets" (since they contain more points than the normal safe sets). However, not all the points of the T^n sets are valid to achieve

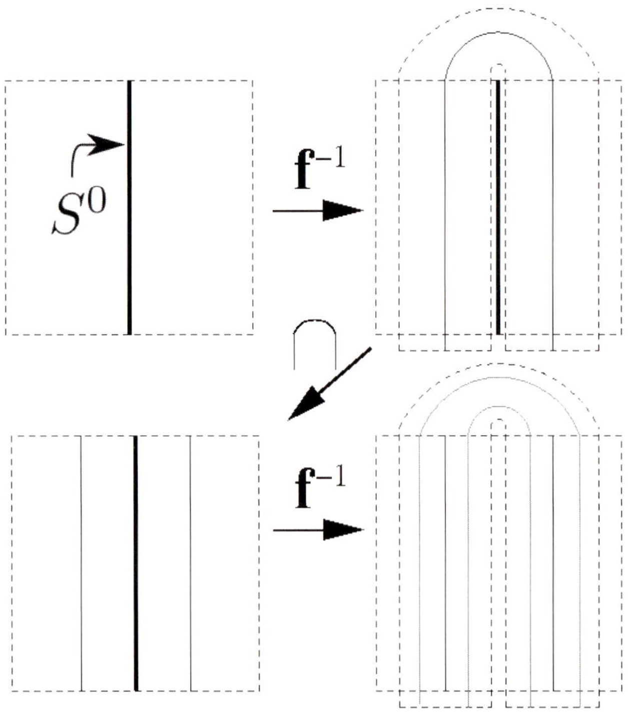

Fig. 4. Basic action to generate inductively the safe sets in a map showing a horseshoe-like behaviour in the phase space. We begin with a vertical segment (S^0) that splits the square in two equal halves and we compute the preimages of that line in Q. The intersections of the kth preimage of S^0 with Q gives the set S^k, that consists of 2^k curves.

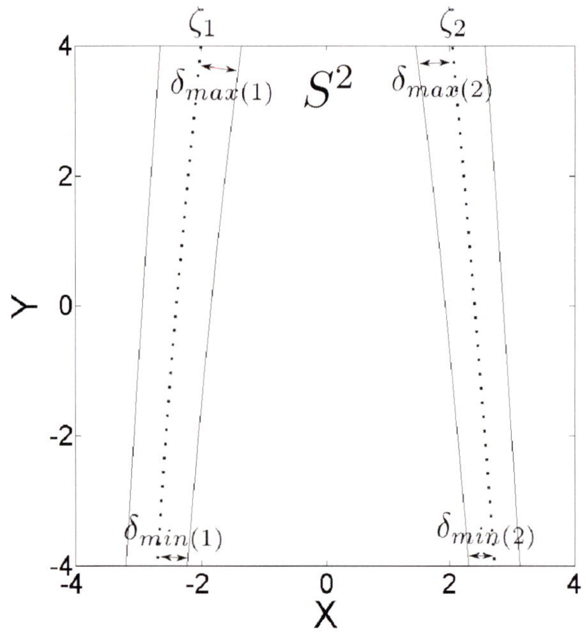

Fig. 5. These are the parameters needed in the partial control technique using the S^2 sets. The curves ζ_1 and ζ_2 are the curves whose points are at the same distance from a curve of each pair of curves of S^2. We can also see that $\delta_{max(i)}$ is the maximum distance from each pair of curves of S^2 to the curves ζ_i as well as $\delta_{min(i)}$ is the minimum distance.

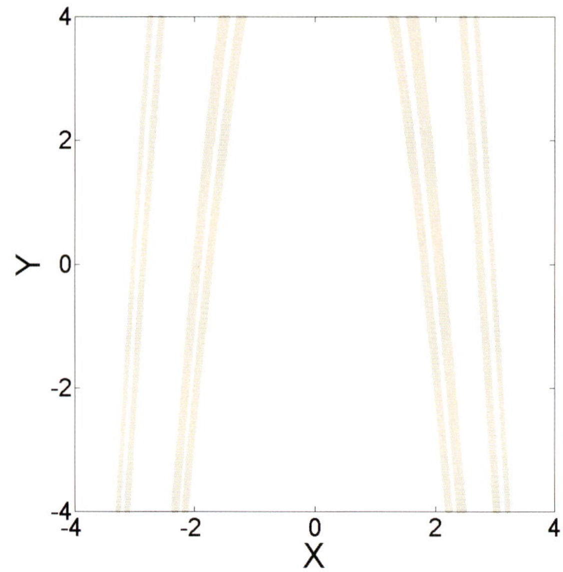

Fig. 6. The escape time set T^3, i.e., the set of points that escape from Q after 3 or more iterations. It consists of four pairs of strips, and it is reminiscent of S^3.

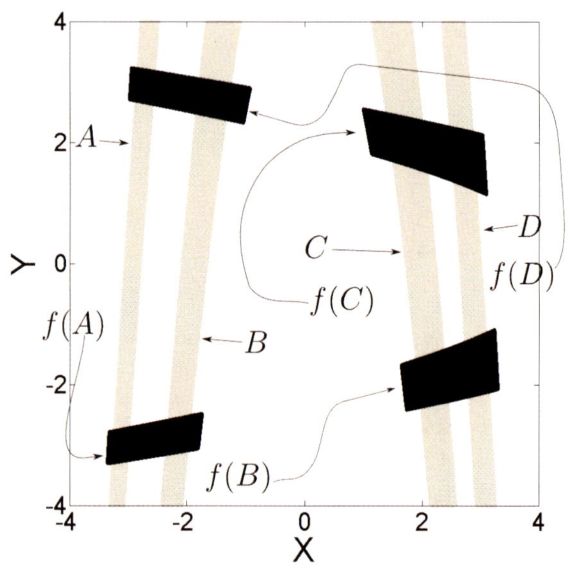

Fig. 7. In this picture we can see in grey the set of points of T^2 (the set of points that stay in the square under 2 or more iterations). We have also plotted in black the images of those points under one iteration. Due to the fact that the images of the strips do not lay between the pairs of strips, it is impossible to use the T^2 sets with a control smaller than the noise.

a control smaller than the noise, as we can see in Fig. 7. In this figure, we can see the forward iterate, in black, of the strips corresponding to the set T^2 (number of points that stay in the square under two iterations or more) which appear colored in grey. We can see in this figure that not all the parts of the forward iterates of the strips lie between some *"pair of strips"*, and this is true for other T^n sets. Thus, if we use the set T^n instead of a safe set in S^n in the partial control strategy it will force us for the worst conditions of noise to apply a control higher than the noise, so it would not be possible to keep trajectories inside Q with $u_0 < \xi_0$.

5. Obtaining Extended Safe Sets from Escape Time Sets

As we said above, clearly the escape time sets T^n are not a good substitute for safe sets S^n in the partial control technique. The next question would be: which points on T^n have to be discarded so that the partial control strategy can be applied with $u_0 \leq \xi_0$? In this section we address this question. We describe here the *maximal extended safe set* E^n_{max}, that is a subset of T^n that can be used in the partial control technique so that trajectories can be kept bounded with $u_0 = \xi_0$. From these sets we can easily define the *extended safe sets* E^n, and for any E^n we show that there is a ξ_0 so that trajectories can be kept inside Q with $u_0 < \xi_0$.

Recall the fixed point p^* shown in Fig. 1. We call $W^s(p^*)$ its stable manifold, and $W^s_L(p^*)$ the vertical curve that is a piece of $W^s(p^*)$ inside Q containing p^*. Consider now the set T^1 and the four vertical curves of $f^{-2}(W^s_L(p^*)) \cap Q$ shown in Fig. 8. We call E^1_{max} the set resulting of "cutting" the two strips of T^1 into two thinner strips as these four curves indicate. The strips of E^1_{max} are mapped as shown in Fig. 8. This is due to the horseshoe mapping and to the fact that points in the stable manifold map into points of the stable manifold under f. This is the limit point of the "good mapping" that we are searching for: the image of each strip of E^1_{max} falls into the space between the pair of strips of E^1_{max}.

Considering this, we define inductively

$$E^{n+1}_{max} = f^{-1}(E^n_{max}) \cap Q = f^{-n}(E^1_{max}) \cap Q. \quad (7)$$

Clearly the set E^n_{max} consists of 2^n strips, that also can be grouped in 2^{n-1} pairs of curves from left to right. Note that by definition it will be contained in T^n. Furthermore, it can be seen that the curves that bound each vertical strip of E^n_{max} are pieces of the stable manifold (since preimages of points of the stable manifold also belong to the stable manifold). These sets will reproduce the good kind of mapping observed for E^1_{max}: The image of each strip of E^n_{max} falls in the space between each pair of strips of E^n_{max}. This is shown for example in Fig. 9 for the set E^2_{max}. Using the geometrical considerations provided above, we can see that by using the sets E^n_{max} instead of S^n in the partial control strategy trajectories can be kept inside Q with $u_0 = \xi_0$.

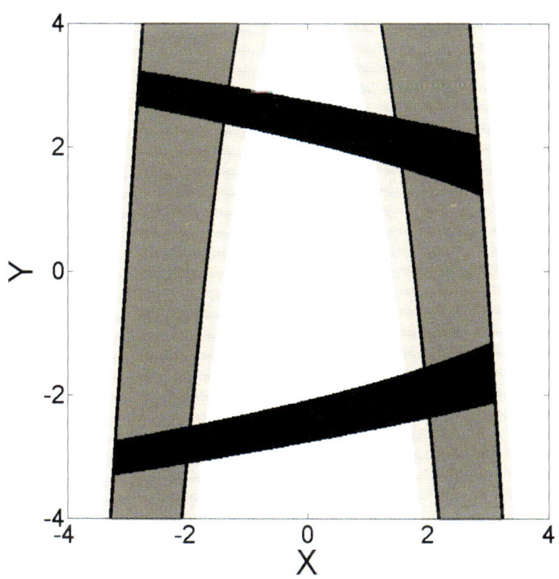

Fig. 8. The set T^1 (light grey) is subdivided using four pieces of the stable manifold of p^* (black curve) so we obtain the maximal extended safe set E^1_{max} (grey). The image of E^1_{max} under f is shown (black), each of its two pieces falls in the space between the pair of strips of E^1_{max}.

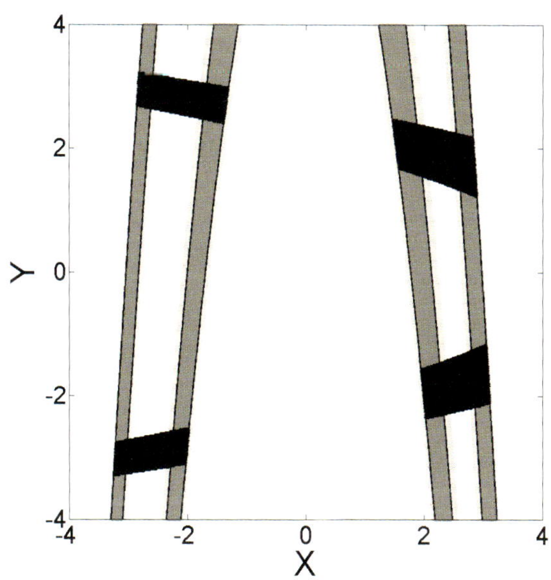

Fig. 9. The set T^2 (light grey) is cut using eight pieces of the stable manifold of p^* (black curve) and gives rise to the maximal extended safe set E^2_{max} (grey). The images of each strip of E^2_{max} under f are shown (black), they fall in the space between each pair of strips of E^2_{max}.

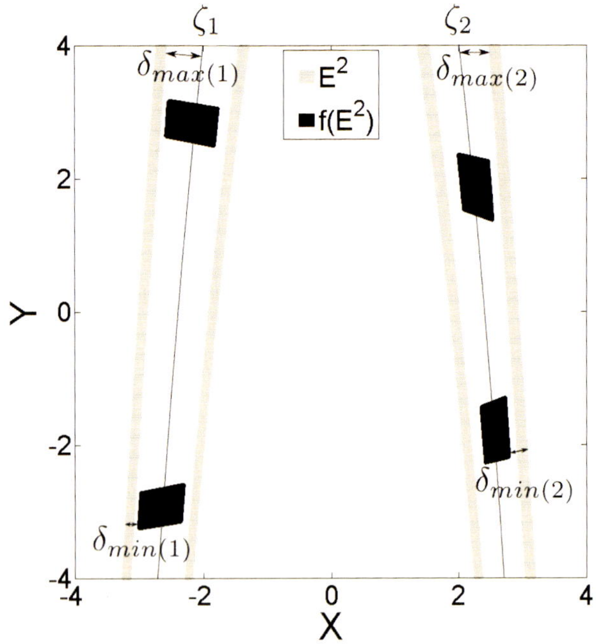

Fig. 10. In this figure we can see an extended safe set E^2 and the parameters needed in the partial control technique using the extended safe set E^2.

With these elements in mind, we can define the *extended safe sets* E^n as follows: An extended safe set E^n is a set of 2^n vertical strips, each of them inside a different strip of E^n_{max}, so that their vertical bounds do not intersect with the vertical bounds of the strips of E^n_{max}. Thus, a extended safe set is obtained when the width of all the strips of the extended safe sets is reduced. If we take zero-width strip we would obtain safe sets as the ones described in Sec. 3. An example of an extended safe set E^2 obtained from E^2_{max} is shown in Fig. 10.

Before concluding this section, we would like to point out that the same procedure that has been carried out with the Hénon map to obtain the extended safe sets could be repeated with the same level of difficulty for any topologically equivalent dynamical system, i.e., a dynamical system acting as a horseshoe on a (topological) square Q. It is important to notice that in order to find the extended safe sets one needs to find the chaotic saddle, a square enclosing it, the escape time sets and the stable manifold of the fixed point of the horseshoe. These can be calculated using time series of the system, and do not require to know exactly the form of the map f. Thus, we consider that this is a first advantage in the use of extended safe sets from the point of view of its applicability.

6. Simulations

Here we show the results that we have achieved by using this new technique. We have carried out different kind of simulations to check that this new technique works as we expected and how it performs for different levels of noise.

We have use as a basis to construct our extended safe sets the T^2 and T^3 escape sets. Of course not all the points inside T^2 or T^3 are valid to keep the condition of a control smaller than the noise. As we know from the 3th step of our strategy, we should remove at least the points that lie beyond the most external overlapping between the stable manifold of the saddle point and the escape time strips. In our simulation we have been a little bit more aggressive and we have removed all the points that had an image on some strip too. Those have been the E^n sets in our simulation.

In our first simulation, we have considered that the maximum level of noise present is $\xi_0 = 0.25$ and the area from where we want to avoid the escape is the square Q, where is located the chaotic saddle. In this situation it is clear that the most effective extended safe set that we can use here is E^3. We have carried out a succesful simulation of 1000 iterations with a control smaller than the noise, as can be seen in Fig. 11.

We have carried out another kind of simulation to compare the technique developed in [23] and the technique proposed here. As we expected, there are some values of the noise in which the S^n are a better strategy in terms of maximum control and

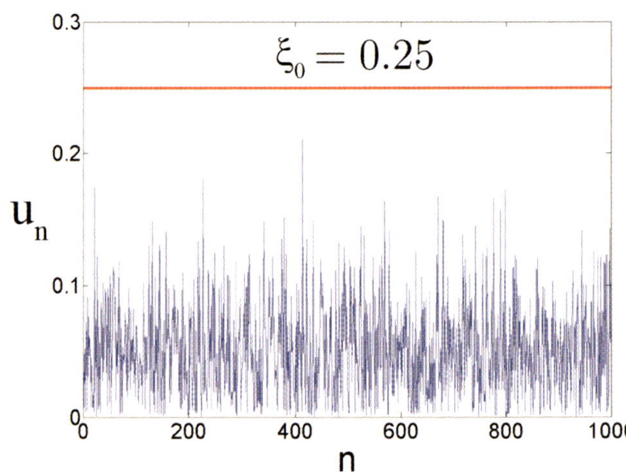

Fig. 11. Here we can see a simulation of 1000 iterations in $x_{n+1} = 6 - 0.4y_n - x_n^2; y_{n+1} = x_n$ where the maximum amplitude of noise is $\xi_0 = 0.25$. As we can check the control applied in each iteration u_0 is smaller than the noise.

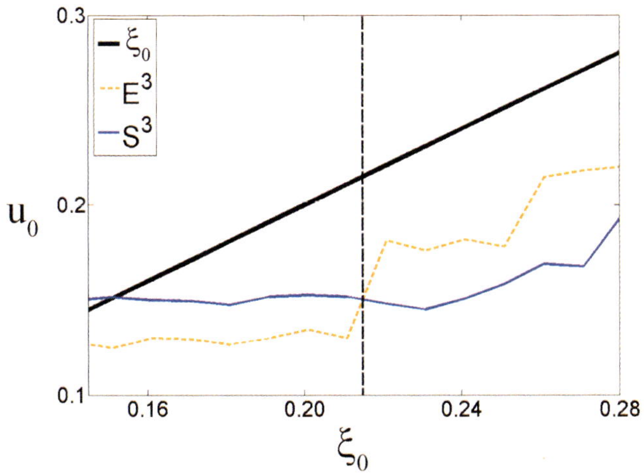

Fig. 12. Here we show a comparison of the safe sets technique and the extended safe sets technique for different values of noise. Here we have carried out series of 2000 iterations and we have computed the maximun control needed for different values of noise. As we can see in the graphic there are some values of noise where the extended safe sets technique is most efficient and others where the safe sets technique is the most efficient. It is also possible to see here that the extended safe set used in the simulation is going to be the optimum for a continuous uniform distribution of noise with a maximun amplitude of roughly 0.19.

others in which the E^n sets are a better solution. This is what we can see in Fig. 12. We can also see that there exists a transition point that indicates the limit where both techniques are more efficient.

7. An Algorithm to Automatically Compute the Safe Sets

The above methodology to find safe sets works in situations in which we have a map that acts like a horseshoe in a topological square Q. But it also provided us the necessary insights to find safe sets in more general settings. We sketch here an algorithm [12] that we have designed in order to find safe sets (and thus apply our partial control technique) in more general situations.

In order to find safe sets in more general settings, one has to think first which is the basic property of a safe set. Given a phase space region Q from which a trajectory escapes under a map f, and where there is a noise ξ bounded by ξ_0, and where it is possible to apply perturbations u also bounded by u_0 before each iteration, a safe set S is a set contained in Q that satisfies the following property:

$$\max_{p\in S, |\xi|\leq\xi_0} d(f(p)+\xi, S) = u_0 < \xi_0, \tag{8}$$

where $d(\cdot,\cdot)$ represents the distance between a point and a set of points.

We can see that the safe sets found for horseshoe maps in the previous section satisfy this condition. For this reason, it is not difficult to see why any set satisfying equation (8) works as a suitable safe set. Assume that p is any point on S. The map takes it to $f(p)$, but it will be deviated by the noise to $q = f(p) + \xi$. Equation (8) implies that no matter which point we consider and the value of ξ, as long as it is bounded by ξ_0, the distance between $q = f(p) + \xi$ and S will be *always* smaller or equal to u_0, and smaller than ξ_0. Thus, with an adequate control u smaller than ξ_0, we can put $f(q) + \xi + u$ on a point of the safe set and this can be repeated forever. Of course, the safe sets for horseshoe maps described previously satisfy this property.

Making use of this idea, we have designed an algorithm [12] that is able to find safe sets in any situation, provided that the phase space region from which trajectories escape Q, the value of the noise ξ_0 and the value of the control u_0 are known. The basic idea of this algorithm is to start with an initial set of points and to take recursively only points that are mapped in a good way (as prescribed by Eq. (8)), until converging to the desired safe set. We have also studied the dynamics of partial control by using the safe sets obtained with this new algorithm [13].

The procedure is then the following: we start with a given set of points $S(0)$ inside the region Q (it is possible to chose as $S(0)$ a grid of points on Q). Then, we pick as points of $S(1)$ only the points $p \in S(0)$ that satisfy the following condition for all ξ such that $|\xi| \leq \xi_0$:

$$d(f(p)+\xi, S(0)) \leq u_0. \tag{9}$$

To obtain S(2), S(3), etc... we apply the general rule: a point $p \in S(n)$ belongs to $S(n+1)$ if it satisfies that for all $|\xi| \leq \xi_0$:

$$d(f(p)+\xi, S(n)) \leq u_0. \tag{10}$$

Thus, we generate iteratively certain sets

$$S(0) \supset S(1) \supset S(2) \supset S(3) \supset \cdots \supset S(n). \tag{11}$$

The resulting set $S(\infty)$, when $n \to \infty$, unless it is not an empty set, is a safe set.

8. Conclusions and Discussion

In this chapter we have shown that it is possible to apply the partial control technique using certain sets, the extended safe sets, that are deeply related with the escape time sets, in a square Q where a horseshoe map exists. The notion of extended safe sets generalizes the notion of safe sets when we use these new sets in the partial control strategy, to keep trajectories bounded in the square, with a control smaller than the noise. The procedure to obtain extended safe sets from escape time sets has been described and it implies the cut of the escapes times using the stable manifold of a saddle fixed point in a horseshoe map (remember that every horseshoe map has associated two fixed points).

From an experimental point of view the use of extended safe sets, being a nonzero measure set, is advantageous. Having an extended surface, it is easier to place trajectories on it without making an error than when dealing with zero measure sets as the safe sets. On the other hand, by construction they are computed from the escape time sets and using the stable manifold of a fixed point as a guide, and this information can be inferred from time series of the system.

Throughout this chapter we have assumed that our control has no errors, that is, that at each iterate we can place the trajectory exactly where we want. However, this is not a critical assumption. As with safe sets, it is possible to keep the condition $u_0 < \xi_0$ using extended safe sets also if we have small control errors [21], i.e., even if at each iteration trajectories are not placed exactly on the extended safe sets. This tolerance to errors depends basically on the value of u_0 needed in absence of errors, the value of ξ_0 and on the expansiveness of the map f, which somehow tells how much are we penalized if we do not apply exactly the required control. To provide an analytical estimate of such tolerance, though, is complicated. However, due to the fact that extended safe sets are "thicker" than the safe sets, we expect that the tolerance for the former is bigger.

We have also described a general algorithm that allows to find safe sets for any nonlinear dynamical system in order to apply the partial control technique. Such safe sets can be found inside the regions from which trajectories escape after having some complex dynamical behavior, and their existence guarantees that trajectories can be kept inside that region with a control smaller than noise. Our algorithm is of a general nature, in the sense that it can be applied for any map or flow. The algorithm eliminates those parts from an initial phase space region that do not hold with the condition of the safe set to finally obtain the desired safe set. We call it Iterative Sculpting Algorithm, as an analogy to removing material as in sculpting a statue.

Finally, we want to emphasize that our analysis reveals the deep relation existing between the escape time sets and these extended safe sets, so we believe that any algorithm implemented in order to detect extended safe sets should make use of this relation: first, searching for the different escape time sets and then discarding the points that are not useful. As we said, this can be helpful both for an experimental detection of safe sets that can allow to obtain partial control with $u_0 < \xi_0$ as well as in generalizations of the partial control technique to dynamical systems in higher dimensions.

Acknowledgments

We acknowledge some fruitful discussions and suggestions with Prof. James A. Yorke. This work was supported by the Spanish Ministry of Science and Innovation under Project No. FIS2009-09898.

References

[1] Aguirre, J., d'Ovidio, F. and Sanjuán, M. A. F. (2004). Controlling chaotic transients: Yorke's game of survival, *Phys. Rev. E* **69**, p. 016203.
[2] Alligood, K. T., Sauer, T. D. and Yorke, J. A. (1997). *Chaos: An Introduction to Dynamical Systems*, 1st edn. (Springer).
[3] Bollt, E. M. (2005). The path towards a longer life: on the invariant sets and the escape time landscape, *Int. J. Bifurcation and Chaos* **15**, p. 1615.
[4] Dhamala, M. and Lai, Y. C. (1999). Controlling transient chaos in deterministic flows with applications to electrical power systems and ecology, *Phys. Rev. E* **59**, p. 1646.
[5] Franaszek, M. (1991). Influence of noise on the mean lifetime of chaotic transients, *Phys. Rev. A* **44**, p. 4065.
[6] In, V., Spano, M. L. and Ding, M. (1998). Mantaining chaos in high dimensions, *Phys. Rev. Lett.* **80**, p. 700.
[7] Kapitaniak, T. and Brindley, J. (1998). Preserving transient chaos, *Phys. Lett. A* **241**, p. 41.
[8] Lai, Y. C. and Tél, T. (2010). *Transient Chaos*, 1st edn. (Springer).
[9] Nusse, H. E. and Yorke, J. A. (1998). *Dynamics: Numerical Explorations*, 2nd edn. (Springer-Verlag, New York).
[10] Ott, E., Grebogi, C. and Yorke, J. A. (1990). Controlling chaos, *Phys. Rev. Lett.* **64**, pp. 1196–1199.
[11] Sabuco, J., Zambrano, S. and Sanjuán, M. A. F. (2010). Partial control of chaotic systems using escape times, *New Journal of Physics* **12**, p. 113038.

[12] Sabuco, J., Zambrano, S., Sanjuán, M. A. F. and Yorke, J. A. (2012a). Finding safety in partially controllable chaotic systems, *Commun. Nonlinear Sci. Numer. Simulat.* **17**, p. 4274–4280.

[13] Sabuco, J., Sanjuán, M. A. F. and Yorke, J. A. (2012b). Dynamics of partial control, *Chaos*, **22**(4), p. 4xxx.

[14] Schwartz, I. B. and Triandaf, I. (1996). Sustaining chaos by using basin boundary saddles, *Phys. Rev. Lett.* **77**, p. 4740.

[15] Tél, T. (1991). Controlling transient chaos, *J. Phys. A: Math. Gen.* **24**, pp. L1359–L1368.

[16] Tél, T. and Gruiz, M. (2006). *Chaotic Dynamics*, 1st edn. (Cambridge University Press, UK).

[17] Wagemakers, A., Zambrano, S. and Sanjuán, M. A. F. (2012). Partial control of transient chaos in electronic circuits, *Int. J. Bifurcation and Chaos*, **22**, p. 1250032.

[18] Yang, W., Ding, M., Mandell, A. J. and Ott, E. (1995). Preserving chaos: Control strategies to preserve complex dymamics with potential relevance to biological disorders, *Phys. Rev. E* **51**, p. 102.

[19] Zambrano, S. and Sanjuán, M. A. F. (2008a). Control of transient chaos using safe sets in simple dynamical systems, in V. Staicu (ed.), *Differential Equations, Chaos and Variational Problems* (Birkhäuser Publishing Ltd), pp. 425–435.

[20] Zambrano, S. and Sanjuán, M. A. F. (2008b). Partial control of a system with fractal basin boundaries, in M. D. Todorov (ed.), *Applications of mathematics in engineering and economics'34: 34th International Conference*, Vol. 1067 (AIP Proceedings), pp. 94–102.

[21] Zambrano, S. and Sanjuán, M. A. F. (2009a). Exploring partial control of chaotic systems, *Phys. Rev. E* **79**, p. 026217.

[22] Zambrano, S. and Sanjuán, M. A. F. (2009b). Partial control of the bouncing ball map, in M. Daniel and S. Rajasekar (eds.), *Nonlinear Dynamics* (Narosa Publishing House, New Delhi), pp. 343–350.

[23] Zambrano, S., Sanjuán, M. A. F. and Yorke, J. A. (2008). Partial control of chaotic systems, *Phys. Rev. E* **77**, p. 055201(R).

CHAPTER 32

PHASE AND COMPLETE SYNCHRONIZATIONS IN TIME-DELAY SYSTEMS*

D. V. SENTHILKUMAR

Potsdam Institute for Climate Impact Research, 14473 Potsdam Germany

M. MANJU SHRII and J. KURTHS*

Institute for Physics, Humboldt University, 12489 Berlin, Germany
Potsdam Institute for Climate Impact Research, 14473 Potsdam Germany
**Institute for Complex Systems and Mathematical Biology,*
University of Aberdeen, United Kingdom

Synchronization is a fundamental nonlinear phenomenon that has been intensively investigated during a couple of decades. Recently, synchronization of time-delay systems with or without delay coupling and even synchronization of low-dimensional dynamical systems described by ordinary differential equations and maps with delay coupling have become an active area of research in view of its potential applications. In this article, we provide an overview of our recent results on phase synchronization in time-delay systems, which usually exhibits hyperchaotic attractors with complex topological properties, noise-enhanced phase and noise-induced complete synchronizations in time-delay systems. Further, we demonstrate the phenomena of delay-enhanced and delay-induced stable synchronous chaos in a delay coupled network of time continuous dynamical system using the framework of master stability formalism (MSF) for the first time.

*We dedicate this work to Professor Leon O. Chua on the occasion of his 75th birthday in honor of his pioneering contribution to nonlinear science.

1. Introduction

Synchronization of chaotic systems driven by common signals have been an area of extensive research since the pioneering works of Fujisaka and Yamada [1] and of Pecora and Carroll [2]. Since the identification of complete (identical) chaotic synchronization, different kinds of chaotic synchronizations have been identified and demonstrated both theoretically and experimentally [3–11]. Recently, synchronization in coupled time-delay systems with or without time-delay coupling has become an active area of research by exploiting the infinite dimensional nature of the underlying systems, which exhibits a large number of positive Lyapunov exponents as a function of the delay time, for potential applications [12–26].

Among the different kinds of synchronization, chaotic phase synchronization (CPS) refers to the coincidence of characteristic time scales of interacting chaotic dynamical systems, while their amplitudes remain chaotic and often uncorrelated [3, 4]. CPS plays a crucial role in understanding a large class of weakly interacting nonlinear dynamical systems and has been demonstrated both theoretically and experimentally in a wide variety of natural systems [3, 4]. Despite our substantial understanding of the phenomenon of CPS and its potential applications in low-dimensional systems, only a very few studies on it have been reported on time-delayed systems, which are essentially infinite-dimensional in nature [20–25]. Due to the highly non-phase-coherent chaotic/hyperchaotic attractors with complex topological properties exhibited by these systems in general, it is often impossible to estimate the phase explicitly and to identify CPS in time-delay systems.

Nevertheless, we have introduced a nonlinear transformation to recast the original non-phase-coherent attractors into smeared limit-cycle attractors to enable to estimate the phase explicitly and to identify CPS in time-delay model systems for the first time in the literature [20]. This study has initiated further investigations in identifying and understanding the mechanism of phase synchronization transition in coupled time-delay systems [20–25]. In particular, phase synchronization in an array of time-delay systems along with their mechanism of transition to global phase synchronization has been reported [23], first experimental demonstration of phase synchronization in coupled time-delay systems with threshold nonlinearity has also been given [26].

Further, the effect of noise on synchronization of time-delay systems have not yet been well understood despite of enormous investigations on the effect of noise in low-dimensional systems. We have also investigated the phenomenon of noise-enhanced phase synchronization in coupled time-delay systems and noise-induced complete synchronization time-delay systems driven by a common Gaussian noise. Furthermore, we demonstrate stable synchronous chaos in a delay coupled network of time continuous dynamical system using the framework of master stability formalism (MSF) for the first time. By exploiting the MSF, we show that delayed coupling in an arbitrary network is capable of both inducing synchronization where there is no synchronization at all and enhancing synchronization to a large parameter space, which even the conjugate coupling without delay is incapable of.

In this chapter, we will provide a brief review of our recent results on synchronization in coupled time-delay systems and synchronization of an arbitrary network with delay coupling using the framework of the master stability formalism. In particular, in Sec. 2 we will discuss briefly the drawbacks in estimating phase in time-delay systems usually exhibiting highly non-phase-coherent hyperchaotic attractors with complex topological properties. We will provide a brief note on the different approaches that we have used in estimating phase and subsequently identifying chaotic phase synchronization in coupled time-delay systems in Sec. 3. Further, we will demonstrate the existence of chaotic phase synchronization in paradigmatic models such as a piecewise linear time-delay systems and Mackey-Glass systems in Sec. 4.

In Sec. 5, we will demonstrate the phenomenon of noise-enhanced phase synchronization in coupled time-delay systems and noise-induced complete synchronization time-delay systems driven by a common Gaussian noise. Further, we will show the phenomenon of delay-enhanced and delay induced synchronizations in an arbitrary network by exploiting the framework of master stability formalism in Sec. 6. Finally, we summarize our results in Sec. 7.

2. Chaotic Phase Synchronization and Time-delay Systems

A scalar time-delay systems with single constant delay is, in general, represented as

$$\dot{x}(t) = -ax(t) + bf(x(t-\tau)), \tag{1}$$

where a and b are positive constants, τ is a time-delay and $f(x)$ is an appropriate nonlinear function. For suitable, choice of the above parameters and nonlinear function, the above system can exhibit hyperchaotic attractors with complex topological properties, which is characterized by multiple positive Lyapunov exponents. Specific details of the paradigmatic models of Mackey-Glass, Ikeda and a couple of piecewise linear time-delay systems will be discussed in the next sections. More details can be found in [5].

CPS has been studied extensively during the last decade in various nonlinear dynamical systems as discussed in the introduction. We usually encounter with the terminologies phase-coherent and non-phase-coherent chaotic attractors while studying CPS. If the flow of a dynamical system has a proper rotation around a fixed reference point, then the corresponding attractor is termed as phase-coherent attractor. In contrast, if the flow does not have a proper rotation around a fixed reference point then the corresponding attractor is called as non-phase-coherent attractor. While methods have been well established in the literature to identify phase and to study CPS in phase-coherent chaotic attractors, methods to identify phase of non-phase-coherent chaotic attractors have not yet been well established.

Nevertheless, the methods available in the literature [3, 4] to calculate the phase of chaotic attractors are restricted to phase-coherent chaotic attractors, while the others to non-phase-coherent chaotic attractors of low-dimensional systems. It is to be noted that these conventional methods available so far in the literature to identify phase of the phase-coherent/non-phase-coherent attractors cannot be used in the case of time-delay systems in general, as such systems will very often exhibit more complicated attractors with more than one positive Lyapunov exponents. Correspondingly methods to calculate the phase of non-phase-coherent hyperchaotic attractors of time-delay systems are not readily available. The most promising approach available in the literature to calculate the phase of non-phase-coherent attractors is based on the concept of curvature [27], but this is often restricted to low-dimensional systems. However, we find that this procedure does not work in the case of nonlinear time-delay systems in general, where very often the attractor is non-phase-coherent and high-dimensional. Hence defining and estimating phase from the hyperchaotic attractors of the time-delay systems itself is a challenging task and so specialized techniques/tools have to be identified to introduce the notion of phase in such systems.

In the following, we introduce different approaches that we have employed to define/estimate phase and CPS in time-delay systems. At first, we have introduced a nonlinear transformation involving time-delay variable that transforms the non-phase-coherent attractors into phase-coherent attractors. After this transformation of the original non-phase-coherent attractor, the transformed attractor allows one to use the conventional techniques. Next, we have used the recurrence based indices for the first time in time-delay systems to identify the onset of PS and subsequent transition to generalized synchronization (GS). Recently, recurrence based indices are shown to be excellent quantifiers [28] of basic kinds of synchronization including CPS in low dimensional systems and even in the case of noisy, non-stationary data. Further, the existence of CPS in coupled time-delay systems can also be identified using the framework of the concept of localized sets [29]. The transition is also confirmed by the changes in the spectrum of Lyapunov exponents of the coupled time-delay system. Further, we find that all these approaches are in good agreement with the indication of onset of CPS.

3. Estimates of Phase and CPS

In this section, we account briefly on the different measures that we have employed in identifying CPS in coupled time-delay systems.

3.1. *Transformation of the original attractor*

We have introduced a nonlinear transformation to rescale the original non-phase-coherent chaotic attractor into smeared limit cycle like attractor with a single center of rotation. The transformation is performed by introducing a new state variable [20, 24],

$$z(t+\tau) = x(t)x(t+\hat{\tau})/x(t+\tau), \qquad (2)$$

where $\hat{\tau}$ is the optimal value of time delay to be chosen in order to avoid any additional center of rotation. This functional form of the transformation (along with a delay time $\hat{\tau}$) has been identified by generalizing the transformation used in the case of chaotic attractors in the Lorenz systems [3]. Now, the projected trajectory in the new state space $(x(t+\tau), z(t+\tau))$ will result in a smeared limit cycle like an attractor with a single fixed center of rotation. Conventional approaches for estimating phase of phase-coherent attractors can now be applied to the transformed attractor in the new state space to estimate the phase explicitly and to identify CPS in time-delay systems with non-phase-coherent attractors. The other estimates that we have employed to identify CPS does not involve estimation of the phase explicitly but instead provides a qualitative and quantitative confirmation of existence of CPS.

It is to be noted that the above transformation (2) can be applied to the non-phase-coherent attractors of any time-delay system in general, except for the fact that the optimal value of $\hat{\tau}$ should be chosen for each system appropriately through trial and error by requiring the geometrical structure of the transformed attractor to have a fixed center of rotation. We have adopted here a geometric approach for the selection of $\hat{\tau}$ and look for an optimum transform which leads to a phase-coherent structure. This is indeed demonstrated for the attractor of piecewise linear and Mackey-Glass time-delay systems in the next section [20,24]. The main point that we want to emphasize here is that even for highly non-phase-coherent hyperchaotic attractors of time-delay systems, there is every possibility to identify suitable transformations of the type (2) to unfold the attractor and to identify phase as we will demonstrate for two paradigmatic time-delay systems. However, formulating a more generalized transformation to include non-phase-coherent attractors of a large class of time-delay systems remains an open problem.

3.2. *Recurrence measures*

The complex synchronization phenomena in the coupled time-delay systems can also be analyzed by means of methods based on recurrence plots [28,30]. These methods help to identify and quantify CPS particularly in non-phase coherent attractors. For this purpose, the generalized autocorrelation function $P(t)$ has been introduced in [28, 30] as

$$P(t) = \frac{1}{N-t}\sum_{i=1}^{N-t}\Theta(\epsilon - \|x_i - x_{i+t}\|), \qquad (3)$$

where Θ is the Heaviside function, x_i is the ith data corresponding to either the drive variable or the response variable and ϵ is a predefined threshold. $\|\cdot\|$ is the Euclidean norm and N is the number of data points. $P(t)$ can be considered as a statistical measure about how often ϕ has increased by 2π or multiples of 2π within the time t in the original space. If two systems are in CPS, their phases increase on average by $K.2\pi$, where K is a natural number, within the same time interval t. The value of K corresponds to the number of cycles when $\|x(t+T) - x(t)\| \sim 0$, or equivalently when $\|x(t+T) - x(t)\| < \epsilon$, where T is the period of the system. Hence, looking at the coincidence of the positions of the maxima of $P(t)$ for both systems, one can qualitatively identify CPS.

A criterion to quantify CPS is the cross correlation coefficient between the drive, $P_1(t)$, and the response, $P_2(t)$, which can be defined as Correlation of Probability of Recurrence (CPR)

$$CPR = \langle \bar{P}_1(t)\bar{P}_2(t)\rangle/\sigma_1\sigma_2, \qquad (4)$$

where $\bar{P}_{1,2}$ means that the mean value has been subtracted and $\sigma_{1,2}$ are the standard deviations of $P_1(t)$ and $P_2(t)$ respectively. If both systems are in CPS, the probability of recurrence is maximal at the same time t and CPR ≈ 1. If they are not in CPS, the maxima do not occur simultaneously and hence one can expect a drift in both the probability of recurrences and low values of CPR.

Further, two indices are defined to quantify GS, which we will use to characterize GS in the next section. The Joint Probability of Recurrence (JPR) [30],

$$JPR = \frac{\frac{1}{N^2}\sum_{i,j}^{N}\Theta(\epsilon_x - \|x_i - x_j\|)\Theta(\epsilon_y - \|y_i - y_j\|) - RR}{1 - RR} \qquad (5)$$

where RR is rate of recurrence, ϵ_x and ϵ_y are thresholds corresponding to the drive and response systems respectively and x_i is the ith data corresponding to the drive variable and y_i is the ith data corresponding to the response variable. RR measures the density of recurrence points and it is fixed as 0.02 [30]. JPR is close to 1 for systems in GS and is small when they are not in GS. The second index depends on the coincidence of the probability of recurrence, which is defined as Similarity of Probability of Recurrence (SPR),

$$SPR = 1 - \langle(\bar{P}_1(t) - \bar{P}_2(t))^2\rangle/\sigma_1\sigma_2. \qquad (6)$$

SPR is of order 1 if both systems are in GS and approximately zero or negative if they evolve independently.

3.3. *Concept of localized sets*

Another interesting framework to identify CPS is the concept of localized sets [29]. This approach provides an easy and efficient way to detect CPS especially in complicated non-phase-coherent attractors. The basic idea of this concept is to define a typical event in one of the systems and then observe the other system whenever this event occurs. These observations give raise to a set D. Depending upon the property of this set D, one can state whether PS exists or not. The coupled systems evolve independently if the sets obtained by observing the corresponding events in the systems spread over the attractor of the systems. On the other hand, if the sets are localized on the attractors then CPS exists between them.

3.4. *Lyapunov exponents*

A transition of a zero Lyapunov exponent of the response system to negative values is used to characterize the onset of phase synchronization in the low-dimensional systems. Chaotic phase synchronization in time-delay systems can also be characterized by the transitions in the Lyapunov exponents of the coupled time-delay systems [20,24].

4. Chaotic Phase Synchronization in Coupled Time-delay Systems

We consider the following unidirectionally coupled drive $x_1(t)$ and response $x_2(t)$ systems,

$$\dot{x}_1(t) = -ax_1(t) + b_1 f(x_1(t-\tau)), \qquad (7a)$$

$$\dot{x}_2(t) = -ax_2(t) + b_2 f(x_2(t-\tau)) + b_3 f(x_1(t-\tau)), \qquad (7b)$$

where b_1, b_2 and b_3 are constants, $a > 0$, τ is the delay time and $f(x)$ is an appropriate nonlinear function. In the following, we demonstrate the emergence of CPS in paradigmatic models of a piecewise linear and Mackey-Glass time-delay systems.

4.1. *CPS in coupled piecewise linear time-delay systems*

The nonlinear function $f(x)$ in the above coupled time-delay systems corresponding to the piecewise linear time-delay systems, which we have studied in detail in [31,32], is an odd piecewise linear function defined as

$$f(x) = \begin{cases} 0, & x \le -4/3 \\ -1.5x - 2, & -4/3 < x \le -0.8 \\ x, & -0.8 < x \le 0.8 \\ -1.5x + 2, & 0.8 < x \le 4/3 \\ 0, & x > 4/3 \end{cases} \qquad (8)$$

We have chosen the value of the parameters as $a = 1.0, b_1 = 1.2, b_2 = 1.1$ and $\tau = 15$. For this parametric choice, in the absence of coupling, the drive $x_1(t)$ and the response $x_2(t)$ systems evolve independently, which exhibit hyperchaotic attractors (Fig. 1) for the chosen parameter values [20]. The original and the transformed hyperchaotic attractors of the piecewise linear time-delay system are shown in Figs. 1(a) and 1(b), respectively. Open circles in Fig. 1(b) correspond to the Poincaré points of the smeared limit-cycle-like attractor. We find the optimal value of $\hat{\tau}$ for the transformation specified by Eq. (2) for the attractor (Fig. 1(a)) of the piecewise linear time-delay system to be 1.6. It is to be noted that on closer examination of the transformed attractor (Fig. 1(b)) in the vicinity of the common center, it does not have any closed loop, unlike the

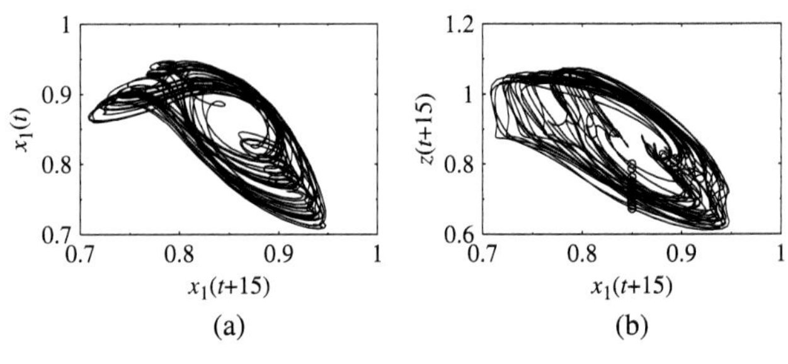

Fig. 1. (a) The non-phase coherent hyperchaotic attractor of the uncoupled piecewise linear time-delay system (7a) for the parameter values $a = 1.0, b = 1.2$ and $\tau = 15$. and (b) Transformed attractor in the $x_1(t+\tau)$ and $z(t+\tau)$ space. Here the Poincaré points are represented as open circles.

case of the original attractor, even though the trajectories show sharp turns in some regime of the phase space.

4.1.1. *CPS from the transformed attractor*

Now, the phase of the transformed attractor can be defined based on the Poincaré section technique [3]. Phases, $\phi_1^z(t)$ and $\phi_2^z(t)$, of the drive $x_1(t)$ and the response $x_2(t)$ systems, respectively, are calculated from the state variables $z_1(t+\tau)$ and $z_2(t+\tau)$. The phase differences ($\Delta\phi = \phi_1^z(t) - \phi_2^z(t)$) between the drive and the response systems are shown in Fig. 2 for different values of the coupling strength b_3. The phase difference $\Delta\phi$ between the coupled piecewise linear time-delay systems for $b_3 = 0.0$ (uncoupled) increases monotonically as a function of time confirming that both systems are in an asynchronous state (also nonidentical) in the absence of coupling between them. For the values of $b_3 = 1.0$ and 1.3, the phase slips in the corresponding phase difference $\Delta\phi$ show that the systems are in a transition state. The strong boundedness of the phase difference is obtained for $b_3 > 1.382$ and it becomes zero for the value of the coupling strength $b_3 = 1.5$, showing a high quality CPS [20].

4.1.2. *CPS from recurrence quantification analysis*

We estimate the recurrence based measures discussed in Sec. 3.2 from 5000 data points after sufficient transients with the integration step $h = 0.01$ and sampling rate $\Delta t = 100$. The generalized autocorrelation functions $P_1(t)$ and $P_2(t)$ (Fig. 3(a)) for the coupling $b_3 = 0.6$ show that the maxima of both systems do not occur simultaneously and there exists a drift between them,

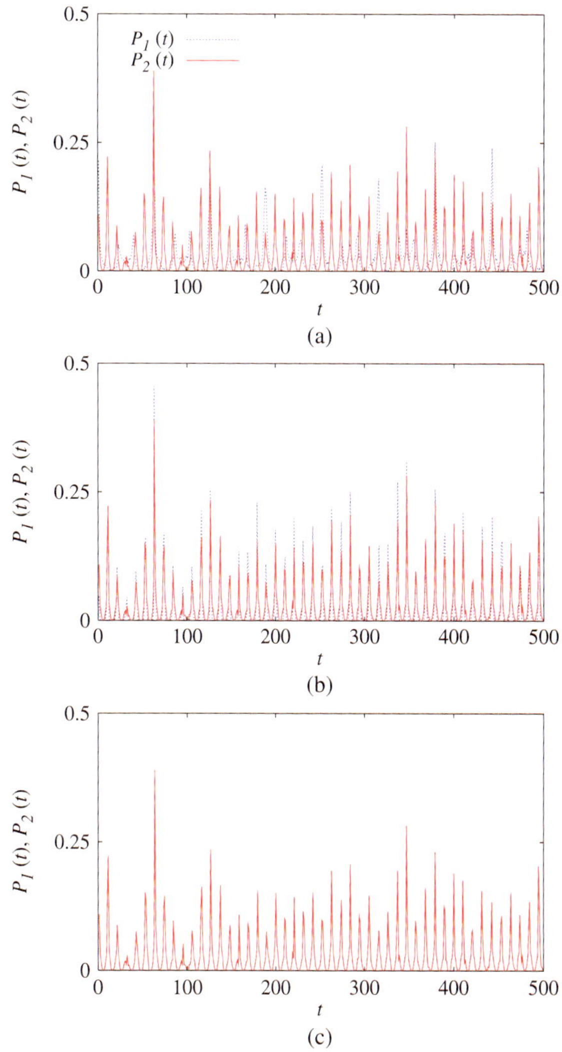

Fig. 3. Generalized autocorrelation functions of both the drive $P_1(t)$ and the response $P_2(t)$ systems. (a) Non-phase synchronization for $b_3 = 0.6$. (b) Phase synchronization for $b_3 = 1.5$. (c) Generalized synchronization for $b_3 = 2.3$.

so there is no synchronization at all. This is also reflected in the rather low value of CPR = 0.381. For $b_3 \in (0.78, 1.381)$, from Fig. 4 we observe the first substantial increase of recurrence reaching CPR $\approx 0.5 - 0.6$. Looking into the details of the generalized correlation functions $P(t)$, we find that now the main oscillatory dynamics becomes locked, i.e. the main maxima of P_1 and P_2 coincide. For $b_3 \in (1.382, 2.2)$ CPR reaches almost 1 as seen in Fig. 4, while now the positions of all maxima of P_1 and P_2 are also in agreement and this is in accordance with strongly bounded nature of phase differences. This is a strong indication for CPS. Note, however that the heights of the peaks are clearly different (Fig. 3(b)). The differences in the peak heights indicate that there is no strong interrelation

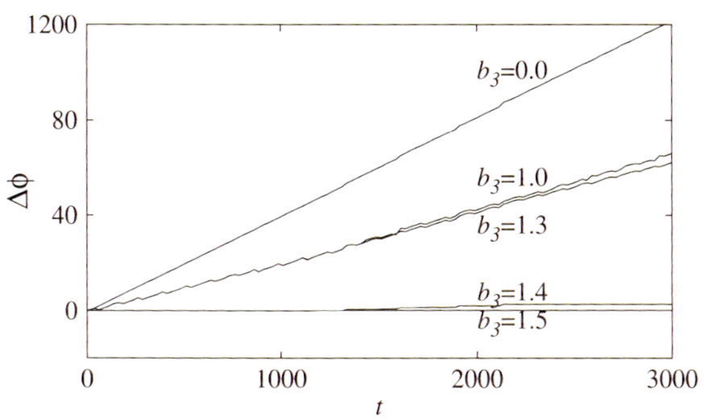

Fig. 2. Phase differences ($\Delta\phi = \phi_1^z(t) - \phi_2^z(t)$) between the coupled piecewise linear time-delay systems for different values of the coupling strength $b_3 = 0.0, 1.0, 1.3, 1.4$ and 1.5.

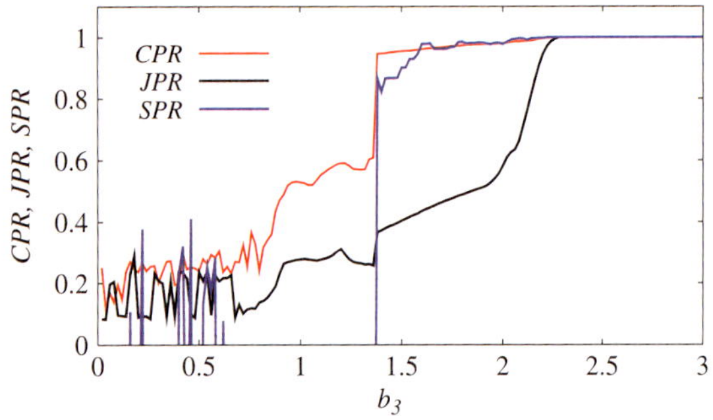

Fig. 4. Indices CPR, JPR and SPR as a function of coupling strength $b_3 \in (0, 3)$.

in the amplitudes. Further increase of the coupling (here $b_3 = 2.21$) leads to the coincidence of both the positions and the heights of the peaks (Fig. 3(c)) referring to GS in systems (7). This is also confirmed from the maximal values of the indices JPR = 1 and SPR = 1, which is due to the strong correlation in the amplitudes of both systems.

4.1.3. *CPS from the framework of localized sets*

We have also confirmed the existence of CPS in coupled piecewise linear time-delay systems using the concept of localized set. We have defined the event in the attractors of both the drive and response system as Poincaré sections, which are shown as open squares in Fig. 5. The sets obtained by observing the response system whenever the defined event occurs in the drive system and that corresponds to drive system whenever the defined event occurs in the response system are shown as filled circles. The observed sets are distributed over both the drive and response attractors as shown in Figs. 5(a) and 5(b), respectively, for the value of the coupling strength $b_3 = 0.1$, at which both the systems evolve independently. For the value of $b_3 = 1.5$, the observed sets are localized on both the attractors as shown in Figs. 5(c) and 5(d) confirming the existence of CPS between both the drive and response systems.

4.1.4. *CPS from spectrum of Lyapunov exponents*

The transition from non-synchronization to CPS is also characterized by changes in the Lyapunov exponents of the coupled time-delay systems (7). The spectrum of the eight largest Lyapunov exponents of the coupled systems is shown in Fig. 6. From this figure one can find that all the positive

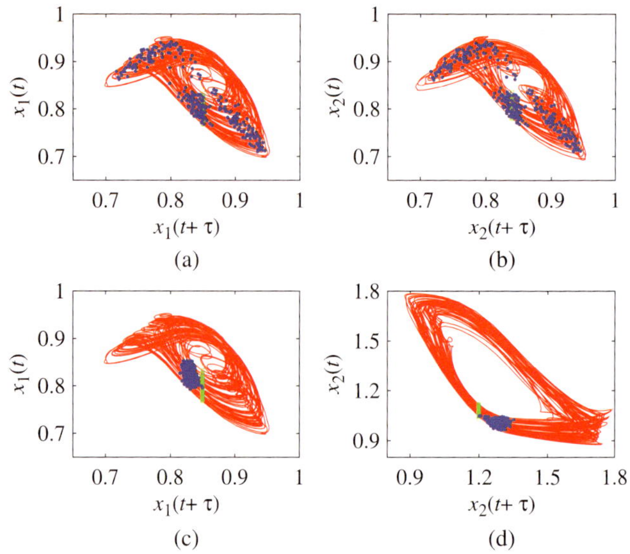

Fig. 5. (a) and (c) Attractor of the drive system, (b) and (d) attractor of the response system. The open squares indicate the events in the corresponding attractors. In (a) and (b) the sets spread over the attractor and hence there is no CPS for the value of the coupling strength $b_3 = 0.1$ and, in (c) and (d) the sets are localized confirming the existence of CPS for $b_3 = 1.5$.

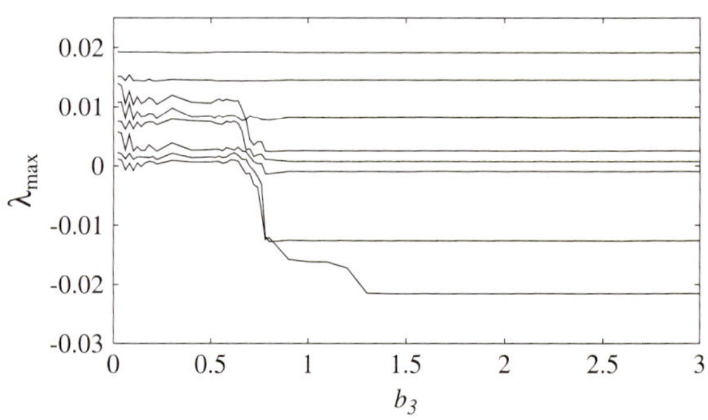

Fig. 6. Spectrum of first eight largest Lyapunov exponents of the coupled systems (7) as a function of coupling strength $b_3 \in (0, 3)$.

Lyapunov exponents, except the largest one ($\lambda_{max}^{(2)}$), corresponding to the response system suddenly become negative at the value of the coupling strength $b_3 = 0.78$ which is an indication of the onset of transition regime. One may also note that at this value of b_3 already one of the Lyapunov exponents of the response system attains negative saturation while another one reaches negative saturation slightly above $b_3 = 0.78$. This is a strong indication that in this rather complex attractor the amplitudes become somewhat interrelated already at the transition to CPS (as in the funnel attractor [27] of the Rössler system). Also the third positive Lyapunov exponent of the response system gradually becomes more negative from $b_3 = 0.78$ and reaches its saturation value at $b_3 = 1.381$ confirming the onset of CPS (which is also indicated by the transition of the indices of CPR and SPR in Fig. 4 in the range of $b_3 \in (0.78, 1.381)$). It is interesting to note that the Lyapunov exponents of the response system $\lambda_i^{(2)}$ (other than $\lambda_{max}^{(2)}$) are changing already at the early stage of CPS ($b_3 \in (0.78, 1.381)$), where the complete CPS is not yet attained. This has also been observed for the onset of CPS in phase-coherent and non-phase-coherent oscillators without time-delay [33–35].

4.2. *CPS in coupled Mackey-Glass time-delay systems*

The nonlinear function $f(x)$ in the coupled time-delay systems, Eq. (7), corresponding to the Mackey-Glass time-delay systems is represented as

$$f(x) = x(t-\tau)/(1.0 + x(t-\tau)^{10}). \tag{9}$$

We have chosen the parameter values as $a = 0.1$, $b_1 = 0.2$, $b_2 = 0.205$ and $\tau = 20$. The non-phase-coherent chaotic attractor of the uncoupled Mackey-Glass system for the above choice of parameters is shown in Fig. 7(a). The transformed attractor, for the optimal value of the delay time $\hat{\tau} = 8.0$ in Eq. (2), is depicted in Fig. 7(b). The Poincaré points are shown as open circles in the Fig. 7(b) from which the instantaneous phase $\phi_1^z(t)$ is calculated using the Poincaré section technique. The phase differences $\Delta\phi = \phi_1^z(t) - \phi_2^z(t)$ between the coupled Mackey-Glass systems for the values of the coupling strength $b_3 = 0.04, 0.08, 0.11, 0.12$ and 0.3 are shown in Fig. 8. For the value of the coupling strength $b_3 = 0.3$, there exists a strong boundedness in the phase difference indicating the existence of CPS [21].

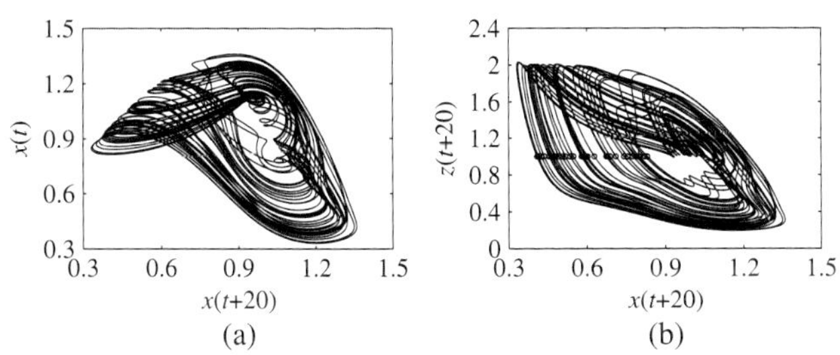

Fig. 7. (a) The non-phase coherent chaotic attractor of the Mackey-Glass time-delay system for the parameter values $a = 0.1, b_1 = 0.2$ and $\tau = 20$. (b) Transformed attractor in the $x_1(t+\tau)$ and $z(t+\tau)$ space along with the Poincaré points represented as open circles.

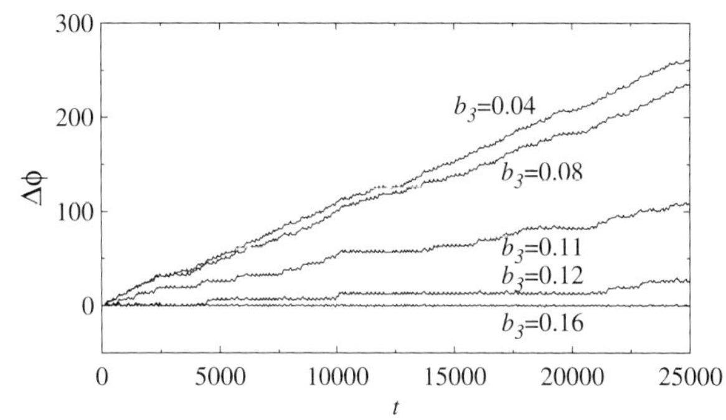

Fig. 8. Phase differences ($\Delta\phi = \phi_1^z(t) - \phi_2^z(t)$) between the coupled Mackey-Glass systems for different values of the coupling strength $b_3 = 0.04, 0.08, 0.11, 0.12$ and 0.3.

The existence of CPS from the original non-phase-coherent chaotic attractors of the Mackey-Glass systems is now investigated using the recurrence quantification measures defined in the Sec. 3.2. We have estimated these measures again using a set of 5000 data points, and the same integration step and the sampling rate as used in Sec. 4.1.2. The maxima of generalized autocorrelations of both the drive $P_1(t)$ and the response $P_2(t)$ systems (Fig. 9(a)) do not occur simultaneously for $b_3 = 0.1$, which indicates the independent evolution of both the systems without any correlation and this is also reflected in the rather low value of CPR $= 0.4$. For $b_3 = 0.3$, the maxima of both $P_1(t)$ and $P_2(t)$ are in good agreement (Fig. 9(b)) and this shows the strongly bounded phase difference. It is to be noted that even though both the maxima coincide, the heights of the peaks are clearly of different magnitudes contributing to the fact that there is no strong correlation in the amplitudes of both the systems indicating CPS. Both the positions and the peaks are in coincidence (Fig. 9(c)) for the value of coupling strength $b_3 = 0.9$ in accordance with the strong correlation in the amplitudes of both systems corresponding to GS. This is also reflected in the maximal values of both JPR=1 and SPR = 1. The spectra of CPR, JPR and SPR are shown in Fig. 10. The onset of CPS is shown by the first substantial increase of the index CPR at $b_3 = 0.11$ and the transition regime is shown by the successive plateaus of CPR in the range $b_3 \in (0.12, 0.23)$. The maximal values of CPR for $b_3 > 0.23$ indeed confirm the existence of high quality CPS.

As discussed in the case of piecewise linear time-delay systems, we have also confirmed the existence of CPS using the concept of localized sets. The event is chosen as Poincaré sections, which are plotted as open squares in Fig. 11. Observed

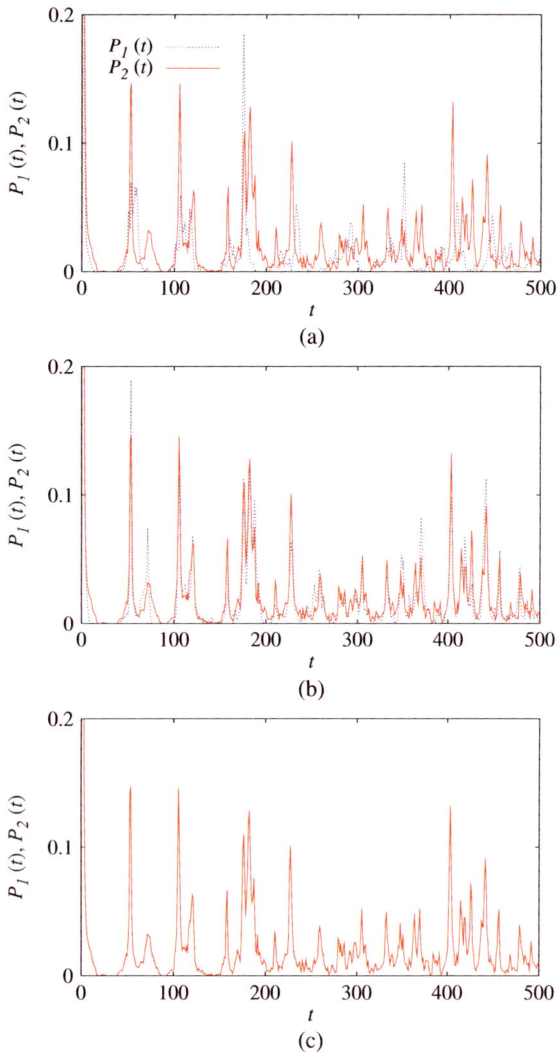

Fig. 9. Generalized autocorrelation functions of both drive systems (7a and 9), $P_1(t)$, and the response system (7b and 9), $P_2(t)$. (a) Non-phase synchronization for $b_3 = 0.1$. (b) Phase synchronization for $b_3 = 0.3$. (c) Generalized synchronization for $b_3 = 0.9$.

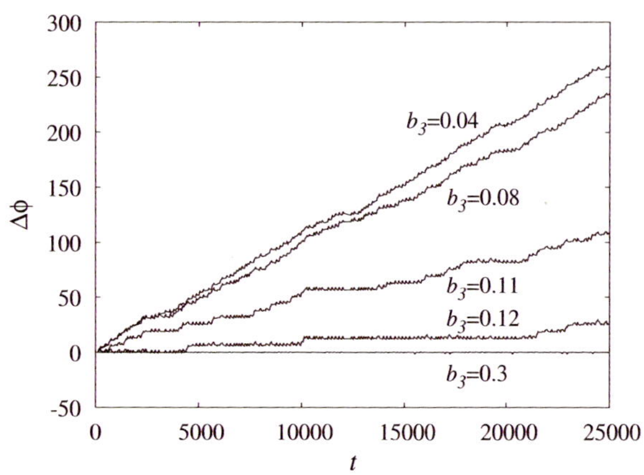

Fig. 10. Indices CPR, JPR and SPR as a function of coupling strength $b_3 \in (0, 1)$.

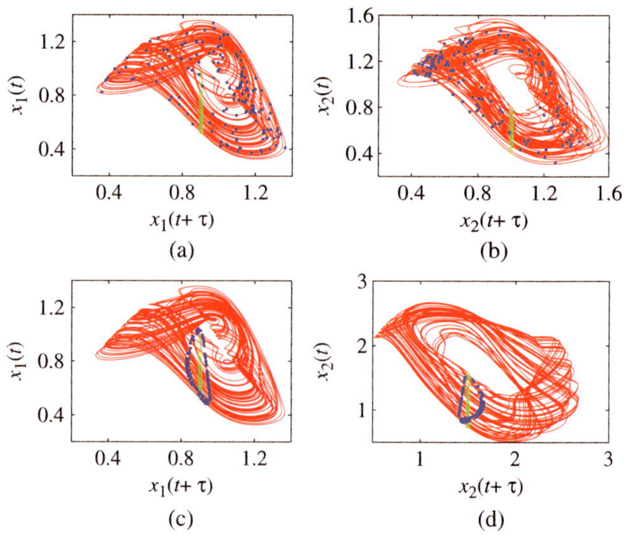

Fig. 11. (a) and (c) Attractor of the drive system, (b) and (d) attractor of the response system of the coupled Mackey-Glass systems. The open squares indicate the events in the corresponding attractors. In (a) and (b) the sets spread over the attractor and hence there is no CPS for the value of the coupling strength $b_3 = 0.04$ and, in (c) and (d) the sets are localized confirming the existence of CPS for $b_3 = 0.3$.

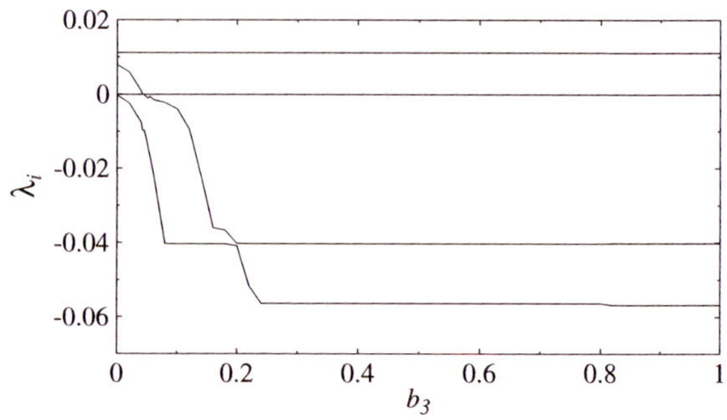

Fig. 12. Spectrum of first four largest Lyapunov exponents of the coupled Mackey-Glass systems as a function of coupling strength $b_3 \in (0, 1)$.

sets from drive/response attractors collected during the Poincaré sections in response/drive attractors are depicted as closed circles. When both the drive and the response systems evolve independently, the observed sets are spread all over the attractors of both systems as shown in Figs. 11(a) and 11(b) for the value of $b_3 = 0.04$. The observed sets are localized on the attractor during the phase synchronized state as shown in Figs. 11(c) and 11(d) for the value of coupling strength $b_3 = 1.5$, confirming the existence of CPS.

The onset of CPS is also characterized by the changes in the spectrum of Lyapunov exponents of the coupled Mackey-Glass systems. The spectrum of the first four largest Lyapunov exponents of the coupled systems is shown in Fig. 12. The zero

Lyapunov exponent of the response system $x_2(t)$ already becomes negative as soon as the coupling is introduced and the onset of CPS is indicated by the negative saturation of the zero Lyapunov exponent at $b_3 = 0.11$. The positive Lyapunov exponent of the response system becomes gradually negative in the transition regime ($b_3 \in (0.12, 0.23)$) and it reaches its negative saturation at $b_3 = 0.23$ at which high quality CPS exists. The transition of the positive Lyapunov exponent to negativity in this rather complex attractor is again a firm indication of some degree of correlation in the amplitudes of both systems even before the onset of CPS. As noted earlier, this behaviour of negative transition of a positive Lyapunov exponent of the response system before CPS has also been observed in [33–35].

4.3. *Experimental confirmation of CPS in time-delay systems*

Experimental confirmation of CPS has already been discussed in the Chapter by Lakshmanan and Murali in the Sec. 8.2 of this book.

5. Effect of Noise on Synchronization in Time-delay Systems

Effects of noise on synchronization have also been studied extensively [3, 4, 36]. The phenomenon of "bubbling", i.e., noise-induced desynchronization due to local instability of the synchronization manifold [37] is an example of a non-constructive effect of noise. Nevertheless, noise can also induce and enhance synchronization in coupled/uncoupled oscillators [38–40]. The influence of stochastic forces is also of great importance in various applications [41,42]. For instance, stochastic process in the central nervous system is essential for the expression of spatiotemporal patterns [43]. Hence investigating noise effects on CPS of delay dynamical systems will provide a better understanding on the stochastic process mediating various natural phenomena including the information processing in the central nervous system. In this section, we will illustrate the phenomenon of noise-enhanced phase synchronization in coupled time-delay systems with threshold nonlinearity. Further, we will show that a common Gaussian noise can drive two independent time-delay systems to evolve in complete synchrony without establishing any explicit coupling between them.

5.1. *Noise-enhanced phase synchronization in time-delayed systems*

Here, we will investigate the phenomenon of noise-enhanced PS in coupled time-delay systems along with its mechanism. As PS is the coincidence of characteristic time scales of interacting dynamical systems, it is cruical and interesting to understand the interplay of noise and the multiple time scales in achieving PS in such systems for various applications. In particular, we will consider the piecewise linear time-delay system with a threshold nonlinearity with parameter mismatches and demonstrate that both an independent and a common additive noise can enhance PS. Interestingly, multiple time scales of a time-delay system are clearly identified from the return times of the flow to a Poincaré section. We find noise-enhanced PS in both basic coupling configurations: unidirectional and bidirectional. On the one hand the response system adjust its time scales to that of the drive in achieving PS in an unidirectional coupling configuration. On the other hand, both subsystems adjust their rhythms to the main time scale of the uncoupled system to achieve PS in bidirectionally coupled systems. Further, we will show that phase slips are accompanied by small and large return times, which are not present in the noise-free system, due to the unlocked unstable periodic orbits present in the coupled system.

We consider the following coupled piecewise linear time-delay system

$$\begin{aligned}\dot{x} &= \omega_x(-x(t) + bf[x(t-\tau)]) \\ &\quad + \varepsilon_x(y(t) - x(t)) + \sigma\xi_x(t), \\ \dot{y} &= \omega_y(-y(t) + bf[y(t-\tau)]) \\ &\quad + \varepsilon_y(x(t) - y(t)) + \sigma\xi_y(t), \end{aligned} \tag{10}$$

where $\omega_x = 1.0$ and $\omega_y = 1.1$ contributes to the frequency mismatch, $b = 1.2$ is the nonlinear parameter, $\tau = 1.33$ the delay, $\varepsilon_{x,y}$ the coupling strength, $\xi_{x,y}(t)$ an added Gaussian noise with zero mean and unit variance, and σ the noise intensity. When $\varepsilon = \varepsilon_x = \varepsilon_y$, then the systems are bidirectionally coupled, whereas $\varepsilon_x = 0$ corresponds to a unidirectional coupling configuration. When $\xi_x = \xi_y$, the Gaussian noise is common otherwise it is independent. The function $f(x(t-\tau))$ is taken to be a symmetric piecewise linear function defined by

$$f(x) = A_{x,y}f^* - Bx. \tag{11a}$$

Here

$$f^* = \begin{cases} -x^* & x < -x^*, \\ x & -x^* \le x \le x^*, \\ x^* & x > x^*, \end{cases} \tag{11b}$$

where x^* is a controllable threshold value, $A_{x,y}$ and B are parameters. The parameter values are fixed in accordance with the values of the circuit elements [26].

5.1.1. *Noise-enhanced CPS in unidirectionally coupled time-delay systems*

Now, we will demonstrate the results for unidirectional coupling. For the drive system $A_x = 5.2$ and for the response system $A_y = 5.0$, while B is fixed as $B = 3.5$. It is to be noted that the results presented here remain invariant irrespective of the values of the parameters $A_{x,y}, B$ and b, which are chosen such that the individual systems exhibit chaotic behavior. We use the Poincaré section, chosen at $x(t-\tau) = 0.8$ and $x(t) < 0.8$ on the attractor, technique to estimate phase and phase difference. In the absence of the coupling, both systems evolve independently with different frequencies due to the frequency mismatch $\omega_x \neq \omega_y$. The unidirectionally coupled, for $\varepsilon_x = 0$ and $\varepsilon = \varepsilon_y$, piecewise linear time-delay system exhibit PS for appropriate values of the coupling strength without noise $\sigma = 0$. In particular, the coupled systems are at the onset of PS at $\varepsilon = 0.95$ and in prefect phase synchrony for $\varepsilon = 0.98$ [44].

The mean frequency difference $\Delta\Omega = \Omega_y - \Omega_x$ as a function of $\varepsilon \in (0, 1)$ for both a common and an independent noise with $\sigma = 0.006$, represented as connected open circles, are shown in Figs. 13(a) and 13(b), respectively. The solid line corresponds to the mean frequency difference in the absence of noise, $\sigma = 0$. Smaller values of the mean frequency difference $\Delta\Omega$, below the onset of exact PS at $\varepsilon_{PS} = 0.95$, for the connected circles compared to that of the solid line clearly illustrate the phenomenon of noise-enhanced PS for both a common and an independent noise.

To demonstrate this phenomenon more clearly, we analyse the duration of PS epochs t_{epoch} as σ is increased. We choose $\varepsilon = 0.88$ well below ε_{PS} and $\xi_x = \xi_y$, that is, a common noise. The instantaneous phase difference $\Delta\phi$ for different σ is depicted in Fig. 13(c). In the absence of noise, $\sigma = 0$, the average phase difference displays several PS epochs, whereas longest possible PS epochs are observed for $\sigma = 0.006$, which is indeed confirmed by the maximal value of the average duration of PS epochs $\langle t_{epoch}\rangle$ shown in Fig. 13(d) ($\langle t_{epoch}\rangle$ is estimated by averaging over 10 different realization, each of duration $t = 2{,}000{,}000$). Note that, as we have fixed the coupling strength well below the onset of perfect PS,

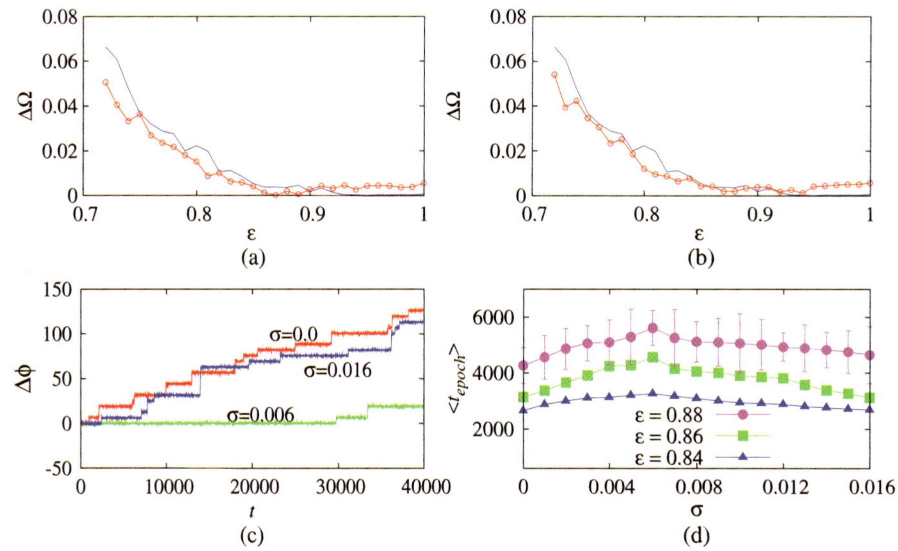

Fig. 13. Frequency difference $\Delta\Omega$ vs coupling strength ε with a common noise (a) and with an independent noise (b) for unidirectional coupling. Solid line corresponds to $\sigma = 0$ and open circles to $\sigma = 0.006$. (c) Phase difference for different noise intensity σ of a common noise. (d) Average duration of PS epochs vs σ of a common noise for different ε. The standard deviation is shown with error bars for $\varepsilon = 0.88$.

we did not achieve perfect PS with noise. However, we have confirmed that if the coupling strength is chosen very close to the onset of perfect PS, noise can induce perfect PS in the coupled time-delay systems. Further increase in the noise intensity results in shortening of the duration of PS epochs as depicted for $\sigma = 0.016$ in Fig. 13(c) and lower values of the average duration of PS epochs $\langle t_{epoch} \rangle$ (see Fig. 13(d)). Hence, increasing σ from zero increases the average duration of PS epochs attaining a maximum for appropriate σ, attributing to noise-enhanced PS, and then $\langle t_{epoch} \rangle$ decreases for larger σ, like the resonance phenomenon due to the competing effects of the noise and the mutual locking of time scales of the coupled system as we will discuss below. This behavior is also observed for the other values of the coupling strength as illustrated in Fig. 13(d). We have also obtained qualitatively similar results for $\xi_x \neq \xi_y$, an independent noise.

Now, we investigate the return times T_k to a Poincaŕe section to understand the constructive role of noise in increasing the duration of PS epochs t_{epoch}. Periodic orbits are manifested by almost vanishing $\Delta X_k = |x_{t+k} - x_t|$ at the Poincaŕe section for every k returns with the corresponding return times T_k. PS is characterized by mutual phase locking of unstable periodic orbits (UPOs) of the coupled chaotic systems and the phase slips are manifested by a few pairs of unlocked UPOs of the coupled system similar to periodically driven chaotic oscillators [3, 4, 38–40, 45, 46]. Noise is capable of preventing the systems from following unlocked UPOs for a long time and also forcing the systems to follow them for a long time resulting in both short and long return times, which do not happen in noise-free systems [38]. This is illustrated in Fig. 14 for $\varepsilon = 0.88$ and for the small noise intensity $\sigma = 0.002$. A phase synchronized epoch along with the accompanying phase slips at its both ends is depicted in Figs. 14(a) and 14(d). Note that both these figures are the same and shown here to compare the region of phase slips with the other figures. The difference between the same variables ΔX_2 and ΔY_1 for every second and first returns, with return times T_2 and T_1, to the Poincaŕe section is shown in Figs. 14(b) and 14(c), respectively, along with their corresponding period-2 and period-1 UPOs in the insets. It is to be noted that the large fluctuation in the difference ΔX_2 and ΔY_1 is due to the oscillations with different (multiple) time scale return to the Poincaŕe section at different instant in contrast to the mutually coupled systems

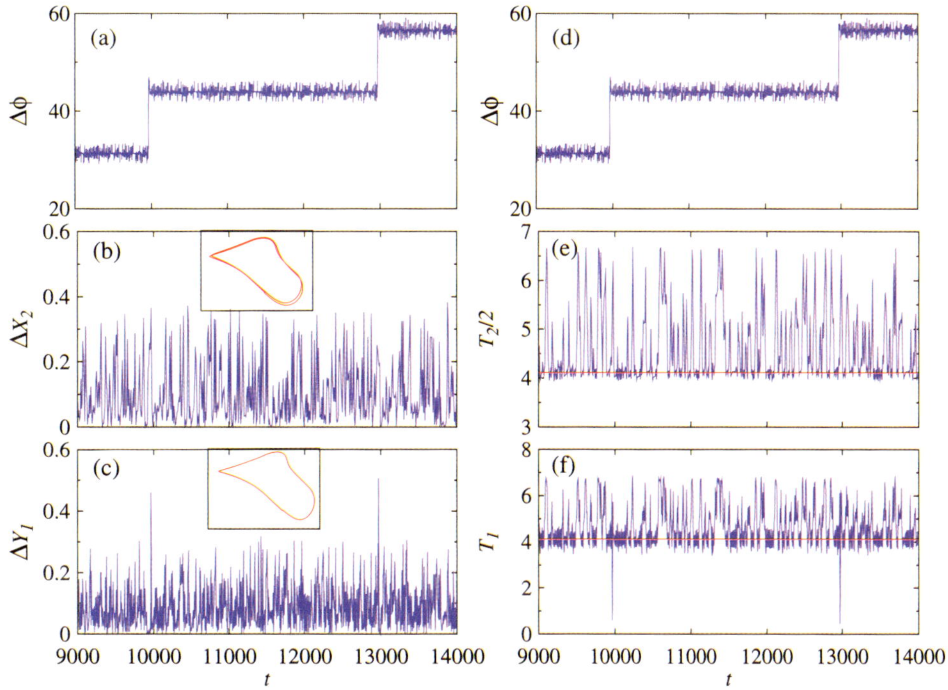

Fig. 14. (a) Phase slips induced by unlocked UPOs are illustrated in (a) and (d). Difference between x and y variables, ΔX_2 and ΔY_1, to the Poincaré section is shown in (b) and (c), respectively, with the insets illustrating period-2 and period-1 unlocked UPOs around $t \approx 9950$ and their corresponding return times T_2 and T_1 in (e) and (f), respectively.

(see below). The phase slip at $t \approx 9950$ is indeed caused by these pairs of unlocked UPOs, which is followed by the system for sufficiently long time to a phase slip to occur. Nevertheless, one can also find other pairs of unlocked UPOs at the phase slip while the majority of the UPOs of the coupled systems are mutually locked. Those pairs with small and large return times (see Fig. 14(f)), corresponding to a large mismatch in their time scales, require a large value of coupling strength for a fixed noise intensity to become locked. On contrary, a large value of the noise intensity, less than a certain threshold for a fixed coupling strength, can also change their time scales to become locked and consequently increasing the duration of PS epochs t_{epoch} resulting in noise-enhanced PS as shown in Fig. 13.

Multiple time scales of the individual system are clearly identified from their return times (Figs. 14(e) and 14(f)). The principle time scale, around which majority of the return times are fluctuating, is indicated by a horizontal line at 4.3, while the other two pronounced time scales are seen at 5.3 and 6.6. It is to be noted that the time scales of the drive system remain unaffected (see Fig. 14(e)), while most of the time scales of the response system become entrained to that of the drive system except those return times with a large difference in the time scales of unlocked UPOs manifesting phase slips (compare Figs. 14(d) and 14(f)).

5.1.2. *Noise-enhanced CPS in bidirectionally coupled time-delay systems*

In the following, we will demonstrate the phenomenon of noise-enhanced PS in bidirectionally coupled time-delay systems where both subsystems adjust their time scales of oscillation to their principle time scale in achieving PS. Now, we fix the same parameters as above except for $A_x = A_y = 5.2$, that is, parameter mismatch only in the frequencies. Because of the frequency mismatch, both systems evolve independently for $\varepsilon = 0.0$, while they are in perfect PS for $\varepsilon = 0.5$ [44]. The average frequency difference $\Delta\Omega$ is depicted in Figs. 15(a) and 15(b) in the presence of a common and an independent noise with $\sigma = 0.04$, indicated by connected open circles, respectively. The solid line corresponds to $\Delta\Omega$ without any noise. $\Delta\Omega$ acquire small values in the presence of noise rather than without noise in almost the entire range of the coupling strength below $\varepsilon_{PS} = 0.5$, which clearly indicates the phenomenon of noise-enhanced PS in bidirectionally coupled systems. The phase difference $\Delta\phi$ for different values of the noise intensity σ with a common Gaussian noise and $\varepsilon = 0.45$ are shown in Fig. 15(c). The average duration of PS epochs $\langle t_{epoch} \rangle$ for three different values of the coupling strength ε is illustrated in Fig. 15(d). From these figures, it is observed that the duration of PS

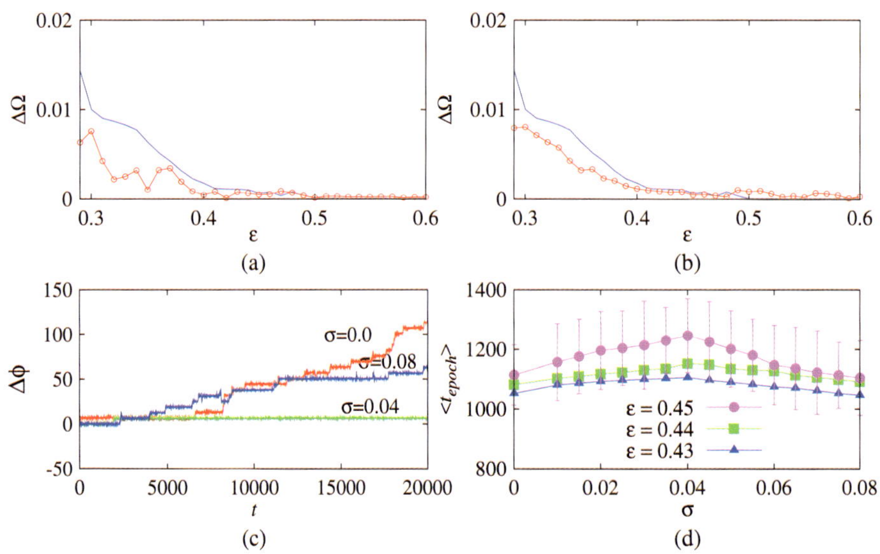

Fig. 15. Bidirectional coupling: Frequency difference $\Delta\Omega$ vs coupling strength ε with a common noise (a) and with an independent noise (b). Solid line corresponds to $\sigma = 0$ and open circles to $\sigma = 0.04$. (c) Phase difference for different noise intensity σ of a common noise. (d) Average duration of PS epochs vs σ of a common noise for different ε. The standard deviation is shown with error bars for $\varepsilon = 0.45$.

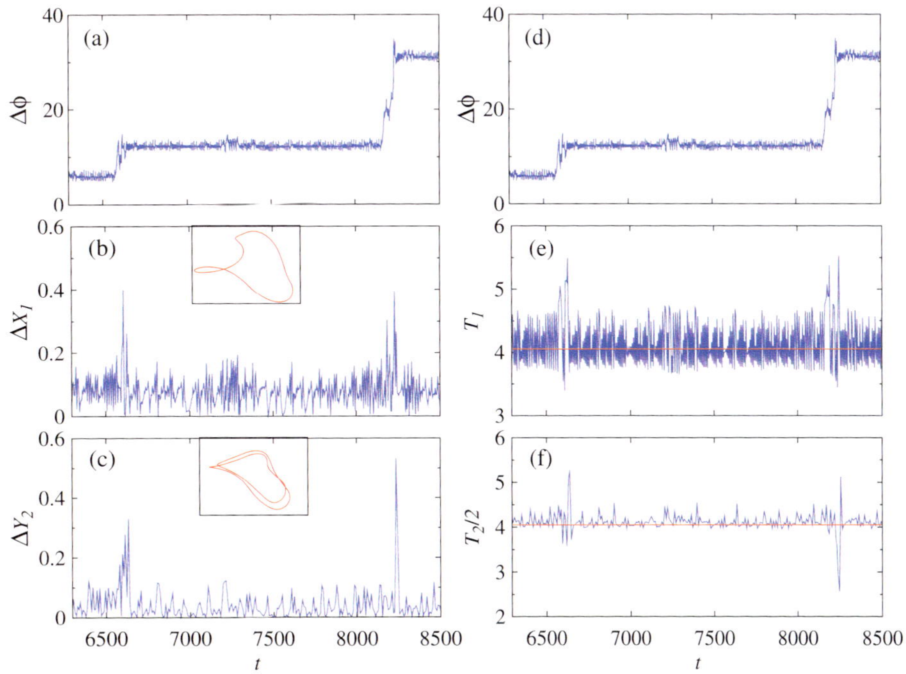

Fig. 16. (a) Phase slips induced by unlocked UPOs are illustrated in (a) and (d). Difference between x and y variables, ΔX_1 and ΔY_2, on the Poincaré section is shown in (b) and (c), respectively, with the insets illustrating period-2 and period-1 unlocked UPOs around $t \approx 6600$ and their corresponding return times T_1 and T_2 in (e) and (f), respectively.

epochs increases with noise intensity and reaches a maximum for an optimal value of the noise intensity with further increase in it resulting in decreasing in the value of $\langle t_{epoch} \rangle$, similar to the phenomenon of resonance due to the competition between the noise and the mutual locking of UPOs. Similar results are obtained for other analyzed values of the coupling strength and for an independent noise.

A phase synchronized epoch with its accompanying phase slips is illustrated in Figs. 16(a) and 16(d) for $\varepsilon = 0.45$ and $\sigma = 0.01$. Difference between the same variables, ΔX_1 and ΔY_2, at every first and second returns to the Poincaré section are shown in Figs. 16(b) and 16(c), respectively. The phase slip around $t \approx 6600$ is occurred between the unlocked period-2 and period-1 UPOs depicted in the insets. One can also find other pairs of unlocked UPOs at this phase slip. Being a bidirectionally coupled system, the subsystems adjust their time scales to their principle time scale, indicated by the horizontal line at 4.05, around which the majority of the return times fluctuate attributing to the majority of locked UPOs. The small and large return times in both subsystems at $t \approx 6600$ and $t \approx 8250$ in Figs. 16(e) and 16(f) is due to the large difference in the time scales of the pair of unlocked UPOs. As discussed for the unidirectional coupling, it requires a large coupling strength or a large noise less than a threshold value to induce changes in their time scales such that the UPOs now become locked. The latter is indeed demonstrated in Fig. 15(c), for the noise intensity $\sigma = 0.04$, and in Fig. 15(d) illustrating the increase in the duration of PS epochs resulting in noise-enhanced PS for fixed values of the coupling strength.

5.2. *Noise-induced synchronization in time-delayed systems*

In this section, we will demonstrate the emergence of synchronization between two piecewise linear time-delay systems without any coupling between them but instead driven by a common Gaussian noise in analogy with the investigation of noise-induced synchronization [39]. To be specific, we consider the following identical time-delay systems [47]

$$\dot{x_1}(t) = -ax_1(t) + bf(x_1(t-\tau)) + \sqrt{2}\sigma\xi(t), \quad (12)$$

$$\dot{x_2}(t) = -ax_2(t) + bf(x_2(t-\tau)) + \sqrt{2}\sigma\xi(t), \quad (13)$$

where a, b are positive constants and τ is the delay time, σ is the noise intensity, $\xi(t)$ is the Gaussian noise and $f(x)$ is of the same form as in Eq. (8).

The parameters are fixed as $a = 1.0$, $b = 1.2$ and $\tau = 20$. On increasing the noise intensity, σ, we found that complete synchronization is established between two piecewise linear time-delay systems driven by a common Gaussian noise and without any explicit coupling between them. To quantify the degree of synchronization, we define the synchronization error as follows

$$\eta = \left\langle \sqrt{(x_1(t) - x_2(t))^2 + (x_1(t-\tau) - x_2(t-\tau))^2} \right\rangle, \tag{14}$$

where $\langle \cdot \rangle$ denotes the time average and we have averaged the error for $N = 200,000$ time steps in our simulations. The synchronization error, η, is illustrated in Fig. 17 as a function of noise intensity σ. It is clear from this figure that the degree of synchronization error decreases for increasing of the noise intensity σ and for $\sigma > 0.006$, η approaches zero indicating that the two time-delay systems driven by a common Gaussian noise oscillate in complete synchrony with each other.

We have also estimated the synchronization error as functions of both the noise intensity σ and the time-delay τ to investigate whether a common noise is capable of inducing synchronization in a wide range of delay times. The synchronization error is depicted in Fig. 18 in the range of $\tau \in (2, 30)$ and $\sigma \in (0, 0.01)$. This figure portrays that synchronization is achieved even for feeble values of noise intensity for small values of delay time, $\tau \in (2, 3)$ and the critical value of σ required for synchronization increases with the delay time in the range $\tau \in (3, 7)$ and for $\tau > 7$, the two time-delay systems are synchronized for $\sigma > 0.006$ as represented in Fig. 17. This confirms that complete synchronization can also be induced in time-delay systems that are driven by a common noise in a wide range of time-delay without any explicit couplings between them.

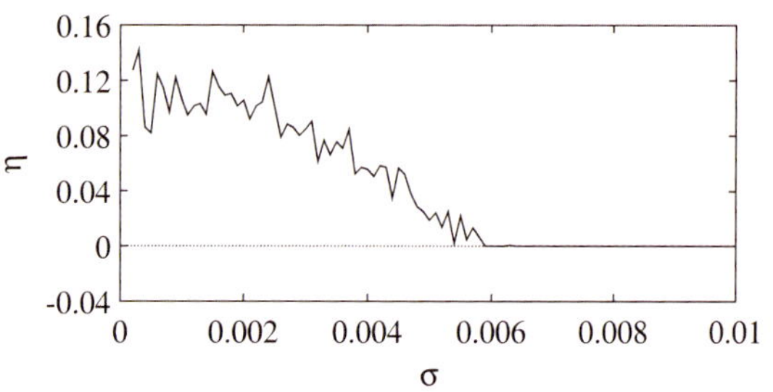

Fig. 17. Synchronization error, η, as a function of noise intensity in the range of $\sigma \in (0, 0.01)$.

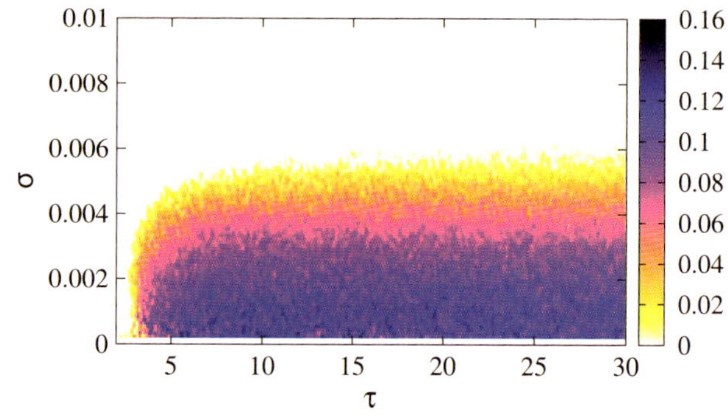

Fig. 18. Synchronization error, η, as a function of both the time-delay and the noise intensity in the range of $\tau \in (2, 30)$ and $\sigma \in (0, 0.01)$.

6. Delay Effects on Synchronization in Complex Networks

Recent investigations on synchronization is focused on nontrivial effects of delay coupling due to a finite propagation time of information transfer in communication channels [48–55]. Delay coupling enhances the synchronizability of networks and interestingly leads to the emergence of a wide range of new collective behavior [48, 49]. Further, it has also been shown that connection delays can actually be conductive to synchronization, so that it is possible for the delayed system to synchronize where the undelayed system does not [48].

Pecora and Carroll introduced an elegant approach, the master stability function (MSF), to investigate the stability properties of a network of oscillators, which allows one to separate the local dynamics of the individual oscillators from the coupling matrix characterizing the topology of the underlying network [56]. Despite of the flurry of independent research activities on both synchronization of delay coupled networks [48–50] and in analysing the stability of synchronization of a network of oscillators by exploiting the MSF [56–58], only a very few recent studies have been dealt with both MSF and delay coupled networks in probing the stability properties of the latter [51–55]. However, these reports limited their analysis either to phase oscillators or to maps except in [51], which reported the enhancement of neural synchrony by delay in two diffusively coupled Hindmarsh-Rose (HR) neurons and in Ref. [54], which analysed optically coupled semiconductor lasers modeled by Lang-Kobayshi equations using MSF. Nevertheless, the authors of [51] admitted that their analysis could not establish the stable synchronized states in the master stability surface of the network of delay coupled HR neurons. Hence, the master stability formalism has not yet been successfully exploited in understanding the stability properties of synchronization transitions in a delay coupled network synthesized from even simple low-dimensional dynamical systems. Further, the effect of delay on the synchronizability of such a network has not yet been investigated using the MSF.

6.1. *Master stability formalism for delay coupled networks*

We consider an arbitrary delay coupled network with N identical chaotic units in accordance with the MSF [56] as follows:

$$\dot{x}_i(t) = f[x_i(t)] - \sigma \sum_{j=1}^{N} g_{ij} h[x_j(t-\tau)], \qquad (15)$$

where $x_i \in \mathbb{R}^m, i = 1, \ldots, N$, f is a nonlinear function describing the dynamics of the individual units, σ is the overall coupling strength, $G = g_{ij}$ is a $N \times N$ matrix, which determines the topology of an arbitrary network, h is a coupling function and τ is the coupling delay. The invariant synchronization manifold is assured by the zero row sum, that is $\sum_{j=1}^{N} g_{ij} = 0 \, \forall \, i$, of each row of the connectivity matrix G. At first, we consider the arbitrary network as undirected such that G is symmetric with real eigenvalues and then we extend our analysis to directed networks, in which case G is asymmetric leading to complex eigenvalues. It has also to be noted that our results remains unaffected in either cases, as the imaginary parts of the complex eigenvalues has only a rotational effect [59], as we will discuss below in detail. If G is diagonalizable, then the stability of the synchronization manifold is governed by the eigenvalues of G and the following block diagonalized variational equation for the transverse modes

$$\dot{\xi}(t) = Df[x(t)]\xi(t) - \varepsilon Dh[x(t-\tau)]\xi(t-\tau). \qquad (16)$$

Here, $x(t)$ is the dynamics within the synchronization manifold, $\xi(t)$ is the perturbation transverse to the synchronization manifold, which could be complex, $Df[x(t)]$ is the Jacobian evaluated on the synchronization manifold and $Dh[x(t-\tau)]$ is an $m \times m$ matrix that determines which components of the oscillator are coupled. The normalized coupling parameter $\varepsilon = \sigma\gamma_k$, where γ_k is an eigenvalue of the coupling matrix G, $k = 0, 1, 2, \ldots, N-1$. It is to be noted that the eigenvalue γ_k is more generic because of the arbitrary G and it could be the eigenvalue corresponding to any specific topology determined by G. Here, $k = 0$ corresponds to the synchronization manifold with $\gamma_0 = 0$, while all the other eigenvalues for $k > 0$ correspond to the transverse modes determining the stability of the synchronization manifold $x(t)$. Thus, the synchronized state is stable for a given network topology, characterized by G, when all the eigenvalues of G for $k > 0$ are negative (or when all the transverse Lyapunov exponents of (16) are negative). Considering the fact that G could have complex eigenvalues arising from asymmetric couplings [59], the generic variational equation, that is the MSF, can be written as

$$\dot{\xi}(t) = Df[x(t)]\xi(t) + (\alpha + i\beta)Dh[x(t-\tau)]\xi(t-\tau), \qquad (17)$$

where, α and β are the real and imaginary parts of the eigenvalue. Separating $\xi(t)$ into the real part $\xi_r(t)$ and the imaginary part $\xi_i(t)$, we get

$$\begin{aligned}\dot{\xi}_r(t) &= Df[x(t)]\xi_r(t) + \alpha Dh[x(t-\tau)]\xi_r(t-\tau) \\ &\quad - \beta Dh[x(t-\tau)]\xi_i(t-\tau), \qquad (18a)\\ \dot{\xi}_i(t) &= Df[x(t)]\xi_i(t) + \alpha Dh[x(t-\tau)]\xi_i(t-\tau) \\ &\quad + \beta Dh[x(t-\tau)]\xi_r(t-\tau). \qquad (18b)\end{aligned}$$

The maximal transverse Lyapunov exponent (λ_{max}) estimated from Eqs. (18) yields a complex surface in the (α, β) plane, namely the master stability surface of the considered network of oscillators. For a given value of the coupling strength σ and the delay time τ, one can locate a point in the complex surface and the sign of λ_{max} at that point determines the stability of the synchronization manifold.

6.2. *Delay-enhanced synchronization in complex networks*

Recently, Huang *et al.* [57] classified the typical behaviors of MSFs into four categories by examining all possible one-component linear coupling configurations in several prototype three-dimensional systems. It is to be noted that the x-component coupling in most of the dynamical systems tends the network to remain synchronized with a single crossing of the maximal transverse Lyapunov exponent with the zero-axis from positive. According to their classification, the λ_{max} of class Γ_2 possesses two successive crossings with the zero-axis in a finite interval of the normalized coupling strength $\varepsilon \in (\varepsilon_1, \varepsilon_2)$ indicating a finite range of the stable synchronized state. Dynamical systems with different component couplings fall under different classes [57].

To demonstrate the existence of the extended regime of the stable synchronous state of a delay coupled network, it would be convenient if the original stable synchronized regime of the network without delay is limited within a finite range of the normalized coupling parameter ε. Therefore, we stick to the appropriate component couplings that come under the class Γ_2 for the following paradigmatic models that have been widely exploited in synchronization studies. $x \to x$ coupled Rössler systems governed by the equation of motions [56,57,60]

$$\begin{aligned} \dot{x} &= -y - z, \\ \dot{y} &= x + 0.2y, \\ \dot{z} &= 0.2 + z(x - 9.0), \end{aligned}$$

$y \to x$ coupled Lorenz systems [57,61] represented as

$$\begin{aligned} \dot{x} &= 10.0(x - y), \\ \dot{y} &= x(28.0 - z) - y, \\ \dot{z} &= xy - 2.0z, \end{aligned}$$

$y \to x$ coupled HR neurons described by the equation of motions [51,62]

$$\begin{aligned} \dot{x} &= y - x^3 + 3.0x^2 - z + 3.2, \\ \dot{y} &= 1.0 - 5.0x^2 - y, \\ \dot{z} &= 0.06(4.0(x + 1.6) - z), \end{aligned}$$

and $z \to x$ coupled Chua's circuit system given by the equation of motions [63–65]

$$\begin{aligned} \dot{x} &= 10.0(y - x + f(x)), \\ \dot{y} &= x - y + z, \\ \dot{z} &= -14.87y. \end{aligned}$$

Here, the piecewise linear function $f(x)$ is given by

$$f(x) = \begin{cases} -bx - a + b & x > 1, \\ -ax & |x| < 1, \\ -bx + a - b & x < -1, \end{cases} \tag{19}$$

where $a = -1.27$ and $b = -0.68$.

Details on the physical meaning of the components of the above dynamical systems can be found in their corresponding references. Now, we demonstrate our results by estimating the largest transverse Lyapunov exponent of the MSF (16) and (18) in both real axis and complex (α, β) planes, respectively [66].

6.2.1. *Delay-enhanced synchronization in a network of Rössler system*

The largest transverse Lyapunov exponent λ_{max} of the MSF (16) of the delay coupled network with the Rössler system as nodes is depicted in Fig. 19(a) for three different values of the coupling delay τ. In the absence of the coupling delay, $\tau = 0$, the stable synchronous state exists only in the range of $\varepsilon \in (0.18, 4.61)$ as indicated by the sign of the λ_{max} (see Fig. 19(a)). As the coupling delay is increased, the stable synchronous regime is gradually extended in the ε parameter space up to a certain threshold. Further increase in it leads to decrease in the regime of a stable synchronous oscillations of the delay coupled network (see Fig. 20). In particular, for $\tau = 0.24$ in the coupling, λ_{max} has a second stable regime for $\varepsilon > 7.38$ and for $\tau = 0.4$ there exists only one single stable synchronous regime but extended in the range of $\varepsilon \in (0.18, 7.38)$, which is almost twice compared to that without delay [66].

In order to understand the effect of delay coupling in the network of Rössler systems more clearly, we have shown λ_{max} as a function of $\varepsilon \in (0, 6)$ and the coupling delay $\tau \in (0, 1)$ in Fig. 20(a). Contours in this figure correspond to different values of λ_{max}. It is evident from this figure that as the coupling delay is increased from zero, the stable synchronous regime increases slowly towards the larger coupling strength up to a certain threshold as indicated by the contour corresponding to $\lambda_{max} = 0$. This confirms the extension of the interval of the stable synchronous state induced by the coupling delay. Further increase in the delay leads to decrease in the regime of a stable synchronous state as indicated by the contour of $\lambda_{max} = 0$. Closer examination of the contour corresponding to $\lambda_{max} = 0$ in the range of coupling delay $\tau \in (0.1, 0.2)$ clearly indicates that the stable synchronous state occurs even for smaller values of ε in a narrow range in the presence of coupling delay. Nevertheless, the networks of Lorenz systems, HR neurons and Chua's circuit systems display more pronounced extended regime of stable synchronous state towards smaller coupling strengths (see Fig. 19), which is indeed attained for larger ε without delay in the network.

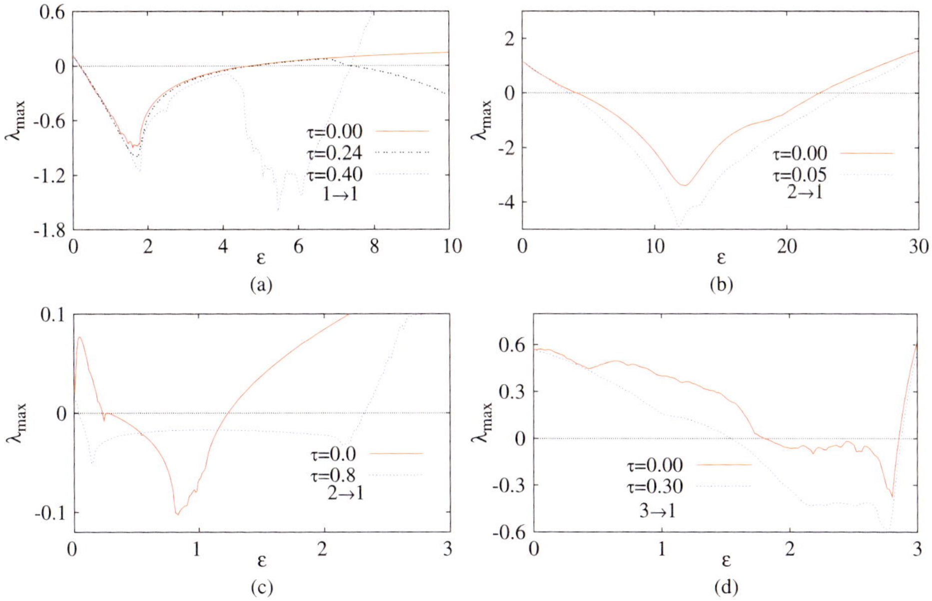

Fig. 19. The largest transverse Lyapunov exponents λ_{max} of the master stability function (16) of a delay coupled network as a function of the normalized coupling parameter ε for different values of the coupling delay τ. The notation $i \to j$ indicates the coupling as being from ith component of one oscillator to the jth component of another one. (a) Rössler system as nodes with $\tau = 0.0, 0.24$ and 0.4. (b) Lorenz system as nodes with $\tau = 0.0$ and 0.05. (c) HR neurons as nodes with $\tau = 0.0$ and 0.8. (d) Chua's circuit system as nodes with $\tau = 0.0$ and 0.3.

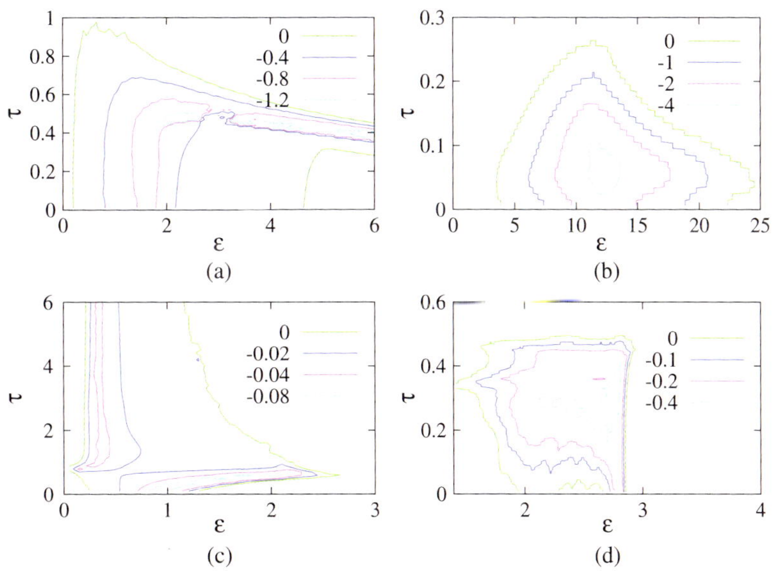

Fig. 20. Contours of the largest transverse Lyapunov exponent λ_{max} in the real plane of the master stability function (16) as a function of ε and the coupling delay τ. (a) $x \to x$ coupled Rössler systems. (b) $y \to x$ coupled Lorenz systems. (c) $y \to x$ coupled Hindmarsh-Rose neurons. (d) $z \to x$ coupled Chua's circuit systems. Contours correspond to different values of λ_{max} encompassing the stable synchronous states.

The stability issue of the synchronized state of the arbitrary network (15) is reduced to that of the block diagonalized variational equation, (16), for the transverse modes such that the sign of the largest Lyapunov exponent determines the stability. As the form of each block is the same with only the scalar multiplier $\varepsilon = \sigma\gamma_k$ differing for each [56] and hence it is straightforward to obtain a scaling relation between the eigenmodes of the connectivity matrix G. Further, one can estimate the maximal number of synchronized oscillators, N_{max}, of the network from the scaling relation in analogy

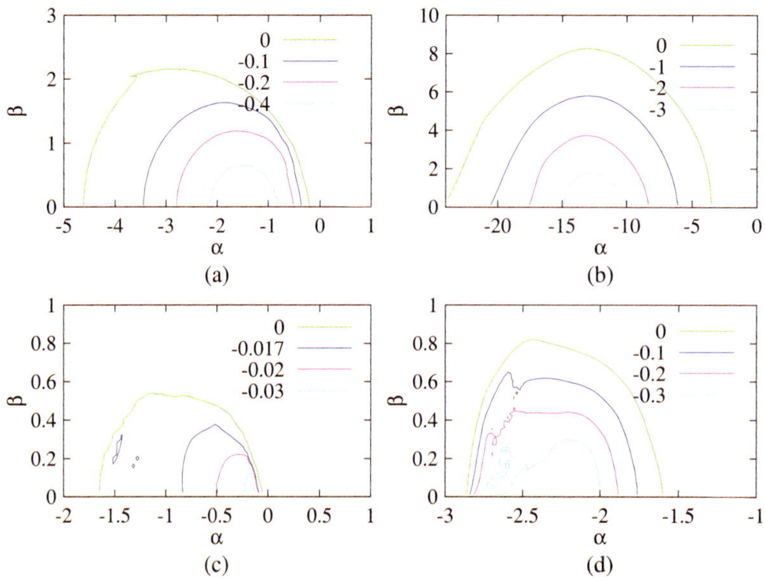

Fig. 21. Contours of the largest transverse Lyapunov exponent λ_{max} in the complex plane of the MSF (18). (a) $x \to x$ coupled Rössler systems with $\tau = 0.24$. (b) $y \to x$ coupled Lorenz systems with $\tau = 0.05$. (c) $y \to x$ coupled Hindmarsh-Rose neurons systems with $\tau = 4.0$. (d) $z \to x$ coupled Chua's circuit systems with $\tau = 0.3$. Contours encompasses the stable synchronous states.

with the one derived for a lattice [58]. However, it is to be noted that the scaling relation facilitates the estimation of N_{max} only from the explicit knowledge of the eigenvalues γ_k in terms of trigonometric functions and hence the estimation of N_{max} is restricted rather to regular networks, whereas for more complex networks it remains an open problem. As N_{max} in the regular networks depends solely on ε at the two successive zero crossings of λ_{max}, namely ε_1 and ε_2. Increase in the interval between ε_1 and ε_2 reflects in increase in the N_{max} [56, 58]. As a consequence, the extended regime of a stable synchronized state as a function of ε in the delay coupled network of Rössler systems facilitates an increase in the N_{max} when compared to the same network without delay [66].

The largest transverse Lyapunov exponent of the MSF (18) for $x \to x$ coupled Rössler systems is illustrated in the complex (α, β) plane in Fig. 21(a) for the value of the coupling delay $\tau = 0.24$. The contours encompasses the stable synchronous regimes corresponding to the negative values of λ_{max}. To the best of our knowledge, this is the first result showing stable synchronous states using the MSF of a delay coupled network of time continuous chaotic units. For $(\alpha, \beta) = (0, 0)$ in Eq. (18), then $\lambda_{max} > 0$ as the nodes evolve independently. Equations (18a) and (18b) become exactly the same as Eq. (16) for $\alpha < 0$ when $\beta = 0$ and the effect of coupling delay for this case is clearly illustrated in Figs. 19 and 20, and explained in appropriate places. The network does not synchronize at all for $\alpha > 0$. Increasing β in Eq. (18) for fixed α rotates the transverse modes as can be seen in Fig. 21 and hence the complex surface defining stability is symmetric along the real axis. Further, λ_{max} becomes positive for larger (or smaller) values of β beyond a certain threshold even if $\alpha < 0$ as depicted in Fig. 21 attributing to the fact that a large imaginary coupling can destabilize the system. In short, varying the real term in $\alpha + i\beta$ has the effect of damping the transverse perturbations to the synchronization manifold, whereas changes in the imaginary terms have a rotational effect between different transverse modes [59].

6.2.2. *Delay-enhanced synchronization in a network of Lorenz system*

Figure 19(b) illustrates λ_{max} of the master stability function (16) of a delay coupled network with the Lorenz systems as its nodes. λ_{max} depicts a finite regime of a stable synchronous chaos for the $y \to x$ coupled Lorenz systems in the range of $\varepsilon \in (4.17, 22.53)$ in the absence of delay coupling. On the other hand, the regime of a stable synchronized state is extended in both directions of the coupling strength in the range of $\varepsilon \in (3.51, 24.14)$ for $\tau = 0.05$

corroborating the delay induced extended regime of a stable synchronous states. As a consequence, the number of synchronized oscillators in the delay coupled network is increased beyond that of the same network without delay. To elucidate this phenomenon of delay enhanced synchrony more clearly, λ_{max} is illustrated as contours in Fig. 20(b) as a function of $\varepsilon \in (0, 25)$ and $\tau \in (0, 0.3)$. The contour for $\lambda_{max} = 0$ confirms the above phenomenon up to a certain threshold value of τ and then the stable regime decreases with further increase in τ. The master stability surface of $y \rightarrow x$ coupled Lorenz systems is depicted in Fig. 21(b) for the coupling delay $\tau = 0.05$. Isocline lines of λ_{max} attributes to the stable synchronous state.

6.2.3. *Delay-enhanced synchronization in a network of Hindmarsh-Rose neurons*

For the case of $y \rightarrow x$ coupled Hindmarsh-Rose neurons, λ_{max} is depicted in Fig. 19(c). There exists a stable synchronized regime in the interval $\varepsilon \in (0.28, 1.23)$ for $\tau = 0.0$, whereas for $\tau = 0.8$ λ_{max} indeed shows a large regime of a stable synchronized state in the range $\varepsilon \in (0.036, 2.31)$ extended for both smaller and larger values of ε. It is a strong evidence that stable synchronization can occur even for smaller values of the coupling strength in the presence of the coupling delay, which is obtained only for sufficiently larger values of ε without delay in the same network. N_{max} can also be increased proportionally as discussed above. Further, the isocline line corresponding to $\lambda_{max} = 0$ as a function of $\varepsilon \in (0, 3)$ and $\tau \in (0, 6)$ in Fig. 20(c) also confirms the extended regime of a stable synchronized state, compared to that for $\tau = 0$, almost in the entire range of the coupling delay we have considered. Contours of λ_{max} estimated from (18) for $y \rightarrow x$ coupled HR neurons depicted in Fig. 21(c) clearly delineate the stable synchronous regimes in the complex plane of the MSF for the coupling delay $\tau = 4.0$.

6.2.4. *Delay-enhanced synchronization in a network of Chua's circuit systems*

Figure 19(d) depicts λ_{max} of $z \rightarrow x$ coupled network of Chua's circuit systems illustrating stable synchronization in the interval $\varepsilon \in (1.84, 2.85)$ in the absence of delay in the coupling. On the other hand for appropriate delay in the coupling, that is for $\tau = 0.3$, λ_{max} in Fig. 19(d) confirms the phenomenon of delay enhanced synchrony by extending the stable synchronization to the interval $\varepsilon \in (1.52, 2.88)$. The contour for $\lambda_{max} = 0$ in Fig. 20(d) confirms the above phenomenon. Further, the isocline line of λ_{max} (see Fig. 21(d)) estimated from (18) for $z \rightarrow x$ coupled network of Chua's circuit systems in the complex plane also corroborates the existence of stable synchronization.

Thus, we have demonstrated the existence of an extended regime of stable synchronized states at both smaller and larger values of the coupling strength in an arbitrary network induced by the coupling delay using the MSF. The generic nature of our results is also exhibited using three paradigmatic models that have been widely exploited in the literature for synchronization studies.

6.3. *Delay-induced synchronization in complex networks*

In this section, we demonstrate the phenomenon of delay-induced synchronization, that is inducing synchronization where there is no synchronization at all wihtout time-delay, in an arbitrary network again using the master stability formalism. In order to demonstrate the phenomenon of delay-induced synchronization, we consider the different component coupling that fall under the class Γ_0, according to the classification by Huang *et al.* [57], which does not exhibit any synchronization at all. In the following we demonstrate delay-induced synchronization in an arbitrary network of $x \rightarrow y$ and $y \rightarrow x$ coupled Rössler systems and $z \rightarrow y$ coupled Chua's circuit systems [67].

6.3.1. *Delay-induced synchronization in a network of Rössler systems*

Now, we consider the network of $x(t-\tau) \rightarrow y(t-\tau)$ coupled Rössler systems and the largest transverse Lyapunov exponent, estimated from the MSF (16), depicted in Fig. 22(a) for $\tau = 0$ indicates that the network is in an asynchronous state as $\lambda_{max} > 0$ in the entire range of ε. In contrast, in the presence of a delay under the same coupling, the network of Rössler systems is entrained to synchronous oscillations as confirmed by the sign of λ_{max} in Fig. 22(a) for $\tau = 1.8$ attributing to the phenomenon of delay-induced synchronization in the network. The effect of delay coupling can be better understood from

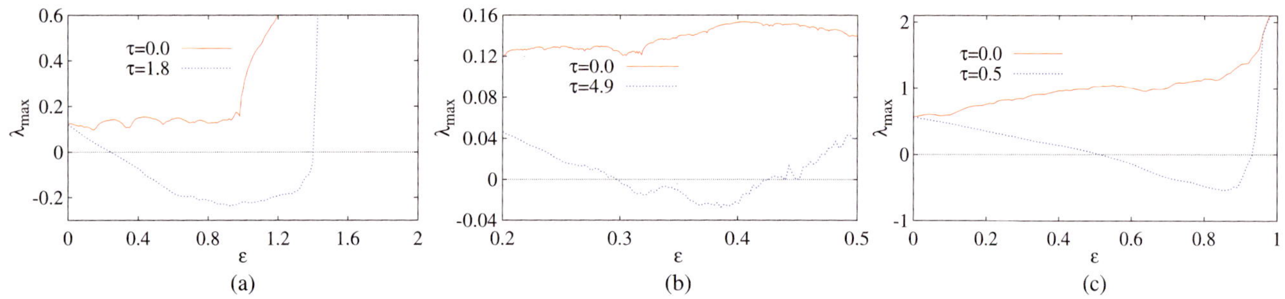

Fig. 22. The largest transverse Lyapunov exponents λ_{max} of the MSF (16) of the delay-coupled network as a function of the normalized coupling parameter ε for different values of the delay τ in the coupling. (a) $x(t-\tau) \to y(t-\tau)$ coupled Rössler system for $\tau = 0.0$ and 1.8. (b) $y(t-\tau) \to x(t-\tau)$ coupled Rössler system for $\tau = 0.0$ and 4.9. (c) $z(t-\tau) \to y(t-\tau)$ coupled Chua's circuit systems for $\tau = 0.0$ and 0.5.

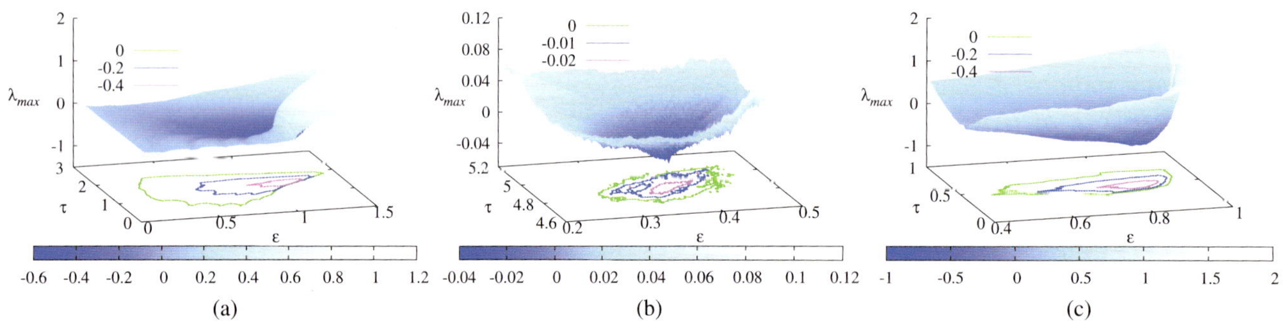

Fig. 23. Surface of the largest transverse Lyapunov exponents λ_{max} as a function of the coupling delay τ and the normalized coupling parameter ε. An increase in the color density corresponds to a decrease in the value of λ_{max} and the contours encompasses the stable synchronized regime corresponding to $\lambda_{max} < 0$. (a) $x(t-\tau) \to y(t-\tau)$ coupled Rössler system, (b) $y(t-\tau) \to x(t-\tau)$ coupled Rössler system. (c) $z(t-\tau) \to y(t-\tau)$ coupled Chua's circuit systems.

Fig. 23(a), where λ_{max} is shown as a surface in the (τ, ε) plane. An increase in the color intensity corresponds to a decrease in λ_{max} and the isoclinic lines encompasses the stable synchronous regime corresponding to $\lambda_{max} < 0$. It is evident from Fig. 23(a) that without delay in the coupling the network of Rössler systems does not exhibit any synchronous oscillation. On the other hand, as τ is increased the network attains synchronization after an appreciable delay and the range of stable synchronization increases with it up to a certain threshold. Further increase in the delay results in a decrease in the spread of stable synchronization and finally reaches asynchronization for further larger delays. The corresponding master stability surface, i.e. the surface of λ_{max} in the complex (α, β) plane, estimated from the MSF (18) is illustrated in Fig. 24(a) for the coupling delay $\tau = 1.8$, in which the stable regimes are displayed by the contours of $\lambda_{max} < 0$.

A similar result is also observed for another choice of the delayed coupling, namely $y(t-\tau) \to x(t-\tau)$. The sign of λ_{max} (see Fig. 22(b)) as a function of ε for both with and without delay in the coupling asserts the phenomenon of delay-induced synchronization in the network of Rössler systems. The surface of λ_{max} in Fig. 23(b) and its contours for $\lambda_{max} < 0$ clearly illustrate the above phenomenon. Further, the complex (α, β) plane of the MSF in Fig. 24(b) for $\tau = 4.9$ with $\lambda_{max} < 0$ confirms the same.

6.3.2. *Delay-induced synchronization in a network of Chua's circuit systems*

Next, we consider the network of $z(t-\tau) \to y(t-\tau)$ coupled Chua's circuit systems. Without any delay in the coupling λ_{max} remains positive in Fig. 22(c) attributing to asynchronization, whereas for the value of delay $\tau = 0.5$, $\lambda_{max} < 0$ in the range $\varepsilon \in (0.5, 0.93)$ corroborating the phenomenon of delay-induced synchronization. Further, this phenomenon is also confirmed from the contours of $\lambda_{max} < 0$ in the surface plot of λ_{max} depicted in Fig. 23(c) as

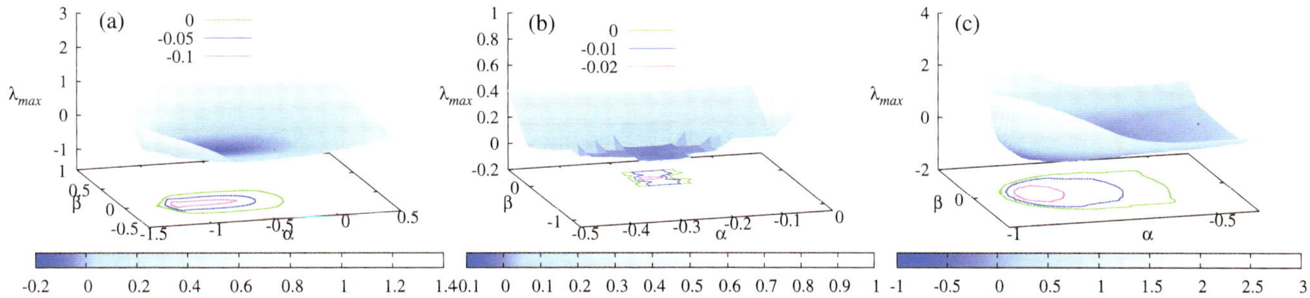

Fig. 24. Master stability surface of the delay-coupled network. An increase in the color density corresponds to a decrease in the value of λ_{max} and the contours encompasses the stable synchronized regime corresponding to $\lambda_{max} < 0$. (a) $x(t-\tau) \to y(t-\tau)$ coupled Rössler system for $\tau = 1.8$. (b) $y(t-\tau) \to x(t-\tau)$ coupled Rössler system for $\tau = 4.9$. (c) $z(t-\tau) \to y(t-\tau)$ coupled Chua's circuit systems for $\tau = 0.5$.

a function of τ and ε. The corresponding master stability surface, estimated from the MSF (18) is illustrated in Fig. 24(c) for the coupling delay $\tau = 0.5$, in which the stable regimes are displayed by the contours of $\lambda_{max} < 0$.

By exploiting the MSF in investigating the delay effects in an arbitrary delay-coupled network, we have demonstrated the phenonemon of delay-induced synchronization in paradigmatic models of Rössler and Chua's circuit systems.

7. Summary and Conclusions

In this chapter, we have demonstrated the existence of CPS in prototype time-delay systems such as a piecewise-linear time-delay system, the Mackey-Glass and experimentally in a piecewise-linear time-delay with threshold nonlinearity possessing highly non-phase-coherent chaotic attractors. We have shown that there is a typical transition from a non-synchronized state to CPS and then to generalized synchronization using different frameworks. Further, we have demonstrated the phenomenon of noise-enhanced PS in both unidirectionally and bidirectionally coupled time-delay systems. It is to be noted that the time scales of the drive system remain unaffected, while most of the time scales of the response system are entrained to that of the drive system in unidirectionally coupled time-delay systems, whereas both systems adjust their rhythms to their principal time scale for mutual coupling. In particular, the return time clearly indicates the multiple time scales of the delay system. We have also demonstrated the phenomenon of noise-induced synchronization in time-delay systems driven by a common Gaussian noise. In addition, We have investigated the phenomenon of delay-enhanced and delay-induced stable synchronous state of an arbitrary delay coupled network using the MSF. We have also shown that there always exist an extended regime of a stable synchronous state as a function of coupling strength for appropriate coupling delays, which cannot be observed without any delay in the coupling. Further, the stable synchronous state is achieved even at smaller values of the coupling strength with delay, which is actually attained at larger coupling strength without delay. The extended regime of synchronization also suggests that the number of synchronized nodes in the delay coupled networks can be increased proportionally beyond the size instability of the same networks without delay.

Acknowledgements

D. V. S and J. K. acknowledges the support from EU under project No. 240763 PHOCUS(FP7-ICT-2009-C) and IRTG 1740 (DFG and FAPESP).

References

[1] H. Fujisaka and T. Yamada, Stability Theory of Synchronized Motion in Coupled-Oscillator Systems, *Prog. Theor. Phys.* **69**, 32–47, (1983).

[2] L. M. Pecora and T. L. Carroll, Synchronization in chaotic systems, *Phys. Rev. Lett.* **64**, 821–824, (1990).

[3] A. S. Pikovsky, M. G. Rosenblum and J. Kurths, *Synchronization — A Unified Approach to Nonlinear Science.* (Cambridge University Press, Cambridge, 2001).

[4] S. Boccaletti, J. Kurths, G. Osipov, D. L. Valladares and C. S. Zhou, The synchronization of chaotic systems, *Phys. Rep.* **366**, 1–101 (2002).

[5] M. Lakshmanan and D. V. Senthilkumar, *Dynamics of Nonlinear Time-delay Systems.* (Springer, Berlin, 2010).

[6] L. O. Chua, L. Kocarev, K. Eckert and M. Itoh, Experimental Chaos Synchronization in Chua's Circuit, *Int. J. Bifurcation and Chaos* **2**, 705–708, (1992).
[7] U. Parlitz, L. O. Chua, L. Kocarev, K. S. Halle and A. Shang, Transmission of Digital Signals by Chaotic Synchronization, *Int. J. Bifurcation and Chaos* **2**, 973–977, (1992).
[8] C. -W. Wu and L. O. Chua, A unified framework for synchronization and control of dynamical systems, *Int. J. Bifurcation and Chaos* **4**, 979–998, (1994).
[9] V. N. Belykh, N. N. Verichev, L. Kocarev and L. O. Chua, On Chaotic Synchronization in a Linear Array of Chua's Circuits, *J. Circuits Systems and Computers* **3**, 579–589, (1993).
[10] T. Kapitaniak, L. O. Chua and G.-Q. Zhong, Experimental synchronization of chaos using continuous control, *Int. J. Bifurcation and Chaos* **4**, 483–488, (1994).
[11] L. O. Chua, M. Itoh, L. Kocarev and K. Eckert, Chaos Synchronization in Chua's Circuit, *J. Circuits Systems and Computers* **3**, 93–108, (1993).
[12] Y. Maistrenko, V. L. Maistrenko and L. O. Chua, Cycles of Chaotic Intervals in a Time-delayed Chua's Circuit, *Int. J. Bifurcation and Chaos* **3**, 1557–1572, (1993).
[13] A. N. Sharkovsky, Y. Maistrenko, P. Deregel and L. O. Chua, Dry turbulence from a time-delayed Chua's circuit, *J. Circuits Systems and Computers* **3**, 645–668, (1993).
[14] E. M. Shahverdiev, S. Sivaprakasam and K. A. Shore, Lag times and parameter mismatches in synchronization of unidirectionally coupled chaotic external cavity semiconductor lasers, *Phys. Rev. E* **66**, 037202(1–4), (2002).
[15] S. Boccaletti, L. M. Pecora and A. Pelaez, Unifying framework for synchronization of coupled dynamical systems, *Phys. Rev. E* **63**, 066219(1–4), (2001).
[16] K. M. Cuomo and A. V. Oppenheim, Circuit implementation of synchronized chaos with applications to communications, *Phys. Rev. Lett.* **71**, 65–68, (1993).
[17] S. Hayes, C. Grebogi, E. Ott and A. Mark, Experimental Control of Chaos for Communication, *Phys. Rev. Lett.* **73**, 1781–1784, (1994).
[18] J. Garcia-Ojalvo and R. Roy, Spatiotemporal Communication with Synchronized Optical Chaos, *Phys. Rev. Lett.* **86**, 5204–5208, (2001).
[19] G. D. VanWiggeren, and R. Roy, Communication with Dynamically Fluctuating States of Light Polarization, *Phys. Rev. Lett.* **88**, 097903(1–4), (2002).
[20] D. V. Senthilkumar, M. Lakshmanan, and J. Kurths, Phase synchronization in time-delay systems, *Phys. Rev. E* **74**, 035205(R), (2006).
[21] D. V. Senthilkumar, M. Lakshmanan, and J. Kurths, Transition from phase to generalized synchronization in time-delay systems, *Chaos* **18**, 023118(1–12) (2007).
[22] D. V. Senthilkumar, M. Lakshmanan, and J. Kurths, Phase synchronization in unidirectionally coupled Ikeda time-delay systems, *Eur. Phys. J. Special Topics* **164**, 35–44, (2008).
[23] Suresh, R. D. V. Senthilkumar, M. Lakshmanan, and J. Kurths, Global phase synchronization in an array of time-delay systems, *Phys. Rev. E* **82**, 016215(1–10) (2010).
[24] D. V. Senthilkumar, J. Kurths and M. Lakshmanan, Stability of synchronization in coupled time-delay systems using Krasovskii-Lyapunov theory, *Phys. Rev. E* **79**, 066208(1–4), (2009).
[25] K. Srinivasan, D. V. Senthilkumar, K. Murali, M. Lakshmanan, and J. Kurths, Synchronization transitions in coupled time-delay electronic circuits with a threshold nonlinearity, *Chaos* **21**, 023119 (1–11), (2011).
[26] D. V. Senthilkumar, K. Srinivasan, K. Murali, M. Lakshmanan, and J. Kurths, Experimental confirmation of chaotic phase synchronization in coupled time-delayed electronic circuits, *Phys. Rev. E* **82**, 065201(R), (2010).
[27] G. V. Osipov, B. Hu, C. Zhou, M. V. Ivanchenko and J. Kurths, Three Types of Transitions to Phase Synchronization in Coupled Chaotic Oscillators, *Phys. Rev. Lett.* **91**, 024101(1–4), (2003).
[28] N. Marwan, M. C. Romano, M. Thiel and J. Kurths, Recurrence plots for the analysis of complex systems, *Phys. Rep.* **438**, 237–329, (2007).
[29] T. Pereira, M. S. Baptista and J. Kurths, General framework for phase synchronization through localized sets, *Phys. Rev. E* **75**, 026216(1–12), (2007).
[30] M. C. Romano, M. Thiel, J. Kurths, I. Z. Kiss and J. L.Hudson, Detection of synchronization for non-phase-coherent and non-stationary data, *Phys. Rev. Lett.* **71**, 466–472, (2005).
[31] D. V. Senthilkumar and M. Lakshmanan, Bifurcation and Chaos in Time Delayed Piecewise Linear Dynamical Systems, *Int. J. Bifurcation and Chaos* **15**, 2895–2912, (2005).
[32] D. V. Senthilkumar and M. Lakshmanan, Transition from anticipatory to lag via complete synchronization in time-delay systems, *Phys. Rev. E* **71**, 016211(1–10), (2005).
[33] M. G. Rosenblum, A. S. Pikovsky and J. Kurths, Phase Synchronization of Chaotic Oscillators, *Phys. Rev. Lett.* **76**, 1804(1–4), (1996).
[34] S. Guan, C. H. Lai and G. W. Wei, Phase synchronization between two essentially different chaotic systems, *Phys. Rev. E* **72**, 016205(1–8), (2005).
[35] B. Hu, G. V. Osipov, H. L. Yang and J. Kurths, Oscillatory and rotatory synchronization of chaotic autonomous phase systems, *Phys. Rev. E* **67**, 066216(1–8), (2003).

[36] A. Arenas, A. Díaz-Guilera, J. Kurths, Y. Moreno and C. S. Zhou Synchronization in complex networks, *Phys. Rep.* **469**, 93–153, (2008).

[37] P. Ashwin, J. Buescu and I. Stewart Bubbling of attractors and synchronisation of chaotic oscillators, *Phys. Lett. A* **193**, 126–139, (1994).

[38] C. Zhou, J. Kurths, I. Z. Kiss and J. L.Hudson, Noise-Enhanced Phase Synchronization of Chaotic Oscillators, *Phys. Rev. Lett.* **89**, 014101(1–4), (2002).

[39] C. Zhou and J. Kurths, Noise-Induced Phase Synchronization and Synchronization Transitions in Chaotic Oscillators, *Phys. Rev. Lett.* **88**, 230602 (1–4), (2002).

[40] K. H. Nagai and H. Kori, Noise-induced synchronization of a large population of globally coupled nonidentical oscillators, *Phys. Rev. E* **81**, 065202(R), (2010).

[41] V. S. Anishchenko, V. V. Astachov, A. B. Neiman, T. E. Vadivasova and L. Schimansky-Geier, *Nonlinear Dynamics of Chaotic and Stochastic Systems.* (Springer, Berlin, 2002).

[42] W. Horsthemke and R. Lefever, *Noise-Induced Transitions: Theory and Applications in Physics, Chemistry and Biology.* (Springer, Berlin, 1984).

[43] G. Deco, V. Jirsa, A. R. McIntosch, O. Sporns and R. Kötter, Key role of coupling, delay, and noise in resting brain fluctuations, *Proc. Natl Acad. Sci. USA* **106**, 10302–10307, (2009).

[44] D. V. Senthilkumar, M. Manju Shrii, and J. Kurths, Noise-enhanced phase synchronization in time-delay systems, *Phys. Rev. E* **85**, 026218(1–6) (2012).

[45] A. Pikovsky, G. Osipov, M. Rosenblum, M. Zaks and J. Kurths, Attractor-Repeller Collision and Eyelet Intermittency at the Transition to Phase Synchronization, *Phys. Rev. Lett.* **79**, 47–50, (1997).

[46] M. A. Zaks, E. H. Park, M. Rosenblum and J. Kurths, Alternating Locking Ratios in Imperfect Phase Synchronization, *Phys. Rev. Lett.* **82**, 4228–4231, (1999).

[47] D. V. Senthilkumar and J. Kurths, Characteristics and synchronization of time-delay systems driven by a common noise, *Eur. Phys. J. Special Topics* **187**, 87–93, (2010).

[48] F. M. Atay, J. Jost and A. Wende, Delays, Connection Topology, and Synchronization of Coupled Chaotic Maps *Phys. Rev. Lett.* **92**, 144101(1–4) (2004).

[49] C. Masoller, and A. C. Martí, Random Delays and the Synchronization of Chaotic Maps *Phys. Rev. Lett.* **94**, 134102(1–4) (2005).

[50] I. Fischer, R. Vicente, J. M. Buldú, C. R. Mirasso, M. C. Torrent, and J. García-Ojalvo, Zero-Lag Long-Range Synchronization via Dynamical Relaying, *Phys. Rev. Lett.* **97**, 123902(1–4) (2006).

[51] M. Dhamala, V. K. Jirsa and M. Ding, Enhancement of Neural Synchrony by Time Delay, *Phys. Rev. Lett.* **92**, 074104(1–4) (2004).

[52] W. Kinzel, A. Englert, G. Reents, M. Zigzag and I. Kanter, Synchronization of networks of chaotic units with time-delayed couplings, *Phys. Rev. E* **79**, 056207(1–4) (2009).

[53] C. -U. Choe, T. Dahms, P. Hövel and E. Schöll, Controlling synchrony by delay coupling in networks: From in-phase to splay and cluster states, *Phys. Rev. E* **81**, 025205(R) (2010).

[54] V. Flunkert, S. Yanchuk, T. Dahms and E. Schöll, Synchronizing Distant Nodes: A Universal Classification of Networks, *Phys. Rev. Lett.* **105**, 254101 (1–4) (2010).

[55] A. Englert, S. Heiligenthal, W. Kinzel and I. Kanter, Synchronization of chaotic networks with time-delayed couplings: An analytic study, *Phys. Rev. E* **83**, 046222(1–4) (2011).

[56] L. M. Pecora and T. L. Carroll, Master Stability Functions for Synchronized Coupled Systems, *Phys. Rev. Lett.* **80**, 2109(1–4) (1998).

[57] L. Huang, Q. Chen, Y. -C. Lai and L. M. Pecora, Generic behavior of master-stability functions in coupled nonlinear dynamical systems, *Phys. Rev. E* **80**, 036204(1–11) (2009).

[58] J. F. Heagy, L. M. Pecora and T. L. Carroll, Short Wavelength Bifurcations and Size Instabilities in Coupled Oscillator Systems, *Phys. Rev. Lett.* **74**, 4185(1–4) (1995).

[59] K. S. Fink, G. Johnson, T. L. Carroll, D. Mar and L. M. Pecora, Three coupled oscillators as a universal probe of synchronization stability in coupled oscillator arrays, *Phys. Rev. E* **61**, 5080(1–11) (2000).

[60] O. E. Rössler, An equation for continuous chaos, *Phys. Lett. A* **57**, 397–398 (1976).

[61] E. N. Lorenz, Deterministic Nonperiodic Flow, *J. Atmos. Sci.* **20**, 130–141 (1963).

[62] J. L. Hindmarsh and R. M. Rose, A Model of Neuronal Bursting Using Three Coupled First Order Differential Equations, *Proc. R. Soc. London, Ser. B* **221**, 87–102 (1984).

[63] T. Matsumoto, L. O. Chua and M. Komuro, The Double Scroll, *IEEE Trans. Circuits Syst.* **CAS 32**, 797–818 (1985).

[64] L. O. Chua, C. A. Desoer and E. S. Kuh, *Linear and Nonlinear Circuits.* (McGraw Hill, Singapore, 1987).

[65] R. N. Madhan Ed., *Chua's Circuit: A Paradigm for Chaos.* (World Scientific, Singapore, 1996).

[66] M. Manju Shrii, D. V. Senthilkumar and J. Kurths, Delay coupling enhances synchronization in complex networks, *EuroPhys. Lett.* **98**, 10003(1–6) (2012).

[67] M. Manju Shrii, D. V. Senthilkumar and J. Kurths, Delay-induced synchrony in complex networks with conjugate coupling, (submitted).

CHAPTER 33

SYMBOLIC DYNAMICS AND SPIRAL STRUCTURES DUE TO THE SADDLE-FOCUS BIFURCATIONS

ANDREY SHILNIKOV

Neuroscience Institute and Department of Mathematics and Statistics,
Georgia State University, Atlanta 30303, USA
ashilnikov@gsu.edu

LEONID SHILNIKOV
(17/12/1934–26/12/2011)

Institute for Applied Mathematics and Cybernetics,
Nizhniy Novgorod, 603006 Russia
diffequ@unn.ac.ru

ROBERTO BARRIO

Departamento de Matemática Aplicada and IUMA,
University of Zaragoza, E-50009. Spain
rbarrio@unizar.es

We examine spiral structures in bi-parametric diagrams of dissipative systems with strange attractors. We show that the organizing center for spiral structures in the Rössler model with the saddle-focus equilibria is related to the change of the structure of the attractor transitioning between the spiral and screw-like types. The structure skeleton is formed by saddle-node bifurcation curves originating from a codimension-two Belyakov point corresponding to the transition to the saddle-focus from a simple saddle. A new computational technique based on the symbolic kneading invariant description for examining dynamical chaos and parametric chaos in systems with Lorenz-like attractors is proposed and tested. This technique uncovers the stunning complexity and universality of spiral structures in the iconic Lorenz equations.

1. Introduction

Over recent years, a great deal of experimental studies and modeling simulations have been directed toward the identification of various dynamical and structural invariants to serve as key signatures uniting often diverse nonlinear systems into a single class.

One such class of low order dissipative systems has been identified to possess one common, easily recognizable pattern involving spiral structures in a biparametric phase space [2, 12, 13, 15]. Such patterns turn out to be ubiquitously alike in both time-discrete [30] and time-continuous systems [6, 20, 21].

Despite the overwhelming number of studies reporting the occurrence of spiral structures, there is still little known about the fine construction details and underlying bifurcation scenarios for these patterns. In this paper we study the genesis of the spiral structures in several low order systems and reveal the generality of underlying global bifurcations. We will start with the Rössler model and demonstrate that such parametric patterns are the key feature of systems with homoclinic connections involving saddle-foci meeting a single Shilnikov condition [44, 45]. The occurrence of this bifurcation causing complex dynamics is common for a plethora of dissipative systems, describing (electro)chemical reactions [32], population dynamics [26, 28], electronic circuits [12, 19, 40], including the Chua circuit [17, 18]. The other group is made of models with the Lorenz attractor. Here, under consideration is the iconic Lorenz equation.

2. Spiral Hubs in Rössler-like Models

One of the most paradigmatic examples of low-dimensional deterministic chaos is the canonical Rössler system [33]:

$$\dot{x} = -(y + z), \quad \dot{y} = x + ay,$$
$$\dot{z} = b + z(x - c),$$

with two bifurcation parameters a and c (we fix $b = 0.2$). For $c^2 > 4ab$ the model has two equilibrium states, $P_{1,2}(ap_\pm, -p_\pm, p_\pm)$, where $p_\pm = (c \pm \sqrt{c^2 - 4ab})/2a$. This classical model exhibits the spiral and screw chaotic attractors after a period doubling cascade followed by the Shilnikov bifurcations of the saddle-focus P_2.

Bi-parametric screening of the Rössler model unveils a stunning universality of the periodicity

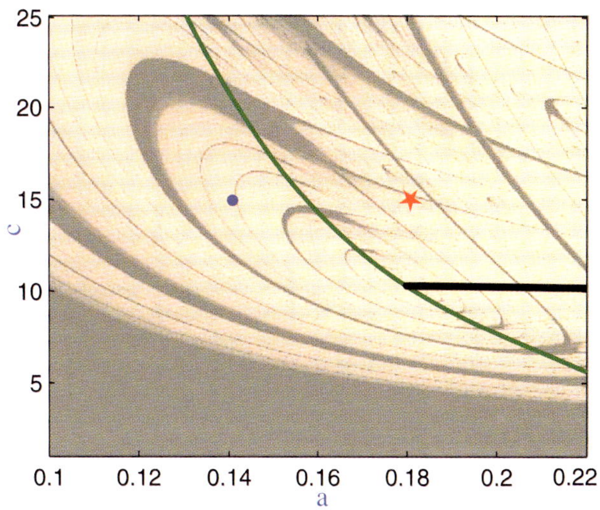

Fig. 1. Spiral structures in the biparametric bifurcation diagram of the Rössler model: light colors are associated with chaotic behavior, whereas dark colors correspond to regular, simple dynamics. The thick dark curve corresponds to the saddle-focus homoclinic bifurcation, and the medium green boundary separates the existence regions of the spiral (at •) and screw-shaped (at ⋆) attractors. Courtesy of [8].

hubs in the bifurcation diagrams shown in Fig. 1 [8]. The diagram is built on a dense grid of 1000×1000 points in the $(a\,c)$-parameter plane. Solutions of the model were integrated using the high precision ODE solver TIDES [1]. The color is related with the Lyapunov exponents, where dark and light colors discriminate between the regions of regular and chaotic dynamics corresponding to a zero and positive maximal Lyapunov exponent λ_1, respectively. The figure reveals the characteristic spiral patterns due to variations of the Lyapunov exponents.

The chaotic-regular regions spiral around a F[ocal] point [12, 13] at $(a, c) = (0.1798, 10.3084)$. This F-point terminates the bifurcation curve (black) corresponding to the formation of a homoclinic loop of the saddle-focus, P_2, in the phase space of the Rössler model [10]. Another curve (medium green) passes through the F-point: crossing it rightward the chaotic attractor in the phase space of the model changes the topological structure from spiral to screw-shaped. This curve has been identified from the examination of 1D Poincaré return maps (see Fig. 2) evaluated on a grid of 1000×1000 points.

The *spiral attractor* in the Rössler model at $a = 0.14$ (the point labeled by "•" in Fig. 1) generates a 1D unimodal map shown in the left panel of Fig. 2. In the case of the *screw attractor* at $a = 0.18$ (the point labeled by "⋆" in the diagram), the corresponding map shown in the right panel in Fig. 2 has

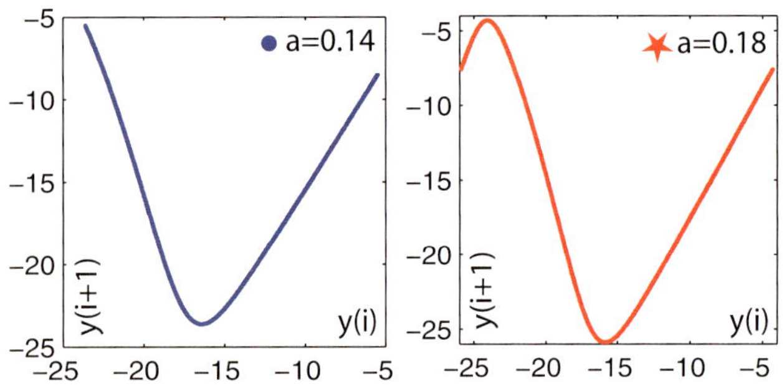

Fig. 2. 1D Poincaré return maps for the spiral and screw-shaped (resp.) chaotic attractors in the Rössler model at $a = 0.14$ and 0.18 for $c = 15$. Courtesy of [8].

a bimodal graph with two critical points. The addition of the second critical point in the map is indicative of a change of the topology of the attractors. This criteria was used to locate the corresponding boundary (medium green line) that separates the existence regions of the attractors of both types in the bifurcation diagram in Fig. 1 [8]. This boundary passes right through the F-point.

The dark bifurcation curve in Fig. 1 corresponds to a formation of the homoclinic orbits to the saddle-focus, P_2, of topological type (1,2), i.e. with 1D stable and 2D unstable manifolds, in the Rössler model. Depending on the magnitudes of the characteristic exponents of the saddle-focus, the homoclinic bifurcation can give rise to the onset of either rich complex or trivial dynamics in the system [22,44,45]. The cases under consideration meet the Shilnikov conditions and hence the existence of a single homoclinic orbit implies chaotic dynamics in the models within the parameter range in the presented diagrams. The magnification of the corresponding bifurcation curve in the diagram (Fig. 1) reveals that what appears to be a single bifurcation curve has actually two branches (Fig. 3). This curve has a U-shape with the turn at the F-point. Figure 3 shows the evolution of the homoclinic loops of the saddle-focus toward a simple saddle along the bifurcation curve. While its low branch corresponds to a primary loop, the upper branch corresponds to the double-pulse homoclinic loops. This suggests that the F-point actually might correspond to a heteroclinic cycle involving two equilibria, i.e. it is similar to T-points considered in the next section (actual study of the phase portrait for the parameter values corresponding to the F-point is yet to be done for the Rösler model).

Figure 4 outlines the structure of the bifurcation unfolding around the spiral hub [6, 8]. The picture depicts a number of the identified folded and cusp-shaped saddle-node bifurcation curves of periodic orbits originating from a codimension-two point, labeled as B[elyakov], toward the spiral hub in the (a, c)-parameter plane for the Rössler model. Note that none of these curves is actually a spiral — the overall spiral structure must be supported by homoclinic bifurcations. At this B-point, the saddle with real characteristic exponents becomes a saddle-focus for smaller values of the parameter a. The unfolding of this bifurcation is known [11] to contain bundles of countably many curves

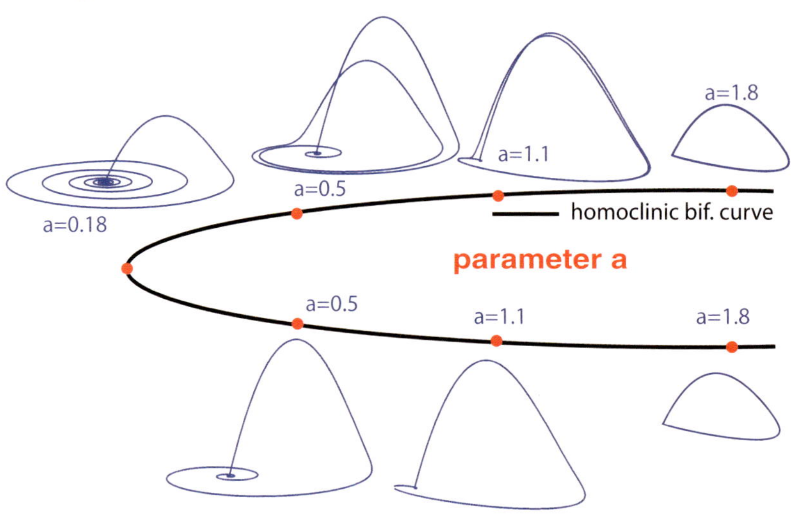

Fig. 3. Transformations of homoclinic orbits of the saddle-focus, P_2, in the Rössler system. In the sketch, the F-point is the turn point for two branches of the U-shaped bifurcation curve corresponding to the primary and secondary (with an additional round) homoclinic loops, which are sampled at the indicated points. Courtesy of [8].

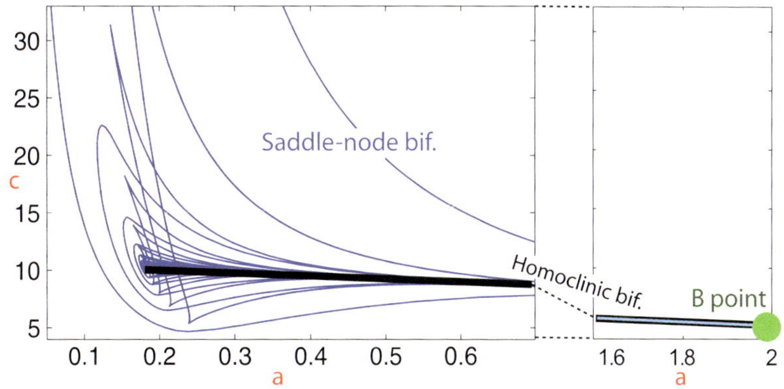

Fig. 4. Spiral structure is outlined with folded and cusp-shaped curves corresponding to saddle-node bifurcation of periodic orbits and originating from the codimension-two homoclinic B[elyakov]-point. Courtesy of [8].

corresponding to saddle-node and period doubling bifurcations of periodic orbits, as well as to various secondary homoclinic bifurcations of the saddle-focus. Indeed, both B- and F-points together globally determine the structure of the (a, c)-bifurcation portrait of the Rössler model [8].

3. Spiral Structures in Systems with the Lorenz Attractor

Dynamical systems theory is aimed to create purely abstract approaches followed by a further development of applicable tools allowing for the search and identification of basic invariants employed to unify diverse nonlinear systems with complex dynamics into a class. One such computationally justified approach for studying complex dynamics capitalizes on the concept of sensitivity of deterministic chaos. Sensitivity of chaotic trajectories can be quantified in terms of the divergence rate expressed through the largest Lyapunov characteristic exponent. The approach has been proven to work out exceptionally well for most systems with chaotic dynamics, like the Rössler model above. However, it has failed, in general, to deliver the desired answers and insights into intrinsic bifurcations of strange attractors in Lorenz-like models.

A strange attractor in the Lorenz equation from hydrodynamics has become a de-facto proof of deterministic chaos and the butterfly-shaped image of the iconic Lorenz attractor, shown in Fig. 5, has become the trademark of Chaos theory and dynamical systems. This theory elaborates on complex trajectory behaviors in nonlinear systems from mathematics, physics, life sciences, finance, etc. Universality of the methods along with bifurcation tools has made them spread wide and deep across all other disciplines.

3.1. *The Lorenz equation*

The Lorenz equation [29] is a system of three differential equations:

$$\dot{x} = -\sigma(x-y), \quad \dot{y} = r\,x - y - xz,$$
$$\dot{z} = -\frac{8}{3}z + xy, \tag{1}$$

with two positive bifurcation parameters: σ being the Prandtl number quantifying the viscosity of the fluid, and r being a Reynolds number used to characterize the fluid dynamics. Equation (1) is Z_2-symmetric: $(x, y, z) \leftrightarrow (-x, -y, z)$ [47].

Figure 6 (courtesy of [39]) exhibits the primary bifurcation curves (left panel) in the parameter plane, and depicts en route fragments of the formation of the Lorenz attractor on the pathway, $\sigma = 10$ [2, 27]. For $r < 1$, Eq. (1) has a single stable equilibrium state at the origin. This equilibrium state undergoes a pitch-fork bifurcation at $r = 1$, so that for $r > 1$ the origin becomes a saddle, O, of the topological type (2,1). This implies that the characteristic exponents of the linearized equation at O can be ordered as follows: $\lambda_3 < \lambda_2 < 0 < \lambda_1$. This implies also that the saddle has a pair of 1D separatrices (due to λ_1) leaving the saddle as $t \to +\infty$, and a 2D stable manifold containing the leading (due to λ_2) invariant z-axis. After the pitch-fork bifurcation, the outgoing separatrices of the saddle tend to two symmetric attractors — equilibrium states, $O_{1,2}(x = y = \pm\sqrt{\frac{8}{3}(r-1)}, z = r-1)$ (Fig. 6(a)).

A homoclinic butterfly bifurcation occurs in the Lorenz equation when both 1D unstable

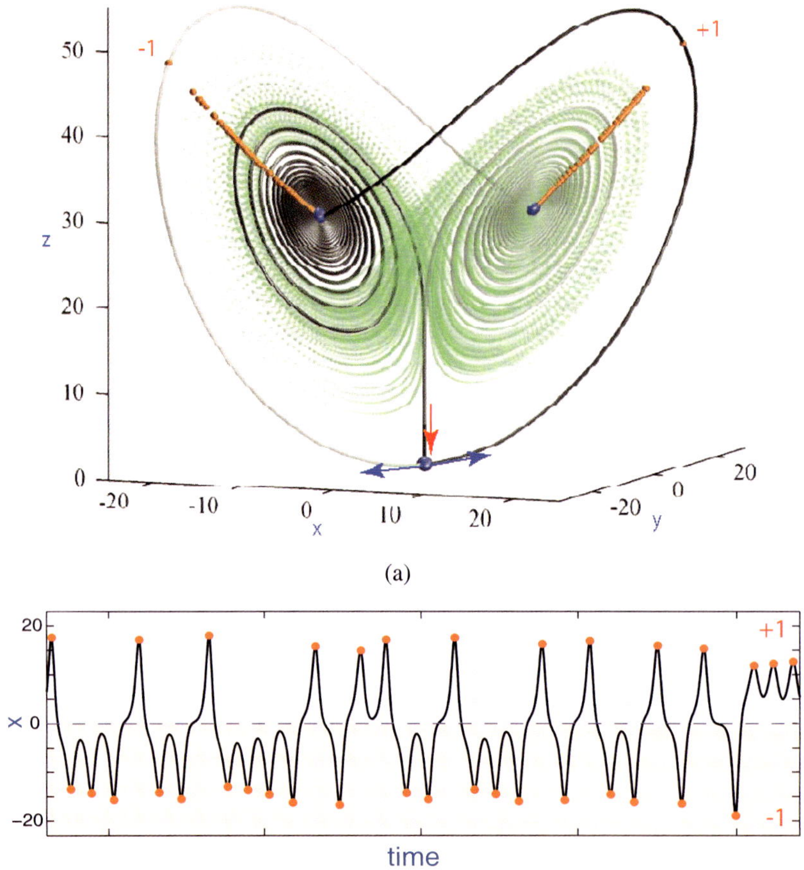

(b)

Fig. 5. (a) Heteroclinic connection (in dark color) between three saddles (blue spheres) overlaid with the Lorenz attractor (green light color) on the background in the 3D phase space of the Lorenz equation at the primary T-point ($r = 30.38$, $\sigma = 10.2$). Orange spheres on the butterfly wings indicating the turning points around the saddle-foci define the kneading sequence entries, $\{\pm 1\}$. (b) Time evolution of the x-coordinate of the right separatrix at other parameter values.

separatrices of the saddle come back to it along the z-axis (Fig. 6(b)). Bifurcations of the separatrices of the saddle at the origin are crucial overall for the Lorenz attractor. In virtue of the symmetry of the Lorenz equation, homoclinic loops come in pairs, and therefore constitute bifurcations of codimension-one, in general. The homoclinic butterfly bifurcation can be continued in the form of a continuous curve, l_1, in the (r, σ)-parameter plane of the Lorenz equation.

The very first homoclinic butterfly made of two separatrices looping a single round about the equilibrium states $O_{1,2}$, causes a homoclinic explosion in the phase space of the Lorenz equation; namely, this bifurcation gives rise to an enumerable set of saddle periodic orbits that further form the skeleton of the Lorenz chaotic set (yet to become the attractor). For the explosion to happen, the so-called *saddle value* $S = \lambda_1 + \lambda_2$, which is the sum of the leading characteristic exponents of the saddle, must be positive at the origin. Otherwise, the unfolding of the homoclinic butterfly produces a single symmetric figure-8 periodic orbit in the aftermath of a gluing bifurcation through which two stable periodic orbits merge into one after flowing into the homoclinic butterfly. In the case of the Lorenz equation with $S > 0$ at the origin, such saddle orbits demarcate a threshold of the "interior" of the chaotic unstable set (Fig. 6(c)).

After the homoclinic butterfly bifurcation, between $r_1 < r < r_2$, the separatrices of the saddle switch targets: now the right/left separatrix tends the left/right focus $O_{2/1}$. For the unstable chaotic set to become the Lorenz attractor, it must become invariant, i.e. must not lose trajectories running away to the stable foci $O_{1,2}$. This happens on another bifurcation curve, l_2, in the parameter space (Fig. 6(c)). After $r \geq r_2$, the Lorenz attractor

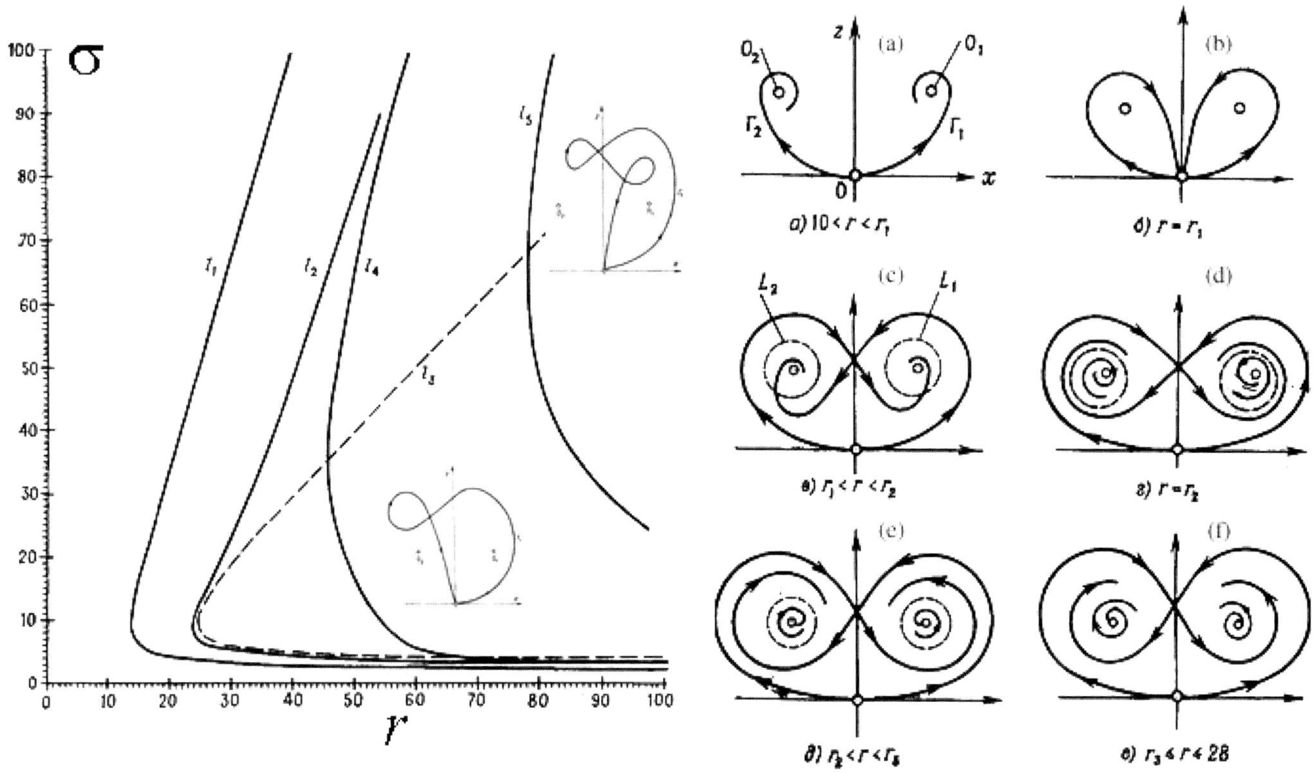

Fig. 6. (Left) The (r, σ)-parameter plane depicting the primary bifurcation curves and the stages of the formation of the Lorenz attractor. l_1 corresponds to the primary homoclinic butterfly; l_2 corresponds to the formation of the Lorenz attractor; l_3 is a supercritical Andronov-Hopf bifurcation curve; curves, l_4 and l_5, correspond to the homoclinic loops (in insets) with kneadings $\{+1, -1\}$ and $\{+1, -1, -1\}$, respectively. Courtesy of [39].

is shielded away from the stable equilibrium states by the 2D stable manifolds of the two "threshold" saddle orbits (Fig. 6(d)) that simultaneously emerge from each separatrix loop after crossing the curve l_1.

As r is increased to r_3 these periodic orbits collapse into the stable equilibria. The equilibrium states become saddle-foci of the (1,2)-type through a subcritical Andronov-Hopf bifurcation. The topological (1,2)-type means that each saddle-focus has 2D unstable and 1D stable manifolds; the latter is formed by two incoming separatrices. Some properties of the eigenvalues of the saddle-foci can be revealed without evaluating their characteristic exponents explicitly. Let $\lambda_1 < 0$ denote the stable exponent of $O_{1,2}$, and $\lambda_{2,3}$ stand for a complex conjugate pair such that $\text{Re}\lambda_{2,3} > 0$. Observe that Eqs. (1) have a constant negative divergence given by $[-\sigma - 1 - 8/3]$, which equals $\sum_{i=1}^{3} \lambda_i$. This implies $\lambda_1 + \text{Re}\lambda_{2,3} < 0$, i.e. the complex conjugate pair is closer to the imaginary axis than the real eigenvalue, and hence the saddle-foci meet the Shilnikov condition [43]. Therefore, if the saddle-focus has a homoclinic loop, the bifurcation causes the emergence of countably many periodic orbits near the saddle-focus. Those periodic orbits constantly undergo saddle-node and period doubling bifurcations as the parameters are varied. Moreover, since the Lorenz equation has the negative divergence, saddle node bifurcations give rise to the onset of stable periodic orbits. Like in the case of the Rössler attractor, a single Shilnikov saddle-focus bifurcation can cause the emergence of spiral or screw-like attractors. However, a strange attractor due to the Shilnikov saddle-focus in a 3D system with a negative divergence is not a genuine chaotic one in the sense that it contains stable periodic orbits within, which may have large or narrow attraction basins in the phase space.

Because of that, such a chaotic attractor is called quasi-attractor thus referring to the existence of stable periodic orbits along with structurally unstable or non-transverse homoclinic orbits in the attractor [3, 40, 41]. Note that in higher dimension, one may have genuine strange attractors with a Shilnikov loop without stable periodic orbits (the so-called pseudo-hyperbolic attractors [40–42, 48]).

The Lorenz attractor is non-hyperbolic because it includes the origin — the saddle equilibrium state of the topological type (2,1) while all other saddle periodic orbits are of the type (2,2), i.e. have 2D stable and unstable manifolds crossing transversally in the 3D phase space. In order for the Lorenz attractor to be strange and chaotic with no stable orbits, it may include neither [homoclinic] saddle-foci, $O_{1,2}$, nor contain only structurally stable homoclinic orbits due to transverse intersections of the manifolds of saddle periodic orbits.

After the stable equilibria are gone through the Andronov-Hopf bifurcation, the Lorenz equation exhibits the strange attractor of the iconic butterfly shape. The wings of the butterfly have symmetric eyes containing the saddle-foci are isolated from the trajectories of the Lorenz attractor. This attractor is structurally unstable [3, 25] as it undergoes bifurcations constantly as the parameters of the equation are varied. The primary cause of structural and dynamical instability of chaos in the Lorenz equation and similar models is the singularity at the origin — the saddle with two one-dimensional outgoing separatrices. Both separatrices fill in densely two spatially symmetric, $[(x, y, z) \leftrightarrow (-x, -y, z)]$, wings of the Lorenz attractor in the 3D phase space (see Fig. 5). The Lorenz attractor undergoes a bifurcation when the separatrices of the saddle change the flip-flop pattern of switching between the butterfly wings centered around the saddle-foci. At the change, the separatrices comes back to the saddle thereby causing additional homoclinic explosions in the Lorenz attractor [2, 27].

The time progression of the "right" (or symmetrical "left") separatrix can be described geometrically and categorized in terms of the number of flip-flops around the equilibria $O_{1,2}$ in the 3D phase space of the Lorenz model (Fig. 5). An alternative way is to follow the time-evolution of the x-coordinate of the separatrix, which is shown in Panel B of Fig. 5. The sign-alternation of the x-coordinate suggests the introduction of a $\{\pm 1\}$-based alphabet to be employed for the symbolic description of the separatrix. Namely, whenever the right separatrix turns around O_1 or O_2, we write down $+1$ or -1, respectively. For example, the time series shown in Panel B generates the following kneading sequence starting with $\{+1, -1, -1, -1, +1, +1, +1, -1, \ldots\}$.

In what follows we demonstrate the new computational toolkit for the analysis of chaos in the Lorenz-like models. The toolkit is inspired by the idea of kneading invariants introduced by Milnor and Thurston [31]. A kneading invariant is a value that is intended to describe quantitatively the state of a complex system that admits a symbolic description using two symbols, here $+1$ and -1. The kneading invariant is supposed to depend monotonically on the parameter controlling the system so that any two systems can be compared and differentiated, or ordered in terms of $>$ and $<$, by the kneadings. Two systems with the same kneading are topologically conjugate [3, 25].

For the symmetric system with the Lorenz attractor, the kneading invariant is assigned to quantify the symbolic description of the separatrix. It reflects respectively and quantitatively a qualitative change in the separatrix behavior as the parameter of the system is changed. So, kneadings can serve as moduli of the topological equivalence to compare or contrast between any two Lorenz attractors or, equivalently, any two Lorenz-like systems.

In virtue of the spatial symmetry of the Lorenz equation, the kneading invariant for either separatrix is defined as a formal series:

$$P(q) = \sum_{n=0}^{\infty} \kappa_n \, q^n. \tag{2}$$

By setting $q \in (0, 1)$, the series (2) is made convergent. The sequence comprised with only $+1$ corresponds to the right separatrix (Fig. 5) tending to either an equilibrium state or a periodic orbit with $x(t) > 0$; the corresponding kneading invariant is given by $P_{\max} = 1/(1 - q)$. When the right separatrix converges to a ω-limit set with $x(t) < 0$ for all times; the corresponding kneading invariant is given $P_{min} = 1 - q/(1 - q)$ because the first $+1$ is followed by an infinite sequence of -1s. Thus, $[P_{\min}, P_{\max}]$ yields the range of the kneading invariant values.

In this computational study of the Lorenz equation we consider the very first 50 entries in the kneading sequences for the right separatrix of the saddle in order to generate the biparametric mapping: $(r, \sigma) \to P_{50}(q)$ with some appropriately chosen value of q. The mapping can be colorized using a non-linear spectrum ranging between to $P^{(50)}_{\min}$ and $P^{(50)}_{\max}$, respectively. In the mapping, a color is associated with a level curve corresponding to the particular value of the kneading invariant that remains constant for the given points in the parameter plane.

Figure 7 represents the kneading-based color mapping for the Lorenz equation in the parameter plane foliated by such kneading level curves on a dense grid of [1000×1000] points. In it, a wide window of a solid color corresponds to a constant

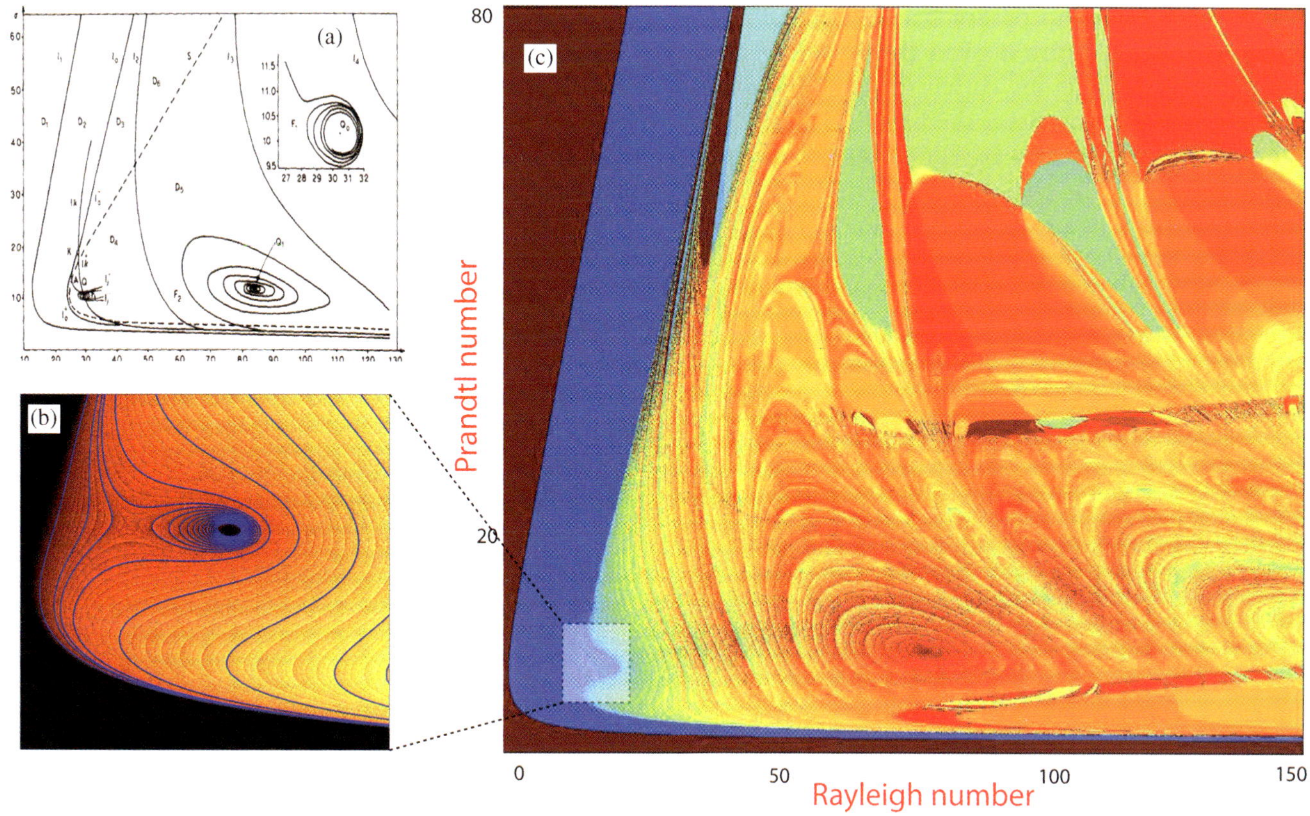

Fig. 7. (a) Bifurcation diagram of the Lorenz equation depicting the two detected T-points and primary homoclinic bifurcation curves (courtesy [15]). (c) The (r, σ)-biparametric sweeping revealing numerous T-points within a fractal mapping of the Lorenz equation in the kneading-based bifurcation diagram. Solid-color regions correspond to constant kneadings: dark (brown) region corresponding to simple Morse-Smale dynamics; lighter (blue) stripe corresponding to the chaotic saddle set after the homoclinic explosion due to the primary homoclinic butterfly. Bright (reddish) region maps the complex chaotic dynamics in the Lorenz equation globally organized by various T-points alternating with saddles in the parameter plane. (b) Zoom of the vicinity of the primary T-point at ($r = 30.4$, $\sigma = 10.2$) to which the homoclinic bifurcation curve converges. Data for curves (purple) are courtesy of Yu. Kuznetsov.

kneading. This implies that the behavior of the separatrix does not change within the windows. It could be due to the existence of robust trivial attractors such as equilibrium states or periodic orbits, to which the separatrix converges as time progress. The boundary between two solid-color regions corresponds to a bifurcation of the trivial attractor, which is associated with a jump between two constant values of the kneading invariant. For example, the jump from $P^{(50)}_{\max}$ to $P^{(50)}_{\min}$ in the kneading value occurs on the borderline between two regions, brown and light blue, corresponds to the formation of the homoclinic butterfly of the saddle. Since, for $r > 1$ there is a single Andronov-Hopf bifurcation for equilibria of the Lorenz equations, all other borderlines between the regions of solid colors should be associated with bifurcations of stable periodic orbits: period-doubling or saddle-nodes. A quick review of the kneading definition (2) suggests that the kneading value does not vary after a period-doubling bifurcation, because the kneading sequence will inherit the same coding; say original $[+1, -1, -1]$ will be repeated twice in the sequence $[+1, -1, -1, +1, -1, -1]$ for a new attractor of double period and so forth. The same is true also for the pitch-fork bifurcations, because continuous transitioning between stable symmetric and asymmetric periodic orbits is undetectable by the kneading technique. We can hypothesize that the borderlines may be only associated with generic saddle-node bifurcations, after which the separatrices must switch to another attractor with a distinct kneading invariant.

What the proposed kneading technique does extraordinary well, compared to the biparametric sweeping based on the finite-time maximal Lyapunov exponent approach, and what we have developed it for, is the detection of bifurcations of the complex sets, see the parametric chaos in Fig. 7. Panel (b) of this figure depicts the kneading mapping at the left-bottom corner of the bifurcation diagram in panel (c). Here, the black (blue in Panel (c)) region corresponds to the unstable complex dynamics, while the region of the existence of the Lorenz attractor is presented in yellow-reddish colors. Dark (purple) curves singled out in panel (b) are bifurcation curves of the simplest homoclinic orbits with the shortest admissible kneadings. One can see from this panel that the diagram is nicely foliated by the kneading level curves of gradually progressing color from red to yellow. This indicates that the Lorenz attractor, while being structurally unstable and sensitive to parameter variations, persists the pseudo-hyperbolic property here because the foliation remains uniform, and transverse to the classical pathway $\sigma = 10$ (Fig. 6). This nice foliation starts breaking around a saddle point after which one partial bifurcation curve spirals onto a singular point. Far from this point, the curve corresponds to the formation of the homoclinic loop with the kneading $\{1, -1, -1, -1\}$, i.e. the right separatrix makes one excursion around the saddle-focus O_1, followed by three revolutions around the saddle-focus O_2, and then return to the saddle at the origin. While moving along the spiraling curve, the separatrix of the saddle makes progressively more turns around O_2. With each turn, the number of the scrolls around the spiral point increments progressively as well. Due to this feature the point $Q_1(r = 30.4, \sigma = 10.2)$ is called a Terminal or T-point, as the bifurcation curve spirals onto it so that the separatrix makes infinitely many turns. The limit case corresponds to the following symbolic sequence $\{+1, -1, -1, -1, \cdots\}$, or $\{+1, (-1)^\infty\}$. In virtue of the symmetry of the Lorenz equation, the T-point corresponds to a closed heteroclinic connection involving all three saddle-equilibria, see Fig. 5. The right (left) separatrix of the saddle merging with the incoming separatrix of the saddle-focus $O_2\,(O_1)$, makes the bifurcation increase the codimension to two, while intersections of the 2D unstable manifolds of the saddle-foci, with the 2D stable manifold of the saddle at the origin remain transverse. Breaking the 1D heteroclinic connection can give rise to a primary homoclinic orbit to the saddle-focus and to a heteroclinic connection between both saddle-foci. The corresponding bifurcation curves bound a sector originating from the T-point in the parameter plane as both r and σ are increased [15, 23]. Bykov [15] proved that near the primary T-point there are countably many alike T-points that are located within the sector. Each new T-point produces other non-overlapping sectors within self-similar structures scaled like fractals. Panel A shows two such identified T-points: primary and another secondary located at ($r = 85$, $\sigma = 11.9$). The primary codimension-two T-point in the Lorenz equation was originally discovered in [5].

As soon as the saddle-foci and their bifurcations become involved in the dynamics of the Lorenz attractor, it loses the purity of the genuine chaotic attractor that used to have neither stable periodic orbits nor non-transverse homoclinic trajectories, and transforms into a quasi-chaotic attractor with stable orbits and tangent homoclinics. The idea of non-transversality or tangency was employed in [14, 16] to numerically identify the second bifurcation curve, l_K, in addition to the curve, l_2, that bounds the existence region of the Lorenz attractor in the parameter plane, see Fig. 8. Note that l_K crosses the initial boundary, l_2. This means that above the intersection point, crossing l_2 rightwards does not guarantee that the basin of the Lorenz attractor will necessarily be isolated from the stable foci, $O_{1,2}$ (Fig. 6). This means too that the

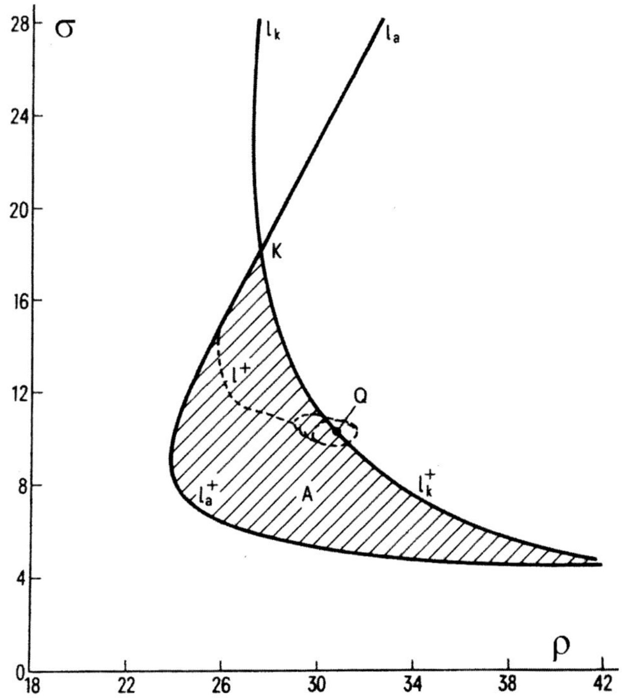

Fig. 8. The (r, σ)-bifurcation diagram showing the existence region (shaded) of the Lorenz attractor. The second bifurcation curve, l_k, passing through the primary T-point, $Q(r = 30.4,\ \sigma = 10.2)$, crosses the first boundary, l_a (l_2 in Fig. 6), thus closing the existence region. Courtesy of [16].

separatrices of the saddle will demonstrate some chaotic transient behaviors prior to them converging to $O_{1,2}$, see [16, 34–36, 38] for full details.

The other feature of the the boundary, l_K, is that it passes through the primary T-point, thereby separating the existence region of the Lorenz attractor from bifurcations of the saddle-foci, and consequently from all secondary T-points. So, for the parameter values to the right of the second boundary, the chaotic dynamics still persist in the Lorenz equation, and moreover it becomes more wider and less predictable [40–42, 48]. The indication to the saying is panel C of Fig. 7 showing chaos, turbulence in the parameter space, with fractal explosions in the forms of multiple spiral structures — "tornado eyes" centered around various T-points. Note that basins of spiraling T-points are separated by corresponding saddles in the parameter plane. One can spot self-similar smaller-scale spiral structures within large-scale ones and so forth. This richness of the structure of the parameter plane indicates that the chaotic attractor, which results from the synergy of the former Lorenz attractor amplified by chaos induced by the Shilnikov saddle-foci, becomes wild, in both dynamical and bifurcation senses.

Figure 9 shows a fragment of the typical biparametric sweeping of the Lorenz equation which is based on the evaluation of the largest Lyapunov exponent spectrum for the separatrices of the saddle calculated over a finite time interval. The same technique was used to identify the spiral structures in the Rössler model with the Shilnikov saddle-focus [6], and was employed in [24] for the study of a wild Lorenz attractor in the 3D Henon map, as well as to carry out a comprehensive three-parametric study of chaotic and regular regions of the Lorenz model [7,9,10]. In contrast to the Rössler model, the Lyapunov-exponent-based biparametric screening reveals no signs of indication of such structures within regions of deterministic chaos — including the Lorenz and quasi-attractors, where variations of the Lyapunov exponents are not significant enough to identify and differentiate between any fine patterns. These chaotic, red-colored regions in Fig. 9 are associated with positive values of the largest Lyapunov exponent, while the dark region corresponds to simple Morse-Smale dynamics with attractor having the largest Lyapunov exponent $L_{\max} \leq 0$ for stable equilibria or periodic orbits, respectively.

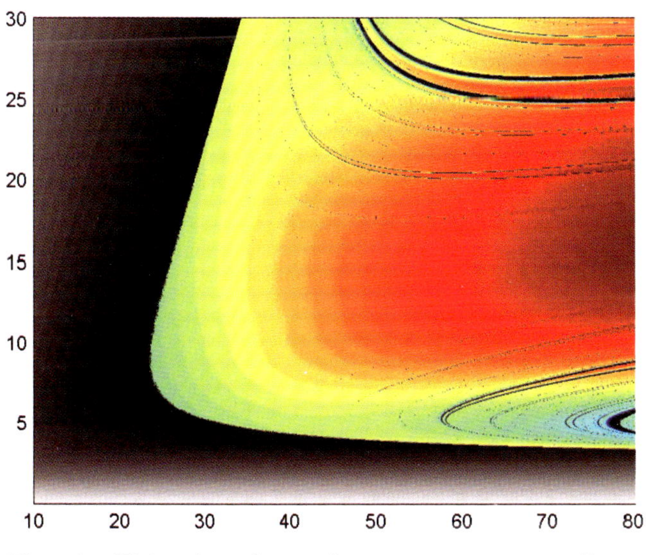

Fig. 9. Finite-time largest-Lyapunov exponent, $L_{\max}$, sweeping of the Lorenz equation shows no sign of spiral structures in the (r, σ)-parameter plane. The dark region corresponds to trivial attractors, where $L_{\max} \leq 0$, while the red color indicates the largest value of the maximal exponent, $L_{\max} > 0$, in chaotic regions right after the saddle-foci become involved.

4. Conclusions

We have examined two formation mechanisms of spiral structures in biparametric mappings of systems with the Shilnikov saddle-focus and with the Lorenz attractor. The feature of the spiral hubs in the Rössler model is that the F[ocal]-point gives rise to the alternation of the topological structure of the chaotic attractor transitioning between the spiral and screw-like types, as well as terminates the primary homoclinic curves of the saddle-focus equilibrium state influencing the forward-time dynamics of the model. The findings let us hypothesize about universality of the structure of the spiral hubs in similar systems with chaotic attractors due to homoclinics of the Shilnikov saddle-focus. The hub structures are outlined by homoclinic and saddle-node bifurcation curves originating from a codimension-two Belyakov point corresponding to the transition to the saddle-focus from a simple saddle.

For thorough explorations of the dynamics of Lorenz-like models we have proposed the algorithmically easy, though powerful toolkit based on the straight-forward evaluation of the kneading invariants derived through symbolic description of a single trajectory — an unstable separatrix of the saddle singularity of the model. We have demonstrated that the presence of multiple spiral T-points, saddles and accompanying fractal structures in the parameter plane is the key signature for systems with the Lorenz attractor. The T-points partition and organize globally the bifurcation structures of the parameter space of a model with the Lorenz attractor. We point out that no other techniques, including approaches based on the Lyapunov exponents for example, can reveal the discovered parametric chaos with such stunning clarity and beauty.

The method should be beneficial for detailed studies of other systems admitting a reasonable symbolic description. It bears an education aspect: the kneading-based screening can be used for in-class presentation to reveal the magic and fascinating complexity of low-order models in the cross-disciplinary field of nonlinear dynamics. The bi-parametric mapping technique is easily adopted for parallel multicore GPU platforms allowing for ultra-fast simulations of models in questions. Additional implementation of high-precision computations for long transients will thoroughly reveal multi-layer complexity of self-similarity of fractal patterns of T-point vortices. In future research we intend to enhance the toolkits for exploration of high-dimensional systems, symmetric and asymmetric [37], admitting symbolic descriptions with more symbols and additional kneading invariants [49].

Acknowledgments

This work is supported by the Spanish Research projects MTM2009-10767 and MTM2012-31883 (to R.B.), and by NSF grant DMS-1009591, GSU Brain & Behaviors program, and MESRF "Attracting leading scientists to Russian universities" project 14.740.11.0919 (to A.S). We thank Dima Turaev for useful comments, Yu. Kuznetsov for sharing the data used in Fig. 7, Fernando Blesa and Sergio Serrano for sharing the data of Figs. 1, 2, 3 and 4, as well as Dane Allen, Bryce Chung and Joe Youker for careful proofreading the manuscript.

References

[1] Abad, A., Barrio, R., Blesa, F. and Rodríguez, M. (2013). Algorithm 924: TIDES, a Taylor series Integrator for Differential Equations, *ACM Transactions on Mathematical Software* **29**(1).

[2] Afraimovich, V., Bykov, V. V. and Shilnikov, L. P. (1977). The origin and structure of the Lorenz attractor, *Sov. Phys. Dokl.* **22**, pp. 253–255.

[3] Afraimovich, V., Bykov, V. V. and Shilnikov, L. P. (1983). On structurally unstable attracting limit sets of Lorenz attractor type, *Trans. Moscow Math. Soc.* **44**, 2, pp. 153–216.

[4] Afraimovich, V. and Shilnikov, L. (1983). Strange attractors and quasiattractors, in, *Dynamics and Turbulence, Pitman, NY*, pp. 1–51.

[5] Andreichikov, I. and Yudovich, V. (1972). Self-oscillating regimes branching from Poiseuille flow in a two-dimensional channel. *Dokl. Akad. Nauk SSSR* **17**, pp. 120–124.

[6] Barrio, R., Blesa, F. and Serrano, S. (2009). Qualitative analysis of the Rössler equations: bifurcations of limit cycles and chaotic attractors, *Phys. D* **238**, 13, pp. 1087–1100.

[7] Barrio, R., Blesa, F. and Serrano, S. (2012). Behavior patterns in multiparametric dynamical systems: Lorenz model, *Internat. J. Bifur. Chaos* **22**(6), p. 1230019 (14 pages).

[8] Barrio, R., Blesa, F., Serrano, S. and Shilnikov, A. (2011b). Global organization of spiral structures in bi-parameter space of dissipative systems with Shilnikov saddle-foci, *Phys. Rev. E* **84**, 035201.

[9] Barrio, R. and Serrano, S. (2007). A three-parametric study of the Lorenz model, *Phys. D* **229**, 1, pp. 43–51.

[10] Barrio, R. and Serrano, S. (2009). Bounds for the chaotic region in the Lorenz model, *Phys. D* **238**, 16, pp. 1615–1624.

[11] Belyakov, L. A. (1980). The bifurcation set in a system with a homoclinic saddle curve, *Mat. Zametki* **28**, 6, pp. 911–922, 962.

[12] Bonatto, C. and Gallas, J. A. C. (2008). Periodicity hub and nested spirals in the phase diagram of a simple resistive circuit, *Phys. Rev. Lett.* **101**, 5, p. 054101.

[13] Bonatto, C., Garreau, J. C. and Gallas, J. A. C. (2005). Self-similarities in the frequency-amplitude space of a loss-modulated CO_2 laser, *Phys. Rev. Lett.* **95**, 14, p. 143905.

[14] Bykov, V. and Shilnikov, A. (1989). Boundaries of the domain of existence of a Lorenz attractor, *Methods in qualitative theory and bifurcation theory (in Russian)* **1**, pp. 151–159.

[15] Bykov, V. V. (1993). The bifurcations of separatrix contours and chaos, *Phys. D* **62**, 1-4, pp. 290–299.

[16] Bykov, V. V. and Shilnikov, A. L. (1992). On the boundaries of the domain of existence of the Lorenz attractor, *Selecta Math. Soviet.* **11**, 4, pp. 375–382.

[17] Chua, L. (1994). Chuas circuit : An overview ten years later,, *J. Circuits, Systems and Computers* **4**, pp. 117–159.

[18] Chua, L. (2007). Chua circuit, *Scholarpedia* **2**(10), p. 1488.

[19] Freire, J. G. and Gallas, J. A. C. (2010). Non-shilnikov cascades of spikes and hubs in a

semiconductor laser with optoelectronic feedback, *Phys. Rev. E* **82**, 3, p. 037202.

[20] Gallas, J. A. C. (2010). The structure of infinite periodic and chaotic hub cascades in phase diagrams of simple autonomous flows, *Internat. J. Bifur. Chaos* **20**, 2, pp. 197–211.

[21] Gaspard, P., Kapral, R. and Nicolis, G. (1984). Bifurcation phenomena near homoclinic systems: A two-parameter analysis, *J. Statist. Phys.* **35**, pp. 697–727.

[22] Gaspard, P. and Nicolis, G. (1983). What can we learn from homoclinic orbits in chaotic dynamics? *J. Statist. Phys.* **31**, 3, pp. 499–518.

[23] Glendinning, P. and Sparrow, C. (1984). Local and global behavior near homoclinic orbits, *J. Statist. Phys.* **35**, 5–6, pp. 645–696.

[24] Gonchenko, S., Ovsyannikov, I., Simo, C. and Turaev, D. (2005). Three-dimensional Hénon-like maps and wild Lorenz-like attractors, *Internat. J. Bifur. Chaos* **15**(11), pp. 3493–3508.

[25] Guckenheimer, J. and Williams, R. F. (1979). Structural stability of Lorenz attractors, *Inst. Hautes Études Sci. Publ. Math.* **50**, 50, pp. 59–72.

[26] Hastings, A. and Powell, T. (1991). Chaos in a three-species food chain, *Ecol.* **72**, pp. 896–903.

[27] Kaplan, J. L. and Yorke, J. A. (1979). Preturbulence: a regime observed in a fluid flow model of Lorenz, *Comm. Math. Phys.* **67**, 2, pp. 93–108.

[28] Kuznetsov, Y. A., De Feo, O. and Rinaldi, S. (2001). Belyakov homoclinic bifurcations in a tritrophic food chain model, *SIAM J. Appl. Math.* **62**, 2, pp. 462–487.

[29] Lorenz, E. (1963). Deterministic nonperiodic flow, *J. Atmospheric Sci.* **20**, pp. 130–141.

[30] Lorenz, E. N. (2008). Compound windows of the Hénon-map, *Phys. D* **237**, 13, pp. 1689–1704.

[31] Milnor, J. and Thurston, W. (1988). On iterated maps of the interval, *Lecture Notes in Math.* **1342**, pp. 465–563.

[32] Nascimento, M. A., Gallas, J. A. C. and Varela, H. (2011). Self-organized distribution of periodicity and chaos in an electrochemical oscillator, *Phys. Chem. Chem. Phys.* **13**, pp. 441–446.

[33] Rössler, O. E. (1976). An equation for continuous chaos, *Phys. Lett. A* **57**, 5, pp. 397–398.

[34] Shilnikov, A. (1986). Bifurcations and chaos in the Marioka-Shimizu model. Part I. *Methods in qualitative theory and bifurcation theory (in Russian)* **1**, pp. 180–193.

[35] Shilnikov, A. (1989). Bifurcations and chaos in the Marioka-Shimizu model. Part II. *Methods in qualitative theory and bifurcation theory (in Russian)* **1**, pp. 130–198.

[36] Shilnikov, A. (1991). Bifurcation and chaos in the Morioka-Shimizu system, *Selecta Math. Soviet.* **10**(2), pp. 105–117.

[37] Shilnikov, A. and Shilnikov, L. (1991). On the nonsymmetric lorenz model, *Internat. J. Bifur. Chaos* **1**, pp. 773–776.

[38] Shilnikov, A. L., Shilnikov, L. P. and Turaev, D. V. (1993). Normal forms and Lorenz attractors, *Internat. J. Bifur. Chaos* **3**, 5, pp. 1123–1139.

[39] Shilnikov, L. (1980). Bifurcation theory and the Lorenz model. *Appendix to Russian edition of "The Hopf Bifurcation and Its Applications."* Eds. J. Marsden and M. McCraken., pp. 317–335.

[40] Shilnikov, L. (1994). Chua's circuit: Rigorous results and future problems, *Internat. J. Bifur. Chaos* **4**(3), pp. 489–518.

[41] Shilnikov, L. (1997). Mathematical problems of nonlinear dynamics: A tutorial. visions of nonlinear mechanics in the 21st century, *Journal of the Franklin Institute.* **334**(5–6), pp. 793–864.

[42] Shilnikov, L. (2002). Bifurcations and strange attractors, *Proc. Int. Congress of Mathematicians, Beijing (China) (Invited Lectures).* **3**, pp. 349–372.

[43] Shilnikov, L. and Shilnikov, A. (2007). Shilnikov bifurcation, *Scholarpedia* **2**(8), p. 1891.

[44] Shilnikov, L. P. (1965). A case of the existence of a countable number of periodic motions, *Sov. Math. Dokl.* **6**, p. 163.

[45] Shilnikov, L. P., Shilnikov, A. L., Turaev, D. and Chua, L. O. (2001). *Methods of qualitative theory in nonlinear dynamics. Part II* (World Scientific Publishing Co. Inc.).

[46] Shilnkov, A. (1993). On bifurcations of the Lorenz attractor in the Shimizu-Morioka model, *Physica D* **62**(1–4), pp. 338–346.

[47] Sparrow, C. (1982). *The Lorenz equations: bifurcations, chaos, and strange attractors*, Applied Mathematical Sciences, Vol. 41 (Springer-Verlag, New York).

[48] Turaev, D. and Shilnikov, L. (1998). An example of a wild strange attractor, *Sbornik. Math.* **189**(2), pp. 291–314.

[49] Barrio, R., Shilnikov, A. and Shilnikov, L. P. (2012). Kneadings, symbolic dynamics, and painting Lorenz chaos. A Tutorial, *Internat. J. Bifur. Chaos* **22**(4), 1230016.

CHAPTER 34

DYNAMICS OF PERIODICALLY FORCED MASS POINT ON CONSTRAINED SURFACE WITH CHANGING CURVATURE

YOSHISUKE UEDA

Waseda University, Emeritus Professor of Kyoto University

The author has long been engaged in the study of nonlinear dynamics. Recently he has developed great interest in the relationship between nonlinear characteristics and outcomes (steady states) arising from nonlinear systems. This dedicatory study in honor of Professor Leon O. Chua deals with the basic behavior of a mass point in a rotating external force around the vertical axis, which is constrained on a surface of revolution opening upward around the same vertical axis. We observed jump phenomena in the amplitude or the altitude of the horizontal circular motion under variations of the angular velocity of the external force almost-stationarily, and obtained the dependence of the jump phenomena on the curvature of the surface of revolution, i.e., discovered an interesting qualitative difference, which is separated by parabolic surface of revolution.

1. Problem Setting and Equation Derivation

1.1. *nth order surface of revolution*

In this study, the dynamics are determined by a downward gravity force and a horizontal force rotating around the axis of revolution of the surface. The fundamental behavior of this system is a periodical circular motion, as the total force is balanced by the damping force, externally-applied force, gravitational force, and constraining force.

Let us take the vertical Z-axis upward, and generate a curve of the constrained surface by $Z = X^n$, i.e., surface of revolution $Z = \sqrt{X^2 + Y^2}^n$ (from now on, we call the surface nth order surface of revolution). Of course, the (X, Y)-plane is horizontal.

1.2. *Derivation of motion equation*

A motion of a constrained mass point m is described by the following equation

$$\left.\begin{aligned} m\frac{d^2X}{dt^2} &= -\delta\frac{dX}{dt} + F_x(t) + R_x(t) \\ m\frac{d^2Y}{dt^2} &= -\delta\frac{dY}{dt} + F_y(t) + R_y(t) \\ m\frac{d^2Z}{dt^2} &= -\delta\frac{dZ}{dt} - mg + R_z(t) \end{aligned}\right\} \tag{1}$$

where δ represents the damping coefficient, mg the gravitational force, $F_x(t), F_y(t)$, the X- and Y-components of the sinusoidal periodic external force, $R_x(t), R_y(t), R_z(t)$ the constrained force by the nth order revolution surface.[a]

1.3. *Normalized equations of motion*

In order to carry out a feasible experimental study, a parameter ℓ is introduced to regulate the actual size of the nth order constrained surface, via

$$Z/\ell = (\sqrt{X^2 + Y^2}/\ell)^n \tag{2}$$

By replacing variables X, Y and Z with ℓx, ℓy and ℓz, and the constrained force $R_*(t)/m\ell$ with $\Lambda(t)N_*(t)$ ($* = x, y, z$), and using Eq. (2), one can represent Eq. (1) as

$$\left.\begin{aligned} \frac{dx}{dt} &= u, & \frac{du}{dt} &= -ku + f_x(t) + \Lambda(t)N_x(t) \\ \frac{dy}{dt} &= v, & \frac{dv}{dt} &= -kv + f_y(t) + \Lambda(t)N_y(t) \\ \frac{dz}{dt} &= w, & \frac{dw}{dt} &= -kw - \frac{g}{\ell} + \Lambda(t)N_z(t) \end{aligned}\right\} \tag{3}$$

where

$$k = \frac{\delta}{m}, \quad f_x = \frac{F_x}{m\ell} = F\cos\omega t,$$
$$f_y = \frac{F_y}{m\ell} = F\sin\omega t \tag{4}$$

Here, $\Lambda(t)$ is an unknown function about the magnitude of the constrained force, $N_x(t), N_y(t), N_z(t) (= R_*(t)/m\ell)$ represent directional components of the constrained force, which are given via partial derivatives of the following constrained condition:

$$\varphi = z - r^n = 0, \quad r^2 = x^2 + y^2 \tag{5}$$

that is,

$$\left.\begin{aligned} N_x(t) &= -nxr^{n-2} \\ N_y(t) &= -nyr^{n-2} \\ N_z(t) &= 1 \end{aligned}\right\} \tag{6}$$

1.4. *Determination of unknown function $\Lambda(t)$*

Derivatives of constrained Eq. (5) are expressed by

$$\frac{d\varphi}{dt} = w - nr^{n-2}(xu + yv) = 0 \tag{7}$$

$$\frac{d^2\varphi}{dt^2} = \frac{dw}{dt} - nr^{n-2}\left(x\frac{du}{dt} + y\frac{dv}{dt} + u^2 + v^2\right)$$
$$-n(n-2)r^{n-4}(xu + yv)^2 = 0 \tag{8}$$

[a] Here, as a mass point, a small iron ball with mass m is considered, where the rolling resistance of a small ball is neglected, while the dissipative force proportional to the velocity of the mass point is taken into account. Constrained surfaces of revolution will be made with nonmagnetic material, and an external force is applied by rotating the magnetic field which is generated by spacially symmetrically arranged three coils injected with three phase alternative currents having variable frequencies.

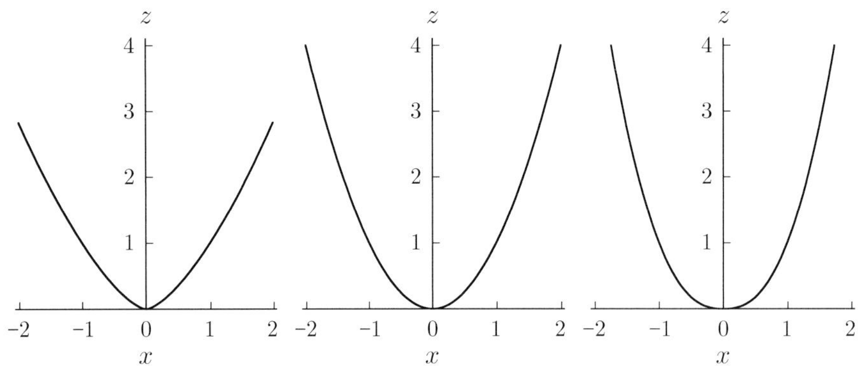

Fig. 1. Generating curves of constrained surface represented by Eq. (5).

By substituting Eq. (3) into Eq. (8), and taking into consideration from Eq. (4) to Eq. (7), the unknown function $\Lambda(t)$ in Eq. (3) is given by

$$\Lambda(t) = \frac{1}{1+n^2r^{2n-2}}\left[\frac{g}{\ell} + nr^{n-2}(xf_x + yf_y + u^2 + v^2) + n(n-2)r^{n-4}(xu+yv)^2\right] \tag{9}$$

Figure 1 shows some examples of generating curves of constrained surface represented by Eq. (5).

2. Derivation of an Autonomous System Approximating the Original Non-autonomous System

In order to examine the general behavior of the horizontal circular motion in which the fundamental component with period $2\pi/\omega$ predominates over all other components. The approximate solution of Eq. (3) is given by

$$\left.\begin{aligned} x(t) &= a(t)\cos\omega t - b(t)\sin\omega t \\ y(t) &= b(t)\cos\omega t + a(t)\sin\omega t \\ z(t) &= c(t) \end{aligned}\right\} \tag{10}$$

where the amplitudes $a(t)$, $b(t)$ and $c(t)$ are slowly varying functions of time t as compared with $\cos\omega t$ and $\sin\omega t$. This implies that their derivatives are sufficiently small quantities. Also, the damping coefficient k is regarded as a small quantity.

The well-known normal means, such as Harmonic Balance Method and Averaging Principle, leads to the following autonomous equation:

$$\left.\begin{aligned} \frac{da}{d\tau} &= \frac{1}{2\omega^2}\left[-k\omega a + \left(\omega^2 - \frac{ng}{\ell}r^{n-2}\right)b\right] \\ &= f(a,b) \\ \frac{db}{d\tau} &= \frac{1}{2\omega^2}\left[-\left(\omega^2 - \frac{ng}{\ell}r^{n-2}\right)a - k\omega b - F\right] \\ &= g(a,b) \\ 0 &= -\frac{g}{\ell} + \Lambda \end{aligned}\right\} \tag{11}$$

where

$$\tau = \omega t, \quad r^2 = a^2 + b^2,$$
$$\Lambda = \frac{\frac{g}{\ell} + nr^{n-2}(Fa + \omega^2 r^2)}{1 + n^2 r^{2n-2}} \tag{12}$$

The autonomous equation (11) turns out to be two-dimensional with respect to $a(t)$ and $b(t)$, and the variable $c(t)$ is determined by Eq. (5). The third equation in Eq. (11) comes from the third equation (3) under the above assumptions.

It is well known that a two-dimensional autonomous equation of a dissipative system has two types of steady states. One is equilibrium points and the other is limit cycles. However, Eq. (11) has only equilibrium points but not limit cycles. This matter originates from the Bendixson (Negative) Theorem on the existence of periodic solutions represented by closed curves.

2.1. *Equilibrium points and frequency response curves*

By putting the right-hand side of Eq. (11) to zero, equilibrium point of the Eq. (11) can be found, as

$$a_0 = -\frac{r_0^2}{F}\left(\omega^2 - \frac{ng}{\ell}r_0^{n-2}\right), \qquad b_0 = -\frac{r_0^2}{F}k\omega \tag{13}$$

and the vertical component c_0

$$c_0 = \sqrt{a_0^2 + b_0^2}^{\,n} \tag{14}$$

where r_0^2 in Eq. (13) is determined from the following relation:

$$\left[k^2\omega^2 + \left(\omega^2 - \frac{ng}{\ell}{r_0}^{n-2}\right)^2\right]{r_0}^2 = F^2 \tag{15}$$

It is noted that Eq. (15) represents frequency response (or resonance) characteristics of the steady-state amplitude (or radius) r_0 of a horizontal circular motion, and c_0 is the height of the circular periodic motion with equi-angular velocity.

2.2. *Stability condition for equilibrium points*

We shall determine stability conditions for the equilibrium point represented by Eq. (13). Let (ξ, η) be a small variation from the equilibrium point (a_0, b_0) and then discuss whether the variation approaches zero or not as the time t increases. The following variational equation is obtained by applying common manipulations:

$$\left.\begin{aligned} \frac{d\xi}{d\tau} &= \left(\frac{\partial f}{\partial a}\right)_0 \xi + \left(\frac{\partial f}{\partial b}\right)_0 \eta \\ \frac{d\eta}{d\tau} &= \left(\frac{\partial g}{\partial a}\right)_0 \xi + \left(\frac{\partial g}{\partial b}\right)_0 \eta \end{aligned}\right\} \tag{16}$$

where left (, right)$_0$ stands for a partial derivative, such as $\partial f/\partial a$, at the equilibrium point $a = a_0$? $b = b_0$.

An application of the well-known Routh-Hurwitz criterion gives the following stability condition:

$$\left.\begin{aligned} p &= -\left(\frac{\partial f}{\partial a}\right)_0 - \left(\frac{\partial g}{\partial b}\right)_0 = \frac{k}{\omega} > 0 \\ q &= \left(\frac{\partial f}{\partial a}\right)_0\left(\frac{\partial g}{\partial b}\right)_0 - \left(\frac{\partial f}{\partial b}\right)_0\left(\frac{\partial g}{\partial a}\right)_0 \\ &= \frac{1}{4\omega^4}\left[\omega^4 + \left(k^2 - \frac{g}{\ell}n^2{r_0}^{n-2}\right)\omega^2 \right. \\ &\quad \left. + \frac{g^2}{\ell^2}n^2(n-1){r_0}^{2n-4}\right] > 0 \end{aligned}\right\} \tag{17}$$

The first relation of the condition (17) answers correctly the Bendixson (Negative) Theorem for its conclusion with a positive value of k. The second relation gives vertical tangency of the frequency response (resonance) characteristics represented by Eq. (15).

3. Horizontal Circular Motion with Equi-angular Velocity

In the preceding section, circular motion representing steady states is obtained as

$$\left.\begin{aligned} x_0(t) &= a_0\cos\omega t - b_0\sin\omega t \\ y_0(t) &= b_0\cos\omega t + a_0\sin\omega t \\ z_0(t) &= c_0 \end{aligned}\right\} \tag{18}$$

By inspection it is easy to see that this steady-state solution (18) is an exact solution of Eq. (3), but non-stationary representative point of Eq. (11) shows approximate solutions which express the transient states.

The solutions represented by Eq. (18) is rewritten in terms of the amplitude r_0 and the phase φ_0 as follows:

$$\left.\begin{aligned} x_0(t) &= r_0\cos(\omega t - \varphi_0) \\ y_0(t) &= r_0\sin(\omega t - \varphi_0) \\ {r_0}^2 &= {a_0}^2 + {b_0}^2 \\ \varphi_0 &= \arctan\left(\frac{-b_0}{a_0}\right) \end{aligned}\right\} \tag{19}$$

4. Frequency Response Curves and Phase Portraits

Here, how curvature of nonlinearity determines outcomes is shown by numerical examples with $n = 2$. The following parameter values are used:

$$\frac{g}{\ell} = 100, \quad k = 0.1, \quad F = 5 \tag{20}$$

These values are given by considering the actual sizes of the revolution surfaces. That is, when the radius of the constrained surfaces is about 9.8[cm], depth also takes about 9.8[cm]. Constants of the external force F and angular velocity ω are used as the so-called control parameters, which are selected according to our purpose. For the amplitude of the external force F, only one case is chosen as $F=5$, but for other cases it will be reconsidered. It should also be added that the choice of the value k is difficult and sometimes controversial, which will not be further discussed here.

4.1. *Response curves and their dependency on the curvature n*

Now, examples of response (or resonance) characteristics (r_0, ω) with stability limit are shown for three curvatures with $n=2, 3/2$, and $5/2$. Also, phase characteristics (φ_0, ω) and bifurcation sets on the (F, ω)-plane are given.

Figure 2 shows the (r_0, ω) and (φ_0, ω) characteristics for the parabolically constrained surface with $n = 2$. In this parabolic case, r_0 is a single-valued function of ω, therefore no jump phenomenon is observed. The response curve (r_0, ω) shows a single sharp peak (drawn in logarithmic scale), and the phase φ_0 changes abruptly from 0 to π with an increase of ω at the resonant point $\omega \simeq 10\sqrt{2}$ (see Eq. (15)). This feature is strikingly similar to the well-known characteristics observed in linear series of LRC-circuits injected with sinusoidal periodic signals of varying frequencies.

Figures 3 and 4 show the (r_0, ω) and (φ_0, ω) characteristics and bifurcation sets on the (F, ω)-plane for the constrained revolution surfaces with order $n=3/2$ and $n=5/2$, respectively. In both cases, r_0 and φ_0 are three-valued in some intervals on ω. In order to make clear boundaries of the intervals, subsidiary vertical lines are added in both figures. On these boundaries, r_0 and φ_0 have vertical tangencies, which are determined as intersections of the characteristic curves with stability limit, given by $q = 0$ (see the second equation in Eq. (17)). Jump phenomena occur at the bifurcation points. In the three-valued intervals, midpoints represent saddle, while upper resonant and lower non-resonant states are equilibrium points (which are stable nodes or foci).

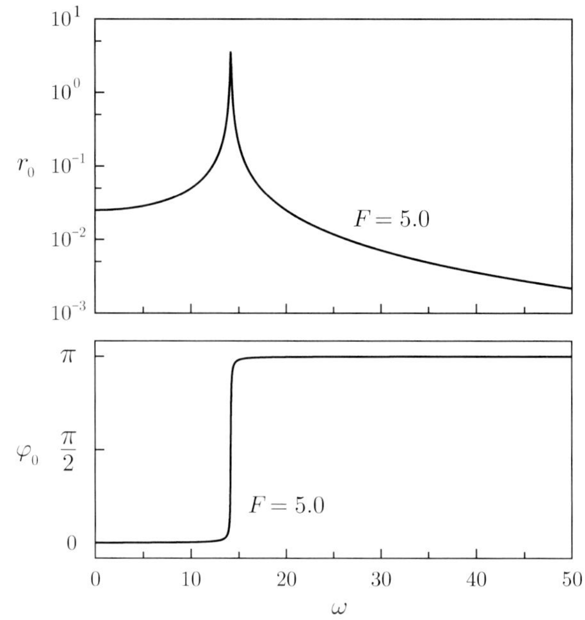

Fig. 2. Frequency response characteristics for the parabolically constrained surface.

Jump-up phenomena from non-resonant to resonant states occur for increasing ω in Fig. 3, but for decreasing ω in Fig. 4. Jump-down phenomena from resonant to non-resonant states take place in the reverse variation of ω. In Fig. 3, the response characteristic r_0 leans to left, and in Fig. 4, to right.[b] These features are similar to the response characteristics observed in periodically forced systems having soft and hard types of restoring terms of second order equations.

4.2. *Phase portraits*

Phase portraits for the two-dimensional autonomous system (11) are shown in Figs. 5 and 6, which correspond to the cases $(n = 3/2)$ and $(n = 5/2)$, respectively. In both figures, point R represents a stable equilibrium point expressing resonant state, stable point N non-resonant states, and D saddle point. The insets (ω-branches) and the outsets (α-branches) of the saddle points D are also shown. In these cases, insets form basin boundaries, or separatrixes between resonant and non-resonant states. Basins for non-resonant states are shown with oblique lines. These phase portraits are qualitatively in almost mirror symmetry relations with respect to $n=2$ mutually. These facts seem to be an interesting nature of nonlinear dynamics.

[b]It should also be added that characteristics (r_0, ω) lean to left for $n = 1.9, 1.8, \ldots$, and to right for $n = 2.1, 2.2, \ldots$.

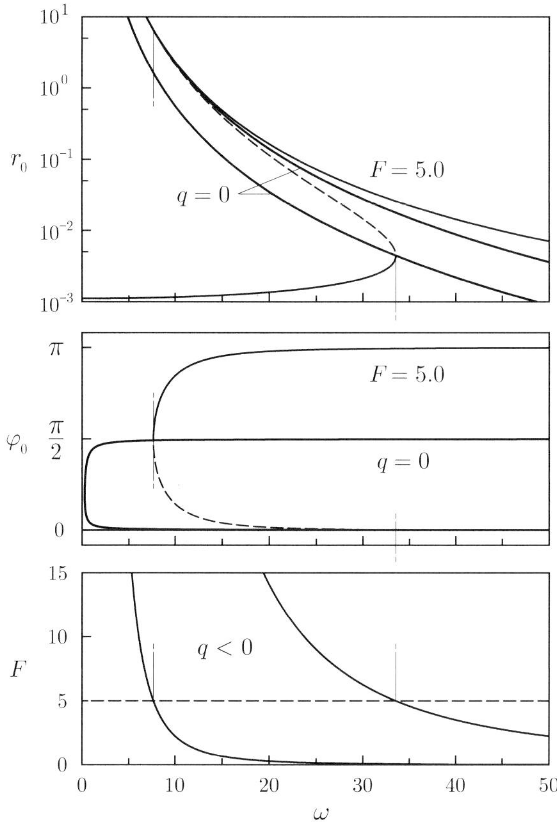

Fig. 3. Frequency response characteristics for the 3/2 order constrained surface.

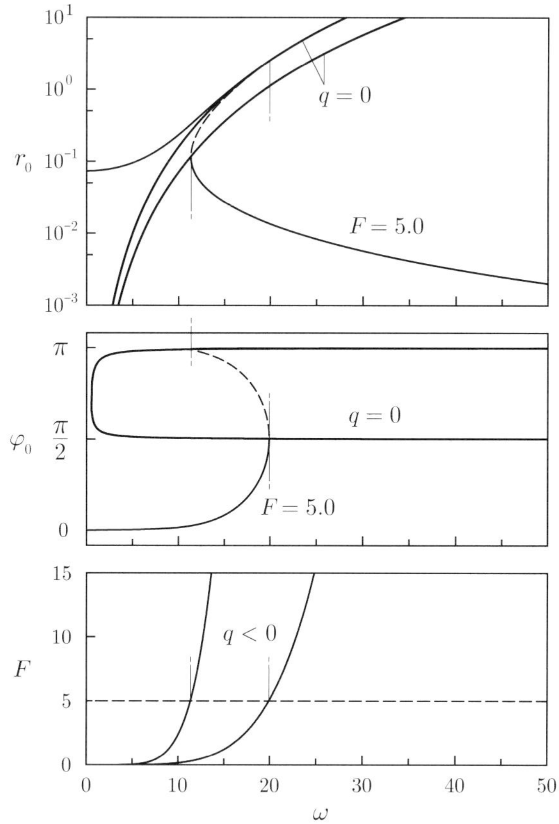

Fig. 4. Frequency response characteristics for the 5/2 order constrained surface.

5. Discussions about the Jumping Mechanism

Interpretation of the above observed jumping phenomena is attempted from a physical standpoint. For that purpose, Eq. (11) is rewritten in the following form:

$$\left.\begin{aligned} \frac{d(a^2+b^2)}{d\tau} &= \frac{1}{\omega^2}\left(-k\omega r^2 - bF\right) \\ &= N(a,b) \\ \frac{d}{d\tau}\left(-\frac{b}{a}\right) &= \frac{1}{2\omega^2 a^2}\left(\frac{g}{\ell}nr^n - \omega^2 r^2 - Fa\right) \\ &= T(a,b) \end{aligned}\right\} \tag{21}$$

where

$$r^2 = a^2 + b^2 \tag{22}$$

The first equation of Eq. (21) determines the derivative of the normal component (radius direction) of the motion of the representative point, and the second equation of Eq. (21) tangential component to the circular trajectory of the motion (see the fourth equation of Eq. (19)).

5.1. *Decomposition of operating force*

When the system is in the steady state, the mass point continuing the horizontal circular motion represented by Eq. (18) or Eq. (19) is supplied with power from the rotating external force. In order to explain the situation, the force operating on the mass point is decomposed into tangential component $F \sin \varphi_0$ and radial component $F \cos \varphi_0$. Let us call the former $F_a = F \sin \varphi_0$ active component and the latter $F_r = F \cos \varphi_0$ reactive component. Both components are on the horizontal plane.

5.2. *Jumping-up phenomena*

Jumping-up phenomenon in Fig. 3 occurs when the angular velocity ω is increased, while in Fig. 4 decreased. These phenomena occur under the mechanism that the saddle point D approaches a non-resonant point N (point N moves a little almost horizontally), and these coalesce into a node-saddle,

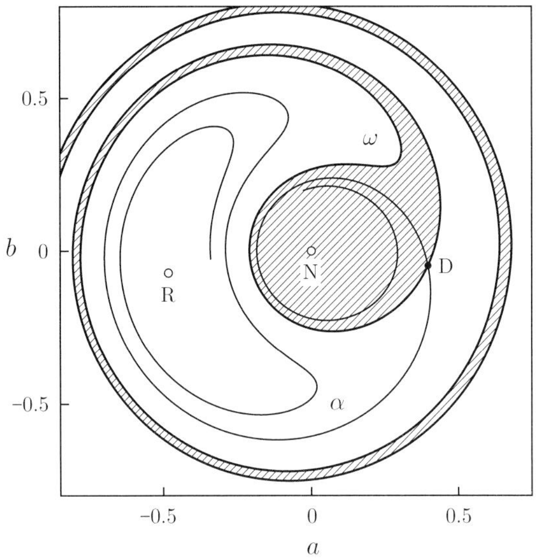

Fig. 5. Phase portrait for Eq. (11) with $n = 3/2$, $F = 5$, $\omega = 15$.

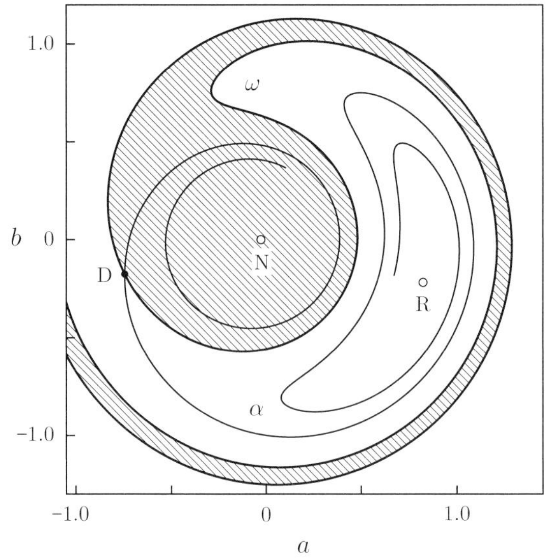

Fig. 6. Phase portrait for Eq. (11) with $n = 5/2$, $F = 5$, $\omega = 15$.

a higher-order equilibrium point, and then disappears. The amplitude characteristic (r_0, ω) and the phase characteristic (φ_0, ω) together demonstrate these states clearly. The well-known fold bifurcation occurs. By inspecting the movement of the representative point of Eq. (11), from (disappeared) *ND* point, representative point takes off to the right (almost on and along the a-axis) in the case of Fig. 3, while to the left in the case of Fig. 4. Moreover, phase changes from 0 to π and from π to 0. Note that the direction φ_0 is taken as clockwise here.

The above implies that the failure ($N(a, b)$ becomes positive) of the first equation of Eq. (21) predominates over the second equation of Eq. (21). This fact implies that the power supplied by the active component to the mass overflows the power consumed by dissipation. In other words, jumping-up phenomena occur via the loss of power balance between supply and consumption.

5.3. *Jumping-down phenomena*

Jumping-down phenomenon in Fig. 3 occur when the angular velocity ω is decreased, while in Fig. 4 increased. These phenomena occur under the mechanism that the saddle point D and the stable equilibrium point R approach each other by moving downward, and they coalesce into a node-saddle, a higher-order equilibrium point, almost on the b-axis, and then disappears. The well-known fold bifurcation occurs. By inspecting the movement of the representative point of Eq. (11), from (disappeared) *RD* point, representative point takes off to the right (almost horizontally) in the case of Fig. 3, while to the left in the case of Fig. 4. Moreover, phase changes from $\pi/2$ to 0 and from $\pi/2$ to π.

The above implies that the failure ($T(a, b)$ becomes non-zero) of the second equation of Eq. (21) predominates over the first equation of Eq. (21). This fact implies that the supplied reactive force to the mass loses the balance with the sum of the centrifugal force and the radial component of the constrained force. In other words, jumping-down phenomena occur via the loss of force balance among reactive, centrifugal and (horizontal component of) constrained forces.

6. Problems Descendent from This Study

Experimental simulation studies on Eq. (11) revealed preceding facts. Though not mentioned there, many complicated motions are observed in systems described by Eq. (3) when the amplitude F takes large values, some of which seems to be chaotic motions. These are left as open problems.

Simulation study for the case of a conic constrained surface of revolution ($n = 1$) showed peculiar behavior. In this case, bottom (origin of the (x, y, z)-space) becomes singular or a capture point, and interesting intermittently bursting motions are observed. No further investigation has been done, however. Several motions that caught our attention may be hidden behind the system.

Introduction of asymmetry into constrained surfaces and/or periodically rotating force will bring unexpected motions, however these are left as further open problems.

7. Conclusion

The qualitative and quantitative behaviors of a mass point in a rotating external force and constrained on a surface of revolution upward has been studied. The basic behavior of this system is periodic motion. Jump phenomena in the amplitude of the periodic circular motion under variation of the angular velocity of the external force almost-stationarily have been observed, and qualitative difference of the jump phenomena on the curvature of the surface of revolution is disclosed, which is separated by a parabolic surface of revolution given by $n = 2$. Moreover, mechanisms of jumping-up and -down phenomena are also clarified by the loss of power balance and reactive force balance, respectively.

Acknowledgments

In the preparation of this dedicatory study in honor of Professor Leon O. Chua, the author was greatly supported by Professor Shin'ichi Oishi, Dean of the Graduate School of Fundamental Science and Engineering, Waseda University, who provided him splendid research environment. He was also aided by his graduated students Mr. Kenta Hosoi and Mr. Takuma Kawamura of the Future University — Hakodate, Dr. Hirofumi Ohta of the Kyoto University and Professor Ralph Abraham of the University of California Santa Cruz, who gave him valuable suggestions and good advice of all kinds. Acknowledgment must also be made to Professor Guanrong (Ron) Chen of the City University of Hong Kong, and Professor Andrew Adamatzky of the University of the West of England at Bristol, for their excellent cooperation with great patience.

As to the references of this study, there are too many good textbooks on Classical Physics and Nonlinear Dynamics or Oscillations. It seems very difficult for the author to refer to a long list of textbooks among many. The author especially appreciates many great precursors in the fields.

CHAPTER 35

SOLITONS FOR DESCRIBING 3-D PHYSICAL REALITY: THE CURRENT FRONTIER

PAUL J. WERBOS*

National Science Foundation, Arlington, Virginia 22230, USA

This chapter begins with a review and assessment of four key frontiers for the fields of memristors, neural networks and chaos: (1) use of learning architectures to expand the possible markets for dense memristor chips, crucial to applications such as power grid intelligent enough to improve the economics of renewable energy; (2) advanced modeling and hardware to build such chips, from graphene to magnetoresistance; (3) second generation quantum computing; and (4) development and use of chaotic soliton models to rebuild the foundations of physics. Current more detailed reviews of the first three frontiers are summarized. The bulk of the paper discusses how a "soliton" model of the electron can fill outstanding gaps in basic physics, such as explanation of the mass and radius of the electron without a need for renormalization. It reviews the two soliton models most widely discussed in physics today, the Skyrme model and the BPS monopole, and proposes alternative Lagrangians starting from the bosonic sector of electroweak theory which may have the required properties.

1. Introduction and the Larger Context

There are many exciting new frontiers opening up for advanced research involving memristors, chaos and beyond. This introduction will provide a very broad review of four of these frontiers — learning based architectures for memristor systems, new physical hardware, the injection of quantum computing capabilities into these systems, and extension of the underlying understanding of physics. After the introduction, this paper will focus on the fourth of these frontiers. It will specify three new dynamical systems, describe their properties, their potential use in providing a new model of the electron, and outline the challenges ahead.

*This chapter represents the views of the author, not the official views of NSF; however, it does constitute work by a federal employee on government time.

The most important new frontier for practical applications of memristors today is the greater use of learning in order to make greater use (and find greater application) of all kinds of massively parallel hardware [1]. Traditional systems for neural networks and machine learning had difficulties in meeting the real needs of applications with a high degree of complexity, or with a mix of continuous and discrete variables; however, we now understand the principles which allow us to overcome those limitations [2, 3].

There is a clear four step roadmap [3] for developing more powerful general-purpose learning systems by Eq. (1) developing more general systems for "vector intelligence" by better integrating fundamental principles for optimal decision-making and prediction; Eq. (2) moving up a level by injecting new approaches to handling complexity in space (as a few systems now begin to do, such as some breakthrough systems which have recently broken world records in image recognition [4, 5]); Eq. (3) handling skill learning and complexity over time in a more principled way (as no working systems really do quite yet); and finally Eq. (4) adding capabilities for brain-like stochastic search [6] which exist at the highest levels of the mammal brain.

Even the fulfillment of this roadmap in technology will not end up replicating the fullest capabilities of the human mind [7], but it will enable many important technological capabilities. For example, the current White House policy statement on the smart grid [8] states that "NSF is currently supporting research to develop a '4th generation intelligent grid' that would use intelligent system-wide optimization to allow up to 80% of electricity to come from renewable sources and 80% of cars to be pluggable electric vehicles (PEV) without compromising reliability, and at minimum cost to the Nation (Werbos 2011)." [9]. In paying for electricity, the world now pays on the order of 10 cents per kilowatt hour (kwh) for 20,000 terawatt hours per year (IEA/OECD data on total electricity in 2008). That amounts to $2 trillion per year. The most cost-effective sources of solar power and wind can supply electricity at about that price, or less, in quantity in the US, and at lower cost in most other places. (By contrast, the latest estimates I hear from industry sources say that the total cost for new nuclear power plants, exploiting the new expedited safety reviews, will still end up costing 13.5 cents per kwh, and present greater problems outside the US.)

This would seem to say that we can become independent of fossil fuel, avoid military conflicts over fossil fuel, and avoid the large-scale environmental threats associated with fossil fuel or nuclear proliferation, simply by shifting over to wind and solar, at a small cost or small profit in the end. But in the real world, the ten cents cost which gets paid to generators is not the full cost of using wind and solar. Use of wind or solar on the real world power grid involves huge new costs in stabilizing the power grid (an operation conducted in real time), and in backup and storage. Yet wind and solar farms contain a lot of switchable power electronics, which could convert these energy sources into assets rather than liabilities, if only we could generate the right control signals to make use of this control authority in real time (every two second rather than every 5 to 15 minutes). At present, the most advanced system for optimal power flow in practical use is the system recently developed by Marija Ilic of Carnegie Mellon, which makes approximate calculations every 15 minutes, and lacks full measures to anticipate areas of future power scarcity, but has shown major savings over older systems in the areas where it has been tested. But with cellular types of neural networks and new cellular hardware, Liang *et al.* [10] have illustrated how it is possible to perform the same kind of task, with a system designed for anticipatory optimization [2], in real time. The sheer speed due to new computing hardware is a crucial factor here. If solar farms and wind farms can be converted from being a net cost to a net asset in stabilizing the grid, the benefits could clearly be on the order of trillions of dollars per year — and perhaps large enough to overcome the obstacles which block today's efforts to solve the urgent problems of the world in that sector.

Of course, the development of new computer hardware to exploit memristor capability and learning together, to provide ever greater computational horsepower, is another important frontier being funded by NSF, a frontier which needs to be advanced in lockstep with the advancement of learning architectures [1]. This frontier requires the use of very fundamental dynamical systems and quantum physics [11] to exploit the utmost in what we can do with hardware. There are many possible physical platforms for memristors, as other chapters in this book will probably review more systematically. In the short term, many of us pay special attention to the stories of transistors now reaching hundreds of gigahertz from IBM, using graphene, and from Intel, using InAs, but many believe that systems using magnetic spin effects or quantum tunnel junctions could do better (in memory and density as well as speed) once the physical problems are better understood and overcome. This in turn requires very complex new mathematical tools, especially for exploiting multibody coherence effects [11]. At these frontiers of physical hardware, it becomes both essential and difficult to account for quantum effects and noise in multibody physics, and to develop a better understanding of the flow of heat (phonons). The better understanding of many-body effects may also allow breakthroughs in the efficiency of solar cells [12] and thermoelectric chips.

Using traditional methods from quantum electrodynamics (QED [13, 14]), it is extremely challenging to understand these effects for physically interesting systems. For example, to model a system with a million electrons, we are asked, in principle, to solve a partial differential equation (the Schrodinger equation) over a space of three million dimensions. Quantum optics has been very effective in understanding and controlling many-body effects such as coherence (lasers) and entanglement, in part because it uses mathematical methods like P, Q and W mapping [15, 16] to convert these systems in high dimensions into traditional partial differential equations (PDE) over three dimensions [15, 16]. To the extent that phonons are bosons, we can apply an extended version of the same mappings (e.g., see [17]) to handle phonons. The emerging field of quantum phononics, borrowing methods from quantum optics, may be of enormous importance to developing advanced memristor hardware. The P, Q and W mappings need to be used together with other mathematical methods, of course, but they bring us back to the need to develop a better understanding of partial differential equations (PDE) over three dimensions, the main theme of this chapter after the introduction. These new tools also provide a pathway to solving serious fundamental problems in the rigorous aximatic formulation of quantum field theory in general [17].

The mapping from many body effects to PDE in 3 dimensions, in order to allow useful large-scale simulations, applications and technologies, is also an important emerging part of empirical nuclear physics [18].

Can we envision even more advanced information technologies, going beyond the capabilities of the mammal brain and the densest possible classical massively parallel computing? Can we inject some kind of quantum computing capability into these kinds of systems? This is a very controversial area, of course, but it makes sense for some of us to explore that third frontier while others work on the more immediate ones.

In my view, we cannot make full use of quantum computing effects until we develop a new, more complete understanding of what quantum mechanics really tells us, and how it really works.

In my view [19, 20], quantum mechanics is ultimately more about the direction of the flow of time than it is about systems over a million dimensions. It is very hard for some people to imagine the possibility of technology which could allow us to affect the direction in time of the flows of information and of heat, but this makes more sense to me, at least, than metaphysical theories of physics which assume that there is no physical reality out there at all [19, 20].

This third frontier leads to a number of immediate empirical opportunities which I have already discussed elsewhere [20]. But it also leads to a kind of fourth frontier, a more scientific frontier. For me, this frontier began with a "simple" question: can we use nonlinear system dynamics in three dimensions to give us a better understanding of what an electron really is, which would then let us use methods like those of quantum optics and empirical nuclear physics [15–18] to model systems of many electrons? Yet in exploring that question, I recently learned that this mathematics is inextricably connected as well to a major high-risk high-potential technology project in Russia [18, Sec. 11.5], aimed at developing methods for the complete conversion of matter to energy. Putin has recently noted that Russia's new project in nuclear technology, if successful, would be as far beyond fission and fusion as those are beyond gunpowder. Yet Manton emphasizes that better understanding and control over coherence effects will be essential to mastering that technology. It now seems that our "simple" question about the electron is important both to the physical assumptions and to the coherence issues which will determine the outcome of that effort.

2. The Current Frontier

Mathematical analysis by itself cannot deduce which dynamical system actually represents the true laws of the universe we live in. However, it can try to develop several credible possibilities, consistent with what we know form experiment.

At this moment in time, the key challenge in the fourth frontier is to understand what kind of stable "solitons" exist in any of three field theory systems. The first is:

$$L = L_W + L_B + L_{\varphi l} + L_{\varphi r} - V, \tag{1}$$

where the first two terms are the usual terms in electroweak theory (EWT) for the usual W and B fields (as on page 56 of Taylor [21]):

$$L_W = -\frac{1}{4} F_{\mu\nu} \cdot F^{\mu\nu}, \quad \text{for } F_{\mu\nu} = \partial_\mu W_\nu - \partial_\nu W_\mu + g W_\mu \wedge W_\nu \tag{2}$$

$$L_B = -\frac{1}{4}(\partial_\mu B_\nu - \partial_\nu B_\mu)(\partial^\mu B^\nu - \partial^\nu B^\mu), \tag{3}$$

where the new "Higgs field" φ is a two-by-two complex matrix which transforms like a spinor/contravariant spinor under Lorentz transforms (i.e. a twistor [22]), and where

$$L_{\varphi l} = c_l \left| \left(\partial_\mu - \frac{i}{2} g \tau \cdot W_\mu - \frac{i}{2} g' B_\mu \right) \varphi \right|^2 \tag{4}$$

$$L_{\varphi r} = c_r \left| \left(\partial_\mu - \frac{i}{2} g' B_\mu \right) \varphi^H \right|^2 \tag{5}$$

$$V = \lambda \left| I - \varphi^H \varphi \right|^2 \tag{6}$$

where c_l, c_r and λ are parameters, and where I define:

$$|M|^2 = Tr(M^H M), \tag{7}$$

where the superscript H is Schwinger's notation for the Hermitian conjugate.

A second possibility is a simple variation of the first, in which Eqs. (4) and (5) are replaced by:

$$L_{\varphi l} = c_l Re \left(\left(\varphi^H \tau^\mu \left(\partial_\mu - \frac{i}{2} g \tau \cdot W_\mu - \frac{i}{2} g' B_\mu \right) \varphi \right) \right) \tag{8}$$

$$L_{\varphi r} = c_r Re \left(\varphi \tau^\mu \left(\partial_\mu - \frac{i}{2} g' B_\mu \right) \varphi^H \right) \tag{9}$$

It is relatively simple how to translate this kind of Lagrangian density into the actual PDE, the Lagrange-Euler equations, and to work out the energy density, the Hamiltonian, which is the starting point for most practical work. The concise review in [14] is quite adequate, but a more extensive discussion may be found in [23, 24] and in texts on classical field theory.

In these models, B is a real covariant vector, and W is a "triplet," a set of three real vectors.

(Thus the wedge in Eq. (2) is simply one way of writing the usual cross product of two vectors in three dimensions; the cross product operates across the three unwritten indices, not across the subscript μ.) So far as I know, the only truly stable "solitons" ever discovered so far, for PDE obeying special relativity over three dimensions, are: Eq. (1) a model of a big star [25] with second order derivatives in the Lagrangian; Eq. (2) models like the Skyrme model [18, 26, 27] which include a field U which is strictly defined as an object in a "topologically interesting space." Some researchers have argued [27] that it is impossible to have truly stable solitons in any field theory, like the two Lagrangians above, in which the fields are defined as ordinary real vectors or spinors without apriori topological constraints, and in which no second order derivatives appear. If truly stable "solitons" are proven to exist in either of the systems above, this by itself will be a fundamental advance in our knowledge. These systems would also seem to enforce the quantization of spin, which is just as important to what we see in nature as the quantization of electric charge.

A more radical possibility, obviously capable of generating two charges, is a closer relative of the system studied by Prasad and Sommerfield, to be discussed in Sec. 4.

My strong conjecture is that one of these Lagrangians, or an extension of one of them, will contain a stable soliton which matches what we know about the electron. In other words, it can explain what an electron is. The challenge before us here is to prove that (and, of course, to find the extensions if necessary).

The remainder of this chapter will address three questions: (1) why do we need a new explanation of what an electron is? (2) how can we meet this challenge, building on the existing literature [18, 26, 27], includuing other "soliton" models? (3) where could success in this challenge lead us?

3. Why Seek A New Model of the Electron?

Most graduate schools now teach us that we already have a perfectly good model of an electron, the model given in quantum electrodynamics (QED) [13,14]. It is often said that QED is the most perfectly proven model of physics ever developed, with predictions tested across a wide range of phenomena in electronics and photonics, accurate to sixteen decimal places. This is not quite the whole story [19, 21], but it is certainly a good starting point. "Maxwell's Laws" are now seen as one part of QED. Physicists have developed many other ways to try to formulate quantum field theories, but for the practical, empirical challenges of electronics and photonics, the canonical version of QED developed by Schwinger *et al.* [13,14], based on operator fields, is still the starting point.

According to QED, the electron is a perfect point particle, with zero radius — unlike the proton or the neutron, which have a measurable radius of about one femtometer, which is also called a "Fermi length." That is a 10^{-15} meters, a millionth of a nanometer, or one hundred-thousandth of an Angstrom unit; in other words, it is about a hundred-thousandth of the size of an atom. But so far as we know today, the radius of the electron is exactly zero, just as QED assumes; recent experiments [28] claim to show an upper bound of 10^{-22} meters for the radius of the electron.

Why would we want a better model than that?

There are at least three reasons.

To begin with, QED is difficult to calculate with when we have millions of electrons in a system. There is some hope that a new model would allow new approaches to many-body computation, like those which have begun to emerge in empirical nuclear physics [18]. This may seem impossible, since the electron is a fermion; however, we now know that compound systems made up of bosons can have emergent statistics which match those of a fermion [26, 27]. Still, the possibility of faster computation for systems of electrons has yet to be proven, and is far from obvious. First we must show we can model the properties of a single electron as proposed here.

Second, a new approach is essential to someday getting back to Einstein's core vision of modeling reality through well-defined continuous PDE. That is what really motivates me here [19, 20], but the details are beyond the scope of this paper.

Third, there are certain problems with the QED model of the electron which people have basically repressed in their minds. These involve the radius of the electron.

We know very well that the electron has a charge which generates a Coulomb field. We know how to calculate the energy of that Coulomb field. If we assume that the charge is located at the origin, we can easily compute the total mass-energy in the Coulomb field, taking the integral of the Coulomb energy from some radius r on out to infinity. At a certain radius, r, the total mass-energy of the Coulomb field equals the total observed mass-energy of the electron. That is called the Thompson radius of the electron. It is just under three femtometers — much, much larger than the "upper limit" of 10^{-22} meters!

How does QED explain this paradox — the paradox of having an electron which has a mass much less than the mass of of one part of its field?

In essence, QED does assume that the Coulomb field of the electron has infinite positive energy, as we get when we send the radius r to zero. This by itself would lead to absurd predictions, since we know the electron does not behave as if it has infinite mass. However, "to fix the predictions," QED assumes an infinite negative offset, δM, located at exactly at the center of the electron. This is commonly presented as "part of the renormalization." People tend to think "this is just part of how we do the calculations." If we ask where this infinite negative energy comes from, perhaps they think subconsciously: "This is just a gift from God or from Schwinger to make our predictions work out." But in fact, what we use to make predictions is part of the theory itself, when it makes our predictions very different from what they otherwise would be.

In summary, the paradox of mass-energy in the Coulomb field greater than the mass of the electron itself gives us only two choices of possible explanation (leaving aside anachronistic alternatives like abandoning special relativity). As in QED, we may assume the presence of negative mass-energy inside every electron, which would call for us to understand much better how negative energy can occur in nature. It would be very exciting if negative energy really did exist, if we could understand it better. But here I am advocating more attention to a different approach to explanation, until and unless we find some credible empirical evidence for negative energy. I am proposing that we model the electron as a "soliton," as a pattern of fields in which other fields besides the Coulomb field exist, and affect the overall energy even beyond three femtometers, and in which energy density is strictly positive definite.

How could we reconcile this notion of a "bigger electron" with the data we now have available?

In fact, the electron may have more than one "radius," depending on how we probe it. Years ago, there was great disturbance when people discovered that the "radius of the proton" is very different depending on how one probes the proton [27].

In general, it seems that there are three ways that we could probe the electron, to determine its radius. First, we can do highly precise measurements [28] of exactly where the center of electric charge is located. Those are the experiments for which there is an upper bound of 10^{-22} meters. However, Prasad and Sommerfield [29] have shown that "soliton" solutions do exist for a field theory similar in spirit to Sec. 2, in which the origin of magnetic charge is all concentrated in exactly one point (radius zero), even as overall energy and other fields are smoothly distributed over a larger radius. The Lagrangians of Sec. 2 are a kind of extension of that earlier work, in which I hope that electric charge is quantized and localized in the same way that magnetic charge is in the earlier work.

It is also possible to probe the electron through high-energy collisions with objects like protons and neutrons. However, those collisions are strongly influenced by the complex physics of the neutron and the proton, which makes them less relevant here.

In order to probe whether there is "something else" of great importance as big as three femtometers, the obvious probe to use is high energy electron-electron (Moller) scattering. In fact, recent work on high-energy spin-polarized Moller scattering has shown major deviations from QED, which people are working to explain by invoking electroweak theory (EWT) [21]. According to the standard model of physics, QED itself should really be seen as one part of the larger theory, EWT. The approach proposed here is really just a variation of that approach. In fact, the equations of Sec. 2 are very close to the standard equations of electroweak theory. The only difference between these systems and the usual bosonic part of EWT is that I assume a Higgs field φ which is a twistor instead of a two-spinor. Yet in discussing how he arrived at EWT, Weinberg has often noted that he had several possible choices for a Higgs field which would have done equally well in performing the only empirical task assigned to it — trying to explain boson masses. The small change here would not undo the empirical success of the theory — but if generates electrons as solitons, it may eliminate the need for anything but bosons. Because of the P, Q and W mappings which work for bosons [15–17], it results in a physical theory which does not require renormalization in order to be mathematically well-defined. Masses emerge directly from integrating the Hamiltonian — including the mass of the electron itself, which is not explained at all by QED or renormalized traditional EWT.

4. Details of the Challenge and Prior Work

4.1. *Types of stability: from solitons to chaoitons*

For practical purposes, the goal in Sec. 2 is to find a set S of solutions to the system in Eq. (1) through 7 (or a variation of that system) which have the following two properties. First, each member of the set S consists of a set of field values $W(\underline{\mathbf{x}})$, $\mathrm{B}(\underline{\mathbf{x}})$ and $\varphi(\underline{\mathbf{x}})$ across all points $\underline{\mathbf{x}}$ in three dimensional space for some time t, such that these field values represent a static equilibrium state of the Lagrange-Euler equations. Second, each member of the set has the same total energy H. Any small perturbation of any member of the set should lead to a new state which has a higher energy, or is another member of the set, or is "equivalent to one of the members of the set." This last option can be important when a small change leads to a small apparent oscillation which can be "gauged away" or the like.

Many other concepts of stability have appeared in the literature on "solitons" in three dimensions.

Many of these concepts are either too restrictive or too loose to represent what we need to satisfy, for an adequate model of the electron. The formal concept of soliton in pure mathematics [30] would require that an electron would keep moving in the same direction, with the same momentum, after a collision. The concept of solitary wave [31] is too loose, and has generally been applied to the study of nonrelativistic systems in any case.

In the 1990's [32], I proposed that we study a class of objects which I called "chaotic solitons" or "chaoitons," which would require strong stability but also allow for the possibility of stable oscillations or even stable localized chaotic patterns. But in the case where energy density is positive definite, I showed that strong energy minimization (as discussed above) is sufficient to prove that a general set of dynamic solutions S constitute a chaoiton. The work of recent years suggests that we do not need to consider chaos or oscillation to model an electron; thus it is enough to seek a static chaoiton here.

Unfortunately, a large part of the prior work on "solitons" in physics assumes a similar but weaker concept of stability which would not be enough to plausibly model an electron. For example, many physicists rely on the concept of "Q stability," in which the states in S are minima of energy only with respect to perturbations which leave the charge Q unchanged [27]. It is known that there are solutions of the simple nonlinear Klein-Gordon equation which meet this criterion, but provide a kind of "ridge" of solutions of ever lower localized energy; in practice, one of these "solitons," when bouncing around a complex universe, would normally leak away its energy and dissipate to nothing. Similar situations can arise when the energy H meets the first order conditions for a minimum of energy, but not the requirement that the Hessian of energy be strictly positive except for perturbations within the set S.

4.2. *From Skyrme to Bogomolnyi, Prasad and Sommerfield (BPS)*

Two types of "soliton" model have attracted great interest in empirical nuclear physics, and in research which aims at a grand unification of EWT and strong nuclear forces: the Skyrme model, and the Bogomolnyi-Prasad-Sommerfield (BPS) monopole. The seminal paper of Prasad and Sommerfield describing that monopole [29] has a citation index of more than 2,000 in Google Scholar. Both types of "soliton" are discussed in great detail in [18, 26, 27].

Both models share a rather surprising property with Eq. (1) through 7 (and with traditional EWT): they require solutions which do ***not*** go to zero as one goes to a radius r infinitely far away from the center of the soliton. If you look closely at Eqs. (1) and (6), for example, you will see that a solution cannot possess finite energy unless the field φ goes to a unitary matrix as $r \to \infty$. The Skyrme model involves a similar matrix field U which is required to be unitary at all points in space. The system studied by BPS was previously discussed by 'tHooft, who described it as:

$$\mathrm{L} = \frac{1}{4}G^a_{\mu\nu}G^a_{\mu\nu} - \frac{1}{2}(D_\mu Q^a)^2 - \frac{1}{8}\lambda(Q^2 - F^2)^2, \tag{10}$$

where the underlying (real) fields are Q^a and A^a_μ for $a = 1, 2, 3$ and $\mu = 0, 1, 2, 3$, where λ and F are parameters, and where he used the definitions:

$$\begin{aligned} G^a_{\mu\nu} &= \partial_\mu A^a_\nu - \partial_\nu A^a_\mu + e\varepsilon_{abc}A^b_\mu A^c_\nu \\ D_\mu Q^a &= \partial_\mu Q^a + e\varepsilon_{abc}A^b_\mu Q^c \end{aligned} \tag{11}$$

From the last term in Eq. (10), it is clear that the three component real vector Q must go to a vector of length F as $r \to \infty$, not to zero. Note that Q is what some would call an "isovector" or a "scalar" or "just a triplet of scalars"; its three components do not change, in the field equations, when we perform a Lorentz transformation.

It is conceivable that someone will someday discover a relativistic field theory without second derivatives in the Lagrangian, in which chaoitons exist which obey the traditional boundary condition that all fields go to zero as $r \to \infty$. However, I do not see any sign in the literature that anyone knows how to do this yet, or that models of that kind would find it easy to explain the remarkable quantization of charge which we observe across all of physics. Thus for now, the approach based on nonzero boundary conditions seems more promising.

The Skyrme model has generally been used only to model protons and neutrons and complex nuclei as solitons. There have been a very few efforts to use it as a model of the electron, without success. Even as a model of the proton and neutron, it does not provide a way to explain why deep inelastic scattering shows that the proton and neutron do seem to be bound states of more fundamental components, such as quarks. This is one of the reasons why I propose that we start from the BPS monopole, rather than the skyrmion, as the main starting point in trying to model the electron. An additional reason, of course, is that we can stay close to EWT, which is our best description today of the behavior of electromagnetic fields and related phenomena. Nevertheless, if the "solitons" of the two systems in Sec. 2 do not have the desired strong stability, it is possible that the addition of a "Skyrme" term in φ or another simpler "fixer" term [33] could solve the problem. It is not necessary to impose the requirement that φ be strictly unitary, in order to do this. In fact, it may not even be necessary to require that φ be unitary asymptotically; one can imagine an alternative Higgs term more in the spirit of:

$$V(\varphi) = \lambda_1(1 - |Tr(\varphi)|^2)^2 + \lambda_2|\det\varphi|^2 \tag{12}$$

The Skyrme model and the BPS monopole both exploit the nonzero boundary conditions in order to create a "topological charge" which is strictly quantized and impossible to change, for reasons of topology [18, 26, 27]. But the BPS monopole is not suitable as a model of the electron, for several reasons.

In the first place, it has only one quantized charge, which is generally interpreted as magnetic charge, not electric charge; magnetic charge is actually a promising idea under serious consideration in grand unification physics, but without electric charge we do not have an electron. In any case, the exact relation between the A field of that model and the ordinary A, W and B fields is not at all clear. Secondly, though the BPS solution does also allow for some electric charge, discovered in the work of Julia and Zee, it does not actually ***quantize*** the electric charge; if electrons were BPS solitons, they would presumably vary continuously all over the map in how much charge they possess. And finally, the family of solutions discovered by Prasad and Sommerfield are a continuous family, with individual members defined by the amount of electric charge and by a continuous parameter C. In the limit as λ goes to zero, they compute the mass-energy of the "soliton" as exactly:

$$M = (C/\alpha)\cosh^2\gamma \tag{13}$$

where α and γ are parameters of the field theory itself. From Eq. (14), it would appear that we have a "ridge" of solutions of different energy, which can dissipate away to zero in practice, even if the magnetic charge itself is fixed and localized. Or to put it another way, the small perturbations which result from differentiating the Prasad-Sommerfield solutions with respect to C would be able to reduce the total energy.

Adding a simple fixer term may be able to solve this problem, but that may or may not be unnecessary when we move to a more realistic model in any case.

It would be very useful and instructive to analyze another possible model, which is a kind of hybrid between Eqs. (10)–(12) and the systems of Sec. 2:

$$L = L_W + L_B - \frac{1}{2}L_{\varphi l} - \frac{1}{2}L_{Qr} - V, \tag{14}$$

where the first two terms are defined as in Eqs. (2) and (3), where I have replaced the twistor φ by a set of two three isovectors, Q and φ, where

$$L_{\varphi l} = c_l\left|(\partial_\mu - \frac{i}{2}g\tau\cdot W_\mu - \frac{i}{2}g'B_\mu)\varphi\right|^2 \tag{15}$$

$$L_{Qr} = c_r\left|(\partial_\mu - \frac{i}{2}g'B_\mu)Q\right|^2 \tag{16}$$

$$V = \lambda_1(Q^2 - F_1^2) + \lambda_2(\varphi^2 - F_2^2) \tag{17}$$

Manton [18, Sec. 11.5] has shown that EWT itself does not possess topological solitons, because the nonzero boundary condition does not impose enough constraint on the usual version of the Higgs field as $r \to \infty$.

Section 2 attempts to solve these problems through a unique new integration of key features.

First, of course, it is a slight variation of EWT itself, such that the connection to electromagnetism and to the W and B fields is quite explicit.

Second, by borrowing the asymmetric left-hand versus right-hand coupling from EWT, it allows coupling to two different components of the fields, one on the left and one on the right, which makes it possible to have two topological charges. This allows quantization of the electric charge, as well as the magnetic charge, at the same time — crucial to modeling the electron, while not disrupting the potential value later on to grand unification physics. Of course, the violation of parity is a fundamental property of EWT and of the empirical work which led up to EWT.

Third, the field φ has eight degrees of freedom in the center of the soliton, but the boundary conditions reduce that to four asymptotically — two on the left, and two on the right, exactly what we need to create the kind of topology which Manton [18] and the others call for.

Even if better models are developed someday which can describe the electron, these fundamental properties appear essential, one way or another.

5. Beyond the Fourth Frontier: Connections to the Rest of Physics

If and when someone fully meets the challenge described in Sec. 2, it will open the door to two kinds of follow-on. First, there are technological connections discussed in earlier sections. Second will be the challenge of unifying a new view of the electron with larger questions in physics. Until we actually have a new model of the electron in hand, we cannot envision all the details, but we know that the connections will be important.

First of all, we may well find that there are several possible variations of these models, which yield solitons (and excited states) with different mixes of electric and magnetic charge and mass. That would give us some choice, in addressing further empirical issues. With this family of models, it is not necessary to inject back-door assumptions like infinite negative mass, or like renormalization in general, to make them mathematically meaningful; however, in quantizing these bosonic field theories, we do have a choice between P, Q and W mappings [15–17], if and when we really need a field operator representation. As in quantum optics and in skyrmion research [18], we can get very far using semiclassical methods alone.

Second, these solitons could be used as a kind of alternative quark model, based on magnetic and electric charge rather than the more complex properties assumed today, as proposed years ago by Julian Schwinger [34]. This would fit with the startling empirical findings of Macgregor [35] of Lawrence Livermore, who found evidence that simple electromagnetic coupling (alpha) seems to be decisive in determining the masses and lifetimes of all the known elementary particles. Though I strongly disagree with Comay's proposed Lagrangian for electromagnetism [36], his concept of the nucleon as a bound state of four objects, core and quarks, all defined my electric and magnetic charge and mass rather than the usual properties, and some of the associated empirical arguments, may be worth thinking about further, in conjunction with the other sources cited in [34].

Finally, the unification of such classical field models with general relativity would be quite straightforward, using the recipe given by Moshe Carmelli [37]. This should also be true for the alternative model of gravity by Moffat, which currently appears more promising in view of empirical data in 2012 suggesting a low density of dark matter.

Of course, this would still leave open the mysteries related to superweak interactions (CP) violation, dark energy, dark matter, and perhaps some other interesting phenomena beyond today's standard model of physics — but it would be a great step forwards in any case.

References

[1] P. Werbos, Memristors for More than Just Memory: How to Use Learning to Expand Applications. In R. Kozma, R. Pino, and G. Pazienza (eds), *Advances in Neuromorphic Memristor Science and Applications*, Springer, 2012.

[2] F. L. Lewis and D. R. Liu, *Reinforcement Learning and Approximate Dynamic Programming for Feedback Control*, Wiley, 2012.

[3] P. Werbos, Intelligence in the Brain: A theory of how it works and how to build it, *Neural Networks*, Vol. 22, Issue 3, April 2009, pp. 200–212.

[4] K. Kavukcuoglu, P. Sermanet, Y.-L. Boureau, K. Gregor, M. Mathieu and Y. LeCun: Learning Convolutional Feature Hierachies for Visual Recognition, *Advances in Neural Information Processing Systems (NIPS 2010), 2010.*

[5] Y. LeCun, K. Kavukvuoglu and C. Farabet: Convolutional Networks and Applications in Vision, *Proc. International Symposium on Circuits and Systems (ISCAS'10), IEEE, 2010.*

[6] P. Werbos, Brain-Like Stochastic Search: A Research Challenge and Funding Opportunity, plenary talk at IEEE Conference on Evolutoinary Computing 1999. Extended version available. http://arxiv.org/abs/1006.0385

[7] P. Werbos, Neural Networks and the Experience and Cultivation of Mind, Neural Networks, 2012.

[8] A POLICY FRAMEWORK FOR THE 21st CENTURY GRID: *Enabling Our Secure Energy Future* http://www.whitehouse.gov/sites/default/files/microsites/ostp/nstc-smart-grid-june2011.pdf

[9] P. Werbos, Computation Intelligence for the Smart Grid — History, Challenges and Opportunities. *IEEE Computational Intelligence Magazine*, August 2011.

[10] J. Liang, R. G. Harley and G. K. Venayagamoorthy, Adaptive Critic Design Based Dynamic Optimal Power Flow Controller for a Smart Grid, in Proc. of Computational Intelligence Applications In Smart Grid (CIASG), IEEE 2011.

[11] J. P. Dowling, G. Klimeck, and P. Werbos, *Final Report on ECCS/NSF Workshop on Quantum, Molecular and High Performance Modeling and Simulation for Devices and Systems (QMHP)*: http://arxiv.org/abs/0709.3310

[12] M. O. Scully, Quantum photocell: Using Quantum Coherence to Reduce Radiative Recombination and Increase Efficiency, *Phys. Rev. Lett.* Vol. 104, 207701, 2010. See also M. O. Scully, K. R. Chapin, K. E. Dorfman, M. Kim and A. Svidzinsky, Quantum Heat Engine Power can be Increased by Noise-induced Coherence, *PNAS*, Vol. 108, No. 37, pp. 15097–15100, 2011.

[13] J. Schwinger, (ed.), *Selected Papers on Quantum Electrodynamics*, Dover, 1958.

[14] F. Mandl and G. Shaw, *Quantum Field Theory*, revised edition. Wiley, 1993.

[15] L. Mandel and E. Wolf, *Optical Coherence and Quantum Optics*. Cambridge University Press, 1995, chapter 11.

[16] D. F. Walls and G. F. Milburn, Quantum Optics (Springer, New York, 1994).

[17] P. Werbos, Classical ODE and PDE Which Obey Quantum Dynamics, *Int'l J. Bifurcation and Chaos*, Vol. 12, No. 10, October 2002, pp. 2031–2049. Slightly Updated as Quant-ph0309031. Also see K. E. Cahill and R. J. Glauber, Ordered Expansions

in Boson Amplitude Operators, *Phys Rev*, Vol. 177, No. 5, pp. 1857–1881, January 1969, and P. Werbos, Soliton-based Approach to Axiomatic Quantization of Classical Field Theory, June 2012, http://www.scribd.com/doc/95547363/How-to-Quantize-June 2012

[18] N. Manton and P. Sutcliffe, Topological Solitons, Cambridge U., 2007 edition.

[19] P. Werbos, Bell's Theorem, Many Worlds and Backwards-Time Physics: Not Just a Matter of Interpretation, *Int'l J Theoretical Physics*, April 2, 2008 (epub date), arXiv: 0801.1234

[20] P. Werbos, A Three Step Program for Return to Reality in Physics, *Global Journal of Mathematical Sciences, International Federation of Nonlinear Analysts*, forthcoming. http://vixra.org/abs/1112.0091

[21] J. C. Taylor, *Gauge Theories of Weak Interactions*. Cambridge U. Press, 1976.

[22] R. Penrose and W. Rindler, *Spinors and Space-Time*, Cambridge U. Press, Corrected Edition, 1987.

[23] D. Lovelock and H. Rund, *Tensors, Differential Forms and Variational Principles*, Wiley, 1975.

[24] R. Adler, M. Bazin and M. Schiffer, *Introduction to General Relativity*, McGraw-Hill, 1965.

[25] G. Hobart, Nonlinear Field Equilibria, *Proc. Phys. Soc.*, Vol. 85, 1965.

[26] R. Rajaraman, *Solitons and Instantons: An Introduction to Solitons and Instantons in Quantum Field Theory*. Elsevier, 2nd edition, 1989.

[27] V. G. Makhankov, Yu. P. Rybakov and V. I. Sanyuk, *The Skyrme Model*, Springer-Verlag, 1994.

[28] H. Dehmelt, A Single Atomic Particle Forever Floating at Rest in Free Space: New Value for Electron Radius, *Physica Scripta* T22: 102–110. 1988. Bibcode: 1988PhST...22..102D. doi: 10.1088/0031-8949/198

[29] M. K. Prasad and C. M. Sommerfield, Exact classical solution for the 't Hoft monopole and the Julia-Zeedyon, Phys. Rev. Lett., Vol. 35, No. 12, 7600762, Sept. 1975.

[30] G. Eilenberger, *Solitons*, Springer-Verlag, 1983.

[31] G. B. Whitham, *Linear and Nonlinear Waves*, Wiley 1974.

[32] P. Werbos, Chaotic Solitons in Conservative Systems: Can They Exist? *Chaos, Solitons and Fractals*, Vol. 3, No. 3, pp. 321–326, 1993; P. Werbos, Can 'Soliton' Attractors Exist in Realistic 3+1−D Conservative Systems? *Chaos, Solitons and Fractals*, Vol. 10, No. 11, pp. 1917–1946, July 1999.

[33] P. Werbos, *Stability Theory and the Foundations of Physics: A Challenge to Real Analysis and Numerical Calculation* http://arxiv.org/abs/0807.2182

[34] P. Werbos, Schwinger's Magnetic Model of Matter: Can It Help Us With Grand Unification? http://arxiv.org/abs/0707.2520

[35] M. Macgregor and H. MacGregor, *Power of α: Electron Elementary Particle Generation With α-quantized Lifetimes and...*, World Scientific, 2007

[36] E. Comay, The Regular Charge-Monopole Theory and Strong Interactions, *Electronic J. of Theoretical Physics* (EJTP), Vol. 9, No. 26, pp. 93–118, 2012.

[37] M. Carmelli, *Classical Fields: General Relativity and Gauge Theory*, Wiley, 1982.

CHAPTER 36

THERMAL SOLITONS IN 1D AND 2D ANHARMONIC LATTICES — SOLECTRONS AND THE ORGANIZATION OF NON-LINEAR FLUCTUATIONS IN LONG-LIVING DYNAMICAL STRUCTURES

M. G. VELARDE*, W. EBELING† and A. P. CHETVERIKOV‡

Instituto Pluridisciplinar, Universidad Complutense de Madrid, Paseo Juan XXIII, 1, Madrid-28040, Spain
**mgvelarde@pluri.ucm.es*
†ebeling@physik.hu-berlin.de
‡chetverikovap@info.sgu.ru

We study the thermal excitation of intrinsic localized modes in the form of solitons in 1d and 2d anharmonic lattices at moderately high temperatures. Such finite-amplitude fluctuations form long-living dynamical structures with life-time in the pico-second range thus surviving a relatively long time in comparison to other thermal fluctuations. Further we discuss the influence of such long-living fluctuations on the dynamics of added excess free electrons. The atomic lattice units are treated as quasi-classical objects interacting by Morse forces and stochastically moving according to Langevin equations. In 2d the atoms are initially organized in a triangular lattice. The electron distributions are in a first estimate represented by equilibrium adiabatic distributions in the actual polarization fields. Computer simulations show that in 2d systems such excitations are moving with supersonic velocities along lattice rows oriented with the cristallographic axes. By following the electron distributions we have also been able to study the excitations of solectron type (electron-soliton dynamic bound states) and estimate their life times.

1. Introduction

In the linear approximation the fluctuations in condensed matter are mainly determined by the spectral distribution of phonons and other infinitessimal modes [1–5]. Nonlinear finite-amplitude fluctuations are less explored. There are however several interesting cases which are well studied [6–29]. Nonlinear fluctuations are of interest in view of their potential for possible novel electronic properties of materials, because of the strong interactions of the state of the atoms with embedded electrons. There is also the claim that nonlinear excitations may be of relevance for the transport phenomena observed in copper oxide layers [30, 31]. Furthermore experiments have shown that the collective vibrations of the atomic lattice of certain superconducting copper oxides behave in a manner that is hard to explain unless one assumes that motions characteristic of the presence of "stripes" are shaking the ion lattice [32–36].

The electron distributions are in a first estimate represented by equilibrium adiabatic distributions in the actual polarization fields offered by the classical lattice dynamics. Here, in particular, we show the existence of strong dynamical excitations forming stripes which are soliton-like and have a Toda-like shape [37, 38]. These entities are moving with supersonic velocities along the cristallographic axes of the lattice over about 10–100 atomic distances and have a lifetime of a few picoseconds.

The dynamics of the electrons is here not studied in detail, instead we are using simple adiabatic approximations assuming that the electron density suitably follows the running lattice soliton compressions. This is just a first approximation which we feel is enough for the purpose of this communication.

2. Solitonic Excitations in Morse Lattices

The Hamiltonian of the atomic lattice system which we consider here, e.g. in 1d, is

$$H_a = \frac{m}{2}\sum_n v_n^2 + \frac{1}{2}\sum_{n,j} V(r_n, r_j). \tag{1}$$

The subscripts locate lattice sites and the summations run from 1 to N with periodic boundary conditions. The characteristic length determining the repulsion between the particles in the lattice is the equilibrium distance $\sigma = L/N$ for a lattice of length L. We assume that the lattice particles repel each other with exponentially repulsive forces and attract each other with weak dispersion forces. We limit ourselves to nearest-neighbors only using their relative distance $r = |r_n - r_j|$. For the interaction we use the Morse potential

$$V = D\{\exp[-2b(r-\sigma)] - 2\exp[-b(r-\sigma)]\}, \tag{2}$$

where b is a stiffness parameter. For 2d systems appropriate modification of the potential is to be done in the long-range part, in order to avoid unphysical cumulative interaction effects. We manage this by imposing a suitable cut-off at 1.5σ.

We take all N units with equal masses m considered as lattice atoms with no internal dynamics. The electrons occupy some volume in the 3d space surrounding the 1d lattice. As we want to account for heating for the *heavy* lattice particles we assume that they obey classical Langevin dynamics appropriately augmenting (1). We include a phenomenological damping γ and some external noise source. For illustration in the computer simulations we shall use $N = 200$. Then in the presence of random forces (hence at non zero temperature) and also external forces the evolution of atomic lattice units is given by the Langevin equations

$$\frac{d}{dt}v_j + \frac{1}{m}\frac{\partial H_a}{\partial x_j} = -\gamma v_j + \sqrt{2D_v}\,\xi_j(t), \tag{3}$$

where $v_j(t)$ denotes velocity of unit j. The stochastic forces $\sqrt{2D_v}\,\xi_j(t)$ model a surrounding heat bath (Gaussian white noise). The parameter γ describes the standard friction frequency acting from the bath. The validity of an Einstein relation is assumed $D_v = k_B T\gamma/m$, where T denotes absolute temperature and k_B is Boltzmann's constant. For convenience, we use the inverse frequency of oscillations around the potential minimum ω_0^{-1} as the time unit. The unit temperature is 2D. Typical values of the parameters for biomolecules are in the range $\sigma \simeq 1 - 5A$; $b \simeq 1 - 5(A)^{-1}$; $D \simeq 0.1 - 0.5eV$. This means that $b\sigma \simeq 1 - 25$ while $1/\omega_0$ is in the range of $0.1 - 0.5ps$ [11, 15, 39–43].

As the energy unit we use the oscillation energy $m\omega_0^2\sigma^2 = 2D(b\sigma)^2$ (recall that in particular we use $b\sigma = 1$). Accordingly, the region where the anharmonic nonlinearity plays a major role is $0.75 < C_v/k_B < 0.95$. In the 1d case the corresponding dimensionless temperatures are in the range $T \simeq 0.1 - 1$. Thus solitons are to be expected stable in the range $T_{sol}^M \simeq 0.2D - 2.0D$ or in electron-volts $T_{sol}^M \simeq 0.02$–02 eV. Clearly for biological macromolecules this estimated range includes physiological temperatures (ca. 300 K).

Let us now add *excess free* electrons, as a form of doping, and consider their interaction with the lattice units [44]. Hence consider that one or several non-interacting electrons are embedded into the atomic lattice. The lattice atoms generate — in the field of charges — a *polarization* potential. Let us assume that the electron is located at position $\mathbf{r}$ and several atoms are located at nearby positions $\mathbf{r}_j$. Let h be a characteristic distance of polarization and U_e

the maximal polarization energy. For the potential generated by the atom number j at the origin we assume

$$U_j(r_j) = -U_e\left[\frac{h^4}{((r-r_j)^2+h^2)^2} - \frac{h^4}{(r_0^2+h^2)^2}\right], \tag{4}$$

$$U_j(r) = 0 \quad \text{if } |r - r_j| > r_0 \; r_0 = 1.5\sigma. \tag{5}$$

We here truncate the polarization potential at the cut-off distance $r = 1.5\sigma$. The distribution of the total electrical potential generated at place $\mathbf{r}$ by all the atoms is then given by

$$V_e(\mathbf{r}) = \sum_j U_j(\mathbf{r} - \mathbf{r_j}). \tag{6}$$

We are interested in the *dynamic* phenomena initiated by solitonic excitations in the lattice and neglect feedback effects. However we are aware that both phenomena, the local compression by a static process (polaron formation) and by a running compression hence the consecuences of soliton excitation are intimately connected [5, 45, 46]. The choice $h \simeq 0.3\sigma$ provides shallow minima at the location of the lattice atoms with significantly deep local minima at the location of a compression.

In a thermal system the lattice units perform quite complex motions, we may expect therefore a rather complex structure of the polarization field acting on the added excess free electrons. A snapshot of the polarization field landscape created by the lattice atoms is presented in Fig. 1. The potential energy has been given in units of the binding energy U_e (4). Taking into account the energy unit $2D(b\sigma)^2$ this means the scale is given by the ratio $\eta = \frac{U_e}{2Db^2\sigma^2} = \frac{1}{2b^2\sigma^2}\frac{U_e}{D}$. For $b\sigma = 1$ the energy scale is therefore $\eta = \frac{U_e}{2D}$. Thus for calculating any physical quantity the actual value taken by η is significant.

Worth noting is that the potential $U(x)$ is time-dependent and changes quickly so that the distribution of the electrons tries to follow it as fast as possible and hence the added excess free electrons are "slaved" accordingly, thus permitting an *adiabatic* approximation. We have a situation similar to that described for the free electron statistics in semiconductors [47, 48]. When the electron density due to doping is sufficiently low, so that the electrons are still nondegenerated we may approximate the Fermi statistics by the Boltzmann statistics. This is often a rather good approximation connecting in a simple way the distribution with the landscapes of the local polarization potential. Then, we take

$$n(x) = \frac{\exp[-\beta U(x)]}{\int dx' \exp[-\beta U(x')]}, \tag{7}$$

with $\beta = 1/k_BT$. Here x denotes the linear coordinate along the lattice. An example of the estimated density is shown in Fig. 2 (not normalized and given in logarithmic scale). The (relatively high) peaks correspond to the enhanced probability of a soliton to meet and trap an electron. This defines the solectron, i.e. an electron "surfing" on a soliton for about 10–50 time units (i.e. a few picoseconds) eventually getting off it and finding another soliton to surf-on and so on, in a kind of "promiscous" quantum dynamics evolution [49, 50]. For $T = 0.1$ we observe several rather stable running excitations (diagonal *stripes*) with velocity around $1.2v_{sound}$. For $T = 1$ (not shown in the figure) one can observe many

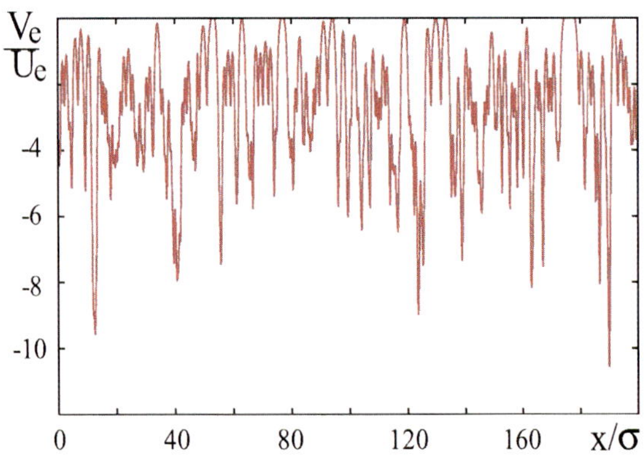

Fig. 1. Morse lattice. Snapshot of the local atomic lattice field potential landscape $U(x)$ acting on electrons in 1d. Many weak excitations and a few very strong solitons clearly appear. For units see main text. Parameter values: $V_e = U, N = 200, h = 0.3\sigma, T = 1$ and $b\sigma = 1$.

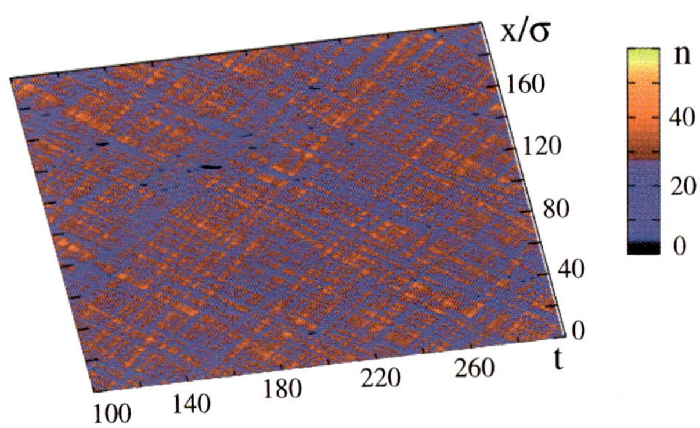

Fig. 2. Morse lattice. Probability distribution of an electron in a heated anharmonic lattice in the *adiabatic* approximation according to local equilibrium distributions. Parameter values: $N = 200, h = 0.3\sigma, T = 0.1, \sigma = 1$ and $b\sigma = 1$.

weak and only a few very stable and high level excitations moving with *supersonic* velocity $1.4v_{sound}$. The probabilities estimated from the Boltzmann distribution are strongly concentrated at the places of polarization potential minima. This means that most of the electrons are concentrated near to solitonic compressions. We wonder about the possible role of such solectrons in electron transfer (ET) in biomolecules or molecular wires found at long distances [39, 40, 51, 52].

3. Solitonic Excitations in 2d Layers

For convenience in the 2d lattice dynamics we use complex coordinates $Z = x + iy$, where x and y are Cartesian coordinates. Then the Langevin equations for the lattice units are

$$\frac{d^2 Z_n}{dt^2} = \sum_k F_{nk}(|Z_{nk}|) z_{nk} + \left[-\gamma \frac{dZ_n}{dt} + \sqrt{2D_v}(\xi_{nx} + i\xi_{ny}) \right], \qquad (8)$$

where an index n identifies a particle among all N particles of the ensemble, again γ is a friction coefficient, D_v defines the intensity of stochastic forces, $\xi_{nx,y}$ denotes statistically independent generators of the Gaussian noise, $Z_{nk} = Z_n - Z_k$. Further $z_{nk} = (Z_n - Z_k)/|Z_n - Z_k|$ is a unit vector defining the direction of the interaction force F_{nk}, corresponding to the Morse potential, between the n-th and the k-th particles. To have dimensionless variables we consider the spatial coordinates normalized to the length σ used in the Morse potential. As earlier noted, time is scaled by the inverse frequency of linear oscillations near the minimum of the Morse potential well, ω_0^{-1} and energy is scaled with $2D$, where D is the depth of the Morse potential well. Further the (dimensionless) parameter b defines the lattice stiffness, i.e. the strength of the repulsion between particles. The interaction force F_{nk} is given by

$$F_{nk} = F_{nk}(|Z_{nk}|) = -\frac{dV(r)}{dr}|_{r=|Z_{nk}|}. \qquad (9)$$

In view of the above only those lattice units with coordinates Z_k, satisfying the condition $|Z_n - Z_k| < 1.5\sigma$, are taken into account in the sum in Eq. (8). In computer simulations the interaction of particles is considered to take place inside a rectangular cell $L_x \cdot L_y$ with periodic boundary conditions and $L_{x,y}$, depending on the symmetry of the initial distribution of units and their number N. For illustration we consider a distribution corresponding to the minimum of potential energy for an equilibrium state of a *triangular* lattice $10\sigma \cdot 10\sqrt{3/2}\sigma$ for $N = 100$ or $20\sigma \cdot 20\sqrt{3/2}\sigma$ for $N = 400$. The positions on a triangular lattice at zero temperature ("cold" lattice) are used as initial conditions. Then the lattice is heated by the stochastic source (white noise) to the temperature $T = mD_v/k_B\gamma$. This corresponds to the mean kinetic energy of a particle $< T_{kin} >$ reaching the value T. The obtained values of Z_n and $V_n = dZ_n/dt$ are subsequently used as new initial values $Z_n(0)$ and $V_n(0)$ for the lattice at corresponding new temperature while setting $D_v = 0$. Notice that by varying $Z_n(0)$ and $V_n(0)$ it is possible to specify a localized excitation in the lattice. Using data about trajectories of particles $Z_n(t)$ and the evolution of velocities $V_n(t)$ we can calculate the mean kinetic and potential energies of the lattice particles, the temperature of the ensemble, and the particle distribution $\rho(Z, t)$ using a Boltzmann-like approximation.

In our computer simulations we used $\gamma = 0.0001 - 0.001$ and checked carefully that the assumed value does not significantly influence the results. In this case the amplitudes of the stochastic forces are small for the temperature values considered. Therefore the (Langevin) noise and the friction act in our computer simulations only as small albeit significant perturbations to the deterministic Newtonian dynamics.

We restrict here consideration to an isolated system (without influence from the bath), and follow the time evolution of quasi-1d excitations running along special lattice rows (similar as in the 1d-case) which may be analytically described as *effective* or *adapted* Toda solitons [24, 37, 38, 53–55]. These soliton-type excitations run in just one characteristic row which is embedded into the 2d layer in the form

$$x_{n+1}(t) - x_n(t) = \sigma - \frac{1}{b_{eff}} ln \left[1 + \frac{\sinh^2 \kappa}{\cosh^2(\kappa n - \beta t)} \right], \qquad (10)$$

where b_{eff} is the dimensionless lattice stiffness. Further the parameter κ is defined by the energy of the soliton and β is the reciprocal characteristic time

$$\beta = 1/\tau = \omega_0 \sinh \kappa. \qquad (11)$$

Such analytical formulae provide a solution of the equation of motion in 1d-lattices of atoms interacting by Toda forces. It appears that the

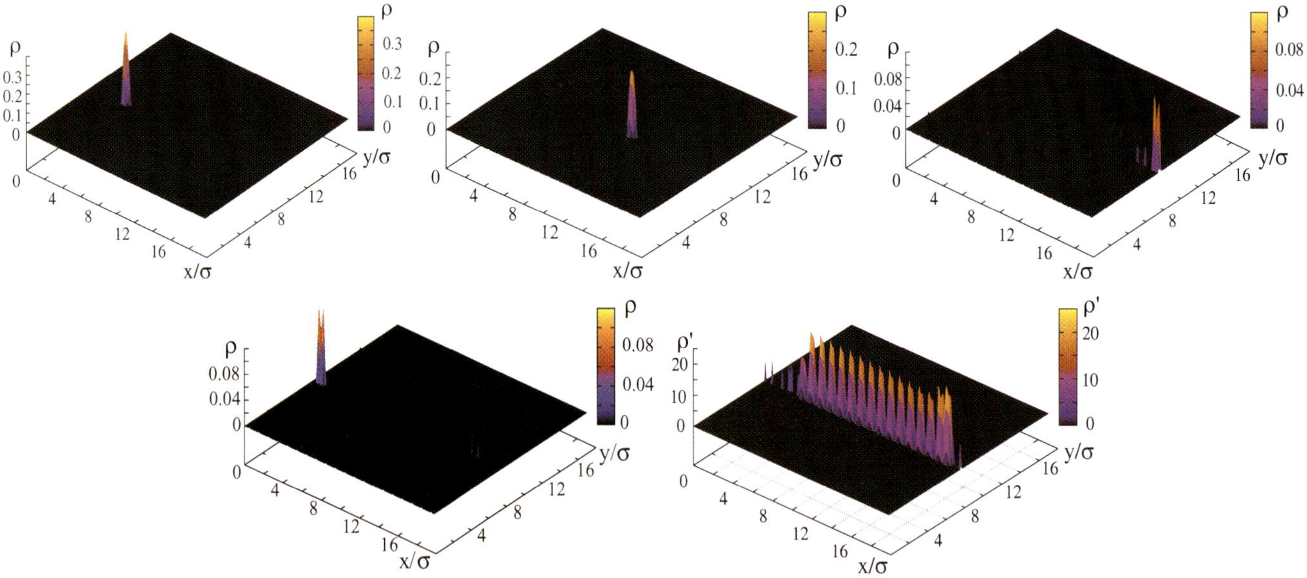

Fig. 3. Morse 2d-lattice. Quasi-1d solitonic excitation running in a 2d triangular Morse lattice along a chosen row. The initial condition is an "optimized Toda profile" that survives practically unaltered for a long time; $b\sigma = 4$. The figure shows four snapshots in sequence and the last panel is a composite cumulating pictures along the complete time sequence.

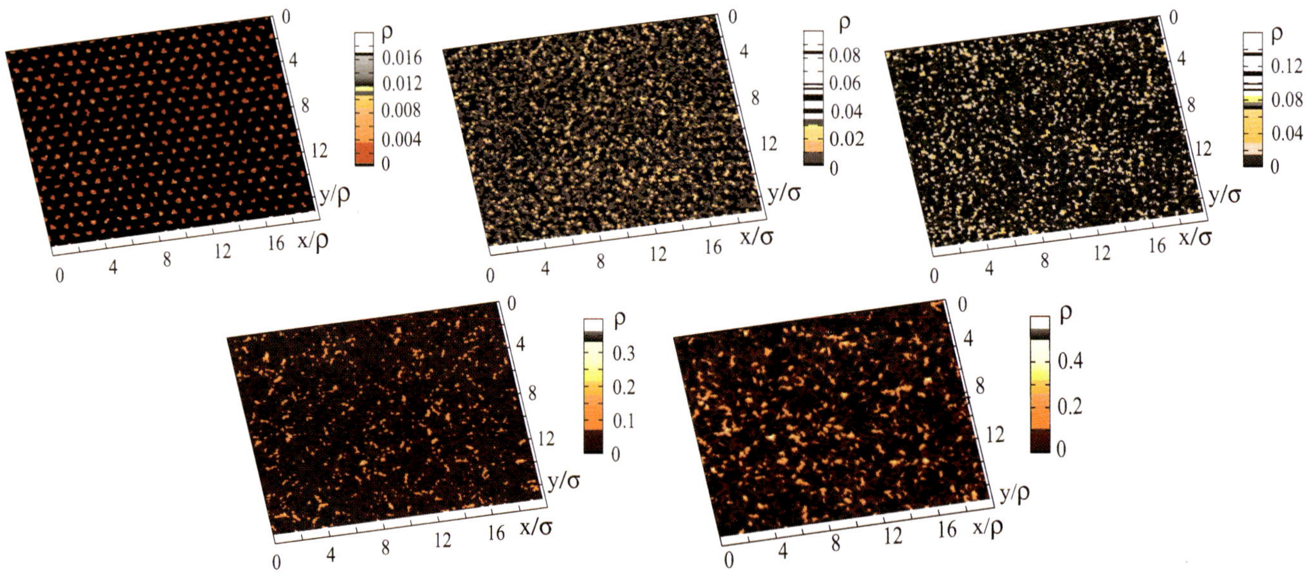

Fig. 4. Morse 2d lattice. Thermal excitation of long-living solitons at different temperatures in a triangular lattice. Upper panels (left to right): $T = 0.1$, $T = 0,5$, $T = 1$; lower panels (left to right): $T = 2$, $T = 5$; with $b\sigma = 4$ ($\gamma = 0.04$ in the first four figures, $\gamma = 0.2$ in the last figure, noise is not switched off for simulation procedure.) Note that the picture shows the long time evolution of the highest peaks in a cumulative representation of the amplitude-filtered density peaks for several time intervals.

Toda formulae may also be used for an approximate description of soliton-like excitations in Morse lattices [54, 55]. The idea is to use the free constants as tunable quantities, in particular by playing with the stiffness b_{eff}. A very good approximation for the solitons in 1d Morse lattices is obtained with the choice $b_{eff} = 3b$ were b is the Morse stiffness. For excitations in 2d-systems we shall use (10) albeit treating the effective stiffnes as a free adjustable parameter. We fixed in one row the initial coordinates and momenta according to (10) and optimized the value of b_{eff} giving the most lasting unaltered structures. In the particular case $b_{eff} = 1.5b$, the excitation remains localized and soliton-like for quite a long time (see Figs. 3–5). By studying numerous other examples we come to the

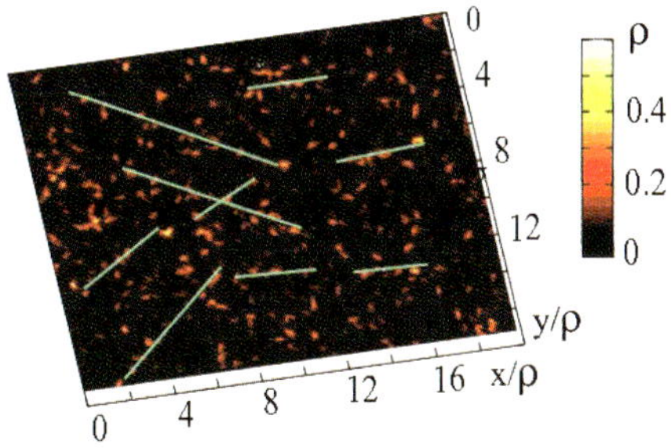

Fig. 5. Morse 2d lattice. Snapshot of the local soliton traces like *stripes* we identify in Fig. 4, last panel, $T = 5$. The electrons are expected to follow this intrinsic localized mode.

conclusion that the quasi-one-dimensional "adapted Toda solitons" with stiffness $b_{eff} = 1.5b$ which are running along rows aligned with the main crystallographic directions are localized excitations in triangular lattices with particular long lasting high stability. (We profit to alert the reader about an unfortunate typo in [54], where rather than $1.5b$ appears $6b$ which is not correct). We expect that this property holds also for other symmetries, as e.g. in $NaCl$ crystals.

Following the Boltzmann-Gibbs approach we may expect that, when heating the lattice, in principle, all possible static and dynamic excitations may spontaneously appear with some non-vanishing probability. The problem which arises is however that the probability may be quite low to find structures with high energy and, further, to identify them in the sea of fluctuations. We have observed quite many localized soliton-like modes which live some 10–20 time units, most of them oriented along the three basic crystallographic directions of the triangular lattice. We may conclude that many excitations we see in the heated system have the character of the localized quasi-one-dimensional modes described in the previous Section.

At low temperature (Fig. 4, $T = 0.1$) small oscillations of lattice atom-particles near equilibrium states are observed. Then at a higher temperature (Fig. 4, $T = 0.5$) particles are able to leave equilibrium states traveling for quite long distances though elongations of bonds are small. At a still higher temperatures (Fig. 4, $T = 1, 2$ and 5; see also Fig. 5) some tracks of short-life-time excitation modes are observed. Both front profile and front velocity, show these excitations as soliton-like.

4. Conclusion and Outlook

In conclusion we feel that intrinsic localized modes behaving as soliton-like excitations in 1d and 2d anharmonic lattices are worth exploring further for they may very well be of significance to experiment. Indeed, on the one hand, in 1d crystal lattices, there is the possibility, yet to be fully explored, of soliton-mediated supersonic (Angstroms per picosecond), fast electron transfer (ET) over long distances (thirty nm and beyond) which only recently has been explored [26, 39, 40, 43, 52]. On the other hand, in 2d systems, our computer simulations show that thermal soliton-like excitations may exist at least up to temperatures of the order of 100 K with life times of several picoseconds. Some of them are organized in stripes reminiscent of those seen in several experimental studies [32–36]. Another process, which we observed in our 2d computer simulations is reminiscent of percolation [53–56], which bears qualitative similarity to dynamical structures observed in cuprates [27].

Acknowledgments

The authors acknowledge fruitful discussions with E. Brandas, L. Brizhik, L. Cruzeiro, S. Larsson, H. Linde, G. Röpke, M. Russell and G. Vinogradov. Correspondence with M. E. Manley and D. M. Newns is also gratefully acknowledged. This research was supported by the Spanish *Ministerio de Ciencia e Innovacion*, under Grants EXPLORA-FIS2009-06585-E and MAT2011-26221.

References

[1] Ashcroft, N. W. & Mermin, N. D. [1976] *Solid State Physics* (Thomson Learning, New York).
[2] Kaganov, M. I. & Lifshitz, I. M. [1979] *Quasiparticles* (Mir, Moscow).
[3] Alexandrov, A. S. (ed.) [2007] *Polarons in Advanced Materials* (Springer, Dordrecht).
[4] Velarde, M. G. [2010] "From *polaron* to *solectron*: The addition of nonlinear elasticity to quantum mechanics and its possible effect upon electric transport", *J. Comput. Appl. Maths.* **233**, 1432–1445 (and references therein).
[5] Cantu Ros, O. G., Cruzeiro, L., Velarde, M. G. & Ebeling, W. [2011] "On the possibility of electric transport mediated by long living intrinsic localized solectron modes", *Eur. Phys. J. B* **80**, 545–554 (and references therein).
[6] Choquard, P. [1967] *The anharmonic crystal* (Benjamin, New York).
[7] Dancz, J. & Rice, S. A. [1977] "Large amplitude vibrationall motion in a one-dimensional chain: Coherent state representation", *J. Chem. Phys.* **67**, 1418–1426.

[8] Holian, B. L. & Straub, G. K. [1978] "Molecular dynamics of shock waves in one-dimensional chains", *Phys. Rev. B* **18**, 1593–1608.
[9] Straub, G. K., Holian, B. L. & Petschek, R. G. [1979] "Molecular dynamics of shock waves in one-dimensional chains. II. Thermalization", *Phys. Rev. B* **19**, 4049–4055.
[10] Rolfe, T. J., Rice, S. A. & Dancz, J. [1979] "A numerical study of large amplitude motion on a chain of coupled nonlinear oscillators", *J. Chem. Phys.* **70**, 26–33.
[11] Christiansen, A. L. & Scott, A. C. (eds.) [1983] *Davydov's Soliton Revisited. Self-trapping of Vibrational Energy in Protein* (Plenum Press, New York).
[12] Lomdahl, P. S. & Kerr, W. C. [1985] "Do Davydov solitons exist at 300K?", *Phys. Rev. Lett.* **55**, 1235–1238.
[13] Sievers, A. J. & Takeno, S. [1988] "Intrinsic localized modes in anharmonic crystals", *Phys. Rev. Lett.* **61**, 970–973.
[14] Burlakov, V. M., Kiselev, S. A. & Pyrkov, V. N. [1990] "Computer simulation of intrinsic localized modes in one-dimensional and two-dimensional anharmonic lattices", *Phys. Rev. B* **42**, 4921–4927.
[15] Scott, A. C. [1992] "Davydov's soliton", *Phys. Rep.* **217**, 1–67 (and references therein).
[16] Wallis, R. F., Franchini, A. & Bortolani, V. [1994] "Localized modes in inhomogeneous one-dimensional anharmonic lattices", *Phys. Rev. B* **50**, 9851–9859.
[17] Franchini, A., Bortolani, V. & Wallis, R. F. [1996] "Intrinsic localized modes in the bulk and at the surface of anharmonic dratimic chains", *Phys. Rev. B.* **53**, 5420–5429.
[18] Zolotaryuk, A. V., Spatschek, K. H. & Savin, A. V. [1996] "Supersonic mechanisms for charge and energy transfers in anharmonic molecular chains", *Phys. Rev. B* **54**, 266–277.
[19] Bortolani, V., Franchini, A. & Wallis, R. F. [1997] "Theory of kink and domain-wall excitation in translationally invariant lattices with quartic anharmonicity", *Phys. Rev. B* **56**, 8047–8053.
[20] Kalosakas, G., Aubry, S. & Tsironis, G. P. [1998] "Polaron solutions and normal-mode analysis in the semiclassical Holstein model", *Phys. Rev. B* **58**, 3094–3104.
[21] Kalosakas, G., Rasmussen, K. O. & Bishop, A. R. [2003] "Nonlinear excitations in DNA: polarons and bubbles", *Synthetic Metals* **141**, 93–97.
[22] Marin, J. L., Russell, F. M. & Eilbeck, J. C. [2001] "Breathers in cuprate-like lattices", *Phys. Lett. A* **281**, 21–25.
[23] Chetverikov, A. P., Ebeling, W. & Velarde, M. G. [2005] "Thermodynamics and phase transitions in dissipative and active Morse chains", *Eur. Phys. J. B* **44**, 509–519.
[24] Chetverikov, A. P., Ebeling, W. & Velarde, M. G. [2006] "Nonlinear excitations and electric transport in dissipative Morse-Toda lattices", *Eur. Phys. J. B* **51**, 87–99.
[25] Velarde, M. G., Ebeling, W. & Chetverikov, A. P. [2005] "On the possibility of electric conduction mediated by dissipative solitons", *Int. J. Bifurcation Chaos* **15**, 245–251.
[26] Velarde, M. G., Ebeling, W., Hennig, D. & Neissner, C. [2006] "On soliton-mediated fast electric conduction in a nonlinear lattice with Morse interactions", *Int. J. Bifurcation Chaos* **16**, 1035–1039.
[27] Pasupathy, A. N., Pushp, A., Gomes, K. K., Parker, C. V., Wen, J., Xu, Z., Gu, G., Ono, S., Ando, Y. & Yazdani, A. [2008] "Electronic origin of the inhomogeneous pairing interaction in the High-T_c superconductor $Bi_2Sr_2CaCu_2O_{8+\delta}$", *Science* **320**, 196–201.
[28] Manley, M. E., Sievers, A. J., Lynn, J. W., Kiselev, S. A., Agladze, N. I., Chen, Y., Llobet, A. & Alatas, A. [2009] "Intrinsic localized modes observed in high-temperature vibrational spectrum of NaI", *Phys. Rev. B.* **79**, 134305-1-5.
[29] Manley, M. E., Abernathy, N., Agladze, N. I., & Sievers, A. J. [2011] "Symmetry-breaking dynamical pattern and localization observed in the equilibrium vibrational spectrum of NaI", *Nature Sci. Reports*, ISSN (online) 2045–2322 (DOI:10.1038/srep00004).
[30] Newns, D. M. & Tsuei, C. C. [2007] "Fluctuating Cu-O-Cu bond model of high-temperature superconductivity", *Nature Phys.* **3**, 184–191.
[31] Nistor, R. A., Martyna, G. J., Newns, D. M., Tsuei, C. C. & Muser, M. H. [2011] "*Ab initio* theory of the pseudogap in cuprate superconductors driven by C4 symmetry breaking", *Phys. Rev. B* **83**, 144503-1-16.
[32] Emery, V. J., Kivelson, S. A. & Tranquada, J. M. [1999] "Stripe phases in high-temperature superconductors", *Proc. Natl. Acad. Sci. USA* **96**, 8814–8817.
[33] Reznik, D., Pintschovius, L., Ito, M., Likubo, S., Sato, M., Goka, H., Fujita, M., Yamada, K., Gu, G. D. & Tranquada, J. M. [2006] "Electron-phonon coupling reflecting dynamic charge inhomogeneity in copper oxide superconductors", *Nature* **440**, 1170–1173.
[34] Zaanen, J. [2006] "Quantum stripe search", *Nature* **440**, 1118–119.
[35] Moler, K. A. [2010] "High-temperature superconductivity — How the Cuprates hide their shape", *Nature* **468**, 643–664.
[36] Parker, C. V. [2010] "On the electrical stripes in Cuprates", *Nature* **468**, 677–620.
[37] Toda, M. [1989] *Theory of Nonlinear Lattices* (2nd. ed.) (Springer, Berlin).
[38] Chetverikov, A. P., Ebeling, W. & Velarde, M. G. [2006] "Dissipative solitons and complex currents

in active lattices", *Int. J. Bifurcation Chaos* **16**, 1613–1632.

[39] Gray, H. B. & Winkler, J. R. [2003] "Electron tunneling through proteins", *Quart. Rev. Biophys.* **36**, 341–372.

[40] Gray, H. B. & Winkler, J. R. [2005] "Long-range electron transfer", *Proc. Natl. Acad. Sci. (USA)* **102**, 3534–3539.

[41] Hennig, D., Neissner, C., Velarde, M. G. & Ebeling, W. [2006] "Effect of anharmonicity on charge transport in hydrogen-bonded systems", *Phys. Rev. B* **73**, 024306-1-10.

[42] Hennig, D., Chetverikov, A. P., Velarde, M. G. & Ebeling, W. [2007] "Electron capture and transport mediated by lattice solitons", *Phys. Rev. E* **76**, 046602-1-9.

[43] Velarde, M. G., Chetverikov, A. P., Ebeling, W., Hennig, D. & Kozak, J. J. [2010] "On the mathematical modeling of soliton-mediated long-range electron transfer", *Int. J. Bifurcation Chaos* **20**, 185–194.

[44] Chetverikov, A. P., Ebeling, W., Röpke, G. & Velarde, M. G. [2007] "Anharmonic excitations, time correlations and electric conductivity", *Contrib. Plasma Phys.* **47**, 465–478.

[45] Landau, L. D. [1933] "Über die Bewegung der Elektronen im Kristallgitter", *Phys. Z. Sowjet.* **3**, 664 [1965, English transl.: "Electron motion in crystal lattices", in *Collected Papers of L. D. Landau*, ed. D. fer Haar (Gordon & Breach, New York), pp. 76–68].

[46] Pekar, S. I. [1954] *Untersuchungen über die Elektronentheorie* (Akademie Verlag, Berlin).

[47] Blakemore, J. S. [1962] *Semiconductor Statistics* (Pergamon Press, Oxford).

[48] Blakemore, J. S. [1985] *Solid State Physics* (Cambridge University Press, Cambridge).

[49] Velarde, M. G., Ebeling, W., Chetverikov, A. P. & Hennig, D. [2008] "Electron trapping by solitons: classical *versus* quantum mechanical approach", *Int. J. Bifurcation Chaos* **18**, 521–526.

[50] Velarde, M. G., Ebeling, W. & Chetverikov, A. P. [2008] "Thermal solitons and solectrons in 1D anharmonic lattices up to physiological temperatures", *Int. J. Bifurcation Chaos* **18**, 3815–3823.

[51] Larsson, S. & Klimkans, A. [1999] "Tunneling transfer in molecular wires", *J. Mol. Struct. (Theochem.)* **464**, 59–65.

[52] Slinker, J. D., Muren, N. B., Renfrew, S. E. & Barton, J. K. [2011] "DNA charge transport over 34 nm", *Nature Chem.* **3**, 228–233.

[53] Chetverikov, A. P., Ebeling, W., Röpke, G. & Velarde, M. G. [2011] "Hopping transfer and stochastic dynamics of electrons in two-dimensional plasmas", *Contrib. Plasma Phys.* **51**, 814–829.

[54] Chetverikov, A. P., Ebeling, W. & Velarde, M. G. [2011] "Localized nonlinear, soliton-like waves in two-dimensional anharmonic lattices", *Wave Motion* **48**, 753–760.

[55] Chetverikov, A. P., Ebeling, W. & Velarde, M. G. [2011] "Properties of nano-scale soliton-like excitations in two-dimensional lattice layers", *Physica D* **240**, 1954–1959.

[56] Chetverikov, A. P., Ebeling, W. & Velarde, M. G. [2011] "Soliton-like excitations and solectrons in two-dimensional nonlinear lattices", *Eur. Phys. J. B* **80**, 137–145.

CHAPTER 37

GLOBAL OPTIMIZATIONS BY INTERMITTENT DIFFUSION

SHUI-NEE CHOW

School of Mathematics, Georgia Institute of Technology, Atlanta, GA 30332, USA

TZI-SHENG YANG

Department of Mathematics, Tunghai University, Taichung 40704, Taiwan

HAO-MIN ZHOU

School of Mathematics, Georgia Institute of Technology, Atlanta, GA 30332, USA

We propose an intermittent diffusion(ID) method to find global minimizers of a given scalar function $g : \mathbb{R}^n \to \mathbb{R}$. The main idea is to add intermittent, instead of continuously diminishing, random perturbations to the gradient flow generated by g, so that the trajectories can quickly escape from the trap of one minimizer and then approach others. During this process, the associated Fokker-Planck equation alternates its type between hyperbolic and parabolic. We prove that by using the ID method one can find, with probability arbitrarily close to 1, a good approximation to the global minimizers in a finite time T provided T is sufficiently large. The convergence rate in probability follows a geometric series. We also prove that for any given finite set of minimizers, a trajectory of the ID method visits an arbitrary small neighborhood of each minimizer with positive probability. Numerical simulations show that the proposed method achieves clear improvements in terms of the time and the frequencies of visiting the global minimizers over some existing global optimization algorithms for many testing problems.

1. Introduction

Global optimization is a classical issue appeared in numerous research fields and applications, such as inverse problems, protein folding, optimal design and digital image processing. In this paper, we will investigate an alternative strategy based stochastic differential equations to find global minimizers for a general objective functional

$$\min_x g(x), \quad x \in \Omega \tag{1}$$

where Ω is an admissible set for $x \in \mathbb{R}^n$. The objective functional $g(x)$ is often defined by an energy functional or a cost functional in many applications. For examples, $g(x)$ could be the double-well potential energy in a 2-phase composite material problem, or the quality factor in an optical spectrometer design, or the total distance in optimal path planning in unmanned vehicle navigation.

As one of the oldest applied mathematics problems, finding minimizers for (1) has been intensively studied and numerous research results having been reported. We refer to some books for more details and references in this subject [10, 11, 25, 27, 37, 42]. Despite the existence of an extensive literature, finding global minimizers for a general g, if it is all possible, can still be extremely challenging in practice, especially when the dimension of x is large and the level sets of g are complicated. One of the most notorious difficulties that every global optimization method has to face is to show how to escape from traps of local minimizers. This becomes more serious if the local search methods, such as gradient descent based flows, are used, because the gradient flows are guaranteed to stuck in basins of (possible local) minimizers. Therefore, finding good initial guess becomes vital in many applications. For examples, the inverse media scattering methods proposed in [7,8] can be viewed as gradient flows for minimizing a regularized data fitting objective, and it is crucial to have good starting points.

To overcome the challenges posed by local minima, many strategies have been proposed. Based on whether randomness being involved in algorithms, they are often classified into two categories: deterministic or stochastic methods. For instance, the well-known branch and bound method and its variations [29, 31, 36] and the cutting angle methods [4,5] are deterministic strategies, while Monte-Carlo sampling [35, 41], genetic algorithms [21], basin-hopping [43] and stochastic tunneling methods [30] are random in nature.

Among the existing global optimization methods, the Metropolis random walks [34] and simulated annealing method [12, 28] are stochastic methods for finding global optimal solutions for a broad range of discrete and continuous problems. In simulated annealing, iterative random sampling procedures are used to generate new admissible states. We note that new states with higher $g(x)$ values are acceptable with a positive probability which is associated with a control parameter called cooling rate or temperature. When the cooling rate is high, the probability to accept a state with a higher $g(x)$ value is large. This is the key idea which allows one to move out of the traps of local minima. In fact, many other stochastic based methods share similar ideas.

The basic idea of simulated annealing is intuitive. It has been broadly used in many problems with remarkable successes. In particular, there are many examples that simulated annealing can give reasonable good approximations to the global minimizers while other methods fail to provide anything close to the global minima. However, it is also well-known that the original simulated annealing may not be very efficient in many applications, due to the requirement of a slow decay of the cooling rate. To improve the efficiency, Szu and Hartley [39] proposed the "fast simulated annealing", which generates a new state according to the Cauchy density with unbounded variance. Later, Ingber [26] generalized this idea and suggested a "non-local generating" of a new state in Metropolis algorithm, which is called "very fast simulated annealing". The study in [39] provides a heuristic sufficient condition so that the state-generating is infinite often in time (i.o.t.), i.e., with probability one that any state x in $\mathbb{R}^n$ will be generated for infinitely many times. More references and analysis can be found in [18, 22, 24, 44].

To further speed up the convergence of simulated annealing, many hybrid strategies have been proposed in recent years. For example, Zhang proposed and compared several hybrid strategies that repeatedly use simulated annealing followed by local gradient search methods [46]. Many other hybrid methods [5, 6, 45] are also studied around the same time. The main motivation for the hybrid methods is to use simulated annealing procedures to break the local traps and use local gradient search to find the minima. Numerical experiments reported in those studies suggest that alternating simulated annealing and deterministic local search methods can significantly improve the efficiency in many challenging problems.

The original simulated annealing method does not have to explicitly use the gradient information in searching for the next samples. Later, some efforts have been devoted to use simulated annealing ideas together with the gradient flow for global optimizations, especially for objective functions g

that are continuously depending on the state variables x. For example, the studies in [1] and [17], suggest to implement the idea of Metropolis algorithm by running the stochastic differential equation:

$$dx(t,\omega) = -\nabla g(x(t,\omega))dt + \sigma(t)dW(t), \quad t \in [0,\infty], \tag{2}$$

where $W(t)$ is the Brownian motion in $\mathbb{R}^n$, ω is a random event (a random path) in the Wiener space of $W(t)$, and the cooling rate $\sigma(t)$ is continuously decreasing to zero. For convenience, we call this implementation the *continuous diminishing diffusion* (CDD) method in this paper. Obviously, unlike the classical simulated annealing methods, the CDD method always accept a new state without checking whether the new state has lower $g(x)$ value or not. It is proved in [13,17] that if $\sigma(t)$ is given by

$$\sigma(t) = c/\sqrt{\log t}, \tag{3}$$

for large $c > 0$, the solution of (2) converges weakly to a distribution concentrated at the global minima of g.

Inspired by the simulated annealing, the CDD method and some recent developments in random dynamical systems, we propose an alternating strategy, called intermittent diffusion (ID), to find the global optimal solution. The main idea is to combine the advantages of gradient flow or other efficient local search algorithms, which may quickly lead to local minimizers, and stochastic perturbations that can promote the trajectories out of the traps of local minima. More precisely, we consider $\sigma(t)$ as a piecewise smooth function (we use piecewise constants in this paper) of t with alternating positive and zero values. When $\sigma(t) = 0$, (2) corresponds to the gradient descent algorithm,

$$\dot{x}(t) = -\nabla g(x(t)). \tag{4}$$

When $\sigma(t) > 0$ the trajectory of (2) may move away from any local minimum with a probability controlled by $\sigma(t)$. Then we repeat such a process multiple times.

This ID procedure is similar to the hybrid methods reported in [45,46], except that we use continuous random perturbations instead of the classical simulated annealing. A perturbed new state is always accepted without checking whether it decreases or increases the value of $g(x)$. In this manner, it is the same as the CDD method. On the other hand, we have a discontinuous cooling rate, which is different from the CDD method. When the cooling rate is zero, any trajectory of the gradient descent flow (or other efficient local search methods) will move towards a local minimizer. When the cooling rate is positive, the trajectories are disturbed to get out of the local traps. Compared to the CDD or other simulated annealing based methods, one finds in the ID procedure global minimizers on its way but not necessarily settles at a global minimizer in the end of the computation. Therefore, the cooling rate, if positive, does not have to go to zero. In fact, it can be relatively large so that it may promote the trajectories out of the traps of local minima quicker. In our experiments, we set the cooling rate as piecewise constants with randomly selected positive values.

The idea of alternating random and deterministic procedures, or more generally alternating local search with global search strategies, is quite common in global optimization methods. For example, the iterated local search method repeatedly starts a local search algorithm with a random initial state or an initial state determined by meta-heuristics (see the review [32] and references therein). The reactive research optimization uses machine learning techniques in search heuristics to update parameters in the model for more accurate computations [9]. Many of them use randomness to get out of the local traps, and use deterministic procedures to find local minimizers. Although each alternating strategy has its own advantages and limitations, their performances all indicate that repetitive computations often improve the speed of convergence to the global minima in variety applications.

An interesting viewpoint for our method can be explained by looking at the probability distribution of the trajectories $x(t,\omega)$ of (2). Its density function $p(t,x)$, which represents the probability density of the trajectories reaching x at time t, satisfies the Fokker-Planck equation,

$$p_t = \nabla \cdot (\nabla g(x)p) + \frac{1}{2}\sigma(t)^2 \Delta p. \tag{5}$$

When $\sigma(t)$ is positive, this is a parabolic equation. When $\sigma(t) = 0$, Eq. (5) becomes a hyperbolic equation. For this reason, we call this method the *intermittent diffusion* (ID) method.

By selecting $\sigma(t)$ as a discontinuous function, (5) becomes degenerate. However, it possesses some interesting properties that are not shared by the standard Fokker-Planck equations for regular diffusion processes. More precisely, in the ID method, when $\sigma(t) > 0$, it is the standard diffusion process that gives a positive probability for trajectories to go everywhere. When $\sigma(t) = 0$, the diffusion term drops out from (5), and the equation becomes a hyperbolic equation with the coefficient $\nabla g(x)$ that asymptotically compresses p toward point distributions (Dirac delta functions) at minimizers. The compression speed is quicker if $\nabla g(x)$ has larger magnitude. This indicates that probability density function p will cluster around the minima, which is actually consistent with the gradient descent flow that every initial state goes to a minimum point. Similar to the hybrid methods, our examples indicate that the ID method speeds up the convergence for many test problems, including a recent study on the shortest path problem [15]. The results suggest that repetitive implementations of such diffusion-compression process help to cluster the probability density function toward the global minima in a quick pace.

Based on the Fokker-Planck equation, we shall prove theoretically that with probability arbitrarily close to 1, the ID method can find the global minima in a finite time T provided T is reasonably large. Furthermore, we show that if the probability of reaching the basin of global minima in one segment (the process of one random diffusion period followed by one gradient flow to a minimum in the ID method) is $\alpha \in (0, 1)$, where α does not have to be close to 1, then the convergence of the ID method is controlled by $(1-\alpha)^N$, where N is the number of segments in each ID implementation. This is a significant improvement over the methods with continuous cooling rate. In addition, we prove that for any given finite sequence of (local or global) minimizers, with a positive probability the trajectory of the ID method visits an arbitrarily small neighborhood of each member of the sequence. Our experiments also show that within finite time intervals, the frequency that the ID method reaches the global minimizer is higher than the CDD method in the tested examples.

We reported partial results of this study in a technical report [16]. The current version is arranged as follows. In the next section, we present the ID method and give two simple examples to illustrate how it is used. A theoretical study is given in Sec. 3. And we show more numerical examples and comparisons in Sec. 4.

2. The Intermittent Diffusion Algorithm

In this section, we present the ID algorithm. In contrast to the existing work on diffusion for global optimizations [13, 17], which gradually decreases the diffusion strength, we employ the deterministic property of (2) that with $\sigma = 0$ the ω-limit set of a trajectory is a minimizer of $g(x)$ [20]. Similar to the hybrid methods for global optimizations, the idea is: (i) to allow the trajectories randomly move away from a small neighborhood of a local minimizer and approach a stable manifold (basin of attraction) of other local minimizers by setting $\sigma > 0$; (ii) if the Hessian matrix of $g(x)$ at any local minimum is negative definite, (the minimizer is called a hyperbolic minimizer in the dynamical system literature), the trajectories reach small neighborhoods of local minimizers within finite time once σ is set to 0. We can realize (i) and (ii) by computing the stochastic perturbed gradient flow (2) with a discontinuous diffusion $\sigma(x,t)$ defined by,

$$\sigma(x,t) = \sum_{j=1}^{N} \sigma_j I_{[S_j,T_j]}(t), \tag{6}$$

where $0 = S_1 < T_1 < S_2 < T_2 \cdots < S_N < T_N < S_{N+1} = T$, and $I_{[S_j,T_j]}(t)$ is the characteristic function of interval $[S_j, T_j]$.

The discontinuous function $\sigma(x,t)$ can be temporally and spatially dependent. In the paper, we focus on the situation where $\sigma(x,t)$ is only time dependent. It "turns off" the diffusion when $t \in [T_j, S_{j+1}]$, so that Eq. (2) becomes a gradient flow. This allows the state to approach a local minimizer. We remark that we use gradient flows to compute the local minima in this paper. However, the gradient flows can be replaced by other more efficient local search algorithms. This is especially necessary when the convergence of gradient flows is slow.

On the other hand, when $\sigma(x,t)$ takes positive values, Eq. (2) "turn on" the diffusion and the trajectories do not stay near the stationary points. It has been shown that in many scenarios, the trajectories eventually escape the traps of the stationary points provided sufficient noise is added. We use this property of diffusion to promote the trajectories getting out of those traps. For this purpose, we do not want to add noise in a decreasing cooling rate. In our numerical experiments, we set $\sigma(x,t)$ to random positive constants for simplicity. The intervals $[S_j, T_j]$ are also picked with random lengths, i.e., $T_j - S_j$ is a random positive number, to help the trajectories move away from one stable manifold to another. In Fig. 1, we illustrate the sample function σ that we used in our simulations. For convenience,

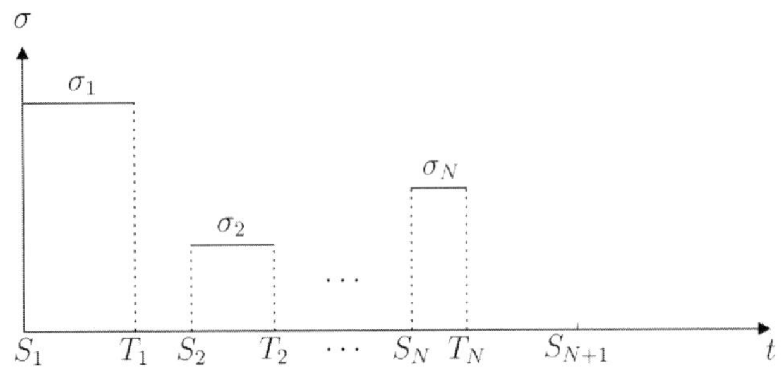

Fig. 1. A sample function of the discontinuous diffusion coefficient σ.

we call the trajectories in the time interval $[S_j, S_{j+1}]$ one *segment* of the (random) dynamical systems.

Following this idea, we present the ID algorithm:

(1) Set α as the scale for diffusion strength, and γ the scale for diffusion time.
(2) Generate a random initial state $x_0 \in \mathbb{R}^n$, and set the optimal state $X_{opt} = x_0$.
(3) Generate two positive random numbers d, t within $[0, 1]$ by uniform distribution and let $\sigma := \alpha d$, and $T := \gamma t$.
(4) Compute the stochastic equation for $t \in [0, T]$

$$dx(t,\omega) = -\nabla g(x(t,\omega)dt + \sigma dW(t), \quad x(0,\omega) = X_{opt}, \tag{7}$$

and record the final state $x_T := x(T, \omega)$.
(5) Compute the solution for the following system until a convergence criterion is satisfied,

$$\dot{x}(t) = -\nabla g(x(t)), \quad x(0) = x_T, \tag{8}$$

and record the final state as X_i. If $g(X_i) < g(X_{opt})$, set $X_{opt} = X_i$. This finishes one segment of ID method.
(6) Repeat Step 2 to Step 2 for N times to obtain N segments of the trajectories, which obtains up to N local minimizers $X_1, \ldots, X_N$. For large enough N, we take X_{opt} as an approximation of the global minimizer.

Remark 37.1.

(1) *Different schemes may be used to solve* (7) *and* (8)*. For example, one can use Euler-Maruyama scheme for* (7) *and a Runge-Kutta scheme for* (8)*.*
(2) *The convergence criterion in Step 2 can be set in different ways. In this paper, we use a simple way to stop the iteration if the absolute value of the difference between two successive iterates is less than a prescribed tolerance $\epsilon > 0$. Different convergence criteria may lead to different computation time intervals $[T_i, S_{i+1}]$ in each computation.*
(3) *Although we only consider constant σ_i in this paper, it can be functions of (t, x) in general. Ideally, $\sigma_i(t, x)$ shall be used to prevent repetitive visits to the same local minima and take the most efficient route to the global minimizer. But how to select $\sigma_i(t, x)$ in practice is still under investigation.*

It is observed in our numerical experiments that the stronger (in a relative sense) noise usually promotes the trajectories to escape the stable manifolds quicker. Furthermore, unlike the CDD method in which the trajectories settle down near the global minimizer asymptotically, the ID method finds the best approximation "on its way". And it does not require the stopping state as the global optimizer. Therefore, the ID method often visits the global minima earlier and more frequently than the CDD in our test examples.

3. An Analysis of the ID Method

In this section, we consider some theoretical properties of the ID algorithm. For convenience, we focus our analysis on the case where σ is taken as piecewise constant functions as shown in Fig. 1. The admissible set Ω is the entire space $\mathbb{R}^n$. To simply our analysis, we only consider the objective function $g(x) \in C^2$ satisfying the following conditions.

(H1) There exist constants $N, \lambda > 0$ such that $D^2 g(x) \geq \lambda I_n$ for $|x| \geq N$, where I_n is the identity matrix.
(H2) Every minimizer of $g(x)$ is hyperbolic, i.e., the Hessian matrix of $g(x)$ at any of its minimizers is negative definite.

Roughly speaking, condition (H1) states that $g(x)$ behaves like $|x|^\gamma$ with $\gamma > 2$ at $|x| = \infty$. Furthermore, it can be verify that condition (H1) implies that all minimizers are in a bounded domain and

$$\int_{\mathbb{R}^n} e^{-\frac{g(x)}{\sigma}} dx < \infty,$$

for any constant $\sigma > 0$. We want to remark that conditions (H1) is considered for technical reasons, it can be relaxed for practical problems. For example, if we search for a global minimizer in a bounded domain, we can simply replace $g(x)$ by an at least quadratically growing penalty function for large $|x|$. This does not change the properties of $g(x)$ in the bounded domain, but make the objective function satisfy the growth condition as required in our theoretical analysis. The hyperbolicity requirements (H2) for the minimizers are also natural, because only those minimizers are stable and observable in practice.

Theorem 1. *Consider a gradient flow*

$$\dot{x}(t) = -\nabla g(x(t)),$$

and its ID process generated by

$$dx(t,\omega) = -\nabla g(x(t,\omega))dt + \sigma(t)dW(t),$$

where $\sigma(t) = \sum_{j=1}^{N} \sigma_j I_{[S_j,T_j]}(t)$. *Let* Q *be the set of global minimizers,* U *a small neighborhood of* Q, *and* X_{opt} *the optimal solution obtained by the ID process. Then for any given* $\varepsilon > 0$, *there exist* $\tau > 0, \sigma_0 > 0$ *and* $N_0 > 0$ *such that if* $T_i - S_i > \tau$, $\sigma_i < \sigma_0$ *(for* $i = 1, \ldots, N$*), and* $N > N_0$,

$$\mathbb{P}(X_{opt} \in U) \geq 1 - \varepsilon. \tag{9}$$

Proof. There are two different cases for the global minimum set Q: it has (1) Lebesgue measure 0, or (2) a positive Lebesgue measure. For instance, if Q contains one global minimum point, or finite number of global minimum points, then its measure is 0. If the global minima form a region in the space, its measure is positive. We give the proof for a simple case assuming there is one global minimum point $\{x^*\}$ of hyperbolic type. Since the Hessian matrix of x^* is negative definite, we can always take $U = B(x^*, \gamma)$, a ball centered at x^* with radius γ for small enough γ. For other situations including multiple global minima, a similar proof can be given with minor modifications and we will omit it in this paper.

Let us consider one segment for $t \in [S_i, S_{i+1}]$ of the ID method first. Since the trajectory rests on a small neighborhood of a local minimizer at the end of each gradient descent period, we may assume $x(S_i) \in B(q_i, \gamma_i)$, where q_i is a minimizer, and $B(q_i, \gamma_i)$ a small neighborhood (ball) centered at q_i with radius γ_i. The location of $x(S_i)$ is a random variable for different realizations, and we denote the distribution of $x(S_i)$ as $p_0^i(x)$, which is a continuous function compactly supported in $B(q_i, \gamma_i)$. This is because $p_0^i(x)$ is the solution at $t = S_i$ of

$$p_t = \nabla \cdot (\nabla g(x)p),$$

with a smooth initial condition (given by the final distribution of previous segment).

For $t \in [S_i, T_i]$, (7) is a diffusion process. We denote the probability density function of the trajectory $x(t,\omega)$ visiting x at time $t \in [S_i, T_i]$ by $p_{\sigma_i}(t - S_i, x; q_i)$, which depends on perturbation strength σ_i. It is well known that p_{σ_i} satisfies the Fokker-Planck equation given by

$$(p_{\sigma_i})_t = \nabla \cdot (\nabla g(x)p_{\sigma_i}) + \frac{1}{2}\sigma_i^2 \Delta p_{\sigma_i}, \tag{10}$$

with initial data $p_0^i(x)$. As the solution of this linear convection diffusion equation, $p_{\sigma_i} \in C^\infty$ continuously depends on $t > 0$ and $\sigma_i > 0$.

In the classical diffusion theory, it is straightforward to verify that (10) has an equilibrium solution which is given by the famous Gibbs distribution,

$$\bar{p}_{\sigma_i}(x) = Ae^{-\frac{2g(x)}{\sigma_i^2}}, \tag{11}$$

where A is a constant defined by

$$A = \left(\int_{\mathbb{R}^n} e^{-\frac{2g(x)}{\sigma_i^2}} dx\right)^{-1}. \tag{12}$$

It is used to normalize p_{σ_i} to be a probability density function. According to [17] and the references cited therein, $\bar{p}_{\sigma_i}(x)$ weakly converges to a point distribution concentrated at x^* as $\sigma_i \to 0$. This implies that

$$\lim_{\sigma_i \to 0} \int_U \bar{p}_{\sigma_i}(x)dx = 1. \tag{13}$$

We remark here that if Q has positive Lebesgue measure, then $\bar{p}_{\sigma_i}(x)$ weakly converges to a uniform distribution on Q as σ_i goes to zero.

In the past two decades, there are intensive studies on the convergence of p_{σ_i} to its steady state $\bar{p}_{\sigma_i}$ [2,33,38]. By Holley-Stroock perturbation lemma [23], it can be shown that with conditions (H2), p_{σ_i} converges to the equilibrium solution $\bar{p}_{\sigma_i}$ exponentially in the relative entropy sense, i.e.

$$H(p_{\sigma_i}|\bar{p}_{\sigma_i}) \leq e^{-2\lambda t} H(p_0^i|\bar{p}_{\sigma_i}),$$

where λ is a constant, depending the structure of g, in the logarithmic Sobolev inequality for $\bar{p}_{\sigma_i}$, and H is the relative entropy defined as

$$H(p|q) = \int p \log\left(\frac{p}{q}\right) dx,$$

for any two probability density functions p and q. Using the Csiszar-Kullback inequality, which states

$$\|p - q\|_{L^1}^2 \leq 2H(p|q),$$

we obtain

$$\|p_{\sigma_i} - \bar{p}_{\sigma_i}\|_{L^1}^2 \leq 2e^{-2\lambda t} H(p_0^i|\bar{p}_{\sigma_i}). \tag{14}$$

Because the initial distribution p_0^i on the right hand side has compact support depending only on $g(x)$ in a bounded region, this estimate is uniform. That is

$$\|p_{\sigma_i} - \bar{p}_{\sigma_i}\|_{L^1}^2 \leq Ce^{-2\lambda t}, \tag{15}$$

where C is constant independent of the initial data p_0^i.

Let us define the attraction set of q_i as

$$K(q_i) = \{x : (8) \text{ with } x \text{ as the initial state converges to a point in } B(q_i, \gamma)\}.$$

Since the Hessian matrix of $g(x)$ has negative eigenvalues near of q_i, we find $B(q_i, \gamma) \subseteq K(q_i)$ provided γ is small enough. Thus, for any given $\alpha \in (0, 1)$ there exists σ_0 such that for $\sigma_i < \sigma_0$, (13) implies

$$\int_U \bar{p}_{\sigma_i}(x)dx > \alpha + (1-\alpha)/2. \tag{16}$$

Also, there exists a $\tau(\alpha)$ such as for $T_i - S_i > \tau(\alpha)$, (15) gives us

$$\|p_{\sigma_i}(T_i - S_i, x) - \bar{p}_{\sigma_i}(x)\|_{L^1} < (1-\alpha)/2. \tag{17}$$

Therefore, by (16) and (17), we have

$$\begin{aligned}\int_U p_{\sigma_i}(T_i - S_i, x)dx &\geq \int_U \bar{p}_{\sigma_i}(x)dx \\ &\quad - \int_U |p_{\sigma_i}(T_i - S_i, x) - \bar{p}_{\sigma_i}(x)\,|\,dx > \alpha. \end{aligned} \tag{18}$$

This implies that

$$\begin{aligned}&\mathbb{P}(x(S_{i+1}) \in U \,|\, x(S_i) \in B(q_i, \gamma_i)) \\ &\quad = \mathbb{P}(x(T_i) \in K(x^*) \,|\, x(S_i) \in B(q_i, \gamma_i)) > \alpha, \end{aligned} \tag{19}$$

due to the fact that the ID method is deterministic on $[T_i, S_{i+1}]$.

Let Θ be the set consisting of local minimizers of g, then

$$\begin{aligned}\mathbb{P}(x(S_{i+1}) \in U^c) &= \sum_{q_i \in \Theta} \mathbb{P}(x(S_{i+1}) \in U^c |\, x(S_i) \in B(q_i, \gamma_i)) \\ &\quad \times \mathbb{P}(x(S_i) \in B(q_i, \gamma_i)), \end{aligned} \tag{20}$$

where U^c is the complement of U. By estimate (19), we have

$$\begin{aligned}&\mathbb{P}(x(S_{i+1}) \in U^c \,|\, x(S_i) \in B(q_i, \gamma_i)) \\ &\quad < (1-\alpha) \quad \text{for all } q_i \in \Theta, \end{aligned} \tag{21}$$

provided $(T_i - S_i) > \tau$ and $\sigma_i < \sigma_0$. This implies

$$\begin{aligned}\mathbb{P}(x(S_{i+1}) \in U^c) &< (1-\alpha) \sum_{q_i \in \Theta} \mathbb{P}(x(S_i) \in B(q_i, \gamma_i)) \\ &< (1-\alpha). \end{aligned} \tag{22}$$

The estimate (22) can be made for all segments $i = 1, 2, \ldots, N$ uniformly. We remark that the estimate (22) is independent of which neighborhood $B(q_i, \gamma_i)$ the $x(S_i)$ will lie in. Since $x(S_i)$ must lie in some $B(q_i, \gamma_i)$, if we repeat it for all intervals $[S_j, S_{j+1}]$, $j = 1, \ldots, N$, with the same fixed α, then the probability that the solution X_{opt} of the ID method does not belong to U satisfies,

$$\begin{aligned}\mathbb{P}(X_{opt} \in U^c) &= \mathbb{P}(\cap_{i=0}^N \{x(S_i) \in U^c\}) \\ &< (1-\alpha)^N. \end{aligned}$$

For any given $\varepsilon \in (0, 1)$, there exists $N_0 > 0$, such as for $N \geq N_0$,

$$\mathbb{P}(X_{opt} \in U^c) < (1-\alpha)^N < \varepsilon,$$

which is equivalent to (9). □

It it easy to obtain the following statement from the proof.

Corollary 37.1. *Assume the probability of attaining an approximation in U, a small neighborhood of the global minimizer x^*, in one segment of the ID process is $\alpha \in (0, 1)$, and there are N segments, then the best approximation X_{opt} obtained by the ID method satisfies*

$$\mathbb{P}(X_{opt} \in U) \geq 1 - (1-\alpha)^N. \tag{23}$$

Furthermore, if we pick $\alpha \geq 1/2$ in the proof, it implies that $x(S_{i+1})$ has larger probability in U than in the neighborhoods of other local minimizers.

We summarize this extension in the following theorem.

Theorem 2. *Consider the ID process for the gradient flow stated in Theorem 1. If $0 < \gamma \ll 1$, then*

$$\mathbb{P}(x(S_{i+1}, \omega) \in U) \geq \mathbb{P}(x(S_{i+1}, \omega) \in B(q_j, \gamma)),$$

where q_j is any local minimizer, provided $T_i - S_i$ is large and σ_i small enough.

We remark that the ID method indicates the convergence is monotone with respect to the number of segments N. And the convergence rate is a geometric series based on the factor α. The following theorem reveals more properties of the ID method.

Theorem 3. *Consider the ID method for the randomly perturbed gradient flow* (2) *and suppose that $g \in C^2$ satisfies* (H2) *and the following conditions*:

(H3) $\lim_{|x| \to \infty} g(x) = \infty$.
(H4) *∇g has a Lipschtiz constant L on $\mathbb{R}^n$.*

Given $0 < \epsilon, T > 0$ and any finite sequence $\{p_i\}_{i=1 \cdots N}$ of local minimizers of g, there exist a

$\Omega' \subset \Omega$ with $\mathbb{P}(\Omega') > 0$ such that for $\omega \in \Omega'$ the trajectory of the ID process visiting each $B(p_i, \epsilon)$ within $[0, T]$.

For convenience of following discussion, we denote $f := -\nabla g$, and $(\Omega, \mathscr{F}, \mathbb{P})$ the Wiener space generated by Brownian motion $W(t)$. To prove this theorem, we need some lemmas for the counterpart equation of (2):

$$dx = f(x)dt + \sigma dW, \quad x(0) = x_0, \tag{24}$$

where σ is a constant. Denoted by $\phi(t, \omega)x_0$, $\omega \in \Omega$ the solution of (24).

Lemma 1. Assume (H3)–(H4) hold. The solution of (24) is a global solution for all $\omega \in \Omega$ (i.e., $\phi(t, \omega)x_0$ is defined for $t \in [0, \infty)$).

Proof. Let $v := x - \sigma W$, we have

$$\frac{dv}{dt} = f(v + \sigma W). \tag{25}$$

Then, the time-derivative of g along trajectories $v(t)$ is given by

$$\begin{aligned}
\frac{dg(v(t))}{dt} &= -f(v) \cdot f(x) &(26)\\
&= -(f(v) - f(x) + f(x)) \cdot f(x) &(27)\\
&\leq -|f(x)|^2 + |f(x)||f(x) - f(v)| &(28)\\
&\leq -|f(x)|^2 + L|f(x)||\sigma W| &(29)\\
&\leq (L|\sigma W(t)|)^2/4, &(30)
\end{aligned}$$

which implies

$$g(v(t)) \leq g(v(0)) + \int_0^t (L|\sigma W(s)|)^2/4 := M(t).$$

Therefore $v(t)$ is bounded by $K_{M(t)}$ which is a bounded set. It follows that $v(t)$ is defined for $t \in [0, \infty)$, and so is $x(t)$. □

Let $\dot{\psi}(t)$ be a piecewise continuous function. We consider the following nonautonomous equation:

$$dy = f(y)dt + \sigma\dot{\psi}dt, \quad y(0) = x_0. \tag{31}$$

Denoted by $S(t, \dot{\psi})x_0$ the solution of (31).

Lemma 2. Assume (H3)–(H4) hold. For any piecewise continuous $\dot{\psi}$, the solution of (31) is a global solution (i.e., $S(t, \dot{\psi})x_0$ is defined for $t\ [0, \infty)$).

Proof. Let $\psi(t) := \int_0^t \dot{\psi}(s)ds$, which is a continuous function of t. The proof is similar to that for Lemma 1 with $W(t)$ replaced by $\psi(t)$. □

Lemma 3. Assume (H3)–(H4) hold. Given $T > 0$, $\epsilon > 0$, x_0 and $q \in \mathbb{R}^n$, there exists a piecewise continuous $\dot{\psi}(t)$ such that solution of (31) satisfies the boundary condition $S(T, \dot{\psi})x_0 \in B(q, \epsilon)$.

Proof. We may assume $\sigma = 1$ for simplicity and $\epsilon < 1$. Let $N \in \mathbb{N}$ and

$$\begin{aligned}
0 := t_0 < t_1 < \cdots < t_N := T,&\\
t_i - t_{i-1} = h := T/N &\quad (32)\\
q_i := x_0 + i((p - x_0)/N), \quad 1 \leq i \leq N. &\quad (33)
\end{aligned}$$

Let V be the tube centered at the line segment from x_0 to q and with radius one, i.e.,

$$V := \bigcup_{t \in [0,1]} B((1 - t)x_0 + tq, 1)$$

Let $C := \sup_{x \in V} |f(x)| < \infty$. We are going to construct $\psi(t)$ on each $(t_i, t_i + 1)$. According to Lemma 2 the solution are defined for $t \in (t_i, t_i + 1)$. For simplicity we set $\psi(t_i) = 0$ at each t_i.

(i) On (t_0, t_1):

Denote $q_0 := x_0$. Let

$$\psi(t) := (q_1 - q_0)t/h, \ t \in (t_0, t_1)$$

then $q_0 + \psi(t) \in V$, $t \in (t_0, t_1)$. Therefore,

$$\sup_{s \in (t_0, t_1)} |f(q_0 + \psi(s))| \leq C.$$

From (31) we have

$$y(t) - (q_0 + \psi(t)) = \int_{t_0}^t f(y(s)) - f(q_0 + \psi(s))ds + \int_{t_0}^t f(q_0 + \psi(s))ds,$$

which implies

$$\begin{aligned}
&|y(t) - (q_0 + \psi(t))|\\
&\quad \leq \int_{t_0}^t |f(y(s)) - f(q_0 + \psi(s))|ds + \int_{t_0}^t |f(q_0 + \psi(s))|ds,\\
&\quad \leq \int_{t_0}^t L|y(s) - (q_0 + \psi(s))|ds + hC.
\end{aligned}$$

By Gronwall inequality, we have

$$|y(t_1) - (q_0 + \psi(t_1))| \leq hCe^{Lh} < \epsilon/2,$$

provided h small. Denote by $\tilde{q}_1 := y(t_1) \in V$.

(ii) On (t_{i-1}, t_i), $2 \le i \le N$:
To define $\psi(t)$ on each (t_{i-1}, t_i), we repeat the following process for $2 \le i \le N$. Let

$$\psi := (q_i - \tilde{q}_{i-1})(t - t_{i-1})/h, \quad t \in (t_{i-1}, t_i)$$

then $\tilde{q}_{i-1} + \psi(t) \in V$, $t \in (t_{i-1}, t_i)$. Therefore,

$$\sup_{s \in (t_{i-1}, t_i)} |f(\tilde{q}_{i-1} + \psi(s))| \le C.$$

Consider (31) with $t \in (t_{i-1}, t_i)$ and $y(t_{i-1}) = \tilde{q}_{i-1}$, then we have

$$\begin{aligned} y(t) - (\tilde{q}_{i-1} + \psi(t)) &= \int_{t_{i-1}}^{t} f(y(s)) - f(\tilde{q}_{i-1} + \psi(s))ds \\ &\quad + \int_{t_{i-1}}^{t} f(\tilde{q}_{i-1} + \psi(s))ds, \end{aligned}$$

which implies

$$\begin{aligned} |y(t) - (\tilde{q}_{i-1} + \psi(t))| &\le \int_{t_{i-1}}^{t} |f(y(s)) - f(\tilde{q}_{i-1} + \psi(s))|ds \\ &\quad + \int_{t_{i-1}}^{t} |f(\tilde{q}_{i-1} + \psi(s))|ds, \\ &\le \int_{t_{i-1}}^{t} L|y(s) - (\tilde{q}_{i-1} + \psi(s))|ds + hC. \end{aligned}$$

By Gronwall inequality, we have

$$|y(t_i) - (\tilde{q}_{i-1} + \psi(t_i))| \le hCe^{Lh} < \epsilon/2,$$

provided h small. Denote by $\tilde{q}_i := y(t_i)$.
At the final stage $i = N$, we have

$$|y(T) - (\tilde{q}_{i-1} + \psi(T))| = |y(T) - q| < \epsilon/2$$

Thus we have $|S(T, \dot{\psi})x_0 - q| < \epsilon$. The proof is complete. □

We will prove that with positive probability the solution $\phi(t, \omega)p$ of (24) can be approximated by solution $S(t, \dot{\psi})p$ of (31) in any finite time interval and any initial point p.

Lemma 4. Assume (H3)–(H4) hold. Given a piecewise continuous $\dot{\psi}$, $T > 0$ and $\gamma > 0$, the set A_γ defined by

$$A_\gamma := \left\{ \omega : \sup_{p \in \mathbb{R}^n} \sup_{t \in [0,T]} |S(t, \dot{\psi})p - \phi(t, \omega)p| < \gamma \right\}, \tag{34}$$

has $\mathbb{P}(A_\gamma) > 0$.

Proof. According to Lemma 1 and Lemma 2, $S(t, \dot{\psi})p$ and $\phi(t, \omega)p$ are defined for $t \in [0, T]$. From (24) and (31) we have

$$\begin{aligned} d(x - y) = (f(x) - f(y))dt - \sigma\dot{\psi}(t)dt \\ + \sigma dW(t). \end{aligned} \tag{35}$$

We may assume $\psi(0) = 0$ since it doesn't affect the solution of (31). Let $\tilde{W}(t) := W(t) - \psi(t)$, $t \in [0, T]$, then we have

$$d\tilde{W} = -\dot{\psi}(t)dt + dW. \tag{36}$$

By Girsanov's Theorem ([19]), $\tilde{W}$ is a Brownian motion under probability $\tilde{\mathbb{P}}$, where

$$\frac{d\tilde{\mathbb{P}}}{d\mathbb{P}} = \exp\left(\int_0^T \dot{\psi}(s) \cdot dW(s) - \frac{1}{2}|\dot{\psi}(s)|^2 ds \right) > 0. \tag{37}$$

Let $\tilde{\Lambda}_\varepsilon := \{\omega : \sup_{t \in [0,T]} |\tilde{W}(t, \omega)| < \varepsilon\}$ and $\Lambda_\varepsilon := \{\omega : \sup_{t \in [0,T]} |W(t, \omega)| < \varepsilon\}$, then $\tilde{\mathbb{P}}(\tilde{\Lambda}_\varepsilon) > 0$, which implies $\mathbb{P}(\Lambda_\varepsilon) > 0$ by (37). It follows that for $\omega \in \Lambda_\varepsilon$

$$\begin{aligned} |\phi(t, \omega)p - S(t, \dot{\psi})p| &\le \int_0^t |f(\phi(s, \omega)p) - f(S(s, \dot{\psi})p)|ds + \sigma|\tilde{W}(t)| \\ &\le \int_0^t L|\phi(s, \omega)p - S(s, \dot{\psi})p)|ds + \sigma\varepsilon. \end{aligned} \tag{38}$$

By Gronwall inequality, we have

$$\sup_{p \in \mathbb{R}^n} \sup_{t \in [0,T]} |S(t, \dot{\psi})p - \phi(t, \omega)p| < \sigma\varepsilon e^{LT}, \quad \omega \in \Lambda_\varepsilon.$$

The assertion follows by choosing ε small enough. □

Proof of Theorem 3:
We still denote by $\phi(t, \omega)$ the flow for (2) with piecewise constant diffusion $\sigma(t)$ shown in Fig. 1. Let q_0 be a initial point. We may assume $B(p_i, \epsilon)$ lying in the stable manifold of p_i for $i = 1, \ldots, N$.

Let $q_i \in \mathbb{R}^n$. For each $1 \le i \le N$, by Lemma 3, there exists a piecewise continuous $\dot{\psi}_i(t), t \in [S_i, T_i]$ such that the solution of

$$dy^{(i)} = f(y^{(i)})dt + \sigma(t)\dot{\psi}_i dt, \quad y^{(i)}(S_i) = q_{i-1}, \tag{39}$$

satisfying

$$y^{(i)}(T) \in B(p_i, \epsilon/2).$$

Since $\sigma(t) = 0$ and $y^{(i)}(T_i)$ lies in the stable manifold of p_i, it follows that

$$y^{(i)}(t) \in B(p_i, \epsilon/2), \; t \in [T_i, S_{i+1}],$$

Selecting $q_i = y^{(i-1)}(S_{i-1})$, $2 \le i \le N$, and let $\dot{\psi}(t)$ be the piecewise continuous function defined by

$$\dot{\psi}(t) = \begin{cases} \dot{\psi}_i(t), & t \in [S_i, T_i], \ 1 \le i \le N, \\ 0, & \text{otherwise.} \end{cases} \tag{40}$$

Then we have

$$S(t, \dot{\psi})q_0 \in B(p_i, \epsilon/2),$$
$$t \in [T_i, S_{i+1}], \quad 1 \le i \le N,$$

By Lemma 4, the set

$$\left\{\omega : \sup_{t\in[0,T]} |S(t, \dot{\psi})q_0 - \phi(t, \omega)q_0| < \epsilon/2\right\}, \tag{41}$$

has positive probability, which implies the set

$$\Omega' := \{\omega : \phi(t, \omega)q_0 \in B(p_i, \epsilon_i), \ t \in [T_i, S_{i+1}], \ 1 \le i \le N\}, \tag{42}$$

has $\mathbb{P}(\Omega') > 0$. The proof is complete.

4. Numerical Examples

In this section, we demonstrate the ID method on four standard test examples for global optimization methods. They are designed to have many local minima which are similar to the global minima, and the local mimima can easily trap the solutions. In all the experiments, we set the parameters $\alpha = 10$, $\gamma = 10$ and $N = 10$.

Problem 1. *Penalized Shubert Function.* Let

$$g_1(x) = \sum_{i=1}^{5} i\cos((i+1)x + 1) \tag{43}$$

be the standard Shubert function. Then the penalized Shubert function is defined by

$$g(x) := g_1(x) + u(x, 10, 100, 2) \tag{44}$$

where $u(x, a, k, m)$ is called penalization function defined by

$$u(x, a, k, m) := \begin{cases} k(x-a)^m, & x > a \\ 0 & -a \le x \le a \\ k(-x-a)^m & x < -a \end{cases}. \tag{45}$$

As shown in Fig. 2, the function g has 19 local minima in the region $\{x : |x| < 10\}$ and three of them are global minima. One realization of the ID method gives the circled points, which is the convergent points of the gradient flow after the diffusion

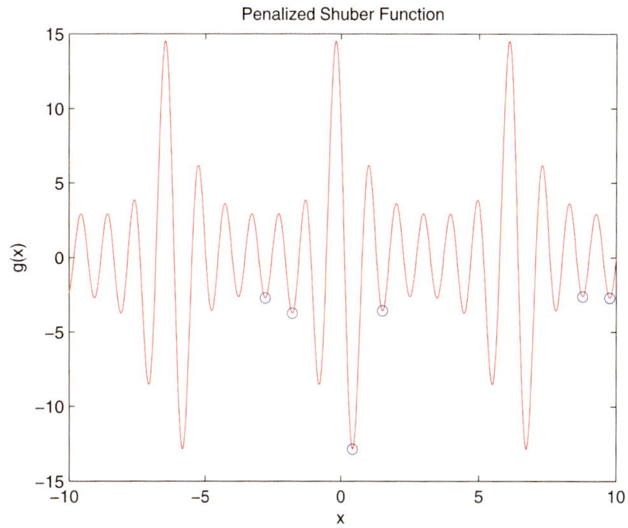

Fig. 2. Numerical experiment for Problem 3. The blue circles mark the convergent points at the non-diffusion step.

is turned off. It finds the global minimum in the middle.

Problem 2. *Two-Dimensional Penalized Shubert Function* is defined by

$$g(x, y) = \left\{\sum_{i=1}^{5} i\cos((i+1)x + 1)\right\} \times \left\{\sum_{i=1}^{5} i\cos((i+1)y + 1)\right\} + u(x, 10, 100, 2) + u(y, 10, 100, 2). \tag{46}$$

It has 760 local minima in the region $\{(x, y) : |x| < 10, |y| < 10\}$, and 18 of them are global minima as shown in Fig. 3. It also shows that an arbitrary realization of ID algorithm gives the circled points. Among them, it finds three global minima.

Problem 3. *Two-Dimensional Penalized Shubert Function with Parameter* β is defined by

$$g(x, y) = \left\{\sum_{i=1}^{5} i\cos((i+1)x + 1)\right\} \times \left\{\sum_{i=1}^{5} i\cos((i+1)y + 1)\right\} + u(x, 10, 100, 2) + u(y, 10, 100, 2) + \beta[(x - 6.0835)^2 + (y + 5.8581)^2], \tag{47}$$

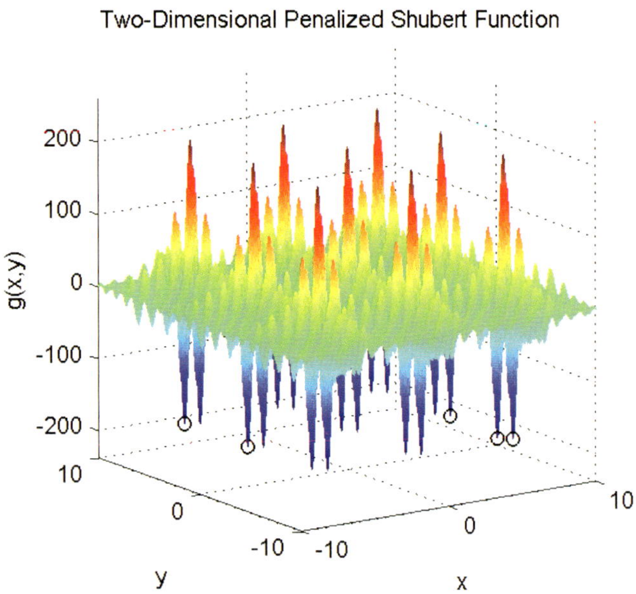

Fig. 3. Numerical experiment for Problem 4. The black circles mark the convergent points at the non-diffusion step.

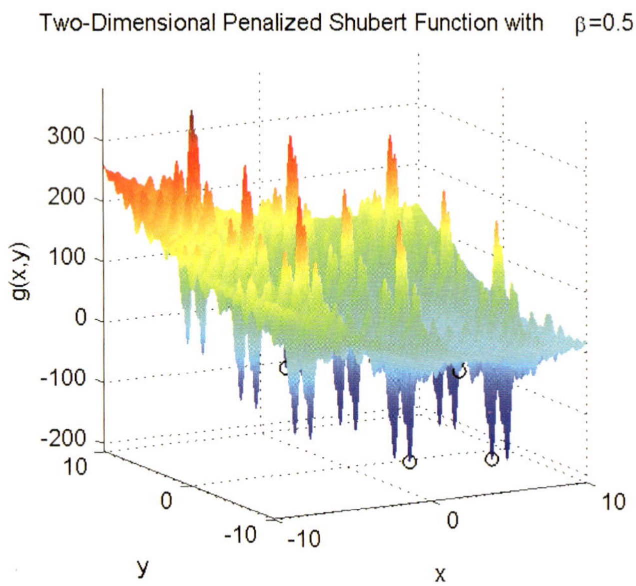

Fig. 4. Numerical experiment for Problem 5. The black circles mark the convergent points at the non-diffusion step.

where $\beta > 0$ is a constant. The point $(x, y) = (6.0835, -5.8581)$ is a global minimizer of g with $\beta = 0$. The function behaves roughly the same with the function considered in Problem 2, but has only one unique global minimizer at $(x, y) = (6.0853, -5.8581)$ as shown in Fig. 4. We show the results of convergent points (circled points) of an arbitrary ID realization. Similar to the previous problems, the ID method is able to find the global minima.

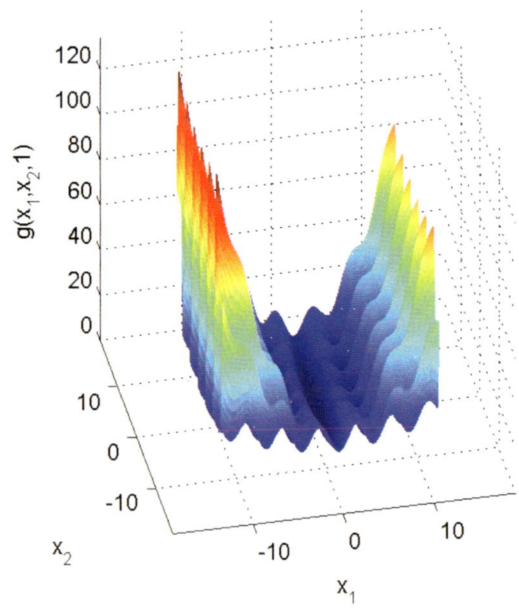

Fig. 5. The projection plot of the graph of $y = g(x_1, x_2, x_3)$ on the space $x_3 = 1$ for Problem 6.

Problem 4. *A Multi-Dimensional Function.* Let

$$g(x) = (\pi/n)\left\{k\sin^2(\pi y_1) + \sum_{i=1}^{n-1}(y_i - A)^2[1 + k\sin^2(\pi y_{i+1})] + (y_n - A)^2\right\}, \tag{48}$$

where

$$x = (x_1, \dots, x_n) \in \mathbb{R}^n,$$
$$y_i = 1 + (x_i - 1)/4, \quad i = 1, \dots, n,$$
$$k = 10, \quad A = 1.$$

The function has roughly 5^n local minima in the region $\{(x, y) : |x| < 10, |y| < 10\}$ and a unique global minimum located at

$$x_i = 1, \; i = 1, \dots, n.$$

For $n = 3$, Fig. 5 is the projection plot of the graph of $y = g(x_1, x_2, x_3)$ on the space $x_3 = 1$, which illustrates the global minimizers is surrounded by many local minimizers. The 10 convergent points visited in a sequential order by an arbitrary ID realization for $n = 3$ and $n = 4$ are given in Table 1 respectively.

The ID method finds the global minima for both cases. And the frequencies of visiting the global minimum are obviously higher than that of any other local minima.

Table 1. The minimizers, listed in a sequential order, found by one ID realization for $n = 3$ and $n = 4$ respectively.

i	n = 3	n = 4
1	$(-6.96, 5, -7)$	$(-6.96, 8.99, 8.99, 4.99)$
2	$(1, 1, 1)$	$(1, 1, 1, 1)$
3	$(1, 1, 1)$	$(1, 1, 1, 1)$
4	$(9, 5, -3)$	$(-2.98, -2.98, -2.98, -2.98)$
5	$(-10.92, -7, 5)$	$(-8.96, -2.99, 1, 1)$
6	$(-3, 1, 1)$	$(-2.98, 4.98, -2.98, 1)$
7	$(-3, 1, 1)$	$(-6.96, 1, 1, 1)$
8	$(1, 1, 1)$	$(1, 1, 1, 1)$
9	$(9, -7, 1)$	$(1, 1, 1, 1)$
10	$(5, 9, 5)$	$(1, 1, 1, 1)$

We also compare the ID method with the CDD method studied in [17]. We run both methods on Problem 3 for 100 independent realizations with the final stopping time $T = 300$ respectively. In each realization, we use the same initial point and Brownian motion path for both methods. We record the number of visits that the ID and the CDD reach a small neighborhood of the unique global minimizer respectively. Our experiments shows that the average number of visits in the 100 realizations for the ID method is 7.5. In contrast, it is 0 for the CDD method, which indicates that it has not reached the small neighborhood of the global minimum in most of the realizations because of the slow decay in the cooling rate. This is not surprising since the convergence time for the CDD method is exponentially long, while the convergence time at non-diffusion step for the ID method is short because the gradient flow drives the trajectories to a small neighborhood of the minimizers.

We also compare the first time of reaching a small neighborhood of the global minimizer for Problem 3 by both methods. The first time that the ID method and the CDD method enter a square neighborhood of the global minimizer with diameter $\epsilon = 0.001$ and $\epsilon = 0.0001$ are recorded respectively for 30 independent realizations. Figure 6 shows the log plot of the first arriving time for $\epsilon = 0.001$. The ID method takes less time to reach the global minimizer than the CDD method does at most realizations. Figure 7 shows the log plot of the first arriving time for a much smaller neighborhood with $\epsilon = 0.0001$. The the ID method takes less time to hit the global minimizer than the CDD method does for all realizations except one, and the time difference is significant large.

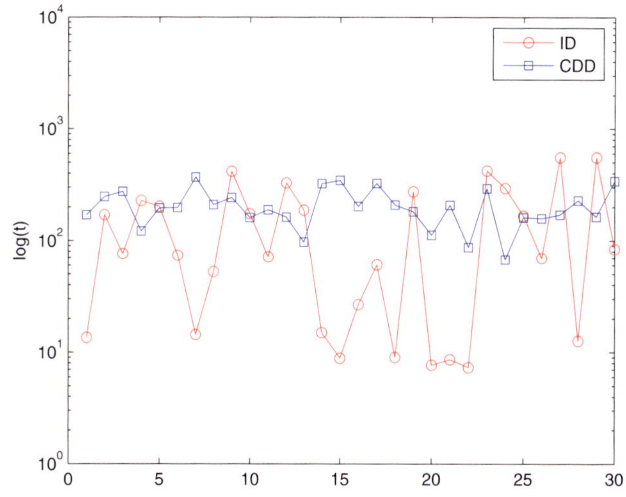

Fig. 6. The log plot of the first time to enter a square neighborhood of the global minimizer with diameter $\epsilon = 0.001$ by ID and CDD for 30 independent realizations.

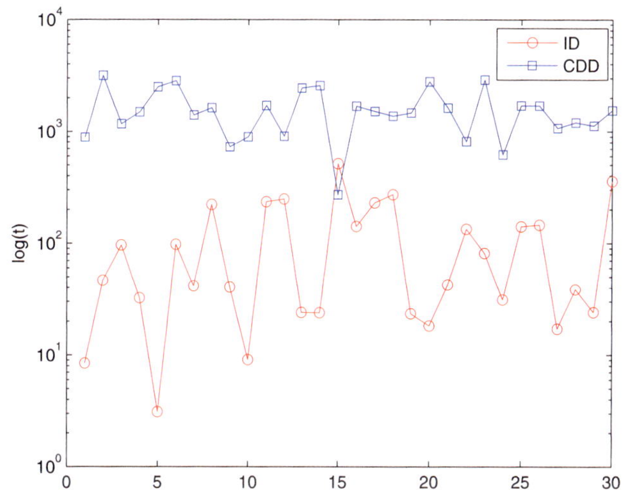

Fig. 7. The log plot of the first time to enter a square neighborhood of the global minimizer with diameter $\epsilon = 0.0001$ by ID and CDD for 30 independent realizations.

Acknowledgment

The authors would like to thank Luca Dieci, Wen Huang, Liangda Huang and Kening Lu for helpful comments and discussions in this project. The project is partially supported by NSF Faculty Early Career Development (CAREER) Award DMS-0645266 and DMS-1042998.

References

[1] F. Aluffi-Pentini, V. Parisi and F. Zirilli, Global optimization and stochastic differential equations, *J. Optmiz. Theory App.*, 47(1), 1–16, 1985.

[2] A. Arnold, P. Markowich, G. Toscani and A. Unterreiter, On convex Sobolev inequalities and the rate of convergence to equilibrium for Fokker-Planck type equations, *Comm. in PDE's*, 26(1 & 2), (2001) pp. 43–100.

[3] L. Arnold, *Random Dynamical Systems*, Springer Verlag, Berlin, 1998.

[4] Bagirov, A. M., and A. M. Rubinov, Cutting angle method and a local search, *Journal of Global Optimization*, 27 (2003), pp. 193–213.

[5] A. Bagirov, A. Rubinov and J. Zhang, Local optimization method with global multidimensional search, *J. of Global Optimization*, 32 (2005), pp. 161–179.

[6] A. M. Bagirov, A. M. Rubinov, and J. Zhang, A multidimensional descent method for global optimization, *Optimization*, Vol. 58, No. 5, 2009, pp. 611–625.

[7] G. Bao and P. J. Li, Inverse medium scattering for the Helmholtz equation at fixed frequency, *Inverse Problems*, 21 (2005), pp. 1621–1641.

[8] G. Bao and P. J. Li, Inverse medium scattering problems for electromagnetic waves, *SIAM J. Appl. Math.*, 65 (2005), pp. 2049–2066.

[9] Battiti, Roberto; Mauro Brunato; Franco Mascia, *Reactive Search and Intelligent Optimization*, Springer-Verlag, 2008.

[10] E. Beale, *Introduction to Optimization*, Wiley, Chichester 1988.

[11] S. Boyd and L. Vandenberghe, *Convex Optimization*, Cambridge University Press, 2004.

[12] V. Cerny, A thermodynamical approach to the travelling salesman problem: an efficient simulation algorithm. *J. Optmiz. Theory App.*, 45(1), 41–51, 1985.

[13] T.-S. Chiang, C.-R. Hwang and S.-J. Sheu, Diffusion for global optimization in $\mathbb{R}^n$, *SIAM J. Control and Optim.*, 25(3), 737–753, 1986.

[14] S.-N. Chow, W. P. Li, Z. X. Liu and H. M. Zhou, A natural order in dynamical systems based on Conley-Markov matrices, accepted by *J. of Differential Equations*.

[15] S.-N. Chow, J. Lu and H. M. Zhou, A fast algorithm for finding shortest path by solving initial value ODE's, *preprint*.

[16] S.-N. Chow, T. S. Yang and H. M. Zhou, Global optimization by intermittent diffusion, *UCLA CAM report*, 2009.

[17] S. Geman and C.-R. Hwang, Diffusion for global optimization, *SIAM J. Control and Optim.*, 24(5), 1031–1043, 1986.

[18] B. Gidas, Metropolis-type Monte Carlo simulation algorithms and simulated annealing, in: J.L. Snell (Ed.), *Topics in Contemporary Probability and Its Applications, in: Probability and Stochastics Series*, CRC Press, Boca Raton, FL, 1995, pp. 159–232.

[19] I. V. Girsanov, On transforming a certain class of stochastic processes by absolutely continuous substitution of measures, *Theory Probab. Appl.*, 5, No. 3 (1960), pp. 285–301.

[20] Jack K. Hale, *Asymptotic behavior of dissipative systems (Mathematical Surveys and Monographs, 25)*, Amer Mathematical Society.

[21] J. Holland, Genetic algorithms and the optimal allocation of trials, *SIAM J. Computing* 2 (1973), pp. 88–105.

[22] R. Holley, S. Kusuoka and D. Stroock, Asymptotics of the spectral gap with applications to the theory of simulated annealing, *J. Func. Anal.*, 83(2), pp. 333–347, 1989.

[23] R. Holley and D. Stroock, Logarithmic Sobolev inequalities and stochastic Ising models, *J. Stat. Phys.*, 46, 5–6, 1159–1194, 1987.

[24] R. Holley and D. Stroock, Simulated annealing via Sobolev inequalities, *Commun. Math. Phys.*, 115(4), 553–569, 1988.

[25] R. Horst and P.M. Pardalos (eds.), *Handbook of Global Optimization*, Kluwer, Dordrecht 1995.

[26] L. Ingber, Very fast simulated re-annealing. *Math. Comput. Modelling*, 12(8), 967–973, 1989.

[27] C. T. Kelley, *Iterative Methods for Optimization*, SIAM, 1999

[28] S. Kirkpatrick, S.; C. D. Gelatt, Jr. and M. P. Vecchi, Optimization by simulated snnealing, *Science*, Vol. 220 No. 4598, pp. 671–680, 1983.

[29] A.H. Land and A.G. Doig, An automated method for solving discrete programming problems, *Econometrica*, 28 (1960), 497–520.

[30] A. Levy and A. Montalvo, The tunneling algorithm for the global minimization of functions, *SIAM J. of Scientific and Statistical Computing*, 6, 1985, pp. 15–29.

[31] J.D. Little, K.C. Murty, D.W. Sweeney and C. Karel, An algorithm for the travelling salesman problem, *Operations Research*, 11 (1963), pp. 972–989.

[32] H. Lourenco and O. C. Martin and T. Sttzle, Iterated Local Search, *Handbook of Metaheuristics*, Springer, 2003, pp. 320–353.

[33] P. Markowich and C. Villani, On the trend to equilibrium for the Fokker-Planck equation: an interplay between physics and functional analysis, *Mat. Contemp.*, 19 (2000), pp. 1–29.

[34] N. Metropolis, A. W. Rosenbluth, M. N. Rosenbluth, A. H. Teller and E. Teller, Equations of state calculations by fast computing machines, *J. Chem. Phys.*, Vol. 21, 1953, pp. 1087–1091.

[35] J. Mockus, Multiextremal problems in design, *Ph.D. Thesis, Nauka*, 1966.

[36] A. Neumaier, Complete search in continuous global optimization and constraint satisfaction, *Acta Numerica* (A. Iserles, ed.), Cambridge University Press, 2004.

[37] J. Nocedal and S. J. Wright *Numerical optimization*, Springer, 1999.

[38] H. Risken, *The Fokker-Planck equation, methods of solution and applications*, Second Edition, Springer-Verlag, 1989.

[39] H. Szu and R. Hartley, Fast simulated annealing, *Phys. Lett. A*, 122(3–4), 157–162. 1987,

[40] Pavlyukevich, Ilya, Simulated annealing for Le'vy-driven jump-diffusions, *Stochastic Process. Appl.* 118 (2008), no. 6, 1071 1105.

[41] A. Torn, Global Optimization as a Combination of Global and Local Search, *Ph.D. Thesis, Abo Akademi University, HHAAA* 13, 1974.

[42] A. Torn and A. Zilinskas, *Global optimization*, Springer Verlag, Berlin, 1989.

[43] D. Wales and J. Doye, Global optimization by basin-hopping and the lowest energy structures of Lennard-Jones clusters containing up to 110 atoms, *J. Phys. Chem. A*, 101, 1997, pp. 5111–5116.

[44] R. L. Yang, Convergence of the simulated annealing algorithm for continuous global optimization, *J. Optim. Theory Appl.*, 104(3), 691–716, 2000.

[45] K. Yiu, Y. Liu and K. Teo, A Hybrid Descent Method for Global Optimization, *J. of Global Optimization*, 28, 2004, pp. 229–238.

[46] J. Zhang, Derivative-free hybrid methods in global optimization and their applications to data mining, protein folding, unfolding, misfolding, *Ph.D Thesis, University of Ballarat*, Australia, 2004.

PART III

MEMRISTORS

CHAPTER 38

HOW WE FOUND THE MISSING MEMRISTOR*

The memristor—the functional equivalent of a synapse—could revolutionize circuit design *By R. Stanley Williams*

IT'S TIME TO STOP SHRINKING. Moore's Law, the semiconductor industry's obsession with the shrinking of transistors and their commensurate steady doubling on a chip about every two years, has been the source of a 50-year technical and economic revolution. Whether this scaling paradigm lasts for five more years or 15, it will eventually come to an end. The emphasis in electronics design will have to shift to devices that are not just increasingly infinitesimal but increasingly capable.

Earlier this year, I and my colleagues at Hewlett-Packard Labs, in Palo Alto, Calif., surprised the electronics community with a fascinating candidate for such a device: the memristor. It had been theorized nearly 40 years ago, but because no one had managed to build one, it had long since become an esoteric curiosity. That all changed on 1 May, when my group published the details of the memristor in *Nature*.

Combined with transistors in a hybrid chip, memristors could radically improve the performance of digital circuits without shrinking transistors. Using transistors more efficiently could in turn give us another decade, at least, of Moore's Law performance improvement, without requiring the costly and increasingly difficult doublings of transistor density on chips. In the end, memristors might even become the cornerstone of new analog circuits that compute using an architecture much like that of the brain

For nearly 150 years, the known fundamental passive circuit elements were limited to the capacitor (discovered in 1745), the resistor (1827), and the inductor (1831). Then, in a brilliant but underappreciated 1971 paper, Leon Chua, a professor of electrical engineering at the University of California, Berkeley, predicted the existence of a fourth fundamental device, which he called a memristor. He proved that memristor behavior could not be duplicated by any circuit built using only the other three elements, which is why the memristor is truly fundamental.

Memristor is a contraction of "memory resistor," because that is exactly its function: to remember its history. A memristor is a two-terminal device whose resistance depends on the magnitude and polarity of the voltage applied to it and the length of time that voltage has been applied. When you turn off the voltage, the memristor remembers its most recent resistance until the next time you turn it on, whether that happens a day later or a year later.

Think of a resistor as a pipe through which water flows. The water is electric charge. The resistor's obstruction of the flow of charge is comparable to the diameter of the pipe: the narrower the pipe, the greater the resistance. For the history of circuit design, resistors have had a fixed pipe diameter. But a memristor is a pipe that changes diameter with the amount and direction of water that flows through it. If water flows through this

THINKING MACHINE: This artist's conception of a memristor shows a stack of multiple crossbar arrays, the fundamental structure of R. Stanley Williams's device. Because memristors behave functionally like synapses, replacing a few transistors in a circuit with memristors could lead to analog circuits that can think like a human brain. IMAGE: BRYAN CHRISTIE DESIGN

*Courtesy of the IEEE.

PICTURING MEMRISTANCE: HP Labs senior fellow R. Stanley Williams [left] and research physicist Duncan Stewart [right] explain the fourth fundamental circuit element. Williams worked with nearly 100 scientists and engineers to find the memristor. PHOTO: PAUL SAKUMA/AP PHOTO

pipe in one direction, it expands (becoming less resistive). But send the water in the opposite direction and the pipe shrinks (becoming more resistive). Further, the memristor remembers its diameter when water last went through. Turn off the flow and the diameter of the pipe "freezes" until the water is turned back on.

That freezing property suits memristors brilliantly for computer memory. The ability to indefinitely store resistance values means that a memristor can be used as a nonvolatile memory. That might not sound like very much, but go ahead and pop the battery out of your laptop, right now—no saving, no quitting, nothing. You'd lose your work, of course. But if your laptop were built using a memory based on memristors, when you popped the battery back in, your screen would return to life with everything exactly as you left it: no lengthy reboot, no half-dozen auto-recovered files.

But the memristor's potential goes far beyond instant-on computers to embrace one of the grandest technology challenges: mimicking the functions of a brain. Within a decade, memristors could let us emulate, instead of merely simulate, networks of neurons and synapses. Many research groups have been working toward a brain in silico: IBM's Blue Brain project, Howard Hughes Medical Institute's Janelia Farm, and Harvard's Center for Brain Science are just three. However, even a mouse brain simulation in real time involves solving an astronomical number of coupled partial differential equations. A digital computer capable of coping with this staggering workload would need to be the size of a small city, and powering it would require several dedicated nuclear power plants.

Memristors can be made extremely small, and they function like synapses. Using them, we will be able to build analog electronic circuits that could fit in a shoebox and function according to the same physical principles as a brain.

A hybrid circuit—containing many connected memristors and transistors—could help us research actual brain function and disorders. Such a circuit might even lead to machines that can recognize patterns the way humans can, in those critical ways computers can't—for example, picking a particular face out of a crowd even if it has changed significantly since our last memory of it.

THE STORY OF THE MEMRISTOR is truly one for the history books. When Leon Chua, now an IEEE Fellow, wrote his seminal paper predicting the memristor, he was a newly minted and rapidly rising professor at UC Berkeley. Chua had been fighting for years against what he considered the arbitrary restriction of electronic circuit theory to linear systems. He was convinced that nonlinear electronics had much more potential than the linear circuits that dominate electronics technology to this day.

Chua discovered a missing link in the pairwise mathematical equations that relate the four circuit quantities—charge, current, voltage, and magnetic flux—to one another. These can be related in six ways. Two are connected through the basic physical laws of electricity and magnetism, and three are related by the known circuit elements: resistors connect voltage and current, inductors connect flux and current, and capacitors connect voltage and charge. But one equation is missing from this group: the relationship between charge moving through a circuit and the magnetic flux surrounded by that circuit—or more subtly, a mathematical doppelgänger defined by Faraday's Law as the time integral of the voltage across the circuit. This distinction is the crux of a raging Internet debate about the legitimacy of our memristor [see sidebar, "Resistance to Memristance"].

Chua's memristor was a purely mathematical construct that had more than one physical realization. What does that mean? Consider a battery and a transformer. Both provide identical voltages—for example, 12 volts of direct current—but they do so by entirely different mechanisms: the battery by a chemical reaction going on inside the cell and the transformer by taking a 110-V ac input, stepping that down to 12 V ac, and then transforming that into 12 V dc. The end result is mathematically identical—both will run an electric shaver or a cellphone, but the physical source of that 12 V is completely different.

Conceptually, it was easy to grasp how electric charge could couple to magnetic flux, but there was no obvious physical interaction between charge and the integral over the voltage.

Chua demonstrated mathematically that his hypothetical device would provide a relationship between flux and charge similar to what a nonlinear resistor provides between voltage and current. In practice, that would mean the device's resistance would vary according to the amount of charge that passed through it. And it would remember that resistance value even after the current was turned off.

He also noticed something else—that this behavior reminded him of the way synapses function in a brain.

Even before Chua had his eureka moment, however, many researchers were reporting what they called "anomalous" current-voltage behavior in the micrometer-scale devices they had built out of unconventional materials, like polymers and metal oxides. But the idiosyncrasies were usually ascribed to some mystery electrochemical reaction, electrical breakdown, or other spurious phenomenon attributed to the high voltages that researchers were applying to their devices.

As it turns out, a great many of these reports were unrecognized examples of memristance. After Chua theorized the memristor out of the mathematical ether, it took another 35 years for us to intentionally build the device at HP Labs, and we only really understood the device about two years ago. So what took us so long?

IT'S ALL ABOUT SCALE. We now know that memristance is an intrinsic property of any electronic circuit. Its existence could have been deduced by Gustav Kirchhoff or by James Clerk Maxwell, if either had considered nonlinear circuits in the 1800s. But the scales at which electronic devices have been built for most of the past two centuries have prevented experimental observation of the effect. It turns out that the influence of memristance obeys an inverse square law: memristance is a million times as important at the nanometer scale as it is at the micrometer scale, and it's essentially unobservable at the millimeter scale and larger. As we build smaller and smaller devices, memristance is becoming more noticeable and in some cases dominant. That's what accounts for all those strange results researchers have described. Memristance has been hidden in plain sight all along. But in spite of all the clues, our finding the memristor was completely serendipitous.

CROSSBAR ARCHITECTURE: A memristor's structure, shown here in a scanning tunneling microscope image, will enable dense, stable computer memories. IMAGE: R. STANLEY WILLIAMS/HP LABS

In 1995, I was recruited to HP Labs to start up a fundamental research group that had been proposed by David Packard. He decided that the company had become large enough to dedicate a research group to long-term projects that would be protected from the immediate needs of the business units. Packard had an altruistic vision that HP should "return knowledge to the well of fundamental science from which HP had been withdrawing for so long." At the same time, he understood that long-term research could be the strategic basis for technologies and inventions that would directly benefit HP in the future. HP gave me a budget and four researchers. But beyond the comment that "molecular-scale electronics" would be interesting and that we should try to have something useful in about 10 years, I was given carte blanche to pursue any topic we wanted. We decided to take on Moore's Law.

At the time, the dot-com bubble was still rapidly inflating its way toward a resounding pop, and the existing semiconductor road map didn't extend past 2010. The critical feature size for the transistors on an integrated circuit was 350 nanometers; we had a long way to go before atomic sizes would become a limitation. And yet, the eventual end of Moore's Law was obvious. Someday semiconductor researchers would have to confront physics-based limits to their relentless descent into the infinitesimal, if for no other reason than that a transistor cannot be smaller than an atom. (Today the smallest components of transistors on integrated circuits are roughly 45 nm wide, or about 220 silicon atoms.)

That's when we started to hang out with Phil Kuekes, the creative force behind the Teramac (tera-operation-per-second multiarchitecture computer)—an experimental supercomputer built at HP Labs primarily from defective parts, just to show it could be done. He gave us the idea to build an architecture that would work even if a substantial number of the individual devices in the circuit were dead on arrival. We didn't know what those devices would be, but our goal was electronics that would keep improving even after the devices got so small that defective ones would become common. We ate a lot of pizza washed down with appropriate amounts of beer and speculated about what this mystery nanodevice would be.

We were designing something that wouldn't even be relevant for another 10 to 15 years. It was possible that by then devices would have shrunk down to the molecular scale envisioned by David Packard or perhaps even *be* molecules. We could think of no better way to anticipate this than by mimicking the Teramac at the nanoscale. We decided that the simplest abstraction of the Teramac architecture was the crossbar, which has since become the de facto standard for nanoscale circuits because of its simplicity, adaptability, and redundancy.

The crossbar is an array of perpendicular wires. Anywhere two wires cross, they are connected by a switch. To connect a horizontal wire to a vertical wire at any point on the grid, you must close the switch between them. Our idea was to open and close these switches by applying voltages to the ends of the wires. Note that a crossbar array is basically a storage

How Memristance Works

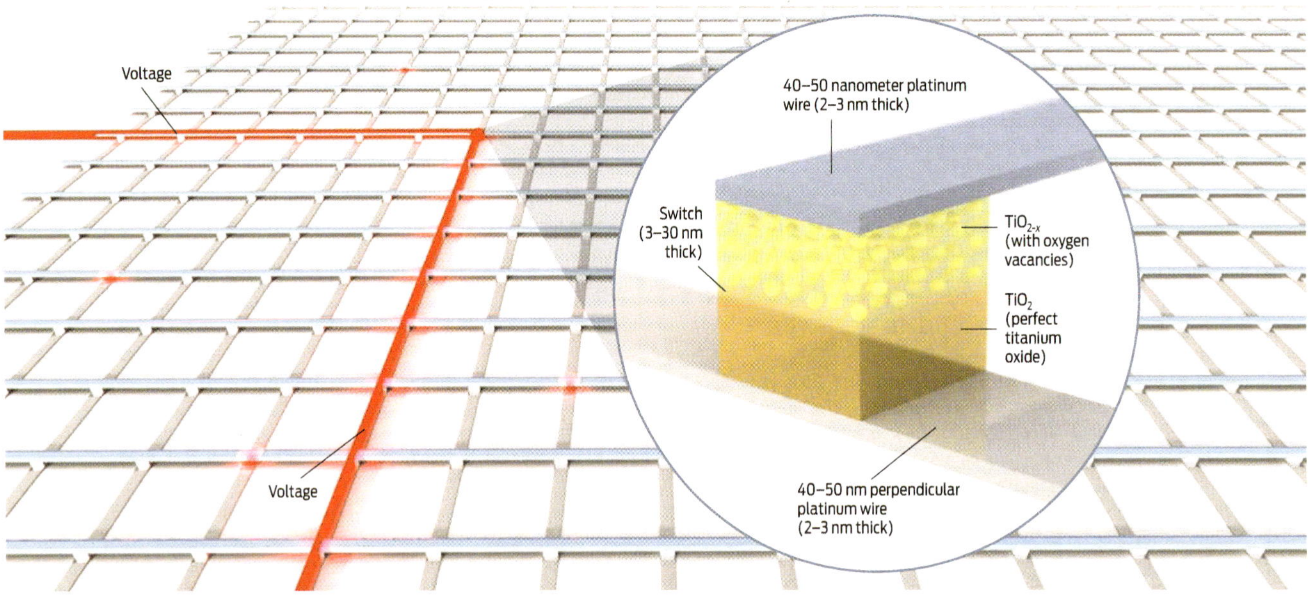

THE CROSSBAR ARCHITECTURE: The crossbar architecture is a fully connected mesh of perpendicular wires. Any two crossing wires are connected by a switch. To close the switch, a positive voltage is applied across the two wires to be connected. To open the switch, the voltage is reversed.

THE SWITCH: A switch is a 40-nanometer cube of titanium dioxide (TiO_2) in two layers: The lower TiO_2 layer has a perfect 2:1 oxygen-to-titanium ratio, making it an insulator. By contrast, the upper TiO_2 layer is missing 0.5 percent of its oxygen (TiO_{2-x}), so x is about 0.05. The vacancies make the TiO_{2-x} material metallic and conductive.

APPLIED MEMRISTANCE: The oxygen deficiencies in the TiO_{2-x} manifest as "bubbles" of oxygen vacancies scattered throughout the upper layer. A positive voltage on the switch repels the (positive) oxygen deficiencies in the metallic upper TiO_{2-x} layer, sending them into the insulating TiO_2 layer below. That causes the boundary between the two materials to move down, increasing the percentage of conducting TiO_{2-x} and thus the conductivity of the entire switch. The more positive voltage is applied, the more conductive the cube becomes.

A negative voltage on the switch attracts the positively charged oxygen bubbles, pulling them out of the TiO_2. The amount of insulating, resistive TiO_2 increases, thereby making the switch as a whole resistive. The more negative voltage is applied, the less conductive the cube becomes.

What makes this switch special—memristive—is that when the voltage is turned off, positive or negative, the oxygen bubbles do not migrate. They stay where they are, which means that the boundary between the two titanium dioxide layers is frozen. That is how the memristor "remembers" how much voltage was last applied.

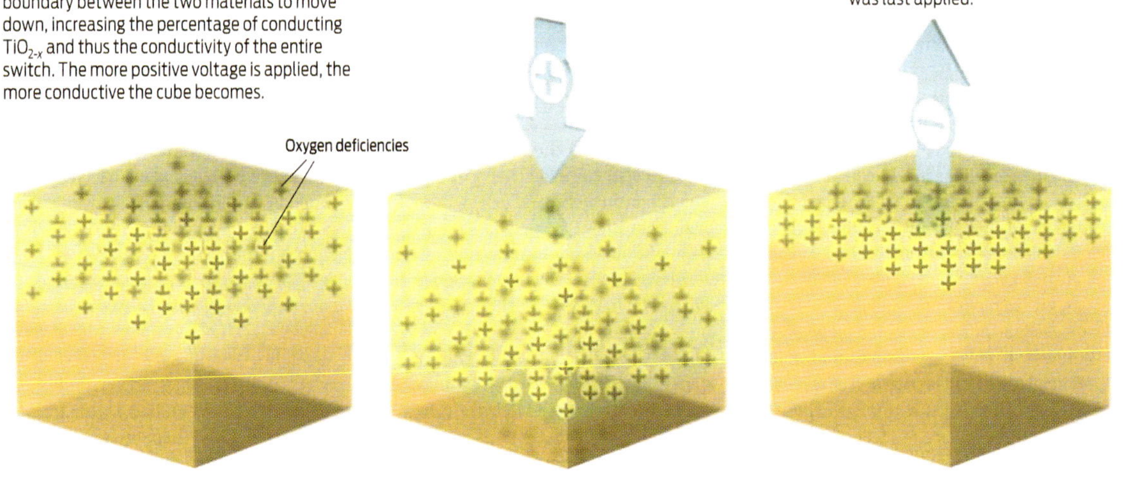

system, with an open switch representing a zero and a closed switch representing a one. You read the data by probing the switch with a small voltage.

Like everything else at the nanoscale, the switches and wires of a crossbar are bound to be plagued by at least some nonfunctional components. These components will be only a few atoms wide, and the second law of thermodynamics ensures that we will not be able to completely specify the position of every atom. However, a crossbar architecture builds in redundancy by allowing you to route around any parts of the circuit that don't work. Because of their simplicity, crossbar arrays have a much higher density of switches than a comparable integrated circuit based on transistors.

But implementing such a storage system was easier said than done. Many research groups were working on such a cross-point memory—and had been since the 1950s. Even after 40 years of research, they had no product on the market. Still, that didn't stop them from trying. That's because the potential for a truly nanoscale crossbar memory is staggering; picture carrying around the entire Library of Congress on a thumb drive.

One of the major impediments for prior crossbar memory research was the small off-to-on resistance ratio of the switches (40 years of research had never produced anything surpassing a factor of 2 or 3). By comparison, modern transistors have an off-to-on resistance ratio of 10 000 to 1. We calculated that to get a high-performance memory, we had to make switches with a resistance ratio of at least 1000 to 1. In other words, in its off state, a switch had to be 1000 times as resistive to the flow of current as it was in its on state. What mechanism could possibly give a nanometer-scale device a three-orders-of-magnitude resistance ratio?

We found the answer in scanning tunneling microscopy (STM), an area of research I had been pursuing for a decade. A tunneling microscope generates atomic-resolution images by scanning a very sharp needle across a surface and measuring the electric current that flows between the atoms at the tip of the needle and the surface the needle is probing. The general rule of thumb in STM is that moving that tip 0.1 nm closer to a surface increases the tunneling current by one order of magnitude.

We needed some similar mechanism by which we could change the effective spacing between two wires in our crossbar by 0.3 nm. If we could do that, we would have the 1000:1 electrical switching ratio we needed.

Our constraints were getting ridiculous. Where would we find a material that could change its physical dimensions like that? That is how we found ourselves in the realm of molecular electronics.

CONCEPTUALLY, OUR DEVICE was like a tiny sandwich. Two platinum electrodes (the intersecting wires of the crossbar junction) functioned as the "bread" on either end of the device. We oxidized the surface of the bottom platinum wire to make an extremely thin layer of platinum dioxide, which is highly conducting. Next, we assembled a dense film, only one molecule thick, of specially designed switching molecules. Over this "monolayer" we deposited a 2- to 3-nm layer of titanium metal, which bonds strongly to the molecules and was intended to glue them together. The final layer was the top platinum electrode.

The molecules were supposed to be the actual switches. We built an enormous number of these devices, experimenting with a wide variety of exotic molecules and configurations, including rotaxanes, special switching molecules designed by James Heath and Fraser Stoddart at the University of California, Los Angeles. The rotaxane is like a bead on a string, and with the right voltage, the bead slides from one end of the string to the other, causing the electrical resistance of the molecule to rise or fall, depending on the direction it moves. Heath and Stoddart's devices used silicon electrodes, and they worked, but not well enough for technological applications: the off-to-on resistance ratio was only a factor of 10, the switching was slow, and the devices tended to switch themselves off after 15 minutes.

Our platinum devices yielded results that were nothing less than frustrating. When a switch worked, it was spectacular: our off-to-on resistance ratios shot past the 1000 mark, the devices switched too fast for us to even measure, and having switched, the device's resistance state remained stable for years (we still have some early devices we test every now and then, and we have never seen a significant change in resistance). But our fantastic results were inconsistent. Worse yet, the success or failure of a device never seemed to depend on the same thing.

We had no physical model for how these devices worked. Instead of rational engineering, we were reduced to performing huge numbers of Edisonian experiments, varying one parameter at a time and attempting to hold all the rest constant. Even our switching molecules were betraying us; it seemed like we could use anything at all. In our desperation, we even turned to long-chain fatty acids—essentially soap—as the molecules in our devices. There's nothing in soap that should switch, and yet some of the soap devices switched phenomenally. We also made control devices with no molecule monolayers at all. None of them switched.

We were frustrated and burned out. Here we were, in late 2002, six years into our research. We had something that worked, but we couldn't figure out *why*, we couldn't model it, and we sure couldn't engineer it. That's when Greg Snider, who had worked with Kuekes on the Teramac, brought me the Chua memristor paper. "I don't know what you guys are building," he told me, "but this is what I want."

To this day, I have no idea how Greg happened to come across that paper. Few people had read it, fewer had understood it, and fewer still had cited it. At that point, the paper was 31 years old and apparently headed for the proverbial dustbin of history. I wish I could say I took one look and yelled, "Eureka!" But in fact, the paper sat on my desk for months before I even tried to read it. When I did study it, I found the concepts and the equations unfamiliar and hard to follow. But I kept at it because something had caught my eye, as it had Greg's: Chua had included a graph that looked suspiciously similar to the experimental data we were collecting.

The graph described the current-voltage (I-V) characteristics that Chua had plotted for his memristor. Chua had called them "pinched-hysteresis loops"; we called our I-V characteristics "bow ties." A pinched hysteresis loop looks like a diagonal infinity symbol with the center at the zero axis, when plotted on a graph of current against voltage. The voltage is first increased from zero to a positive maximum value, then decreased to a minimum negative value and finally returned to zero. The bow ties on our graphs were nearly identical [see graphic, "Bow Ties"].

That's not all. The total change in the resistance we had measured in our devices also depended on how *long* we applied the voltage: the longer we applied a positive voltage, the lower the resistance until it reached a minimum value. And the longer we applied a negative voltage, the higher the resistance became until it reached a maximum limiting value. When

Resistance To Memristance

INTRODUCING A NEW fundamental circuit element earned R. Stanley Williams some grief along with his newfound fame. After the *Nature* article appeared in May, online comments pages boiled over with skepticism. "Is this a hoax?" someone asked on the Wikipedia memristor page on 30 April 2008, the day the news broke, in one of the milder statements of disbelief. Seven months later, the debate continues.

Skeptics argue that the memristor is not a fourth fundamental circuit element but an example of bad science. The crux of their argument rests on two fundamental misunderstandings: first, skeptics overlook the expanded design space that arises from working with nonlinear circuit elements. The second and more profound misunderstanding concerns Leon Chua's mathematical definition of a memristor.

At first, most people—including Williams—assumed that Chua defined memristance strictly as the relationship between electric charge and magnetic flux. However, the actual definition of memristance is more general. Linking electric charge and magnetic flux is one way to satisfy the definition, but it's not the only one. In fact, it turns out you can bypass magnetic interaction altogether.

Chua's general memristance definition has two parts. The first equation defines how the memristor's voltage depends on current and a "state variable"—that is, a quantity that measures some physical property of a device, like the length of a column of mercury in a thermometer. The column's length correlates to the thermometer's temperature, and adding or removing heat makes the column longer or shorter. In Williams's memristor, the state variable is the thickness of the stoichiometric titanium dioxide in the switch; increasing or decreasing that thickness causes the device's resistance to increase or decrease.

The second equation expresses how the changing state variable (the TiO_2's thickness) depends on the amount of charge flowing through the device. In Williams's memristor, the TiO_2's thickness depends on the distribution of the oxygen vacancies throughout the material.

Here is what you need to remember: one, a magnetic interaction is not necessary for memristance. Two, in nonlinear circuit elements, memristance is not the same thing as nonlinear resistance. Three, because no combination of passive devices can reproduce the properties of a memristor, memristance is a fundamental circuit quantity.

Williams himself is sanguine about the memristor's reputation. "A hundred years after Einstein proposed his theory of relativity," he says, shrugging, "you still have some people arguing against it." —*Sally Adee*

we stopped applying the voltage, whatever resistance characterized the device was frozen in place, until we reset it by once again applying a voltage. The loop in the I-V curve is called hysteresis, and this behavior is startlingly similar to how synapses operate: synaptic connections between neurons can be made stronger or weaker depending on the polarity, strength, and length of a chemical or electrical signal. That's not the kind of behavior you find in today's circuits.

Looking at Chua's graphs was maddening. We now had a big clue that memristance had something to do with our switches. But how? Why should our molecular junctions have anything to do with the relationship between charge and magnetic flux? I couldn't make the connection.

Two years went by. Every once in a while I would idly pick up Chua's paper, read it, and each time I understood the concepts a little more. But our experiments were still pretty much trial and error. The best we could do was to make a lot of devices and find the ones that worked.

But our frustration wasn't for nothing: by 2004, we had figured out how to do a little surgery on our little sandwiches. We built a gadget that ripped the tiny devices open so that we could peer inside them and do some forensics. When we pried them apart, the little sandwiches separated at their weakest point: the molecule layer. For the first time, we could get a good look at what was going on inside. We were in for a shock.

WHAT WE HAD was not what we had built. Recall that we had built a sandwich with two platinum electrodes as the bread and filled with three layers: the platinum dioxide, the monolayer film of switching molecules, and the film of titanium.

But that's not what we found. Under the molecular layer, instead of platinum dioxide, there was only pure platinum. Above the molecular layer, instead of titanium, we found an unexpected and unusual layer of titanium dioxide. The titanium had sucked the oxygen right out of the platinum dioxide! The oxygen atoms had somehow migrated through the molecules and been consumed by the titanium. This was especially surprising because the switching molecules had not been significantly perturbed by this event—they were intact and well ordered, which convinced us that they must be doing something important in the device.

The chemical structure of our devices was not at all what we had thought it was. The titanium dioxide—a stable compound found in sunscreen and white paint—was not just regular titanium dioxide. It had split itself up into two chemically different layers. Adjacent to the molecules, the oxide was stoichiometric TiO_2, meaning the ratio of oxygen to titanium was perfect, exactly 2 to 1. But closer to the top platinum electrode, the titanium dioxide was missing a tiny amount of its oxygen, between 2 and 3 percent. We called this oxygen-deficient titanium dioxide TiO_{2-x}, where x is about 0.05.

Because of this misunderstanding, we had been performing the experiment backward. Every time I had tried to create a switching model, I had reversed the switching polarity. In other words, I had predicted that a positive voltage would switch the device off and a negative voltage would switch it on. In fact, exactly the opposite was true.

It was time to get to know titanium dioxide a lot better. They say three weeks in the lab will save you a day in the library every time. In August of 2006 I did a literature search and found about 300 relevant papers on titanium dioxide. I saw that each of the many different communities researching titanium dioxide had its own way of describing the compound. By the end of the month, the pieces had fallen into place. I finally knew how our device worked. I knew why we had a memristor.

The exotic molecule monolayer in the middle of our sandwich had nothing to do with the actual switching. Instead, what it did was control the flow of oxygen from the platinum dioxide into the titanium to produce the fairly uniform layers of TiO_2 and TiO_{2-x}. The key to the switching was this bilayer of the two different titanium dioxide species [see diagram, "How Memristance Works"]. The TiO_2 is electrically insulating (actually a semiconductor), but the TiO_{2-x} is conductive, because its oxygen vacancies are donors of electrons, which makes the vacancies themselves positively charged. The vacancies can be thought of like bubbles in a glass of beer, except that they don't pop—they can be pushed up and down at will in the titanium dioxide material because they are electrically charged.

Now I was able to predict the switching polarity of the device. If a positive voltage is applied to the top electrode of the device, it will repel the (also positive) oxygen vacancies in the TiO_{2-x} layer down into the pure TiO_2 layer. That turns the TiO_2 layer into TiO_{2-x} and makes it conductive, thus turning the device on. A negative voltage has the opposite effect: the vacancies are attracted upward and back out of the TiO_2, and thus the thickness of the TiO_2 layer increases and the device turns off. This switching polarity is what we had been seeing for years but had been unable to explain.

On 20 August 2006, I solved the two most important equations of my career—one equation detailing the relationship between current and voltage for this equivalent circuit, and another equation describing how the application of the voltage causes the vacancies to move—thereby writing down, for the first time, an equation for memristance in terms of the physical properties of a material. This provided a unique insight. Memristance arises in a semiconductor when both electrons and charged dopants are forced to move simultaneously by applying a voltage to the system. The memristance did not actually involve magnetism in this case; the integral over the voltage reflected how far the dopants had moved and thus how much the resistance of the device had changed.

We finally had a model we could use to engineer our switches, which we had by now positively identified as memristors. Now we could use all the theoretical machinery Chua had created to help us design new circuits with our devices.

Triumphantly, I showed the group my results and immediately declared that we had to take the molecule monolayers out of our devices. Skeptical after years of false starts and failed hypotheses, my team reminded me that we had run control samples without molecule layers for every device we had ever made and that those devices had never switched. And getting the recipe right turned out to be tricky indeed. We needed to find the exact amounts of titanium and oxygen to get the two layers to do their respective jobs. By that point we were all getting impatient. In fact, it took so long to get the first working device that in my discouragement I nearly decided to put the molecule layers back in.

A month later, it worked. We not only had working devices, but we were also able to improve and change their characteristics at will.

But here is the real triumph. The resistance of these devices stayed constant whether we turned off the voltage or just read their states (interrogating them with a voltage so small it left the resistance unchanged). The oxygen vacancies didn't roam around; they remained absolutely immobile until we again applied a positive or negative voltage. That's memristance: the devices remembered their current history. We had coaxed Chua's mythical memristor off the page and into being.

Bow Ties

LEON CHUA'S original graph of the hypothetical memristor's behavior is shown at top right; the graph of R. Stanley Williams's experimental results in the *Nature* paper is shown below. The loops map the switching behavior of the device: it begins with a high resistance, and as the voltage increases, the current slowly increases. As charge flows through the device, the resistance drops, and the current increases more rapidly with increasing voltage until the maximum is reached. Then, as the voltage decreases, the current decreases but more slowly, because charge is flowing through the device and the resistance is still dropping. The result is an on-switching loop. When the voltage turns negative, the resistance of the device increases, resulting in an off-switching loop. *—R.S.W.*

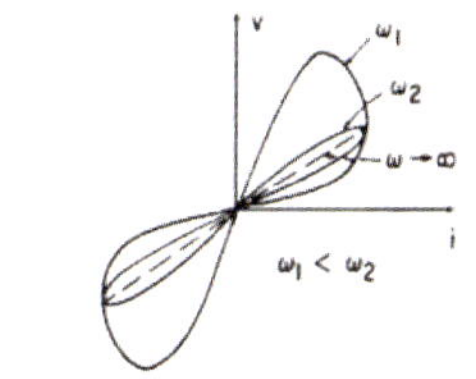

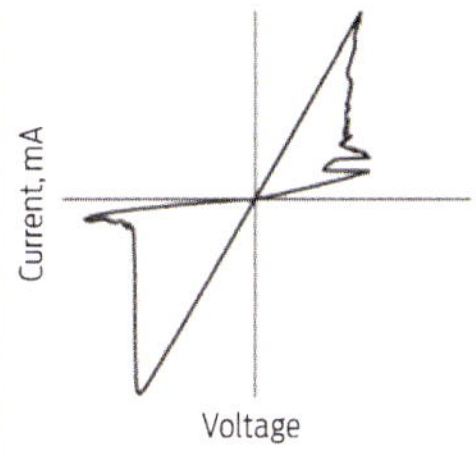

EMULATING THE BEHAVIOR of a single memristor, Chua showed, requires a circuit with at least 15 transistors and other passive elements. The implications are extraordinary: just imagine how many kinds of circuits could be supercharged by replacing a handful of transistors with one single memristor.

The most obvious benefit is to memories. In its initial state, a crossbar memory has only open switches, and no information is stored. But once you start closing switches, you can store vast amounts of information compactly and efficiently. Because memristors remember their state, they can store data indefinitely, using energy only when you toggle or read the state of a switch, unlike the capacitors in conventional DRAM, which will lose their stored charge if the power to the chip is turned off. Furthermore, the wires and switches can be made very small: we should eventually get down to a width of around 4 nm, and then multiple crossbars could be stacked on top of each other to create a ridiculously high density of stored bits.

Greg Snider and I published a paper last year showing that memristors could vastly improve one type of processing circuit, called a field-programmable gate array, or FPGA. By replacing several specific transistors with a crossbar of memristors, we showed that the circuit could be shrunk by nearly a factor of 10 in area and improved in terms of its speed relative to power-consumption performance. Right now, we are testing a prototype of this circuit in our lab.

And memristors are by no means hard to fabricate. The titanium dioxide structure can be made in any semiconductor fab currently in existence. (In fact, our hybrid circuit was built in an HP fab used for making inkjet cartridges.) The primary limitation to manufacturing hybrid chips with memristors is that today only a small number of people on Earth have any idea of how to design circuits containing memristors. I must emphasize here that memristors will never eliminate the need for transistors: passive devices and circuits require active devices like transistors to supply energy.

The potential of the memristor goes far beyond juicing a few FPGAs. I have referred several times to the similarity of memristor behavior to that of synapses. Right now, Greg is designing new circuits that mimic aspects of the brain. The neurons are implemented with transistors, the axons are the nanowires in the crossbar, and the synapses are the memristors at the cross points. A circuit like this could perform real-time data analysis for multiple sensors. Think about it: an intelligent physical infrastructure that could provide structural assessment monitoring for bridges. How much money—and how many lives—could be saved?

I'm convinced that eventually the memristor will change circuit design in the 21st century as radically as the transistor changed it in the 20th. Don't forget that the transistor was lounging around as a mainly academic curiosity for a decade until 1956, when a killer app—the hearing aid—brought it into the marketplace. My guess is that the real killer app for memristors will be invented by a curious student who is now just deciding what EE courses to take next year. ❑

CHAPTER 39

AFTERMATH OF FINDING THE MEMRISTOR

R. STANLEY WILLIAMS

Hewlett-Packard Labs, USA

Observing the response to our paper "The missing memristor found" [1] over the past four years in both the popular press and the scientific literature has been fascinating. A significant part of the scientific process is to vet descriptions of new ideas or objects, and the bigger the potential impact of a concept, the more rigorous that scrutiny should be. However, intertwined with this process are many human issues of desire for recognition and priority of discovery, as well as an often strong bias to reject anything new without actually understanding it. There are a lot of misconceptions about memristors floating around that are difficult to correct with only a few explanatory pages. Real understanding requires a great deal of hard work, and the resources essential to achieve that understanding already exist in the literature. However, for the vast majority of us, skimming over a few papers is completely insufficient; I spent years reading and re-reading Prof. Leon Chua's papers before I started to really get an appreciation of what he was saying. I have several copies of many different papers completely covered with highlighter of many colors and with my scrawled notes — each time I read one of his paper, and I continually refer back to them, I learn something new and my appreciation deepens. Although he has written some wonderful tutorials, most of Prof. Chua's writings are formal and dense with information, and thus can be intimidating; they absolutely require a level of mathematical sophistication to comprehend, but to those who persevere, they are marvels of rigor and, eventually, clarity. Here I will provide a personal guide to the set of papers that I have found most helpful in my research in the hope that this will encourage others to study them and answer the questions they have, and then I will finish with some observations and comments.

During the 1960's, Prof. Chua established the mathematical foundations for nonlinear circuit theory, which was the basis for his classic 1969 textbook *Introduction to Nonlinear Network Theory* as well as a large number of papers in refereed journals. As a result of this work, Prof. Chua made an interesting observation that led to his discovery of the memristor as a mathematical entity, reported in 1971 [2]. For completely linear circuits (which is highly restrictive, since real physical systems will display nonlinearity beyond some operating range), there are only three independent two-terminal passive circuit elements: the resistor R, the capacitor C and the inductor L. However, when he generalized the mathematical relations to be nonlinear, there was another independent differential relationship that in principle coupled the charge q that flowed through a circuit and the flux ϕ in the circuit, $d\phi = M\,dq$, that was mathematically different from the nonlinear resistance that coupled the voltage v to the current i, $dv = Rdi$. He mathematically explored the properties of this new model nonlinear circuit element, and found that it was essentially a resistor with memory — it was a device that changed its resistance depending on the amount of charge that flowed through the device, and thus he called this hypothetical circuit element M a memristor. This conclusion was independent of any physical mechanism that might couple the flux and charge, and none was postulated. Moreover, the memristor definition did not require causality. In other words, the mathematical relationship between flux and charge could be the result of some other cause — any mechanism that led to the constraint embodied by the equation $d\phi = M\,dq$ would lead to a device with the circuit properties of a memristor. This prediction of the properties of a new circuit element from symmetry principles was totally unique and revolutionary, and did not depend on any experimental observation. He published these initial findings essentially as a curiosity — it was not obvious at that time that a physical analog of such a circuit element existed, and thus he called it the "missing element".

In 1976, with his then student Sung Mo Kang, he published a critical generalization of the original memristor concept [3], but this has not been cited with the frequency of the 1971 paper, so fewer people seem to be aware of it. Chua and Kang introduced the fact that a "memristive device" has a state variable (or variables), indicated by w, that describes the physical properties of the device at any time. A memristive system is characterized by two equations, the "quasi-static" conduction equation that relates the voltage across the device to the current through it at any particular time via a generalized resistance,

$$v = R(w, i)i$$

and the dynamical equation, which explicitly asserts that the derivative of the state variable w is a function f of itself and the current through the device,

$$\frac{dw}{dt} = f(w, i).$$

Neither the flux ϕ nor the charge q explicitly appears in either of these two equations, but if $w = q$, $R(w, i) = R(w)$ and $f(w, i) = i$, the two equations reduce to the original definition of a memristor. Furthermore, the quasi-static conduction equation places a requirement on the current-voltage characteristic of the device — if a memristive system is driven with some type of cyclic excitation, such as a sinusoidal current, the plot of the voltage vs. the current will be a Lissajous curve for which the voltage is always equal to zero when the current is zero, and vice versa. Chua called this curve a "pinched hysteresis loop", and it has an important physical interpretation — neither a memristor nor a memristive system stores either charge or energy (like a capacitor, for example), but they do 'remember' their history because of their changing resistance. This 1976 paper showed many other properties of the generalized memristor and also discussed possible examples, but again this was a mathematical exercise that was independent of any physical mechanism known at the time. The importance to real systems is that if one can identify the state variable with a physical property of a device and experimentally determine the dynamical and quasistatic equations, then one has a useful model for the element that can be used for designing a circuit that would utilize the device.

There is another pair of papers that are critical for not only understanding how memristors stand as independent devices, but how to appropriately understand a nonlinear circuit element model, how to construct one from a physics-based mechanism or black-box electrical measurements, and how the model relates to an actual nonlinear circuit [4, 5]. The two papers are best read together; the 1980 paper [4] is mathematically thorough, broad in coverage and filled with deep insights, whereas the 1984 paper is more tutorial and descriptive. I often find myself going back and forth between the two for the complementary viewpoints they express. We learn that no circuit model is an exact equivalent because no physical device can be exactly mimicked by a mathematical equation. A particular physical device may be best described by different models depending on the operating range, with the 'best' model being the simplest one that produces realistic results. There are several properties that a realistic model should have, including well-posedness (no mathematical artifacts that cause an unphysical situation), the capability to be simulated, qualitative similarity to the physical system (e.g. same initial and asymptotic behavior), the ability to predict previously unexplored operating modes, and structural resilience (stability under small perturbations of the model parameters). These concepts are made clear through mathematical definitions and examples. Thus, one needs an appropriate set of models (I think of them as basis functions) that are as complete as possible to describe a real system. Creating a device model is an art that can utilize a wide range of inputs and insights — there is no unique way to define the best possible model; was it useful in enabling a circuit to be designed and did it predict the properties of the circuit to within some desired accuracy? If there is a physical device for which the properties are well described by a particular model, then we can call that device by the name of the model, understanding that a more complete description may require some attribute of a different model. For example, all physical inductors have an intrinsic resistance, which is usually described as a model resistance in series with a model inductance.

The final two papers are both tutorials and are written in a much more informal style [6, 7]. They are very useful for people who just want to get a light overview of memristors without digging into the mathematical details, but also contain significant new insights and are therefore valuable even for people who have mastered the first four papers. However, no one should read only [6, 7] and think that they comprehend the subject — any word description can be misconstrued or misrepresented; the actual definitions are all mathematical. In his 2002 publication [6], Prof. Chua correctly realized that as electronic device dimensions shrink into the nanometer scale, their properties will become more nonlinear and thus the issues for understanding nonlinear circuits are becoming increasingly more relevant and critical. He used a fascinating analogy working up from the 'Laws' of motion postulated by Aristotle, Newton and Einstein to illustrate the necessity of choosing the right variables for a model in the first place and then what happens when the model progresses from an initial linear approximation to a more realistic nonlinear formulation. There follows several completely worked out examples to illustrate nonlinear circuit element modeling from his previous papers, including memristors. In the final paper [7], he describes memristors and memristive systems, and makes the observation that in fact the latter are a relatively straightforward generalization of the former, and recommends that from now on to simplify the nomenclature that both should be called memristors. By creating memristor models for the pinched i–v hysteresis loops of each, he shows that specific physical examples of memristors include several devices that are the subject of contemporary research: bipolar and unipolar resistive switches, often called RRAM or ReRAM; "atomic switches"; spin-transfer torque (STT) RAM devices; and phase-change memory devices; which are based on a wide variety of materials and physical mechanisms [7].

An important issue to understand is that the discovery of the memristor mathematical model does not conflict with nor compete for priority against the various realizations of physical devices that exhibit this circuit behavior. It is complementary, in that it provides a mathematical framework for designing and actually using the devices in circuits. It also provides an important mathematical constraint for those who are interested in the physics of their devices — any mechanism proposed for how the device operates needs to be in agreement with the memristor equations or it is not valid. Thus, researchers who are working on various types of resistive switching devices need not fear the memristor, but rather should embrace it. It is a high-level mathematical model that can be used to predict the circuit behavior of a wide variety of physical devices, it provides a unifying framework to put the circuit properties of all the devices into context, and therefore provides insight into how the various devices may substitute for each other in a wide variety of (especially nonmemory) applications originally developed for a different device.

Another issue is that no matter how careful one tries to be, any word description of a mathematical model will likely be incomplete, just as the model itself is only an approximation of the properties of a physical system. Thus, papers or discussions that argue about the meaning of a particular word or phrase often miss the point, since words can be ambiguous and interpreted (or twisted) in different ways. That is why we use mathematics in science — when a question arises about the specific meaning of a concept, we must go back to the defining equations. This is where Prof. Leon Chua's work stands out — precise, complete, insightful and totally rigorous.

References

[1] D. Strukov, G. S. Snider, D. R. Stewart and R. S. Williams: "The Missing Memristor Found," *Nature* **453**, 80 (2008).

[2] L. O. Chua, "Memristor — missing circuit element," *IEEE Transactions on Circuit Theory*, Vol. CT-18 (5), pp. 507–519, 1971.

[3] Chua, L.O. & Kang, S. M. Memristive devices and systems. *Proc. IEEE* **64**, 209–223 (1976).

[4] Chua, L. O. Device Modeling via basic nonlinear circuit elements. *IEEE Trans. Circ. & Syst.* **CAS-27**, 1014–1044 (1980).

[5] Chua, L. O. Nonlinear Circuits. *IEEE Trans. Circ. & Syst.* **CAS-31**, 69–87 (1984).

[6] Chua, L. O. Nonlinear Circuit Foundations for Nanodevices, Part I: The Four-Element Torus. *Proc. IEEE* **91**, (2003).

[7] Chua, L. Resistance switching memories are memristors. *Appl. Phys. A* **102**, 765–783 (2011).

CHAPTER 40

THE SINGING ARC: THE OLDEST MEMRISTOR?

JEAN-MARC GINOUX* and BRUNO ROSSETTO†

Laboratoire PROTEE, *I.U.T. de Toulon, Université du Sud, BP 20132, F-83957 La Garde Cedex, France*
**ginoux@univ-tln.fr*
†rossetto@univ-tln.fr

On April 30th 2008, the journal *Nature* announced that the missing circuit element, postulated thirty-seven years before by Professor Leon O. Chua has been found. Thus, after the capacitor, the resistor and the inductor, the existence of a fourth fundamental element of electronic circuits called "memristor" was established. In order to point out the importance of such a discovery, the aim of this article is first to propose an overview of the manner with which the three others have been invented during the past centuries. Then, a comparison between the main properties of the singing arc, i.e. a forerunner device of the triode used in Wireless Telegraphy, and that of the memristor will enable to state that the singing arc could be considered as the oldest memristor.

1. Introduction

Contrary to what one might think, it is not by experimenting, but by logical deduction that Professor L. O. Chua was able to postulate the existence of a missing circuit element. In his now famous publication of 1971, [6] considered the three basic building blocks of an electric circuit: the capacitor, the resistor and the inductor as well as the three laws linking the four fundamental circuit variables, namely, the electric *current* i, the *voltage* v, the *charge* q and the *magnetic flux* φ. Then, [6, p. 507] explained that:

> "... by the *axiomatic* definition of the three classical circuits elements, namely, the *resistor* (defined by a relationship between v and i), the *inductor* (defined by a relationship between φ and i), and the *capacitor* (defined by a relationship between q and v). Only one relationship remains undefined, the relationship between φ and q."

He thus concluded from the logical as well as axiomatic points of view, that it is necessary, for the sake of *completeness*, to postulate the existence of a fourth circuit element to which he gave the name *memristor* since it behaves like a nonlinear resistor with memory. The manner with which these three first laws that he calls: "relationships" have been discovered will be briefly recalled in Sec. 2 while the fourth law that he has established as well as the main properties of the memristor will be presented in Sec. 3. Then, it will be shown in Sec. 4 that the singing arc which exhibits analogous characteristic to the missing circuit element can be considered as the oldest memristor.

2. Fundamental Passive Circuits Elements

Anyone who has taken a physics class is familiar with the capacitor, the resistor, the inductor and of course the generator. But the question that one can ask is when and how these devices have been invented. This section aims to answer this question.

2.1. *Le condensatore and Volta's law*

In a way we could say that the history of the *capacitor* begins with the History of Physics, i.e. at the time of the Ancient Greece. Indeed, according to many historians of sciences, Thales of Miletus (624–547 B.C.) who is generally considered as the first physicist, has highlighted the fact that the rubbed *amber* ($\eta\lambda\varepsilon\kappa\tau\varrho\o\nu$ elektron in Greek) has the property of attracting light objects, such as pieces of straw or fabric. At the beginning of the seventeenth century, the English physician, physicist and natural philosopher William Gilbert (1544–1603) extended in his book *De Magnete, Magneticisque Corporibus, et de Magno Magnete Tellure* (On the Magnet and Magnetic Bodies, and on the Great Magnet the Earth) published in 1600, the number of substances capable of "becoming electric" (triboelectric), that is to say, after friction they exhibit the same property as amber. In the second chapter entitled: On the Magnetick Coition, and first on the Attraction of Amber, or more truly, on the Attaching of Bodies to Amber, [16, p. 48] wrote:

> "*Nam non solum succinum, & gagates (vt illi putant) allectant corpuscula.* For it is not only amber and jet (as they suppose) which entice small bodies; but Diamond, Sapphire, Carbuncle, Iris gem, Opal, Amethyst, Vincentina, and Bristolla (an English gem or spar), Beryl, and Crystal do the same."

This list has grown with the glasses of various compositions, such as resins, or sulfur. After the body is rubbed with dry hands, or with paper, leather or fabric, feathers and straw are more or less attracted. But the observed phenomena, did not received spectacular interests before the 1700s. During this period, the electrostatic generator, then called *friction machine*, appeared. The German scientist Otto von Guericke (1602–1686) is considered as the inventor of a primitive form of frictional electrical machine. In 1671, he had the idea to make a sulphur globe that could be rotated and rubbed by hand. In 1707, Isaac Newton (1642–1727) suggested to use a glass globe instead of a sulphur one and in 1709 Francis Hauksbee (1666–1713) improved the basic design, by rotating a glass sphere to be rotated rapidly against a woollen cloth. Thus, in the middle of the eighteenth century, the *Electricty* becomes fashionable. In France, it was widely popularized by the French clergyman and physicist Jean-Antoine Nollet (1700–1770) known as Abbé Nollet. On March 14th 1746, he made one of his most spectacular experiments in the galerie des Glaces at Versailles in the presence of his Majesty the King Louis XV. He electrified a hundred eighty royal guards holding hands with an electrostatic generator (see Fig. 1) positioned at the beginning of the human chain.

The same year in January, Nollet who has been asked to translate a letter send by the Dutch scientist Pieter Musschenbroek (1692–1761), then professor at Leiden, to René Réaumur (1683–1757) and written in Latin, had credited him erroneously for the discovery[1] of a device that he called the

[1]This episode has been clarified by Edmond Hoppe in his History of Physics [20, p. 448].

Fig. 1. Abbé Nollet's friction machine, [25, p. 24].

Fig. 2. The Leyden jar.

"Leiden jar". In fact, this invention was made one year earlier, on October 10th 1745 by a prelate named Ewald Georg von Kleist (1700–1748) who wanted to confine electricity in a bottle. So, he filled the bottle with water closed by a cork pierced by a nail. Holding the bottle with one hand he attached the nail against an electrostatic generator. Then, after having moved away the bottle from the machine, he touched the nail with his other hand and received a violent electric discharge. In fact von Kleist had invented the first capacitor.

However, it's the Italian physicist Alessandro Volta (1745–1827) who gave it this name half a century later. Volta is generally well-known for his invention in 1800 of the electric battery. Ten years before, in 1771, he had improved and popularized the electrophorus: a capacitive generator used to produce electrostatic charge via the process of electrostatic induction. Then, he published his research about the electrophorus in the *Philosophical transactions of the Royal Society of London* in 1782 and they were reproduced in his Complete Works in 1816. In this paper [45, vol. 1, part 2, p. 50] wrote:

> "Questo artificio, voi forse già l'indovinate, consiste a riunire all eletrometro medesimo il *Condensatore*."[2]

Then, [45, vol. 1, part 1, p. 250] exposed the law that Prof. Chua called "relationship" linking the tension v, the charge q and the capacity C of the capacitor:

> "Ciò che abbiam detto comprendersi facilmente che la tensione debbe essere in ragione inversa delle capacita, ci viene poi mostrato nella maniera pi chiara dall esperienza."[3]

This law is generally expressed as:

$$q = Cv \tag{1}$$

We suggest to name this relationship: Volta's law. During the same period Benjamin Franklin (1706–1790) invented the terminology "electric

[2] "This device, you probably already guessed, is to bring together all the same eletrometers under the name *Capacitor*."

[3] "What we have said can easily be understood that the voltage ought to be in inverse proportion to the capacity, as it is shown clearly by experiments."

charge" while Henry Cavendish (1731–1810) introduced the notion of electrostatic capacity of a sphere one inch in diameter that he called "globular inches".

Unfortunately, this series of papers on electrical phenomena, which had appeared in the *Philosophical transactions of the Royal Society of London*, i.e. the bulk of his electrical experiments, did not become known until they were collected and published by James Clerk Maxwell (1831–1879) a century later, in 1879, long after other scientists had been credited with the same results. On March 20th 1800, Volta wrote his first letter to Joseph Banks (1743–1820) in which he described the first true battery which came to be known as the Voltaic Pile (see Fig. 3). This letter was send for publication to the *Philosophical transactions* and read at the *Royal Society of London* on June 26th 1800 (see [44]).

On November 7th and 20th 1801, Volta presented his device at the Institut de France before

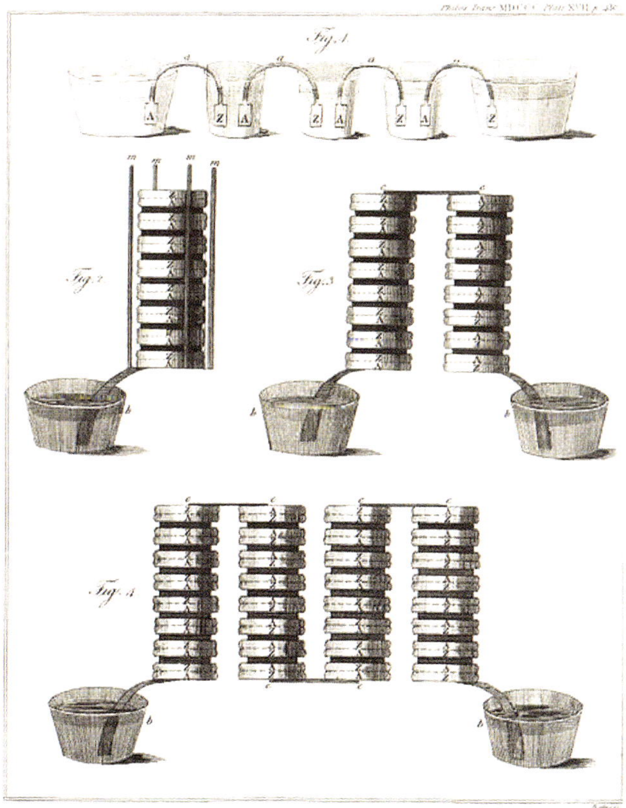

Fig. 3. Voltaic pile, [44, p. 430].

Napoléon who awarded him a gold medal. Although Volta made use of the term "tensione" in his articles introducing thus in a qualitative way the *voltage*, it's the French Physicist and Mathematician André Marie Ampère (1775–1836) who defined accurately the concepts of *current* and *voltage* in his first mémoir of 1820 [1, p. 59] wrote:

> "L'action électromotrice se manifeste par deux sortes d'effets que je crois devoir d'abord distinguer par une définition précise. J'appellerai le premier tension électrique, le second courant électrique[4]."

Nevertheless, it was unknown at that time among most of the physicists, including Ampère, the precise idea of *resistance* and of a possible "relationship" between the *voltage* of a voltaic pile and the *current* it produces in a conductor. For further research on Alessandro Volta see for example [29].

2.2. *Elektrischer Widerstand and Ohm's law*

The German physicist Georg Simon Ohm (1789–1854) started his experiments on electrical currents in 1825 while using a voltaic pile. The year after, he replaced it by copper-bismuth thermoelectric couples and could thus established his famous law [26, 27]:

$$v = Ri \tag{2}$$

Ohm defined thus the concept of "elektrischer Widerstand" (resistance) and of resistivity with great precision. Their measurements, in relative value, became thus accessible. In 1827, he summarized his research in a book entitled: *Die galvanische Kette, mathematisch bearbeitet (The galvanic circuit invesitgated mathematically)* [28] in which he introduced the concept of potential difference (P.D.) that he called "gefälle". [28, p. 19] wrote:

> "... wo man unter dem Ausdrucke "Gefälle" die Differenz solcher Ordinaten zu verstehen hat, die zu zwei um die Längeneinheit von einander entfernten Stellen gehören[5]".

For further research on Georg Ohm see for example [31, 32].

[4] "The electromotive action is characterized by two kinds of effects that I believe I must first distinguish by a precise definition. I will call the first voltage, the second electric current."

[5] "That is to say, the steepness or extent of the fall of potential in equal lengths of different portions of the circuit of like material but of different size, will be directly proportional to the resistance of such portions."

2.3. *Induction and Faraday's law*

Everybody knows the story of Michael Faraday (1791–1867) apprentice bookbinder who attended lectures by the eminent English chemist Humphry Davy (1778–1829) of the Royal Institution. When Davy damaged his eyesight in an accident with nitrogen trichloride he hired Faraday as secretary and then as Chemical assistant. After graduating Faraday became his successor in 1827 and later one of the most famous physicists. On August 29th 1831, Faraday made his first experimental demonstration of electromagnetic induction. He wrapped two wires around opposite sides of an iron-ring coil (see Fig. 4). In one side (A) he sent a current by connecting it to a battery while the opposite side (B) was attached to a galvanometer. At the moment of opening and closing the current, the galvanometer indicated current of opposite sense. He called the first coil an "inductor" with a primary current passing through it. He called the current passing through the second coil and galvanometer the secondary current.

The phenomenon he called "induction" was caused by the change in magnetic flux that occurred when the battery was connected, or disconnected. In 1834, Heinrich Lenz (1804–1865) formulated his famous law which describes "flux through the circuit" and gave the direction of the induced electromotive force and current resulting from electromagnetic induction that [13, p. 125, p. 129, p. 139] had described thus:

> "The power which electricity of tension possesses of causing an opposite electrical state in its vicinity has been expressed by the general term Induction; which, as it has been received into scientific language, may also, with propriety, be used in the same general sense to express the power which electrical currents may possess of inducing any particular state upon matter in their immediate neighborhood, otherwise indifferent. It is with this meaning that I purpose using it in the present paper. [...] For the purpose of avoiding periphrasis, I propose to call this action of the current from the voltaic battery, *volta-electric* induction. [...] but, as a distinction in language is still necessary, I propose to call the agency thus exerted by ordinary magnets, *magneto-electric* or *magnelectric* induction."

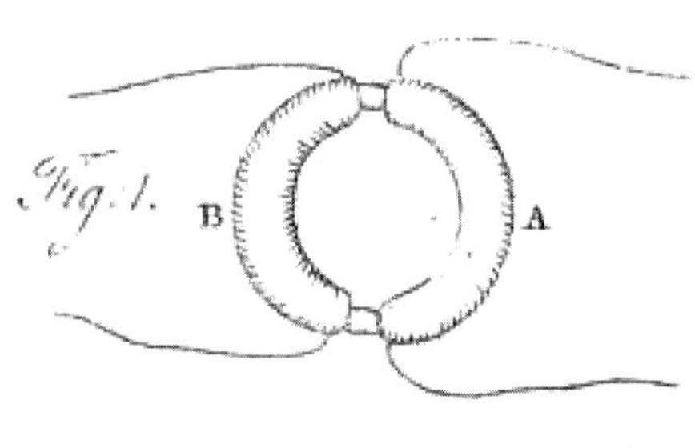

Fig. 4. Faraday's iron-ring coil apparatus, [13, p. 131].

Ten years later, starting from the work of Faraday and Lenz, Franz Neumann (1798–1895) established the mathematical law of induction. Based on the idea suggested by the rule of Lenz that the induced currents are created in the case of the motion, by working against electromagnetic forces, he derived the mathematical law for this phenomenon in the form:

$$e = -\frac{d\varphi}{dt} \tag{3}$$

often called "Faraday's law", although the works of Faraday did not contain any mathematical formula. This law means that the induced electromotive force is proportional to the "magnetic flux" φ cut per unit time, or the change per unit time of the induced "magnetic flux" φ embraced by the circuit. The self-inductance L of a coil traversed by a changing current of intensity i is then related to the "magnetic flux" φ by the formula:

$$\varphi = Li \tag{4}$$

For further research on Michael Faraday see for example [37].

As recalled in the previous sections, for each of the three circuit elements: capacitor (C), resistor (R) and inductor (L), a *linear* relationship, i.e. a *linear* law linking amongst themselves, a different pair of variables (q, v), (v, i) and (φ, i), has been discovered. These three linear laws seem to suggest that circuit theory was dominated by a kind of "linear thinking".

3. The Singing Arc

In fact, in the late nineteenth century, a French engineer Jean-Marie-Anatole Gérard-Lescuyer observed in an electromagnetic device, a nonlinear behavior that he considered then as paradoxical. By coupling a dynamo acting as a generator to a magneto-electric machine he built what was called a series dynamo machine. [15, p. 215] noticed a surprising phenomenon which he described as follows.

> "As soon as the circuit is closed the magneto-electrical machine begins to move; it tends to take a regulated velocity in accordance with the intensity of the current by which it is excited; but suddenly it slackens its speed, stops, and starts again in the opposite direction, to stop again and rotate in the same direction as before. In a word, it receives a regular reciprocating motion which lasts as long as the current that produces it."

He thus reported periodic reversals in the rotation of the magneto-electric machine although the source current was constant. He considered this phenomenon as an "electrodynamical paradox". According to him, an increase in the velocity of the magneto-electric machine gives rise to a current in the opposite direction which reverses the polarity of the inductors and, then reverses its rotation. Many years later, the Dutch physicist Baltahzar Van der Pol (1889–1959) will establish that this nonlinear phenomenon belongs to the class of "relaxation oscillations" in the second Dutch version [40, p. 39] of his eponymous article on "Relaxation-Oscillations" [42]. It has been pointed out by one of us [17] that there was at least four versions of this article:

(i) Over Relaxatietrillingen, *Physica*, 6, 154–157, [39],
(ii) Over Relaxatie-trillingen, *Tijdschrift Nederlandsch Radiogenoot* 3, 25–40, [40],
(iii) Über Relaxationsschwingungen, *Jb. Drahtl. Telegr.* 28, 178–184, [41],
(iv) On relaxation-oscillations, *Philosophical Magazine*, serie 7, (2) 978–992, [42].

The English version [42], the best known of his article, was the last to be published in the last two months of 1926. Van der Pol gave the explanation of Gérard-Lescuyer's electrodynamical paradox by showing that the relation between the potential difference v across the dynamo's terminals and the current intensity i through it may be represented by a nonlinear function. In other words, as we will see further, the tension-current characteristic (v, i) of the dynamo can be modeled by a nonlinear function.

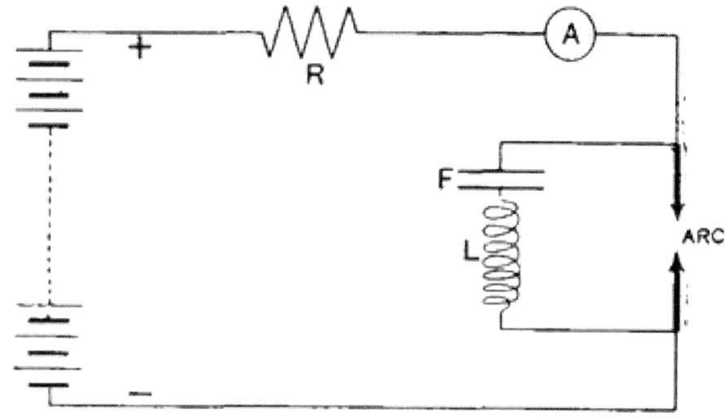

Fig. 5. Singing arc circuit diagram, [10, p. 248].

At the end of the nineteenth century a device, ancestor of the incandescent lamp, called electric arc was used for illumination of lighthouses and cities.[6] It exhibited independently of its low intensity light, a major drawback: the noise generated by the electrical discharge disturbed the residents. In London, the British physicist William Du Bois Duddell (1872–1917), was commissioned in 1899 by the British authorities to solve this problem. He had the idea of combining an oscillatory circuit composed of an inductor L and a capacitor of capacitance C (F in Fig. 5) with the electric arc and then stopped the rustling (see Fig. 5).

Following his investigations, [10, p. 248] reported in an article entitled *Musical Arc*:

> "A direct-current arc of suitable length and current between solid carbons, will give out a musical note if it be shunted with a condenser in series with a self-induction, as in Fig. 5 ..."

In fact, Duddell had designed an oscillatory circuit susceptible to produce sounds and more than that: electromagnetic waves. Thus, this apparatus was used as emitter and receiver for the wireless telegraphy till the advent of the triode. By producing spark, the *musical arc*, or *singing arc*, or *Duddell's arc*, generated electromagnetic waves highlighted by the experiments of [21]. In France, the engineer André Blondel (1863–1938) who was assigned to the Lighthouses and Beacons Service, to conduct research on the electric arc in order to improve the outdated French electrical system, made an extensive analysis of the musical arc [3]

[6]The electrical arc (artificial in contrast to the flash of lightning) is associated with the electrical discharge produced between the ends of two electrodes (e.g. carbon), which also emits light. It is still used today in cinema projectors, plasma and thermal metallurgy in "arc welding" or smelting (arc furnaces).

using Duddell's circuit (Fig. 5). He presented his main contribution on this subject in a fundamental paper entitled: "On the singing arc phenomenon" in which he pointed out the fact that the arc oscillations were "nearly sinusoidal" [3, p. 46]. Although the "audion" was invented in 1907 by Lee de Forest (1873–1961) [14], it was not until the First World War that it was widely distributed for military and commercial purposes under the name of "triode" coined by the British physicist William Eccles (1875–1966) ([12, p. 11]). In April 1919, the French engineer Paul Janet (1863–1937) exhibited an analogy between the *series dynamo machine*, the *singing arc* and the *triode*. [22, p. 764] wrote about Gérard-Lescuyer's experiment with the series dynamo machine:

> "It seemed interesting to me to report unexpected analogies between this experiment and the sustained oscillations so widely used today in wireless telegraphy, for instance, those produced by Duddell's arc or by the three-electrodes lamps used as oscillators. The production and the maintenance of oscillations in all these systems mainly result from the presence, in the oscillating circuit, of something analogous to a negative resistance[7]."

At the end of June 1920, the French engineer Jean-Baptiste Pomey (1861–1943) made use of a cubic function to represent the tension-current characteristic (v, i) of the *singing arc* [30, p. 380]. Three weeks later, [38, p. 703] employed the same type of modeling for the triode. Ten years later, on march 11th 1930, during a conference given at the École Supérieure d'Électricité (today Sup'Elec), [43, p. 300] claimed that the "nearly sinusoidal" oscillations produced by the *singing arc* represent one of the "very nice example of relaxation oscillations".

Thus, in the early 1930s, at least three analogous devices, namely the *series dynamo machine*, the *singing arc* and the *triode*, with nonlinear tension-current characteristic (v, i) were known. Let's focus on the main properties of the *singing arc*.

3.1. *Arc characteristics*

According to the British engineer Mrs Hertha Ayrton (1854–1923) the arc "static characteristic" (static because the voltage and currents are supposed to increase or decrease slowly during the experiments as it is the case when the oscillatory circuit LC is disconnected.), i.e. the tension-current characteristic (v, i) of the *singing arc* may be represented by an hyperbola [2]:

$$e = a + \frac{b}{i} \tag{5}$$

where a and b are constants depending on the conditions of the experiments (supply current, length of the arc, thermal conductivity and nature of the electrodes, as well as on the temperature and pressure of the surrounding gas...). The French physicist Henri Bouasse (1866–1953) plotted some arc "static characteristic" (DiH) and (ASC) as exemplified on Fig. 6:

As noticed by Bouasse, the curve must not be extended to the asymptote $i = 0$ since for small currents the above relation (6) is no longer valid (between O and A in Fig. 6).

3.2. *Arc hysteresis*

In the case of *Duddell's arc*, since an oscillatory circuit LC has been connected to the arc, an alternating current flowing out of this oscillatory circuit superposes itself on the direct current in the arc, and thus turns it into an oscillating current,

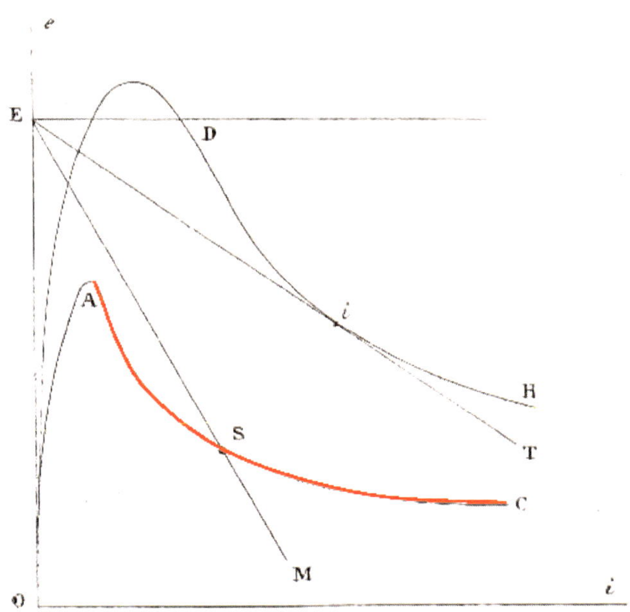

Fig. 6. Singing arc static characteristic, [5, p. 250].

[7]The concept of "negative resistance" was introduced by the German physicist Hans [23, p. 568].

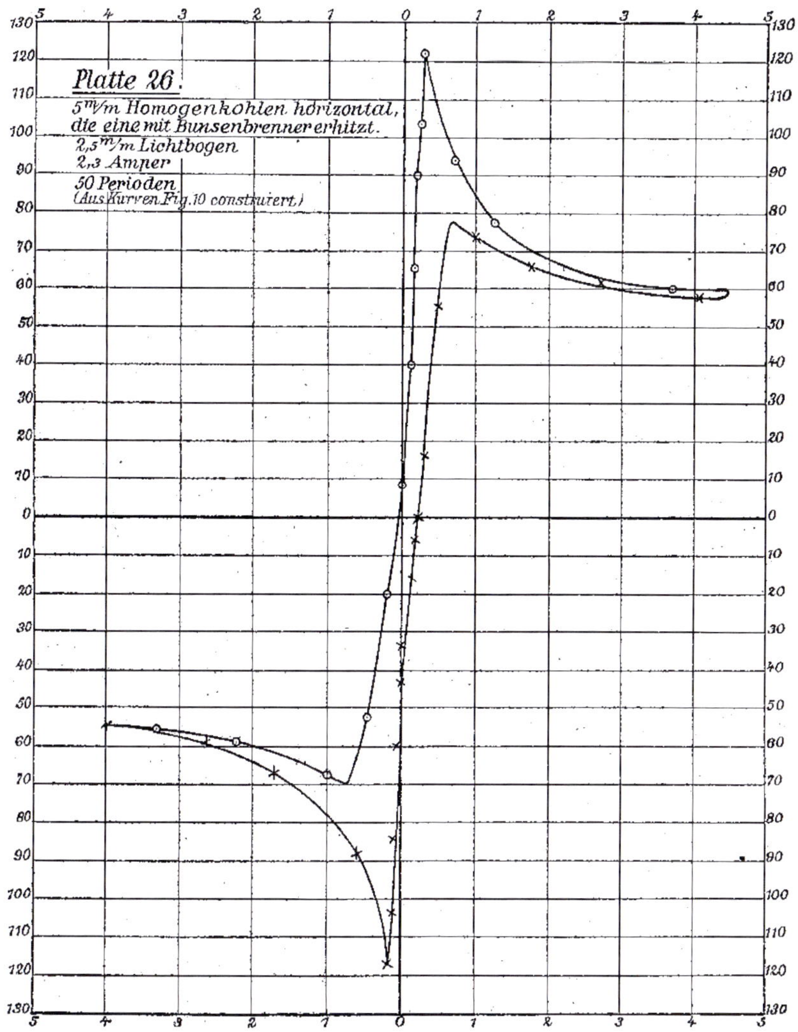

Fig. 7. Arc hysteresis, arc "dynamic characteristic", [35, p. 305].

which lags 180 degrees behind the current in the shunt circuit, i.e., it is increasing while the shunt current is decreasing, and *vice versa*. In 1905, the German physicist Hermann Theodor Simon (1870–1918) highlighted this effect that he called "Lichtbogenhysteresis", i.e. "arc hysteresis" [35, p. 305]. When the voltage and current are varied very rapidly (as it is the case when the oscillatory circuit LC is connected), the curve is called "dynamic characteristic" and the rising voltage curve is different from the falling one (see Fig. 7).

So, when the arc is functioning its "static characteristic" is no longer valid and must be replaced by its "dynamic characteristic". Figure 8 represents a "zoom" of the right top of the Fig. 7 and illustrates with Fig. 9 diagrammatically one form taken by the above curve under these conditions, i.e., when the oscillatory circuit LC is connected. Let's

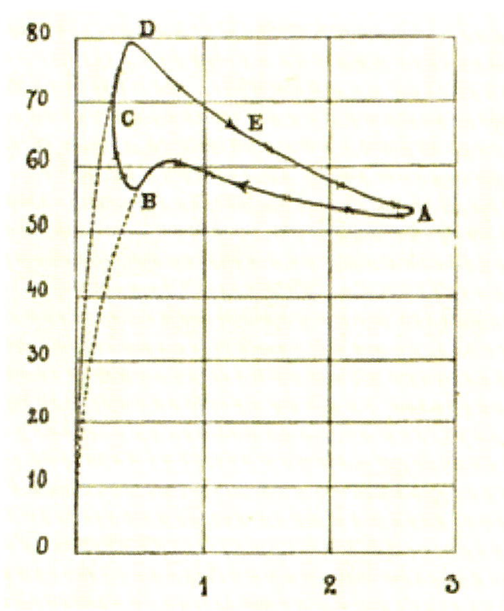

Fig. 8. Arc "dynamic characteristic", [33, p. 150].

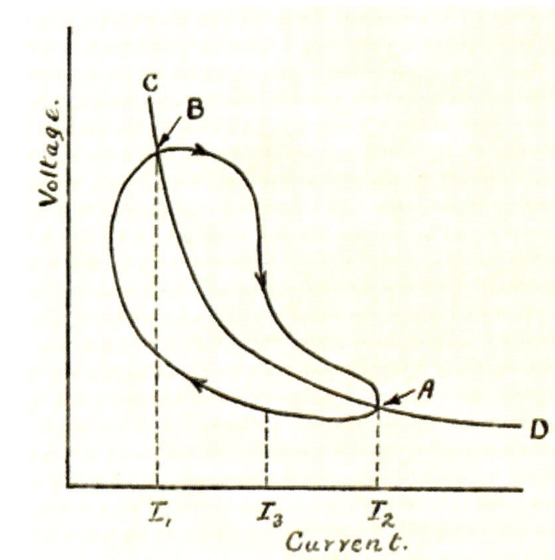

Fig. 9. Arc "dynamic characteristic", Arc "static characteristic" (CD), [9, p. 115].

notice that in this case the "dynamic characteristic" becomes a closed curve. Since both $v(t) > 0$ and $i(t) > 0$ in the dynamic regime, the hysteresis curve in this oscillation did not pass through the origin. This hysteresis loop is therefore not "pinched" [8] as would be the case if either $v(t)$ or $i(t)$ attains the "zero" value during each period.

This phenomenon is closely related to the fact that the number of ions existing between the electrodes depends upon the current and the temperature of the electrodes at the preceding instant, the number increasing as the current and electrode temperature increase. More precisely, "arc hysteresis" transcribes the fact that the temperature is lagging behind the current. Hence with rising current the number of ions is less at a given current value than with decreasing current at the same current value, and the voltage necessary to produce a given current is greater in the first case than in the second for the same reason. Let's consider the point A of Fig. 9, with a large current through the arc, the electrodes will be at a high temperature, and consequently there will be a great supply of ions to maintain the arc. If now the current is suddenly diminished from its value I_2 at this point towards some other lower value, say I_3, the temperature of the electrodes, and therefore also the supply of ions, will not *immediately* decrease to the values they would normally have for the current in question, since the electrodes will require a certain time interval for cooling down. The arc gap will therefore be in a more conducting state than usual, for the current I_3, and therefore the P.D. between its terminals, will be lower. Now, lets' consider the point B of Fig. 9 with a small current I_1 and an impoverished supply of ions. When the current is increased the electrode temperatures do not immediately respond to the change, so that consequent upon this the gap is less ionised than usual, and exhibits therefore a higher apparent resistance or what is the same thing, the P.D. between its terminals is higher than the usual value for the same current. So, we can say that if, the current in the circuit is alternately increased and decreased between two values corresponding to points A and B of the curve, the voltage will vary in opposite manner, that is, respectively decrease and increase. In other words, the ratio dV/dI of corresponding voltage and current variations is negative, which is often expressed in saying that the alternating-current resistance of the arc is negative (since the resistance R of a conductor is precisely equal to the ratio dV/dI of corresponding voltage and current variations).

Moreover, in a note at the French *Comptes Rendus* of the Academy of Sciences in which he established the second-order differential equation of the singing arc oscillations,[8] [4, p. 900] defined thus the "arc hysteresis" phenomenon:

> "Ce mot, bien que formé très correctement au point de vue étymologique, est devenu impropre pour l'arc par le fait qu'il est djà employé en magnétisme pour définir un phénomène de retard sensiblement indépendant de la fréquence, tandis qu'ici le retard va en décroissant rapidement quand la fréquence augmente.[9]"

4. Memristor: The Missing Circuit Element

As previously recalled, starting from the logical as well as axiomatic points of view, [6] was led to postulate the existence of a fourth circuit element to which he gave the name *memristor* since it behaves

[8]It has been stated by one of us [18] that [30, p. 380] had already stated this equation in 1920 by using a cubic function in order to represent the characteristic of the singing arc.

[9]"Although this word is correctly formed on an etymological point of view, it has became inappropriate for the arc because it has been already used in magnetism for defining a phenomenon of delay strictly independent of the frequency, while in this case the delay is decreasing quickly as the frequency increases."

like a nonlinear resistor with memory. By considering the three classical circuit elements, the *capacitor*, the *resistor*, the *inductor* and the "relationships" linking the variables involved, respectively: "Volta's law" $q = Cv$ (Eq. (1)), "Ohm's law" $v = Ri$ (Eq. (2)) and "Faraday's law" $\varphi = Li$ (Eq. (3)), he deduced, for the sake of *completeness*, that only one "relationship", i.e. only one law, remained undefined.

4.1. *Theoretical imperfections*

The philosopher of sciences Dudley [34] distinguished three kinds of scientific problems, first, problems related to scientific domain concerning for example the clarification and extension of the field, then the theoretical problems that require that one accounts, in a more profound manner, for a certain theory of the domain, and finally the *theoretical imperfections*. These, arise eventually when the theory was developed in response to a given theoretical problem. The theory can be revealed as incomplete with respect to the considered domain, be a simplification, or cause problems that Shapere called *incompleteness of the black box*. In fact, these three types of theoretical imperfections are distinguished from each other starting from the domain considered by the theory and the observed correspondence between theory and reality. Thus, if we consider a scientific domain as a whole corpus of information that a theory must account, we can say that a theory can be revealed as incomplete with respect to the considered domain if, on one hand there is no reason to suppose that the theory is not fundamentally correct and, on the other hand there is no reason to suppose that this theory can not be completed so that it can account for an unexplained part of the corpus.

Following this idea of *completeness*, [6] postulated the existence of a fourth circuit element in addition to the three others that he called it *memristor*. Then, he established the "relationship" between the magnetic flux φ and the charge q:

$$\varphi = Mq \tag{6}$$

We suggest to call this relationship: Chua's law. A that time, he also proposed a symbol (see Fig. 10).

4.2. *Memristor DC characteristics*

In order to emphasize a strong analogy between the memristor properties and that of the singing arc let's focus on the interesting properties 3-5

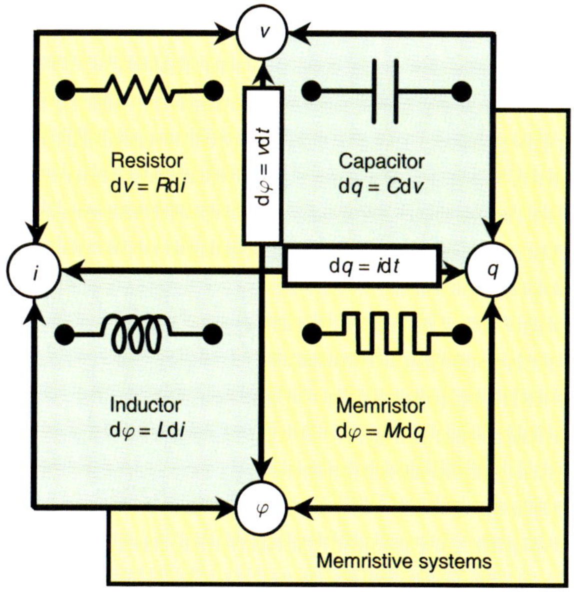

Fig. 10. The four fundamental two-terminal circuit elements, [36, p. 80].

established by Prof. Chua in his second publication on this subject [7, p. 211 and next].

Let's begin with the:

> "*Property 3–DC Characteristics*
> A time-invariant current-controlled memristive one-port under dc operation is equivalent to a time-invariant current-controlled nonlinear resistor if $f(x, I) = 0$ has a unique solution $x = X(I)$ such that for each value of $I \in \mathbb{R}$, the equilibrium point $x = X(I)$ is globally asymptotically stable."

[7, p. 211] then observed that a time invariant memristive one-port under dc operation behaves just like a nonlinear resistor and took as an example the dc characteristics of a short neon tube (Fig. 11).

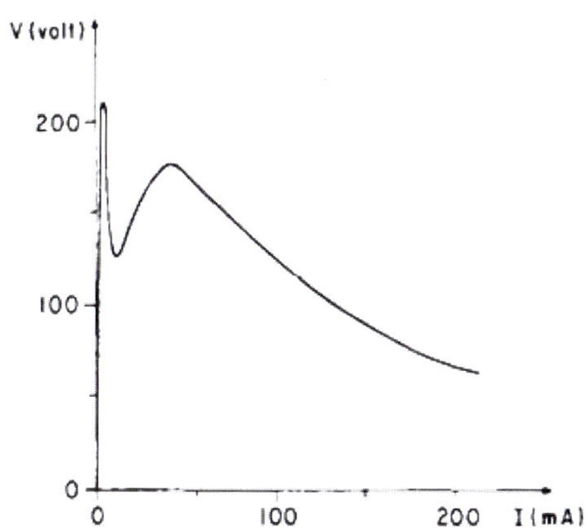

Fig. 11. The dc characteristics of a short neon tube, [7, p. 211].

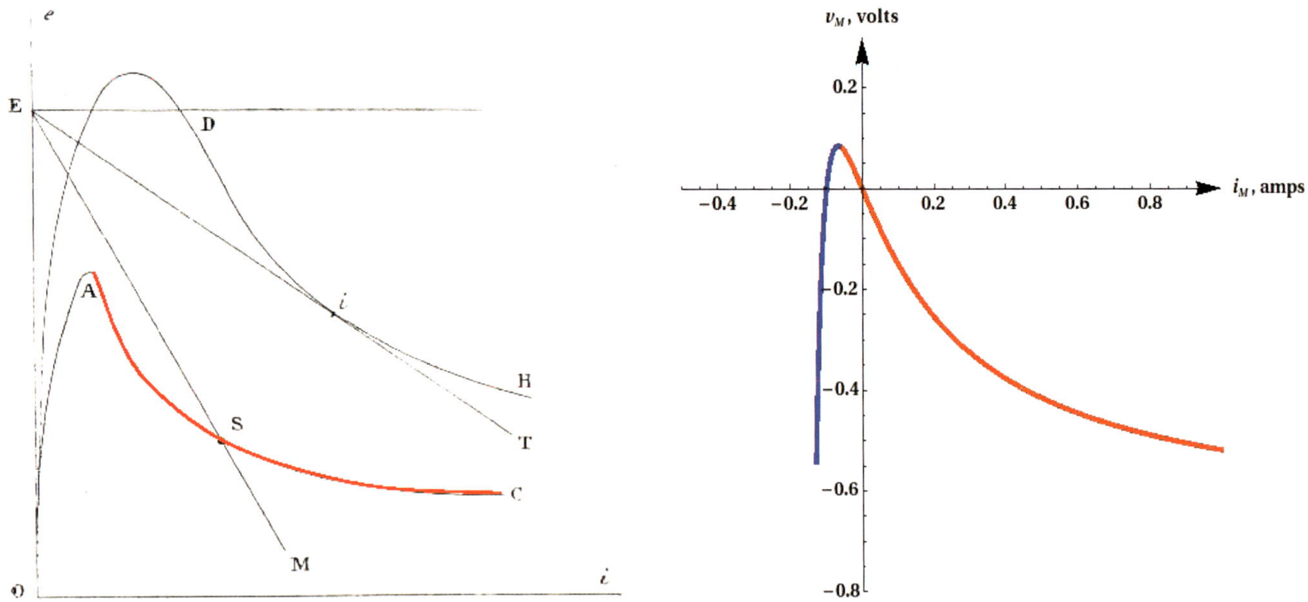

(a) Singing arc, [5, p. 250]

(b) Memristor, [24, p. 1574]

Fig. 12. The DC $v_M - i_M$ curve for the singing arc and for the memristor.

Starting from 25 mA Fig. 6 resembles to Fig. 11. However, such similarity as been already pointed out by [42, p. 992] who stated that the neon tube was another example of relaxation oscillator as well as the *singing arc* and the *triode*.

If we compare now the singing arc static characteristic Fig. 6 with the DC $v_M - i_M$ curve of the memristor (Fig. 12) plotted in a recent publication of [24, p. 1574] the comparison between both curves shape is obvious. Especially if you take into account the legend which specifies that the locally-active region highlighted in red on Fig. 12 begins far from the origin exactly like in the case of the singing arc (Cf. supra).

But, we can go further by comparing the equation of each DC $v_M - i_M$ curve. For the singing arc it has been given by Eq. (5). In their paper [24, p. 1580] gave for the memristor the following equation:

$$v_M = i_M \left(-1 + \frac{i_M^2}{(i_M + \alpha)^2} \right) \beta \tag{7}$$

This Eq. (7) may be written as:

$$v_M = i_M \left(-1 + \frac{1}{\left(1 + \frac{\alpha}{i_M}\right)^2} \right) \beta$$

A Taylor series expansion of the ratio up to order three reads provided that $\alpha/i_M \ll 1$:

$$\frac{1}{\left(1 + \frac{\alpha}{i_M}\right)^2} = 1 - \frac{2\alpha}{i_M} + \frac{12\alpha^2}{i_M^2} + h.o.t.$$

Thus, we find for the DC $v_M - i_M$ curve of the memristor:

$$v_M = -2\alpha\beta + \frac{12\alpha^2}{i_M} + \ldots \tag{8}$$

Since Eqs. (5) and (8) are identical, we can conclude that the singing arc "static characteristic" corresponds exactly to Chua's DC $v_M - i_M$ curve of the memristor.

4.3. *Memristor: An hysteretic system*

According to [7, p. 209] memristive systems are hysteretic. In order to study this effect let's focus on the

> "*Property 4-Double-Valued Lissajous Figure Property* A current-controlled memristive one-port under periodic operation with $i(t) = I cos\omega t$ always gives rise to a $v - i$ Lissajous figure whose voltage u is at most a double-valued function of i

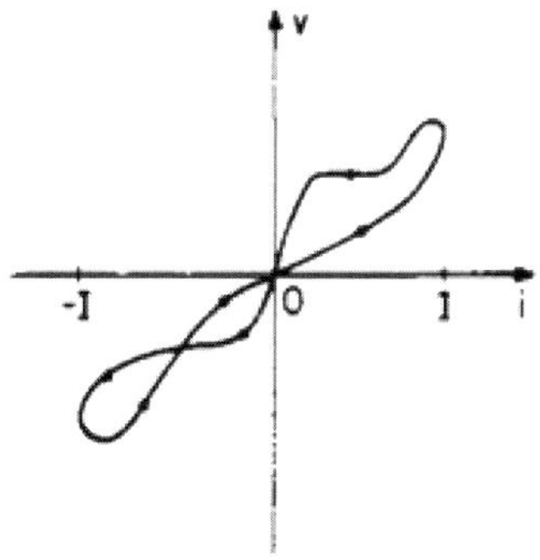

Fig. 13. Hysteretic effect, [7, p. 212].

Precisely, it has been stated for the singing arc "dynamic characteristic" that the rising voltage curve is different from the falling one. By plotting any parallel to the $v - axis$ it is easy to check on Figs. 7–9 that such a curve is double-valued. Moreover, the closed curve used by [7, p. 212] to emphasize this property and plotted on Fig. 13 is analogous to the "dynamic characteristic" presented on Fig. 7.

Property 5 was summarized by [7, p. 2132] by saying that "the $v - i$ curve is odd symmetric with respect to the origin." It is the case of Fig. 6 if we model it with a cubic function.

Then, let's compare now this sentence from Chua's conclusions [7, p. 220]

> "Various generic properties of memristive systems have been derived and shown to coincide with those possessed by many physical devices and systems. Among the various properties of memristive systems, the frequency response of the Lissajous figure is especially interesting. As the excitation frequency increases toward infinity, the Lissajous figure shrinks and tends to a straight line passing through the origin-except for some pathological cases where the bibs[10] stability property is not satisfied. The physical interpretation of this phenomenon is that the system possesses certain inertia and cannot respond as rapidly as the fast variation in the excitation waveform and therefore must settle to some equilibrium state. **This implies that the hysteretic effect of the memristive system decreases as the frequency increases** and hence it eventually degenerates into a purely *resistive* system."

and that of [4, p. 900] in which he claimed that "hysteresis effect is decreasing quickly as the frequency increases".

[10]bounded-input bounded-state.

At last, let's notice that the Duddell's singing arc circuit in Fig. 5 can be proved via Circuit theory principles, to be equivalent to the simplest chaotic circuit as exemplified in [24] paper. Since one more state variable is needed, in addition to L and C in Duddell's circuit and since one nonlinearity is also needed to obtain chaos as in [24] article, and since Duddell's circuit exhibits "hissing"sounds, and, "noise-like" chaotic waveforms as observed with an oscilloscope by [3], it follows that the Duddell's singing arc circuit in Fig. 5 is experimentally a chaotic circuit, and hence the singing arc must be a memristor.

5. Discussion

In this work, we have first recalled the history of the three classical circuit elements namely *capacitor*, *resistor*, *inductor* and the relationships, i.e. *Volta's law*, *Ohm's law* and *Faraday's law* involving the four fundamental circuit variables: the *current*, the *voltage*, the *charge* and the *magnetic flux* in order to highlight the *incompleteness* of the Circuit Theory which lead Professor Chua to postulate the existence of a fourth element he called *memristor*. Then, we focused on the properties of an old nonlinear device called *singing arc*, originally used in Wireless Telegraphy before the invention of the triode. By comparing the *singing arc* to the *memristor* we have highlighted an analogy between some of their properties. The *memristor* DC characteristic curve has exactly the same hyperbolic profile and the same equation as the *singing arc* "static characteristic". Moreover, both *memristor* and *singing arc* are hysteretic systems exhibiting the same kind of closed curve in order to describe the "hysteresis effect". So, we conclude that the *singing arc* could be considered as the oldest *memristor*. We believe that the hysteresis characteristic in Fig. 7 should have intersected each other at the origin, forming a "pinched" hysteresisloop. We therefore conjectured that the slight discrepancy of the "almost pinched" characteristic near the origin is due on the one hand to the imperfection of instruments used in this measurement and, on the other hand to the great variability of the conditions of the experiments.

Acknowledgments

Authors would like to thank Dr. Bharathwaj Muthuswamy for his helpful advices.

References

[1] Ampère, A. M. [1820] "Mémoire sur les effets des courants électriques," *Annales de chimie et de physique*, Vol. 15, 59–74 & 170–218.

[2] Ayrton, H. [1895] "Variation of Potential Difference of the Electric Arc with Current, Size of Carbons, and Distance Apart," *The Electrician*, Vol. XXXIV, 335, 364, 399, 471, 541, 610; 1895, Vol. XXXV, 418, 635, 743; 1895, Vol. XXXVI, 36, 225; 1896, Vol. XXXVI, 539.

[3] Blondel, A. [1905] "Sur les phénomènes de l'arc chantant," *Éclairage Électrique*, **44** (28), 41–58 and 81–104.

[4] Blondel, A. [1926] "Contribution à la théorie de l'arc chantant musical," *Comptes rendus hebdomadaires des séances de l'Académie des sciences*, **82** (12 avril 1926), 900–903.

[5] Bouasse, H. [1924] *Oscillations électriques*, (Librairie Delagrave, Paris).

[6] Chua, L. O. [1971] "Memristor — The Missing Circuit Element," *IEEE Transactions on Circuit Theory* **18**(5), 507–519.

[7] Chua, L. O. & Kang S. M. [1976] "Memristive devices and systems," *Proceedings of the IEEE* **64**, 209–223.

[8] Chua, L. O. [2011] "Resistance switching memories are memristors," *Applied Physics A*, Vol. A 102, 765–783.

[9] Coursey, P. [1919] *Telephony Without Wires*, (The Wireless Press Ltd, London).

[10] Du Bois Duddell, W. [1900] "On Rapid Variations in the Current through the Direct-Current Arc," *Journ. Inst. Elec. Eng.* **30**, pp. 232–283, (December 13th 1900).

[11] Du Bois Duddell, W. [1900] "On Rapid Variations in the Current through the Direct-Current Arc," *Journ. Inst. Elec. Eng.* **46**, 269–273 & 310–313, (December 14th & 21th 1900).

[12] Eccles, W. [1919] "An Investigation of the Internal Action of a Triode Valve," *The Radio Review*, 11–20, 67–77 & 127–134.

[13] Faraday, M. [1832] "Experimental Researches in Electricity," *Phil. Trans. R. Soc. Lond.*, (122) 125–162.

[14] de Forest, L. [1907] "Oscillation-responsive device," *US Patent* n$^\circ$ 836070 (1906) and "Wireless telegraphy," *US Patent* n$^\circ$ 841386 (1907).

[15] Gérard-Lescuyer, J. M. A. [1880] "On an electrodynamical paradox," *Philosophical Magazine*, V, **10**, 215–216.

[16] Gilbert, W. [1600] *De Magnete, Magneticisque Corporibus, et de Magno Magnete Tellure*, translation by Silvanus Phillips Thompson and the Gilbert Club; limited to 250 copies. (Chiswick Press, London).

[17] Ginoux, J. M. [2011] *Analyse mathématiques des phénomènes oscillatoires non linéaires*, (2nd) Ph-D Thesis, Université of Paris VI.

[18] Ginoux, J. M. [2011] *Blondel et les oscillations auto-entretenues*, submitted at *Archive for History of Exact Sciences.*

[19] Hawksbee, F. [1709] *Physico-Mechanical Experiments on Various Subjects*, 2nd Ed., London.

[20] Hoppe, E. [1928] *Histoire de la physique*, translation from German by Henri Besson (Payot, Paris).

[21] Hertz, H. R. [1887] "Ueber sehr schnelle electrische Schwingungen," *Annalen der Physik*, **Vol. 267** (7), 421–448.

[22] Janet, P. [1919] "Sur une analogie électrotechnique des oscillations entretenues," *Comptes Rendus Hebdomadaires de l'Académie des Sciences* **168**, 764–766.

[23] Luggin, H. [1888] "Versuche und Bemerkungen über den galvanischen Lichtbogen," *Centralblatt für Elektrotechnik*, **10**, 567–581.

[24] Muthuswamy, B. & Chua, L. O. [2010] "Simplest chaotic circuit," *International Journal of Bifurcation and Chaos*, Vol. 20, No. 5, 1567–1580.

[25] Nollet, J. F. [1753] *Essai sur l'électricité des corps*, 3rd Ed. (Chez les Frères Guerin, Paris).

[26] Ohm, G. [1826] "Bestimmung des Gesetzes, nach welchem Metalle die Contaktelektricitt leiten, nebst einem Entwurfe zu einer Theorie des Voltaischen Apparates und des Schweiggerschen Multiplicators," *Journal fr Chemie und Physik*, Vol. 46, 137–166.

[27] Ohm, G. [1826] "Versuch einer Theorie der durch galvanische Krfte hervorgebrachten elektroskopischen Erscheinungen," *Annalen der Physik und Chemie*, Vol. 6, 459–469.

[28] Ohm, G. [1827] *Die galvanische Kette, mathematisch bearbeitet*, (Riemman, Berlin).

[29] Pancaldi, G. [2003] *Volta, Science and Culture in the Age of Enlightenment*, (Princeton Univ. Press).

[30] Pomey, J. B. [1920] *Introduction la théorie des courants téléphoniques et de la radiotélégraphie*, (Gauthier-Villars, Paris).

[31] Pourprix, B. [1989] "La mathématisation des phénomènes galvaniques par G. S. Ohm (1825–1827), La mathématisation 1780–1830," *Revue d'Histoire des Sciences*, 42 (1–2) 139–154.

[32] Pourprix, B. [1995] "G. S. Ohm théoricien de l'action contigue," *Archives Internationales d'Histoire des Sciences*, 45 (134) (1995), pp. 30–56.

[33] Ruhmer, E. [1908] *Wireless Telephony*, (Van Nostrand Company, New York).

[34] Shapere, D. [1974], "Scientific Theories and Their Domain," in F. Suppe, The Structure of Scientific Theories, University of Illinois Press, Urbana (IL), pp. 518–565.

[35] Simon, H. Th. [1905] "Über die Dynamik der LichtbogenVorgänge und über Lichtbogenhysteresis," *Physikalische Zeitschrift*, **6**, 297–319.

[36] Strukhov, D. B., Snider, G. S., Stewart, G. R. & Williams R. S. [2008] "The missing memristor found," *Nature*, **453**, 80–83.

[37] Thompson, S. [1901] *Michael Faraday, His Life and Work*, (Cassell and Company, London).

[38] Van der Pol, B. [1920] "A theory of the amplitude of free and forced triode vibrations," *Radio Review* (London), **1**, 701–710 and 754–762.

[39] Van der Pol, B. [1926] "Over Relaxatie-trillingen," *Physica*, **6**, 154–157.

[40] Van der Pol, B. [1926] "Over "Relaxatie-trillingen"," *Tijdschrift van het Nederlandsch Radiogenootschap*, **3**, 25–40.

[41] Van der Pol, B. [1926] "Über "Relaxationsschwingungen"," *Jahrbuch der drahtlosen Telegraphie und Telephonie*, **28**, 178–184.

[42] Van der Pol, B. [1926] "On "relaxation-oscillations"," *The London, Edinburgh, and Dublin Philosophical Magazine and Journal of Science*, **7**(2), 978–992.

[43] Van der Pol, B. [1930] "Oscillations sinusoïdales et de relaxation," *Onde Électrique*, **9**, 245–256 & 293–312.

[44] Volta, A. [1800] "On the Electricity Excited by the Mere Contact of Conducting Substances of Different Kinds. In a Letter from Mr. Alexander Volta, F. R. S. Professor of Natural Philosophy in the University of Pavia, to the Rt. Hon. Sir Joseph Banks, Bart. K. B. P. R. S.," *Phil. Trans. R. Soc. Lond.*, (1800) (90), 403–431.

[45] Volta, A. [1816] *Collezione dell' Opere del Cavaliere Conte Alessandro Volta*, 3 Volumes, Firenze.

CHAPTER 41

TWO CENTURIES OF MEMRISTORS

THEMISTOKLIS PRODROMAKIS

Centre for Bio-Inspired Technology,
Department of Electrical and Electronic Engineering,
Imperial College London, London SW7 2AZ, UK

Nanoscale implementations of memory-resistors are currently showcased as prospective solutions to emerging memory and unconventional computing. Whilst the theoretical background was established exactly 40 years ago, most researchers became aware of the memristor after HP's work on TiO2 thin films. Nonetheless, memristive attributes are not something new. In this chapter we present historical evidence on the electric arc, which we believe is the first ever-reported memristor. The first experiments on the electric arc are ascribed to Sir Humphry Davy, just one year after Volta's discovery of the battery, classifying the memristor as a two hundred year old device.

1. Introduction

The realization of a nanoscale memory-resistor by Hewlett Packard in 2008 [1] came almost 40 years after its theoretical inception [2] and ever since it has precipitated an enormous interest on emerging applications that leverage on the dynamic nature of such devices. Nonetheless, as the research community became acquainted with memristive attributes, it became evident that these unconventional characteristics have been to a great extent observed and documented before. The functional properties of memristors were first documented by Chua [2] and later on Chua and Kang [3], with their main fingerprint being a pinched-hysteresis loop when subjected to a bipolar periodic signal. This particular signature has been explicitly observed in a number of devices for more than one century now while it can be extrapolated for devices that appeared as early as the dawn of nineteenth century.

2. Hysteresis of Memristors Due to Rate-dependent Discharge Mechanisms

Hysteresis is typically noticed in systems/devices that possess certain inertia, causing the value of a physical property to lag behind changes in the mechanism causing it; manifesting memory [4]. Such causes are typically associated with irreversible rate-dependent electro- or thermo-dynamic changes that are contingent on both the present as well as the past environment. Particularly in the case of nanoscale memristors, this inertia can be associated with the displacement of mobile ions or oxygen vacancies [5, 6], the formation and rupture of conductive filaments [7, 8] or even the phase-transitions of an active core [9, 10]. And despite the fact that these mechanisms have a more substantial effect in lamella devices, similar attributes can also be supported by considerably larger ionic systems [11], contingent upon the extend of the stimulating cause, the nature of the pertinent ions and the barrier medium that governs their kinetics.

Essentially, this signature memristor property is quite characteristic for all devices in support of a discharge phenomenon, no matter what the conveying medium is. For example, naturally occurring ionic systems, such as biological ion-channels [12, 13], allow the passive displacement of ions due to an electrochemical gradient along the cells membrane. The extend of this electrochemical gradient sets the probability of an ion-channel being in an open or closed state and in consequence governs the amount of ions that are conveyed through the channel. The intra- and extra-cellular current conduction per channel is governed by distinct rate-limiting dynamics and therefore a pinched-hysteresis loop should inherently appear when an appropriate periodic signal is applied [14]. Such hysteretic effects have been deliberately omitted from past publications as being errors provoked by the employed instrumentation set-ups, while very early investigations only resolved the static characteristics of such systems. Recent studies however, have identified that ion-channels can genuinely undergo a hysteresis in their voltage dependence during normal gating. This is clearly illustrated in Fig. 1 for hyperpolarization-activated, cyclic nucleotide-gated (HCN) ion-channels [15], attributed to the rate-limiting conformational changes of the channels pore during gating charge movement [16].

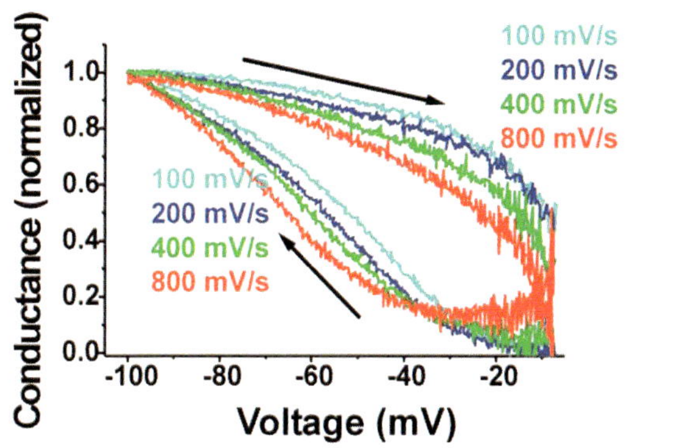

Fig. 1. Conductance of spHCN channels during voltage ramps from −10 mV to −100 mV, and back to −10 mV, using different ramp speeds. Reproduced with permission from [15].

Chemical diffusion and ion migration are considered as the facilitating mechanisms for most functional properties of organic and inorganic systems; both of them being in essence distinct forms of ionic discharge. Hence, almost all naturally occurring and artificially made systems can be classified as being memristive; not a surprise that memristance has also been observed in human blood [17]! Ipso facto, memristors should not be considered as being a novel subject that has just emerged, notwithstanding the tremendous abundance of space for innovation on novel architectures [18], materials [19], models [20] and applications [21, 22] that exploit such effects at the nanoscale, which the scientific community is currently engaging in.

3. Discharge Tubes can be Classified as Memristors

To further prove this, our quest for discovering the earliest artificially made memristor is innately focused on historical devices that support ionic-discharge, particularly when luminous effects accompany this discharge. Over the past century, discharge tubes and low-pressure lamps have been extensively scrutinized, facilitating the first experimental studies on the behavior of atoms and electrons that have been efficiently translated into a multitude of innovations, particularly in lighting. Discharge lamps are recognized as being rather dynamic since the time required for ionization and deionization to take place, depends not only upon the instantaneous current flow, but also upon the current, which has previously flowed through it as well as the rate of change of current. When a potential difference (PD) is applied across a gas-discharge tube that is sufficiently larger than the static ignition potential of the tube, the current flow increases rapidly. However, the deionization time is observed to be substantially larger than the time that it takes to decrease this current flow abruptly, rendering an increase in the tubes resistance as deionization occurs. This dynamic response was especially observed for two types of glow discharge tubes, as depicted in Fig. 2 by Reich and Depp [23] in 1938.

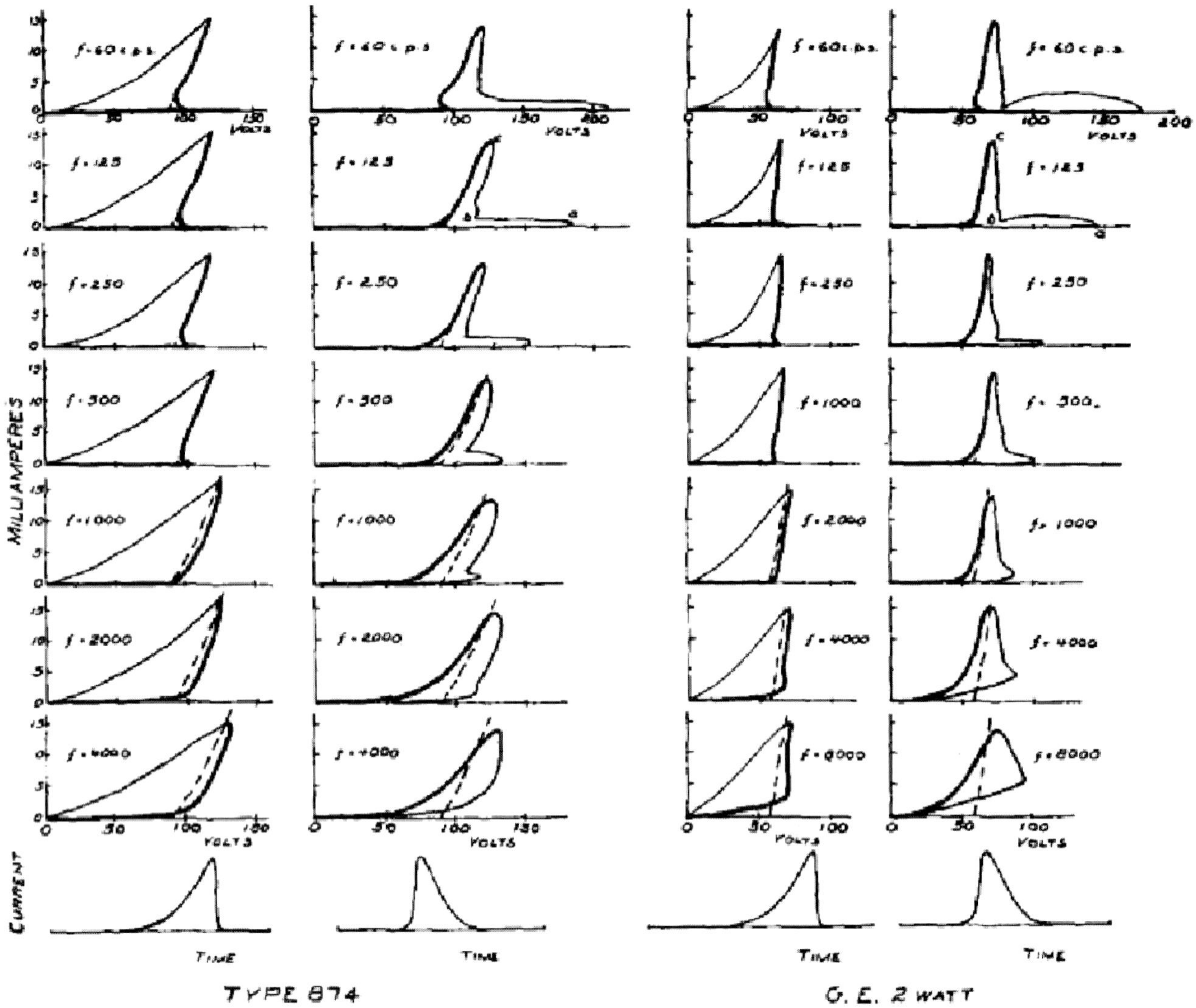

Fig. 2. Dynamic characteristics for two types of tubes; current amplitude constant frequency and current waveform varied; rod of type 874 tube is used as cathode. Reproduced with permission from [23].

This has been one of the earliest known demonstrations on the vacuum tubes capacity to act as variable resistors, whose magnitude can be controlled by means of voltage applied on a control circuit. The resemblance of these results with the I–V characteristics observed for example in the case of nanoscale TiO2 memristors [1] is apparent.

Gas is normally a poor conductor of electricity, however at sufficiently high electric fields the molecules in the vicinity of the electrodes are ionized and are set in motion, allowing the conduction of small currents. These ionized molecules predominantly support the establishment of current percolation paths in a rather similar manner to nanoionic systems [24] that are classified as memristors, including resistive switching elements [25, 26], quantum-point contact switches [27] and polymeric devices [11]. Despite the fact that the intrinsic mechanism differs substantially in the aforementioned systems, they all exhibit a rather similar dynamic response that is often exemplified by a pinched-hysteresis loop [25].

An excellent analysis of the dynamic characteristics of devices exhibiting discharge phenomena was presented by V.J. Francis [28], where alike dynamic characteristics were observed for tungsten filaments, high-pressure mercury-vapour lamps, low-pressure mercury tubes, discharge-tubes and sodium tubes as illustrated in Fig. 3 respectively. The operation of these devices relies upon distinct configurations, with the discharge phenomenon being expressed in rather dissimilar environments. Yet, the cause of the observed hysteresis relies upon the same physical ionic mechanisms that are veritably supported even down to the nanoscale. Francis has correctly attributed the cause of this

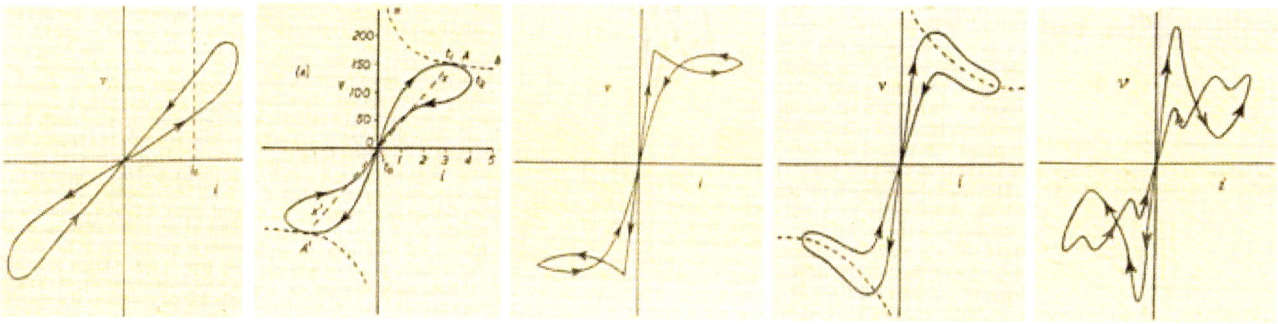

(a) (b) (c) (d) (e)

Fig. 3. Dynamic characteristics of (a) tungsten filament, (b) high-pressure mercury-vapour lamp, (c) low-pressure mercury tube, (d) discharge tube, (e) sodium tube. Reproduced with permission from [28].

dynamic response to rate-limiting effects associated with the establishment and sustaining of ionic discharge. These loops are typically observed when a device is excited by a periodic PD, causing the maxima and minima of this PD to be out of phase with the corresponding maxima and minima of the current flowing through the device. Representative phase plots are reproduced in Fig. 4 that demonstrate this phenomenon for a low-pressure mercury lamp serially connected with (a) a capacitor, b) a resistor and (c) an inductor. It can be observed that in all cases the tubes current and voltage have identical zero crossings in time, implying a type-II [4,29] pinched hysteresis loop.[a]

4. Discharge at Different Length Scales

The larger the distance between the electrodes, i.e. the barrier across which the PD is applied, the larger the PD must be in order to establish discharge. On the other hand, when the separation gap is miniscule, a few volts are sufficient for mobilizing ions across this barrier, which is also the case in nanoscale implementations of memristors. And one should be able to steadily sustain this effect as the barrier width is enlarged, provided that the PD across the device is enhanced appropriately. In fact the effect on the barrier width was thoroughly studied for a predecessor of discharge tubes, also known as the electric arc [30]. The electric arc has monopolized the interest of researchers during the nineteenth century, setting the foundations of the lighting industry. In similar fashion, the functional property of the electric arc depends upon a discharge phenomenon that has been mainly observed for carbon electrodes placed a few millimetres apart with a relatively large PD being applied across them. Quoting De La Rue [31] the stratified discharge in a vacuum tube is simply a magnified form of the electric arc that is in essence a modification of the same phenomenon.

Hertha Ayrton was one of the first eminent female engineers (first female MIEE) that have extensively studied the electric arc [32]. As the electric arc was employed commercially since 1879,[b] Ayrtons work complemented the work of Swan and Edison towards addressing the effects of the electrodes degeneration due to oxygen coming into contact with the carbon electrodes that caused the hissing of the arc, as depicted in Fig. 5. She proved that if air was excluded from the arc, or when nitrogen was introduced in isolation, the hissing did not occur; eventually leading to the development of gas-discharge tubes. Most importantly, she was one of the first noticing that the characteristic curve of an alternating current (AC) arc shows lower voltages when the current is decreasing than when it is increasing that is dependent upon the supply current, the length of the arc, the thermal conductivity of the electrodes as well as the pressure of the surrounding gas. This rate-limiting response was later on exploited by Duddell [33]

[a]It should be noted that all pinched hysteresis loops appearing on Fig. 3 are not self-crossing as in the case of an ideal memristor [2]. This type of hysteresis is typically observed in thermistors and has been classified by Pershin and Di Ventra [4] as well as Biolek [29] as a type II pinched hysteresis loop.

[b]Mosley Street in Newcaste-upon-Tyne was the first street to be lit by electric light, after Joseph Swan installed an incandescent lamp on the 3rd February 1879.

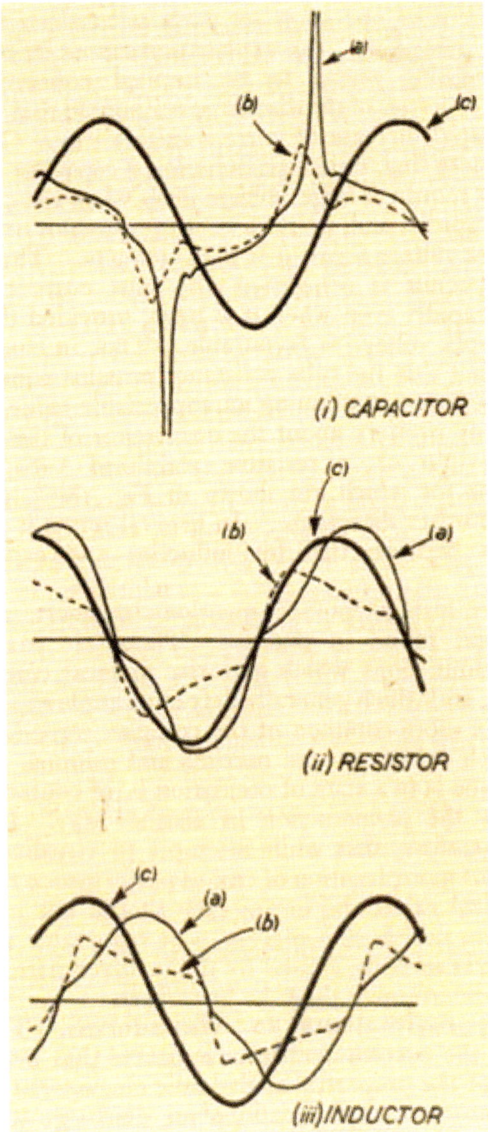

Fig. 4. Phase relations of (a) tube current, (b) tube voltage and (c) mains voltage for a low-pressure mercury lamp, with various stabilizers. Reproduced with permission from [28].

as well, who interpreted it as an oscillating arc; essentially demonstrating hysteresis. Duddell in [36] pointed that if it were possible to obtain perfectly pure carbon electrodes, then the resistance of the arc between them would be very high, so high that it might be impossible to maintain a true arc between them at all. He also expressed the opinion that traces of impurities, such as the vapours of the alkaline earths, are essential to provide the carriers of the electric charges in the vapour column, so as to render it conducting. Although this was mentioned more than a century ago, this statement was rather prophetic when considering that nowadays nanoscale memristors leverage intrinsic defects for establishing distinct conductive states.

Other studies exemplifying the rate-limiting response of the electric arc were carried out towards the fall of the nineteenth century, with Frith *et al.* [37] observing that the current flowing through a hissing arc was oscillatory. At the same time it was demonstrated that when an electric arc was driven with relatively high frequency AC currents, there was no effect on the overall resistance of the arc. Likewise, this was also noted eight years before Frith in a paper by Luggin [38], where he observed that when the current was alternated with sufficient rapidity, the PD would rise and fall in phase with the current; demonstrating a linear resistance.[c] If however the current was alternated at lower frequencies, the electric arc could in fact exhibit a negative resistance. Analogously, it is nowadays shown that nanoscale memristor implementations based on metal-oxides can demonstrate a negative differential resistance [39].

The physical mechanism of discharge in the electric arc is ascribed to thermionic emission, with the term thermionic mainly being used in the eighteen hundreds to describe the process of increasing the energy of ions by heating. The application of relatively large PD on carbon electrodes generate considerable heat that in turn cause the flow of charged carriers across the arcs barrier, allowing large enough current densities to be established. The number of emitted carriers increases in proportion with the temperature of the electrodes, as described by Richardsons law [40], which is also applicable for describing conduction mechanisms in solid-state Schottky barriers [41]. Quite a few scientists have demonstrated that air was unable to insulate when a PD was adequate enough to heat the bounding carbon electrodes, with Blondlot indicating that the leaking-currents obey a non-linear trend and thus do not comply with Ohms law [42].

Thermionic discharge is however applicable to thermally sensitive resistors, also known as thermistors that have been previously classified as memristors [3]. It is rather interesting to note that the first reported thermistor is attributed to the British physicist Michael Faraday in 1833, based

[c]This limiting linear-resistance signature of the memristor was predicted and proved in property 6 of [3]. As the frequency ω tends to a large value, the pinched hysteresis loop must tend to a straight line, whose limiting slope however, is not constant, but depends on the amplitude and waveform of the excitation.

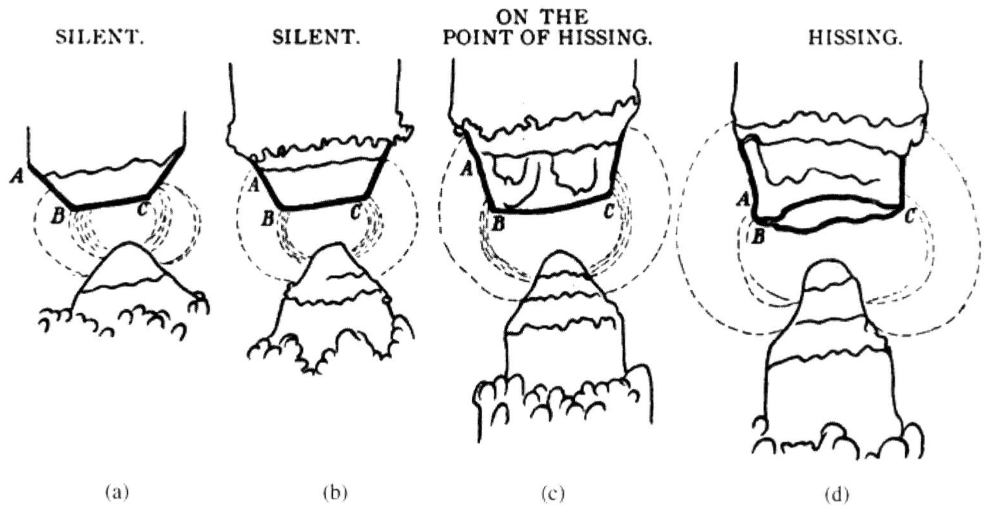

Fig. 5. The Hissing of the Electric Arc, as presented to the IEE in 1899 by Hertha Ayrton [32]. Her conclusion: Air causes hissing and inconstant light when it comes into contact with the carbon crater of the arc lamp. To the best of our knowledge, the noise-like waveforms associated with this hissing phenomenon presented two years later by Duddell in Fig. 7 [33] is the first chaotic phenomenon reported in electrical engineering, and possibly in all man-made devices. Therefore the hissing arc qualifies as the first chaotic electrical circuit, instead of the Van der Pol oscillator [34, 35].

on his study on the semiconducting behavior of Ag_2S. In his book [43], Faraday writes: "I have lately met with an extraordinary case which is in direct contrast with the influence of heat upon metallic bodies. On applying a lamp under the sulphuret between the poles, the conducting power rose rapidly with the heat, and at last the galvanometer needle jumped into a fixed position, and the sulphuret was found conducting in the manner of a metal. On removing the lamp and allowing the heat to fall, the effects were reversed, the needle at first began to vibrate a little, then gradually left its transverse direction, and at last returned to a position very nearly that which it would take when no current was passing through the galvanometer." What was clearly happening is that the PD applied across the poles was insufficient to allow current conduction across the barrier. Thermionic emission was eventually being facilitated by the thermal energy, provided by the lamp, that supplemented the missing energy necessary for exceeding the barrier's potential. Not surprisingly, Ag_2S has recently been employed to facilitate resistive switching in nanoscale elements [44] that demonstrate alike rate-limiting functionality [45] with what was observed almost two centuries ago by Faraday.

Ayrton in the preface of her book [30] notes: "From the first it seemed to me that the fact that the resistance of the material in the gap between the carbons must not only depend upon the current, but that it must depend upon it in many apparently contradictory ways, could not but lead to curious complications in the relation between the PD and the current." But who was the first one to observe this peculiar response, which without argument is the signature property of memristors? No one else but the inventor of the electric arc, Sir Humphry Davy,[d] who was also Faradays mentor.

5. The Oldest Known Memristor

One of the greatest scientific breakthroughs of mankind occurred in 1800, with Alessandro Volta [46] establishing the voltaic pile, comprising a number of alternating discs of zinc and copper with pieces of cardboard soaked in brine between the metals that enabled the establishment of unprecedented PDs. Humphry Davy leveraged the voltaic pile for observing the effect of large currents on: (1) living things [47], (2) chemical decomposition [48] and (3) heating and producing sparks [49]. In his letter printed on Nicholson's Journal

[d]Bonaparte was quite an admirer of Davy's work and although at that time England and France were at war, Napoleon awarded him with The Prix Napoleon de l'Institut. It is interesting to note that Napoleon invited Davy to France for receiving his award, for which he arranged a special visa and covered all the necessary costs, including the cost of a research assistant who was Faraday!

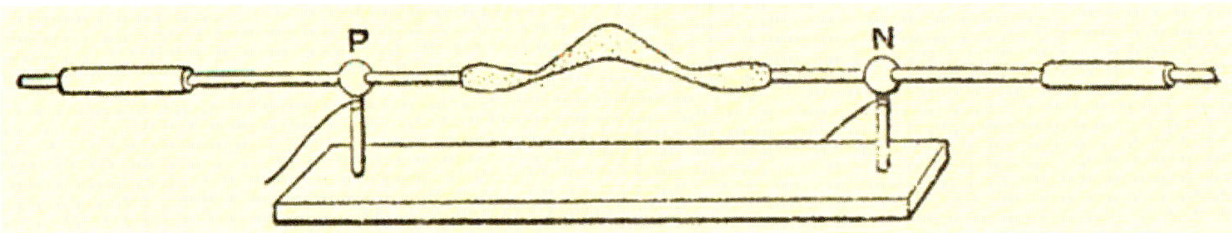

Fig. 6. Horizontal arc, copied from Fig. 18 in Davy's Elements of Chemical Philosophy. Reproduced with permission from [53].

[49] dated September 22nd 1800, Davy writes: "The earlier experimenters on animal electricity noticed the power of well burned charcoal to conduct the common galvanic influence. I have found that this substance possesses the same properties as metallic bodies in producing the shock and spark when made a medium of communication between the ends of the galvanic pile of Signor Volta." Later on, in a lecture before the Royal Institution, given in 1801 [50], Davy describes some experiments on the spark yielded by the pile, stating: "When, instead of the metals, pieces of well-burned charcoal were employed, the spark was still larger and of a vivid whiteness." And again: "The spark is most vivid when the charcoal is hot." Clearly, this is not the definition of the electric arc, but of a spark, since the electric arc requires the establishment of a continuous luminous filament without intermittent gaps.

Davy's failure of acknowledging his discovery of the electric arc from early on was due to the lack of a sufficiently large PD that would have enabled him to initiate and sustain a continuous luminous filament from a de-ionized state. In 1808 Davy requested the purchase of a large galvanic battery, comprising 2000 pairs of copper and zinc plates serially connected, that was installed in the Royal Institution on July 11 of the same year [51]. In his experiment in the Royal Institution in 1808, as recounted in the Philosophical Transactions for 1810 [52], Davy demonstrates a functional electric arc; charcoal, even when ignited to whiteness in oxymuriatic or muriatic acid gases, by the Voltaic battery, effects no change in them; if it has been previously free from hydrogen and moisture by intense ignition in vacuo. This was better comprehended though in his 1812 book Elements of Chemical Philosophy [53], where he writes: "When pieces of charcoal about an inch long and one sixth of an inch in diameter, were brought near each other a bright spark was produced, and more than half the volume of the charcoal became ignited to whiteness, and by withdrawing the points from each other a constant discharge took place through the heated air, in a space equal at least to four inches, producing a most brilliant ascending arch of light, broad, and conical in form in the middle." Figure 6 is reproduced from Davys book and illustrates his experimental set-up; serving as the first schematic representation of an electric arc (arch), published exactly two centuries ago.

This two-terminal device is essentially the same that Davy employed in his early experiments, however the lack of sufficiently large PD prevented the earlier discovery of the electric arc. Yet, Davy had already established the electric arc in late 1800, nearly at the same time that Volta discovered the battery, whose series connections enabled Davy to establish a sustained electric arc, which rightfully qualifies as the earliest known memristor. Although, Davy originally exploited this incandescent device for electrolysis [54] his contributions are far more reaching, as we now know that he set the foundations for lighting but also analogue computing, both in the early days when vacuum tubes were used [55] and hopefully in the future as memristor-based computing emerges.

6. Conclusion

Just as Columbus did not invent America, the memristor is not an invention. Rather it is a description of a basic phenomenon of nature that manifests itself in various dissipative devices, made from different materials, internal structures and architectures. We can therefore expect to find many such devices in the future, both in the nano as well as macroscale. And as we do so, we should not neglect the immense developmental phase that over the past two centuries has matured the electric arc into reliable and robust devices, baring potential solutions towards developing resilient nanoscale memristors. We end this historical narrative by noting that even though the memristor has seen its light of joy only in 2008, and has been recognized as the fourth circuit element [56] along with the resistor, capacitor and inductor, it actually predates the resistor which was formally published by Ohm [57] in 1827, and the inductor, which was formally published by Faraday in 1831 [43].

Acknowledgments

The authors wish to acknowledge the financial support of Dr Wilf Corrigan, the CHIST-ERA ERA-Net, the Lindemann Trust and the Royal Academy of Engineering. We would also like to acknowledge the assistance of Mrs. E.B. Haigh in tracking historical evidence to support this work.

Appendices

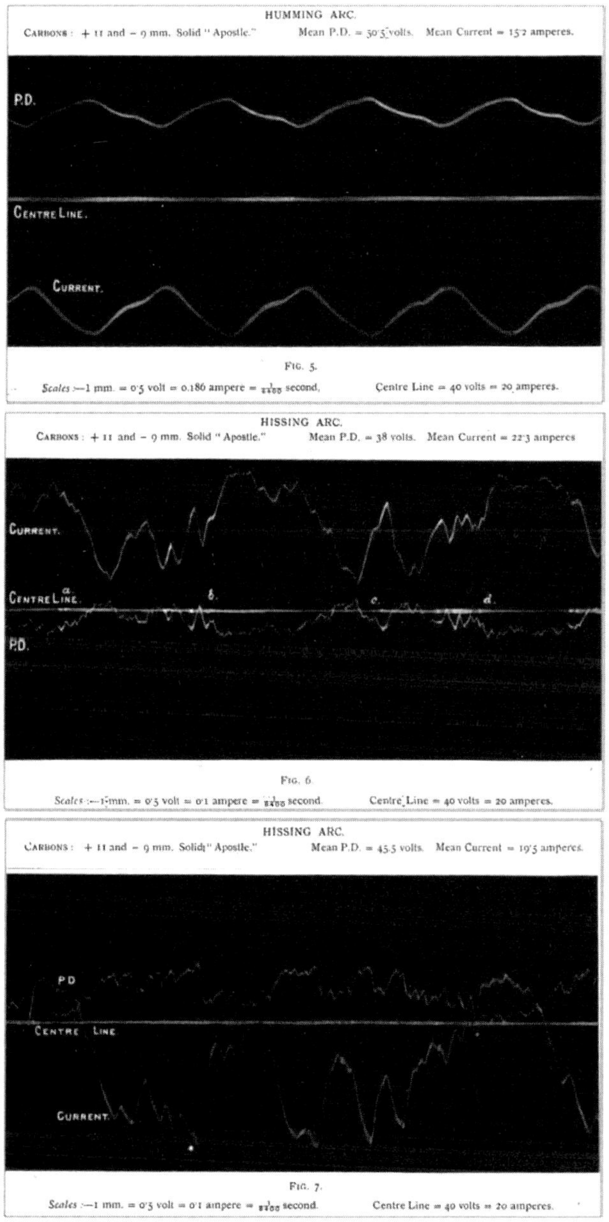

Fig. 7. Demonstration of arc hissing. Quoting Duddell [33] If the current through the humming arc be increased until the arc hisses, the variations in P.D., current, and light change, I find, in a most striking manner from the regular periodic variations of Fig. 5 to the very irregular variations shown in Figs. 6 and 7. Reproduced with permission from [33].

References

[1] D. B. Strukov, G. S. Snider, D. R. Stewart, and R. S. Williams, The missing memristor found, Nature, Vol. 453, No. 7191, pp. 80–83, May 2008.

[2] L. Chua, Memristor — Missing Circuit Element, IEEE Trans. Circuit Theory, Vol. 18, No. 5, pp. 507, 1971.

[3] L. O. Chua and S. M. Kang, Memristive devices and systems, Proceedings of the IEEE, Vol. 64, No. 2, pp. 209–223, 1976.

[4] Y. V. Pershin and M. Di Ventra, Memory effects in complex materials and nanoscale systems, Advances in Physics, Vol. 60, No. 2, pp. 145–227, Apr. 2011.

[5] J. J. Yang, M. D. Pickett, X. Li, D. A. A. Ohlberg, D. R. Stewart, and R. S. Williams, Memristive switching mechanism for metal/oxide/metal nanodevices, Nature Nanotech, Vol. 3, No. 7, pp. 429–433, Jun. 2008.

[6] M.-J. Lee, C. B. Lee, D. Lee, S. R. Lee, M. Chang, J. H. Hur, Y.-B. Kim, C. J. Kim, D. H. Seo, S. Seo, U.-I. Chung, I.-K. Yoo, and K. Kim, A fast, high-endurance and scalable non-volatile memory device made from asymmetric Ta2O5?x/TaO2?x bilayer structures, Nature Materials, Vol. 10, No. 8, pp. 625–630, Jul. 2011.

[7] D.-H. Kwon, K. M. Kim, J. H. Jang, J. M. Jeon, M. H. Lee, G. H. Kim, X.-S. Li, G.-S. Park, B. Lee, S. Han, M. Kim, and C. S. Hwang, Atomic structure of conducting nanofilaments in TiO2 resistive switching memory, Nature Nanotech, Vol. 5, No. 2, pp. 148–153, Jan. 2010.

[8] K. M. Kim, D. S. Jeong, and C. S. Hwang, Nanofilamentary resistive switching in binary oxide system; a review on the present status and outlook, Nanotechnology, Vol. 22, No. 25, p. 254002, May 2011.

[9] M. Wuttig and N. Yamada, Phase-change materials for rewriteable data storage, Nature Materials, Vol. 6, No. 11, pp. 824–832, 2007.

[10] T. Driscoll, H. T. Kim, B. G. Chae, M. Di Ventra, and D. N. Basov, Phase-transition driven memristive system, Appl. Phys. Lett., Vol. 95, No. 4, p. 043503, 2009.

[11] V. Erokhin, T. Berzina, and M. P. Fontana, Hybrid electronic device based on polyaniline-polyethyleneoxide junction, J. Appl. Phys., Vol. 97, No. 6, p. 064501, 2005.

[12] B. Hille, Ion Channels of Excitable Membranes (3rd Edition), 3rd ed. Sinauer Associates Inc 2001-07-01, 2001, p. 814.

[13] J. N. C. Kew and C. H. Davies, Ion channels. Oxford University Press, USA, 2010, p. 562.

[14] L. Chua, V. Sbitnev, and H. Kim, Hodgkin-Huxley Axon is Made of Memristors, International Journal of Bifurcation and Chaos, Vol. 22, No. 3, pp. 1–81, 2012.

[15] R. Mannikko, Hysteresis in the Voltage Dependence of HCN Channels: Conversion between Two Modes Affects Pacemaker Properties, The Journal of General Physiology, Vol. 125, No. 3, pp. 305–326, Feb. 2005.
[16] A. Bruening-Wright, F. Elinder, and H. P. Larsson, Kinetic Relationship between the Voltage Sensor and the Activation Gate in spHCN Channels, The Journal of General Physiology, Vol. 130, No. 1, pp. 71–81, Jul. 2007.
[17] S. P. Kosta, Y. P. Kosta, M. Bhatele, Y. M. Dubey, A. Gaur, S. Kosta, J. Gupta, A. Patel, and B. Patel, Human blood liquid memristor, IJMEI, Vol. 3, No. 1, p. 16, 2011.
[18] H. Yan, H. S. Choe, S. Nam, Y. Hu, S. Das, J. F. Klemic, J. C. Ellenbogen, and C. M. Lieber, Programmable nanowire circuits for nanoprocessors, Nature, Vol. 470, No. 7333, pp. 240–244, Feb. 2011.
[19] A. Chanthbouala, A. Crassous, V. Garcia, K. Bouzehouane, S. Fusil, X. Moya, J. Allibe, B. Dlubak, J. Grollier, S. Xavier, C. Deranlot, A. Moshar, R. Proksch, N. D. Mathur, M. Bibes, and A. Barthlmy, Solid-state memories based on ferroelectric tunnel junctions, pp. 1–11, Jan. 2012.
[20] T. Prodromakis, B. P. Peh, C. Papavassiliou, and C. Toumazou, A Versatile Memristor Model With Nonlinear Dopant Kinetics, IEEE Trans. Electron Devices, Vol. 58, No. 9, pp. 3099–3105. 2011.
[21] J. Borghetti, G. S. Snider, P. J. Kuekes, J. J. Yang, D. R. Stewart, and R. S. Williams, Memristive switches enable stateful logic operations via material implication, Nature, Vol. 464, No. 7290, pp. 873–876, Apr. 2010.
[22] A. Gelencser, T. Prodromakis, C. Toumazou, and T. Roska, A Biomimetic Model of the Outer Plexiform Layer by Incorporating Memristive Devices, Physical Review E. 2012.
[23] H. J. Reich and W. A. Depp, Dynamic Characteristics of Glow Discharge Tubes, J. Appl. Phys., Vol. 9, No. 6, p. 421, 1938.
[24] R. Waser and M. Aono, Nanoionics-based resistive switching memories, Nature Materials, Vol. 6, No. 11, pp. 833–840, 2007.
[25] L. Chua, Resistance switching memories are memristors, Appl. Phys. A, Vol. 102, No. 4, pp. 765–783, Jan. 2011.
[26] R. Waser, R. Dittmann, G. Staikov, and K. Szot, Redox-Based Resistive Switching Memories — Nanoionic Mechanisms, Prospects, and Challenges, Adv. Mater., Vol. 21, No. 25, pp. 2632–2663, Jul. 2009.
[27] T. Hasegawa, A. Nayak, T. Ohno, K. Terabe, T. Tsuruoka, J. K. Gimzewski, and M. Aono, Memristive operations demonstrated by gap-type atomic switches, Appl. Phys. A, Vol. 102, No. 4, pp. 811–815, Feb. 2011.
[28] V. Francis, Fundamentals of discharge tube circuits. London: Methuen, 1948.
[29] D. Biolek, Z. Biolek, and V. BIOLKOV, Pinched hysteretic loops of ideal memristors, memcapacitors and meminductors must be self-crossing, Electronics Letters, Vol. 47, No. 25, pp. 1385–1387, 2011.
[30] H.M. Ayrton, The electric arc. 1902, p. 479.
[31] W. de la Rue and H. W. Muller, Experimental Researches on the Electric Discharge with the Chloride of Silver Battery, Proc. R. Soc. Lond., pp. 563–572, Nov. 1879.
[32] H.M. Ayrton, The hissing of the electric arc, Electrical Engineers, Journal of the Institution of, Vol. 28, No. 140, pp. 400–436, 1899.
[33] W. Duddell, On rapid variations in the current through the direct-current arc, Electrical Engineers, Journal of the Institution of, Vol. 30, No. 148, pp. 232–267, 1901.
[34] B. Van Der Pol and J. Van der Mark, Frequency demultiplication, Nature, Vol. 120, No. 3019, pp. 363–364, 1927.
[35] M. Kennedy and L. Chua, Van der Pol and chaos, Circuits and Systems, IEEE Transactions on, Vol. 33, No. 10, pp. 974–980, 1986.
[36] W. Duddell, On the Resistance and Electromotive Forces of the Electric Arc, Philosophical Transactions of the Royal Society A: Mathematical, Physical and Engineering Sciences, Vol. 203, No. 359, pp. 305–342, Jan. 1904.
[37] J. Frith and C. Rodgers, On the resistance of the electric arc, The London Philosophical Magazine, No. XLI, 1896.
[38] H. Luggin, Centralblatt for Elektrotechnik. 1888.
[39] M. D. Pickett, J. Borghetti, J. J. Yang, G. Medeiros-Ribeiro, and R. S. Williams, Coexistence of Memristance and Negative Differential Resistance in a Nanoscale Metal-Oxide-Metal System, Adv. Mater., Vol. 23, No. 15, pp. 1730–1733, Feb. 2011.
[40] O. W. Richardson, The Emission of electricity from hot bodies, by O.W. Richardson, ... 1916, p. 304.
[41] C. Crowell, The Richardson constant for thermionic emission in Schottky barrier diodes, Solid-State Electronics, Vol. 8, No. 4, pp. 395–399, 1965.
[42] M. R. Blondlot, LECTRICIT: Recherches sur la transmission de llectricit faible tension par Vintermdiaire de lair chaud, COMPTES RENDUS, Vol. CIV, p. 283, Sep. 1887; M. Sapoff and R. M. Oppenheim, Theory and application of self-heated thermistors, Proceedings of the IEEE, Vol. 51, No. 10, pp. 1292–1305, 1963.
[43] M. Faraday, Experimental researches in electricity. 1833.
[44] Z.-M. Liao, C. Hou, Q. Zhao, D.-S. Wang, Y.-D. Li, and D.-P. Yu, Resistive Switching and Metallic-Filament Formation in Ag2S Nanowire Transistors, Small, Vol. 5, No. 21, pp. 2377–2381, Oct. 2009.

[45] A. Nayak, T. Tamura, T. Tsuruoka, K. Terabe, S. Hosaka, T. Hasegawa, and M. Aono, Rate-Limiting Processes Determining the Switching Time in a Ag2S Atomic Switch, J Phys Chem Lett, Vol. 1, No. 3, pp. 604–608, 2010.

[46] A. Volta, On the Electricity Excited by the Mere Contact of Conducting Substances of Different Kinds. In a Letter from Mr. Alexander Volta, FRS Professor of Natural Philosophy in the University of Pavia, to the Rt. Hon. Sir Joseph Banks, Bart. KBPRS, Philosophical Transactions of the Royal Society of London, Vol. 90, No. 2, pp. 403–431, 1800.

[47] H. Davy, Letter to Mr Nicholson, Vol. 4. J. Nat. Phil. Chem. Arts, 1801, p. 527.

[48] H. Davy, Letter to James Watt, Davy Letters. 08-Jan.-1801.

[49] H. Davy, Additional experiments on Galvanic electricity, in a letter to Mr. Nicholson, dated September 22, 1800, Nicholson's) Journal of Natural Philosopy, Chemistry, and the Arts, Vol. 4, pp. 326–328, Sep. 1800.

[50] H. Davy, The journal of the Royal Institution of Great Britain, Vol. I. 1802, p. 166.

[51] J. Flemming, Problems in the Physics of an Electric Lamp, Nature, Vol. 42, No. 1078, pp. 198–201, Jun. 1890.

[52] H. Davy, Philosophical transactions of the Royal Society of London, Vol. XIV. 1810, p. 232.

[53] H. Davy, Elements of Chemical Philosophy. Philadelphia — Bradford and Inskeep, 1812, p. 511.

[54] R. Lamont-Brown, Humphry Davy: life beyond the lamp. The History Press, 2004, p. 224.

[55] N. Metropolis, J. Howlett, and G.-C. Rota, A history of computing in the twentieth century. Academic Pr, 1980, p. 659.

[56] J. M. Tour and T. He, Electronics: the fourth element., Nature, Vol. 453, No. 7191, pp. 42–43, Apr. 2008.

[57] G. S. Ohm, Die Galvanische Kette, Mathematisch Bearbeitet (1827) (German Edition). Kessinger Publishing, LLC, 1827, p. 250.

CHAPTER 42

STATE EQUATIONS FOR ACTIVE CIRCUITS WITH MEMRISTORS

MARTIN HASLER

School of Computer and Communication Sciences,
Ecole Polytechnique Fédérale de Lausanne,
Lausanne, Switzerland

Nonlinear dynamic circuits including memristors are considered. Graph-theoretical criteria are given as sufficient conditions for the existence of global state equations, with the capacitor and memristor charges as state variables (for convenience we do not consider inductors here). They are based on earlier work on circuits without memristors. Global state equations in these charges fail to exist when the charges are dependent or when singular points are present, so-called impasse points. The values of state variables at initial time uniquely determine a solution. Therefore, even for numerical circuit analysis that in general does not use state equations, the determination of a set of state variables is important, and to know whether, with the "canonical candidates" for states, global state equations exist, is the first and most important step.

The graph-theoretical criteria are easy to check directly for simple circuits. We give a few simple examples that illustrate the various cases. Larger circuits have to be examined by combinatorial algorithms.

1. Introduction

Typically, a circuit is described by a set of equations that are partly differential and partly non-differential, i.e. static. Such systems are called *differential-algebraic equations*. Even though for the purpose of numerical analysis it is advantageous to solve algebra-differential systems directly, and most general purpose circuit analysis programs actually do so, the question whether or not it is possible to reduce them to a purely differential form, the *state equations*, is not only of theoretical interest. In fact, differential-algebraic systems that cannot be reduced to a system of first order differential equations, not even locally, possess singular points, where the solution either does not exist (*impasse points*) or is not unique. Of course, this is only a property of the circuit model, and not of the circuit itself, but it indicates a singular behavior of the physical circuit also from a practical point of view, like e.g. jumps from a state to a distant one. Therefore, it is important to have computable criteria that allow excluding the potential presence of singular points.

Circuits are by definition connections of *circuit elements*. The connections are realized by *ports* that carry two variables, a *current* and a *voltage*. While the sum of the currents of connected ports must be zero, the voltages must be identical. The circuit elements themselves are described by equations involving their port variables, and possibly additional, internal variables such as *charges* and *fluxes*. They are called the *constitutive relations* of the elements. The constitutive relations may be linear or nonlinear and the corresponding elements carry the attribute linear or nonlinear, accordingly.

The electronics literature is full of different circuit elements, the most classical being the resistor, the capacitor, the inductor, the transistor, the operational amplifier, controlled sources, etc. This list of elements is redundant in the sense that a large part of them can be obtained by connection of only a few of them, the *basic circuit elements*. Different sets of basic circuit elements can be chosen. Here we favor the set composed of nonlinear locally passive resistors, independent current and voltage sources and linear capacitors as 1-ports, and the ideal operational amplifier as 2-port.

In addition we find it useful to decompose the ideal operational amplifier into two juxtaposed singular 1-ports, the nullator and the norator (Fig. 2). In this way, all basic elements are 1-ports. Inductors, transistors, controlled sources, etc., can all be obtained by connecting these basic elements. Even nonlinear resistors with non-monotonic characteristics, and nonlinear capacitors and inductors have equivalent circuits composed exclusively of the basic circuit elements.

The *state* of a circuit at a time t_0 is the minimal information necessary to determine the solution of the circuit equations for $t > t_0$. It is given by a vector of circuit variables. This vector is not unique,

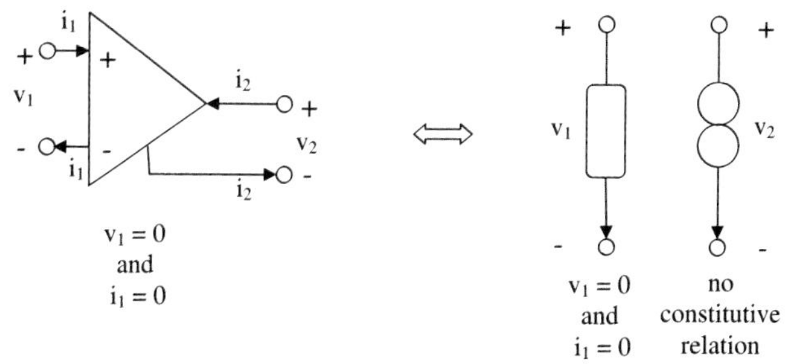

Fig. 2. Equivalent circuit for an ideal operational amplifier: juxtaposition of a nullator and a norator.

various choices can be taken. However, there is a privileged candidate, the vector of capacitor charges and the inductor fluxes. The state equations in this case are of the form

$$
\begin{aligned}
\frac{dq_1}{dt} &= F_1(q_1, \ldots, q_{n_C}, \phi_1, \ldots, \phi_{n_L}, \\
&\qquad \times E_1, \ldots, E_{n_{VS}}, I_1, \ldots, I_{n_{CS}}) \\
&\;\;\vdots \\
\frac{d\phi_{n_L}}{dt} &= F_{n_C+n_L}(q_1, \ldots, q_{n_C}, \phi_1, \ldots, \phi_{n_L}, \\
&\qquad \times E_1, \ldots, E_{n_{VS}}, I_1, \ldots, I_{n_{CS}})
\end{aligned} \tag{1}
$$

where n_C, n_L, n_{CS} and n_{VS} are, respectively, the number of capacitors, inductors, independent voltage sources and independent current sources. The voltages E_k and currents I_k injected into the circuit by the independent sources may depend on time or be constant. Most often, the state equations in this form exist. Two main exceptions can be distinguished:

(1) The charges and fluxes are linearly dependent. In this case, usually a linearly independent subset can be chosen such that a system of state

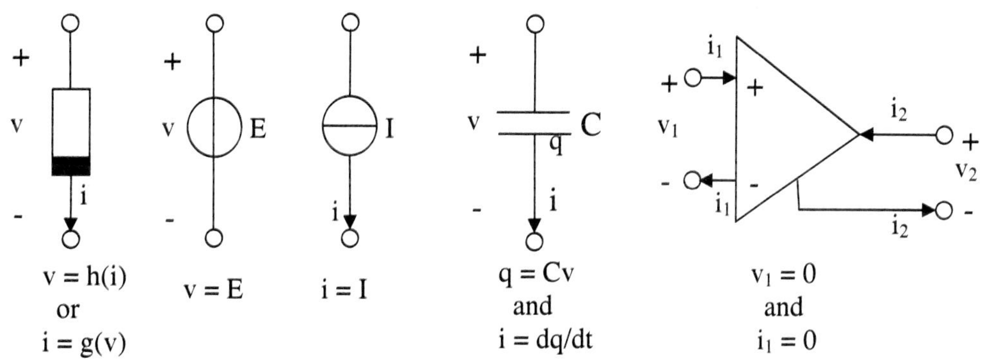

Fig. 1. Classical list of basic circuit elements.

equations of the same form as (1) exists for the reduced set of independent charges and fluxes.

(2) The system of equations in the form (1) does not exist, but it exists in almost all points locally. In this case, the solutions of the equations exist and are uniquely determined by the state variables, as long as they stay away from exceptional points, where they normally break down. This is the phenomenon of impasse points. At other exceptional points, the solution would not be uniquely determined by the variables that constitute the state at the regular points.

It is important to know whether global state equations exist, especially in order to exclude impasse points that would cause problems for numerical simulations, even when the simulations are not based on state equations. The problem is as classical one, but the renewed interest in memristors poses it once more.

2. The Memristor

The memristor is a 1-port described by the constitutive relations [1].

$$\frac{dq}{dt} = i, \frac{d\phi}{dt} = v, \phi = h(q) \text{ (charge controlled) or}$$
$$q = g(\phi) \text{ (flux controlled)} \tag{2}$$

Its symbol is represented in Fig. 3. Is it necessary to enlarge the list of basic circuit elements by adjoining the memristor? The answer is negative, as shows the equivalent circuit of Fig. 4 for a charge controlled memristor. To show the equivalence, let us analyze the 1-port of Fig. 4. The auxiliary currents and voltages i_2, i_3, v_2, v_3 already take Kirchhoff's equations and the constitutive relations of the ideal operational amplifiers into account. The constitutive relations of the capacitors and the resistors become

$$\begin{aligned} q_1 &= C_1 v_3, \frac{dq_1}{dt} = i_1 \\ i_3 &= g(v_3) \\ v_2 &= R_2 i_3 \\ q_2 &= C_2 v_2, \frac{dq_2}{dt} = i_2 \\ v_1 &= R_1 i_2 \end{aligned} \tag{3}$$

Eliminating all currents and voltages internal to the 1-port leads to

$$\begin{aligned} \frac{dq_1}{dt} &= i_1, \frac{dq_2}{dt} = \frac{v_1}{R_1}, \\ q_2 &= C_2 R_2 g\left(\frac{q_1}{C_1}\right) \end{aligned} \tag{4}$$

Setting $q = q_1$ and $\phi = R_1 q_2$ yields indeed the constitutive relations of a charge-controlled memristor:

$$\begin{aligned} &\frac{dq}{dt} = i_1, \frac{d\phi}{dt} = v_1, \\ &\phi = f(q), \text{ with } f(x) = C_2 R_1 R_2 g\left(\frac{x}{C_1}\right) \end{aligned} \tag{5}$$

A similar equivalent circuit can be given for the flux controlled memristor.

The drawback of these equivalent circuits is that they use two capacitors which would suggest two state variables, namely the two charges. But by the very nature of the memristor, the charge and the flux, and therefore in the equivalent circuit the two charges are dependent and therefore there can be only one state variable. For this reason, we prefer to view the memristor as a charge or flux controlled resistor and use the constitutive relations

$$\frac{dq}{dt} = i, \; v = R(q) \cdot i \text{ or } i = G(\phi) \cdot v \tag{6}$$

where R and G are the derivatives of the functions h and g in (2), respectively. Here only a single state variable candidate q or ϕ appears.

3. Existence of Global State Equations

The existence of global state equations is a classical problem that has been solved long ago, and we

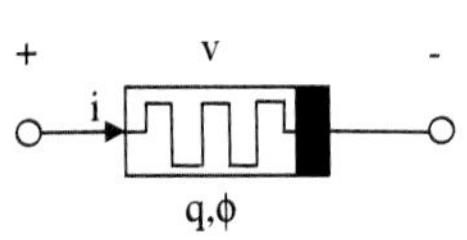

Fig. 3. Memristor symbol.

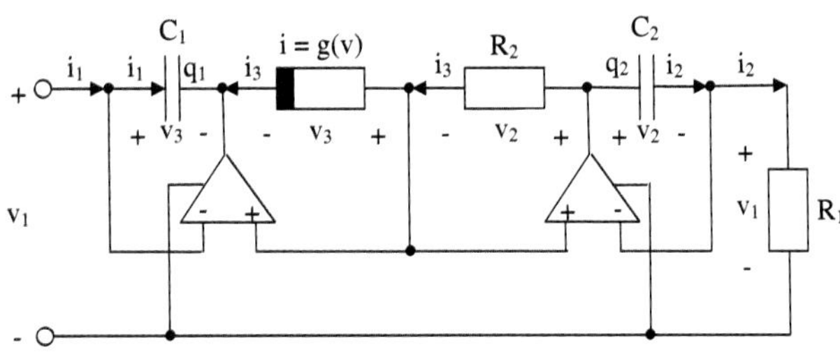

Fig. 4. Equivalent circuit of a charge controlled memristor.

shortly review it here. Let us limit the attention to circuits composed of linear capacitors, linear and nonlinear resistors, independent voltage and current sources, nullators and norators. As remarked above, this is sufficient; the other circuit elements can be replaced by equivalent circuits. Alternatively, what follows can easily be extended to circuits with a richer list of circuit elements. The global state equations for circuits composed of the above list of basic circuit elements are expected to be of the form

$$\begin{aligned} \frac{dq_1}{dt} &= F_1(q_1, \ldots, q_{n_C}, E_1, \ldots, E_{n_{VS}}, \\ &\quad \times I_1, \ldots, I_{n_{CS}}) \\ &\vdots \\ \frac{dq_{n_C}}{dt} &= F_{n_C}(q_1, \ldots, q_{n_C}, E_1, \ldots, E_{n_{VS}}, \\ &\quad \times I_1, \ldots, I_{n_{CS}}) \end{aligned} \tag{7}$$

In principle, we obtain these equations by writing down all circuit equations, comprising Kirchhoff's equations and the constitutive relations of the circuit elements and then eliminating all circuit variables other than the capacitor charges, the voltage source voltages and the current source currents. Since, in general, some of the constitutive relations are nonlinear, the elimination procedure cannot be done explicitly, and even the theoretical possibility to perform this elimination, i.e. the equivalence of equations of the form (7) with the full set of circuit equations is by no means trivial.

If we number the edges of the circuit, which correspond to the ports, starting from the capacitors, and we replace the capacitor charges by their voltages and the charge derivatives by their currents, we get

$$\begin{aligned} i_1 &= F_1(C_1 v_1, \ldots, C_{n_C} v_{n_C}, E_1, \ldots, E_{n_{VS}}, \\ &\quad \times I_1, \ldots, I_{n_{CS}}) \\ &\vdots \\ i_{n_C} &= F_{n_C}(C_1 v_1, \ldots, C_{n_C} v_{n_C}, E_1, \ldots, E_{n_{VS}}, \\ &\quad \times I_1, \ldots, I_{n_{CS}}) \end{aligned} \tag{8}$$

These are also the equations of the *associated resistive circuit* obtained by replacing the capacitors by independent voltage sources ($v_k = \tilde{E}_k$), when all circuit variables except the capacitor voltages and currents, the independent voltage source voltages and the independent current source currents are eliminated. The equations of the associated resistive circuit can be written in this form if it has exactly one solution for any combination of source values, including the voltage sources that replace the capacitors.

Therefore, the existence of global state equations is reduced to the existence and uniqueness of the solution of a resistive circuit, the associated resistive circuit, for arbitrary source values. Thus, a problem of a dynamical system is reduced to a static problem. Note that we use here the term "resistive circuit" in the sense of "static circuit", not just "circuit composed of only resistors", it may contain also sources and active elements.

Now let us also admit passive charge-controlled memristors to the circuit, described by the constitutive relations

$$\frac{dq_k}{dt} = i_k, \; v_k = R_k(q_k) \cdot i_k \tag{9}$$

Passivity means that R_k has nonnegative values. If we number their edges in the circuit graph after the capacitors, from $n_C + 1$ to $n_C + n_M$, where n_M is the number of memristors, then we expect the state equations to have the following form:

$$\begin{aligned} \frac{dq_1}{dt} &= F_1(q_1, \ldots, q_{n_C}, R(q_{n_C+1}), \ldots, \\ &\quad \times R(q_{n_C+n_M}), E_1, \ldots, E_{n_{VS}}, \\ &\quad \times I_1, \ldots, I_{n_{CS}}) \\ &\vdots \\ \frac{dq_{n_C+n_M}}{dt} &= F_{n_C+n_M}(q_1, \ldots, q_{n_C}, R(q_{n_C+1}), \ldots, \\ &\quad \times R(q_{n_C+n_M}), E_1, \ldots, E_{n_{VS}}, \\ &\quad \times I_1, \ldots, I_{n_{CS}}) \end{aligned} \tag{10}$$

Again, we define an associated resistive circuit by replacing the capacitors by independent voltage sources ($v_k = \tilde{E}_k$) and in addition by replacing the memristors by linear resistors ($\tilde{R}_k = R(q_k)$). Actually, for each choice of memristor charges the associated resistive circuit has different linear resistors replacing the memristors. The equations of the associated resistive circuit corresponding to the state equations are:

$$\begin{aligned} i_1 &= F_1(C_1 \tilde{E}_1, \ldots, C_{n_C} \tilde{E}_{n_C}, \tilde{R}_1, \ldots, \\ &\quad \times \tilde{R}_{n_M}, E_1, \ldots, E_{n_{VS}}, I_1, \ldots, I_{n_{CS}}) \\ &\vdots \end{aligned} \tag{11}$$

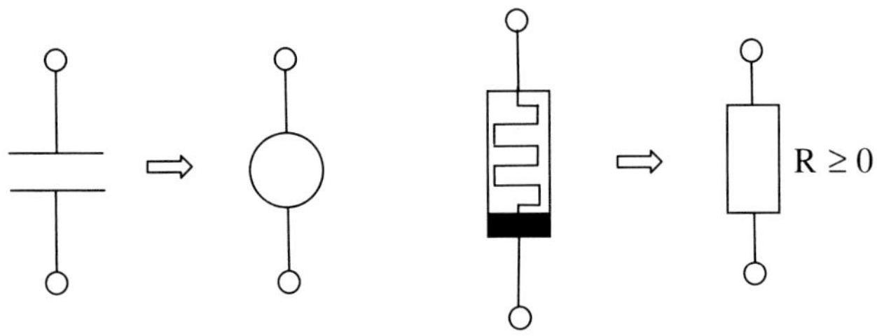

Fig. 5. Element replacements to obtain the associated resistive circuit.

$$i_{n_C+n_M} = F_{n_C+n_M}(C_1\tilde{E}_1, \ldots, C_{n_C}\tilde{E}_{n_C}, \tilde{R}_1, \ldots, \times\tilde{R}_{n_M}, E_1, \ldots, E_{n_{VS}}, I_1, \ldots, I_{n_{CS}})$$

Their existence is linked to the existence and uniqueness of the solution of the associated circuit for any choice of independent voltage and current sources, and any nonnegative value of the resistors replacing the memristors. Hence, we get the following theorem.

Theorem 1. *Consider a circuit composed of linear capacitors, linear and nonlinear resistors, independent voltage and current sources, nullators, norators and passive charge controlled memristors. Consider the associated resistive circuit obtained by replacing the capacitors by voltage sources and the memristors by nonnegative linear resistors (Fig. 5). Global state equations in the capacitor and memristor charges exist for the original circuit, if the associated resistive circuit has exactly one solution for all possible combinations of voltage source voltages (including the replacement sources for the capacitors), current source currents and nonnegative resistances for the linear resistors replacing the memristors.*

Note that resistive circuit does not mean only "circuit composed of resistors", but includes also active and non-reciprocal circuit elements such as the operational amplifier, here decomposed into a nullator-norator pair.

4. Graph Theoretical Criteria for the Existence of Global State Equations

Theorem 1 allows proving the existence of global state equations by proving the existence and uniqueness of the solution of the associated resistive circuit. This problem was the object of intensive research in the 1970's and 1980's. Three main approaches can be distinguished.

(1) Study of transistor circuits directly from positivity properties of the circuit equations [7,10].
(2) Study of nonlinear resistive circuits with controlled sources or operational amplifiers based on "cactus graphs" [8,9].
(3) Study of nonlinear resistive circuits with nullators and norators, based on "conjugate trees" and "uniform partial orientations" [3,4].

We favor here our own approach, because we think it is easy to understand and readily extensible to memristor circuits. Let us give a short summary of the results.

In [4], the notion of *circuit structure* is introduced. A circuit structure is a graph with labeled edges, where the labels are types of circuit elements. For the graphical representation of a circuit structure, each edge is replaced by the usual symbol of the element type. A specific circuit, where all element parameters have specified values, has a certain circuit structure, if the graph and the element types of the circuit are the same as the graph and the edge labels of the circuit structure.

In [4] both nonlinear and linear resistive circuit structures are used. Here, we are only interested the existence and uniqueness of the global state equations and only indirectly in the existence and uniqueness of the solution of the associated nonlinear resistive circuit. Therefore, we shall only introduce the circuit structures used directly in formulating the graph theoretical criteria, namely the linear resistive circuit structures.

The circuit element types admitted as edge labels in a linear resistive circuit structure are

- linear resistor, restricted to a strictly positive and finite resistance
- voltage source
- current source
- nullator
- norator

The symbols are represented in Fig. 6.

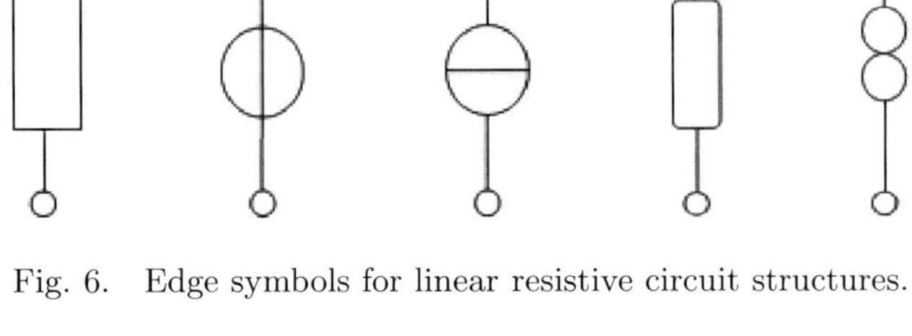

Fig. 6. Edge symbols for linear resistive circuit structures.

There is a relation between the existence and the uniqueness of the solution of a **single** nonlinear resistive circuit and the existence and uniqueness of the solutions of an **infinity** of linear resistive circuits. These linear resistive circuits have in fact all the same linear resistive circuit structure, or they have one among a well-defined finite number of such structures.

Thus, we have to specify, which linear resistive circuit structures are associated with a nonlinear resistive circuit. This association can be partially understood by noting that the difference between any two solutions of the given nonlinear resistive circuit is the solution of a certain linear resistive circuit. Note that we do not perform any linearization, the two solutions can be far apart. Furthermore, in any two solutions, the voltage source voltages and the current source currents must be identical. Therefore, in the linear resistive circuit the sources must be zero. Now, if the linear resistive circuit has exactly one solution, it must be the trivial one, where all currents and voltages vanish. This implies that the two solutions of the nonlinear resistive circuit must be identical. As the two solutions were arbitrarily chosen, the solution of the nonlinear resistive circuit must be unique.

Since we postulate an arbitrary pair of solutions for the nonlinear circuit the linear circuit can have all kinds of resistor values. Therefore, we must consider simultaneously a whole set of resistive circuits. They have all the same linear resistive circuit structure, or slightly different ones. These are the linear resistive circuit structures associated with the nonlinear resistive circuit. They are obtained by taking the circuit graph and by replacing on each edge the circuit element with a label of Fig. 6. On most edges there is only one possible label. Sometimes, two different labels are possible. If there are m such edges, 2^m resistive circuit structures have to be analyzed, in principle. In practice, much less effort is sufficient.

Finally, according to Sec. 3, in order to establish the global existence of state equations, the existence and uniqueness of the solution of the associated nonlinear resistive circuit has to be proved. We can combine this association with the association of the nonlinear resistive circuit with linear resistive circuit structures. We limit this to our basic set of circuit elements plus charge controlled passive memristors. The extension to other circuit elements is rather obvious.

Definition 42.1. The following substitutions of circuit elements (Figs. 1–3) by labels (Fig. 6) edge by edge defines the association of a nonlinear dynamic circuit with a finite number of linear resistive circuit structures.

By combination of results of [4] and Theorem 1, we obtain the following theorem.

Theorem 2. *Consider a nonlinear dynamical memristor circuit composed of elements listed on the left-hand-side of Fig. 7. It possesses global state equations in the capacitor and memristor charges, if all linear resistive circuits that have one of the linear resistive circuit structures resulting from the substitution of Fig. 7 have exactly one solution.*

In order to apply Theorem 2, the graph-theoretical criteria of [4] for linear resistive circuit structures are used. They are reproduced here for convenience. First, we have to introduce two additional notions.

Definition 42.2. A *pair of conjugate trees* (t_1, t_2) in a linear resistive circuit structure is defined by the following conditions:

- t_1 contains all norator, no nullator, all voltage source, no current source and some resistor edges.
- t_2 contains no norator, all nullator, all voltage source, no current source and the same resistor edges as t_1.

A *uniform partial orientation of the resistor edges* is an orientation of part of the resistor edges, such that the following two conditions are satisfied:

- Each orientated resistor edge belongs to at least one uniform loop composed of orientated resistor edges, voltage source edges and norator edges.
- Each orientated resistor edge belongs to at least one uniform cutset composed of orientated resistor edges, current source edges and norator edges.

A *loop* is *uniform* if all orientated edges of the loop are directed in the same way around the loop. A *cutset* is *uniform*, if all orientated edges of the cutset are orientated in the same way across the cutset. A partial orientation of the edges is *trivial* if no edge is orientated.

Theorem 3. *Let S be a linear resistive circuit structure. All circuits with structure S have exactly one solution, if and only if the following two conditions are satisfied.*

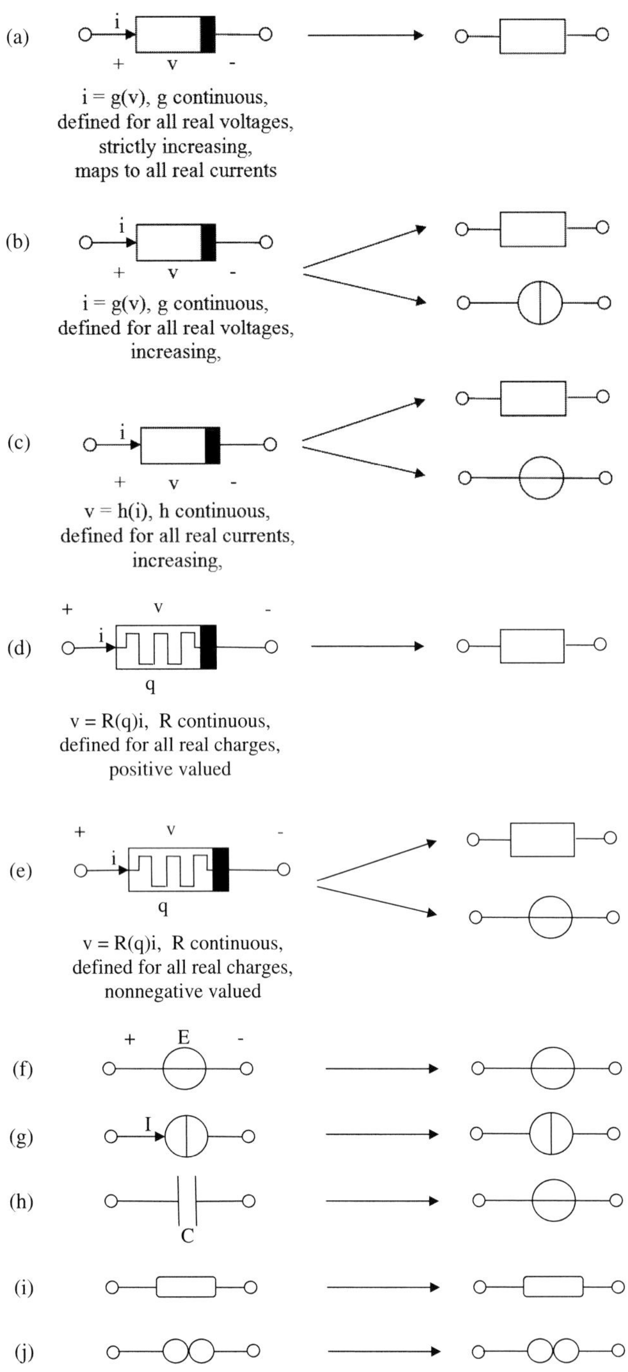

Fig. 7. Classical list of basic circuit elements.

(1) *S has a pair of conjugate trees*
(2) *S has no non-trivial uniform partial orientation of the resistor edges.*

Example 1. Consider the linear resistive circuit structure of Fig. 8. A pair of conjugate trees is represented in Fig. 9 and a uniform partial orientation in Fig. 10.

It follows that in this example, condition b) of Theorem 3 is not satisfied. In fact, a linear resistive

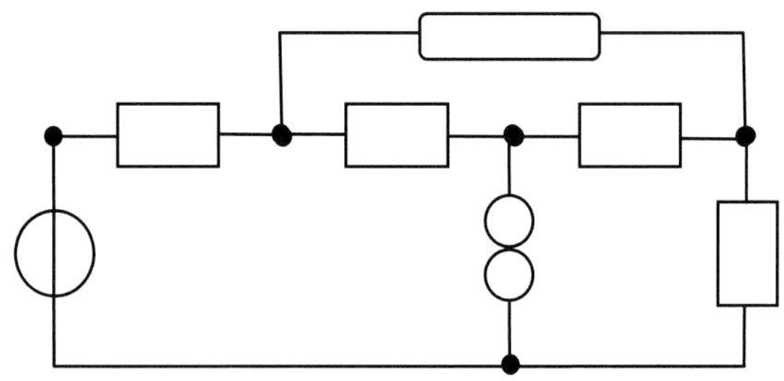

Fig. 8. Linear resistive circuit structure.

circuit with this structure and all equal resistances has no solution if the source voltage is different from 0 and an infinity of solutions if the source voltage vanishes.

We can now combine Theorems 2 and 3 and formulate the main theorem of this paper.

Theorem 4. *Consider a nonlinear dynamical circuit composed of elements listed on the left-hand-side of Fig. 7. It possesses global state equations in the capacitor and memristor charges, if all linear resistive circuit structures resulting from the substitutions of Fig. 7 have a pair of conjugate trees and no non-trivial partial uniform orientation of the resistors edges.*

Example 2. Consider the circuit composed of a voltage source connected in series with a linear resistor and the 1-port of Fig. 4. We suppose that the characteristic of the nonlinear resistor given by the function g is increasing, but not strictly increasing, i.e. there is at least one interval of voltages where the current is constant. Does this circuit have global state equations in the state variables q_1, q_2? In order to answer this question, we apply Theorem 4.

In a first step, we replace the two operational amplifiers by two nullator-norator pairs. Then, we apply the element-wise substitutions of Fig. 7. Only for the nonlinear resistor, two substitutions are possible. Therefore we only have to examine two linear resistive circuit structures. They are given in Fig. 12.

Now, we have to look for a pair of conjugate trees in these structures. It turns out that there is no such pair. This can be seen as follows. In the upper circuit structure neither of the resistor edges pointed to by the dashed arrow can be part of a nullator-resistor-voltage source tree, because each of them forms a loop with a voltage source and a nullator edge. As the resistor edges of the conjugate norator-resistor-voltage source tree must

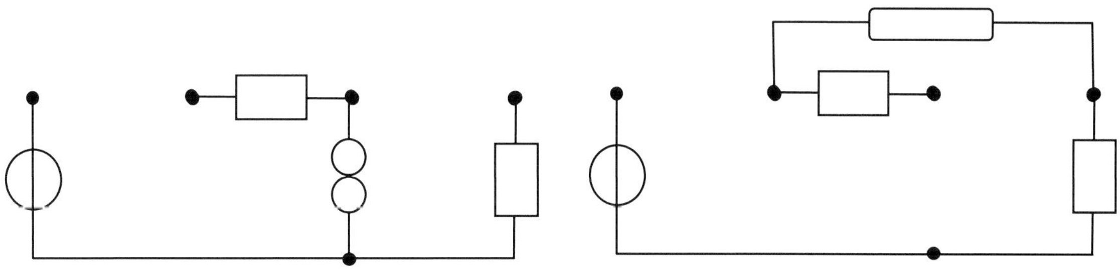

Fig. 9. Pair of conjugate trees for the linear resistive circuit structure of Fig. 8.

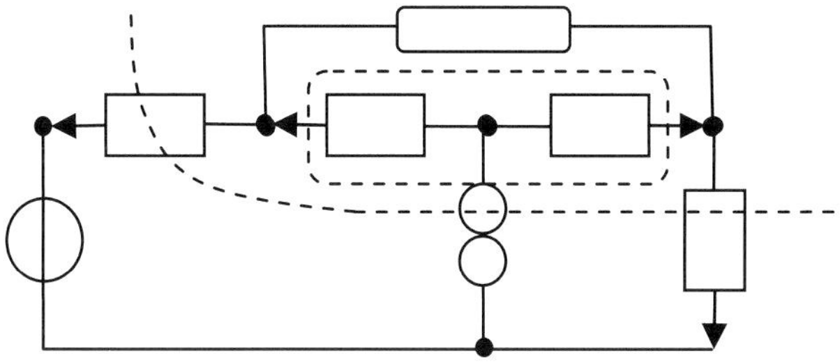

Fig. 10. Uniform partial orientation of the resistor edges of the linear resistive circuit structure of Fig. 8. Actually, all resistor edges are orientated. Clearly, the voltage source, the two resistor and the norator edges form a uniform loop and the two other resistor and the norator edges form another uniform loop. Two uniform cutsets are indicated by dotted lines. Two resistor edges belong to one of them and the other two to the other. Except for the norator, no other edge belongs to them.

be the same, they cannot be part of this tree, either. Therefore, in any pair of conjugate trees, the vertex between the two resistor edges is isolated in the norator-resistor-voltage source tree, which is a contradiction. Therefore, the conditions of Theorem 4 are not satisfied. Even though this is not necessary anymore, for the lower circuit structure a similar reasoning could be applied.

The conclusion of this analysis is that we suspect that there are no global state equations with the two capacitor charges as state variables. Actually, we already know this from the direct analysis

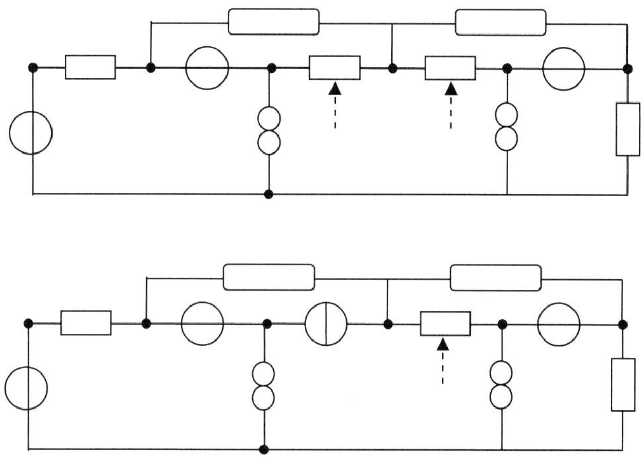

Fig. 12. Linear resistive circuit structures associated with the nonlinear dynamic circuit of Fig. 11.

of the 1-port of Fig. 4, which established the dependence of q_2 on q_1. It established also the equivalence of the 1-port with a charge-controlled memristor.

Example 3. Since Theorem 4 admits memristors directly, we can consider the circuit of Fig. 13 in place of the circuit of Fig. 11. The relation between g(v) and R(q) is, using (5)

$$R(q) = \frac{C_2}{C_1} R_1 R_2 \frac{dg}{dv}\left(\frac{q}{C_1}\right) \tag{12}$$

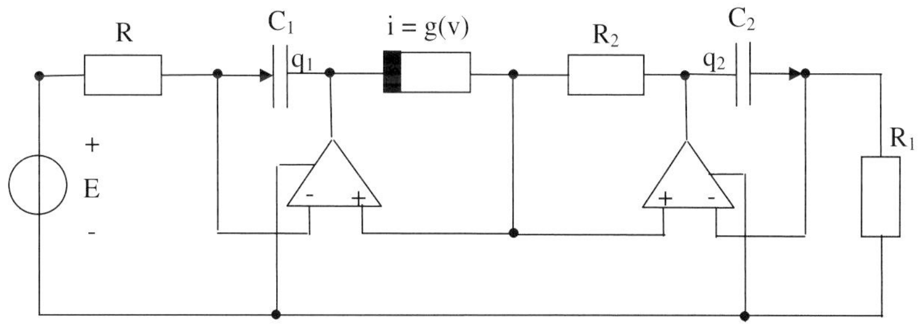

Fig. 11. Circuit containing the 1-port of Fig. 4.

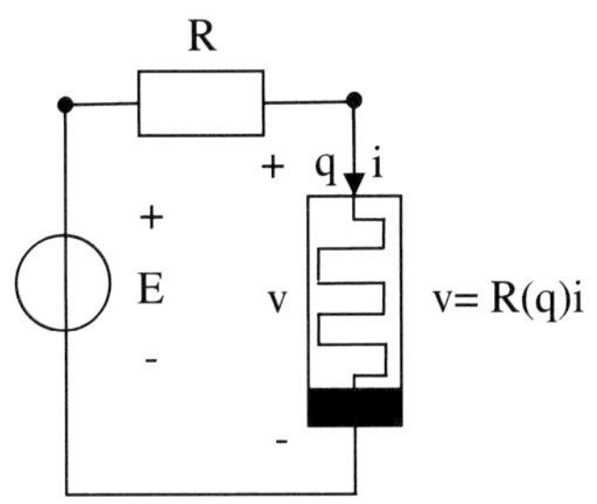

Fig. 13. Memristor circuit equivalent to the circuit of Fig. 11.

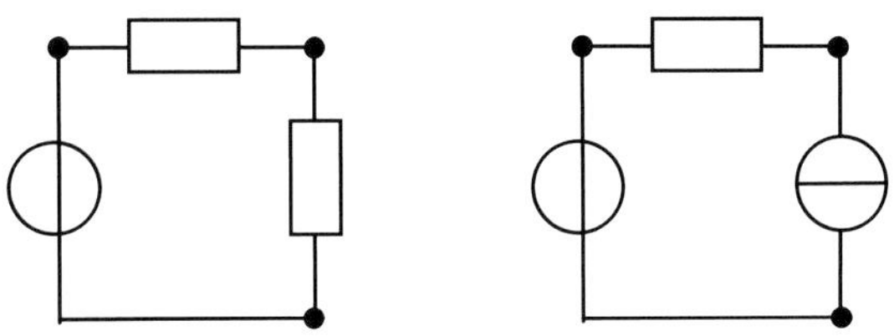

Fig. 14. Linear resistive circuit structures associated with the circuit of Fig. 13.

The associated linear circuit structures are represented in Fig. 14. Clearly, all linear resistive circuits with either of the structures of Fig. 14 have exactly one solution. This can also easily be confirmed by the criteria of Theorem 3. Therefore, the circuit of Fig. 13 has a global state equation in the state variable q. In this simple case, it can be written explicitly:

$$\frac{dq}{dt} = \frac{E}{R + R(q)} \tag{13}$$

This clearly shows the advantage in this context to use the constitutive relations (6) instead of (2), because per memristor there is only one potential state variable.

Example 4. One of the simplest examples of a memristor circuit without global state equations in the memristor charges is represented in Fig. 15. For simplicity, suppose that $R(q)$ is strictly positive for all q. Then there is only a single associated linear resistive circuit structure, which we have already encountered in Example 1. As we have already shown, this structure has a pair of conjugate trees, but also a nontrivial partial uniform orientation of the resistors. Therefore, we suspect that that the circuit of Fig. 15 has no global state equation in the memristor charge.

Again, in this simple case, we can perform a direct analysis. We take a short-cut by using only the auxiliary current i_1, having taken into account

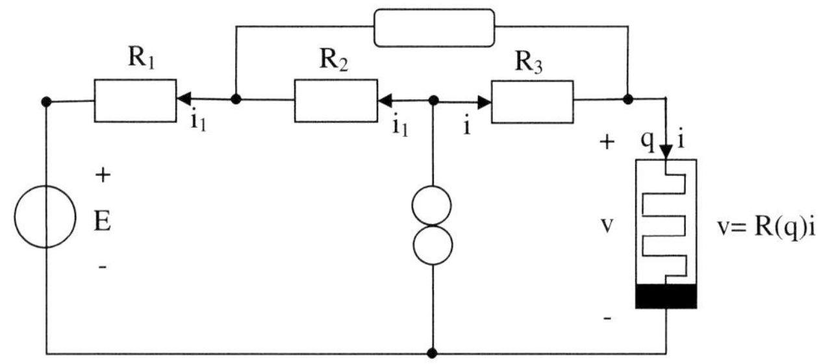

Fig. 15. Simple circuit with a memrister that does not have a global state equation in the memristor charge.

the zero current of the nullator and Kirchhoff's current law. Taking into account the zero voltage of the nullator and Kirchhoff's voltage law, we can write

$$\begin{aligned} E + R_1 i_1 &= R(q)i \\ R_2 i_1 &= R_3 i \end{aligned} \tag{14}$$

Solving for i:

$$i = \frac{E}{R(q) - \frac{R_1 R_3}{R_2}} \tag{15}$$

and thus the state equation should be

$$\frac{dq}{dt} = \frac{E}{R(q) - \frac{R_1 R_3}{R_2}} \tag{16}$$

Clearly, at the values of q where the denominator in the RHS of (16) vanishes, the state equation does not even exist locally, and these values of q do not define solutions. At all other values of q, the state equation exists locally and q determines a solution.

Actually, by discussing the sign of the RHS of (16), one can understand the qualitative behavior of the solutions. Suppose that in the original constitutive relations (2) of the memristor, the function g is of sigmoid type (Fig. 16, left). Its derivative, that appears in the constitutive relations (6), is represented in Fig. 16, right. We suppose that the linear resistances are such that $\mathrm{R_1R_3/R_2}$ is located between the minimum and the maximum of $R(q)$ (Fig. 16, right). This yields the qualitative picture of Fig. 17 of the RHS of (16) as a function of q and the direction of motion of the solutions on the q-axis.

Thus, the following conclusions about the qualitative behavior of the solutions can be drawn;

- Solutions starting from q_1 or q_2 do not exist, because the corresponding memristor currents are infinite. Even if we did accept infinite derivatives of q, the solutions would not exist in q_2 and would not be unique in q_1 (observe the direction

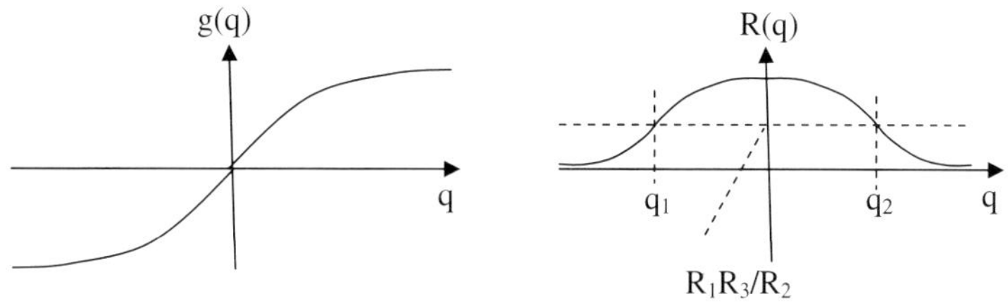

Fig. 16. Characteristics of the memristor.

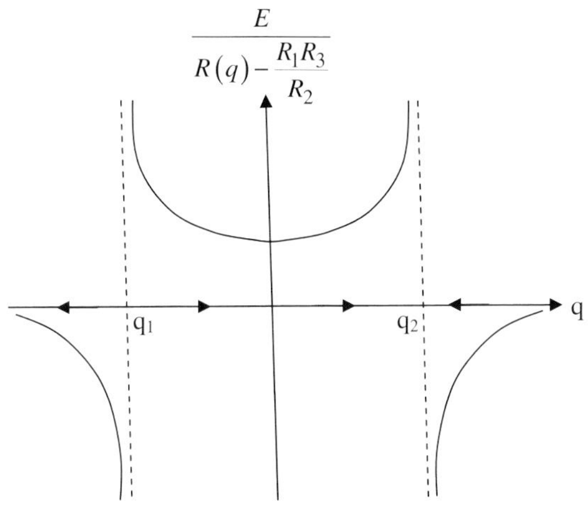

Fig. 17. RHS of (16) as a function of q. Its sign determines the direction of motion of the solutions on the q-axis. It is indicated by arrows.

of motion in their neighborhood). This is the signature of impasse points.

— Any solution starting in the interval $-\infty < q < q_1$ diverges to $-\infty$. Any solution starting in the interval $q_1 < q < q_2$ or $q_2 < q < +\infty$ converges to q_2, but q_2 is not an equilibrium point.

All examples in this paper are small circuits, where direct reasoning can be applied, both as far as the graph-theoretical criteria and as far as the qualitative behavior of the solutions are concerned. The graph-theoretical algorithms for larger circuits have to be analyzed by appropriate combinatorial algorithms. This is discussed in [6], where an algorithm for finding a pair of conjugate trees of polynomial complexity in the circuit size is described. Unfortunately, no algorithm of polynomial complexity in the circuit size has been found so far for excluding or finding partial uniform orientations of the resistors edges. To the best of our knowledge, the same problem persists also for the other two approaches [7–9].

5. Conclusions

We have given graph-theoretic criteria for the existence of global state equations in the capacitor and memristor charges in nonlinear circuits composed of basic circuit elements. Our list of basic circuit elements consists of: linear capacitors, linear resistors, voltage controlled and current controlled nonlinear resistors with increasing characteristics, nonlinear passive charge-controlled memristors, independent voltage and current sources, nullators and norators (a nullator-norator pair is equivalent to an ideal operational amplifier). Other circuit elements commonly used in electronics, such as transistors, inductors, nonlinear resistors with non-increasing characteristics, flux-controlled memristors, etc. have all equivalent circuits in the basic circuit elements.

The graph theoretical criteria give sufficient conditions for the existence of global state equations. One has to show that all linear resistive circuits (they are in general active non-reciprocal circuits, not just circuits composed of resistors and sources) with a certain structure, or with several structures, have exactly one solution. This problem, in turn, is reduced to the two purely graph-theoretical properties of the circuit structur(es). For small circuits, as those of the examples that we have given, the criteria can be verified or disproved by direct reasoning. For larger circuits, combinatorial algorithms have to be used.

The rather theoretical problem of the existence of global state equations is of practical importance. Even in numerical simulations, where usually a mixed differential-algebraic system is solved that can directly be established from the circuit description, it is important to know what the state variables are. The initial values of the state variables define the solutions; they can be chosen independently, while the other circuit variables have to satisfy constraints. Sometimes, the "usual candidates for state variables" are dependent and some of them have to be eliminated to obtain global state equations. We have given a simple example of them, linked to different descriptions of a memristor. Otherwise, we did not pursue here the reduction process, which is a typical and rather involved problem of differential-algebraic systems. Furthermore, there is the danger of impasse points, i.e. special values of circuit variables that fail to determine a solution even if for all other values they determine a solution. We also have given a simple example of a memristor circuit with this singular behavior. Non-detected impasse points can be lead to wrong conclusions of numerical simulations. Impasse points can be eliminated by completing the circuit model with suitable parasitic capacitors. Knowing their presence not only leads to better circuit models, but also indicates singular behavior, such as jumps in the solutions [5].

References

[1] Chua, L. O. (1971). Memristor — the missing circuit element. *IEEE Transactions on Circuit Theory, CT-18*(5), 507–519.

[2] Chua, L. O., & Alexander, G. R. (1971). The effedts of nonlinear reactances on nonlinear networks. *IEEE Trans. on Circuit Theory*, 520.

[3] Fosséprez, M., & Hasler, M. (1990). Resistive circuit topologies that admit several solutions. *Int. J. Circuit Theory and Applications, 18*, 625–638.

[4] Hasler, M. (1986). Non-linear non-reciprocal resistive circuits with a structurally unique solution. *Int. J. Circuit Theory and Applications, 14*, 237–262.

[5] Hasler, M., & Neirynck, J. (1986). *Nonlinear Circuits.* Boston: Artech House.

[6] Hasler, M., Marthy, C., Oberlin, A., & de Werra, D. (1994). A discrete model for studying existence aqnd uniqueness of solutions in nonlinear resistive circuits. *Discrete Applied Mathematics*, 169–184.

[7] Nielson, R. O., & Willson, A. N. (1984). A fundamental result concerning the topology of transistor circuits with multiple equilibria. *Proc. IEEE, 68*, 196–208.

[8] Nishi, T., & Chua, L. O. (1984). Non-linear op-amp circuits: existence and uniqueness of the solution by inspection. *Int. J. Circuit Theory and Applications, 12*, 145–173.

[9] Nishi, T., & Chua, L. O. (1984). Topological criteria for nonlinear resistive circuits containing controlled sources to have a unique solution. *IEEE Trans. Circuits and Systems, CAS-31*, 722–741.

[10] Willson (ed.), A. N. (1974). *Nonlinear Networks.* New York: IEEE Press.

CHAPTER 43

ANALYTICAL ANALYSIS OF MEMRISTIVE NETWORKS

TORSTEN SCHMIDT, WILLI NEUDECK, UTE FELDMANN and RONALD TETZLAFF

Faculty of Electrical and Computer Engineering, Technische Universität Dresden, Dresden, Germany

During the last years, many technical realizations of memristors have been conceptually and practically realized. The theory behind the memristor has been extended as well but mostly for memristors and less for electrical circuits with memristors which are often difficult to analyze due to the nonlinearity and the dynamics of the memristor. The theory of single memristor networks is the very aim of this paper. Based on the Multilinear Theory of Nonlinear Networks and on Volterra Series representation, a general description of single memristorcircuits is obtained.

1. Introduction

The theoretical concept of the memristor, a special kind of resistor with memory, was published first in 1971 by Leon O. Chua [4]. But the scientific interest in memristors remained limited. This changed abruptly by a letter to "Nature" in 2008. Strukov *et al.* [13] observed that certain nanoscale devices with thin semiconductor layers can be described as memristors. Since then, the memristor has a wide attention in the international research community of electrical engineers, physicists and biologists. Recent investigations are mainly focused on the realization of ultra dense VLSI resistive random access memories (ReRAM). The current main areas of research with respect to memristors are depicted in Fig. 1, but it can be assumed that more interesting fields will be opened up in future.

During the last years, many technical realizations of different memristor types have been

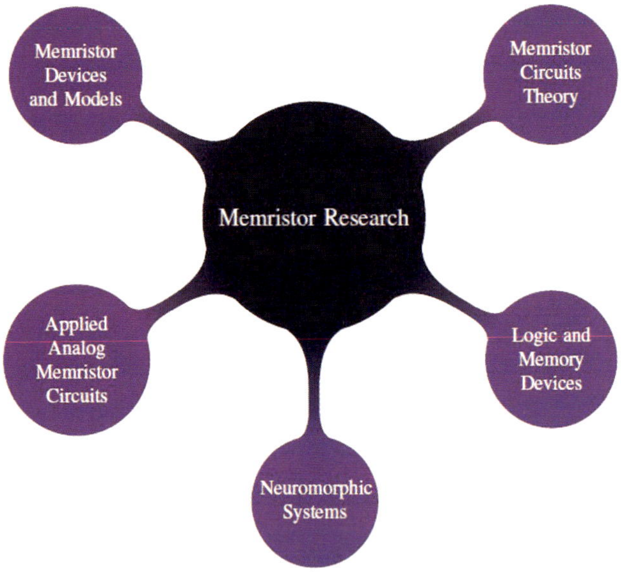

Fig. 1. The current main areas of research with respect to memristors.

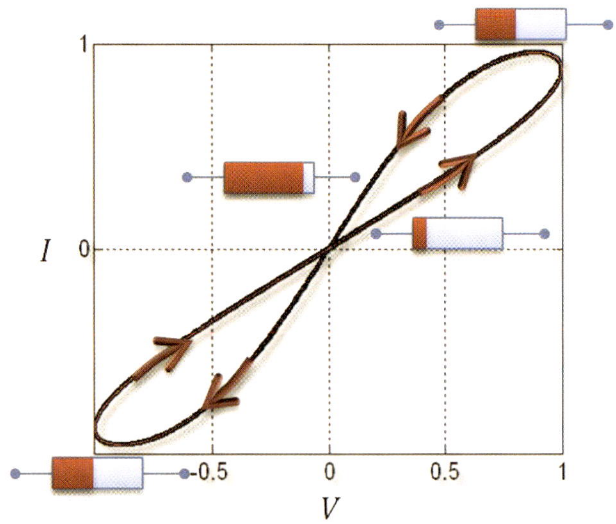

Fig. 2. The pinched hysteresis loop of a so-called HP memristor driven by a periodic source voltage $v(t) = \hat{V}\cos(\omega t)$. The memristor symbols indicate the state of the memristor, i.e. the wider the red region becomes, the higher is the conductance of the device.

conceptually and practically realized [3, 8, 15]. The theory behind the memristor has been extended as well but mostly within the frame of single memristors [3, 7] and less for extended networks which include memristors [6, 11]. Electrical circuits with memristors are often difficult to analyze due to the nonlinearity and the dynamics of the memristor.

The analytical description of memristor networks is the very aim of this chapter which is structured as follows: First, we will give a short introduction to the basic properties of single memristors in general and the HP memristor in particular. In Sec. 3, we then consider the *Multilinear Theory of Nonlinear Networks* [10], which can be applied to simple structured memristor networks. For more complex networks we introduce a method based on a *Volterra Series* representation. By means of this, a general description of single memristor circuits is obtained.

2. The Behavior of Single Memristors

According to Chua's definition [4] a memristor is a device which has a characteristic curve in the $\Phi - Q$ domain where $\Phi(t) = \int_{-\infty}^{t'} v(t')dt'$ denotes the flux and $Q(t) = \int_{-\infty}^{t'} i(t')dt'$ denotes the charge. The characteristic curve may be presented by either of the two possibilities,

$$\Phi = F(Q) \quad or \quad Q = G(\Phi). \tag{1}$$

In the first case, we name the device a charge controlled memristor whereas in the second case, the device is called a flux-controlled memristor. The memristance M and the memductance W are defined by

$$M = \frac{dF(Q)}{dQ} \quad and \quad W = \frac{dG(\Phi)}{d\Phi}, \tag{2}$$

respectively.

In 1976, Chua and Kang generalized the memristor to the concept of memristive systems [5]. The memristance of such a system depends on an internal state vector $\mathbf{x}$. A memristive system is defined by

$$\begin{aligned} v(t) &= M(\mathbf{x}) \cdot i(t) \\ \dot{\mathbf{x}} &= f(\mathbf{x}, i). \end{aligned} \tag{3}$$

A capacitance and inductance can be defined as well with an additional state vector giving the memory related quantities $L(\mathbf{x})$ (meminductance) and $C(\mathbf{x})$ (memcapacitance) [14]. The best investigated type of memristor is the TiO_2-memristor which usually is called the HP memristor for sake of simplicity. Starting with a linear drift model [13], the memristive behavior could be described simply by a series of voltage dividers. A more advanced model takes into account a nonlinear drift and the boundaries of the memristor by means of a window function $F(x)$ (e.g. [9]),

$$\begin{aligned} v(t) &= (R_{ON}x + R_{OFF}(1-x)) \cdot i(t) \\ \dot{x} &= \lambda \cdot i(t) \cdot F(x). \end{aligned} \tag{4}$$

More sophisticated models based on coupled ionic/electronic drift-diffusion [12] and tunneling mechanisms (in particular Simmons tunneling) [16] have been developed. In [7] it has been shown that the dynamical behavior of this element can be determined by calculating a Bernoulli-type differential equation

$$\dot{i}(t) - \frac{\dot{v}(t)}{v(t)} i(t) + \frac{k}{v(t)} [i(t)]^3 = 0. \tag{5}$$

Another approach in [2] provides the analytical treatment of a wider class of memristors, including the HP memristor with nonlinear drift and points out that in general, the dynamics of those systems can be described by a (generalized) Abel-type differential equation.

However, those descriptions are valid only for single memristors. Going further, even simple memristor circuits reveal very interesting properties. In Fig. 3, a circuit consisting of a capacitor, memristor and inductor is depicted. Let's imagine that the capacitor will be initially charged to +20 mV in one case and to −20 mV in another case. In Fig. 4 the quantities show that on the one hand, the initial voltage of +20 mV leads to a smooth relaxation of all quantities. On the contrary to this, the initial voltage of −20 mV leads to a short current spike and a following relaxation.

This simple example shows that memristors can change a circuit behavior dramatically which may be considered to develop a new type of circuit realizations. To describe simple networks with memristors, in the following we will apply the multilinear theory of Schetzen [10].

3. The Multilinear Approach

3.1. *Basics*

According to [10], the multilinear theory is based on the assumption that every (node) voltage and every (branch) current can be expressed as a series of the form

$$V(t) = \sum_{n=0}^{\infty} V_n(t).$$

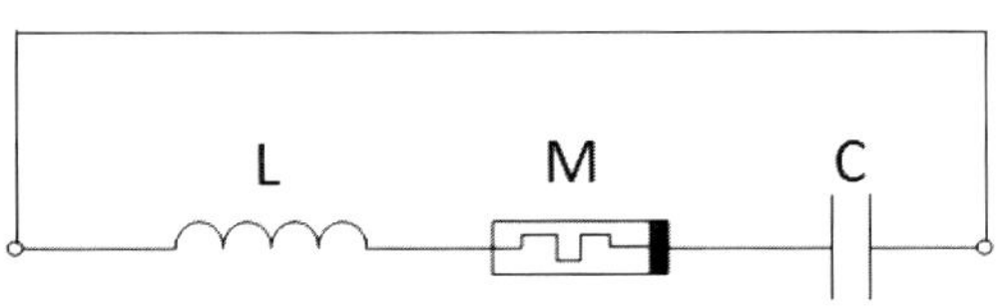

Fig. 3. A simple LCM branch with a pre-charged capacitor is shorted.

The n-th order voltages (or currents) $V_n(t)$ must be defined such that a multiplication of *every source* in the circuit with *the same* constant α will result in the *new* voltage (or current)

$$V(t) = \sum_{n=1}^{\infty} \alpha^n V_n(t). \tag{6}$$

Both of the Kirchhoff laws for current and voltage still hold for each order n [10]. And for each of the basic circuit elements, there are small linear sub-circuits (the "n-th order models") which implement the relationship between $V_n(t)$ and $I_n(t)$. The n-th order model of a circuit element does not depend on any quantity of order n or higher, but it may contain a time-dependent voltage source depending on voltages or currents of any order *lower* than n [10].

Given a nonlinear circuit, the analysis works as follows: One starts by replacing each circuit element with its first order model. The resulting linear circuit is solved, and the time-dependent sources of the second-order models are calculated. Then, the circuit elements are replaced by their second-order models, and the circuit (which is still linear) is solved again. This process may be repeated several times. In the end, the voltages and currents of the different orders are added simply, setting $\alpha = 1$, to get an approximation of the voltages and currents in the real circuit.

A diagram with an overview of the analysis using the multilinear theory of nonlinear networks is shown in Fig. 5. The n-th order models of several basic circuit elements (including the memristor, whose model will be derived in Sec. 3.2) can be found in Table 1.

Using the multilinear theory of nonlinear networks, one will get the same results, which could be obtained by an analysis based on Volterra series [10]. One advantage of the multilinear theory is the fact that the network structure of the circuit is not lost in the process. It is hoped that this might help with the study (or even the design) of nonlinear circuits.

3.2. *Extension to memristors*

Extending the multilinear theory to charge-controlled memristors involves calculations which are similar to those for nonlinear resistors, capacitors, and inductors. First, the relationship between $Q(t)$ and $I_n(t)$ has to be determined. If $I(t)$ is given

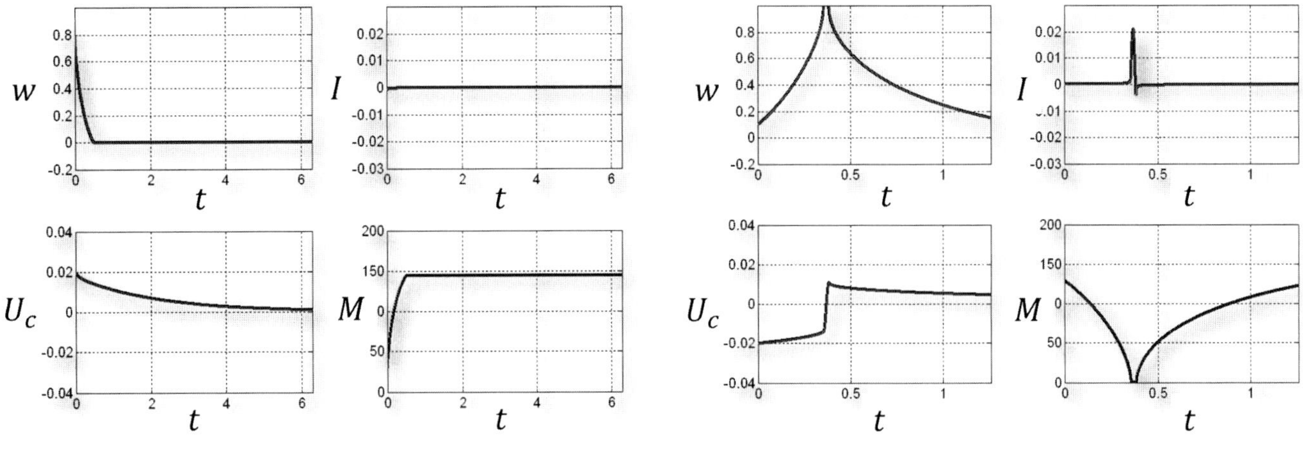

Fig. 4. Applying an initial voltage $U_c = +20\,\mathrm{mV}$, the voltages and currents simply relax to zero (a). On the contrary, with an initial voltage of $-20\,\mathrm{mV}$, the current exhibits a remarkable spike (b).

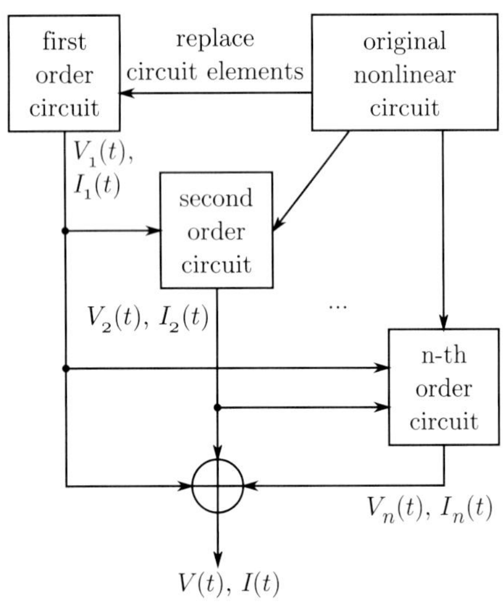

Fig. 5. Scheme of circuit analysis using the multilinear theory of nonlinear networks.

by a series of the form (6), the "charge" of the memristor can be expressed as

$$\begin{aligned} Q(t) &= Q(t_0) + \int_{t_0}^{t} \sum_{n=1}^{\infty} \alpha^n I_n(t')dt' \\ &= Q(t_0) + \sum_{n=1}^{\infty} \alpha^n \int_{t_0}^{t} I_n(t')dt' \qquad (7) \\ &= Q(t_0) + \sum_{n=1}^{\infty} \alpha^n Q_n(t), \end{aligned}$$

where $Q_n(t)$ is the time integral over the n-th order current. For now, $Q(t_0)$ is assumed to be zero.

The incremental memristance has to be represented approximately as a Taylor series (or a suitable polynomial)

$$M(Q) = M_0 + \sum_{k=1}^{\infty} M_k Q^k. \qquad (8)$$

For instance, the simple model of the memristor implementation in [13] has a memristance of this form.

By inserting (7) and (8) into $V(t) = M(Q(t)) \cdot I(t)$, we get

$$\begin{aligned} V(t) &= M_0 \sum_{n=1}^{\infty} \alpha^n I_n(t) \\ &\quad + \sum_{k=1}^{\infty} M_k \left(\sum_{n=1}^{\infty} \alpha^n Q_n(t) \right)^k \cdot \sum_{r=1}^{\infty} \alpha^r I_r(t) \\ &= M_0 \sum_{n=1}^{\infty} \alpha^n I_n(t) + \sum_{k,r=1}^{\infty} \sum_{n_1=1}^{\infty} \cdots \sum_{n_k=1}^{\infty} \\ &\quad \times (M_k Q_{n_1}(t) \cdots Q_{n_k}(t) \alpha^{n_1+\cdots+n_k+r} I_r(t)). \end{aligned}$$

Sorting the summands according to the exponent of α, the n-th order voltage can be obtained

$$\begin{aligned} V_n(t) &= M_0 I_n(t) \\ &\quad + \sum_{r=1}^{n-1} \sum_{k=1}^{n-r} \sum_{\substack{n_1,\ldots,n_k \in \mathbb{N}\setminus\{0\} \\ n_1+\cdots+n_k = n-r}} \\ &\quad \times (M_k Q_{n_1}(t) \cdots Q_{n_k}(t) I_r(t)). \qquad (9) \end{aligned}$$

Table 1. The n-th order models of basic circuit elements.

Circuit element	Replacement by n-th order model	Details
Independent voltage source, $V(t)$ is given [10]	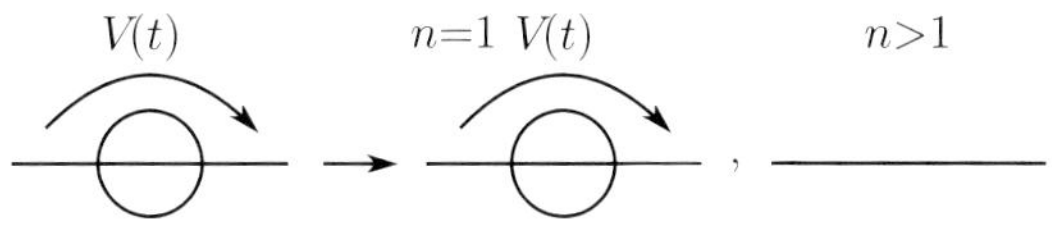	first order: no change higher orders: short circuit
Independent current source, $I(t)$ is given [10]	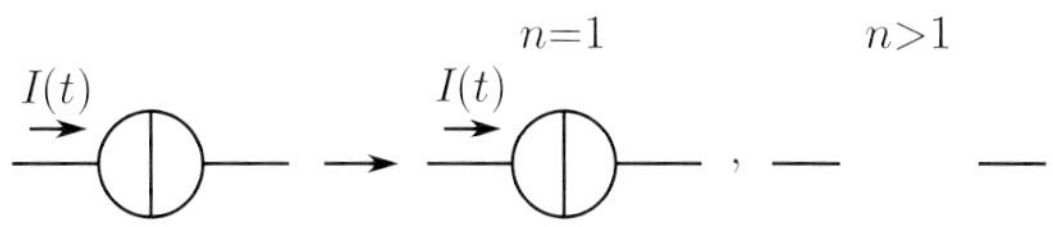	first order: no change higher orders: open circuit
Nonlinear resistor, $V(t) = \sum_{k=1}^{\infty} R_k I^k(t)$ [10]	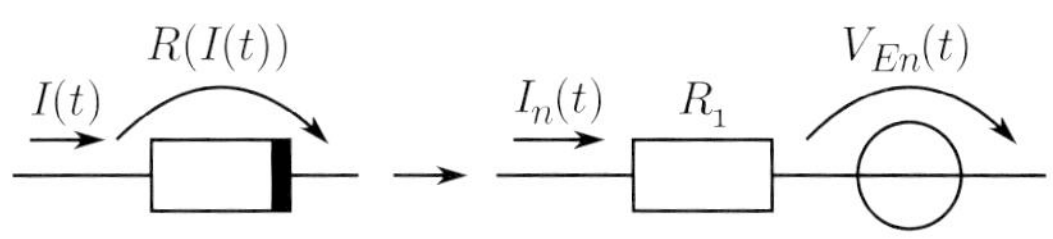	$V_{E1}(t) = 0$ i.e. shorted $V_{E2}(t) = R_2 I_1^2(t)$ $V_{E3}(t) = 2R_2 I_1(t) I_2(t) + R_3 I_1^3(t)$
Nonlinear capacitor, $V(t) = \sum_{k=1}^{\infty} S_k Q^k(t)$ [10]	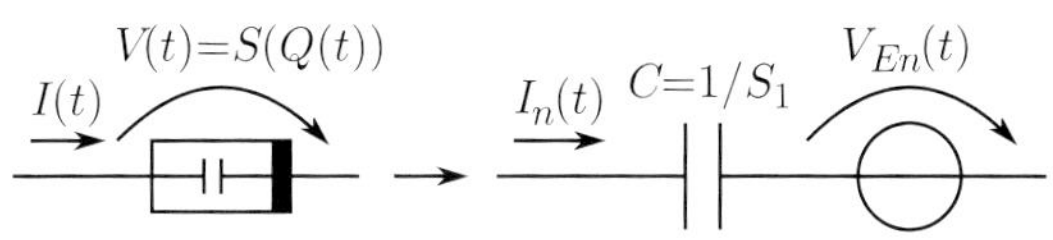	$V_{E1}(t) = 0$ $V_{E2}(t) = S_2 Q_1^2(t)$ $V_{E3}(t) = 2S_2 Q_1(t) Q_2(t) + S_3 Q_1^3(t)$
Nonlinear inductor, $V(t) = \dot{\Phi}(t)$ $\Phi(t) = \sum_{k=1}^{\infty} L_k I^k(t)$ [10]	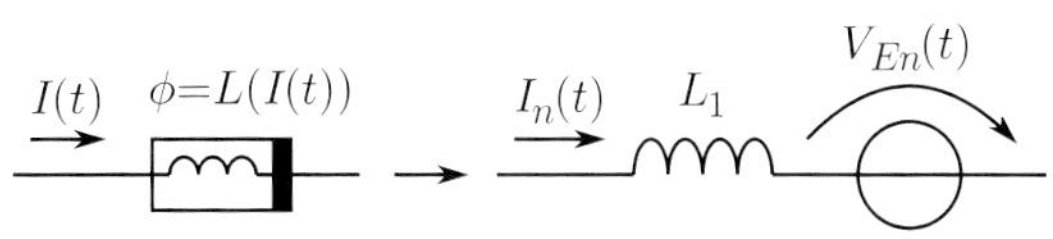	$V_{E1}(t) = 0$ $V_{E2}(t) = \frac{d}{dt} L_2 I_1^2(t)$ $V_{E3}(t) = \frac{d}{dt}(2L_2 I_1(t) I_2(t) + L_3 I_1^3(t))$
Memristor, $V(t) = M(Q(t))I(t)$ $M(Q) = \sum_{k=0}^{\infty} M_k Q^k$ (Section 3.2)	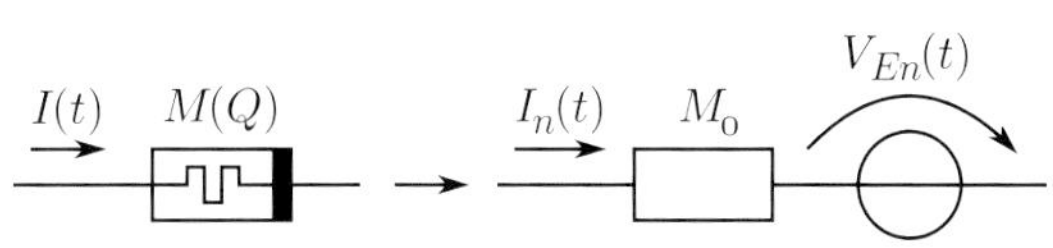	$V_{E1}(t) = 0$ $V_{E2}(t) = M_1 Q_1(t) I_1(t)$ $V_{E3}(t) = M_1 Q_2(t) I_1(t) + M_2 Q_1^2(t) I_1(t)$ $+ M_1 Q_1(t) I_2(t)$

For the first three orders n, we get

$$V_1(t) = M_0 I_1(t),$$

$$V_2(t) = M_0 I_2(t) + \Big(M_1 Q_1(t) I_1(t)\Big),$$

$$V_3(t) = M_0 I_3(t) + \Big(M_1 Q_2(t) I_1(t) + M_2 Q_1^2(t) I_1(t) + M_1 Q_1(t) I_2(t)\Big).$$

It can be seen that $V_n(t)$ is the sum of M_0 times the n-th order current and a term in parentheses, which does not depend on any quantity of order $\geq n$. The n-th order memristor-model therefore is a linear resistor M_0, connected in series with a time-dependent voltage source (see Table 1).

3.3. *CM-circuit*

For the sake of simplicity it is assumed that the capacitor in the CM-circuit (Fig. 6) is linear, and

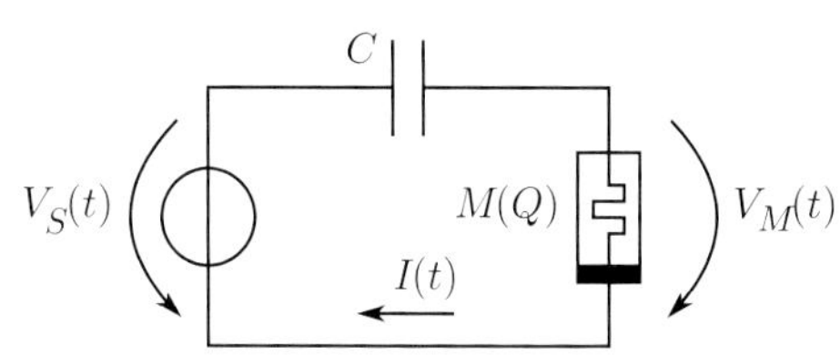

Fig. 6. Schematic diagram of a CM-circuit.

that the memristance is given as

$$M(Q) = M_0 + M_1 Q \quad \text{with } M_0 C > 0.$$

The circuit shall be excited by an independent sinusoidal voltage source

$$V_S(t) = \hat{V}_S \cos(\omega_0 t + \varphi_{V_S}), \tag{10}$$

and we want to calculate the current $I(t)$ through the memristor.

Furthermore, $Q(t) = Q_C(t)$ is assumed where $Q_C(t)$ denotes the charge of the capacitor. Then,

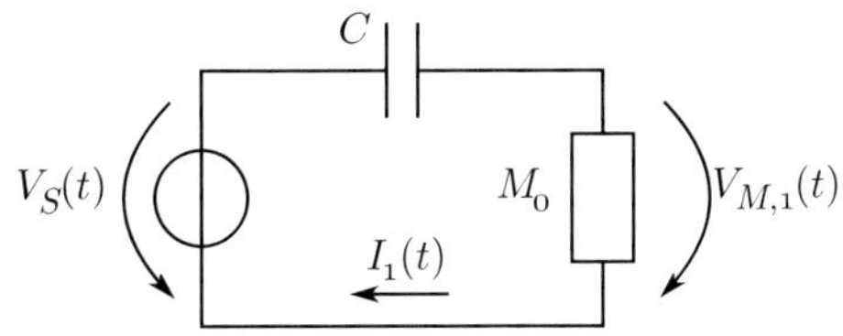

Fig. 7. First order model of the CM-circuit.

the resulting system has an attractive fixed point $Q_{FP} = 0$ for $V_S(t) \equiv 0$. Thus, for sufficiently small $Q(0)$, the circuit with $Q_C(t) = Q(t)$ and no excitation has unique asymptotic behavior. We assume that is true for the excited circuit. Consequently, we disregard the transient response, and calculate the steady state current only. (However, if $Q(0) - Q_C(0)$ *was* nonzero, $Q(t)$ would follow $Q_C(t)$ with a constant offset, i.e. it would act as a "shadow state". The analysis of the system therefore would be slightly different.)

First order circuit: First, the circuit elements in the above mentioned circuit have to be replaced by their individual first order models, which can be taken from Table 1. The resulting first order circuit is shown in Fig. 7.

As the first order circuit is linear, the first order current $I_1(t)$ can be determined directly. For a sinusoidal excitation $V_S(t)$ with the angular frequency ω_0, the current $I_1(t) = \hat{I}_{1,\omega_0} \cos(\omega_0 t + \varphi_{I_1,\omega_0})$ must be sinusoidal, too. Employing the transfer function

$$\underline{G}_1(j\omega) = \frac{\underline{I}_1}{\underline{V}_S} = \frac{1}{M_0 - j\frac{1}{\omega C}},$$

the amplitude $\hat{I}_{1,\omega_0}$ and the phase φ_{I_1,ω_0} of the first order current follow as

$$\hat{I}_{1,\omega_0} = \hat{V}_S |\underline{G}_1(j\omega_0)| = \frac{\hat{V}_S}{\sqrt{M_0^2 + \left(\frac{1}{\omega_0 C}\right)^2}}$$

and

$$\varphi_{I_1,\omega_0} = \varphi_{V_S} + \arg \underline{G}_1(j\omega_0) = \varphi_{V_S} + \arctan \frac{1}{\omega_0 M_0 C}.$$

Integrating $I_1(t)$ with respect to time, we get

$$Q_1(t) = \frac{\hat{I}_{1,\omega_0}}{\omega_0} \cos\left(\omega_0 t + \varphi_{I_1,\omega_0} - \frac{\pi}{2}\right).$$

Second order circuit: The second order circuit contains a new independent voltage source, which is a part of the memristor model. (Fig. 8.) The voltage $V_{E2}(t)$ of this source has to be determined by using the results for the first order circuit.

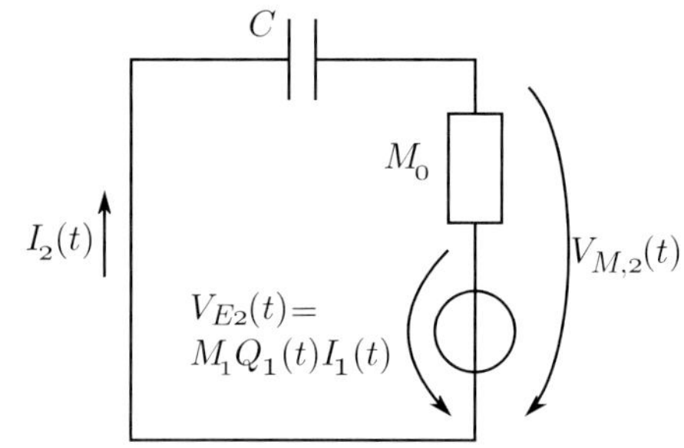

Fig. 8. Second order model.

With the abbreviation $A = \omega_0 t + \varphi_{I_1,\omega_0}$, $V_{E2}(t)$ can be written as

$$\begin{aligned} V_{E2}(t) &= M_1 Q_1(t) I_1(t) \\ &= M1 \frac{\hat{I}^2_{1,\omega_0}}{\omega_0} \cos(A) \cos\left(A - \frac{\pi}{2}\right) \\ &= \frac{M_1 \hat{I}^2_{1,\omega_0}}{2\omega_0} \left[\cos\left(2A - \frac{\pi}{2}\right) + \cos\frac{\pi}{2}\right] \\ &= \hat{V}_{E2,2\omega_0} \cos(2\omega_0 t + \varphi_{V_{E2},2\omega_0}). \end{aligned}$$

The sole excitation of the second order circuit therefore is a sinusoidal voltage with the angular frequency $2\omega_0$, and the amplitude/phase

$$\hat{V}_{E2,2\omega_0} = \frac{M_1 \hat{I}^2_{1,\omega_0}}{2\omega_0}, \quad \varphi_{V_{E2},2\omega_0} = 2\varphi_{I_1,\omega_0} - \frac{\pi}{2}.$$

Again, a transfer function $\underline{G}_2(j\omega) = -\underline{G}_1(j\omega)$ is used to calculate the current $I_2(t)$. The amplitude of the second order current is

$$\hat{I}_{2,2\omega_0} = \frac{\hat{V}_{E2,2\omega_0}}{\sqrt{M_0^2 + \left(\frac{1}{2\omega_0 C}\right)^2}},$$

and it has a phase of

$$\varphi_{I_2,2\omega_0} = \varphi_{V_{E2},2\omega_0} + \arctan \frac{1}{2\omega_0 M_0 C} + \pi.$$

$Q_2(t)$ can be obtained by integrating $I_2(t)$.

Third order circuit: The only difference between the second and the third order circuit is the time dependent voltage source that belongs to the memristor model. Both $V_{E3}(t)$ and $I_3(t)$ can be written as the sum of two sinusoidal components with the angular frequencies ω_0 and $3\omega_0$ respectively.

Results: To get an approximation of the current $I(t)$ in the original circuit, we just add the current

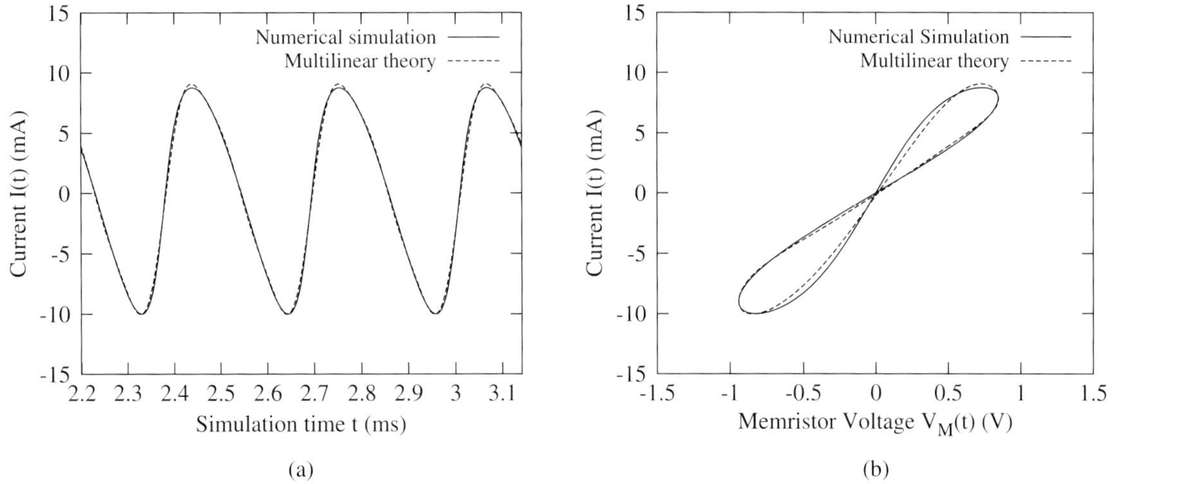

Fig. 9. Comparison of the results obtained by applying the multilinear theory (dashed lines) to those calculated in a numerical simulation of the CM-circuit (solid lines). The diagrams show the current in the circuit (a) and the pinched hysteresis loop of the memristor (b).

of first, second, and third order leading to

$$\begin{aligned} I(t) \approx\ & \hat{I}_{1,\omega_0}\cos(\omega_0 t + \varphi_{I_1,\omega_0}) \\ & + \hat{I}_{2,2\omega_0}\cos(2\omega_0 t + \varphi_{I_2,2\omega_0}) \\ & + \hat{I}_{3,\omega_0}\cos(\omega_0 t + \varphi_{I_3,\omega_0}) \\ & + \hat{I}_{3,3\omega_0}\cos(3\omega_0 t + \varphi_{I_3,3\omega_0}). \end{aligned} \tag{11}$$

A numerical simulation was carried out using $C = 1\,\mu\mathrm{F}$, $M_0 = 100\,\Omega$, $M_1 = 0.1\,\mathrm{G}\Omega(\mathrm{As})^{-1}$, $\omega_0 = 20000\,\mathrm{s}^{-1}$, $\hat{V}_S = 1\,\mathrm{V}$, $\varphi_{V_S} = 0.166667\,\pi$ and $Q(0) = 2.5\,\mu\mathrm{As}$. In Fig. 9 it can be seen, that (after decay of the transient) the approximation obtained by using (11) is very close to a numerical simulation of the example circuit.

4. A General Description of Single Memristor Circuits by Means of the Volterra Series Approach

Although the application of the multilinear theory leads to accurate approximations in several cases, it is desirable to describe circuits with memristors on a more general basis. This can be achieved by introducing the so called single memristor circuits shown in Fig. 10. A general linear two-port which might be represented by its impedance matrix is coupled to a memristor and a given current source. Please note that this specific form is defined without loss of generality, i.e. a voltage source instead of the current source can be described similarly.

In the following, the Volterra Series are introduced and applied to the SMC.

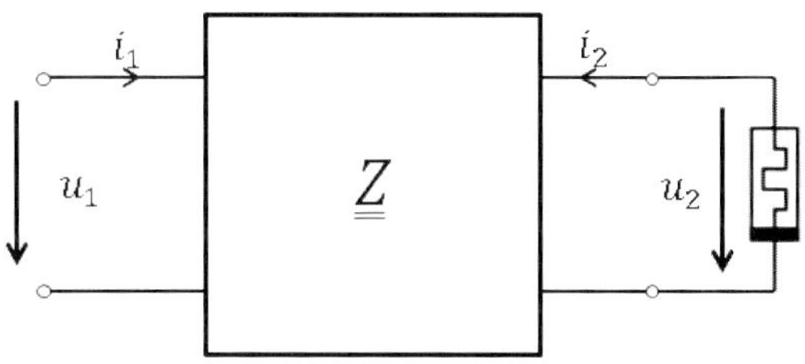

Fig. 10. The general structure of a single memristor circuit consisting of a linear two-port connected to a source and a memristor.

4.1. *The Volterra series approach*

Let's assume a nonlinear system with an input function $x(t)$ and the corresponding output function $y(t)$. The Volterra series relates the input to the output function according to

$$\begin{aligned} y(t) = \sum_{n=1}^{\infty}\int_{-\infty}^{\infty}\cdots\int_{-\infty}^{\infty} h_n(\tau_1,\ldots,\tau_n)x(t-\tau_1)\ldots \\ \times x(t-\tau_n)d\tau_1\ldots d\tau_n. \end{aligned} \tag{12}$$

The index n defines the so-called Volterra order and hence (12) can be imagined as a decomposition of y into the output y_n of homogeneous systems of order n. The functions $h_n(\tau_1,\ldots,\tau_n)$ denote each the Volterra kernel of order n. Once the kernels h_n are known, it is possible to determine the correct output function corresponding for a given input. The kernels in the complex frequency domain are the Laplace transforms of the kernels in the time domain

$$H_n(s_1, \ldots, s_n) = \int_{-\infty}^{\infty} \cdots \int_{-\infty}^{\infty} h_n(\tau_1, \ldots, \tau_n) \times e^{-(s_1\tau_1 + \cdots + s_n\tau_n)} d\tau_1 \ldots d\tau_n. \tag{13}$$

In order to determine the Volterra kernels, we use the harmonic probing method [1]. This method is based on the application of a n-tone input signal $\sum_{k=1}^{n} e^{s_k t}$ in order to determine the Volterra-kernel of order n, e.g. to carry out the procedure for the first order needs an input of $x(t) = e^{st}$. The response can be directly calculated according to (12) as

$$y(t) = \sum_{k=1}^{\infty} H_k(s, \ldots, s) e^{kst}. \tag{14}$$

Using (14) and extracting an equation only for terms with the factor e^{st}, provides the first order kernel $H_1(s)$. The same procedure can be executed for higher order kernels and is explained in detail in [1].

4.2. *Theory of single memristor circuits*

In this section we will apply the Volterra series approach to the general case of a SMC. Without loss of generality, we consider the linear part of the circuit being represented by its impedance matrix $\mathbf{Z}$. If the impedance matrix doesn't exist for the circuit, one can use the admittance matrix and carry out a similar analysis. Also, without loss of generality, we define a current source providing the input signal i_1. The output signal then is the current i_2 flowing through the memristor. Knowing i_2, other circuit variables can be calculated by means of classical circuit theory. In order to describe a wider class of memristors, we use a polynomial model for the memristance

$$M(Q) = \sum_{k=1}^{\infty} m_k Q^k. \tag{15}$$

This approach holds for a charge controlled memristor, but it could be replaced as well by a flux dependent equation. The first term with $k = 0$ was neglected in the sum (15) in order to separate the linear part from the nonlinear part of the circuit. However, if the memristor had a non-vanishing resistance at $Q = 0$, it can be taken into account as a linear resistive one-port in the linear part of the circuit. In this case, we could redefine the function $M(Q)$ as a function $\tilde{M}(Q)$ which already includes this substitution, i.e. we do not need to take care of this in the further analysis. Generally, the state variable Q does not denote the charge but is rather related to the charge and a constant summand, i.e. the time derivatives of both quantities are equal. Equation (15) represents also the HP-type memristor which just includes the first order. Please note that this approach also can include saturation effects. The linear system with its impedance operator fulfills the following dynamic equations

$$\begin{aligned} u_1 &= Z_{11} i_1 + Z_{12} i_2 \\ u_2 &= Z_{21} i_1 + Z_{22} i_2, \end{aligned} \tag{16}$$

where Z_{ij} are the impedance matrix elements. Considering only the second row and replacing U_2 by the product of the current and the memristance, gives

$$Z_{21} i_1 + Z_{22} i_2 = u_2 = -i_2 \sum_{k=1}^{\infty} m_k \left(\int i_2 dt' \right)^k. \tag{17}$$

So far, we have obtained a representation of the system which includes only the input current i_1 and the output current i_2. We now have to derive the Volterra kernels of different orders by applying the previously described methods. For the further calculation it is convenient to separate the different Volterra orders first. This is done by introducing a factor c to the input i_1. As it is obvious from (12), the n-th order output i_{2n} will be multiplied by a factor c^n. Equation (18) will hence become

$$Z_{21} c i_1 + Z_{22} \sum_n c^n i_{2n} = -\sum_n c^n i_{2n} \sum_{k=1}^{\infty} m_k \left(\int \sum_n c^n i_{2n} dt' \right)^k \tag{18}$$

This equation leads to a separation of the Volterra orders by simply sorting orders of c. The next step is to draw equations for each Volterra order out of (18) and afterwards to determine the Volterra kernels by means of an auxiliary n-tone input signal. The first order is represented by factors of the c^1 terms

$$Z_{21} i_1 + Z_{22} i_{21} = 0. \tag{19}$$

If the input is $i_1 = 1$, i.e. an amplitude "1" for the single tone signal $e^{j\omega_1 t}$, for the output we get $i_{21} = H_1(\omega_1)$. Inserting these expressions into (19)

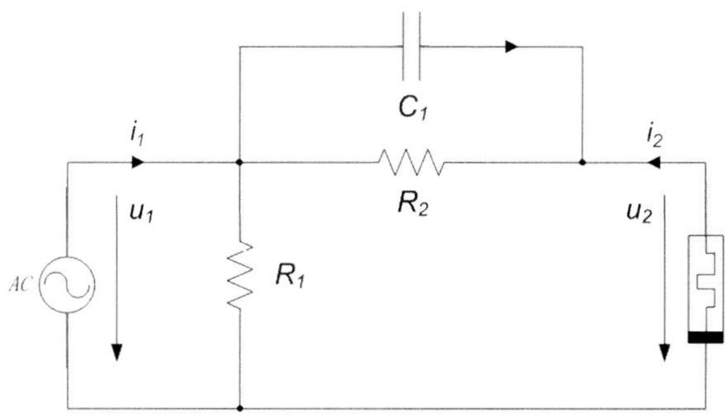

(a)

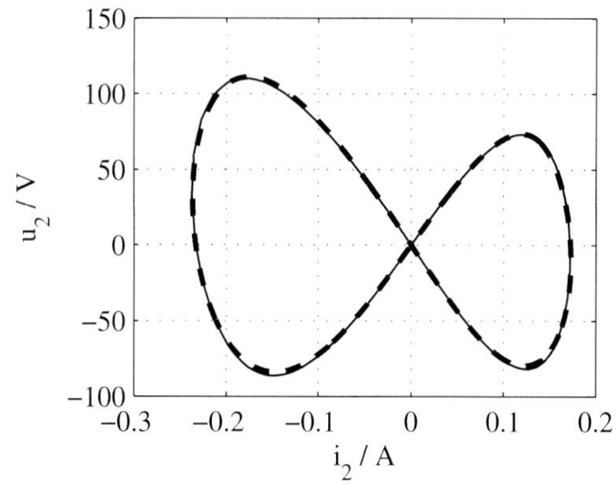

(b)

Fig. 11. (a) A simple circuit with two dynamic devices. Parameters: $\omega = 200\,\text{Hz}$, $C_1 = 1\mu\text{F}$, $R_1 = 1\text{k}\Omega$, $R_2 = 10^{14}\Omega$, $m_1 = 9.1 \cdot 10^5\,\Omega/C$. (b) The Volterra based pinched hysteresis loop (dashed) is in good agreement to the numerically calculated curve. Please note that the linear part of the memristor has been included in the linear part of the network and hence the hysteresis loop is aligned horizontally.

leads to the following equations

$$Z_{21}(\omega_1) + Z_{22}(\omega_1)H_1(\omega_1) = 0. \tag{20}$$

$$\longrightarrow H_1(\omega_1) = -\frac{Z_{21}(\omega_1)}{Z_{22}(\omega_1)}. \tag{21}$$

Collecting the c^2 terms for the second order, yields the equation

$$Z_{22}i_{22} = -i_{21}m_1\left(\int i_{21}(t')dt'\right). \tag{22}$$

Similar to the first order, the amplitude of the response to $e^{\mathrm{j}(\omega_1+\omega_2)t}$ according to the multi tone-probing is $i_{22} = H_2(\omega_1, \omega_2)$. This allows the extraction of the second order kernel

$$\longrightarrow H_2(\omega_1, \omega_2) = -\frac{m_1}{\mathrm{j}\omega_2}\frac{1}{Z_{22}(\omega_1+\omega_2)}H_1(\omega_2)H_1(\omega_1). \tag{23}$$

And for the third order, we obtain in a similar way

$$Z_{22}i_{23} = -\left(i_{21}m_1\int i_{22}(t')dt' + m_2 i_{21}\int i_{21}(t')dt'\int i_{21}(t'')dt'' + m_1 i_{22}\int i_{21}dt'\right) \tag{24}$$

$$\longrightarrow H_3(\omega_1, \omega_2, \omega_3) = -\frac{1}{Z_{22}(\omega_1+\omega_2+\omega_3)}\left[\frac{m_1}{\mathrm{j}(\omega_2+\omega_3)}H_1(\omega_1)H_2(\omega_2,\omega_3) + \frac{m_2}{\omega_2\omega_3}H_1(\omega_1)H_1(\omega_2)H_1(\omega_3) + \frac{m_1}{\mathrm{j}\omega_3}H_2(\omega_1,\omega_2)H_1(\omega_3)\right]. \tag{25}$$

In (23) and (24) antisymmetric kernels have been used to minimize the number of terms. Although the analytical expressions of the kernels become more complex with increasing order, their form remains the same, no matter which circuit is considered.

4.3. *Examples*

Results for the first case are given in Fig. 11. The inner part of the circuit which consists of resistors and a capacitor, can be represented by its impedance matrix $\mathbf{Z}$

$$\mathbf{Z} = \begin{pmatrix} R_1 & R_1 \\ R_1 & R_1 + \dfrac{1}{sC + 1/R_2} \end{pmatrix}. \tag{26}$$

We assume a HP-type memristor which contains the first order memristance $M(Q) = m_1 \cdot Q$. The input signal is given by $i_1 = \hat{i}_1 \sin(\omega_0 t)$. The output $i_2(t)$ is determined first by a numerical simulation, using the Rosenbrock algorithm which is implemented in the ode23s-solver in Matlab, and

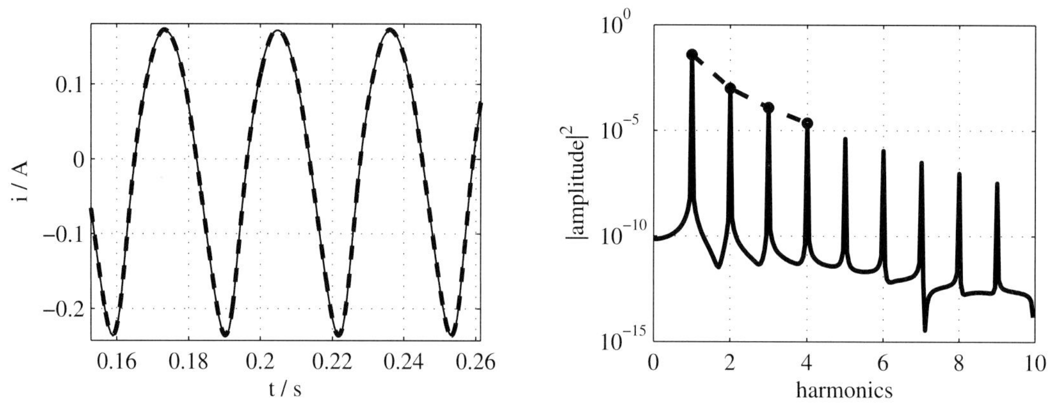

Fig. 12. (a) The comparison of $i(t)$ between the Volterra result (dashed) and the numerical curve (solid). (b) The Volterra series of the fourth order provides good results for the amplitudes up to the fourth harmonic.

second by the Volterra based single memristor circuit approach.

In Fig. 11, the pinched hysteresis loop of the result obtained numerically is compared to the Volterra series based result, which has been determined by taking into account terms up to Volterra order four. It confirms the high approximation accuracy of the Volterra approach. The power spectrum of the output function is presented in Fig. 12. Up to the 4th harmonic, the Volterra amplitudes show a maximal relative error of 0.3% as compared to the numerically calculated amplitudes. In general, we can expect the Volterra ansatz providing good approximations for the higher harmonics up to the highest Volterra order which we take into account in the analytical calculation.

The parameter m_1 is a measure for the strength of the nonlinearity of the system. In the calculation, we used $m_1 = 9.1 \cdot 10^5 \Omega/C$ which is the highest value of m_1 that leads to a steady state output. Increasing the value of m_1 further, forces the memristor into a state of being an energy source. Thus, the current i_2 and the voltage u_2 will finally diverge. Hence we can conclude for the considered circuit that in all cases with steady states, the Volterra approach provides an excellent approximation.

A second example and the corresponding results are given in Figs. 13 and 14. In this case, there are four dynamic devices and we have four independent states. As in the previous example, the time dependent circuit quantities which are calculated by means of the derived Volterra Kernels, are in good agreement with their corresponding numerical values.

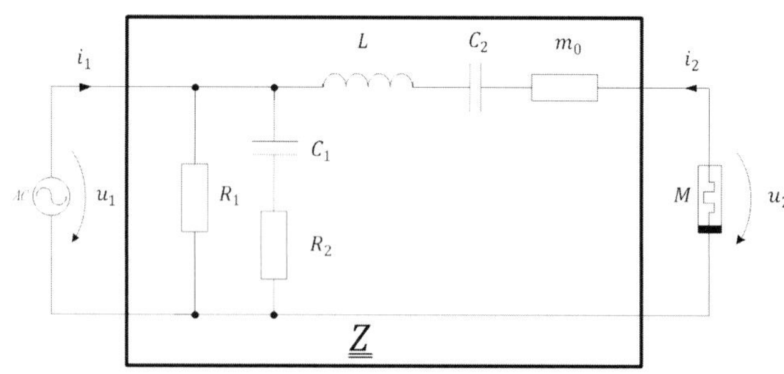

Fig. 13. Another example circuit with three branches and four dynamic devices. Parameters: $\omega = 200\,\text{Hz}$, $L = 5\text{mH}$, $C_1 = 2\mu\text{F}$, $C_2 = 1\mu\text{F}$, $R_1 = 1\text{k}\Omega$, $R_2 = 100\Omega$, $m_0 = 2\text{k}\Omega$, $m_1 = 0.83 \cdot 10^6 \Omega/C$.

5. Conclusions

In this chapter we have introduced two methods for the analysis of circuits with memristors. The multilinear method proved to deliver good results for several circuits with a given structure while the Volterra series based approach describes the behavior of a class of memristor circuits. We introduced the class of SMC and provided an analytical solution for this class in the form of (21)–(25). The derived equations can be adapted easily to other circuit forms (e.g. when the linear system is given by its admittance matrix) and other input-output quantities. Provided a unique steady state, the described Volterra based SMC analysis is quite general and hence gives the possibility to investigate SMC systematically. The influence of circuit parameters (e.g. a capacity value) on the output properties can hence be studied analytically without the need to carry out numerical simulations for

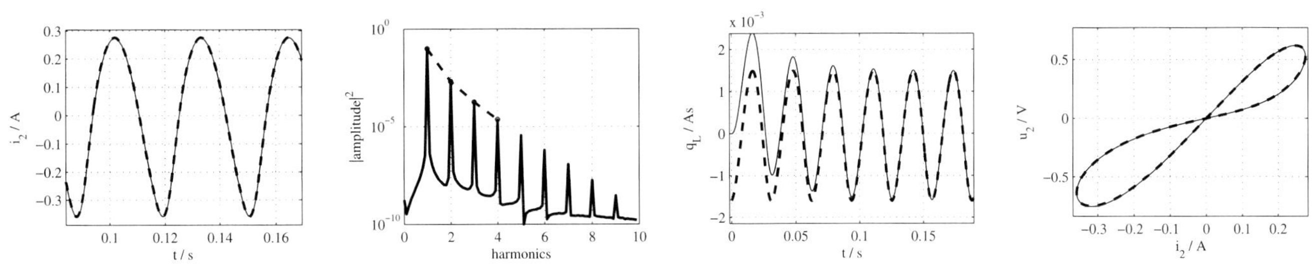

Fig. 14. The comparison between the Volterra result (dashed) and the numerically calculated curves. For all quantities, the Volterra approach delivers good approximations. As expected, the Volterra result does not reproduce the transient behavior of the circuit.

each specific parameter setup. For the future, it is planned to extend the current approach to circuits with an arbitrary number of memristors.

References

[1] Boyd, S., Tang, Y. and Chua, L. (1983). Measuring volterra kernels, *Circuits and Systems, IEEE Transactions on* **30**, 8, pp. 571–577.

[2] Cai, W., Ellinger, F., Tetzlaff, R. and Schmidt, T. (2011a). Abel dynamics of titanium dioxide memristor based on nonlinear ionic drift model, Submitted to em IEEE Transactions on Circuits and Systems, arXiv:1105.2668.

[3] Cai, W., Schmidt, T., Joerges, U. and Ellinger, F. (2011b). A feedback spin-valve memristive system, Submitted to IEEE Transactions on Circuits and Systems, arXiv:1104.5518.

[4] Chua, L. O. (1971). Memristor — the missing circuit element, *IEEE Trans. Circuit Theory* **18**, 5, pp. 507–519, URL http://ieeexplore.ieee.org/xpls/abs_all.jsp?arnumber=1083337.

[5] Chua, L. O. and Kang, S. M. (1976). Memristive devices and systems, *Proc. IEEE* **64**, 2, pp. 209–223, http://ieeexplore.ieee.org/xpls/abs_all.jsp?arnumber=1454361.

[6] Corinto, F., Ascoli, A. and Gilli, M. (2011). Nonlinear dynamics of memristor oscillators, *IEEE Trans. Circuits Syst.* **58**, 6.

[7] Drakakis, E. M., Yaliraki, S. N. and Barahona, M. (2010). Memristors and bernoulli dynamics, in *2010 12th International Workshop on Cellular NanoscaleNetworks and their Applications (CNNA)*, http://ieeexplore.ieee.org/xpls/abs_all.jsp? arnumber=5430324.

[8] Erokhin, V. (2010). Organic memristors: basic principles, in *Proceedings of 2010 IEEE International Symposium on Circuits and Systems (ISCAS)*, pp. 5–8, http://ieeexplore.ieee.org/xpls/abs_all.jsp?arnumber=5537145.

[9] Joglekar, Y. N. and Wolf, S. J. (2009). The elusive memristor: properties of basic electrical circuits, *European Journal of Physics* **30**, 4, pp. 661–675, http://iopscience.iop.orag/0143-0807/30/4/001.

[10] Schetzen, M. (1985). Multilinear theory of nonlinear networks, *Journal of the Franklin Institute* **320**, 5, pp. 221–247, http://linkinghub.elsevier.com/retrieve/pii/0016003285900237.

[11] Schmidt, T., Feldmann, U., Neudeck, W. and Tetzlaff, R. (2011). Analytical approach to single memristor circuits, in *Proc. of the European Conference of Circuits Technology and Devices (ECCTD2011)*, pp. 93–96.

[12] Strukov, D., Borghetti, J. and Williams, R. (2009). Coupled ionic and electronic transport model of thin-film semiconductor memristive behavior, *small* **5**, 9, pp. 1058–1063.

[13] Strukov, D. B., Snider, G. S., Stewart, D. R. and Williams, R. S. (2008). The missing memristor found, *Nature* **453**, 7191, pp. 80–83, http:// www.nature.com/nature/journal/v453/n7191/full/nature06932.html.

[14] Ventra, M. D., Pershin, Y. V. and Chua, L. O. (2009). Circuit elements with memory: Memristors, memcapacitors, and meminductors, *Proc. IEEE* **97**, 10, pp. 1717–1724, http://ieeexplore.ieee.org/xpls/abs_all.jsp?arnumber=5247127.

[15] Wang, X., Chen, Y., Xi, H., Li, H. and Dimitrov, D. (2009). Spintronic memristor through spin-torque-induced magnetization motion, *IEEE Electron Device Lett.* **30**, 3, pp. 294–297, http://ieeexplore.ieee.org/xpls/abs_all.jsp?arnumber=4781542.

[16] Williams, S. (2010). Talk at the second memristor and memristive systems symposium, Berkeley.

CHAPTER 44

HARDWARE MEMRISTOR EMULATORS

ANDREW L. FITCH and HERBERT H. C. IU

School of Electrical, Electronic and Computer Engineering,
The University of Western Australia

CHI K. TSE

Department of Electronic and Information Engineering,
The Hong Kong Polytechnic University

This chapter reviews a number of recent methods to model memristive systems using analog and micro-processor based circuits. As memristors have not yet entered the market in any format, these emulator circuits provide effective platforms for researchers to develop applications for memristive systems.

1. Introduction

This chapter examines a variety of recent methods to model memristive systems using analog and micro-processor based circuits. Since Hewlett-Packard's (HP) announcement in 2008 of the successful fabrication of a memristor [1], researchers have developed a number of circuits to observe memristive behavior either as a stand-alone phenomenon or in a practical application as a component.

The development of functional memristor models is important, if not essential, in the exploration of the properties and the development of applications of this new component. Currently, memristors have not entered the market in any format. It is hoped, although there has been no indication of it happening, that at some time in the future the memristor will be available as a through-hole or surface mount two-terminal device. This will greatly benefit application development and prototyping. Few are prepared to wait for this to happen, and for now, all research into developing applications for memristive systems and knowledge of memristors must be done with circuit-based and mathematical models.

2. The Memristor Modeling Circuits

In this section, a number of proposed memristor modeling circuits (MMC) will be examined.

2.1. *JFET based flux-controlled MMC*

Valsa *et al.* [2] developed a flux controlled MMC in 2010. The circuit is intended to be used for obtaining proof of theoretical concepts regarding memristor properties by observing a memristive response to different input signals. The circuit uses a JFET to provide voltage controlled conductance (G_M) and an operational amplifier based integrator to obtain a voltage representing flux (v_ϕ) which is used to control G_M. In response to an input sinusoidal signal of the appropriate frequency and amplitude, the circuit enables the observation of a characteristic hysteretic loop passing through zero.

Due to the use of an integrator sub-circuit to obtain the voltage corresponding to flux, the circuit's operating region in terms of frequency is limited to signals lower than 5 Hz. This region can be adjusted to work with higher frequency input signals by changing the values of either the resistor R or capacitor C and thereby altering the RC relationship of the integrator, as seen in Fig. 1, but the range of the operating region will still remain at a similar width.

This circuit cannot be used to emulate a memristor in a functional circuit; it is designed for observing memristive behavior in an experimental environment rather than a practical one.

2.2. *Pulse controlled microprocessor based MMC*

Pershin and Ventra [3] developed a programmable analog circuit, also presented in 2010. This circuit approached the design of a suitable memristor emulator from a completely different direction. The approach was taken based on the following concept. The thin film TiO_2 memristors behave as ideal memristors, whereas their behavior is not ideal and considerably more complex [3, 4]. Pershin and Ventra chose to use pulse trains to set the memristor's state and then use low voltage analog signals for actual signal processing. This exploits the threshold behavior observed in memristors where larger signals alter the conductance of the device and much smaller signals can pass through with no or negligible effect [4, 5].

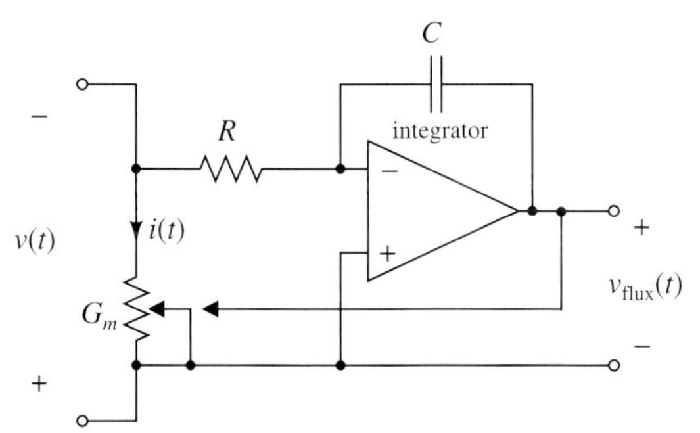

Fig. 1. Basic schematic of memristor modeling circuit (MMC) [2].

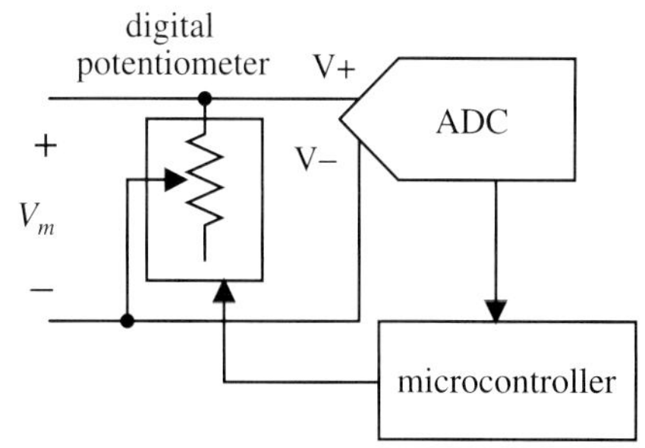

Fig. 2. Block diagram of memristor emulator [3].

The block diagram shown in Fig. 2 shows that the circuit contains a microprocessor, an analog-to-digital converter (ADC) and a digital potentiometer. The circuit operates with the ADC converting the analog signal to its binary equivalent and supplying this to the microprocessor. The microprocessor then sets the digital potentiometer to the resistance value required. According to Pershin and Ventra [3], the microprocessor is coded to process signals according to

$$I(t) = {R_M}^{-1}(x, V_M, t)V_M(t), \tag{1}$$

$$\dot{x} = f(x, V_M, t), \tag{2}$$

where x is a vector representing n internal state variables, R_M is memory resistance and R_M^{-1} is its inverse, being memductance [3]. In Valsa *et al.* [2], it has been pointed out that the disadvantages of using a digital system to emulate an analog component are the limited resolution in the equivalent digital signal and the physical properties of the digital potentiometer. This comment can be justified considering the digital potentiometer used is a 256 position 10 kΩ type [3] which would give steps of 39 Ω. The circuit is also limited by the ADC sampling frequency of 1 kHz, meaning that the operating signals applied to the circuit should be at most 50 Hz and ideally less than 10 Hz [3].

Pershin and Ventra [3] have pointed out that the ADC, microprocessor and digital potentiometer can all be upgraded to provide greater frequency handling capability and stepping resolution, and that the components chosen for the demonstration are good enough to provide proof of theory. The advantages of their method may become more evident when further experimental data describing the operational characteristics and parameters of actual memristors are obtained and published. At that time it is a relatively simple process to upgrade the basic system and alter the coding to suit the data as it is revealed. Furthermore, there will not be just one type of memristors, and in the same sense there are a variety of capacitors in terms of construction materials, capacitance value and performance ratings. It can thus be expected that a number of different memristors and memristive systems will be developed in the near future. Pershin and Ventra's circuit is an adaptable design that can be modified to effectively emulate the many variations in memristor types that are expected to be available.

2.3. *Plug-and-play MMC*

A third MMC has been presented by Sodhi and Gandhi [6], as shown in Fig. 3, which is primarily intended for educational purposes rather than as a means to develop memristor based applications. Nevertheless, the circuit operates effectively up to at least 100 Hz and shows V-I characteristics simulating that of the TiO_2 memristor. The circuit is derived from careful analysis of the equations describing the TiO_2 memristor in the paper by Strukov *et al.* [1], i.e.,

$$v(t) = \left(R_{ON}\frac{w(t)}{D} + R_{OFF}\left(1 - \frac{w(t)}{D}\right)\right)i(t), \quad (3)$$

$$w(t) = \mu_v \frac{w(t)}{D} q(t), \quad (4)$$

$$M(q) = R_{OFF}\left(1 - \frac{\mu R_{ON}}{D^2} q(t)\right), \quad (5)$$

where D is the total width of the TiO_2 film, $w(t)$ is the width of the region of high dopant concentration on the TiO_2 film, R_{OFF} and R_{ON} are the limit values of the memristor resistance for $w(t) = 0$ and $w(t) = D$, and μ_v is the dopant mobility. Here, (5) describes the memristance of the circuit, obtained from equations (3) and (4) and simplified by assuming $R_{ON} \ll R_{OFF}$ [1].

Designed with the same goal as the Valsa *et al.*'s circuit [2], this circuit serves to observe memristive behavior rather than as a circuit for practical implementation. The circuit shows good correspondence with SPICE results, displaying regular pinched hysteretic loops in response to a sinusoidal input.

2.4. *A memristive system for chaotic circuits*

The MMC presented by Muthuswamy and Chua [7] does not conform to the ideal memristor described by Chua in 1971 [8], nor does it follow the HP model [1]. It is more specifically a memristive device as described by Chua and Kang [9]. The equations describing the memristive system are given as [7]

$$V_M = \beta(x^2 - 1)i_M, \quad (6)$$

$$\dot{x} = i_M - \alpha x - i_M x, \quad (7)$$

where x represents the memristor's internal state, α and β are constants. The memristive symbols described by (6) and (7) are given in Fig. 4 and the plot derived from them is given in Fig. 5.

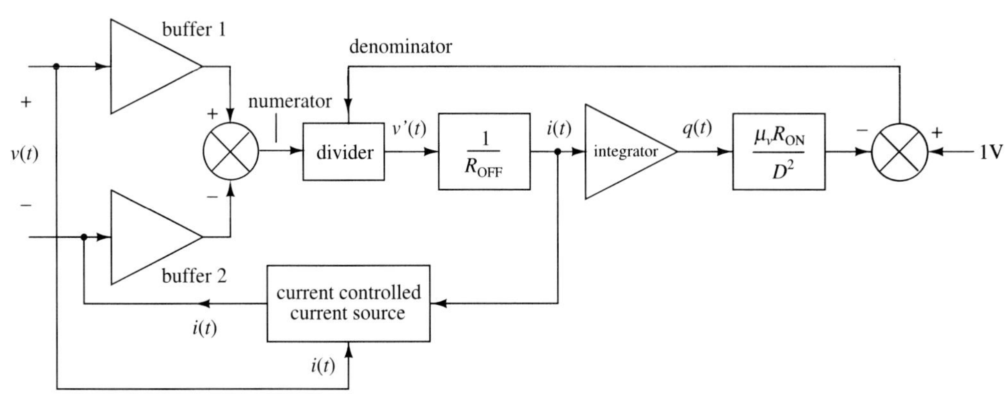

Fig. 3. Block diagram of the plug-and-play circuit [4].

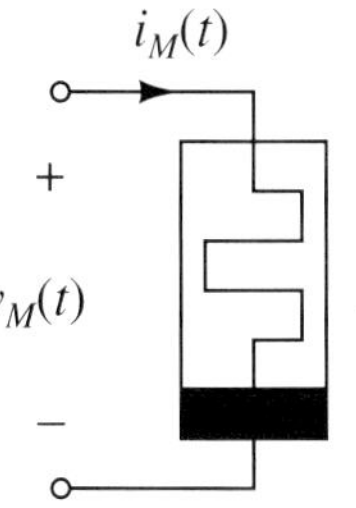

Fig. 4. Memristor circuit symbol.

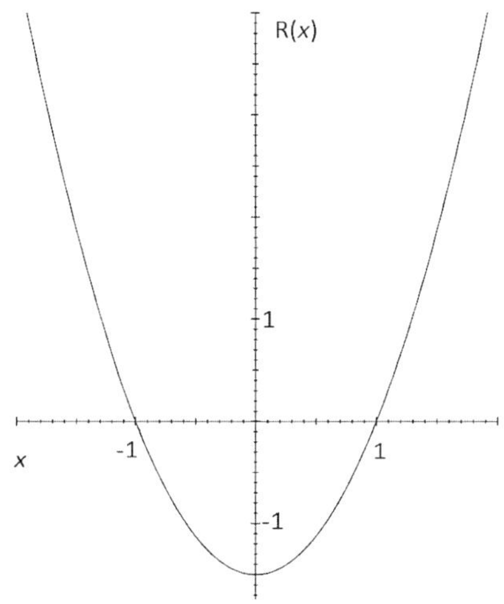

Fig. 5. Plot of the memristive function of the MMC [7].

The paper by Muthuswamy and Chua [7] describes the application of the MMC in a simple chaotic oscillator. For the chaotic oscillator circuit to operate correctly, it requires three independent state variables, which cannot occur in a single loop circuit with an ideal charge controlled type memristor.

2.5. *A light dependent resistor based MMC for practical applications*

A practical MMC design is studied in more detail here as it is a design based on the HP flux controlled memristor model and is able to function in a circuit as a memristive system [10]. This circuit uses a light dependent resistor (LDR) to obtain a controllable conductance representing the memristance. The resistance value of the LDR is controlled by a light emitting diode (LED) which maintains a brightness controlled by a voltage corresponding to the flux of the input signal. The advantage of using the LDR is that it allows the control circuit to be separated from the signal path.

The MMC has 3 operational amplifier stages and a Silonex NSL-32 analog opto-coupler, which contains the LDR and a LED. The input signal takes two paths through the MMC. One path connects directly to the LDR, the other path is connected to the control circuit that determines the resistance value of the LDR. The first operational amplifier stage is a difference inverting buffer that can be modified to adjust the gain of the input signal if necessary. Although this stage could be eliminated from the circuit if the MMC is to be used with a grounded terminal, it is useful to include it. One reason is, as mentioned, to increase or decrease the gain of the input signal as desired. The other reason is that when the MMC is implemented as a two-terminal device, this stage is used as a difference stage to obtain the voltage drop of the signal across the MMC.

The integrated value of the signal is obtained from the second stage. This is a voltage corresponding to flux (ϕ) which will be used to remotely control the resistance value of the LDR. This varying, controlled resistance is the emulation of the memristive function. The third operational amplifier stage serves to introduce an offset voltage and drives the LED that controls the LDR by its brightness. The offset voltage needs to be high enough to ensure the varying output is always above the LED turn-on voltage [10]. This MMC is based on the flux control model proposed by Biolek *et al.* [11].

The block diagram of this circuit is shown in Fig. 6 and an equivalent circuit is shown in Fig. 7. This version of the circuit as shown and described assumes that the output terminal is connected to ground. The full MMC with grounded output terminal is given in Fig. 8. Table 1 gives the values of the components. For the purpose of analysis, it is convenient to consider the circuit as grounded at the output terminal, although further into the chapter the version with two open terminals will be considered. To determine how the resistance of the LDR varies in response to a sinusoidal input,

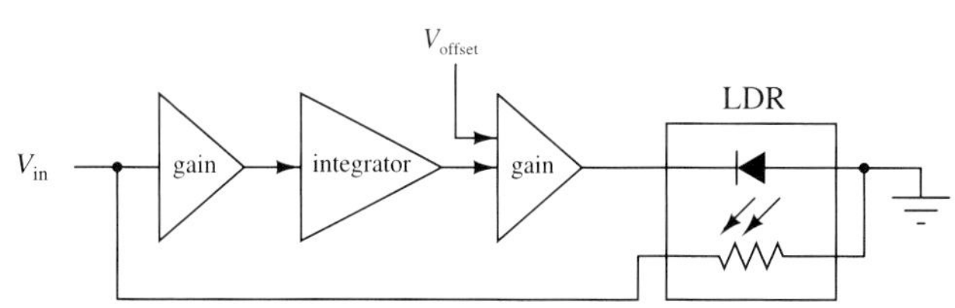

Fig. 6. Basic block diagram of grounded terminal light dependent resistor (LDR) based MMC.

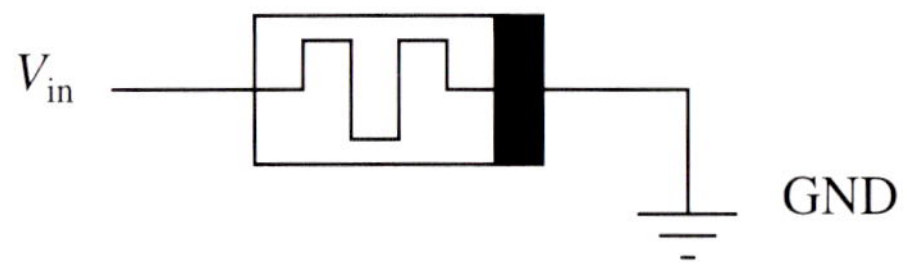

Fig. 7. Equivalent circuit of grounded LDR based MMC.

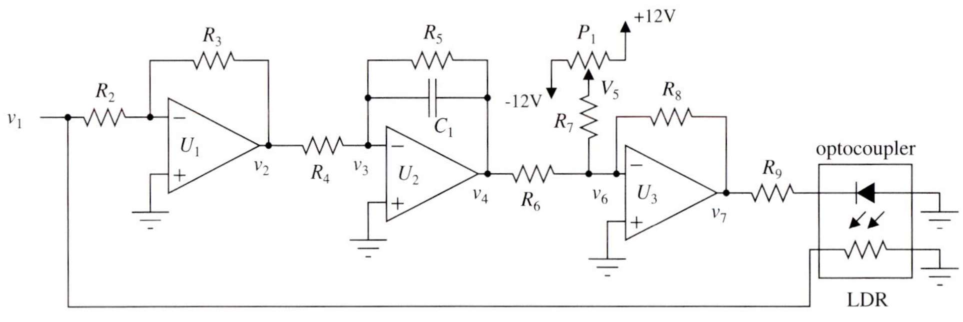

Fig. 8. Circuit implementation of the grounded LDR based MMC.

Table 1. Component values for MMC of Fig. 8.

Components	Values
R_2, R_3	100 kΩ
R_6, R_7, R_8	10 kΩ
R_4	150 kΩ
R_5	680 kΩ
R_9	1 kΩ
R_1	22 Ω
P_1	100 kΩ potentiometer
C_1	220 nF
U_{1-3}	TL074 quad operational amplifier
Opto-coupler	Silonex NSL-32

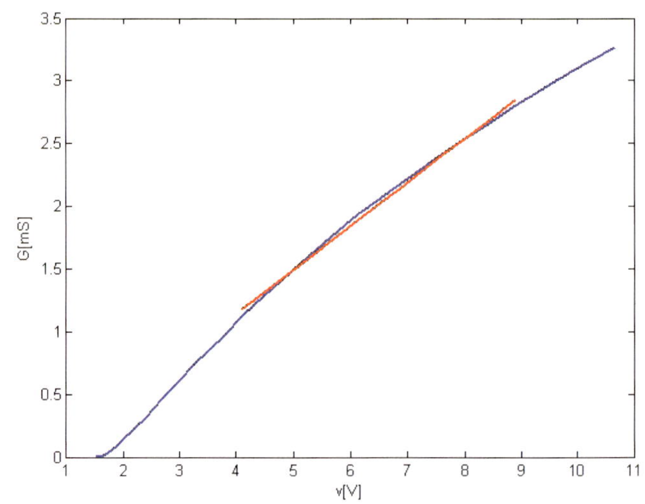

Fig. 9. Measured voltage/conductance characteristics of the Silonex NSL-32 analog opto-coupler.

we consider the equations given in [2] then modify them to suit the current circuit. The following equations describe the MMC described in [10] when the output is grounded.

The first step is to determine the approximate linear operating region of the LDR as a relationship of the voltage across the LED, through a 1 kΩ resistor, and the conductance of the LDR. These data are used to set the offset voltage of the MMC to the center of the linear region.

Figure 9 shows the approximate linear region is between 4 V and 9 V, giving a fitted linear expression of

$$G(v) = 0.3v + 0.11. \tag{8}$$

The MMC is examined and defined using (8) through (14). For input $v_1(t)$,

$$v_2(t) = -\frac{R_3}{R_2} v_1(t). \tag{9}$$

The integrator stage at U_2 realizes $v_1(t)$ as flux (ϕ), R_4 and C_1 establish the integration time constant τ, i.e.,

$$\begin{aligned} v_4(t) &= -\frac{1}{\tau}\int v_2(t)dt \\ &= -\frac{1}{R_4 C_1}\int v_2(t)dt = v_\phi. \end{aligned} \tag{10}$$

The next stage, U_3, sums v_ϕ and the offset voltage v_5, giving

$$v_7(t) = -\frac{R_8}{R_6} v_4(t) - \frac{R_8}{R_7} v_5. \tag{11}$$

We can see that $v_7(t)$ can be obtained by substituting (9) into (10), and then (10) into (11). We can then insert (11) into (8) to obtain an equation describing the memductance of the MMC, i.e.,

$$W(t) = 0.3(0 - v_7(t)) + 0.11. \tag{12}$$

Using a sinusoidal signal at the input gives

$$v_1(t) = V\sin(\omega t), \tag{13}$$

thus allowing $v_7(t)$ to be determined as

$$v_7(t) = \frac{R_8 R_3 V}{R_2 R_4 R_6 C_1}\cos(\omega t) - \frac{R_8}{R_7} v_5. \tag{14}$$

For $V = 1\,\text{V}$, $\omega = 2\pi\,\text{rad/s}$, offset voltage $V_5 = 5.5\,\text{V}$ and the component values given for the schematic in Table 1, Eqs. (12) and (14) can be used to determine that the memductance of the MMC varies between 1.45 mS and 2.07 mS. This gives a memristance range of 483.1 Ω to 689.7 Ω.

This range can be adjusted by a number of methods. The gain of the first operational amplifier stage at U_1 can be altered to give a corresponding change in the value of $v_7(t)$. This can also be adjusted at the summing stage U_3 by changing the values of components R_8 or R_6.

The circuit can be simply modified to operate as a two-terminal (non-grounded) MMC as shown in Fig. 10. The block diagram shows a feedback path to the non-inverting input of the first operational amplifier stage. To configure this stage as a difference circuit, the feedback path is in series with a 100 kΩ resistor and another 100 kΩ resistor is connected to ground at the non-inverting input.

Figure 11 shows the constructed circuit of the memristor emulator. Figures 12 to 14 show the characteristic pinched hysteretic loop expected from an MMC as an indication of memristive behavior. Figure 13 shows the loop narrowing at the higher frequency which is typical memristor behavior, in which a linear response is approached the input frequency is increased. Figure 14 shows the loop range decreases in proportion to a reduction in the amplitude of the input signal. Figures 15 and 16 show the experimental results with a triangle and square input signals at a frequency of 5 Hz and 0.8 V amplitude.

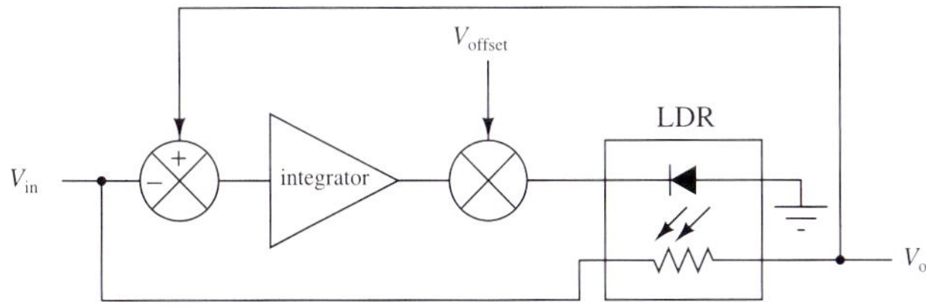

Fig. 10. Modified memristor mimicking circuit block diagram in order to operate as a two-terminal device.

Fig. 11. Laboratory construction of memristor emulator.

Fig. 12. The I-V relationship of LDR memristor at 5 Hz when the sinusoidal input amplitude is 0.8 V. The x-axis scale is 200 mV/division and y-axis scale is 0.4 mA/division.

Fig. 13. The I-V relationship of LDR memristor at 10 Hz when the sinusoidal input amplitude is 0.8 V. The x-axis scale is 200 mV/division and y-axis scale is 0.4 mA/division.

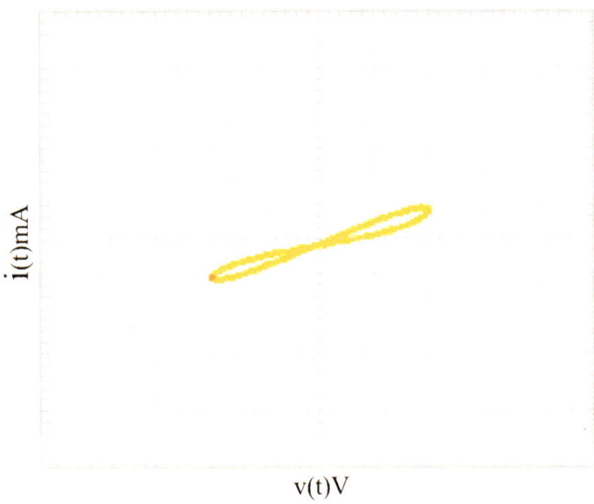

Fig. 14. The I-V relationship of LDR memristor at 5 Hz when the sinusoidal input amplitude is 0.4 V. The x-axis scale is 200 mV/division and y-axis scale is 0.4 mA/division.

Fig. 15. The I-V relationship of LDR memristor with triangular input signal at 5 Hz, the input amplitude is 0.8 V. The x-axis scale is 200 mV/division and y-axis scale is 0.4 mA/division.

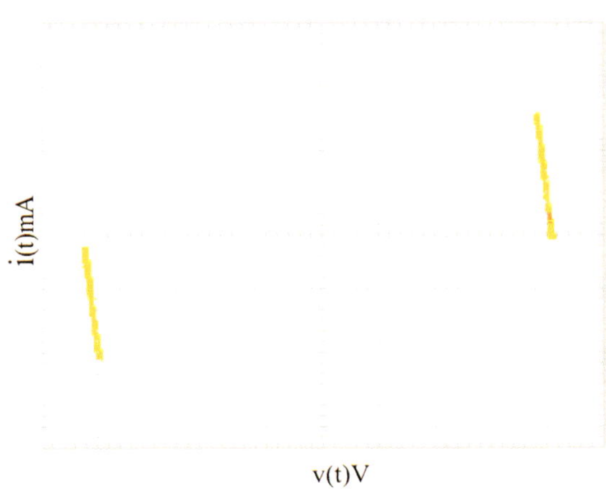

Fig. 16. The I-V relationship of LDR memristor with square-wave input signal at 5 Hz, the input amplitude is 0.8 V. The x-axis scale is 200 mV/division and y-axis scale is 0.4 mA/division.

This MMC is the simplest of the versions reviewed, particularly in terms of the number of components, simplicity of component types, ease of construction and ease of setup. It is interesting as it is designed to be used as a memristive system in actual circuits and may prove to be an effective method to study these types of circuits. The main drawbacks of this circuit, which are not trivial but surmountable, are its limited frequency range for effective operation and its lack of long-term memory which is expected of a true memristive system. The operating frequency range of the circuit can be adjusted to higher or lower rates by changing the *RC* combination in the integrator stage and installing a faster opto-coupler. A sample and hold based analog memory function is being pursued to refine the circuit and bring it closer to a true MMC. In the meantime this circuit is suited to development of applications in analog neural networks and analog signal processing where the long term memory function may be superfluous to the circuit requirements.

3. Conclusion

There is a great sense of anticipation of the capabilities and possibilities that can be achieved using memristor based devices. A number of interesting memristor based chaotic and nonlinear circuits have been developed [12]–[17] and the variety of research approaches into memristor applications for memory systems and analog and neuro-computing is evidenced by the large and constantly growing volume of published results [18]–[28]. In this chapter we attempt to provide an overview of, and some insight into, the effort and the ingenuity being applied to this field of research. The memristor modeling circuits described in this chapter constitute the first steps in the development of this field. It can be seen that the systems presented are novel and carefully designed but these are fledgling solutions and further development will enable more realistic and practical emulators to be built.

References

[1] D.B. Strukov, G.S. Snider, G.R. Stewart and R.S. Williams, "The missing memristor found," *Nature* **453**, 7191, pp. 80–83, (2008).

[2] J. Valsa, D. Biolek and Z. Biolek, "An analog model of the memristor," *International Journal of Numerical Modeling: Electronic Networks, Devices and Fields* DOI:10.1002/jnm.786, (2010).

[3] Y. V. Pershin and M. Di Ventra, "Practical approach to programmable analog circuits with memristors," *IEEE Transactions on Circuits and Systems I: Regular Papers* **57**, 8, pp. 1857–1864, (2010).

[4] J.J. Yang, M.D. Pickett, X. Li, D.A.A. Ohlberg, D.R. Stewart and R.S. Williams, "Memristive switching mechanism for metal/oxide/metal nanodevices," *Nat. Nanotechnol.* **3**, 7, pp. 429–433, (2008).

[5] S. Liu, N. Wu and A. Ignatiev, "Electric-pulse-induced reversible resistance change effect in megnetoresistive films," *Appl. Phys. Lett.* **76**, 19, pp. 2749–2751, (2000).

[6] A. Sodhi and G. Gandhi, "Circuit mimicking TiO_2 memristor: a plug and play kit to understand the fourth passive element," *International Journal of Bifurcation and Chaos* **20**, 8, pp. 2537–2545, (2010).

[7] B. Muthuswamy and L.O. Chua, "Simplest chaotic circuit," *International Journal of Bifurcation and Chaos* **20**, 5, pp. 1567–1580, (2010).

[8] L.O. Chua, "Memristor — The missing circuit element," *IEEE Transactions on Circuit Theory* **18**, 5, pp. 507–513, (1971).

[9] L.O. Chua and S.M. Kang, "Memristive devices and systems," *Proceedings of the IEEE* **64**, 2, pp. 209–223, (1976).

[10] A.L. Fitch, H.H.C. Iu, X.Y. Wang, V. Sreeram and W.G. Qi, "Realization of an analog model of memristor based on light dependent resistor," *Proc. IEEE International Symposium on Circuits and Systems*, pp. 1139–1142, (2012).

[11] D. Biolek, V. Biolkova and Z. Kolka, "Mutators simulating memcapacitors and meminductors," *Proc. IEEE Asia Pacific Conference on Circuits and Systems*, pp. 800–803, (2010).

[12] F. Corinto, A. Ascoli and M. Gilli, "Nonlinear dynamics of memristor oscillators," *IEEE Transactions on Circuits and Systems I: Regular Papers* **58**, 6, pp. 1323–1336, (2011).

[13] B. Muthuswamy, "Implementing memristor based chaotic circuits," *International Journal of Bifurcation and Chaos* **20**, 5, pp. 1335–1350, (2010).

[14] T. Driscoll, Y.V. Pershin, D.N. Basov and M. Di Ventra, "Chaotic memristor," *Applied Physics A* **102**, pp. 885–889, (2011).

[15] H.H.C. Iu, D.S. Yu, A.L. Fitch, V. Sreeram and H. Chen, "Controlling chaos in a memristor based circuit using a Twin-T notch filter," *IEEE Transactions on Circuits and Systems I: Regular Papers* **58**, 6, pp. 1337–1344, (2011).

[16] A.L. Fitch, D.S. Yu, H.H.C. Iu and V. Sreeram, "Hyperchaos in a memristor based modified canonical Chua's circuit," *International Journal of Bifurcation and Chaos* **22**, 6, 1250133, (2012).

[17] I. Petras, "Fractional-order memristor-based Chua's circuit," *IEEE Transactions on Circuits and Systems II: Express Briefs* **57**, 12, pp. 975–979, (2010).

[18] F. Merrikh-Bayat and S.B. Shouraki, "Memristor-based circuits for performing basic arithmetic operations," *Procedia Computer Science*, pp. 128–132, (2011).

[19] S. Shin, K. Kim and S.M. Kang, "Memristor applications for programmable analogic ICs," *IEEE Transactions on Nanotechnology* **10**, 2, pp. 266–274, (2011).

[20] T.A. Wey and W.D. Jemison, An automatic gain control circuit with TiO_2 memristor variable gain amplifier, *2010 IEEE International NEWCAS Conference*, pp. 49–52, (2010).

[21] Y.V. Pershin and M. Di Ventra, "Experimental demonstration of associative memory with memristive neural networks," Available: http://arXiv.org/abs/arXiv:0905.2935, (2009).

[22] S.H. Jo, T. Chang, I. Ebong, B.B. Bhadviya, P. Mazumder and W. Lu, "Nanoscale memristor device as synapse in neuromorphic systems," *Nano Letters* **10**, 4, pp. 1297–1301, (2010).

[23] G.M. Huang, H. Yenpo and L. Peng, "Memristor system properties and its design applications to ciruits such as nonvoltile memristor memories," *Proc. International Conference on Communications, Circuits and Systems*, pp. 805–810, (2010).

[24] F. Mohsin, "A multivalued storage system using memristor," *Proc. International Conference on Computer and Information Technology*, pp. 343–346, (2010).

[25] H. Manem and G.S. Rose, "A read-monitored write circuit for 1T1M multi-level memristor memories," *Proc. IEEE International Symposium on Circuits and Systems*, pp. 2938–2941, (2011).

[26] C.K. Volos, I.M. Kyprianidis and I.N. Stouboulos, "The memristor as an electric synapse-synchronization phenomena," *Proc. International Conference on Digital Signal Processing*, pp. 1–6, (2011).

[27] K. Hyongsuk, M. Sah, C. Yang, T. Roska and L.O. Chua, "Neural synaptic weighting with a pulse-based memristor circuit," *IEEE Transactions on Circuits and Systems I: Regular Papers* **59**, 1, pp. 148–158, (2012).

[28] H. Xu, X. Tian, H. Nie, Q. Li, H. Liu and X. Xu, "On development and expectation of memristor," *Proc. International Conference on Intelligent System Design and Engineering Application*, pp. 1035–1038, (2010).

CHAPTER 45

LEON CHUA'S MEMRISTOR

GUANRONG CHEN*

Department of Electronic Engineering,
City University of Hong Kong, Hong Kong
eegchen@cityu.edu.hk

This essay was published in the **IEEE Circuits and Systems Magazine**, Second Quarter, 2008, page 55.

My simple Google search for the keyword "memristor" today surprisingly resulted in nearly 80,000 items, many referring back to Leon Chua's seminal paper "Memristor — The Missing Circuit Element" published in the *IEEE Transactions on Circuits Theory* (vol. 18, no. 5, pp. 507–519, September 1971). In that paper Leon argued that in addition to the three well-known basic circuit elements, namely resistor, capacitor, and inductor, a fourth element should exist. He named this element the memristor, meaning the "memory resistor" because it can remember the current passing through it even after the current has disappeared. Is that a reality? On April 30, 2008, the Hewlett-Packard research team proudly announced their realization of a memristor prototype, with an official publication entitled "The Missing Memristor Found" in Nature (vol. 453, pp. 80–83, May 1, 2008).

I am puzzled. Perhaps not only I, but everyone except Leon, believed that the set of three basic circuit elements are theoretically self-contained. How could Leon see a gap there so as to introduce the fourth element? Mathematically, there are five well-known relations among the four circuit quantities, charge q, current i, voltage v, and magnetic flux f: two are fundamental physical laws, $dq = idt$ and $df = vdt$, and three are basic circuit equations, $dv = Rdi$, $dq = Cdv$, and $df = Ldi$, where d is the differential operator, t is the time variable, and R, C, and L are resistance, capacitance, and inductance, respectively. These two physical laws and three circuit equations are so familiar to electrical engineers that no one seemed to have any doubt of something missing when teaching students in classrooms. Leon's contribution is to introduce one missing component that satisfies an additional equation, $df = Mdq$, where M is the so-called "memristance". Does this make sense? Well, logically, yes. However, what about physically and technologically?

The HP Media Coverage Report on the "Existence of New Basic Element for Electronic Circuits," released on April 30, 2008, included twelve news reports that appeared the same day in the New York Times, Wall Street Journal, CNET, Bloomberg, EE Times, Crunch Gear, New Scientist,

InfoWorld, and PopSci.com, followed by many more articles in the IEEE Spectrum, Washington Post, and Scientific American, among others, which have convinced me that Leon's memristor is in practice today. The HP scientists are credited for their endeavor to make the idea become a reality.

To some extent, Leon's discovery is similar to that of the Russian chemist Dmitri Mendeleev who created and used a periodic table in 1869 to find many unknown properties and missing elements. Leon postulated in 1971 that there should be one component missing from the table of six possible circuit equations. This component has finally come to existence. Thus, as Leon commented, future electrical engineering textbooks need to be changed due to this discovery. As a side note, in the early 1980s, Leon also postulated several other missing circuit elements, namely, some higher-order nonlinear circuit elements, which are yet to be coined.

The impact of Leon's discovery is promising. As suggested by its name, memristor is useful for building electronic memories. According to HP scientists, the prototype device could lead to extremely dense computer memory chips that consume much less power than the current DRAM memory chips which are rapidly reaching their size limits. Moreover, memristor could handle values other than binary, based on which some future devices could possibly mimic biological functions towards a new generation of brain-like computers. Let us hope that Leon's baby memristor will grow fast and become commercialized soon, so that the nano-scale electronics can be further enabled by the industrialized memristors.

According to Leon, the memristor is not only an invention — it is a basic scientific discovery. The existence of the fourth circuit element along with the sixth circuit equation was predicted in 1971 when Leon was only thirty some years old. This sets a good example for researchers, especially young generations of scientists and engineers, that one should persist when working on fundamental problems in scientific research, even in such a mature field as the classical circuit theory.

Once again, congratulations Leon!